INTERFACIAL ELECTROKINETICS AND ELECTROPHORESIS

SURFACTANT SCIENCE SERIES

ADDITIONAL VOLUMES IN PREPARATION

INTERFACIAL ELECTROKINETICS AND ELECTROPHORESIS

edited by
Ángel V. Delgado
University of Granada
Granada, Spain

CRC Press
Taylor & Francis Group
Boca Raton London New York

CRC Press is an imprint of the
Taylor & Francis Group, an **informa** business

Preface

The phrase *zeta potential* is frequently used in this book. This concept pervades all the chapters that follow, which were written by a significant number of outstanding scientists in the field of colloid and interface science. However, we do not intend to merely add another text to the (short) list of excellent monographs on zeta potential, some authored by contributors to this volume. The aim of this book is to provide the reader with as many tools as he or she might possibly need to solve any problem in the physics and chemistry of interfaces (mainly, but not only) between solid particles and liquids. The intent is to demonstrate that electrokinetic phenomena, associated with forces of any origin causing relative motion of two phases in contact, are extremely interesting from the fundamental point of view, challenging the application of classical laws of physics to apparently simple problems. In addition, they are the basis of a number of techniques that have long proved useful in predicting or explaining many behaviors of systems, in which electrical properties of interfaces play an essential role. In order to apply a proper methodology to a particular problem, the reader must be familiarized with the origin of the processes involved and understand what kind of information can be obtained. For this reason, we have tried to find a balance between the theory and the experimental methods, or between "fundamental" and "applied" electrokinetics, by including many examples of various applications.

The book is divided into five parts. The general aspects of electrokinetics of suspended particles (both theoretical and experimental) are included in the first part. Following an overview of the phenomenology, simple theory, and experimental methods (Delgado and Arroyo), the contributions of Dukhin, Shilov, and Lyklema address the fundamental problem of the role of surface conductivity (in particular, behind the slipping plane) in electrokinetics. Kovačević et al. consider another essential question, i.e., the definition, meaning, and experimental accessibility of different electric potentials, which can be defined in the double layer. The theory of electrophoresis for different geometries or for complex structures is considered in the chapters by Ohshima, Velegol et al., and Kim and Yoon. Experimental aspects of this widespread technique, with particular emphasis on the reproducibility of measurements on well-defined samples of the same nature, and the search for standards of universal validity are the topics dealt with independently by Matijević and Furusawa. Mishchuk and Dukhin focus on the so-called electrokinetic phenomena of the second kind, which appear when electric fields are applied to internally conducting particles, such as ion exchangers. A number of chapters are devoted to the rapidly emerging field of AC electrokinetics, with chapters by Grosse (analysis of relaxation phenomena in the double layer), Shilov et al. (dielectric dispersion and particle electrorotation), and Gimsa (providing many details on the latest techniques based on the application of AC fields to colloid dispersions). Nonspherical particle electro-orientation, and its implications for optical birefrigence in dispersed systems, is discussed by Bellini and Mantegazza. Finally, Keh, Ozaki, and Sasaki consider the sedimentation potential, while the contribution by A. S. Dukhin et al. deals with the promising field of electroacoustic phenomena.

In the second part, "Membranes and Porous Plugs," rigorous treatments of electro-osmosis in porous media of different geometries by Adler et al., and Li, are followed by applications in the characterization of (mainly inorganic) membranes by Ricq et al.

In the third part, "Polymers and Particles of Biological Interest," the latter area is emphasized. Stein gives a review of the electrokinetic properties of polystyrene latex, and Martín-Rodríguez et al. use such particles as substrates for protein adsorption. Lee et al. introduce the modifications of the classical electrokinetic models, required when dealing with particles of complex or nonrigid surfaces, as is common with biological cells. Next, Knippel and Budde focus on the use of electrophoresis as an aid in the diagnosis of human pathologies. Finally, Makino describes a number of methods for significantly modifying the surface of particles by polymer coating, and the implications of these treatments on the interactions of colloids with red blood cells.

The fourth part, "Inorganic Particles," deals mainly with specific inorganic disper-sions. Sprycha describes technological applications of electrokinetics in the ink and print-ing industry, Sondi and Pravdić review the electrical surface properties of clays, and Das discusses electrokinetic properties of a large number of natural and synthetic inorganic compounds.

The final part, "Gas/Liquid and Liquid/Liquid Interfaces: Vesicles," is devoted to dispersed systems of deformable interfaces. The general aspects of gas/liquid electroki-netics are reviewed by Graciaa et al., and Spasic discusses a general model for the electro-viscoelastic properties of liquid/liquid interfaces. The chapters by O, Chibowski, and Wiącek deal with emulsions; the last two contributions focus on the electrokinetic proper-ties of liposomes (De Meulaenaer et al.) and on the effects of applied fields on their stability (Matsumura and Furusawa).

Because of the wide variety of topics, the book should be useful to readers from both academia and industry. Scientists dealing with a variety of processes, such as adsorption, adhesion, corrosion, and many others, will appreciate finding all the information con-tained in one volume. Graduate students with projects in colloid and interface science will find this book an excellent source of references. Finally, practitioners will want this book on their shelves, because electrokinetic techniques are used routinely in many areas of technology, including paint, food, drugs, oil, detergents, dairy, and many other industries.

I am indebted to the series editor, Professor Arthur T. Hubbard, for suggesting that I edit this volume, and to the contributors, who have spared no effort in writing their chapters. The list of authors could be longer, considering the amount and quality of research on the subject of this book, but as with any other publication, there are limita-tions in size and also restrictions with respect to deadlines. I would like to take this opportunity to thank the members of my research group for their support, advice, and patience. Finally, I am also pleased to acknowledge the efficiency and care of the people in charge of this project at Marcel Dekker, Inc.

Ángel V. Delgado

Contents

Contributors

Pierre M. Adler Laboratoire des Milieux Poreux et Fracturés, Institut de Physique du Globe de Paris, Paris, France

John L. Anderson Department of Chemical Engineering, Carnegie Mellon University, Pittsburgh, Pennsylvania

Francisco J. Arroyo Department of Physics, University of Jaén, Jaén, Spain

S. Békri Laboratoire des Milieux Poreux et Fracturés, Institut de Physique du Globe de Paris, Paris, France

Tommaso Bellini Department of Physics, Politecnico di Milano, Milan, Italy

Axel Budde HaSo Tec Hard- and Software Technology GmbH, Rostock, Germany

Emil Chibowski Department of Interfacial Phenomena, Maria-Curie Sklodowska University, Lublin, Poland

Ana Čop Department of Chemistry, Faculty of Science, University of Zagreb, Zagreb, Croatia

Patrice Creux Department of Physics, University of Pau, Pau, France

Kalyan K. Das Tata Research Development and Design Centre, Pune, India

Bruno De Meulenaer Department of Food Technology and Nutrition, Ghent University, Ghent, Belgium

Ángel V. Delgado Department of Applied Physics, University of Granada, Granada, Spain

Edwin Donath Max-Planck-Institute for Colloids and Interfaces, Berlin, Germany

Andrei S. Dukhin Dispersion Technology, Inc., Mount Kisco, New York

Stanislav S. Dukhin   Civil and Environmental Engineering, New Jersey Institute of Technology, Newark, New Jersey

Patrick Fievet   Laboratoire de Chimie des Materiaux et Interfaces, Besançon, France

Kunio Furusawa   Department of Chemistry, University of Tsukuba, Tsukuba, Ibaraki, Japan

Jan Gimsa   Institut für Biologie, Humboldt Universität zu Berlin, Berlin, Germany

Philip J. Goetz   Dispersion Technology, Inc., Mount Kisco, New York

Fernando González-Caballero   Department of Applied Physics, University of Granada, Granada, Spain

Alain Graciaa   Department of Physics, University of Pau, Pau, France

Constantino Grosse   Department of Physics, National University of Tucumán, San Miguel de Tucumán, and Consejo Nacional de Investigaciones Cientificas y Técnicas, Buenos Aires, Argentina

Roque Hidalgo-Álvarez   Department of Applied Physics, University of Granada, Granada, Spain

Jyh-Ping Hsu   Department of Chemical Engineering, National Taiwan University, Taipei, Taiwan

Nikola Kallay   Department of Chemistry, Faculty of Science, University of Zagreb, Zagreb, Croatia

Huan J. Keh   Department of Chemical Engineering, National Taiwan University, Taipei, Taiwan

Jae Young Kim   Department of Chemical Engineering, Pohang University of Science and Technology, Pohang, Korea

Eberhard Knippel   Consultant, Langenfeld, Germany

Davor Kovačević   Department of Chemistry, Faculty of Science, University of Zagreb, Zagreb, Croatia

Jean Lachaise   Department of Physics, University of Pau, Pau, France

Eric Lee   Department of Chemical Engineering, National Taiwan University, Taipei, Taiwan

Dongqing Li[*]   Department of Chemical Engineering, University of Alberta, Edmonton, Alberta, Canada

J. Lyklema   Laboratory for Physical Chemistry and Colloid Science, Wageningen University, Wageningen, The Netherlands

---

[*]*Current affiliation*: Department of Mechanical and Industrial Engineering, University of Toronto, Toronto, Ontario, Canada.

Kimiko Makino Faculty of Pharmaceutical Sciences, Science University of Tokyo, Tokyo, Japan

Francesco Mantegazza Department of Experimental Medicine, Università di Milano-Bicocca, Monza, Italy

Antonio Martín-Rodríguez Department of Applied Physics, University of Granada, Granada, Spain

Egon Matijević Center for Advanced Materials Processing, Clarkson University, Potsdam, New York

Hideo Matsumura Life-Electronics/Photonics, National Institute of Advanced Industrial Science and Technology, Tsukuba, Japan

Nataliya A. Mishchuk Electrochemistry of Disperse Systems, Institute of Colloid Chemistry and Chemistry of Water, Ukrainian National Academy of Sciences, Kiev, Ukraine

O Boen Ho Akzo Nobel Surface Chemistry, Deventer, The Netherlands

Hiroyuki Ohshima Faculty of Pharmaceutical Sciences, Science University of Tokyo, Tokyo, Japan

Juan Luis Ortega-Vinuesa Department of Applied Physics, University of Granada, Granada, Spain

Masataka Ozaki Department of Environmental Science and Graduate School of Integrated Science, Yokohama City University, Yokohama, Japan

Velimir Pravdić Center for Marine and Environmental Research, Ruđer Bošković Institute, Zagreb, Croatia

Laurence Ricq Laboratoire de Chimie des Materiaux et Interfaces, Besançon, France

Hiroshi Sasaki School of Science and Engineering, Waseda University, Tokyo, Japan

Vladimir N. Shilov Institute of Biocolloid Chemistry, Ukrainian National Academy of Sciences, Kiev, Ukraine

Yuri Solomentsev Motorola Advanced Products, Austin, Texas

Ivan Sondi Center for Marine and Environmental Research, Ruđer Bošković Institute, Zagreb, Croatia

Aleksandar M. Spasic Department of Chemical Engineering, Institute for Technology of Nuclear and Other Mineral Raw Materials, Belgrade, Yugoslavia

Ryszard Sprycha Daniel J. Carlick Technical Center, Sun Chemical Corporation, Carlstadt, New Jersey

Hans Nikolaus Stein Chemical Technology, Eindhoven University of Technology, Eindhoven, The Netherlands

Anthony Szymczyk Laboratoire de Chimie des Materiaux et Interfaces, Besançon, France

Jean-François Thovert Laboratoire de Combustion et Détonique, Poitiers, France

Paul Van der Meeren Department of Applied Analytical and Physical Chemistry, Ghent University, Ghent, Belgium

Jan Vanderdeelen Department of Applied Analytical and Physical Chemistry, Ghent University, Ghent, Belgium

Darrell Velegol Department of Chemical Engineering, The Pennsylvania State University, University Park, Pennsylvania

Agnieszka Wiącek Department of Interfacial Phenomena, Maria-Curie Sklodowska University, Lublin, Poland

Fong-Yuh Yen Department of Chemical Engineering, National Taiwan University, Taipei, Taiwan

Byung Jun Yoon Department of Chemical Engineering, Pohang University of Science and Technology, Pohang, Korea

INTERFACIAL ELECTROKINETICS AND ELECTROPHORESIS

1

Electrokinetic Phenomena and Their Experimental Determination: An Overview

ÁNGEL V. DELGADO University of Granada, Granada, Spain

FRANCISCO J. ARROYO University of Jaén, Jaén, Spain

I. INTRODUCTION

Consider a spherical solid particle 1 cm in diameter. Its surface S and volume V are, respectively, $3.14 \times 10^{-4}\,m^2$ and $5.24 \times 10^{-7}\,m^3$, and thus its surface/volume ratio will be $S/V \sim 600\,m^{-1}$. Assume that we divide the particle in a number N of spherical particles of radius 100 nm such that their total volume equals that of the original 1-cm sphere. However, their surface would be $15.7\,m^2$, and $S/V \sim 3 \times 10^7\,m^{-1}$. This simple and well-known example explains that an essential contribution to the properties of a system formed by dispersing the N particles in, say 1 L of water will be connected with the influence of the surfaces and interfaces of the particles. In particular, the electrical state of the surface of the particles may be determinant: if each of them bears a surface potential of 100 mV (about the order of magnitude typical of colloidal particles in aqueous media), the repulsive electrostatic force between two such particles dispersed in water and located at a surface-to-surface distance of 10 nm would be $F^{EL} \sim 2.12 \times 10^{-12}\,N$. This force has to be compared to the strength of other interactions that must or could exist between them. Thus, their gravitational attraction at the same distance would be $F^{G} \sim 6.3 \times 10^{-15}\,N$ (if their density is $10^3\,kg/m^3$); their van der Waals attraction $F^{LW} \sim 8 \times 10^{-13}\,N$ (using typical values of the Hamaker's constant, see Ref. 1). Again, these examples show that in most instances the electrostatic interactions are mainly responsible for the macroscopic properties of the suspensions.

In this context, electrokinetic phenomena and the techniques associated with them demonstrate their importance. They are manifestations of the electrical properties of the interface, and hence deserve attention by themselves. Furthermore, however, they are a valuable (unique, in many cases) source of information on those electrical properties, because of the possibility of their being experimentally determined.

In this chapter, we describe some of the most widely used electrokinetic phenomena and techniques. These include classical electrophoresis, streaming potential and current, and electro-osmosis. Attention will also be paid to other rapidly growing techniques, such as those based on electroacoustic measurements, electrorotation, dielectrophoresis, or low-frequency dielectric dispersion, some of which appear to be suitable for the electrokinetic analysis of suspensions even in the case of very high concentrations of solids, when conventional techniques are inapplicable.

II. CLASSICAL DESCRIPTION OF THE ELECTRICAL DOUBLE LAYER

We will admit as an experimental fact that most solids acquire an electrical surface charge when dispersed in a polar solvent, in particular, in an electrolyte solution. The origins of this charge are diverse [1–4], and include:

1. *Preferential adsorption of ions in solution.* This is the case of ionic surfactant adsorption. The charged entities must have a high affinity for the surface in order to avoid electrostatic repulsion by already adsorbed ions.
2. *Adsorption–desorption of lattice ions.* Silver iodide particles in Ag^+ or I^- solutions are a typical example: the crystal lattice ions can easily find their way into crystal sites and become part of the surface. They are called potential-determining ions (p.d.i.).
3. *Direct dissociation or ionization of surface groups.* This is the mechanism through which most polymer latexes obtain their charge. Thus, acid groups as sulfate and/or carboyxl are responsible for the negative charge of anionic polymer lattices. When the pH is above the pK_a of dissociation of these groups, most of them will be ionized, rendering the surface negative. In the case of oxides, zwitterionic MOH surface groups can generate either positive or negative charge, depending on the pH; H^+ and OH^- would hence be p.d.i.'s for oxides.
4. *Charge-defective lattice: isomorphous substitution.* This is a mechanism typical – almost exclusive, in fact – of charging in clay minerals: a certain number of Si^{4+} and Al^{3+} cations of the ideal structure are substituted by other ions with lower charge and of almost the same size. As a consequence, the crystal would be negatively charged, although this structural charge is compensated for by surface cations, easily exchangeable in solution [5].

    Whatever the mechanism (and there are instances in which more than one of them participates), the net surface charge must be compensated for by ions around the particle so as to maintain the electroneutrality of the system. Both the surface charge and its compensating countercharge in solution form the electrical double layer (EDL in what follows). In spite of the traditional use of the word "double," its structure can be very complex, not fully resolved in many instances, and it may contain three or more layers, extending over varying distances from the solid surface.

    Close to the latter or on the surface itself one can find charges responsible for the surface charge σ_0, the so-called titratable charge. In their immediate vicinity, ions capable of undergoing specific adsorption might be located: their distance to the solid will be of the order of an ionic radius, since it is assumed that they have lost their hydration shell, at least in the direction of the solid surface. Let us call σ_i the surface charge at such a plane of atoms, located at a distance β_i from the solid (see Fig. 1). If, as usual, we assume that the interface has planar geometry, and x is the outward distance normal to it, then we can say that the region between $x = 0$ and $x = \beta_i$ is free of charge, and a capacitor whose plates are the surface and the β_i plane is identified. If C_i is its specific (per unit surface area) capacitance, then

$$\psi_0 - \psi_i = \frac{\sigma_0}{C_i} \tag{1}$$

where ψ_0 is the potential at the solid surface. Let us note that ions responsible for ψ_i will not only have electrostatic interactions with the surface: in fact they often overcome electrical repulsions, and are capable, for instance, of increasing the positive charge of

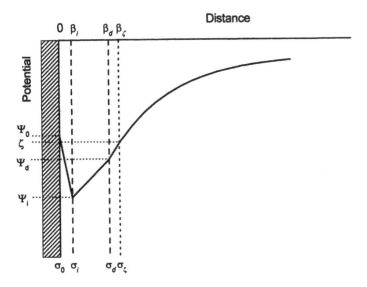

FIG. 1 Schematic representation of the potential distribution at a negatively charged interface.

an already positive surface. It is usual to say that the missing interactions are of a chemical nature, although this is not, strictly speaking, always the case. There is a large variety of situations, ranging from the formation of chemical (covalent) bonds, to weaker interactions including van der Waals attraction, hydrogen bonds, hydrophibic–hydrophilic forces, and so on [3]. Because of the typical lack of information about this inner part of the ionic atmosphere, the treatment is usually not free from more or less realistic hypotheses and assumptions.

Still further from the surface, at a distance $x = \beta_d$ and beyond, ions undergoing only electrostatic interactions with the surface are located, and because of their less intense interactions with the solid, they are also subjected to collisions with solvent molecules, so that they are in fact distributed over a certain distance to the solid. This third layer in the ionic distribution is in fact characterized by a volume charge density $\rho(x)$, although it is of practical use to introduce a surface charge density σ_d, located at $x = \beta_d$, according to

$$\sigma_d = \int_{\beta_d}^{\infty} \rho(x)\,dx \tag{2}$$

for a plane interface, or

$$\sigma_d = \frac{1}{(a + \beta_d)^2} \int_{\beta_d}^{\infty} r^2 \rho(r)\,dr \tag{3}$$

for a spherical interface of radius of a, r being the radial co-ordinate, with origin at the particle center.

If ψ_d is the potential at $x = \beta_d$, another capacitance can be distinguished between β_i and β_d:

$$\psi_i - \psi_d = \frac{\sigma_0 + \sigma_i}{C_2} \tag{4}$$

Now, because of electroneutrality,

$$\sigma_0 = -\sigma_i - \sigma_d \tag{5}$$

and Eq. (4) can also be written:

$$\psi_i - \psi_d = -\frac{\sigma_d}{C_2} \tag{6}$$

In order to become familiar with the nomenclature (unfortunately, there are almost as many criteria as authors, concerning the symbols used for the different potentials), let us mention:

1. The volume ionic distribution extending from $x = \beta_d$ is called the *diffuse layer* or *diffuse part of the double layer*.
2. The region between $x = 0$ and $x = \beta_d$ is often named the *Stern layer* or *inner part of the double layer*, or *dense part of the double layer*.
3. The plane $x = \beta_i$ is the *inner Helmholtz plane* (IHP) and that at $x = \beta_d$ is called the *outer Helmholtz plane* (OHP). That is, the OHP identifies the beginning of the diffuse layer.

The diffuse layer can be described mathematically in a simple way: the equilibrium condition for ions in this layer can be written [6]:

$$-ez_i \nabla \psi - k_B T \nabla \ln n_i = 0, \quad i = 1, \ldots, N \tag{7}$$

where the first term corresponds to the electrostatic force on ions of type i (charge ez_i, number concentration n_i) and the second is the thermodynamic force. Integration of Eq. (7) under the condition $n_i = n_i^0(\infty)$ for $\psi = 0$ leads to the Boltzmann distribution:

$$n_i(\vec{r}) = n_i^0(\infty) \exp[-ez_i \psi(\vec{r})/k_B T], \quad i = 1, \ldots, N \tag{8}$$

where $n_i^0(\infty)$ is the number concentration of ions of type i, far from the particle, k_B is the Boltzmann constant, and T is the absolute temperature. Finally, the Poisson equation will give us the relationship between the potential and ionic concentrations:

$$\nabla^2 \psi(\vec{r}) = -\frac{1}{\varepsilon_{rs}\varepsilon_0} \rho(\vec{r}) = -\frac{1}{\varepsilon_{rs}\varepsilon_0} \sum_{i=1}^{N} ez_i n_i^0(\infty) \exp\left[-\frac{ez_i \psi(\vec{r})}{k_B T}\right], \tag{9}$$

$\varepsilon_{rs}\varepsilon_0$ being the dielectric permittivity of the dispersion medium. Equation (9) (the Poisson–Boltzmann equation) is the starting point of the Gouy–Chapman description of the diffuse layer.

It will be clear that there is no general solution to this partial differential equation, but in certain cases [6, 7]:

1. A flat interface, with low potential. In this case:

$$\psi = \psi_d e^{-\kappa x} \tag{10}$$

where κ^{-1} is the Debye length, and it is clearly a measure of the diffuse layer thickness. Its value is

$$\kappa^{-1} = \left\{ \frac{\varepsilon_{rs}\varepsilon_0 k_B T}{\sum\limits_{i=1}^{N} e^2 z_i^2 n_i^0(\infty)} \right\}^{1/2}$$ (11)

To get an idea of the typical values of κ^{-1}, the following practical formula for a 1-1 electrolyte ($N = 2, z_1 = 1, z_2 = -1$) is useful (water at 25°C is the solvent):

$$\kappa^{-1} = 0.308 c^{-1/2} \text{ nm}$$

if c is the molar concentration of electrolyte ($n_1 = n_2 = 10^3 N_A c$ for a 1-1 electrolyte).

2. A flat interface, in a symmetrical z-valent electrolyte ($z_1 = -z_2 = z$) for arbitrary ψ_d potential:

$$y(x) = 2 \ln \left[\frac{1 + e^{-\kappa x} \tanh(y_d/4)}{1 - e^{-\kappa x} \tanh(y_d/4)} \right]$$ (12)

where y is the dimensionless potential:

$$y = \frac{ze\psi}{k_B T}$$ (13)

and a similar expression can be given for y_d.

3. A spherical interface (radius a) at low potentials (the so-called Debye approximation):

$$\psi(r) = \psi_d \left(\frac{a}{r}\right) e^{-\kappa(r-a)}$$ (14)

whereas numerical solutions or approximate analytical expressions have to be applied in other cases. This is illustrated in Fig. 2, where Eqs (10) and (12) are compared for the flat interface, and in Fig. 3, where the approximate solution, Eq. (12), is compared to the full numerical calculation [8].

III. ELECTROKINETIC POTENTIAL AND ELECTROKINETIC PHENOMENA

Let us assume that an electric field is applied parallel to the solid/solution interface in Fig. 1, and that the solid wall is fixed in our co-ordinate system. From the preceding discussion, it will be clear that the liquid adjacent to the solid has a net electric charge, opposite to that on the surface. Part of the ions in that liquid will likely be strongly attached to the surface by short-range attractive forces, and can be considered immobile (however, it is not rare that their mobility has a value close to that in the bulk solution, see Chapter 2) and the same will be admitted with respect to the liquid in that region. On the other hand, both ions and liquid outside it can be moved by the external field: in fact the electric force will act on the ions (mainly, counterions) and they will drag liquid in their motion. That is, a relative movement between the solid and the liquid will occur: this is the very essence of electrokinetic phenomena. The potential existing at the boundary between the mobile and immobile phases is known as electrokinetic or zeta (ζ) potential. The exact location (distance β_ζ in Fig. 1) of that so-called slipping or shear plane is a matter of investigation (see Chapter 4), and, in fact, even the existence of the latter plane and of the zeta potential itself

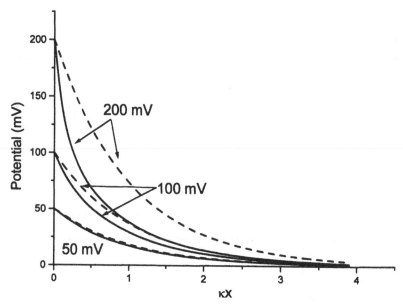

FIG. 2 Potential distribution at a flat interface, calculated by using the Debye–Hückel approximate formula [Eq. (10), dashed lines] and the full calculation [Eq. (12), solid lines], for the values of Ψ_d indicated. Monovalent electrolytes.

is strictly an abstraction [9], since they are based on the assumption that the viscosity of the liquid medium jumps discontinuously from infinity in the Stern layer to a finite value in the diffuse atmosphere. One possible way to overcome, at least formally, this uncertainty is to assume a gradual variation of the viscosity η from the surface to the beginning of the diffuse part [7, 9], but the quantitative, experimental verification of such a variation is not accessible. Since all treatments on electrokinetic phenomena rely in fact on the existence of the zeta potential, we will not pursue this question any more, and admit the model of viscosity jump as one that works reasonably well. This means that electrokinetic techniques will give us information on the zeta potential, wherever it is located. Trying to extract more information is risky and increasingly model dependent. Dukhin [7] gives a clear physical explanation of the reasons:

1. ψ_d and ζ can be considered identical only at low electrolyte concentrations, while at high concentration, when the double-layer thickness is very much reduced, any attempt to obtain information about ψ_d from electrokinetic measurements requires knowledge of the exact position of the shear plane.
2. The liquid in the immediate vicinity of the surface (the stagnant layer) may have (unknown) rheological properties very different to those in the bulk, and the difference between ψ_d and ζ could be related to that fact. Furthermore, even if ψ_d and ζ can be considered identical, the viscosity and dielectric constant of the liquid likely differ from those of the bulk, and this might affect the electrokinetic behavior.
3. It is possible (and likely) that the surface is far from being molecularly flat: if the typical dimension of the roughness is smaller than the double-layer thickness, κ^{-1}, the standard treatment is not affected by this lack of homogeneity. If, on the contrary, κ^{-1} and the characteristic depth of valleys on the surface are compar-

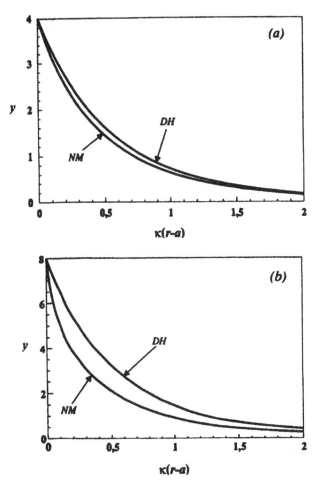

FIG. 3 Dimensionless potential around a spherical particle as a function of the reduced distance to the surface, for monovalent electrolytes. DH: Debye–Hückel approximation; NM: full numerical calculation. In (a), $y_d = 4$, and in (b), $y_d = 8$. (From Ref. 8, with permission of Academic Press.)

able, the discussion of the distinction between ψ_d and ζ (and, for that matter, the interpretation of ψ_d and ζ themselves) can be somewhat meaningless.

In spite of these difficulties, (few) experimental works have indeed approached this core problem of electrokinetics from an experimental point of view. Thus, using a Gouy–Chapman description of the double layer (that is, ignoring the existence of the Stern layer or, in other words, identifying ψ_d and ψ_0), Eversole and coworkers [10, 11] proposed to rewrite Eq. (12) as

$$\ln \tanh\left(\frac{y_\zeta}{4}\right) = \ln \tanh\left(\frac{y_0}{4}\right) - \kappa\beta_\zeta \tag{15}$$

so that measuring the zeta potential as a function of the concentration of an indifferent electrolyte, β_ζ would be the slope of the linear plot of the left-hand side term of Eq. (15) versus κ. In that way, they obtained β_ζ values ranging from 8 to 63 Å. Smith [12] obtained $\beta_\zeta < 5$ Å for rutile and silica, and indistinguishable from zero in AgI suspensions.

Similarly, Lyklema [9] concluded that the identification of ψ_d and ζ in the latter systems is plausible within experimental error.

Furthermore, advances in the theory of electrokinetic phenomena [13–18] have allowed us to relate the observed electrokinetic effects not only to the zeta potential, but also to other parameters of the double layer and to the existence of a Stern layer with ions capable of moving under the action of external fields. This increased number of parameters to be determined require for their estimation a well-planned experimental investigation and often experiments with different electrokinetic techniques.

Coming back to the main topic of the present section, and having discussed the notion of the electrokinetic potential, we will now focus on the different electrokinetic phenomena that can be distinguished by the mobile–immobile phases, the nature of the applied field, and the quantity that must be experimentally determined. In the following sections we will describe them individually in a more thorough way but a brief list of definitions seem to be in order here.

1. *Electrophoresis.* It is the translation of a colloidal particle relative to the liquid under the action of an externally applied field, $\vec{E}_\infty$, constant in time and position independent. Under certain conditions, a linear relationship exists between the steady electrophoretic velocity, $\vec{v}_e$ (attained by the particle just a few microseconds after application of the field), and the applied field:

$$\vec{v}_e = \mu_e \vec{E}_\infty \tag{16}$$

where μ_e is the quantity of interest, known as the electrophoretic mobility.

2. *Sedimentation potential.* It is the potential difference, V_s, sensed by two electrodes placed at a known vertical distance in the suspension, subjected to a gravitational (or equivalently, centrifugal) field, $\vec{g}$. Clearly, the potential will only be generated if the density of the particle ρ_p, and of the dispersion medium, ρ_s, are different.

3. *Electrorotation.* It is the rotational motion of colloidal particles. It is provoked by an applied rotating field, and the quantity of interest is the angular velocity of the particle, $\vec{\Omega}$. It depends on the frequency of the field, ω, and can be zero, positive (i.e., cofield rotation), or negative (counterfield rotation).

4. *Dielectrophoresis.* A nonhomogeneous (spatially varying) electric and harmonically time-alternating field is applied in this case to the particles. They undergo a translational motion towards or away from the high-field region, depending on the value of their induced dipole moment. This translation is known as dielectrophoresis.

5. *Diffusiophoresis.* It is the motion of the suspended particles under the action of an externally applied concentration gradient of the electrolyte solution (or a gradient of solvent composition in nonelectrolyte solution) that constitutes the dispersion medium. The presence of this macroscopic concentration gradient induces a local gradient of electrical potential in the vicinity of the particle, thus provoking a sort of electrophoretic motion. Its reciprocal phenomenon is termed *capillary osmosis*: the concentration gradient-induced electric field sets the liquid in the vicinity of the double layer into motion.

6. *Electro-osmosis.* It is the motion of the liquid adjacent to a charged surface due to an externally imposed electric field. The phenomenon may occur in, for example, flat or cylindrical capillaries, membranes, porous plugs, etc.

7. *Streaming potential and streaming current.* In these phenomena, the motion of the liquid is forced by an applied pressure gradient. The motion of the charged liquid gives rise to an electric current (streaming current) if there is a return path for the charges, or an electrical potential (streaming potential) if the sensing electrodes are connected to a high-input impedance voltmeter (open circuit).

8. *Dielectric dispersion.* The polarization of the EDL, a phenomenon that constitutes a very important contribution to electrokinetic phenomena, determines also the dielectric dispersion of a suspension, i.e., the change with the frequency of an applied AC field of the dielectric permittivity of a suspension of colloidal particles. The phenomenon is dependent on the concentration of particles, their zeta potential, and the ionic composition of the medium, and appears to be very sensitive to most of these quantities.

9. *Electroacoustic phenomena.* They include: ESA (electrokinetic sonic amplitude), which consists in the generation of sound waves under the action of an alternating electric field applied to the suspension, and its reciprocal, CVP (colloid vibration potential), in which a mechanical (ultrasonic) wave is forced to propagate in the system. Electroacoustic phenomena have been very recently studied on a rigorous basis, both experimentally and theoretically.

The list could be made longer, taking the idea of electrokinetics in a wide sense (response of the colloidal system to an external field that reacts differently to particles and liquid). Thus, we could include:

- *Electroviscous effects*: the viscosity of a suspension in its Newtonian range is different from that of the dispersion liquid, and depends on the surface potential of the particles, and on the ionic characteristics of the medium.
- *DC conductivity*: the DC conductivity of the suspension is also dependent on the properties of the solid/liquid interfaces.
- *Electric birefringence*: the torque exerted by an external field on anisotropic particles will provoke their orientation. This affects the refractive index of the suspension. The variation of the refractive index with the frequency of the field, if it is alternating, is related to the double-layer characteristics.

We will now give some ideas about the simplest theoretical explanation for the observed phenomena, as well as on the experimental techniques available. The treatment will not be excessively detailed, to avoid overlapping with the other chapters of this book. In the following sections we will proceed with such an overview.

IV. ELECTROPHORESIS

A. Simple Theory

Assume that we have a spherical particle of radius a in the presence of an electric field that, far from the particle, equals $\vec{E}_\infty$. The particle is considered to be nonconducting and with a dielectric permittivity much smaller than that of the dispersion medium. For the moment, we will also assume that the electrolyte concentration is very low and that a is also very small, so that the following inequality holds between the double-layer thickness [Eq. (11)] and the radius:

$$\kappa^{-1} \gg a \quad \text{or} \quad \kappa a \ll 1 \tag{17}$$

that is, we are in the thick double layer (or Hückel) approximation. Because the ionic atmosphere extends over such long distances, the volume charge density inside it will be very small, and the applied field will hence provoke almost no liquid motion around the particle. As a consequence, the only forces acting on the latter are Stokes' drag ($\vec{F}_s$) and electrostatic ($\vec{F}_E$). Since the particle moves with constant velocity (the electrophoretic velocity, $\vec{v}_e$), the net force must vanish:

$$\vec{F}_s = -6\pi\eta a \vec{v}_e$$
$$\vec{F}_E = Q\vec{E}_\infty \tag{18}$$
$$\vec{F}_s + \vec{F}_E = 0$$

In these equations, η is the viscosity of the dispersion medium, and Q is the total surface charge on the particle. From Eq. (18):

$$\vec{v}_e = \frac{Q}{6\pi\eta a}\vec{E}_\infty \tag{19}$$

Now, remembering [19] that the potential on the surface, under condition (17), is

$$V(a) = \frac{1}{4\pi\varepsilon_{rs}\varepsilon_0}\frac{Q}{a} \tag{20}$$

the identification of $V(a)$ with the zeta potential, ζ, leads to

$$\vec{v}_e = \frac{2}{3}\frac{\varepsilon_{rs}\varepsilon_0}{\eta}\zeta\vec{E}_\infty \tag{21}$$

or the electrophoretic mobility, μ_e:

$$\mu_e = \frac{2}{3}\frac{\varepsilon_{rs}\varepsilon_0}{\eta}\zeta \tag{22}$$

which is the well-known Hückel formula.

Let us now consider the opposite situation, for which an analytical solution exists, i.e., a thin double-layer approximation:

$$\kappa^{-1} \ll a \quad \text{or} \quad \kappa a \gg 1 \tag{23}$$

In this case, the surface charge is screened by double layer ions in a short distance, what means that, as described in Section III, electroneutrality is lost in that region. The field will hence provoke motions of the charged liquid that affect the particle motion itself: the solution is in this case somewhat more complicated, but still achievable.

For the moment, we still assume a spherical particle with a constant surface potential, ζ; another important assumption is that the surface conductivity of the EDL is small enough to have negligible influence on the field-induced potential distribution. The problem is better solved if a reference system is used that is centered in the sphere. Since in the laboratory system the liquid does not move far from the particle, the use of the co-ordinate system fixed to the particle will yield a liquid velocity equal to $-\vec{v}_e$ at long distances.

Figure 4 is a scheme of the situation described. Under the assumption that the double layer is everywhere thin, the tangential velocity of liquid at a distance r from the particle's surface, which is large enough when compared to the EDL thickness but, simultaneously, is small when compared to the particle radius ($a \gg r - a \gg \kappa^{-1}$), can be estimated if it is known for a flat interface, and thus it is much easier than the general problem. In fact, the tangential velocity distribution of the liquid with respect to a solid

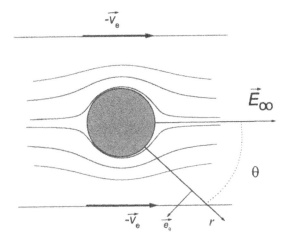

FIG. 4 Scheme of the liquid flow around a negatively charged spherical particle. Far from the interface, the liquid moves with constant velocity $-\vec{v}_e$.

flat interface with a potential equal to ζ, under the action of a tangential field E_t [2, 4, 13] (see also Section IX below) is

$$v_t(x) = -\frac{\varepsilon_r \varepsilon_0}{\eta}[\zeta - \psi(x)]E_t \tag{24}$$

where x is the distance taken outwards of the interface, and $\psi(x)$ is the equilibrium potential distribution in the double layer (in the framework of the linear approximation with respect to an external field, it is assumed that the double layer retains its equilibrium configuration in the presence of a field that is weak enough). Equation (24) describes in fact the x-distribution of electro-osmotic velocity, and for distances exceeding the double-layer thickness, when $\psi(x) \rightarrow 0$, the velocity is called the electro-osmotic slipping velocity, $\vec{v}_s$:

$$\vec{v}_s = -\frac{\varepsilon_{rs} \varepsilon_0}{\eta}\zeta \vec{E}_t \tag{25}$$

Using Eq. (25) outside any portion of the quasiflat double layer of our spherical particle, would allow us to write:

$$\vec{v}_s(\theta) = -\frac{\varepsilon_{rs} \varepsilon_0}{\eta}\zeta \vec{E}_t(\theta) \tag{26}$$

where $\vec{v}_s(\theta)$ is the tangential velocity of the liquid near the particle's surface outside a thin double layer in a position characterized by the angle θ with the field direction, where the tangential field is $\vec{E}_t(\theta)$.

Now, the potential $\delta\psi(\vec{r})$ due to the external field around a nonconducting sphere in a conducting medium is given by [19, 20]

$$\delta\psi(\vec{r}) = -E_\infty r \cos\theta - \frac{E_\infty a^3}{2r^2}\cos\theta \tag{27}$$

(recall that it is assumed that there is no double-layer deformation or relaxation effect) and hence the tangential field:

$$E_t(\theta) = -\frac{1}{r}\frac{\partial \delta\psi}{\partial \theta}\bigg|_{r=a} = -\frac{3}{2}E_\infty \sin\theta \tag{28}$$

Since the net force on the liquid is zero, and no pressure gradient is applied, the liquid flow must have a potential nature [21]:

$$\vec{v}(\vec{r}) = -\nabla\Phi(\vec{r}) \tag{29}$$

where Φ is the velocity potential. Note that Eq. (29) is a solution of Navier–Stokes equation:

$$\eta \nabla \times \nabla \times \vec{v}(\vec{r}) - \nabla p + \vec{f} = 0 \tag{30}$$

for the case $\nabla p = 0$ (no pressure gradient) and $\vec{f} = 0$ (no external body force on the fluid). Furthermore, Eq. (29), together with the condition of incompressibility of the liquid, $\text{div}\,\vec{v} = 0$, leads us to a Laplace equation for the velocity potential:

$$\nabla^2\Phi = 0 \tag{31}$$

The normal velocity of the liquid (i.e., the radial component of $\nabla\Phi$) must be zero at the surface of the particle:

$$\nabla_r\Phi|_{r=a} = 0 \tag{32}$$

Equations (31) and (32) have their equivalent, formally identical versions, for the electric potential:

$$\nabla^2\delta\psi = 0$$
$$\nabla_r\delta\psi|_{r=a} = 0 \tag{33}$$

hence, the velocity potential outside the double layer must have a position dependence similar to that of $\delta\psi$ [Eq. (27)]:

$$\Phi = -\vec{v}_\infty \cdot \vec{r} - \frac{1}{2}\frac{a^3}{r^3}\vec{v}_\infty \cdot \vec{r} \tag{34}$$

where $\vec{v}_\infty$ is the uniform velocity of the liquid far from the particle (like $\vec{E}_\infty$ was the uniform external field). Now, using Eqs (26), (28), and (34):

$$-\frac{1}{r}\frac{\partial\Phi}{\partial\theta}\bigg|_{r=a} = -\frac{3}{2}v_\infty \sin\theta = v_s = -\frac{\varepsilon_{rs}\varepsilon_0}{\eta}\zeta\left[-\frac{3}{2}E_\infty \sin\theta\right] \tag{35}$$

Equations (27), (34), and (35) reflect the geometrical similarity of the distributions of electrical potential and velocity potential with a coefficient of similarity equal to $(-\varepsilon_r\varepsilon_0/\eta)\zeta$ and hence the velocity of liquid far from the particle would be

$$\vec{v}_\infty = -\frac{\varepsilon_{rs}\varepsilon_0}{\eta}\zeta\vec{E}_\infty \tag{36}$$

Equation (36) is known as the Smoluchowski equation for electro-osmosis.

Returning to the laboratory system, the electrophoretic velocity of the particle would be $\vec{v}_e = -\vec{v}_\infty$:

$$\vec{v}_e = \frac{\varepsilon_{rs}\varepsilon_0}{\eta}\zeta\vec{E}_\infty \tag{37a}$$

and from here follows the Smoluchowski formula for the electrophoretic mobility [2]:

$$\mu_e = \frac{\varepsilon_{rs}\varepsilon_0}{\eta}\zeta \tag{37b}$$

From our deduction above, it will be clear that Eq. (37b) is valid whatever its geometry provided that [22, 23]:

1. The disperse particle acquires some surface charge, compensated for by an excess of charge of the opposite sign in the medium.
2. The particle is rigid and of arbitrary shape with uniform surface electrical potential, ζ, with respect to the liquid far from the interface.
3. The particle dimensions are such that the curvature radius of the interfce at any position is much larger than the double-layer thickness.
4. The particle is nonconducting.
5. No surface conductance effects are essential.
6. The dielectric constant and viscosity of the medium are everywhere the same (see, however, Refs 23–26).
7. The applied field, although distorted by the presence of the particle, is vectorially added to the local, equilibrium double-layer field.

B. More Elaborate Treatments

Henry [27] was the first author that eliminated the restriction 3 above, and solved the problem for spheres (also for infinite cylinders) of any radius, a, i.e., of any κa value, although for small zeta potentials, since it is assumed that Eq. (14) holds for the potential distribution in the equilibrium double layer. Restricting ourselves to the case of spheres, Henry's equation for nonconducting particles reads:

$$\mu_e = \frac{2}{3}\frac{\varepsilon_{rs}\varepsilon_0}{\eta}\zeta f(\kappa a) \tag{38}$$

where

$$f(\kappa a) = 1 + \frac{(\kappa a)^2}{16} - 5\frac{(\kappa a)^3}{48} + \cdots \tag{39}$$

and an approximate formula for $f(\kappa a)$ has been given by Ohshima [20].

Another key contribution to the understanding and evaluation of electrophoretic mobility and, in general, of the physics underlying electrokinetic phenomena is due to Overbeek [23, 28]; Booth [29–31] also produced a theory that followed similar lines – for spheres in both cases. These authors first considered that during the electrophoretic migration the double layer loses its original symmetry, and becomes polarized: the nonequilibrium potential distribution is no longer the simple addition of that created by the external field around the nonconducting sphere and that of the equilibrium EDL [32]. The mathematical problem in hand is now much more involved, and until the advent of computers only approximate theories (low ζ, large κa: see Refs 23, 28, and 29–31) were available. The first (numerical) treatments of the problem, valid for arbitrary values of the radius, the zeta potential, or the ionic concentrations, were elaborated by Wiersema et al. [33] and O'Brien and White [34].

This is not the proper place to describe these treatments, so we simply show some results in Fig. 5. The validity of Smoluchowski formula for large κa and low to moderate zeta potentials is clearly observed; it is also evident that Henry's treatment is valid for low ζ irrespective of the double-layer thickness.

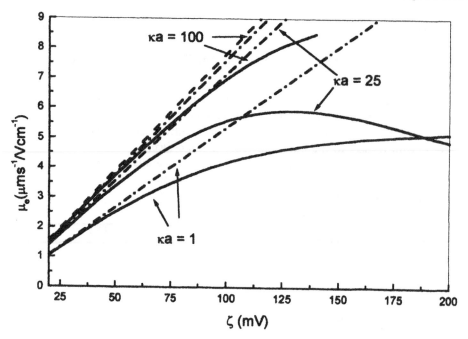

FIG. 5 Electrophoretic mobility vs. zeta potential for spherical particles of radius $a = 100$ nm and for $\kappa a = 1$, 25, and 100 in KCl solutions. Dashed line: Smoluchowski equation; dashed–dotted lines: Henry formula [27]; solid lines: O'Brien and White's theory [34].

C. Experimental Determinations

1. Moving Boundary

Nowadays it is a seldom used method, but we would like to mention it here, mainly because it is very intuitive and also because of its historical role in μ_e determinations [35]. Figure 6 is a simple view of the design proposed in Ref. 36. In essence, electrophoretic mobility determinations involve placing the suspension in a U-shaped tube and on top of it, a liquid without particles with the same electrolyte concentration as the suspension. If a DC field is applied between both ends of the tube, electrophoretic migration provokes a movement of both boundaries towards the electrode with polarity opposite to that of the particles. The method presents certain drawbacks [36]:

- The motion of both boundaries might be different, mainly if the system has low ionic strength or if the suspension is concentrated.
- It is difficult to apply to suspensions with low concentration of particles.
- The limits between the solution and the suspension may become blurred, because of the mixing of electrolyte and particles in the boundaries.

These are some of the reasons why the method ceased to attract the attention of colloid researchers in recent years.

2. Mass Transport Electrophoresis

Since the mass/charge ratio of a colloidal particle is normally much larger than for an ion, a method in which the mass transported by electrophoresis can be measured appears to be

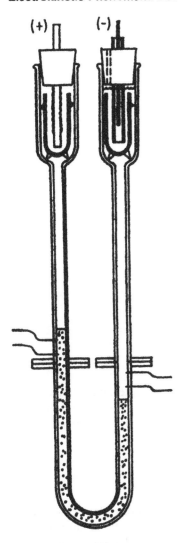

FIG. 6 A moving boundary electrophoresis setup. (From Ref. 36, with permission of Academic Press.)

very suitable for supensions. This is the so-called mass-transport method (Fig. 7), in which interest was lost after the 1980s; but it reached some popularity in the 1960s and 1970s, even at the commercial level, since there were some commercially available devices [37–39]. Some problems can also be mentioned with this method [39]:

- Particles can deposit on the electrodes, and affect the current flow through suspensions.
- Sedimentation of the dispersed phase may prevent its transport from the storage to the measuring chambers (Fig. 7).
- The density of the particles should be as large as possible, compared to that of the medium.

Furthermore, the method works best with concentrated suspensions, for which the electrophoretic mobility may differ from that of dilute systems. The treatment of the

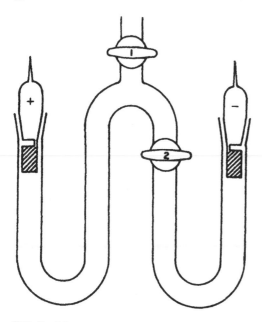

FIG. 7 Mass transport electrophoresis cell, according to Homola and Robertson. (From Ref. 39, with permission of Academic Press.)

mobility of concentrated suspensions is out of the scope of this chapter, but some models have been elaborated, mainly under simplifying assumptions such as a thin double layer or low zeta potentials [40–45]. In any case, no simple equations are available for the mobility–zeta relationship in this case.

3. Microelectrophoresis

This is probably the most widespread method: it is based on the direct observation, with suitable magnifying optics, of individual particles in their electrophoretic motion. In fact, it is not the particle that is seen but its scattering pattern when illuminated in a dark background field, as illustrated in Fig. 8. The method has some advantages, including [24, 46]:

1. The particles are directly observed in their medium.
2. The suspensions to be studied can be (actually, they should be) dilute: if they are not, the view through the microscope would be a sort of fog, where individual particles cannot be identified. However, with dilute systems, the aggregation times are very large, even in the worst conditions, so that velocities can be measured.
3. If the suspension is polydisperse, the observer can identify a (somewhat wide) size range of particle sizes to track.

Its main disadvantage, however, is precisely related to the bias and subjectivity of the observer, who can easily select only a narrow range of velocities, which might be poorly representative of the true average value of the suspension. Hence, some manufacturers (see a few websites in Ref. 47) have modified their designs to include automatic tracking by

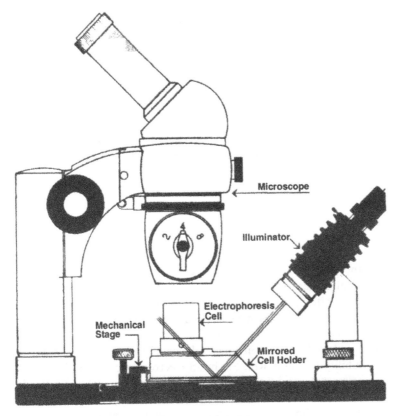

FIG. 8 Typical components of a microelectrophoresis setup. (Courtesy of Zeta-Meter, Inc., USA.)

digital image processing: the observer's eye is substituted by a video camera and a computer.

Another source of error is the location of the so-called *stationary level*: if the electrophoresis channel is cylindrical, electro-osmotic flow in the channel walls will provoke a velocity distribution in the hydrodynamically closed cylinder [2] given by

$$v_L = v_{eo} \left[2\frac{r^2}{R^2} - 1 \right] \tag{40}$$

where v_{eo} is the electro-osmotic liquid velocity close to the wall, R is the capillary radius, and r is the radial distance from the cylinder axis. From Eq. (40), it is clear that $v_L = 0$ if $r = R/\sqrt{2}$, so that the true electrophoretic velocity will be displayed only by particles moving in a cylindrical shell placed at $0.292R$ from the channel wall. It is easy to estimate the uncertainties associated with errors in the measuring position: if $R \sim 2\,\text{mm}$ and the microscope has a focus depth of $\sim 50\,\mu\text{m}$, then an uncertainty $\geq 2\%$ in the velocity will always be present.

These arguments apply also to electrophoresis cells with rectangular or squared cross sections. The electro-osmotic flow profile is in this case more involved, but it can be approximated by [48, 49]

$$v_L(z) = A v_{eo}(z/b)^2 + \Delta v_{eo}(z/b) + (1 - A)v_{eo} \tag{41a}$$

where z is the vertical position of the observation point ($z = 0$ at the channel center), $2b$ is the vertical dimension of the channel, v_{eo} is the average electro-osmotic velocity of the upper and lower cell walls, Δv_{eo} is the difference between the electro-osmotic velocities at the upper and lower cell walls, and

$$A = \left(\frac{2}{3} - \frac{0.420166}{K}\right)^{-1}, \quad K = a/b \tag{41b}$$

a being the horizontal dimension of the channel. Note that Eqs (41) and (42) can be simplified if $a \gg b$, and $\Delta v_{eo} = 0$. A parabolic velocity profile can be obtained:

$$v_L(z) = \frac{v_{eo}}{2}\left[3\left(\frac{z}{b}\right)^2 - 1\right] \tag{42}$$

so that the stationary levels are situated at $z = \pm 0.577b$.

Some authors have suggested (see Chap. 9 for details) that a procedure to avoid this problem would be to cover the cell walls, whatever the geometry, with a layer of uncharged chemical species, for instance, polyacrylamide. However, it is possible that the layer becomes detached from the walls after some usage, and this would mask the electrophoretic velocity, measured at an arbitrary depth with an electro-osmotic contribution, the absence of which can only be ascertained by measuring μ_e of standard, stable particles, which in turn remains an open problem in electrokinetics.

Recently, Minor et al. [50] analyzed the time dependence of both the electro-osmotic flow and electrophoretic mobility (see also Ref. 51) in an electrophoresis cell. They concluded that, for most experimental conditions, the colloidal particle reaches its steady motion after the application of an external field in a much shorter time than electro-osmotic flow does. Hence, if electrophoresis measurements are performed in an alternating field with a frequency much larger than the reciprocal of the characteristic time for steady electro-osmosis ($\tau \sim 10^0$ s), but smaller than that of steady, electrophoresis $\tau \sim 10^{-4}$s, the electro-osmotic flow cannot develop. In such conditions, electro-osmosis is suppressed, and the velocity of the particle is independent of the position in the cell (Fig. 9 is an

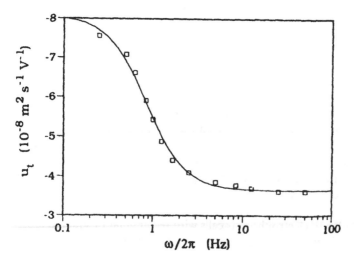

FIG. 9 Electrophoretic mobility of particles as a function of the frequency of the applied field. (From Ref. 50, with permission of Academic Press.)

example). Another way to overcome the electro-osmosis problem is to place both electrodes, providing the external field, inside the cell, completely surrounded by electroneutral solution; since no net external field acts on the charged layer close to the cell walls, the associated electro-osmotic flow will not exist [52].

4. Electrophoretic Light Scattering (ELS)

Since the 1970s, visual microelectrophoresis techniques have been increasingly replaced by automated methods based on the analysis of the (laser) light scattered by electrophoresing particles [53–56]. These are known as electrophoretic light scattering (ELS) methods [57] and have different principles of operation. In some cases, heterodyne optics is used: the light scattered by the moving particles is mixed with a reference beam, and due to the Doppler shift of the scattered light, a beat pattern is produced, the frequency of which, f_e, can be related to the electrophoretic velocity. Such a frequency can be measured by means of a spectrum analyzer or by analysis of the correlation function of the scattered light. Knowledge of f_e leads immediately to that of the velocity since

$$v_e = \frac{2\pi f_e}{E_\infty k \cos \alpha} \tag{43}$$

where k is the modulus of the scattering vector $\vec{k}$ (the difference between the wave vectors of scattered and incident light) and α is the angle between $\vec{k}$ and the direction of electrophoretic migration.

Other instruments are based on the formation of optical fringes in a properly selected (according to the location of the stationary level) volume of the cell: the laser beam is divided into two beams that interfere in that volume, provoking a fringe pattern oriented perpendicular to the direction of electrophoretic motion. When the particles move through the fringes, they will go alternatively through light and darkness, and thus the scattered intensity will fluctuate at a rate depending on the electrophoretic velocity.

More recently, a new technique has become available, also commercially: it is called phase analysis light scattering or PALS [58–60]. It is especially suited for particles moving with very low electrophoretic velocities, i.e., close to the isoelectric point, or in nonpolar solvents, where very low mobilities are attained by the particles. The key of the method is the use of moving fringes of the interference pattern of the laser beams mentioned above: thus, if one of the laser beams is frequency shifted by a well-defined frequency, ω_s, and a colloidal particle is stationary at the probed volume, the scattered light will be a sinusoidal wave with frequency ω_s [58]. The phase difference between scattered and reference beams will be constant in that case. However, if the particle undergoes electrophoretic motion, that phase difference will change with time in such a way that the mean phase change includes information about the electrophoretic motion. Note that if the particle traverses only half a fringe separation (typically below 1 μm), the phase shift would be π radians, and since it is possible to measure much smaller phase differences (well below 0.1 rad), the precision of the method is much higher than in standard laser Doppler velocimetry. Figure 10 is a scheme of the experimental device (a commercial version is available, see Ref. 57). As observed, the frequency shift is produced by Bragg cells driven by a single-side band modulator to produce beams with frequencies of 80 MHz and 80 MHz $+ f_s$ (f_s is the shift frequency, between 1 and 10 kHz). The method is capable of detecting electrophoretic mobilities as low as 10^{-12} m^2 V^{-1} s^{-1}, that is, 10^{-4} μm s^{-1}/V cm^{-1} in practical mobility units.

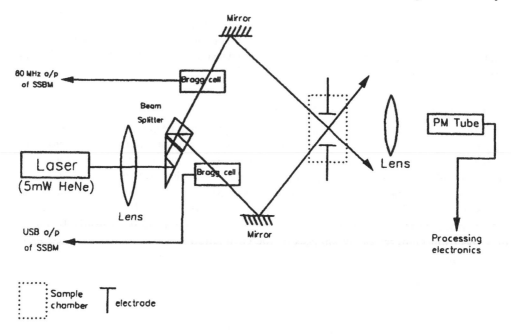

FIG. 10 The PALS (phase analysis light scattering) electrophoresis technique. (From Ref. 58, with permission of Academic Press.)

V. SEDIMENTATION POTENTIAL (DORN EFFECT)

A. Physical Principles

When a colloidal particle has a density different from that of the surrounding liquid, sedimentation (or buoyancy) will take place. The presence of the double layer gives further rise to the generation of an electric field that, summed over all the particles (if their average separation is larger than their size), generates the sedimentation (or flotation) potential. This is the Dorn effect, a simplified theory of which will be described below [7, 61]; the detailed description of these phenomena can be found in Chapter 16 of this volume.

When the particle is falling, the flow of liquid around it will alter the spatial distribution of double-layer charges (Fig. 11): the normal fluxes of counterions (cations) and coions (anions) carried upwards by the liquid are roughly identical, as they are due to the convective motion of electroneutral solution. However, the double layer is enriched in cations (we assume a negative surface charge); hence, their tangential flux will be much larger than that of anions, so the lower pole of the particle will be enriched in anions, which are in very low amounts around the particle surface. By the same token, positive ions will accumulate at the upper end, thus originating a dipole oriented against the gravitational field $\vec{g}$ (the orientation would be in the same direction as $\vec{g}$ for a positive particle). The electric field generated by this dipole will extend beyond the limits of the double layer, since the potential generated will be

$$\psi^{\mathrm{d}} = -\frac{1}{4\pi\varepsilon_{\mathrm{rs}}\varepsilon_0}\,\frac{\vec{d}\cdot\vec{r}}{r^3} \tag{44}$$

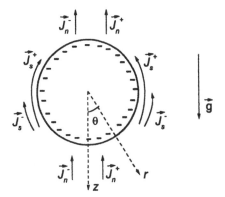

FIG. 11 Diagrammatic illustration of the charge fluxes (normal, $\vec{J}_n^{\pm}$, and tangential $\vec{J}_s^{\pm}$) around a sedimenting particle.

$\vec{d}$ being the dipole moment induced by the sedimentation of the charged particle. If n is the number of particles per unit volume, and they are considered identical, the total field, $\vec{E}_{sed}$ can be calculated as [13]

$$\vec{E}_{sed} = -\frac{n\vec{d}}{\varepsilon_{rs}\varepsilon_0} \tag{45}$$

Thus, obtaining $\vec{d}$ will suffice to calculate the sedimentation potential. This can be done [13] by a continuity equation for the current density on the double-layer limit (which can be assumed to coincide with the solid surface $r = a$ for a thin double layer). At any point (any θ value) of the surface $r = a$, the normal current is given by

$$J_n(\theta) = -K^{\infty}\frac{\partial \psi^d(r,\theta)}{\partial r}\bigg|_{r=a} \tag{46}$$

whereas the corresponding convective surface current I_s will be:

$$I_s = \int_0^{\infty} \rho(x)v_{\theta}(x)\mathrm{d}x \tag{47}$$

where $v_{\theta}(x)$ is the tangential component of the fluid velocity [13]:

$$v_{\theta} = -\frac{3}{2}\frac{U_{sed}x}{a}\sin\theta \tag{48}$$

with $x = r - a$. U_{sed} is the limiting sedimentation velocity of a sphere of radius a, and density ρ_p in a liquid of density ρ_m:

$$U_{sed} = \frac{2}{9}\frac{a^2(\rho_p - \rho_m)}{\eta}g \tag{49}$$

In Eq. (47), $\rho(x)$ is the charge density in the thin double layer, at distance x from the slipping plane; using Poisson's equation for a flat interface:

$$\rho(x) = -\varepsilon_{rs}\varepsilon_0\frac{\mathrm{d}^2\Psi}{\mathrm{d}x^2} \tag{50}$$

Integration of Eq. (47) after substitution of Eqs (48)–(50) leads to

$$I_s = \frac{3\varepsilon_{rs}\varepsilon_0\zeta \sin\theta}{2a} \tag{51}$$

where use has been made of the equality $\Psi(x = 0) = \zeta$.

In steady state, the change in the charge convectively transported from θ to $\theta + \Delta\theta$ must equal the normal charge flow. This boundary condition is equivalent [13] to

$$\nabla_S(I_S) = K^\infty \left.\frac{\partial\psi^d}{\partial r}\right|_a \tag{52}$$

where ∇_s is the surface divergence operator:

$$\nabla_s(\cdot) = \frac{1}{a\sin\theta}\frac{\partial}{\partial\theta}(\sin\theta\cdot)$$

Substitution of Eqs (44) and (51) into Eq. (52) yields the following expression for the induced dipole moment:

$$\vec{d} = 6\pi \frac{\varepsilon_{rs}^2\varepsilon_0^2\zeta\vec{U}_{sed}a}{K^\infty} \tag{53a}$$

and hence, using Eq. (48):

$$\vec{d} = \frac{4}{3}\frac{\pi\varepsilon_{rs}^2\varepsilon_0^2\zeta a^3\vec{g}}{K^\infty\eta}(\rho_p - \rho_m) \tag{53b}$$

The final expression for the sedimentation potential according to Smoluchowski is obtained by using Eqs (45) and (53b), and taking into account that the volume fraction of solids is $\phi = 4\pi a^3 n/3$:

$$\vec{E}_{sed} = -\frac{\varepsilon_{rs}\varepsilon_0\zeta\vec{g}}{K^\infty\eta}(\rho_p - \rho_m)\phi \tag{54}$$

Using reasonable values for the quantities involved ($\zeta = 100\,mV$, $\rho_p - \rho_m = 10^3\,kg\,m^{-3}$, $K^\infty = 150\,\mu S\,cm^{-1}$, $\eta = 0.8904 \times 10^{-3}\,Pa\cdot s$, $\phi = 0.1$), $|\vec{E}_{sed}|$ would be of the order of $\sim 5\,mV\,m^{-1}$. This means that the best signal-to-noise ratio in a sedimentation potential measurement is obtained when the sensing electrodes are sufficiently far apart and the particle concentration is rather large.

Furthemore, the Smoluchowski equation is only valid for particles with thin double layers and negligible surface conductance (low zeta potentials). Hence, the theory was later generalized to arbitrary κa values by Booth [62] for low zeta potentials and was developed for arbitrary ζ by Stigter [63] and Ohshima et al. [64]. Considering the fact that rather concentrated suspensions are often used in sedimentation potential determinations [65], theories have also been generalized to this situation [66]. Finally, Carrique et al. [67] have also analyzed the effect of the presence of a dynamic Stern layer on the sedimentation potential.

Figure 12 shows some examples in which full numerical results are compared to the Smoluchowski equation, Eq. (54). Note that, as expected, the Smoluchowski approach is less valid the higher the zeta potential and the thicker the double layer.

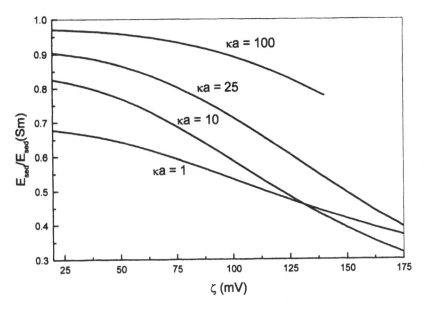

FIG. 12 Full calculation of the sedimentation field (E_{sed}) of spherical particles of 100 nm radius, as compared to Smoluchowski formula [E_{sed}(Sm)], as a function of zeta potential, for the κa values indicated.

B. Experimental Determinations

The reader is referred to Chapter 16 in this book for the latest methods of sedimentation and flotation potential determinations. Here, we will illustrate the principles of measurement with reference to the device used in our laboratory [68], schematically shown in Fig. 13. In the upper tube (A) the suspension to be sedimented is stored, and (B) is the sedimentation cell itself. The sedimentation potential is determined when the optical system (lenses L, and photocells C) detects a uniform particle concentration between both electrodes. The same system is used to determine the particle concentration, n: to that end, the mass M of particles collected in flask F is weighed, and the total sedimentation time, t_s, is recorded. The mass m of particles between the electrodes during measurements of E_{sed} is given by $m = (M/t_s)t_f$, where t_f, the time of flow, is the time needed for the particles to move between the electrodes.

VI. ELECTROROTATION

As shown below, the dielectric behavior of suspensions, associated with the polarizability of the individual particles and their double layers, can be determined by impedance spectroscopy. The change with frequency of the impedance of the suspension can be related to the real and imaginary parts of the dipole moments induced in the colloidal units by the external field: as the frequency of the latter is increased, some of the polarization mechanisms can follow the external field, but others relax and disperse. If data are acquired over a sufficiently wide frequency range, a rich information can be obtained on the internal structure of the particles, and on their double layers.

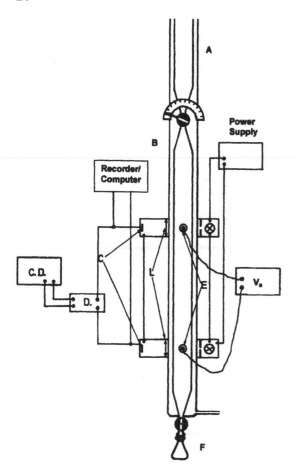

FIG. 13 Experimental setup for sedimentation potential measurements.

An alternative way of performing this analysis is by electrorotation [69–71], consisting in the rotation of the particles under the action of a rotating external field. A classical explanation of the phenomenon can be given as follows: the applied field induces a dipole moment in the particle, which rotates at the same frequency as the field rotates; the dispersion processes provoke a phase shift between the field and the induced dipole and, as a consequence, a torque $\vec{M}$ acts on the particle, given by

$$M = -4\pi\varepsilon_{rs}\varepsilon_0 E_0^2 \mathrm{Im}C_0^* \tag{55}$$

where E_0 is the applied field strength, and C_0^* is the (complex) dipole coefficient, related to the dipole moment $\vec{d}^*$ induced by the external field $\vec{E}$ as follows:

$$\vec{d}^* = 4\pi\varepsilon_{rs}\varepsilon_0 a^3 C_0^* \vec{E} \tag{56}$$

From Eq. (55) it can be shown that the angular velocity of the particle rotation is given by [72, 73]

$$\Omega = -\frac{\varepsilon_{rs}\varepsilon_0 E_0^2}{2\eta}\,\mathrm{Im}(C_0^*) \tag{57}$$

where η is the viscosity of the dispersion medium and a is the particle radius.

This simple model is, however, incomplete: while a good agreement between theory and experiment was found by Arnold et al. [74] for a frequency of the applied field above $\sim 100\,\mathrm{kHz}$, in the range of interest from the electrokinetic point of view (the α-dispersion range, typically between $100\,\mathrm{Hz}$ and $10\,\mathrm{kHz}$) a significant disagreement was found: the particle should rotate in the opposite direction to that of the field rotation, contrary to experimental findings, systematically demonstrating cofield rotation. Grosse and Shilov [72] and later Zimmerman et al. [73] first explained the discrepancy. They used an interesting parallelism with the theory of electrophoresis: in the presence of a static field, the particle experiences electrophoretic motion in spite of the fact that the total force on the particle plus its double layer is zero because of the electroneutrality of the whole. The motion occurs in fact by the electro-osmotic slipping of the liquid with respect to the particle surface. The same idea is applied to obtaining an electro-osmotic component of the electrorotation of the particle. The authors apply these ideas to two different models of EDL polarization: the Schwarz or surface diffusion model (which assumes that counterions can redistribute on the particle surface, but cannot exchange with the bulk electrolyte solution) and the standard model, in which double-layer polarization involves diffuse-layer deformation and subsequent formation of a gradient of electrolyte concentration (concentration polarization) that extends over distances comparable to the particle radius. Figure 14 is a proof of the good agreement between theoretical results and the experimental data of Ref. 74.

The experimental determination of electrorotation velocity is not easy, hence its scarce (but increasing) application in colloid science. Since the frequency of electrorotation depends on the applied field strength [Eq. (57)] and the hydrodynamical resistance, it is possible to select E_0 such that the angular velocity is around or below $1\,\mathrm{s}^{-1}$: this allows one to observe the rotation directly with a video-microscope system [75]. The rotating field is

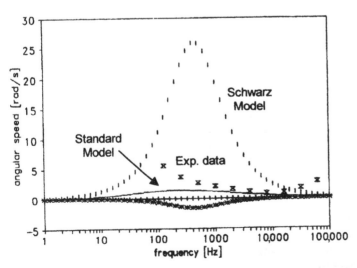

FIG. 14 Electrorotation speed of a 2.65-μm particle in a rotating field of $9580\,\mathrm{V\,m}^{-1}$, as a function of the frequency of the field. The experimental data are compared to the Schwarz model and the standard electrokinetic theory. The crosses and asterisks correspond to the contribution of the electro-osmotic torque, calculated by means of the two models cited. (From Ref. 72, Copyright 1998, with permission of Elsevier Science.)

produced by four needle electrodes driven by square-wave signals with 90° phase shift. Figure 15 is an example of the experimental data obtained in Ref. 75 on red blood cells: note that using a microscopic technique requires particles to be large enough (several micrometers in diameter) and nonspherical.

Recently, the so-called electrorotational light scattering (ERLS) techniques have been introduced (see, e.g., Refs. 76, 77): in this case, electrorotation is combined with dynamic light scattering. By analyzing the autocorrelation function of the light scattered by an ensemble of particles undergoing electrorotation, it is possible to gain information on their rotation frequency. Figure 16 is an example of the modifications, introduced by Gimsa et al. [76], in the optical chamber of a commercial dynamic light-scattering device (N4-MD, Beckman/Coulter) to accommodate the four electrodes producing the electric field.

Let us finally mention that electrorotation has found wide application in the field of biology [78–80], since it has been demonstrated that this phenomenon is very sensitive for monitoring the viability of biological cells [81]. A very interesting example is shown in Fig. 17, where a group of oocysts of *cryptosporidium parvum* (responsible for human infection when present in drinking water) is shown during electrorotation: viable oocysts rotated counterclockwise whereas nonviable ones rotated clockwise.

VII. DIELECTROPHORESIS

In this electrokinetic phenomenon, the applied electric field is spatially nonhomogeneous, and this causes the translation (*dielectrophoresis*) of the polarized particle [69, 82]. Measurements involve microscopically analyzing the motion of the particles as a function of the frequency of the applied field. The particles will move towards or away from the high-field region, depending on the relative directions of their induced dipole moment and

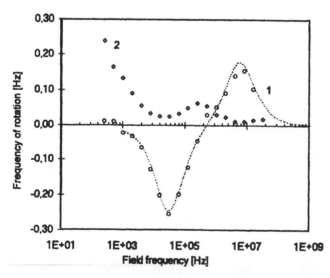

FIG. 15 Electrorotation speed of red blood cells as a function of the frequency of the applied field. 1: Native cells in dilute PBS and sucrose with a conductivity of $2.6\,\mathrm{mS\,cm^{-1}}$ and a voltage of $7.5\,\mathrm{V}$ between the electrodes; 2: fixed red blood cells in a solution with conductivity $2.9\,\mathrm{mS\,cm^{-1}}$ and $15\,\mathrm{V}$ applied. (From Ref. 75, Copyright 1998, with permission of Elsevier Science.)

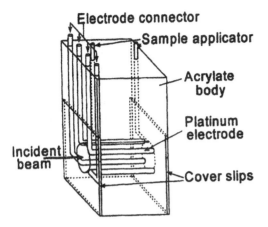

FIG. 16 Cuvette for electrorotational light scattering. The rotating field is perpendicular to the laser beam. (From Ref. 76, Copyright 1995, with permission of Elsevier Science.)

the field. The change in the velocity of motion with the frequency of the alternating field is thus due to the different frequency dependencies (dispersions) of the dielectric constants of the particles (with their double layers) and the medium. This can be made explicit by writing the expression for the dielectrophoretic force $\vec{F}_{\mathrm{DP}}$ acting on the particles [70, 73] in terms of the induced dipole coefficient:

$$\vec{F}_{\mathrm{DP}} = 2\pi\varepsilon_{\mathrm{rs}}\varepsilon_0 a^3 \, \mathrm{Re}(C_0^*)\nabla E_{\mathrm{rms}}^2 \tag{58}$$

where E_{rms} is the root-mean-square amplitude of the applied field. From Eq. (58), it is clear that the main quantities describing the dielectrophoretic phenomena are the induced

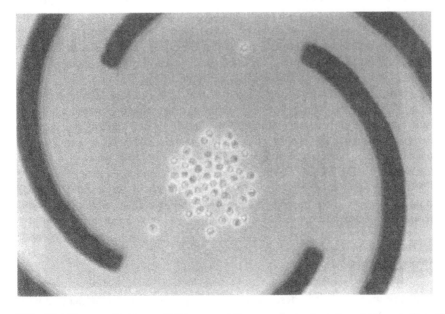

FIG. 17 Picture of oocysts of *Cryptosporidium parvum* undergoing electrorotation. (From Ref. 81, with permission of the Institute of Physics Publishing.)

dipole moment of the particles and the spatial variation of the field. The classical electrokinetic theory was applied to the calculation of C_0^* by Baygents [83]. It must be pointed out that, as in the case of electrorotation, calculation of the particle velocity does not only involve knowledge of the induced dipole moment: a generalized theory [84] should also take into account that the external force acts on both the particle and the surrounding liquid, and that part of the force acting on the latter is transmitted to the former.

Wang and coworkers [85, 86] considered also the related AC electrokinetic phenomenon of *travelling wave dielectrophoresis*. In this, the particle linear motion is induced by a travelling field that is made to displace along the measuring chamber by a periodical set of appropriately phased electrodes.

Experimental determinations of the dielectrophoretic motion in conventional dielectrophoresis have often been based on the changes in the light scattered by the suspensions because of the dielectrophoretic aggregation of the particles. In fact, Khusid and Acrivos [87] have examined the conditions upon which dielectrophoresis is accompanied by formation of chain-like aggregates parallel to the field, or disk-like aggregates with perpendicular orientation. Allsopp et al. [88] have recently proposed impedance measurements for detecting dielectrophoresis.

A precise optical technique (based on PALS and called DPALS by the authors) has been set up by Gimsa's group [70, 89] and is schematically shown in Fig. 18 [70]. A nonhomogeneous electric field is generated in the measuring cell by two peaked electrodes, and the scattered light is phase analyzed in a very similar way to that described for electrophoresis and electrorotation. As observed in Fig. 19 [70], the particle velocity can be measured with an accuracy of the order of $1\,\mu m\,s^{-1}$, and the effect of the frequency of the applied field (above ~ 1 kHz, to avoid unwanted electrode effects) on the translational velocity is clearly appreciable.

Let us mention that dielectrophoresis has also found wide application in manipulation and sorting of particles and biological cells. Together with standard electrophoresis, it is perhaps the electrokinetic phenomenon that is most often used with practical applications in mind. There is a huge variety of experimental methods in order to produce the inhomogeneous field and to visualize or separate the particles. Thus, Green and Morgan [90] use a set of four hyperbolic polynomial electrodes (Fig. 20), with a low field in the central region and high-field zones along the electrode edges. Using fluorescent latex spheres, they could obtain some beautiful results, as shown in Fig. 21.

Particle separation experiments are often performed with interdigitated microelectrode arrays in two [91] or three [92] dimensions. Based on the dielectrophoresis phenomenon, a technique has recently become available for particle or cell separation, namely, dielectrophoresis/gravitational field-flow fractionation (DEP/G-FFF). As is well known, FFF is a technique in which separation is achieved by the relative positions of different particles in a fluid flow. In DEP/G-FFF, the relative positions and velocities of unequal particles or cells are also controlled by the dielectric properties of the colloid and the frequency of the applied field. The method has been applied of course to model polystyrene beads, but, most interestingly, to suspensions of different biological cells [92–96].

VIII. DIFFUSIOPHORESIS

In the previous section we have briefly described a kind of particle motion associated with a nonuniform electric field. More examples can be given of phenomena in which the particles move due to nonhomogeneous fields of different nature (for instance, a tempera-

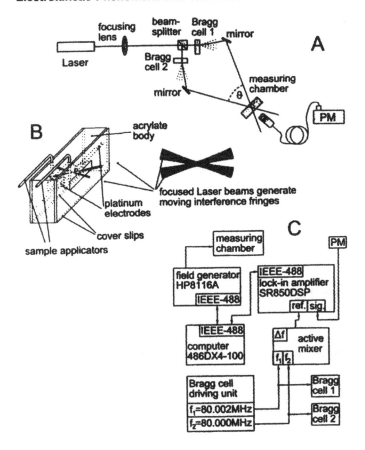

FIG. 18 Experimental setup for dielectrophoretic phase analysis light scattering (DPALS). A: optical setup; B: electrode chamber; C: block diagram of the electronics. (From Ref. 70, Copyright 1999, with permission of Elsevier Science.)

ture gradient or a solute concentration gradient). These kinds of particle locomotions were called phoretic motions by Anderson [97], and one of them is *diffusiophoresis*, i.e., the migration of colloidal particles due to a gradient of solute concentration in the dispersion medium. Derjaguin and his group were the first investigators that described the phenomenon and performed experimental verifications [32, 98].

Let us note that the solute can be either nonionic or ionic, and hence diffusiophoresis can occur even in the absence of an EDL: the particle moves toward or away from the high-concentration region, depending on its long-range interactions with the solute molecules [99–102].

The differential equations of the phenomenon are, in the case of an electrolyte concentration gradient, identical to those of electrophoresis, with the only difference being the appropriate boundary conditions far from the particle surface [103]. Also, as in electrophoresis, the simplest mathematical treatments concern the case of low ζ and flat geometry [98] that were then generalized to spherical particles with low ζ and thin double layer, or low ζ and finite double layer thickness [100]. Prieve and Roman [103], using a methodology similar to that of O'Brien and White [34], found a numerical solution to the problem for arbitrary values of ζ and κa, assuming a symmetrical electrolyte solution. The

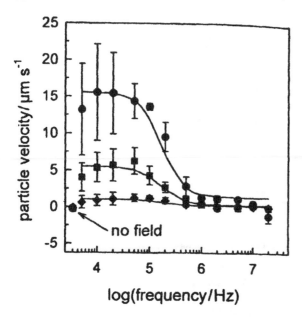

FIG. 19 Dielectrophoretic velocity of latex particles in an external conductivity of $0.75\,\mathrm{mS\,cm^{-1}}$, as a function of the frequency of the field, for different driving voltages (●: 16 V; ■: 8 V; ◆: 4 V). (From Ref. 70, Copyright 1999, with permission of Elsevier Science.)

thin double layer ($\kappa a \gg 1$) expression for the diffusiophoretic velocity, $\vec{v}_{\mathrm{dp}}$, is to first order in $1/\kappa a$ given by [100, 103]

$$\vec{v}_{\mathrm{dp}} = \vec{v}_{\mathrm{CD}}\left\{1 + \frac{G_{\mathrm{CD}}}{\kappa a}\right\} + \vec{v}_{\mathrm{EO}}\left\{1 + \frac{G_{\mathrm{ED}}}{\kappa a}\right\} \tag{59}$$

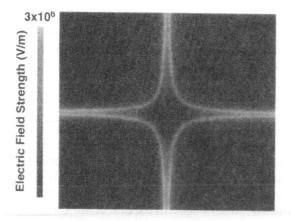

FIG. 20 Electric field strength 100 nm above the surface of a polynomial electrode, with 10 V peak-to-peak applied between adjacent electrodes. The highest fields are located on the electrode edges (brightest regions). (From Ref. 90. Copyright 1999, reprinted with permission of the American Chemical Society.)

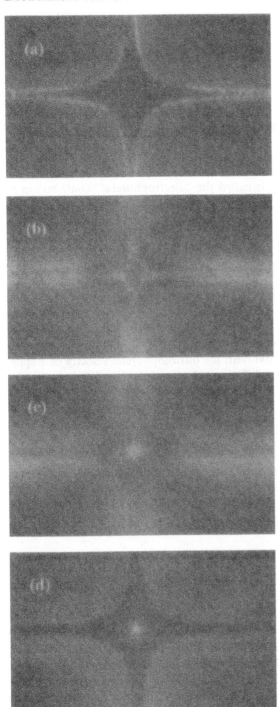

FIG. 21 Video images of latex particles collecting on the polynomial electrodes shown in Fig. 20. (a): Particles on the electrode edges, for an applied field of frequency 500 kHz; (b–d): pictures obtained at 3 s intervals after switching the field frequency to 5 MHz—note how the particles are forced to go toward the low-field region. (From Ref. 90. Copyright 1999, reprinted with permission of the American Chemical Society.)

where

$$\vec{v}_{CD} = \frac{2\varepsilon_{rs}\varepsilon_0}{\eta}\left(\frac{k_B T}{ze}\right)^2 [-\ln(1-\gamma^2)]\nabla \ln n_\infty \tag{60}$$

is called the "chemophoretic" contribution to diffusiophoresis. Here, z is the common valency of the ions ($z_+ = -z_- = z$), $\gamma = \tanh(ze\zeta/4k_B T)$, and ∇n_∞ is the constant electrolyte concentration gradient that is imposed in the solution. In Eq. (59), G_{CD} is

$$G_{CD} = -\frac{21}{2} + O(\tilde{\zeta}) \tag{61}$$

The second term needed to calculate $\vec{v}_{dp}$ is called the "electrophoretic" contribution to diffusiophoresis, since it is given by

$$\vec{v}_{EO} = \frac{\varepsilon_{rs}\varepsilon_0\zeta}{\eta}\frac{\beta k_B T}{ze}\nabla \ln n_\infty \tag{62}$$

with

$$\beta = \frac{m_- - m_+}{m_- + m_+} \tag{63}$$

and $m_+(m_-)$ is the dimensionless mobility of cations (anions). Note that in an electrolyte like KCl ($m_+ \approx m_-$) this term is irrelevant for the calculation of $\vec{v}_{dp}$. Let us also point out that, for typical values of the quantities involved, the diffusiophoretic velocity of a spherical particle is comparable to the average diffusion velocity of the ions [103], so that for having a measurable $|\vec{v}_{dp}|$ ($\approx 1\,\mu m\,s^{-1}$), $|\nabla \ln n_\infty|$ must be $\approx 30\,cm^{-1}$.

Let us finally recall that a gradient of mixed solvent composition can also lead to diffusiophoresis. Kosmulski and Matijević [104] proposed calling the phenomenon *solvophoresis* in this case. In their experimental set up, the gradient of solvent composition was created by carefully placing a layer of alcohol on water: although they have complete miscibility, a stable boundary layer with spatially varying composition could be observed. Aqueous polystyrene latexes were used as the aqueous phase, and ethanol was the light phase placed on it. The shift of the turbid/clear boundary with time was recorded, and an example is shown in Fig. 22.

IX. ELECTRO-OSMOSIS

A. Theory

We now return, so to speak, to more classical electrokinetic phenomena. If a flat piece of solid is placed in contact with an aqueous solution, the formation of the double layer will lead to the appearance of a diffuse region, enriched in counterions and depleted in coions, so that a nonzero charge density $\rho(x)$ will exist along the double layer extension. Here, x is the dimension perpendicular to the solid/liquid interface. If an external field $\vec{E}_\infty$ is now applied parallel to the latter, the electrical force on the charge region will set the ions into motion, and they will drag liquid with them: this liquid flow is called electro-osmotic and the electrokinetic phenomenon is electro-osmosis.

The velocity of the liquid at sufficiently large (see below) distance from the interface, $\vec{v}_{eo}$, is easy to find for a flat double layer [2, 13]; with reference to Fig. 23: since a steady velocity distribution is reached soon after the application of the external field, the net force (electric and viscous) on any double layer volume element of thickness dx should be zero:

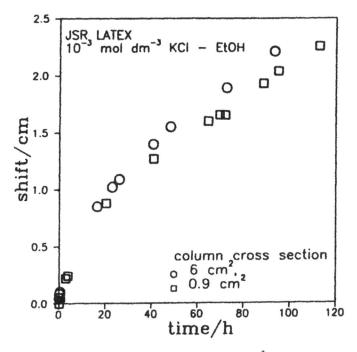

FIG. 22 Solvophoresis of latex particles in 10^{-3} M KCl against ethanol for two different cross-sectional areas of the experimental cell. (From Ref. 104, with permission of Academic Press.)

$$\rho(x)E_\infty \, dx + \left[\eta\left(\frac{dv}{dx}\right)_{x+dx} - \eta\left(\frac{dv}{dx}\right)_x\right] = 0 \tag{64}$$

or

$$\rho(x)E_\infty + \eta\frac{d^2v}{dx^2} = 0 \tag{65}$$

Furthermore, the charge density can be related to the equilibrium potential distribution, $\psi(x)$, by Eq. (9) and consideration that in our problem $\nabla^2 \equiv d^2/dx^2$. Hence:

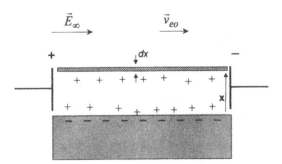

FIG. 23 Scheme of the electro-osmotic flow of the charged liquid in the double layer at a plane interface.

$$\frac{1}{\varepsilon_{rs}\varepsilon_0} \frac{d^2\psi(x)}{dx^2} = \eta \frac{d^2 v}{dx^2} \tag{66}$$

which easily integrates to

$$v(x) = -\frac{\varepsilon_{rs}\varepsilon_0}{\eta} E_\infty [\zeta - \psi(x)] \tag{67}$$

with the conditions:

$$\psi(x = 0) = \zeta$$

$$\left.\frac{d\psi}{dx}\right|_{x \to \infty} = 0 \tag{68}$$

$$\left.\frac{dv(x)}{dx}\right|_{x \to \infty} = 0$$

Far from the particle, $\psi \to 0$, and $v(x)$ is the electro-osmotic velocity:

$$\vec{v}_{eo} = -\frac{\varepsilon_{rs}\varepsilon_0}{\eta} \zeta \vec{E}_\infty \tag{69}$$

which is the Smoluchowski formula for electro-osmosis [identical to Eq. (36), or Eq. (37a) for the electrophoretic mobility except for the obvious change of sign].

According to Dukhin and Derjaguin [13], Eq. (69) is also applicable to electro-osmotic flow through systems closer to the actual laboratory situation, for instance to flat capillaries (formed between identical plane-parallel surfaces) or to cylindrical capillaries, provided that:

1. The separation between opposite walls (or the cylindrical capillary radius) R is much smaller than the linear dimension L of the capillary.
2. The double-layer thickness at any point of the walls is small as compared to R.

In the case of cylindrical capillaries of radius R, the correction to the Smoluchowski equation if point 2 is not fulfilled is exactly solvable for the case of low ζ potentials (Debye–Hückel approximation), as shown in, e.g., Refs 105 and 106:

$$\langle \vec{v}_{eo} \rangle = \frac{2}{R^2} \int_0^R v(r) r dr = -\frac{2\varepsilon_{rs}\varepsilon_0 \vec{E}_\infty \zeta}{\eta R^2} F_1(\kappa R) \tag{70}$$

where $\langle \vec{v}_{eo} \rangle$ is the average velocity of the electro-osmotic flow, and

$$F_1(\kappa R) = \int_0^R r\left(1 - \frac{I_0(\kappa r)}{I_0(\kappa R)}\right) dr = R^2\left(\frac{1}{2} - \frac{I_1(\kappa R)}{\kappa R I_0(\kappa R)}\right) \tag{71}$$

Here, I_0 and I_1 are, respectively, the zeroth-order and first-order modified Bessel functions of the first kind [107], and r is the cylindrical radial co-ordinate.

Although $\vec{v}_{eo}$ (or $\langle \vec{v}_{eo} \rangle$) can sometimes be measured, experimental determinations refer mainly to the fluid flow rate:

$$Q = \pi R^2 \langle \vec{v}_{eo} \rangle \tag{72}$$

However, particularly in the case of porous plugs, where electro-osmosis (and streaming potential/current) find larger applicability, it is preferred to use a measurable quantity, the current I transported by the charged liquid. The total current consists of two contributions [106], namely: (1) that due to the bulk conductivity of the electrolyte, and (2) the surface conductivity, due to current through the double layer, including the convective current due to the electro-osmotic flow itself. Thus, one can write:

$$I = \pi R^2 K^\infty E_\infty + 2\pi R k_s E_\infty \tag{73}$$

where k_s is the specific surface conductivity.

Using Eqs (72) and (73):

$$\frac{Q}{I} = \frac{\varepsilon_{rs}\varepsilon_0\zeta}{\eta[K^\infty + 2k_s/R]}F_1(\kappa R) \tag{74a}$$

so that by simultaneously measuring Q and I, an estimation of ζ and k_s can be reached [106].

We can now consider how these equations can be applied to porous plugs. Smoluchowski himself [35] demonstrated that, in the absence of surface conductance, and with thin double layers everywhere in the pores, Eq. (74a) can be applied in the form:

$$\frac{Q}{I} = \frac{\varepsilon_{rs}\varepsilon_0\zeta}{\eta K^\infty} \tag{74b}$$

However, if surface conductance is not negligible, its correction in the case of complex systems such as porous plugs is not rigorously possible, although some approaches are available. The reader is referred to Refs 2, 3, 13, and 108 for details about different approaches, which usually involve simulating the plug as a number of parallel capillaries or spherical cells with dimensions such that the hydrodynamic behavior of the plug is equivalent to that of the proposed association of capillaries.

B. Electro-osmotic Measurements

There exists a variety of procedures for the measurement of electro-osmosis either in capillaries or in porous plugs [2, 13]; in the authors' laboratory the design shown in Fig. 24 [109] was used for porous plugs of crushed materials. As observed, the plug is contained between perforated platinum electrodes. Since the electro-osmotic flow rate is typically quite low ($\sim 10^{-2}\,\text{cm}^3\,\text{s}^{-1}$), a more precise determination can be achieved if to the electro-osmotic flow a Poisseuille flow provoked by a solution pressure head is superimposed. The experiment is run for both directions of the current, and the flow is determined by the time dependence of the position of a bubble created in the horizontal calibrated capillary: the electro-osmotic flow will be one-half the difference between the total flows measured in both directions of the current.

X. STREAMING POTENTIAL AND STREAMING CURRENT

A. Physical Principles

Assume that we have a cylindrical capillary of radius R and length L filled with an electrolyte solution. For the moment, we will also suppose that the concentration of electrolyte and the radius are such that $\kappa R \gg 1$, that is, the double layer is thin. If the surface charge on the capillary is negative, an excess of positive counterions will exist in the vicinity of the walls (see Fig. 25).

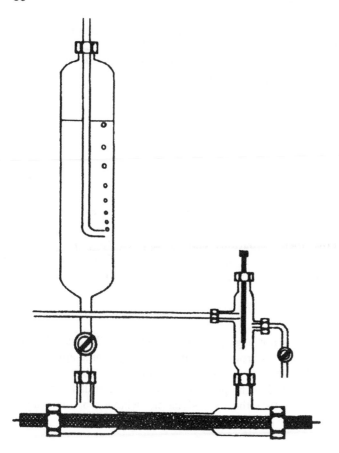

FIG. 24 Setup for electro-osmotic flow measurements. The porous plug is contained in between the perforated electrodes. The horizontal calibrated capillary is used for flow rate measurements. (From Ref. 109, with permission of Academic Press.)

Let us further assume that a hydrostatic pressure gradient $\nabla p = -(\Delta p/L)\hat{z}$ is applied to the system. This will tend to accumulate ions (and hence also the positive counterions) at the low-pressure side of the capillary. If the circuit is not electrically closed (like in Fig. 25, where a high-input impedance voltmeter is connected to the electrodes E and E′), a potential difference will be produced that will prevent further transport of positive charge. In this stationary situation, the potential measured, V_s, is the streaming potential. We will follow Hunter's derivation of V_s [2]. The liquid velocity distribution in the cylinder is well known:

$$v_z(r) = \Delta p \frac{R^2 - r^2}{4\eta L} \tag{75}$$

and since the liquid bears a net charge density $\rho(r)$, its motion brings about an electric current:

$$I_{str} = \int_0^R 2\pi r v_z(r)\rho(r)\,dr \tag{76}$$

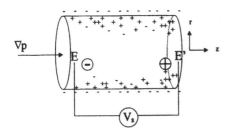

FIG. 25 Scheme of the generation of the streaming potential V_s between the electrodes E–E' in a cylindrical capillary with negatively charged walls.

However, by the thin-double-layer hypothesis, $\rho(r)$ is zero except in a very narrow cylindrical shell close to the wall, so that the liquid velocity profile can be considered linear (not parabolic) without error:

$$v_z(r) \cong \frac{\Delta pR}{2\eta L}(R - r) = \frac{\Delta pR}{2\eta L}x \tag{77}$$

with $x \equiv R - r$. Hence [2]

$$I_{\text{str}} = -\frac{\pi R^2 \Delta p}{\eta L}\int_R^0 x\rho(x)\,dx \tag{78}$$

and $\rho(x)$ can be calculated from $\psi(x)$ by Eq. (9). Integrating, one easily obtains:

$$I_{\text{str}} = -\frac{\varepsilon_{\text{rs}}\varepsilon_0 \pi R^2}{\eta}\frac{\Delta p}{L}\zeta \tag{79}$$

In steady state, this current (the streaming current) must be compensated for by a conduction current, produced by the streaming potential V_s:

$$I_c = K^\infty \pi R^2 \frac{V_s}{L} \tag{80}$$

and the condition $I_{\text{str}} + I_c = 0$ leads to

$$\frac{V_s}{\Delta p} = \frac{\varepsilon_{\text{rs}}\varepsilon_0}{\eta K^\infty}\zeta \tag{81}$$

an expression due to Smoluchowski. As in the case of electro-osmosis, if surface conductance is appreciable, the surface conductivity k_s must also enter the calculation:

$$\frac{V_s}{\Delta p} = \frac{\varepsilon_{\text{rs}}\varepsilon_0}{\eta(K^\infty + 2k_s/R)}\zeta \tag{82}$$

We can now ask ourselves how do we know that surface conductance exists, and how to correct for it. If streaming potential measurements are performed on a given capillary or porous plug for increasing electrolyte concentrations, then if the electrolyte ions are indifferent to the interface at hand (i.e., they do not undergo specific, nonelectrostatic interactions with the surface), V_s should decrease with concentration. This must be so because ζ must decrease (double-layer compression) and K^∞ clearly increases. If this is the observed behavior, then most likely k_s in Eq. (82) can be neglected. However, most often the case is that a maximum in the ζ-concentration plot is observed: at low ionic strengths, the denominator on the right-hand side of Eq. (82) is underestimated if k_s is not considered, and so is ζ. Only when the concentration is high enough ($K^\infty \gg 2k_s/R$), is the normal decreasing trend recovered.

Alternatively, experiments may be performed on capillaries of different radii, or on plugs of different compactions (i.e., porosities): if the apparent zeta potentials [obtained from Eq. (82) under the assumption $k_s = 0$] change with the capillary or (equivalent) pore radius, again surface conductance corrections will be necessary.

The correction eventually needed for surface conduction may be done in several ways. One of them is to measure the conductivity K^* of the liquid in the capillary or plug, by connecting the electrodes directly to a conductivity bridge. The measured conductivity approximately accounts for K^∞ and k_s:

$$\frac{V_s}{\Delta p} \simeq \frac{\varepsilon_{rs}\varepsilon_0}{\eta K^*} \zeta \tag{83}$$

This is equivalent to measuring the resistance R_{plug} of the plug or capillary immersed in the working solution, and also the resistance R_∞ when a solution of high electrolyte concentration is used. In this case [2]:

$$\frac{V_s}{\Delta p} \simeq \frac{\varepsilon_{rs}\varepsilon_0}{\eta K^\infty} \zeta \frac{R_{plug}}{R_\infty} \tag{84}$$

The corrections needed to account for the effect of capillary radius are the same as mentioned above for electro-osmosis. Thus, according to Rice and Whitehead [105], if ζ is low:

$$\frac{V_s}{\Delta p} = \frac{\varepsilon_{rs}\varepsilon_0}{\eta K^\infty} \zeta f(\kappa R, \beta) \tag{85a}$$

where

$$f(\kappa R, \beta) = \frac{1 - \dfrac{2I_1(\kappa R)}{\kappa R I_0(\kappa R)}}{1 - \beta \left[1 - \dfrac{2I_1(\kappa R)}{\kappa R I_0(\kappa R)} - \dfrac{I_1^2(\kappa R)}{I_0^2(\kappa R)} \right]} \tag{85b}$$

$$\beta = \frac{(\varepsilon_{rs}\varepsilon_0\kappa\zeta)^2}{\eta K^\infty} \tag{85c}$$

Note that in Eq. (85a) the zeta potential must be low, and the surface conductance is neglected: the conductivity in the capillary is considered to be equal to K^∞ throughout its whole section. This may be corrected by using K^* instead of K^∞ in Eqs (85a)–(85c) (see also Ref. 110 for a treatment of the problem, valid for any zeta potential or double-layer thickness).

Also, similarly to the case of electro-osmosis, the previous equations are applicable not only to single capillaries but also to bundles of capillaries and to plugs formed with colloidal particles. The latter can be simulated as a set of parallel capillaries, with dimensions related to the porosity experimentally determined for the plug [111], but, as mentioned above, rigorous treatments have also been proposed [65, 108, 112] in which the electrokinetic effects in membranes or concentrated arrays of colloidal particles are analyzed using different hydrodynamic cell models. It must be stressed that the treatment mentioned requires that the liquid flow be characterized by a low Reynolds number (creeping flow) and that the plug is homogeneous, conditions that usually apply in most experimental situations.

As above mentioned, an alternative determination to streaming potential is streaming current: in this case the probing electrodes are connected to a current meter, and Eq. (79) is used to obtain ζ. Since no conduction currents are now needed to be considered, corrections for surface conductance are not necessary either. Let us mention, however, that measured currents are rather low, so that some experimental difficulties must be overcome [2, 113].

B. Measurements

There are presently some commercial instruments for V_s determinations [114], mainly based on similar procedures. Figure 26 is a drawing of the system used in our laboratory until recently: it is designed for measurements on porous plugs, and in fact the same cell and perforated electrodes described for electro-osmosis are used. Nitrogen is used to drive the liquid through the plug in both directions (so the homogeneity and isotropy of the plug is demonstrated), and the streaming voltage is measured directly in a voltmeter (values of even hundreds of millivolts are typical). A different system was described by van der Hoven and Bijsterbosch [115], as schematically shown in Fig. 27, and a computer-controlled version of the latter is described in Ref. 116.

XI. LOW-FREQUENCY DIELECTRIC DISPERSION OF SUSPENSIONS

A. Mechanisms of Dielectric Relaxation in Colloidal Particles

We have previously mentioned two groups of electrokinetic phenomena (electrorotation and dielectrophoresis), whereby the dipole moment induced in a colloidal particle by an external field is probed by determining the motion of *individual* particles in the presence of

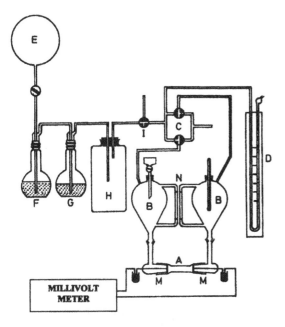

FIG. 26 Experimental setup for streaming potential measurements. The nitrogen gas in E forces the solution through the porous plug A, contained between the perforated electrodes M. (From Ref. 68.)

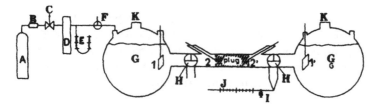

FIG. 27 Four-electrode streaming potential setup according to Van den Hoven and Bijsterbosch. (From Ref. 115, Copyright 1987, with permission of Elsevier Science.)

rotating or inhomogeneous fields. The dipole moment and its relaxations can also be observed if a *collective* response of the suspension to an external AC field is determined, namely, the dielectric constant or the conductivity of the system, as a function of the frequency of the field. Our aim in this section is to derive expressions for either of the two macroscopic quantities in terms of the microscopic individual dipole moments of the particles, which we consider to be identical and spherical (radius a).

Let us for the moment think in terms of the time domain and then we will shift to the more familiar frequency domain. Assume (Fig. 28) that a step electric field $\vec{E}_\infty$ is applied, at $t = 0$, to a colloidal particle, negatively charged, suspended in a large volume of electrolyte solution of dielectric permittivity $\varepsilon'_{rs}\varepsilon_0{}^*$ and conductivity K^∞. As is most often the case, the particle dielectric permittivity $\varepsilon'_{rp}\varepsilon_0$ is much smaller than that of the liquid.

A very short time after the field is applied, ionic motions toward and around the particle have not started, and only orientation of dipoles in both the particle and the liquid medium can occur (Fig. 28a): because $\varepsilon'_{rp} \ll \varepsilon'_{rs}$, a dipole $\vec{d}^{(1)}$ is formed, which is oriented opposite to the field [51]:

$$\vec{d}^{(1)} = 4\pi\varepsilon'_{rs}\varepsilon_0 a^3 \frac{\varepsilon'_{rp} - \varepsilon'_{rs}}{\varepsilon'_{rp} + 2\varepsilon'_{rs}} \vec{E}_\infty \tag{86}$$

When the time after switching on the field is of the order of the Maxwell–Wagner relaxation time, τ_{MW}, which for the case of a nonconducting sphere with low surface conductivity (low ζ) is

$$\tau_{MW} \cong \frac{\varepsilon'_{rs}\varepsilon_0}{K^\infty} \tag{87}$$

ionic migrations in both the solution and the double layer begin. A new dipole, $\vec{d}^{(2)}$ (Fig. 28b), is now formed that superimposes to $\vec{d}^{(1)}$:

$$\vec{d}^{(2)} = 4\pi\varepsilon'_{rs}\varepsilon_0 a^3 \frac{2Du - 1}{2Du + 2} \vec{E}_\infty \tag{88}$$

where Du is the Dukhin number [4], relating surface and bulk conductivities:

$$Du = \frac{K_s}{K^\infty a} \tag{89}$$

If Du $\ll$ 1, normal fluxes brought about by the field (accumulation of cations at the left pole and depletion at the right) cannot be compensated for by the double-layer conductance, so $\vec{d}^{(2)}$ is opposed to the field. On the contrary, for Du $\gg$ 1, the cations brought

---

*We use hereafter a prime (') in the real part of the dielectric constant of any of the phases involved.

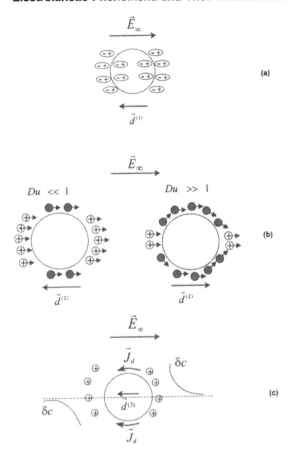

FIG. 28 Schematic representation of the dipoles induced in the double layer by the action of an external field. (a) Dipoles associated with bound charges at the solid/liquid interface; (b) dipoles originated by charge fluxes for low (left) and high (right) surface conductance; (c) concentration polarization of the double layer. (From Ref. 51, with permission from Academic Press.)

normally to the particle are transported at a faster rate tangentially to it (high surface conductivity), and they accumulate at the right and are depleted at the left; $\vec{d}^{(2)}$ is in this case parallel to the field.

The last fundamental time scale is related to the fact that, on the right side of the particle, normal outward fluxes of cations from the double layer find inward fluxes of anions brought normally from the bulk by the field. As a consequence, an increase in neutral electrolyte concentration (Fig. 28c) is produced, and, by the same reasoning, a decrease will occur at the left side. A gradient of neutral electrolyte concentration is thus produced around the particle, with a characteristic time:

$$\tau_{VD} \cong \frac{a^2}{2D_{eff}} \tag{90}$$

where D_{eff} is the effective diffusion coefficient:

$$D_{eff} = \frac{2D^+ D^-}{D^+ + D^-} \tag{91}$$

As a consequence, the double layer will be compressed at the right side (higher electrolyte concentration) of the negative particle, and expanded at the left, so a third dipole $\vec{d}^{(3)}$ is formed that points against the field. Also, the concentration gradient gives rise to diffusion fluxes $\vec{J}_D$ (Fig. 28c) from right to left, which oppose the tangential fluxes provoked by the field.

When these processes are reinterpreted in the frequency domain, an electric field $\vec{E} = \vec{E}_0 e^{-i\omega t}$ is assumed to be acting on the particle, and the phase difference between the induced dipole moment and the field can be interpreted by considering that the latter is a complex quantity, $\vec{d}^*$:

$$\vec{d}^* = 4\pi\varepsilon_{rs}\varepsilon_0 a^3 (c_1 + ic_2)\vec{E} \tag{92}$$

where c_1 and c_2 are, respectively, the real and imaginary parts of the induced dipole coefficient [cf. Eq. (56)]. Figure 29 shows some calculations of c_1 and c_2 as a function of ω for different ζ values and $\kappa a = 10$.

The current density, $\vec{J}$, through a suspension with volume fraction ϕ (with ϕ small enough for any double-layer overlapping to be negligible) will have two components in the presence of the AC field, namely, the conductive, $\vec{J}_{DC}$, and the displacement $\vec{J}_D = \partial\vec{D}/\partial t$ currents:

$$\vec{J} = \vec{J}_{DC} - i\omega\varepsilon_r^*\varepsilon_0\vec{E} \tag{93}$$

where $\varepsilon_r^*\varepsilon_0$ is the complex dielectric permittivity of the suspension. The corresponding complex conductivity will be

$$K^* = K_{DC} - i\omega\varepsilon_r^*\varepsilon_0 \tag{94}$$

K_{DC} being the constant field (DC) conductivity. Based on the condition of low ϕ, a linear dependence between suspension quantities and volume fraction can be assumed:

$$K^* = K_s^* + \phi\Delta K^* \tag{95}$$

$$K_{DC} = K_\infty + \phi\Delta K_{DC} \tag{96}$$

$$\varepsilon_r^* = \varepsilon_{rs}' + \phi\Delta\varepsilon_r^* \tag{97}$$

Here ΔK^*, ΔK_{DC}, and $\Delta\varepsilon_r^*$ are the so-called increments of complex conductivity, DC conductivity, and complex dielectric constant, respectively. They represent the role of the particles and their double layers on the overall conductive and dielectric properties of the suspension. Thus,

$$\Delta\varepsilon_r^* = \Delta\varepsilon_r' + i\Delta\varepsilon_r'' \tag{98}$$

$$\Delta K^* = \Delta K_{DC} - i\omega\varepsilon_0\Delta\varepsilon_r^* = \Delta K_{DC} + \omega\varepsilon_0\Delta\varepsilon_r'' - i\omega\varepsilon_0\Delta\varepsilon_r' \tag{99}$$

where the double prime ($''$) denotes the imaginary part of any complex quantity, and K_s^* is the complex conductivity of the solution:

$$K_s^* = K_\infty - i\omega\varepsilon_{rs}'\varepsilon_0 \tag{100}$$

Note that if a finite current out of phase with the field is measured at low frequencies, Eq. (99) indicates that this can be interpreted macroscopically as a large real part of the dielectric constant of the suspension. In our problem, the largest out-of-phase currents must come for the slowest processes, i.e., the diffusion fluxes originated by concentration polarization. At low frequencies, a high dielectric constant is thus expected for the suspension. As the frequency increases, the slow processes cannot follow the field: as a consequence, they are frozen and the dielectric constant decreases. This is the α (or volume

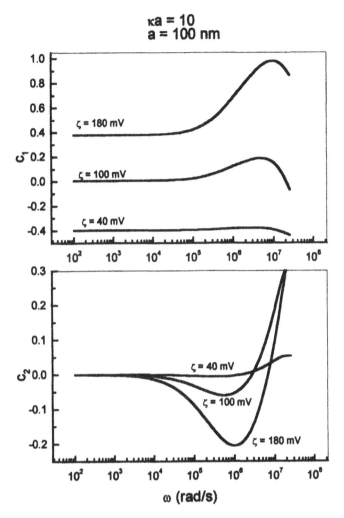

FIG. 29 Real (c_1) and imaginary (c_2) components of the induced dipole coefficient as a function of the frequency of the field, for different zeta potentials. The particle radius and κa values are indicated in the figure.

diffusion) relaxation of the suspension, which will occur for $\omega \sim 1/\tau_{VD}$. At still higher frequencies (typically in the megahertz range), the Maxwell–Wagner relaxation ($\omega \sim 1/\tau_{MW}$) will be observable: for those frequencies, ions cannot rearrange back and forth around the particle as fast as required by the field.

It only remains to relate the dielectric constant ε_r^* to the complex dipole moment. This can be done easily in terms of the complex conductivity [13, 117–121]:

$$K^* = K_s^*(1 + 3\phi C_0^*) \tag{101}$$

so that

$$\Delta K^* = 3C_0^* K_s^* \tag{102}$$

Hence

$$\text{Re}(\Delta\varepsilon_r^*) = \Delta\varepsilon_r' = 3\varepsilon_{rs}'\left[c_1(\omega) - \frac{K^\infty}{\omega\varepsilon_0\varepsilon_{rs}'}c_2(\omega)\right] \tag{103}$$

$$\text{Im}(\Delta\varepsilon_r^*) = \Delta\varepsilon_r'' = 3\varepsilon_{rs}'\left[\frac{K^\infty}{\omega\varepsilon_0\varepsilon_{rs}'}[c_1(\omega) - c_1(0)] + c_2(\omega)\right] \tag{104}$$

where use has been made of the identity $\Delta K_{DC} = 3c_1(0)K^\infty$, $c_1(0)$ being the $\omega \to 0$ limit of $c_1(\omega)$.

Summarizing, the main experimentally accessible quantities, ε_r' (real part of the dielectric constant of the suspension) and K (real part of the electrical conductivity of the suspension), are given by

$$\varepsilon_r' = \varepsilon_{rs}' + 3\phi\varepsilon_{rs}'\left[c_1 - \frac{K^\infty}{\omega\varepsilon_0\varepsilon_{rs}'}c_2\right]$$

$$K = K^\infty + 3\phi K^\infty\left[c_1 + \frac{\omega\varepsilon_0\varepsilon_{rs}'}{K^\infty}c_2\right] \tag{105}$$

The key to the solution of the problem lies in the calculation of $c_1(\omega)$ and $c_2(\omega)$. There is no simple expression for these quantities that are extremely sensitive to the structure and dynamics of the double layer (this, in turn, points to the merit of low-frequency dielectric dispersion – LFDD – as an electrokinetic technique). Probably, the first theoretical treatment is the one due to Schwarz [122], which considered only surface diffusion of counterions (it is, as mentioned before, a so-called *surface diffusion model*) without exchange with diffuse layer ions. An extensive treatment of the subject, with an account of diffuse atmospheric polarization, was first given by Dukhin and Shilov [117], although their treatment was limited to particles with thin double layers. A full numerical treatment of LFDD in suspensions is due to DeLacey and White [121], and comparison with this numerical model allowed one to show that the thin-double-layer approximations [117, 123, 124] worked reasonably well in a wider than expected range of values of both ζ and κa [125]. Figure 30 is an example of the calculations of $\Delta\varepsilon_r'$ following the DeLacey and White procedure. Note that: (1) $\Delta\varepsilon_r'$ is very sensitive to the zeta potential, (2) at low frequencies $\Delta\varepsilon_r'$ can be very high, and (3) the relaxation of the dielectric constant takes place in the low-kilohertz frequency range, in accordance with Eq. (90).

The significant differences found between theory and experiment for this and other electrokinetic phenomena prompted a number of authors to reconsider one of the main hypotheses of the so-called standard electrokinetic model: the assumption that ions in the Stern layer are absolute immobile. Zukoski and Saville [126] were among the first in modifying, in a quantitative manner, the electrokinetic theories to account for the possibility of lateral transport of ions in the inner part of the double layer [the so-called dynamic stern layer (DSL) model]. The improvements achieved were checked against DC conductivity and electrophoretic mobility data. Later, the theory was also elaborated for the analysis of the dielectric constant of suspensions in the DSL scenario [127–129]. Mangelsdorf and White [15–17] performed a slightly different study of the problem, and the ability of this theory to reach a better similarity between theory and experiment has also been demonstrated [130, 131]. Let us finally mention that Kijlstra and coworkers [132–134] also were successful in modifying Fixman's equations [123, 124] to include ionic motions in the stagnant (immobile fluid) layer adjacent to the solid.

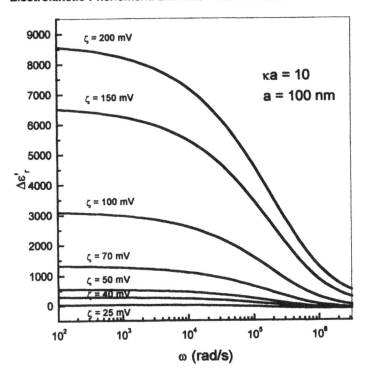

FIG. 30 Real part of the dielectric increment for suspensions of spherical particles as a function of the frequency of the external field. Different zeta potentials as indicated.

B. Measurement of the Dielectric Constant of Suspensions

One of the most usual techniques for measuring the dielectric constant (or, recall its equivalence, the conductivity) of suspensions as a function of the frequency of the applied field is based on the use of a conductivity cell connected to an impedance analyzer. This technique has been widely employed since it was first proposed by Fricke and Curtis [135], and in all cases, the distance between electrodes can be changed (see, e.g., Refs 130, 131, and 136–138). Figure 31 is, for example, a scheme of the one used in our laboratory; experimental details of the method followed are given in Ref. 139. The need for variable electrode separation comes from the problem of electrode polarization at low frequencies, since it can be shown that the electrode impedance dominates over that of the sample at sufficiently low frequencies. The use of different electrode separations stems from the idea that the effects of electrode polarization do not depend on the distance. Grosse and Tirado [140] have recently introduced a method (the *quadrupole method*) in which the correction for electrode polarization is optimized by measuring the cell impedance in four situations (for the whole frequency range of interest, and for each electrode separation): (1) with the cables coming from the impedance analyzer in short circuit (*short* correction), (2) with the cables disconnected from the cell (*open*), (3) with the cell filled with an electrolyte solution of known conductivity and dielectric constant, and (4) with the cell filled with the suspension to be analyzed. Figure 32 shows data obtained in our laboratory with a suspension of polystyrene particles, using their method; the improvement is such that in many cases measurements can be performed down to one order of magnitude lower in frequency than achieved by the classical separation method.

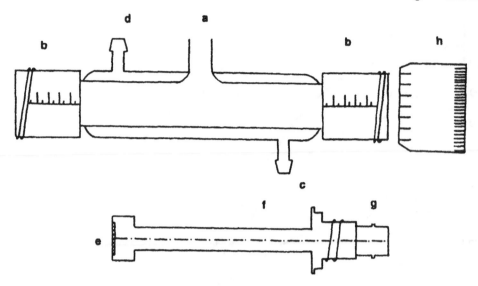

FIG. 31 Schematic representation of the conductivity cell used for LFDD measurements. The electrodes e are supported by Teflon pieces f, and their position is varied by means of the micrometer screws b–h.

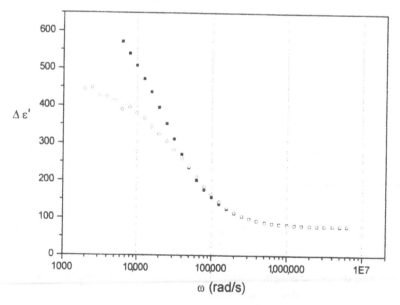

FIG. 32 Dielectric increment of a suspension of latex spheres in 5×10^{-4} M NaCl solution as a function of frequency. $\bigcirc$: quadrupole technique for correction of electrode effects; $\blacksquare$: standard correction method.

This is not, however, the only possible design. The four-electrode method has also been employed with success [132–134, 141]: in this case, since the electrode functions are separated (the sensors and current supplying electrodes are different), the polarization is not the main problem, but the electronics of the measurements are rather more complicated.

XII. ELECTROACOUSTIC PHENOMENA

Recall from Section V that when an external field (such as gravity) produces the deformation (polarization) of the EDL, a dipole is generated in each particle of the suspension. The superposition of the fields of individual particles is sensed as a macroscopic dipolar field in the system (Dorn effect). Let us now assume that, instead of a constant field, it is a sound wave that passes through the suspension: a relative motion between the particles and the surrounding liquid (which are assumed to have different densities) is hence produced. This relative motion provokes again a distortion of the ionic atmosphere, and as a consequence, an alternating electric field. This simple idea is the core of the group of electrokinetic phenomena known as *electroacoustic phenomena* [142–147]. The one described above, in which a pressure wave produces an AC field, is called the *colloid vibration potential* (CVP); the reciprocal phenomenon also occurs: a sound wave is produced by application of an AC electric field to the suspension. The latter electrokinetic phenomenon and technique are called ESA (*electrokinetic sonic amplitude*).

Enderby [148] and Booth and Enderby [149] first obtained an expression for the CVP of colloidal spheres under the following assumptions:

1. The suspension is dilute, i.e., no hydrodynamic or electrical interactions exist between the particles.
2. Particles are nonconducting.
3. The oscillatory boundary layer thickness is much larger than the EDL thickness.
4. The zeta potential is low.
5. The double layer is thin.

Their expression for CVP (see Ref. 143 for a summary of the derivation) can be written (the sensing electrodes are assumed to be at a distance equal to half the wavelength of sound in the liquid):

$$\text{CVP} = \frac{2\Delta p \phi}{K^\infty} \frac{\rho_p - \rho_m}{\rho_m} \frac{\varepsilon_{rs}\varepsilon_0}{\eta} \zeta \qquad (106)$$

where Δp is the pressure amplitude, ρ_p (ρ_m) is the density of the particle (medium), and the other symbols have already been defined in this chapter.

It must be mentioned here that one of the most promising potential applications of these methods is their usefulness with concentrated systems (high-volume fractions of solids, ϕ), since the effect to be measured is also in this case a collective one. The generalization of the theory to concentrated suspensions requires the use of a cell model to account for particle–particle interactions. After some approaches using different cell models [65, 150], recent results seem to suggest that the Shilov–Zharkikh cell model [44] gives the best description of the phenomenon (Ref. 147; see also Chapter 17 in this volume).

In the case of the ESA phenomenon, O'Brien [146] showed that the ESA signal in the presence of an AC field of frequency ω can be found from

$$\text{ESA}(\omega) = A(\omega)\phi\frac{\rho_p - \rho_m}{\rho_m}\langle\mu_d^*\rangle \tag{107}$$

where A is an instrument factor, and $\langle\mu_d^*\rangle$ is the particle-averaged dynamic mobility, a complex quantity defined as the factor of proportionality between the electrophoretic velocity and the AC field:

$$\begin{aligned}\vec{v}_e &= \mu_d^*\vec{E}_0 e^{i\omega t} \\ \mu_d^* &= \mu e^{i\theta}\end{aligned} \tag{108}$$

The magnitude, μ, of μ_d^* is the particle velocity amplitude per unit field strength, and its phase θ measures the lag between the particle motion and the applied field, due to the inertia of the particle: as shown in Fig. 33, both μ and θ are functions of the frequency and the zeta potential (they also depend on the particle size). For a spherical particle with a thin double layer [145]:

$$\mu_d^* = \frac{2}{3}\frac{\varepsilon_{rs}\varepsilon_0\zeta}{\eta}G(\omega a^2/\nu)[1 + f] \tag{109}$$

where $\nu = \eta/\rho_m$ is the kinematic viscosity of the liquid, and the function G reads [145, 151]:

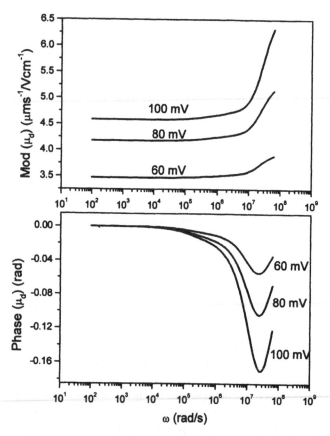

FIG. 33 Modulus (top) and phase (bottom) of the dynamic mobility of spherical particles of radius 100 nm as a function of the frequency of the field and the zeta potential. In all cases, $\kappa a = 10$.

$$G(x) = \left\{ 1 - \frac{ix[3 + 2\Delta\rho/\rho_m]}{9\left[1 + (1 - i)\sqrt{x/2}\right]} \right\}^{-1} \tag{110}$$

with $\Delta\rho \equiv \rho_p - \rho_m$. The function f depends on the Dukhin number (Du) as follows:

$$f = \frac{1 - i\omega' - (2\mathrm{Du} - i\omega'\varepsilon_p/\varepsilon_{rs})}{2(1 - i\omega') + (2\mathrm{Du} - i\omega'\varepsilon_p/\varepsilon_{rs})} \tag{111}$$

where $\omega' = \omega/\omega_{MW} = \omega\tau_{MW}$ [see Eq. (87)].

The link between CVP and ESA comes from the flux–force relationships of non-equilibrium thermodynamics. Thus, the macroscopic (average) current, $\vec{J}$, and particle velocity, $\vec{v}$, are related to the pressure gradient ∇p and external field $\vec{E}$ by [151]

$$\vec{J} = \alpha\nabla p + K^*\vec{E}$$
$$\vec{v} = \beta\nabla p + \mu_d^*\vec{E} \tag{112}$$

where α, β, K^*, and μ_d^* are the transport coefficients that satisfy the relationship:

$$\alpha = \phi\frac{\Delta\rho}{\rho_m}\mu_d^* \tag{113}$$

Equation (113) may be considered as a generalization of Onsager's reciprocity relation (originally derived for stationary processes) to dynamic (time-varying) processes. When measuring CVP we are in open-circuit conditions [$\vec{J} = 0$ in Eq. (112)]; hence, the voltage drop per unit pressure drop will equal:

$$\mathrm{CVP} = \frac{\Delta V}{\Delta p} = \frac{\alpha}{K^*} = \phi\frac{\Delta\rho}{\rho_m}\frac{\mu_d^*}{K^*} \tag{114}$$

and [151]:

$$\mathrm{ESA} \propto \mathrm{CVP} \times K^* \tag{115}$$

where K^* is the complex conductivity of the suspension.

Experimental measurements have increased during the last 10 years, mainly because there are commercial instruments devised for both kinds of determinations [152]. The ESA effect is determined with the device schematically shown in Fig. 34 [153, 154]. A radio-frequency voltage pulse is sent through the sample across two parallel-plate electrodes attached to a pair of glass blocks (delay lines). The sound wave travels along the glass blocks and is sensed by a piezoelectric transducer placed at the right delay rod. Figure 34 also shows the signal produced by the transducer as a function of time: the first signal is not of ESA origin, but is called "cross-talk" [154], as it is an electrical signal radiated from the voltage pulse and detected by the transducer. The second and third oscillations, on the contrary, are actual ESA pulses coming from the right and left electrodes, respectively. The first of them is the one that the system uses to measure the dynamic mobility and zeta potential.

We will not go into the details of the CVP (or CVI, *colloid vibration intensity*) experimental device, as this is fully described in Chapter 17 of this volume.

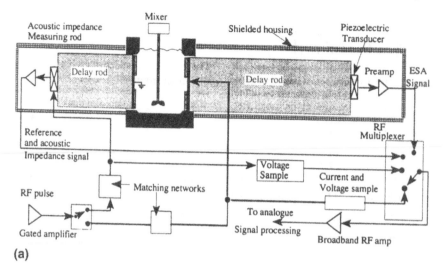

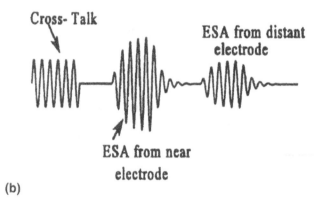

FIG. 34 (a) Schematic diagram of the ESA setup; (b) the signal from the ESA transducer. (From Ref. 153, copyright 1998, with permission of Elsevier Science.)

XIII. CONCLUSION

We have tried to give a (necessarily brief) overview of the electrokinetic phenomena and their associated techniques: some are classical and are among the oldest experimental determinations in our science. Others are more recently discovered, explained, and implemented. In any case, we hope that the reader will have found the starting information needed to choose a technique suitable for any particular problem. A much more detailed description of most of the phenomena described above will be found in the chapters that follow.

ACKNOWLEDGMENTS

The assistance of members of our research group is gratefully acknowledged, as well as financial support by CICYT, Spain (MAT98-0940). Thanks are also due to Professor V.N. Shilov, of the Ukrainian Academy of Sciences for clarifying discussions.

REFERENCES

1. RJ Hunter. Foundations of Colloid Science. Oxford: Oxford University Press, 1987, vol. I.
2. RJ Hunter. Zeta Potential in Colloid Science. New York: Academic Press, 1981.
3. J Lyklema. In: TF Tadros, ed. Solid/Liquid Dispersions. London: Academic Press, 1987, ch. 3.
4. J Lyklema. Fundamentals of Interface and Colloid Science. vol. II. Solid–Liquid Interfaces. New York: Academic Press, 1995.
5. HJ Van Olphen. An Introduction to Clay Colloidal Chemistry. New York: Wiley, 1977.
6. WB Russel, DA Saville, WR Schowalter. Colloidal Dispersions. Cambridge: Cambridge University Press, 1989.
7. SS Dukhin. In: E Matijević, ed. Surface and Colloid Science. New York: Wiley, 1974, vol. 7, ch. 1.
8. JJ López García, AA Moya, J Horno, A Delgado, F González-Caballero. J Colloid Interface Sci 183: 124, 1996.
9. J Lyklema. J Colloid Interface Sci 58: 242, 1977.
10. WG Eversole, PH Lahr. J Chem Phys 9: 530, 1941.
11. WG Eversole, PH Lahr, WW Boardman. J Chem Phys 9: 758, 1941.
12. AL Smith. J Colloid Interface Sci 55: 525, 1976.
13. SS Dukhin, BV Derjaguin. In: E Matijević, ed. Surface and Colloid Science. New York: Wiley, 1974, vol. 7, ch. 2.
14. CF Zukoski IV, DA Saville. J Colloid Interface Sci 114: 32, 1986.
15. CS Mangelsdorf, LR White. J Chem Soc, Faraday Trans 86: 2859, 1990.
16. CS Mangelsdorf, LR White. J Chem Soc, Faraday Trans 94: 2441, 1998.
17. CS Mangelsdorf, LR White. J Chem Soc, Faraday Trans 94: 2583, 1998.
18. J Lyklema, M Minor. Colloids Surfaces A 140: 33, 1998.
19. WKH Panofski, M Phillips. Classical Electricity and Magnetism. Reading: Addison-Wesley, 1975.
20. H Ohshima. In: H Ohshima, K Furusawa, eds. Electrical Phenomena at Interfaces. New York: Marcel Deker, 1998, ch. 2.
21. LD Landau, EM Lifshitz. Fluid Mechanics. Oxford: Pergamon Press, 1959.
22. FA Morrison. J Colloid Interface Sci 34: 210, 1970.
23. JThG Overbeek. Adv Colloid Interface Sci 3: 97, 1950.
24. AM James. In: RJ Good, RS Stromberg, eds. Surface and Colloid Science: New York: Plenum Press, 1979, vol. 9.
25. J Lyklema, JThG Overbeek. J Colloid Interface Sci 16: 501, 1961.
26. RJ Hunter. J Colloid Interface Sci 22: 231, 1966.
27. DC Henry. Proc Roy Soc 133A: 106, 1931.
28. JTG Overbeek. Kolloid Beih 54: 287, 1943.
29. F Booth. Trans Faraday Soc 44: 955, 1948.
30. F Booth. Nature 161: 83, 1948.
31. F Booth. Proc Roy Soc 203A: 514, 1950.
32. BV Derjaguin, SS Dukhin. In: E Matijević, ed. Surface and Colloid Science. New York: Wiley, 1974, vol. 7, ch. 3.
33. PH Wiersema, AL Loeb, JTG Overbeek. J Colloid Interface Sci 22: 78, 1966.
34. RW O'Brien, LR White. J Chem Soc, Faraday Trans 2 74: 1607, 1978.
35. JTG Overbeek. In: HR Kruyt, ed. Colloid Science. Amsterdam: Elsevier, 1952, vol. 1.
36. RP Tison. J Colloid Interface Sci 60: 519, 1977.
37. RP Long, S Ross. J Colloid Interface Sci 20: 438, 1965.
38. RP Tison. J Colloid Interface Sci 52: 611, 1975.
39. A Homola, AA Robertson. J Colloid Interface Sci 51: 202, 1975.
40. S Levine, GH Neale. J Colloid Interface Sci 47: 520, 1974.
41. S Levine, GH Neale. J Colloid Interface Sci 49: 330, 1974.

42. LD Reed, FA Morrison. J Colloid Interface Sci 54: 117, 1976.
43. JL Anderson. J Colloid Interface Sci 82: 248, 1981.
44. NI Zharkikh, VN Shilov. Colloid J 43: 865, 1981.
45. AS Dukhin, VN Shilov, Yu Borkovskaya. Langmuir 15: 3452, 1999.
46. AL Smith. In: GD Parfitt, ed. Dispersions of Powders in Liquids. London: Applied Science, 1981.
47. www.cad-inst.com; www.lavallab.com; www.zeta-meter.com; www.webzero.co.uk
48. K Oka, K Furusawa. In: H Ohshima, K Furusawa, eds. Electrical Phenomena at Interfaces. New York: Marcel Dekker, 1998, ch. 8.
49. S Komagata. Res Electrotech Lab (Jpn) 348: March, 1933.
50. M Minor, AJ van der Linde, HP Leeuwen, J Lyklema. J Colloid Interface Sci 189: 370, 1997.
51. VN Shilov, AV Delgado, F González-Caballero, J Horno, JJ López-Garcia, C Grosse. J Colloid Interface Sci 232: 141, 2000.
52. EE Uzgiris. Rev Sci Instrum 45: 74, 1974.
53. R Wave, WH Flyare. J Colloid Interface Sci 39: 670, 1972.
54. EE Uzgiris. Optics Comm 6: 55, 1972.
55. E Malher, D Martin, C Duvivier. Studia Biophys 90: 33, 1982.
56. E Malher, D Martin, C Duvivier, B Volochine, JF Stolz. Biorheology 19: 647, 1982.
57. www.beckman.com/coulter; www.bic.com; www.malvern.co.uk
58. JF Miller, K Schätzel, B Vincent. J Colloid Interface Sci 143: 532, 1991.
59. F Manerwatson, W Tscharnuter, J Miller. Colloids Surfaces A 140: 53, 1988.
60. JF Miller, O Velev, SCC Wu, HJ Ploehn. J Colloid Interface Sci 174: 490, 1995.
61. M Ozaki, H Sasaki. In: H Ohshima, K Furusawa, eds. Electrical Phenomena at Interfaces. New York: Marcel Dekker, 1998, ch. 10.
62. F Booth. Proc Roy Soc Lond, Ser A 203: 514, 1950.
63. D Stigter. J Phys Chem 84: 2758, 1980.
64. H Ohshima, TW Healy, LR White. J Chem Soc, Faraday Trans 2 74: 1607, 1978.
65. S Levine, G Neale, N Epstein. J Colloid Interface Sci 57: 424, 1976.
66. H Ohshima. J Colloid Interface Sci 208: 295, 1998.
67. F Carrique, FJ Arroyo, AV Delgado. J Colloid Interface Sci 227: 212, 2000.
68. F González-Caballero. A Study on Streaming and Sedimentation Potentials. Application to the Electrokinetic Properties of Quartz and Fluorite. PhD thesis, University of Granada, Spain, 1974.
69. J Gimsa. Colloids Surfaces A 149: 451, 1999.
70. P Eppmann, B Prüger, J Gimsa. Colloids Surfaces A 149: 443, 1999.
71. B Prüger, P Eppmann, J Gimsa. Colloids Surfaces A 136: 199, 1998.
72. C Grosse, VN Shilov. Colloids Surfaces A 140: 199, 1998.
73. V Zimmerman, C Grosse, VN Shilov. Colloids Surfaces A 159: 299, 1999.
74. WN Arnold, HP Schwan, U Zimmermann. J Phys Chem 91: 5093, 1987.
75. B Neu, R Georgieva, H Bäunmler, VN Shilov, E Kooppal, E Donath. Colloids Surfaces A 140: 325, 1998.
76. J Gimsa, B Prüger, P Eppmann, E Donath. Colloids Surfaces A 98: 243, 1995.
77. B Prüger, P Eppmann, J Gimsa. Colloids Surfaces A 136: 199, 1998.
78. R Pethig. IOP Conf Ser 118: 13, 1991.
79. AD Goater, R Pethig. Parasitology 117: 5177, 1998.
80. KL Chan, H Morgan, E Morgan, IT Camerson, MR Thomas. Biochim Biophys Acta-Molec Basis Disease 1500: 313, 2000.
81. AD Goater, JPH Burt, R Pethig. J Phys D: Appl Phys 30: L65, 1997.
82. HA Pohl. Dielectrophoresis. Cambridge: Cambridge University Press, 1978.
83. JC Baygents. Colloids Surfaces A 92: 67, 1994.
84. O Shramko, V Shilov, T Simonova. Colloids Surfaces A 140: 385, 1998.
85. XB Wang, Y Huang, FF Becker, PRC Gascoyne. J Phys D: Appl Phys 27: 1571, 1994.

86. XB Wang, MP Hughes, Y Huang, FF Becker, PRC Gascoyne. Biochim Biophys Acta-Gen Subj 1243: 185, 1995.
87. B Khusid, A Acrivos. Phys Rev E 54: 5428, 1996.
88. DWE Allsopp, KR Milner, AP Brown, WB Betts. J Phys D: Appl Phys 32: 1066, 1999.
89. J Gimsa, P Eppmann, B Prüger. Biophys J 73: 3309, 1997.
90. NC Green, H Morgan. J Phys Chem B 103: 41, 1999.
91. J Yang, Y Huang, X Wuang, FF Becker, PRC Gascoyne. Anal Chem 71: 911, 1999.
92. J Suehiro, R Pethig. J Phys D: Appl Phys 31: 3298, 1998.
93. WH Arnold. Inst Phys Conf Ser 163: 63, 1999.
94. H Morgan, MP Hughes, NG Green. Biophys J 77: 516, 1999.
95. J Cheng, EL Sheldon, L Wu, MJ Heller, JP O'Connell. Anal Chem 70: 2321, 1998.
96. R Pethig. Crit Rev Biotechnol 16: 331, 1996.
97. JL Anderson. Annu Rev Fluid Mech 21: 61, 1989.
98. BV Derjaguin, SS Dukhin, AA Korotkova. Kolloidn Zh 23: 53, 1961.
99. JL Anderson, ME Lowell, DC Prieve. J Fluid Mech 117: 107, 1982.
100. DC Prieve, JL Anderson, JP Ebel, ME Lowell. J Fluid Mech 148: 247, 1984.
101. HJ Keh, TY Huang. J Colloid Interface Sci 160: 354, 1993.
102. HJ Keh, TY Huang. Colloids Surfaces A 92: 51, 1994.
103. DC Prieve, R Roman. J Chem Soc, Faraday Trans 2 83: 1287, 1987.
104. M Kosmulski, E Matijević. J Colloid Interface Sci 150: 291, 1992.
105. CL Rice, P Whitehead. J Phys Chem 69: 4017, 1965.
106. S Arulanandam, D Li. J Colloid Interface Sci 225: 421, 2000.
107. M Abramowitz, IA Stegun. Handbook of Mathematical Functions. New York: Dover, 1972.
108. RW O'Brien. J Colloid Interface Sci 110: 477, 1986.
109. R Hidalgo-Álvarez, F González-Caballero, JM Bruque, G Pardo. J Colloid Interface Sci 82: 45, 1981.
110. A Szymczyk, B Aoubrza, P Fievet, J Pagetti. J Colloid Interface Sci 216: 285, 1999.
111. MY Chang, AA Robertson. Can J Chem Eng 45: 67, 1967.
112. RW O'Brien, WT Perrins. J Colloid Interface Sci 99: 20, 1984.
113. C Werner, H Korber, R Zimmermann, SS Dukhin, HJ Jacobasch. J Colloid Interface Sci 208: 329, 1998.
114. www.anton-paar.com
115. TJJ Van der Hoeven, BH Bijsterbosch. Colloids Surfaces A 22: 187, 1987.
116. O El Gholabzouri, MA Cabrerizo, R Hidalgo-Álvarez. J Colloid Interface Sci 199: 38, 1998.
117. SS Dukhin, VN Shilov. Dielectric Phenomena and the Double Layer in Disperse Systems and Polyelectrolytes. Jerusalem: Wiley, 1974.
118. SS Dukhin. Adv Colloid Interface Sci 44: 1, 1993.
119. RW O'Brien. J Colloid Interface Sci 81: 234, 1980.
120. RW O'Brien. Adv Colloid Interface Sci 16: 281, 1982.
121. EHB DeLacey, LR White. J Chem Soc, Faraday Trans 2 77: 2007, 1981.
122. G Schwarz. J Phys Chem 66: 2636, 1962.
123. M Fixman. J Chem Phys 72: 5177, 1980.
124. M Fixman. J Chem Phys 78: 1483, 1983.
125. F Carrique, L Zurita, AV Delgado. J Colloid Interface Sci 170: 176, 1995.
126. CF Zukoski, DA Saville. J Colloid Interface Sci 114: 45, 1986.
127. LA Rosen, DA Saville. J Colloid Interface Sci 140: 82, 1990.
128. LA Rosen, DA Saville. J Colloid Interface Sci 149: 542, 1992.
129. LA Rosen, JC Baygents, DA Saville. J Chem Phys 98: 4183, 1993.
130. FJ Arroyo, F Carrique, T Bellini, AV Delgado. J Colloid Interface Sci 210: 194, 1999.
131. FJ Arroyo, F Carrique, AV Delgado. J Colloid Interface Sci 217: 411, 1999.
132. J Kijlstra. Dielectric Relaxation in Colloids. PhD thesis, University of Wageningen, The Netherlands, 1992.
133. J Kijlstra, HP van Leeuwen, J Lyklema. Langmuir 9: 1625, 1993.

134. J Kijlstra, HP van Leeuwen, J Lyklema. J Chem Soc, Faraday Trans 88: 3441, 1992.
135. H Fricke, HJ Curtis. J Phys Chem 41: 729, 1937.
136. MM Springer. Dielectric Relaxation of Dilute Polystyrene Lattices. PhD thesis, University of Wageningen, The Netherlands, 1979.
137. K Lim, EI Frances. J Colloid Interface Sci 110: 201, 1986.
138. C Grosse, AJ Hill, KR Foster. J Colloid Interface Sci 127: 167, 1989.
139. F Carrique, L Zurita, AV Delgado. Acta Polym 45: 115, 1994.
140. C Grosse, MC Tirado. Mat Res Soc Symp Proc 430: 287, 1996.
141. DF Myers, DA Saville. J Colloid Interface Sci 131: 448, 1989.
142. RJ Hunter. Introduction to Modern Colloid Science. Oxford: Oxford Science Publications, 1993, ch. 8.
143. S Takeda, N Tobori, H Sugawara, K Furusawa. In: H Ohshima, K Furusawa, eds. Electrical Phenomena at Interfaces. Fundamentals, Measurements, and Applications. New York: Marcel Dekker, 1998, ch. 13.
144. E Kissa. Dispersions. Characterization, Testing, and Measurement. New York: Marcel Dekker, 1999, ch. 13.
145. RW O'Brien. J Fluid Mech 190: 71, 1988.
146. RW O'Brien. J Fluid Mech 212: 81, 1990.
147. A Dukhin, H Ohshima, VN Shilov, PJ Goetz. Langmuir 15: 3445, 1999.
148. J Enderby. Proc Phys Soc 207A: 321, 1951.
149. F Booth, J Enderby. Proc Phys Soc 208A: 351, 1952.
150. S Kuwabara. J Phys Soc Jpn 14: 527, 1959.
151. RW O'Brien, BR Midmore, A Lamb, RJ Hunter. Faraday Disc Chem Soc 90: 301, 1990.
152. www.dispersion.com; www.matec.com
153. RJ Hunter. Colloids Surfaces A 141: 37, 1998.
154. RW O'Brien, DW Cannon, WN Rowlands. J Colloid Interface Sci 173: 406, 1995.

2

Nonequilibrium Electric Surface Phenomena and Extended Electrokinetic Characterization of Particles

STANISLAV S. DUKHIN   New Jersey Institute of Technology, Newark, New Jersey

VLADIMIR N. SHILOV   Institute of Biocolloid Chemistry, Ukrainian National Academy of Sciences, Kiev, Ukraine

I. INTRODUCTION

Three branches of colloid science—the theory of the electrical double layer, electrokinetic phenomena, and the electric surface forces—have always been developed in close connection with each other and traditionally on the basis of the concept of complete equilibrium (thermodynamic concept) of the double layer (DL). However, the DL transforms from its equilibrium state into a nonequilibrium one under any effect causing internal ion flow [1, 2], e.g., under the influence of an external electric field. The transition from the thermodynamic concept to the electrochemical macrokinetics concept [3, 4] not only causes these three branches of colloid science to be transformed, but also to be supplemented by new branches.

The theory of the nonequilibrium DL [1, 5–7] has developed in parallel with that of the equilibrium DL, and, in fact, during the last decades many colloid electrochemical phenomena, which are caused by the nonequilibrium state of the DL, have been identified. They can be called nonequilibrium electric surface phenomena (NESP) [8]. In addition to the theory of equilibrium surface forces, a theory of nonequilibrium surface forces is also being developed [9], together with the dynamics of DL interaction [10–13].

Experimental studies of NESP contribute much more than electrokinetic measurements to surface characterization. The new information that is obtained is richer and more precise when a set of electric surface phenomena is studied [2–4]. For a long time, the particle charge was characterized by the electrokinetic charge σ_ζ, i.e., the charge calculated from the measured value of the electrokinetic potential ζ. This procedure was subject to fair criticism [2] because σ_ζ is frequently one order less than the charge of the DL measured by other methods. The DL disequilibration enables the surface conductivity and mobile charge σ_m to be measured [2], demonstrating that σ_m can be significantly larger than σ_ζ. Values of σ_m close to the titratable charge, σ_t, have very often been obtained in experimental researches [8, 14]. This is the reason for using σ_m to characterize the surface charge.

In Refs 2 and 4, the theory of DL disequilibration was incorporated into the models of NESP and electrokinetic phenomena. In particular, a method, known as "thin DL approximation" [15], was elaborated [5–7], and, on the basis of such approximation, the

55

mobile charge of a particle could be calculated, and its surface charge characterized. A variety of integrated electrokinetic investigations were furthermore described in Ref. 8 and their usefulness for particle characterization was discussed. It became clear that the interpretation of experimental data should be based on their comparison with the various existing static and dynamic DL models [14, 16].

The fundamental concept of the surface conductivity on solids was elaborated in detail by Lyklema [17] including a broad review of the preceding investigations. We shall follow this key work using similar terms in equations as well as a similar nomenclature.

Recent developments in the application of NESP to the extended electrokinetic characterization of colloids will be considered in this chapter.

II. GENERAL

A. General Characteristics of the Nonequilibrium DL of Colloid Particles

1. Induced Dipole Moment

The DL changes its equilibrium state into a nonequilibrium one under any effect causing ion flow inside it, since lateral ion flows disturb its equilibrium structure. If these effects are steady, the resulting flow of ions will also reach a steady state, and instead of the conditions of equality of flows to zero, which are valid for the equilibrium layer ($\vec{j}^{\pm} = 0$), it is the condition of equality of the divergence of flows to zero that is satisfied [18]:

$$\nabla \cdot \vec{j}^{\pm} = 0 \tag{1}$$

where $\vec{j}^{+}(\vec{j}^{-})$ is the flux of coions (counterions) at any point of the ionic atmosphere. In order to illustrate the qualitative distinction between the equilibrium and nonequilibrium DL, we will consider a special case of the polarization, provoked by an external stationary electric field, of the DL of a spherical particle with homogeneously charged surface. The equilibrium DL is spherically symmetric and this implies the absence of a dipole moment (Fig. 1). Under the external electric field, a tangential flow of the ions appears, which redistributes them over the surface: the DL is deformed and polarized, diverting from its original spherically symmetrical structure. The tangential flows in the DL are locked through the electrolyte volume, and the stationary ion exchange between the DL and the adjacent electrolyte volume is developed only by the occurrence of a spatial current distribution and an electric field which is beyond the limits of the DL and attains high values in a region with linear dimensions close to the particle size.

The electric field beyond the limits of the DL is called a long-range field, as distinct from the short-range field of the equilibrium DL; the latter is localized in its interior if its thickness κ^{-1} is much smaller than the particle size a, i.e., when the condition:

$$\kappa a \gg 1 \tag{2}$$

is satisfied. The stationary tangential current of ions of the DL is maintained owing to the charge supply from the volume to the left hemisphere of the particle (see Fig. 1). Correspondingly, the field lines of the long-range electric field arising near one hemispherical surface run into the volume and approach the other hemisphere from the volume where the electric field outlets are distributed. Therefore, the long-range field is set up by a dipole characterizing the polarization of the DL. Thus, the nonequilibrium DL of a spherical particle may be described as a superposition of the symmetrical DL and the induced dipole (Fig. 1) [1, 2, 6].

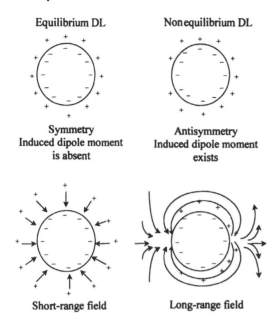

Equilibrium DL   Nonequilibrium DL

Symmetry
Induced dipole moment
is absent

Antisymmetry
Induced dipole moment
exists

Short-range field   Long-range field

FIG. 1 A comparison of the characteristics of the equilibrium and nonequilibrium double layers.

2. Concentration Polarization of the DL of Colloid Particles

Surface flows of ions within the DL carry mainly counterions, while electromigration flows within the electrolyte volume consist of carriers of both signs in equivalent amounts. The difference between the electromigration flows of anions (coions, j^+) and cations (counterions, j^-) in the dispersion medium and at the particle surface gives rise to a polarization gradient of electrolyte concentration along with the polarization field [2, 4, 6, 19]. The electrolyte concentration outside the DL changes within a distance of the order of the particle size and, in accordance with the principle of local equilibrium, the structure of the regions of a much smaller size is considered to be at equilibrium. Thus, the DL can be taken to be in a state of local equilibrium with the electrolyte concentration beyond its limits if the inequality given by Eq. (2) is satisfied. This means that the normal components of diffusion and electromigration flows are almost completely mutually canceled out and the Boltzmann distribution of ion concentrations, which exactly holds in equilibrium DL, holds approximately in the thin polarized DL. Let us note that the increase of electrolyte concentration along the particle surface provokes a compression of the diffuse part of the DL, thus resulting in a decrease of the Stern potential at a constant value of the surface charge; then, however, the potential jump in the DL will vary along the particle surface.

Overbeek [1] and Booth [20] first formulated the problem of polarization of the diffuse part of the DL of spherical colloid particles in an external electric field using a series expansion in the zeta potential, ζ. The method of Overbeek and Booth is too cumbersome to be applied to more complex models such as DL polarization of nonspherical particles and of spherical ones in an alternating field. In Ref. 6, the same results were obtained in a more appropriate analytical form. The theory was then generalized to alternating electric fields [7], which made it possible to predict the effect known as *low-frequency dielectric dispersion* or LFDD [4, 21]. The next step was the generalization of the theory to the case of LFDD of polyelectrolyte solutions with macroions of linear structure, and also to the analysis of the electro-orientation effect [19, 22–24].

B. General Characteristics of NESP

During the last few decades, the use of the ideas of electrochemical macrokinetics has ensured a qualitatively new level of development in colloid electrochemistry: the electrochemical macrokinetics of colloids, which included a number of advances (Fig. 2) justifying that NESP have become an essential part of many major branches of modern colloid science:

Electrochemical thermodynamics of colloids	Electrochemical macrokinetics of colloids
1. Interface electrochemistry	*1. Nonequilibrium double layer*
• Equilibrium double layer (DL)	• Induced dipole of particle
2. Classical electrokinetic phenomena	*2a. Classical electrokinetic phenomena at DL disequilibration*
• Electroosmosis	• Influence of DL disequilibration on electrophoresis, electroosmosis, streaming potential, Dorn effect, vibration potential, ESA
• Electrophoresis	
• Streaming potential	
• Dorn Effect	*2b. New electrokinetic phenomena*
• Vibration potential	• Diffusiophoresis
• ESA	• Capillary osmosis
	• Reverse capillary osmosis
	• Electrochemical theory of desalting by means of reverse osmosis
	• Electrodiffusiophoresis
	• Anisotropy of electrophoresis
	3. Nonequilibrium electrosurface phenomena
	3a. Linear
	• Surface conduction
	• Low frequency dielectric dispersion
	3b. Nonlinear
	• Electroorientation in a.c. field and electrooptic phenomena
	• Electrorotation in rotating field
	• Steady electrorotation in d.c. field
	• Dipolophoresis
	• Aperiodic electrophoretic drift
	• Nonlinear electrophoresis at large Peclet numbers
	• Electroviscous effects
4. Equilibrium surface forces	*4. Nonequilibrium surface forces*
• DLVO theory	• Diffusiophoretic interaction
	• Electrocoagulation
	• Dynamics of colloid particle interaction and slow coagulation
	• Theory of electrofiltration
	• Microfiltration

FIG. 2 A comparison of the electrochemical thermodynamics and macrokinetics of colloids.

1. A number of new phenomena were discovered.
2. DL disequilibration was incorporated into the theory of classical electrokinetic phenomena.
3. Based on the theory of the nonequilibrium DL, the dynamics of particle interactions became the subject of study to incorporate substantial refinements into the theory of slow coagulation.

Some fields where these advances were particularly noticeable are:

- *Membrane phenomena.* The thermodynamics of irreversible processes enables the interrelation between different NESP to be established. In this way, the relationship between reverse osmosis on charged membranes, with pore size commensurable with the diffuse layer thickness, and capillary osmosis can be determined. Therefore, it is possible to apply the methods of the theory of NESP to the theory of reverse osmosis and to substantiate its newer version, namely, low-pressure reverse osmosis [25].
- *Transport phenomena and separation.* NESP cause the transport of disperse particles under the influence of solution concentration gradients (diffusiophoresis [26, 27] and electrodiffusiophoresis [28]), and the inhomogeneity of alternating electric and concentration fields within the diffusion layer arising on passing alternating electric current (aperiodical electrodiffusiophoresis). These phenomena offer new opportunities for the control of coating processes [28] and for separation processes. Typical examples of baromembrane processes are reverse osmosis, ultrafiltration, and microfiltration. Among electromembrane processes, electrodialysis is especially well known. However, with the help of an electric field one can certainly perform more than just a demineralization process. Electromembrane separation processes of colloid and semicolloid systems and of macromolecular solutions may naturally be referred to as electrofiltration. Electrofiltration of a colloid is an electromembrane analog of microfiltration, and electrofiltration of macromolecular solutions and semicolloids is the analog of ultrafiltration. Also, the whole variety of NESP and equilibrium electric surface phenomena manifests itself in the electrofiltration processes [29].
- *Nonequilibrium surface forces.* Nonequilibrium surface forces relate to the nonequilibrium DL in the same way as equilibrium surface forces (described by the DLVO theory) relate to the equilibrium DL. These forces are conditioned by the nonequilibrium DL, caused by electric current [30], liquid flow [31] and phase development [32]. The theory of reversible and irreversible electrocoagulation [29, 30] is based on the nonequilibrium surface forces. The DL, polarized by the liquid flow, can influence the particle deposition during flotation [33, 34] and filtration. Here, electrokinetic phenomena – diffusiophoresis and dipolophoresis – are important. Similar conditions are brought about by phase transitions (dissolution; crystallization in electrolyte) and exchange processes which allow vital activity of the cell [32].
- *Disequilibration of the inner part of the DL and its manifestation in electrokinetic phenomena and particle interaction.* Most investigations have been concerned with the effects caused by the concentration polarization of the diffuse part of the DL. During the last few years, interest in the Stern layer disequilibration has grown [11, 12, 16, 35–37]. The effects of the Stern layer conductance on electrokinetic transport and on the low-frequency dielectric properties are being inves-

tigated. The problem of Stern-layer disequilibration is also important in connection with its manifestation in the dynamics of colloid particle interaction.

C. Electrokinetic Phenomena

Electrokinetic phenomena arise as a DL response to different external forces, which cause ion fluxes and liquid flux within it. Since these fluxes extend outside the DL and consequently become measurable, electrokinetic phenomena are a valuable source of information about charge and potential in the vicinity of a surface. Electrokinetic phenomena are typically second-order phenomena – from the point of view of the thermodynamics of irreversible processes, they are described by nondiagonal terms of the matrix of kinetic coefficients – thermodynamic forces of a certain kind create fluxes of another type. For example, in electro-osmosis and electrophoresis an electric force leads to a mechanical motion, and in streaming current (potential) an applied mechanical force produces an electric current (potential).

There is an essential difference in the definitions of NESP and electrokinetic phenomena: Some NESP *are not second-order phenomena*, and consequently cannot be qualified as electrokinetic phenomena. Furthermore, the DL disequilibration is often negligible, for instance, in the case of electrokinetic phenomena in straight capillaries of arbitrary radius, and also for weakly charged surfaces of large radius of curvature (this means large electrokinetic radius, $\kappa a \gg 1$, and small Du number, see below). Thus, at least in these important regimes of electrokinetic phenomena, they cannot be qualified as NESP.

D. Coupling and Decoupling Electrokinetic Phenomena and Surface Conduction

The DL response to external electric fields causes also surface conduction (a first-order phenomenon). In this case, the force (electric field) and the flux (surface current) are conjugate. In addition, when the applied field is alternating, the surface current will also be alternating, and its out-of-phase component causes another interesting phenomenon known as *dielectric dispersion*. Obviously, both surface conduction and dielectric dispersion are related to electrokinetic phenomena [17], but being first order they cannot be strictly considered as electrokinetic phenomena themselves.

Let us consider another example: the electro-osmotic slip or streaming current can occur independently of surface conduction in straight capillaries. However, the coupling of streaming current and surface conduction manifests itself in streaming potential. On the other hand, there is usually no surface conduction independent of electro-osmotic transport. Correspondingly, the Bikerman equation [38, 39] for the specific surface conductivity of the diffuse layer, $K^{\sigma d}$, consists of two terms. The first one, K_m^σ takes into account the transfer of diffuse-layer ions relative to the liquid due to migration in the electric field, whereas the second term K_{eo}^σ covers the convective charge transfer by electro-osmotic flow. When the dimensionless electrokinetic potential $\tilde{\zeta}$ is high enough, i.e., $z\tilde{\zeta} \gg 1$, only one ionic species, the counterion, has to be considered. The Bikerman equation thus reduces to

$$K^{\sigma d} = \frac{K}{\kappa} \exp(\tilde{\zeta}/2)\left(1 + \frac{3m}{z^2}\right) \tag{3}$$

where K is the bulk conductivity, $\tilde{\zeta} = e\zeta/kT$ is the dimensionless zeta potential, z is the counterion valence, and m is the dimensionless ionic mobility. Here, the second term represents the electro-osmotic role in the charge transport in the diffuse layer. Very often, $m = 0.15$ [1, 2].

Decoupling of electrokinetic phenomena and surface current is easy in the case of straight capillaries, because it is then possible to determine independently the streaming current (and calculate ζ from it) and the surface conduction (from which K^σ can be calculated). These are two independent measurements that yield independent information about the DL, namely ζ and K^σ. This is the necessary condition for the quantitative characterization of a particle surface. Such a decoupling is more difficult for other electrokinetic systems, including the simplest model system, namely, a dilute suspension of monodisperse spherical particles. Actually, the surface conductivity can be measured with sufficient accuracy only if its role is commensurate with the bulk conductivity K [2]:

$$\frac{K^\sigma}{Ka} \approx 1 \tag{4}$$

This dimensionless ratio has been assigned the name of "Du" [17]. Since the surface current is sufficiently strong it influences the tangential electric field distribution within the DL and correspondingly the electro-osmotic slip. In other words, the stream of charge influences the stream of matter [2] and their coupling has not been disregarded by the condition (4). This influence leads to DL polarization [1–4], which cannot be ignored under the condition (4).

The rigorous theories of Overbeek and Booth [1, 20] for DL polarization and its influences on electrophoresis may be compared with Henry's theory [40], which accounts for the surface current influence, while disregarding DL polarization. The former theories are more exact and general. This leads to the conclusion that, whenever the surface conduction must be taken into consideration, DL polarization is also an essential factor. The decoupling is possible [2] in the limiting case:

$$\mathrm{Du} = \frac{K^\sigma}{Ka} \ll 1 \tag{5}$$

which is the second necessary condition for applicability of the Smoluchowski theory. This leads us to distinguish two approaches to the theory of electrokinetic phenomena [2]:

1. Equilibrium DL and electrokinetic phenomena (Du $\ll$ 1).
2. Nonequilibrium DL and NESP (Du $\sim$ 1).

Under the condition (4), an electrokinetic phenomenon is at the same time a nonequilibrium electric surface phenomenon. Hence the term *NESP* can be used, as it is done above to specify the second approach. Although the latter yields a more exact and profound understanding of electrokinetic phenomena, the first approach completely retains its independent significance, since under definite conditions DL polarization is very weakly manifested in electrokinetics. The DL polarization extremely complicates the mechanisms of electrokinetic phenomena.

In the classical regime of electrokinetic phenomena, when the Smoluchowski theory is valid, only the ζ potential can be measured. The nonequilibrium state of the DL provides an opportunity to measure in parallel ζ and K^σ. Both the diffuse and Stern layers can be characterized with NESP, because both ζ and K^σ are measurable. NESP advantages will be considered with respect to electrokinetic surface characterization in Section V.

It must be mentioned that LFDD, although in Lyklema's classification, is not strictly an electrokinetic phenomenon, but presents the characteristic feature of being fully determined by the mobile charge in the DL, and thus provides an opportunity of measuring surface conductivity even in the first interpretation.

III. ELECTROKINETIC SURFACE TRANSPORT MODELS AND THEORIES OF ELECTROKINETIC PHENOMENA AND NESP

The exact laws of statistical physics, electrodynamics, and hydrodynamics are the foundation of the theories for DL, electrokinetics, and colloid stability. However, model considerations are used in such aspects as space dependence of local parameters like dielectric constant, solution viscosity, and ionic mobility, and in formulation of the boundary conditions. In electrokinetics, the central notion is that of *electrokinetic potential* ζ, introduced through the use of the *slipping plane* model. Unfortunately, there is no unambiguous way of determining the position of the slipping plane. We will consider this issue in the following paragraphs.

A. Electrokinetic Potential and Slipping Plane

Tangential liquid flow along an immobile charged surface can be caused by applied mechanical force or by an external electric field (electro-osmotic slip). Experience and recent molecular dynamic simulations have shown that in such tangential motion usually a very thin layer of fluid (with thickness d^{ek}) remains adhered to the surface and is called a hydrodynamically stagnant layer (SL). The space charge for $x > d^{ek}$ is electrokinetically active, and the particle (if spherical) behaves hydrodynamically as it has a radius $a + d^{ek}$. Its surfce potential with respect to the bulk is the potential at the slipping plane, and is known as electrokinetic or ζ potential. The countercharge part beyond that plane is called the electrokinetic charge σ_ζ. It can be expressed through the measured ζ with the use of Gouy–Chapman theory, which yields the space charge distribution within the diffuse part of the DL. For water, the SL is not more than a few molecular layers thick [17] but it can no longer be ignored in electrokinetics, because a substantial fraction of countercharge is located in that layer. The σ_ζ computed from ζ seldom amounts to more than 3–5 μC/cm^2, whereas the surface charge σ_0 may be 10 times as high, even on nonporous surfaces [17].

B. Standard and Nonstandard Electrokinetic Models

There is no convective ion flux behind the slipping plane, according to its definition. However, this definition does not exclude the possibility of ion fluxes behind it under the action of a tangential electric field (ion-migration fluxes). The majority of theories disregard this possibility, thus implying the absence of tangential ionic transport behind the slipping plane under any conditions. As a result, a real particle with its liquid SL is replaced by a virtual solid particle with radius equal to that of the particle radius plus d^{ek} (rigid particle ζ model [17]).

   A single DL characteristic is present in this model, namely, ζ. If possible, measurements of both electrokinetic phenomena and surface charge by titration will allow one to evaluate its usefulness, and this has been done for some systems (see Section III.C). However, the interpretation of integrated electrokinetic measurements for other systems is impossible using a rigid particle ζ model, that hence cannot be considered as universal.

   All classical electrokinetic theories [1, 18, 41] are based on the assumptions of the standard electrokinetic model (SEM), in which it is also emphasized that the notions of slipping plane and electrokinetic potential are associated with a surface that is smooth at the molecular level. Consideration of the roughness of the interface in electrokinetics is the single difference between the SEM and rigid particle ζ models. However, this difference is very important because porous or rough surfaces are quite abundant. Hence, a nonstandard electrokinetic model (NSEM) was proposed [14] to discriminate between those surfaces that fulfill the assumptions of the SEM, and those that must be described with an NSEM model.

C. Experimental Verification of the Standard Electrokinetic Model

The SEM is based on a number of untested assumptions. Clearly, it cannot be universal, i.e., it is not suitable for the description of all systems. The reuqirement that the surface should be homogeneous, smooth, and impervious can be satisfied by properly selecting the system. Systems that meet these requirements and have been used for checking the standard model, are emulsion droplets in water [42], silver iodide crystals [43], partially fused molecularly smooth quartz [44], mica, etc. The titratable and electrokinetic charges were measured for these systems. As these investigations were performed in broad ranges of electrolyte concentrations, the small difference between the charges, originating from independent phenomena, point to the correctness of the model for the slipping plane (electrokinetic potential).

D. Dynamic Stern Layer

Smoluchowski and Bikerman paid no attention to the Stern-layer role in electrokinetics. Later this approach was introduced into the rigid particle ζ model or SEM. In parallel, another approach has been developing for many decades. In particular, the notion of quasiequilibrium between inner and outer parts of the DL and the bulk was introduced in the first version of the theory [5]. Some additional equations, namely, Eqs (100) and (101) in Ref. 19, were obtained later on. The estimated moderate Stern-layer polarization was justified by the intuitive estimation that the surface ion diffusivity is small in comparison with the bulk value. If so, the ionic flux in a Stern layer must be small in comparison with the tangential ionic flux in the diffuse layer and can be neglected. In this respect, the Zukoski and Saville estimation [45] of the surface diffusivity of the adsorbed ion as comparable to the bulk diffusivity was unexpected and attracted attention [46].

E. Conduction Behind the Slipping Plane

On the basis of experimental results, Bikerman [39] arrived at the conclusion that gels represent a much greater obstacle to the motion of a liquid than to the passage of a current. The same conclusion is valid for charged surfaces covered with an adsorbed layer of macromolecules, or even for uncharged (or charged) surfaces with an adsorbed layer of polyelectrolyte. A charge fixed on the surface of a particle or on segments of an adsorbed macroion is compensated for by mobile counterions partly distributed inside the gel and partly in the diffuse layer beyond it.

What we originally called anomalous surface conductivity [2] is caused by ion flow along that portion of the diffuse layer that is confined between the rough surface of the particle and the slipping plane. However, this term should now be discarded for a number of reasons. The term anomalous was proposed a long time ago, when there was a shortage of experimental data regarding the conductivity of the SL. The term additional surface conductivity $K_r^{\sigma i}$ was used later [14]. Subscript "r" denotes the roughness role in the additional conductivity rise. There is a potential drop across the portion of the diffuse layer between the rough surface and the slipping plane, corresponding to the difference between effective values of Stern and electrokinetic potentials ($\Psi_{ef} - \zeta_{ef}$) (see Section VIII.B).

Recently, Lyklema [17] gave convincing evidence for the stagnant-layer conductivity (SLC):

1. The fact that many experimental K^σ values exceed $K^{\sigma d}$.
2. The observation that ζ potentials obtained for a given system and fixed conditions, but by different techniques, may be substantially different, but can be harmonized if $K^{\sigma i}$ contributions are introduced.
3. For some systems, computed ζ potentials depend in a bizarre way on the electrolyte concentration unless the interpretation of the experiments is improved to take account of a nonzero $K^{\sigma i}$.

Important results regarding SLC were obtained [47, 48], and recently, great attention has been paid to SLC [49–53].

F. Theories of Electrokinetic Phenomena and NESP

Of the various possibilities, we will only deal with those cases for which the usefulness of electrokinetic characterization has already been substantiated. Thus, in the megahertz region, an induced dipole moment (i.d.m.) is produced by Maxwell–Wagner (MW) polarization [54, 55] at the expense of the space charge arising in the electrolyte layer adjacent to the particle surface whose thickness is of the order of κ^{-1}. The time of MW polarization can be estimated by the Einstein equation, i.e.,

$$\tau_{MW} = \frac{\kappa^{-2}}{2D} \tag{6}$$

where D is the ion diffusion coefficient.

This well-known equation was obtained in the MW theory (see, e.g., Ref. 2), which expressed the electric conductance of a suspension in terms of the electric conductances of the particle and of the medium, and volume fraction. O'Konski [56] pointed out that in the absence of particle conductance, but in the presence of surface conductivity, the MW polarization mechanism is also operating and showed that the equation for i.d.m. is also valid in this case if the conductance of the particle is replaced by $2k^\sigma/a$. Thus, the surface conductance of spherical particles can be determined from the dependence of the measured suspension conductance on the volume fraction. The contribution of Hunter and O'Brien and their collaborators [57, 58], who conducted systematic investigations of MW dispersion as the method of determining specific surface conductivity, was significant.

The process of i.d.m. formation under the effect of direct current is complicated by the establishment of a gradient of neutral electrolyte concentration at distances of the order of the particle radius [7] for which the required time τ_c exceeds τ_{MW} under the condition (2). The value of τ_c is

$$\tau_c = a^2/2D \tag{7}$$

While nonequilibrium surface phenomena in the MW relaxation region can be characterized as true electric or electrohydrodynamic phenomena, the stationary or low-frequency nonequilibrium surface phenomena are diffusive–electric or diffusive–

electrohydrodynamic phenomena. Overbeek [1] was the first to develop the theory of DC electrophoresis in terms of the diffusive–electrohydrodynamic phenomenon.

The usefulness of the thin DL approximation [5, 59] lies in the fact that it enables us to obtain expressions for different nonstandard electrokinetic models. The first of such expressions became known as the Dukhin–Semenikhin equation [60], and it accounts for the role of the conductivity behind the slipping plane. O'Brien and White [61] later elaborated a numerical procedure in order to obtain the electrophoretic mobility as a function of zeta potential, particle radius, and valences, concentrations, and mobilities of the ions in solution. In this theory, the authors identified the Ψ_{ef} and ζ potentials in the Dukhin–Semenikhin equation and thus have neglected one of its main advantages, namely, the possibility of describing electrophoresis within the framework of the nonstandard electrokinetic model. A decade later, Midmore et al. [62] extended O'Brien and White's theory, and eliminated such simplification.

If the frequency of the alternating field is not too high (smaller than the MW characteristic frequency), a local equilibrium between the DL and the adjoining volume of electrolyte has time to set up during a cycle. This allowed [7] generalization of the theory of thin DL polarization for the case of an alternating field of frequency $\omega < 2\pi/\tau_c$. The polarization of a particle turns out to be associated with the concentration gradient along the outer boundary of the DL, which periodically varies in time with some delay with respect to the applied field. This phase delay is mathematically expressed by the fact that the dipole moment d_0^* is a complex quantity.

Because of the presence of conduction, the dispersion medium (electrolyte) polarizes also with a substantial phase delay, so that its permittivity ε^* is also complex:

$$\varepsilon^* = \varepsilon' - i\frac{K}{\omega} \tag{8}$$

where ε' is the permittivity of the medium. The real part, $\Delta\varepsilon$, of the variation $\Delta\varepsilon^*$ in the permittivity of a dispersion medium caused by the introduction of particles with concentration N can be inferred from the MW theory [54, 55] as

$$\Delta\varepsilon = \text{Re}\{\Delta\varepsilon^*\} = 4\pi N(\text{Re}\{\varepsilon^*\}\text{Re}\{C_o^*\} - \text{Im}\{\varepsilon^*\}\text{Im}\{C_o^*\}) \tag{9}$$

where C_o^* is the induced dipole moment coefficient, defined, for a spherical particle of radius a, by

$$\vec{d_o^*} = C_o^* a^3 \vec{E} \tag{10}$$

$\vec{E}$ being the applied field.

As a result of multiplying the imaginary parts of the particle dipole coefficient and ε^*, $\Delta\varepsilon$ substantially grows as ω decreases, because of the unlimited growth of the imaginary part of ε^* according to Eq. (9). Lyklema [17] simplified the equation derived in [4, 7] for the LFDD of a suspension as follows:

$$\frac{\Delta\varepsilon(\omega)}{\varepsilon'} = \frac{9}{16}\varphi(\kappa a)^2\left(\frac{2\text{Du}^d}{1 + 2\text{Du}^d}\right)^2\frac{1}{(1 + \omega\tau_c)(1 + \sqrt{\omega\tau_c})} \tag{11}$$

where φ is the volume fraction, and Du^d can be obtained by means of Eq. (5) and Bikerman's expression for the surface conductivity of the diffuse layer [39].

The theory has been confirmed by other theoretical models [36, 45, 63, 64] and also by experimental determinations of the dielectric constant of suspensions [36]. It was extended to disperse systems with arbitrary values of κa [65]. In all these studies, the influence of the salt diffusion in the bulk electrolyte solution adjoining the particle on its DL polarization occurs as the main mechanism of LFDD. Saville called it the volume diffusion mechanism (VDM).

Recently, the system of equations of the thin-DL approximation was numerically solved [66]. It was concluded that in many cases the theoretical model [5, 6] provided LFDD estimations sufficiently good to be used for the interpretation of experimental results. On the other hand, theoretical predictions for conductivity do not properly reproduce numerical results for highly charged particles even under condition (2). The numerical calculation showed that the charged region surrounding the particles extends further than in equilibrium. This result is in agreement with the first version of the theory [5]: initially, it was assumed that the deviation from electroneutrality was small. Later, this was replaced by the approximate condition of complete electroneutrality.

The investigations of Saville's group [35, 45] attracted attention to the dynamic Stern layer in electrokinetics and LFDD. Initially, Schwarz [67] proposed lateral transport of counterions as the main mechanism of LFDD. Counterions were supposed by Schwarz to be capable of moving due to electromigration and diffusion along the surface, but to be unable to exchange with the adjacent electrolyte (surface diffusion model, SDM). In contradistinction, free exchange of ions between the diffuse part of the DL and the adjacent electrolyte was stated [5, 7] to be a direct consequence of the standard model since it is precisely that free exchange that determined the local equilibrium between a thin DL and its adjacent electrolyte. This means that the Schwarz model of counterions' "displacement without exchange" may well apply to the Stern layer only (not to the diffuse part). More recently, Kijlstra et al. [68], Rosen et al. (69), and Razilov et al. [70, 71] developed models in which the Stern layer is supposed to be in equilibrium with the neighboring diffuse layer.

Shilov and Dukhin [21] modified the Schwarz theory by incorporating the necessary screening by the field-induced diffuse layer, in the absence of an equilibrium DL (i.e., in the isoelectric point). Lyklema et al. [36] modified the theory for situations of nonzero Stern potential, still under the restriction of small values of Du, where concentration polarization is negligible. More recently, Razilov et al. [72, 73] extended this theory for moderate values of Du [Du = $O(1)$], by studying the mutual influence of polarization of the Stern layer and the diffuse layer. It is noteworthy that this case is important for surface conductivity measurements and for its use in surface characterization.

Hence, a theory describing polarization of the Stern layer is only available for limiting situations. An intermediate situation has been the object of a study by Hinch et al. [74], who took into account retarded surface-site reactions and showed their effect on LFDD. However, in their model surface ions were not allowed to displace along the surface.

Minor [75] derived the most general equation. It comprises the equations of all preceding theories that can be obtained in some extreme cases. However, the use of this theory for experimental interpretation is difficult because the characteristic times needed for tangential redistribution of ions in the Stern layer and for establishing equilibrium between it and the diffuse layer are two additional unknown values used in his theory.

Delgado et al. [76] proved that one can separate the dispersion curves of the SDM and VDM by varying the concentration of particles in the suspension. it has been found that VDM is much more strongly dependent than SDM on particle concentration. The reason is that, when the particle concentration increases, the characteristic length for the propagation of volume diffusion processes decreases together with the decrease in free electrolyte volume, whereas the characteristic length for the surface diffusion remains constant. Correspondingly, when the particle concentration is raised, the relaxation time of the VDM effect must decrease, whereas it must remain constant for the SDM mechanism. These authors demonstrated that VDM prevailed in the systems studied. The relative importance of VDM and SDM must depend on the specific system considered. The results of Barchini and Saville [77] could be explained [76] as a superposition of surface and volume diffusion of ions.

The binary-electrolyte approach is employed even in suspensions with essentially nonbinary media with respect to the LFDD theory. The influence of small concentrations of "additional" ions, i.e., the ions that are present in addition to the binary electrolyte, on LFDD is considered in Ref. 78. These additional counterions can undergo significant adsorption oscillations that can increase to a large extent the dielectric increment of a suspension. This phenomenon provides another opportunity for surface characterization.

G. Effect of Sol Concentration on Electrokinetic Phenomena and on NESP

When sols are not dilute, particle–particle interaction influences the electrophoretic mobility. The interaction has a three-fold nature: overlapping of hydrodynamic, electric, and ion diffusion fields created by neighboring particles. There are several pure hydrodynamic cell models that are successfully applied for solving complicated hydrodynamic problems in concentrated systems. Electrokinetic cell models are the results of some generalization of hydrodynamic cell models. There are many ways to perform this generalization and (correspondingly) many ways to create different electrokinetic cell models. The well-known Onsager relationship [79] and Smoluchowski law, both of which are valid no matter the suspension concentration are two criteria which determine a proper choice of a cell model. The electrokinetic Levine–Neale cell model [80] for the limiting case corresponding to the conditions (2) and (5) does not agree simultaneously with the exact Smoluchowski law and the Onsager relationship.

In order to overcome this disagreement, a cell model [81] was formulated from the very beginning on the basis of the following requirements: fulfillment of the Onsager relationship; reduction to the Smoluchowski law at small Du, and agreement with the well-known asymptotic laws available for dilute suspensions. Thus, A. Dukhin et al. [82, 83] have elaborated the vibration potential theory for concentrated systems with use of the Shilov–Zharkih cell model. An experimental test on silica Ludox Tm (30 nm) and rutile R-746 Dupont (about 30 nm) confirmed the theory [83] for particle concentrations up to 45% (v/v).

H. Extended Electrokinetic Characterization of Polydisperse Colloids

The sensitivity of DL polarization to particle size causes a dimension dependence of the electrophoretic mobility. There are commercial devices for the measurement of distributions in electrophoretic mobility and in particle radius for a polydisperse suspension. Mathematical algorithms for calculating the electrokinetic and effective Stern potentials were developed [84], and their extension to concentrated suspensions has been accomplished as well [83]. Thus, there is an opportunity for the extended characterization of polydisperse systems with respect to electrokinetic potential and surface conductivity by means of the vibration potential measurements, for instance.

IV. GENERALIZED STANDARD ELECTROKINETIC MODEL AND ELECTROKINETIC CONSISTENCY TEST

A. Generalized Standard Electrokinetic Model

Recently, Lyklema et al. [85] proved that "stagnant layers are general properties of liquids adajcent to solid walls." This is the result of molecular dynamics (MD) simulation. At the same time they studied the ionic mobilities in the SL with MD computation and found for monovalent counter ions a value of 0.96 for the ion mobility ratio between the SL and the bulk. New experiments on the conductivity and streaming potential of polystyrene latex plugs [86, 87] illustrated the aforesaid results of MD. It is shown that, for this system, accounting for the SLC is crucial to converting the streaming potentials into ζ potentials. For nonpenetrable surfaces, lateral mobilities of monovalent ions in the Stern layer are not much lower than those in the bulk.

The collected evidence, comprising the results described in Section IV.D and the newest results [85–87], strongly suggests that "at least in a number of systems conduction behind the shear plane does take place" [86]. This understanding of the universal mechanism [85] of mobile SL ions characterizes a new standard for electrokinetic transport modeling. Correspondingly, a generalized standard electrokinetic model (GSEM) can be proposed. Because of its more general character, the applicability range of GSEM is wider. Its main features include: (1) conduction behind the shear plane (contrary to SEM assumptions), (2) there is no demand on surface smoothness on the molecular level (another important difference from SEM), and (3) the DL is described by at least two parameters, namely, ζ and $K^{\sigma i}$, the situation $K^{\sigma i} = 0$ corresponding to SEM.

B. Incorporation of the Generalized Standard Electrokinetic Model into the Theory

In view of the success of the Gouy–Stern theory in interpreting static DL properties, it is consistent to do the same for the surface conductivity [17], i.e.,

$$K^\sigma = K^{\sigma i} + K^{\sigma d} \tag{12}$$

As a first approximation, $K^{\sigma i}$ is determined by the ions in the SL and $K^{\sigma d}$ by the diffuse part. The assumption is made that the surface charge is immobile. The assumption that the inner layer contains only one type of ion (counterions, either or not specifically adsorbed) with mobility u_i and surface charge density σ_i yields:

$$K^{\sigma i} = \sigma_i u_i \tag{13}$$

A special notation:

$$\Theta = K^{\sigma i}/K^{\sigma d} \tag{14}$$

is introduced to generalize electrokinetic theories to account for SLC. This generalization is achieved [17] by means of a more general expression for K^σ, namely, Eq. (12). This leads to the generalization of the equation for Du number [17]:

$$\mathrm{Du}(\zeta, \Theta) = \frac{1 + \Theta + 3m/z^2}{\kappa a} \exp(\tilde{\zeta}/2) - 1 \tag{15}$$

This equation with $\Theta = 0$:

$$\mathrm{Du}(\zeta, 0) \cong \mathrm{Du}(\zeta) \tag{16}$$

follows from the Bikerman equation for $K^{\sigma d}$ and corresponds to the rigid particle ζ model.

The electrophoretic mobility is a function of ζ and Du. The LFDD depends on Du according to Eq. (11). The GSEM in electrophoresis and LFDD theories is accounted for by substituting into Eq. (11) the more general expression, Eq. (15).

C. Reconsideration of Conditions for ζ Determination with Use of the Smoluchowski Theory

The majority of published ζ potential values are determined with the use of one electrokinetic technique and the Smoluchowski equation. Measurement of ζ enables the surface conductivity $K^{\sigma d}(\zeta)$ to be calculated according to the Bikerman equation, and the condition for applicability of the Smoluchowski equation (5) to be specified:

$$Du(\zeta) \ll 1 \tag{17}$$

However, the stronger the conduction behind the slipping plane, the narrower is the area of Smoluchowski theory applicability. Under actual experimental conditions even $\Theta = 3$–5 is possible. Thus, a large quantitative difference between the rigorous condition:

$$Du(\zeta, \Theta) \ll 1 \tag{18}$$

and the simplified condition (17), which is not sufficient if so large a ζ value can exist. This means that *most ζ values reported in the literature should be revised, especially those obtained at low salt concentrations*, which correspond to larger Du values.

For some systems and for some conditions, Du can be evaluated and the condition (17) can be examined using the surface charge measured by titration, the Gouy–Chapman theory, and the Bikerman equation. If Smoluchowski theory applicability is justified in this way, one electrokinetic technique is sufficient for ζ measurement. But in the absence of independent information, the general principle of justification for application of the Smoluchowski theory is the use of two experimental techniques. However, for some systems (sufficiently large particles, sufficiently low ζ, sufficiently high electrolyte concentration) this complication of the experimental procedure can be avoided, because for such systems and even for Θ of about 5–10, the difference between Du (ζ) and Du (ζ, Θ) is not large. A criterion to ensure that only one electrokinetic technique is enough can be given:

$$Du(\zeta) \ll 1/10 \tag{19}$$

D. Electrokinetic Consistency Test and Current State of Electrokinetic Theory

Attempts to interpret the results of integrated electrokinetic investigations leads to formulation of the notion of electrokinetic nonconsistency, because the investigation of the different phenomena often gives different values for the electrokinetic potential of the same system. This contradiction is eliminated by proper use of a SLC model.

All electokinetic theories can be classified into two groups with respect to the electrokinetic consistency test:

1. Theories for which the electrokinetic consistency test is possible for systems with or without SLC:
   Monodisperse spherical particles with thin DL. The equations for colloid conductivity, LFDD, and electrophoresis were generalized with the use of Du (ζ, Θ) by Lyklema [17] and were aplied to the interpretation of experimental data for

(Stober) silica and hematite. This approach provided electrokinetic consistency and yielded ζ values. These theories can be recommended for rigorous investigation with the electrokinetic consistency test for spherical particles. If monodisperse isometric particles can be modeled as spherical ones, this is an appropriate approach as well. A good agreement has been shown to exist between the equations derived in Refs 17 and 88 for electrophoresis in the presence of SLC.

The results of some modern theories developed for the thin-DL case, accounting for DL polarization, have to be refined by means of the substitution in the final equation of Du (ζ, Θ) for Du (ζ). In electrophoresis, such a refinement will lead to an increase in ζ, as compared to published values, calculated without SLC. On the other hand, in the case of conductivity of suspensions and their LFDD, this refinement leads to values of ζ lower than those obtained without considering SLC. Interestingly, this provides the electrokinetic consistency, because ζ determination in the framework of the rigid particle ζ model causes zeta potential underevaluation in electrophoresis, and overestimtion in LFDD.

2. Theories that make the electrokinetic consistency test possible only for systems without SLC. The rigid particle ζ model is used in these theories. There is no objection regarding the application of these important theories in the case of particles with low SLC. However, the electrokinetic consistency test will not be possible when they are applied to particles with an essentially high SLC. They are, nevertheless, applicable to spherical particles with thick DL, and to spheroidal and cylindrical particles with thin DL. The most basic assumption is that ζ potentials are fully defined by the nature of the colloid, particle, or pore size, its surface charge, the pH and electrolyte concentration in the solution, and the nature of the electrolyte and the solvent. Stated otherwise, for a colloidal surface with all these parameters fixed, ζ is a well-defined characteristic. So, if two investigators find different values for ζ of a given system under fixed conditions there are only two options. Either the systems were different, e.g., because of minor specific adsorption of impurities, or the conversion of the experimental signal into ζ was flawed [17].

V. EXTENDED ELECTROKINETIC SURFACE CHARACTERIZATION

A. Electrokinetic Potential and Stern Potential

The present state of thinking is that the slip-plane position is close, or for practical reasons identical, to the outer Helmholtz plane (OHP) and correspondingly the difference between Stern and electrokinetic potential is not large [17]. As the identification of electrokinetic potential and Stern potential is an approximation, it perhaps cannot be considered as universal. The subdivision of the solution side of the DL into two parts by the OHP is rather pragmatic, though somewhat artificial, because all complications regarding finite ion size, specific adsorption, discrete charge, and surface heterogeneity reside within the Stern layer. As a result, its thickness d is left unspecified for the moment, and it can only be said that d is less than 1 nm, with a large uncertainty in its exact value [17, 89].

In addition, there is no unambiguous way of determining the position of the slipping plane. Apart from the fact that this notion is already an abstraction from reality, the positioning requires a detailed knowledge of potential–distance distribution and this requires a model [89]. *A priori* it is not obvious why the OHP (a static feature) should be identical to the slipping plane (dynamically determined, see Ref. 89): "However, for practical purposes the outcome might be that the two are close enough to consider them as the same" [89], i.e., Ψ_d and ζ are close. At least, the difference between these potentials cannot be large at low electrolyte concentration. The diffuse layer is thick at low concentration and this makes rather unimportant the possible difference between the distances of OHP and slipping plane to the wall.

B. Generalized Standard Electrokinetic Model and Extended Electrical Surface Characterization.

The existing trend to evaluate the Stern potential with the use of ζ deserves attention. A new opportunity arises with SLC measurement. As a first approximation, σ_{ek} characterizes the diffuse layer, and SLC characterizes the Stern layer (with the correction for difference in ion mobilities of Stern layer ions with respect to those in the bulk). This correction is sometimes small and can be neglected. For example, it was found that for K^+ ions on (Stober) silica, the aforesaid ratio is 0.96, and 0.7 for Cl^- [90]. Recent studies [86] on sulfate–polystyrene lattices yields a value of 0.89 for this ratio for the monovalent ions studied. An even larger value, namely 0.98, for this ratio follows from the measurements of streaming potentials and conductivities of latex plugs [87]. Thus, with this ratio value of near unity, combined measurements, comprising two electrical surface phenomena, enable us to determine *the charge distribution between diffuse and Stern layers*, which can be called *the extended electrical surface characterization.* This new level in surface characterization is possible as well, if the difference in mobilities of adsorbed and bulk ions is not negligible. Three measurements have then to be fulfilled to determine the charge distribution between Stern and diffuse layers and the adsorbed counterion mobility.

C. Mobile Charge and Surface Charge

In many investigations, the convective transport of the surface charge, i.e., the term K_{eo} in the Bikerman equation, is disregarded, a hypothesis that is justified due to the small value of the electrokinetic charge. This simplification leads to the possibility of mobile charge determination [2] for the diffuse part of the DL:

$$\sigma_{md} = K^{\sigma d}/u_i \tag{20}$$

The bulk value of counterion mobility u_i can be used within the diffuse part of DL. In the case of monovalent ions, Eq. (20) can be transformed into the equation for the entire mobile charge:

$$\sigma_m = (K^{\sigma d} + K^{\sigma i})/u_i = K^\sigma/u_i \tag{21}$$

because there is little difference in mobility values for adsorbed and free counterions. Equation (21) is more useful because K^σ is a measurable quantity, and because the information about σ_m enables us to estimate the surface charge:

$$\sigma_o = -\sigma_m \tag{22}$$

Thus, the recent information [85–87] about the small difference in mobilities for adsorbed and free monovalent counterions proved the possibility of surface charge determination by means of surface conductivity measurement. Earlier, this method was proposed [2] with the assumption that the ion mobility change within the DL is not large.

Recently, Grosse et al. [91] considered a theoretical model that relates both the high- and the low-frequency relaxation to a single parameter of the particle surface: its surface conductivity. Correspondingly, the dielectric properties of suspensions of polystyrene particles were measured within a unique broad-frequency range. This work and the subsequent research [92] are important for the fundamentals of DL investigation.

D.   Extended Electrokinetic Surface Characterization of Flat Solid Surfaces

Probably, the most important systems for electrokinetic investigations with flat surfaces are polymeric materials. There exists a large variety of materials suitable for the preparation of flat surfaces. Flat polymer samples can, for example, be prepared as thin, smooth, and well-defined layers through techniques like spin coating, solution casting, adsorption, or plasma deposition on top of plane carriers. These flat surfaces are further easily assessible for complementary surface analysis to be compared with the resutls of electrokinetic studies, e.g., surface spectroscopies, microscopic investigations, contact-angle measurements, and ellipsometry.

An experimental setup has been developed and applied to the combined determination of the electrokinetic potential and the surface conductivity of flat surfaces [93]. The key feature of the new device (designated as a microslit electrokinetic setup) is the variability of the distances between two parallel flat sample surfaces (10 mm × 20 mm) forming a slit channel. The setup allows one to decrease this distance down to about 1 μm keeping the surfaces parallel. In consequence, streaming-potential measurements can be performed at a given solid/liquid interface both under conditions where surface conductivity is negligible and under conditions where surface conductivity significantly contributes to the total channel conductivity. The zeta potential is calculated at different channel geometries based on streaming potential and channel conductivity and, alternatively, based ons streaming-current measurements and the dimensions of the cross-section of the slit channel. In a series of measurements, a plasma-deposited fluoropolymer (PDFP) layer on top of a glass carrier, and an adsorbed layer of the blood protein fibrinogen on top of the PDFP layer, were characterized by zeta potential and surface conductivity measurements in different aqueous electrolyte solutions (KCl, KOH, HCl). The contribution of the diffuse layer to the surface conductivity was calculated from the zeta potential, and compared with the experimentally determined surface conductivity. The hydrodynamically mobile charge contributes only about 10% or less to K^σ in all cases although the polymer surface was found to be smooth on the namometer scale. The resulting high SLC is attributed to both the high specific mobilities of the ions accumulated in the inner layer (OH^- and H_3O^+) and to the contribution of these surface charge-creating species to the surface conductivity.

After adsorption of fibrinogen onto the PDFP surface, the SLC increased by about an order of magnitude. The latter fact is assumed to be caused by the presence of mobile ions in the interfacial volume of the adsorbed protein layer. In addition to the electrochemical characterization of the adsorbed protein layer, its hydrodynamic thickness was determined by means of liquid flow measurements with the microslit electrokinetic setup. The obtained value of 48 ± 5 nm correlates well with the protein dimensions given in the literature and is of the order of magnitude of the optical layer extension determined by ellipsometry.

E. Extended Characterization of Nonspherical Particles Based on Electro-orientation and Electrorotation Phenomena

The polarizability of nonspherical particles is anisotropic, i.e., polarizabilities along the long and the short axes differ. The result is that the i.d.m. becomes anisotropic, and the action of an external field on it generates a torque, which orients the particle. A distribution in the orientation of particles is established as a result of the orienting effect of the field and the disorientation effect of the rotational diffusion. Anisotropy of orientations under the effect of the field results in a change in optical properties of the system (electro-optical phenomena [94]) and in the anisotropy of electrical conduction. Investigations of time and field dependence of these effects allow one to determine K^σ. The theory of the polarization of nonspherical particles and their electro-orientation was experimentally confirmed [95] and used to determine K^σ from data on suspensions of synthetic diamonds and polygorskite. Good agreement was observed with K^σ values measured on the basis of orienting conductimetric effects, which demonstrates the correctness of the theory and the experiment. An important achievement is a generalization of these methods over polydisperse systems. In the latter case it was shown [95] that the effect of polydispersity is essential and that its neglect results in substantial errors.

Systematic measurements of electrophoretic mobility, dielectric dispersion, and electric birefringence [96] have been performed as a function of the concentration of nonionic surfactant and salt added to aqueous suspensions of charged rod-like PTFE (Teflon) particles. Such an extended characterization provides more reliable information.

The relaxation of the DL polarization in the case of rotating external electric fields causes a phase lag between the latter and the symmetry axis of the field-induced charge distribution, giving rise to torques on both the particle and its adjacent liquid, and causing a slow rotation of the particle (electrorotation effect). It is shown [97] that the main mechanism determining electrorotation at low frequencies is the rotational component of electro-osmotic slip velocity. Hence, there is a perspective of electrokinetic potential measurement with the use of this new NESP.

VI. EXTENDED ELECTROKINETIC CHARACTERIZATION OF NEUTRAL POLYMER ADSORPTION

When neutral polymer are adosrbed at an interface, they may influence the ζ potential through the adsorption characteristics of the ions present and/or by shifting the position of the plane of shear away from the particle surface [98]. In a more elaborate study [99], researchers were able to distinguish very clearly the effect of the polymer on ion adsorption from its effect on the shear plane. The initial effect is to move the isoelectric point to more positive values with little effect on the maximum value of ζ [99]. The authors attribute this to the adsorption of the polymer with a horizontal configuration on the surfce (called trains). Only at higher levels of adsorbed polymer is there a pronounced formation of loops, which force the shear plane out from the surface.

By analogy with the hydrodynamic layer thickness, one can define an electrokinetic thickness d_{ek} in terms of the effective shift in the plane of shear. The theory for its dependence has been worked out for electro-osmosis and electrophoresis [100, 101] and for streaming current [102]. The effect of poly(ethylene oxide) on the ζ potentials of quartz, antimony sulfide, and silver iodide [103] was investigated. The results were interpreted in terms of the apparent shift in the position of the slipping plane, by neglecting the water flow and the ion flux behind it. This corresponds to the extreme case of an impermeable adsorbed layer. This simple approximation is used in many works, and the conditions of its applicability are established in Ref. 104.

The electro-osmotic slip along a flat charged surface with an adsorbed neutral polymer was also considered. The change in electro-osmotic velocity is characterized by a change in the electrokinetic potential. Its dimensionless values after polymer adsorption, $\tilde{\zeta}_\Gamma$, and before adsorption, $\tilde{\zeta}_0$, are interconnected by means of the equation [104]:

$$\tilde{\zeta}_\Gamma = \tilde{\zeta}_0 \frac{1 + \int_0^{\kappa LM} (f_0/\tilde{\zeta}_0) \sinh y \cdot dy}{\cosh(\kappa LM)} \tag{23}$$

where $f_0(y)$ is the potential distribution in the DL as a function of the distance to the solid surface, L is the thickness of the adsorbed polymer layer, and

$$M = \sqrt{n_0 f_g / \eta} \cdot \kappa^{-2} \tag{24}$$

where n_0 is the segment concentration, which is considered as independent of y; f_g is the coefficient of the hydrodynamic resistance per segment, approximately given by $f_g = 6\pi\eta a$, where a characterizes the linear dimension of a segment.

The exact analytical result of the integration is given in Ref. 104, and it is a rather complicated equation that simplifies to

$$\tilde{\zeta}_\Gamma = 4\tanh(\tilde{\zeta}_0/4)\left[1 + \frac{1}{M-1}\right]e^{-\kappa L} \tag{25}$$

with the condition:

$$\kappa L > 0.5 \tag{26}$$

If M is such that

$$M \gg 1 \tag{27}$$

then Eq. (25) transforms into the well-known equation for the case of an impermeable adsorbed layer:

$$\tilde{\zeta}_\Gamma = 4\tanh(\tilde{\zeta}_0/4)e^{-\kappa L} \tag{28}$$

The results of calculations performed with the more general Eq. (23), and its simplified versions Eqs (25) and (28) are given in Fig. 3. Comparison of the curves 1, 2, and 3 at two values of M leads to three conclusions. First, the simple Eq. (25) is valid under condition (26) because the difference between curves 1 and 2 is negligible in this case. Second, the condition of adsorbed layer impermeability can be specified as

$$M > 10 \tag{29}$$

because under this condition the difference between curves 1, 2, and 3 is almost negligible. Third, the adsorption layer permeability leads to the essential correction of Eq. (28), which disregards the permeability, because the difference between the curves for different M values is large. This difference will be even larger with decreasing M value, which cannot be shown with the use of Eq. [25], because it is not valid for very small M values. For this case, the more general Eq. (14) in Ref. 104 has to be used.

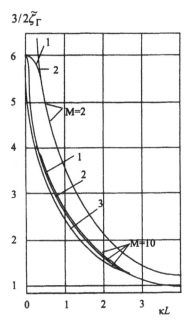

$3/2\tilde{\zeta}_\Gamma$

FIG. 3 Electrokinetic potential decrease after nonionic polymer adsorption as a function of κL for $M = 2$ and $M = 10$; L is the adsorption layer thickness. The calculation results are according to Eq. (23) (curve 1); Eq. (25) (curve 2); Eq. (28) (curve 3).

The generalization of Eq. (28) by taking into account [105] the deviation of the spherical DL from the flat geometry is given by the second term on the right-hand side (r.h.s.) of the following equation:

$$\tilde{\zeta}_\Gamma = f_0(L) + \frac{1}{\kappa L_1} \sinh \frac{f_0(L)}{2} \left[\tanh^2 \frac{\tilde{\zeta}_0}{4} - \tanh^2 \frac{f_0}{4} - 2\kappa L \right] \tag{30}$$

where $L_1 = a + L$.

The thickness of the adsorbed polymer layer L was calculated [106] without taking into account the second term on the r.h.s. of Eq. (30). We obtained an electrokinetic thickness of 5.5 and 11 nm for polymer concentrations of 10^{-4} and 10^{-3} M, respectively, by the application of Eq. (30) to the experimental results of Ref. 107. These values of thickness are nearly twice as small as those calculated in Ref. 105 by neglecting the second term in Eq. (30). It is noteworthy that this smaller thickness agrees better with those measured by other methods [107].

This approach has to be completed by considering changes in the inner part of the DL [98]. The measurement of polymer adsorption effects on the surface conductivity offers the opportunity of a more precise characterization of the neutral polymer adsorption. The sensitivity of the surface conductivity to the latter has been demonstrated [105]: poly(vinyl alcohol) adsorption leads to a substantial reduction in the dilute suspension conductivity, measured for polygorskite and antimony sulfide [105]. This effect was assessed by the changes in the values of the Du number on the basis of Eq. (12), and the assumption that the adsorbed polymer layer was a SL. Thus, there is a promising perspective for the investigation of neutral polymer adsorption combining electrokinetic and surface conductivity measurements. The latter will provide information about the SL charge, i.e., about the change in ion adsorption and, correspondingly, in the diffuse part of the DL. The incorporation of surface conductivity determination in the electrokinetic characterization of polymer adsorption is simplified owing to the small difference in the mobilities of adsorbed and free ions, demonstrated for monovalent ions [86]. This enables us to obtain σ_i changes by means of the determination of $K^{\sigma i}$ changes caused by polymer adsorption. Another simplification relates to the $K^{\sigma i}$ dependence on the coverage (the surface concentration of the adsorbed segments of macromolecule), because this dependence may be linear.

The microslit electrokinetic setup (see Section V.D) offers a unique opportunity for the investigation of the neutral polymer adsorption layer by means of extended electrokinetic investigation. The derivation of the equation for surface conductivity changes due to polymer adsorption is simplified in this case.

The importance of surface conductivity measurements for the electrokinetic characterization of the polyelectrolyte adsorption is emphasized in Ref. 108 as well: adsorbed polyelectrolyte modifies essentially the space distribution of mobile ions in the diffuse layer, simultaneously screening the surface charge and the fixed charge of the polyelectrolyte, the latter being distributed over the whole polymer layer. For the characteristic situation when the adsorbed polymer molecules are charged opposite to the surface, the conditions determining concentration polarization may essentially (even qualitatively) change. In particular, it may occur that the surface conductivity K^{σ} is large, but the fractions of ions of both signs contributing to K^{σ} are comparable. This is contrary to the usual case (see Section II.A.2), when surface flows of ions within the DL carry mainly counterions. The description of the corrections for surface conductivity to the electrophoretic mobility of particles with adsorbed polyelectrolyte was made possible by the theory described in Ref. 108, which was experimentally demonstrated to be valid for arbitrary contributions of ions of both signs to K^{σ}.

VII. INVESTIGATION OF THE DYNAMIC STERN LAYER OF LIPOSOMES BY UTILIZING THE ISOELECTRIC AND ISOCONDUCTING POINTS

A. Surface Conductivity in Vicinity of Isoelectric Point

The investigation of the conduction behind the slipping plane is complicated because of the conduction through the diffuse part of the DL. This complication can be eliminated by working in the vicinity of the isoelectric point (i.e.p.), i.e., at zero conductivity in the diffuse layer. Such an investigation, recently proposed [17] has been already accomplished [109, 110]. The surface conductivity of certain divalent metal ions adsorbed on negatively charged liposomes was determined by combining measurements of both electrophoresis and conductivity of dilute liposome suspensions with volume fraction $\varphi \approx 0.1$. The surface mobility of adsorbed Be^{2+}, Ca^{2+}, and La^{3+} counterions manifests in a rather large surface conductivity at the i.e.p. of the liposomes [109, 110]. If the adsorbed counterions in the Stern layer were immobile, the liposome particles at the i.e.p. would behave like uncharged particles. Therefore, the conductivity K^* of the suspensions should be smaller than that of the pure solution: $K^*/K = 1 - 3\varphi/2$. However, the experimentally determined data of this ratio around the i.e.p. is slightly larger than unity for Ca^{2+} and substantially larger than unity for La^{3+}. This means that there exists surface current around the liposome particles that must originate from the mobility of counterions in the Stern layer. It must be noted that these large values of the conductivity ratio were detected at frequencies much more lower than the reciprocal time of charging of the liposome mebrane, so that the current through the membrane must be negligibly small.

B. Evaluation of the Mobility of Adsorbed Counterions

In addition to the qualitative conclusion of the preceding section, a method of experimental discrimination between low and high surface mobilities of adsorbed counterions through the use of i.e.p. and isoconducting point (i.c.p.) determinations is elaborated and verified in experiments with NaCl, $BeSo_4^-$, and $LaCl_3$ electrolytes [109] and CsCl and $CaCl_2$ electrolytes [110]. Under the condition of the i.e.p., the charge density of the Stern layer σ_i is equal to the surface charge $-\sigma_o$, because of the electroneutrality in the DL. In this case Du is described as follows:

$$\text{Du} = \frac{K^\sigma}{Ka} = \frac{\sigma_i D^\sigma}{2azFDC} \tag{31}$$

where D^σ is the diffusivity of the adsorbed counterions, F is the Faraday constant, and C is the concentration of the z-valent symmetric electrolyte solution.

Setting $\sigma_i = \sigma_o$, and explicitly writing Du_e for Du at the i.e.p. in Eq. (31), the ratio of the diffusivity in the Stern layer to that in the bulk is expressed as

$$\frac{D^\sigma}{D} = \frac{2azFC_e\text{Du}_e}{\sigma_o} \tag{32}$$

where C_e is the electrolyte concentration at the i.e.p.

This equation can be used for the determination of the adsorbed counterion mobility if there is independent information regarding surface charge. Unfortunately, this information is not available for liposomes. The problem is solved in Ref. 110 by the measurement of two concentration dependencies, namely, that in electrolytes with strongly adsorbed counterions and in an indifferent electrolyte. Cesium ions have very low affinity for binding to egg PC liposome [110]. Therefore, Du associated with Cs^+ counterions can be approximated by a diffuse-type DL:

$$\text{Du} = \frac{2}{\kappa a}\left(1 + \frac{3m}{z^2}\right)\left[\cosh\frac{z\tilde{\Psi}_d}{2} - 1\right] \approx 1.45\frac{\exp\left(\frac{z\tilde{\Psi}_d}{2}\right)}{\kappa a} \tag{33}$$

The diffuse-layer charge density for monovalent ions is expressed by the Gouy–Chapman theory:

$$|\sigma_d| = |\sigma_o| = \frac{4FC}{\kappa}\sinh\frac{\tilde{\Psi}_d}{2} \approx \frac{2FC}{\kappa}\exp\left(\frac{\tilde{\Psi}_d}{2}\right) \tag{34}$$

Comparing Eq. (34) with Eq. (33), Du is expressed as follows:

$$\text{Du} \approx 1.45\frac{|\sigma_o|}{2FCa} \tag{35}$$

At the i.c.p. (Du $= 1/2$ at $C = C_c$), the surface charge density is calculated as

$$|\sigma_o| \approx FaC_c/1.45 \tag{36}$$

Therefore, we can estimate the surface charge density of the particles of radius a by the determination of the i.c.p. C_c of nonadsorbing ions, e.g., Cs^+. Combining Eq. (36) with Eq. (32), we can obtain a useful equation:

$$D^\sigma/D \approx 2.9zC_e Du_e/C_c \tag{37}$$

Taking C_c of Cs^+ as $2 \times 10^{-3}\,M$, the following estimation of the diffusivity is obtained [110]:

1. In the case of Ca^{2+}: $C_e = 6 \times 10^{-4}\,M$; $Du_e = 0.5$. Using these values, we obtain $D^\sigma/D = 0.87$.
2. In the case of La^{3+}: $C_e = 5 \times 10^{-6}\,M$; $Du_e = 3.5$. Using these values, the result is $D^\sigma/D = 0.07$.

These data show the same trend as that established for polymer lattices [87], namely, the larger the ion valency the smaller is its surface mobility.

VIII. ELECTROKINETIC POTENTIAL INTERPRETATION AND SURFACE ROUGHNESS

A. Stagnant Layer Near a Rough Surface

Bikerman [39] attributed the SL phenomenon to surface roughness. Lyklema et al. [85] established that the SLs are general properties of liquids adjacent to solid walls, and in particular adjacent to a molecularly smooth one. This important conclusion does not mean that the surface roughness does not affect the electrokinetic slip. The liquid within valleys of a rough surface is hydrodynamically immobilized [2]. Hence, there are two mechanisms of liquid immobilization near a solid–liquid interface, namely, the universal mechanism [85] and that caused by the surface roughness. The first mechanism is a generic property, while the second is important for sufficiently rough surfaces only.

Thus, the solid–liquid interfaces may be classified into two groups. For the first group the roughness role in electrokinetics is negligible, and for the second the roughness cannot be disregarded. For the first group, the SL perhaps almost coincides with the Stern layer, and its thickness d_{ek} is less than 1 nm.

The introduction of the thickness d_r for the SL, caused by the surface roughness, is useful at least for a qualitative analysis. The quantification of d_r has to be done with account for the statistical description of the roughness. Different models exist for roughness and, correspondingly, different mean d_r values will be predicted. As the first crude approximation, the roughness role in electrokinetics can be neglected, if

$$d_r < d_{ek} < 1\,nm \tag{38}$$

This condition is not as strict as that assumed in the SEM, namely, that the surface has to be molecularly smooth. The important conclusion follows that the GSEM (Section V.D) may have a wider applicability than SEM. However, there is no proof that surfaces for which:

$$d_r > d_{ek} < 1\,nm \tag{39}$$

do not exist; 1 nm is considered as the maximum thickness for the Stern layer. The term molecular condenser is well known for Stern layers and corresponds to a Stern layer thickness of less than 1 nm. For this case of a thin Stern layer, the condition (38) is violated even at a roughness scale that is not large. In this case, the discrimination between the SL that is a generic property [85] of the liquid adjacent to a solid wall, and the SL arising due to the surface roughness, is important.

B. Effective Slipping Plane and Effective Potentials for a Rough Surface

A simplification in electrokinetic theory is the assumption that the distribution of potential, electric charge density, and hydrodynamic field can be considered as functions of the distance to the flat surface. For a rough surface this simplification is wrong [2]. It means that the applicability of the notion of a flat slipping surface is questionable under the condition (39). Even if it is so, this does not mean that the Smoluchowski theory is not useful in this case, but rather that the convenient notions of slipping plane and electrokinetic potential are rather crude approximations [2].

There are common features in liquid slip along a smooth surface with adsorbed uncharged polymer and along a rough clean surface. In both cases, a hydrodynamically SL arises due to the viscous resistance enhancement near a surface caused either by adsorbed macromolecules or by roughness. In both cases the slipping plane would "move" outward. Although the notion of a flat slipping surface is an oversimplification under condition (39), this approach corresponds to the current state of electrokinetic theories.

This approach enables us to model the hydrodynamic slip along a rough surface as a slip along a flat surface with SL thickness d_r. The effective slipping plane position and corresponding roughness cannot be ignored if a significant fraction of the countercharge resides inside the SL. At a given roughness scale, the larger the surface charge density and the higher the electrolyte concentration (the smaller the diffuse-layer thickness), the larger is the countercharge part located behind the slipping plane. A simple calculation shows that this charge can exceed the electrokinetic charge by an order of magnitude. If d_r exceeds 1 nm, the charge behind the slipping plane cannot be considered as a Stern layer, because its thickness is usually less than 1 nm. This is the part of the diffuse layer located within the hydrodynamically immobile liquid.

Thus, if condition (38) holds, the slipping plane, characterized by its position d_{ek}, is the boundary between Stern and diffuse layers. On the other hand, under condition (39) the slipping plane, characterized in this case by d_r, is the boundary between the hydrodynamically mobile and immobile parts of the diffuse layer. It is obvious that the potential of this boundary differs from the Stern potential because this is a boundary between two parts of the diffuse layer. In this case, a measured electrokinetic potential cannot be interpreted as a Stern potential. Moreover, the question arises of what the Stern potential means for a rough surface: if the local surface properties are invariant along it, the Stern layer thickness is invariant as well, at least as a crude approximation. Its outer boundary will then reproduce the geometry of the rough surface. This means that the notion of the OHP has to be revised for rough surfaces under condition (39). In addition, the boundary condition for the Poisson–Boltzmann equation is formulated in terms of the outer boundary of the Stern layer. Hence, the consideration of different models for a rough surface and correspondingly for the outer boundary of the Stern layer is necessary to quantify the diffuse layer. As this task is extremely difficult, the simplified approach inherent in the current state of electrokinetic theory was used [2], namely, an effective flat outer boundary for the Stern layer was considered, and its corresponding potential was assumed to be the effective Stern potential, Ψ_{ef}.

C. Surface Conductivity and Roughness

The solution of the Poisson–Boltzmann equation, assuming a flat effective outer boundary of the Stern layer, enable us to [2] to express the diffuse layer charge near a rough surface through Ψ_{ef} by means of the well-known Gouy–Chapman theory, and requires us to generalize the Bikerman equation for surface conductivity:

$$K^{\sigma d} = K_m^{\sigma d}(\Psi_{ef}) + K_{eo}^{\sigma d}(\zeta_{ef}) \tag{40}$$

The contribution $K_m^{\sigma d}$ is caused by the movement of charges with respect to the liquid by ion migration in the electric field, and the term $K_{eo}^{\sigma d}$ is caused by the charge movement together with the moving liquid. The entire diffuse layer charge is involved in migration, i.e., $K_m^{\sigma d}$ depends on Ψ_{ef}. The part of the diffuse layer charge beyond the effective slipping plane is involved in electro-osmosis, i.e., $K_{eo}^{\sigma d}$ depends on ζ_{ef}.

The electro-osmotic component is small in comparison with the migration component even if Stern and electrokinetic potentials are comparable. Since an effective Stern potential near a rough surface exceeds the electrokinetic potential, the necessity of accounting for the electro-osmotic term in Eq. (40) is questionable.

D. Mobile Charge

Equation (21) is more exact if condition (39) holds, because neglecting the second term in Eq. (40) can be justified. It can be emphasized that Eq. (21) can be derived without the introduction of the effective Stern potential. This means that the uncertainty associated with the introduction of a flat outer boundary for the Stern layer does not diminish the exactness of Eq. (21).

Thus, the rather rough surface corresponding to the condition (39) can be character-ized by the values of the mobile charge (surface charge) and by the effective values of Stern and electrokinetic potentials. The characterization of a rough surface by its mobile charge is more exact than that based on the use of effective potentials, because the evaluation of the uncertainty arising at the averaging procedure, which is the basis for the notions of effective values of potentials, is an unsolved and extremely difficult task. The evaluation of the SL thickness for a rough surface can be accomplished with use of the equation:

$$d_\Gamma = \kappa^{-1} \ln \frac{\tanh \tilde{\Psi}_{ef}/4}{\tanh \tilde{\zeta}_{ef}/4} \tag{41}$$

The term *generalized standard electrokinetic model* was proposed in Section IV.A for systems satisfying the condition (38). The term *nonstandard electrokinetic model* may be proposed for the systems that satisfy the condition (39).

The difference between Stern and electrokinetic potentials was accounted for in the initial version of the thin DL polarization theory [5]. Correspondingly, both potentials are present in the theories for electrophoresis [60] and LFDD [21], and hence they in fact correspond to the NSEM model. A number of investigations detailed in Refs 8 and 14 show how those theoretical treatments can be used for the estimation of K^σ, Ψ_{ef}, and ζ.

The values of σ_m obtained in this way proved to be very close to σ_t. Moreover, the low values of σ_ζ as compared to σ_m established in these investigations confirm that K^σ and σ_m are the most important characteristics of these surfaces [2]. In some of the papers cited above, a thickness much larger than 1 nm was determined for d_r. Later, there were com-ments in the literature that these large values of SL thickness are meaningless, since d_r

should not exceed 1 nm [17, 90]. However, it must be taken into account that the extension of the SL can be larger in rough than in flat surfaces.

Recently, it was shown [111] that small protrusions on the particle surface with dimensions much smaller than the particle dimension, but of the order of κ^{-1}, may have a dramatic influence on the DLVO interaction energy. Correspondingly, there is great interest in the modification of DLVO theory in the case of rough surfaces [112]. This supports the conclusion that the generalization of electrokinetic theory to cover rough surfaces is very important.

IX. SUMMARY

Both MD simulations and experimental investigations by Lyklema and coworkers [85] proved that "stagnant layers are general properties of liquids adjacent to a solid wall," and "at least in a number of systems, conduction behind the shear plane does take place." This characterizes a new standard for electrokinetic transport modeling, that can be called the generalized standard electrokinetic model (GSEM). Its range of applicability is wider, since, unlike the standard model (SEM), the existence of SLC is considered, and there is no requirements concerning surface smoothness on the molecular level. Thus, the DL is described by at least two parameters (ζ and $K^{\sigma i}$) in the GSEM, and only one, ζ, is needed in SEM.

Electrokinetic consistency tests are possible for systems with or without SLC in the framework of theories that take into account the existence of SLC. Inconsistencies are unavoidable when the Du number is not small if SLC is disregarded.

The mobile and surface charges are approximately equal if the electrokinetic charge, σ_ζ, is small in comparison with the titratable one, and the adsorbed counterion mobility approximates that in the bulk. The second condition is satisfied by monovalent ions, according to recent measurements of the adsorbed counterion mobility. The first condition is satisfied very well in many systems, because σ_ζ is often very small in comparison with σ_t. For these systems, surface conductivity measurements yield nearly exact values for the surface charge. In general, the surface conductivity yields at least an estimation of the surface charge, because the electro-osmotic component of the surface conductivity is always rather small according to the Bikerman equation.

For smooth surfaces without adsorbed macromolecules, σ_ζ characterizes the diffuse layer, and the SLC characterizes the Stern layer. Combined measurements comprising two electrical surface phenomena allow determination of the charge distribution between diffuse and Stern layers. This new level in surface characterization is more precise, if the difference in mobilities for adsorbed and bulk counterions is small.

In addition to circular capillaries and dilute suspensions of monodisperse spherical particles the measurement of K^σ is now possible in polydisperse suspensions of different volume fractions (CVP techniques), and for particles with shape approximated by a spheroid. For flat surfaces, both ζ and $K^{\sigma i}$ can be measured (microslit electrokinetic setup). The integrated measurement in the vicinities of the i.e.p. and i.c.p. is the method for the determination of the adsorbed counterion mobilities. Furthermore, the surface conductivity measurements provide additional information about macromolecular adsorption.

There are two mechanisms of liquid immobilization near a solid–liquid interface, namely, the universal mechanism and that caused by the surface roughness. The first mechanism is a generic property, while the second one is important for sufficiently rough surfaces only. The notions of slipping plane, OHP, and Stern and electrokinetic potentials are oversimplifications in the second case because the shape of boundaries between the Stern layer and diffuse layer, as well as between stagnant and mobile liquid is extremely complicated. The notions of OHP and slipping plane can perhaps be introduced in this case as a very crude approximation. Correspondingly, the interpretation of effective values of Stern and electrokinetic potentials is not clear at large roughness. In contradistinction, the mobile charge can provide more precise evaluation of the surface charge in the case of a rough surface, because the role of the electro-osmotic component of the surface conductivity is small.

The roughness determines the boundary of applicability of the GSEM, and it is the main source of uncertainty for electrokinetic potential interpretation. It is very important to clarify whether the roughness can be disregarded in electrokinetics (with the exception for the extreme case of very rough surfaces) or the notions of slipping plane and electrokinetic potential are oversimplifications that can cause a substantial decrease in precision. The roughness role in electrokinetics decreases with decreasing electrolyte concentration and surface charge.

The role of all the generalizations of the DL model described in electrophoresis can also be applied to the Dorn effect (or sedimentation potential), and vice versa, consideration about their role in the Dorn effect are pertinent for electrophoresis. This statement is a direct consequence of the Onsager symmetry relationships applied to electrokinetics [79].

ABBREVIATIONS

DL	double layer
DLVO	Derjaguin, Landau, Verwey, Overbeek
GSEM	generalized standard electrokinetic model
i.c.p.	isoconducting point
i.d.m.	induced dipole moment
i.e.p.	isoelectric point
LFDD	low-frequency dielectric dispersion
NESP	nonequilibrium electric surface phenomena
MD	molecular dynamics
MW	Maxwell–Wagner
NSEM	nonstandard electrokinetic model
OHP	outer Helmholtz plane
PDFP	plasma-deposited fluoropolymer
SEM	standard electrokinetic model
SL	stagnant layer
SLC	lateral conductivity of stagnant layer

REFERENCES

1. JTG Overbeek. Kolloidchem Beih 59:287, 1943.
2. SS Dukhin, BV Derjaguin. In: E Matijevic, ed. Surface and Colloid Science, vol. 7. New York and Toronto: Wiley, 1974.
3. BV Derjaguin, SS Dukhin, VN Shilov. Adv Colloid Interface Sci 13:141, 1980.
4. SS Dukhin, VN Shilov. Dielectric Phenomena and the Double Layer in Disperse Systems and Polyelectrolytes. New York and Toronto: John Wiley, 1974.

5. SS Dukhin. Diffusion–electrical theory of electrophoresis. Proceedings of the XXth International Congress of Pure and Applied Chemistry, Moscow, 1965, A72, pp 68–70; SS Dukhin. In: BV Derjaguin, ed. Izsledovaniya v oblasti povernostnykh sil. Moscow: Nauka, 1967, pp 335–352; In: BV Derjaguin, ed. Research in Surfaces Forces, vol. III. New York: Plenum Press, 1971, pp 312–341.
6. SS Dukhin, VN Shilov. Kolloidn Zh 31:706, 1969.
7. VN Shilov, SS Dukhin. Kolloidn Zh 32:117, 1970.
8. SS Dukhin. Adv Colloid Interface Sci 44:1, 1993.
9. SS Dukhin. In: BV Derjaguin, ed. Research in Surface Forces. New York: Consult Bureau, 1963, vol. 1, pp 27–34.
10. J Lyklema. Mater Sci Forum 1:25, 1988.
11. SS Dukhin, J Lyklema. Langmuir 3:94, 1987.
12. SS Dukhin, J Lyklema. Faraday Disc Chem Soc 90:261, 1990.
13. J Kijlstra, J Lyklema, SS Dukhin, SYu Shulepov. Kolloidn Zh 54:92, 1992.
14. SS Dukhin. Adv Colloid Interface Sci 61:17, 1995.
15. RJ Hunter. Foundation of Colloid Science. Oxford: Oxford University Press, 1989, vol. 11, ch. 13.
16. VF Zukoski, DA Saville. J Colloid Interface Sci 114:32 and 45, 1986.
17. J Lyklema. Fundamentals of Interface and Colloid Science. New York and London: Academic Press, 1995, Vol. 11, ch. 4.
18. JTG Overbeek. Adv Colloid Sci 3:97, 1950.
19. SS Dukhin, VN Shilov. Adv Colloid Interface Sci 13:153, 1980.
20. F Booth. Trans Faraday Soc 44:955, 1948.
21. VN Shilov, SS Dukhin. Kolloidn Zh 32:293, 1970.
22. SS Dukhin. In: E Matijevic, ed. Surface and Colloid Science, vol. 3, 1970, pp 82–137.
23. VN Shilov, Yu Ya Eremova. Kolloidn Zh 57:255, 1995.
24. C Grosse, S Pedrosa, VN Shilov. J Colloid Interface Sci 220:31, 1999.
25. SS Dukhin, MP Sidorova, AE Jaroschuk. Membrane Chemistry and Reverse Osmosis. Leningrad: Khimija, 1991 (in Russian).
26. BV Derjaguin, SS Dukhin, AA Korotkova. Kolloidn Zh 23:409, 1961.
27. DC Prieve, RJ Roman. J Chem Soc, Faraday Trans 2 83:1287, 1987.
28. ZR Ulberg, SS Dukhin. Prog Org Coat 18:1–50, 1990.
29. SS Dukhin, VR Estrela-Lopis, EK Zholkovsky. Electric Surface Phenomena and Electrofiltration. Kiev: Naukova Dumka, 1985 (in Russian).
30. SS Dukhin. Croat Chem Acta 53:167, 1987.
31. SS Dukhin. In: BG Goodrich, AI Rusanov, eds. Modern Theory of Capillarity. Berlin: Akademischer Verlag, 1981, pp 83–107.
32. SS Dukhin. Croat Chem Acta 60:395, 1987.
33. BV Derjaguin, SS Dukhin, NN Rulov. In: E Matijevic, ed. Surface and Colloid Science, vol. 13. New York: Wiley, 1984, pp 71–114.
34. SS Dukhin, NN Rulov, DD Dimitrov. Coagulation and Thin Film Dynamics. Kiev: Naukova Dumka, 1986 (in Russian).
35. LA Rosen, DA Saville. Langmuir 7:36, 1991.
36. J Lyklema, SS Dukhin, VN Shilov. J Electroanal Chem Interfacial Electrochem 143:1, 1983.
37. VE Shubin, RJ Hunter, RW O'Brien. J Colloid Interface Sci 159:174, 1993.
38. JJ Bikerman. J Phys Chem 46:725, 1942.
39. JS Bikerman. Physical Surfaces. New York: Academic Press, 1970, ch. 8.
40. DC Henry. Trans Faraday Soc 44:1021, 1948.
41. M Smoluchowski. In: W. Graetz. Handbuch der Electrizitat und der Magnetismus. vol. 2. Leipzig: Barth, 1921, p 366.
42. DA Haydon, FH Taylor. Phil R Soc, London, Ser A 253:255, 1960.
43. BH Bijsterbosch, J Lyklema. Adv Colloid Interface Sci 9:147, 1978.
44. VM Muller, IP Sergeeva, VD Sobolev, NV Churaev. Kolloidn Zh 48:718, 1986.

45. CF Zukoski, DA Saville. J Colloid Interface Sci 107:322, 1985.
46. BR Midmore, RW O'Brien. J Colloid Interface Sci 123:486, 1988.
47. AG Van der Put, BH Bijsterbosch. J Colloid Interface Sci 75:512, 1980.
48. R Hidalgo-Alvarez, JA Molon, FJ de las Nieves, BH Bijsterbosch. J Colloid Interface Sci 149:23, 1992.
49. JM Peula-Garcia, R Hidalgo-Alvarez, FJ de las Nieves. Colloids Surfaces 127:19, 1997.
50. X Wu, TGM van de Ven. Langmuir 12:3859, 1998.
51. EM Egorova, AS Dukhin, IE Svetlova. Biochim Biophys Acta 1104:102, 1992.
52. EM Egorova. Colloids Surfaces 131:7, 1998.
53. EM Egorova. Colloids Surfaces 131:19, 1998.
54. JC Maxwell. Electricity and Magnetism. Oxford: Clarendon Press, 1892, p 313.
55. KW Wagner. Isolierstoffe der Electrotechnik. Berlin: Springer, 1924.
56. CT O'Konski. J Phys Chem 64:605, 1960.
57. RW O'Brien. J Colloid Interface Sci 81:234, 1981; 113:81, 1986.
58. BR Midmore, RJ Hunter. J Colloid Interface Sci 122:521, 1988.
59. M Fixman. J Chem Phys 72:5177, 1980.
60. SS Dukhin, NM Semenikhin. Kolloidn Zh 32:366, 1970.
61. RW O'Brien, LR White. J Chem Soc, Faraday Trans 2:1607, 1978.
62. BR Midmore, D Diggins, RJ Hunter. J Colloid Interface Sci 129:153, 1989.
63. RW O'Brien. Adv Colloid Interface Sci 16:281, 1982.
64. M Fixman. J Chem Phys 78:1483, 1983.
65. EHB De Lacey, LR White. J Chem Soc, Faraday Trans 2 74:2007, 1981.
66. SE Pedrosa, C Grosse. J Colloid Interface Sci 219:37, 1999.
67. G Schwarz. J Phys Chem 66:2636, 1962.
68. J Kijlstra, HP van Leeuwen, J Lyklema. J Chem Soc, Faraday Trans 2 88:3441, 1992.
69. LA Rosen, JC Baygents, DA Saville. J Chem Phys 98:4183, 1993.
70. IA Razilov, G Pendze, SS Dukhin. Kolloidn Zh 56:612, 1994.
71. IA Razilov, G Pendze, J Lyklema, SS Dukhin. Kolloidn Zh 56:736, 1994.
72. IA Razilov, SS Dukhin. Colloid J 57:364, 1995.
73. IA Razilov, F González-Caballero, AV Delgado, SS Dukhin. Kolloidn Zh 58:222, 1996.
74. EJ Hinch, JD Sherwood, WC Chew, PN Sen. J Chem Soc, Faraday Trans 2 80:535, 1984.
75. M Minor. Electrodynamics of Colloids. PhD thesis, Wageningen University, 1998.
76. AV Delgado, FJ Arroyo, F González-Caballero, VN Shilov, YB Borkovskaya. Colloids Surfaces 140:139, 1998.
77. R Barchini, DA Saville. J Colloid Interface Sci 173:86, 1995.
78. IA Razilov, NI Zharkikh, F Gonzálex-Caballero, AV Delgado, SS Dukhin. Colloids Surfaces 121:173, 1997.
79. HR Kruyt. Colloid Science. vol. 1. New York: Elsevier, 1952.
80. S Levine, GH Neale. J Colloid Interface Sci 47:520, 1974.
81. VN Shilov, NI Zharkih, YuB Borkovskaya. Colloid J 43:434, 1981.
82. AS Dukhin, V Shilov, Yu Borkovskaya. Langmuir 15:3452, 1999.
83. AS Dukhin, VN Shilov, H Ohshima, PJ Goetz. Langmuir, in press (2000).
84. AS Dukhin, TGM van de Ven. J Colloid Interface Sci 165:9, 1994.
85. J Lyklema, S Rovillard, J de Coninck. Langmuir 14:5659, 1998.
86. J Lykelma, M Minor. Colloids Surface 140:33, 1998.
87. M Lobbus, HP van Leeuwen, J Lyklema. Colloids Surface 161:103, 2000.
88. TS Simonova, VN Shilov. Kolloidn Zh 48:379, 1986.
89. J Kijlstra. Double Layer Relaxation in Colloids. PhD thesis. Wageningen University, 1992.
90. J Lyklema, HP van Leeuwen, M Minor. Adv Colloid Interface Sci 83:33, 1999.
91. C Grosse, M Tirado, W Pieper, R Pottel. J Colloid Interface Sci 205:26, 1998.
92. JJ Lopez-Garcia, J Horno, F González-Caballero, C Grosse, AV Delgado. J Colloid Interface Sci, in press (2000).

93. C Werner, H Korber, R Zimmermann, S Dukhin, H-J Jacobach. J Colloid Interface Sci 208:329, 1998.
94. SP Stoylov. Colloid Electro-Optics. London: Academic Press, 1991.
95. VV Voitilov, SA Kakorin, AA Trusov. Kolloidn Zh 48:139, 1986.
96. AV Delgago, F Carrique, FJ Arroyo, T Bellini, F Mantegazza, ME Giardini, V Degiorgio. Colloids Surfaces 140:157, 1998.
97. C Grosse, VN Shilov. Colloids Surfaces 140:199, 1998.
98. GJ Fleer, LK Koopal, J Lyklema. Kolloid Z-Z Polym 250:689, 1972.
99. LK Koopal, J Lyklema. Disc Faraday Soc 59:230, 1975.
100. R Varoqui. Nouveau J 6:187, 1982.
101. PG de Gennes. CR Acad Sci Paris 197:883, 1983.
102. MA Cohen Stuart, FHWH Waajen, SS Dukhin. Colloid Polym J 262:423, 1984.
103. BV Eremenko, BE Platonov, AA Baran, ZA Sergienko. Kolloidn Zh 38:618, 1976.
104. SS Dukhin, NM Semenikhin, VA Bichko. In: BV Derjaguin, ed. Poverhnostnye sily v tonkih plenkah. Moskva: Nauka, 1979, pp 85–93.
105. SS Dukhin, LM Dudkina. Kolloidn Zh 40:232, 1978.
106. SS Dukhin, BV Derjaguin, NM Semenikhin. Dokl AN SSSR 192:367, 1970.
107. MJ Garvey, TF Tadros, B Vincent. J Colloid Interface Sci 55:441, 1976.
108. E Donath, D Walter, VN Shilov, E Knippel, A Budde, K Lowack, CA Helm, H Mohwald. Langmuir 13:5294, 1997.
109. SV Verbich, SS Dukhin, H Matsumura. J Disp Sci Technol 20:83, 1999.
110. H Matsumura, SV Verbich, SS Dukhin. Colloids Surfaces 159:271, 1999.
111. S Bhattacharjee, C-H Ko, M Elimelech. Langmuir 14:3365, 1998.
112. JY Walz. Adv Colloid Interface Sci 74:119, 1998.

3

The Role of Surface Conduction in the Development of Electrokinetics

J. LYKLEMA   Wageningen University, Wageningen, The Netherlands

I. INTRODUCTION

Surface conduction is an electrokinetic phenomenon in its own right, but it also plays an auxiliary role in conjunction with other electrokinetic processes. The reason is that upon tangential motion along charged interfaces the excess conductivity of the mobile ions in the electric double layer contributes to flow of charges and, indirectly, to fluid motion.

Historically, recognizing the role of surface conduction in the interpretation of electrokinetic observations has twice contributed to an improvement in the meaning of electrokinetic or ζ potentials. First, in the 1930s it has led to the insight that ζ does not depend on the size and shape of the object to be studied. Second, in the 1990s it resolved the problem that ζ was sometimes found to depend on the electrokinetic method used. Thus, understanding surface conduction has greatly contributed to the definiteness of ζ as a real characteristic of charged surfces, only determined by the electrical properties of the interface.

This is, in a nutshell, the theme of the present contribution.

II. SURFACE CONDUCTION

The phenomenon of surface conduction has already been recognized by von Smoluchowski in 1905 [1]. It is the excess conduction tangential to a charged surface. We shall assume this surface to be flat ($\kappa a \gg 1$), unless otherwise specified. Our symbol is K^σ; the SI units are $A\,V^{-1}$ or S. The defining equation is

$$j^\sigma = K^\sigma E \tag{1}$$

where j^σ is the excess current density, tangential to the surface (in $C\,s^{-1}\,m^{-1} = A\,m^{-1}$) under the influence of the electric field parallel to the surface, E. Equation (1) is the two-dimensional analog of Ohm's law. It is noted that j^σ is a current per unit length, i.e., a linear current density as opposed to the bulk current density in $A\,m^{-2}$, occurring in Ohm's law. In reality, double layers have a finite thickness but for the phenomenological description this spatial excess is assigned to one line, just as in the Gibbs convention for adsorption, where surface excesses can be formally assigned to a certain dividing plane. Only upon elaboration does one have to specify this plane.

Let x be the distance in the double layer from the surface. The tangential current $j(x)$ is different at each position x. Let us call the bulk current density $j(\infty)$, then the excess in each infinitesimal layer dx is $[j(x) - j(\infty)]dx$. Integration over the entire double layer yields:

$$j^\sigma = \int_0^\infty [j(x) - j(\infty)]dx \tag{2}$$

Expressing all currents in terms of concentrations, $c_i(x)$ and mobilities, $u_i(x)$:

$$j^\sigma = \int_0^\infty \sum_i [c_i(x) - c_i(\infty)]|z_i|F\, u_i(x)E\,dx \tag{3}$$

where F is the Faraday and i indicates the ionic species. Hence, from Eqs (1) and (3):

$$K^\sigma = F \sum_i |z_i| \int_0^\infty [c_i(x) - c_i(\infty)]u_i(x)\,dx \tag{4}$$

It follows from Eq. (4) that the surface excess conductivity can be computed if we know the ionic distribution and the ionic mobilities. Experience has shown that, in the diffuse part of the double layer, to a good approximation $u_i(x) = u_i(\infty)$. So far that double layer part $u_i(x)$ can be taken out of the integral. For the inner, or Stern, layer this is not usually the case. In fact, studying the tangential mobility in this part will be one of the themes of this chapter.

Anticipating further analysis, we mention the following elaboration of Eq. (4), given by Bikerman for a diffuse double layer [2, 3]:

$$K^{\sigma d} = \sqrt{8\varepsilon_0\varepsilon c RT}\left\{\frac{u_+}{A-1} - \frac{u_-}{A+1} + \frac{4\varepsilon_0\varepsilon RT}{\eta z F}\frac{1}{A^2-1}\right\} \tag{5}$$

where ε_0 is the dielectric permittivity of the solvent, c is the electrolyte concentration, η is the viscosity of the solvent, z is the valency of the (symmetrical) electrolyte, u_+ and u_- are the mobilities of cation and anion, respectively, and

$$A = \coth\left(-\frac{zF\zeta}{4RT}\right) \tag{6}$$

Equation (5) can be converted into the following expression if $u_+ = u_- = |z|FD/RT$, where $D = D_+ = D_-$ is the ionic diffusion coefficient:

$$K^{\sigma d} = \frac{4F^2 c z^2 D}{RT\kappa}\left(1 + \frac{3m}{z^2}\right)\left[\cosh\left(\frac{zF\zeta}{2RT}\right) - 1\right] \tag{7}$$

where

$$m = \left(\frac{RT}{F}\right)^2 \frac{2\varepsilon_0\varepsilon}{3\eta D} \tag{8}$$

and κ^{-1} is the Debye length.

For other elaborations, see Ref. 4. Equation (5) and its variants apply for a purely diffuse double layer, obeying Poisson–Boltzmann statistics, behaving ideally with respect to ε, η, and u in the sense that these physical quantities are assumed to be homogeneous and identical to those in the bulk.

Bikerman realized that application of the tangential field E also leads to electro-osmosis, that is, fluid motion induced by the moving ions. The term $3m/z^2$ accounts for this feature.

Bikerman's model applies to the outer, or diffuse, part of the current leading to a simplified double layer picture which is still generally accepted, viz., that of a double layer of which a part resides in the *stagnant layer*, that is, a thin layer of solvent molecules which, upon tangential motion, remain adhered to the solid surface, whereas the other part is fully mobile. The stagnant layer is separated from the mobile part of the double layer by an (infinitesimally thin) *slip plane*. The potential at this plane is the *electrokinetic potential* ζ, occurring in Eqs (6) and (7). So, phenomenologically the picture is that of a solid, including the stagnant layer, which behaves as if it were inert and which is electro-kinetically characterized by the potential ζ.

Below we shall revisit this model.

III. DISCOVERY OF SURFACE CONDUCTION AND FIRST EVALUATION

At the beginning of the twentieth century it was not obvious what potentials could be found in a double layer, let alone what ζ meant physically. One of the problems was that at that time ζ potentials, measured by different electrokinetic techniques, and all interpreted on the level of Helmholtz–Smoluchowski equations, were sometimes found to be different. This observation gave rise to a debate on the physical meaning: is, under given conditions of surface charge, electrolyte concentration, pH, etc., the ζ potential a unique character-istic of a charged surface? Even Verwey and Overbeek, in their famous book [5], consid-ered ζ as insufficiently established and decided to go for the surface potential ψ^o as the interaction-determining double-layer parameter. This decision has played its role in the quantitative interpretation of the so-called Schulze–Hardy rule, an old point of contention that we shall not discuss here.

More in line with the present theme was the equally troublesome observation that some electrokinetic measurements with capillaries gave rise to radius-dependent ζ poten-tials if surface conduction was not accounted for. So, the debate whether or not ζ poten-tials were radius dependent was resolved when it was discovered that surface conduction should be properly accounted for [6]. Let us illustrate this for the streaming potential E_{str} of a capillary of radius a and electrokinetic potential ζ. The expression on the Helmholtz–Smoluchowski level, including surface conduction, is

$$E_{str} = \frac{\varepsilon_o \varepsilon \zeta \Delta p}{\eta(K^L + 2K^\sigma/a)} \tag{9}$$

and that without surface conduction

$$E_{str} = \frac{\varepsilon_o \varepsilon \zeta \Delta p}{\eta K^L} \tag{10}$$

In these equations Δp is the pressure difference applied across the capillary and K^L is the bulk conductivity. The denominator accounts for the back-conduction of accumulated charge. In the case of negligible surface conduction ($2K^\sigma/a \ll K^L$) all conduction takes place through the bulk of the lumen which extends to the slip lane and has the same conductivity K^L everywhere.

At this stage it is helpful to introduce the *dimensionless surface conductivity ratio*

$$Du = \frac{K^\sigma}{aK^L} \tag{11}$$

which I have dubbed the *Dukhin number* after S.S. Dukhin who explicitly introduced this parameter. (See Ref. 7, where references to older work can be found.) Dukhin himself called it *Rel*. Expressions for what virtually is the same quotient can also be found in other literature under different names and symbols, for instance in the O'Brien–White heory [8]. That the ratio K^σ/aK^L plays an important role in electrokinetic phenomena was already recognized by Bikerman [9] long ago. The Dukhin number enters the equation for the streaming potential, Eq. (9), explicitly:

$$E_{str} = \frac{\varepsilon_0 \varepsilon \zeta \Delta p}{\eta K^L (1 + 2Du)} \tag{12}$$

The Dukhin number also occurs in the interpretation of other electrokinetic phenomena, like electrophoresis, even though on the Helmholtz–Smoluchowski level the surface conductivity does not enter the equation. The reason is that high surface conductivity tends to "contract" field lines around the moving particle.

With Eq. (9) or (12) in mind, the "historical" query regarding the radius dependence of ζ can be solved; because of Du in the denominator of Eq. (12), E_{str} depends on a, but ζ does not. If surface conduction is significant, but ignored in the interpretation, i.e., if Eq. (10) is used instead of Eq. (9) or (12), ζ seemingly depends on a. The smaller a, the more the computed ζ is underestimated. Quantitatively, ζ and K^σ can both be found, if the streaming potential is measured at more than one radius. For instance, one could plot $\varepsilon_0 \varepsilon \Delta p / E_{str} \eta$ as a function of a^{-1}, which should be linear. From the slope and intercept $2K^\sigma/\zeta$ and K^L/ζ are obtained. Measuring K^L is rarely a problem, so ζ and K^σ can be found and the linearity ascertained. Such types of analysis have already been carried out long ago by Rutgers [10].

Anticipating the following sections, it is also noted that ζ is underestimated if K^σ is taken into account but if a too low value for this parameter is substituted. This takes us to the quantitative evaluation.

IV.   QUANTITATIVE ASPECTS. CONDUCTION AND MOBILITIES IN THE STAGNANT LAYER

In a chapter on electrokinetic phenomena in Kruyt's *Colloid Science*, Overbeek addresses the state of the art of K^σ measurements up to about 1951 [11]. He tabulated data available at that time, exclusively referring to glass surfaces (with that material well-defined capillaries can be manufactured). At that time data for sols were not available. The results demonstrated a wide spread, which partly reflected the state of the art at that time. Data ranged from less than 10^{-9} S to $\sim 10^{-6}$ S, depending on the electrolyte concentration, the pH (which was not always controlled), and the nature and pretreatment of the glass. Overbeek compared these data with the predictions of Bikerman's equation (5) and concluded that most of these data exceeded Bikerman's $K^{\sigma d}$ by factors of 10–10^3. According to Eq. (7), $K^{\sigma d}$ increases with salt c_{salt} and ζ. However, in practice, high ζ values are found at low c_{salt}. As a result, there is a window of practically relevant $\zeta(c_{salt})$ data, leading to theoretical $K^{\sigma d}$ values between 0 and 10^{-9} S [4]. Even allowing for some experimental uncertainty in the data, it is obvious that Eq. (7) substantially underestimates K^σ. Later experiments confirmed this. The notion of an additional source of surface conduction, besides that in the diffuse part of the double layer, forces itself on the investigator. Some referred to this phenomenon as "anomalous surface conduction." However, nowadays it is known that this additional conductivity is the rule rather than the exception. It finds its origin in conduction in the stagnant part of the double layer. Therefore, we shall not use the term "anomalous." As this stagnant layer more or less coincides with the Stern layer, one also speaks of *dynamic Stern layers*.

Quantitatively, the total surface conductivity consists of two contributions, $K^{\sigma i}$, caused by the stagnant layer, and $K^{\sigma d}$ by the double layer part beyond it. For the latter we can safely substitute Eq. (7). Hence, assuming additivity:

$$K^{\sigma} = K^{\sigma i} + K^{\sigma d} \tag{13}$$

The computation of $K^{\sigma i}$ is now at issue. Some detailed elaborations are available [12–15] for this. Most of these are multiparameter theories, requiring an adsorption isotherm equation for the counterions in the Stern layer. Below we shall employ a more simple model, assuming that stagnant layers usually contain only one type of counterion (i) and that the charge, attributed to these ions (σ_i^i), can be assessed with reasonable confidence without invoking an adsorption isotherm equation. This is possible if simultaneously surface charge and electrokinetic charge densities are available. Let the tangential mobilities of these ions be u_i^i, then we have simply

$$K^{\sigma i} = \sigma_i^i u_i^i \tag{14}$$

Of course, the stagnant layer exhibits no electro-osmosis.

Before discussing alternative methods of determining K^{σ} and finding data for u_i^i, let us formulate three considerations that simplify the analysis without significant loss of precision.

1. For practical purposes the stagnant layer may be identified with the Stern layer. We realize that both layers are abstractions from reality. Hydrodynamically it is unrealistic to have a thin layer, a few molecular cross-sections wide and having an infinitely high viscosity, separated jump-like from the bulk with normal viscosity over an infinitesimally thin slip layer. Nor do deviations from the ideal Poisson–Boltzmann statistics drop stepwise from finite to zero at the outer Helmholtz plane; rather, they decrease more gradually. However, both in fluid dynamics and double-layer electrochemistry experience has shown that idealization in terms of step functions works very well, even if the precise thicknesses of these two layers are not known; in fact, there is no unambiguous way of locating the step. The literature, of course, contains many examples of better double-layer theories (some are reviewed in Ref. 16), but the academic satisfaction of such improvements has to be paid for by advanced mathematics, which is beyond the daily scope of the average colloid scientist. Likewise, theories to account for a more gradual transition of viscosity, like that based on the viscoelectric effect [17], cover only part of the slip phenomena.

2. Important colloid phenomena such as stability and rheology only involve the diffuse part of the double layer, and a variety of experiments have corroborated the correctness of the underlying theory. Such experiments include thickness measurements in thin fluid films, and direct measurement of the interaction force in the so-called surface-force apparatus. The only questions remaining are: (1) does interaction take place at constant charge or constant potential, or perhaps something between these, and (2) if interaction is controlled by a certain potential, or charge, what is then its value? As we are dealing with the diffuse part, this potential is that of the outer Helmholtz plane, i.e., the potential ψ^d from where the diffuse part of the double layer starts. In many experi-

ments where ζ potentials are also available ψ^d and ζ appear to be identical or close, suggesting that the part that is electrokinetically active is about the same as the diffuse part. This identification is commonplace and acceptable for many practical purposes. However, the issue deserves more attention. It follows from the insight of substantial conduction in the stagnant layer that the mostly used Helmholtz–Smoluchowski equation for the electrophoretic mobility underestimates the ζ potential. Because of this, the interpretations of many excellent experimental studies about interaction have to be revisited. It may be added that finite mobility of ions in the inner layer also has direct consequences for the mechanism of interaction in the DLVO range [18]. So, as far as the definitions and values of ψd or σ^d are regarded, there still is a problem. However, when the properties of the inner layer are at issue, the difficulties are alleviated. This is our second rule.

3. The quantitative argument behind this rule is that both in electrostatics and electrokinetics the contribution of the diffuse part is often minor. In electrostatics the contribution of the diffuse part is often minor. In electrostatics, surface charges (σ^o) on model colloids such as oxides, latex, and vesicles can readily reach values of 10–$40\,\mu\mathrm{C\,cm}^{-2}$, whereas diffuse charges (σ^d), obtained from, say stability and Gouy–Chapman theory, rarely exceed $5\,\mu\mathrm{C\,cm}^{-2}$. As a result, in the electroneutrality balance:

$$\sigma^o + \sigma^i + \sigma^d = 0 \tag{15}$$

no large errors in σ^i are made if σ^d is not accurately known, and therefore:

$$\sigma^o + \sigma^i + \sigma^{ek} \approx 0 \tag{16}$$

also is a good approximation, even if we are not one hundred percent certain of the value of ζ. Like σ^d, σ^{ek} rarely exceeds $5\,\mu\mathrm{C\,cm}^{-2}$.

In surface conduction we have already seen that often $K^{\sigma i}$ strongly exceeds $K^{\sigma d}$. The consequences are similar as for electrostatics: from Eq. (13) we can confidently establish $K^{\sigma i}$ from K^σ by subtracting $K^{\sigma d}$, obtained from Eq. (7) even if we do not know ζ well.

Obviously, all of this breaks down when we are close to the point of zero charge, in which case the inner layer is sparsely (or not at all) inhabited by ions. However, then we have no problem because $K^\sigma \approx K^{\sigma d}$ and the Helmholtz–Smoluchowski equation for electrophoresis applies. We shall not consider this lower limit.

The conclusion of the foregoing considerations is that we can obtain tangential mobilities of ions (u_i^i) in the stagnant layer according to the following scheme:

1. Measure K^σ (see below).
2. Measure σ^o by titration or otherwise.
3. Estimate ζ and/or $\sigma^{ek} \approx \sigma^d$ from electrokinetics.
4. Find $K^{\sigma i}$ by subtracting $K^{\sigma d}$ from K^σ, Eqs (7) and (13).
5. Find σ^i by subtracting σ^d from σ^o.
6. Use Eq. (14) to find u_i^i.

The advantages of this procedure are that it does not require an adsorption isotherm equation. Neither do we have to specify the thickness of the stagnant layer.

Regarding step 1, we have already given one method of measuring K^σ in Section III. It applies to capillaries of differing sizes. Let us now add two more recent ones.

1. *Kijlstra method* [19]. In Section III we discussed how incorporating surface conduction in Eq. (9) for the streaming potential "harmonized" ζ potentials in that the spurious radius dependence could be eliminated. More experiments are needed (E_{str} as a function of a), but then more information is obtainable (ζ and K^σ). More recently, another disharmony was discovered in that ζ potentials obtained by different electrokinetic experi-

ments yielded different values, even if the standard theory of O'Brien and White [8] is used. This theory automatically accounts for surface conduction, although only in the diffuse part of the double layer because dynamic Stern layers [14, 15] had not been considered.

Basically, neglect of $K^{\sigma i}$ (i.e., underestimating K^σ) works through in different directions for different groups of electrokinetic experiments. Typically, ζ potentials from streaming potentials and electrophoresis are *under*estimated, whereas those from dielectric spectroscopy are *over*estimated. Based on this idea, Kijlstra et al. considered K^σ as an adjustable parameter, of which the real value was such that it harmonized ζ potentials. The experiments involved electrophoresis and dielectric spectroscopy of silica and hematite sols. The advantage of this "second harmonization" is that now K^σ data also become available for dilute sols.

2. *Minor method* [20]. In this approach the conductivity of poly(styrene sulfonate) latex plugs was measured directly as a function of the bulk conductivity K^L. The last-mentioned quantity was modified by changing the electrolyte concentration. Resulting graphs of K(plug) versus K^L are straight lines, obeying the following equation, derived by O'Brien and Perrins [21]:

$$K(\text{plug}) = [1 + 3\varphi f(0)]K^L - \frac{6\varphi f(0)}{a}K^\sigma \qquad (17)$$

where φ is the volume fraction and $f(0) = f(\text{Du} = 0)$ accounts for the packing of the plug. The product $\varphi f(0)$ is obtainable from the slope; hence, K^σ can be obtained. By this method K^σ data are now also obtainable for concentrated sols.

More recently these two methods, or combinations thereof, have been amended and extended to a variety of systems, including bacterial cell walls [22], porous silicas [23], and liposome vesicles [24]. As a result of these studies we now have an interesting set of u_i^i data available; they are collected in Table 1, where the ratio:

$$R = u_i^i / u_i^L \qquad (18)$$

TABLE 1 Ratios R for the Ionic Mobility u_i^i in the Stagnant Layer with Respect to the Bulk Mobility u_i^L

System	Counterion	R	Reference
Silica (Stöber)	K^+	0.96	19, 25
Silica (Monosphere-100)	K^+	0.7	23
Silica (Monosphere-100)	Mg^{2+}	0.7	23
Hematite	Cl^-	0.7	19, 25
Poly(styrene sulfate) latex	H^+, Li^+, Na^+, K^+	0.85	20
Poly(styrene sulfate) latex + adsorbed PEO layer		~0.6	26
Bacterial cell walls	Na^+	0.2–0.5[a]	22
Liposome vesicles	Na^+	~1.0	24
Liposome vesicles	$Ca^{2+}, Cd^{2+}, Cu^{2+}$	~0.6	24
Liposome vesicles	Cs^+	~1.0	27, 28[b]
Liposome vesicles	Ca^{2+}	~0.8	27, 28[b]
Liposome vesicles	La^{3+}	~0.07	27, 28[b]

[a] Depending on species.
[b] Obtained by a different method exploiting the difference between the isoelectric and isoconducting points.

is given for a variety of systems. From this table the following is concluded:

1. For nonporous surfaces and monovalent countercations R is unity or close to it.
2. Porous surfaces (Monosphere, bacterial cell walls) have lower R values for the same ions, probably because these ions have to make detours.
3. Chloride ions adsorb specifically on hematites; also, for that system R is lower than for monovalent counterions.
4. For bivalent counterions R is much lower than for monovalent ones.
5. For a given substrate and valency no ion specificity is observed [compare the bivalent counterions on vesicles in Ref. 24 and the monovalent ones on poly(styrene sulfate) latices]. This nonspecificity even applies to the proton, for which the conduction mechanism is very different from that of other ions.
6. Adsorbed polymers, also forcing tangentially moving ions to make detours, also lower R.

Regarding the consistence of the data, those from Refs. 27 and 28, stemming from a completely different group, using other liposomes, and a very different technique, agree very well with those from Ref. 24, which is gratifying.

The inescapable conclusion is that, in the immobilized (stagnant) layer, ions can move with up to the same mobility as in the bulk. What is the reason for this apparent paradox?

V. AN INTERIM BALANCE: WHAT DO WE KNOW ABOUT STAGNANT LAYERS?

At this stage it is appropriate to review the information we have collected on stagnant layers:

1. Stagnant layers are hydrodynamically immobilized.
2. They have been observed for all solid–aqueous solution interfaces carrying an electric double layer.
3. They have been observed on positive and negative surfaces.
4. They exist both adjacent to hydrophilic and hydrophobic surfaces.
5. From electrokinetics it is not known whether they also occur at water–air or at solid–oil interfaces.
6. Ions in those layers move almost unimpeded.

From this list some inferences can be drawn. First, from point 3 it may be concluded that it is very likely that stagnant layers also occur on *uncharged* surfaces; however, for those conditions their presence cannot be established electrokinetically. From the observation that such layers occur both on hydrophilic (oxides) and hydrophobic (silver iodide, polystyrene latices) surfaces, it is inferred that the solid–water interaction is not the reason for their existence. Rather it is the *stacking of liquid molecules against a hard wall*. When this is true, stagnant layers must also occur at solid–oil interfaces, although these cannot be measured electrokinetically because σ^i is very small. By the same token, stagnant fluid layers will be absent at water–air interfaces. Liquid–liquid interfaces form an intermediate case that deserves further study.

Perhaps the most enigmatic feature is that, although the water in stagnant layers is hydrodynamically immobilized, the embedded ions move almost unimpeded. On further inspection, this feature does not stand on its own; a similar situation arises with gels. For these systems the fluid is immobilized, although the diffusion coefficients of dissolved molecules do not differ greatly from those in the bulk. Because of this analogy we may classify the behavior of stagnant layers as that of a *two-dimensional* (2D) *gel*.

VI. MOLECULAR DYNAMICS (MD) INTERPRETATION

Thanks to a constructive co-operation with the Modélisation Moléculaire group of the University of Mons, Belgium, it was possible to subject liquid flow tangential to solid surfaces by MD simulation. By this technique the general properties of stagnant layers could be reproduced and given a molecular interpretation. The first results have been published [29] and we shall now summarize some of these.

The simulations involved a large number of molecules, viz., 40,000, of which 32,000 were in the liquid and 8000 constituted the solid. The liquid consisted of 31,500 solvent molecules ("water") and 500 solute molecules ("ions"). Interactions are for the moment only accounted for in terms of Lennard–Jones (L–J) pair energies. This is of course not representative for electrostatics, but as explained above, stagnancy and slip are general features, so a model without charges should already capture the main physical features. The ions were chosen slightly larger than the water molecules and were stronger attracted by the wall. The L–J interactions with water were chosen in such a way that they represented water well in previous simulations.

In the first part of the simulations, water and ions were randomly mixed and the system left to equilibrate. Upon this equilibration the ions enriched the layer adjacent to the surface. The phenomenon of stacking of solvent molecules against a hard wall was confirmed: for water the density distribution $\rho(z)$ exhibited the familiar maxima and minima, petering out beyond a few oscillations. The counterions had their first maximum slightly beyond that of the water, because they had larger radii.

The logical physical step is to identify both the Stern layer and the stagnant layer with the first, and main, maximum in $\rho(z)$. For the Stern layer this is a relatively new idea; mostly the interpretation has a more electrostatic nature and runs in terms of dipolar and quadrupolar interactions, ignoring the fact that such layers are also formed with apolar solvents [30]. In fluid dynamics, stagnant layers do not play an important role because their contributions to the fluid flux is relatively small; in Poiseuille flow through capillaries the major part of it goes through the center. It is now also understood that the charge on the surface has no significant influence on the thickness, and that this thickness is not sharply defined; it is statistically defined. As a measure of the quality of the step-function abstraction the amplitudes of the second and following oscillations in $\rho(z)$ may be taken. Certainly the first maximum and minimum dominate the stacking behavior of the fluid molecules.

In the second part of the simulations a tangential force was applied to the ions and their trajectories could be established. Regarding their qualitative physical behavior, one of the main findings was that ions in the stagnant layer can move in that layer, or escape from it to the diffuse part, from where they either may return or be exchanged against another ion to keep the distribution intact. Ions in the bulk were much more subject to thermal noise than those in the stagnant layer, apparently because the molecular stacking in that layer forced them to stay there (or to escape completely). More recent MD experiments, in which electrical fields are explicitly included, support this observation [31].

The excursions that ions can make into the diffuse part help to explain why for monovalent ions $R \sim 1$. When ions can escape and return easily, and travel a substantial fraction of their time in the diffuse part, where bulk properties prevail, their averaged mobilities are dominated by those in the bulk. This also explains the lack of specificity. Bivalent ions are more strongly attracted by the solid, so it is more difficult for these to be "short circuited," hence their lower R. Still another phenomenon is accounted for, viz., that of electro-osmosis. Although no tangential external force is applied to the solvent, molecules do move with the field, simply because they are entrained by the ions; sites left by an ion are mostly replenished by solvent molecules.

All of this can be made quantitative. By choosing reasonable values for the molecular parameters it is possible to mimic experimental mobilities. Perhaps the most essential distinction is that between the *individual* and *collective displacement* of water molecules. The former follows from the trajectories and, as with ions, a bulk-like displacement rate can be easily simulated. The collective mobility follows from the viscosity, which can be obtained from the pressure tensor correlation function [29], which in adjacent layers is anisotropic. Finding this viscosity requires simulation of very large numbers of molecules, a condition which is met in our case. It transpired that the collective mobility of solvent molecules was so low that for practical purposes the first layer behaved as stagnant. In this way, not only is the physics of stagnancy explained, but at the same time the 2D-gel behavior is also accounted for. In conclusion, MD simulations are very rewarding in solving the mysteries of electrokinetics and we feel that the potentialities are not yet exhausted [31].

REFERENCES

1. M von Smoluchowski. Phys Z 6:529, 1906.
2. JJ Bikerman. Z Phys Chem A 163:378, 1933.
3. JJ Bikerman. Kolloid-Z 72:100, 1935.
4. J Lyklema. Fundamentals of Interface and Colloid Science (FICS). vol. II. Academic Press, London, San Diego, New York, Boston, Sydney, Tokyo, Toronto, 1995, Section 4.3f.
5. EJW Verwey, JTG Overbeek. Theory of the Stability of Lyophobic Colloids. Elsevier, Amsterdam, New York, 1948, p 49.
6. AJ Rutgers, M de Smet, W Rigole. J Colloid Sci 14:330, 1959.
7. SS Dukhin. Adv Colloid Interface Sci 44:1, 1993.
8. RW O'Brien, LR White. J Chem Soc, Faraday Trans 2 78:1483, 1983.
9. JJ Bikerman. Trans Faraday Soc 36:154, 1940.
10. AJ Rutgers. Trans Faraday Soc 36:69, 1940.
11. JTG Overbeek. In: Colloid Science, vol. I. Irreversible Systems. (HR Krugt, ed.) Elsevier, Amsterdam, Houston, London, New York, 1952, ch. V, section 12.
12. CF Zukoski, DA Saville. J Colloid Interface Sci 114:32, 1986.
13. LA Rosen, JC Baygents, DA Saville. J Chem Phys 98:4183, 1993.
14. CS Mangelsdorf, LR White. J Chem Soc, Faraday Trans 86:2859, 1990.
15. CS Mangelsdorf, LR White. J Chem Soc, Faraday Trans 94:2583, 1998.
16. J Lyklema. Fundamentals of Interface and Colloid Science (FICS). vol. II. Academic Press, London, San Diego, New York, Boston, Sydney, Tokyo, Toronto, 1995, section 3.6.
17. J Lyklema. Fundamentals of Interface and Colloid Science (FICS). vol. II. Academic Press, London, San Diego, New York, Boston, Sydney, Tokyo, Toronto, 1995, section 4.4.
18. J Lyklema, HP van Leeuwen, M Minor. Adv Colloid Interface Sci 83:33, 1991.
19. J Kijlstra, HP van Leeuwen, J Lyklema. Langmuir 9:1625, 1993.
20. M Minor, AJ van der Linde, HP van Leeuwen, J Lyklema. J Colloid Interface Sci 189:370, 1997.
21. RW O'Brien, WT Perrins. J Colloid Interface Sci 99:20, 1984.

22. A van der Wal, M Minor, W Norde, AJB Zehnder, J Lyklema. J Colloid Interface Sci 186:71, 1997.
23. M Löbbus, HP van Leeuwen, J Lyklema. Colloids Surf. A. Physicochem. Eng. Aspects 161:103, 2000.
24. R Barchini, HP van Leeuwen, J Lyklema. Langmuir 16:8238, 2000.
25. J Lyklema. Fundamentals of Interface and Colloid Science (FICS). vol. II. Academic Press, (see p. 96), 1995, section 4.6f.
26. M Minor, HP van Leeuwen, J Lyklema. Langmuir 15:6677, 1999.
27. H Matsumura, SV Verbich, SS Dukhin. Colloids Surfaces 159:271, 1999.
28. SV Verbich, SS Dukhin, H Matsumura. J Disp Sci Technol 20:83, 1999.
29. J Lyklema, S Rovillard, J de Coninck. Langmuir 14:5659, 1998.
30. J Lyklema. Fundamentals of Interface and Colloid Science (FICS). vol. II. Academic Press, (see p. 96), 1995, sections 3.6c, 3.6d.
31. J de Coninck, J Lyklema, in course of preparation.

4

Evaluation and Usage of Electrostatic Potentials in the Interfacial Layer

DAVOR KOVAČEVIĆ, ANA ČOP, and NIKOLA KALLAY   University of Zagreb, Zagreb, Croatia

I. INTRODUCTION

In the past decades the phenomena at the solid/liquid interface have been often interpreted in terms of the surface complexation model (SCM), sometimes called the site binding model [1, 2]. Most of the related experimental studies concern adsorption equilibrium and development of the surface charge. In addition, electrokinetic measurements are often reported. Adsorption of ionic species, i.e., binding of ionic groups, is markedly affected by the electrostatic potentials at the interface to which bound charged species are exposed. Interfacial electrostatic potentials could be calculated from the equilibrium parameters determined from adsorption and electrokinetic studies. In some cases they could also be measured. This chapter is devoted to the electrostatic potentials at the interface and will concern their definition, meaning, evaluation, and also their use in understanding the processes influenced by electrostatic interactions.

II. SURFACE REACTIONS

The SCM takes into account interactions of ions from the bulk of the solution with specified active surface groups. It does not consider adsorption just as accumulation of species at the interface, but rather as an interfacial chemical reaction, i.e., formation of surface complexes. The concept of surface complexation includes different possible reactions. The simplest mechanism was proposed by Bolt and van Riemsdijk [3] and is called the "1-pK model." It assumes charge separation among surface atoms and just one reaction, i.e., protonation of negative sites. The classical 2-pK model, introduced by Parks [4], assumes amphoteric surface OH groups that undergo protonation and deprotonation, resulting in positively and negatively charged surface groups, respectively. The more general approach, called the MUSIC model (*multisite complexation model*), proposed by Hiemstra and coworkers [5–7], considers different types of active sites at the surface. This model was recently modified using the Pauling concept of charge distribution and resulted in the CD (*charge distribution*)–MUSIC model [8, 9].

A. 1-pK Model

According to the 1-pK model [2, 3], the surface consists of partially charged atomic groups the charge of which depends on the type of metal oxide and on the co-ordination number of the central metal ion. Negative groups bind, while positive groups release the proton. In the case of, e.g., hematite, the protonation of surface $MOH^{-1/2}$ groups results in the formation of positive surface sites, $MOH_2^{+1/2}$:

$$MOH^{-1/2} + H^+ \rightarrow MOH_2^{+1/2} \qquad K = \exp(\phi_0 F/RT)\frac{\Gamma_{(MOH_2^{+1/2})}}{a_{(H^+)}\Gamma_{(MOH^{-1/2})}} \qquad (1)$$

where M denotes an iron atom at the surface, Γ is the surface concentration of surface species, K is the (thermodynamic) equilibrium constant, and ϕ_0 is the potential that affects the state of surface species. According to the 1-pK model, only one reaction is responsible for surface charging, and contrary to the 2-pK model no neutral active surface groups are assumed to exist.

B. 2-pK Model

According to the 2-pK model [2, 4], surface charging of metal oxides is a consequence of protonation (p) and deprotonation (d) of amphoteric surface MOH groups:

$$MOH + H^+ \rightarrow MOH_2^+ \qquad K_p = \exp(\phi_0 F/RT)\frac{\Gamma_{(MOH_2^+)}}{a_{(H^+)}\Gamma_{(MOH)}} \qquad (2)$$

$$MOH \rightarrow MO^- + H^+ \qquad K_d = \exp(-\phi_0 F/RT)\frac{a_{(H^+)}\Gamma_{(MO^-)}}{\Gamma_{(MOH)}} \qquad (3)$$

where K_p and K_d are the corresponding equilibrium constants. In commonly accepted approximation, both positive and negative surface species (MOH_2^+ and MO^-) are exposed to the same electrostatic potential denoted as ϕ_0. In the literature this potential is sometimes called the potential of the inner plane of the compact (Helmholtz) layer, but more often just the "surface potential." Although the term surface potential has a general meaning it will be used here for the potential ϕ_0 defined as the potential that affects the state of ionic surface groups produced by interactions with potential determining ions (p.d.i.). For metal oxides, potential determining ions are H^+ and OH^-, since they are directly or indirectly responsible for interactions with surface-active groups and because they directly determine the surface potential ϕ_0.

C. Counterion Association

Charged surface groups may bind ions of opposite charge – counterions. Counterion association will be described here on the basis of the 2-pK concept as

$$MOH_2^+ + A^- \rightarrow MOH_2^+ \cdot A^- \qquad K_A = \exp(-\phi_\beta F/RT)\frac{\Gamma_{(MOH_2^+ \cdot A^-)}}{a_{(A^-)}\Gamma_{(MOH_2^+)}} \qquad (4)$$

$$MO^- + C^+ \rightarrow MO^- \cdot C^+ \qquad K_C = \exp(\phi_\beta F/RT)\frac{\Gamma_{(MO^- \cdot C^+)}}{a_{(C^+)}\Gamma_{(MO^-)}} \qquad (5)$$

where A^- and C^+ denote anions and cations, respectively, and K_A and K_C are the respective equilibrium constants. The state of associated counterions is affected by electrostatic potential ϕ_β, sometimes called the potential of the outer plane of the compact layer. In this chapter, for ϕ_β, the term association potential will be used. Counterion association is governed by electrostatic Coulombic interactions. This process takes place in "neutral" electrolyte solutions consisting of ions that do not undergo chemical bonding with surface sites. The counterion association is enabled by the local electrostatic field created by oppositely charged surface groups and by the overall electrostatic field due to the presence of other ionic groups at the interface. In reality, the local electric field alone is not sufficient to bind counterions of a neutral electrolyte so that significant association occurs only at high values of the association potential ϕ_β. The effect of the ϕ_β potential on counterion association equilibrium is expressed through the exponential term in Eqs (4) and (5).

Another approach to counterion association is based on the Bjerrum concept originally developed for ionic association in the bulk of the solution. Statistical distribution of counterions around the oppositely charged central ion is considered by Boltzmann statistics including the electrostatic Coulombic term. The probabililty of finding the counterion at a certain distance has a minimum at the so-called critical distance d_{crit} so that integration between the minimum possible separation d_{min} and d_{crit} provides the fraction of associated pairs, and consequently, the equilibrium constant of the ion-pairing process. This equilibrium constant is higher for higher d_{crit} values, i.e., for ions of higher charges and for media of lower permittivity. The equilibrium constant also depends on the minimum possible separation, which is directly related to the ionic size. Relatively large ions ($d_{min} > d_{crit}$) cannot approach each other close enough to overcome the critical distance and become associated. The Bjerrum theory is an excellent explanation for complete dissociation of strong electrolytes. At the interface the situation is somehow different [10–12]. Half of the space is occupied by the solid phase and is thus a "forbidden zone." Second, the electrostatic potential is not only due to the Coulombic potential of the central ion but also due to the presence of other ions at the interface. The consequence is that critical distance depends on the angle, i.e., the "critical plane" is not a sphere. As demonstrated in Fig. 1, the cross-section looks like an ellipse, the extension of which is higher for higher surface potentials. This extension allows the association, since at high enough surface potentials the critical distance could exceed the minimum separation.

Accordingly, the association space and surface association equilibrium constant would increase with the surface potential, in agreement with experimental findings. The Bjerrum model applied to interfaces, introducing the association space concept, would explain the effect of the medium permittivity and the so-called lyotropic effect. Smaller counterions would be characterized by lower d_{min} and consequently higher association equilibrium constant. However, the effective size of ions is not a clear concept, e.g., the lithium ion is effectively larger than the cesium ion owing to the higher degree of hydration. If the hydration shell is destroyed in the association process the opposite order in the lyotropic series could be found.

In conclusion one may say that the association space concept does not contradict the common approach described by Eqs (4) and (5). In both cases the association is predicted to be more pronounced at higher potentials. The specificity of a counterion is characterized by its effective size (minimum separation) in the association space model (ASM), while in the common approach it is reflected in the value of the association equilibrium constant. The common approach would predict a low degree of association at low potentials while the ASM results in the complete absence of counterion association for poten-

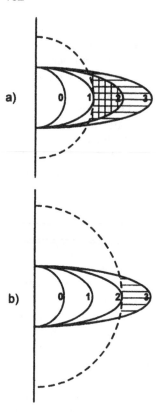

FIG. 1 Schematic presentation of critical boundaries of counterion's association space around a fixed central surface charged group for gradually increasing surface potentials [10–12]. Dashed lines are minimum distances between centers of central ion and counterion for relatively small (a) and large (b) counterions. The numbers represent the values of surface potentials at an arbitrary scale. The shadowed area is the association space: for relative potential 2, vertical shading, and for relative potential 3, horizontal shading. (From Ref. 12.)

tials lower than the critical one, i.e., the potential at which the critical distance becomes larger than the minimum separation.

D. Specific Adsorption

The above considerations dealt with metal oxides in aqueous solutions of "neutral" electrolytes. Potential-determining ions (H^+ and OH^- in the case of metal oxides) are responsible for the formation of surface charge while other ions do not chemically react with surface groups but could become associated with oppositely charged surface groups due to electrostatic (Coulombic) interactions. However, several molecules and ionic groups may be bound "chemically" to the surface. Such processes are commonly called specific adsorption. Specific adsorption will be demonstrated here on the examples of adsorption of salicylic acid and cadmium on ferric oxides.

The SCM considers the interaction of specific surface groups with defined species from the bulk of the solution. Therefore, one should at first analyze the association and dissociation equilibria of the adsorbent in the liquid medium. The interpretation of adsorption and electrokinetic data [13] showed that the binding of singly charges species (HL^-) is responsible for adsorption of salicylic acid on hematite [14]:

$$MOH + HL^- \rightarrow ML^- + H_2O \qquad K_{(HL^-)} = \exp(-\phi_a F/RT)\frac{\Gamma_{(ML^-)}}{\Gamma_{(MOH)}a_{(HL^-)}} \qquad (6)$$

Adsorption of cadmium on goethite is described by the following mechanism [15]:

$$MO^- + CdOH^+ \rightarrow MO^- \cdot CdOH^+ \qquad K_{a(1,1)} = \exp(\phi_a F/RT)\frac{\Gamma_{(MO^- \cdot CdOH^+)}}{\Gamma_{(MO^-)}a_{(CdOH^+)}} \qquad (7)$$

$CdOH^+$ being the species that actually adsorb. The above equilibria are influenced by the electrostatic potential affecting the state of bound ionic species. This potential will be called the adsorption potential and denoted by ϕ_a.

III. STRUCTURE OF THE ELECTRICAL INTERFACIAL LAYER

Several models of the electrical interfacial layer (EIL) are described in the literature [2,12,13] and all of them may be considered as a simplification of the general scheme presented in Fig. 2. At first, we will distinguish between "planes" and "layers." For a

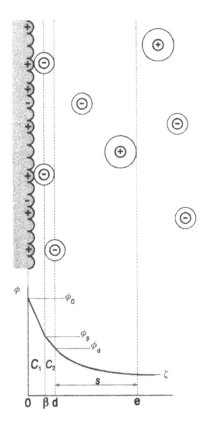

FIG. 2 Schematic presentation of the general model of the electrical interfacial layer.

simple case of a metal oxide in aqueous solution of a "neutral" electrolyte four equipotential planes are postulated: (1) the 0-plane, characterized by ϕ_0 potential, in which surface groups formed by interactions with potential determining ions are located; (2) the β-plane, characterized by ϕ_β potential, in which centers of associated counterions are located; (3) the d-plane (ϕ_d potential), which is the onset of the diffuse layer; and (4) the electrokinetic slipping or shear plane (e-plane), characterized by ζ potential, which is located within the diffuse layer close to the d-plane, and being the hypothetical border between mobile and immobile parts of the diffuse layer. These four planes divide the interfacial layer into four layers. The inner compact layer is the space between 0- and β-planes. The borders of the outer compact layer are β- and d-planes, while the diffuse part of the interfacial layer is divided into two parts, immobile (between d- and e-planes) and mobile (between e-plane and infinity). In practice, this last layer extends up to several Debye–Hückel lengths.

The general EIL model can be reduced to more simple models as follows. The triple layer model (TLM), proposed by Davis, James, and Leckie [16–18], can be derived from the general model by taking the slipping plane to be identical to the onset of the diffuse layer, so that $\phi_d = \zeta$. In other words, the assumption while using the Davis–James–Leckie TLM (DJL-TLM) is that the distance between the onset of the diffuse layer and the slipping (shear) plane equals zero. Accordingly, this model includes three layers as presented in Fig. 3.

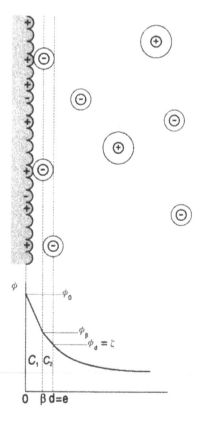

FIG. 3 Schematic presentation of the triple layer model of the electrical interfacial layer as proposed by Davis, James, and Leckie [16–18], with assumption $\phi_d = \zeta$.

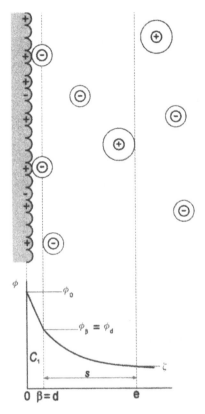

FIG. 4 Schematic presentation of the triple layer model of the electrical interfacial layer, with assumption $\phi_\beta = \phi_d$.

Another version of the TLM assumes that the onset of the diffuse layer is at the β-plane and consequently that $\phi_\beta = \phi_d$. Such an approach results in a slipping-plane separation between 5 and 20 Å. This version of TLM is presented in Fig. 4.

The two versions of the TLM, as presented in Figs 3 and 4, cannot be experimentally distinguished since the interpretation of experimental data requires only $|\phi_0| > |\phi_\beta| > |\zeta|$, which can be achieved by both versions.

More reduced models are those of Helmholtz and Gouy–Chapman. The latter one may be derived from the general model by taking the d-plane as identical to the 0- and β-planes, while the classical Helmholtz model could be obtained formally taking the β- and d-planet as identical and by "compressing" the diffuse layer. However, the experimental data cannot be successfully interpreted by these two oversimplified concepts so that they should be incorporated in the more general scheme: the Helmholtz model for the inner compact layer and the Gouy–Chapman theory for the diffuse part of the interfacial layer.

The above schemes of EIL models consider the simple case of a metal oxide in a neutral electrolyte solution. The situation changes when specific adsorption takes place. In such a case, another "surface" reaction is present: binding of adsorbed ions with certain surface groups. Also, the plane in which this adsorption takes place should be introduced. In doing so two main possibilities appear: the introduction of a new a-plane or the assumption that adsorbed ions are located in one of the already postulated planes (0-, β-, or d-plane).

According to the SCM, the total surface concentration of active surface sites in the interfacial layer (Γ_{tot}) is the sum of all contributions. In the absence of specific adsorption:

$$\Gamma_{tot} = \Gamma_{(MOH)} + \Gamma_{(MOH_2^+)} + \Gamma_{(MO^-)} + \Gamma_{(MO^- \cdot C^+)} + \Gamma_{(MOH_2^+ \cdot A^-)} \tag{8}$$

The surface charge densities in the 0-plane (σ_0) and in the β-plane (σ_β) are given by

$$\sigma_0 = F(\Gamma_{(MOH_2^+)} + \Gamma_{(MOH_2^+ \cdot A^-)} - \Gamma_{(MO^-)} - \Gamma_{(MO^- \cdot C^+)}) \tag{9}$$

$$\sigma_\beta = F(\Gamma_{(MO^- \cdot C^+)} - \Gamma_{(MOH_2^+ \cdot A^-)}) \tag{10}$$

Surface charge density in the diffuse layer (σ_d) is equal in magnitude, but different in sign, to the net charge bound to the surface (σ_s):

$$\sigma_s = -\sigma_d = \sigma_0 + \sigma_\beta = F(\Gamma_{(MOH_2^+)} - \Gamma_{(MO^-)}) \tag{11}$$

As commonly accepted, the potential drop between the 0- and d-planes can be calculated on the basis of the constant capacitance (C_1) of the Helmholtz layer:

$$C_1 = \frac{\sigma_0}{\phi_0 - \phi_\beta} \tag{12}$$

The possible potential drop in the region between the β- and d-planes depends on the capacitance of the second capacitor (C_2) as

$$C_2 = \frac{\sigma_\beta}{\phi_\beta - \phi_d} \tag{13}$$

The surface charge density in the diffuse layer (σ_d) is, according to the Gouy–Chapman theory, defined as

$$\sigma_d = -\sqrt{8RT\varepsilon I_c} \, \sinh \frac{\phi_d F}{2RT} \tag{14}$$

where I_c is the ionic strength. The potential at the onset of the diffuse layer is accordingly defined as

$$\phi_d = \frac{2RT}{F} \, \text{arcsinh} \, \frac{-\sigma_d}{\sqrt{8RT\varepsilon I_c}} \tag{15}$$

IV. ZERO-CHARGE CONDITION

The zero-charge condition at the surface is commonly described by two quantities, i.e., the isoelectric point (i.e.p.) and the point of zero charge (p.z.c.). The i.e.p. corresponds to the condition at which $\zeta = 0$, and consequently to the zero effective charge: $\sigma_s = -\sigma_d = 0$, and $\phi_d = 0$. For oxides it is given by the characteristic pH value (pH_{iep}) which could be experimentally obtained by common electrokinetic methods such as electrophoresis, electro-osmosis, by measuring the streaming potential or current, and on the basis of the sedimentation potential [2, 19, 20]. The i.e.p. of conductive surfaces, as in the case of metals, could be obtained by adhesion methods, showing that metallic surfaces have pH_{iep} values close to those of the corresponding oxides [21, 22].

The p.z.c. is defined through the consumption of potential-determining ions from the bulk of the solution. In the case of metal oxides it could be obtained by pH titration of the suspension at different ionic strengths, by the pH-shift method, or by using the mass titration method [23]. In the case of metal oxides the p.z.c. corresponds to the pH at which

$$\Gamma_{(MOH_2^+)} + \Gamma_{(MOH_2^+ \cdot A^-)} = \Gamma_{(MO^-)} + \Gamma_{(MO^- \cdot C^+)} \tag{16}$$

In the absence of specific adsorption, and in the case of negligible [$\Gamma_{(MOH_2^+ \cdot A^-)} = 0$; $\Gamma_{(MO^- \cdot C^+)} = 0$] or symmetric [$\Gamma_{(MOH_2^+ \cdot A^-)} = \Gamma_{(MO^- \cdot C^+)}$] counterion association, the p.z.c. corresponds to $\sigma_0 = 0$. In these systems all other interfacial potentials (ϕ_0, ϕ_β, ϕ_d) are equal to zero and also $\Gamma_{(MOH_2^+)} = \Gamma_{(MO^-)}$. According to Eqs (2) and (3), the p.z.c. is related to protonation and deprotonation equilibrium constants by

$$pH_{pzc} = 0.5 \log(K_p/K_d) \tag{17}$$

There are two additional possibilities to define the zero-charge condition. The first of them is the point of zero potential (p.z.p.) at which $\phi_0 = 0$, and the second one is characterized by $\sigma_0 = 0$. The p.z.p. is of special interest since it is available experimentally. One would need to construct the proper metal oxide electrode and measure its potential with respect to the reference electrode. The problem is that one cannot simply obtain the absolute values of the surface potential, but, at low concentrations of a neutral electrolyte, one may assume that the p.z.p. corresponds to the i.e.p. so that zero value of the surface potential could be found. Once the system is calibrated one may follow the change in p.z.p. as a function of concentration of specifically adsorbable ions. Such data may be extremely useful to conclude on the EIL structure and on the mechanism of binding of ions that chemically adsorb at the surface.

V. POTENTIALS IN THE ELECTRICAL INTERFACIAL LAYER

A. Electrokinetic Potential

The electrokinetic ζ potential could be evaluated from electrokinetic phenomena such as electrophoresis, etc. The strict physical meaning of this quantity is given by the equation used for its evaluation from the measured data, e.g., by the equation that relates ζ potential and the electrophoretic mobility of the particles. However, the intention is to obtain the quantity which would correspond to the hypothetical slipping or shear plane. The importance of ζ potential lies in the fact that it is a measurable quantity that can be obtained independently. As will be described later, the only other potential that can be measured is the surface potential ϕ_0. The methods that can be used for evaluation of ζ potential are extensively explained in other chapters so that here only the relation of ζ potential with other electrostatic potentials in the electrical interfacial layer will be described.

The measured ζ potential could be the starting point in calculation of other electrostatic potentials. The direct connection between ζ potential, which is the potential somewhere in the diffuse layer, and the ϕ_d potential, which is characteristic of the onset of the diffuse layer, is given by the Gouy–Chapman theory. The relationship between the potential at the onset of the diffuse layer (ϕ_d) and the potential at a distance x from the onset of the diffuse layer (ϕ_x) is given by

$$\phi_d = 2RTF^{-1} \ln\left[\frac{\exp(-x\kappa) + \tanh(F\phi_x/4RT)}{\exp(-x\kappa) - \tanh(F\phi_x/4RT)}\right] \tag{18}$$

with κ being the Debye–Hückel reciprocal length:

$$\kappa = \sqrt{\frac{2F^2 I_c}{\varepsilon RT}} \tag{19}$$

where I_c is the ionic strength, and permittivity ε is the product of the relative value for the medium ε_r, and the permittivity of the vacuum ε_0. It is obvious that calculated potentials depend on the value of permittivity, i.e., on the relative permittivity of the medium. In the calculations the value of the relative permittivity of bulk water is commonly used, which is questionable since the interfacial water may have different properties, e.g., lower permittivity. According to Eq. (18), the difference between ζ potential and ϕ_d is determined by the slipping-plane separation distance so that the use of the bulk water permittivity will lead to the apparent value of the separation distance.

The relationship between ζ potential and ϕ_d potential is extensively discussed in the literature. Some authors [24–26] believe that the slipping-plane separation distance is negligible, so that $\zeta \approx \phi_d$, while others [27–35] suggest that such an approximation would cause considerable error, especially in systems at high ionic strengths.

In the discussion regarding the slipping-plane separation it is worth noting that Eversole and coworkers [28, 29] suggested a method for its evaluation by interpreting ζ-potential dependency on the ionic strength on the basis of the Gouy–Chapman theory. The Gouy–Chapman relationship (Eq. 18) can be linearized as

$$\ln \tanh\left(\frac{F\zeta}{4RT}\right) = \ln \tan\left(\frac{F\phi_d}{4RT}\right) - \kappa s \tag{20}$$

with the slope providing the slipping-plane separation (s). Such a procedure assumes constant potential ϕ_d (at constant activity of potential-determining ions) despite increase in the electrolyte concentration. However, knowing that ϕ_d decreases with ionic strength, mainly due to counterion association, no linearity could be expected and the slope of the tangent at a certain point should be always higher than the slipping-plane separation.

Therefore, the Eversole method determines only the upper possible limit of the value of the slipping-plane separation s. Anothe possibility for evaluation of s is based on the comparison of adsorption data with electrokinetic measurements. By using this approach [27], the results obtained with an iminodiacetic acid/hematite system yielded a value of $s = 16\,\text{Å}$. For other systems, such as hematite/water, hematite/water–methanol, and hematite/water–ethanol with neutral electrolytes [30, 31], a similar value was obtained ($s = 12 \pm 3\,\text{Å}$). By modeling the data obtained by potentiometric measurements, Healy and White [32] obtained a slipping-plane separation (at higher ionic strengths) of $20\,\text{Å}$. The slipping-plane distance can be calculated also from the pH dependence of the ζ potential in the vicinity of the i.e.p. at different ionic strengths [33]; for chromium hydroxide the slipping-plane distance was thus determined to be 17–$20\,\text{Å}$. Harding and Healy [34] interpreted the electrokinetic measurements of latex particles at low ionic strengths and obtained a slipping-plane separation of $20\,\text{Å}$. Using electrokinetic measurements of latex particles, Chow and Takamura [35] estimated the slipping-plane separation to be $6\,\text{Å}$. The results on the adsorption of organic acids on metal oxides were interpreted by means of simultaneous analysis of adsorption and electrokinetic measurements [14, 15, 36,

37] and a slipping-plane separation of $15 \pm 5 \text{ Å}$ was obtained. All the results mentioned suggest that the slipping-plane separation should not be neglected, and that its value is somewhere between 5 and 20 Å. When neglecting the slippping-plane separation one is forced to introduce the potential drop between the β-plane in which counterions are associated and the onset of the diffuse layer and use the DJL–TLM with the outer-layer capacitor. The reason for such a refinement is that adsorption data require higher potentials to which adsorbed species are exposed with respect to the measured ζ potential. At present, one cannot experimentally distinguish between these two different approaches so that both concepts, introduction of the second capacitor and the slipping-plane separation, are in use.

The practical use of the ζ potential could be either on the quantitative or semiquantitative level [19, 20]. In the latter case one can, e.g., conclude on the sign of charge of particles and predict the conditions required for colloid stability. On the quantitative level one may use the electrokinetic data [14, 15, 36, 37] to calculate ϕ_d potential, which is then useful in the interpretation of the ionic adsorption equilibria and the mutual particle interactions. Electrokinetic potentials may be converted into electrokinetic charge (density), which provides information regarding the effective charge of the particle moving in the electric field.

The i.e.p. is an important characteristic of the solid phase in contact with the liquid. This makes electrokinetic methods popular tools in interfacial chemistry. In the case of specific adsorption the i.e.p. moves from its original value. Specifically adsorbed anions would change the i.e.p. to lower, and cations to higher, pH values. The p.z.c. moves in the opposite direction with respect to the i.e.p. These effects may be used to detect the presence of specific adsorption. Classical electrokinetic methods cannot be used for conductive metallic surfaces, with the exception of metals that can be prepared in colloidal form so that electrophoresis can be applied. For other cases one may apply adhesion methods [21, 22], which show that i.e.p. values of metals are approximately equal to those of the corresponding oxides. The adhesion method was found to be useful for characterization of the oxide layer at the metal surface and it is thus applicable in studying corrosion processes [38].

B. ϕ_d Potential

The potential at the onset of the diffuse layer (ϕ_d) cannot be obtained by direct measurements. It is related to both the potential affecting the state of the bound ions in the outer plane of the compact layer (ϕ_β or ϕ_a) and to the experimental ζ potential. In the case of neutral electrolytes, according to the DJL–TLM presented in Fig. 3, $|\phi_d| < |\phi_\beta|$ and $\phi_d \approx \zeta$. On the other hand, according to the second version of the TLM presented in Fig. 4, $\phi_d \approx \phi_\beta$ and $|\phi_d| > |\zeta|$. Considering these two approaches one may conclude that $|\phi_\beta| \geq |\phi_d| \geq |\zeta|$. Also, it is clear that the $\phi_d(\text{pH})$ function should follow the experimental ζ-potential dependency on pH, being somehow higher in magnitude. The difference should be more pronounced at higher ionic strengths. With respect to the surface potential (ϕ_0), the potential at the onset of the diffuse layer (ϕ_d) should be lower in magnitude. For further discussion, we shall use the second version of the TLM (Fig. 4) with the assumption that $\phi_\beta = \phi_d$ and $|\phi_d| \geq |\zeta|$, i.e., $s \geq 0$. In such a case:

$$\phi_d = \phi_\beta = \phi_0 - \frac{\sigma_0}{C} \tag{21}$$

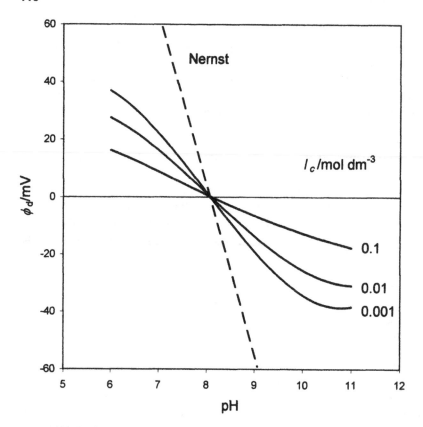

FIG. 5 Effect of ionic strength on the ϕ_d(pH) function calculated for γ-Al$_2$O$_3$/NaCl(aq) system on the basis of the SCM. Dashed line represents the Nernstian slope. (Data from Ref. 39.)

which means that the ϕ_d(pH) function follows the ϕ_0(pH) function up to a certain extent. As will be shown in Section V.D, ϕ_0 decreases with pH. This function is approximately linear, with a slope lower than the Nernstian one. The second term in Eq. (21) depends on the surface charge density in the 0-plane, which results in the leveling of the ϕ_d(pH) function. At high electrolyte concentration the magnitude of σ_0 increases, mainly due to counterion association, so that ϕ_d is significantly reduced with respect to ϕ_0. In Figs 5 and 6 the effect of pH [39] and ionic strength [40] on the ϕ_d potential is presented.

In Fig. 7 the effect of pH on the ϕ_d potential in the case of a salicylic acid/hematite system at two different ionic strengths is presented [36]. For comparison, the ζ(pH) function is also displayed.

The significance of the potential at the onset of the diffuse layer, ϕ_d, lies in the fact that it determines the equilibrium in the diffuse layer. This potential represents the overall (net) charge of colloid particles and thus determines their aggregation kinetics. According to the above discussion, the electrostatic energy barrier between two interacting particles will be reduced by addition of electrolytes for two reasons. At first the association of counterions will decrease the ϕ_d potential, and second the diffuse layer will be more "compressed," which means that ions will be distributed more closely to the d-plane. In a 1:1 electrolyte systems all ions may be considered to behave similarly in

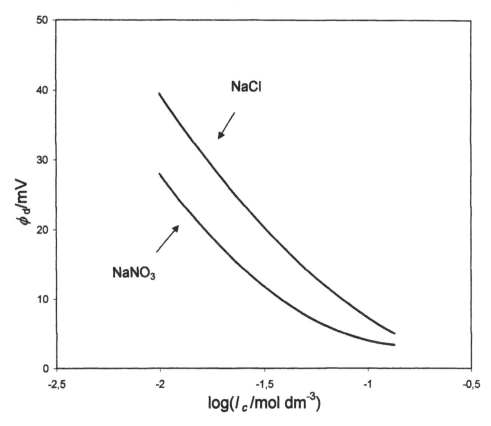

FIG. 6 Effect of ionic strength on the ϕ_d potential at pH 4, calculated for α-Fe_2O_3/NaCl(aq) and α-Fe_2O_3/NaNO$_3$(aq) systems. (Data from Ref. 40.)

the diffuse layer so that no difference may be expected between effects of, e.g., sodium nitrate and potassium perchlorate. However, the surface association will depend on the nature of the ions, so that counterions characterized by higher association equilibrium constants will reduce ϕ_d to a greater extent and will promote the coagulation process more efficiently. According to the SCM this effect will be more pronounced for counterions of higher charge number through their association equilibrium constant and the exponential term in the equilibrium relationship. Association of cations with negative surface sites is described by

$$MO^- + C^{z+} \rightarrow MO^- \cdot C^{z+} \qquad K = \exp(z\phi_\beta F/RT)\frac{\Gamma(MO^- \cdot C^{z+})}{a(C^{z+})\Gamma(MO^-)} \qquad (22)$$

Another approach, based on the Bjerrum concept, would suggest that ions of higher charge exhibit extended association space and consequently higher values of association equilibrium constants and more pronounced association. Accordingly, both theoretical approaches would predict that counterions of higher charge number reduce ϕ_d potential more efficiently so that the critical coagulation concentration of these ions will be lower, which is known as the Schulze–Hardy rule [2,12].

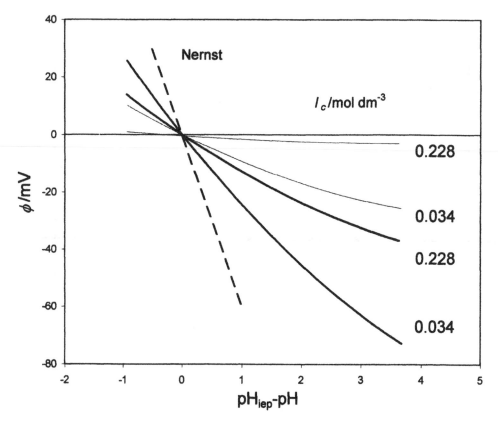

FIG. 7 Effect of ionic strength on ϕ_d (heavy lines) and ζ potential (thin lines) calculated for α-Fe_2O_3/salicylic acid system. Dashed line represents the Nernstian slope. (Data from Ref. 36.)

C. ϕ_β and ϕ_a Potentials

The potential affecting the state of associated counterions in the interfacial layer (ϕ_β) is defined by Eqs (4) and (5). This potential cannot be obtained by direct measurements but rather indirectly by interpreting the adsorption equilibrium data. The problem is that in most of cases adsorption of counterions is not measured directly but rather deduced from experimental σ_0(pH) functions. According to the DJK–TLM, as presented in Fig. 3, the ϕ_β potential is higher in magnitude than the ϕ_d potential. The difference between these two potentials is proportional to the charge density in the β-plane, and the proportionality constant is the reciprocal capacity of the assumed capacitor with β- and d-planes (C_2). High C_2 values would correspond to closer proximity of the β- and d-planes and result in $\phi_\beta \approx \phi_d$. Within the second simplification of the general scheme of the EIL (Fig. 4) $C_2 \rightarrow \infty$ and $\phi_\beta = \phi_d$, so that the ϕ_β potential could be deduced from electrokinetic measurements via Eq. (18).

A similar situation applies for the "adsorption potential" ϕ_a. If specifically adsorbed ions are not located in the 0-plane, one may relate ϕ_a to the ϕ_0 potential, introducing the capacitor C_1 and the approximation $\phi_a \approx \phi_d$.

The interesting feature of the ϕ_β and ϕ_a potentials is the "charge reversal." If the concentration of specifically adsorbed ions is high with respect to σ_0 or if the adsorbed ions

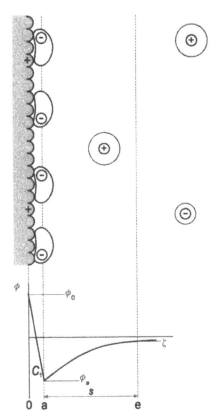

FIG. 8 Schematic presentation of electric interfacial layer in the case of "charge reversal" due to the specific adsorption of anions. (From Ref. 13.)

are of higher charge number, the signs of potentials ϕ_0 and ϕ_β (or ϕ_a) could be opposite. Such a situation is demonstrated in Fig. 8. Again, ϕ_β (ϕ_a) are close (or equal) to the potential at the onset of diffuse layer ϕ_d.

D. ϕ_0 Potential

Potential ϕ_0 will be defined here by Eqs (2) and (3) as the potential that determines the equilibrium involving interactions of surface groups with potential determining ions. For metal oxides, according to the 2-pK model, Eqs (2) and (3), the following equation could be derived:

$$\phi_0 = \frac{RT}{2F} \ln \frac{K_p}{K_d} - \frac{RT}{2F} \ln \frac{\Gamma_{(MOH_2^+)}}{\Gamma_{(MO^-)}} + \frac{RT}{F} \ln a_{(H^+)} \tag{23}$$

$$= \frac{RT \ln 10}{F} (pH_{pzc} - pH) - \frac{RT \ln 10}{2F} \log \frac{\Gamma_{(MOH_2^+)}}{\Gamma_{(MO^-)}}$$

while the 1-pK model, Eq. (1), results in

$$\phi_0 = \frac{RT}{F} \ln K - \frac{RT}{F} \ln \frac{\Gamma_{(MOH_2^{+1/2})}}{\Gamma_{(MOH^{-1/2})}} + \frac{RT}{F} \ln a_{(H^+)} \tag{24}$$

$$= \frac{RT \ln 10}{F}(pH_{pzc} - pH) - \frac{RT \ln 10}{F} \log \frac{\Gamma_{(MOH_2^{+1/2})}}{\Gamma_{(MOH^{-1/2})}}$$

In both cases the slope of the $\phi_0(pH)$ function should be lower than the Nernstian one, given by $RT \ln 10/F$. Deviation from the Nernst equation is due to the pH dependence of the ratio of surface concentrations of positive and negative groups. Below the p.z.c. the positive groups prevail while at higher pH values their concentration is reduced. Negative groups exhibit the opposite trend. Since the surface charge is given by the difference between surface concentrations of positive and negative groups, and the deviation from the Nernst equation by their ratio, one can conclude that the surface will be closer to Nernstian at higher values of $\Gamma(MOH_2^+)$ and $\Gamma(MO^-)$. This condition may be achieved with systems characterized by higher values of protonation and deprotonation equilibrium constants and low values of the counterion association equilibrium constants. Also, the higher value of the total concentration of active surface sites (Γ_{tot}) would lead to the potential being closer to Nernstian. However, Γ_{tot} is limited by the surface structure and could have significantly higher values only if surface reactions take place in a layer

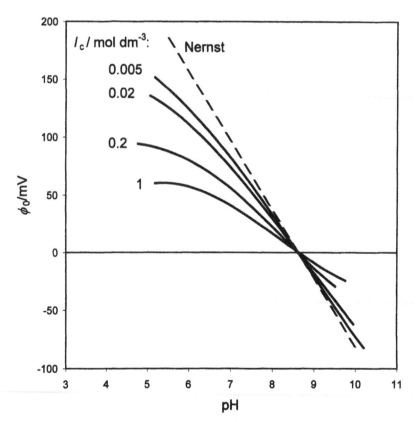

FIG. 9 Effect of ionic strength on the $\phi_0(pH)$ function calculated for α-Fe$_2$O$_3$/KNO$_3$(aq) system; $\log K_p = 6.4$, $\log K_d = -10.5$, $\Gamma_{tot} = 1 \times 10^{-5}$ mol m^{-2}. Dashed line represents Nernstian slope. (Data from Ref. 1.)

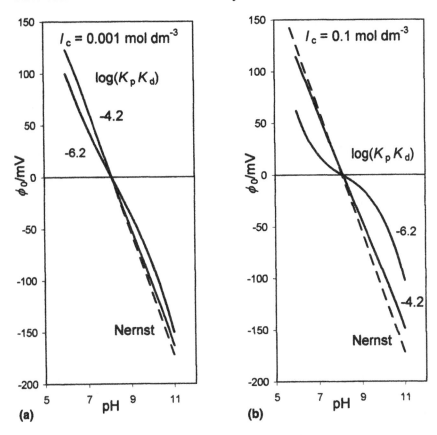

FIG. 10 Effect of magnitudes of K_p and K_d values on the ϕ_0(pH) function for γ-Al$_2$O$_3$/electrolyte system. Dashed line represents Nernstian slope. (Data from Ref. 39.)

but not in the plane, e.g., in the case of a porous surface or, as in the case of glass, when a thin "gel" layer is formed.

The ϕ_0(pH) function could be analyzed quantitatively by numerical simulation [1, 24, 26, 39, 41–43] based on the SCM using the equilibrium parameters obtained by interpretation of the adsorption measurements, i.e., of the σ_0(pH) function. Such an example is presented in Fig. 9, which shows that the ϕ_0(pH) function is approximately linear and that deviation from the Nernst equation is more pronounced at higher ionic strengths [1].

Depending on the values of the equilibrium parameters, a significant deviation from linearity in the region around the p.z.c. could be observed. According to Fig. 10, this deviation is significant for low values of the equilibrium constants K_p and K_d, and is more pronounced at higher ionic strengths [39].

The above theoretical speculations may be tested, since the ϕ_0 potential can also be obtained by direct measurements. For that purpose two approaches are in use: (i) application of field-effect transistors (FETs) [44–47] and (ii) measurements of the electrode potential of a metal covered by an oxide layer [48–55]. In both cases the measured quantity is the total potential drop in the circuit. Assuming that the only potential that depends on pH is the potential drop within the interfacial layer one may conclude that the measured potential difference is equal to the relative value of the surface potential. The absolute scale of the surface potential can be obtained by locating its zero value at the p.z.c. and/or i.e.p. The procedure is correct in the absence of specific adsorption and in the case of negligible or symmetric counterion association. In this case these two zero points coincide.

The origin of the surface potential may be a redox process involving ions from the solid phase [48–50] or the surface complexation as described by Eq. (1) or Eqs (2) and (3) [51–55]. Redox equilibrium will be explained by means of the example of manganese dioxide [48]. The following reactions may take place at the surface:

$$MnO_2 + H^+ + e^- \rightarrow MnOOH \tag{25}$$

$$MnO_2 + 4H^+ + 2e^- \rightarrow Mn^{2+} + 2H_2O \tag{26}$$

Since both reactions involve H^+ ions, their equilibrium should depend on pH. In the case of reaction (25) the following equation could be derived:

$$E = E^\circ + \frac{RT \ln 10}{F} \log a_{(H^+)} = E^\circ - \frac{RT \ln 10}{F} pH \tag{27}$$

where E° is the standard value of the potential. In the second case, the potential for reaction (26) is equal to

$$E = E^\circ + \frac{RT \ln 10}{2F} \log \frac{a_{(H^+)}^4}{a_{(Mn^{2+})}} = E^\circ - \frac{RT \ln 10}{2F} \log a_{(Mn^{2+})} - \frac{2RT \ln 10}{F} pH \tag{28}$$

According to the above analysis, the slope of the function $E(pH)$ should be either $RT \ln 10/F$ or $2RT \ln 10/F$, depending on the reaction being responsible for the equilibrium at the surface. Experimental results show that in the acidic region (pH < 6) the slope is between -90 and $-100\,mV$ which corresponds to mechanism (26) and Eq. (28). The slope is lower in magnitude than $120\,mV$, which may be due to slight changes in the local equilibrium concentration of Mn^{2+} ions. In the basic region (pH > 7) the slope is found to be $-60\,mV$, in accordance with the Nernst equation [Eq. (27)] for equilibrium (25). As expected, in this region perfect agreement with Nernstian behavior is obtained, because no dissolved species are involved in the reaction so that only the activity of H^+ ions is responsible for the equilibrium.

Surface complexation is a more complicated mechanism with respect to the redox process. In the simplest case of an oxide in contact with "neutral electrolyte solution" four surface reactions take place, Eqs (2)–(5). Each reaction influences others through surface charge and consequently surface potential. They also compete for active surface sites. It may be concluded that surface complexation is responsible for the development of surface potential in the absence of redox processes. According to the SCM, the slope of the $\phi_0(pH)$ function should be either close to Nernstian ($-59.2\,mV$ at 25°C) or lower, this decrease being more pronounced at higher ionic strengths. Also, depending on the nature of the surface sites, a minimum of the slope can be observed in the zero-charge region. In some cases it is not simple to conclude on the nature of the process responsible for the development of the surface potential. For example, in the case of a redox process involving the same number of H^+ ions and electrons the function $\phi_0(pH)$ will be linear, with slope $RT \ln 10/F$. Similar observations may be found in the case of the surface complexation mechanism. However, the latter case can be recognized through the sensitivity on addition of electrolyte.

In the literature one can find several examples of "open-circuit potential" measurements with metal oxide electrodes. Some of them are presented in Fig. 11. Electrodes made by Penners et al. [52], by depositing hematite on platinum wire, showed behavior

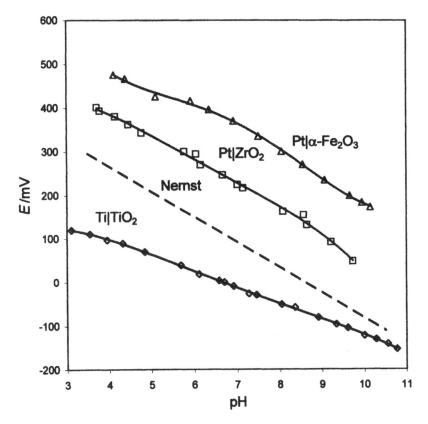

FIG. 11 "Open-circuit potential" of different metal oxide electrodes: (Δ) Pt|α-Fe$_2$O$_3$|5 × 10^{-3} mol dm^{-3} KCl, 20°C (data from Ref. 52); ($\Diamond$) Ti|TiO$_2$|2 × 10^{-3} mol dm^{-3} KNO$_3$, 25°C (data from Ref. 53.); ($\square$) Pt|ZrO$_2$|1 × 10^{-3} mol dm^{-3} KNO$_3$ or NaClO$_4$, ~ 25°C (data from Ref. 54). Dashed line represents the Nernstian slope.

close to Nernstian in the pH range 6–10 in 5×10^{-3} mol dm^{-3} KCl solution. In the acidic region below pH 6 the magnitude of the slope decreased significantly. Avena et al. [53] prepared a Ti/TiO$_2$ electrode and showed that surface potential changed linearly with pH with a slope of −39 mV, independent of ionic strength. Again, in the acidic region below pH 5 the magnitude of the slope was decreased. This finding was confirmed by the interpretation of adsorption data [56]. In the case of a ZrO$_2$ electrode, prepared by Ardizzone and Radaelli [54], the results were sensitive to thermal treatment. However, the electrode prepared at 400°C showed reversible and fast response with a slope of −59 mV. Kallay and Čakara [55] prepared an ice electrode and found the slope to be lower in magnitude than the Nernstian one (from −40 to −46 mV at 0°C) with the maximum of its magnitude at the i.e.p. of ice at pH ≈ 4. The important requirement for these electrodes is nonporosity of the oxide layer since in the opposite case the electrode would behave as the oxide electrode of the second kind, the potential of which would depend on pH through the solubility.

These results correspond to the surface complexation mechanism. However, in the case of manganese dioxide [48], iridium dioxide [49], and palladium oxide [50] the redox process was found to be responsible for the development of the surface potential.

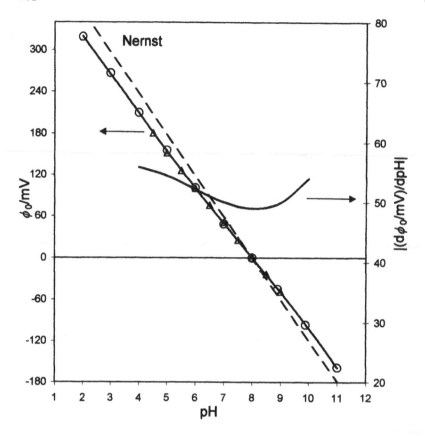

FIG. 12 Measurements of surface potential of γ-Al$_2$O$_3$ by FET device in ($\bigcirc$) 0.1 mol dm^{-3} NaCl; ($\triangle$) 0.1 mol dm^{-3} NaH$_2$PO$_4$/Na$_2$HPO$_4$. Dashed line represents the Nernstian slope. (Data from Ref. 47.)

There are numerous reports, e.g., Refs 44–47, of direct measurement of ϕ_0 using FETs. One purpose of the work was to prepare an ion-selective FET electrode [44]. Here, we will focus on the results that contribute to our knowledge of the surface complexation equilibria. Most of the reports deal with SiO$_2$ and Al$_2$O$_3$ surfaces exposed to aqueous electrolyte solutions [45–47]. Cichos and Geidel [46] examined the effect of electrolytes on the surface potential of silica and alumina. In the case of silica, the minimum in the magnitude of the slope was found at pH $\approx$ 7, and the region of lower slope significantly extended at higher concentrations of electrolytes. Alumina showed a very good linearity of the ϕ_0(pH) function with a slope independent of electrolyte concentration. Bousse et al. [47] examined in detail the deviation of the alumina surface from Nernstian behavior and found a minimum magnitude of the slope ($|d\phi_0|/dpH$) at pH 8, which is the p.z.c. of alumina. The results were successfully intrepreted by the SCM. Figure 12 demonstrates the pH dependence of the surface potential and the slope for alumina.

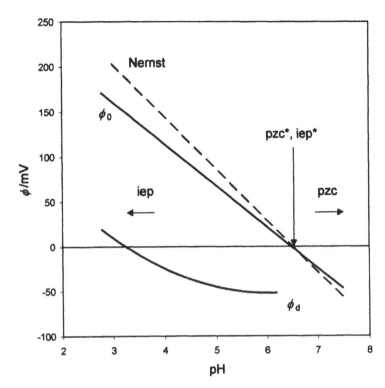

FIG. 13 Effect of specific adsorption on ϕ_0(pH) and ϕ_d(pH) functions for α-Fe$_2$O$_3$/salicylic acid system. Asterisk represents values for pure hematite in absence of specific adsorption. (Data from Ref. 14.)

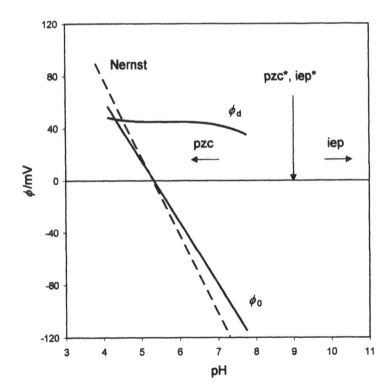

FIG. 14 Effect of specific adsorption on ϕ_0(pH) and ϕ_d(pH) functions for Cd^{2+}/goethite system. Asterisk represents values for pure goethite in absence of specific adsorption. (Data from Ref. 15.)

VI. THE SHIFT OF "ZERO POINTS" IN THE CASE OF SPECIFIC ADSORPTION

In the simplest case of metal oxides in aqueous solutions of neutral electrolytes (with ions that do not show affinity towards adsorption) all four "zero points" coincide so that $pH_{iep} = pH_{pzc} = pH_{pzp} = pH(\sigma_0 = 0)$. Any specific adsorption will shift these "zero points" differently. For example, specific adsorption of anions will shift the i.e.p. to lower pH values, but the p.z.c. in the opposite direction. Such a finding is in accordance with all variation of the SCM. However, the possible shift in pH_{pzp} would depend on the mechanism of binding of these ions and on the structure of the EIL so that this information may be used to study these systems. This hypothesis is worth examining theoretically, because ϕ_0 is a measurable quantity.

In the case of specific adsorption, when some ions are chemically bound to certain surface groups, Eqs (2) and (3) are still valid. However, specific adsorption will change the zero point and may also indirectly affect the ratio of surface concentrations of positive and negative surface groups. Figure 13 represents the results obtained by the analysis of adsorption and electrokinetic data for a salicylic acid/hematite system [14]. It is obvious that the i.e.p. shifts towards lower pH values after adsorption of anions while the p.z.c. shifts in the opposite direction. The point of zero potential has the same value as the i.e.p. and p.z.c. before adsorption. In the case of cation adsorption, presented in Fig. 14 for the $Cd^{2+}/$ goethite system [15], it is shown that the i.e.p. and p.z.c. shift in expected different directions, and that the p.z.p. is shifted in the same direction as the p.z.c.

REFERENCES

1. M Kosmulski, R Sprycha, J Szczypa. In: N Kallay, ed. Interfacial Dynamics. New York: Marcell Dekker, 2000, p. 163.
2. J Lyklema. Fundamentals of Interface and Colloid Science. vol. II. London: Academic Press, 1995.
3. GH Bolt, WH van Riemsdijk. In: GH Bolt, ed. Soil Chemistry. B. Physico-chemical Models. Amsterdam: Elsevier, 1982.
4. GA Parks. Chem Rev 65:177, 1965.
5. T Hiemstra, WH van Riemsdijk, GH Bolt. J Colloid Interface Sci 133:91, 1989.
6. T Hiemstra, JCM de Wit, WH van Riemsdijk. J Colloid Interface Sci 133:105, 1989.
7. T Hiemstra, WH van Riemsdijk. Colloids Surfaces 59:7, 1991.
8. T Hiemstra, WH van Riemsdijk. J Colloid Interface Sci 179:488, 1996.
9. P Venema, T Hiemstra, WH van Riemsdijk. J Colloid Interface Sci 183:515, 1996.
10. N Kallay, M Tomić. Langmuir 4:559, 1988.
11. M Tomić, N Kallay. Langmuir 4:565, 1988.
12. N Kallay, R Sprycha, M Tomić, S Žalac, Ž Torbić. Croat Chem Acta 63:467, 1990.
13. N Kallay, D Kovačević, A Čop. In: N Kallay, ed. Interfacial Dynamics. New York: Marcel Dekker, 2000, p. 249.
14. D Kovačević, I Kobal, N Kallay. Croat Chem Acta 71:1139, 1998.
15. D Kovačević, A Pohlmeier, G Özbas, HD Narres, N Kallay. Progr Colloid Polym Sci 112:183, 1999.
16. JA Davis, RO James, JO Leckie. J Colloid Interface Sci 63:480, 1978.
17. JA Davis, JO Leckie. J Colloid Interface Sci 67:90, 1978.
18. JA Davis, RO James, JO Leckie. J Colloid Interface Sci 74:32, 1980.

19. RJ Hunter. Zeta Potentials in Colloid Science. London: Academic Press, 1981.
20. N Kallay, V Hlady, J Jednačak-Bišćan, S Milonjić. In: BW Rossiter, RC Beatzold, eds. Investigation of Surfaces and Interfaces. Part A of vol. IX of Physical Methods in Chemistry. New York: Wiley-Interscience, 1993, p. 73.
21. N Kallay, Ž Torbić, E Barouch, J Jednačak. J Colloid Interface Sci 118:431, 1987.
22. N Kallay, Ž Torbić, M Golić, E Matijević. J Phys Chem 95:7028, 1991.
23. T Preočanin, N Kallay. Croat Chem Acta 71:1117, 1998.
24. R Sprycha. J Colloid Interface Sci 102:173, 1984.
25. RJ Hunter, HJL Wright. J Colloid Interface Sci 37:564, 1971.
26. R Sprycha, J Szczypa. J Colloid Interface Sci 102:288, 1984.
27. R Torres, N Kallay, E Matijević. Langmuir 4:706, 1988.
28. WG Eversole, PH Lahr. J Chem Phys 9:530, 1941.
29. WG Eversole, WW Boardman. J Chem Phys 9:798, 1941.
30. P Hesleitner, N Kallay, E Matijević. Langmuir 7:178, 1991.
31. P Hesleitner, N Kallay, E Matijević. Langmuir 7:1554, 1991.
32. TW Healy, LR White. Adv Colloid Interface Sci 9:303, 1978.
33. R Sprycha, E Matijević. Langmuir 5:479, 1989.
34. IH Harding, TW Healy. J Colloid Interface Sci 107:382, 1985.
35. RS Chow, K Takamura. J Colloid Interface Sci 125:226, 1988.
36. D Kovačević, N Kallay, I Antol, A Pohlmeier, H Lewandovski, HD Narres. Colloids Surfaces 140:261, 1998.
37. D Kovačević, A Pohlmeier, G Özbas, HD Narres, MJ Schwuger, N Kallay. Colloids Surfaces 166:225, 2000.
38. N Kallay, D Kovačević, I Dedić, V Tomašić. Corrosion (NACE) 50:588, 1994.
39. R Sprycha. J Colloid Interface Sci 127:12, 1989.
40. M Čolić, DW Fuerstenau, N Kallay, E Matijević. Colloids Surfaces 59:169, 1991.
41. R Sprycha. J Colloid Interface Sci 127:1, 1989.
42. R Sprycha. J Colloid Interface Sci 110:278, 1986.
43. LK Koopal, WH Van Riemsdijk, MG Roffey. J Colloid Interface Sci 118:117, 1987.
44. JN Zemel. Anal Chem 47:255A, 1975.
45. JF Schenk. J Colloid Interface Sci 61:569, 1977.
46. C Chicos, T Geidel. Colloid Polym Sci 261:947, 1983.
47. L Bousse, NF de Rooij, P Bergveld. Surface Sci 135:479, 1983.
48. I Tari, T Hirai. Electrochim Acta 26:1657, 1981.
49. S Ardizzone, A Carugati, S Trasatti. J Electroanal Chem 126:287, 1981.
50. E Kinoshita, F Ingman, G Edwall, S Glab. Electrochim Acta 31:29, 1986.
51. S Ardizzone, S Trasatti. Adv Colloid Interface Sci 64:173, 1996.
52. NHG Penners, LK Koopal, J Lyklema. Colloids Surfaces 21:457, 1986.
53. MJ Avena, OR Camara, and CP De Pauli. Colloids Surfaces 69:217, 1993.
54. S Ardizzone, M Radaelli. J Electroanal Chem 269:461, 1989.
55. N Kallay, D Čakara. J Colloid Interface Sci 232:81, 2000.
56. N Kallay, D Babić, E Matijević. Colloids Surfaces 19:457, 1986.

5

Electrophoresis of Charged Particles and Drops

HIROYUKI OHSHIMA Science University of Tokyo, Tokyo, Japan

I. INTRODUCTION

The motion of charged colloidal particles in an applied electric field, which is called electrophoresis, depends on their zeta potential or electric charge [1–7]. In this chapter we derive equations relating the electrophoretic mobility (the electrophoretic velocity per unit applied electric field) to the zeta potential or electric charges of various types of colloidal particles (hard particles, liquid drops, and soft particles) in a liquid containing an electrolyte in an applied electric field both for dilute and concentrated suspensions. We deal mainly with static electrophoresis, i.e., electrophoresis of colloidal particles when a static electric field is applied. We also consider dynamic electrophoresis and other electrokinetic phenomena such as sedimentation potential and conductivity of suspensions of colloidal particles with particular emphasis on an Onsager relationship holding between electrophoretic mobility and sedimentation potential.

II. SMOLUCHOWSKI'S EQUATION

The electrophoretic mobility μ of a spherical particle moving with a velocity U in an electrolyte solution in an applied electric field E is given by the ratio U/E, where $U = |U|$ and $E = |E|$. The most widely employed formula relating the electrophoretic mobility μ of a colloidal particle to its zeta potential ζ is Smoluchowski's formula (8),

$$\mu = \frac{\varepsilon_r \varepsilon_0}{\eta} \zeta \tag{1}$$

Here, ε_r and η are, respectively, the relative permittivity and the viscosity of the electrolyte solution, and ε_0 is the permittivity of a vacuum. The zeta potential ζ is defined as the potential at the plane where the liquid velocity u relative to the particle ($u \rightarrow -U$ far from the particle) is 0. This plane is called the slipping plane or shear plane. The slipping plane does not necessarily coincide with the particle surface. Only if the slipping plane is located at the particle surface, does the zeta potential ζ become equal to the surface potential ψ_0. In the following we treat the case where $\zeta = \psi_0$.

Smoluchowski's equation (1) is readily derived from the condition of balance between electric and viscous forces acting on the particle, as described below. Both the liquid velocity u and electric potential ψ decay (nearly exponentially) over the distance of the order of the Debye length $1/\kappa$ from the particle surface (Fig. 1), κ being the Debye–Hückel parameter. For a general electrolyte composed of N ionic mobile species of valence z_i and bulk concentration (number density) n_i^∞, κ is defined by

$$\kappa = \left(\frac{1}{\varepsilon_r \varepsilon_0 kT} \sum_{i=1}^{N} z_i e^2 n_i^\infty \right)^{1/2} \tag{2}$$

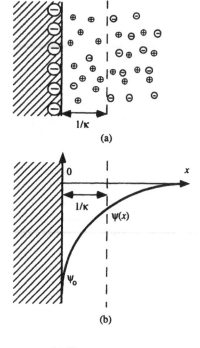

(a)

(b)

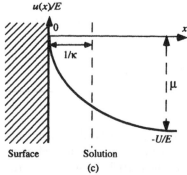

Surface   Solution

(c)

FIG. 1 Distribution of ions (a), potential $\psi(x)$ (b), and liquid velocity $u(x)$ (c) near the surface of a solid particle moving with a velocity U in an applied electric field E; x is the distance measured from the particle surface, ψ_0 is the surface potential of the particle, which is approximately equal to the zeta potential ζ, $\mu = U/E (U = |U|$ and $E = |E|)$ is the electrophoretic mobility, and $1/\kappa$ is the Debye length.

where e is the elementary electric charge, k is Boltzmann's constant, and T is the absolute temperature. Let σ be the charge density of the particle surface. The magnitude of the electric force F_e per unit area acting on the particle surface is then

$$F_e = \sigma E \tag{3}$$

Since the electric potential ψ varies from ψ_0 to zero with a decay constant κ^{-1} (Fig. 1), the electric field at the particle surface is approximately given by $\psi_0/(\kappa^{-1})$. This field, which is generated by σ, is equal to $\sigma/\varepsilon_r\varepsilon_0$. Thus, we have $\psi_0/(\kappa^{-1}) = \sigma/\varepsilon_r\varepsilon_0$, viz.,

$$\psi_0 = \frac{\sigma}{\varepsilon_r\varepsilon_0\kappa} \tag{4}$$

The magnitude of the viscous force F_h acting on the particle surface per unit area, on the other hand, is given by the product of the viscosity η and the velocity gradient at the particle surface, the magnitude of which, in turn, is approximately given by $U/(\kappa^{-1})$. Thus,

$$F_h = \eta U/(\kappa^{-1}) \tag{5}$$

In the stationary state, these two forces must be equal: $F_e = F_h$. Thus, if the surface potential ψ_0 is identified as the zeta potential ζ, then Smoluchowski's mobility formula (1) follows from Eqs (3)–(5).

Smoluchowski's equation (1) has been derived on the basis of some approximations. This formula, however, is the correct limiting mobility equation for very large particles and is valid irrespective of the shape of the particle provided that the dimension of the particle is much larger than the Debye length $1/\kappa$, and thus the particle surface can be considered to be locally planar. For a sphere with radius a, this condition is expressed by $\kappa a \gg 1$.

III. HENRY'S EQUATION

The electrophoretic mobility of very small spheres ($\kappa a \ll 1$) is given by Hückel's equation [9]:

$$\mu = \frac{2\varepsilon_r\varepsilon_0}{3\eta}\zeta \tag{6}$$

The difference between Smoluchowski's equation (1) and Hückel's equation (6) by a factor of 2/3 can be explained as follows. Consider a spherical colloidal particle of radius a. The origin of the spherical polar co-ordinate system (r, θ, ϕ) is held fixed at the center of the particle. The applied electric field is distorted in the presence of a colloidal particle in such a way that the applied field becomes parallel to the particle surface. The potential of the applied electric field, which is $-Er\cos\theta$ in the absence of the particle, is distorted to become

$$-E\left(r + \frac{a^3}{2r^2}\right)\cos\theta \tag{7}$$

where the second term corresponds to the distortion of the applied electric field due to the presence of the particle. It follows from Eq. (7) that the potential of the applied field near the particle surface $r \approx a$ is larger than the original undistorted field by a factor of 3/2. The electrophoretic mobility is determined mainly by electrolyte ions in the double layer (of thickness $1/\kappa$). As is seen in Fig. 2, for thick double layers ($\kappa a \ll 1$) most electrolyte ions in the double layer experience an undistorted original field. For thin double layers ($\kappa a \gg 1$),

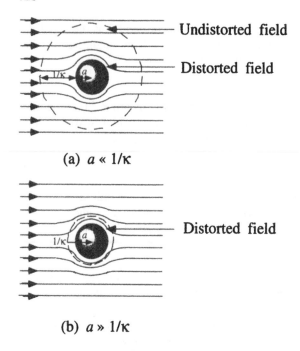

(a) $a \ll 1/\kappa$

(b) $a \gg 1/\kappa$

FIG. 2 The applied electric field is distorted so as to be parallel to the particle surface. For thick double layers ($\kappa a \ll 1$) (a) most electrolyte ions in the double layer experience an undistorted original field. For thin double layers ($\kappa a \gg 1$) (b), most electrolyte ions in the double layer experience a distorted field.

on the other hand, most electrolyte ions in the double layer experience a distorted field. This is the reason why Smoluchowski's equation (1) differs from the Hückel equation (6) by 3/2.

Henry [10] derived the mobility equations for spheres of radius a and an infinitely long cylinder of radius a, which are applicable for low ζ and any value of κa. Henry's equation for the electrophoretic mobility μ of a spherical colloidal particle of radius a with a zeta potential ζ is expressed as

$$\mu = \frac{\varepsilon_r \varepsilon_0}{\eta} \zeta f(\kappa a) \tag{8}$$

with

$$f(\kappa a) = 1 - e^{\kappa a}\{5E_7(\kappa a) - 2E_5(\kappa a)\} \tag{9}$$

where $E_n(\kappa a)$ is the exponential integral of order n and $f(\kappa a)$ is called Henry's function. As $\kappa a \to \infty$, $f(\kappa a) \to 1$ and Eq. (8) tends to Smoluchowski's equation (1), while if $\kappa a \to 0$, then $f(\kappa a) \to 2/3$ and Eq. (8) becomes Hückel's equation (6). Ohshima [11] has derived the following simple approximate formula for Henry's function $f(\kappa a)$ with relative errors less than 1%:

$$f(\kappa a) = \frac{2}{3}\left[1 + \cfrac{1}{2\left(1 + \cfrac{2.5}{\kappa a\{1 + 2\exp(-\kappa a)\}}\right)^3}\right] \tag{10}$$

For the case of a cylindrical particle, the electrophoretic mobility depends on the orientation of the particle with respect to the applied electric field. When the cylinder is oriented parallel to the applied electric field, its electrophoretic mobility μ is given by Smoluchowski's equation (1), viz.,

$$\mu_{//} = \frac{\varepsilon_r \varepsilon_0}{\eta} \zeta \tag{11}$$

If, on the other hand, the cylinder is oriented perpendicularly to the applied field, then the mobility depends not only on ζ but also on the value of κa. Henry [10] showed that $f(\kappa a)$ for a cylindrical particle of radius a oriented perpendicularly to the applied field is given by [see Eq. (3) in Ref. 12].

$$\mu_\perp = \frac{\varepsilon_r \varepsilon_0}{\eta} \zeta f(\kappa a) \tag{12}$$

with

$$f(\kappa a) = 1 - \frac{4(\kappa a)^4}{K_0(\kappa a)} \int_{\kappa a}^\infty \frac{K_0(t)}{t^5} dt + \frac{(\kappa a)^2}{K_0(\kappa a)} \int_{\kappa a}^\infty \frac{K_0(t)}{t^3} dt \tag{13}$$

where $K_0(x)$ is the zero-order modified Bessel function of the second kind. Ohshima [13] obtained an approximate formula for Henry's function for a cylinder:

$$f(\kappa a) = \frac{1}{2} \left[1 + \frac{2}{\left(1 + \dfrac{2.55}{\kappa a \{1 + \exp(-\kappa a)\}} \right)^2} \right] \tag{14}$$

the relative error being less than 1%. As $\kappa a \to \infty$, $f(\kappa a) \to 1$ and Eq. (14) gives Smoluchowski's equation (1), while if $\kappa a \to 0$, then $f(\kappa a) \to 1/2$. For a cylindrical particle oriented at an arbitrary angle between its axis and the applied electric field, its electrophoretic mobility averaged over a random distribution of orientation is given by [14, 15]

$$\mu_{av} = \frac{1}{3}\mu_{//} + \frac{2}{3}\mu_\perp \tag{15}$$

where $\mu_{//}$ is the mobility for parallel orientation given by Smoluchowski's formula (1) and $\mu_\perp$ is the mobility for perpendicular orientation given by Eq. (14).

Figure 3 compares Henry's equations for a sphere and a cylinder.

IV. ACCURATE MOBILITY EXPRESSION

Henry's mobility equation (8) assumes that the double-layer potential distribution around a spherical particle remains unchanged during electrophoresis. For high zeta potentials, the double layer is no longer spherically symmetrical. This effect is called the relaxation effect (Fig. 4). Equation (8) does not take into account the relaxation effect and thus this equation is correct to the first order of zeta potential ζ. Full electrokinetic equations determining electrophoretic mobility of spherical particles with arbitrary values of zeta potentials were derived independently by Overbeek [16] and Booth [17]. Wiersema et al. [18] solved the equations numerically using an electronic computer. The computer calculation of the electrophoretic mobility was considerably improved by O'Brien and White [19].

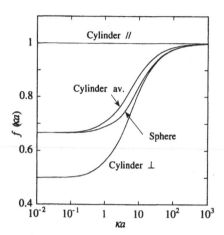

FIG. 3 Henry's function $f(\kappa a)$ for a spherical particle and a cylindrical particle of radius a as a function of scaled radius κa. The cylinder is oriented parallel ($\mu_{//}$) or perpendicularly ($\mu_{\perp}$) to the applied field; $\mu_{av} = (\mu_{//} + 2\mu_{\perp})/3$ [Eq. (15)].

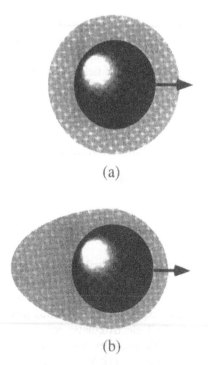

FIG. 4 The relaxation effect. The electrical diffuse double layer around a spherical particle is spherical for low zeta potentials (a), but is no longer spherical for high zeta potentials (b).

Approximate analytic mobility expressions, on the other hand, have been proposed by several authors [16, 17, 20–22].

Ohshima et al. [22] showed that analytic approximate solutions to electrokinetic equations are most easily derived by introducing the electrochemical potential of ionic species, which are

$$\eta \nabla \times \nabla \times \boldsymbol{u} + \nabla p + \rho_{el} \nabla \psi = 0 \tag{16}$$

$$\nabla \cdot \boldsymbol{u} = 0 \tag{17}$$

$$v_i = \boldsymbol{u} - \frac{1}{\lambda_i} \nabla \mu_i \tag{18}$$

$$\nabla \cdot (n_i v_i) = 0 \tag{19}$$

$$\rho_{el}(\boldsymbol{r}) = \sum_{i=1}^{N} z_i e n_i(\boldsymbol{r}) \tag{20}$$

$$\mu_i(\boldsymbol{r}) = \mu_i^{\infty} + z_i e \psi(\boldsymbol{r}) + kT \ln n_i(\boldsymbol{r}) \tag{21}$$

$$\Delta \psi(\boldsymbol{r}) = -\frac{\rho_{el}(\boldsymbol{r})}{\varepsilon_r \varepsilon_o} \tag{22}$$

where $p(\boldsymbol{r})$ is the pressure, $\rho_{el}(\boldsymbol{r})$ is the charge density resulting from the mobile charged ionic species given by Eq. (20), $\psi(\boldsymbol{r})$ is the electric potential, $\mu_i(\boldsymbol{r})$ and $n_i(\boldsymbol{r})$ are, respectively, the electrochemical potential and concentration (the number density) of the ith ionic species, and μ_i^{∞} is a constant term in $\mu_i(\boldsymbol{r})$. Equations (16) and (17) are the Navier-Stokes equation and the equation of continuity for an incompressible flow. Equation (18) expresses that the flow $v_i(\boldsymbol{r})$ of the ith ionic species is caused by the liquid flow $\boldsymbol{u}(\boldsymbol{r})$ and the gradient of the electrochemical potential $\mu_i(\boldsymbol{r})$, given by Eq. (21). Equation (19) is the continuity equation for the ith ionic species, and Eq. (22) is Poisson's equation. Ohshima et al. [22] derived an accurate analytic approximate expression for the elctrophoretic mobility of a spherical colloidal particle of radius a and zeta potential ζ in a symmetrical electrolyte of valence z and bulk concentration n with a relative error of less than 1% for $10 \le \kappa a \le \infty$, which is

$$
\begin{aligned}
E_m = \text{sgn}(\zeta) \Bigg[& \frac{3}{2}\bar{\zeta} - \frac{3F}{1+F}H + \frac{1}{\kappa a}\Bigg\{-18\left(t + \frac{t^3}{9}\right)K + \frac{15F}{1+F}\left(t + \frac{7t^2}{20} + \frac{t^3}{9}\right) \\
& - 6(1+3\tilde{m})(1 - e^{-\tilde{\zeta}/2})G + \frac{12F}{(1+F)^2}H + \frac{9\tilde{\zeta}}{1+F}(\tilde{m}G + mH) \\
& - \frac{36F}{1+F}\left(\tilde{m}G^2 + \frac{m}{1+F}H^2\right)\Bigg\}\Bigg]
\end{aligned}
\tag{23}
$$

with

$$E_m = \frac{3\eta ze}{2\varepsilon_r \varepsilon_o kT}\mu \tag{24}$$

$$\bar{\zeta} = \frac{ze|\zeta|}{kT} \tag{25}$$

$$F = \frac{2}{\kappa a}(1 + 3m)(e^{\tilde{\zeta}/2} - 1) \tag{26}$$

$$G = \ln \frac{1 + e^{-\tilde{\zeta}/2}}{2} \tag{27}$$

$$H = \ln \frac{1 + e^{\tilde{\zeta}/2}}{2} \tag{28}$$

$$K = 1 - \frac{25}{3(\kappa a + 10)} \exp\left[-\frac{\kappa a}{6(\kappa a - 6)}\tilde{\zeta}\right] \tag{29}$$

$$t = \tanh(\tilde{\zeta}/4) \tag{30}$$

$$m = \frac{2\varepsilon_r\varepsilon_o kT}{3\eta z^2 e^2}\lambda \tag{31}$$

$$\tilde{m} = \frac{2\varepsilon_r\varepsilon_o kT}{3\eta z^2 e^2}\tilde{\lambda} \tag{32}$$

where E_m is the scaled electrophoretic mobility; sgn $(\zeta) = +1$ if $\zeta > 0$ and -1 if $\zeta < 0$; $\tilde{\zeta}$ is the magnitude of the scaled zeta potential; λ and $\tilde{\lambda}$ are, respectively, the ionic drag coefficients of counterions and coions; and m and $\tilde{m}$ are the corresponding dimensionless quantities. The drag coefficient λ of an ionic species is further related to the limiting conductance Λ° of that ionic species by

$$\lambda = \frac{N_A e^2 |z|}{\Lambda^\circ} \tag{33}$$

where N_A is Avogadro's number. For K^+ and Cl^- ions, for example, $m = 0.176$ ($\Lambda^\circ = 73.5 \times 10^{-4}$ m^2 Ω^{-1}mol^{-1}) and $m = 0.169$ ($\Lambda^\circ = 76.3 \times 10^{-4}$ m^2 Ω^{-1}mol^{-1}), respectively.

In Fig. 5 we plot the mobility–zeta potential relationship for KCl for several values of κa. The $\kappa a = \infty$ and $\kappa a = 0$, which are both given by straight lines, correspond to Smoluchowski's equation (1) and Hückel's equation (6), respectively. It is seen that there is a mobility maximum, which is due to the relaxation effect. That is, as the zeta potential increases, the tangential flow of counterions in the double layer along the particle surface increases. This surface current tends to equalize the potential around the surface and hence retards the motion of the particle. According to the results of the computer calculation by O'Brien and White [19], there is a mobility maximum, when mobility is plotted as a function of ζ for $\kappa a > 3$. It is also seen that the mobility tends to a limiting value as $\zeta \to \infty$, which is given by

$$E_m \to \text{sgn}(\zeta)3\ln 2 + O\left(\frac{1}{\kappa a}\right) \tag{34}$$

V. CONCENTRATED SUSPENSION

So far we have discussed electrokinetic processes in a dilute suspension of colloidal particles. For concentrated suspensions, hydrodynamic and electrostatic interactions between

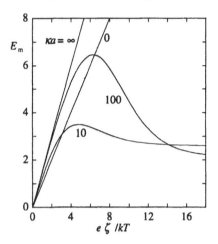

FIG. 5 The reduced electrophoretic mobility E_m of a spherical colloidal particle of radius a in a KCl solution at 25°C as a function of reduced zeta potential $e\zeta/kT$ for various values of κa; E_m is defined by Eq. (24).

particles become important. The simplest but most effective way to take into account the interparticle interactions is to employ Kuwabara's cell model [23]. In this model each particle of radius a is considered to be surrounded by a concentric spherical shell of an electrolyte solution, having an outer radius b such that the particle/solution volume ratio in a unit cell is equal to the particle volume fraction ϕ throughout the entire system (Fig. 6), viz.,

$$\phi = (a/b)^3 \tag{35}$$

and the fluid vorticity is zero at the outer surface of the cell.

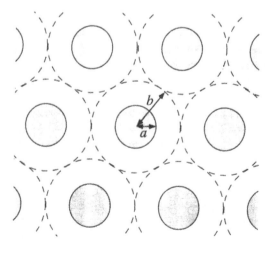

FIG. 6 Spherical particles of radius a in concentrated suspensions in the cell model [23]. Each sphere is surrounded by a virtual shell of outer radius b. The particle volume fraction ϕ is given by $\phi = (a/b)^3$ and the porosity is $\varepsilon = 1 - \phi$.

Levine and Neale [24] derived a mobility expression applicable for low zeta potentials. Kozak and Davis [25, 26] also developed a more general theory for the electrokinetics of concentrated suspensions and porous media, which is applicable to all zeta-potential values but ignores double-layer overlapping. For low ζ potentials, Ohshima [27] derived a simple approximate formula with relative errors of less than 4%. The result is

$$\mu = \frac{\varepsilon_r \varepsilon_0 \zeta}{\eta} f(\kappa a, \phi) \tag{36}$$

with

$$f(\kappa a, \phi) = \frac{2}{3}\left[1 + \frac{1}{2(1 + 2.5/\kappa a[1 + 2\exp(-\kappa a)])^3}\right]M_1 + M_2 \tag{37}$$

where

$$M_1 = 1 - \frac{3}{(\kappa a)^2}\frac{\phi}{1-\phi}(1 + \kappa a Q) - \frac{(\kappa a)^2}{3(1-\phi)P}\left[\phi^{1/3} + \frac{1}{\phi^{2/3}} - \frac{9}{5\phi^{1/3}} - \frac{\phi^{4/3}}{5}\right] \tag{38}$$

$$M_2 = \frac{2(\kappa a)^2}{9P}\frac{1 + \phi/2}{1-\phi}\left[\phi^{1/3} + \frac{1}{\phi^{2/3}} - \frac{9}{5\phi^{1/3}} - \frac{\phi^{4/3}}{5}\right] \tag{39}$$

$$P = \cosh[\kappa(b-a)] - \frac{\sinh[\kappa(b-a)]}{\kappa b}$$

$$= \cosh[\kappa a(\phi^{-1/3} - 1)] - \frac{\phi^{1/3}}{\kappa a}\sinh[\kappa a(\phi^{-1/3} - 1)] \tag{40}$$

$$Q = \frac{1 - \kappa b \cdot \tanh[\kappa(b-a)]}{\tanh[\kappa(b-a)] - \kappa b} = \frac{1 - \kappa a \phi^{-1/3} \cdot \tanh[\kappa a(\phi^{-1/3} - 1)]}{\tanh[\kappa a(\phi^{-1/3} - 1)] - \kappa a \phi^{-1/3}} \tag{41}$$

In Fig. 7 we plot $f(\kappa a, \phi)$ given by Eq. (37), which is Henry's function for a concentrated suspension. As $\phi \to 0$, Eq. (37) reduces to Eq. (10). It is seen from Fig. 7 that, at the limit of thin double layers around the particles ($\kappa a \to \infty$), the electrophoretic mobility is described by Smoluchowski's formula (1) for a single particle with infinitesimally thin electrical double layers. In other words, in the large κa limit the mobility in concentrated suspensions does not depend on the particle volume fraction ϕ. Further, it is seen that as κa decreases and/or the particle volume fraction ϕ increases (the porosity ε decreases), the mobility rapidly decreases because of interparticle interactions.

Kozak and Davis [28] extended the theory of Levine and Neale [24] to electro-osmosis in an array of circular cylinders, where the electro-osmotic velocity U of liquid flowing perpendicularly to the cylinders in an applied electric field E is considered. Ohshima [29] derived an approximate formula for the electro-osmotic velocity for this system with relative errors of less than 5.6%:

$$U = \frac{\varepsilon_r \varepsilon_0 \zeta}{\eta}\left[\frac{1}{2}\left\{1 + \frac{1}{(1 + 2.55/\kappa a[1 + \exp(-\kappa a)])^2}\right\}N_1(\kappa a, \varepsilon) + N_2(\kappa a, \varepsilon)\right]E \tag{42}$$

where $U = |U|$, $E = |E|$, and

$$N_1(\kappa a, \varepsilon) = 1 + \frac{2(1-\varepsilon)}{\varepsilon}\frac{1}{\kappa a}\frac{K_1(\kappa b)I_1(\kappa a) - I_1(\kappa b)K_1(\kappa a)}{K_1(\kappa b)I_0(\kappa a) + I_1(\kappa b)K_0(\kappa a)} - \frac{2}{2-\varepsilon}N_2(\kappa a, \varepsilon) \tag{43}$$

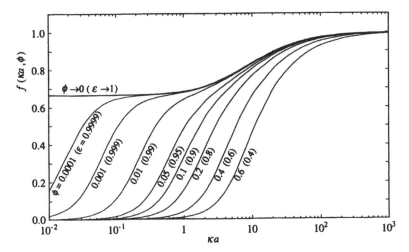

FIG. 7 Henry's function $f(\kappa a, \phi)$ for concentrated suspensions as a function of scaled radius a and particle volume fraction ϕ (or porosity $\varepsilon = 1 - \phi$). (From Refs. 24 and 27.)

$$N_2(\kappa a, \varepsilon) = \frac{(2 - \varepsilon)(1 - \varepsilon)^{1/2}}{4\varepsilon} \frac{\kappa a}{K_1(\kappa b)I_0(\kappa a) + I_1(\kappa b)K_0(\kappa a)} \left\{ 1 - \frac{3}{4(1 - \varepsilon)} - \frac{1 - \varepsilon}{4} \frac{\ln(1 - \varepsilon)}{2(1 - \varepsilon)} \right\}$$

(44)

Note that in the limit $\varepsilon \to 1$, U/E reduces to $\mu_\perp$ given by Eq. (12).

VI. CHARGED DROPS

The electrophoretic velocity of liquid drops is faster than that of rigid particles because the flow velocity of the surrounding liquid need not become zero at the drop surface and is conveyed into the drop interior [30–36]. The electrophoretic mobility of a drop thus depends on the viscosity of the drop. Here, we treat the example of mercury drops. In this instance the drop surface is always equipotential so that it is relatively easy to solve the electrokinetic equations for this case.

For the case where ζ is low and κa is large, the following approximate equation was derived by Levine and O'Brien [31], viz.,

$$\mu = \frac{\varepsilon_r \varepsilon_0}{\eta} \zeta \frac{\eta}{3\eta_d + 2\eta} \kappa a$$

(45)

where η_d is the viscosity of the drop. Ohshima et al. [32] extended the theory of Levine and O'Brien [31] and derived:

$$\mu = \frac{\varepsilon_r \varepsilon_0}{\eta} \zeta \left[\frac{\eta}{3\eta_d + 2\eta} \kappa a + \frac{3\eta_d + \eta}{3\eta_d + 2\eta} + 2e^{\kappa a} E_5(\kappa a) - \frac{15\eta_d}{3\eta_d + 2\eta} e^{\kappa a} E_7(\kappa a) \right]$$

(46)

which is applicable for low ζ and all values of κa. The first term on the right agrees with that of Eq. (45). For $\zeta_d \to \infty$, Eq. (46) reduces to Henry's equation (8) for a rigid sphere. Ohshima [35] has shown that Eq. (46) is approximated well by

$$\mu = \frac{2\varepsilon_r\varepsilon_o}{3\eta}\zeta\left[1 + \frac{1}{2\left(1 + \frac{1.86}{\kappa a}\right)^3}\right]\left[\frac{\eta}{3\eta_d + 2\eta} + \frac{3(\eta_d + \eta)}{3\eta_d + 2\eta}\right] \tag{47}$$

with a maximum relative error of less than 1%.

Levich and Frumkin (see Ref. 30) derived an approximate analytic mobility equation applicable for arbitrary values of ζ, and Ohshima et al. [32] derived a more accurate mobility expression correct to the order of $1/\kappa a$. The leading term of their expression is

$$\mu = \frac{2\varepsilon_r\varepsilon_o kT}{3\eta ze}\frac{\text{sgn}(\zeta)}{\left(\frac{\eta_d}{\eta} + \frac{2}{3} + D\right)}\left(\frac{\kappa a}{2}\right)\left(e^{\tilde\zeta/2} - e^{-\tilde\zeta/2}\right) \tag{48}$$

with

$$D = m(1 - e^{\tilde\zeta/2})^2 + \tilde m(1 - e^{-\tilde\zeta/2})^2 \tag{49}$$

where $\tilde\zeta, F, m,$ and $\tilde m$ are already given In Eqs (25), (26), (31), and (32). We find that at the limit of large κa:

$$\frac{\mu(\text{mercury})}{\mu(\text{rigid})} = O(\kappa a) \tag{50}$$

That is, at this limit the mobility of mercury drops is larger than that of rigid particles by an order of κa. It is also to be noted that mercury drops behave like rigid particles at very high ζ (solidification effect) [30, 32].

For a concentrated suspension of mercury drops with the particle volume fraction ϕ, Ohshima [36] derived the following mobility formula:

$$\mu = \frac{\varepsilon_r\varepsilon_o\zeta}{\eta}\left[\frac{2}{3}\left\{1 + \frac{2}{2(1 + 1.86/\kappa a)^3}\right\}M_1 + M_2\right] \tag{51}$$

with

$$M_1 = \frac{\eta}{3\eta_d + 2\eta}\kappa aQ + \frac{3(\eta_d + \eta)}{3\eta_d + 2\eta} - \frac{3}{(\kappa a)^2}\frac{\phi}{1 - \phi}(1 + \kappa aQ)$$

$$- \frac{(\kappa a)^2}{(1 - \phi)P}\left[\frac{\eta_d\phi^{1/3}}{3\eta_d + 2\eta} + \frac{\eta_d + \eta}{(3\eta_d + 2\eta)\phi^{2/3}} - \frac{3}{5\phi^{1/3}} - \frac{(\eta_d - \eta)\phi^{4/3}}{5(3\eta_d + 2\eta)}\right] \tag{52}$$

$$M_2 = \frac{2(\kappa a)^2}{3P}\frac{1 + \phi/2}{1 - \phi}\left[\frac{\eta_d\phi^{1/3}}{3\eta_d + 2\eta} + \frac{\eta_d + \eta}{(3\eta_d + 2\eta)\phi^{2/3}} - \frac{3}{5\phi^{1/3}} - \frac{(\eta_d - \eta)\phi^{4/3}}{5(3\eta_d + 2\eta)}\right] \tag{53}$$

where P and Q are given by Eqs. (40) and (41).

VII. SURFACE POTENTIAL/SURFACE CHARGE DENSITY RELATIONSHIP

If zeta potential ζ is identified as surface potential ψ_o, then one can calculate the surface charge density σ of particles or drops from their surface potential ψ_o. Numerical tables of ψ_o/σ relationships for a sphere [37] and approximate analytic expressions for a sphere [38–

40] and a cylinder [38, 41] are available. In this section we give accurate analytical relationships between surface potential and surface charge density for various cases [38–41].

For a spherical particle in a 1:1 electrolyte (e.g., NaCl) of concentration n:

$$\sigma = \frac{2\varepsilon_r\varepsilon_0\kappa kT}{e}\sinh\left(\frac{e\psi_0}{2kT}\right)\left[1 + \frac{1}{\kappa a}\frac{2}{\cosh^2(4e\psi_0/kT)} + \frac{1}{(\kappa a)^2}\frac{8\ln[\cosh(e\psi_0/4kT)]}{\sinh^2(e\psi_0/2kT)}\right]^{1/2}$$

(54)

where

$$\kappa = (2ne^2/\varepsilon_r\varepsilon_0 kT)^{1/2}$$

(55)

is the Debye–Hückel parameter for a 1:1 electrolyte. The relative error of Eq. (54) is less than 1% for $0.5 \le \kappa a < \infty$. For a spherical particle in a 2:1 electrolyte (e.g., CaCl$_2$) of concentration n:

$$\sigma = \frac{\varepsilon_r\varepsilon_0\kappa kT}{e}pq\left[1 + \frac{4}{\kappa a}\frac{(3-p)q-3}{(pq)^2} + \frac{4}{(\kappa a)^2(pq)^2}\left\{6\ln\left(\frac{q+1}{2}\right) + \ln(1-p)\right\}\right]^{1/2}$$

(56)

where

$$\kappa = \left(\frac{6\pi e^2}{\varepsilon_r\varepsilon_0 kT}\right)^{1/2}$$

(57)

is the Debye–Hückel parameter for a 2:1 electrolyte, and

$$p = 1 - \exp(-e\psi_0/kT)$$

(58)

$$q = \left[\frac{2}{3}\exp\left(\frac{e\psi_0}{KT}\right) + \frac{1}{3}\right]^{1/2}$$

(59)

The relative error of Eq. (56) is 1% for $0.5 \le \kappa a < \infty$. For a spherical particle in a mixed solution of 1:1 electrolyte of concentration n_1 and 2:1 electrolyte of concentration n_2:

$$\sigma = \frac{\varepsilon_r\varepsilon_0\kappa kT}{e}\left[pt + \frac{2}{\kappa apt}\left\{(3-p)t - 3 - \frac{3^{1/2}(1-\eta)}{2\eta^{1/2}}\ln\left(\frac{\{1+(\eta/3)^{1/2}\}\{t-(\eta/3)^{1/2}\}}{\{1-(\eta/3)^{1/2}\}\{t+(\eta/3)^{1/2}\}}\right)\right\}\right]$$

(60)

where

$$\kappa = \left[\frac{2(n_1+3n_2)e^2}{\varepsilon_r\varepsilon_0 kT}\right]^{1/2}$$

(61)

is the Debye–Hückel parameter for a mixed solutions of 1:1 and 2:1 electrolytes, p is given by Eq. (58), and

$$t = \left[\left(1-\frac{\eta}{3}\right)\exp\left(\frac{e\psi_0}{kT}\right) + \frac{\eta}{3}\right]^{1/2}$$

(62)

$$\eta = \frac{3n_2}{n_1+3n_2}$$

(63)

The relative error of Eq. (60) is less than 1% for $5 \le \kappa a \le \infty$. Note that, for low potentials, Eqs (54), (56) and (60) all reduce to

$$\sigma = \varepsilon_r \varepsilon_0 \kappa \psi_o \left(1 + \frac{1}{\kappa a}\right) \tag{64}$$

which holds irrespective of the type of electrolyte.

For a cylindrical particle in a 1:1 electrolyte of concentration n:

$$\sigma = \frac{2\varepsilon_r \varepsilon_0 \kappa kT}{e} \sinh\left(\frac{e\psi_o}{2kT}\right) \left[1 + \left\{\left(\frac{K_1(\kappa a)}{K_0(\kappa a)}\right)^2 - 1\right\} \frac{1}{\cosh^2(e\psi_o/4kT)}\right]^{1/2} \tag{65}$$

where κ is given by Eq. (55), and $K_n(x)$ is the nth-order modified Bessel function of the second kind. For a spherical particle in a 2:1 electrolyte of concentration n:

$$\sigma = \frac{\varepsilon_r \varepsilon_0 \kappa kT}{e} pq \left[1 + 2\left\{\left(\frac{K_1(\kappa a)}{K_0(\kappa a)}\right)^2 - 1\right\} \frac{(3-p)q - 3}{(pq)^2}\right]^{1/2} \tag{66}$$

where κ, p, and q are, respectively, given by Eqs (57)–(59). For a cylindrical particle in a mixed solution of 1:1 electrolyte of concentration n_1 and 2:1 electrolyte of concentration n_2

$$\sigma = \frac{\varepsilon_r \varepsilon_0 \kappa kT}{e} pt \left[1 + \frac{2}{(pt)^2}\left\{\left(\frac{K_1(\kappa a)}{K_0(\kappa a)}\right)^2 - 1\right\}\right.$$

$$\left. \times \left\{(3-p)t - 3 - \frac{3^{1/2}(1-\eta)}{2\eta^{1/2}} \ln\left(\frac{\{1 + (\eta/3)^{1/2}\}\{t - (\eta/3)^{1/2}\}}{\{1 - (\eta/3)^{1/2}\}\{t + (\eta/3)^{1/2}\}}\right)\right\}\right]^{1/2} \tag{67}$$

where κ, p, and t are, respectively, given by Eqs (61), (58), and (62). For low potentials, Eqs (65)–(67) all reduce to

$$\sigma = \varepsilon_r \varepsilon_0 \kappa \psi_o \frac{K_1(\kappa a)}{K_0(\kappa a)} \tag{68}$$

It must be noted here that in the above σ/ψ_o relationships the electrolyte concentration n (and also n_1 and n_2) is given in units of m^{-3}. If the electrolyte concentration is given in units of M (mol/liter), then n must be replaced by $1000 N_A n$, N_A being Avogadro's number.

VIII. ONSAGER'S RELATIONSHIP

The basic equations governing electrophoresis also describe other electrokinetic phenomena, e.g., the electric conductivity and sedimentation of a suspension of colloidal particles. In particular, Onsager's relationship holds between electrophoretic mobility μ and sedimentation potential E_{SED}:

$$E_{SED} = -\frac{\phi(\rho_p - \rho_o)}{K^\infty} \mu \boldsymbol{g} \tag{69}$$

which is correct to the first order of the particle volume fraction ϕ. In Eq. (69), ρ_p and ρ_o are, respectively, the mass densities of the particle and the electrolyte solution, $\boldsymbol{g}$ is the gravity, and K^∞ is the conductivity of the electrolyte solution in the absence of the particles, defined by

$$K^\infty = \sum_{t=1}^{N} z_i^2 e^2 n_i^\infty / \lambda_i \tag{70}$$

Equation (69) was originally derived by de Groot et al. [42] on the basis of irreversible thermodynamics, and later a direct proof was given by Ohshima et al. [43].

For a concentrated suspension with particle volume fraction ϕ, Ohshima [44] derived the following Onsager relationship on the basis of Kuwabara's cell model [23]:

$$E_{\text{SED}} = -\frac{\phi(1-\phi)}{(1+\phi/2)} \frac{(\rho_p - \rho_o)}{K^\infty} \mu \boldsymbol{g} \tag{71}$$

where μ is the electrophoretic mobility of particles in concentrated suspensions. For $\kappa a \to \infty$, Eq. (72) holds for any ζ. For arbitrary values of κa, Eq. (71) holds for low ζ. Later Ohshima [45] generalized Eq. (71) to obtain:

$$E_{\text{SED}} = -\frac{\phi(\rho_p - \rho_o)\mu}{K^\infty} \left(\frac{K^*}{K^\infty}\right) \boldsymbol{g} \tag{72}$$

where K^* is the electrical conductivity of the suspension. For concentrated suspensions of particles or drops of radius a with low zeta potentials, Ohshima [36, 46] derived:

$$\frac{K^*}{K^\infty} = \frac{1-\phi}{1+\phi/2} \left(1 - 3\phi(\frac{e\zeta}{kT})L(\kappa a, \phi)\frac{\displaystyle\sum_{i=1}^{N} z_i^3 n_i^\infty / \lambda_i}{\displaystyle\sum_{i=1}^{N} z_i^2 n_i^\infty / \lambda_i}\right) \tag{73}$$

with

$$L(\kappa a, \phi) = \frac{(1 + \kappa a Q)}{(\kappa a)^2(1-\phi)(1+\phi/2)}\left\{1 + \frac{1}{2(1+\delta/\kappa a)^3}\right\} \tag{74}$$

where Q is defined by Eq. (41) and $\delta = 2.5/(1 + z\exp(-\kappa a))$ for rigid particles and $\delta = 1.86$ for mercury drops.

IX. SOFT PARTICLES (POLYELECTROLYTE-COATED PARTICLES)

In this section we consider soft particles, i.e., colloidal particles covered with a polyelectrolyte layer (Fig. 8). For such particles one must consider the potential distribution and the liquid flow distribution not only outside but also inside the surface charge layer. As is seen in Fig. 9, the potential deep inside the polyelectrolyte layer is almost equal to the Donnan potential, provided that the surface layer is much thicker than the Debye length $1/\kappa$. A number of theoretical studies have been made [47–57] on the basis of the model of Debye and Bueche [58]. This model assumes that the polymer segments are regarded as resistance centers distributed in the polyelectrolyte layer, exerting frictional forces on the liquid flowing in the polyelectrolyte layer. That is, a friction term $-\gamma\boldsymbol{u}$, γ being a frictional coefficient, is incorporated into electrokinetic equations for the liquid flow $\boldsymbol{u}$ in the polyelectrolyte layer. The Navier–Stokes equation for the liquid flow inside the surface charge layer is thus given by

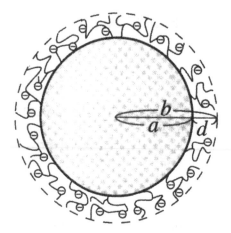

FIG. 8 A soft particle. The hard particle core of radius a is covered with a layer of polyelectrolytes of thickness d; $b = a + d$.

$$\eta \nabla \times \nabla \times \boldsymbol{u} + \gamma \boldsymbol{u} + \nabla p + \rho_{\text{el}} \nabla \psi = 0 \tag{75}$$

Ohshima [55–57] has presented a theory for electrophoresis of a soft particle, which unites two different electrophoresis theories of hard particles and of spherical polyelectrolytes (Fig. 10). For a spherical soft particle consisting of a hard particle core of radius a covered with a layer of polyelectrolytes of thickness d, in which dissociated groups of valence Z are distributed with a uniform density N, Ohshima obtained [55–57]:

$$\mu = \frac{\varepsilon_r \varepsilon_0}{\eta} \frac{\psi_0 / \kappa_m + \psi_{\text{DON}} / \lambda}{1 / \kappa_m + 1 / \lambda} f\left(\frac{d}{a}\right) + \frac{ZeN}{\eta \lambda^2} \tag{76}$$

with

$$f\left(\frac{d}{a}\right) = \frac{2}{3}\left[1 + \frac{1}{2(1 + d/a)^3}\right] \tag{77}$$

$$\lambda = (\gamma / \eta)^{1/2} \tag{78}$$

$$\psi_0 = \frac{kT}{ze}\left(\ln\left[\frac{ZN}{2zn} + \left\{\left(\frac{ZN}{2zn}\right)^2 + 1\right\}^{1/2}\right] + \frac{2zn}{ZN}\left[1 - \left\{\left(\frac{ZN}{2zn}\right)^2 + 1\right\}^{1/2}\right]\right) \tag{79}$$

$$\psi_{\text{DON}} = \frac{kT}{ze}\ln\left[\frac{ZN}{2zn} + \left\{\left(\frac{ZN}{2zn}\right)^2 + 1\right\}^{1/2}\right] \tag{80}$$

$$\kappa_m = \kappa\left[1 + \left(\frac{ZN}{2zn}\right)^2\right]^{1/4} \tag{81}$$

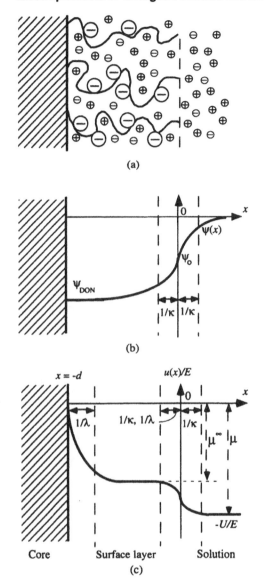

(a)

(b)

(c)

FIG. 9 Distribution of ions (a), potential $\psi(x)$ (b), and liquid velocity $u(x)$ (c) across the surface layer of a soft particle moving with a velocity U in an applied electric field E; ψ_0 is the surface potential and ψ_{DON} is the Donnan potential.

where ψ_0 is the surface potential (i.e., the potential at the boundary of the surface charge layer and the surrounding solution), ψ_{DON} is the Donnan potential of the surface charge layer, and κ_m is the effective Debye–Hückel parameter of the surface charge layer that involves the contribution of the fixed charges ZeN.

Equation (76) is applicable for the case where the relaxation effect is negligible and $\lambda a \gg 1$, $\kappa a \gg 1$, $\lambda d \gg 1$, and $\kappa d \gg 1$. In Fig. 11, we plot the function $f(d/a)$, which varies from 2/3 to 1, as d/a increases. For $d \ll a$ [$f(d/a) \approx 1$], the polyelectrolyte layer can be regarded as planar and Eq. (76) becomes

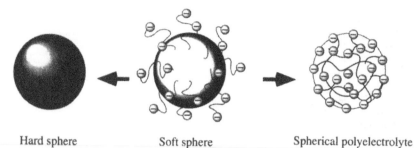

Hard sphere   Soft sphere   Spherical polyelectrolyte

FIG. 10 A soft sphere becomes a hard sphere in the absence of the polyelectrolyte layer and tends to a spherical polyelectrolyte in the absence of the particle core.

$$\mu = \frac{\varepsilon_r \varepsilon_o}{\eta} \frac{\psi_o/\kappa_m + \psi_{DON}/\lambda}{1/\kappa_m + 1/\lambda} + \frac{ZeN}{\eta \lambda^2} \tag{82}$$

while for $d \gg a$ [$f(d/a) \approx 2/3$], the soft particle behaves like a spherical polyelectrolyte with no particle core. In the limit $a \to 0$, the particle core vanishes and the particle becomes a spherical polyelectrolyte (a porous charged sphere). For low potentials, in particular, Eq. (76) tends to

$$\mu = \frac{ZeN}{\eta \lambda^2} \left[1 + \frac{2}{3} \left(\frac{\lambda}{\kappa} \right)^2 \frac{1 + \lambda/2\kappa}{1 + \lambda/\kappa} \right] \tag{83}$$

which agrees with the Hermans–Fujita equation for spherical polyelectrolytes [59]. The function $f(d/a)$ corresponds to Henry's function $f(\kappa a)$ [see Eq. (10)] for hard particles.

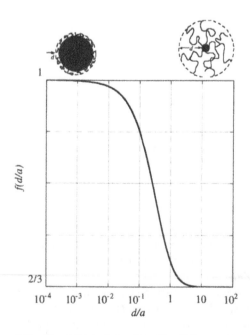

FIG. 11 $f(d/a)$, defined by Eq. (77), as a function of d/a. (From Ref. 55.)

That is, the thickness of the polyelectrolyte layer d plays essentially the same role as the thickness of the electrical double layer $1/\kappa$.

Equation (76) does not involve the thickness of the surface charge layer or the position of the slipping plane. This implies that the zeta potential and the slipping plane both lose their meaning. At the limit of $\lambda \to \infty$, Eq. (76) becomes the Smoluchowski equation (1), as should be expected. On theo ther hand, at the limit of high electrolyte concentrations, Eq. (76) attains a nonzero value, viz.,

$$\mu \to \mu^{\infty} \equiv \frac{ZeN}{\eta\lambda^2} \tag{84}$$

This is the most remarkable difference between Eq. (76) and Smoluchowski's equation (1). This can be understood by noting the velocity distribution of liquid flow within the surface charge layer (Fig. 9). That is, in the midregion of the surface charge layer the liquid velocity is shown to have an almost constant value independent of the position, and this constant value gives μ^{∞}. It can also be shown that the parameter $1/\lambda$ is the distance between the slipping plane and the region where the liquid velocity is almost constant.

Equation (76) [or Eq. (82)] involves two parameters, ZeN and $1/\lambda$, the latter of which can be considered to characterize the "softness" of the polyelectrolyte layer, because at the limit of $1/\lambda \to 0$, the particle becomes rigid. Experimentally, these parameters may be determined from a plot of measured mobility values of a soft particle as a function of electrolyte concentration by a curve-fitting procedure [60–72]. Note that N and n are given in units of m^{-3}. If one uses the units of M, then N and n must be replaced by $1000N_A N$ and $N_A n$, where N_A is Avogadro's number.

For cylindrical soft particles, Ohshima [73] derived the following mobility expressions:

$$\mu_{//} = \frac{\varepsilon_r \varepsilon_0}{\eta} \frac{\psi_0/\kappa_m + \psi_{DON}/\lambda}{1/\kappa_m + 1/\lambda} + \frac{ZeN}{\eta\lambda^2} \tag{85}$$

$$\mu_{\perp} = \frac{\varepsilon_r \varepsilon_0}{\eta} \frac{\psi_0/\kappa_m + \psi_{DON}/\lambda}{1/\kappa_m + 1/\lambda} f\left(\frac{d}{a}\right) + \frac{ZeN}{\eta\lambda^2} \tag{86}$$

with

$$f\left(\frac{d}{a}\right) = \frac{1}{2}\left[1 + \frac{1}{(1+d/a)^2}\right] \tag{87}$$

The above theory assumes that the fixed-charges are uniformly distributed in the polyelectrolyte layer and that the relative permittivity in the polyelectrolyte layer takes the same values as that in the bulk solution phase. These effects are discussed in Refs 74–76. The case where the polyelectrolyte layer is not fully ion-penetratable is considered in Ref. 77.

So far the polyelectrolyte layer has been assumed to have a definite thickness with a uniform segment density distribution. When, however, the polyelectrolyte layer has originated from polymer adsorption, there are some cases in which the effect of the segment density distribution of adsorbed polymers becomes important. Varoqui [78] considered the case where electrically neutral polymers are adsorbed with an exponential segment density distribution on to the particle surface with a charge density σ_0. He assumed that frictional coefficient is expressed as

$$\tau \exp(-x/d) \tag{88}$$

where x is the distance measured from the particle surface, τ is a constant, and d is the average thickness of the polymer layer, and he derived the following mobility expression:

$$\mu = \frac{\sigma_0}{\eta\kappa} \frac{[\Gamma(\kappa d + 1)]^2}{I_0(2\lambda d)} \sum_{n=0}^{\infty} \frac{(\lambda d)^{2n}}{[\Gamma(\kappa d + n + 1)]^2},$$ (89)

with

$$\lambda = (\tau/\eta)^{1/2}$$ (90)

where $\Gamma(z)$ is the gamma function and $I_0(z)$ is the zero-order modified Bessel function.

Ohshima [79] extended Varoqui's theory [78] to the case where adsorbed polymers are charged with a density

$$\rho_{\text{fix}}(x) = \rho_0 \exp(-x/d)$$ (91)

In Eq. (91) ρ_0 corresponds to the average value of the fixed charge density within the polyelectrolyte layer (Fig. 12). The mobility expression thus obtained is

$$\mu = \frac{\sigma_0}{\eta\kappa} \frac{[\Gamma(\kappa d + 1)]^2}{I_0(2\lambda d)} \sum_{n=0}^{\infty} \frac{(\lambda d)^{2n}}{[\Gamma(\kappa d + n + 1)]^2}$$

$$+ \frac{\rho_0}{\eta\lambda^2} \frac{(\kappa d)^2}{(\kappa d)^2 - 1} \left[1 - \frac{1}{I_0(2\lambda d)} - \frac{1}{I_0(2\lambda d)} \frac{[\Gamma(\kappa d)]^2}{\kappa d} \sum_{n=0}^{\infty} \frac{(\lambda d)^{2n+2}}{[\Gamma(\kappa d + n + 1)]^2} \right]$$ (92)

where the first term on the right arises from the charged particle core [which agrees with Eq. (89)] and the second from the charged polymer layer.

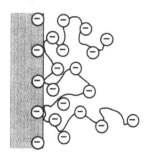

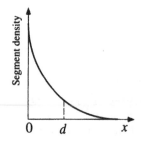

FIG. 12   Schematic representation of the surface of a particle adsorbed by polyelectrolytes (upper) and the segment density distribution (lower).

X. DYNAMIC ELECTROPHORESIS

When a suspension of colloidal particles is in an oscillating electric field, the electrophoretic mobility of the particles depends on the frequency ω of the applied field. O'Brien [80], Babchin and coworkers [81, 82], Sawatzky and Babchin [83], and Fixman [84] proposed approximate theories for the dynamic mobility and derived approximate mobility formulas. James et al. [85] compared these approximations. Mangelsdorf and White [86, 87] obtained the full electrokinetic equations governing the dynamic electrophoresis of spherical colloidal particles, as well as their numerical computer solutions, and an approximate expression for the dynamic mobility for low zeta potentials. For the case where $\kappa a \gg 1$ and the dynamic relaxation effect is neglected, O'Brien's formula [80] reads:

$$\mu(\omega) = \frac{\varepsilon_r \varepsilon_o \zeta}{\eta} \frac{1 - i\gamma a}{H(a) - \Gamma} \tag{93}$$

with

$$\gamma = \sqrt{\frac{i\omega \rho_o}{\eta}} = (i + 1)\sqrt{\frac{\omega \rho_o}{2\eta}} \tag{94}$$

$$H(a) = 1 - i\gamma a - \frac{(\gamma a)^2}{3} \tag{95}$$

$$\Gamma = \frac{2(\gamma a)^2(\rho_p - \rho_o)}{9\rho_o} \tag{96}$$

where ρ_p and ρ_o are, respectively, the mass densities of the particle and the electrolyte solution. When $\omega = 0$, Eq. (89) becomes Smoluchowski's equation (1). For arbitrary κa and low zeta potentials, the mobility is expressed by [83, 87, 88]

$$\mu(\omega) = \frac{\varepsilon_r \varepsilon_o}{\eta} \zeta \frac{\kappa^2}{\gamma^2 [H(a) - \Gamma]}$$
$$\times \left[e^{\beta a} \{ E_5(\beta a) - i\gamma a E_4(\beta a) - \frac{(\gamma a)^2}{3} E_3(\beta a) \} - H(a) e^{\beta a} E_5(\beta a) \right. \tag{97}$$
$$\left. + \frac{2}{3} \left(\frac{\gamma}{\kappa} \right)^2 \left(1 - \frac{i\gamma \kappa a}{\beta} \right) \right]$$

with

$$\beta = \kappa - i\gamma \tag{98}$$

Equation (97), which is applicable for all κa at zero particle permittivity and low zeta potentials, reduces to Henry's equation (8) in the static limit ($\omega \to 0$) and to O'Brien formula (93) for $\kappa a \to \infty$.

The ratio $\mu(\omega)/\mu_s$, where $\mu_s = \varepsilon_r \varepsilon_o \zeta/\eta$ is Smoluchwski's mobility formula [Eq. (1)], can be interpreted as a dynamic Henry function. It can be shown that Eq. (93) is well approximated by [88]

$$\mu(\omega) = \frac{2\varepsilon_r \varepsilon_o \zeta}{3\eta} \left[1 + \frac{1}{2(1 + 2.5/\kappa a\{1 + 2\exp(-\kappa a)\})^3} \right] \frac{1 - i\gamma a(1 + 1/\kappa a)}{[H(a) - \Gamma](1 - i\gamma/\kappa)} \tag{99}$$

from which the following simple expression for the dynamic/static mobility ratio is derived:

$$\frac{\mu(\omega)}{\mu(0)} = \frac{1 - i\gamma a(1 + 1/\kappa a)}{[H(a) - \Gamma](1 - i\gamma/\kappa)} \tag{100}$$

Ohshima [89] derived expressions for the dynamic mobility formulas for cylindrical particles. The dynamic electrophoretic mobility $\mu_\perp(\omega)$ of a cylinder oriented perpendicularly to the applied field is given by

$$\mu_\perp(\omega) = \frac{\varepsilon_r\varepsilon_0\zeta}{2\eta}\left[1 + \frac{1}{(1 + 2.55/\kappa a\{1 + \exp(-\kappa a)\})^2}\right]\frac{1}{1 + (\gamma/\kappa)^2}$$
$$\times \frac{H_1^{(1)}(\gamma a)/\{\gamma a H_0^{(1)}(\gamma a)\} + K_1(\kappa a)/\{\kappa a K_0(\kappa a)\}}{H_1^{(1)}(\gamma a)/\{\gamma a H_0^{(1)}(\gamma a)\} - (1 + \Delta\rho/2\rho_0)/2} \tag{101}$$

When the cylinder is parallel to the applied field, the mobility is given by

$$\mu_{//}(\omega) = \frac{\varepsilon_r\varepsilon_0\zeta}{\eta}\frac{1}{1 + (\gamma/\kappa)^2}\frac{H_1^{(1)}(\gamma a)/\{\gamma a H_0^{(1)}(\gamma a)\} + K_1(\kappa a)/\{\kappa a K_0(\kappa a)\}}{H_1^{(1)}(\gamma a)/\{\gamma a H_0^{(1)}(\gamma a)\} - (1 + \Delta\rho/\rho_0)/2} \tag{102}$$

For a cylindrical particle oriented at an arbitrary angle between its axis and the applied electric field, its electrophoretic mobility averaged over a random distribution of orientation is given by Eq. (15) as in the static case.

An approximate expression for the dynamic electrophoretic mobility of spherical colloidal particles in concentrated suspensions in an oscillating field has been derived by Ohshima [90], which depends on the frequency ω of the applied field and the particle volume fraction ϕ. Ohshima and Dukhin [91] derived an Onsager relationship between dynamic electrophoretic mobility and colloid vibration potential (or current), and Dukhin et al. [92–94] presented a general theory of dynamic electrophoresis in concentrated suspensions.

REFERENCES

1. SS Dukhin, BV Derjaguin. In: E Matijvić, ed. Surface and Colloid Science, vol 7. New York: John Wiley, 1974, ch. 3.
2. RJ. Hunter. Zeta Potential in Colloid Science. New York: Academic Press, 1981.
3. TGM van de Ven. Colloid Hydrodynamics. New York: Academic Press, 1989.
4. RW O'Brien. In: RJ Hunter, ed. Foundations of Colloid Science, vol. 2. Oxford University Press, 1989, ch. 13.
5. SS Dukhin. Adv Colloid Interface Sci 44: 1, 1993.
6. J Lyklema. Fundamentals of Interface and Colloid Science, vol. II. Solid–Liquid Interfaces. New York: Academic Press, 1995.
7. H Ohshima. In: H Ohshima, K Furusawa, eds. Electrical Phenomena at Interfaces, Fundamentals, Measurements, and Applications. 2nd ed. New York: Marcel Dekker, 1998, ch. 2.
8. M von Smoluchowski. Z Phys Chem 92: 129, 1918.
9. E Hückel. Phys Z 25: 204, 1924.
10. DC Henry. Proc R Soc London, Ser A 133: 106, 1931.
11. H Ohshima. J Colloid Interface Sci 168: 269, 1994.
12. WPJT van der Drift, A de Keizer, JTG Overbeek. J Colloid Interface Sci 71: 67, 1979.
13. H Ohshima. J Colloid Interface Sci. 180: 299 (1996).

14. A de Keizer, WPJT van der Drift, JTG Overbeek. Biophys Chem 3: 107, 1975.
15. D Stigter. J Phys Chem 82: 1424, 1978.
16. JTG Overbeek. Kolloid-Beihefte 54: 287, 1943.
17. F Booth. Proc R Soc London, Ser A 203: 514, 1950.
18. PH Wiersema, AL Loeb, JTG Overbeek. J Colloid Interface Sci 22: 78, 1966.
19. RW O'Brien, LR White. J Chem Soc, Faraday Trans 2 74: 1607, 1978.
20. SS Dukhin, NM Semenikhin. Kolloid Zh 32: 360, 1970.
21. RW O'Brien, RJ Hunter. Can J Chem 59:1878, 1981.
22. H Ohshima, TW Healy, LR White. J Chem Soc, Faraday Trans 2 79: 1613, 1983.
23. S Kuwabara. J Phys Soc Japan 14: 527, 1959.
24. S Levine, GH Neale. J Colloid Interface Sci 47: 520, 1974.
25. MW Kozak, EJ Davis. J Colloid Interface Sci 127: 497, 1989.
26. MW Kozak, EJ Davis. J Colloid Interface Sci 129: 166, 1989.
27. H Ohshima. J Colloid Interface Sci 188: 481, 1997.
28. MW Kozak, EJ Davis, J Colloid Interface Sci 112: 403, 1986.
29. H Ohshima, J Colloid Interface Sci 210: 397, 1999.
30. VG Levich. Physicochemical Hydrodynamics. Englewood Cliffs, NJ: Prentice Hall, 1962.
31. S Levine, RN O'Brien. J Colloid Interface Sci 43: 616, 1973.
32. H Ohshima, TW Healy, LR White. J Chem Soc, Faraday Trans 2 80: 1643, 1984.
33. JC Baygents, DA Saville. J Chem Soc, Faraday Trans 87: 1883, 1991.
34. JC Baygents, DA Saville. J Colloid Interface Sci 146: 9, 1991.
35. H Ohshima. J Colloid Interface Sci 189: 376, 1997.
36. H Ohshima. J Colloid Interface Sci 218: 535, 1999.
37. AL Loeb, JTG Overbeek, PH Wiersema. The Electrical Double Layer Around a Spherical Colloid Particle. Cambridge, MA: MIT Press, 1961.
38. H Ohshima, TW Healy, LR White. J Colloid Interface Sci 90: 17, 1982.
39. H Ohshima. J Colloid Interface Sci 171: 525, 1995.
40. H Ohshima. In: H Ohshima, K Furusawa, eds. Electrical Phenomena at Interfaces, Fundamentals, Measurements, and Applications. 2nd ed. New York: Marcel Dekker, 1998, ch. 1.
41. H Ohshima, J. Colloid Interface Sci. 200: 291 (1998).
42. SR de Groot, P Mazur, JTG Overbeek J Chem Phys 20: 1825, 1952.
43. H Ohshima, TW Healy, LR White, RW O'Brien. J Chem Soc, Faraday Trans 2 80: 1299, 1984.
44. H Ohshima. J Colloid Interface Sci 208: 295, 1998.
45. H Ohshima. Adv Colloid Interface Sci. 88:1, 2000.
46. H Ohshima. J Colloid Interface Sci 212: 443, 1999.
47. E Donath, V Pastuschenko. Bioelectrochem Bioenerg 6: 543, 1979.
48. IS Jones. J Colloid Interface Sci 68: 451, 1979.
49. RW Wunderlich. J Colloid Interface Sci 88: 385, 1982.
50. S Levine, M Levine, KA Sharp, DE Brooks. Biophys J 42: 127, 1983.
51. KA Sharp, DE Brooks. Biophys J 47: 563, 1985.
52. H Ohshima, T Kondo. Colloid Polym Sci 264: 1080, 1986.
53. H Ohshima, T Kondo. J Colloid Interface Sci 116: 305, 1987.
54. H Ohshima, T Kondo. J Colloid Interface Sci 130: 281, 1989.
55. H Ohshima. J Colloid Interface Sci 163: 474, 1994.
56. H Ohshima. Adv Colloid Interface Sci 62: 443, 1995.
57. H Ohshima. Colloids Surfaces A: Physicochem Eng Aspects 103: 249, 1995.
58. P Debye, A Bueche. J Chem Phys 16: 573, 1948.
59. JJ Hermans, H Fujita, Koninkl Ned Akad Wetenschap Proc B 58: 182, 1955.
60. H Ohshima, K Makino, T Kato, K Fujimoto, T Kondo, H Kawaguchi. J Colloid Interface Sci 159: 512, 1993.
61. K Makino, T Taki, M Ogura, S Handa, M Nakajima, T Kondo, H Ohshima. Biophys Chem 47: 261, 1993.

62. H Ohshima, K Makino, T Kato, K Fujimoto, T Kondo, H Kawaguchi. J Colloid Interface Sci 159: 512, 1993.
63. K Makino, T Taki, M Ogura, S Handa, M Nakajima, T Kondo, H Ohshima. Biophys Chem 47: 251, 1993.
64. K Makino, M Ikekita, T Kondo, S Tanuma, H Ohshima. Colloid Polym Sci 272: 487, 1994.
65. T Mazda, K Makino, H Ohshima. Colloids Surfaces B; Biointerfaces 5: 75, 1995; Colloids Surfaces B: Biointerfaces 10: 303, 1998.
66. CC Ho, T Kondo, N Muramastu, H Ohshima. J Colloid Interface Sci 178: 442, 1996.
67. S Takashima, H Morisaki. Colloids Surfaces B: Biointerfaces 9: 205, 1997.
68. R Bos, HC van der Mei, HJ Busscher. Biophys Chem 74: 251, 1998.
69. T Kondo. In: H Ohshima, K Furusawa, eds. Electrical Phenomena at Interfaces, Fundamentals, Measurements, and Applications. 2nd ed. New York: Marcel Dekker, 1998, ch. 30.
70. K Makino. In: H Ohshima, K Furusawa, eds. Electrical Phenomena at Interfaces, Fundamentals, Measurements, and Applications. 2nd ed. New York: Marcel Dekker, 1998, ch. 31.
71. A Larsson, M Rasmusson, H Ohshima. Carbohydr Res 317: 223, 1999.
72. H Morisaki, S Nagai, H Ohshima. Microbiology 145: 2797, 1999.
73. H Ohshima. Colloid Polym Sci 275: 480, 1997.
74. H Ohshima, T Kondo. Biophys Chem 39: 191, 1991.
75. JP Hsu, WC Hsu, YI Chang. Colloid Polym Sci 265: 911, 1987.
76. JP Hsu, YP Fan, J. Colloid Interface Sci. 172: 230 1995.
77. H Ohshima, K Makino. Colloids Surfaces A: Physicochem Eng Aspects 109: 71, 1996.
78. R Varoqui. Nouv J Chim 6: 187, 1982.
79. H Ohshima. J Colloid Interface Sci 185: 269, 1997.
80. RW O'Brien. J Fluid Mech 190: 71, 1988.
81. AJ Babchin, RS Chow, RP Sawatzky. Adv Colloid Interface Sci 30: 111, 1989.
82. AJ Babchin, RP Sawatzky, RS Chow, EE Isaacs, H Huang. Proceedings of International Symposium on Surface Charge Characterization. 21st Annual Meeting of Fine Particle Society, San Diego, CA, p 49, 1990.
83. RP Sawatzky, AJ Babchin. J Fluid Mech 246: 321, 1993.
84. M Fixman. J Chem Phys 78: 1483, 1983.
85. RO James, J Texter, PJ Scales. Langmuir 7: 1993, 1991.
86. CS Mangelsdorf, LR White. J Chem Soc, Faraday Trans 88: 3567, 1992.
87. CS Mangelsdorf, LR White. J Colloid Interface Sci 160: 275, 1993.
88. H Ohshima. J Colloid Interface Sci 179: 431, 1996.
89. H Ohshima. J Colloid Interface Sci 185: 131, 1997.
90. H Ohshima. J Colloid Interface Sci 195: 137, 1997.
91. H Ohshima, AS Dukhin. J Colloid Interface Sci 212: 449, 1999.
92. AS Dukhin, H Ohshima, VN Shilov, PJ Goetz. Langmuir 15: 3445, 1999.
93. AS Dukhin, VN Shilov, YuB Borkovskaya. Langmuir 15: 3452, 1999.
94. AS Dukhin, VN Shilov, H Ohshima, PJ Goetz. Langmuir 15: 6692, 1999.

6

Electrophoresis of Complex and Interacting Particles

DARRELL VELEGOL The Pennsylvania State University, University Park, Pennsylvania

JOHN L. ANDERSON Carnegie Mellon University, Pittsburgh, Pennsylvania

YURI SOLOMENTSEV Motorola Advanced Products, Austin, Texas

I. INTRODUCTION

A. Classical Electrophoresis

Most classical works on electrophoresis are for individual particles that are uniformly charged [1]. In practice these conditions are often not met, and so the purpose of this chapter is to summarize recent theoretical and experimental results for *nonuniformly charged* particles and *interacting* particles (see Fig. 1). Several of the theoretical results might seem counterintuitive at first; however, many are supported by experimental data [2–5]. This empowers the electrokinetic results in this chapter to be used for a variety of important new analytical applications, including measurements of colloidal forces and charge distributions on individual particles [4, 6]. A new experimental technique to determine forces between coagulated particles, "differential electrophoresis," has been developed using the concepts presented here [5].

Electrokinetic theory is based on the equations of electrostatics and hydrodynamics. These equations have been laid out clearly by previous authors [7–10]. In general, these equations can be quite difficult to solve, especially for complex problems like those in this chapter; therefore, we focus on the electrophoresis of particles with thin, unpolarized electrical double layers (EDLs). The starting point for the discussion is the Smoluchowski result [11], which relates the zeta (ζ) potential of a particle to its translational electrophoretic velocity (U) and angular electrophoretic velocity (Ω):

$$U = \frac{\varepsilon \zeta \mathbf{E}_\infty}{\eta}, \qquad \Omega = 0 \tag{1}$$

where ε is the permittivity and η the viscosity of the fluid, and $\mathbf{E}_\infty$ is the uniform applied electric field in the suspension. This equation was developed for spheres, but Morrison extended it to arbitrary shapes and pointed out that the result contains four key assumptions [12]:

1.   The particle is rigid and nonconducting, and it has a negligible electrical permittivity compared to the surrounding fluid which is conductive.

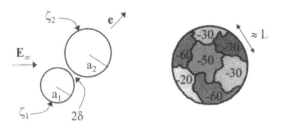

FIG. 1 Example of interacting and nonuniformly charged particles. Left: particles 1 and 2 with radii (a_i), zeta potentials (ζ_i), and gap (2δ). Right: schematic example of a particle with a nonuniform ζ potential (units mV) on each "patch."

2. The EDL around the particle is infinitesimally thin (i.e., $\kappa R \to \infty$ for the whole particle, where R is the local curvature of the particle and κ^{-1} is the Debye length).
3. The particle is in an unbounded fluid.
4. The ζ potential over the surface of the particle is uniform.

The first assumption is satisfied for a broad range of particles. It is in fact usual for the permittivity of the particle to be much less than that of the fluid. For polystyrene latex particles, the dielectric constant is 2.5, whereas for water it is 80 at room temperature. With regard to conductivity, even conducting particles (e.g., gold) often behave as non-conducting particles since the transfer of charge across the fluid–solid interface is usually zero or negligible. While we will consider only rigid particles in this chapter, it is also useful to note that many fluid particles behave as rigid particles hydrodynamically [13, 14].

Equation (1) is applicable for a *sphere* not only when E_∞ is uniform, but also when E_∞ varies spatially and the field used in Eq. (1) is evaluated at the position of the center of the particle [15]. This is not true for nonspherical particles, however. For spheroidal particles there is an angular velocity which is proportional to ∇E_∞, and the translational velocity depends on both E_∞ and $\nabla^2 E_\infty$ [16].

The assumption concerning "thin" EDLs has been corrected in two important ways. Shortly after Eq. (1) was developed, the result for infinitely thick double layers (i.e., $\kappa R \to 0$) was obtained by Hückel for arbitrary ζ potentials [17]. Later, Henry developed the equation for spheres with arbitrary κa, where a is the sphere radius [18]; however, this equation requires that the ζ potential on the sphere be small relative to the thermal electrical potential (kT/e). In practice, the Henry result usually works well for $Ze|\zeta|/kT < 2$, where Z is the valence of the symmetric electrolyte. Yoon and Kim [19] extended Henry's analysis to spheroids.

The second correction to Eq. (1) for thin EDLs is for "concentration polarization," which involves a significant conduction of ions within the EDL. Dukhin and Derjaguin [20] clarified the understanding of "thin double layers," showing that "thin" means

$$\frac{\kappa a}{\cosh(Ze\zeta/2kT)} \gg 1 \qquad (2)$$

where κ^{-1} is the Debye length and a is the particle radius. Following the work of Wiersema et al. [21], O'Brien and White [7] developed a numerical scheme to calculate electrophoretic mobility for spheres with arbitrary κa and ζ potential, showing that for finite double layers, particles reach a maximum electrophoretic mobility at a particular ζ potential. If the ζ potential on a particle is higher than this, then the mobility of the particle actually decreases because of the increasing significance of concentration polarization of the double layer.

Another cause for thick double layers – at least over part of the surface – is surface roughness [22]*. If the radius of curvature is small for the local surface roughness, then so is κR. However, the scarce data on surface roughness indicates that this is not ordinarily the case. Finally, finite double layers have been considered for interacting spheres [23, 24]; in general, the differences from the results for thin EDLs for interacting particles are not usually large, and so we will not discuss finite or thick double layers in this chapter.

B. Scope of This Chapter

This chapter focuses on the third and fourth assumptions above. We consider particles with infinitesimal EDLs, but an arbitrary distribution of ζ potential on the surface, and also particles that interact with other particles or with boundaries. The applied electric field (E_∞) is assumed to be uniform on the length scale of the particles. Here, we report advances that have occurred since a previous review of the topic [10] and emphasize two key points:

1. More than just the average (monopole) ζ potential of particles can be determined from electrophoresis measurements; for example, the second (dipole) moment of the ζ distribution of the particle's surface, which is often critical in understanding the stability and other behavior of colloidal suspensions, is also measurable.
2. The electrophoresis of interacting particles can be used to probe other colloidal phenomena (e.g., surface forces) by providing a controlled means to exert forces between the particles.

II. SINGLE PARTICLES WITH NONUNIFORMLY CHARGED SURFACES

A. Spheres

The importance of charge nonuniformity on electrophoresis was addressed by Teubner [9], and in his seminal paper he gave integrals that account for the charge nonuniformity. However, it is easier to understand the role of charge nonuniformity in terms of moments of the ζ potential distribution [15]. The distribution of ζ potential can be expressed in terms of the monopole moment (i.e., the average), the dipole moment, the quadrupole moment, etc.; and the electrokinetic equations can be solved to give the particle's translational and angular velocities in terms of these moments. Interestingly, the electrophoretic velocities of a sphere depend only on the monopole $(\langle\zeta\rangle)$, dipole (P_1) and quadrupole (P_2) [15]:

$$U = \frac{\varepsilon}{\eta}\left[\langle\zeta\rangle I - \frac{1}{2}P_2\right]\cdot E_\infty \tag{3}$$

$$\Omega = \frac{\frac{9}{4}\varepsilon}{\eta a}P_1 \times E_\infty \tag{4}$$

---

*Particle roughness was seen to consist of 30-nm hemispheres on a 10-μm colloidal sphere; however, the SEM required that the particle was dried, and the effects of this on producing roughness are unknown.

where I is the identity tensor, and the moments are defined by integrals over the surface of the sphere:

$$\langle \zeta \rangle = \frac{1}{4\pi} \int_0^{2\pi} \int_0^{\pi} \zeta \sin\theta d\theta d\phi \tag{5}$$

$$P_1 = \frac{1}{4\pi} \int_0^{2\pi} \int_0^{\pi} \zeta n \sin\theta d\theta d\phi \tag{6}$$

$$P_2 = \frac{1}{4\pi} \int_0^{2\pi} \int_0^{\pi} \zeta(3nn - I) \sin\theta d\theta d\phi \tag{7}$$

where n is the unit normal pointing away from the surface; for a sphere, this is the unit vector in the r-co-ordinate direction. In an electrophoresis experiment, one usually aims to measure $\langle \zeta \rangle$. Equation (3) shows that in fact typical electrophoresis experiments measure both the monopole and the quadrupole. Indeed, the movement due to the quadrupole need not even be in the same direction as the electric field (E_∞), depending on the orientation of the particle relative to the applied field.

An important type of charge nonuniformity is achieved by *random* charging. It is not always appreciated that a random placement of charges on a particle's surface implies a *nonuniform* charge distribution. Velegol et al. [25] recently studied the electrophoresis of spheres that are randomly charged. They developed a model in which the particle has a number (N) of finite-size patches, with each patch on the particle having a ζ potential from a Gaussian distribution with an average ζ potential ($\langle \zeta \rangle$) and standard deviation (σ_ζ). The length scale (L) of the patch is arbitrary (see Fig. 1); if one considers the entire sphere, then $N = 1$, but we could also select a patch size that, say, corresponds to the area of interaction between two particles that nearly touch as would exist in a secondary potential energy minimum. Since in an electrophoresis experiment the measurable quantities are the particle translational and angular velocities, Eq. (4) can be used to obtain the dipole moment (P_1) on an individual particle. In practice the magnitude of the dipole moment ($P_1 = \sqrt{P_1 \cdot P_1}$) is easier to measure. After measuring the angular velocity of many particles and finding the ensemble average dipole ($\langle P_1 \rangle$), one would like to interpret the result in terms of the standard deviation of the ζ potential on the particle surface. Using statistical concepts, Velegol et al. [25] related the ensemble average of P_1 (e.g., over many spheres) and its sphere-to-sphere standard deviation to the number (N) of patches and the standard deviation (σ_ζ) of ζ potential for each patch:

$$\langle P_1 \rangle = \frac{4}{\sqrt{6\pi}} \frac{\sigma_\zeta}{\sqrt{N}} = 0.921 \frac{\sigma_\zeta}{\sqrt{N}} \tag{8}$$

$$\sigma_{P_1} = \sqrt{1 - \frac{16}{6\pi}} \frac{\sigma_\zeta}{\sqrt{N}} = 0.389 \frac{\sigma_\zeta}{\sqrt{N}} \tag{9}$$

The average dipole moment of the sphere, which is measurable, is proportional to the standard deviation of the ζ potential. Note that, for a constant $\langle P_1 \rangle$ (since this is a measurable property), $\sigma_\zeta \sim \sqrt{N}$ for a random distribution of the ζ potential, which implies that $\sigma_\zeta \sim L^{-1}$.

To gain intuition into the above relations, consider a set of spherical particles with $a = 1$ μm in an aqueous solution with $\kappa^{-1} = 1$ nm. Suppose the measured average translational mobility is $\langle |U| \rangle / E_\infty = 5.0 \times 10^{-8}$ m^2/V·s and the angular mobility is $\langle |\Omega| \rangle / E_\infty =$

1.0×10^{-3} m/V·s, then using Eq. (4) with Ω as the maximum angular velocity gives $\langle P_1 \rangle = 0.63$ mV. The next step is to obtain a standard deviation of ζ potential on a pertinent length scale (L). From the theory of colloidal forces, the spheres interact most at the regions near contact over a length scale (L). From the theory of colloidal forces, the spheres interact most at the regions near contact over a length scale $L \approx \sqrt{\kappa^{-1}a}$, which, here, is 32 nm. If we consider round patches on the sphere, the ratio of the sphere area to the patch area gives $N \approx 4000$. From Eq. (8) we have $\sigma_\zeta = 43$ mV for each patch. This is a very large standard deviation on this length scale, which is important to the interaction between two nearly touching particles; however, as shown in the next paragraph, it is highly unlikely that this variation would be detected in translational electrophoresis measurements. Theoretical work has shown that such large standard deviations of ζ in the region of contact between two particles can cause instability in colloidal suspensions, even though the average ζ potential indicates that the suspension should be stable [26, 27].

Knowing the value of σ_ζ from the angular velocity, we can also obtain the expected quadrupole contribution to the translational velocity. If we define the z co-ordinate axis by $E_\infty = E_\infty i_z$, then we are interested in the quantity $P_2 = \sqrt{(P_2 \cdot i_z) \cdot (P_2 \cdot i_z)}$, especially its average and standard deviation. Statistical analysis gives [25]

$$\langle P_2 \rangle = \frac{4 + 3 \ln 3}{\sqrt{10\pi}} \frac{\sigma_\zeta}{\sqrt{N}} = 1.302 \frac{\sigma_\zeta}{\sqrt{N}} \tag{10}$$

$$\sigma_{P_2} = \sqrt{2 - \frac{(4 + 3 \ln 3)^2}{10\pi}} \frac{\sigma_\zeta}{\sqrt{N}} = 0.553 \frac{\sigma_\zeta}{\sqrt{N}} \tag{11}$$

The average magnitude of the quadrupole is directly related to the standard deviation of ζ (and thus, to the dipole), and so in the example above, $\langle P_2 \rangle = 0.89$ mV and $\langle \zeta \rangle = -71$ mV. So whereas the dipole is the leading order term in the angular velocity and therefore readily measurable, the quadrupole would likely go completely undetected in this experiment because the translational mobility would be masked by $\langle \zeta \rangle$.

The results presented above are summarized as follows:

- Random charging of the surface of a particle leads to a *nonuniform* rather than uniform distribution of ζ potential.
- The results above hold for any random distribution of ζ potential (Poisson, uniform, etc.) if the number of patches is sufficiently large ($N > 30$) and are reasonable approximations for smaller N. This is by the central limit theorem of statistics.
- The dipole and quadrupole are measurable but related quantities, both proportional to $\sigma_\zeta/\sqrt{N}$. Thus, the standard deviation of ζ potential on any given patch (σ_ζ) of the particle's surface depends inversely on the length scale (L) of the patch.
- Whereas translational electrophoresis can usually be expected to give a good approximation for the average ζ potential, rotational electrophoresis provides a means to obtain $\sigma_\zeta/\sqrt{N}$.

The results given above for infinitesimal EDLs have been extended [25] to thick EDLs by using Yoon's results for arbitrary EDL thickness [28]. Yoon's solutions are cast in terms of the monopole, dipole, and quadrupole, making the results above readily usable with only a minor modification of the three precoefficients that account for arbitrary κa. Although the extension to arbitrary κa requires low ζ potentials on the particles, this is often satisfied for small particles.

B. Spheroids

Fair and Anderson [29] developed expressions for the electrophoretic mobility of an ellipsoid with an arbitrary variation of ζ potential on its surface, assuming unpolarized, thin EDLs. Since other shapes can be approximated by spheroids (e.g., disks, doublets, slender bodies), this is a broadly applicable class of solutions. They found

$$U = \frac{\varepsilon}{3\eta V_p} \int_{S^+} \int n \cdot r(I - nn)\zeta dS \cdot G \cdot E_\infty \tag{12}$$

$$\Omega = \frac{\varepsilon}{\eta V_p} H \cdot \int_{S^+} \int n \cdot rr \times (I - nn)\zeta dS \cdot G \cdot E_\infty \tag{13}$$

where $V_p = \frac{4}{3}\pi b_1 b_2 b_3$ is the volume of an ellipsoid (semiaxes b_1, b_2, b_3), n is the unit normal to the surface, and r is the position on the surface. The surface S^+ denotes the outer edge of the double layer, which for infinitesimal EDLs is *geometrically* the same as the true surface of the particle. For a uniform ζ potential, Eqs (12) and (13) reduce to the Smoluchowski result given by Eq. (1).

The tensors G and H are geometric only. For a sphere of radius a, $G = \frac{3}{2}I$ and $H = I/2a^2$, so that Eqs (3)–(7) are recovered. For a spheroid with axes (b, b, c), expressions are given in the original paper [29] for G and H in terms of the aspect ratio $\alpha = b/c$:

$$H = \frac{1}{b^2 + c^2}(I - ee) + \frac{1}{2b^2}ee \tag{14}$$

$$G = G_\parallel ee + G_\perp(I - ee) \tag{15}$$

The unit vector (e) is along the c axis. The G coefficients are well approximated by the following:

prolate spheroid $(\alpha < 1)$

$$G_\parallel \approx 1 + 0.3\alpha + 0.2\alpha^2, \qquad G_\perp \approx 2 - 0.62\alpha + 0.12\alpha^2 \tag{16}$$

oblate spheroid $(\alpha > 1)$

$$G_\parallel \approx 0.863 + 0.637\alpha, \qquad G_\perp \approx 1 + 0.75/\alpha - 0.25/\alpha^2 \tag{17}$$

These approximations are accurate at the limits (i.e., small α, large α, and $\alpha = 1$) and are within 1% of the correct result for all α.

An interesting example of the electrophoresis of complex particles is that of proteins [30]. Chae and Lenhoff developed the method of Teubner [9] to analyze the translational and rotational mobility of proteins in free solution, accounting for complicated protein shapes and charge distributions (including charges interior to the proteins). They used their method to analyze experimental results for ribonuclease A, showing good agreement under conditions when the protein was charged. This work was motivated by the desire to model globular proteins realistically, accounting for the distributed charge within the molecule.

C. Using Spheroid Results to Approximate a Disk (Clay Platelets)

The importance of Eqs (12) and (13) can be demonstrated by approximating other geometries. For example, the limit $\alpha \to 0$ is a rod, while the limit $\alpha \to \infty$ is a flat plate. In this section we will approximate a disk as an oblate spheroid. Kaolinite clay particles have a disk-like shape and are believed to be nonuniformly charged. The surface chemistry of kaolinite is well referenced in the literature [31, 32] and it suggests that the faces and edge of the disk have different ζ potentials. Indeed, even the two basal surfaces might have different ζ potentials because of their different surface chemistries [33]. In order to estimate the charge nonuniformity on the clay particles, one can use electrophoresis. Since the typical face diameter of clay platelets* is about 1 μm, and the thickness is about 0.1 μm, kaolinite particles can be approximated by an oblate spheroid as shown in Fig. 2.

Choosing an "equivalent spheroid" to approximate a disk is a subtle process. Three parameters must be fixed: the spheroid's major axis $(2b)$ and minor axis $(2c)$, and the local surface charge density $\rho_s(x)$. Several constraints are available to determine these parameters, including the total surface area of the disk, the total charge on the disk (or perhaps the total charges on any region of the disk), the volume of the disk, the hydrodynamic resistance of the disk parallel to its axis (e), and the hydrodynamic resistance of the disk perpendicular to e. We choose to match the surface area and the particle volume, which allows determination of the two semiaxes of the spheroid.

The surface area (S) and volume (V) of a spheroid with axes (b, b, c) are given by

$$S = \begin{cases} 2\pi b^2 + \dfrac{\pi c^2}{e}\ln\left(\dfrac{1+e}{1-e}\right), & e = \sqrt{1 - \dfrac{c^2}{b^2}} & \text{if } b > c \text{ (oblate)} \\[2ex] 4\pi b^2 & & \text{if } b = c \\[2ex] 2\pi b^2 + 2\pi bc\dfrac{\sin^{-1}e}{e}, & e = \sqrt{1 - \dfrac{b^2}{c^2}} & \text{if } b < c \text{ (prolate)} \end{cases} \tag{18}$$

$$V = \frac{4}{3}\pi b^2 c \tag{19}$$

The eccentricity (e) is always less than or equal to unity. In approximating the disk with the oblate spheroid, we must simultaneously solve for the semiaxes b and c as well as the edge region (x_0). These are obtained from the disk diameter (D) and the disk thickness (2δ) by maintaining the areas of the faces and the edge. The results of these calculations are given in Table 1.

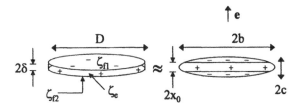

FIG. 2 Approximation of a disk by a spheroid, as a model for clay platelets.

---

*These particles are especially suitable for the coating of paper.

TABLE 1 Approximating a Disk by a Spheroid.[a] The Disk Diameter (D) and Thickness (2δ) are Related to the Major Semiaxis (b) and Minor Semiaxis (c) of the Oblate Spheroid. The "Edge" Region has a Thickness $2x_0$ Corresponding to Edge of Disk

$D/2\delta$	$D/2b$	b/c	x_0/c
1	0.6595	2.3238	0.7593
2	0.7871	2.7341	0.6332
3	0.8386	3.3908	0.5626
4	0.8675	4.0843	0.5146
5	0.8866	4.7832	0.4786
6	0.9003	5.4810	0.4500
7	0.9108	6.1761	0.4265
10	0.9316	8.2453	0.3744
20	0.9606	15.0411	0.2841
50	0.9823	35.1716	0.1900
100	0.9907	68.5688	0.1375
1000	0.9990	668.6515	0.0446

[a]See Fig. 2.

In practice the size of the clay particles is estimated from sedimentation experiments and is given as an "equivalent diameter" (d_e), which is the diameter of a sphere that would have the same hydrodynamic settling velocity as the disk. We approximate the settling velocity U_s) of a disk using the spheroid, as[*]

$$U_s = \frac{d_e^2 \Delta \rho g}{18\eta} = \left[\frac{1}{3}M_{\parallel} + \frac{2}{3}M_{\perp}\right]\frac{4}{3}\pi c^3 \alpha^2 \Delta \rho g \tag{20}$$

where $M_{\parallel}$ and $M_{\perp}$ are the Stokes' mobility coefficients for a spheroid moving parallel or perpendicular to the e vector [34]. For oblate spheroids:

$$M_{\parallel} = \frac{1}{6\pi\eta}\frac{3}{8b}\left[\frac{2\alpha^{-1}}{1-\alpha^{-2}} + \frac{2(1-2\alpha^{-2})}{(1-\alpha^{-2})^{3/2}}\tan^{-1}\sqrt{\alpha^2-1}\right] \tag{21}$$

$$M_{\perp} = \frac{1}{6\pi\eta}\frac{3}{8b}\left[\frac{-\alpha^{-1}}{1-\alpha^{-2}} + \frac{3-2\alpha^{-2}}{(1-\alpha^{-2})^{3/2}}\sin^{-1}\sqrt{1-\alpha^{-2}}\right] \tag{22}$$

where $\alpha = b/c > 1$. From Eq. (20) we have

$$\frac{b}{d_e} = \left[\frac{2\alpha}{16\pi\eta b(M_{\parallel} + 2M_{\perp})}\right] \tag{23}$$

The charges are distributed on the disk-like spheroid as follows:

---

[*]The particle size of kaolinite is often expressed as an equivalent spherical diameter (d_e), which is the diameter of a sphere that has the same sedimentation velocity as the actual particle. Particles larger than $d_e = 5$ μm are called "booklets," and are made of several plate-like particles. In paper coating, particles < 2 μm are desirable.

$$
\begin{aligned}
0 \le |z| \le z_0 : & \quad \zeta = \zeta_e \\
z_0 < z \le 1 : & \quad \zeta = \zeta_{f1} \\
-1 \le z < -z_0 : & \quad \zeta = \zeta_{f2}
\end{aligned}
\tag{24}
$$

where $z(= x/c)$ is the dimensionless position along the c semiaxis.

The expressions from the previous section can be simplified, and the mobility of a spheroid with an *axisymmetric* charge distribution can be described as [29].

$$
U = \frac{\varepsilon}{\eta} [m_\perp (I - ee) + m_\parallel ee] \cdot E_\infty
\tag{25}
$$

$$
\Omega = \frac{\varepsilon}{\eta c} m_r e \times E_\infty
\tag{26}
$$

where $m_\parallel$ and $m_\perp$ are electrophoretic translation mobility coefficients that depend on even moments of ζ potential, and m_r is an electrophoretic rotation mobility coefficient that depends on the odd moments of ζ potential. They can be written as

$$
m_\parallel = \frac{\zeta_{f1} + \zeta_{f2}}{2} + \left[\zeta_e - \frac{\zeta_{f1} + \zeta_{f2}}{2} \right] f_\parallel(\varepsilon, z_0)
\tag{27}
$$

$$
m_\perp = \frac{\zeta_{f1} + \zeta_{f2}}{2} + \left[\zeta_e - \frac{\zeta_{f1} + \zeta_{f2}}{2} \right] f_\perp(\varepsilon, z_0)
\tag{28}
$$

$$
m_r = (\zeta_{f1} - \zeta_{f2}) f_r(\varepsilon, z_0)
\tag{29}
$$

As the disk becomes infinitesimally thin (i.e., $D/\delta \to \infty$), then $f_\parallel \to 1, f_\perp \to 0$, and $f_r \to 0$. Values for the f functions are listed in Table 2.

D. Using Spheroids to Approximate Doublets

A colloidal doublet can be approximated as a prolate spheroid. Such an equivalence was made by Nir and Acrivos [35] using the hydrodynamic characteristics of each geometry as the criteria for matching. They matched the angular velocities of a doublet and a spheroid undergoing Jeffery orbits, finding that $c/b = 1.982$ for the equivalent axis ratio. For electrophoresis we can do a similar calculation. Once again we have a variety of parameters to choose from, but we choose to equate the total surface charge (or equivalently here, the surface area) and the electrophoretic angular velocity of a doublet and a spher-

TABLE 2 Electrophoretic Mobility Coefficients for Oblate Spheroid Equivalent Model of a Disk[a]

$D/2\delta$	$f_\parallel$	$f_\perp$	f_r
2	0.9344	0.5606	0.07963
3	0.9260	0.5051	0.06195
5	0.9110	0.4145	0.03893
10	0.9107	0.3261	0.01617
20	0.9191	0.2520	0.000589
30	0.9261	0.2146	0.000310
50	0.9358	0.1736	0.000132

[a] See Eqs (27) and (28).

oid. Fair and Anderson [29] showed that for a spheroid with an axisymmetric charge distribution:

$$\Omega \equiv e \times \frac{de}{dt} = \frac{9\varepsilon}{4\eta c}\beta_1 \times E_\infty \qquad (30)$$

$$\beta_1 = \frac{G_\perp}{3}\int\limits_{-1}^{1}\zeta(z)z\,\frac{1 + z^2\left(\dfrac{\alpha^2 - 1}{\alpha^2 + 1}\right)}{1 + z^2(\alpha^2 - 1)}dze \qquad (31)$$

where z is a dimensionless position ($-1 \leq z \leq 1$) along the c axis and e is the unit vector giving the orientation of that axis; $\alpha = b/c$ and $G_\perp$ is given by Eqs (16) and (17).

We want to find the semiaxes, b and c, that make a spheroid equivalent electrophoretically to a *rigid* doublet comprising two spheres of equal radius (a) but different zeta potential (see Fig. 3). The angular velocity of a doublet composed of two spheres (1 and 2, each uniformly charged) is [36]

$$\Omega \equiv e \times \frac{de}{dt} = \frac{\varepsilon(\zeta_2 - \zeta_1)}{2\eta a}Ne \times E_\infty \qquad (32)$$

where e is the unit vector along the line between centers (from sphere 1 to 2). For a rigid doublet of equal size spheres, $N = 0.6400$ [36].

We will let the top half of the spheroid have a ζ potential ζ_2) and the bottom half ζ_1, as shown in Fig. 3. Solving the integral in Eq. (31) leads to

$$N = \frac{3G_\perp a}{4c}\left(\frac{\alpha^2 - 1 + 2\alpha^2\ln\alpha}{\alpha^4 - 1}\right) \qquad (33)$$

We thus need to solve for two parameters ($b < c$) from Eq. (18), with $S = 8\pi a^2$ (for two spheres in a doublet) and Eq. (33) with $N = 0.64$. This calculation reveals that $b/a = 0.99993$ and $c/a = 2.39192$, and hence $c/b = 2.392$ (compared to 1.982 obtained by Nir and Acrivos [35] for equivalence based solely on hydrodynamic behavior). That is, if we have a spheroid with these semiaxes and the zeta potentials as shown above, the spheroid will have the same total charge and electrophoretic angular velocity as a doublet in which the spheres have radius a.

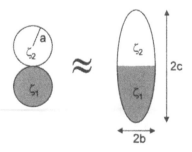

FIG. 3 Approximating a doublet with a spheroid. The equivalent axis ratios are $b/a = 0.99993$ and $c/a = 2.39192$.

III. PARTICLE–PARTICLE AND PARTICLE–BOUNDARY INTERACTIONS

A. Results for Two Interacting Spheres

The electrophoresis of two interacting but *force-free* and *couple-free* particles is complicated by the fact that we no longer have a potential flow field outside the EDL. This is because the flow field produced by particle 1 exerts a force on particle 2, and vice versa. Methods that have been used to solve this problem are reflection calculations [37, 38], numerical calculations [39–43], and semianalytical results for small gaps between the spheres [44]. Results have also been obtained for thin but polarized EDLs, but we will not discuss those results here [23]. A significant difference between the interactions in electrophoresis versus seidmentation is that electrophoretic interactions are much shorter range; the flow fields driven by sedimentation and other body forces decay as $1/r$, where r is the distance from the moving particle, whereas in electrophoresis the flow fields decay as $1/r^3$. The paper by Zeng et al. [43] summarizes many of the reflection and numerical results in the literature and puts them into convenient empirical forms for a broad range of gaps and size ratios of spheres.

Chen and Keh [37] developed the first reflection results and numerical solutions for the electrophoretic mobility of two spheres. They gave the translational and rotational velocities of two force-free, couple-free particles as a function of the ζ potential on the spheres (assumed uniform on each sphere) and the gap between the spheres. They put their calculations in a form developed by Fair and Anderson [36]:

$$U_1 = \frac{\varepsilon\zeta_1}{\eta}E_\infty + \frac{\varepsilon(\zeta_2 - \zeta_1)}{\eta}\left[M_{12}^p ee + M_{12}^n(I - ee)\right]\cdot E_\infty \tag{34}$$

$$U_2 = \frac{\varepsilon\zeta_2}{\eta}E_\infty - \frac{\varepsilon(\zeta_2 - \zeta_1)}{\eta}\left[M_{21}^p ee + M_{21}^n(I - ee)\right]\cdot E_\infty \tag{35}$$

$$\Omega_1 = \frac{\varepsilon(\zeta_2 - \zeta_1)}{\eta L}N_{12}e \times E_\infty \tag{36}$$

$$\Omega_2 = -\frac{\varepsilon(\zeta_2 - \zeta_1)}{\eta L}N_{21}e \times E_\infty \tag{37}$$

The "p" and "n" superscripts on the elctrophoretic mobility coefficients (M and N) represent the motion parallel and perpendicular to the center-to-center axis (e = unit vector from particle 1 to 2, where 2 is the larger particle). For large separations the method of reflections gives [37]

$$M_{12}^p = \frac{a_2^3}{R^3} + \tfrac{13}{2}\frac{a_1^3 a_2^3}{R^6} + O(R^{-8}), \qquad M_{21}^p = \frac{a_1^3}{R^3} + \tfrac{13}{2}\frac{a_1^3 a_2^3}{R^6} + O(R^{-8}) \tag{38}$$

$$M_{12}^n = -\tfrac{1}{2}\frac{a_2^3}{R^3} - \tfrac{1}{4}\frac{a_1^3 a_2^3}{R^6} + O(R^{-8}), \qquad M_{21}^n = -\tfrac{1}{2}\frac{a_1^3}{R^3} - \tfrac{1}{4}\frac{a_1^3 a_2^3}{R^6} + O(R^{-8}) \tag{39}$$

$$N_{12} = \tfrac{15}{4}\frac{a_1^2 a_2^4}{R^6} + O(R^{-8}), \qquad N_{21} = \tfrac{15}{4}\frac{a_1^4 a_2^2}{R^6} + O(R^{-8}) \tag{40}$$

where R is the center-to-center distance between the two particles. These results show the sharp decay of the interaction, which is $O(R^{-3})$.

Near-field results have also been obtained, both for the case when the electric field is oriented parallel [45] and perpendicular [44, 46] to the line between the particles. Velegol et al. [44] showed that in fact some of the coefficients become singular as the gap between the spheres goes to zero, a fact not appreciated in the work of Keh and coworkers. The lubrication result for electrophoresis along the line of centers is [45]

$$k_E \equiv \frac{U_2 - U_1}{U_2^\infty - U_1^\infty} = \frac{\dfrac{c_1^E(1+\beta)^3}{\beta^2}\lambda}{1 - \dfrac{(1+7\beta+\beta^2)}{5\beta}\lambda \ln\left[\dfrac{(1+\beta)^2}{\beta}\lambda\right] + c_2^E\dfrac{(1+\beta)^2}{\beta}\lambda} \tag{41}$$

where $\lambda = 2\delta/(a_1 + a_2)$, 2δ is the gap between the particles, $\beta = a_1/a_2$ (where a_1 is the smaller particle), and U_i^∞ is the velocity of the particle when the separation is large (given by Smoluchowski's equation). The coefficient c_1^E is 1.50553 for $\beta = 1$, 0.84548 for $\beta = 0.5$, and 0.23567 for $\beta = 0.2$. The coefficient c_2^E is 2.24 for $\beta = 1$ and 1.84 for $\beta = 0.5$ (difficult to calculate for other β). For $\beta = 1$ we know [3]

$$M_{12}^p = \frac{1 - k_E}{2} \tag{42}$$

For the case when e is perpendicular to E_∞, electrophoretic motion is still described by an analytical result, albeit more complicated [44]. A solution for N was obtained only for $\beta = 1$. The coefficients were determined using lubrication theory. For $\beta = 1$ we have $M_{12}^n = M_{21}^n$ and $N_{12} = -N_{21}$, and

$$M_{12}^n = 1 + y_{11}^a\left[-3.00\pi a\lambda^{\sqrt{2}/2-1} + I_{22}^{\text{outer}}\right] + y_{12}^a\left[3.00\pi a\lambda^{\sqrt{2}/2-1} + I_{21}^{\text{outer}}\right] \\ + y_{11}^b\left[3.00\pi a^2\lambda^{\sqrt{2}/2-1} + K_{22}^{\text{outer}}\right] + y_{12}^b\left[-3.00\pi a^2\lambda^{\sqrt{2}/2-1} + K_{21}^{\text{outer}}\right] \tag{43}$$

$$\frac{N_{12}}{L} = -y_{11}^b\left[-3.00\pi a\lambda^{\sqrt{2}/2-1} + I_{22}^{\text{outer}}\right] - y_{12}^b\left[3.00\pi a\lambda^{\sqrt{2}/2-1} + I_{21}^{\text{outer}}\right] \\ - y_{11}^c\left[3.00\pi a\lambda^{\sqrt{2}/2-1} + K_{22}^{\text{outer}}\right] + y_{12}^c\left[-3.00\pi a\lambda^{\sqrt{2}/2-1} + K_{21}^{\text{outer}}\right] \tag{44}$$

for $\lambda = \delta/a$ (2δ is the gap; a is the radius of the spheres) and

$$I_{21}^{\text{outer}} = -5.48901a, \quad I_{22}^{\text{outer}} = -8.16940a \tag{45}$$

$$K_{21}^{\text{outer}} = 11.97955a^2, \quad K_{22}^{\text{outer}} = -14.96163a^2 \tag{46}$$

$$6\pi a y_{11}^a = \frac{0.89056\left(\ln\dfrac{1}{2\lambda}\right)^2 + 5.77196\left(\ln\dfrac{1}{2\lambda}\right) + 7.06897}{\left(\ln\dfrac{1}{2\lambda}\right)^2 + 6.04250\left(\ln\dfrac{1}{2\lambda}\right) + 6.32549} \tag{47}$$

$$6\pi a y_{12}^a = \frac{0.48951\left(\ln\dfrac{1}{2\lambda}\right)^2 + 2.80545\left(\ln\dfrac{1}{2\lambda}\right) + 1.98174}{\left(\ln\dfrac{1}{2\lambda}\right)^2 + 6.04250\left(\ln\dfrac{1}{2\lambda}\right) + 6.32549} \tag{48}$$

$$6\pi a^2 y_{11}^b = \frac{0.20052\left(\ln\frac{1}{2\lambda}\right)^2 + 0.29918\left(\ln\frac{1}{2\lambda}\right) - 1.18857}{\left(\ln\frac{1}{2\lambda}\right)^2 + 6.04250\left(\ln\frac{1}{2\lambda}\right) + 6.32549} \tag{49}$$

$$6\pi a^2 y_{12}^b = \frac{-0.20052\left(\ln\frac{1}{2\lambda}\right)^2 - 1.39080\left(\ln\frac{1}{2\lambda}\right) - 0.28208}{\left(\ln\frac{1}{2\lambda}\right)^2 + 6.04250\left(\ln\frac{1}{2\lambda}\right) + 6.32549} \tag{50}$$

$$6\pi a^3 y_{11}^c = \frac{0.20052\left(\ln\frac{1}{2\lambda}\right)^2 + 4.20672\left(\ln\frac{1}{2\lambda}\right) + 6.96083}{\left(\ln\frac{1}{2\lambda}\right)^2 + 6.04250\left(\ln\frac{1}{2\lambda}\right) + 6.32549} \tag{51}$$

$$6\pi a^3 y_{12}^c = \frac{0.20052\left(\ln\frac{1}{2\lambda}\right)^2 - 0.79328\left(\ln\frac{1}{2\lambda}\right) + 0.22486}{\left(\ln\frac{1}{2\lambda}\right)^2 + 6.04250\left(\ln\frac{1}{2\lambda}\right) + 6.32549} \tag{52}$$

The $\lambda^{\sqrt{2}/2-1}$ dependencies of the coefficients are interesting and not initially intuitive; seldom do such irrational powers appear. It says that, as the particles come closer together, their angular velocities *increase*. This is opposite to the behavior of sedimentation. An intuitive explanation for this behavior is that since electrophoresis exerts a force on the diffuse charge region of the thin double layers, as the particles come closer together they can "push off" each other more efficiently.

The motivation for the above calculations was to obtain an effective angular velocity of two spheres that are *freely rotating* (i.e., close but not touching) and yet behave as a single doublet because colloidal forces acting along the line between centers hold the particles at a fixed separation (see the "freely rotating" case of Fig. 4). This would conceivably happen for spheres that are in a DLVO secondary energy minimum. The coefficient for the angular velocity of two freely rotating spheres is [3]

$$N = 1 - M_{12}^n - M_{21}^n \tag{53}$$

where N is defined by Eq. (32) with the radius a replaced by $(a + \delta)$. In addition, we wanted to know the effect on N if the ζ potential in the gap between the spheres was not uniform. We found that if on sphere 1 the ζ potential is kT/e outside the gap region but zero in the gap, while on sphere 2 the ζ potential is $-kT/e$ on the entire surface, the change in the angular velocity is less than 15%. This calculation was done for a dimensionles gap $2\delta/2a = 0.01$; the nonuniformity in zeta potential becomes more important for smaller gaps. Finally, a calculation was done to examine the effect of finite, overlapping double layers. The EDLs are still thin relative to the particle size (eg., $\kappa a = 100$), but can be large relative to the gap dimension (eg., $\kappa\delta = 3$). Under this condition the infinitesimal EDL result, given by Eqs (43)–(53), predicts the angular velocity to within 5%.

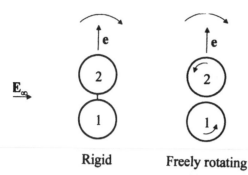

Rigid   Freely rotating

FIG. 4 Rigid and freely rotating doublets. Experimentally, one can see only the whole doublet rotating. For rigid doublets, each sphere rotates with the same angular velocity as the *e* vector. For freely rotating doublets, the spheres may rotate relative to each other, and they both rotate in the opposite direction as the *e* vector.

B. Rigid Doublets

In order to find the electrophoretic mobility of two spheres, Fair and Anderson [36] used the Lorentz reciprocal theorem and the results from Keh and Yang [41, 42] (see also Ref. 43). In the freely rotating case, the spheres can rotate force and torque free. However, for rigid doublets, the constraints are

$$F_1 + F_2 = 0, \, T_1 + T_2 = 0 \tag{54}$$

$$\Omega_1 = \Omega_2 = \Omega \equiv e \times \frac{de}{dt} \tag{55}$$

The sum of the forces or torques on a doublet must be zero since the doublet is not accelerating; however, this is not true for each particle. This constraint is most important, as it slows the rotation. The M and N coefficients from above have been calculated for rigid doublets, and the results vary slowly with the gap [3–5]. We list here only the results for two spheres in a rigid doublet configuration with no gap between them ($\delta = 0$), since the dependence on δ is weak when $\delta \ll a$:

$$U_0 = \frac{\varepsilon \zeta_1}{\eta} E_\infty + \frac{\varepsilon(\zeta_2 - \zeta_1)}{\eta} [M^p ee + M^n(I - ee)] \cdot E_\infty \tag{56}$$

$$\Omega_1 = \Omega_2 = \Omega = \frac{\varepsilon(\zeta_2 - \zeta_1)}{\zeta L} Ne \times E_\infty \tag{57}$$

where the Ms and N are given in Table 3 [36, 42, 43]; r_{01} is the distance from the center of particle 1 to the center of hydrodynamic stress (point 0) of the doublet, and U_0 is the velocity of point 0. For equal size spheres $r_{01} = a$, the midpoint between the centers of the particles, but this is not true for the case of unequal size spheres [36].

C. Estimates for Rigid and Freely Rotating Doublets

Measurements of the coefficient N can be used to probe the rigidity of a colloidal doublet because the value of N depends on whether or not the surfaces of the particles in the gap region are able to move parallel to each other. Here, we provide an approximate analysis

TABLE 3 Mobility Coefficients for Rigid Doublets[a]

$a_1/a_2 \leq 1$	M^p	M^n	N	$r_{01}/(a_1 + a_2)$
0.2	0.98238	0.7684	0.1560	0.9648
0.5	0.8429	0.3470	0.4880	0.7920
1	0.5000	0.1800	0.6400	0.5000

[a] See Eqs (56) and (57).

of the dynamics of a doublet to illustrate this dependence. The approximate analyses in this section and Section III.E emphasize that, by using simple ideas of connectivity and zero- or first-order approximations for particle interactions, one can often gain valuable qualitative and semiquantitative insights into the behavior of groups or clusters of particles without resorting to solving fully the electrokinetic equations for the entire system.

Consider the orientation of the doublet as shown in Fig. 4, with the forces and translational velocities in the direction of the electric field while the torques and angular velocities are directed perpendicular to the plane of the paper. If the particles are free to rotate such that each achieves mechanical equilibrium with the surrounding fluid, then the total force and torque on each particle must be zero:

$$F_i = 0, \; T_i = 0 \tag{58}$$

Because the double layers are very thin relative to the radius of the particles ($\kappa a_i \gg 1$), the force and torque on the particles can be calculated at the outer edge of the double layer, and hence the forces are all hydrodynamic in nature (i.e., electrical forces are zero at the edge of the EDL to first order in the electric field, although the electric field is not) [10]. If electrostatic and hydrodynamic interactions between the particles are neglected, then each particle would move (without rotation) at the velocity U_{i0} given by Eq. (1). The angular velocity of the doublet is defined by $\Omega = e \times (U_2 - U_1)/L$, so N is defined by

$$\Omega = \frac{U_2 - U_1}{L} = N \frac{U_{20} - U_{10}}{L} \tag{59}$$

Since we assume here that $U_i = U_{i0}$, then $N = 1$ in the absence of interactions. Effects of particle interactions have been determined by a "method of reflections" technique [37]. Using only the first nonzero reflections of the disturbances to the fluid velocity and electric field caused by each sphere, corrections to the velocity of each sphere are determined and then Eq. (59) gives the following for freely rotating doublets:

$$N = N_{\mathrm{fr}} = 1 + \frac{1}{2}(1 + \beta^3)\left(\frac{a_2}{L}\right)^3 + O\left(\frac{a_2}{L}\right)^6 \tag{60}$$

where L is the center-to-center separation of the particles and is essentially equal to $a_2(1 + \beta)$ for a doublet held together by colloidal forces, and $\beta = a_1/a_2$ where particle 2 is defined to be the larger particle. The above expression is the result of two interactions between the spheres; the motion of sphere 2 creates a disturbance to the fluid velocity leading to a term $(1/2)(a_2/L)^3$, while sphere 1 disturbs the electric field and contributes $(1/2)(a_1/L)^3$ to the value of N_{fr}. Note that $\Omega_i = 0$ for both spheres until terms of $O(a_2/L)^8$ are included. To the order given in Eq. (60) the two spheres move relative to each other *without rotating*; hence, their surfaces *slide* by one another unless there is a mechanical restraint. Chen and Keh [37] give the expansions for the sphere interactions to $O(a_2/L)^{12}$, from which N_{fr} could be determined. Note that $N_{\mathrm{fr}} > 1$ for the case of free rotation.

Now consider the rigid-body case. Because the two particles act as one, due to forces between them directed both along the line of centers and tangent to the surfaces, mechanical equilibrium is achieved when the *total* force and torque on the doublet are zero [36]:

$$F_1 + F_2 = 0 \tag{61}$$

$$T_1 + T_2 + \mathrm{L}\, F_2 = 0 \tag{62}$$

$$\Omega_1 = \Omega_2 = \Omega \tag{63}$$

Equation (63) is the important one, as it forces each particle to rotate with the doublet. Neglecting hydrodynamic interactions between the spheres, the force and torque on each sphere are

$$F_i = -6\pi\eta a_i(U_i - U_i^{\mathrm{f}}) \tag{64}$$

$$T_i = -8\pi\eta a_i^3(\Omega_i - \Omega_i^{\mathrm{f}}) \tag{65}$$

where U_i^{f} and Ω_i^{f} are the velocities for the freely rotating case. Combining Eqs (59)–(62) we have the following for rigid-body rotation, correct to $O(a_2/L)$ [7]:

$$N = N_{\mathrm{rb}} = \left[1 + \frac{4}{3}\left(\frac{a_2}{L}\right)^2 \frac{(1+\beta)(1+\beta^3)}{\beta}\right]^{-1} N_{\mathrm{fr}} \tag{66}$$

The above equation indicates that $N < 1$ for all separations L down to the minimum value of contact, $a_2(1 + \beta)$.

The essence of the rigid-body constraint is Eq. (63). The coupling between the two spheres *prevents their surfaces from moving tangentially relative to each other*, thus imparting an angular velocity to each of $O(a_2/L)^2$. This is why $N_{\mathrm{rb}} < 1$ for all values of separation. In addition to providing insight into the essential hydrodynamic difference between the freely rotating and rigid-body conditions for a doublet, the above analysis results in reasonably accurate approximations for N. For $L/a_2 > 1.05(1 + \beta)$, Eq. (60) is accurate to within 8%, while Eq. (66) is accurate to within 30%.

D. Slender Bodies

Particles can aggregate into chains, and the electrophoresis of these chains can be approximated by slender-body theory. An example is shown in Fig. 5. Semianalytical formulas have been derived [47–49] for the motion of long, slender particles whose charge varies along the contour. The results are valid when $\kappa b_0/\ln(\lambda) \gg 1$ where b_0 is the characteristic radius of the cross-section of the particle and λ is the ratio of contour length to $2b_0$. The Lorentz reciprocal theorem for Stokes flow was used to evaluate the translational and angular velocities of the slender body based on a distribution of stokeslets, which must be computed from hydrodynamic theory based on the geometry.

An interesting example is a straight cylinder of diameter $2b$ and length $2L$ with the slenderness parameter $\varepsilon = b/L \ll 1$. Define the first three moments of the ζ-potential distribution as

$$p_0 = \frac{1}{2}\int\limits_{-1}^{+1} \zeta\, ds \qquad p_1 = \frac{1}{2}\int\limits_{-1}^{+1} s\zeta\, ds \qquad p_2 = \frac{1}{2}\int\limits_{-1}^{+1} (3s^2 - 1)\zeta\, ds \tag{67}$$

a

b   c

FIG. 5 Chains of particles. These can be approximated as "slender bodies": (a) long chain that can be well approximated as a "slender body", (b) and (c) chains with the same set of particles, but which will have different electrophoretic velocities. Whereas the chain in (c) will move at the average mobility of the two types of spheres, the chain in (b) will move at a mobility close to that of the lighter-shaded sphere since these are the "exposed" particles.

where s is the dimensionless position along the axis ($s = \pm 1$ represents the ends of the cylinder). Note that p_0 is the area-averaged zeta potential (ζ_{ave}) on the surface. Using the formulation of Eqs (25) and (26) and the results of ref. 47, the mobilities are

$$m_{\parallel} = \zeta_{ave} + \frac{5/6}{3 - 2\ln(4\varepsilon)}p_2, \; m_{\perp} = \zeta_{ave} + \frac{5/6}{1 - 2\ln(4\varepsilon)}p_2, \; m_r = 3p_1 \tag{68}$$

where $\zeta_{ave} = p_0$, the area-average zeta potential of the cylinder. These expressions are valid when $\kappa b \gg 1$ and $\varepsilon \ll 1$. This example illustrates three interesting facets of nonuniformly charged particles in electric fields: (1) the particle rotates into alignment with its dipole moment (p_1), (2) the mobility is different along the axis compared to perpendicular to it (also note that the mobility is *greater* for the *perpendicular* motion if p_2 has the same sign as ζ_{ave}), and (3) *neutral* particles can have a finite mobility. Concerning the third observation, the local charge per surface area is proportional to ζ for small ζ, so the particle would be neutral if $\zeta_{ave} = 0$; however, a finite p_2 would lead to translation in the electric field.

The effect of hydrodynamic shielding on electrophoretic mobility can be illustrated by considering a cylinder having N segments each of which has its own zeta potential ζ_i, as shown in Fig. 6. In this example the length of each segment equals the cylinder's diameter, so the slenderness parameter ε equals N^{-1}. Note that this geometry is a reasonable approximation for a chain of spherical particles. Here, we consider a straight cylinder, but we could generalize the results to a curved chain using the basic relations derived in Ref. 47. For the straight cylinder we have

$$m_{\parallel} = \sum_{i=1}^{N} A_{\parallel}^{i}\zeta_i, \, m_{\perp} = \sum_{i=1}^{n} A_{\perp}^{i}\zeta_i, \; m_r = \sum_{i=1}^{N} A_r^{i}\zeta_i \tag{69}$$

In the case of *translational* motion parallel and perpendicular to the axis of the cylinder, $\sum(A^i) = 1$ because Smoluchowski's equation must hold if all segments have the same zeta potential, and $A^{N+1-i} = A^i$ because of symmetry. For rotation we have $\sum(A_r^i) = 0$ and $A_r^{N+1-i} = -A_r^i$. The general formulas for the coefficients are given below:

$$A_{\parallel}^{i} = \{2[(1 - i\varepsilon)\ln(1 - i\varepsilon) - \ln[1 - (i - 1)\varepsilon] - i\varepsilon\ln(4i\varepsilon)] +$$
$$\varepsilon[6 - 4\ln(2\varepsilon) - 2(1 - i)\ln[4\varepsilon(i - 1)(1 - (i - 1)\varepsilon)]]\}\{6 - 4\ln(4\varepsilon)\}^{-1} \tag{70}$$

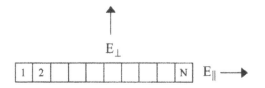

CONFIGURATION	$m_\parallel$	$m_\perp$	m_r
	0	0	-1.165
	0	0	-0.314
	0.105	0.180	-0.203
	-0.105	-0.180	0.203

FIG. 6 Modeling a straight chain of equal size particles by a cylinder of $N = 10$ segments. The darker spheres have a zeta potential of $+\zeta_0$ and the lighter spheres have $-\zeta_0$. The mobility coefficients (m) are expressed in units of ζ_0.

$$A_\perp^i = \{2[(1 - i\varepsilon)\ln(1 - i\varepsilon) - \ln[1 - (i - 1)\varepsilon] - i\varepsilon\ln(4i\varepsilon)] +$$
$$\varepsilon[6 - 4\ln(2\varepsilon) - 2(1 - i)\ln[4\varepsilon(i - 1)(1 - (i - 1)\varepsilon)] - 4]\}[2 - 4\ln(4\varepsilon)]^{-1} \tag{71}$$

$$A_r^i = \frac{C}{2}[(2\varepsilon i - 1)^2 - (2\varepsilon(i - 1) - 1)^2] + \frac{D}{4}\{-2\varepsilon i - 1)^2 + (2\varepsilon(i - 1) - 1)^2 +$$
$$((2\varepsilon i - 1)^2 - 1)\ln[4\varepsilon i(1 - \varepsilon i)] - [(2\varepsilon(i - 1) - 1)^2 - 1]\ln[4\varepsilon(i - 1)(1 - \varepsilon(i - 1))]\} \tag{72}$$

$$C = \left(\frac{9}{2}\right)\frac{1 + 2\ln(2\varepsilon)}{-5 + 6\ln(4\varepsilon)}, \quad D = \left(\frac{9}{2}\right)\frac{1}{-5 + 6\ln(4\varepsilon)} \tag{73}$$

Note that Eq. (72) corrects a typographical error in Eq. (1.3) of Ref. 48.

As an example consider a chain of $N = 10$ segments ($= \varepsilon^{-1}$). The A coefficients are given in Table 4. Consider the case where five of the spheres have $\zeta = +\zeta_0$ and five have $\zeta = -\zeta_0$; thus, the chain is neutral. Figure 6 shows that the relative placement of the charged spheres has a significant effect on the electrophoretic mobility; furthermore, this is a case where a *neutral* chain has a significant electrophoretic mobility. The ends of the chain ($i = 1$ and 10) are weighted the most because they are hydrodynamically exposed.

Han and Yang [50] extended the slender body approach [47] to rod-like particles with low ζ potentials but arbitrary κb. The results were then used to compute the orientation distribution of rod-like particles during capillary electrophoresis. The analysis uses singularity solutions of the linearized Poisson–Boltzmann equation as the base set of the electrostatic part of the problem. They considered two cases that allow analytical expres-

TABLE 4 A coefficients Defined in Eq. (69) and Computed
from Eqs (70)–(73) with $\varepsilon = 0.1 = N^{-1}$ and $N = 10^a$

Position of segment i	$A_{\parallel}^i$	$A_{\perp}^i$	A_r^i
1	0.1280	0.1478	−0.2492
2	0.1016	0.1028	−0.1544
3	0.0934	0.0888	−0.1016
4	0.0893	0.0818	−0.0584
5	0.0876	0.0788	−0.0191

[a] Note that $A^{11-i} = A^i$ for the translational coefficients and $A_r^{11-i} = -A_r^i$ for rotation.

sions for rotational mobility of the particle. As expected the result for an infinitesimal double layer coincides with Eq. (67). For very thick double layers relative to the radius of the fiber, the factor 1/2 in Eq. (67) is replaced by 1/4. Similar expressions were obtained by Chen and Koch [51]

E. Rigid Networks of Particles

Insight into the effects of variations in ζ among spherical particles in a cluster can be gained simply by applying basic concepts of mechanics and neglecting disturbances in the velocity and electric fields caused by each particle. Consider a group of M spherical particles in an unbounded fluid, each of which has zeta potential ζ_i and radius a_i. The electric field far from the cluster is E_∞. The double layers on each particle are assumed to be thin and unpolarized; thus, in the absence of interactions between the particles, each would move according to the Smoluchowski equation with $\zeta = \zeta_i$ for each.

Now consider the case where the M spherical particles are *connected* by rigid rods of negligible hydrodynamic resistance, as shown in Fig. 4. The cluster of M particles moves as a rigid body with translational velocity U_o (i.e., the velocity of the center "o" of the group) and angular velocity Ω. Using Stokes' law (remembering that hydrodynamic and electrostatic interactions among the particles are neglected) we calculate the net force on each particle to be

$$F_i = -6\pi\eta a_i [U_o + \Omega \times r_{oi} - U_{i0}] \tag{74}$$

where U_{i0} is given by Eq. (1), and r_{oi} is the vector from the center of the cluster (point 'o') to the center of particle i. Point 'o' is determined such that there is no coupling between the translational and angular velocites [52]; since hydrodynamic interactions are neglected in this analysis, point 'o' is given by

$$\sum_{i=1}^{M} a_i r_{oi} = 0 \tag{75}$$

The velocity of the cluster is determined by summing all the forces to zero: $\sum F_i = 0$, so that

$$U_0 = \frac{\varepsilon}{\eta} \frac{\sum a_i \zeta_i}{\sum a_i} E_\infty \tag{76}$$

Consider a cluster that is "neutral." For low zeta potentials (about $2kT/e$ or less), the charge per area on the particle surface is proportional to ζ_i; thus, a neutral cluster would have $\sum a_i^2 \zeta_i = 0$. However, from (76) we see that U_0 is not necessarily zero unless all the particles are neutral. So here we have another example of how a neutral body (the M connected particles) has a finite electrophoretic mobility even though it is neutral. Allowing for hydrodynamic and electrostatic interactions among the particles does not change the essence of this result.

The connected cluster of M particles will also rotate in an electric field. The angular velocity is found by requiring the hydrodynamic torque (T) on the cluster to be zero:

$$T = \sum_{i=1}^{M} [r_{oi} \times F_i - 8\pi\eta a_i^3 \Omega] = 0 \tag{77}$$

Because the cluster is force free, the torque is independent of the placement of the origin point 'o'. From the above equations we have the following for the angular velocity:

$$\Omega \cdot \sum_{i=1}^{M} a_i \left[r_{oi}^2 I - r_{oi} r_{oi} + \frac{4}{3} a_i^2 I \right] = \frac{\varepsilon}{\eta} \sum_{i=1}^{M} a_i \zeta_i r_{oi} \times E_\infty \tag{78}$$

Note that "alignment" of the cluster with the electric field occurs when the right-hand size of Eq. (78) equals zero. If all the spheres have the same zeta potential, then the right-hand side automatically equals zero because of Eq. (75) so that $\Omega = 0$ for any orientation.

As an example, consider $M = 2$ as shown in Fig. 4; L equals the center-to-center separation and $\beta = a_1/a_2 \leq 1$ (i.e., particle '2' is always the larger sphere). The unit vector e gives the direction of the line of centers:

$$e = \frac{r_{o2} - r_{o1}}{|r_{o2} - r_{o1}|} \tag{79}$$

The center (o) is defined along e according to Eq. (75); thus, $r_{o2} = \beta L/(1 + \beta)$ and $r_{o1} = L - r_{o2}$. The angular velocity of the vector e is then given by

$$\Omega = N \frac{\varepsilon}{\eta L} (\zeta_2 - \zeta_1) e \times E_\infty \tag{80}$$

where the dimensionless coefficient N is given by Eq. (66) with $N_{fr} = 1$ (which is the case if hydrodynamic and electrostatic interactions between the particles are neglected, as assumed in this section). Note that no rotation occurs about the e vector for this axisymmetric case.

The above discussion neglects hydrodynamic and electrostatic interactions among the particles, except for the rigid connectedness that makes the cluster a rigid body. The disturbance to the electric field caused by particle j and felt by particle i goes as r_{ij}^{-3}, which is a weak interaction. On the other hand, the leading interaction of the velocity fields is of order r_{ij}^{-1}, so neglect of the hydrodynamic interactions is a more serious assumption. However, inclusion of these interactions would not change the basic structure of the above equations, and the results of the simplified analysis presented above are still conceptually correct and semiquantitative.

F. Sphere Approaching a Flat Wall

Electrophoretic deposition is a process that is used to produce dense monolayer or multi-layer films of particles from nanometer to micrometer sizes [53]. An electric field is used to transport the particles to the surface of the electrode, which is usually a thin metallic film on a solid substrate. The particles become loosely deposited such that they remain near the electrode surface, but are mobile in the plane parallel to the electrode. Two-dimensional particle aggregation occurs during continued application of the field [54–56]. A theory based on electroosmotic convection about the deposited particles has been validated for the aggregation of two [57, 58] and three particles [59] under direct current (dc) electric field conditions.

Consider the electrophoresis of a spherical particle *toward* the electrode (a constant potential surface). The effect of the wall on the electrophoretic velocity is determined by the dimensionless separation $\lambda = a/z$ where z is the distance of the center of the particle from the wall. The following equation is valid over a broad range of separations [60]:

$$\frac{U}{U_0} = F^{(e)}(\lambda) = 1 - \frac{5\lambda^3}{8} + \frac{\lambda^5}{4} - \frac{5\lambda^6}{8} + O(\lambda^8) \tag{81}$$

where U_0 is given by Smoluchowski's equation (1). The mobility as λ approaches unity has also been calculated [61, 62].

Particles often experience a gravitational force as well. Using superscript (s) for sedimentation, we have the following for the wall effect on settling of the particle toward the electrode [59].

$$\frac{U^{(s)}}{U_0^{(s)}} = F^{(s)}(\lambda) = \left[\frac{1 - 0.700\lambda}{1 + 0.554\lambda}\right] F^{(e)}(\lambda); \; U_0^{(s)} = \frac{2}{9} \frac{a^2 \Delta \rho g}{\eta} \tag{82}$$

The wall slows the sedimentation more than it does electrophoresis for a given separation, and as the particle approaches contact, the hindrance is about a factor of 5 greater on sedimentation than on electrophoresis. This means that the position z of the particle can be controlled within limits by applying an electric field to counterbalance gravitational settling.

If a spherical particle is stationary near a surface, say, due to colloidal forces that prevent it from sticking to the wall, then the particle acts as an electro-osmotic pump as shown in Fig. 7. In this figure the particle has a negative ζ potential, and the electric field acts on the positive space charge of the double layer to force fluid upward on the sides of the particle. To conserve mass the fluid must flow parallel to the surface toward the particle. The streamlines shown in the figure were computed by solving the electrostatic equation for the electrical potential and then the Stokes' equations for the velocity field (or the stream function of the velocity field) [59]. The aggregation of particles under dc fields is primarily due to entrainment of neighboring particles in the electro-osmotic flow generated by each deposited particle [57, 58]. The strength of the electro-osmotic flow at any distance from the deposited particle depends on the particle size, so larger particles entrain smaller particles, and clusters of particles entrain other particles and clusters [56,59]. In this way the deposited particles are consolidated on the surface, leaving bare areas for other particles to deposit. The film is stabilized at the end of some time period, once a dense layer (or multilayer) is formed, by adjusting experimental conditions and forcing the particles to adhere to the surface.

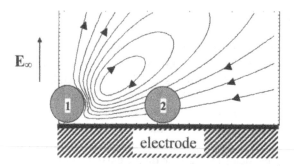

FIG. 7 Streamlines generated by electro-osmosis about particle 1 [59]. In this example, the particles are negatively charged and the electrode is positive. There is a small gap of fluid between the particle and the electrode, thus imparting two-dimensional mobility to the particles. The electric field generates an electrokinetic slip velocity about particle 1, pushing fluid upward along that particle's surface and drawing fluid laterally in toward the particle to conserve mass. A neighboring particle (2) would be convected toward particle 1 and vice versa, as particle 2 also generates an electro-osmotic flow field.

IV. EXPERIMENTAL RESULTS

Few experimental results exist concerning the electrophoresis of complex and interacting particles. In many cases, even though the experiments were on complex particles, the analysis was done as if the particles were uniformly charged, giving an "average" charge. The first experiments done to study the electrophoresis of colloidal doublets were by Fair and Anderson [2]. They found results quantitatively and qualitatively consistent with their theory for rigid doublets [36]. In these experiments they observed both the translational and rotational motion of the doublets; this appears to be the first reported rotation of dipolar particles by electrophoresis (versus rotation by dielectric polarization). The results for the translational mobility of colloidal doublets compared to within a few per cent of theoretical predictions for mobility parallel to the axis of the doublet. The mobility of the doublet perpendicular to its axis was difficult to measure since the doublet rotated.

Fair and Anderson's measurements of the angular velocity of colloidal doublets were qualitatively in agreement with theory. They allowed two particles with different zeta potentials [1.1-μm carboxylate-coated polystyrene (PS) latex and 2.5-μm amidine-coated PS latex] to aggregate by Brownian motion. The resulting doublets had a dipole moment of zeta potential, and therefore rotated in an applied electric field. Due to the effects of Brownian rotation, they were not able to obtain quantitative agreement with theory. The use of video microscopy enabled Velegol et al. [3] to confirm the theory precisely. An example of doublet rotation trajectories is shown in Fig. 8.

It has also been shown that clusters of particles that have an area-average ζ potential of nearly zero can have a finite electrophoretic mobility, as discussed previously [32]. The particle trajectories were viewed with a video camera on a MKII Microelectrophoresis Apparatus (Rank Brothers). Two types of particles were used: a carboxylate-coated PS latex (4.42 μm) and an amidine-coated PS latex (2.50 μm). At 10 mM NaCl, the carboxylate particle had $\zeta = -47$ mV and the amidine particle had $\zeta = +53$ mV. Clusters of three or more particles had measured mobilities different from that which would be expected based on area-averaged zeta potential. For example, a cluster of three amidines and one carboxylate that was expected from an area-averaged zeta potential to have a mobility of

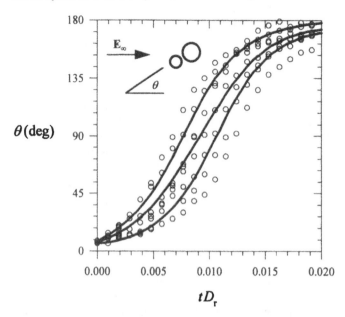

FIG. 8 Typical rotation data for a single doublet. The data are for a doublet formed from an amidine particle ($2a = 2.51 \pm 0.12$ μm, $\zeta = +74 \pm 11$ mV) and a sulfate particle ($2a = 2.75 \pm 0.09$ μm, $\zeta = -113 \pm 5$ mV). For this doublet $N = 0.57 \pm 0.04$, whereas the theoretical value for a rigid doublet with this size ratio is $N = 0.64^{+0.00}_{-0.02}$. The center line is $\cos^{-1}\langle \cos\theta(t)\rangle$, obtained by solving the orientation probability equation, and the two outer lines represent approximately the region of one standard deviation; i.e., they represent $\cos^{-1}[\langle\cos\theta(t)\rangle \pm \sigma_{\cos}]$ where $\sigma^2_{\cos} = \langle\cos^2\theta(t)\rangle - [\langle\cos\theta(t)\rangle]^2$. Roughly two-thirds of the data should (and do) fall between these lines. (From Ref. 3. Reproduced with permission from *Langmuir*.)

0.169 μm/cm / V·s in fact had a mobility of $+2.312$ μm cm / V·s. This reinforces the theoretical calculations presented here showing that the area-averaged ζ potential is not the relevant property for the electrophoretic mobility of complex particles, rather it is a hydrodynamic-based average of the ζ potential that determines the motion.

V. SUMMARY AND FUTURE DIRECTIONS

In this chapter we have demonstrated that nonuniformly charged particles and clusters of particles behave in ways unpredictable from simply applying Smoluchowki's equation assuming an area average of the ζ potential. Geometry and hydrodynamics are important to the electrophoretic motion of complex bodies. The rich electrokinetic behavior of complex particles in electric fields should be viewed as an opportunity, and the curious dynamics such as rotation of dipolar particles and translation of neutral particles or a connected cluster of particles could be exploited in separations and synthesis of colloidal material.

Electrophoretic theory for nonuniformly charged or interacting particles is developed to the point where it can be used to test various parameters important in understanding colloidal forces and stability. Historically, this has meant only the zeta potential, but more recently "differential electrophoresis" has been used to measure the normal and tangential forces between colloidal particles [5]. Initial results have shown large attractive forces between polymer colloids that have no surfactant on them. Tangential forces, which are not predicted by the classical theory for colloidal forces, have also been found [6]. These tangential forces, measured between PS latex particles and silica particles, showed interesting time-dependent effects over short (but nonBrownian) time scales.

Recent measurements show that nominally "homogeneous" colloidal particles can actually have a nonuniform charge distribution on them [25, 63], thus supporting the models developed in Section II.A. Since many researchers have shown that charge non-uniformity on colloidal particles can be a cominant factor in interparticle forces [26, 27, 64–66], these experiments provide one explanation for why the classical DLVO theory [67] of colloidal forces often fails to predict the behavior of colloidal suspensions. An impor-tant consequence of this work could be the identification of particles with a small charge nonuniformity. Thus, just as translational electrophoresis has been used for years to measure the "average" zeta potential, now rotational electrophoresis can be used to measure the standard deviation of zeta potential on individual particles or even between particles of the same nominal specifications. By identifying manufacturing techniques that produce more homogeneous particles, downstream manufacturing design and costs could be greatly reduced – even if initial manufacturing costs were slightly higher. In addition, in some cases a certain charge distribution might be desirable, and electrophoretic rotation experiments could determine when this exists.

ACKNOWLEDGMENTS

The authors thank Yashodhara Pawar and Lena Asavathiratham for work on the calcula-tions involving a spheroid model for clay particles. D.V. thanks the National Science Foundation for CAREER grant CTS 9984443 and the Petroleum Research Fund for G Grant 35400. J.L.A. acknowledges support under NSF grant CTS9814064.

REFERENCES

1. RJ Hunter. Zeta Potential in Colloid Science. New York: Academic Press, 1981.
2. MC Fair, JL Anderson. Langmuir, 8: 2850, 1992.
3. D Velegol, JL Anderson, S Garoff. Langmuir 12: 675–685, 1996.
4. D Velegol, JL Anderson, S Garoff. Langmuir 12: 4103, 1996.
5. JL Anderson, D Velegol, S Garoff. Langmuir 16: 3372, 2000.
6. D Velegol, S Catana, JL Anderson, S Garoff. Phys Rev Lett 83: 1243, 1999.
7. RW O'Brien, LR White. J Chem Soc, Faraday Trans 2 74: 1607, 1978.
8. DA Saville, Ann Rev Fluid Mech 9: 321, 1980.
9. M Teubner. J Chem Phys 76: 5564, 1982.
10. JL Anderson. Ann Rev Fluid Mech 21: 61, 1989.
11. M Smoluchowski. In: W Graetz, ed. Handbuch der Elektrizitat und des Magnetismus. Leipzig: Barth, 1921.
12. FA Morrison Jr. J Colloid Int Sci, 34: 210, 1970.
13. J Blawzdziewicz, V Cristini, M Loewenberg. J Colloid Int Sci 211: 355, 1999.
14. DA Edwards, H Brenner, DT Wasan. Interfacial Transport Processes and Rheology. Boston: Butterworth-Heinemann, 1991.
15. JL Anderson. J Colloid Int Sci 105: 45, 1985.
16. Y. Solomentsev, JL Anderson. Ind Eng Chem Res 34: 3231, 1995.

17. E Hückel. Phys Zeitschr 25: 204, 1924.
18. DC Henry. Proc Roy Soc Lond Ser A 133: 106, 1931.
19. BJ Yoon, S Kim. J Colloid Int Sci. 128: 275, 1989.
20. SS Dukhin, BV Derjaguin. In: E Matijevic ed. Surface and Colloid Science. Vol. 7. New York: Wiley, 1974.
21. PH Wiersema, AL Loeb, JThG Overbeek. J Colloid Int Sci 22: 78, 1966.
22. L Suresh, JY Walz. J Colloid Int Sci 183: 199, 1996.
23. HJ Keh, FR Yang. J Colloid Int Sci 158: 199, 1993.
24. AA Shugai, SL Carnie, DYC Chan, JL Anderson. J Colloid Int Sci 191: 357, 1997.
25. D Velegol, JD Feick, LR Collins, J Colloid Int Sci 230:114, 2000.
26. WJC Holt, DYC Chan. Langmuir 13: 1577, 1997.
27. JM Stankovich. Electrical Double Layer Interactions Computed Using the Poisson–Boltzmann Theory. PhD thesis, University of Melbourne, Australia. 1998.
28. BJ Yoon. J Colloid Int Sci 142: 575, 1991.
29. MC Fair, JL Anderson. J Colloid Int Sci 127: 388, 1989.
30. K-S Chae, AM Lenhoff. Biophys J 68: 1120, 1995.
31. WB Jepson. Phil Trans Roy Soc Lond 311: 191, 1984.
32. Y Pawar. Electrophoresis of Heterogeneously-Charged Colloidal Particles. PhD thesis, Carnegie Mellon University, 1993.
33. A Weiss, J Russow. Proceedings of 203rd International Clay Conference. (Th Rosenquist and P Graff-Peterson, eds.), Pergamon Press, Oxford (1963).
34. J Happel, H Brenner. Low Reynolds Number Hydrodynamics. Boston: Kluwer, 1991.
35. A Nir, A Acrivos. J Fluid Mech 59: 209, 1973.
36. MC Fair, JL Anderson. Int J Multiphase Flow 16: 663, 1990. (Corrigenda in Int J Multiphase Flow 16: 1131, 1990.)
37. SB Chen, HJ Keh. AIChE J 34: 1075, 1988.
38. SC Nichols, M Loewenberg, RH Davis. J Colloid Int Sci 176: 342, 1995.
39. HJ Keh, SB Chen. J Colloid Int Sci 130: 542, 1989.
40. HJ Keh, SB Chen. J Colloid Int Sci 130: 556, 1989.
41. HJ Keh, FR Yang. J Colloid Int Sci 139: 105, 1990.
42. HJ Keh, FR Yang. J Colloid Int Sci 145: 362, 1991.
43. S Zeng, AZ Zinchenko, RH Davis. J Colloid Int Sci 209: 282, 1999.
44. D Velegol, Y Solomentsev, JL Anderson, SL Carnie. Colloids Surfaces A 140: 59, 1998.
45. M Loewenberg, RH Davis. J Fluid Mech 288: 103, 1995.
46. Y Solomentsev, D Velegol, JL Anderson. Phys Fluids 9: 1209, 1997.
47. Y Solomentsev, JL Anderson. J Fluid Mech 279: 197, 1994.
48. JL Anderson, Yu Solomentsev. In: KS Schmitz, ed. Macro-Ion Characterization: From Dilute Solutions to Complex Fluids. Am Chem Soc Symp Ser, no. 548, 1994.
49. JL Anderson. J Electrostat 34: 189, 1993.
50. SP Han, S-M Yang. J Colloid Interface Sci 177: 132, 1996.
51. SB Chen, DL Koch. J Colloid Int Sci 180: 466, 1996.
52. J Happel, H. Brenner, Low Reynolds Number Hydrodynamics. 2nd ed. Leyden: Noordhoff, 1973.
53. P Sarkar PS Nicholson. J Am Ceram Soc 79: 1987, 1996.
54. M Giersig, P Mulvaney. J Phys Chem 97: 6334, 1993; Langmuir 9: 3408, 1993.
55. M Trau, DA Saville, IA Aksay. Science, 272: 706, 1996; Langmuir 13: 6375, 1997.
56. M Bohmer. Langmuir 12: 5747, 1997.
57. SA Guelcher, Y Solomentsev, JL Anderson. Powder Technol. 110: 90, 2000.
58. Y Solomentsev, SA Guelcher, M Bevan, JL Anderson. Langmuir 16: 9208, 2000.
59. Y Solomentsev, M Bohmer, JL Anderson. Langmuir 13: 6058, 1997.
60. H Keh, JL Anderson. J Fluid Mech 153: 417, 1985.
61. HJ Keh, LC Lien. J Fluid Mech 224: 305, 1991.
62. HJ Keh, JS Jan. J Colloid Int Sci 183: 458, 1996.

63. JD Feick, D Velegol. "Measurements of charge nonuniformity on polystyrene latex particles," submitted (2001).
64. J Czarnecki. Adv Colloid Int Sci 24: 283, 1986.
65. SJ Miklavic, DYC Chan, LR White, TW Healy. J Phys Chem 98: 9022, 1994.
66. ML Grant, DA Saville. J Colloid Int Sci 171: 35, 1995.
67. WB Russel, DA Saville, WR Schowalter. Colloidal Dispersions. New York: Cambridge University Press, 1989 (with corrections 1991).

7
Electrophoresis Theory of Nonspherical Particles

JAE YOUNG KIM and BYUNG JUN YOON Pohang University of Science and Technology, Pohang, Korea

I. INTRODUCTION

The electrokinetic or zeta potential of a charged particle is commonly determined by using electrophoresis measurements. In capillary electrophoresis the velocity of a charged particle is measured either by direct observation through a microscope or by using other optical techniques such as laser light scattering. The measured velocity of the particle comprises both electrophoretic and nonelectrophoretic contributions. The electrophoretic contribution is the part that depends linearly on the electrokinetic potential of the particle and the applied electric field. The nonelectrophoretic contribution is attributable to several factors such as electro-osmotic flow of the surrounding fluid and other nonelectrostatic forces applied to the particle during experiments.

To determine the electrokinetic potential of a charged particle, one must isolate the electrophoretic contribution from the observed particle velocity. The electrokinetic potential is then estimated using an appropriate relationship between the electrokinetic potential and the electrophoretic velocity of the particle. The isolation of the electrophoretic contribution and the estimation of the electrokinetic potential therefrom both require mathematical model analyses for the motion of a charged particle in an electrophoresis experiment. The mathematical model consists of a set of partial differential equations: the conservation equations for mass, momentum, and ionic species, and the Poisson equation. When the mathematical model is simple, the governing differential equations can be solved analytically and a simple explicit relationship between the electrokinetic potential and the electrophoretic mobility can be derived. The most famous model is due to Henry [1], who considered spherical particles with small and uniform electrokinetic potentials. As the particle shape or the electrical surface condition becomes more complicated, the analysis becomes more involved and explicit relationships between the electrokinetic potential and the electrophoretic mobility may not become available.

Electrophoresis measurement furnishes us the value of the electrokinetic potential as a single parameter that characterizes the electrical property of the particle. Although we may interpret the furnished value as one of the lumped electrical properties of the particle, it is worth while to explore possible connections between the electrophoretic measurement data and the surface charge or potential distribution of the particle. For homogeneous spherical particles the electrical surface condition may be uniform so that a single parameter suffices to represent their electrical properties. On the other hand, for nonspherical particles the electrical surface condition is not uniform and the electrical property must assume the form of surface distribution. In electrophoresis analysis for such nonspherical particles it is natural to ask the following questions. First, how do we interpret physically the meaning of the lumped electrokinetic potential obtained from the electrophoretic measurement? Second, is it possible to determine the electrokinetic potential distribution? For spherical particles with arbitrary electrokinetic potential distributions the answers to these questions are partly available [2, 3]. For Brownian spheres with small electrokinetic potentials it has shown that the lumped value from the electrophoresis measurement is nothing but an area average (or monopole moment) of the electrokinetic potential distribution. In addition, the dipole moment induces the electrophoretic rotation and the quadrupole moment contributes to the electrophoretic translation. When spherical particles are nonBrownian, we thus may be able to determine the dipole and quadrupole moments of the electrokinetic potential distribution by measuring the orientation and the rotational velocity of the sphere.

When an electric field is applied to a suspension of charged nonspherical particles, each particle experiences electrostatic, hydrodynamic, gravitational, and Brownian forces. Since each of these forces affects the dynamics of the particle in a very complicated way, the determination and proper interpretation of the electrokinetic potential for nonspherical particles is a challenging task. In particular, a continuous change in the particle orientation during the course of the electrophoresis experiment complicates the analysis. During the electrophoresis of nonspherical particles the particle orientation is affected by both electrophoretic and nonelectrophoretic contributions. Therefore, the isolation of the electrophoretic contribution and the determination of the electrokinetic potential require more careful analyses for nonspherical particles.

Previous studies on the electrophoresis theory for nonspherical particles are mostly concerned with cylindrical and spheroidal particles. Henry's work [1] for infinite cylinders with small and uniform electrokinetic potentials was later extended by including double-layer relaxation [4] and by considering finite cylinders [5]. Electrophoresis of spheroidal particles with small and uniform electrokinetic potentials has been analyzed under various conditions: in quiescent fluid [6], with gravitational settling [7], and in electro-osmotic bulk flow [8]. Electrophoresis of disks with small and uniform electrokinetic potentials has also been analyzed [9]. The aforementioned works all assume that the double-layer thickness is arbitrary. When the double layer is very thin compared to the particle size, the electrophoresis analysis may become more tractable. Utilizing the assumption of thin double layers, electrophoretic motion has been analyzed for the following systems: a spheroid with a polarized double layer [10, 11], a spheroid with a nonuniform electrokinetic potential [12], a dumbbell-like particle which consists of two spheres with unequal electrokinetic potentials [14], and a spheroid in nonhomogeneous electric field [15]. When the particle shape is highly irregular, the set of governing partial differential equations must be solved numerically. For this purpose, boundary element methods have been developed and utilized for studying the electrophoretic mobility of biomolecules [16–20].

The outline of this chapter is as follows. Section II summarizes the general theory for the dynamics of charged dielectric particles in electrophoresis experiments. By following the general scheme developed by Teubner [21], force and torque balance equations are obtained to derive the relationship between the velocity and the electrical condition of the particle. The effects of external flow and gravitational settling are also included in the balance equations. For the further development of the theory we introduce the assumption that the double-layer potentials are small. Sections III and IV deal with the applications of the theory developed in Section II. Section III discusses the electrophoresis of slightly deformed spheres. Extending the earlier work for spheres [3], we determine the electro-phoretic mobility tensors of slightly deformed spheres in terms of the multipole moments of the shape function and the electrokinetic potential distribution of the particle. Section IV deals with the electrophoresis of spheroidal particles, including electro-osmotic bulk flow and gravitational settling. By performing the trajectory analyses for spheroidal par-ticles under such circumstances the effects of nonelectrophoretic contributions on the apparent motion of the particle are discussed. Section V concludes this chapter by sum-marizing major findings and further comments.

II. THEORY

A. Force and Torque Balances

Consider a charged dielectric particle undergoing a rigid-body motion through an electro-lyte solution in a capillary under the influence of an electric field. The translational and rotational motion of the particle is governed by Newton's laws of motion. The force and torque exerted on the particle consist of hydrodynamic, electrostatic, gravitational, and stochastic Brownian contributions. Based on the quasisteady-state approximation, Brownian force and torque are commonly excluded in the balance equations for the electrophoresis analysis. However, the Brownian rotation must be properly incorporated in the determination of the orientation distribution for nonspherical particles. Neglecting the particle inertia, the force and torque balance equations are given by

$$0 = \mathbf{F}^H + \mathbf{F}^E + (m_p - m_f)\mathbf{g} \tag{1a}$$

$$0 = \mathbf{T}^H + \mathbf{T}^E + (m_p\mathbf{x}_m - m_f\mathbf{x}_b) \times \mathbf{g} \tag{1b}$$

Here, the superscripts H and E denote hydrodynamic and electrostatic contributions, respectively. The last term in the force balance equation is the gravity force on the particle, where $m_p - m_f$ denotes the difference between the masses of the particle and displaced fluid. In the torque balance equation the last term appears only when the center of mass $\mathbf{x}_m$ and the center of buoyancy $\mathbf{x}_b$ are different. The hydrodynamic force and torque are defined by

$$\mathbf{F}^H = \int_S (\sigma \cdot \mathbf{n})dS \tag{2a}$$

$$\mathbf{T}^H = \int_S (\mathbf{x} - \mathbf{x}_c) \times (\sigma \cdot \mathbf{n})dS \tag{2b}$$

and the electrostatic force and torque are defined by

$$\mathbf{F}^E = \int_S (\sigma^M \cdot \mathbf{n})dS, \tag{3a}$$

$$\mathbf{T}^E = \int_S (\mathbf{x} - \mathbf{x}_c) \times (\sigma^M \cdot \mathbf{n})dS \tag{3b}$$

Here, $\mathbf{x}_c$ denotes the hydrodynamic center of the particle, σ the hydrodynamic stress tensor, and σ^M the Maxwell stress tensor. The integration is over the particle surface S, for which the outward unit normal vector is denoted by $\mathbf{n}$.

B. Governing Equations

Since the hydrodynamic condition of the electrolyte solution around a charged particle affects the electrostatic condition of the solution, the hydrodynamic and Maxwell stresses cannot be determined independently. The governing equations for the coupled hydrodynamics and electrostatics are the Stokes equations:

$$-\nabla p + \mu \nabla^2 \mathbf{u} + \rho \mathbf{E} = 0, \quad \nabla \cdot \mathbf{u} = 0 \tag{4a, b}$$

for the velocity $\mathbf{u}$ and the pressure p, and the Poisson equation:

$$\nabla^2 \Psi = -\frac{\rho}{\epsilon} \tag{5}$$

for the electric field $\mathbf{E} = -\nabla \Psi$ and the charge density ρ. Here, ϵ is the dielectric constant of the electrolyte solution. The charge density ρ of the electrolyte solution is given by

$$\rho = e \sum_i z_i n_i \tag{6}$$

where e is the elementary charge, z_i and n_i are valence and number density of the ith ionic species. The number density n_i is governed by the conservation equation:

$$\nabla \cdot \mathbf{J}_i = 0 \tag{7}$$

$$\mathbf{J}_i = n_i \mathbf{u} - ez_i n_i w_i \nabla \Psi - kT w_i \nabla n_i \tag{8}$$

in which the ion flux $\mathbf{J}_i$ is governed by bulk flow, electromigration, and diffusion. The symbol w_i is the ion mobility and kT is the Boltzmann energy. For the Stokes equations the boundary condition at the particle surface is the condition of the rigid-body motion, and the fluid velocity approaches an undisturbed electro-osmotic flow field, $\mathbf{V}$, far from the particle:

$$\mathbf{u}|_S = \mathbf{U} + \mathbf{\Omega} \times (\mathbf{x} - \mathbf{x}_c) \tag{9a}$$

$$\mathbf{u}|_\infty = \mathbf{V} \tag{9b}$$

For the Poisson equation the boundary condition at the particle surface is the distribution of the surface potential Ψ_S, and the potential approaches an undisturbed linear field far from the particle:

$$\Psi|_S = \Psi_S(\mathbf{x}_S) \tag{10a}$$

$$\Psi|_\infty = -\mathbf{E}^\infty \cdot \mathbf{x} \tag{10b}$$

The hydrodynamic stress σ and the Maxwell stress σ^M are defined by

$$\sigma = -p\delta + \mu(\nabla \mathbf{u} + \nabla \mathbf{u}^\dagger) \tag{11}$$

$$\sigma^M = \epsilon \mathbf{E}\mathbf{E} - \frac{1}{2}\epsilon E^2 \delta \tag{12}$$

Here, δ is a dyadic unit and μ is the viscosity of the fluid.

C. Hydrodynamic Force and Torque

Hydrodynamic force and torque can be conveniently determined after decomposing Eqs (4a, b) and (9a, b) into two subproblems:

$$-\nabla p_1 + \mu \nabla^2 \mathbf{u}_1 = 0, \quad \nabla \cdot \mathbf{u}_1 = 0 \tag{13a, b}$$

$$\mathbf{u}_1|_S = \mathbf{U} + \mathbf{\Omega} \times (\mathbf{x} - \mathbf{x}_c), \quad \mathbf{u}_1|_\infty = \mathbf{V} \tag{14a, b}$$

$$-\nabla p_2 + \mu \nabla^2 \mathbf{u}_2 + \rho \mathbf{E} = 0, \quad \nabla \cdot \mathbf{u}_2 = 0 \tag{15a, b}$$

$$\mathbf{u}_2|_S = 0, \quad \mathbf{u}_2|_\infty = 0 \tag{16a, b}$$

The first problem is a classical Stokes problem for a uncharged particle undergoing rigid-body motion in an arbitrary flow field. Various solution methods are available for several classes of nonspherical particles [22, 23]. The resulting hydrodynamic force and torque exerted on the particle depend linearly on $\mathbf{U}$, $\mathbf{\Omega}$, and $\mathbf{V}$, so that

$$\mathbf{F}_1^H = \mathbf{F}_1^H(\mathbf{U}, \mathbf{\Omega}, \mathbf{V}) \tag{17a}$$

$$\mathbf{T}_1^H = \mathbf{T}_1^H(\mathbf{U}, \mathbf{\Omega}, \mathbf{V}) \tag{17b}$$

The second problem describes the fluid motion due to an electro-osmotic flow around a stationary particle in the absence of bulk flow. The solutions of the second problem are difficult to obtain for nonspherical particles. Fortunately, however, the hydrodynamic force and torque can be determined directly using the reciprocal theorem [21]. The resulting hydrodynamic force and torque are given by

$$\mathbf{F}_2^H = \int_V \rho \mathbf{u}^F \cdot \mathbf{E} dV \tag{18a}$$

$$\mathbf{T}_2^H = \int_V \rho \mathbf{u}^T \cdot \mathbf{E} dV \tag{18b}$$

Here, the velocity fields, $\mathbf{u}^F$ and $\mathbf{u}^T$, are dyadic Stokes fields around the particle in the absence of bulk flow under the boundary conditions:

$$\mathbf{u}^F|_S = \boldsymbol{\delta} \tag{19a}$$

$$\mathbf{u}^T|_S = -\boldsymbol{\epsilon} \cdot (\mathbf{x} - \mathbf{x}_c) \tag{19b}$$

Here, $\boldsymbol{\epsilon}$ is unit isotropic triadic. The expressions for $\mathbf{u}^F$ and $\mathbf{u}^T$ for spherical and ellipsoidal particles are available [21, 33].

D. Total Force and Torque

Before we obtain the total force and torque balance equations it is useful to recast the expressions for the electrostatic force and torque into the volume integrals. Using the divergence theorem, Eqs (3a, b) can be rewritten as

$$\mathbf{F}^E = -\int_V \rho \mathbf{E} dV \tag{20a}$$

$$\mathbf{T}^E = -\int_V (\mathbf{x} - \mathbf{x}_c) \times \rho \mathbf{E} dV \tag{20b}$$

Here, V denotes the fluid volume outside the particle. Combining Eqs. (17a,b), (18a,b), and (20a,b), the most general forms for the total force and torque balance equations are given by

$$0 = \mathbf{F}_1^H(\mathbf{U}, \boldsymbol{\Omega}, \mathbf{V}) + \int_V \rho[\mathbf{u}^F - \boldsymbol{\delta}] \cdot \mathbf{E} dV + (m_p - m_f)\mathbf{g} \tag{21a}$$

$$0 = \mathbf{T}_1^H(\mathbf{U}, \boldsymbol{\Omega}, \mathbf{V}) + \int_V \rho[\mathbf{u}^T \cdot \mathbf{E} - (\mathbf{x} - \mathbf{x}_c) \times \mathbf{E}] dV + (m_p \mathbf{x}_m - m_f \mathbf{x}_b) \times \mathbf{g} \tag{21b}$$

These are a set of two equations for two unknowns, $\mathbf{U}$ and $\boldsymbol{\Omega}$. By solving these equations the electrophoretic motion of the particle can be determined. To solve Eqs (21a,b) we need three sets of information: $(\mathbf{F}_1^H, \mathbf{T}_1^H)$, $(\mathbf{u}^F, \mathbf{u}^T)$, and $(\rho, \mathbf{E})$. The first and the second sets require the solutions of the Stokes equations for uncharged particles. The Stokes solutions for several classes of nonspherical particles are available and various solution methods are developed for particles of arbitrary shape. The third sets requires the solutions of the Poisson equation coupled with the ion conservation equations. The determination of ρ and $\mathbf{E}$ for nonspherical particles under general conditions is difficult. Simplifying assumptions are commonly introduced for further analysis for ρ and $\mathbf{E}$.

When electro-osmotic bulk flow and gravitational settling are not present, Eqs (21a, b) simplify to

$$-\mathbf{F}_1^H(\mathbf{U}, \boldsymbol{\Omega}) = \mu \mathbf{A} \cdot \mathbf{U} + \mu \mathbf{B}^\dagger \cdot \boldsymbol{\Omega} = \int_V \rho[\mathbf{u}^F - \boldsymbol{\delta}] \cdot \mathbf{E} dV \tag{22a}$$

$$-\mathbf{T}_1^H(\mathbf{U}, \boldsymbol{\Omega}) = \mu \mathbf{B} \cdot \mathbf{U} + \mu \mathbf{C} \cdot \boldsymbol{\Omega} = \int_V \rho[\mathbf{u}^T \cdot \mathbf{E}(\mathbf{x} - \mathbf{x}_c) \times \mathbf{E}] dV \tag{22b}$$

Since the Stokes equations are linear, $(\mathbf{F}_1^H, \mathbf{F}_1^T)$ is linear with respect to $(\mathbf{U}, \boldsymbol{\Omega})$. The tensor coefficients $\mathbf{A}, \mathbf{B}, \mathbf{C}$ are the hydrodynamic resistance tensors, which depend only on the shape of the particle. When the applied electric field is small, the volume integrals in Eqs (22a, b) are linear to $\mathbf{E}^\infty$. By inverting Eqs (22a, b) we can then obtain a linear relationship between $(\mathbf{U}, \boldsymbol{\Omega})$ and $\mathbf{E}^\infty$. The tensor coefficient of $\mathbf{E}^\infty$ in such a relationship is the electrophoretic mobility tensor. The electrophoretic mobility tensor contains information on the geometry and the electrical condition of the particle. Most electrophoresis theories developed earlier deal with Eqs (22a,b) rather than with Eqs (21a, b).

E. Equilibrium Double Layer in Weak Field

Without bulk flow and an external electric field the ion conservation equations integrate to

$$n_i = n_i^\infty \exp\left(-\frac{ez_i \psi}{kT}\right) \tag{23}$$

which is the Boltzmann distribution. Here, n_i^∞ is the bulk number density and ψ is the double-layer potential in the absence of external electric field. Substitution of Eq. (23) into the Poisson equation yields the Poisson–Boltzmann equation:

$$\nabla^2 \psi = -\frac{e}{\epsilon} \sum_i z_i n_i^\infty \exp\left(-\frac{ez_i \psi}{kT}\right) \tag{24}$$

The Poisson–Boltzmann equation cannot be solved analytically in general. However, when the double-layer potential is small ($e\psi/kT \ll 1$), we can linearize the equation to obtain

$$\nabla^2 \psi = \kappa^2 \psi \tag{25}$$

where

$$\kappa^2 = \frac{e^2}{\epsilon kT} \sum_i z_i^2 n_i^\infty \tag{26}$$

The electric potential Ψ, which appears in Eq. (5), comprises the potentials due to the double layer and the externally applied electric field. Under general conditions a distinction between the two is not clear. However, when the externally applied field is small compared with the fields that occur in the double layer, we can assume additivity of the potentials:

$$\Psi = \psi + \Phi \tag{27}$$

Here, Φ denotes the perturbation due to the applied field. When the double-layer potential is small, the equations for ψ and Φ are Eq. (25) and

$$\nabla^2 \Phi = 0 \tag{28}$$

with the boundary conditions:

$$\psi|_\infty = 0, \quad \psi|_S = \zeta(\mathbf{x}_s) \tag{29a, b}$$

$$\Phi|_\infty = -\mathbf{E}^\infty \cdot \mathbf{x}, \quad \frac{\partial \Phi}{\partial n}|_S = 0 \tag{30a, b}$$

Here, ζ denotes the equilibrium electrokinetic potential. The boundary condition for Φ on the particle surface implies that the dielectric constant of the particle is much smaller than that of the electrolyte solution. Such an assumption for the electrostatic boundary condition is somewhat restrictive. However, our electrophoresis theory is more general than the assumption on which the theory is based, for the electrophoretic dynamics of the particle is independent of the boundary condition for Φ on the particle surface [24]. Since the exact value of Φ on the particle surface has no effect on the electrophoresis analysis, we usually set $\Psi_S = \zeta$, i.e., the electric potential on the particle surface remains equal to the equilibrium electrokinetic potential. Under the condition of weak applied field and the additivity of potentials, the electrostatic body force $\rho\mathbf{E}$ simplifies to

$$\rho\mathbf{E} = \epsilon\nabla^2(\psi + \Phi)\nabla(\psi + \Phi) \simeq \epsilon\nabla^2\psi\nabla\Phi = \epsilon\kappa^2\psi\nabla\Phi \tag{31}$$

Here, we neglect the term $\nabla^2\Phi\nabla\Phi$, which is quadratic in $\mathbf{E}^\infty$. According to Eq. (31), ρ and $\mathbf{E}$ are decoupled. The charge density ρ is entirely governed by the equilibrium double layer, and the double-layer, and the double-layer potential does not affect the electric field $\mathbf{E}$.

F. Electro-osmotic Bulk Flow

During the electrophoresis experiment the electro-osmotic flow is developed in the capillary because of the motion of counterions near the inner surface of the capillary. The electrolyte solution in the capillary is either supplied afresh continuously from a reservoir or confined in a closed cell assembly. In the former setup (open capillary system) the electro-osmotic flow results in a uniform velocity profile across the capillary so that this flow does not alter the particle motion significantly. However, in the latter setup (closed capillary system) the condition of no net flow across the capillary results in a nonuniform velocity profile, which profoundly affects the velocity and orientation of nonspherical particles. In practice, closed capillary systems are favored because of their simple setup.

Consider a straight closed capillary of uniform cross-section. The capillary is placed horizontally with the electric field applied along the horizontal axis of capillary. Constant electric field is applied along the x axis ($\mathbf{E}^\infty = E^\infty \mathbf{e}_x$) and the gravity acts along the negative z axis ($\mathbf{g} = -g\mathbf{e}_z$). For the capillary of either circular of slit type the fully developed electro-osmotic flow field is given by a one-dimensional parabolic flow field:

$$\mathbf{V} = (\alpha_0 + \alpha_y y^2 + \alpha_z z^2)\mathbf{e}_x \tag{32}$$

The electro-osmotic bulk flow about the center of the particle $\mathbf{x}_c$ is then given by

$$\begin{aligned}\mathbf{V} = &(\alpha_0 + \alpha_y y_c^2 + \alpha_z z_c^2)\mathbf{e}_x \\ &+ 2\alpha_y y_c(y - y_c)\mathbf{e}_x + 2\alpha_z z_c(z - z_c)\mathbf{e}_x \\ &+ \alpha_y(y - y_c)^2\mathbf{e}_x + \alpha_z(z - z_c)^2\mathbf{e}_x\end{aligned} \tag{33}$$

In capillary electrophoresis the thickness of the double layer formed at the inner wall of the capillary is usually much smaller than the gap distance of the capillary. The electro-osmotic velocity of the fluid near the cell wall is then given by the Smoluchowski formula:

$$V_e = -\frac{\epsilon \zeta_w E^\infty}{\mu} \tag{34}$$

Here, ζ_w denotes the electrokinetic potential of the cell wall. The coefficients in Eq. (32) are then given by

$$\alpha_0 = -\frac{V_e}{2}, \quad \alpha_y = \frac{3V_e}{2h^2}, \quad \alpha_z = 0$$

for a slit of thickness $2h$, and

$$\alpha_0 = -V_e, \quad \alpha_y = \alpha_z = \frac{2V_e}{h^2}$$

for a circular cylinder of radius h.

III. ELECTROPHORESIS OF SLIGHTLY DEFORMED SPHERE

A. Particle Shape and Electrokinetic Potential Distribution

When the shape of nonspherical particles deviates only slightly from that of a sphere, their electrophoretic motion can be analyzed using perturbation methods. The surface of a slightly deformed sphere is represented by

$$r = a[1 + \beta f(\vartheta, \varphi)] \tag{35}$$

Here, β is a small parameter and (r, ϑ, φ) are the spherical polar co-ordinates having their origin at the center of the undeformed sphere with radius a. The shape function f is an arbitrary function of ϑ and φ. The function f is typically expressed in terms of spherical harmonics. However, in this work we use the multipole expansion form:

$$f(\vartheta, \varphi) = M^f + \mathbf{D}^f \cdot \mathbf{n} + \mathbf{Q}^f : \mathbf{nn} \tag{36}$$

Only the multipole moments up to the quadrupole moment are included in our analysis. The first three moments of f are defined by

$$M^f = \frac{1}{4\pi}\int_\Omega f \, d\Omega \tag{37a}$$

$$\mathbf{D}^f = \frac{3}{4\pi} \int_\Omega \mathbf{n} f \, d\Omega \tag{37b}$$

$$\mathbf{Q}^f = \frac{5}{8\pi} \int_\Omega (\mathbf{n}\mathbf{n} - \boldsymbol{\delta}) f \, d\Omega \tag{37c}$$

Here, Ω denotes the solid angle and $\mathbf{n}$ the outward unit normal to the sphere. Note that the quadrupole moment $\mathbf{Q}^f$ is symmetric and traceless. Although we consider only up to the quadrupole moments of f, by varying the components of $\mathbf{D}^f$ and $\mathbf{Q}^f$ a wide variety of particle shapes can be described.

The electrical surface condition of nonspherical particles is usually nonuniform. For such particles their electrokinetic potentials can be also expressed as functions of ϑ and φ. Following Eq. (36), the function ζ is expressed as

$$\zeta(\vartheta, \varphi) = M^\zeta + \mathbf{D}^\zeta \cdot \mathbf{n} + \mathbf{Q}^\zeta : \mathbf{n}\mathbf{n} \tag{38}$$

Following Eqs (37a–c), the multipole moments of ζ are defined accordingly. Note that the multipole moments of ζ are defined not over the particle surface but over the solid angle. Hence, the monopole moment M^ζ is different from the area average of ζ over the particle surface. The relationship between M^ζ and the area average of ζ, which is defined by

$$\zeta_{av} = \frac{\int_S \zeta \, dS}{\int_S dS} \tag{39}$$

is given by

$$\zeta_{av} = \frac{M^\zeta}{1 + 2\beta M^f + O(\beta^2)} \left[= 1 + 2\beta \left(M^f + \frac{\mathbf{D}^\zeta \cdot \mathbf{D}^f}{3M^\zeta} + \frac{2\mathbf{Q}^\zeta : \mathbf{Q}^f}{15M^\zeta} \right) + O(\beta^2) \right] \tag{40}$$

B. Force and Torque Balances

For a small parameter β any field variable q can be expanded by

$$q = q^{(0)} + \beta q^{(1)} + \beta^2 q^{(2)} + \cdots \tag{41}$$

Using Eqs (22a) and (31), the total force balance equation in the absence of electro-osmotic bulk flow and gravitational settling is given by

$$-\mathbf{F}_1^H = \epsilon\kappa^2 \int_V (\mathbf{u}^{F(0)} - \boldsymbol{\delta}) \cdot \psi^{(0)} \nabla \Phi^{(0)} dV + \beta\epsilon\kappa^2 \int_V \mathbf{u}^{F(1)} \cdot \psi^{(0)} \nabla \Phi^{(0)} dV$$

$$+ \beta\epsilon\kappa^2 \int_V (\mathbf{u}^{(F0)} - \boldsymbol{\delta}) \cdot (\psi^{(1)} \nabla \Phi^{(0)} + \psi^{(0)}) \nabla \Phi^{(1)} dV + O(\beta^2) \tag{42}$$

The torque balance equation can be obtained in a similar way. The integrations in Eq. (42) are over the volume outside the undeformed sphere. The first volume integral corresponds to the result obtained for an undeformed sphere [3]. The next three integrals require the solutions for $\mathbf{u}^{F(1)}$, $\psi^{(1)}$, and $\Phi^{(1)}$. Since the volume integrals in Eq. (42) are all linear with respect to $\mathbf{E}^\infty$, we can symbolically denote the sum of three integrals as $\pi a \epsilon \mathbf{I}^F \cdot \mathbf{E}^\infty$. For slightly deformed spheres the hydrodynamic resistance tensor $\mathbf{B}$ vanishes, and Eq (22a, b) simplify to

$$\mu \mathbf{A} \cdot \mathbf{U} = \pi a \epsilon \mathbf{I}^F \cdot \mathbf{E}^\infty + O(\beta^2) \tag{43a}$$

$$\mu \mathbf{C} \cdot \mathbf{\Omega} = \pi a^2 \epsilon \mathbf{I}^T \cdot \mathbf{E}^\infty + O(\beta^2) \tag{43b}$$

Since the two equations are decoupled, $\mathbf{U}$ and $\mathbf{\Omega}$ can be determined independently. The components of $\mathbf{A}$ and $\mathbf{C}$ for slightly deformed spheres are given by

$$\mathbf{A} = 6\pi a \left[(1 + \beta M^f)\delta - \frac{\beta}{5}\mathbf{Q}^f \right] + O(\beta^2) \tag{44a}$$

$$\mathbf{C} = 8\pi a^3 \left[(1 + 3\beta M^f)\delta - \frac{3\beta}{5}\mathbf{Q}^f \right] + O(\beta^2) \tag{44b}$$

Inverting Eqs (43a, b), we obtain

$$\mathbf{U} = \frac{\epsilon \zeta_0}{\mu} \mathbf{M} \cdot \mathbf{E}^\infty \tag{45a}$$

$$\mathbf{\Omega} = \frac{\epsilon \zeta_0}{\mu a} \mathbf{N} \cdot \mathbf{E}^\infty \tag{45b}$$

The electrophoretic mobility tensors $\mathbf{M}$ and $\mathbf{N}$ are normalized to the Smoluchowski limit. Here, ζ_0 denotes the characteristic electrokinetic potential of the particle.

C. Hydrodynamics

The solutions for $\mathbf{u}^F$ and $\mathbf{u}^T$ for slightly deformed spheres are readily available [25]. When a particle translates with a velocity $\mathbf{U}$ in the Stokes flow, the Stokes solution $\mathbf{u}$ is linear with respect to $\mathbf{U}$ and we can write $\mathbf{u} = \mathbf{u}^F \cdot \mathbf{U}$. Expanding $\mathbf{u}$ about $r = a$, we obtain

$$\mathbf{u}|_S = \mathbf{u}^{(0)}|_{r=a} + \beta \mathbf{u}^{(1)}|_{r=a} + a\beta f \frac{\partial \mathbf{u}^{(0)}}{\partial r}\bigg|_{r=a} + O(\beta^2) \tag{46}$$

Since $\mathbf{u}|_S = \mathbf{U}$, the boundary conditions for $\mathbf{u}^{(0)}$ and $\mathbf{u}^{(1)}$ are given by

$$\mathbf{u}^{(0)}|_{r=a} = \mathbf{U} \tag{47a}$$

$$\mathbf{u}^{(1)}|_{r=a} = -af \frac{\partial \mathbf{u}^{(0)}}{\partial r}\bigg|_{r=a} \tag{47b}$$

The zeroth-order solution is the well-known solution for a translating sphere, which is given by

$$\mathbf{u}^{F(0)} = \delta\left(\frac{3}{4}\lambda + \frac{1}{4}\lambda^3\right) + \mathbf{n}\mathbf{n}\left(\frac{3}{4}\lambda - \frac{3}{4}\lambda^3\right) \tag{48}$$

where $\lambda = a/r$. The first-order solution $\mathbf{u}^{(F(1)}$ is given by $\sum_{k=0}^{2} \mathbf{u}^{F(1,k)}$, where $\mathbf{u}^{F(1,k)}$ are

$$\mathbf{u}^{F(1,0)} = M^f \delta\left(\frac{3}{4}\lambda^3 + \frac{3}{4}\lambda\right) + M^f \mathbf{n}\mathbf{n}\left(-\frac{9}{4}\lambda^3 + \frac{3}{4}\lambda\right) \tag{49a}$$

$$\mathbf{u}^{F(1,1)} = \mathbf{D}^f \cdot \mathbf{n}\delta\left(\frac{3}{4}\lambda^4 + \frac{3}{4}\lambda^2\right) + \mathbf{D}^f \cdot \mathbf{n}\mathbf{n}\mathbf{n}\left(-\frac{15}{4}\lambda^4 + \frac{9}{4}\lambda^2\right)$$
$$+ (\mathbf{n}\mathbf{D}^f + \mathbf{D}^f\mathbf{n})\left(\frac{3}{4}\lambda^4 - \frac{3}{4}\lambda^2\right) \tag{49b}$$

$$\mathbf{u}^{F(1,2)} = \mathbf{Q}^f\left(-\frac{6}{20}\lambda^5 - \frac{1}{20}\lambda^3 - \frac{3}{20}\lambda\right) + \mathbf{n}\mathbf{Q}^f \cdot \mathbf{n}\left(\frac{3}{2}\lambda^5 - \frac{17}{20}\lambda^3 - \frac{3}{20}\lambda\right)$$

$$+ \mathbf{Q}^f \cdot \mathbf{nn}\left(\frac{3}{2}\lambda^5 - \lambda^3\right) + \mathbf{Q}^f : \mathbf{nn}\delta\left(\frac{3}{4}\lambda^5 + \frac{1}{4}\lambda^3\right) \tag{49c}$$

$$+ \mathbf{Q}^f : \mathbf{nnnn}\left(-\frac{21}{4}\lambda^5 + \frac{15}{4}\lambda^3\right) - \frac{1}{2}\lambda^3(\boldsymbol{\epsilon} \cdot \mathbf{n}) \cdot \mathbf{Q}^f \cdot (\boldsymbol{\epsilon} \cdot \mathbf{n})$$

The corresponding solutions for $\mathbf{u}^T$ are also obtained using a similar method.

D. Electrostatics

The solutions for $\psi^{(1)}$ and $\Phi^{(1)}$ are determined by solving Eqs (25) and (28). We first consider the double-layer potential. Expanding ψ about $r = a$, we obtain

$$\psi|_S = \psi^{(0)}|_{r=a} + \beta\psi^{(1)}|_{r=a} + a\beta f\frac{\partial\psi^{(0)}}{\partial r}\bigg|_{r=a} + O(\beta^2) \tag{50}$$

The boundary condition, Eq. (29b), suggests that the boundary conditions for $\psi^{(0)}$ and $\psi^{(1)}$ are given by

$$\psi^{(0)}|_{r=a} = \zeta \tag{51a}$$

$$\psi^{(1)}|_{r=a} = -af\frac{\partial\psi^{(0)}}{\partial r}\bigg|_{r=a} \tag{51b}$$

The zeroth-order solution, in terms of the multipole moments of ζ over the solid angle, is given by [3]

$$\psi^{(0)} = M^\zeta\lambda e^{-\kappa(r-a)} + \mathbf{D}^\zeta \cdot \mathbf{n}\lambda^2 e^{-\kappa(r-a)}\frac{K_1(\kappa r)}{K_1(\kappa a)} + \mathbf{Q}^\zeta : \mathbf{nn}\lambda^3 e^{-\kappa(r-a)}\frac{K_2(\kappa r)}{K_2(\kappa a)} \tag{52}$$

where the polynomial function K_n is defined by

$$K_n(x) = \sum_{s=0}^{n}\frac{2^s n!(2n-s)!}{s!(2n)!(n-2)!}x^s$$

The first-order solution can be obtained in a similar manner. Setting the right-hand side of Eq. (51b) to Θ, the multipole expansion form of the solution for $\psi^{(1)}$ is given by

$$\psi^{(1)} = M^\Theta\lambda e^{-\kappa(r-a)} + \mathbf{D}^\Theta \cdot \mathbf{n}\lambda^2 e^{-\kappa(r-a)}\frac{K_1(\kappa r)}{K_1(\kappa a)} + \mathbf{Q}^\Theta : \mathbf{nn}\lambda^3 e^{-\kappa(r-a)}\frac{K_2(\kappa r)}{K_2(\kappa a)} + \cdots \tag{53}$$

Here, the multipole moments of Θ are given by

$$M^\Theta = M^\zeta M^f K_1 + \mathbf{D}^\zeta \cdot \mathbf{D}^f\frac{1}{3}A_1 + \mathbf{Q}^\zeta : \mathbf{Q}^f\frac{2}{15}A_2 \tag{54a}$$

$$\mathbf{D}^\Theta = M^\zeta\mathbf{D}^f K_1 + \left(\mathbf{D}^\zeta M^f + \frac{2}{5}\mathbf{D}^\zeta \cdot \mathbf{Q}^f\right)A_1 + \mathbf{Q}^\zeta \cdot \mathbf{D}^f\frac{2}{5}A_2 \tag{54b}$$

$$\mathbf{Q}^\Theta = M^\zeta\mathbf{Q}^f K_1 - \frac{1}{3}\mathbf{D}^\zeta \cdot \mathbf{D}^f\delta A_1 + \frac{1}{2}(\mathbf{D}^\zeta\mathbf{D}^f + \mathbf{D}^f\mathbf{D}^\zeta)A_1$$
$$+ \mathbf{Q}^\zeta M^f A_2 + \frac{2}{7}(\mathbf{Q}^\zeta \cdot \mathbf{Q}^f + \mathbf{Q}^f \cdot \mathbf{Q}^\zeta - \frac{2}{3}\mathbf{Q}^\zeta : \mathbf{Q}^f\delta)A_2 \tag{54c}$$

where $A_1 = (3K_2 - K_1)/K_1$, $A_2 = (5K_3 - 2K_2)/K_2$, and $K_n = K_n(\kappa a)$. Although the solution for $\psi^{(1)}$ contains the multipole moments higher than the quadrupole moment, only the multipole moments up to the quadrupole moment contribute to the electrophoretic mobility.

Now we consider the potential Φ. Expanding $\partial\Phi/\partial n$ about $r = a$ we obtain

$$\left.\frac{\partial\Phi}{\partial n}\right|_S = \left.\frac{\partial\Phi^{(0)}}{\partial n}\right|_{r=a} + \beta\left.\frac{\partial\Phi^{(1)}}{\partial n}\right|_{r=a} - a\beta(\nabla f \cdot \nabla\Phi^{(0)})|_{r=a} + a\beta f\frac{\partial}{\partial r}\left(\frac{\partial\Phi^{(0)}}{\partial n}\right)\Big|_{r=a} + O(\beta^2)$$

(55)

The boundary condition, Eq. (30b), suggests that the boundary conditions for $\Phi^{(0)}$ and $\Phi^{(1)}$ are given by

$$\left.\frac{\partial\Phi^{(0)}}{\partial n}\right|_{r=a} = 0$$

(56a)

$$\left.\frac{\partial\Phi^{(1)}}{\partial n}\right|_{r=a} = -af\frac{\partial}{\partial r}\left(\frac{\partial\Phi^{(0)}}{\partial n}\right)\Big|_{r=a} + a(\nabla\Phi^{(0)} \cdot \nabla f)|_{r=a}$$

(56b)

In addition, Eq. (30a) furnishes the boundary conditions:

$$\Phi^{(0)}|_\infty = -\mathbf{E}^\infty \cdot \mathbf{x}$$

(57a)

$$\Phi^{(1)}|_\infty = 0$$

(57b)

The zeroth-order solution is given by

$$\Phi^{(0)} = -\left(r + \frac{a}{2}\lambda^2\right)\mathbf{E}^\infty \cdot \mathbf{n}$$

(58)

The first-order solution is also obtained as a multipole expansion form. Setting the right-hand side of Eq. (56b) to Π, the solution for $\Phi^{(1)}$ is given by

$$\Phi^{(1)} = -M^\Pi a\lambda - \mathbf{D}^\Pi \cdot \mathbf{n}\frac{a}{2}\lambda^2 - \mathbf{Q}^\Pi : \mathbf{nn}\frac{a}{3}\lambda^3 - \mathbf{O}^\Pi : \mathbf{nnn}\frac{a}{4}\lambda^4$$

(59)

Here, the multipole moments of Π are given by

$$M^\Pi = 0,$$

(60a)

$$\mathbf{D}^\Pi = 3\mathbf{E}^\infty M^f - \frac{3}{5}\mathbf{E}^\infty \cdot \mathbf{Q}^f$$

(60b)

$$\mathbf{Q}^\Pi = \frac{9}{4}(\mathbf{D}^f\mathbf{E}^\infty + \mathbf{E}^\infty\mathbf{D}^f) - \frac{3}{2}\mathbf{D}^f \cdot \mathbf{E}^\infty\delta$$

(60c)

$$O^\Pi_{ijk} = 2(E^\infty_i Q^f_{jk} + E^\infty_j Q^f_{ki} + E^\infty_k(Q^f_{ij}) - \frac{4}{5}E^\infty_l(Q^f_{ll}\delta_{jk} + Q^f_{lj}\delta_{ki} + Q^f_{lk}\delta_{ij})$$

(60d)

E. Electrophoretic Mobility and Examples

Substituting the multipole expansion solutions for $\mathbf{u}^F$, ψ, and Φ into Eq. (42), we determine the expression for $\mathbf{I}^F$. The volume integrals are first evaluated by performing the angular integrations over the solid angle. Subsequent integrations over the radial co-ordinate result in an explicit expression for the volume integrals. After a few steps of algebraic manipulations the expression for $\mathbf{I}^F$ is obtained by

$$
\begin{aligned}
\mathbf{I}^F =\ & M^\zeta \delta H_1 + \mathbf{Q}^\zeta H_2 \\
& + \beta M^\zeta M^f \delta (K_1 H_1 + H_3) + \beta M^\zeta \mathbf{Q}^f (K_1 H_2 + H_4) \\
& + \beta \mathbf{D}^\zeta \cdot \mathbf{D}^f \delta \left[\frac{1}{3} A_1 (H_1 - H_2) + H_5 \right] + \beta (\mathbf{D}^\zeta \mathbf{D}^f + \mathbf{D}^f \mathbf{D}^\zeta) \left(\frac{1}{2} A_1 H_2 + H_6 \right) \\
& + \beta \mathbf{Q}^\zeta M^f (A_2 H_2 + H_7) + \beta \mathbf{Q}^\zeta \cdot \mathbf{Q}^f \left(\frac{2}{7} A_2 H_2 + H_8 \right) \\
& + \beta \mathbf{Q}^f \cdot \mathbf{Q}^\zeta \left(\frac{2}{7} A_2 H_2 + H_9 \right) + \beta \mathbf{Q}^f : \mathbf{Q}^\zeta \delta \left[\left(\frac{2}{15} H_1 - \frac{4}{21} H_2 \right) A_2 + H_{10} \right]
\end{aligned}
\tag{61}
$$

Here, H_n are functions of κa and their definitions are listed in Table 1. The corresponding expression for $\mathbf{I}^T$ can be obtained in a similar way. Substituting the expressions for $\mathbf{I}^F$ and $\mathbf{I}^T$ into Eqs (43a, b), the electrophoretic mobility tensors $\mathbf{M}$ and $\mathbf{N}$ can be determined. For spherical particles the expressions for $\mathbf{M}$ and $\mathbf{N}$ reduce to the results previously obtained [3]. When $\beta = 0$, only M^ζ and $\mathbf{Q}^\zeta$ contribute to $\mathbf{M}$, and only $\mathbf{D}^\zeta$ contributes to $\mathbf{N}$.

To highlight the main feature of the theory we consider the electrophoretic motion of axisymmetric particles with axisymmetric electrokinetic potential distributions. When the particle shape is axisymmetric, the dipole moment of f is parallel to the axis of revolution, and the quadrupole moment of f is diagonal and traceless. When we denote the unit directional vector along the axis of revolution as $\mathbf{d}$, $\mathbf{D}^f$ and $\mathbf{Q}^f$ can be expressed as $\mathbf{D}^f = D^f \mathbf{d}$ and $\mathbf{Q}^f = Q^f \mathbf{dd} - Q^f/2(\delta - \mathbf{dd})$, respectively. The components of $\mathbf{D}^\zeta$ and $\mathbf{Q}^\zeta$ follow

TABLE 1 Definitions of the Functions $H_n(x)$ in Eq. (61)

$H_1(x) = 1 + x + (12 - x^2)E_5(x)$

$H_2(x) = \dfrac{1}{1 + x + x^2/3}\left[-5 + x - \dfrac{2}{5}x^2 + (30 - x^2)E_7(x) \right]$

$H_3(x) = 6[-2 + (8 - x^2)E_5(x)]$

$H_4(x) = \dfrac{6}{5}[2 - (8 - x^2)E_5(x)]$

$H_5(x) = \dfrac{3}{5(1 + x)}[25 - 5x - 4x^2 - 15(10 - x^2)E_7(x)]$

$H_6(x) = \dfrac{3}{10(1 + x)}[25 - 5x + 2x^2 - 5(30 - x^2)E_7(x)]$

$H_7(x) = \dfrac{2}{5(1 + x + x^2/3)}[-50 + 10x + x^3 + 15(20 - x^2)E_7(x)]$

$H_8(x) = \dfrac{1}{175(1 + x + x^2/3)}[525 - 105x - 43x^2 + 35x^3 - 5(630 - 113x^2 + 3x^4)E_7(x)]$

$H_9(x) = \dfrac{1}{175(1 + x + x^2/3)}[175 - 35x - 71x^2 + 35x^3 - 5(210 - 127x^2 + 3x^4)E_7(x)]$

$H_{10}(x) = \dfrac{1}{175(1 + x + x^2/3)}[-95x^2 - 35x^3 + 5(60x^2 - 5x^4)E_7(x)]$

$E_n(x) = \displaystyle\int_1^\infty t^{-n} e^{x(1-t)}\,dt$

---

the same characteristics. Under such conditions for f and ζ, the second-order tensor $\mathbf{I}^F$ is diagonal and the electrophoretic mobility tensors $\mathbf{M}$ and $\mathbf{N}$ can be represented by

$$\mathbf{M} = M_\parallel \mathbf{dd} + M_\perp (\boldsymbol{\delta} - \mathbf{dd}) \qquad (62a)$$

$$\mathbf{N} = N\boldsymbol{\epsilon} \cdot \mathbf{d} \qquad (62b)$$

The symbols $M_\parallel$ and $M_\perp$ represent the components parallel and perpendicular to the axis of revolution, respectively. Combining Eqs (45b) and (62b), we can show that $\mathbf{N} \cdot \mathbf{E}^\infty = -N\mathbf{d} \times \mathbf{E}^\infty$. Thus, $\boldsymbol{\Omega}$ is perpendicular to $\mathbf{d}$ and there exists no electrophoretic rotation about the axis of revolution. In addition, when the axis of revolution $\mathbf{d}$ aligns parallel to the external electrical field, there is no electrophoretic rotation. When axisymmetric particles possess fore–aft symmetry (e.g., spheroidal particles), $N = 0$ and the electrophoretic rotation is not present.

As an illustrative example the electrophoretic motion is analyzed for an axisymmetric particle which lacks fore–aft symmetry. The orientation of the particle relative to the applied electric field is described with the aid of Euler angles [26] defined in Fig. 1. Affixing the body co-ordinates $\tilde{x}\tilde{y}\tilde{z}$ to the particle, the first Euler angle ϕ measures the angle between the x and $\tilde{x}$ axes after rotating the body co-ordinates about the $\tilde{z}$ the axis. We set the axis of revolution as the $\tilde{z}$ axis. The second Euler angle θ measures the angle between the z and $\tilde{z}$ axes after rotating the body co-ordinates about the $\tilde{x}$ axis. The last Euler angle that measures the rotation about the $\tilde{z}$ axis is not necessary for the analysis of axisymmetric particles. In terms of Euler angles the unit directional vector $\mathbf{d}$ is given by

$$\mathbf{d} = \sin\theta \sin\phi \, \mathbf{e}_x - \sin\theta \cos\phi \, \mathbf{e}_y + \cos\theta \, \mathbf{e}_z \qquad (63)$$

When $\mathbf{E}^\infty = E^\infty \mathbf{e}_x$, Eqs (45a, b) furnish the equations for the time rate of change for $\mathbf{x}_c$, θ, and ϕ. After normalizing the length and times scales to a and $a\mu/\epsilon\zeta_0 E^\infty$, respectively, the set of trajectory equations are given by

$$\dot{X} = (M_\parallel - M_\perp)\sin^2\theta \sin^2\phi + M_\perp \qquad (64a)$$

$$\dot{Y} = -\frac{1}{2}(M_\parallel - M_\perp)\sin^2\theta \sin 2\phi \qquad (64b)$$

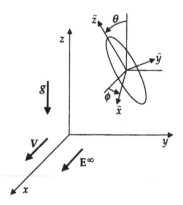

FIG. 1 Co-ordinate systems for the electrophoretic analysis of nonspherical particles. The co-ordinate system xyz is fixed in the fluid, while the co-ordinate system $\tilde{x}\tilde{y}\tilde{z}$ forms body co-ordinates affixed to the particle with the $\tilde{z}$ axis as its axis of revolution.

$$\dot{Z} = \frac{1}{2}(M_\parallel - M_\perp)\sin 2\theta \sin \phi \tag{64c}$$

$$\dot{\theta} = -N\cos\theta\sin\phi \tag{64d}$$

$$\dot{\phi} = -N\cos\phi/\sin\theta \tag{64e}$$

We can integrate Eqs (64d–e) to obtain

$$\tan\theta = \frac{C}{\cos\phi} \tag{65}$$

where C is the integration constant. The rotational motion is not periodic. According to Eqs (64d–e) and (65), $\dot{Y} = \dot{Z} = \dot{\theta} = \dot{\phi} = 0$ when $\theta = \phi = \pi/2$. Thus, an axisymmetric particle at arbitrary orientation eventually aligns parallel to the applied electric field and thereafter the particle translates along the applied electric field without any lateral motion.

A typical trajectory of an axisymmetric particle which lacks fore–aft symmetry is shown in Fig. 2. The particle shape is given by $f = -2\cos\vartheta + \cos^2\vartheta$, for which $M^f = 2/3$, $D^f = -2$, and $Q^f = 4/3$. Assuming $\zeta_0 = 1$, the electrokinetic potential distribution is given by $\zeta = 4/3 + \cos\vartheta - \cos^2\vartheta$, for which $M^\zeta = 1$, $D^\zeta = 1$, and $Q^\zeta = -2/3$. The initial orientation of the particle is $(\theta = \pi/6, \phi = \pi/2)$, and we set $\beta = 0.2$ and $\kappa a = 1$. The alignment of the particle along the electric field occurs over a relatively short time. The electrophoretic mobility components $M_\perp$ and $M_\parallel$ for the same particle at arbitrary κa are shown in Fig. 3. The results for spherical particles with the same electrokinetic potential distribution are also shown. As κa increases, the variation of ζ affects the electrophoretic mobility more strongly and thus the difference between $M_\perp$ and $M_\parallel$ becomes more pronounced. For an ensemble of axisymmetric particles with a uniform orientation distribution their average electrophoretic mobility is given by

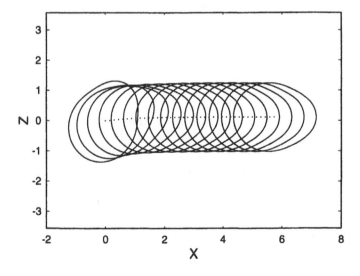

FIG. 2 Trajectory and snapshots for the electrophoretic motion of an axisymmetric particle which lacks fore–aft symmetry. The electrokinetic potential distribution is given by $\zeta = \zeta_0(4/3 + \cos\vartheta - \cos^2\vartheta)$.

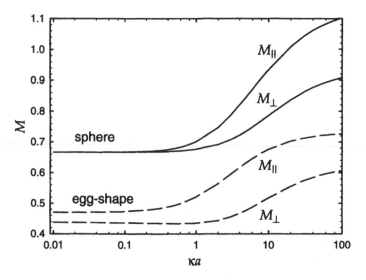

FIG. 3 Normalized electrophoretic mobility components $M_\parallel$ and $M_\perp$ for spheres and the axisymmetric particles considered in Fig. 2. The electrokinetic potential distributions for both particles are identical and equal to that used in Fig. 2.

$$M_{av} = \frac{1}{3}M_\parallel + \frac{2}{3}M_\perp \tag{66}$$

As shown earlier [3], the application of Eq. (66) to $M_\parallel$ and $M_\perp$ for spherical particles considered in Fig. 3 results in the classical result of Henry. However, in practice, when random Brownian rotation is not strong enough, spherical particles with nonuniform electrokinetic potential distributions may prefer to align along the electric field. For such spherical particles the electrophoresis measurement only furnishes the value of $M_\parallel$. Thus, proper interpretation of the electrophoresis measurement of such particles requires a careful analysis.

IV. ELECTROPHORESIS OF SPHEROID

A. Force and Torque Balances

Many nonspherical particles can be modeled as either prolate or oblate spheroidal particles. The solutions of the Stokes equations for ellipsoidal particles can be obtained using various methods. The expressions for $(\mathbf{u}^F, \mathbf{u}^T)$ and $(\mathbf{F}_1^H, \mathbf{T}_1^H)$ are available in terms of either ellipsoidal harmonics or singularity solutions [23]. For an ellipsoidal particle with the semiaxes $a \geq b \geq c$ the solutions for $(\mathbf{F}_1^H, \mathbf{T}_1^H)$, which are commonly known as the Faxen relations, are given by

$$\mathbf{F}_1^H = \mu\mathbf{A} \cdot \int_E f_1 \left[1 + \frac{c^2}{2}q^2\nabla^2\right][\mathbf{V} - \mathbf{U}]dA, \tag{67a}$$

$$\mathbf{T}_1^H = \mu\mathbf{C} \cdot \int_E f_2\left[\frac{1}{2} \times \mathbf{V} - \mathbf{\Omega}\right]dA + \mu\mathbf{H} : \int_E f_2\left[1 + \frac{c^2}{6}q^2\nabla^2\right]\frac{1}{2}\left[\nabla\mathbf{V} + \nabla\mathbf{V}^\dagger\right]dA \tag{67b}$$

where

$$q = \sqrt{1 - \tilde{x}^2/a_e^2 - \tilde{y}^2/b_c^2}, \quad a_e = \sqrt{a^2 - c^2}, \quad b_e = \sqrt{b^2 - c^2}$$

$$f_n = \frac{(2n-1)q^{2n-3}}{2\pi a_e b_e}$$

Here, the integration is over the fundamental focal ellipse $E(\tilde{x}, \tilde{y})$. Corresponding expressions for prolate and oblate spheroids can be obtained by setting $b = c$ and $a = b$, respectively. The expressions for the material tensors $\mathbf{A}, \mathbf{C}$, and $\mathbf{H}$, which only depend on the shape of the spheroid, are readily available [23].

Substituting Eqs (67a, b) into Eqs (21a, b), the translational and angular velocities of an ellipsoid are given by

$$\mathbf{U} = \int_E f_1 \left[1 + \frac{c^2}{2} q^2 \nabla^2 \right] \mathbf{V} dA + \frac{1}{\mu} \mathbf{A}^{-1} \cdot \int_V \rho[\mathbf{u}^F - \boldsymbol{\delta}] \cdot \mathbf{E} dV + \frac{(m_p - m_f)}{\mu} \mathbf{A}^{-1} \cdot \mathbf{g} \quad (68a)$$

$$\boldsymbol{\Omega} = \int_E f_2 \left[\frac{1}{2} \times \mathbf{V} \right] dA + \mathbf{C}^{-1} \cdot \mathbf{H} : \int_E f_2 \left[1 + \frac{c^2}{6} q^2 \nabla^2 \right] \frac{1}{2} \left[\nabla \mathbf{V} + \nabla \mathbf{V}^\dagger \right] dA$$

$$+ \frac{1}{\mu} \mathbf{C}^{-1} \cdot \int_V \rho[\mathbf{u}^T \cdot \mathbf{E} - (\mathbf{x} - \mathbf{x}_c) \times \mathbf{E}] dV \quad (68b)$$

For homogeneous ellipsoidal particles the center of mass and the center of buoyancy are identical so that the torque due to gravity is not included in Eq. (68b). The surface integrals over E correspond to nonelectrophoretic contributions due to electro-osmotic bulk flow, while the volume integrals over V correspond to electrophoretic contributions. Since the electrophoretic contribution in Eq. (68a) is linear with respect to $\mathbf{E}^\infty$, we represent it by introducing the electrophoretic mobility tensor $\mathbf{M}$:

$$\frac{1}{\mu} \mathbf{A}^{-1} \cdot \int_V \rho[\mathbf{u}^F - \boldsymbol{\delta}] \cdot \mathbf{E} dV = \frac{\epsilon \zeta_0}{\mu} \mathbf{M} \cdot \mathbf{E}^\infty \quad (69)$$

Here, the electrophoretic mobility tensor $\mathbf{M}$ is normalized to the Smoluchowski limit. For ellipsoidal particles $\mathbf{M}$ is diagonal. The evaluation of $\mathbf{M}$ requires the solutions for $(\rho, \mathbf{E})$.

B. Electrostatics

To determine $(\rho, \mathbf{E})$ for ellipsoidal particles the solutions for the Laplace equation and the linearized Poisson–Boltzmann equation are required. The solutions for the Laplace equation for ellipsoidal particles are readily available in terms of ellipsoidal harmonics. The solutions for the linearized Poisson–Boltzmann equation are available only for spheroidal particles in terms of spheroidal wave functions [6]. We first introduce the prolate and oblate spheroidal co-ordinate systems. The equation for the prolate spheroid is

$$\frac{\tilde{x}^2 + \tilde{y}^2}{c^2} + \frac{\tilde{z}^2}{a^2} = 1 \quad (70)$$

We choose the $\tilde{z}$ axis as the axis of symmetry. The prolate spheroidal co-ordinates (ξ, η, ω) are defined by

$$\tilde{x} = k\sqrt{\xi^2 - 1}\sqrt{1 - \eta^2} \cos \omega, \quad (1 \le \xi < \infty) \quad (71a)$$

$$\tilde{y} = k\sqrt{\xi^2 - 1}\sqrt{1 - \eta^2}\sin\omega, \quad (-1 \le \eta \le 1) \tag{71b}$$

$$\tilde{z} = k\xi\eta, \quad (0 \le \omega \le 2\pi) \tag{71c}$$

where $k = \sqrt{a^2 - c^2}$. The radial co-ordinate of the point on the spheroid surface, ξ_0, is a/k. The equation for the oblate spheroid is

$$\frac{\tilde{x}^2 + \tilde{y}^2}{a^2} + \frac{\tilde{z}^2}{c^2} = 1 \tag{72}$$

The oblate spheroidal co-ordinates (ξ, η, ω) are defined by

$$\tilde{x} = k\sqrt{1 + \eta^2}\sqrt{1 - \eta^2}\cos\omega, \quad (0 \le \xi < \infty) \tag{73a}$$

$$\tilde{y} = k\sqrt{1 + \eta^2}\sqrt{1 - \eta^2}\sin\omega, \quad (-1 \le \eta \le 1) \tag{73b}$$

$$\tilde{z} = k\xi\eta, \quad (0 \le \omega \le 2\pi) \tag{73c}$$

The radial co-ordinate of the point on the spheroid surface is c/k.

The general solution of the linearized Poisson–Boltzmann equation in spheroidal co-ordinates is given by

$$\psi(\xi, \eta, \phi) = R_n^m(\kappa k, \xi)S_n^m(\kappa k, \eta)\begin{pmatrix} \cos m\omega \\ \sin m\omega \end{pmatrix} \tag{74}$$

Since spheroidal particles are axisymmetric, the electrokinetic potential distribution for homogeneous spheroids may be axisymmetric as well. In such cases $m = 0$, and the solution can be written as

$$\psi(\xi, \eta) = \sum_n a_n R_n(\kappa k, \xi)S_n(\kappa k, \eta) \tag{75}$$

The radical function $R_n(\kappa k, \xi)$ is expanded in modified spherical Bessel functions of the first and third kinds. The angular function $S_n(\kappa k, \eta)$, which is given by

$$S_{2n} = \sum_{r=0} d_{2r}^{2n} P_{2r}(\eta) \tag{76a}$$

$$S_{2n+1} = \sum_{r=0} d_{2r+1}^{2n+1} P_{2r+1}(\eta) \tag{76b}$$

is expanded in Legendre functions of the first kind. Here, the coefficient d_r^n is a function of κk. Note that S_{2n} and S_{2n+1} are even and odd functions of η, respectively. When the electrokinetic potential distribution is centrosymmetric, only S_{2n} is required for the solution. The unknown coefficient a_n in Eq. (75) can be determined using the orthogonality property of $S_n(\kappa k, \eta)$. When the electrokinetic potential is uniformly constant ($\zeta = \zeta_0$), the solution is given by

$$\psi = \zeta_0 \sum_{r=0}^{\infty} \frac{d_d^{2r}}{N_{2r}} \frac{R_{2r}(\kappa k, \xi)}{R_{2r}(k, \xi_0)} S_{2r}(\kappa k, \eta) \tag{77}$$

where

$$N_{2r} = \sum_{p=0}^{\infty} \frac{\{d_{2p}^{2r}\}^2}{4p + 1}$$

When the electrokinetic charge density is uniformly constant $(-\epsilon \partial \psi / \partial n = \sigma)$, the solution is given by

$$\psi = -\frac{\sigma \kappa}{\epsilon} \sum_{r=0}^{\infty} \frac{\beta_{2r}}{N_{2r} \dfrac{R_{2r}(\kappa k, \xi)}{R'_{2r}(\kappa k, \xi_0)} S_{2r}(\kappa k, \eta)} \qquad (78)$$

where

$$\beta_{2r} = 2 \int_0^1 \sqrt{\xi_0^2 \pm \eta^2} \, S_{2r}(\kappa k, \eta) d\eta$$

The prime sign at the radial function represents the differentiation with respect to the radial co-ordinate. In the expression for β_{2r} the positive sign is for prolate spheroids and the negative sign is for oblate spheroids.

C. Electrophoretic Mobility

The volume integral on the left-hand side of Eq. (69) can be evaluated numerically to determine the electrophoretic mobility tensor $\mathbf{M}$ for spheroidal particles. Utilizing the axisymmetry of a spheroid, we determine the mobility components parallel and perpendicular to the axis of revolution. The values of $M_{\parallel}$ and $M_{\perp}$ for prolate and oblate spheroids $(c/a = 0.1, 0.2, 0.4, 0.6, 0.8)$ are shown in Figs 4 and 5, respectively. The results for spheres and infinite cylinders are shown as thick lines. The solid lines are for spheroids with uniform electrokinetic potentials, while the dotted lines are for spheroids with uniform electrokinetic charge densities. For the latter cases the characteristic electrokinetic potential ζ_0 is defined as the potential averaged over the particle surface. When the characteristic

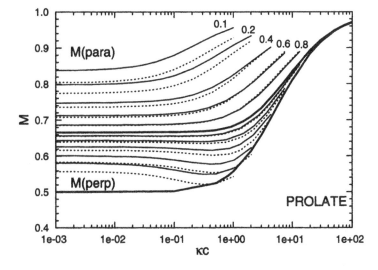

FIG. 4 Normalized electrophoretic mobility components $M_{\parallel}$ and $M_{\perp}$ for prolate spheroids. The solid lines are for spheroids with a uniform potential, the dotted lines are for spheroids with a uniform charge density. Starting from the curve right next to Henry's curve for spheres, the values of c/a are 0.8, 0.6, 0.4, 0.2, and 0.1. The thick line at the bottom is the curve for an infinite cylinder.

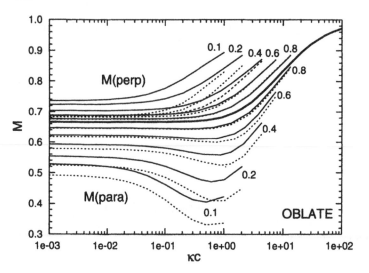

FIG. 5 Normalized electrophoretic mobility components $M_\parallel$ and $M_\perp$ for oblate spheroids. The solid lines are for spheroids with a uniform potential, the dotted lines are for spheroids with a uniform charge density. Starting from the curve adjoining Henry's curve for spheres, the values of c/a are 0.8, 0.6, 0.4, 0.2, and 0.1.

electrokinetic potentials are identical, the electrophoretic mobility of a spheroid with a uniform surface potential is larger than that with a uniform surface charge density.

When the double layer is very thick compared to the particle size, the double-layer potential approaches the solution of the Laplace equation. At this limit one can readily determine $M_\parallel$ and $M_\perp$ for spheroids. The results for prolate spheroids are given by

$$M_\parallel = \frac{1+e^2}{2e^2} - \frac{1}{e\ln\left(\frac{1+e}{1+e}\right)} \tag{79a}$$

$$M_\perp = \frac{3e^2-1}{4e^2} + \frac{1}{2e\ln\left(\frac{1+e}{1-e}\right)} \tag{79b}$$

and the results for oblate spheroids are given by

$$M_\parallel = \frac{2e^2-1}{2e^2} + \frac{\sqrt{1-e^2}}{2e\cot^{-1}\left(\frac{\sqrt{1-e^2}}{e}\right)} \tag{80a}$$

$$M_\perp = \frac{1+2e^2}{4e^2} - \frac{\sqrt{1-e^2}}{4e\cot^{-1}\left(\frac{\sqrt{1-e^2}}{e}\right)} \tag{80b}$$

These results are for spheroids with uniform electrokinetic potential distributions. For an ensemble of spheroidal particles with a uniform orientation distribution the average electrophoretic mobility can be determined by using Eq. (66). The average mobility curves obtained from the results in Figs 4 and 5 deviate only slightly from the curve for spherical particles. It is interesting to note that the substitution of Eqs (79a, b) or (80a, b) into Eq. (66) results in $M_{\mathrm{av}} = 2/3$, the exact result for spheres.

D. Electrophoretic Trajectory Analysis

Nonelectrophoretic contributions such as electro-osmotic bulk flow and gravitational settling strongly affect the apparent motion of spheroidal particles in electrophoresis experiments. To assess the importance of such nonelectrophoretic contributions to the motion of spheroidal particles we perform trajectory analysis. The translational and rotational motion of spheroidal particles under such conditions can be studied by integrating Eqs (68a, b). We assume that the electrokinetic potential distribution is uniformly constant. Starting from Eqs (68a, b), the equations for the rate of change of $\mathbf{x}_c$, θ, and ϕ are given by

$$
\begin{aligned}
\dot{X} = {} & A + BY^2 + CZ^2 + \frac{B}{3}q_y + \frac{C}{3}q_z + (M_\| - M_\perp)\sin^2\theta\sin^2\phi + M_\perp \\
& - \frac{aG}{2}(A_\|^{-1} - A_\perp^{-1})\sin 2\theta\sin\phi
\end{aligned}
\tag{81a}
$$

$$
\dot{Y} = -\frac{1}{2}(M_\| - M_\perp)\sin^2\theta\sin 2\phi + \frac{aG}{2}(A_\|^{-1} - A_\perp^{-1})\sin 2\theta\cos\phi
\tag{81b}
$$

$$
\dot{Z} = \frac{1}{2}(M_\| - M_\perp)\sin 2\theta\sin\phi - aG[(A_\|^{-1} - A_\perp^{-1})\cos^2\theta + A_\perp^{-1}]
\tag{81c}
$$

$$
\dot{\theta} = -\frac{1}{2}\left(\frac{r^2 - 1}{r^2 + 1}\right)BY\sin 2\theta\sin 2\phi + \frac{2}{r^2 + 1}CZ(r^2\cos^2\theta + \sin^2\theta)\sin\phi
\tag{81d}
$$

$$
\dot{\phi} = -\frac{2}{r^2 + 1}BY(r^2\cos^2\phi + \sin^2\phi) + \frac{2r^2}{r^2 + 1}CZ\cot\theta\cos\phi
\tag{81e}
$$

where

$$
q_y = 1 - e^2 + e^2\sin^2\theta\cos^2\phi, \quad q_z = 1 - e^2\sin^2\theta
$$

for a prolate spheroid, and

$$
q_y = 1 - e^2\sin^2\theta\cos^2\phi, \quad q_z = 1 - e^2\cos^2\theta
$$

for an oblate spheroid. Here, $e = \sqrt{1 - c^2/a^2}$ is the eccentricity of the spheroid and r is the aspect ratio of the spheroid, which is defined by the ratio of the lengths of the semiaxes of the symmetry axis and transverse axis. For prolate spheroids $r > 1$, while for oblate spheroids $0 < r < 1$. The definitions for the co-ordinates systems and the Euler angles are identical to those shown in Fig. 1. In Eqs (81a–e) the length and time scales are normalized. The length scale is the major semiaxis of the spheroid ($\mathbf{X} = \mathbf{x}_c/a$) and the time scale is $a\mu/\epsilon\zeta_0 E^\infty$. The dimensionless parameter G is $(m_{\mathrm{p}} - m_{\mathrm{f}})g/a\epsilon\zeta_0 E^\infty$. The dimensionless constants A, B, and C in Eq. (81a) characterize the electro-osmotic bulk flow developed in the capillary. Their expressions are given by

$$A = \frac{1}{2}\left(\frac{\zeta_w}{\zeta_0}\right), \quad B = -\frac{3}{2}\left(\frac{\zeta_w}{\zeta_0}\right)\left(\frac{a}{h}\right)^2, \quad C = 0$$

for a slit type, and

$$A = \frac{\zeta_w}{\zeta_0}, \quad B = C = -2\left(\frac{\zeta_w}{\zeta_0}\right)\left(\frac{a}{h}\right)^2$$

for a circular cylinder.

Compared with Eqs (64a–c), previously derived for axisymmetric particles, electrophoretic contributions in the translational velocity are identical. The rotational motion is entirely determined by the electro-osmotic bulk flow, since the electrophoretic rotation is not present for spheroidal particles with uniform electrokinetic potential distributions. According to Eq. (81a), the motion of the spheroid along the direction of applied electric field is governed by four contributions: fluid velocity at the center of the spheroid $(A + BY^2 + CZ^2)$, additional contribution due to a quadratic part of the flow field (q_y and q_z terms), electrophoretic contribution ($M_\parallel$ and $M_\perp$ terms), and gravitational settling. The second contribution is linear with respect to B or C; thus it is usually negligible when a/h is small. The lateral motion of the spheroid can be significant when the difference between $M_\parallel$ and $M_\perp$ is not small. When all orientations are equally probable the observables in electrophoresis experiments are the quantities averaged over the Euler angles θ and ϕ. Averaging Eqs. (81a–c), we obtain

$$< \dot{X} > = A + BY^2 + CZ^2 + \frac{(B+C)}{9}q_a + \frac{1}{3}[M_\parallel + 2M_\perp] \tag{82a}$$

$$< \dot{Y} > = 0 \tag{82b}$$

$$< \dot{Z} > = -\frac{aG}{3}[A_\parallel^{-1} + 2A_\perp^{-1}] \tag{82c}$$

Here, $q_a = 3 - 2e^2$ for a prolate spheroid and $q_a = 3 - e^2$ for an oblate spheroid. On average there is no lateral motion in the Y direction. The averaged motion in the X direction consists of three parts: electro-osmotic bulk flow contribution, quadratic flow contribution, and electrophoretic contribution. Gravitational settling does not affect the averaged motion in the X direction. When we further average Eq. (82a) over the cross-section of the capillary, the first part in Eq. (82a) drops out. Since the second part in Eq. (82a) is usually very small, the area averaged form of Eq. (82a) reduces to Eq. (66).

Typical trajectories of prolate spheroids initially placed in a capillary of slit type are shown in Fig. 6. To illustrate the role of electro-osmotic bulk flow on the apparent motion of spheroids the gravitational settling effects are neglected. A prolate spheroid has a major semiaxis of 2 μm and its aspect ratio is 10. The applied electric field is 500 V/m. The zeta potential of the spheroids is set to 20 mV. The capillary has dimensions of 1×3 cm and an inner gap distance of 0.6 mm. The range of the dimensionless coordinate Y is then $-150 \le Y \le 150$. The absolute value of the zeta potential of the cell wall is set to 100 mV. The double-layer thickness is 10 nm. The initial orientation is fixed at $\theta = 0.5\pi$ and $\phi = 0.3\pi$, thus there is no motion lateral in the Z direction. Although the trajectories depart significantly from a straight path along the X axis, they occur over a relatively long time. The magnitude of lateral motion depends strongly on the initial position of the spheroid. The shear rate of the electro-osmotic flow increases linearly as Y increases from the center to the cell wall. Accordingly, a spheroid placed at large Y rotates faster than

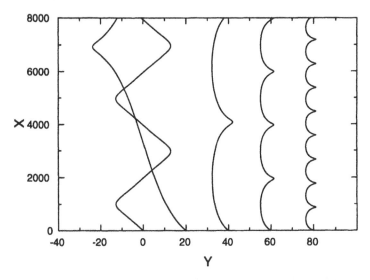

FIG. 6 Trajectories for the electrophoretic motion of prolate spheroids in electro-osmotic flow developed in a slit-type capillary. In all cases the initial orientation of the spheroids is $\theta = 0.5\pi$ and $\phi = 0.3\pi$.

that at small Y. Since a spheroid at large Y translates slowly, trajectories at large Y show waves of high frequency and small amplitude. On the other hand, trajectories at small Y, as long as they do not cross over the center plane of the cell, show waves of low frequency and large amplitude.

In a simple method for isolating the electrophoretic contribution from the observed particle velocity, one usually measures the velocity of the particle at the stationary level where the electro-osmotic bulk flow is zero. As long as the particle remains at the stationary level during the electrophoresis experiment, the electrophoretic velocity can be determined by dividing its travel distance in the X direction, $X(\tau)$, by the travel time, τ. For nonspherical particles, however, $X(\tau)/\tau$ depends on τ because of the lateral motion and orientation effects. The plots for $X(\tau)/\tau$ for spheroids studied in Fig. 6 are shown in Fig. 7. As τ increases, the calculation of the electrophoretic velocity deviates from the instantaneous velocity, the value at $\tau = 0$, for a given initial orientation. In fact, the distribution of $X(\tau)/\tau$ over the orientation space (θ, ϕ) varies along τ. Consequently, the averaged electrophoretic velocity of randomly oriented spheroids at $\tau = 0$ can be different from that at nonzero τ.

V. CONCLUDING REMARKS

To understand the meaning of the electrokinetic potential properly and to utilize the electrophoretic measurement data to find their correlations with the surface charge or potential distribution of the particle we must devise a suitable mathematical model which incorporates every detail of the electrophoresis experiment. Although such models are too complex to analyze mathematically, over the last several decades many efforts have been made to improve the electrophoresis theory so as to make it applicable to a wider range of experimental systems. This chapter reports one of such efforts to furnish a more

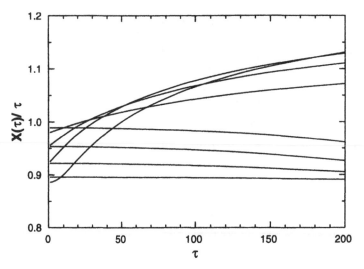

FIG. 7 Normalized electrophoretic velocity obtained by dividing $X(\tau)$ by τ. Spheroids are initially placed at the stationary level. Cases with eight different initial orientations are shown.

sound basis for the interpretation of the electrophoresis measurement data, in particular for nonspherical particles.

We address two main issues: (1) how to characterize effectively the shape and the electrokinetic potential distribution for the electrophoresis analysis for nonspherical particles, and (2) how to deal with nonelectrophoretic contributions that affect the apparent motion of nonspherical particles. When the shape of nonspherical particles does not deviate significantly from that of spheres, their electrophoretic mobility can be conveniently determined in terms of the multipole moments. For axisymmetric particles with axisymmetric electrokinetic potential distributions the explicit expressions for their electrophoretic mobility tensors are derived. Based on the electrophoretic trajectory analysis it is shown that axisymmetric particles may prefer to align parallel to the applied electric field. As for the second issue, we discuss the electrophoretic motion of heavy spheroidal particles suspended in electro-osmotic flow developed in a capillary. The electro-osmotic flow results in the lateral migration of the particle and strongly affects the particle orientation. Therefore, to analyze the electrophoresis measurement of such particles the interplay of electro-osmotic flow and the electrophoretic motion must be carefully examined.

ACKNOWLEDGMENTS

This work was supported by the Korea Science and Engineering Foundation through the Automation Research Center at the Pohang University of Science and Technology and also by the Ministry of Education under the BK21 program.

REFERENCES

1. DC Henry. Proc Roy Soc A 133: 106, 1931.
2. JL Anderson. J Colloid Interface Sci 105: 45, 1985.

3. BJ Yoon. J Colloid Interface Sci 142: 575, 1991.
4. D Stigter. J Phys Chem 82: 1417, 1978.
5. JD Sherwood. J Chem Soc, Faraday Trans 2 78: 1091, 1982.
6. BJ Yoon, S Kim. J Colloid Interface Sci 128: 275, 1989.
7. BJ Yoon. J Colloid Interface Sci 187: 459, 1997.
8. BJ Yoon. J Colloid Interface Sci 214: 215, 1999.
9. JD Sherwood, HA Stone. Phys Fluids 7: 697, 1995.
10. RW O'Brien, DN Ward. J Colloid Interface Sci 121: 402, 1988.
11. H Keh, TY Huang. J Colloid Interface Sci 160: 354, 1993.
12. MC Fair, JL Anderson. J Colloid Interface Sci 127: 388, 1989.
13. MC Fair, JL Anderson. Int J Multiphase Flow 16: 663, 1990.
14. Y Solomentsev, JL Anderson. J Fluid Mech 279: 197, 1994.
15. Y Solomentsev, JL Anderson. Ind Eng Chem Res 34: 3231, 1995.
16. KS Chae, AM Lenhoff. Biophys J 68: 1120, 1995.
17. SA Allison, VT Tran. Biophys J 68: 2261, 1995.
18. SA Allison. Macromolecules 29: 7391, 1996.
19. SA Allison, M Potter, JA McCammon. Biophys J 73: 133, 1997.
20. SA Allison, S Mazur. Biopolymers 46: 359, 1998.
21. M Teubner. J Chem Phys 76: 5564, 1982.
22. J Happel, H Brenner. Low Reynolds Number Hydrydynamics. The Hague: Martinus Nijhoff, 1983.
23. S Kim, SJ Karrila. Microhydrodynamics: Principles and Selected Applications. Boston, MA: Butterworth Heinemann, 1991.
24. RW O'Brien, LR White. J Chem Soc, Farday Trans 2 74: 1607, 1978.
25. H Brenner. Chem Eng Sci 19: 519, 1964.
26. H Goldsten. Classical Mechanics. 2nd ed. Reading, MA: Addison-Wesley, 1980.

8

A Critical Review of Electrokinetics of Monodispersed Colloids

EGON MATIJEVIĆ Clarkson University, Potsdam, New York

I. INTRODUCTION

There are few techniques used more extensively than electrokinetics in studies involving colloid dispersions. The information obtained from these measurements is invaluable in the characterization of fine particles and in the interpretation of various phenomena observed with such materials.

Two aspects regarding electrokinetics need to be clearly distinguished. The first refers to the methods of mobility measurements, and the second to the interpretation of data, paticularly in terms of calculated ζ potentials. Both these parts are fraught with considerable difficulties. Indeed, there is good reason to believe that many reported experimental data, and even more so the values of the corresponding electrokinetic potentials, are of doubtful quality. These problems are extensively dealt with in various chapters of this volume.

Another matter of concern is a common oversimplification of the significance of the mobility, and especially of the zeta potential, in the explanation of colloid phenomena, such as dispersion stability and particle adhesion, or of some processes represented, for example, by the retention in paper processing, etc. Certainly, the particle charge plays an important role in all mentioned aspects of fine particles, and many others, but the relationships of surface potential to the observed effects are neither straightforward nor easily quantified. Thus, it is necessary to evaluate carefully each and every case and refrain from unjustified generalizations.

There is one useful aspect of electrokinetics, not always fully recognized, which deals with the great sensitivity of the surface charge to extremely small amounts of contaminants. The latter may not be assayed even with quite sophisticated analytical techniques, yet are readily detected by mobility measurements.

A related application is in the use of mobility measurements to follow interactions of solutes with colloid particles. There is a twofold purpose of such studies: one is to ascertain that adsorption took place, and the other is intentionally to modify the particle charge. Numerous reports show that often minute amounts of surfactants, chelating agents, and other ionic species can greatly affect the magnitude of the surface potential or even reverse the sign of its charge. However, the latter effects are not the subject of this article.

It is obvious that the use of well-defined and reproducible fine particles offers an essential advantage in the proper evaluation of electrokinetic effects. Monodispersed colloids, which are now available in a variety of chemical compositions, sizes, shapes, and structures [1–4] are ideally suited for such studies. The purpose of this review is to summarize the published results attained with uniform inorganic dispersions, and to illustrate some phenomena and discrepancies. Thus, the chapter focuses on the experimental evidence and not on possible pitfalls in the evaluation of the mobility data in terms of the calculated ζ potentials.

II. REPRODUCIBILITY OF MOBILITY MEASUREMENTS

It would be of considerable practical and theoretical interest to have a generally accepted standard dispersion for electrokinetic evaluations. Originally, polymer latexes appeared as a possible choice for this purpose, but they proved to be elusive in terms of reproducibility of the results. Monodispersed spherical inorganic colloids, especially metal oxides, seem to be an alternative possibility. An additional advantage of such particles is the ease with which the sign and the magnitude of the surface potential can be varied by simple adjustment of the pH.

One carefully tested dispersion consists of uniform spherical amorphous chromium hydroxide particles, which are exceedingly simple to prepare [5] and have a convenient isoelectric point (i.e.p.), making it easy to carry out studies with negatively and positively charged colloids of the same chemical composition (6–9).

Figure 1 illustrates data obtained with such colloids using two different instruments (PenKem System 3000 and DELSA, Coulter Electronics) [7]. The consistency of the mobilities is excellent, especially if one considers that the principles on which the two instruments are based differ.

Figure 2 displays a series of mobility curves of monodispersed chromium hydroxide dispersions in the presence of different concentrations of Na_2SO_4, which resulted in fairly constant values over a broad pH range at sufficiently high sulfate concentrations [8]. In view of these results, it was suggested that such dispersions could be used as an electrokinetic "standard," but no acceptance of this proposal has been noted.

III. DETECTION OF IMPURITIES

As outlined in the introduction, one important use of electrokinetics is to detect surface contaminations. A few example are offered here.

Figure 3 shows two mobility curves for amorphous spherical chromium hydroxide particles as a function of the pH. The freshly prepared dispersion has an i.e.p. at pH 8.5. The second set of data is for the same system kept for 3 days in alkaline solutions contained in glass tubes. The significant shift in the mobility is due to a minute amount of silicates leached from the container and adsorbed on the particle [7].

The mobilities of uniform spheres of hematite (α-Fe_2O_3) change with repeated rinsings with water, displayed by different curves in Fig. 4 [10], until the surface is purified as given by the solid line. This colloidal hematite was prepared by aging $FeCl_3$ solutions at elevated temperatures [11]. As a result some chloride was incorporated in the particles, which was removed from the surface layers by extensive washings. Since these anions are occluded in the bulk of the precipitated solids and slowly diffuse towards the surface, it takes a large number of washings to eliminate this contaminant.

The final example deals with monodispersed amorphous aluminum hydrous oxide particles, which were obtained by forced hydrolysis of aluminum sulfate solution [12]. The

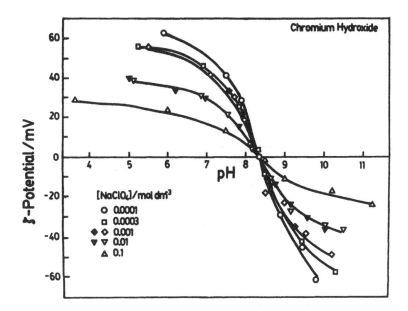

FIG. 1 ζ potentials of monodispersed spherical amorphous chromium hydroxide particles as a function of pH in the presence of different concentrations of sodium perchlorate. Open and solid symbols represent data taken with the Pen-Kem 3000 and DELSA (Coulter Electronic) instruments, respectively.

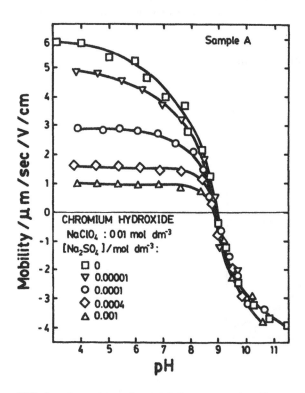

FIG. 2 Electrophoretic mobilities of monodispersed spherical amorphous chromium hydroxide particles in the presence of 0.01 mol dm^{-3} NaClO$_4$ as a function of pH at different concentrations of Na$_2$SO$_4$: none ($\square$); 1×10^{-5} ($\triangledown$); 1×10^{-4} ($\bigcirc$); 4×10^{-4} ($\diamond$); 1×10^{-3} mol dm^{-3} ($\triangle$).

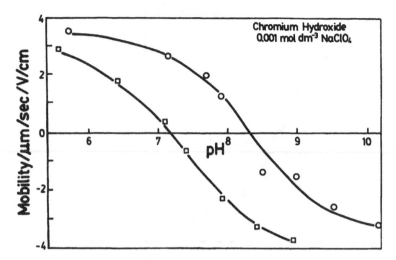

FIG. 3 Electrophoretic mobilities of monodispersed $Cr(OH)_3$ particles as a function of pH in 1×10^{-3} mol dm^{-3} NaClO$_4$. Purified sample (○) and sample stored for 3 days in alkaline solutions in a glass container (□).

dispersions so attained showed an i.e.p. at pH 7.2 (Fig. 5). However, on washing with a basic aqueous solution the curve shifted, yielding an i.e.p. at pH 9.3. Obviously, this change was due to elimination of the sulfate ions, which were retained during the precipitation of the particles.

It should be noted that in all the cases described the contaminants were present in quantities too small to be detected by other techniques.

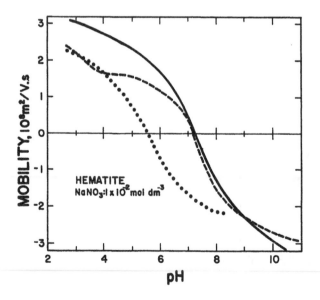

FIG. 4 Electrophoretic mobilities of spherical hematite (α-Fe$_2$O$_3$) particles (0.12 μm in diameter) as a function of pH in 1×10^{-2} mol dm^{-3} NaNO$_3$: completely cleaned (35 washing cycles) (————), after 20 washing cycles (– – – –), and freshly prepared (•••••).

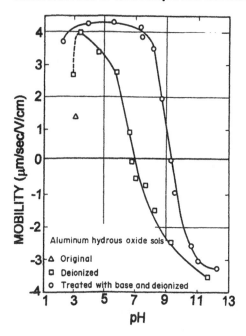

FIG. 5 Electrophoretic mobilities of spherical amorphous aluminum hydrous oxide particles prepared by aging a 1×10^{-3} mol dm^{-3} solution of $Al_2(SO_4)_3$ at 98°C for 24 h and treated as follows: $\triangle$, original sol at room temperature; $\square$, sol freed of the original electrolyte by centrifugation and redispersed in dilute NaOH solutions; $\bigcirc$, pH of the sol adjusted to 9.7 by NaOH, followed by deionization, and redispersion at various pH values.

IV. SURVEY OF PUBLISHED DATA

Apparently, the only comprehensive review of electrokinetic results for inorganic oxides and hydroxides was published by Parks in 1965 [13].[*] The tabulated i.e.p. showed considerable variations, which depended on the method of preparation and treatments of individual dispersions. For example, reported data for aluminum (hydrous) oxides were as low as 3.5 and as high as 11, and for iron (hydrous) oxide the values ranged from 1.9 to 9! Furlong and Parfitt [14] listed the i.e.p. for TiO_2 from 2.7 to 6.0. Needless to say, discrepancies in the measured mobilities or in calculated ζ potentials at the same pH values for the same compounds varied just as much.

In contrast to the cited review, which was based on measurements mostly obtained with ill-defined particles, here an attempt is made to summarize electrokinetic data published with dispersions consisting of well-defined colloids of narrow size distributions and of simple and composite natures. In doing so one eliminates at least the possible effects of polydispersity and of other conditions inherent in less carefully prepared systems. Every effort is made to compare results using dispersions presumably obtained by the same or

---

[*] After this article was written, Professor M. Kosmulski (Lublin, Poland) informed the author that his elaborate tabulation of electrokinetic properties of metal (hydrous) oxides will appear in the book titled *Chemical Properties of Material Surfaces* to be published by Marcel Dekker Inc. in 2001.

similar procedures in different laboratories. Preference is given to mobility measurements, whenever possible, in order to eliminate uncertainties arising from calculations of the ζ potentials by individual investigators, which may be based on different assumptions. Data for a few compounds will be presented in some detail, while the references to other, less frequently studied materials, are given in Table 1 (see later).

A. Titanium Dioxide

Information on the electrokinetic behavior of titanium dioxide is of particular interest, because dispersions of this material, consisting of "monodispersed" spheres, have been prepared by chemical reactions in aerosols and by precipitation in solutions. In the former method, droplets of titanium(IV) alkoxides are contacted with water vapor, which process rapidly yields TiO_2 spheres with the elimination of alcohol [15]. These particles are very pure, since no ionic species, which could be incorporated as contaminants, are present in the reacting system. The transmission electron micrograph in Fig. 6 illustrates one such product.

Figure 7 compares electrophoretic data for several TiO_2 dispersions obtained by the hydrolysis of Ti(IV) ethoxide using the aerosol process [15–18]. While the i.e.p. is quite consistent, the mobilities at a given pH and at the same ionic strength, using the same electrolyte (KNO_3), differ significantly. It should be noted that the solid lines refer to particles which were converted into anatase [18] by mild heat treatment, but they retained their spherical shape. Furthermore, different designs of aerosol generators were used in the preparation of TiO_2 [15, 16], and the measurements were carried out with different instruments (Rank Mark VIII and DELSA). None of these experimental specifics should account for the observed discrepancies in the reported mobilities.

0.5 μm

FIG. 6 Transmission electron micrograph of TiO_2 particles prepared by the aerosol technique using droplets of Ti(IV) ethoxide interacted with water vapor.

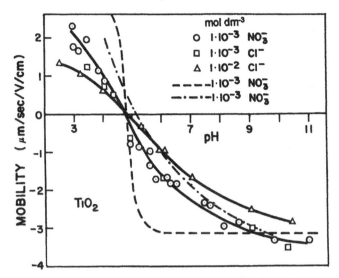

FIG. 7 Electrophoretic mobilities of spherical TiO_2 particles prepared by the aerosol method from Ti(IV) ethoxide droplets and converted into anatase at 25°C as a function of pH in the presence of different anions at constant ionic strength $\mu = 1 \times 10^{-3}$ mol dm^{-3} (NO$_3^-$: O, Cl : □) and at $\mu \times 10^{-2}$ mol dm^{-3} (Cl; △), with Na$^+$ as the cation [18]. (−·−·−·) TiO_2 particles obtained by the same technique without subsequent treatment [15]. (- - - -) Particles obtained in a different aerosol generator [16].

Figure 8 compares ζ potentials of a "high purity" spherical colloidal titania (anatase) prepared by solution precipitation using Ti(IV) ethoxide [19], with ellipsoidal rutile particles, obtained by the hydrolysis of $TiCl_4$, and thoroughly washed [20]. Again the i.e.p. is reasonably close and not much different from the value obtained with the "aerosol" samples. However, the ζ potentials at pH > i.e.p. are substantially lower at the same ionic strength for the sample precipitated in $TiCl_4$ solution.

Finally, Fig. 9 displays ζ potentials of the aerosol TiO_2 (anatase) sample in 1×10^{-1} mol dm^{-3} aqueous KCl solution and in 30% ethanol/water mixtures in the presence of different alkali chlorides [21]. There is a considerable effect of alcohol on the surface potential, which becomes less pronounced as the ionic strength decreases. Similar results were obtained with methanol/water mixtures [21]. This example points to another combination of parameters which may greatly affect the electrokinetic properties of a monodispersed system.

B. Iron Oxides

One significant aspect of iron oxides is the availability of "monodispersed" particles of the same chemical composition, but in a variety of shapes. Thus, hematite (α-Fe_2O_3) was obtained as spheres, ellipsoids, rods, cubes, platelets, and some other more complex morphologies [11, 22–28]. In addition, uniform particles of different composition, such as akageneite (β-FeOOH), have also been produced [11]. Obviously, it should, in principle, be possible to investigate the effects of the shape and structure of particles on their electrokinetic properties.

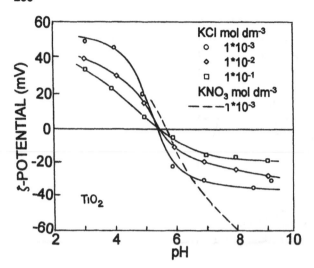

FIG. 8 ζ potentials of spherical TiO$_2$ particles prepared by precipitation in Ti(IV) ethoxide solutions at three different ionic strengths ($\square$, $\bigcirc$, $\diamondsuit$) [19], and of needle-type rutile obtained by precipitation in TiCl$_4$ solutions (– – – –) [20].

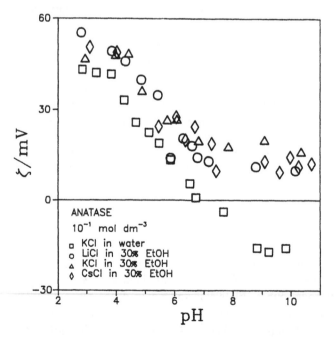

FIG. 9 ζ potentials of spherical (aerosol) anatase particles in 1×10^{-1} mol dm^{-3} alkali chlorides as a function of pH in water and 30% ethanol [21].

Figure 10 compares the mobilities as a function of the pH of hematite particles (α-Fe$_2$O$_3$) of different shapes and of needle-like akageneite (β-FeOOH) [11]. It is surprising that, despite the variation in the preparation procedures and particle morphologies, relatively little difference is noted in their electrophoretic behavior. Indeed, considering the possible experimental errors it would not be justified to draw any conclusions regarding the effect of the shapes on the electrokinetics of these dispersions. The data that follow will further support this statement.

Figure 11 displays four sets of mobility data from different laboratories, obtained with nearly spherical hematite particles of comparable size, prepared by essentially the same procedure. Two solid symbols refer to the i.e.p. taken from plots of ζ potential as a function of the pH. Obviously the value obtained by Kandori et al. [23] must be erroneous for some reason. The results in Fig. 11 are both surprising and disturbing, because it is not easy to account for the significant differences in the behaviors of the supposedly identical, or at least very similar, dispersions. It should be noted that measurements were carried out with different instruments, which should not be responsible for the observed discrepancies.

No less inconsistency is observed with ellipsoidal hematite, synthesized by using the same procedure. Figure 12 shows that neither the i.e.p. nor the mobilities at the same pH agree as reported for these dispersions.

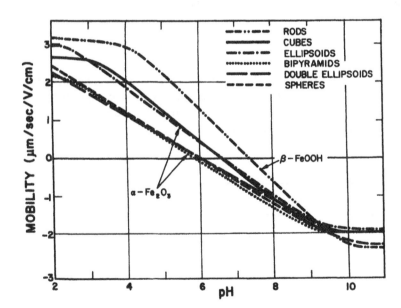

FIG. 10 Electrophoretic mobilities of different iron(III) hydrous oxide particles as function of pH. β-FeOOH sols: rods, formed from solution of 0.09 mol dm^{-3} in FeCl$_3$ and 0.01 mol dm^{-3} in HCl aged 24 h at 100°C (– – • • – – • •); α-Fe$_2$O$_3$ sols containing different shaped particles formed in solutions as given below: cubes, 0.09 mol dm^{-3} in FeCl$_3$ and 0.01 mol dm^{-3} in HCl aged 24 h at 150°C (————); ellipsoids, 0.018 mol dm^{-3} in Fe(No$_3$)$_3$ and 0.05 mol dm^{-3} in HNO$_3$ aged 24 h at100°C (– • – •); bipyrimidal, 0.018 mol dm^{-3} in Fe(ClO$_4$)$_3$ and 0.05 mol dm^{-3} in HClO$_4$ aged 3 days at 100°C (• • • • •); double ellipsoids, 0.018 mol dm^{-3} in FeCl$_3$ and 0.05 mol dm^{-3} in HCl aged 1 week at 100°C (— — —), and spheres, 0.0315 mol dm^{-3} in FeCl$_3$ and 0.005 mol dm^{-3} in HCl aged 2 weeks at 100°C (- - - - -).

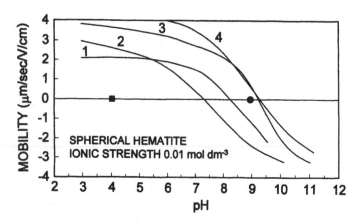

FIG. 11 Electrophoretic mobilities of nearly spherical hematite (α-Fe$_2$O$_3$) particles prepared in different laboratories: 1. Regazzoni et al. [29], $r = 50$ nm; 2. Hesleitner et al. [10], $r = 120$ nm; 3. Shudel et al. [30], $r = \sim 100$ nm; 4. Penners et al. [31], $r = 40$ nm. ●, Hesleitner et al. [32], $r = 96$ nm; ■, Kandori et al. [23], $r = 800$ nm.

Finally, Fig. 13 compares data obtained with needle-type β-FeOOH particles of different length from two laboratories. While the i.e.p. is reasonably close for all samples, the mobilities differ dramatically above and below this pH value.

C. Aluminum (Hydrous) Oxides

Fewer cases are available with well-defined aluminum (hydrous) oxides than with the above described systems. However, in view of the importance of these compounds in a large number of applications, Fig. 14 summarizes the best reported data. There is less fluctuations in the value of the i.e.p. than, for example, with iron oxides, but still little reproducibility in the overall mobilities is noted.

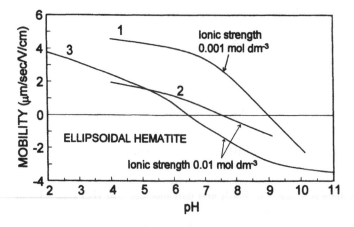

FIG. 12 Electrophoretic mobilities of ellipsoidal hematite (α-Fe$_2$O$_3$): 1. Haq and Matijević [33], $\sim 400/100$ nm; 2. Delgado and González-Caballero [34], $\sim 180/70$ nm; 3. Garg and Matijević [35], $\sim 600/150$ nm.

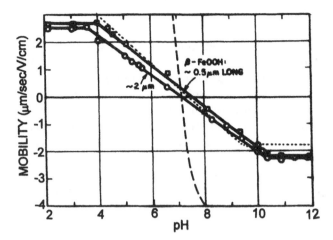

FIG. 13 Electrophoretic mobilities of needle-type alkageneite (β-FeOOH) particles: ($\square$,$\bigcirc$) two different rod sizes [36]; ($\cdots\cdots$) retrace of data shown in Fig. 10; ($---$) data from Ref. 37.

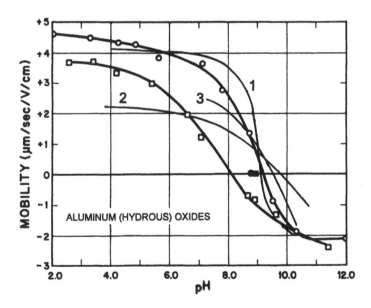

FIG. 14 Electrophoretic mobilities of aluminum (hydrous) oxide particles. 1. The same purified dispersion as in Fig. 5 [12]. 2. Amorphous spherical aluminum hydroxide particles (prepared as in 1) of $r \sim 260$ nm at $\mu = 1 \times 10^{-2}$ mol dm^{-3} [38]. 3. Spherical aluminum hydroxide particles prepared by the aerosol technique (Al sec-butoxide droplets hydrolyzed with water vapor); $r = 350$ nm, $\mu = 1 \times 10^{-3}$ mol dm^{-3} [39]. $\square$ and $\bigcirc$, α-AlOOH particles prepared by forced hydrolysis of Al(ClO$_4$)$_3$ and AlCl$_3$, respectively [40]. Isoelectric points from plots of ζ potentials: $\bullet$, particles prepared as in 1 and 2 [41]; $\blacksquare$, γ-alumina, $r = 10$ nm [20].

D. Silica

Monodispersed silica has been prepared over a broad range of modal sizes, i.e., from several nanometers to several micrometers. The exceedingly uniform particles are usually obtained by the hydrolysis of tetraethyl orthosilicate (TEOS), following the process developed by Stöber et al. [42]. Figure 15 displays mobility data determined by microelectrophoresis of two such dispersions with particles of somewhat different size [43, 44]. The curve given by triangles was essentially produced with silica particles of radius $r = 400$ nm by Hsu et al. [45]. The full circle shows the i.e.p. from plots of the ζ potential as a function of the pH for silica of $r = 350$ nm at an ionic strength of 1×10^{-3} mol dm^{-3} [43, 46]. In another study of Stöber silica of $r = 90$ nm, the i.e.p. was reported at pH ~ 3 [47]. Included also is the curve for nanosized Ludox HS silica, as determined by the moving boundary technique [48], which yields a much lower i.e.p at pH $<$ 2.

It is interesting that a finely dispersed quartz (Sikron) of 14 μm in radius had an i.e.p. at pH 4 [49]. Obviously, the difference in the i.e.p. between the very small (nanosized) and larger colloidal particles is rather significant.

E. Other Compounds

Electrokinetic data for a number of inorganic monodispersed colloids of various compositions have been reported, but mostly in small enough numbers of examples to justify a comparison. For this reason, Table 1 lists only the i.e.p. for a few such dispersions with some indication of their morphological properties. This kind of presentation was also opted because some results were extracted from plots of mobilities, while others were from ζ potentials. For details the reader should consult the indicated references.

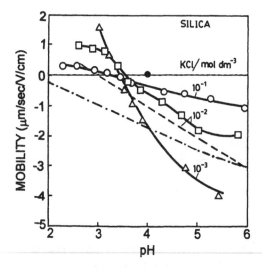

FIG. 15 Electrophoretic mobilities of spherical silica particles prepared by the Stöber method, $r =$ 150 nm, at three ionic strengths (○, □, △) [43] and of $r = 100$ nm (– – –) [44]. ●, i.e.p. of a silica ($r = 300$ nm) from plots of ζ potentials as a function of pH [46]; (– • – •) Ludox HS silica, $r = 15$ nm [48].

TABLE 1 Isoelectric Points of Some Monodispersed Inorganic Colloids

Compound	Particle shape and size	i.e.p.	Reference
MnO, MnO$_2$	Spheres (0.5–2.0 μm)	6.0	50
MnO$_2$	Spheres, monoclinic (0.5 μm, ~ 3 μm)	4.0	51
Co$_3$O$_4$	Cubes	5.5	52
Co(OH)$_2$	Not given	10.5	53
NiO	Spherical (~ 2 μm)	9.5	54
NiO	Platelets (100 nm)	12.7	55
Ni(OH)$_2$	Platelets (100 nm)	9.3	55
CuO	Ellipsoids (100 nm long)	7.5	56
CuO	Ellipsoids (1 μm long)	5.0	57
Cu(OH)$_2$	Needles (500 nm long)	9.5–10.5	58
ZrO$_2$	Spheres (400 nm)	4.0	59
ZrO$_2$	Hollow spheres (170 μm)	4.5	60
Zr (hydrous) oxide	Spheres (20 μm)	8	61
Zr (hydrous) oxide	Spheres (0.1–1 μm)	5.9	62
SnO$_2$	Spheroids (50 nm)	4.2	63
Sn(OH)$_4$	Spheroids (~ 200 nm)	4.5	33
Sb$_2$O$_4$	Prizmatic crystals (2 μm)	7.5	64
Y(OH)(CO$_3$)	Spheres (~ 150 nm)	7.5	65, 66
		8.6	67
Y$_2$O$_3$	Spheres (~ 120 nm)	8.5	68
Y$_2$O$_3$		9.1	67
Y$_2$O$_3$	Hollow spheres	8.5	68
Eu(OH)CO$_3$	Spheres (~ 150 nm)	7.5	69
Gd(OH)CO$_3$	Spheres (~ 150 nm)	7.5	69
CeO$_2$	Spheres (~ 100 nm)	6.0	70
	Calcined	5.2	
ZnS	Spheres (~ 300 nm)	5.5	71, 72
	Spheres	3.0	73
CdS	Spheres (400 nm)	3.5	74
BaSO$_4$	Spheroids (200 μm)	6.5	75

V. COMPOSITE PARTICLES

Electrokinetics may yield useful information with regard to internally or externally composite colloids, as illustrated below.

A. Internally Composite Particles

It has been demonstrated on a number of examples that uniform particles of internally mixed composition can be obtained by precipitation in solutions containing two or more different metal salts [76]. Depending on the method of preparation the resulting colloids can be either internally homogeneous or inhomogeneous. As a rule, if the process is slow the latter case prevails. As a result the composition of particles varies from the center to the periphery. Figure 16 illustrates this effect on the mixed cadmium/nickel basic phosphate system. Uniform spherical particles were obtained by aging at 80°C solutions containing metal sulfate salts, phosphoric acid, urea, and sodium dodecyl sulfate [77]. The ratio of the two metals changes as the particles grow, with Cd in excess, until the solute content is exhausted. It is to be expected that the surface charge characteristics should change accordingly.

Indeed, electrokinetic measurements are indicative of the surface composition, as shown in the example of monodispersed spherical silica/alumina particles, prepared by the cohydrolysis of mixed alkoxides [TEOS and Al(t-OBu)$_4$] in different molar ratios [78]. Although in all cases TEOS was in excess in the reacting system and silicon dioxide in the resulting particles, the mobilities clearly indicated varying surface compositions (Fig. 17). Thus, particles precipitated in a solution of molar ratio [Si]/[Al] = 4 had an i.e.p. of 9.2, typical of alumina, while those prepared in a solution of [Si]/[Al] = 8, behaved as pure silica. Finally, the solids produced in systems of [Si]/[Al] = 6 showed the i.e.p. corresponding to an aluminosilicate surface.

In another example, spherical nickel ferrites were prepared with different [Ni]/[Fe] ratios, yet all mobilities as a function of the pH yielded a single curve (Fig. 18) [79]. Since

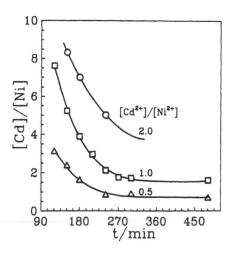

FIG. 16 Internal composition of mixed nickel basic and cadmium phosphate particles, precipitated in solutions of initial molar ratios [Cd^{2+}]/Ni^{2+}] = 2.0, 1.0, and 0.5, respectively, as a function of aging time at 80°C [77].

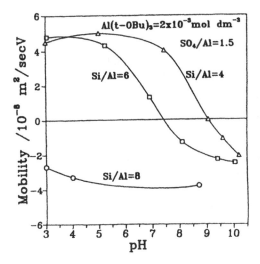

FIG. 17 Electrophoretic mobilities as a function of the pH of dispersions prepared from solutions of initial concentration ratio [TEOS]/[Al(t-OBu)$_3$] of 4 ($\triangle$), 6 ($\square$), and 8 ($\bigcirc$) at a constant concentration of Al(t-OBu)$_3$ = 2×10^{-3} mol dm^{-3} and the ratio [SO$_4$]/[Al] = 1.5, all aged at 98°C for 24 h [78].

the i.e.p. of NiO and Fe$_2$O$_3$ are quite different, one would expect that the electrokinetic behavior of these ferrites should systematically shift with the change in the chemical composition of the particles. However, the experimental evidence indicates that the surface potentials are independent of the [Ni]/[Al] ratio in these ferrites, which leads to the conclusion that the surface chemical composition is essentially the same, while that of the bulk varies.

In both these examples, electrophoretic mobility measurements offered information on the chemical properties of particle surfaces in a rapid manner, which would be difficult or time consuming to achieve by other more sophisticated techniques.

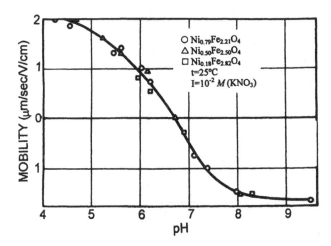

FIG. 18 Electrophoretic mobilities of nickel ferrite particles of various chemical composition as a function of the solution pH [79].

B. Coated Particles

Covering particles with a shell of different chemical compositions is a useful method to achieve some desirable surface or morphological properties of colloids, which cannot be obtained directly. For example, by coating ellipsoids of hematite (α-Fe$_2$O$_3$) with a silica layer, the resulting dispersion will react as the latter, yet the particles will be of anisometric shape, which cannot be produced with silica by the usual preparation procedures [80].

Recently, it was shown that inorganic particles can be covered with either inorganic or organic layers, but also that organic colloids can be enveloped with inorganic shells. Indeed, the rather diverse cores are readily coated with material of the same chemical composition. For example, latex, silica, and hematite have been covered with yttrium basic carbonate, which can be converted into yttrium oxide, using essentially the same procedure [44, 66, 81].

Electrokinetic measurements are again useful in confirming that such coating has taken place, since the surface charge characteristics of cores and shells are usually different. Some example will illustrate such effects.

Figure 19 shows electrophoretic mobilities of spherical hematite particles, of ovalbumin, and of the same cores covered with a uniform layer of this protein. The coated iron oxide has not only the same i.e.p., but also the mobilities as a function of the pH are essentially identical, within experimental error, with those of the pure ovalbumin. Consequently, the composite particles will exhibit the properties of the protein in terms of the surface potential and interactions [82].

In another example, hematite particles were covered with Sn(OH)$_4$ [33]. Again, the coated particles behaved just as the shell material prepared separately, as displayed in Fig. 20.

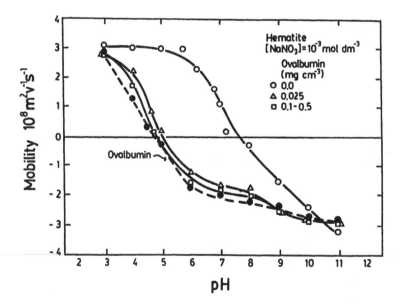

FIG. 19   Electrophoretic mobilities as a function of the pH of spherical hematite (α-Fe$_2$O$_3$) particles ($\bigcirc$), of ovalbumin (1 wt%) ($\bullet$), and of hematite coated with this protein ($\square$, $\triangle$).

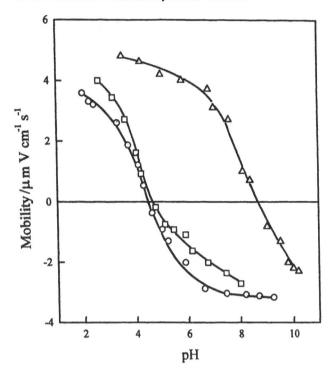

FIG. 20 Electrophoretic mobilities as a function of the pH of hematite cores ($\triangle$), of Sn(OH)$_4$ ($\bigcirc$), and of hematite coated with Sn(OH)$_4$ ($\square$).

Analogous behavior was documented with a number of other systems, such as hematite coated with chromium hydroxide [36], polystyrene latex coated with yttrium basic carbonate [68], manganese carbonate coated with nickel carbonate, [85], or titania coated with polyurea by the aerosol method [16], to mention a few.

VI. CONCLUSIONS

The results of this survey are certainly disturbing. Despite the fact that the comparisons of the published electrokinetic mobility data refer to reasonably well defined dispersions, the discrepancies are outside expected experimental errors. The causes of the disagreements are difficult to pinpoint. While they may be due in some cases to inaccurate measurements, it is more likely that the differences are caused by minor inconsistencies in the surface chemistry of presumably identical dispersed particles. It is especially important to recognize how a little contamination may affect the electrokinetic behaviors in a significant manner.

These results also cause concern regarding interpretations of different colloid phenomena, especially in terms of generalizations based on surface potentials. In most cases the latter are identified with the ζ potentials, which are calculated from mobility measurements using various assumptions. Consequently, assigning quantitative significance to these numbers is even less justified.

In view of these findings, it is essential to select a "standard" dispersion that should be evaluated in a number of laboratories with different instrumentation in order to arrive at some meaningful comparisons. These tests should be carried out with one sample distributed to different locations and again with the dispersions of the same material prepared independently by exactly the same procedure in these laboratories.

On a positive note, the usefulness of electrokinetics is demonstrated when employed as a tool to detect changes in surface charge characteristics, caused by minor contaminations inherent in the system or by additives intentionally introduced into a dispersion. In such instances the absolute values of mobilities (i.e., ζ potentials) are not as essential as are the relative changes in these parameters. Indeed, the shifts in the magnitude of the potentials or the reversal of the sign of the surface charge may be indicative, at least qualitatively, of some changes of particles behavior in adsorption, adhesion, or other processes, as well as their stability when dispersed in liquids.

Electrokinetic measurements remain to be an indispensable tool, as stated in Section I, but the results must be treated judiciously.

REFERENCES

1. E Matijević. Langmuir 10: 8, 1994.
2. Fine Particles, Synthesis, Characterization, and Mechanisms of Growth (T Sugimoto, ed.), Marcel Dekker, 2000.
3. E Matijević. Chem Mater 5: 412, 1993.
4. E Matijević. Annu Rev Mater Sci 15: 483, 1985.
5. R Demchak, E Matijević. J Colloid Interface Sci 31: 257, 1969.
6. E Matijević, AD Lindsay, S Kratohvil, ME Jones, RI Larson, NW Cayey. J Colloid Interface Sci 36: 273, 1971.
7. R Sprycha, E Matijević. Langmuir 5: 479, 1989.
8. R Sprycha, E Matijević. Colloids Surfaces 47: 195, 1990.
9. JJ Spitzer, LJ Danielson, LG Hepler. Colloids Surfaces 3: 321, 1981.
10. P Hesleitner, D Babić, N Kallay, E Matijević. Langmuir 3: 815, 1987.
11. E Matijević, P Scheiner. J Colloid Interface Sci 63: 509, 1978.
12. R Brace, E Matijević. J Inorg Nucl Chem 35: 3691, 1973.
13. GA Parks. Chem Rev 65: 177, 1965.
14. DN Furlong, GD Parfitt. J Colloid Interface Sci 65: 548, 1978.
15. M Visca, E Matijević. J Colloid Interface Sci 68: 308, 1979.
16. FC Mayville, RE Partch, E Matijević. J Colloid Interface Sci 120; 135, 1987.
17. E Matijević, Q Zhong, RE Partch. Aerosol Sci Technol 22: 171, 1995.
18. P Gherardi, E Matijević. J Colloid Interface Sci 109: 57, 1986.
19. EA Barringer, H Kent Brown. Langmuir 1: 420, 1985.
20. GR Wiese, TW Healy. J Colloid Interface Sci 51: 427, 1975.
21. M Kosmulski, E Matijević. Colloids Surfaces 64: 57, 1992.
22. HG Peners, LK Koopal. Colloids Surfaces 19: 337, 1986.
23. K Kandori, A Yasukawa, T Ishikawa. J Colloid Interface Sci 180: 446, 1996.
24. S Hamada, E Matijević. J Colloid Interface Sci 84: 274, 1981.
25. S Hamada, E Matijević. J Chem Soc, Faraday Trans 1 78; 2147, 1982.
26. M Ozaki, S Kratohvil, E Matijević. J Colloid Interface Sci 102: 146, 1984.
27. M Ozaki, E Matijević. J Colloid Interface Sci 107: 199, 1985.
28. M Ozaki, N Ookoshi, E Matijević. J Colloid Interface Sci 137: 546, 1990.
29. AE Regazzoni, MA Blesa, AJG Maroto. J Colloid Interface Sci 122: 315, 1988.

30. M Schudel, SH Behrens, H Holthoff, R Kreutzschmar, M Borkovec. J Colloid Interface Sci 196: 241, 1997.
31. HG Penners, LK Koopal, J Lyklema. Colloids Surfaces 21: 457, 1986.
32. P Hesleitner, N Kallay, E Matijević. Langmuir 7: 178, 1991.
33. I ul Haq, E Matijević. Progr Colloid Polym Sci 109: 185, 1998.
34. AV Delgado, F Gonzáles-Caballero. Croat Chem Acta 71: 1087, 1998.
35. A Garg, E Matijević. Langmuir 4: 38, 1988.
36. E Matijević, Y Kitazawa. Colloid Polymer Sci 261: 527, 1983.
37. SB Kanungo. J Colloid Interface Sci 162: 86, 1994.
38. H Sasaki, E Matijević, E Barouch. J Colloid Interface Sci 76:319, 1980.
39. BJ Ingebrethsen, E Matijević. J Aerosol Sci 11:271, 1980.
40. WB Scott, E Matijević. J Colloid Interface Sci 66: 447, 1978.
41. FK Hansen, E Matijević. J Chem Soc, Faraday Trans I 76: 1240, 1980.
42. W Stöber, A Fink, E Bohn. J Colloid Interface Sci 26: 62, 1968.
43. M Kosmulski, E Matijević. Langmuir 7: 2066, 1991.
44. H Giesche, E Matijević. J Mater Res 9: 436, 1994.
45. WP Hsu, R Yu, E Matijević. J Colloid Interface Sci 156: 56, 1993.
46. M. Kosmulski and E. Matijević. Langmuir 8: 1060 (1992).
47. HA Kettlson, R Pelton, MA Brook. Langmuir 12: 1134, 1976.
48. LH Allen, E Matijević. J Colloid Interface Sci 31: 287, 1969.
49. D Bauer, A Fuchs, W Jaeger, E Killmann, K Lunkwitz, R Rehmet, S Schwarz. Colloid Surf. 156: 291, 1999.
50. I ul Haq and E Matijević. J Colloid Interface Sci 192: 104, 1997.
51. S Hamada, Y Kudo, J Okada, H Kano. J Colloid Interface Sci 118: 356, 1987.
52. T Sugimoto, E Matijević. J Inorg Nucl Chem 41: 165, 1979.
53. RO James, TW Healy. J Colloid Interface Sci 40: 53, 1972.
54. I ul Haq, E Matijević, K Akhtar. Chem Mater 9: 2659, 1997.
55. L Durand-Keklikian, I ul Haq, E Matijević. Colloids Surfaces A 92: 267, 1994.
56. SH Lee, YS Her, E Matijević. J Colloid Interface Sci 186: 193, 1997.
57. P McFadyen, E Matijević. J Inorg Nucl Chem 35:1883, 1973.
58. L Durand-Keklikian, E Matijević. Colloid Polym Sci 268: 1151, 1990.
59. B Aiken, WP Hsu, E Matijević. J Mater Sci 25: 1886, 1990.
60. N Kawahashi, C Persson, E Matijević. J Mater Chem 1:577, 1991.
61. LA Pérez-Maqueda, E Matijević. J Mater Res 12: 3286, 1997.
62. MA Blesa, AJG Maroto, SI Passaggio, NE Figliolia, G Rigotti. J Mater Sci 20: 4601, 1985.
63. M Ocaña, E Matijević. J Mater Res 5: 1083, 1990.
64. A Koliadima, LA Pérez-Maqueda, E Matijević. Langmuir 13: 3733, 1997.
65. B Aiken, AP Hsu, E Matijević. J Am Ceram Soc 71: 845, 1988.
66. N Kawahashi, E Matijević. J Colloid Interface Sci 138: 534, 1990.
67. R Sprycha, J Jablonski, E Matijević. J Colloid Interface Sci 149: 561, 1992.
68. N Kawahashi, E Matijević. J Colloid Interface Sci 143: 103, 1991.
69. E Matijević, WP Hsu. J Colloid Interface Sci 118: 506, 1987.
70. WP Hsu, L Rönnquist, E Matijević. Langmuir 4: 31, 1988.
71. JDG Durán, A Ontiveros, E Chibowski, F Gonzalés-Caballero. J Colloid Interface Sci 214: 53, 1999.
72. JDG Durán, MC Guindo, AV Delgado. J Colloid Interface Sci 173: 436, 1995.
73. D Murphy Wilhelmy, E Matijević. J Chem Soc, Faraday Trans I 80: 563, 1984.
74. E Matijević, D Murphy Wilhelmy. J Colloid Interface Sci 86: 476, 1982.
75. G Pozarnsky, E Matijević. J Adhesion 63: 53, 1997.
76. E Matijević. In: E Pelizzetti, ed. Fine Particles Science and Technology. Vol. 12. NATO ASI, Series 3. Kluwer Academic, 1996, pp 1–16.
77. J Quibén, E Matijević. Colloids Surfaces 82:237, 1994.
78. S Nishikawa, E Matijević. J Colloid Interface Sci 165: 141, 1994.

79. AE Regazzoni, E Matijević. Corrosion 38: 212, 1982.
80. M Ohmori, E Matijević. J Colloid Interface Sci 150: 594, 1992.
81. B Aiken, E Matijević. J Colloid Interface Sci 126: 645, 1988.
82. JE Johnson, E Matijević. Colloid Polym Sci 270: 364, 1992.

9

Towards a Standard for Electrophoretic Measurements

KUNIO FURUSAWA University of Tsukuba, Tsukuba, Ibaraki, Japan

I. INTRODUCTION

Zeta-potential measurements can provide valuable information necessary for preparing stable colloidal suspensions in many applications, including food preparation, agriculture, pharmaceuticals, the paper industry, ceramics paints, coatings, photographic emulsions, etc. The concept of zeta potential is also very important in such diverse processes as environmental transport of nutrients, sol–gel synthesis, mineral recovery, wastewater treatment, corrosion, and many more. The historical prominence of zeta potential, ζ, in colloid and surface science has been due to its experimental accessibility via measurement of the electrophoretic mobility, μ which is the terminal velocity of the particle, v, per unit field strengtht, E:

$$v = \mu E \tag{1}$$

In an externally applied electric field, ζ can be calculated from the mobility using the simple Smoluchowski relationship [1]:

$$\mu = \frac{\varepsilon \zeta}{\eta} \tag{2}$$

where ε and η are the dielectric constant and viscosity of the medium, respectively. At present, a continuing problem with the measurement of the electrophoretic mobility, particularly where quantitative information is required, is the lack of universally recognized standards for interlaboratory comparison and calibration of the various commercial instruments. This problem is compounded by the observed variability between measurements on similar materials obtained from different laboratories or users [2, 3]. There are many possible instrumental origins of this variability, including improper alignment of measurement optics, incorrect determination of the cell position, dirty cell walls, and deteriorated electrodes. Variations may also occur due to sample-related problems such as contamination, chemical and dispersion instability, errors in pH measurement, etc. The research community has long recognized the need for mobility standards to address these and related problems [4, 5].

The selection of a reference material is not a trivial matter: colloidal systems are thermodynamically unstable, difficult to prepare reproducibly, subject to contamination, and often chemically unstable over long periods. As a consequence of their high surface-to-volume ratio, finely dispersed systems are particularly subject to physical and chemical changes by processes such as aging, flocculation, dissolution, photochemical degradation, and Ostwald rippening. Latex particles are often suggested as model colloids and mobility standards [6–8], primarily because they can be prepared as monodisperse spheres. Inorganic materials such as sulfated chromium hydroxide [9] and hematite (α-Fe_2O_3) [10] have also been proposed as standards.

In this chapter, the results of the collaborative studies of zeta potential measurements conducted in Japan to seek a suitable standard sample will be introduced first [11, 12]. The measurements were carried out, in addition to those at Tsukuba University, at nine university laboratories in Japan using different electrophoretic apparatus, which included the Rank Brother M-2, PEN KEM-501, PEM KEM-3000, and Malvern Zetasizer-F.

In Section III, based on the above collaborative studies, a new electrophoretic technique, using a reference sample as a standard, is introduced [13, 14]. The most familiar method for determining the zeta potential is electrophoresis, which consists of setting up a potential gradient in the solution. The electrophoretic migration of colloid particles is always superimposed on the electro-osmotic liquid flow from the cell wall, and the apparent particle velocity observed coincides with the true electrophoretic mobility only at two special cell depths which are known as the stationary levels. Unfortunately, however, the velocity gradient of the liquid at the stationary levels is usually large, and thus the observed velocity of the particles changes rapidly with the cell depth, so that the errors in electrophoretic mobility measurements may be substantial. However, if the electrophoretic measurements are carried out using some reference sample as a standard, the electrophoretic mobility of the unknown sample can be determined at any cell depth by substracting the mobility of the reference particles at the same level, because the liquid flow velocity induced by the electro-osmotic effect of the cell wall comes to the same value for both kinds of particles under the same experimental conditions. So, if the electrophoretic mobility is detected at one-half depth in the cell, the real electrophoretic mobility is also given at the maximum of the parabolic velocity as a function of the cell depth. At that point, slight errors in focusing are less important than the usual measurements at the stationary levels.

Section IV reports on the synthesis of a new reference colloidal dispersion stabilized only sterically in an aqueous medium, without any electrostatic repulsion [14, 15]. Special attention is given to the synthesis of the reference particles, which are covered with polymer layers as densely as possible [14] and display a zero zeta potential under a variety of solution conditions. It becomes apparent that a dense adsorption layer of hydroxypropyl cellulose (HPC), with a lower critical solution temperature (LCST), formed on latex particles with a low surface charge density at temperatures higher (60–80°C) than the LCST plays a role in completely shielding the electrostatic effect arising from the surface charge on the bare particles [15, 16]. Here, the experimental method and the merit of a new electrophoretic technique based on the use of a reference sample will be explained.

In Section V, a recent joint work [17] in the USA to select and evaluate a colloid material for use as a mobility standard in electrophoretic light scattering will be described. A published report summarizing this work concluded that an acidic suspension of well-aged phosphated goethite (α-FeOOH) would be a suitable reference material for the determination of electrophoretic mobility. This recommendation was based on such criteria as adjustable mobility over a wide pH range, low solubility, temporal stability, kinetic stability, sample reproducibility, availability, and sufficient light scattering, among others.

II. COLLABORATIVE STUDY OF ZETA POTENTIAL MEASUREMENTS IN JAPAN USING A REFERENCE SAMPLE

In 1970, a group of Japanese surface and colloid chemists, who had engaged in the study of electrokinetic phenomena and/or colloid stability, formed a committee under the Division of Surface Chemistry in the Japan Oil Chemist's Society. This group measured, compared, and discussed zeta potentials for such samples as titanium oxide, microcapsules, silica, and some polymer lattices or silver iodide. Table 1 indicates some examples of simultaneous measurements of zeta potential for four samples. All these measurements have been carried out by the microelectrophoretic technique using the respective electrophoretic apparatus belonging to each member laboratory. Usually, the mobility measurements were performed at a constant field strength of 4–5 V cm^{-1}, using a rectangular glass (or quartz) cell, and the mobility was determined at the respective stationary levels in the cell. The zeta potential was calculated by the Smoluchowski equation (2). As can be seen from Table 1, the collaborative results for titanium oxide and microcapsule dispersions do not agree with each other and, in particular, the data for the microcapsule dispersion indicated a large deviation within some members. On the other hand, the ζ potential measured for polystyrene lattices, which were prepared in a surfactant-free system by the Kotera–Furusawa–Takeda method [6], and silver iodide sol prepared by the usual way, were very similar among those tested and were reproducible. Furthermore, it is known from the data for titanium oxide that the zeta potential determined by the streaming potential method is slightly lower than the value obtained by the electrophoretic (EP) technique.

At the next stage of the collaborative study, simultaneous measurements of ζ potential and critical flocculation concentration (c.f.c) were conducted for two standard latex suspensions (Samples 1 and 2) so as to assess the reliability of each of these measurement

TABLE 1 Results of Collaborative Studies of Zeta-Potential Measurements for Various Samples

1. *Titanium oxide*						
Laboratory	A	B	C	D	E	F
Method	EP	SP	SP	SP	EP	SP
ζ (mV)	−24.8	−14.8	−8	−21.0	−35.72	−25
2. *Polystyrene lattices (EP)*						
Laboratory	A	B	C	D		
ζ (mV)	−48	−42.5–46	−47	−42.3		
3. *Microcapsule (EP)*						
Laboratory	A	B	C	D		
ζ (mV)	−27.0	−26.1	−70	−35.4		
4. *AgI sol (EP)*						
Laboratory	A	B	C	D		
ζ (mV) (pA g 3)	+50			+66.4		
ζ (mV) (pI 4)		−47.5–51	−45			
5. *AgI sol (EP)*						
Laboratory	A	B	C			
U (μm cm^{-1} V^{-1} s^{-1})	−3.675	−3.8	−3.67			
ζ (mV)	−44.5	−46	−44.4			

techniques and find means for their improvement [12]. Sample 1 was a negative styrene/ styrene sulfonate copolymer latex which was prepared according to the Juang and Krieger method [18]. Sample 2 was an amphoteric polymer latex synthesized by the Homola method [19]. These latex samples were employed after purification by ion-exchange treatment and dialysis against distilled water.

The zeta potential measurements were carried out at nine laboratories using different electrophoretic apparatus, which included the Rank Brother M-2, Pen Kem-501, Pen Kem-3000 and Malvern Zetasizer-F. As can be seen from Fig. 1 and Table 2, the data from all the laboratories showed fairly good agreement and displayed possible means for their improvement:

1. Elevated potential supply (Pen Kem-3000 and Malvern Zetasizer-F) was found to increase the slope of the zeta potential versus pH curve near the isoelectric point of Sample 2.

2. Differences between the zeta-potential values reported from each laboratory indicate a constant deviation over the whole pH range, i.e., data G show relatively high values, while data A indicate low values over the whole pH range. This tendency suggests that the difference between data G and data A must be

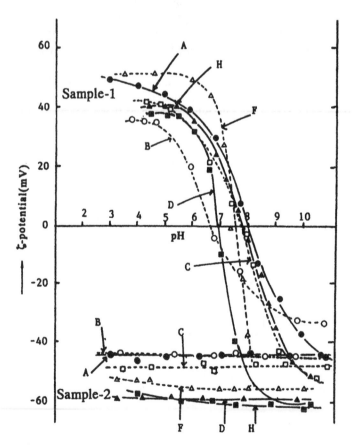

FIG. 1 Zeta potential versus pH curves for Samples 1 and 2. A, Rank Brother M-2; B, PEM KEM-500; C, Rank Brothers M-2; D, E, Laser–Doppler method; E, PEN KEM-3000; F, Malvern Zetasizer-F.

TABLE 2 Zeta-Potentials of Sample 1 and Isoelectric Points (pH^0) of Sample 2

Laboratory		A	B	C	D	E	F	G	H	I
Sample 1 (ζ, mV)	pH = 4	−44.0	−45.0	−48.0	−57.7	−58.0	−53.0	−62.0	−59.0	−44.5
	pH = 10	−44.0	−45.0	−48.0	−60.5	−58.0	−54.0	−65.0	−62.0	−46.0
Sample 2	pH^0	8.0	6.6	7.8	6.8	7.1	7.6	8.2	7.8	6.5

based on the same reason, which can be eliminated completely by using a definite standard simple.

The c.f.c. values for KNO_3, $Mg(NO_3)_2$, and $La(NO_3)_3$ for Sample 1 were determined simultaneously by a static method, and the results were compared with those determined by a kinetic method and dynamic light scattering (LS). These results for the c.f.c. and some related data are presented in Table 3. The results obtained from the four laboratories by a static method agreed fairly well, and the order of the magnitudes of the c.f.c. values determined by the LS technique and the kinetic method was static < LS technique < kinetic method. Furthermore, it was found that in determining the c.f.c., the solid concentration in the medium was an important factor.

III. ELECTROPHORETIC MEASUREMENTS USING A REFERENCE SAMPLE AS A STANDARD

From the above collaborative measurements on the zeta potential, it was realized that in determination of the zeta potential of an unknown sample, electrophoretic measurements using the Sample 1 in the above section as a standard are very useful for obtaining reliable data. Zeta-potential measurements using (PSSNa) lattices as a standard were conducted. Sample 1 ($2a = 420$ nm, $\sigma_0 = 7.0 \ \mu C \ cm^{-2}$) was made by incorporating small amounts of an ionic comonomer, sodium p-vinylbenzene sulfonate, in a polystyrene chain according to Juang and Krieyer [18].

The electrophoretic mobility measurements in this work were carried out by using a Zeecom IP-120B zeta-potential analyzer (Chiba, Japan). The apparatus performs automatic

TABLE 3 Critical Flocculation Concentration (c.f.c) for Sample 1 and Some Related Data

Laboratory	KNO_3 c.f.c (M)	ζ^a (mV)	$Mg(NO_3)_2$ c.f.c (M)	ζ^a (mV)	$La(NO_3)_3$ c.f.c (M)	ζ^a (mV)
A	1.1×10^{-1}	−40	4.2×10^{-3}	−35	1.8×10^{-4}	−19
B	1.01×10^{-1}	−39.4	4.6×10^{-3}	−20.9	2.2×10^{-4}	−3.7
C^b	2.0×10^{-1}	−28	2.0×10^{-2}	−23	3.0×10^{-4}	−11
D	0.7×10^{-1}	−41	5.4×10^{-3}	−35	1.9×10^{-4}	−14
A^c			5.6×10^{-2}	−15		
D^d			1.05×10^{-3}			

[a] Zeta-potential at c.f.c.
[b] Data measured by diluted dispersed system.
[c] Data measured by kinetic method.
[d] Data measured by dynamic light scattering technique.

tracking of the particles in the center of a 10 cm long electrophoretic cell [14]. The particle velocities were measured by frequently changing the direction of the field to minimize possible errors from cell leakage. At least 50 particles were counted in each measurement.

The apparent electrophoretic mobility (U_{app}) of an unknown colloid sample is always the sum of two contributions, one of which is the real electrophoretic mobility U_{el} and the other is the liquid flow velocity induced by the electro-osmotic effect U_{osm}) of the cell wall, which changes as a parabolic function of the cell depth h:

$$U_{app} = U_{el} + U_{osm} \tag{3}$$

$$U_{osm} = \frac{U_o}{2}\left(3\frac{h^2}{b^2} - 1\right) \tag{4}$$

where b is the half-thickness of the cell and U_o is the electro-osmotic flow at the cell wall ($h = b$). Similarly, the apparent velocity of the reference sample (U'_{app}) is also the sum of the true electrophoretic mobility (U'_{el}) and the electro-osmotic flow velocity (U'_{osm}), i.e.,

$$U'_{app} = U'_{el} + U'_{osm} \tag{5}$$

Under the same experimental conditions using a finite electrophoretic cell, so that, $U_{osm} = U'_{osm}$, the following relation holds, from Eqs (3) and (5):

$$U_{el} = U'_{el} = U_{app} - U'_{app} \tag{6}$$

Equation (6) indicates that if U'_{el} is known exactly, the U_{el} value of any unknown sample can be determined from the difference between the two apparent mobilities at any cell depth. So, if the particle mobility of the unknown sample is determined at the one-half depth in the cell, the actual electrophoretic mobility is given at the maximum of the particle velocity as a function of the cell depth:

$$U_{el} - U'_{el} = U_{app}(max) - U'_{app}(max) \tag{7}$$

Figure 2 shows an example indicating the electrophoretic mobility profiles obtained experimentally for the reference sample (Sample 1) and an unknown sample (SM lattices) along the cell depth in 10^{-3} M KCl solution at 25°C. SM lattices employed as an unknown sample were prepared by copolymerization of styrene with 5% methacrylic acid at 70°C. It is apparent that both profiles indicate reasonable parabolic curves and that the curve for the reference lattices shows a constant mobility at the two stationary levels. Furthermore, the difference between the two apparent mobilities at the cell center agrees well with the velocity of the SM lattices at the stationary level.

Figure 3 shows the zeta potential versus pH curves for the SM lattices which have been determined from the maximum mobilities using Sample 1 as a standard. In Figure 3, the same relation obtained from the velocities of the SM lattices at the stationary level are also indicated. As can be seen, both curves agree fairly well over the whole pH range.

All these results indicate that, if we have a reliable colloid sample whose zeta-potential is exactly determined, the zeta-potential of an unknown sample can be determined precisely from the measurements of apparent electrophoretic mobility at the cell center. In that case, slight errors in focusing, i.e., errors due to the depth of view field are less important, since the velocity gradient near the level of observation is very small.

According to Eq. (4), the zeta potential of the cell wall in contact with aqueous solutions can be measured by means of the plane interface technique, which involves

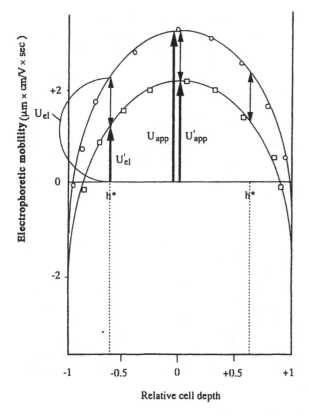

FIG. 2 Examples of electrophoretic mobility profiles of PSSNa lattices (U') and SM lattices (U); h^*, stationary level. (□) PSSNa lattices; (○) SM lattices (10^{-3} M KCl, 25°C).

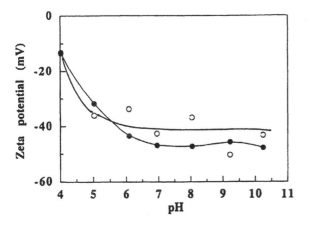

FIG. 3 Zeta-potential versus pH curves of unknown sample (SM lattices) determined by the maximum velocity of reference sample (●) and the usual method (○).

establishing the liquid flow velocity–depth profile using reference (polystyrene latex) particles). The electro-osmotic velocity (U_o) obtained by extrapolation of the velocity profile to the cell wall permits calculation of the zeta potential of the cell wall–solution interface. The plane interface technique for measuring ζ at the air–aqueous solution interface presented by Huddleston and Smith [20] is unique in its use of an open cell, and the zeta-potential measurement of various solid–solution and air–solution interfaces [21], including a dissimilar cell system, has been extensively conducted [22].

Here, we would like to emphasize again that the determination of the zeta-potential of the cell wall can also be possible from the maximum velocity of the reference sample, instead of from the usual plane interface procedure [20]. According to Eq. (3), $U'_{app} = U'_{el} - U'_0/2$ at $h = 0$, i.e., from the measured apparent velocity of the reference sample at the cell center ($U'_0/2$), the zeta potential of the cell wall can be quickly determined, if U'_{el} is previously known.

Figure 4 shows some examples of apparent flow velocity profiles of a standard latex sample (Sample 1) at various pH values in which both boundaries refer to the glass–solution interface. A symmetrical parabola was obtained at all pH conditions where the surface charge of glass is consistent with both sides.

In Figure 5, open circles refer to the zeta potential of the cell wall–solution interfaces which were determined from the maximum velocity of the reference sample. On the other hand, full circles in the same figure indicate the results which were obtained by extrapolation of the liquid flow velocity at the cell wall. It was found that both zeta potential series determined by the two methods agreed very well with each other over a wide range of pH, and it was realized that our new procedure using U'_0 was also useful for determining the zeta potential of the solid–solution interface.

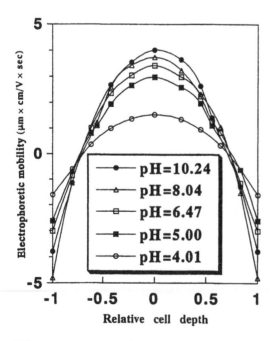

FIG. 4 Apparent flow velocity profile of standard latex sample at various values of pH (10^{-3} M KCl, 25°C).

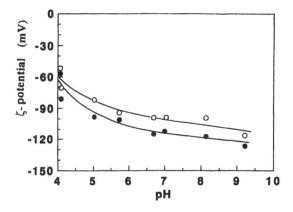

FIG. 5 Zeta-potential versus pH curves of cell wall/solution interface. (○) Determined by the maximum velocity of PSSNa lattices; (●) determined by the plane interface technique.

IV. SYNTHESIS OF A REFERENCE SAMPLE WITHOUT ANY ELECTROSTATIC EFFECT

It will be clearly realized from Section III that, if the electrophoretic measurements are carried out by using a reference sample without any electrostatic effect as a standard, the electrophoretic mobility of an unknown sample could be determined by just subtracting the mobility of the new reference sample at the cell center. Also, the zeta potential of the cell wall can be quickly determined from the measured electrophoretic mobility at the cell center, because U'_{el} of this sample is zero. As our final goal, syntheses of new reference particles that are sterically stabilized in an aqueous medium without any electrostatic effect were studied under various experimental conditions.

It has been reported that the adsorption behavior of hydroxypropyl cellulose (HPC) with a LCST depends significantly on the adsorption temperature [23]. In Fig. 6, the molecular structure of HPC and the molecular weights of HPC-H, HPC-M, and HPC-L are indicated. The maximum adsorption (A_m) of HPC at LCST is several times larger than at room temperature. Furthermore, the high A_m values obtained at the LCST are maintained for a long period at room temperature, and the dense adsorption layer of HPC shows a strong protective action against flocculation of the particles [24]. These results suggest that the dense (or thick) adsorption layer of HPC plays a role in the synthesis of a

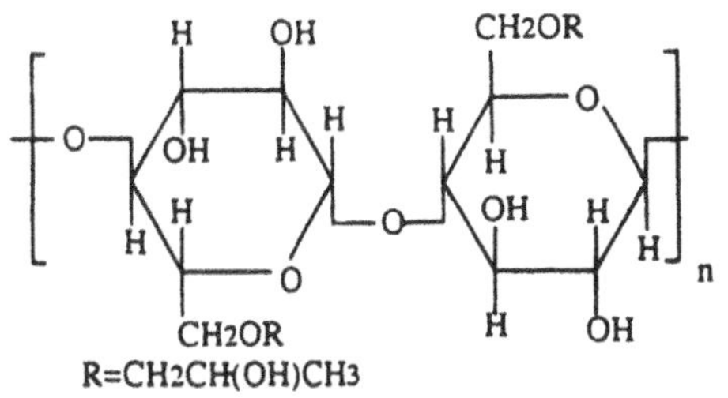

FIG. 6 Molecular structure of hydroxypropylcellulose (HPC). Molecular weights–HPC-H: 925,000; HPC-M: 303,000; HPC-L: 53,000.

new reference sample for determination of the zeta potential of other colloid systems. Special attention has been paid to the synthesis of reference particles that are covered with an HPC layer as densely as possible and display a zero zeta potential under a variety of solution conditions.

Concerning the adsorption of HPC to the latex surface, it has been established that this behavior is remarkably influenced by the surface nature of the original lattices; the lower the surface charge density of the particles (i.e., the stronger the hydrophobic nature of the surface), the higher the adsorption amount [25]. For the sake of clarity, Fig. 7 shows adsorption isotherms of HPC-M on the three kinds of lattices (STL, STSL, and SAL-H) at neutral pH. Latex STSL has a higher surface charge ($\sigma_0 = -9.2 \ \mu C \ cm^{-2}$) than STL, and the charge consists entirely of sulfate groups. The charge density of latex SAL-H ($\sigma_0 = -18 \ \mu C \ cm^{-2}$) is far higher than that of STSL, although it consists of acrylamide groups and carboxyl groups. From Fig. 7, it is apparent that the higher the surface charge, the lower the amount of adsorption. The same trend can be detected in Fig. 8, where the amounts of $K_2S_2O_8$ (KPS) used as an initiator in the latex polymerization are plotted against the saturated amounts of adsorbed HPC for the respective latex surfaces. The surface charge density (σ_0) becomes higher with increasing KPS concentration under the same polymerization conditions.

The adsorption treatments were carried out as follows. The fresh polystyrene lattices prepared by the Kotera–Furusawa–Takeda method [6] were treated with HPC-M (or -H and -L) solutions (0.05–0.08 wt%) at the LCST (45–50°C) for 2 h, and the amounts of HPC adsorbed on the latex surfaces were determined calorimetrically from depletion of the solution [26]. To complete the HPC adsorption, one portion of the dispersion was then heated in an oil bath at 50–80°C by rotating the adsorption tubes for another 20 h. After that, the residual HPC remaining in the medium was completely removed by repeated centrifugation–decantation–redispersion cycles.

In Fig. 9, the experimental results of the second adsorption (heating) process under different regimes are indicated, where the residual amounts of HPC dissolved in each solution are plotted against the duration of the heating process as the temperature is raised to the various upper limits (50–80°C) and subsequently reduced to room temperature (25°C). As can be seen from Fig. 9, as the temperature is raised, the residual con-

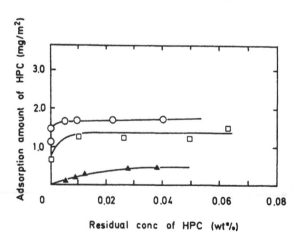

FIG. 7 Adsorption isotherms of HPC-M on three different lattices (48°C, pH 5–8). (○) STL ($\sigma_0 = -1.5 \ \mu C \ cm^{-2}$); (□) STSL ($\sigma_0 = -9.2 \ \mu C \ cm^{-2}$); (▲) SAL-H ($\sigma_0 = -18.0 \ mC \ cm^{-2}$).

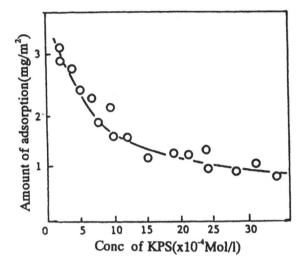

FIG. 8 Relationship between the concentration of KPS and the saturated amount of HPC-M at the LCST.

centration of HPC gradually decreases over time and attains final equilibrium values. Furthermore, the final concentrations of HPC maintain the same values after the temperature in each system is reduced to 25°C. These results indicate that raising the medium temperature contributes to a reduction in the concentration of HPC remaining in the

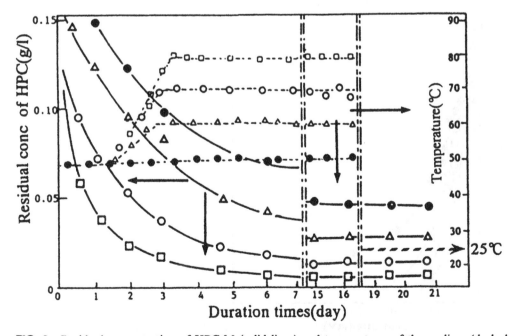

FIG. 9 Residual concentration of HPC-M (solid lines) and temperatures of the medium (dashed lines) as a function of time elapsed during temperature raising and reducing. Heating temperature of the medium: (●) 50°C; (△) 60°C; (○) 70°C; (□) 80°C. Latex sample: diameter 401 nm, $\sigma_o = 0.5\ \mu C$ cm^{-2}.

solution and that the thick (or dense) adsorbed layers of HPC built up at elevated temperatures are maintained on the latex surface after the temperatures are reduced to 25°C.

Figure 10 shows the relationship between the zeta potentials and the adsorption amounts of the HPC-coated lattices, which were removed from the adsorption vessels after various time intervals during the heating process. It is apparent that the negative ζ-potential value of the latex suspension decreases with increasing HPC adsorption amounts and finally attains a real zero when the adsorption amount exceeds 3.0 mg m^{-2} as seen in Fig. 10. Also, Fig. 9 indicates that such a high adsorption amount can be achieved easily with the medium at a high temperature according to the solvency of the medium; i.e., less solvency leads to greater adsorption on the latex surface. It is expected that the adsorption layer of HPC formed at the high medium temperature is so dense that the permittivity of the layer will be nearly zero and brings about a zero zeta-potential of the composite [27].

Now, according to Eqs (3) and (5), the apparent electrophoretic velocity of the new reference samples indicates directly the liquid flow velocity ($U_{ref} = U_{osm}$), because the reference particles are suspended without any electrostatic effects. This fact allows us to determine the U_{el} value of the unknown sample by substracting the mobility of the reference particles from the U_{app} of the unknown sample at the same level, i.e.,

$$U_{el} = U_{app}(\text{max}) - U_{ref}(\text{max}) \qquad (8)$$

Figure 11 shows examples of the electrophoretic mobility profiles obtained by using the new reference sample and unknown sample (SM lattices) along the cell depth in a 10^{-3} M KCl solution. It is apparent that both profiles indicate parabolic curves and that the

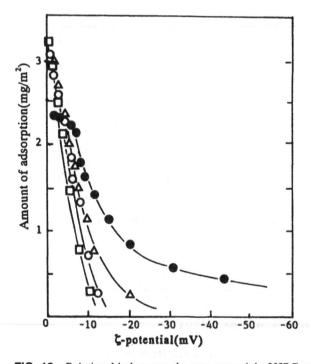

FIG. 10 Relationship between the zeta-potential of HPC-coated lattices and the amount of adsorption of HPC. Adsorption temperature: (●) 50°C; (△) 60°C; (○) 70°C; (□) 80°C. Latex sample: diameter 380 nm, $\sigma_0 = 0.6$ μC cm^{-2}.

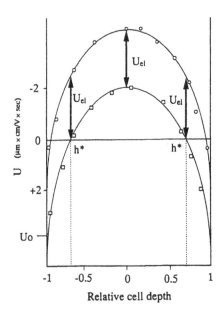

FIG. 11 Electrophoretic mobility profile obtained by using the new reference sample (□) and an unknown sample (○) in 10^{-3} M KCl at 25°C.

curve for the new reference sample shows zero mobility at the two stationary levels. Furthermore, the difference between the two apparent mobilities at the cell center agrees with the velocity of the SM lattices at the stationary levels. All these results indicate that the new reference sample synthesized here is suspended without any electrostatic effects and serves as a good standard in the determination of the zeta potential of the other colloid systems.

Figures 12 and 13 show the ζ values for amphoteric lattices and negative AgI sol determined by using the new reference sample and the usual technique. It was found that the zeta potential series determined by the two methods agree very well with each other over a wide range of pH and KI concentrations.

For extensive application of the reference sample, the stability of the particles is manifested in a few specific examples. In Fig. 14, the mobility of the reference sample at the stationary level is plotted against pH and electrolyte concentration where each sample has been incubated for 24 h at 25°C. It is evident that, over the entire ranges, all series show nearly zero mobility within experimental error and the zero mobility is especially held in the case of the reference sample treated with HPC-M. These results induce us to use the reference sample in many fields of colloid science and technology, because the zeta potential in the new technique can be given at the maximum of the parabolic velocity as a function of the cell depth. In that way, errors in determining the zeta potential are minimized and the measurements of ζ are exact over wide fields of colloid systems.

The real advantage of this new method may lie in the determination of the electrical properties of a solid surface. According to Eq. (4), $U_{osm} (= U_{ref}) = -U_o/2$ at $h = 0$; i.e., from the apparent velocity of the new reference sample measured at the cell center, the zeta potential of a cell wall can be directly determined from the U_o value. Figure 15 shows ζ versus KCl concentration and pH curves of the cell wall measured directly with the usual

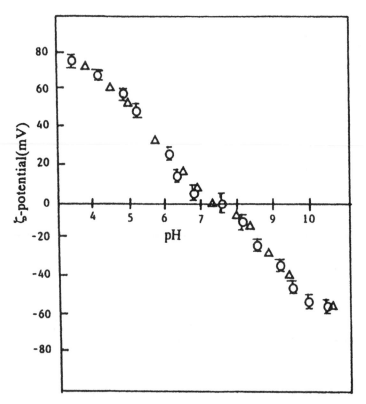

FIG. 12 Determination of the zeta-potential of amphoteric lattices in 10^{-4} M KCl: (○) new technique; (△) usual technique.

plane interface technique and with our new technique using the electrically neutral sample. As can be seen, the two zeta-potential series determined by the two methods agree very well with each other. Our new methods will be more convenient than the conventional plane interface technique or streaming potential measurements.

V. STANDARD REFERENCE MATERIAL FOR MOBILITY MEASUREMENT BY ELECTROPHORETIC LIGHT SCATTERING

Over the past few years, electrophoretic light scattering (ELS) has largely replaced such traditional techniques of mobility measurement as microelectrophoresis and moving-boundary methods. ELS provides rapid, reproducible, and relevant mobility measurements on dilute suspensions of colloids. Furthermore, an important feature of ELS is that it can determine the mobility of very small particles, such as proteins, polyelectrolytes, micelles, and liposomes. In spite of this, a continuing problem in the utilization of ELS, particularly where quantitative information is required, is the lack of universally recognized standards for interlaboratory comparison and calibration of the various commercial instruments.

Tejedor-Tejedor and Anderson [28] first suggested phosphated geothite (α-FeOOH) as a standard material for electrophoresis, and recent joint research supported by the

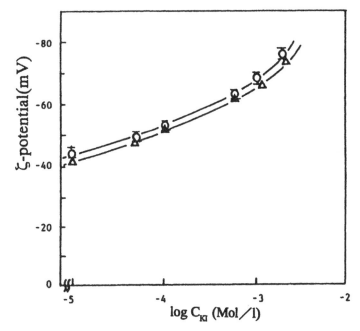

FIG. 13 Determination of the zeta-potential of AgI sol in various KI concentrations: (○) new technique; (△) usual technique.

Department of Energy as part of the Ceramic Technology Project [29], and involving researchers of four other university laboratories, suggested again that phosphated goethite would serve as an absolute standard zeta potential and mobility measurement by ELS. This standard reference material (SRM) is a positive electrophoretic mobility standard containing 500 mg dm^{-3} microcrystalline goethite (α-FeOOH) and 100 mmol g^{-1} phosphate in 0.05 mol dm^{-3} sodium perchlorate electrolyte solution at pH 2.5. This SRM is intended for the calibration and evaluation of ELS instruments and must be diluted prior to use. A certified mobility value (2.53 μm cm V^{-1} s^{-1}) was determined from a statistical analysis of round robin data from five laboratories according to NIST guidelines. The following is the summary of the joint work to recognize why this material would be suitable as a reference for determination of electrophoretic mobility [17].

A. Synthesis Procedure for SRM (Microcrystal Goethite Suspension)

Microcrystalline goethite was synthesized by precipitation and aging of a hydrolyzed ferric nitrate solution (final Fe/ON =2.9) in several batches according to the procedure of Atkinson et al. [30]. The final aged precipitate was washed repeatedly with deionized water, freeze dried and stored in an air-tight polyethylene bottle until needed. The powder obtained from several batches yields a BET single-point surface area of 84 ± 3 m^2 g^{-1}. The resulting particles were acicular with approximate dimensions of 60 nm × 20 nm as determined from electron microscope images.

The SRM suspension was prepared in the following manner. Freeze-dried powder (10 g) was dispersed in 2 L of 0.05 mol L^{-1} sodium perchlorate solution containing 100 mmol potassium dihydrogen phosphate per gram of powder at pH 2.5. This suspension

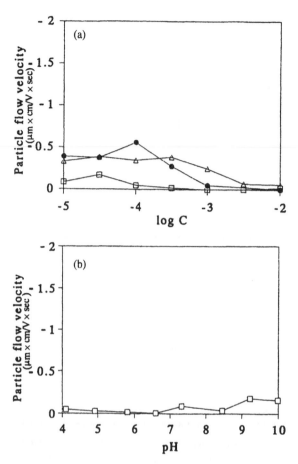

FIG. 14 Electrophoretic mobilities at stationary level of the new reference samples treated by the different HPC molecules: (a) electrophoretic mobilities versus KCl concentration curves; (b) electrophoretic mobilities versus pH curves. (●) HPC-H; (□) HPC-M; (△) HPC-L.

was aged for 60 days and treated by ultrasonication during the first 10-day period. Following aging, the suspension was diluted by a factor of 10 with 0.05 mol sodium perchlorate solution, ultrasonicated, and aged for an additional 19 days. From this suspension, 40 cm^3 aliquots were transferred into high-density polyethylene bottles with a unidisperse system and aged for 1 month. This final 500 mgL^{-1} goethite suspension constituted the SRM, from which 10 mL aliquots were diluted with deionized water to a volume of 100 mL prior to analysis. Smaller test suspensions were initially prepared from the freeze-dried powder. In this case, the suspensions were made up at a solids concentration of 500 mg L^{-1}, and either KNO_3 or $NaNO_3$ was used as supporting electrolyte. The suspension pH was adjusted as required in the test samples.

B. Characterization and Aging Studies of SRM

The long-term mobility for orthophosphate goethite at pH 2.5 is shown in Fig. 16. This suspension exhibited excellent stability with respect to mobility, pH, and visible appearance over the 1-year period of this study. Also, scatter in the mobility measurements was

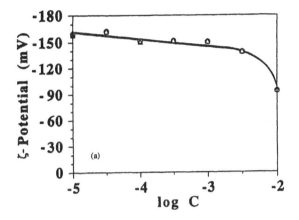

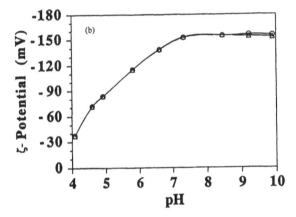

FIG. 15 Zeta-potential versus KCl concentration (a) and pH (b) curves of the cell wall–solution interface: (○) determined by maximum velocity of HPC-coated lattices; (△) determined by the plane interface technique.

typically low, with only two out of 49 sets of test measurements exceeding a sample standard deviation of -0.05 μm cm V^{-1} s^{-1}. After an initial increase over the first 60 days of aging, the mobility reached a plateau value (2.4–2.5 μm cm V^{-1} s^{-1}).

The stability of the pH 2.5 orthophosphated goethite (SRM) after dilution from the 500 mg dm^{-3} stock suspension is shown in Fig. 17. The mean and sample deviation of each measurement set is plotted as a function of time up to 6 h following dilution. The suspension was hand shaken prior to each set of measurements, but otherwise untreated during the period following the initial preparation. A linear regression of these data shows no apparent trend. The slope is nearly equal to zero and the y-intercept value (2.54 μm cm V^{-1} s^{-1}) is virtually identical to the collective mean (2.53 μm cm V^{-1} s^{-1}). Measurements made on this suspension after 24 h yield a mean value of 2.50 ± 0.02 μm cm V^{-1} s^{-1}. These results clearly demonstrate that the prepared (diluted) SRM suspension is sufficiently stable to provide a practical analysis window (approximately 1 day).

Figure 17 (inset) also shows a typical intensity-weighted mobility distribution as measured by the Malvern ZetaSizer 111 instrument. Note that the signal-to-noise ratio is quite large, while the full-width at half-height is about 0.3 μm cm V^{-1} s^{-1}.

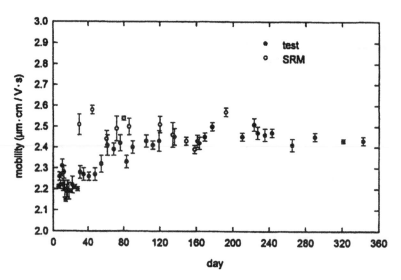

FIG. 16 Mobility of test batch (●) and SRM (○) as a function of aging time. Samples were aged in the concentrated form and diluted by a factor of 10 for analysis.

A further consideration in developing a suitable mobility standard is the effect of temperature variations expected under normal ambient conditions. As shown in Fig. 18, a linear temperature dependence was found for the mobility of the SRM over the range 18–28°C. The mobility varied by only 002 μm cmn V^{-1} s^{-1} per °C over this range, and correlated well with the fluidity $(1/\eta)$ dependence on temperature as predicted from Eq. (2).

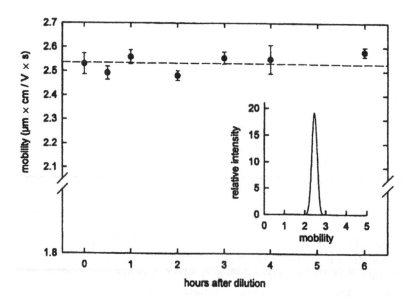

FIG. 17 Mobility of SRM as a function of time following dilution from the stock suspension. Inset shows a typical intensity-weighted mobility distribution for SRM measured on the Malvern ZetaSizer III system.

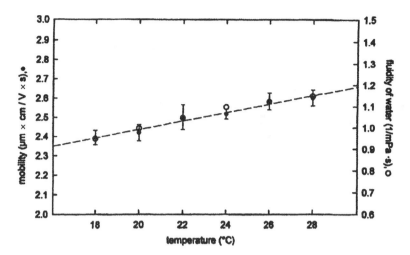

FIG. 18 Temperature dependence of mobility for SRM. Also shown is the fluidity of water ($1/\eta$) as a function of temperature.

C. Interlaboratories and Different Instrument Evaluation of SRM

In order to determine the reproducibility between different laboratories and instruments, four groups along with NIST (using three well-established commercial instruments) participated in round robin laboratory testing of the present SRM. The participants, in addition to NIST, included the University of Florida (DELSA 440), the University of Wisconsin (SYSTEM 3000), Pen-Kem Incorporated (SYSTEM 3000), and Coulter Corporation (DELSA 440). Randomly selected sample suspensions were sent to these laboratories with specific instructions for preparation and analysis within a fixed time.

Figure 19 shows the entire set of mobility data from all participating laboratories, including NIST, plotted versus bottle number. The dotted lines in this graph represent the mean and the sample standard deviation. The certified mobility value and uncertainty of the electrophoretic mobility for SRM have been determined as 2.53 ± 0.12 μm cm V^{-1} s^{-1}. As seen, the variation between laboratories and instruments was found not to be statistically significant. This demonstrates that, if properly analyzed according to a specific set of instructions under normal laboratory conditions, the SRM will generate consistent and repeatable mobility values on the instruments employed in this study. Stability during transport was also a consideration. Round robin results suggested that the transport of samples does not present a serious problem. In order to test this conclusion further, several samples used in the round robin were returned to NIST and reanalyzed in NIST. These samples yielded mobility values within the certified uncertainty.

Perhaps the single most pervasive problem affecting mobility analysis is contamination during sample preparation. This is particularly true of oxides where the surface sites tend to form complexes with a variety of ionic solution species. Of particular nuisance are sulfates, phosphates, and organic surfactants. Other contributing factors are the relatively low concentration of solids and high specific surface area common to colloidal suspensions; low levels of contaminants can have a relatively large impact on particle mobility in these systems. The water source used in the cleaning and dilution of suspensions should be deionized and not simply distilled. For an accurate measurement

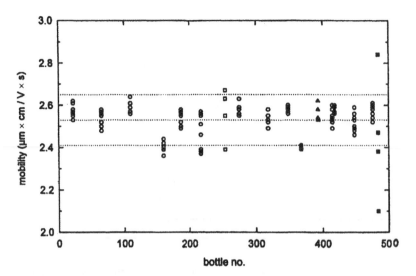

FIG. 19 Round robin results for each of five laboratories, including NIST homogeneity data, plotted as mobility versus bottle number. The dotted lines represent the unweighted mean and standard uncertainty for all measurements

of SRM, glassware, pipettes, and pH electrodes must be thoroughly clean of potential contaminants.

Preconditioning of the electrophoretic cell appears to reduce measurement variation and drifting of values. The origin of this effect is not entirely known, and may involve an equilibrium with the cell walls. In the case of the present SRM, the effect may be due to deposition of the positively charged particles on the negatively charged walls. Regardless of this, preconditioning with the sample for a short period just prior to analysis, followed by a fresh introduction of sample, is the recommended procedure.

D.   Protein and Polyelectrolyte as a Electrophoretic Mobility Standard

One of the features of ELS is that it determines the mobility of very small particles such as proteins, polyelectrolytes, micelles, and liposomes. At present, definite standard samples are necessary to check the accuracy of the observed mobility. Protein or polyelectrolyte might be suitable for this purpose, because many values have been reported that were measured by using the Tiselius electrophoretic instrument. In addition, reagents of high purity are commercially available, and the mobility of protein and polyelectrolyte can be measured by using an ELS instrument. On the other hand, in the case of inorganic pigments, mobility fluctuates, depending on the methods of sample synthesis and preparation. In addition, aggregation and sedimentation often occur during sample preparation.

A comparison of the mobilities of three protein and polyelectrolyte systems is shown in Table 4. The mobility of polyelectrolyte of small hydrodynamic radius was measured at relatively low concentration and the results agreed well with reference values. The isoelectric point of a protein was also measured and this agreed with the reference value.

In conclusion, some protein and polyelectrolyte systems might be suitable as electrophoretic mobility standards, especially in ELS.

TABLE 4 Mobility of Protein and Polyelectrolyte

Sample	MW ($\times 10^3$)	Solvent	Concentration (mg mL^{-1})	E (V cm^{-1})	R_h (nm)	Mobility (10^{-4} cm^2 V^{-1} s^{-1})	
RCA–BSA	66	1	1.8	39.3	5.3	-2.12 ± 0.17^a	-2.58
NaPSS	177	2	4.0	87.0	3.6	-4.58 ± 0.39^a	-4.50
NaPSS	400	2	4.0	75.8	7.0	-4.77 ± 0.17^a	-4.50

[a] Measured by electrophoretic light scattering instrument.
E: electric field; R_h: hydrodynamic radius; RCA-BSA: reduced carboxyamidomethyl bovine serum albumin; NaPSS: sodium poly(styrene sulfate).
Solvent 1: 1 mg/mL^{-1} sodium dodecyl sulfate, 10 mM sodium phosphate buffer (pH 7.0).
Solvent 2: 10 mM NaCl.
The uncertainty, $\pm$, is the standard deviation calculated from the results of a least squares fitting for a flow profile with six data points.
Source: Ref. 31.

REFERENCES

1. M von Smoluchowski. Z Phys Chem 92: 129, 1918.
2. RW O'Brien, LR White. J Chem Soc, Faraday Trans. 2 74: 1607, 1978.
3. GA Park. Chem Ref 65: 177, 1965.
4. J Wang, RE Riman, DJ Shanefield. In: MJ Cima, ed. Ceramic Transactions. Vol. 26. Westerville, OH; American Ceramic Society, 1992, p 240.
5. WA Zeltner, J Wang, OO Omatete, MA Janney, MI Tejedor-Tejedor, MA Anderson, RE Riman, DJ Shanefield, JH Adair. In: JH Adair, JA Casey, S Venigalla eds. Characterization Techniques for the Solid–Solution Interface. Westerville, OH: American Ceramic Society, 1993, p 87.
6. A Kotera, K Furusawa, Y Takeda. Kolloid Z Z Polym 239: 677, 1970.
7. K Furusawa, W Norde, J Lyklema. Kolloid Z Z Polym 250: 908, 1972.
8. HJ Van den Hul, JW Vanderhoff. J Electroanal Chem 37: 161, 1972.
9. R Sprycha, E Matijevic. Colloids Surfaces 47: 195, 1990.
10. A Ben-Taleb, P Vera, AV Delgado, V Gallardo. Mater Chem Phys 37: 68, 1994.
11. Japanese Surface and Colloid Chemical Group (Japan Oil Chemist's Society). Yukagaku 25: 239, 1975.
12. K Furusawa, S Usui, M Ozaki, K Konno, A Kitahara. Yukagaku 37: 632, 1988.
13. K Furusawa, Q Chen, N Tobori. J Colloid Interface Sci 137: 456, 1990.
14. K Furusawa, K Uchiyama. Colloids Surfaces 140: 217, 1998.
15. K Furusawa, T Tagawa. Colloid Polym Sci 263: 1, 1985.
16. K Furusawa, Y Kimura, T Tagawa. ACS Symp Ser 240: 131, 1984.
17. VA Hackley, RS Premachandran, SG Malghan, SB Schiller. Colloids Surfaces A 98: 209, 1995.
18. MS Juang, IM Krieger. J Polym Sci 14: 2089, 1976.
19. A Homola, RO James. J Colloid Interface Sci 59: 123, 1977.
20. H Sasaki, A Muramatsu, H Arakatsu, S Usui. J Colloid Interface Sci 142: 266, 1991; RW Huddleston and AL Smith. In R J Akers, ed. Foams. New York: NY; Academic 1976, p 163.
21. S Usui, Y Imamura, H Sasaki. J Colloid Interface Sci 118; 335, 1987.
22. NL Burns. PhD thesis, Royal Institute of Technology, Stockholm, Sweden, 1996.
23. K Furusawa, Y Kimura, T Tagawa. ACS Symp Ser 240: 131, 1984.
24. T Tagawa, S Yamashita, K Furusawa. Kobunshi Ronbunshu 40: 273, 1983.
25. K Furusawa, T Tagawa. Colloid Polym Sci 263: 1, 1985.

26. M Dubois, KE Gilles, JK Hamilton, PA Rebers, F Smith. Anal Chem 28: 350, 1956.
27. H Ohshima, T Kondo. J Colloid Interface Sci 116: 456, 1990.
28. MI Tejedor-Tejedor, MA Anderson. Langmuir 6: 602, 1990.
29. Ceramic Technology Project. Department of Energy contract no. DE-ACo5-40R21400, 1992.
30. RJ Atkinson, AM Posner, JP Quirk. J Inorg Nucl Chem 30: 2371, 1968.
31. K Oka. In: H Dhsima and K Furusawa, eds. Electrical Phenomena at Interfaces. New York: NY; Marcel Dekker Inc., 1998, p 167.

10

Electrokinetic Phenomena of the Second Kind

NATALIYA A. MISHCHUK Institute of Colloid Chemistry and Chemistry of Water, Ukrainian National Academy of Sciences, Kiev, Ukraine

STANISLAV S. DUKHIN New Jersey Institute of Technology, Newark, New Jersey

I. INTRODUCTION

Classical electrokinetic phenomena (electrokinetic phenomena of this first kind) are caused by the electrical double layer existing at electrified interfaces. Both the velocity of liquid along the charged immobile interface (electro-osmosis) and the velocity of a particle mobile in liquid (electrophoresis) are functions of the surface potential and strength of the external electric field.

According to Smoluchowski, at low voltage, the velocity of electrokinetic phenomena is a linear function of the strength of the electric field. At larger voltage, deviation from the Smoluchowski law can be considerable. However, the relative difference between velocities that have been obtained experimentally and those calculated on the basis of the Smoluchowski equation never exceed several units [1–3]. The situation changed essentially during the last decade. New phenomena, called electrokinetic phenomena of the second kind, were predicted theoretically and discovered experimentally [4–9].

The name highlights a new mechanism underlying these phenomena. If the standard electrokinetic phenomena are caused by a charge of the usual (or primary) electrical double layer, the new phenomena are caused by the secondary double layer, which arises in strong electric fields behind the quasiequilibrium primary double layer and contains an induced space charge. The large density and thickness of the induced charge create an extremely high growth of velocities that exceed the velocities of earlier-known phenomena by a factor of 10 or more. The new phenomena are specially pronounced for particles with ion type of conductivity, although they are also intense for metals and other materials with electron-type conductivity, and semiconductors with hole-type conductivity.

The secondary double layer appears to be due to concentration polarization. This process plays an important role in the electrochemistry of colloids and especially in the electrochemical macrokinetics of disperse systems [3] and polarization of membranes [10, 11]. However a large induced space charge behind the primary double layer and, as a consequence, the new electrokinetic phenomena, may appear as a result of concentration polarization only when certain surface and volume characteristics of contacting phases are present. The necessary conditions are as follows: (1) electric current through the interface provided by current carriers with the same sign of charge (ions, electrons, or holes); (2) curved interface; (3) high conductivity of solid phase K_i in comparison with the conductivity of liquid K_e; and (4) the strength of the electric field and particle size satisfying the condition $2Ea \gg \Phi_{cr}$, where Φ_{cr} is some critical potential, the value of which depends on the type of particle conductivity (it equals about 100 mV for an ion-exchange particle and about 1.5 V for a particle with electron or hole-type conductivity), E is the applied field strength, and a is the particle radius.

II. ELECTRO-OSMOSIS OF THE SECOND KIND

A. General Notion of Concentration Polarization of Flat Ideally Selective Membrane

The theory of concentration polarization of a flat ion-exchange membrane was developed for two cases: weak polarization at low electric field (sublimiting regime) and strong one at high electric field (overlimiting regime).

1. Low Electric Field

The concentration polarization is maximum in the case of high or ideal selectivity of a membrane. The electric current passing through an ideally selective membrane placed in an aqueous solution of electrolyte is brought about only by the migration of counterions. The transport properties of the membrane and the electrolyte solution are not identical, therefore the continuity condition for cation and anion fluxes during the transition from one medium to another cannot be satisfied only by means of their migration in the electric field.

Electrolyte concentration decreases near the side of a membrane where counterions enter it from electrolyte. The coions move in the opposite direction, away from the membrane surface. This leads to a decrease in coion concentration because the withdrawal of coions is not compensated for by their intake from the membrane. In low fields, electroneutrality is preserved. The decrease in counterion concentration is equal to the decrease in coion concentration. This means that there is a diffusion flux from the electrolyte to the membrane surface, and thus a diffusion layer is formed. The diffusion flux j_D is proportional to the decrease in concentration of ΔC and inversely proportional to the thickness of the diffusion layer δ, which, at low electric field, coincides with the thickness of the region of concentration polarization L:

$$j_D = D\Delta C/L \tag{1}$$

where D is the diffusion coefficient of ions (for simplification we suppose that diffusion coefficients for the anion and cation coincide: $D^+ = D^- = D$).

The concentration decrease in the region of concentration polarization (Fig. 1, curve 1) may be described by the following expression [4], in which $\tilde{C}$ is the dimensionless concentration, C/C_0:

$$\tilde{C}(x) = 1 - \frac{i(L-x)}{2FDC_0} \tag{2}$$

where i is the current density, x is the distance from the membrane surface, F is Faraday's constant, and C_0 is the bulk concentration of electrolyte. The highest concentration

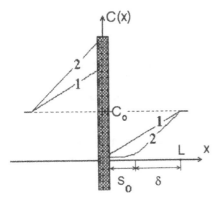

FIG. 1 Polarization of a flat ion-selective membrane at low (curves 1) and high (curves 2) electric field.

decrease is $\Delta C = C_0$. Therefore, the highest diffusion flux Eq. (1) is equal to $j_D^{\text{lim}} = DC_0/L$. It is called the limiting diffusion flux.

Owing to concentration polarization, the growth of the electric field is accompanied by a decrease in local conductivity. The increase in electric current then stops. This phenomenon is known as the limiting current. One can show that the density of the limiting current is equal to $2Fj_D^{\text{lim}}$ or

$$i_{\text{lim}} = \frac{2FDC_0}{L} \tag{3}$$

The potential drop $\tilde{\phi}(x) = F\phi(x)/RT$ and the charge density $\tilde{\sigma}(x) = \sigma(x)/C_0$ in the region of concentration polarization may be expressed as [4]

$$\tilde{\phi}(x) = \ln\frac{C(x)}{C_0} = \ln\left(1 - \frac{i}{i_{\text{lim}}}\frac{(L-x)}{L}\right) \tag{4}$$

$$\tilde{\sigma}(x) = \frac{\left(\dfrac{i}{i_{\text{lim}}}\right)^2}{\left(1 - \dfrac{i}{i_{\text{lim}}}\dfrac{L-x}{L}\right)^2}\left(\frac{\kappa^{-1}}{L}\right)^4 \tag{5}$$

where R is the gas constant, T is the absolute temperature and $\kappa^{-1} = \kappa^{-1}(C_0)$ is the thickness of the electrical double layer (Debye length).

The expressions presented above are correct for the condition $i < i_{\text{lim}}$ (sublimiting regime). One can see that if $i \to i_{\text{lim}}$ and $x \to 0$, the potential drop ϕ and the charge density σ are close to infinity. This inaccuracy was eliminated by the theory of strong concentration polarization, which takes into account the deviation from electroneutrality. The peculiarities of this deviation were analyzed for flat [10–13] and curved [4–7] interfaces.

2. High Electric Field

The main feature of the profiles of ion concentration at strong electric field [10–13] is that one can distinguish three different regions (Fig. 1, curve 2). The first is a large region with linear variation of ion concentration and local electroneutrality. This is the diffusion layer (or the convective-diffusion layer, CDL, in the presence of the liquid flow) with thickness

δ. The second region, absent at low voltage, is a region of approximately constant low electrolyte concentration C_S with thickness S_0. Here, the concentration of counterions is much higher than that of coions, which is very low. This is an induced *space* charge layer (SCL) because it is not directly connected with the membrane surface. It is precisely this charge that causes the electro-osmosis of the second kind. The third region is the quasi-equilibrium electrical double layer (DL). Since the DL is very thin in comparison with the thickness of the SCL and CDL, it is not shown in the figure.

At the opposite side of the membrane, the counterions come out of the ion-exchange membrane and coions move to its surface. As a result, the electrolyte concentration increases, the electroneutrality is preserved with quite high acuracy, and the space charge region is absent.

Taking into account that the conductivity of the SCL is 100–1000 times lower than the conductivity of the electrolyte [10–13] and the conductivity of a membrane is usually high, the potential drop V on the membrane with three layers of concentration polarization (DL, SCL, and CDL) takes place mainly in the SCL. An increase in the electric field leads to the growth of S_0 and, as a result, $V \to \Phi$, where Φ is the potential drop in the SCL.

The formation of the SCL, the thickness of which increases with the growth of the applied potential difference, reduces the thickness of the diffusion layer. At low voltage, the thickness of the region of concentration polarization is equal to the thickness of the diffusion layer $L = \delta$, while at high voltage $L = S_0 + \delta$. Thus, $\delta < L$. The thickness of the DL, κ^{-1}, is usually considerably lower than the thickness of the diffusion layer, δ, and can be neglected in the presented evaluation. As a result, the current through the membrane in a strong electric field preserves its diffusive nature and increases due to the growth of the SCL S_0 and the decrease in thickness of the diffusion layer δ (Fig. 1):

$$i = \frac{2FDC_0}{\delta} \tag{6}$$

For $\delta < L$, the current density satisfies the conditition $i > i_{\text{lim}}$ and this regime is called overlimiting. It is exactly this regime in which the appearance of electrokinetic phenomena of the second kind is possible.

The current growth caused by the emergence of the SCL for a flat membrane is not very large. In the absence of additional factors (thermoconvection [14] or water dissociation [15, 16]), it makes several tens of percents [10–13]. Traditionally, the name "overlimiting regime" combines all three factors mentioned above which lead to the growth of current over i_{lim} [Eq. (3)]. However, for electrokinetic phenomena of the second kind, only the appearance of an induced space charge and the decrease in thickness of the diffusion layer are important. Moreover, it will be shown below (Section V) that under conditions of electrokinetic phenomena of the second kind the role of other factors diminishes.

The theory of concentration polarization of a flat ion-exchange membrane under the overlimiting regime was developed numerically [10, 11] and analytically [12, 13]. The first analytical description [12] was used by us to develop the theory of electro-osmosis of the second kind near a curved interface [4–7]. However, due to rather complicated expressions [12], the description of electro-osmotic velocity was also difficult. The present chapter offers a simpler version of this theory, the correctness of which is proved with the help of analytical theory [13] developed after the first publications on electro-osmosis of the second kind [4–7].

The substantial simplification of the theory is possible under the condition $I = i/i_{\text{lim}} = L/\delta \gg 1$, when the region of an induced space charge is large and can be described in the following way. On the basis of numerical calculations [10, 11] and analytical solution [13], we suppose that the concentration of coions is low and that the current inside the SCL is provided only by counterions with concentration $C_s(x)$

$$I = \frac{i}{i_{\text{lim}}} = \frac{F^2}{RT} DC_s(x)E(x)\frac{L}{2FDC_0} = \frac{1}{2}\frac{FL}{RT}\frac{C_s(x)}{C_0}E(x) \tag{7}$$

where $E(x)$ is the local value of the strength of the electric field. Simultaneously, the charge density can be described by the Poisson equation:

$$\frac{d^2\phi(x)}{dx^2} = \frac{dE(x)}{dx} = -\frac{4\pi}{\varepsilon}FC_s(x) \tag{8}$$

The equation for $E(x)$ can be obtained as a combination of Eqs (7) and (8) as

$$\left(\frac{F}{RT}\right)^2 E(\tilde{x})\frac{dE(\tilde{x})}{d\tilde{x}} = -\frac{I}{\kappa^{-2}} \tag{9}$$

where the normalized distance $\tilde{x} = x/S_0$ is introduced, and the approximation $L \approx S_0$, correct for $I \gg 1$, is used. The solution of Eq. (9) is

$$\left(\frac{FE(x)}{RT}\right)^2 = \frac{2I}{\kappa^{-2}}(1 - x) + \text{const} \tag{10}$$

where the constant can be obtained by the asymptotic method [13]. However, as the analysis of a more accurate description of an electric field can show [13], the first term on the right-hand side of Eq. (10) describes almost the whole region of the induced space charge with 100% exactness. The inaccuracy appears only near the external boundary of the SCL, $x = S_0$. Taking into account that the size of this nonexact region is about several percent units of the thickness of the SCL, the constant in the first approximation can be neglected ($const = 0$). This allows one to simplify the description of the field and potential distribution and obtain an expression for electro-osmosis of the second kind in a form considerably simpler than it was done in Ref. 4. Thus, the strength of electric field can be written:

$$\frac{FE(\tilde{x})}{RT} = \frac{2\sqrt{I}}{\kappa^{-1}}\sqrt{1 - \tilde{x}} = E^*\sqrt{1 - \tilde{x}} \tag{11}$$

and the potential drop as

$$\tilde{\phi}(x) = \frac{F\phi(x)}{RT} = \phi_i + \int_0^x E(x)dx = \phi_i + \frac{4}{3}\frac{\sqrt{I}S_0}{\kappa^{-1}}[(1 - \tilde{x})^{3/2} - 1]$$
$$= \phi_i + \phi^*[(1 - \tilde{x})^{3/2} - 1] \tag{12}$$

where ϕ_i is the constant of integration that plays the part of an internal potential of ion-exchange material and depends on polarization processes [4]. The constants E^* and ϕ^* are given by

$$E^* = \frac{2\sqrt{I}}{\kappa^{-1}}; \quad \phi^* = \frac{4}{3}\frac{\sqrt{I}S_0}{\kappa^{-1}}$$

The full potential drop in the SCL is equal to

$$\tilde{\Phi} = \frac{F\Phi}{RT} = \tilde{\phi}(1) - \phi_i = \frac{4}{3}\frac{S_0^{3/2}}{\kappa^{-1}\delta^{1/2}} = \frac{4}{3}\frac{\delta}{\kappa^{-1}}I^{3/2} \tag{13}$$

The local value of the induced space charge can be defined according to Eqs (7) and (11) as

$$\tilde{C}_s(x) = \frac{C_s(x)}{C_0} = \frac{I}{\dfrac{1}{2}\dfrac{FL}{RT}E(x)} = \frac{\kappa^{-1}\sqrt{I}}{L\sqrt{1-\tilde{x}}} = \frac{C_s^*}{\sqrt{1-\tilde{x}}} \tag{14}$$

with $C_s^* = \kappa^{-1}\sqrt{I}/L$. Due to the simplification in Eq. (10), the strength of the electric field, the potential drop, and density of the induced space charge in Eqs (11), (12), and (14) depend on certain parameters E^*, ϕ^*, C_s^* and are functions of distance. This fact is very important for the development of the theory of electro-osmosis of the second kind. It allows one to obtain every function, necessary for this theory, in terms of angular and distance multipliers and, as a result, to use the local flat approximation.

B. General Notion of Polarization of a Spherical Particle with Ionic Conductivity

In the case of a flat membrane, the regions of concentration polarization near the opposite sides of a membrane have different properties, although they depend on each other due to the current through the membrane. The picture is much more complicated for a spherical particle where the transition from the layer with decreasing electrolyte concentration to the layer with an increasing one should be accomplished in the presence of electro-osmosis.

At low electric field, the pictures of concentration polarization for spherical conductive and insulating particles are almost antisymmetrical [3]: the regions of decrease and increase in electrolyte concentration each occupy half the particle surface. Similarly to the flat ion-exchange membrane, the increase in electric field leads to the appearance of an induced space charge near one side of a particle and, as a result, to the asymmetry of the investigated processes.

The scheme of particle polarization under conditions of electro-osmosis of the second kind is shown in Fig. 2. This is a cation-exchange particle for which the counterions are cations. The particle is surrounded by the electrical DL (which cannot be shown on the scale of the figure) and the region of concentration polarization that includes the induced space layer (S_0) and diffusion layer (δ). Near the surface of the hemisphere $0 \le \theta \le \pi/2$, counterions move to and through the conductive particle. Here, the concentration of electrolyte decreases and an induced space charge appears. Near the opposite part of the particle surface, similarly to the membrane, the concentration of electrolyte increases and deviation from electroneutrality is very small. Contrary to the polarization of a flat membrane, near a conducting particle there are both the normal component of electric field and the one that is tangential to the particle surface. The normal component provides current through the particle and forms an induced space charge. The tangential component affects the induced space charge and sets in motion not only the charge but also the liquid where it exists. Hence, electro-osmosis of the second kind appears. Near the opposite side of the particle (without induced space charge), the electro-osmotic flow caused by the nonequilibrium electrical DL (electro-osmosis of the first kind or classical electro-osmosis) exists.

The general picture of particle polarization is not only the result of the asymmetrical appearance of an induced space charge but also the manifestation of asymmetrical hydrodynamic flow. The movement of liquid affects the thickness of the CDL. Earlier it has been established that at weak concentration polarization the thickness of the CDL near the spherical particle δ is inversely proportional to the velocity of liquid [17–20]. Taking into account that the velocity of electro-osmosis of the second kind near one side of a particle is

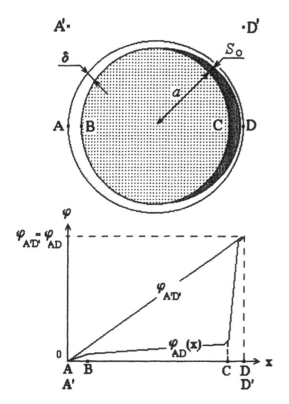

FIG. 2 Scheme of polarization of a spherical particle and potential distribution $\varphi_{A'D'}(x)$ in electrolyte and $\varphi_{AD}(x)$ across the granule under conditions of electro-osmosis of the second kind.

considerably larger than the velocity of classical electro-osmosis near the opposite side, the thickness of the CDL changes considerably along the particle surface (Fig. 2). In turn, the change in δ strongly affects the current density through the particle [Eq. (6)], resulting in the change in induced space charge, distribution of electric field, etc.

The density and thickness of an induced space charge and, correspondingly, the velocity of electrokinetic phenomena of the second kind grow when the strength of the electric field and particle size increase because the potential drop in the region of concentration polarization depends on these parameters. This can be shown by the simplest qualitative analysis of potential drop φ_{AD} in the particle with the region of concentration polarization (between points A and D) and potential drop $\varphi_{A'D'}$ in "free" electrolyte (between points A' and D').

$$\varphi_{AD} = \varphi_{AB} + \varphi_{BC} + \varphi_{CD} \tag{15}$$

where φ_{AB} is the potential drop across the region of concentration polarization with increasing electrolyte concentration, φ_{BC} is the potential drop across the section of a particle, and φ_{CD} is the potential drop across the region of concentration polarization with decreasing electrolyte concentration, which includes the SCL and CDL.

Similarly to the potential drop in a membrane, φ_{CD} is mainly defined by potential drop Φ on the SCL. The potential drop φ_{AB} in the region of the increased concentration of electrolyte satisfies the condition $\varphi_{AB} \ll \varphi_{CD}$. Finally, the value of potential drop φ_{BC}

depends on the correlation between the conductivities of electrolyte K_e and particle K_i. Under condition:

$$K_e \ll K_i \tag{16}$$

the potential drop φ_{BC} is relatively small and Eq. (15) can be rewritten as

$$\varphi_{AD} = \varphi_{CD} \approx \Phi \tag{17}$$

The potential drop $\varphi_{A'D'}$ between points A′ and D′ located at the same distance from each other as points A and D, i.e., at a distance approximately equal to $2a$, is about φ_{AD} and equals $2aE$ because electric field E is homogeneous far from a particle:

$$\varphi_{AD} \approx \varphi_{A'D'} \approx 2aE \tag{18}$$

On comparison of Eqs (17) and (18), the conclusion is reached that

$$\Phi \approx 2aE \tag{19}$$

The evaluation was carried out only for $\theta = 0$ (see Fig. 2). Equation (18) shows the maximum value of potential drop on the SCL, which decreases with the growth of angle θ. However, this evaluation is important because it demonstrates the role of particle size for the potential drop on the SCL and is useful for the development of a relatively simple theory of electrokinetic phenomena. As will be shown in the next section, this potential drop plays the role of "electrokinetic" potential, similar to ζ potential in classical electrokinetic phenomena. However, the new "electrokinetic potential" Φ is independent of the surface charge of the particle, typical for the classical electrokinetic phenomena and, moreover, is a linear function of both particle size and strength of the external electric field. Therefore, it can reach values which considerable exceed the ζ potential (e.g., for $a = 50\ \mu m$ and $a = 500\ \mu m$ at $E = 20$ V/cm, the new "electrokinetic potential" Φ reaches the value of 200 and 2000 mV, respectively). This means that electrokinetic phenomena of the second kind strongly depend on the particle size and can reach large velocities even at low voltage.

It is necessary to stress that a small value of potential drop φ_{BC} means that the internal potential of ion-exchange particle ϕ_i [see Eq. (12)] under condition (16) is approximately constant. Thus, the condition (16) not only corresponds to the case when the potential drop on the SCL is maximum but also allows one to simplify the model of electro-osmosis of the second kind because $\partial\phi_i/\partial\theta \approx 0$.

C. Local Flat Model of Electro-osmosis of the Second Kind

The theory of electrokinetic phenomena of the second kind was developed for a thin polarization layer that satisfies the condition $L \ll a$ [4–7]. This condition allows us to use a local flat model that considerably simplifies the mathematical description of the processes investigated. The velocity of electro-osmosis of the second kind can be obtained on the basis of the Navier–Stokes' equation:

$$\eta\Delta\vec{V} = \sigma\nabla\phi + \nabla p \tag{20}$$

where η is the liquid viscosity, $\vec{V}$ is the liquid velocity, σ is the charge density (in the electrical DL for classical electro-osmosis and in the SCL for electro-osmosis of the second kind), ϕ is the potential distribution, and p is the pressure.

For a thin nonpolarized DL, the right-hand side of this equation can be reduced only to the first term, equivalent to the well-known Smoluchowski equation for electro-osmosis, and can be used not only for the traditional electrical double layer but also for the SCL. However, the process of polarization of the DL and the appearance of the SCL change the pressure near the particle surface, leading to the appearance of a velocity component related to this pressure that is absent in the Smoluchowski equation. The scheme developed earlier for the polarization of thin double layer [3, 18–20] can also be used to solve the problem discussed. In order to obtain the distribution of pressure inside the polarized layer, it is necessary to apply the divergence operator to Eq. (20) and use the equation of contintuity of liquid flow $\nabla \cdot \vec{V} = 0$. This procedure results in

$$\Delta p = -\nabla \cdot (\sigma \nabla \phi) \tag{21}$$

Neglecting the tangential derivatives in comparison with the normal ones and integrating this equation over the polarized layer (the SCL in the case investigated) allows us to obtain the second term of Eq. (20) in the following form:

$$\frac{\partial}{\partial \theta} p(x) = \frac{1}{2} \frac{\partial}{\partial \theta} \left(\frac{\partial \phi}{\partial x} \right)^2 \tag{22}$$

As a result, the equation for electro-osmosis can be written as a sum of two terms:

$$\frac{\partial^2 \tilde{V}_\theta}{\partial \tilde{x}^2} = -\frac{\partial^2 \tilde{\phi}}{\partial \tilde{x}^2} \frac{\partial \tilde{\phi}}{\partial \theta} + \frac{1}{2} \frac{\partial}{\partial \theta} \left(\frac{\partial \tilde{\phi}}{\partial \tilde{x}} \right)^2 \tag{23}$$

where $\tilde{x} = x/a$; $\tilde{V}_\theta = V_\theta \dfrac{4\pi\eta a}{\varepsilon (RT/F)^2}$. Both terms in Eq. (23) will be calculated independently. With Eq. (12), the first term can be rewritten as

$$\frac{\partial^2 \tilde{V}_{\theta 1}}{\partial \tilde{x}^2} = -\frac{3}{4} \tilde{\phi}^* \frac{\partial \tilde{\phi}^*}{\partial \theta} \left[1 - \tilde{x} - (1 - \tilde{x})^{-1/2} \right] \tag{24}$$

After integration with the boundary conditions:

$$\frac{\partial}{\partial \tilde{x}} \tilde{V}_{\theta 1}(\tilde{x} = 1) = 0 \text{ and } \tilde{V}_{\theta 1}(\tilde{x} = 0) = 0 \tag{25}$$

the "Smoluchowski" component of velocity at $\tilde{x} = 1$ equals:

$$\tilde{V}_{\theta 1}(\tilde{x} = 1) = -\frac{7}{8} \tilde{\phi}^* \frac{\partial \tilde{\phi}^*}{\partial \theta} \tag{26}$$

Substitution of Eq. (12) into Eq. (23) leads to the differential equation:

$$\frac{\partial^2 \tilde{V}_{\theta 2}}{\partial \tilde{x}^2} = \frac{1}{2} \frac{\partial}{\partial \theta} \left(\frac{\partial \tilde{\phi}}{\partial \tilde{x}} \right)^2 = \frac{9}{4} \tilde{\phi}^* \frac{\partial \tilde{\phi}^*}{\partial \theta} (1 - \tilde{x}) \tag{27}$$

After integration with boundary conditions equivalent to conditions (25), the "pressure" component of electroosmotic velocity at $\tilde{x} = 1$ equals:

$$\tilde{V}_{\theta 2}(\tilde{x} = 1) = -\frac{3}{8} \tilde{\phi}^* \frac{\partial \tilde{\phi}^*}{\partial \theta} \tag{28}$$

One can see that the component of velocity caused by pressure has the same sign as the "Smoluchowski" component. This result is similar to the polarization of nonconducting particles with large surface conductivity (regime of large Peclet numbers) when both components of electro-osmotic velocity have the same sign and the pressure component is rather large [21]. The coincidence of the directions of both electro-osmosis components can be proved by simple qualitative considerations. In the case of a cationite (cation-exchange particle), the cations move along the line of the electric field and enter the particle near the pole $\theta = 0$. The pressure is maximum near this pole because $d\phi/dx$ reaches its maximum here and the pressure gradually decreases with the increase in θ, i.e., $V_{\theta 2}$ is positive. The counterions entering the space charge move partially inside the particle and partially along its surface, i.e., in the direction of increasing θ. Thus, the Smoluchowski component of the electro-osmotic slip, i.e., $V_{\theta 1}$, $V_{\theta 1}$, is positive too. As a result, the directions of both components coincide. Since the direction of liquid slip corresponds to increasing θ, the direction of electrophoretic velocity corresponds to decreasing θ because the directions of electrophoresis and electro-osmotic slip are reversed. This result is reasonable because it is shown that the negative particle (cationite) moves in the direction opposite to the electric field.

Let us consider an anionite in an electric field of the same direction. The anions (in this case, counterions) enter the field near the pole $\theta = \pi$ and produce a space charge in its vicinity. Thus, the pressure is maximum near this pole and decreases with the decrease in θ, i.e., $V_{\theta 2}$ is negative. The anions move along the particle surface opposite to the electric field, i.e., in the direction of decreasing θ, so $V_{\theta 1}$ is negative. The directions of both components coincide again. Since the direction of electrophoresis is opposite to the one of electro-osmosis, the anionite particles move towards the external field. The anionite particle, being a positively charged particle, moves along the lines of the electric field, in agreement with the definition of the sign of the electric field, namely, that its direction is the direction of motion of a positive charge.

The total electro-osmotic velocity is equal to the sum of Eqs (26) and (28):

$$\tilde{V}_\theta(\tilde{x} = 1) = \tilde{V}_{\theta 1}(\tilde{x} = 1) + \tilde{V}_{\theta 2}(\tilde{x} = 1) = -\frac{5}{4}\tilde{\phi}^* \frac{\partial \tilde{\phi}^*}{\partial \theta} \tag{29}$$

Taking into account that the maximum potential drop in the region of the induced space charge Φ is evaluated by Eq. (9) as $2aE$, and the gradient of this potential along the particle surface $\frac{\partial \phi^*}{a\partial \theta}$ takes values between E and $2E$, the conclusion can be reached that the velocity [Eq. (29)] is between the two limiting values:

$$\frac{5}{8}\tilde{\Phi}^2 |\tilde{V}_\theta| < \frac{5}{4}\tilde{\Phi}^2 \quad \text{or} \quad \frac{5}{4}\left(\frac{\varepsilon}{4\pi\eta}\Phi E\right) < |V_\theta| < \frac{5}{2}\left(\frac{\varepsilon}{4\pi\eta}\Phi E\right) \tag{30}$$

To compare this result with the Smoluchowski expression, one should remember that for a nonpolarized particle the maximum value of the potential gradient equals $\frac{\partial \phi^*}{a\partial \theta}\big|_{\theta=\pi/2} = \frac{3}{2}E$. On the basis of this expression and Eq. (30), it can be concluded that the "effective electrokinetic potential" ζ_{eff} of electro-osmosis of the second kind takes values between:

$$\frac{5}{6}\Phi < \zeta_{\text{eff}} < \frac{5}{3}\Phi \tag{31}$$

Roughly speaking, the potential drop in the region of the induced space charge Φ plays the role of electrokinetic potential. Thus, accounting for Eqs (19) and (31), the velocity of electro-osmosis can be described by the expression $V \sim \zeta_{\text{eff}}E$, identical to the one of Smoluchowski where

$$\zeta_{\text{eff}} \approx \Phi \approx 2Ea \tag{32}$$

Thus, the velocity of electro-osmosis of the second kind is proportional to particle size a and squared electric field E^2.

Approximately the same value for the effective electrokinetic potential was obtained in the first theory of electro-osmosis of the second kind [3]. However, as is seen from Eqs (26) and (28), about two-thirds of this velocity is due to the induced space charge and the other one-third is the result of the change in pressure caused by polarization processes. In order to obtain a more exact value for the coefficient preceding Φ in Eq. (31), a very complicated problem of potential distribution around particle should be solved.

For the purposes of simplification, only the velocity component caused by induced space charge has been analyzed above. The two other components, created by the polarized electrical DL and CDL are considerably smaller. As was shown in Ref. 3, the velocity component caused by the SCL is higher than the other components under condition $\Phi >$ 250 mV or $\tilde{\Phi} > 10$. This condition is equivalent to condition $\Phi \gg \zeta$ because the electrokinetic potential of different ionite particles ζ is about 50–100 mV.

It is necessary to stress that, although for derivation of Eq. (29) the condition $S(\theta) \gg \delta(\theta)$ was used, the numerical analysis of velocity under the opposite conditions showed that Eq. (29) could also be used under condition $S(\theta) \sim \delta(\theta)$ and even $S(\theta) < \delta(\theta)$ with a high degree of exactness. This is a result of the similarity between the Smoluchowski formula and Eq. (29). With decrease in voltage, the effective potential ζ_{eff} turns smoothly into the classical electrokinetic potential ζ.

D. Potential Distribution Around an Ion-Exchange Particle Under Conditions of Electro-osmosis of the Second Kind

The description of the potential distribution around a polarized particle requires solving the system of equations (Poisson, Navier–Stokes, convective-diffusion equations) together with a set of boundary conditions, and the equations of continuity of electrical and hydrodynamic fluxes. These equations and conditions include not only all those factors that are traditionally used to describe concentration polarization of nonconducting particles but also the strong deviation from electroneutrality behind the DL and the strongly asymmetrical picture of hydrodynamic flow, considerably complicating the problem investigated. However, even in the case of nonconducting particles, theoretical difficulties in developing the theory have been overcome with strong restrictions to generality, for instance, by imposing conditions of small differentials of concentration or local electroneutrality which apply to weak fields [22–24] or strong fields with several additional restrictions [21].

The attempts to obtain an analytical solution for potential distribution under conditions of electro-osmosis of the second kind are still in development. However, even the numerical solution of this problem is accompanied by serious difficulties and has been achieved only for a definite set of parameters [25, 26].

The results of these calculations are shown in Fig. 3 (curves 1–3). The potential distribution in the range of angles of $0 < \theta < 70°$ is close to $\Phi \cos \theta$ and displays more complicated dependencies for larger angles. Moreover, the deviations from this law for different curves vary.

Despite the roughness of the approximation of potential distribution, it allows us to evaluate the possible behavior of angular distribution of velocity [Eq. (27)] as

$$\tilde{V}_\theta(\tilde{x} = 1) \sim -\frac{5}{4}\tilde{\Phi}^2 \sin\theta \cos\theta \tag{33}$$

This means that the maximum velocity of electro-osmosis of the second kind is localized at $\theta \sim 45°$ instead of $\theta = 90°$ found for classical electro-osmosis. The absolute

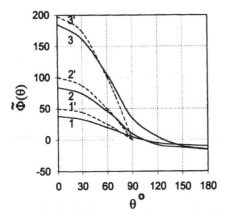

FIG. 3 Potential distribution around a particle at fixed potential drop $\tilde{\Phi}$ between points A and D (see Fig. 2): $\tilde{\Phi} = 50$ (curves 1, 1'); 100 (curves 2, 2'); 200 (curves 3, 3'). Curves 1–3 correspond to numerical calculations; curves 1'–3' refer to the approximation $\tilde{\Phi}(\theta) = \tilde{\Phi} \cos \theta$.

value of velocity and the position of the maximum velocity are qualitatively in good agreement with the experimental data (see Section II.E).

The change in sign of potential in Fig. 3 corresponds to the change in current direction. With the growth of voltage, the distribution of potential becomes more asymmetrical. This tendency is qualitatively similar to the distribution of the field near a strongly polarized nonconducting particle [21].

E. Experimental Investigation of Electro-osmosis of the Second Kind

The existence of electro-osmosis of the second kind was proved by means of the direct observation of liquid near a cationite (cation-exchange resin) surface [9]. The investigated particle of cationite (granule) was placed on the bottom of an electrophoresis cell and fixed by a special glass. After switching on the electric field, the liquid movement in the meridional direction appeared. The visualization of the electro-osmotic movement of liquid was undertaken using dispersed particles (with particle size equal to 2–5 µm) the electro-migration velocity of which was considerably less than the velocity of the electro-osmotic flow (the ζ potential of particles is considerably less than the potential drop in the granule $\Phi = 2Ea$).

The observed picture of electro-osmotic flow depends on time. The investigated processes can be divided into three qualitatively different stages: (I) initial stage, (II) developed electro-osmosis; and (III) damping of electro-osmotic slip (Fig. 4a, b).

I.   During the initial stage, the SCL and CDL are formed and the electro-osmotic movement of liquid along the particle surface appears. The normal component of the liquid stream provides continuity of liquid flow and, together with the tangential stream, forms closed electro-osmotic whirls that cover approximately half the granule surface. An asymmetrical picture is caused by the rough change in the tangential component of electro-osmotic velocity along the granule surface. Electro-osmosis of the second kind is many times larger than the classical electro-osmosis that occurs near the opposite side of the particle. Due to this considerable difference between electro-osmotic velocities, the streamlines of electro-osmotic flow of the second kind form the closed whirls.

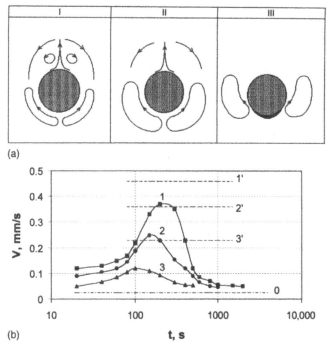

(a)

(b)

FIG. 4 (a) Three stages of electro-osmotic slip near a cation-exchange particle (KU-2-8). (b) Experimental electro-osmotic velocity as a function of time, counted from the moment of switching on of the electric field for $C_0 = 10^{-4}$ mol/L, $E = 10$ V/cm, and $a = 0.42$ mm (curve 1), 0.33 mm (curve 2), and 0.21 mm (curve 3). Curve 0 – electro-osmotic velocity according to Smoluchowski theory for dimensionless electrokinetic potential $\bar{\zeta} = 2.5$; curves $1'$–$3'$: theoretical values according to stationary model of electro-osmosis of the second kind. (Adapted from Ref. 9.)

II. Electro-osmotic movement develops in a few minutes after the beginning of the process. The whirls cover approximately three-quarters of the particle surface. The velocity $V(t)$ reaches the maximum value V_{max}. The most probable mechanism of the growth in velocity is water desalination near the side of the particle with an induced space charge, improving condition (16). This statement was checked by the additional measurement of electro-osmotic velocity under the special nonstationary regime [27]. The growth of velocity can be also explained by the nonstationarity of polarization and hydrodynamic flow. A certain transition time is necessary to reach the stationary distribution of ions and electric field inside and around the granule and to overcome the inertial properties of liquid and reach its stationary velocity.

III. Damping of the electro-osmotic stream starts within 7–15 min. Diminution of electro-osmotic velocity accompanies the extension of the region between whirls and the decrease of whirl size. This is caused by the sedimentation of small particles to visualize the flow on the granule [27], forming a layer of 2–5 μm during the above-mentioned time. The sediment overflows the region of an induced space charge and, as a consequence of the sediment electrical and hydrodynamic resistance, the electro-osmotic flow is damped. Velocity damping depends on the concentration of small particles. At greater concentration, the negative influence of the sediment increases. However, for good visualization, the concentration should be rather large. As a result, maximum measured velocities (Fig. 4b, curves 1–3) are lower than the theoretically predicted ones (curves $1'$–$3'$). At the same time, they are considerably higher than the velocities (curve 0) calculated for electrokinetic potential $\tilde{\zeta} \approx 2.5$ (obtained by the measurement of electrophoretic mobility under conditions that provide linear electrophoresis) according to the Smoluchowski theory.

The tangential component V_θ of electro-osmotic velocity for different granule sizes, electric fields, and electrolyte concentrations is shown in Figs 5–7. The measurements were carried out a distance from the granule surface $x = 50$ µm and $\theta \approx 30°–50°$. Figure 5 shows that V_θ increases linearly with increase in the granule size, although the experimental increase in velocity is slightly less than the theoretical one. It should be stressed that such linear dependence $V_\theta(a)$ exists at any time (e.g., at 30 or 120 s from the moment of switching on the voltage). The experimental dependence on the strength of the electric field $V_\theta(E)$ also corroborates qualitatively the theoretical one (Fig. 6). The experimental dependence $V_\theta(E)$ has the form of $V_\theta(E) \sim E^{2.2}$, whereas the theoretical dependence is $V_\theta(E) \sim^2$. The experimental values agree well with theory if the strength of the electric field satisfies the condition $2FEa/RT > 50$. Unfortunately, the applicability of the theory at stronger electric fields cannot be analyzed owing to a warming up of the electrolyte and thermoconvection in the cell. Since the condition (16) can be strictly satisfied only at very low concentration of electrolyte, the velocity of electro-osmosis strongly depends on electrolyte concentration [9]. The measured velocities increase with the decrease in concentration and reach saturation at an electrolyte concentration of $10^{-5} – 10^{-6}$ mol/L. Such saturation can be probably caused by two other factors (see Section V). One of them is the expansion of the SCL and tangential drift of induced space charge that increases at low electrolyte concentration. The second factor is water dissociation, which also decreases with electrolyte concentration.

Figure 7 demonstrates the angular dependence of the electro-osmotic velocity. For $E = 15$ V/cm. the maximum of electro-osmotic velocity (curve 1) corresponds to the maximum of the theoretical angular function $\sin\theta \cos\theta$ (curve 1). For lower electric field (curves 2 and 3), the maximum is at larger angles (60°–70°). Taking into account that for classical electrokinetic phenomena electro-osmosis is maximum at 90°, curves 2 and 3 presented in Fig. 7 show a tendency of transition between electro-osmosis of the second kind and the classical one (see curve 3, calculated according to Smoluchowski's

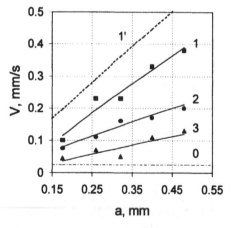

FIG. 5 Electro-osmotic velocity as a function of the particle radius a for $C_0 = 10^{-4}$ mol/L and $E = 10$ V/cm. Curve 1 – $V_{eo,max}$; curve 2 – $V_{eo}(t = 120$ s); curve 3 – $V_{eo}(t = 30$ s); curve 1' – theoretical dependence; curve 0 – electro-osmotic velocity according to Smoluchowski theory for dimensionless electrokinetic potential $\tilde{\zeta} = 2.5$. (Adapted from Ref. 9.)

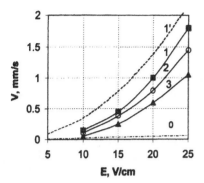

FIG. 6 Electro-osmotic velocity as a function of electric field strength for $C_0 = 10^{-4}$ mol/L: $a = 0.31$ mm (curve 1); 0.28 mm (curve 2); 0.26 mm (curve 3). Curve $1'$ – theoretical dependence for $a = 0.31$ mm; curve 0 – electro-osmotic velocity according to Smoluchowski theory for dimensionless electrokinetic potential $\tilde{\zeta} = 2.5$. (Adapted from Ref. 9.)

theory). For example, at $E = 5$ V/cm the dimensionless potential drop on the particle equals 15 and, according to Ref. 4, the component of velocity related to the DL and CDL is about one-half of the component related to the SCL. As a result, experimental curves 2 and 3, being a sum of all components, reach maximum between 45° and 90°.

All the results described above were obtained for cation-exchange granules. The velocity of electro-osmosis near anion-exchange granules is three to four times lower than for cation-exchange ones. This effect can be explained by the influence of two processes. The first of them is water dissociation caused by the catalytic influence of surface groups or by creation of bipolar contact (Section V). The second one is caused by the formation of a space charge by macroions of impurities that play the role of counterions but have lower mobility than ions of the used electrolyte or form an immobile charged sediment that diminishes the current through the particle and causes hydrodynamic resistance to electro-osmotic flow of liquid.

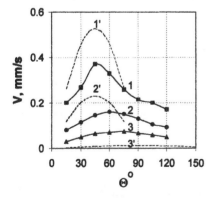

FIG. 7 Tangential component of electro-osmotic velocity as a function of angle for $a = 0.42$ mm and $C_0 = 10^{-4}$ mol/L at $E = 15$ V/cm (curves 1, $1'$); $E = 10$ V/cm (curves 2, $2'$) and $E = 5$ V/cm (curves 3, $3'$). 1–3 – experimental curves; $1'$, $2'$ – theoretical curve for electro-osmosis of the second kind, and $3'$ – theoretical curve calculated according to Smoluchowski theory for $\tilde{\zeta} = 2.5$. (Adapted from Refs. 9 and 25.)

III. ELECTROPHORESIS OF THE SECOND KIND

When a particle is suspended in an electrolyte, the electro-osmotic movement of liquid is transmitted to the particle. This results in the appearance of electrophoresis independently of its nature: either classical electrophoresis or electrophoresis of the second kind. However, each type of electrophoresis is closely related to the peculiarities of electro-osmosis causing the movement of a suspended particle. In the particular case of electrophoresis of the second kind, its velocity should be defined in terms of effective electrokinetic potential, $\zeta_{eff} = 2Ea$. This means that the most interesting results can be obtained for large particles, when the measurement of electrophoresis of the second kind is complicated by rapid particle sedimentation. Indeed, because of this obstacle, electrophoresis of the second kind was not discovered earlier. Since the traditional scheme of an electrophoretic cell was not suitable for large particles, several new methods were proposed [8, 28].

Electrophoresis of the second kind was also called "superfast electrophoresis" to demonstrate that its velocity considerably exceeds that of the classical one.

A. Electrophoresis of Ion-Exchange Particles and Fibers

Generally, the electrophoretic velocity can be obtained by integration of electro-osmotic velocity around the particle. The expression for linear electrophoresis takes the following form:

$$\tilde{V}_{ef} = \frac{4}{3\pi} \int_0^\pi \tilde{V}_{eo}(\theta) \sin\theta d\theta \tag{34}$$

We will also use this expression to describe electrophoresis of the second kind. However, it is necessary to stress that, due to another angle distribution of hydrodynamic flow, the coefficient before the integral can differ a little.

In the investigated case of electrokinetic phenomena of the second kind only the evaluation of electro-osmotic velocity [Eq. (33)] for the angles $0 < \theta < 70°$ is available. However, taking into account that the velocity for this range of angles is maximum and the velocity of electro-osmosis near the opposite side of the particle is negligibly small, we can evaluate the velocity of electrophoresis as

$$\tilde{V}_{ef} \approx \frac{4}{3\pi} \int_0^{\pi/2} \tilde{V}_{eo}(\theta) \sin\theta d\theta \approx \frac{5}{3\pi} \tilde{\Phi}^2 \int_0^{\pi/2} \sin^2\theta \cos\theta d\theta = -\frac{5}{9\pi}\tilde{\Phi}^2 \approx -\frac{2}{7}\tilde{V}_{eo}^{max} \tag{35}$$

As is seen from Fig. 3 for the numerical calculation of potential distribution and from Fig. 7 for the experimental velocity of electro-osmosis, the angular dependence changes with the field. This results in the appearance of an additional multiplier in the expression for the velocity of electrophoresis:

$$\tilde{V}_{ef} \approx -\frac{5}{9\pi}\tilde{\Phi}^2 f(\tilde{\Phi}) \tag{36}$$

Thus, the velocity of electrophoresis differs from the squared function of the strength of the electric field. Due to the enlargement of the region of electro-osmosis of the second kind (Figs 3 and 7), it can be claimed that the velocity of electrophoresis for this range of field should increase as

$$\tilde{V}_{ef} \sim \tilde{\Phi}^{2+k} \tag{37}$$

where $k > 0$. However, at larger electric fields, the relative role of the tangential processes increases (Section V), leading to slower growth in the thickness of the induced space charge and, correspondingly, slower growth in velocity. This means that with the growth of applied voltage, the parameter k not only can decrease, but also even change its sign.

Figure 8 shows the electrophoretic velocity of different cation and anion exchangers (mean diameter of 750 µm) as a function of the strength of the external field [8]. The cation-exchange particles move towards the anode and the anion-exchange particles move towards the cathode. When gradients are in the range 75–300 V/cm, the absolute values of the electrophoretic velocity of different ion-exchanger particles considerably exceed the velocity of classical electrophoresis. The difference between various plots of $V(E)$ is due to the disparity in the conductivity of the particles. For example, the conductivity of particles of KU-2-8 is higher than the conductivity of those that correspond to KB-4. The shape of the studied particles also plays an important part. For instance, the EDE-10-P particles have a plate-like shape, whereas the AB-17 particles are spherical. The smaller hydrodynamic resistance of the former is responsible for electrophoretic velocity higher than that of the latter. It can be seen that the experimental values (curves 1–5) are lower than the theoretical ones (curve $1'$, $1''$). The cause of the discrepancy will be discussed below.

A detailed investigation of the dependence of electrophoretic velocity on the particle size and voltage was carried out [8] for particles of KU-2-8, which have high conductivity and selectivity and satisfy the conditions of electrokinetic phenomena of the second kind better than other ion exchangers. The data for electrophoretic mobility as a plot of $\log U = \log(V/E)$ against $\log E$ are shown in Fig. 9. It is seen that the small particles (1–10 µm) move with almost the same mobility when the field strength is below 20 V/cm. In this range of the field strength and particle size, the electrophoresis follows the classical patter. When the external field increases, the larger particles begin to move faster. For smaller particles, a higher E is required for electrophoresis of the second kind. This correlates with the condition according to which electrokinetic phenomena of the second kind occur when the potential drop across the space charge region reaches 100–200 mV, i.e., when the effective electrokinetic potential ζ_{eff} becomes higher than the "classical" ζ potential.

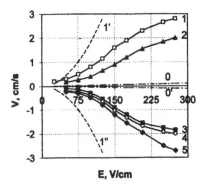

FIG. 8 Electrophoretic velocity as a function of electric field strength for different types of particles: cation exchangers KU-2-8 (curve 1) and KB-4 (curve 2); anion exchangers AN-1 (curve 3), AB-17 (curve 4), and EDE-10P (curve 5); $2a = 750$ µm. Curves 0, $0'$ – electrophoretic velocity according to Smoluchowski theory for $\tilde{\zeta} = 2.5$ and $\tilde{\zeta} = -2.5$; 1, $1'$ – theoretical curves for electrophoresis of the second kind for cation and anion exchangers with ideal selectivity. (Adapted from Refs. 8 and 25.)

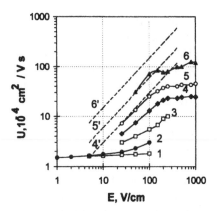

FIG. 9 Electrophoretic mobility as a function of electric field strength for different sizes of particles KU-2-8: $2a = 1$ μm (curve 1); 10 μm (curve 2); 50 μm (curve 3); 100 μm (curves 4, 4′); 200 μm (curves 5, 5′); 500 μm (curves 6, 6′). Curves 1–6: experimental, and 4′–6′: theoretical dependences. (Adapted from Refs. 8 and 25.)

For particles with diameters $2a = 10$–500 μm in fields of 25–200 V/cm, the dependences $U(E)$ are close to linear with almost the same slope (about 45°), demonstrating the second-order dependence of electrophoretic velocity on the external field. For larger fields (200–1000 V/cm), the curve reaches saturation. As a consequence, the first-order dependence of electrophoretic velocity on the voltage occurs. This could be explained by the negative value of parameter k in Eq. (37), because the tangential drift of charge becomes important and the particles move in the regime of large Reynolds numbers (Section V). Both factors cause a decrease in velocity.

Similar experimental investigations were also carried out for ion exchange fibers [8]. The trends for cylindrically shaped particles are similar to those for spherical particles. In weak electric field, the mobility of fibers is almost independent of voltage, while with the increase in applied voltage the electrophoretic mobility starts to rise. For particles and fibers with the same size in the direction of the field (diameter of spherical particles and length of fibers), the slope of the curves and the value of electrophoretic mobility are almost equal. The measurement for nonconducting fibers showed that their electrophoretic mobility at high electric field is lower by a factor of several tens than that for conducting fibers. Moreover, their velocity is independent of the fiber length.

The occurrence of electrophoresis of the second kind is associated with the high conductivity of the particles [Eq. (16)]. The conductivity of ion-exchange resins depends on the degree of cross-linkage, the temperature, the type of electrolyte solution, and other factors. Particularly interesting is the strong dependence of the conductivity on pH that affects the dissociation of carboxyl groups and, as a result, changes not only the conductivity but also the sign of ions that provide current through the ion exchanger [29]. This allows one to undertake analysis of the role of condition (16) the results of which, obtained in Ref. 30, are shown in Fig. 10. The theoretical curve 2′ in Fig. 10 was calculated according to expression (56), evaluating the role of nonzero potential drop in a particle caused by its finite internal electrical resistance. For numerical calculations, the experimental dependence of internal conductivity of ion-exchange particle KB-4 on pH [29] was used. As for KU–2-8, it is known that its conductivity is almost constant over a rather large pH interval that is reflected by curve 1. Unfortunately, experimental data for its internal conductivity at very low and high pH are unknown and therefore a theoretical curve cannot be drawn.

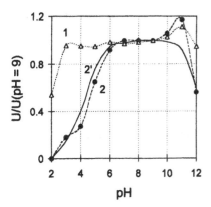

FIG. 10 Normalized electrophoretic mobility of particles as a function of pH for cation exchangers KU-2-8 (curve 1) and KB-4 (curves 2, 2′) at $2a = 750$ μm and $E = 150$ V/cm. Curve 1, 2 – experiment; curve 2 – theory. (Adapted from Refs. 8 and 30.)

To obtain a general notion of the difference between the velocities of electrokinetic phenomena of the second kind and the classical velocities, it is useful to show the obtained experimental results for electrophoresis and electro-osmosis in the same figure and compare them with Smoluchowski's velocity for ζ potential, measured at low voltage (Fig. 11). As one can see, experimental (curves 1 and 2) and theoretical (curves 1′ and 2′) values of velocity for electrokinetic phenomena of the second kind almost coincide at low voltage. The difference between the classical velocity curves 1″ and 2″) and the velocity of electrokinetic phenomena of the second kind (1, 1′ or 2, 2′) is small at low voltage and grows with its increase. At 1000 V/cm, this difference reaches almost three orders of magnitude according to theoretical predictions (curves 2′ and 2″) and about two orders according to experimental measurements (curve 2) and the theoretical approximation of Smoluchowski's law for high voltage (curve 2″).

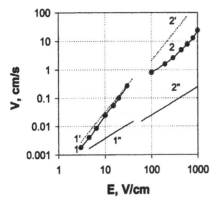

FIG. 11 Electro-osmotic (curves 1, 1′, 1″) and electrophoretic (curves 2, 2′, 2″) velocities for cation exchanger KU-2-8 placed in distilled water at $2a = 600$ μm. Curves 1, 2 – experimental values; curves 1′, 2′ – theoretical ones for electrokinetic phenomena of the second kind. Cuve 1″, 2″ – linear velocities of electro-osmosis at $\theta = \pi/2$ and electrophoresis, calculated for $\tilde{\zeta} = 2.5$. (Adapted from Ref. 25.)

Since the theoretical evaluation of velocity is based on the use of very approximated distribution of electric field around a particle, the deviation between curves 2 and 2' is rather large. In addition, the deviation can be explained by the influence of numerous factors that accompany polarization of a particle in a strong electric field (Section V), particularly by the negative action of tangential processes and by braking of particles caused by the influence of walls and the large Reynolds number.

B. Electrophoresis of the Second Kind for Particles with Electron and Hole-Types of Conductivity

The theory of polarization processes and electrokinetic phenomena of the second kind for both metallic and semiconducting (with hole-type conductivity) particles is closely linked to the main features of polarization processes for ion-exchange particles. In all cases these are unipolar conductors. The current through an ionite particle is ensured by ions with a charge opposite to that of ionized surface groups of the ionite material (i.e., by counterions). In metallic or other particles (e.g., carbon) with electron type of conductivity the current is caused by the electromigration of electrons. In semiconducting particles, the current is caused by the electromigration of holes. Both electron and hole types of conductivity are related to oxidation/reduction reactions on the particle surface that complicate concentration polarization and change the properties of electrokinetic phenomena.

In spite of the similarity of polarization processes that take place for ion exchangers and metallic or semiconducting particles, there are also essential distinctions between them since induced space charges of opposite sign can arise on opposite sides of the particles [5]. The sign of the space charge coincides with the sign of the ions transported from the electrolyte to the conducting particle surface. However, on the opposite sides of a metal or semiconducting particle, different electrochemical reactions take place; as a consequence, ions of opposite sign are transported to these surfaces. Close to the part of the surface where electron transport is accomplished from the metal to the cations continually transported from the volume of electrolyte, a cationic space charge appears. An anionic space charge is formed near the part of the surface where the anion draws of electrons from the metal surface. Two regions of the induced space charge are also formed inside the particle. The sign of these charges is opposite to the sign of induced space charges outside of the particle. The influence of an external electric field on the induced space charges localized near the particle surface leads to the appearance of electro-osmosis and electrophoresis of the second kind.

Generally, the magnitude of the induced space charge formed on the opposite sides of the particle and electro-osmosis provoked by these charges are not symmetrical because of a number of reasons. One of such reasons is the different type of ions that are oxidized or reduced on the particle surface, the presence of different admixtures in solution which affect the oxidation–reduction potential of ions, and the process of dissociation of water, as well as the surface properties of metallic or carbon particles, etc. All factors modify the density and the thickness of induced space charges.

The asymmetry of cationic and anionic induced space charges leads to asymmetry in the electro-osmotic slip: higher induced space charge causes more intensive hydrodynamic flow. Indeed, the essential asymmetry of the electro-osmotic flow near the opposite sides of particles suspended in electrolyte allows the movement of particles in electric field. Full symmetry would have led to mutual compensation of forces affecting the particles and, consequently, to the immobility of the particle. On the other hand, the considerable predominance in the magnitude of the induced space charge of one sign over the induced charge of the opposite sign leads to the prevalence of forces, which act on the particle from one of its sides. This determines the direction and the velocity of the particle movement in strong electric fields.

Electrokinetic phenomena of the second kind require the flow of electric current through the particles. Through an ion-exchanger particle, current flows at arbitrarialy small voltage. For particles with electron-type conductivity, electric current through a particle at low voltage is impossible [31, 32]. It can be realized only at a high voltage sufficient for electrochemical oxidation–reduction reactions at the particle/solution interface. Every ion that transfers its charge to the particle surface has its equilibrium (cathode or anode) potential Φ_c and Φ_a. Current flows through the metallic particle–solution system only if the potential drop at the interface Φ exceeds the sum of the two potential drops Φ_a and Φ_c. This sum is called the voltage of decomposition of the electrolyte, i.e., $\Phi_d = \Phi_c + \Phi_a$. For example, the decomposition voltage of water is 1.23 V.

Moreover, a current should flow along the surface of the particle. Taking into account that the angular dependence of potential drop on the polarized layer is close to $\cos \theta$ (Fig. 3), the necessary condition of current at a given angle θ is

$$\Phi \cos \theta \geq \Phi_d \qquad (38)$$

A similar condition was found for polarization processes in period fields [33]. Because a portion of the total potential drop on metallic particles is used for decomposition of the electrolyte, these materials undergo electrophoresis of the second kind at higher voltages than those for ion-exchange particles, and their velocity in the first case should be lower than in the second one by a factor $(2Ea - \Phi_d)^2/(2Ea)^2$. Thus, if the velocity of electrophoresis of ion-type conducting particles in strong electric fields is described by Eqs. (36 and 37), the electrophoretic velocity of particles with electron-type conductivity can be roughly estimated by

$$\tilde{V}_{ef} \approx \frac{5}{9\pi}(\tilde{\Phi} - \tilde{\Phi}_d)^2 f(\tilde{\Phi}) \qquad (39)$$

This equation is valid when an induced space charge is formed only near one side of a particle. For electronic conducting particles, the regions of concentration polarization near the opposite side of a particle contain an induced space charge of the opposite sign. As mentioned above, this leads to electro-osmotic flow in the opposite direction and partial compensation of the forces affecting the particle suspended in the electrolyte. As a result, the electrophoretic velocity of an electron-conducting particle is not only smaller than the velocity of an ion-conducting particle but also smaller than the velocity, calculated according to Eq. (39). To analyze the extent of this decrease, it was supposed [28, 34] that the velocities of electro-osmosis near opposite sides of a particle are related to certain potentials $\Phi_{1,2}$, the sum of which $\Phi_1 + \Phi_2$ is equal to $\tilde{\Phi} - \tilde{\Phi}_d$. At $f_1(\Phi) \approx f_2(\Phi) \approx 1$, the velocity of electrophoresis can be evaluated as $\tilde{V}_{ef} \sim \int_0^{\pi} (\tilde{V}_{eo1} - \tilde{V}_{eo2}) \sin \theta d\theta \sim (\Phi_1^2 - \Phi_2^2)$, or by introducing the parameter $\gamma = \Phi_1/(\Phi_1 + \Phi_2)$, as

$$\tilde{V}_{ef} \sim (\Phi - \Phi_d)^2 (2\gamma - 1) \qquad (40)$$

Thus, for the antisymmetric picture ($\gamma = 1/2$) the velocity is equal to zero. When an induced space charge is formed only near one side of a particle $\gamma = 1$), the expression for electrophoretic velocity corresponds to Eq. (39). For $1/2 < \gamma < 1$, the velocity takes intermediate values and finally, at $\gamma < 1/2$, the velocity changes its sign.

An optimum case of the manifestation of such asymmetry could be the direct investigation of electro-osmosis of the second kind. However, as is shown above, the study of electrokinetic phenomena of the second kind for metallic or semiconductive particles can be performed only in much stronger electric fields than those required for ion-exchange particles. The method developed for the visualization of electro-osmotic flow [9] cannot be used to study electro-osmosis due to intensive thermoconvection in a cell.

The experimental investigation of electrophoresis of the second kind was undertaken for a particle with both electron-type [28, 34] and hole-type [35] conductivities. Several experimental curves from Refs 28 and 34 are shown in Fig. 12. The electrophoretic velocity of metallic (Al/Mg alloy) and activated carbon particles substantially exceeds the electrophoretic velocities expected for nonconducting particles, although it is lower than the velocity of an ionite particle. According to our model, this is due to the influence of the appearance of an induced space charge near the opposite side of the particle. In the case of an activated carbon particle, a certain decrease in velocity can be caused by the porosity of the particle, which leads to additional ion-type conductivity and creates the possibility of transport of both anions and cations through the particle. As a result, the induced space charge and, correspondingly, the velocity of electrophoresis decrease. Curve 5 for $\gamma = 3/5$ demonstrates the possible values of particle velocity according to the theoretical conception.

The upward curvature that can be seen in Fig. 12 indicates that the electrophoretic velocities of different particles are proportional to the value of the electric field raised to a power larger than 1 but lower than 2. Taking into account that the velocities obtained are considerably higher than the possible values of linear velocity, it can be suggested that these rather large values are caused by the induced space charge. More detailed analysis of these curves is impossible because of the nonsphericity of the particles (Al/Mg alloy is fiber shaped, while graphite particles are plate-like) and unknown distribution of charges near the opposite sides of the particles, which strongly depend on reactions on the particle surface.

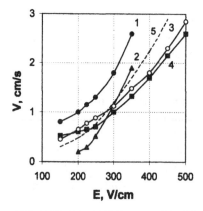

FIG. 12 Electrophoretic velocity as a function of electric field strength for a cationite particle KU-2-8 with size $2a = 400$ μm (curve 1); graphite particles with size $2a = 400$ μm (curve 2); Al/Mg alloy with size $2a = 300$–500 μm (curve 3); activated carbon particles with size $2a = 350$–450 μm (curve 4). The measurements were carried out in distilled water. Curve 5 shows theoretical evaluation of velocity for $\gamma = 4/5$. (Adapted from Refs. 8 and 34.)

IV. UNLIMITED GROWTH OF CURRENT UNDER CONDITIONS OF ELECTRO-OSMOSIS OF THE SECOND KIND

As mentioned above (Section II), electrokinetic phenomena exist due to the appearance of the induced space layer that decreases the thickness of the diffusion layer, thus causing the growth of current over its limiting value. Another factor, important for the growth of current, is a decrease in thickness of the diffusion layer caused by hydrodynamic flow [36]. The mutual influence of both factors intensifies the current through a flat ion-selective membrane several times [37] and leads to the strong growth of current under conditions of electro-osmosis of the second kind near a spherical particle [4] or an ion-exchange membrane with heterogeneous conductivity [38–41]. The latter effect was called "disappearance of phenomenon of limiting current" [4, 7].

The explanation of the phenomenon is very simple. Under conditions of electro-osmosis of the second kind, the velocity of liquid movement V is proportional to the square of the electric field strength E^2. Taking into account that the thickness of the CDL near a spherical particle depends on the Peclet number Pe $= aV/D$ or the liquid velocity V [17–20] as $\delta = \sqrt{aD/V}$, one can see that electro-osmosis of the second kind reduces the thickness of CDL as E^{-1}:

$$\delta \sim \frac{1}{E} \tag{41}$$

Current through a spherical particle is related to the process of polarization and preserves its diffusion nature similarly to current through a flat interface. However, the thickness of the CDL [Eq. (41)] decreases, causing the growth of current. Considering Eqs (6) and (41) it is clear that in the case of a spherical surface the current density is a linear function of the applied field:

$$i = \frac{2FDC_0}{\delta} \sim E \tag{42}$$

Thus, a linear rapid growth of current for a spherical interface occurs instead of the rather small overlimiting current for a flat one. Similar results are possible for a pair of spherical [7, 42, 43] or cylindrical [43, 44] particles and for a flat ion-selective membrane with heterogeneous conductivity [38–41].

It is necessary to stress that Eq. (41) was obtained for weak concentration polarization, when the angular distribution of velocity is close to sinusoidal. In our more complicated case, Eq. (41) describes only a qualitative dependence of current on electric field strength.

A. Current Through Two Spherical Particles

A special scheme for an experimental cell has been developed to investigate the concentration polarization between two curved surfaces where nonhomogeneity of the electric field is caused by the geometry of intersurface space [35, 36]. The cell is connected to the source of current in the regime of desalination in the central chamber between ion-selective granules (Fig. 13a). For maximum manifestation of concentration polarization and electro-osmosis of the second kind, the electrode chambers should be filled with an electrolyte of high conductivity, whereas in the central chamber it is necessary to use an electrolyte with low conductivity. Under such conditions, the potential drop in the electrode chambers can be neglected. Due to the high conductivity of the ion-exchange material, the

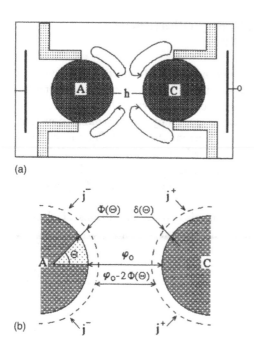

(a)

(b)

FIG. 13 (a) Scheme of experimental cell for measurement of current through a pair of ion-exchange granules; closed circles show electro-osmotic flow of liquid. (b) Scheme of ion fluxes and potential drop between a pair of ion-exchange granules.

potential drop in the granules can also be ignored. Thus, the potential drop ϕ_0 between the electrodes takes place totally in the space between the two granules. Hence, a theoretical analysis of polarization related to the intergranule region can be undertaken. The potential distribution in the discussed case is simpler than in the case of a single particle (Section II) because it takes place in a closed space and the analysis of polarization should be carried out only at one side of each granule.

The potential drop ϕ_0 is identical between each symmetrical point on the surface of the granules (Fig. 13b). However, with the increase of angle θ, the distance between symmetrical points and, correspondingly, the resistance of the liquid grow. As a result, the current density changes along the surface of the granule from its maximum value on the line connecting the centers of the granules to zero far from this line. At high voltage, the induced charge layer appears and the potential drop Φ on this layer in an angular function. Thus, there appear the induced space charge and tangential electric field that cause electro-osmosis of the second kind. If the distance between granules is much less than their radius, the strength of the electric field behind the boundaries of the diffusion layer [with thickness $\delta(\theta)$ and $S(\theta \ll \delta(\theta)$] can be evaluated as the ratio between $\phi_0 - 2\Phi(\theta)$ and $h(\theta) - 2\delta(\theta)$. Expressing the intensity of current outside the diffusion layers in these terms and representing the current density as a limiting current for given $\delta(\theta)$, we obtain the following formula for the angular dependence of the potential drop:

$$\tilde{\Phi}(\theta) = \frac{\tilde{\phi}_0}{2} + 1 - \frac{h(\theta)}{2\delta(\theta)} \qquad (43)$$

where

$$h(\theta) = h_0 + 2a(1 - \cos\theta); \quad \tilde{\phi}_0 = \frac{F\phi_0}{RT} \tag{44}$$

The angle dependence of the electro-osmotic velocity of the second kind is obtained [7] on the basis of Eq. (29):

$$V \sim \Phi(\theta)\frac{\partial\Phi\theta)}{\partial\theta} \approx \frac{m}{2}\frac{D\tilde{\phi}_0}{\delta}\sin\theta \tag{45}$$

under the restrictions:

$$\frac{h(\theta)}{\delta(\theta)} \ll \tilde{\phi}_0; \quad \frac{\partial}{\partial\theta}\ln\delta(\theta) \ll \frac{\partial}{\partial\theta}\ln h(\theta) \tag{46}$$

A solution of the convection–diffusion equation [7] shows that the electro-osmotic mixing of liquid between granules determines the thickness of the CDL and the density of current as

$$\delta = \frac{10a}{\tilde{\phi}_0} \quad \text{and} \quad i = \frac{2FD_0C_0}{10a}\tilde{\phi}_0 \tag{47}$$

Thus, the density of current is proportional to the value of applied voltage ϕ_0. In the case when the thickness (d) of walls that limit the electrode chamber is small ($d \ll a$), the walls do not affect electro-osmosis. Consequently, the current can be calculated as

$$I \approx iS_g \tag{48}$$

where $S_g = 2\pi a^2$ is half the surface of a spherical particle.

The experimental measurements of current were carried out for different combinations of flat and spherical interfaces (membrane/membrane; membrane/granule; granule/granule). The scheme of the experimental cell with flat and spherical interfaces was similar to the one presented in Fig. 13a with flat membranes replacing one or both granules. The mathematical description of the processes in the cell with a granule and a membrane is quite complicated. While in the case of two granules every granule is responsible for the structure of its own CDL, in the case of a granule and a membrane, the electro-osmotic flow of the granule affects the CDL of the membrane. Due to the peculiarities of the electro-osmotic flow between a granule and a membrane, the thickness of the CDL of the membrane noticeably changes along its surface [35], and its average value considerably exceeds the thickness of the CDL of the granule. The area of the membrane surface was 10 times larger than the area of the granule surface. This allowed one to analyze the 10-fold difference between the density of current through the curved and flat surfaces.

The measured values of current are shown in Fig. 14. The current through two membranes (curve 1) is quite close to the theoretical value (curve 1'). Other things being equal, the current through two granules (curve 4 in Fig. 14) is considerably higher than the one through two membranes (curve 1). Nevertheless, the values of current are lower than they are supposed to be according to the theoretical prediction (curve 4') based on Eq. (48). This discrepancy can be explained by the action of two factors: the property of the anion-selective material and the peculiarity of the experimental cell.

The first factor is related to the difference between polarization of cation- and anion-selective materials: the velocity of electro-osmosis of anion-exchange particles is considerably lower than in cation exchangers (Section II). Thus, the thickness of the CDL is larger and the possible density of current for an anion-exchange granule is lower than the density

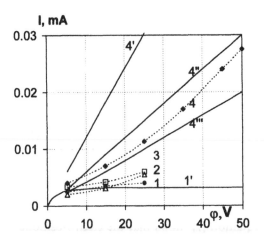

FIG. 14 Current as a function of applied voltage in different experimental schemes: 1 – two ion-exchange flat membranes; 1′ – theoretical model of limiting current; 2 – anion-exchange granule/cation-exchange membrane, 3 – cation-exchange granule/anion-exchange membrane; 4 – two ion-exchange granules; 4′– theoretical dependence without complicating factors; 4″ – theoretical dependency with account of peculiarities of electro-osmosis near anion-exchange granule; 4‴ – theoretical dependency with account of a stagnant region.

for the cation exchanger. As a result, this limits the measured current. The calculations with account of the actual velocity for an anion-exchange particle are shown as curve 4″.

The second factor is the deterioration of liquid movement caused by the special tubes for fixation of particles (Fig. 13a). As a result, the electro-osmotic flow cannot reduce the thickness of the CDL in the region which becomes considerably larger than the thickness predicted for a "free" particle. Thus, this part of the granule surface is lost for the high density of current. The rough evaluation of this factor [35] is shown by curve 4‴ in Fig. 14. On the basis of this, it can be stated that the experimental results are in good agreement with the theoretical prediction.

The values of current for two versions of the granule/membrane system (Fig. 14, curves 2 and 3) are close to the ones that are true for current through two membranes (Fig. 14, curve 1). Current through the anion-exchange granule and cation-exchange membrane (curve 2) is close to the current through the opposite versions of the granule and the membrane (curve 3).

The change in pH during measurement of current was less than 0.3–0.5. Thus, contribution of H^+ and OH^- ions in fluxes through granules and membranes can be neglected (see Section V.B).

B. Current Through Two Ion-Exchange Fibers

The models for flat and curved surfaces presented above are based on the notion of the large conductivity of ion-exchange materials and relatively low conductivity of liquid. This corresponds to many systems where the investigation of current through an ion-exchange material is both of scientific and practical interest. At the same time, it is well known that, due to the low values of the diffusion coefficient, the conductivity of ion-exchange materials cannot be indefinitely large and depends considerably on the conductivity of the liquid that is used [29]. Therefore, a special experimental scheme for investigation of the combination of external and internal transport processes was created (Fig. 15).

Two cation- and anion-exchange fibers are set in parallel. On one side, their ends are inserted into electrode chambers, while on the other side the ends are held with the aid of a nonconducting material to fix the distance between the fibers. The fibers are connected to the power source in the regime of desalination. The potential difference ϕ_0, established by the external source, is distributed along and between the fibers. Owing to the action of the electric field, the cations and anions move to the fibers and through them to the electrode chambers. As a result, regions of low electrolyte concentration in the space between fibers near their surfaces appear. In a strong electric field, the thickness of the region of concentration polarization is defined by electro-osmosis of the second kind, whereas in a weak electric field, when electro-osmosis of the second kind is absent or minimal, it is defined by the velocity of the external hydrodynamic flow or is equal to half the distance between the fibers in the absence of the external flow.

The curved surface of fibers is similar to the surface of spherical granules. Thus, electro-osmotic slip of the second kind appears and leads to significant compression of the CDL: $\delta \ll h/2$. Its thickness can be evaluated by Eq. (47) and the current can be described as

$$I_0 = \frac{FDC_0}{5a}\tilde{\phi}_0 \pi aL \tag{49}$$

where a is the radius of the fiber, and L is its length in the desalination chamber. The current [Eq. (49)] is lower by a factor $h/10a$ than the one that takes place without concentration polarization and is greater by $\tilde{\phi}_0 h/10a$ than in the case of concentration polarization without electro-osmotic slip of the second kind [43, 44].

However, with such intense external diffusion transport (i.e., the transport of ions outside the fibers), the internal diffusion transport of ions becomes a new limiting factor. If the conductivity of the fibers is not very large, they are not able to carry away all the current that can be provided by the ion fluxes to the fiber surface.

The theory of concentration polarization of fibers with arbitrary conductivity was developed in Refs 43 and 44. For the purposes of simplification of the theory, a few approximations were used. The object of the study is long fibers, placed in an electrolyte with low conductivity, with a short distance h between them ($h \ll L$). This allows one to describe the characteristics of concentration polarization between the fibers by the solution of a local one-dimensional problem, and to divide the current into two components: the current in the electrolyte that is perpendicular to the fiber surface and the current in the

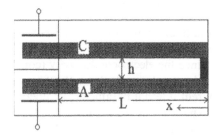

FIG. 15 Scheme of an experimental cell for investigation of polarization of ion-exchange fibers. C, A: cation- and anion-exchange fibers; h: distance between fibers; L: length of fibers inside the desalination chamber; x: distance to the free end of the fibers.

fiber that is directed along the longitudinal axis. It is assumed that the properties of the fibers are similar and, consequently, the potential difference is distributed uniformly between them.

The normal component of current density is determined by the thickness of the CDL $\delta(x)$ specified by the electro-osmotic flow that depends on x. The current along the fiber is replenished by the normal flux of ions with displacement from $x = 0$ to $x = L$. Thus, longitudinal current at the free end of the fiber is equal to zero and reaches its maximum value at $x = L$.

With account of all the above-mentioned assumptions, the distribution of potential along the fiber was obtained [43, 44] in the following form:

$$\tilde{\phi}(x) = \frac{\tilde{\phi}_0}{2} \frac{\exp(-\sqrt{b}\tilde{x}) + \exp(\sqrt{b}\tilde{x})}{\exp(-\sqrt{b}) + \exp(\sqrt{b})} \tag{50}$$

where $b = \frac{2C_0 D_0 L^2}{5 D_i C_i a^2}$ is a dimensionless parameter that determines the relative characteristics of the electrolyte and the fiber; D_1, C_1 are the diffusion coefficient and the concentration of the current carriers (counterions) inside the fibers. In the limiting case $b \ll 1$, Eq. (50) can be rewritten as $\tilde{\phi} = \tilde{\phi}_0/2$. Hence, due to the large internal conductivity of the fibers, they are isopotential. In the opposite limiting case $b \gg 1$, when the conductivity of the fibers is low, the potential changes considerably along the fibers. Its value near the free end of the fiber is significantly lower than the value $\tilde{\phi}_0/2$ that is given by the power source. It follows from Eq. (50) that the potential decreases more rapidly, the larger the parameter b, i.e., the higher electrolyte concentration and the thinner and longer the fiber. However, this means that the lower the current normal to the fiber surface, the less productively the fiber surface is utilized.

Considering the potential distribution along the fibers and its corresponding influence on electro-osmosis of the second kind and local value of current density, the measured current takes the value:

$$I_1 = I(\tilde{x} = 1) = \frac{F D_1 C_1}{2L} \pi a^2 \, th(\sqrt{b})\sqrt{b}\tilde{\phi}_0 \tag{51}$$

Under the assumption $th(\sqrt{b}) \approx \sqrt{b} \ll 1$, Eq. (51) leads to $I_1 = \pi F D_0 C_0 L \tilde{\phi}_0/5$, which coincides with Eq. (49). This means that the current in the system is independent of the coefficient of diffusion and the concentration of the current carriers in the fiber. In the opposite case, $b \gg 1$, the role of internal conductivity is substantial.

The experimental data for current through the ion-exchange fibers at $C_0 = 0.001$ mol/L are shown in Fig. 16. The theoretical curves are calculated for three values of internal conductivity, because the internal electroconductivity of the ion-exchange material strongly depends on the concentration of electrolyte in which it is placed [29]. However, in the case investigated, the fibers are placed in liquid with different concentrations of electrolyte because in the electrode chamber its concentration is considerably higher than in the central chamber between the fibers. Since the electroconductivity (concentration of ions and their diffusion coefficients) could be changed along the fibers, numerical calculations were carried out for several different internal parameters of fibers, which, according to different papers, are close to our experimental conditions.

It can be observed that the current is higher than in the case of the granules, this being explained by the larger surface area of the fiber. However, the latter is approximately 300 times larger than that of the granule, while the current, for instance, at 50 V, is only 15 times higher. This considerable difference can be caused by the lower internal conductivity

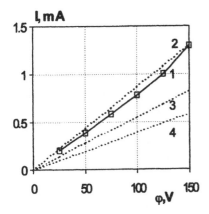

FIG. 16 Current through two ion-exchange fibers as a function of applied voltage : curve 1 – experiment; curves 2 – 4 theoretical curves for $C_i = 1$ mol/L at different values of diffusion coefficient inside the fibers: $D_i = 10^{-8}$ cm^2/s (2), 2×10^{-8} cm^2/s (3), and 4×10^{-8} cm^2/s (4).

of the fibers. Nevertheless, the most important factor is the radius of the fibers because both the resistance of the fibers and the parameter b are inversely proportional to the area of their cross-section. Thus, there is a possibility of current increase due to the larger surface area of the fibers ($\pi a L$ for the fibers is noticeably larger than the surface area $2\pi a^2$ for the granules), although, at the same time, there is a limiting factor related to the different area of their cross-sections.

C. Current Through Flat Heterogeneous Membrane

Electro-osmosis of the second kind also appears near a flat interface with heterogeneous conductivity of different sections [38–41]. Due to the nonuniform conductivity, the lines of electric current are curved. As a result, the electric field has two components: normal V_n, and tangential V_t, to the surface. Thus, the conditions for electro-osmosis of the second kind exist. The normal component of the field V_n creates the region of induced space charge and the action of the tangential component V_t on the induce space charge leads to the formation of electro-osmotic whirls. For equal sizes of sections with high and low conductivity, the symmetry of processes occurring in the region near the surface resembles those at the surface of a spherical particle. This means that, while solving the problem of formation of electro-osmotic whirls and their effect on current, the basic ideas of the theory of electro-osmosis of the second kind at the surface of an ion-exchange granule can be used.

It should be noted that the same size of sections with significantly different conductivity is optimum for the appearance of the comparable values of the normal and tangential components of an electric field. With a greater size of conducting sections, the tangential components of the electrical field is insignificant. With a greater size of nonconducting sections, the additional advantage in the value of the tangential component of the electric field directly at the membrane surface is accompanied by a growth in the total resistance of the membrane and, correspondingly, by a decrease in the normal component of the electric field that negatively affects current through the membrane and the value of the induced space charge.

Another necessary condition for strong electro-osmotic flow is a rather strong electric field, needed to form a wide region of induced space charge. If the thickness of the induced space charge, S_0 is much higher than the thickness of the electrical double layer, κ^{-1}, and is close to the characterstic dimension d of the sections with different conductivity, the action of the tangential component of the electric field on the induced space charge is maximum. It was exactly this case that was described analytically [38, 41]. A more general case was analyzed with the help of numerical methods [39].

It is necessary to emphasize that the peculiarities of concentration polarization of a heterogeneous membrane depend on the characteristics of external hydrodynamic flows. It was shown that, under laminar conditions, the increase in current due to electro-osmotic convection is small [38, 41]. However, when electro-osmotic convection is combined with turbulent pulsation, the current growth above the limiting value can be significant.

V. FACTORS AFFECTING VELOCITY OF ELECTRO-OSMOSIS AND ELECTROPHORESIS OF THE SECOND KIND

The theory of electrokinetic phenomena of the second kind that was presented above does not take into account many processes, which can complicate concentration polarization and reduce the experimental velocity in comparison with the theoretical one. A few such processes are described below.

A. Tangential Drift of Induced Space Charge

The theory of electro-osmosis of the second kind was developed with approximation of thin CDLs and SCLs. However, the increase in applied electric field leads to different behaviors of these layers: the thickness of the CDL decreases [Eq. (41)] and that of the SCL increases as $S_0 \sim \Phi^{1/3}$ [5]. Thus, the condition $S_0 \ll a$ limits the applicability of the theory for high electric field.

In addition to this, the condition shows the strength of an electric field above which the velocity of electrokinetic phenomena decreases. Indeed, to simplify the developed theory, the tangential flux of counterions has been neglected. Due to this neglect, the calculated value of density and thickness of an induced space charge are higher than the real ones. This means that the theoretical values of electrokinetic velocity are also higher than the real ones. On the basis of analysis of expressions for tangential and normal components of ion fluxes [5], the following critical value of the electric field strength is obtained:

$$\frac{FE_{\text{crl}}}{RT} = (\kappa a)^{4/5}/a \tag{52}$$

For example, for $a = 0.025$ cm and $C_0 = 10^{-5}$ mol/L, the critical value of the field is 750 V/cm. This means that at 400–500 V/cm the tangential processes become important.

B. Water Dissociation

According to condition (16), electrokinetic phenomena of the second kind exist at low electrolyte concentration. This means that, due to concentration polarization, the density of counterions in the region of an induced space charge may reach the value of 10^{-7} mol/L corresponding to local equilibrium of the reaction of water dissociation $C_H C_{\text{OH}} = 10^{-14}$ mol^2/L [45]. Hydrogen ions and hydroxyl ions cause an increase in electrical conductivity in the region of an induced space charge and the current through the interface. As a result, the space decreases [45, 46].

In the case of flat surfaces, water dissociation increases with electric field [45], while in the case of curved surfaces, the opposite effect takes place [46, 47]. Owing to electro-osmosis of the second kind, the density of the induced space charge increases and water dissociation decreases. The critical electrical strength E_{cr2} may be calculated according to the following expressions [47, 48]:

$$\sqrt{C_H C_{OH}} \approx C_s \text{ or } \frac{FE_{cr2}}{RT} \approx \left(\sqrt{C_H C_{OH}/C_0}\right)^3 (\kappa a)^2/2a \tag{53}$$

The theoretical conclusion about weak water dissociation under conditions of electrokinetic phenomena of the second kind has been corroborated experimentally with the help of a specially developed method based on the change in color of a particle in response to the change in pH [47, 48].

Water dissociation caused by the low density of an electrolyte in the region of an induced space charge exists independently of the nature of a particle material. However, there is a possibility of existence of another mechanism of water dissociation, i.e., the one caused by the catalysis of dissociation caused by surface groups of the anion-exchange material [15, 16]. This leads to an additional decrease in the induced space charge and, as a result, to lower velocity of electro-osmosis and electrophoresis for anion-exchange particles as compared to the cation-exchange ones. The second process that decreases the velocity of electro-osmosis is the concentration of ionized impurities (anions) in the region of induced space charge [49]. The sediment of macroions forms a bipolar contact with the material of an anion exchanger, similar to bipolar membranes, usually causing the process of water dissociation [50].

C. Heat Effects

Heat processes caused by electric current play an important part in the concentration polarization of ion-exchange membranes. Local warming of liquid in the region of low electrolyte concentration leads to an increase in ion-diffusion coefficients [51] and to thermal convection [14]. These factors stimulate the change in ion flows and raise the concentration near the membrane. As a result, the induced space charge near the membrane decreases or disappears.

Another conclusion follows from the analysis of heat processes near the curved surface under conditions of electro-osmosis of the second kind [52]. Taking into account the local warming and heat transfer caused by heat conductivity and electro-osmotic movement of the liquid, the temperature rise in the region of induced space charge, ΔT, may be described by the expression:

$$\Delta T = \frac{FDC_0\tilde{\Phi}^2}{30K_T a\sqrt{m}}\left[\frac{(9\tilde{\Phi})^{1/3}\kappa^{-2/3}a^{1/3}}{4m^{1/6}} + \left(\frac{K_T}{DC_p}\right)^{1/2}\frac{a\sqrt{m}}{\tilde{\Phi}}\right] \tag{54}$$

where K_T is the heat conductivity, C_p is the specific heat capacity, and the parameter m is equal to 0.2 for an aqueous electrolyte solution. For the large interval of parameters of the investigated system, $\Delta T = 2°–5°C$. This temperature increase leads to thermal convection along the particle surface with velocity:

$$V_T = \beta g \Delta T (S_0 + \partial_T)^2/2\nu \tag{55}$$

where β is the thermal expansion coefficient, g is the acceleration due to gravity, $\delta_T = (K_T /DC_p)^{1/2}(a/\sqrt{m}\tilde{\Phi})$ is the thickness of the thermoconvective layer, and ν is the kinematic viscosity. The value of velocity may be equal to approximately $10^{-4} - 10^{-3}$ cm/s. Thus, the velocity of thermal convection is two or three orders of magnitude lower than the velocity of electro-osmosis of the second kind. This means that the thermal convection in the SCL cannot significantly affect the induced space charge and change the characteristics of electrokinetic phenomena of the second kind. However, the general thermal convection in the experimental cell can diminish the accuracy of velocity measurements.

D. High Concentration of Electrolyte

Condition (16) is completely correct only in special cases because the conductivity of a particle depends on the electrolyte concentration [29]. If K_i is close to K_e, the electroosmosis and electrophoresis velocity can be still described by Eq. (29), but the dimensionless potential drop $\bar{\Phi}$ in the SCL is less than $2FEa/RT$. In this case, the change in electrolyte concentration leads to a change in velocity [30], approximately given by

$$V(C_2) = V(C_1)\frac{1 + 2\sqrt{m}K_e(C_2)/K_i(C_2)}{1 + 2\sqrt{m}K_e(C_1)/K_i(C_1)} \tag{56}$$

where the subscripts 1 and 2 refer to two different solutions of electrolyte.

For example, if $K_e(C_1)/K_i(C_1) = 0.001$ and $K_e(C_2)/K_i(C_2) = 0.2$, the change in velocity is $V(C_2)/V(C_1) \approx 0.85$.

E. Large Reynolds Numbers and Influence of Walls of Experimental Cell

At high electric fields, the electrophoretic velocity of large particles corresponds to the regime of large Reynolds numbers, $Re = aV/\nu > 1$. For example, at $E = 1000$ V/cm, an ion-exchange particle with radius $a = 100$ μm moves with velocity 4.5 cm/s ($Re = 4.5$), while a particle with radius $a = 250$ μm moves at 10 cm/s ($Re = 25$). At the same time, theoretical evaluation of electrophoretic velocity yields values almost 10 times larger.

This fact can be explained by the influence of the two following factors. The first one is that, according to the theoretical and experimental investigations in the field of hydrodynamics [53, 54], the increase in particle velocity over $Re > 1$ leads to the growth of hydrodynamic resistance to particle movement. This means a deviation of the experimental velocity from the value theoretically predicted on the basis of the hydrodynamics of low Reynolds number.

The second factor is related to the movement of a particle near the solid walls of the experimental cell. In order to reduce thermal convection, a distance between walls of about 1–1.5 mm was taken. According to the theory developed for low Reynolds' numbers [54, 55], the braking of a particle caused by the presence of walls was in this case about 5–10%. It is obvious that, in the cell used for electrophoretic measurements, the two factors strengthen each other, leading to a greater decrease in velocity than the one caused by the influence of separate factors.

The role of walls in electro-osmosis was small because measurements were carried out near the equator of a large particle, far from walls.

VI. APPLICATIONS

Owing to their considerable intensity, the new electrokinetic phenomena may stimulate the development of the electrotechnology of disperse systems. Electrophoresis of the second kind can be applied to separate particles that have different conductivity and size [56], while electro-osmosis of the second kind may be of use in intensification of electrodialysis [7, 42, 43] and electrofiltration [27]. Other interesting possibilities are connected with investigations of a nonstationary induced space charge and its manifestation in different technologies [42, 57]. In this respect, the development of further research in electro-osmosis and electrophoresis of the second kind is of great interest.

VII. CONCLUSION

Strong concentration polarization of conducting particles leads to the appearance of a large induced space charge behind the electrical DL, which considerably changes the main characteristics of electrokinetic phenomena and current through a particle. The velocity of electrokinetic phenomena of the second kind strongly depends on the size of a particle and for large particles exceeds the velocity of the classical phenomena by several dozen times. The correlation between experimental results and theoretical predictions proves the correctness of the developed model for strong concentration polarization of conductive particles and electrokinetic phenomena of the second kind. Owing to the very complicated problem of potential distribution around a particle, the existing theory has a qualitative character and can be regarded as only the first level of theory. The second level, leading to a more accurate theory, is currently under development.

LIST OF SYMBOLS AND ABBREVIATIONS

a	radius of a particle or a fiber
C_0	electrolyte concentration
C_p	specific heat capacity
C_S	density of induced space charge
CDL	convective-diffusion layer
D	diffusion coefficient
DL	(electrical) double layer
E	electric field strength behind the region of concentration polarization
E_S	electric field strnegth in the region of induced space charge
F	Faraday constant
h	thickness of a channel
i	density of current
i_{lim}	limiting current
K_e	conductivity of electrolyte
K_i	internal conductivity of particle or fiber
L	thickness of the region of concentration polarization
p	pressure
R	gas constant
S_0	thickness of the region of induced space charge
SCL	induced space charge layer
T	temperature
U	electrophoretic mobility of a solid particle
V_{ef}	electrophoretic velocity of a solid particle
V_{eo}	electro-osmotic velocity of liquid

V_T	thermoconvective velocity of liquid
x	distance from a granule surface
δ	thickness of convective-diffusion layer
δ_T	thickness of thermoconvective layer
η	dynamic viscosity of liquid
θ	angle, calculated from the direction of external electric field
κ^{-1}	thickness of Debye layer
σ	charge density
Φ	potential drop in the region of induced space charge
ϕ_i	internal potential of polarized particle
ϕ_0	potential drop between the membranes

REFERENCES

1. SS Dukhin, BV Derjaguin. In: E Matievich, ed. Electrokinetic phenomena. Surface and Colloid Science. Vol. 7. New York, Toronto: John Wiley and Sons, 1974.
2. J Lyklema. Fundamental of Interface and Colloid Science. vol. 2. London: Academic Press, 1995, ch. 4.
3. SS Dukhin. Adv Colloid Interface Sci 44: 1, 1993.
4. SS Dukhin, NA Mischuk. Kolloidn Zh 51: 659, 1989.
5. SS Dukhin, NA Mishchuk. Kolloidn Zh 52: 452, 497, 1990.
6. SS Dukhin. Adv Colloid Interface Sci 35: 173, 1991.
7. SS Dukhin, NA Mishchuk. J Membr Sci 79: 199, 1993.
8. AA Baran, Ya Babich, AA Tarovsky, NA Mishchuk. Colloids Surfaces A 68: 141, 1992.
9. NA Mishchuk, PV Takhistov. Colloids Surfaces A 95: 119, 1995.
10. I Rubinstein, L Shtilman. J Chem Soc, Faraday Trans 2 75: 231, 1979.
11. I Rubinstein. J Chem Soc, Faraday Trans 2 77: 1595, 1981.
12. VV Nikonenko, VI Zabolotski, NP Gnusin. Electrokhimiya 25: 180, 1989.
13. AV Listovnichij. Electrokhimiya 51: 1651, 1989.
14. VK Varentsov, MV Pevnitskaja. Izvest Sib Ot Akad Nauk SSSR, Ser Khim 4: 134, 1973.
15. R Simons. Electrochim Acta 30: 275, 1985.
16. Y Tanaka, M Seno. J Chem Soc, Faraday Trans 82: 2065, 1986.
17. SS Dukhin. In: AI Rusanov, FC Goodrich, eds. Modern Kapillaritatstheorie. Berlin: Akademie-Verlag, 1981, p 83.
18. SS Dukhin. Surface Forces in Thin Films and Disperse Systems. Moskow: Nauka, 1967, p 364.
19. SS Dukhin, BV Derjaguin. Electrophoresis. Moscow: Nauka: 1976, p 132.
20. SS Dukhin. In: BV Derjaguin, ed. Research in Surface Forces. vol. 3. NY-L. Consult. Bureau, 1971.
21. NA Mishchuk, SS Dukhin. Kolloidn Zh 50: 1111, 1988.
22. JTG Overbeek. Kolloid Beih 54: 287, 1943.
23. J Lyklema, SS Dukhin, VN Shylov. J Electroanal Chem 143:4, 1983.
24. RJ Hunter. Foundation of Colloid Science. vol. 11. Oxford: Oxford University Press, 1989, ch. 12.
25. NA Mishchuk. Electrokinetic phenomena at strong concentration polarization of interface. Thesis. ICCWC, Kiev, 1996, 370 pp.
26. NA Mishchuk, PV Takhistov. Ukr Khim Zh 57: 240, 1991.
27. NA Mishchuk, PV Takhistov. Khim Tehnol Wody 14: 25, 1992.
28. NA Mishchuk, S Barany, AA Tarovsky, F Madai. Colloids Surfaces 140: 43, 1998.

29. NP Gnusin, VD Grebenyuk. Electrochemistry of Granulated Ioniates. Kiev: Naukova Dumka, 1972, p 178.
30. Ya Babich, A Baran, NA Mishchuk. Ukr Khim Zh 56: 1034–1037, 1990.
31. HR Kruyt. Colloid Science. vol. 1. New York: Elsevier, 1952.
32. DD Dukhin, NV Shilov. Adv Colloid Interface Sci 13: 153, 1980.
33. VA Murtsovkin, GI Mantrov. Kolloidn Zh 54: 105, 1992.
34. A Baran, N Mishchuk, D Prieve. J Colloid Interface Sci 207: 240, 1998.
35. S Barany, Ya Babich, F Madai. Kolloidn Zh 60: 726, 1998.
36. AA Sonin, RF Probstein. Desalination 5: 293, 1989.
37. SS Dukhin. Khim Tehnol Wody 11: 675, 1989.
38. NA Mishchuk, SS Dukhin. Khim Tehnol Wody 13: 963, 1991.
39. I Rubinstein. Phys Fluids A 3 (1): 2301, 1991.
40. VI Zabolotsky VV Nikonenko. Electrokhimiya 32: 246, 1996.
41. NA Mishchuk. Colloids Surfaces A 140: 75, 1998.
42. NA Mishchuk. Desalination 117: 283, 1998.
43. NA Mishchuk, F Gonzalez-Caballero, PV Takhistov. Colloids Surfaces A 181, 131, 2001.
44. NA Mishchuk, PV Takhistov. Khim Tehnol Wody 15: 707, 1993.
45. AV Listovnichij. Dopov Acad Nauk Ukr Ser B 8: 39, 1988 (in Russian).
46. NA Mishchuk. Khim Tehnol Wody 11: 1067, 1989.
47. PV Takhistov, AV Listovnichij, NA Mishchuk. Khim Tehnol Wody 12: 1070, 1990.
48. NA Mishchuk. Colloids Surfaces, A 159, 467, 1999.
49. SS Dukhin, NA Mishchuk, NA Klimenko, VN Goncharuk. Khim Tehnol Wody 11: 885, 1989.
50. EK Zholkovslij, VI Koval'chuk. Electrokhimiya 24: 74, 1988.
51. NA Mishchuk. Kolloidn Zh 53: 402, 1991.
52. NA Mishchuk. Khim Tehnol Wody 13: 212, 1991.
53. C. Ossen. Neuere Methoden und Ergebnisse in der Hydrodynamik. Leipzig: Academiesche Verlagsgesellschaft, 1927.
54. J Happel, H Brenner. Low Reynolds Number Hydrodynamics. Englewood Cliffs: Prentice Hall, 1965.
55. WL Haberman, RM Sayre. David Taylor Model Basin Report No. 1143, Washington, DC: US Navy Dept., 1958.
56. SS Dukhin, AA Tarovsky, NA Mishchuk. Ya Babich, AA Baran, AA Bajchenko. USSR Patent no. 1610643, 1991.
57. NA Mishchuk, LK Koopal, F Gonzalez-Caballero. Colloids Surfaces, A 176, 195, 2001.

11

Relaxation Mechanisms of Homogeneous Particles and Cells Suspended in Aqueous Electrolyte Solutions

CONSTANTINO GROSSE National University of Tucumán, San Miguel de Tucumán, and Consejo Nacional de Investigaciones Científicas y Técnicas, Buenos Aires, Argentina

I. INTRODUCTION

Dielectric spectroscopy is a powerful tool for studying all kinds of colloidal suspensions: macromolecules [1–5], micelles [6–8], latex [9–18], lyposomes [19], and cells [20–27]. Among these systems, polystyrene particles and cell suspensions stand out as being the simplest, which still exhibit all the main relaxations. This makes them ideal study subjects for analyzing the response of colloidal suspensions to an applied electric field. Such an analysis constitutes the purpose of this work.

We start with some very basic concepts such as the meaning of a dielectric measurement, the definition of dielectric properties of homogeneous and inhomogeneous materials, and the elementary frequency response. We then define the simplest possible models of homogeneous particle and cell suspensions and discuss their dielectric behavior. This presentation if undertaken in two steps: in the time domain, discussing the physics of the different mechanisms, and in the frequency domain, calculating the parameters of each relaxation. The aim is to show, as clearly as possible, why these simple models present such a complicated dielectric behavior, and how can the measurement of this behavior be used to determine different parameters of the system.

We purposefully ignore many complications found in real suspensions such as the effects of bound water, static membrane conductance, frequency dependence of the membrane dielectric properties, deviations from perfect spherical geometry, convection effects in the electrolyte solution, complex interfaces such as the cell wall, anomalous surface conductivity, high concentrations of particles, different ion valences, more than two types of ions, and nonlinear effects [28–42]. While their inclusion may be necessary in some cases, the interpretation of the dielectric spectra always requires a thorough understanding of the main mechanisms, which we shall consider here.

II. DIELECTRIC MEASUREMENT

The dielectric properties of a material can be defined on the basis of the following idealized experiment. A measurement cell is made with the sample placed between the parallel plates of a capacitor (area A, separation d), Fig. 1. The measurement cell is connected to a voltage source:

$$V(t) = V_0 \text{Re}(e^{i\omega t})$$

and the amplitude and phase of the current are measured:

$$I(t) = I_0(\omega)\text{Re}\left\{e^{i[\omega t - \phi(\omega)]}\right\} \tag{1}$$

The measurement cell is now represented as a parallel circuit made of an ideal capacitor $C(\omega)$ and an ideal resistor $R(\omega)$, having both the same dimensions as the cell. Fig. 1. The term "ideal" means that the admittance of the resistor has no imaginary part while the admittance of the capacitor is purely imaginary:

$$Y^*(\omega) = 1/R(\omega) + i\omega C(\omega)$$

where the asterisk denotes a complex quantity. Solving either the real or the imaginary parts of the equation:

$$I^*(t) = Y^*(\omega)V^*(t)$$

leads to

$$1/R(\omega) = I_0(\omega)\cos[\phi(\omega)]/V_0$$

$$\omega C(\omega) = -I_0(\omega)\sin[\phi(\omega)]/V_0$$

The conductivity and the absolute permittivity of the sample are finally defined as

$$\sigma(\omega) = d/R(\omega)A$$

$$\varepsilon(\omega) = C(\omega)d/A$$

so that the admittance of the measurement cell becomes

$$Y^*(\omega) = [\sigma(\omega) + i\omega\varepsilon(\omega)]A/d$$

In this expression, the term in square brackets is the complex conductivity of the material:

$$K^*(\omega) = \sigma(\omega) + i\omega\varepsilon(\omega) \tag{2}$$

We shall use this representation of the dielectric properties in all that follows.

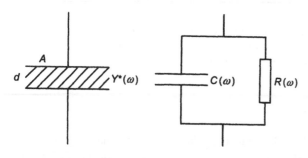

FIG. 1 Measurement cell and its equivalent representation.

III. DIELECTRIC PROPERTIES OF A SUSPENSION

A colloidal suspension is made of particles, which we shall assume for simplicity to be spherical and identical, suspended in a continuous medium, usually an electrolyte solution. The complex conductivity of this system depends in a rather complex fashion on many parameters such as the dielectric properties of the intervening media, the size and structure of the particles, the volume fraction they occupy, and the way in which they are distributed in space. The major part of this chapter will be devoted to the determination of the function $K_s^*(\omega)$. In this section we shall only deduce the most basic relations.

The dielectric properties of a suspension can be defined in the following way. We consider two macroscopic spherical samples with the same radius ρ, one made of the suspension, and the other of some homogeneous material, Fig. 2. We now choose the complex conductivity of this material in such a way that, when immersed in the same external medium and acted upon by the same external field, the total field outside the two spheres is also the same. The complex conductivity obtained in this fashion is defined as being the complex conductivity of the suspension: $K_s^*(\omega)$.

Since, by hypothesis, the dielectric behavior of the two spheres must always be the same, we are free to choose the configuration which leads to the result in the simplest possible way. We therefore consider that both spheres are immersed in the same electrolyte solution in which the particles are suspended, which is characterized by a complex conductivity $K_e^*(\omega)$.

A uniform electric field is now applied to both systems:

$$\vec{E}(r, \theta, t) = E \, \text{Re}\big(e^{i\omega t}\big)[\cos(\theta)\hat{a}_r - \sin(\theta)\hat{a}_\theta]$$

where r and θ are spherical co-ordinates, $\hat{a}_r$ and $\hat{a}_\theta$ are the corresponding unit vectors, and E is the amaplitude of the field. The resulting potential outside the sphere made of the suspension (left half of Fig. 2) is determined by the sum of the contributions of each of its particles:

$$U_e^*(r, \theta, \omega) = -Er\cos(\theta) + N \, d^*(\omega)a^3 E \frac{\cos(\theta)}{r^2} \tag{3}$$

where N is the total number of particles, a is the radius of a particle, and $d^*(\omega)$ is its dipolar coefficient (actually this expression is rigorously valid only far from the sphere since the particles are not located at its center). This coefficient is a nondimensional quantity, which is related to, but different from, the dipole moment of a particle.

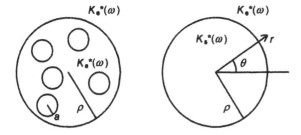

FIG. 2 Configurations used to define the complex conductivity of a suspension.

In order to obtain the potential for the homogeneous sphere (right half of Fig. 2) it is necessary to solve Laplace's equation:

$$\nabla^2 U^*(r, \theta, \omega) = 0$$

in spherical co-ordinates and taking into account the axial symmetry of the problem. The appropriate solutions inside and outside the sphere are

$$U_s^*(r, \theta, \omega) = A^*(\omega)E\, r\cos(\theta)$$

$$U_e^*(r, \theta, \omega) = -E\, r\cos(\theta) + B^*(\omega)\rho^3 E\,\frac{\cos(\theta)}{r^2}$$

where the coefficients $A^*(\omega)$ and $B^*(\omega)$ are determined using the following boundary conditions on the surface of the sphere of radius ρ:

1. The potential $U^*(r, \theta, \omega)$ must be continuous:

    $$U_s^*(\rho, \theta, \omega) = U_e^*(\rho, \theta, \omega)$$

2. The radial component of the complex current density must be continuous:

    $$-K_s^*(\omega)\frac{\partial U_s^*(r, \theta, \omega)}{\partial r}\bigg|_\rho = -K_e^*(\omega)\frac{\partial U_e^*(r, \theta, \omega)}{\partial r}\bigg|_\rho \tag{4}$$

This last condition follows from the following two conditions:

a. The continuity equation: the difference between the radial components of the current densities on both sides of the interface is equal to the time derivative of the field-induced surface charge density:

$$-\sigma_e(\omega)\frac{\partial U_e^*(r, \theta, \omega)}{\partial r}\bigg|_\rho + \sigma_s(\omega)\frac{\partial U_s^*(r, \theta, \omega)}{\partial r}\bigg|_\rho = \frac{\partial \kappa^*(\theta, \omega)}{\partial t} = -i\omega\kappa^*(\theta, \omega)$$

b. The Gauss equation: the difference between the radial components of the electric displacements on both sides of the interface is equal to the field-induced surface charge density:

$$-\varepsilon_e(\omega)\frac{\partial U_e^*(r, \theta, \omega)}{\partial r}\bigg|_\rho + \varepsilon_s(\omega)\frac{\partial U_s^*(r, \theta, \omega)}{\partial r}\bigg|_\rho = \kappa^*(\theta, \omega)$$

Multiplying this equation by $i\omega$ and adding the result to the preceding equation leads to Eq. (4).

The resulting expressions for the potentials are

$$U_s^*(r, \theta, \omega) = -\frac{3K_e^*(\omega)}{K_s^*(\omega) + 2K_e^*(\omega)}E\, r\,\cos(\theta)$$

$$U_e^*(r, \theta, \omega) = -E\, r\,\cos(\theta) + \frac{K_s^*(\omega) - K_e^*(\omega)}{K_s^*(\omega) + 2K_e^*(\omega)}\rho^3 E\,\frac{\cos(\theta)}{r^2} \tag{5}$$

Equating the expressions for the potential outside the left and the right spheres in Fig. 2, Eqs (3) and (5), leads to the following result for the complex conductivity of the suspension:

$$K_s^*(\omega) = K_e^*(\omega)\frac{1 + 2vd^*(\omega)}{1 - vd^*(\omega)} \tag{6}$$

where

$$v = Na^3/\rho^3$$

is the volume fraction occupied by the particles in the suspension.

This expression, which has a very general character, shows that the central problem that must be solved in order to determine the dielectric properties of a suspension is the calculation of the dipolar coefficient of a single suspended particle. The assumptions used in its derivation (the particles are spherical, identical, and have identical dipole coefficients) can usually be removed using an appropriate averaging process.

In practice, the calculation of the dipolar coefficient $d^*(\omega)$ is only possible for small values of the volume concentration v, since the interactions among neighboring particles may then be neglected. For low concentrations, Eq. (6) reduces to

$$K_s^*(\omega) = K_e^*(\omega)[1 + 3vd^*(\omega)] \tag{7}$$

which can be separated into real and imaginary parts using Eq. (2):

$$\sigma_s(\omega) = \sigma_e(\omega) + 3v\sigma_e(\omega)\left\{ \text{Re}[d^*(\omega)] - \frac{\omega\varepsilon_e(\omega)}{\sigma_e(\omega)}\text{Im}[d^*(\omega)] \right\} \tag{8}$$

$$\varepsilon_s(\omega) = \varepsilon_e(\omega) + 3v\varepsilon_e(\omega)\left\{ \text{Re}[d^*(\omega)] + \frac{\sigma_e(\omega)}{\omega\varepsilon_e(\omega)}\text{Im}[d^*(\omega)] \right\} \tag{9}$$

These are the fundamental equations for the calculation of the dielectric properties of dilute suspensions. They show that the suspension conductivity and permittivity depend on both the in- and the out-of-phase parts of the dipolar coefficient, and that the out-of-phase part has a major bearing on the conductivity at high frequencies and on the permittivity at low frequencies.

While the first addend inside the braces in Eq. (8) is proportional to the in-phase part of the conduction current density, the second term is proportional to the in-phase part of the displacement current density:

$$J_D^*(\omega)e^{i\omega t} \propto \frac{\partial}{\partial t}\left[\varepsilon_e(\omega)d^*(\omega)e^{i\omega t}\right] = i\omega\varepsilon_e(\omega)d^*(\omega)e^{i\omega t}$$

Separating the first and last terms of this expression into real and imaginary parts leads to

$$\text{Re}[J_D^*(\omega)] \propto -\omega\varepsilon_e(\omega)\text{Im}[d^*(\omega)]$$

As for Eq. (9), the first addend inside the braces is proportional to the in-phase part of the electric displacement, while the second term is proportional to the in-phase part of the field-induced charge density:

$$\kappa^*(\omega)e^{i\omega t} \propto \int \sigma_e(\omega)d^*(\omega)e^{i\omega t}dt = \sigma_e(\omega)d^*(\omega)\frac{e^{i\omega t}}{i\omega}$$

Separating the first and last terms of this expression into real and imaginary parts leads to

$$\text{Re}[\kappa^*(\omega)] \propto \frac{\sigma_e(\omega)}{\omega}\text{Im}[d^*(\omega)]$$

The expressions for the DC conductivity and the limiting high-frequency permittivity are particularly simple:

$$\sigma_s(0) = \sigma_e(0) + 3v\sigma_e(0)d(0) \tag{10}$$

$$\varepsilon_{s\infty} = \varepsilon_{e\infty} + 3v\varepsilon_{e\infty}d_{\infty} \tag{11}$$

where the general condition:

$$\text{Im}[d^*(0)] = \text{Im}[d_{\infty}^*] = 0$$

was used. Expressions (10) and (11) show that $\sigma_s(0)$ and $\varepsilon_{s\infty}$ can be obtained by solving just the DC and the limiting high-frequency problems. On the other hand, in order to calculate $\varepsilon_s(0)$ and $\sigma_{s\infty}$, the general frequency dependent solution must be obtained and the appropriate limit taken (this is not the case if an alternative formalism based on the calculation of the stored and the dissipated energy is used [43, 44]).

IV. MAXWELL MIXTURE FORMULA

The dielectric properties of any isotropic spherical particle can always be represented by an equivalent homogeneous particle with radius a and complex conductivity $K_p^*(\omega)$. Neglecting any interactions among neighboring particles, its dipolar coefficient can be deduced from Eq. (5) writing $K_p^*(\omega)$ instead of $K_s^*(\omega)$ and a instead of ρ:

$$d^*(\omega) = \frac{K_p^*(\omega) - K_e^*(\omega)}{K_p^*(\omega) + 2K_e^*(\omega)} \tag{12}$$

The form of this expression shows that the real part of the dipolar coefficient is always confined inside the following bounds:

$$-1/2 \le \text{Re}[d^*(\omega)] \le 1$$

The limit $\text{Re}[d^*(\omega)] = -1/2$ corresponds, at low frequencies, to the case when the conductivity of the particle is much lower than the conductivity of the electrolyte solution: $\sigma_p(\omega) \ll \sigma_e(\omega)$, while $\text{Re}[d^*(\omega)] = 1$ applies to the opposite case $\sigma_p(\omega) \gg \sigma_e(\omega)$.

Combining Eqs (6) and (12), leads to the Maxwell mixture formula:

$$\frac{K_s^*(\omega) - K_e^*(\omega)}{K_s^*(\omega) + 2K_e^*(\omega)} = v\frac{K_p^*(\omega) - K_e^*(\omega)}{K_p^*(\omega) + 2K_e^*(\omega)} \tag{13}$$

which was obtained by Maxwell for the DC case [45] and extended by Wagner for alternating fields [46].

The validity of Eq. (13) is limited by the assumptions used for its derivation: the particles are spherical, isotropic, and identical, and the interactions among neighbors may be neglected. These conditions are met in two situations for which Eq. (13) is rigorous:

1. When the concentration of particles is low, in which case Eq. (13) reduces to

$$K_s^*(\omega) = K_e^*(\omega)\left[1 + 3v\frac{K_p^*(\omega) - K_e^*(\omega)}{K_p^*(\omega) + 2K_e^*(\omega)}\right] \tag{14}$$

This expression is often used instead of eq. (7) in order to calculate the dielectric properties of suspensions. Expression (7) has, nevertheless, a more general character (as will be shown in the description of the low-frequency dispersion). Furthermore, the dielectric behavior of the dipolar coefficient has often a simpler interpretation than that of the equivalent complex conductivity of the particle.

2. When there is just a single particle in the center of the sphere of radius ρ. While this configuration is not a suitable representation of a suspension, it can be used as a model for nonhomogeneous particles with central symmetry (a particle with a membrane,

for example). Therefore, the dielectric behavior of a particle made of a central sphere of radius a_1 and complex conductivity $K_1^*(\omega)$, surrounded by a shell of external radius a_2 and complex conductivity $K_2^*(\omega)$, Fig. 3, is exactly the same as that of a homogeneous particle of radius a_2 and complex conductivity $K_p^*(\omega)$ given by

$$K_p^*(\omega) = K_2^*(\omega) \frac{2(1 - v)K_2^*(\omega) + (1 + 2v)K_1^*(\omega)}{(2 + v)K_2^*(\omega) + (1 - v)K_1^*(\omega)} \tag{15}$$

where

$$v = a_1^3/a_2^3$$

Equation (15), which can also be written in the form of Eq. (13), is rigorously valid for any value of the volume fraction : $0 \le v \le 1$.

V. SINGLE TIME CONSTANT RELAXATION

The simplest form of the frequency dependence of the dielectric properties of a medium is called a single time constant, or Debye type, relaxation [47]. This behavior is only found in a few simple systems such as pure polar liquids, or dilute suspensions of spherical particles when the conductivity and permittivity of both media are frequency independent and at least one of them is conductive.

In this last case, the analytic expression for the relaxation can be obtained from Eq. (14) with the following substitutions:

$$K_e^*(\omega) = \sigma_e + i\omega\varepsilon_e$$

$$K_p^*(\omega) = \sigma_p + i\omega\varepsilon_p$$

This leads to

$$K_s^*(\omega) = \sigma_s(0) + i\omega\left[\varepsilon_{s\infty} + \frac{\Delta\varepsilon_s}{1 + i\omega\tau_s}\right] \tag{16}$$

where

$$\sigma_s(0) = \sigma_e\left[1 + 3v\frac{\sigma_p - \sigma_e}{\sigma_p + 2\sigma_e}\right] \tag{17}$$

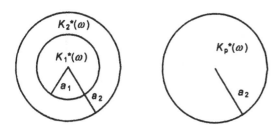

FIG. 3 Inhomogeneous particle with central symmetry and its dielectric equivalent.

is the DC conductivity,

$$\varepsilon_{s\infty} = \varepsilon_e\left[1 + 3v\frac{\varepsilon_p - \varepsilon_e}{\varepsilon_p + 2\varepsilon_e}\right] \tag{18}$$

is the limiting high frequency permittivity,

$$\tau_s = \frac{\varepsilon_p + 2\varepsilon_e}{\sigma_p + 2\sigma_e} \tag{19}$$

is the relaxation time, and

$$\Delta\varepsilon_s = 9v\frac{(\varepsilon_p\sigma_e - \varepsilon_e\sigma_p)^2}{(\varepsilon_p + 2\varepsilon_e)(\sigma_p + 2\sigma_e)^2} \tag{20}$$

is the relaxation amplitude. If the full form of the Maxwell mixture formula, Eq. (13), were to be used instead of Eq. (14), Eqs (17)–(20) would have changed, but Eq. (16) would still hold.

The meaning of the different parameters can be appreciated by combining Eqs (2) and (16), in order to obtain the conductivity and permittivity of the suspension:

$$\sigma_s(\omega) = \sigma_s(0) + \frac{\omega^2\Delta\varepsilon_s\tau_s}{1 + \omega^2\tau_s^2} = \sigma_s(0) + \frac{\Delta\varepsilon_s}{\tau_s} - \frac{\Delta\varepsilon_s/\tau_s}{1 + \omega^2\tau_s^2}$$

$$\varepsilon_s(\omega) = \varepsilon_{s\infty} + \frac{\Delta\varepsilon_s}{1 + \omega^2\tau_s^2}$$

The frequency dependence of these expressions is represented in Fig. 4, where

$$\varepsilon_s(0) = \varepsilon_{s\infty} + \Delta\varepsilon_s$$

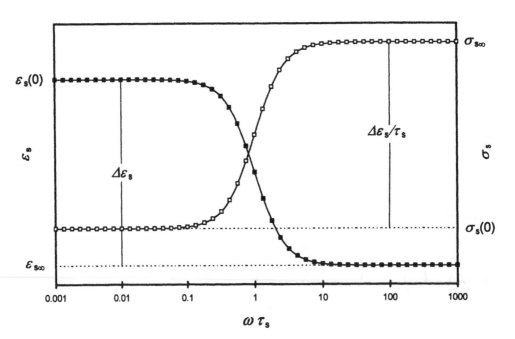

FIG. 4 Parameters that characterize a single time constant relaxation.

is the DC permittivity, and

$$\sigma_{s\infty} = \sigma_s(0) + \Delta\varepsilon_s/\tau_s \tag{21}$$

is the limiting high-frequency conductivity.

The dielectric behavior expressed in Eq. (16) is determined by some mechanism of charge redistribution or polarization in the system, and by the corresponding time. In the case just considered, this mechanissm is the accumulation of field-induced electric charge on the surface of the particles whereas, for a polar liquid, it is the reorientation of the polar molecules. The relaxation time is a measure of the time it takes for the charge to build up or for the molecules to reorient.

Real systems such as cell suspensions, present dielectric spectra that are much more complex. They can be usually decomposed, nevertheless, in a superposition of single time constant relaxations, in which each characteristic time is mainly associated with a single relaxation process. The number of terms in the superposition can be calculated in the following way [48]. For a system made of particles with dielectric properties characterized by N_p single time constant relaxations, immersed in a medium with N_e relaxation terms, the number of terms which characterize the suspension is

$$N_s = N_p + 2N_e + 1 \tag{22}$$

This number is reduced by one if both media are insulating. Among these N_s relaxation times, N_e coincide with the N_e relaxation times of the external medium:

$$\tau_{sj} = \tau_{ej}$$

while their original amplitudes are only modified by the volume fraction occupied by the particles:

$$\Delta\varepsilon_{sj} = \Delta\varepsilon_{ej}(1 - 3v/2) \tag{23}$$

These results also hold for the full form of the Maxwell mixture formula, Eq. (13), except for Eq. (23), which changes to

$$\Delta\varepsilon_{sj} = \Delta\varepsilon_{ej}\frac{2(1 - v)}{2 + v}$$

VI. CONSIDERED SYSTEMS

In the following sections we shall analyze the dielectric behavior of dilute suspensions of homogeneous particles and cells. These are the simplest systems that still exhibit the main relaxation processes usually found in colloidal suspensions. The dielectric properties of the components of these systems can be represented as follows.

A. Electrolyte Solution

We consider an aqueous electrolyte solution with complex conductivity:

$$K_e^*(\omega) = \sigma_e(0) + i\omega\left[\varepsilon_{e\infty} + \frac{\varepsilon_e(0) - \varepsilon_{e\infty}}{1 + i\omega\tau_w}\right] \tag{24}$$

where the dielectric behavior corresponds to a single time constant process with the relaxation time τ_w of water. This is usually a sufficiently good approximation [49].

The frequency dependence of the conductivity and permittivity of the electrolyte solution, Eqs (2) and (24), is represented in Fig. 5 using the following parameters:

$$\varepsilon_e(0) = 80\varepsilon_0 \tag{25}$$

$$\varepsilon_{e\infty} = 4\varepsilon_0 \tag{26}$$

$$\sigma_e(0) = 0.01 \ \text{S/m} \tag{27}$$

$$\varepsilon_0 = 8.85 \times 10^{-12} \ \text{Farad/m} \tag{28}$$

$$\tau_w = 9 \times 10^{-12} \ \text{s} \tag{29}$$

As can be seen, the conductivity strongly increases with frequency, rising from the static value $\sigma_e(0) = 0.01$ S/m, which is solely due to ion movement, to a value of the order of 75 S/m. This increment is due to the reorientation of the permanent dipole moments of the water molecules under the action of the applied field. At high frequencies their angular velocity is in phase with the field, leading to a high displacement current. At low frequencies it is the polarization which is in phase with the field, leading to a decrease in conductivity and to a corresponding increment of the permittivity.

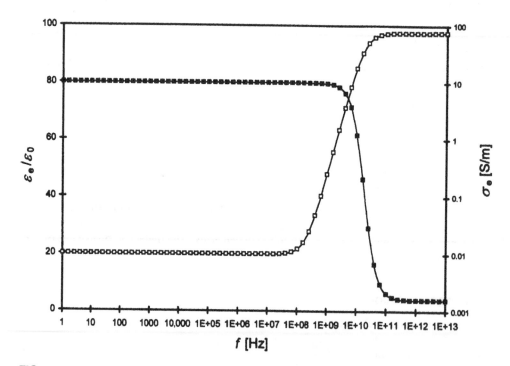

FIG. 5 Permittivity and conductivity of the electrolyte solution. The relaxation frequencies of the conductivity and permittivity spectra seem to differ due to the logarithmic scale used for the conductivity.

B. Homogeneous Particles

From the dielectric standpoint, a homogeneous particle such as latex, for example, is just an insulating sphere surrounded by a thin layer that has an enhanced conductivity. The origin of this layer is mainly in the hydrophilic head groups of molecules that make the particle and dissociate on its surface, and in the adsorption of ions from the electrolyte solution. The net charge that the particle so acquires does not contribute directly to the surface conductivity, since it is fixed, but indirectly by means of the radial field it produces that attracts counterions from the bulk electrolyte solution.

The particle interior is characterized by a complex conductivity:

$$K_i^*(\omega) = i\omega\varepsilon_i$$

where it is assumed that the particle is perfectly insulating and that its permittivity ε_i does not depend on frequency. The dielectric properties of the layer with enhanced ion concentration are represented by a surface conductivity λ, determined as the additional conductivity multiplied by the layer thickness [50], and assumed to be frequency independent. All these are usually acceptable approximations.

The dipolar coefficient of a suspended particle can be calculated with the help of Eq. (12). In order to do this, it is first necessary to determine the complex conductivity of an equivalent particle: a homogeneous particle with the same radius a and the same dielectric properties as the original particle together with its surface layer. This can be done by considering that the particle is surrounded by a thin layer of thickness d, which has an enhanced static conductivity, and using Eq. (15) with the following substitutions:

$$v = a^3/(a+d)^3 \cong 1 - 3d/a$$

$$K_1^*(\omega) = i\omega\varepsilon_i$$

$$K_2^*(\omega) = \lambda/d + \sigma_e(\omega) + i\omega\varepsilon_e(\omega)$$

The result obtained for $d \ll a$ is

$$K_p^*(\omega) = 2\lambda/a + i\omega\varepsilon_i \tag{30}$$

Therefore, the insulating particle surrounded by a layer with surface conductivity λ, behaves as if it had a bulk conductivity equal to $2\lambda/a$.

The dielectric properties of the homogeneous particle, which is dielectrically equivalent to the original particle together with its surface layer, are represented in Fig. 6 using the following parameters:

$$\varepsilon_i = 2\varepsilon_0 \tag{31}$$

$$a = 5 \times 10^{-6} \text{ m} \tag{32}$$

$$\lambda = 0; \quad 1.25 \times 10^{-8}; \quad 5 \times 10^{-8}; \quad 20 \times 10^{-8} \text{ S} \tag{33}$$

As can be seen, the frequency dependence of the dielectric properties of the equivalent particle are extremely simple: a constant permittivity value, which does not depend on the surface conductivity, and a constant conductivity value that is proportional to the surface conductivity.

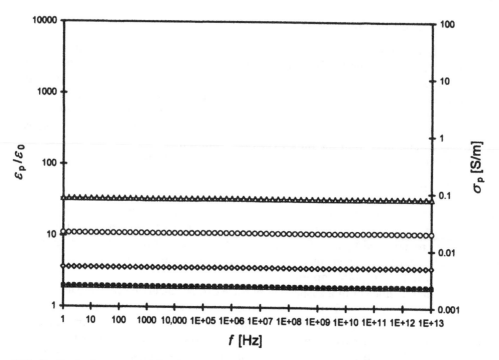

FIG. 6 Equivalent permittivity (black symbols) and conductivity (white symbols) of the homogeneous particle model, Eq. (30). Squares: $\lambda = 0$; diamonds: $\lambda = 1.25 \times 10^{-8}$ S; circles: $\lambda = 5 \times 10^{-8}$ S; triangles: $\lambda = 20 \times 10^{-8}$ S.

C. Cells

From the dielectric standpoint, the most basic representation of a cell is a conducting sphere surrounded by an insulating membrane. The whole cell is further surrounded by a thin layer that has an enhanced conductivity. This layer is due, among other causes, to the net charge which the interior of a living cell must have in order to establish its membrane potential. The radial field of this charge attracts counterions for the bulk electrolyte solution. A radial field is also created by hydrophilic head groups of the lipid molecules that make the membrane and dissociate on the outer cell boundary. Another contribution is the layer of glycocalix or wall, surrounding some cells, which is rich in fixed charges that attract ions from the electrolyte solution.

The inner part of the cell is characterized by a complex conductivity:

$$K_i^*(\omega) = \sigma_i(0) + i\omega \left[\varepsilon_{i\infty} + \frac{\varepsilon_i(0) - \varepsilon_{i\infty}}{1 + i\omega\tau_w} \right]$$

where it was assumed that the cell interior is mainly composed of water so that its relaxation properties are those of water, except for changes in the static and the limiting high-frequency permittivities that are due to the presence of other substances such as lipids.

The dielectric properties of the membrane are characterized by

$$K_m^*(\omega) = i\omega\varepsilon_m$$

where it is assumed that it is perfectly insulating and that its permittivity ε_m does not depend on frequency. This is usually a good approximation at least for low frequencies, less so at high frequencies (gigahertz region) where the permittivity usually shows a frequency dependence. The dielectric properties of the layer with enhanced conductivity are represented by a surface conductivity λ, assumed to be frequency independent.

Finally, the geometric parameters of the cell are its radius a and the thickness of the membrane h. In order to simplify the discussion, we shall consider that the cell membrane is thin ($h \ll R$).

The dipolar coefficient of a suspended cell can be calculated with the help of Eq. (12). In order to do this, it is first necessary to determine the characteristic parameters of an equivalent particle: a homogeneous particle with the same size and the same dielectric properties as the whole cell. Its complex conductivity $K_p^*(\omega)$ can be determined using Eq. (15) for the system made of the internal part of the cell and the cell membrane:

$$v = (a - h)^3/a^3 \cong 1 - 3h/a$$
$$K_1^*(\omega) = K_i^*(\omega)$$
$$K_2^*(\omega) = i\omega\varepsilon_m$$

and adding to the result the contribution of the surface conductivity. This leads to

$$
K_p^*(\omega) = i\omega\left[\varepsilon_i(\omega) + \frac{\varepsilon_m a/h}{1 + i\omega\dfrac{\varepsilon_m a}{\sigma_i(\omega)h}}\right] + \frac{2\lambda}{a} =
$$

$$
= \frac{\left(\dfrac{\omega\varepsilon_m a}{h}\right)^2 \Big/ \sigma_i(\omega)}{1 + \omega^2\left[\dfrac{\varepsilon_m a}{\sigma_i(\omega)h}\right]^2} + \frac{2\lambda}{a} + i\omega\left\{\varepsilon_i(\omega) + \frac{\varepsilon_m a/h}{1 + \omega^2\left[\dfrac{\varepsilon_m a}{\sigma_i(\omega)h}\right]^2}\right\}
$$

(34)

At high frequencies, the second addend inside the braces becomes negligible. The conducting particle surrounded by an insulating membrane behaves then as if the membrane did not exist (it is electrically shorted). On the other hand, at low frequencies, the conductivity of the equivalent particle reduces to the contribution of the surface conductivity, while its permittivity attains a very large value determined mainly by the properties of the membrane.

The dielectric properties of the homogeneous particle, which is dielectrically equivalent to the cell together with its surface layer, are represented in Fig. 7 using the following parameters:

$$\varepsilon_i(0) = 50\varepsilon_0 \tag{35}$$

$$\varepsilon_{i\infty} = 4\varepsilon_0 \tag{36}$$

$$\sigma_i(0) = 1 \text{ S/m} \tag{37}$$

$$\varepsilon_m = 6\varepsilon_0 \tag{38}$$

$$a = 5 \times 10^{-6} \text{ m} \tag{39}$$

$$h = 10^{-8} \text{ m} \tag{40}$$

$$\lambda = 0; \quad 1.25 \times 10^{-8}; \quad 5 \times 10^{-8}; \quad 20 \times 10^{-8} \text{ S} \tag{41}$$

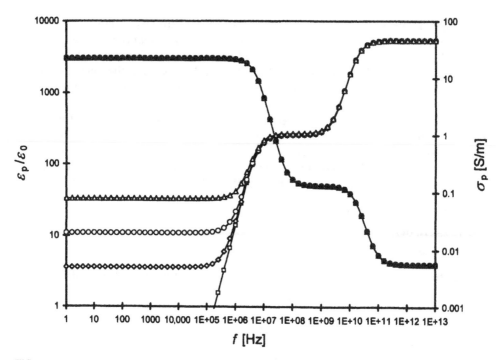

FIG. 7 Equivalent permittivity (black symbols) and conductivity (white symbols) of the cell model, Eq. (34). Squares: $\lambda = 0$; diamonds $\lambda = 1.25 \times 10^{-8}$ S; circles: $\lambda = 5 \times 10^{-8}$ S; triangles; $\lambda = 20 \times 10^{-8}$ S. The relaxation frequencies of the conductivity and permittivity spectra seem to differ due to the logarithmic scales.

which are reasonable values chosen to produce a spectrum in which all the main relaxations are clearly visible.

It is apparent that the cell model used leads to two single time constant relaxations, in accordance with Eq. (22) with $N_p = 1$ (cell interior), $N_e = 0$ (cell membrane), and one conductive medium (cell interior). The surface layer does not add any relaxation term because its influence is calculated at the limit $d \to 0$ [Eq. (15) with $\nu \to 1$].

VII. DIELECTRIC PROPERTIES OF THE CONSIDERED SYSTEMS

The dipolar coefficient of a suspended particle can be obtained using Eqs (12) and (24), together with either Eq. (30) (for homogeneous particles) or Eq. (34) (for cells). The real and imaginary parts of $d^*(\omega)$, calculated using the same parameters as in Figs 5 and either 6 (for homogeneous particles) or 7 (for cells), are represented in Figs 8 and 9.

While the dielectric properties of the homogeneous particle, which is equivalent to the real one together with its surface conductivity, are frequency independent (Fig. 6), its dipolar coefficient varies strongly with frequency in two distinct regions (Fig. 8). The high-frequency dependence occurs at frequencies higher than the range corresponding to the dipolar relaxation of water, Fig. 5, and is independent of the surface conductivity. On the other hand, the lower frequency dependence varies strongly with the surface conductivity.

For a suspended cell, the dielectric properties of the equivalent particle vary with frequency in two regions (Fig. 7), while its dipolar coefficient shows three relaxations (Fig.

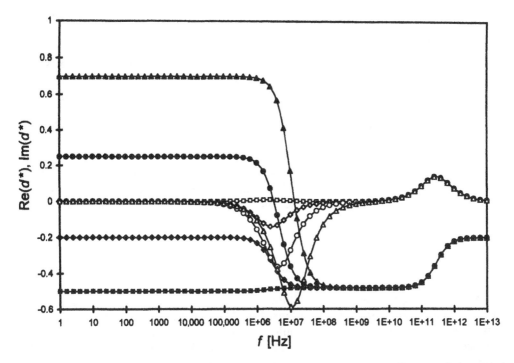

FIG. 8 Real (black symbols) and imaginary (white symbols) parts of the dipolar coefficient for the homogeneous particle model. Squares: $\lambda = 0$; diamonds: $\lambda = 1.25 \times 10^{-8}$ S; circles: $\lambda = 5 \times 10^{-8}$ S; triangles: $\lambda = 20 \times 10^{-8}$ S.

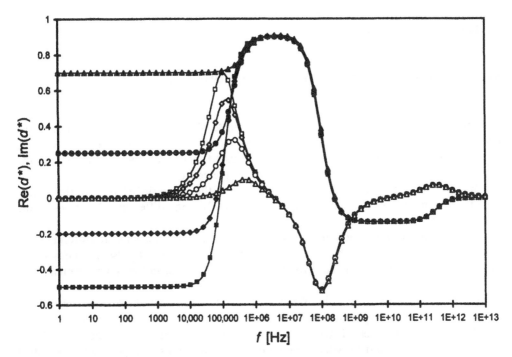

FIG. 9 Real (black symbols) and imaginary (white symbols) parts of the dipolar coefficient for the cell model. Squares: $\lambda = 0$; diamonds: $\lambda = 1.25 \times 10^{-8}$ S; circles: $\lambda = 5 \times 10^{-8}$ S; triangles: $\lambda = 20 \times 10^{-8}$ S.

9). The higher frequency one is very similar to that of a homogeneous particles (Fig. 8). The middle-frequency relaxation occurs at frequencies that are roughly a decade higher for cells than for homogeneous particles, while its parameters are almost independent of the surface conductivity. On the other hand, this conductivity has a very strong bearing on the low-frequency behavior.

The frequency dependencies of the dipolar coefficients of homogeneous particles and cells will be discussed qualitatively in Section VIII and quantitatively in Sections IX–XI.

The complex conductivity of the suspension can finally be obtained combining the expression for the dipolar coefficient $d^*(\omega)$ with Eq. (7). The frequency dependence of the conductivity and permittivity, Eqs (8) and (9), calculated using the same parameters as in Figs 5 and either 6 (homogeneous particles) or 7 (cells), and a volume fraction $v = 0.1$, are represented in Figs 10 and 11.

In agreement with Eq. (22), the spectra for suspended homogeneous particles, Fig. 10, consist of a superposition of three single time constant relaxations: $N_p = 0, N_e = 1$, and two conductive media (the third relaxation is barely visible at frequencies of the order of 10^{12} Hz).

As for cell suspensions, Fig. 11, although the spectra should consist of a superposition of five single time constant relaxations, Eq. (22) with $N_p = 2, N_e = 1$, and two conductive media, they actually contain only four terms (the fourth is barely visible at frequencies of the order of 10^{12} Hz). The reason for this is the assumption that the relaxation times of the electrolyte solution and of the cell interior have the same value τ_w.

The expressions for the characteristic parameters of these relaxations are in general extremely complicated, so that a physical interpretation of the mechanisms involved is very difficult. Fortunately, in many cases of practical interest, the relaxation times differ widely. The relaxation processes then become independent of one another, leading to a simple formulation and interpretation of the characteristic parameters. This is the situation that will be considered in what follows.

VIII. TIME DOMAIN DESCRIPTION OF THE HIGH-FREQUENCY RELAXATIONS

There are two ways to analyze the dielectric response of a system: in the time and in the frequency domains. In the first approach, it is considered that a uniform field is suddenly applied to the system and its evolution is studied as a function of time. In the second, the system is studied under the action of an alternating electric field.

We shall first present a qualitative description of the dielectric behavior of the system in the time domain. Next, we shall quantitatively analyze the different relaxations in the frequency domain.

When a uniform electric field is suddenly applied to the suspension, different phenomena occur in succession. The first is an extremely strong current pulse that lasts a very short time: less than 10^{-15} s. It is due to all the electronic clouds in the system, which move in response to the perturbation. This current comes to a halt as soon as the orbitals attain equilibrium with the external field. What remains is a polarization state of the system, determined by the limiting high-frequency permittivities of the component media: $\varepsilon_{j\infty}$.

At this stage, the field distribution around the homogeneous particle and cell are represented in Figs 12 and 13, which correspond to suspensions with parameters specified in Eqs (25)–(29) and either (31)–(33) (homogeneous particles) or (35)–(41) (cells). These and the following figures have been drawn in such a way that the field line density is

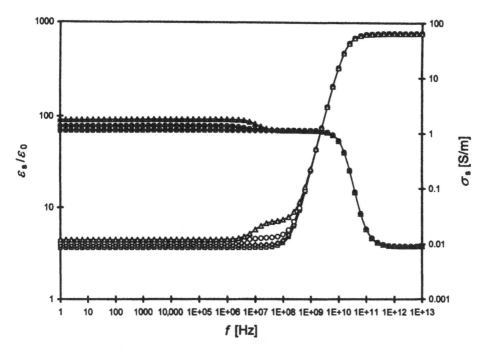

FIG. 10 Permittivity (black symbols) and conductivity (white symbols) of the homogeneous particle suspension, Eqs (7) and (12), for a volume fraction $v = 0.1$. Squares: $\lambda = 0$; diamonds: $\lambda = 1.25 \times 10^{-8}$ S; circles: $\lambda = 5 \times 10^{-8}$ S; triangles: $\lambda = 20 \times 10^{-8}$ S. The relaxation frequencies of the conductivity and permittivity spectra seem to differ due to the logarithmic scales.

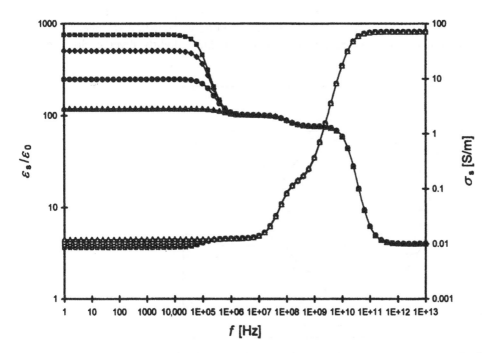

FIG. 11 Permittivity (black symbols) and conductivity (white symbols) of the cell suspension, Eqs (7) and (12), for a volume fraction $v = 0.1$. Squares: $\lambda = 0$; diamonds: $\lambda = 1.25 \times 10^{-8}$ S; circles: $\lambda = 5 \times 10^{-8}$ S; triangles: $\lambda = 20 \times 10^{-8}$ S. The relaxation frequencies of the conductivity and permittivity spectra seem to differ due to the logarithmic scales.

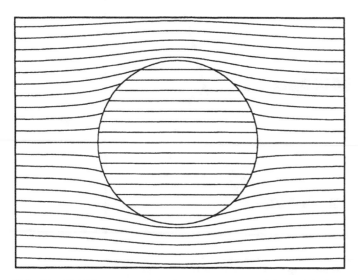

FIG. 12   Field lines around the homogeneous particle model after the electronic clouds attain equilibrium. Also, lines of the in-phase field for frequencies above the gamma relaxation. The field distribution solely depends on the high-frequency permittivities of the component media: $\varepsilon_{e\infty}$ and ε_i.

strictly proportional everywhere to the field strength [51]. The fields inside the suspended particles have been calculated using the expressions in Appendix 1.

   The field is slightly perturbed by the presence of the homogeneous particle, Fig. 12, since the limiting high-frequency permittivity of the electrolyte solution is assumed to be larger than the permittivity of the particle. Correspondingly, the field is stronger inside the particle because of its lower permittivity. The field lines slightly diverge close to the

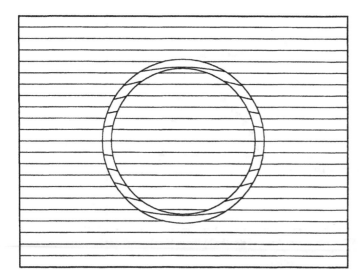

FIG. 13   Field lines around the cell model after the electronic clouds attain equilibrium. Also, lines of the in-phase field for frequencies above the gamma relaxation. The field distribution solely depends on the high-frequency permittivities of the component media: $\varepsilon_{e\infty}$, ε_m, and $\varepsilon_{i\infty}$.

particle, which corresponds to a negative value of the real part of the high-frequency dipolar coefficient, Fig. 8.

However, the field is not perturbed by the presence of the cell, Fig. 13, since the high-frequency permittivities of the electrolyte solution and of the cell interior are assumed to be the same (in general, these values need not be exactly equal to one another so that the field could be slightly deformed). The field is weaker inside the membrane because of its higher permittivity, but this difference does not modify the field outside the cell due to the extremely low volume occupied by the membrane (its thickness has been exaggerated in Fig. 13 and in the following figures). The unperturbed field lines outside the cell correspond to a zero value of the real part of the high-frequency dipolar coefficient, Fig. 9.

Actually, after the first sudden burst, the current does not stop but continues to flow at a much lower rate than before, being mainly due to the permanent dipole moments of the water molecules. Without an applied field, these moments are randomly oriented so that when the field is applied, they find themselves in a nonequilibrium configuration. In order to attain equilibrium they have to rotate, meaning that, on average, the positive and negative charges that make the dipole moments move in opposite directions, contributing to a current. This current is much weaker than the previous one since there are fewer charges involved, and because their movement is much slower due to the viscous-type drag acting on the reorienting molecules. However, the current lasts much longer, due to the lower speed of the charges and because they move through longer distances. When equilibrium is attained (with almost, but not totally, random dipole orientations) this current stops, and what remains is a new polarization state determined by the static permittivities of the component media $\varepsilon_j(0)$.

At this stage, the field distributions around the homogeneous particle and cell are represented in Figs 14 and 15. The field lines strongly diverge close to the homogeneous particle because the permittivity of the particle's interior is much lower than the static permittivity of the electrolyte solution. Correspondingly, the field inside the particle is much stronger than outside. This outer field configuration corresponds to a negative value of the dipolar coefficient, Fig. 8. The field lines around a suspended cells are much less deformed since the cell's inner static permittivity is lower than, but close to, the static permittivity of the electrolyte solution. Correspondingly, the dipolar coefficient value is negative but close to zero, Fig. 9.

Again, the current flow does not completely stop but continues at a much lower rate, being due now to the movement of ions. This current is much weaker than before mainly because there are so much less ions than water molecules. However, it lasts as long as there is a field, since it is a conduction current (ions are free) in contrast with the previous two displacement currents (charges which make permanent dipole moments and electronic clouds are bound).

Since it is assumed that the homogeneous particle is insulating, ionic current can only exist in the external medium. Considering that the field direction is towards the right, this current is directed towards the bulk electrolyte solution on the right-hand side of the particle. Therefore, a negative surface charge starts to build up on its right side while positive charge builds up on its left side. This charge distribution produces inside the particle a field in the same direction as the applied field, increasing the field strength. Outside the particle, it produces a dipolar field that opposes the applied field on the symmetry axis, bending the field lines away from the particle. This process goes on until the normal component of the current density reduces to zero on the surface of the particle. At this stage, which is represented in Fig. 16, the field-induced surface charge density stops changing with time, and the system attains a stable state with a field line configuration

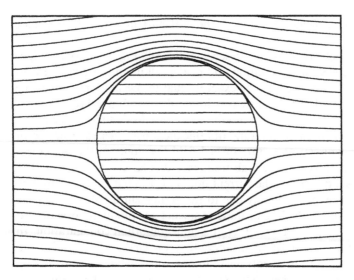

FIG. 14 Field lines around the homogeneous particle model after the reorientation of the water molecules occurs. Also, lines of the in-phase field for frequencies above the delta and below the gamma relaxations. The field distribution solely depends on the low-frequency permittivities of the component media: $\varepsilon_e(0)$ and ε_i.

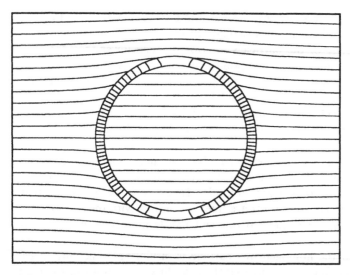

FIG. 15 Field lines around the cell model after the reorientation of the water molecules occurs. Also, lines of the in-phase field for frequencies above the delta and below the gamma relaxations. The field distribution solely depends on the low-frequency permittivities of the component media: $\varepsilon_e(0)$, ε_m, and $\varepsilon_i(0)$.

corresponding to an insulating particle inside a conducting medium: the dipolar coefficient is $d = -1/2$, Fig. 8. The field-induced charges, which are responsible for the slight difference between Figs 14 and 16, could not be drawn in this last figure, in view of the very low value of the charge density (Appendix 1).

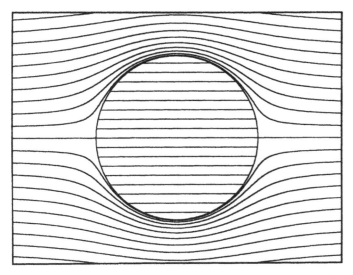

FIG. 16 Field lines around the homogeneous particle model after the buildup of the field-induced surface charge is completed. Also, lines of the in-phase field for frequencies below the delta relaxations. The field distribution solely depends on the condition that, at low frequencies, the particle is insulating while the electrolyte solution is conducting.

The above description corresponds to the case when there is no surface conductivity. Its presence modifies the current flow since the enhanced conductivity layer leads to the appearance of a surface current around the particle. At short times, this current transfers positive (negative) charges to the right (left) side of the particle. The field of this charge distribution modifies the current flow around the particle until the final configuration is reached in which the total current leaving the particle from its right side is equal to the surface current traversing the particle's equator towards its right side. Therefore, the state when the surface charge distribution stops changing with time is attained with a nonvanishing radial component of the field on the particle's surface. Figure 17 represents the field distribution for a very high surface conductivity: $\lambda = 20 \times 10^{-8}$ S. The field lines in the electrolyte solution correspond now to a highly conducting particle inside a medium with low conductivity: the dipolar coefficient strongly increases while the field inside the particle diminishes due to the shielding provided by the conductive layer. This behavior illustrates the strong dependence of the low-frequency dipolar coefficient on the surface conductivity, Fig. 8.

As for the suspended cell, the field intensities inside and outside the membrane represented in Fig. 15 are comparable, whereas the conductivity inside the cell is assumed to be much higher than outside it, Eqs (27) and (37). Therefore, the current density arriving from inside to the right-hand side of the cell is much higher than the one leaving this side towards the electrolyte solution. As a result, positive charge starts to build up at the right side of the cell and negative charge on its left side. These charges produce inside the cell a field in the opposite direction to the applied field, which diminishes the current. This process goes on until the currents arriving and leaving each side of the cell membrane become identical.

At this stage, which is represented in Fig. 18, the total charge distribution around the cell stops changing with time, while the charge densities on the inner and the outer sides of the membrane continue to increase, both at approximately the same rate. This does not

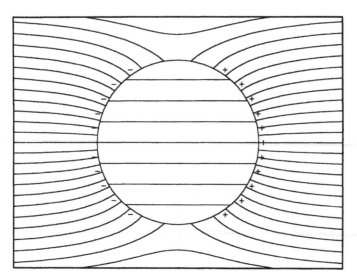

FIG. 17 Same as Fig. 16 but for a strong surface conductivity $\lambda = 20 \times 10^{-8}$ S. The field distribution in the external medium solely depends on the ratio between the particle's equivalent conductivity $2\lambda/a$ and the low-frequency electrolyte solution conductivity $\sigma_e(0)$.

mean, however, that the fields inside the cell and in the electrolyte solution become constant. On the contrary, due to the finite thickness of the membrane, they continue to change but at a much lower rate than before.

The reason for this (Appendix 2) is that the charge distributions built on both sides of the membrane create uniform fields inside the cell, and dipolar fields outside. Since a

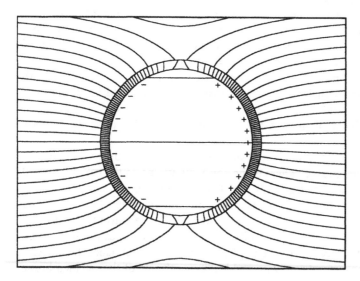

FIG. 18 Field lines around the cell model after the current density arriving at one side of the membrane becomes equal to the one leaving from the other side. Also, lines of the in-phase field for frequencies above the beta and below the delta relaxations. The field distribution solely depends on the low-frequency conductivities of the component media: $\sigma_e(0)$ and $\sigma_i(0)$.

dipolar field decreases as r^{-3}, the field on the outer side of the membrane, produced by a surface charge density located on its inner side, is weaker than the field that this same charge density would create if it were located on the outer side. On the contrary, since a uniform field does not vary with distance, the field on the inner side of the membrane is the same for charge densities located on its inner or outer sides. Because of this, the charge density on the inner side of the membrane increases at a slightly higher rate than on the outer side. This has two consequences: first, the field inside the cell progressively decreases until it vanishes, and second, the field outside the cell changes until it corresponds to that of an insulating particle inside a conducting medium.

At this stage, which is represented in Fig. 19, the situation is finally stable since the charge distributions on each side of the membrane attain constant values: the inner one because there is no current flow inside, and the outer one because there is no radial component of the current flow outside.

This description corresponds to the case when there is no surface conductivity. At short times, its presence leads to the appearance of a surface current around the cell, because of which part of the radial current arriving at, or leaving the cell, does not contribute to the buildup of a charge density on the outer side of the membrane. This is why the situation when the total charge distribution around the cell stops changing with time is attained with field lines converging towards the cell slightly stronger than in Fig. 18. The difference, which corresponds to a slight increase of the dipolar coefficient, is nevertheless too weak to be visible in the figure (it is also practically undetectable in Fig. 9). This insensitivity to the value of the surface conductivity is not a general feature, but is due to the choice of a very high internal conductivity as compared to the conductivity of the electrolyte solution.

At longer times, the field inside the cell slowly diminishes until the final stable situation represented in Fig. 20 is attained. This figure differs from the corresponding

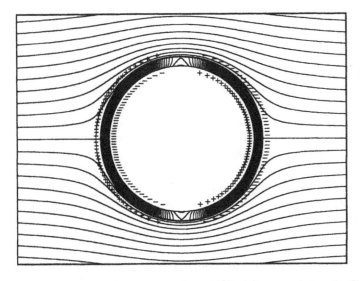

FIG. 19 Field lines around the cell model after the charge densities on both sides of the cell membrane have finished changing with time. Also, lines of the in-phase field for frequencies below the beta relaxation. The field distribution solely depends on the condition that the low-frequency conductivities inside and outside the cell are different from zero whereas the membrane is insulating.

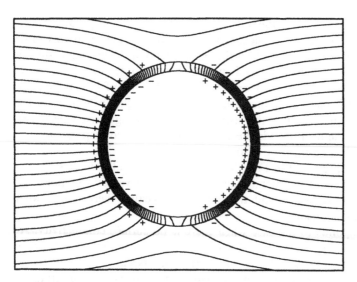

FIG. 20 Same as Fig. 19 but for a strong surface conductivity $\lambda = 20 \times 10^{-8}$ S. The field distribution solely depends on the ratio between the particle's equivalent conductivity $2\lambda/a$ and the low-frequency electrolyte solution conductivity $\sigma_e(0)$.

Fig. 19 in two aspects: first, the external field corresponds now to a conductive particle in a conductive medium (the radial component of the field does not vanish on the surface since it leads to a surface current flowing along it), and second, the charge accumulated on each side of the membrane as well as the field strength inside it are now smaller, since the conducting layer partially shields the cell.

IX. GAMMA RELAXATIONS

We shall now quantitatively analyze in the frequency domain the relaxations that we have qualitatively described, starting with the relaxation associated with the reorientation of water molecules.

The gamma relaxation is found in the giga- to tera-hertz range. It is actually composed of two relaxations, which are due to the frequency dependence of the dielectric properties of the electrolyte solution and to the difference between the permittivities of the two media.

At these high frequencies, the field distribution inside the system is determined exclusively by the permittivities of the components. This does not mean that there are no currents inside the system. On the contrary, these currents are quite strong but they lead to negligible charge densities on the interfaces, since their magnitudes are proportional to the period of the field: $1/\omega$, which is very small.

Therefore, the frequency dependence of the dipolar coefficient can be calculated considering that the components that make the suspension are characterized by the following complex conductivities:

$$K_e^*(\omega) = i\omega \left[\varepsilon_{e\infty} + \frac{\varepsilon_e(0) - \varepsilon_{e\infty}}{1 + i\omega\tau_w} \right] \qquad (42)$$

for the electrolyte solution,

$$K_p^*(\omega) = i\omega\varepsilon_i \qquad (43)$$

for the homogeneous particles, and

$$K_p^*(\omega) = i\omega\left[\varepsilon_{i\infty} + \frac{\varepsilon_i(0) - \varepsilon_{i\infty}}{1 + i\omega\tau_w}\right] \qquad (44)$$

for the cells. This last expression corresponds to the cell model without the membrane, because at these high frequencies it is electrically shorted, and its permittivity does not contribute either, due to its very small volume (less than 1% of the volume of the cell).

Combining Eqs (12) and (42) with either Eq. (43) or (44) leads to the dipolar coefficient that is characterized by a single time constant relaxation:

$$d^*(\omega) = d_{\gamma H} + \frac{d_{\gamma L} - d_{\gamma H}}{1 + i\omega\tau_w \dfrac{\varepsilon_i + 2\varepsilon_{e\infty}}{\varepsilon_i + 2\varepsilon_e(0)}} \qquad (45)$$

for homogeneous particles, and

$$d^*(\omega) = d_{\gamma H} + \frac{d_{\gamma L} - d_{\gamma H}}{1 + i\omega\tau_w \dfrac{\varepsilon_{i\infty} + 2\varepsilon_{e\infty}}{\varepsilon_i(0) + 2\varepsilon_e(0)}} \qquad (46)$$

for cells. In these expressions, the lower index γ denotes the considered relaxation, whereas the indexes L and H indicate the limiting low- and high-frequency values of the dipolar coefficient corresponding to this relaxation, Figs 21 and 22.

Because of the factors multiplying τ_w in Eqs (45) and (46), the relaxation time of the dipolar coefficient is much smaller than the relaxation time of water. This means that, in the time domain, the field configuration around the homogeneous particle (cell) represented in Fig. 12 (13) transforms to the one represented in Fig. 14 (15) before the reorientation of the water molecules is completed. Afterwards this reorientation continues, without further changes in the field distribution (Appendix 3).

At frequencies above this relaxation, the dipolar coefficient for homogeneous particles has the value:

$$d_{\gamma H} = \frac{\varepsilon_i - \varepsilon_{e\infty}}{\varepsilon_i + 2\varepsilon_{e\infty}} \qquad (47)$$

which is rather small and negative in the considered case, since it is assumed that the particle's frequency-independent permittivity $\varepsilon_i = 2\varepsilon_0$ is lower than the limiting high-frequency permittivity of the electrolyte solution: $\varepsilon_{e\infty} = 4\varepsilon_0$. The corresponding value for cells is

$$d_{\gamma H} = \frac{\varepsilon_{i\infty} - \varepsilon_{e\infty}}{\varepsilon_{i\infty} + 2\varepsilon_{e\infty}} \qquad (48)$$

which is zero in the considered case, since it is assumed that the high-frequency permittivities of the cell interior and the electrolyte solution are the same: $\varepsilon_{i\infty} = \varepsilon_{e\infty} = 4\varepsilon_0$.

At frequencies below this relaxation, the dipolar coefficient for homogeneous particles is

$$d_{\gamma L} = \frac{\varepsilon_i - \varepsilon_e(0)}{\varepsilon_i + 2\varepsilon_e(0)}$$

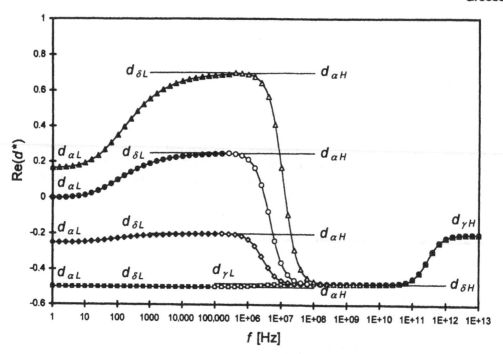

FIG. 21 Real part of the dipolar coefficient of the homogeneous particle model calculated as a superposition of individual relaxations, each one of which extends from a low frequency (d_{jL}) to a high frequency (d_{jH}) value. Squares: $\lambda = 0$; diamonds: $\lambda = 1.25 \times 10^{-8}$ S; circles: $\lambda = 5 \times 10^{-8}$ S; triangles: $\lambda = 20 \times 10^{-8}$ S.

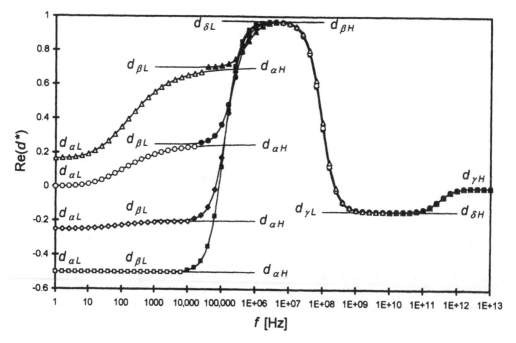

FIG. 22 Same as Fig. 21 but for the cell model.

which is much more negative than in Eq. (47), since $\varepsilon_e(0) \gg \varepsilon_{e\infty}$ (consequently, the field lines in Fig. 14 strongly diverge away from the particle). As for cells:

$$d_{\gamma L} = \frac{\varepsilon_i(0) - \varepsilon_e(0)}{\varepsilon_i(0) + 2\varepsilon_e(0)}$$

which is relatively small and negative, in view of the assumption that $\varepsilon_i(0) < \varepsilon_e(0)$.

Combining the expressions obtained for the dipolar coefficient with Eqs (7) and (24) leads to the complex conductivity of the suspension, which consists of the sum of two terms. The corresponding relaxation parameters for homogeneous particle suspensions are

$$\tau_{\gamma 1} = \tau_w \tag{49}$$

$$\Delta \varepsilon_{\gamma 1} = \Delta \varepsilon_e (1 - 3\nu/2) \tag{50}$$

$$\tau_{\gamma 2} = \tau_w \frac{\varepsilon_i + 2\varepsilon_{e\infty}}{\varepsilon_i + 2\varepsilon_e(0)} \tag{51}$$

$$\Delta \varepsilon_{\gamma 2} = 9\nu \frac{\Delta \varepsilon_e \varepsilon_i^2}{2(\varepsilon_i + 2\varepsilon_{e\infty})[\varepsilon_i + 2\varepsilon_e(0)]} \tag{52}$$

where

$$\Delta \varepsilon_e = \varepsilon_e(0) - \varepsilon_{e\infty}$$

For cell suspensions, the relaxation parameters are

$$\tau_{\gamma 1} = \tau_w \tag{53}$$

$$\Delta \varepsilon_{\gamma 1} = \Delta \varepsilon_e \left(1 + 3\nu \frac{\Delta \varepsilon_i - \Delta \varepsilon_e}{\Delta \varepsilon_i + 2\Delta \varepsilon_e} \right) \tag{54}$$

$$\tau_{\gamma 2} = \tau_w \frac{\varepsilon_{i\infty} + 2\varepsilon_{e\infty}}{\varepsilon_i(0) + 2\varepsilon_e(0)} \tag{55}$$

$$\Delta \varepsilon_{\gamma 2} = 9\nu \frac{(\Delta \varepsilon_i \varepsilon_{e\infty} - \Delta \varepsilon_e \varepsilon_{i\infty})^2}{(\Delta \varepsilon_i + 2\Delta \varepsilon_e)(\varepsilon_{i\infty} + 2\varepsilon_{e\infty})[\varepsilon_i(0) + 2\varepsilon_e(0)]} \tag{56}$$

where

$$\Delta \varepsilon_i = \varepsilon_i(0) - \varepsilon_{i\infty}$$

The relaxation time of the main term coincides with the relaxation time of water. It constitutes the only relaxation whose amplitude is not proportional to the volume fraction of the particles. The second low-amplitude term has the same relaxation frequency as the dipolar coefficient, which is typically more than a decade higher.

Above the gamma relaxation, the permittivity of the system is low because of the low value of the permittivity of the electrolyte solution (and cell interior). On the other hand, the conductivity is very high because the reorientation of the polar water molecules leads to a large displacement current that is much larger than the conduction current due to ionic movement. Below it, the permittivity of the system strongly increases because of the increase in the permittivity of the electrolyte solution (and cell interior). Correspondingly, the conductivity becomes much smaller, since the reorientation velocity of the polar molecules shifts out of phase with the field so that it does not contribute any more to the conductivity.

X. DELTA RELAXATION

The delta relaxation is found in the high megahertz range. In the frequency range of this relaxation, the boundary condition across the interface (or membrane) [analogous to Eq. (4) but applicable to the particle or cell] changes from the continuity of the electric displacement (high frequencies) to the continuity of the current density (low frequencies). Since these frequencies are below the relaxation frequency of water, the permittivity and conductivity of the component media can now be considered as being frequency independent.

Therefore, the frequency dependence of the dipolar coefficient can be calculated considering that the components that make the suspension are characterized by the following complex conductivities:

$$K_e^*(\omega) = \sigma_e(0) + i\omega\varepsilon_e(0) \tag{57}$$

for the electrolyte solution, Eq. (30) for homogeneous particles, and

$$K_p^*(\omega) = \sigma_i(0) + 2\lambda/a + i\omega\varepsilon_i(0) \tag{58}$$

for cells. This last expression corresponds to the high-frequency limit of Eq. (34), written using the static value of $\varepsilon_i(\omega)$.

Combining Eqs (12) and (57) with either Eq. (30) or (58) leads to

$$d^*(\omega) = d_{\delta H} + \frac{d_{\delta L} - d_{\delta H}}{1 + i\omega\dfrac{\varepsilon_i + 2\varepsilon_e(0)}{2\lambda/a + 2\sigma_e(0)}} \tag{59}$$

for homogeneous particles, and to:

$$d^*(\omega) = d_{\delta H} + \frac{d_{\delta L} - d_{\delta H}}{1 + i\omega\dfrac{\varepsilon_i(0) + 2\varepsilon_e(0)}{\sigma_i(0) + 2\lambda/a + 2\sigma_e(0)}} \tag{60}$$

for cells. The frequency dependence of the real parts of these expressions are represented in Figs 21 and 22.

At frequencies higher than the delta relaxation, the dipolar coefficient for homogeneous particles has the value:

$$d_{\delta H} = \frac{\varepsilon_i - \varepsilon_e(0)}{\varepsilon_i + 2\varepsilon_e(0)} = d_{\gamma L}$$

and the corresponding distribution of field lines is represented in Fig. 14. At frequencies lower than this relaxation, the dipolar coefficient usually increases, depending on the value of the surface conductivity:

$$d_{\delta L} = \frac{2\lambda/a - \sigma_e(0)}{2\lambda/a + 2\sigma_e(0)} \tag{61}$$

The corresponding distribution of field lines is represented in Fig. 17, while Fig. 16 illustrates the situation when $\lambda = 0$. In this last case, the dipolar coefficient decreases, attaining the minimum possible value of $-1/2$.

The dipolar coefficient for cells has the following value at frequencies higher than the delta relaxation:

$$d_{\delta H} = \frac{\varepsilon_i(0) - \varepsilon_e(0)}{\varepsilon_i(0) + 2\varepsilon_e(0)} = d_{\gamma L}$$

and the corresponding field line distribution is represented in Fig. 15. At frequencies lower than this relaxation the dipolar coefficient strongly increases and becomes positive:

$$d_{\delta L} = \frac{\sigma_i(0) + 2\lambda/a - \sigma_e(0)}{\sigma_i(0) + 2\lambda/a + 2\sigma_e(0)}$$

This happens because it is assumed that the conductivity of the cell interior is much higher than that of the electrolyte solution, Eqs (27) and (37) (the surface conductivity has almost no bearing on the dipolar coefficient except when the conductivity of the cell interior is low or for very small cells). The field distribution in and around the cell is represented in Fig. 18.

While in Figs 14 and 15 the field lines are determined by the permittivities of the components, they only depend on their conductivities in Figs 16–18. This occurs because when the frequency is lowered, the currents start to build up appreciable charge densities on the interfaces. As a result, part of the current density shifts out of phase with the field, and correspondingly part of the charge density shifts in phase.

Combining the expressions obtained for the dipolar coefficient with Eqs (7) and (57) leads to the complex conductivity of the suspension. It consists of a single term, with the same relaxation time as the dipolar coefficient, since the dielectric properties of the electrolyte solution are frequency independent at these frequencies. The relaxation parameters for homogeneous particle suspensions are

$$\tau_\delta = \frac{\varepsilon_i + 2\varepsilon_e(0)}{2\lambda/a + 2\sigma_e(0)} \tag{62}$$

$$\Delta\varepsilon_\delta = 9v \frac{[\varepsilon_i\sigma_e(0) - \varepsilon_e(0)2\lambda/a]^2}{[\varepsilon_i + 2\varepsilon_e(0)][2\lambda/a + 2\sigma_e(0)]^2} \tag{63}$$

while, for cell suspensions, they become

$$\tau_\delta = \frac{\varepsilon_i(0) + 2\varepsilon_e(0)}{\sigma_i(0) + 2\lambda/a + 2\sigma_e(0)} \tag{64}$$

$$\Delta\varepsilon_\delta = 9v \frac{\{\varepsilon_i(0)\sigma_e(0) - \varepsilon_e(0)[\sigma_i(0) + 2\lambda/a]\}^2}{[\varepsilon_i(0) + 2\varepsilon_e(0)][\sigma_i(0) + 2\lambda/a + 2\sigma_e(0)]^2} \tag{65}$$

These parameters only depend on the radius a through the surface conductivity terms, becoming totally independent of a when $\lambda = 0$. For cells, the delta relaxation parameters are almost independent of their size and of the surface conductivity (except for very small cells with low internal conductivity). Correspondingly, the delta relaxation time for cells is mainly dependent on the conductivity of the cell interior, which explains the higher relaxation frequency for cells than for homogeneous particles (Figs 22 and 21, respectively).

The permittivity of the suspension increases because the field is partially excluded from part of the volume of the system. This process corresponds to the Maxwell-Wagner relaxation of the homogeneous particle, or the internal part of the cell, together with their conductive layers, immersed in the electrolyte solution [Eqs (62)–(65) are analogous to Eqs

(19) and (20)]. The cell membrane is still shorted at these frequencies, so that it does not contribute to the permittivity value.

XI. BETA RELAXATION

This relaxation is usually found in the low megahertz range and is due to the variation with frequency of the dielectric properties of the cell, Eq. (34) (there is no beta relaxation in homogeneous particle suspensions). At low frequencies the membrane ceases to be shorted, and finally determines a new boundary condition between the internal medium and the electrolyte solution: there can be no current flow among these media. For $\lambda = 0$ this condition reduces to the requirement that the normal component of the current density on the external boundary of the membrane must vanish. This can be seen in Fig. 19, where the field lines are represented for frequencies below the beta relaxation.

The field distribution around the cell at frequencies above and below this relaxation only depends on the intervening conductivities. Because of this, the frequency dependence of the dipolar coefficient can be calculated considering a suspension of homogeneous particles characterized by

$$K_p^*(\omega) = 2\lambda/a + i\omega \frac{\varepsilon_m a/h}{1 + i\omega \dfrac{\varepsilon_m a}{\sigma_i(0)h}} \tag{66}$$

(where the last addend represents the frequency dependence of the conductivity of the equivalent particle), in a medium with

$$K_e^*(\omega) = \sigma_e(0)$$

Combining these expressions with Eq. (12) leads to

$$d^*(\omega) = d_{\beta H} + \frac{d_{\beta L} - d_{\beta H}}{1 + i\omega \dfrac{\varepsilon_m a}{h}\left[\dfrac{1}{\sigma_i(0)} + \dfrac{1/2}{\lambda/a + \sigma_e(0)}\right]} \tag{67}$$

The frequency dependence of the real part of this expression is represented in Fig. 22.

At frequencies higher than the beta relaxation, the dipolar coefficient has the value:

$$d_{\beta H} = \frac{\sigma_i(0) + 2\lambda/a - \sigma_e(0)}{\sigma_i(0) + 2\lambda/a + 2\sigma_e(0)} = d_{\delta L}$$

and the corresponding field line distribution around the cell is represented in Fig. 18.

At frequencies lower than this relaxation, the dipolar coefficient strongly decreases:

$$d_{\beta L} = \frac{2\lambda/a - \sigma_e(0)}{2\lambda/a + 2\sigma_e(0)} \tag{68}$$

because the real part of Eq. (66) reduces to $\sigma_p(0) = 2/\lambda/a$, which is the equivalent conductivity of an insulating particle surrounded by a conducting layer. The field distribution around the cell is represented in Figs 19 and 20. In the first, the dipolar coefficient has the value $-1/2$, which corresponds to an insulating particle in a conductive medium.

Combining the expression obtained for the dipolar coefficient with Eqs (7) and (57) leads to the complex conductivity of the suspension. It consists of a single term with the same relaxation time as the dipolar coefficient. The relaxation parameters are

$$\tau_\beta = \frac{\varepsilon_m a}{h}\left[\frac{1}{\sigma_i(0)} + \frac{1/2}{\lambda/a + \sigma_e(0)}\right] \tag{69}$$

$$\Delta\varepsilon_\beta = \frac{9v}{4}\frac{\varepsilon_m a/h}{\left[1 + \frac{\lambda/a}{\sigma_e(0)}\right]^2} \tag{70}$$

which are both proportional to a/h when $\lambda = 0$.

The permittivity of the suspension strongly increases, because of the high surface capacity of the cell membrane that becomes fully charged at low frequencies. This increment is diminished by the surface conductivity that partially shields the cell.

The relaxation time can be interpreted as a product of a resistance multiplied by a capacity. This capacity is in series with two resistances: that of the internal medium and that of the external medium plus the surface layer (the coefficient $1/2$ is due to the spherical geometry of the system).

XII. TIME DOMAIN DESCRIPTION OF THE LOW-FREQUENCY DISPERSION

While the results obtained so far are in qualitative agreement with high-frequency spectra of real suspensions, they do not reproduce the low-frequency dielectric behavior, which is generally characterized by an additional high-amplitude dispersion. This dispersion, which cannot be deduced from Eqs (7) and (12), will be described in the following sections.

The configurations represented in Figs 16 and 19 do not evolve with time, since they correspond to stable states. On the other hand, Figs 17 and 20 represent configurations that are nonstable in the usual case when the surface conductivity is mostly due to ions of a single sign. We shall now describe how these configurations evolve with time.

We first note that the field line distribution outside a homogeneous particle at frequencies below the delta relaxation, Fig. 17, is exactly the same as the corresponding distribution for cells below the beta relaxation, Fig. 20. This happens because, at low frequencies, the dipolar coefficients only depend on the conductivities of both the electrolyte solution and the surface layer, Eqs (61) and (68). Because of this, the phenomena that occur in the electrolyte solution at low frequencies are almost the same for homogeneous particles and for cells, as long as their surface layers have the same properties. Therefore, in what follows, we shall limit the discussion to a single case: a suspension of homogeneous particles. We shall also choose, for definiteness, that the particles have a negative charge, so that the ions that make the surface conductivity, or counterions, are positive. We begin with the situation represented in Fig. 23, which corresponds to a very high surface conductivity: $\lambda = 20 \times 10^{-8}$ S, the same as in Fig. 17.

The total current that enters the left side of the space represented in Fig. 23 (and abandons it from the right side) flows entirely within the electrolyte solution. On the other hand, the total current that traverses the vertical plane located at the center of this figure is mostly confined to the surface layer. In a stable situation these currents must have the same value (or there would be a build-up of charge somewhere inside the space represented in the figure).

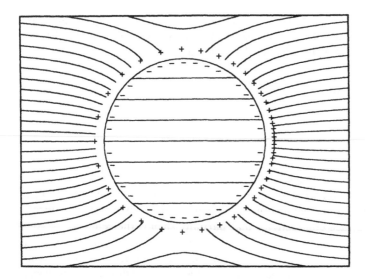

FIG. 23 Initial field-line distribution for the homogeneous particle model partially shielded by a high-conductivity layer of counterions ($\lambda = 20 \times 10^{-8}$ S). The separation between the counterions and the particle's surface represents the thickness of the double layer. The radial field lines have been omitted.

The current transported by each ion type is proportional to its concentration. However, the relative concentrations of positive and negative ions in the bulk electrolyte solution and in the surface layer are different: close to the particle there are almost exclusively positive ions. This means that the fraction of the total current transported by each ion type in the bulk electrolyte solution is different from the corresponding fraction transported near the particle.

If, for instance, the valences and diffusion coefficients of both types of ions were the same, the flow of positive ions entering from the left side the space represented in Fig. 23 would be the same as the flow of negative ions leaving the left side of this same space. On the other hand, the flow of positive ions traversing from left to right the vertical plane located at the center of this figure would be much larger than the flow of negative ions traversing this same plane from right to left.

This means that the number of positive ions arriving per unit time towards the left side of the particle (or leaving it from its right side) is smaller than the number of positive ions being transferred per unit time along the particle's surface from its left to its right side. On the other hand, the number of negative ions arriving per unit time towards the right side of the particle (or leaving it from its left side) is larger than the number of negative ions being transferred per unit time from the right to the left side of the particle. Therefore, the concentration of ions of both signs increases on the right-hand side of the particle and decreases on its left side. This process is often referred to as the build up of "neutral salt clouds."

Although these changes in the ion densities cannot modify in a direct way the electric field, since the ion clouds are neutral, they do modify the ion flows. This happens because the current densities depend not only on the electric field but also on the concentration gradients. The stable situation, which is finally reached, is represented in Figs 24 and 25, which correspond to positive and negative ions, respectively.

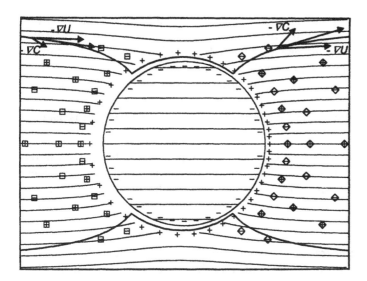

FIG. 24 Final field and ion distribution for the homogenous particle model surrounded by a high-conductivity layer of counterions ($\lambda = 20 \times 10^{-8}$ S) and for a totally symmetric electrolyte: $D^+ = D^-$. Signs inside a diamond represent ions added to, whereas signs inside a square represent ions subtracted from, the normal electrolyte solution concentration. The vectors show how the electric potential and the concentration gradients combine to determine the flow of counterions (thick lines). The counterion layer is not centered because of the dependence of the Debye screening length on the ion concentration (see text). The radial field lines have been omitted.

In both these figures, the electric field distribution corresponds to a particle with an equivalent conductivity which is higher (but not by much) than the conductivity of the electrolyte solution. Positive ions move from the left side of Fig. 24 pulled toward the particle by the electric field and by the concentration gradient, which tends to curve the lines of ion flow toward the symmetry axis. Therefore, the motion of these ions is as if they

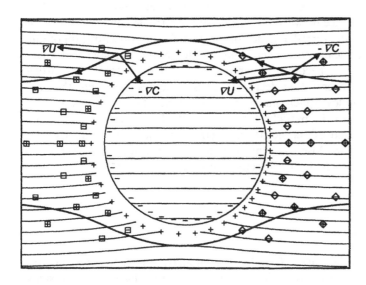

FIG. 25 Same as Fig. 24 but for coions.

were following field lines corresponding to a highly conductive particle in a much less conductive electrolyte solution (except that they move along the surface of the particle and not right through it). Negative ions moving from the right side of Fig. 25 are pulled toward the particle by the electric field but are repelled by the concentration gradient. Their motion is as if they were following field lines corresponding to an insulating particle in an electrolyte solution.

The final dipolar coefficient of the particle is positive, but rather weak. It is made of two contributions:

1. *Fast part*. This dipole coefficient is positive (the dipole points to the right in the same direction as the applied field) and is due to the redistribution to counterions in the conducting layer. This is the coefficient of the particle just before the build up of neutral clouds begins. In Fig. 23 it is represented by the nonuniform distribution of positive ions within the counterion layer.

2. *Slow part*. The buildup of neutral density clouds is a process that demands a long time, since ions must be transferred from one side of the particle to the other, but these clouds, being neutral, cannot contribute directly to a dipole moment. They are responsible, nevertheless, for its change by means of two mechanisms:

    a. A decrement of the fast part of the dipolar coefficient due to the tangential component of the concentration gradient, which tends to decrease the redistribution of counterions in the conducting layer. The distribution of positive ions within this layer appearing in Figs 24 and 25 is more uniform than in Fig. 23 due to this mechanism.

    b. The second mechanism is related to the thickness of the counterion layer, which is of the order of the Debye screening length $1/\chi$, where

$$\chi = \sqrt{\frac{(z^+ e)^2 C^+ (z^- e)^2 C^-}{kT\varepsilon_e(0)}}$$

In this expression $z^\pm$ are the valences of the two ion types, and $C^\pm$ are their concentrations. The thickness of the counterion layer is represented in Figs 23–25 by the distance between the surface of the particle and the positive ion distribution. On the right-hand side of the particle the concentration of ions increases, which leads to a decrement of the counterion layer thickness. Correspondingly, it becomes larger on the left-hand side of the particle where the ion concentration decreases. These changes mean that the positively charged hollow sphere made of the counterion layer shift towards the left with respect to the particle, Figs 24 and 25. Therefore, the charge distribution composed of the negatively charged particle surrounded by the positive counterions loses its central symmetry, creating a dipole moment that points to the left, in the direction opposite to the applied field.

XIII. ALPHA DISPERSION

This high-amplitude dispersion is usually found in the kilohertz range. It occurs because the conducting layer surrounding the particle is made of ions that have mainly a single sign. This asymmetry is unimportant at high frequencies, but plays a decisive role at very low frequencies. The conducting layer permits the transfer of counterions from one side of the particle to the other, but excludes the flow of coions.

There is no universally accepted model for this dispersion. The first one, due to Schwartz [52], was widely used but is now superseded by a series of more realistic models [53–59]. We shall use here the model of Shilov and Dukhin [53], which was the first one to provide an analytical solution to the whole equation set that determines the ion movement around the particle.

This model was developed for charged insulating particles characterized by their ζ potential, immersed in an electrolyte solution characterized by its frequency-independent permittivity, its viscosity, the concentration $C = C^+ = C^-$ and valence $z = z^+ = z^-$ of its two types of ions, and their diffusion coefficients $D^\pm$. In a framework of the so-called standard model that was used, the ζ potential coincides with the equilibrium electrical potential on the surface of the particle.

In terms of these parameters, the conductivity of the electrolyte solution is

$$\sigma_e(0) = \sigma_e^+(0) + \sigma_e^-(0)$$

$$\sigma_e^\pm(0) = \frac{Cz^2 e^2 D^\pm}{kT} \rightarrow \frac{Cz^2 e^2 D}{kT} = \frac{\sigma_e(0)}{2}$$

where k is the Boltzmann constant, T is the absolute temperature, and Einstein's equation relating the ion mobilities and diffusion coefficients was used. In this, and in the following expressions, the right-hand term following the arrow corresponds to the simplest case of a fully symmetric electrolyte: $D^+ = D^- = D$.

The surface conductivity is not considered as an independent parameter, but is rather calculated as a function of the ζ potential:

$$\lambda = \lambda^+ + \lambda^-$$

$$\lambda^\pm = \frac{2\sigma_e^\pm(0)}{\chi}\left[e^{\mp\frac{ze\zeta}{2kT}} - 1 \right] \rightarrow \frac{\sigma_e(0)}{\chi}\left[e^{\mp\frac{ze\zeta}{2kT}} - 1 \right] \tag{71}$$

This last expression is only valid when the radius of the particle is much larger than the Debye screening length:

$$\chi a \gg 1 \tag{72}$$

The low-frequency dielectric behavior of the system is calculated by solving the Poisson and Navier–Stokes equations in the electrolyte solution, together with the equations for the ion flows, including conduction, diffusion, and convection terms. For the sake of simplicity, we shall neglect here the contribution of convection to the flows of ions. This simplification introduces an error of the order of 15% in the dispersion amplitude [60], but greatly simplifies the results and makes them independent of the fluid viscosity. The expression for the dipolar coefficient, obtained under the assumption (72), has a frequency dependence that does not correspond to a single time constant relaxation, being substantially broader:

$$d^*(\omega) = d_{\alpha H} + \frac{d_{\alpha L} - d_{\alpha H}}{1 + i\frac{W^2 S}{1+W+iW}} \tag{73}$$

where

$$S = \frac{2(R+2)}{(R^+ + 2)(R^- + 2)} \rightarrow \frac{R+2}{2\left[R\left(1 - \frac{2}{\chi a}\right) + 1\right]}$$

$$R^{\pm} = \frac{2\lambda^{\pm}}{a\sigma_e^{\pm}(0)} \rightarrow \frac{4}{\chi a}\left[e^{\mp\frac{ze\zeta}{2kT}} - 1\right]$$

$$R = \frac{2\lambda}{a\sigma_e(0)} = R^+ \frac{D^+}{D^+ - D^-} + R^- \frac{D^-}{D^+ + D^-} \rightarrow \frac{R^+}{2} + \frac{R^-}{2}$$

$$W = \sqrt{\omega\frac{a^2(D^+ + D^-)}{4D^+D^-}} \rightarrow \sqrt{\omega\frac{a^2}{2D}}$$

The frequency dependence of the real part of Eq. (73) is represented in Figs 21 and 22.

At frequencies higher than the α dispersion, the dipolar coefficient has the value:

$$d_{\alpha H} = \frac{R-1}{R+2} = d_{\delta L}$$

which is associated with the fast part of the dipole moment, and corresponds to an insulating particle surrounded by a conducting layer with surface conductivity λ and immersed in an electrolyte solution with conductivity $\sigma_e(0)$. The corresponding field line distribution around the particle is represented in Figs 17 and 23.

At frequencies lower than this dispersion, the dipolar coefficient decreases due to the slow part of the dipole moment, which is associated with the buildup of neutral ion clouds around the particle and is always negative:

$$\begin{aligned}d_{\alpha L} = d_{\alpha H} &- \frac{3D^+D^-}{(D^+ + D^-)^2}\frac{(R^+ - R^-)^2}{(R+2)(R^+ + 2)(R^- + 2)} = \frac{R^+ - 1}{R^+ + 2}\frac{D^+}{D^+ + D^-} \\ &+ \frac{R^- - 1}{R^- + 2}\frac{D^-}{D^+ + D^-} \rightarrow d_{\alpha H} - \frac{3R}{4(R+2)}\frac{R + 8/\chi a}{R(1 - 2/\chi a) + 1}\end{aligned} \qquad (74)$$

The corresponding field line distribution around the particle is represented in Figs 24 and 25, which differ from Fig. 23 in the following aspects (all these figures correspond to the same surface conductivity $\lambda = 20 \times 10^{-8}$ S and in Figs 24 and 25 it was further assumed that $D^+ = D^-$). The counterion distribution around the particle is more uniform in Figs 24 and 25, which means that it gives rise to a weaker dipolar field than in Fig. 17. It also leads to a stronger uniform field inside the particle. The counterion layer is furthermore shifted to the left. This does not further increase the field inside the particle, but produces outside it the field of a dipole oriented opposite to the applied field, weakening even more the resulting dipolar field.

Combining the expression obtained for the dipolar coefficient with Eqs (7) and (57) leads to the complex conductivity of the suspension. The dispersion amplitude has two terms corresponding to the real and imaginary parts in Eq. (9). Neglecting the first, as usually done in view of the low frequencies corresponding to this dispersion, it reduces to

$$\Delta\varepsilon_\alpha = \frac{9v\varepsilon_e(0)}{4}\chi^2 a^2 \frac{(R^+ - R^-)^2}{(R^+ + 2)^2(R^- + 2)^2} \rightarrow \frac{9v\varepsilon_e(0)}{16}\chi^2 a^2 \frac{R(R + 8/\chi a)}{[R(1 - 2/\chi a) + 1]^2} \tag{75}$$

A relaxation time cannot be strictly defined since this is not a single time constant process. A characteristic time, calculated as the reciprocal of the angular frequency at which the permittivity increment is half the value of $\Delta\varepsilon_\alpha$ is, approximately:

$$\tau_\alpha = \frac{a^2(D^+ + D^-)}{4D^+ D^-} \rightarrow \frac{a^2}{2D} \tag{76}$$

The above results simplify in the usual case when the surface conductivity is not too low: $R^+ \gg R^-, R^- \ll 1$, so that

$$R^+ = R\frac{D^+ + D^-}{D^+} \rightarrow 2R$$

With these approximations, the limiting low-frequency value of the dipolar coefficient, Eq. (74), becomes

$$d_{\alpha L} = \frac{R - 1}{R + 2} - \frac{3R^2 D^-}{2(R + 2)[(R + 2)D^+ + RD^-]} = \frac{1}{4}\frac{R\frac{2(2D^+ - D^-)}{D^+ - D^-} - 2}{R + 1} \tag{77}$$

$$\rightarrow \frac{R - 1}{R + 2} - \frac{3R^2}{4(R + 2)(R + 1)} = \frac{1}{4}\frac{R - 2}{R + 1}$$

while the dispersion amplitude simplifies to

$$\Delta\varepsilon_\alpha = \frac{9v\varepsilon_e(0)}{16}\chi^2 a^2 \left[\frac{R}{R + \frac{2D^+}{D^+ + D^-}}\right]^2 \rightarrow \frac{9v\varepsilon_e(0)}{16}\chi^2 a^2 \left(\frac{R}{R + 1}\right)^2 \tag{78}$$

This expression shows that while the value of $\Delta\varepsilon_\alpha$ increases with the charge of the particle or its surface conductivity, it is bounded:

$$\Delta\varepsilon_\alpha \leq \frac{9v\varepsilon_e(0)}{16}\chi^2 a^2$$

XIV. DIELECTRIC SPECTRUM

The relaxation parameters deduced in the preceding sections can be used now to obtain the permittivity and conductivity expressions for the suspension, valid over the whole frequency range.

In doing this, it should be noted that although the individual treatment of each relaxation leads to limiting values of the dipole coefficient which couple among neighboring relaxations, Fig. 21:

$$d_{\gamma L} = d_{\delta H}, d_{\delta L} = d_{\alpha H}$$

and Fig. 22:

$$d_{\gamma L} = d_{\delta H}, d_{\delta L} = d_{\beta H}, d_{\beta L} = d_{\alpha H}$$

this does not occur with the limiting values of the conductivity or permittivity calculated using Eqs (10) and (11). Therefore, the permittivity increment of the delta relaxation for cell suspensions, for example, cannot be calculated as a difference between the limiting high-frequency permittivities of the beta and delta relaxations:

$$\Delta\varepsilon_\delta \neq 3v\varepsilon_e(0)(d_{\beta H} - d_{\delta H})$$

Analogously, the conductivity increment of this same relaxation cannot be calculated as a difference between the limiting low frequency conductivities of the gamma and delta relaxations:

$$\Delta\varepsilon_\delta/\tau_\delta \neq 3v\sigma_e(0)(d_{\gamma L} - d_{\delta L})$$

This means that Eqs (10) and (11) only provide the correct values for the DC conductivity and the limiting the high-frequency permittivity when used with the full expression of $d^*(\omega)$: an expression that comprises all the relaxations of the system. They also provide approximate values for these parameters when used with expressions of the dipolar coefficient corresponding to the lowest and the highest frequency relaxations, respectively [Eqs (73) and (45)]. Nevertheless, when used with an expression of the dipolar coefficient corresponding to any given relaxation, Eqs (10) and (11) provide parameters that would only be correct if the system did not have any other relaxations (Appendix 4).

The approximate expressions for the permittivity of suspensions of homogeneous particles and cells, valid over the whole frequency range are, respectively:

$$\varepsilon_s(\omega) = \varepsilon_{e\infty} + 3v\varepsilon_{e\infty}d_{\gamma H} +$$

$$+\frac{\Delta\varepsilon_{\gamma 1}}{1+\omega^2\tau_{\gamma 1}{}^2} + \frac{\Delta\varepsilon_{\gamma 2}}{1+\omega^2\tau_{\gamma 2}{}^2} + \frac{\Delta\varepsilon_\delta}{1+\omega^2\tau_\delta{}^2} + \frac{\Delta\varepsilon_\alpha(1+W)}{(1+W)^2+W^2(1+WS)^2} \tag{79}$$

$$\varepsilon_s(\omega) = \varepsilon_{e\infty} + 3v\varepsilon_{e\infty}d_{\gamma H} +$$

$$+\frac{\Delta\varepsilon_{\gamma 1}}{1+\omega^2\tau_{\gamma 1}{}^2} + \frac{\Delta\varepsilon_{\gamma 2}}{1+\omega^2\tau_{\gamma 2}{}^2} + \frac{\Delta\varepsilon_\delta}{1+\omega^2\tau_\delta{}^2} + \frac{\Delta\varepsilon_\beta}{1+\omega^2\tau_\beta{}^2} + \frac{\Delta\varepsilon_\alpha(1+W)}{(1+W)^2+W^2(1+WS)^2} \tag{80}$$

The first two addends in these expressions represent the limiting permittivity of the suspensions at high frequencies, Eqs (11) and (47) or (48). The following addends correspond to the different single time constant relaxations, and are expressed as functions of their relaxation times, Eqs (49), (51), and (62), or (53), (55), (64), and (69), and amplitudes, Eqs (50), (52), and (63), or (54), (56), (65), and (70). The last addends correspond to the low-frequency dispersion term, and were obtained by combining Eq. (73), written as:

$$d^*(\omega) = d_{\alpha H} + (d_{\alpha L} - d_{\alpha H})\left[\frac{(1+W)^2 + W^2(1+WS)}{(1+W)^2 + W^2(1+WS)^2} - i\frac{SW^2(1+W)}{(1+W)^2 + W^2(1+WS)^2}\right] \tag{81}$$

with Eq. (9). Since the frequencies corresponding to this relaxation are much lower than $\sigma_e(0)/\varepsilon_e(0)$ [this quotient is of the order of the characteristic frequency of the delta relaxation, Eqs (62) or (64)] the first addend inside the braces in Eq. (9) was neglected.

The approximate expressions for the conductivity of suspensions of homogeneous particles and of cells, valid over the whole frequency range are, respectively:

$$\sigma_s(\omega) = \sigma_e(0) + 3v\sigma_e(0)d_{\alpha L} +$$

$$+\frac{\omega^2 \Delta\varepsilon_{\gamma 1}\tau_{\gamma 1}}{1 + \omega^2\tau_{\gamma 1}^2} + \frac{\omega^2 \Delta\varepsilon_{\gamma 2}\tau_{\gamma 2}}{1 + \omega^2\tau_{\gamma 2}^2} + \frac{\omega^2 \Delta\varepsilon_{\delta}\tau_{\delta}}{1 + \omega^2\tau_{\delta}^2} + \frac{\omega\Delta\varepsilon_{\alpha}W(1 + WS)}{(1 + W)^2 + W^2(1 + WS)^2} \tag{82}$$

$$\sigma_s(\omega) = \sigma_e(0) + 3v\sigma_e(0)d_{\alpha L} +$$

$$+\frac{\omega^2 \Delta\varepsilon_{\gamma 1}\tau_{\gamma 1}}{1 + \omega^2\tau_{\gamma 1}^2} + \frac{\omega^2 \Delta\varepsilon_{\gamma 2}\tau_{\gamma 2}}{1 + \omega^2\tau_{\gamma 2}^2} + \frac{\omega^2 \Delta\varepsilon_{\delta}\tau_{\delta}}{1 + \omega^2\tau_{\delta}^2} + \frac{\omega^2 \Delta\varepsilon_{\beta}\tau_{\beta}}{1 + \omega^2\tau_{\beta}^2} + \frac{\omega\Delta\varepsilon_{\alpha}W(1 + WS)}{(1 + W)^2 + W^2(1 + WS)^2}$$

$$\tag{83}$$

The first two addends in these expressions represent the DC conductivity of the suspensions and were obtained using Eqs (10) and (81). The following addends correspond to the different single time constant relaxations, and were obtained using Eq. (21), which relates the permittivity and conductivity increments. The last addend corresponds to the low-frequency dispersion term, and was obtained combining Eqs (8) and (81). In view of the low values of the frequencies corresponding to the alpha dispersion, the second addend inside the braces in Eq. (8) was neglected (actually, for very high frequencies, this second addend becomes dominant and finally diverges, which is due to the use in this model of a hypothesis of local equilibrium which ceases to be valid).

The permittivity and conductivity spectra for homogeneous particle and for cell suspensions, Eqs (79), (82) and (80), (83), are represented in Figs 26 and 27.

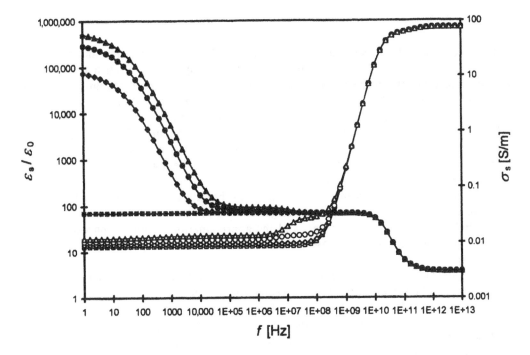

FIG. 26 Permittivity (black symbols) and conductivity (white symbols) of the homogeneous particle suspensions with a volume fraction $v = 0.1$. Squares: $\lambda = 0$; diamonds: $\lambda = 1.25 \times 10^{-8}$ S; circles: $\lambda = 5 \times 10^{-8}$ S; triangles: $\lambda = 20 \times 10^{-8}$ S. The relaxation frequencies of the conductivity and permittivity spectra seem to differ due to the logarithmic scales.

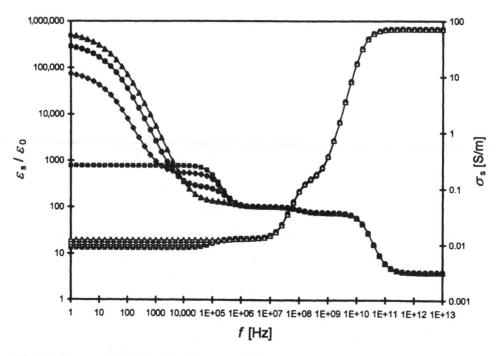

FIG. 27   Same as Fig. 26 but for cell suspensions.

XV.   CONCLUSION

In the preceding sections, we examined individually the main relaxations that make the dielectric spectrum of homogeneous particle and cell suspensions. Such a treatment is approximate, and only valid when the relaxation times of the different mechanisms greatly differ (this was purposely the case in the systems considered). The validity of this approximation can be appreciated comparing the exact dipolar coefficient spectra [Eqs (12) and either (30) or (34)] with their approximate forms [either Eqs (45) and (59) or Eqs (46), (60), and (67)], Figs 28 and 29. The low-frequency dispersion has not been included in this comparison since there is no single model that comprises the high-frequency relaxations together with the low-frequency dispersion.

As can be seen, the agreement is very good for homogeneous particles and generally good for cells, less so in the frequency range between the delta and the beta relaxations. This is indeed the most critical range, and the approximation becomes unacceptable for smaller cells. Therefore, in many cases, these two relaxations must be treated together as a whole [61].

Our approximate results clearly show which is the information on the system parameters that can be extracted from the interpretation of the dielectric spectra.

1.   *Gamma relaxation*, either Eqs (49) and (50) or (53) and (54).

The characteristic frequency of this relaxation is essentially fixed, since it depends neither on the particle size nor on its dielectric properties. Its parameters give no information about the dielectric properties of the homogeneous particle but they make it possible to determine the permittivity change in the cell interior. This determination is usually difficult because the cells only contribute a small fraction to the total relaxation amplitude, and because of the presence of bound water.

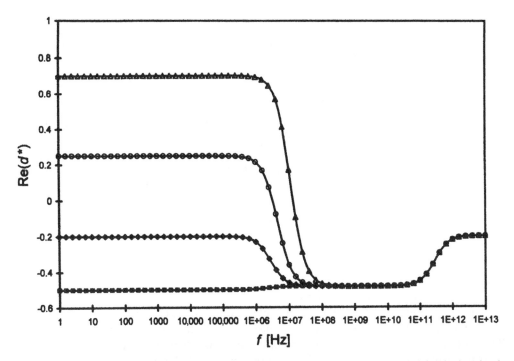

FIG. 28 Real part of the dipolar coefficient of the homogeneous particle model calculated using approximate expressions corresponding to the individual relaxations (symbols), compared to the exact solution (full lines). Squares: $\lambda = 0$; diamonds: $\lambda = 1.25 \times 10^{-8}$ S; circles: $\lambda = 5 \times 10^{-8}$ S; triangles: $\lambda = 20 \times 10^{-8}$ S.

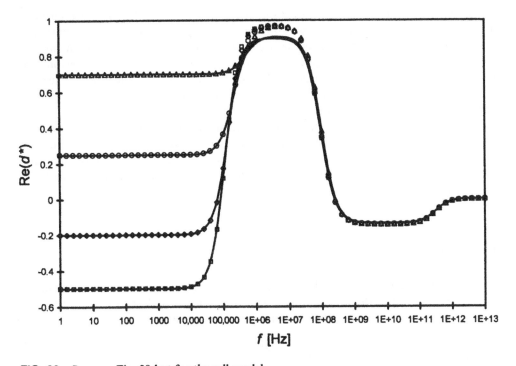

FIG. 29 Same as Fig. 28 but for the cell model.

The interpretation of the second low-amplitude relaxation [either Eqs (51) and (52) or (55) and (56)] is even more problematic, mainly because in the terahertz region the electrolyte solution starts to show deviations from the single time constant relaxation behavior, Eq. (24) [62].

2. *Delta relaxation*, either Eqs (62) and (63) or (86) and (65).

For homogeneous particles, the characteristic frequency of this relaxation mostly depends on the surface conductivity and on the particle's size. Its parameters provide the best information about the surface conductivity.

As for cells, the characteristic frequency of this relaxation mostly depends on the conductivity of the cell interior and is almost independent of the cell size. Its parameters provide information about the conductivity of the cell interior and its surface conductivity. The interpretation is often subject to great uncertainties because of the neighboring beta relaxation which has a much greater amplitude. A possible solution to this problem is in the measurement of the electrorotation speed [60, 63], which is proportional to the imaginary part of the dipolar coefficient. As shown in Fig. 9, the imaginary part is comparable in size for the delta and beta relaxations.

3. *Beta relaxation*, Eqs (69) and (70).

This relaxation provides the most straightforward interpretation mainly due to its large amplitude and because it is the only relaxation which depends on properties of the cell membrane. Its relaxation time and amplitude are almost proportional to the cell radius so that the determination of the membrane capacity strongly improves for large cells. For very small cells, the alpha, beta, and delta relaxations come close to one another, making any meaningful interpretations almost impossible. Other difficulties, which are present even for large cells, have their origin in the conductivity of the cell membrane and in the diffusion effects close to its surface [30, 34].

4. *Alpha dispersion*, Eqs (75) and (76).

This dispersion is related to a series of important properties of the suspended particle, mainly its charge or ζ potential and its surface conductivity. For homogeneous particles, the main difficulty is of an experimental nature: low-frequency dielectric measurements are extremely difficult to perform due to electrode polarization and because suspensions are almost totally resistive at low frequencies [$\phi(\omega) \ll 1$ in Eq. (1)]. As for cell suspensions, there is a further difficulty related to the description of the surface conductivity which, for real cells, can strongly deviate from Eq. (71).

These caveats notwithstanding, dielectric measurements provide rich information about the properties of suspended particles, information that is often inaccessible by any other technique. Meaningful interpretations generally require precise measurements over a broad frequency range and always a good understanding of the different relaxation mechanisms.

APPENDIX 1

The fields inside a suspended particle and the field-induced charge densities on the interfaces can be expressed in the following way as a function of the dipolar coefficient.

For *homogeneous particles*, the potential in the two media has the form:

$$U_e^*(r, \theta, \omega) = -Er\cos(\theta) + d^*(\omega)a^3 E\frac{\cos(\theta)}{r^2}$$

$$U_i^*(r, \theta, \omega) = -A_i^*(\omega)Er\cos(\theta)$$

The condition of the continuity of the potential across the boundary determines the field inside the particle:

$$A_i^*(\omega) = 1 - d^*(\omega)$$

while the Gauss equation (discontinuity of the normal component of the electric displacement) determines the surface charge density:

$$\kappa_e^*(\omega)/E = \varepsilon_e(\omega)[1 + 2d^*(\omega)] - \varepsilon_i A_i^*(\omega) = \varepsilon_e(\omega) - \varepsilon_i + d^*(\omega)[\varepsilon_i + 2\varepsilon_e(\omega)]$$

For *cells*, the potential in the three media has the form:

$$U_e^*(r, \theta, \omega) = -Er\cos(\theta) + d^*(\omega)a^3 E\frac{\cos(\theta)}{r^2}$$

$$U_m^*(r, \theta, \omega) = -A_m^*(\omega)Er\cos(\theta) + B_m^*(\omega)Ea^3\frac{\cos(\theta)}{r^2}$$

$$U_i^*(r, \theta, \omega) = -A_i^*(\omega)Er\cos(\theta)$$

The continuity of the potential across the inner and outer sides of the membrane leads to the following equations:

$$-A_m^*(\omega) + B_m^*(\omega) = -1 + d^*(\omega)$$

$$-A_i^*(\omega)(1 - 3g) = -A_m^*(\omega)(1 - 3g) + B_m^*(\omega)$$

where:

$$1 - 3g = (1 - h/a)^3$$

(it is worth noting that this last expression does not correspond to a series expansion, so that it is not restricted to the assumption that the membrane thickness is small as compared to the cell size). The continuity of the normal component of the complex current density across the inner side of the membrane provides the third equation:

$$K_i^*(\omega)A_i^*(\omega)(1 - 3g) = K_m^*(\omega)[A_m^*(\omega)(1 - 3g) + 2B_m^*(\omega)]$$

These expressions lead to the following results:

$$A_i^*(\omega) = \frac{K_m^*(\omega)[1 - d^*(\omega)]}{K_m^*(\omega) + g[K_i^*(\omega) - K_m^*(\omega)]}$$

$$A_m^*(\omega) = \frac{[K_i^*(\omega) + 2K_m^*(\omega)][1 - d^*(\omega)]/3}{K_m^*(\omega) + g[K_i^*(\omega) - K_m^*(\omega)]}$$

$$B_m^*(\omega) = \frac{[K_i^*(\omega) - K_m^*(\omega)][1 - d^*(\omega)](1 - 3g)/3}{K_m^*(\omega) + g[K_i^*(\omega) - K_m^*(\omega)]}$$

The field-induced surface charge densities can be finally obtained using the Gauss equation on both sides of the membrane:

$$\kappa_i^*(\omega)/E = \varepsilon_m[A_m^*(\omega) + 2B_m^*(\omega)/(1 - 3g)] - \varepsilon_i(\omega)A_i^*(\omega) = \frac{[K_i^*(\omega)\varepsilon_m - K_m^*(\omega)\varepsilon_i(\omega)][1 - d^*(\omega)]}{K_m^*(\omega) + g[K_i^*(\omega) - K_m^*(\omega)]}$$

$$\kappa_e^*(\omega)/E = \varepsilon_e(\omega)[1 + 2d^*(\omega)] - \varepsilon_m[A_m^*(\omega) + 2B_m^*(\omega)]$$

$$= 3\varepsilon_e(\omega) - \frac{\{K_i^*(\omega)\varepsilon_m + 2K_m^*(\omega)\varepsilon_e(\omega) + 2g[K_i^*(\omega) - K_m^*(\omega)][\varepsilon_e(\omega) - \varepsilon_m]\}[1 - d^*(\omega)]}{K_m^*(\omega) + g[K_i^*(\omega) - K_m^*(\omega)]}$$

APPENDIX 2

In order to clarify the charging process of the cell membrane, we shall first consider a spherical surface in air, with radius a and a surface charge density:

$$\kappa(\theta) = \kappa_0 \cos(\theta)$$

This configuration produces the following internal and external potentials:

$$U_i(r, \theta) = \frac{\kappa_0}{3\varepsilon_0} r \cos(\theta)$$

$$U_e(r, \theta) = \frac{\kappa_0 a^3}{3\varepsilon_0} \frac{\cos(\theta)}{r^2}$$

Therefore, the fields just inside and outside the poles of two concentric surface charge densities with radii $a - h$ and a, immersed in a uniform field E are

$$E_i(a - h, 0) = E_i(a - h, \pi) = E - \frac{\kappa_i + \kappa_e}{3\varepsilon_0}$$

$$E_e(a, 0) = E_e(a, \pi) = E + \frac{2\kappa_i}{3\varepsilon_0}\left(1 + \frac{3h}{a}\right) + \frac{2\kappa_e}{3\varepsilon_0}$$

where it is assumed that $h \ll a$.

Considering that these charge densities are created by the fields on both sides of an insulating membrane, they must obey the following continuity equations:

$$\frac{d\kappa_i(t)}{dt} = \sigma_i(0)\left[E - \frac{\kappa_i(t) + \kappa_e(t)}{3\varepsilon_0}\right]$$

$$\frac{d\kappa_e(t)}{dt} = -\sigma_e(0)\left[\frac{E + 2\kappa_i(t)}{3\varepsilon_0}\left(1 - \frac{3h}{a}\right) + \frac{2\kappa_e(t)}{3\varepsilon_0}\right]$$

To first order in h/a, the solutions of these equations are

$$\kappa_i(t) = k_1 e^{-t/\tau_1} + k_2\left[1 + \frac{h}{a}\frac{6\sigma_e(0)}{\sigma_i(0) + 2\sigma_e(0)}\right]e^{-t/\tau_2} + \frac{3\varepsilon_0 Ea}{2h} \tag{84}$$

$$\kappa_e(t) = k_1\frac{2\sigma_e(0)}{\sigma_i(0)}\left[1 - \frac{h}{a}\frac{3\sigma_i(0)}{\sigma_i(0) + 2\sigma_e(0)}\right]e^{-t/\tau_1} - k_2 e^{-t/\tau_2} - \frac{3\varepsilon_0 Ea}{2h}\left(1 - \frac{2h}{a}\right) \tag{85}$$

where k_1 and k_2 are coefficients determined by the initial $\kappa_i(0)$ and $\kappa_e(0)$ values, and

$$\tau_1 = \frac{3\varepsilon_0}{\sigma_i(0) + 2\sigma_e(0)}\left\{1 + \frac{h}{a}\frac{6\sigma_i(0)\sigma_e(0)}{[\sigma_i(0) + 2\sigma_e(0)]^2}\right\}$$

$$\tau_2 = \frac{\varepsilon_0[\sigma_i(0) + 2\sigma_e(0)]a}{2\sigma_i(0)\sigma_e(0)h}\left\{1 - \frac{h}{a}\frac{6\sigma_i(0)\sigma_e(0)}{[\sigma_i(0) + 2\sigma_e(0)]^2}\right\}$$

The first, very short relaxation time corresponds to the delta relaxation. During the whole beta relaxation process ($t \gg \tau_1$), Eqs (84) and (85) show that the current flow just inside the membrane is slightly larger than just outside it. Correspondingly, at the end of this process, the total charge density on the outer side of the membrane is slightly smaller than on its inner side:

$$\kappa_{i\infty} = \frac{3\varepsilon_0 E a}{2h}$$

$$\kappa_{e\infty} = -\frac{3\varepsilon_0 E a}{2h}\left(1 - \frac{2h}{a}\right)$$

This qualitative relationship is necessary since $\kappa_{e\infty}$, together with E, must produce inside the cell the same field strength as $\kappa_{i\infty}$. Nevertheless, this smaller charge density produces outside a dipole field stronger than the inner distribution since its radius is larger. Together, the two distributions lead to a dipolar coefficient equal to $-1/2$.

APPENDIX 3

The reason why the relaxation time $\tau_{\gamma2}$ of the dipolar coefficient is shorter than the relaxation time τ_w of water is due to the presence of a spherical boundary around the supended particle. To simplify the discussion, we shall consider the simplest system with this kind of boundary: a spherical drop of water in air.

When a step field is applied to this system, the polarization $P(t)$ increases at a rate that is approximately proportional to the difference between its equilibrium and its present values:

$$\frac{dP(t)}{dt} = \frac{[\varepsilon_i(0) - \varepsilon_0]E_i(t) - P(t)}{\tau_w} \tag{86}$$

where $E_i(t)$ is the field inside the system. For a sphere, $E_i(t)$ is related to the external field E by

$$E_i(t) = E - \frac{P(t)}{3\varepsilon_0} \tag{87}$$

which means that the internal field decreases as the polarization increases. Therefore, the difference between the two terms on the right-hand side of Eq. (86) changes more rapidly than if $E_i(t)$ were constant, which leads to a faster rate of change for the polarization. Combining Eqs (86) and (87) leads to

$$\frac{dP(t)}{dt} = \frac{[\varepsilon_i(0) - \varepsilon_0]E}{\tau_w} - \frac{[\varepsilon_i(0) + 2\varepsilon_0]P(t)}{3\varepsilon_0 \tau_w}$$

This expression shows that the polarization of a spherical sample has a relaxation time:

$$\tau = \frac{3\varepsilon_0}{\varepsilon_i(0) + 2\varepsilon_0}\tau_w$$

which is shorter than the relaxation time of a continuous sample filling the space between the plates of the condenser to which the step voltage is applied.

APPENDIX 4

We consider, for sake of simplicity, that the particles in a suspension have dipolar coefficients characterized by just two relaxation processes:

$$d^*(\omega) = \frac{\Delta_L}{1 + i\omega\tau_L} + \frac{\Delta_H}{1 + i\omega\tau_H} + d_\infty \tag{88}$$

We further assume that the high-frequency relaxation occurs at frequencies that are low compared to the dipolar relaxation of water ($\tau_H \gg \tau_w$). Under these conditions, the permittivity of the suspension can be calculated using Eq. (9), considering that the dielectric parameters of the electrolyte solution are frequency independent: $K_e^*(\omega) = \sigma_e(0) + i\omega\varepsilon_e(0)$. This leads to

$$\varepsilon_s(\omega) = \varepsilon_e(0) + 3v\varepsilon_e(0) \left\{ \frac{\Delta_L \left[1 - \frac{\sigma_e(0)\tau_L}{\varepsilon_e(0)} \right]}{1 + \omega^2\tau_L^2} + \frac{\Delta_H \left[1 - \frac{\sigma_e(0)\tau_H}{\varepsilon_e(0)} \right]}{1 + \omega^2\tau_H^2} + d_\infty \right\} \tag{89}$$

If the two processes are considered separately, the expression for the dipolar coefficient corresponding to just the high-frequency relaxation term reduces to

$$d^* = \frac{\Delta_H}{1 + i\omega\tau_H} + d_\infty \tag{90}$$

so that the permittivity expression becomes

$$\varepsilon_s(\omega) = \varepsilon_e(0) + 3v\varepsilon_e(0) \left\{ \frac{\Delta_H \left[1 - \frac{\sigma_e(0)\tau_H}{\varepsilon_e(0)} \right]}{1 + \omega^2\tau_H^2} + d_\infty \right\} \tag{91}$$

The corresponding expressions for just the low-frequency relaxation term are

$$d^*(\omega) = \frac{\Delta_L}{1 + i\omega\tau_L} + \Delta_H + d_\infty \tag{92}$$

$$\varepsilon_s(\omega) = \varepsilon_e(0) + 3v\varepsilon_e(0) \left\{ \frac{\Delta_L \left[1 - \frac{\sigma_e(0)\tau_L}{\varepsilon_e(0)} \right]}{1 + \omega^2\tau_L^2} + \Delta_H + d_\infty \right\} \tag{93}$$

As can be seen in Fig. 30, the full expression for the dipolar coefficnet, Eq. (88), is equal at high frequencies to the high-frequency term, Eq. (90) and, at low frequencies, to the low-frequency term, Eq. (92). On the other hand, Fig. 31 shows that although the full permittivity expression, Eq. (89), is equal at high frequencies to the high-frequency term, Eq. (91), it differs at low frequencies from the low-frequency term, Eq. (93).

A similar situation arises with the conductivity spectrum. Combining Eqs (8) and (88) leads to the following expression for the conductivity of the suspension:

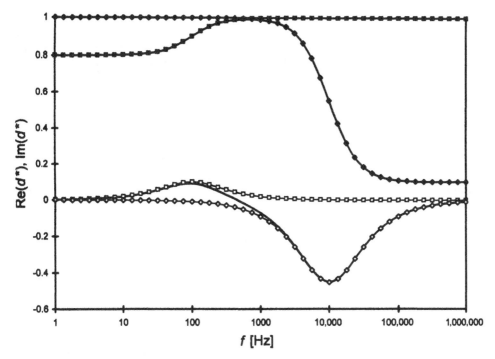

FIG. 30 Real (black symbols) an imaginary (white symbols) parts of the dipolar coefficient for a hypothetical suspension, Eq. (88), characterized by the following parameters: $\Delta_L = -0.2$; $\Delta_H = 0.9$; $d_\infty = 0.1$; $2\pi\tau_L = 0.01$; $2\pi\tau_H = 0.0001$. Squares: low-frequency term; diamonds: high-frequency term; thick lines: full expression.

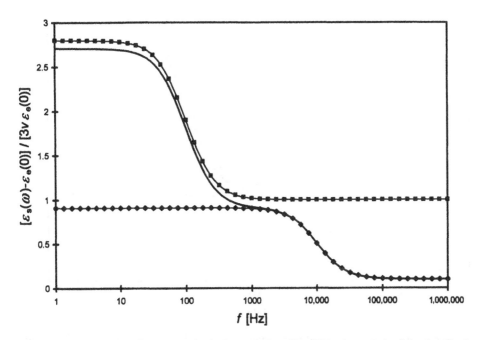

FIG. 31 Permittivity of the hypothetical suspension, Eq. (89), characterized by the dipolar coefficient parameters specified in Fig. 30 and by $\varepsilon_e(0) = 80\varepsilon_0$ and $\sigma_e(0) = \varepsilon_e(0)/\sqrt{\tau_L\tau_H}$. Squares: permittivity corresponding just to the low-frequency terms, Eq. (93); diamonds: same but for just the high-frequency term, Eq. (91); thick line: full permittivity expression.

$$\sigma_s(\omega) = \sigma_e(0) + 3v\sigma_e(0)\left\{\Delta_L + \Delta_H + d_\infty + \frac{\Delta_L\omega^2\tau_L^2\left[\frac{\varepsilon_e(0)}{\sigma_e(0)\tau_L} - 1\right]}{1 + \omega^2\tau_L^2} + \frac{\Delta_H\omega^2\tau_H^2\left[\frac{\varepsilon_e(0)}{\sigma_e(0)\tau_H} - 1\right]}{1 + \omega^2\tau_H^2}\right\}$$

(94)

Considering separately the two relaxation terms, Eq. (90) for the high-frequency process leads to

$$\sigma_s(\omega) = \sigma_e(0) + 3v\sigma_e(0)\left\{\Delta_H + d_\infty + \frac{\Delta_H\omega^2\tau_H^2\left[\frac{\varepsilon_e(0)}{\sigma_e(0)\tau_H} - 1\right]}{1 + \omega^2\tau_H^2}\right\}$$

(95)

while the conductivity expression calculated using the low-frequency term, Eq. (92), results in

$$\sigma_s(\omega) = \sigma_e(0) + 3v\sigma_e(0)\left\{\Delta_L + \Delta_H + d_\infty + \frac{\Delta_L\omega^2\tau_L^2\left[\frac{\varepsilon_e(0)}{\sigma_e(0)\tau_L} - 1\right]}{1 + \omega^2\tau_L^2}\right\}$$

(96)

Again, Fig. 32 shows that the full conductivity expression, Eq. (94), only coincides at low frequencies with the low-frequency term, Eq. (96), but does not coincide at high

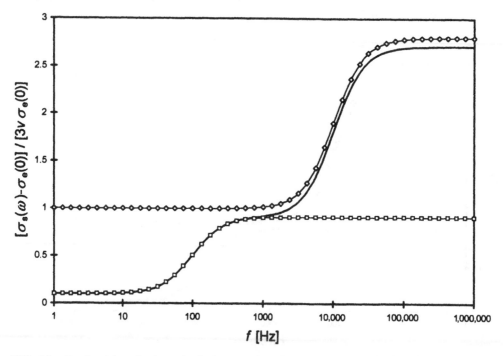

FIG. 32 Conductivity of a hypothetical suspension, Eq. (94), characterized by the dipolar coefficient represented in Fig. 33 and by $\varepsilon_e(0) = 80\varepsilon_0$ and $\sigma_e(0) = \varepsilon_e(0)/\sqrt{\tau_L\tau_H}$. Squares: conductivity corresponding just to the low-frequency term, Eq. (96); diamonds: same but for just the high-frequency term, Eq. (95); thick line: full conductivity expression.

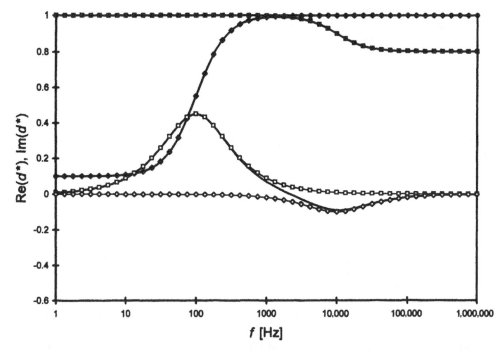

FIG. 33 Real (black symbols) an imaginary (white symbols) parts of the dipolar coefficient for a hypothetical suspension, Eq. (88), characterized by the following parameters: $\Delta_L = -0.9$; $\Delta_H = 0.2$; $d_\infty = 0.8$; $2\pi\tau_L = 0.01$; $2\pi\tau_H = 0.0001$. Squares: low-frequency term; diamonds: high-frequency terms; thick lines: full expression.

frequencies with the high-frequency term, Eq. (95). This figure was drawn using the dipolar coefficient spectrum represented in Fig. 33, since there is no single set of parameters for which the differences being discussed are clearly visible both in the permittivity and the conductivity plots.

Figures 31 and 32 permit us to conclude that, in order to construct the full permittivity (conductivity) spectrum from the component relaxation terms, the highest (lowest) frequency term must be used together with just the frequency-dependent part of the remaining terms.

REFERENCES

1. G Schwarz. Adv Mol Relaxation Processes 3: 281–295, 1972.
2. F Van der Touw, M Mandel. J Phys Chem 2: 231–241, 1974.
3. S Takashima, A Minakata, Dielectr Lit 37: 602–653, 1975.
4. M Sakamoto, H Kanda, R Hayakawa, Y Wada. Biopolymers 15: 879–892, 1976.
5. S Takashima, C Gabriel, RJ Sheppard, EH Grant. Biophysics 46: 29–34, 1984.
6. EAS Cavell. J Colloid Interface Sci 62: 495–502, 1977.
7. U Kaatze. Progr Colloid Polym Sci 67: 117–130, 1980.
8. RM Hill, J Cooper. J Colloid Interface Sci 174: 24–31, 1995.
9. HP Schwan, G Schwarz, J Maczuk, H Pauly. J Phys Chem 66: 2626–2635, 1962.
10. C Ballario, A Bonincontro, C Cametti. J Colloid Interface Sci 72: 304–313, 1979.

11. S Sasaki, A Ishikawa, T Hanai. Biophys Chem 14: 45–53, 1981.
12. MM Springer, A Korteweg, J Lyklema. Electroanal Chem 153: 55–66, 1983.
13. KH Lim, EI Franses. J Colloid Interface Sci 110: 201–213, 1986.
14. DE Dunstan, LR White. J Colloid Interface Sci 152: 308–313, 1992.
15. F Carrique, L Zurita, AV Delgado. J Colloid Interface Sci 166: 128–132, 1994.
16. MR Gittings, DA Saville. Langmuir 11: 798–800, 1995.
17. G Blum, H Maier, F Sauer, HP Schwan. J Phys Chem 99: 780–789, 1995.
18. C Grosse, M Tirado, W Pieper, R. Pottel. J Colloid Interface Sci 205: 26–41, 1998.
19. SA Barker, DQM Craig, RM Hill, KMG Taylor. J Colloid Interface Sci 166: 66–72, 1994.
20. HP Schwan. Advances in Biological and Medical Physics. vol. 5. New York: Academic Press, 1957.
21. R Pethig. Dielectric and Electronic Properties of Biological Materials. Chichester: John Wiley, 1979.
22. MA Stuchly, SS Stuchly. J Microwave Power 15: 19–26, 1980.
23. HP Schwan. Blut 46: 185–197, 1983.
24. HP Schwan. In Chiabrera, Nicolini, Schwan, eds. Interactions Between Electromagnetic Fields and Cells. New York: Plenum, 1985, pp 75–97.
25. A Irimajiri, K Asami, T Ichinowatari, Y Kinoshita. Biochim Biophys Acta 896: 203–213, 1987.
26. KR Foster, HP Schwan. Dielectric properties of tissues and biological materials: a critical review. CRC Critical Reviews in Biomedical Engineering, vol. 17. Boca Raton, Florida: CRC Press, 1989.
27. S Takashima. Electrical Properties of Biopolymers and Membranes. Bristol: Adam Hilger, 1989.
28. S Takashima, A Casaleggio, F Giuliano, M Morando, P Arrigo, S Ridella. S. Biophysics 49: 1003–1008, 1986.
29. U Kaatze. Phys Med Biol 35: 1663–1681, 1990.
30. C Grosse, HP Schwan. Biophys J 63: 1632–1642, 1992.
31. H Fricke. Phys Rev 24: 575–587, 1924.
32. C Grosse, S Pedrosa, VN Shilov. J Colloid Interface Sci 220: 31–41, 1999.
33. EM Trukhan. Sov Phys–Solid State 4: 2560–2570, 1963.
34. A Garcia, R Barchini, C Grosse. J Phys D: Appl Phys 18; 1891–1896, 1985.
35. DAG Bruggeman. Ann Phys 24: 636–664, 1935.
36. T Hanai. In: P Sherman, ed. Emulsion Science. London: Academic Press, 1968, pp 353–478.
37. EHB DeLacey, LR White. J Chem Soc, Faraday Trans 2 77: 2007–2039, 1981.
38. HP Schwan. In Chiabrera, Nicolini, Schwan, eds. Interactions Between Electromagnetic Fields and Cells. New York: Plenum, 1985, pp 371–389.
39. U Zimmermann. Rev Physiol Biochem Pharmacol 105: 175–256, 1986.
40. R Barchini, DA Saville. J Colloid Interface Sci 173: 86–91, 1995.
41. AV Delgado, FJ Arroyo, F Gonzalez-Caballero, VN Shilov, YB Borkovskaya. Colloids Surfaces A 140: 139–149, 1998.
42. M Minor, A van der Wal, J Lyklema. In: E. Pelizetti, ed. Fine Particles Science and Technology. Dordrecht, The Netherlands: Kluwer Academic, 1996, pp 225–238.
43. C Grosse. Ferroelectrics 86: 171–179, 1988.
44. C Grosse, VN Shilov. J Colloid Interface Sci 193: 178–182, 1997.
45. JC Maxwell. Electricity and Magnetism. vol. 1. Oxford: Clarendon Press, 1892.
46. KW Wagner. Arch Elektrotech 2: 371–387, 1914.
47. P Debye. Polar Molecules. New York: Dover Publications, 1929.
48. C Grosse, R Barchini, C Halloy, R Pottel. J Phys D: Appl Phys 19: 1957–1964, 1986.
49. U Kaatze, V Uhlendorf. Z Phys Chem Neue Folge 126: 151–164, 1981.
50. CT O'Konski. J Phys Chem 64: 605–619, 1960.
51. C Grosse, M Tirado. IEEE Trans Educ 39: 69–76, 1996.
52. G Schwarz. J Phys Chem 66: 2636–2642, 1962.

53. SS Dukhin, VN Shilov. Dielectric Phenomena and the Double Layer in Disperse Systems and Polyelectrolytes. Jerusalem: Kerter Publishing House, 1974.
54. WC Chew, PN Sen. J Chem Phys 77; 4683–4693, 1982.
55. M Fixman. J Chem Phys 72: 5177–5186, 1980.
56. RW O'Brien. Adv Colloid Interface Sci 16: 281–320, 1982.
57. J Lyklema, SS Dukhin, VN Shilov. Electroanal Chem 143: 1–21, 1983.
58. C Grosse, KR Foster. J Phys Chem 91: 3073–3076, 1987.
59. C Grosse. J Phys Chem 92: 3905-3910, 1988.
60. C Grosse, VN Shilov. J Phys Chem 100: 1771–1778, 1996.
61. H Pauly, HP Schwan. Z Naturforsch 14b: 125–131, 1959.
62. NE Hill, WE Vaughan, AH Price, M Davies. Dielectric Properties and Molecular Behaviour. London: Van Nostrand Reinhold, 1969.
63. WM Arnold, U Zimmermann. J Electro 21: 151–191, 1988.

12

Suspensions in an Alternating External Electric Field: Dielectric and Electrorotation Spectroscopies

VLADIMIR N. SHILOV   Institute of Biocolloid Chemistry, Ukrainian National Academy of Sciences, Kiev, Ukraine

ÁNGEL V. DELGADO and FERNANDO GONZÁLEZ-CABALLERO   University of Granada, Granada, Spain

CONSTANTINO GROSSE   National University of Tucumán, San Miguel de Tucumán, and Consejo Nacional de Investigaciones Científicas y Técnicas, Buenos Aires, Argentina

EDWIN DONATH   Max-Planck-Institute for Colloids and Interfaces, Berlin, Germany

I. FREQUENCY DEPENDENCE OF INDUCED DIPOLE MOMENT OF DISPERSED PARTICLES, AND ADMITTANCE OF SUSPENSIONS: GENERAL FEATURES

The phenomena that result from the action of external electric fields on dispersed systems are currently used in many methods of characterization of dispersed particles, as well as in technology. The response of a dispersed system to an applied electric field may have different physical natures. The most important ones for the characterization of suspended particles are the purely electrical and the electromechanical responses.

The measurable values associated with the purely electrical response to an harmonic electric field are the absolute permittivity ε and conductivity k of the system. In order to simplify the following considerations, it is convenient to represent the time dependence of the harmonic external field $E(t)$, having a frequency ω, by means of the complex multiplier $e^{i\omega t}$:

$$E(t) = E \cdot e^{i\omega t} \tag{1}$$

The well-known advantage of this representation is that all the field-induced magnitudes $X(a, \zeta, \ldots, t)$ also have this same time dependence (in the linear approximation with respect to the external field):

$$X(a, \zeta, \ldots, t) = X(a, \zeta, \ldots, \omega) \cdot e^{i\omega t} \tag{2}$$

If any of the magnitudes X had a relaxation at a frequency close to ω, it would be out of phase with respect to the external field. In the representation used, this corresponds to the appearance of an imaginary part of $X(a, \zeta, \ldots, \omega)$ and also to a frequency dependence of $X(a, \zeta, \ldots, \omega)$. Here, ζ is the electrokinetic or zeta potential, and a is the particle dimension.

Correspondingly, the full electric response of the disperse system can be character-ized by the value of its complex permittivity:

$$\varepsilon^*(\omega) = \varepsilon(\omega) - \frac{ik(\omega)}{\omega} \tag{3}$$

It should be emphasized that the permittivity $\varepsilon(\omega)$ and conductivity $k(\omega)$ are real functions of the frequency.

The contribution $\delta\varepsilon^*$ of the dispersed particles to the complex permittivity of a dilute suspension is determined by the superposition of the long-range electric fields of the polarized particles together with their double layers:

$$\delta\varepsilon^*(\omega) = \varepsilon^*(\omega) - \varepsilon_m^*(\omega) = 4\pi\varepsilon_m^*(\omega)\frac{1}{V}\sum_i \frac{d_i^*}{E} \tag{4}$$

Here, d_i^* is the effective induced dipole moment of particle i (summation index) and V is the volume of the suspension. The function:

$$\varepsilon_m^*(\omega) = \varepsilon_m - \frac{ik_m}{\omega} \tag{5}$$

represents the complex permittivity of the dispersion medium. The effective induced dipole moment d_i^* appears in the expression for the electric potential distribution σ_e^*, valid in the electroneutral electrolyte solution outside the double layer of the particle:

$$\phi_e^* = -Er\cos\theta + \frac{d_i^*}{r^2}\cos\theta \tag{6}$$

The first term on the right-hand side of this expression is the potential of the homogeneous external field, and θ is the angle between the radius vector of the point and the direction of the field. The second term in Eq. (6) represents the long-range potential distortion pro-duced by all the polarized inhomogeneities inside a closed surface situated in the electro-neutral solution, and enclosing the particle together with its double layer. It must be emphasized that the value of d_i^* characterizes the effective induced dipole moment of the whole system, including the particle and the surrounding layer of solution bearing the diffuse part of its electric double layer.

Equation (4) can be written in the following more familiar form (see, e.g., Ref. 1):

$$\delta\varepsilon^* = 4\pi n\varepsilon_m^* \frac{d^*}{E} \tag{7}$$

where d^* is the effective induced dipole moment averaged over all the particles and n is the number of particles per unit volume of the suspension. The value of d^* only coincides with the effective induced dipole moment of every particle in the case that all the particles are identical.

The contribution of the particles to the permittivity $\delta\varepsilon$ and conductivity δk of the suspension can be expressed through the real and imaginary parts of Eq. (7):

$$\delta\varepsilon + \mathrm{Re}\delta\varepsilon^* = 4\pi n\left(\varepsilon_m \mathrm{Re}\frac{d^*}{E} + \frac{k_m}{\omega}\mathrm{Im}\frac{d^*}{E}\right) \tag{8}$$

$$\delta k = -\omega\,\mathrm{Im}\,\delta\varepsilon^* = 4\pi m\left(k_m \mathrm{Re}\frac{d^*}{E} - \omega\varepsilon_m\mathrm{Im}\frac{d^*}{E}\right) \tag{9}$$

The ratio of effective induced dipole moment to the external field strength

$$\gamma_i^* = \frac{d_i^*}{E}; \quad \gamma^* = \frac{d^*}{E} \tag{10}$$

is denoted as the polarizability, rather than the effective polarizability, for conciseness, and following the usage in other works [1, 3, 4] dealing with the double-layer polarization.

If the frequency ω is far away from any polarization dispersion range, the time variations of the induced dipole moment d^* are almost in phase with the time variation of external field E and hence, in accordance with Eq. (10), the polarizabilities γ_i^* and γ^* are both almost purely real magnitudes. Inside a dispersion frequency range, the induced dipole moment is out of phase with respect to the external field, so that both γ_i^* and γ^* are essentially complex.

It can be seen from Eqs (7) and (8) that both the real and imaginary parts of the complex polarizability contribute to the permittivity $\delta\varepsilon$ and the conductivity δk. However, in order to answer the question on which of the contributions [$\mathrm{Re}(\gamma^*)$ or $\mathrm{Im}(\gamma^*)$] is the most important, and under which conditions, it is necessary to introduce an important concept dealing with a point in the frequency scale. This point corresponds to the reciprocal of the relaxation time of the electrolyte solution:

$$\omega_{\mathrm{el}} \equiv \frac{1}{\tau_{\mathrm{el}}} \equiv \frac{k_{\mathrm{m}}}{\varepsilon_{\mathrm{m}}} \tag{11}$$

The value of τ_{el} characterizes the time required for the screening of charges and electric field perturbations in the electrolyte solution. Its role in the time dimension is similar to that of the Debye screening length $R_{\mathrm{D}} \equiv 1/\chi$ in the length dimension.

The angular frequency ω_{el}, Eq. (11), divides the frequency scale into two ranges (see Fig. 1):

$$\omega < \omega_{\mathrm{el}} \tag{12}$$

and

$$\omega > \omega_{\mathrm{el}} \tag{13}$$

In the left-hand range, where the frequencies are low with respect to ω_{el}, Eqs (5) and (11) show that the absolute value of the imaginary part of the complex permittivity of the electrolyte solution, $\varepsilon_{\mathrm{m}}^*$, exceeds the real part. Therefore, in this frequency range the major part of the total current is made of conduction current (determined by ion movement), whereas the displacement current (related to the polarization of the medium) contributes with a smaller part. At very low frequencies, where Eq. (12) becomes a strong inequality, the conduction current represents the overwhelming part of the total current.

Let us consider that the polarizability of the dispersed particles is characterized by two relaxation processes, one with characteristic frequency ω_1, far in the left-side range:

$$\omega_1 \equiv \frac{1}{\tau_1} \ll \omega_{\mathrm{el}} \equiv \frac{1}{\tau_{\mathrm{el}}} \tag{14}$$

and the other with characteristic frequency ω_{h} far in the right-side range:

$$\omega_{\mathrm{h}} \equiv \frac{1}{\tau_{\mathrm{h}}} \gg \omega_{\mathrm{el}} \equiv \frac{1}{\tau_{\mathrm{el}}} \tag{15}$$

The particle is made of a nonconducting medium so that, besides these two dispersion bands, it will necessarily exhibit the usual Maxwell-Wagner dispersion with the well-known polarizability expression:

$$\gamma_{\mathrm{MW}} = a^3 \frac{\varepsilon_{\mathrm{pef}}^* - \varepsilon_{\mathrm{m}}^*}{2\varepsilon_{\mathrm{m}}^* + \varepsilon_{\mathrm{pef}}^*} \tag{16}$$

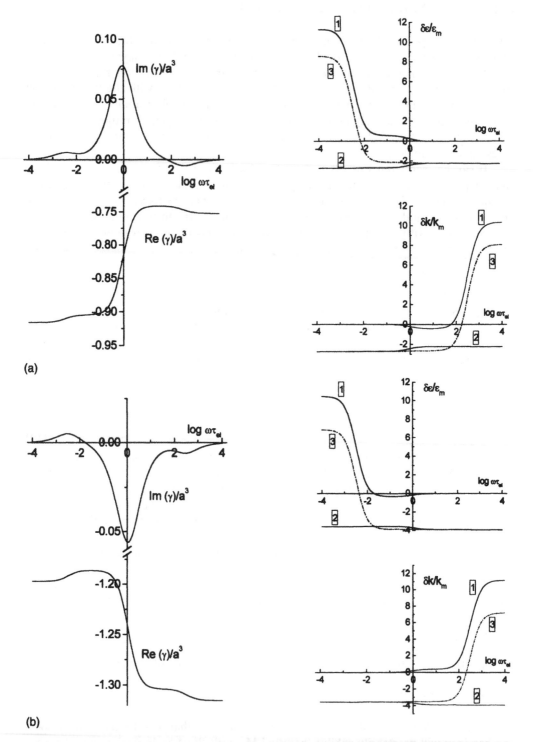

FIG. 1 Left characteristic features of the spectrum of particle polarizability [Re(γ)/a^3 and Re(γ)/a^3] for three relaxation processes of particle polarization (see text). Right: associated spectra of the specific dielectric increment ($\delta\varepsilon/\varepsilon_m$) and conductivity increment ($\delta k/k_m$) of suspensions. (a): $\varepsilon_p/\varepsilon_m = 1/2 > k_{pef}/k_m$; (b): $\varepsilon_p/\varepsilon_m = 1/40 < k_{pef}/k_m$. Parameters: $\varepsilon_m = 80\varepsilon_0$; $k_m = 2.4 \times 10^{-3}$ S/m; $k_p = 0$; $k_{per}/k_m = 2\lambda/ak_m = 0.2$; $\omega_l = 10^{-2.5}\omega_{el}$; $\omega_h = 10^{2.5}\omega_{el}$.

Here, the effective complex permittivity of the particle:

$$\varepsilon_{pef}^* = \varepsilon_p - \frac{i}{\omega}\frac{2\lambda}{a} \tag{17}$$

includes not only the value of the inner permittivity ε_p, but also the contribution of the surface conductivity λ, by means of an equivalent inner conductivity $k_{pef} = \frac{2\lambda}{a}$ [2].

The real and imaginary parts of γ_{MW} vary with frequency in the central part of Fig. 1, since the characteristic frequency of the Maxwell–Wagner relaxation is almost always close to $1/\tau_{el}$. The low- and high-frequency limits of γ_{MW} have the following values:

$$\gamma_{MW}|_l = a^3 \frac{k_{pef} - k_m}{2k_m + k_{pef}} \tag{18}$$

$$\gamma_{MW}|_h = a^3 \frac{\varepsilon_p - \varepsilon_m}{2\varepsilon_m + \varepsilon_p} \tag{19}$$

The characteristic frequency of the Maxwell–Wagner dispersion for nonconducting particles with a surface conductivity that is not very high, practically coincides with ω_{el}. Therefore, assuming that the strong inequalities (14) and (15) hold, we may consider $\gamma_{MW}|_l$ as being the high-frequency limit of the polarizability for the low-frequency dispersion, and $\gamma_{MW}|_h$ as the low-frequency limit of the polarizability for the high-frequency dispersion. Considering, for simplicity, that simple Debye type functions describe the frequency dependence of both the low- and high-frequency relaxation processes:

$$\gamma^*|_l = \gamma_{MW}|_l + \frac{G_l}{1 + i\omega\tau_l} \tag{20}$$

$$\gamma^*|_h = \gamma_{MW}|_h + \frac{G_h \cdot i\omega\tau_h}{1 + i\omega\tau_h} \tag{21}$$

where G_i and G_h are, respectively, the amplitudes of the low- and high-frequency dispersions.

Under these assumptions, the full frequency dependence of the polarizability is described by the following expression:

$$\gamma^* = \frac{G_l}{1 + i\omega\tau_l} + a^3 \frac{\varepsilon_{pef}^* - \varepsilon_m^*}{2\varepsilon_m^* + \varepsilon_{pef}^*} + \frac{G_h \cdot i\omega\tau_h}{1 + i\omega\tau_h} \tag{22}$$

Plots of the frequency dependence of both the real and imaginary parts of γ^*, of the contribution of the particles to the permittivity ($\delta\varepsilon$) and conductivity (δk^*) of the suspension (curves labelled 3), and of the components of these magnitudes ($\varepsilon_m \text{Re}\frac{d^*}{E}$ (curve 2) and $\frac{k}{\omega}\text{Im}\frac{d^*}{E}$ (curve 1) for $\delta\varepsilon^*$; $k_m \text{Re}\frac{d^*}{E}$ (curve 2) and $\omega\varepsilon_m\text{Im}\frac{d^*}{E}$ (curve 1) for δk^*), calculating using Eqs (8)–(10) and (22), are represented in Fig. 1. It can be seen that in the left-hand frequency range $\left(\omega \ll \frac{k_m}{\varepsilon_m}\right)$, the relative contribution of the particles to the permittivity is very large and surpasses by many times their relative contribution to the conductivity. This corresponds to a well-known feature of the LFDD (low-frequency dielectric dispersion) phenomenon. The physical mechanism of this dispersion, first described in Ref. 3, consists in the conversion of the conductive current into a very large (at low frequencies) electrostatic induction, due to the action of the out-of-phase field created by the slow [4] polarization of the particles.

In the right-side range ($\omega \ll k_m/\varepsilon_m$), Fig. 1 shows a very large variation in the conductivity and a comparatively small variation in the permittivity of the suspension, caused by the high-frequency dispersion of the induced dipole moment of the particles. This high-frequency conductivity dispersion (HFCD) is in a certain sense the symmetrical phenomenon of LFDD. Due to the low-frequency relaxation process, the out-of-phase dipole moment produces an out-of-phase conductive current, which is indistinguishable phenomenologically from an in-phase displacement. This gives rise to a very high permittivity at low frequencies. Analogously, due to the high-frequency relaxation process, the out-of-phase dipole moment produces an out-of-phase polarization, which is indistinguishable phenomenologically from a conductive current, since its time derivative is in phase with the field. This gives raise to a very high conductivity at high frequencies.

Both at very low and very high frequencies, the major influence of the presence of the particles on the suspension properties is determined by the imaginary part of the complex polarizability:

$$\delta\varepsilon \approx 4\pi n \frac{k_m}{\omega} \mathrm{Im}\gamma^* \quad \text{for} \quad \omega \cong \omega_i \ll \omega_{el} \tag{23}$$

$$\delta k \approx 4\pi n \omega \varepsilon_m \mathrm{Im}\gamma^* \quad \text{for} \quad \omega \cong \omega_h \gg \omega_{el} \tag{24}$$

It is finally worth noting that the usual relaxation type of the frequency dependence of the admittance components – i.e., a monotonic decrease (increase) of the permittivity (conductivity) with increasing frequency – requires, in view of Eqs (8), (9), (23), and (24), $\mathrm{Im}(\gamma^*)$ to be positive (negative) for the low-frequency (high-frequency) dispersion:

$$\mathrm{Im}\gamma^* > 0 \quad \text{for} \quad \omega \cong \omega_i \ll \omega_{el} \tag{25}$$

$$\mathrm{Im}\gamma^* < 0 \quad \text{for} \quad \omega \cong \omega_h \gg \omega_{el} \tag{26}$$

As for the vicinity of ω_{el}, the contributions of both $\mathrm{Re}(\gamma^*)$ and $\mathrm{Im}(\gamma^*)$ to the components of the complex admittance have comparable values, so that both signs of $\mathrm{Im}(\gamma^*)$ are possible for the usual behavior of these components. The sign of $\mathrm{Im}(\gamma^*)$ in the Maxwell–Wagner dispersion depends on the ratios of the conductivities and dielectric permittivities of the two intervening media. If the particle-to-medium permittivity ratio $\varepsilon_p/\varepsilon_m$ is greater (smaller) than the corresponding conductivity ratio $\frac{k_{pef}}{k_m} = \left(k_p + \frac{2\lambda}{k_m}\right)\Big/ k_m$, the sign of $\mathrm{Im}(\gamma^*)$ is negative (positive) (cf. Fig. 1a and Fig. 1b). Another interesting feature of the Maxwell–Wagner dispersion is that, in spite of a very large absolute value and sign variations of $\mathrm{Im}(\gamma^*)$, the frequency dependence of both components of the admittance is very small and always of the usual type. This occurs because of an almost complete compensation of the contributions of $\mathrm{Im}(\gamma^*)$ and $\mathrm{Re}(\gamma^*)$ to both the permittivity [Eq. (8)] and conductivity [Eq. (9)] amplitudes.

II. GENERAL FEATURES OF ELECTROROTATION SPECTRA: DISCREPANCIES BETWEEN THE TRADITIONAL THEORY AND LOW-FREQUENCY EXPERIMENTAL RESULTS

A widely used and investigated kind of electromechanical response of dispersed particles to an external electric field is electrophoresis. While there is no direct relation between this phenomenon and the admittance of a suspension in the general case, the theory of the volume diffusion mechanism of the LFDD [1, 3] established that the diffuse part of the electrical double layer (EDL), which is responsible for electrophoresis, has also a very important role in LFDD (the so-called α-dispersion). There is, however, another kind of electromechanical response that is a close relative of the phenomenon of dispersion of the admittance of a suspension. It is the phenomenon of electrorotation, which we will consider here.

Electrorotation is best described as the slow (of the order of units of rad/s) rotation of suspended particles under the influence of an external electric field that rotates with an angular velocity in the range 10^4–10^9 rad/s (or 10^3–10^8 Hz). Electrorotation is caused by relaxation process taking place during the polarization of the particle, the bulk solution, and the interface. If the frequency of the rotating field becomes comparable to the reciprocal of the characteristic time of at least one of the relaxation processes contributing to the polarization of the particle, an angle forms between the induced dipole moment and the external field. The interaction of the electric field $\vec{E}$ with the nonparallel dipole moment $\vec{d}_i^*$ of the particle results in a torque $\vec{M}_i$ [5]:

$$\vec{M}_i = 4\pi\varepsilon_m[\vec{d}_i^* \times \vec{E}] \tag{27}$$

where the factor relates the induced dipole moment to its effective value $\vec{d}_i^*$, which appears in Eq. (6).

The angle between the induced dipole moment and the external field vectors, which arises in a rotating field, is equal to the phase difference between the complex induced dipole moment and external alternating field. Therefore, the torque [Eq. (27)] can be expressed in accordance with Eq. (10) by means of the imaginary part of the complex polarizability:

$$M_i = -4\pi\varepsilon_2 \mathrm{Im}\gamma_i^* \tag{28}$$

In the framework of the traditional approach of the electrorotation theory [5–15], this torque M is the only cause for the field-induced angular speed Ω_{Mi} of electrorotation of a spherical particle:

$$\Omega_{\mathrm{Mi}} = \frac{M_i}{8\pi\eta a^3} = -\frac{\varepsilon_m E^2 \mathrm{Im}\gamma_i^*}{2\eta a^3} \tag{29}$$

where η is the viscosity of the bulk solution.

The negative sign in this formula shows that for positive values of $\mathrm{Im}(\gamma_i^*)$ (the induced dipole moment leads in phase the external field so that the induced dipole moment vector $\vec{d}_i^*$ rotates ahead of vector $\vec{E}$), the torque $\vec{M}_i$ tends to rotate the particle in the direction opposite to the field rotation, causing the so-called counterfield electrorotation. Correspondingly, if $\mathrm{Im}(\gamma_i^*)$ is negative, the induced dipole moment vector lags the rotating field, leading to cofield electrorotation.

The rotational velocity of colloidal particles is usually recorded as a function of the frequency of the applied rotating field, leading to electrorotation spectra. The above considerations show a close relationship [5,7] between these spectra and those corresponding to the permittivity of the suspension, since both are determined [in the framework of the applicability of Eq. (29), see below] by the same characteristic: the effective induced dipole moment of a polarized particle and its surrounding double layer. This line of reasoning is in good agreement with many experimental findings obtained at high enough frequencies of field rotation [5, 7–15].

However, even using the traditional approach based on Eq. (29), there are some essential differences in the character of the information that can be deduced from admittance and electrorotation spectroscopies. The characteristic feature of electrorotation spectra is that they reflect the frequency dependence of $Im(\gamma_i^*)$ only, while admittance spectra depend on both $Im(\gamma_i^*)$ and $Re(\gamma_i^*)$. Furthermore, Figs 1a and 1b show that in the frequency range of the Maxwell–Wagner dispersion, the contributions of $Im(\gamma_i^*)$ and $Re(\gamma_i^*)$ almost fully compensate each other in the case of nonconducting particles. Therefore, electrorotation spectra appear to be much more sensitive to the particle parameters that determine the Maxwell–Wagner polarization as, for example, the permittivity of nonconducting particles suspended in electrolyte solution, than admittance spectra. If the particle is inhomogeneous with regard to its conductivity and dielectric parameters, several minima and maxima can be observed in an electrorotation spectrum, each one corresponding to a specific dispersion process.

Another distinguishing feature of electrorotation spectra is that they are determined by the induced dipole moment of a single rotating particle, while the permittivity of a suspension is a macroscopic property that depends on the average induced dipole moment of the suspended particles. This is why the electrorotation technique may be considered as a dielectric spectroscopy technique applicable at the single-particle level. It has been shown that the obtaining of electrorotation spectra is especially suitable for studying the biological cell membrane and the dialectric properties of the cell interior [8, 10, 11, 13–18].

It is important to note that while Eqs (8) and (9) are always valid for sufficiently diluted suspensions, there are some important cases when relation (29) is not applicable for the description of particle electrorotation. The incompleteness of the traditional approach based on this equation becomes apparent by comparing measurements of low-frequency electrorotation with the most general features of the permittivity dispersion corresponding to the lowest frequency range (see Figs 1a and 1b). The generally measured behavior of $\delta\varepsilon(\omega)$ consists of a monotonic decrease with increasing frequency. This behavior corresponds (see Eq. (25)] to a positive sign of $Im(\gamma_i^*)$ and hence, according to Eq. (29), to a counterfield peak of the low-frequency electrorotation. On the other hand, recent measurements of electrorotation at the lowest attainable frequencies [Eq. (25)] clearly show a cofield peak of electrorotation. Such a behavior was first reported [5] in 1987 by Arnold et al. for polymer latex particles suspended in electrolyte solutions, i.e., precisely for the systems for which the low-frequency admittance dispersion was extensively studied [19–21], always showing the usual behavior (decreasing permittivity and increasing conductivity with increasing frequency). This strange behavior of the low-frequency electrorotation was later confirmed for many different systems [12, 15, 22, 23]. The observations of cofield electrotation together with the usual frequqency dependence of the admittance components at low frequencies clearly pointed at an essential restriction in the applicability of Eq. (29).

The physical cause of this restriction, clarified in Ref. 24, is the following. When the medium surrounding the particle is an electrolyte solution, a significant part of the field-induced charges are located in the external Debye atmosphere, i.e., within the hydrodynamically mobile liquid. Therefore, a significant part of the total torque is applied to the liquid within the Debye atmosphere rather than directly to the particle, inducing a rotation of the Debye atmosphere with respect to the particle's surface. Consequently, besides using Eq. (29) to determine the joint rotation of the particle together with its Debye atmosphere, it is necessary to determine the relative rotation of the particle with respect to its Debye atmosphere. Instead of being determined by the total induced dipole moment, this relative motion is produced by torques applied to the particle and to its double layer. These torques are equal in absolute value but opposite in sign so that the total torque is zero.

This new component of electrorotation constitutes a rotational analog of electrophoresis: a rotation of the particle with respect to its double layer without the action of a total torque just as in electrophoresis, where there is a relative translation of the particle with respect to its double layer without the action of a total force.

The effect on the electrorotation spectra of this relative rotation of a particle and its double layer is especially important when a relatively small total torque is the result of two large opposite torques applied to the inner and outer parts of the double layer. This is precisely the case for both known mechanisms of the LFDD: the concentration polarization (volume diffusion mechanism of double-layer polarization) and the Schwarz mechanism (surface diffusion mechanism of double-layer polarization).

Quantitatively, the contribution of this new factor to electrorotation is up to now only considered [24, 25] for the limiting case of a very thin souble layer, i.e., when the Debye screening length $R_D = 1/\chi$ is very small as compared to the radius a of the spherical particle:

$$\chi a = \frac{1}{R_d} \gg 1 \tag{30}$$

When this condition is fulfilled, the angular velocity Ω_S can be expressed in terms of the rotational component of the electro-osmotic slip velocity of the outer surface of the double layer with respect to the surface of the particle. The local value of this velocity, v_S, at any given point of the quasiflat [due to Eq. (30)] particle surface is expressed as a function of the tangential component of the electric field, E_S, by the well-known Smoluchowski formula:

$$v_S = -\frac{\varepsilon_m \zeta}{\eta} E_S \tag{31}$$

Here it is important to note that the local value of the ζ potential includes both the equilibrium (field-independent) component ζ_0 and the field-induced part $\delta\zeta$:

$$\zeta = \zeta_0 + \delta\zeta \tag{32}$$

The rotational component of the electro-osmotic slip arises as a result of the tangential component of the electric field acting on the field-induced component of the zeta potential, $\delta\zeta$, whenever the symmetry axis of the E_S distribution forms a finite angle with that of the $\delta\zeta$ distribution. This angle arises due to the relaxation of the double-layer polarization, which causes a phase difference between the E_S and $\delta\zeta$ distributions.

The distribution of ζ along the surface may be represented in the usual form:

$$\zeta \equiv \zeta_0 + \delta\zeta = \zeta_0 + K^* E a \cos\theta \tag{33}$$

where K^* is a field-independent quantity, characteristic of the polarization of the particle surface. It can be considered as the complex amplitude of the field-induced variation of the ζ potential. The distribution of E_S follows from Eqs (6) and (10):

$$E_S = \frac{1}{r}\frac{\partial\phi}{\partial\theta}\bigg|_{r=a} = -\left(1 - \frac{\gamma_i^*}{a^3}\right) E \sin\theta \tag{34}$$

Equations (33) and (34) show that the axes of symmetry of $\delta\zeta$ and E_S rotate synchronously with $E = E_0 e^{i\omega t}$ and are directed, respectively, as $K^* E a = K^* E_0 a e^{i\omega t}$ and

$-\left(1 - \frac{\gamma_i^*}{a^3}\right)E = -\left(1 - \frac{\gamma_i^*}{a^3}\right)E_0 e^{i\omega t}$. Therefore, they form finite phase angles both with E and with one another.

There are two pairs of mutually orthogonal components of the rotating axes:

$$Ea\,\text{Im}\,K^*, \quad -E\,\text{Re}\left(1 - \frac{\gamma_i^*}{a^3}\right) = -E\left(1 - \text{Re}\frac{\gamma_i^*}{a^3}\right)$$

and

$$Ea\,\text{Re}\,K^*, \quad -E\,\text{Im}\left(1 - \frac{\gamma_i^*}{a^3}\right) = \text{Re}\frac{\gamma_i^*}{a^3}$$

Both pairs contribute, in accordance with Eq. (31), to the tangential movement of the outer surface of the double layer with respect to the particle's surface, which leads to rotating electro-osmotic slip velocity v^S (in a plane containing the center of the particle and the rotating field vector):

$$v_S^R = -\frac{\varepsilon_m a}{2\eta} E^2 \left\{ \frac{\text{Im}\gamma^*}{a^3} \text{Re}\,K^* + \left(1 - \frac{\text{Re}\gamma^*}{a^3}\right) \text{Im}\,K^* \right\} \tag{35}$$

The factor $\frac{1}{2}$ in this expression appears because of the difference between the rotation of the outer surface of the double layer as a whole and the general angle dependence of the local v_S value [24, 25].

The angular velocity of the particle with respect to the liquid at the outer surface of the thin double layer is, Eq. (35):

$$\Omega_S = -\frac{v_S^R}{a} = \frac{\varepsilon_m}{2\eta} E^2 \left\{ \frac{\text{Im}\gamma^*}{a^3} \text{Re}\,K^* + \left(1 - \frac{\text{Re}\gamma^*}{a^3}\right) \text{Im}\,K^* \right\} \tag{36}$$

The total angular velocity of electrorotation, observed experimentally, is represented by the superposition of two contributions:

$$\Omega = \Omega_M + \Omega_S \tag{37}$$

- The traditional Ω_M caused by the action of the total torque acting on the particle together with its double layer and expressed in Eqs (27) and (28). It is a function of just one characteristic magnitude of the polarized particle: the imaginary part of its polarizability.
- The new Ω_S, caused by the rotational electro-osmotic slip of the particle with respect to the double layer Eqs (36) and (33), which depends on two characteristic magnitudes of the polarized particle: its polarizability and the complex amplitude of the field-induced variation of the ζ potential. This new contribution explains the very strong quantitative and qualitative discrepancies between admittance and electrorotation spectra observed at very low frequencies.

We will now discuss the admittance and electrorotation spectra for two models of double-layer polarization: the volume diffusion model, based on the so-called standard model of the kinetic properties of the double layer, and the surface diffusion model, based on the concept of a diffusion-controlled polarization of the dense part of the double layer, and without any exchange (or with a hindered exchange) between the dense and diffuse parts of the double layer.

We will use these two models of the double-layer kinetics together with a single model for the volumetric properties of the disperse particle and electrolyte solution. The suspended particle is represented by an insulating sphere of radius a, while the surrounding electrolyte solution is characterized by its viscosity ζ, absolute permittivity ε_m, the valencies of its ions $z^{\pm}$, their diffusion coefficients $D^{\pm}$, and their concentrations far from the particle $C^{\pm}(\infty)$. We will consider that there are only two types of ions and that there are no ion pairs.

III. LOW-FREQUENCY DISPERSION OF INDUCED DIPOLE MOMENT AND AMPLITUDE OF FIELD-INDUCED ZETA POTENTIAL IN FRAMEWORK OF THE STANDARD MODEL OF THE THIN EDL

A. Induced Dipole Moment

According to the standard model [26], which was used in the great majority of works on the theory of double-layer polarization, e.g., Refs 27–34, we consider that the particle has a uniform surface charge density σ in the dense part of the EDL. We assume, as is usually done, that the viscosity, the diffusion coefficients, and the permittivity have constant values everywhere outside the charged surface of the particle. These assumptions are equivalent to the equality of the potential change across the diffuse part of the double layer and the ζ potential. We further assume that the ions belonging to the dense part of the double layer are immobile. This makes it possible to identify the surface of the particle with the boundary between the condensed and the diffuse parts of the double layer, the surface potential with the potential at this boundary, and the surface charge with the charge inside this boundary.

The problem of double-layer polarization consists in finding the solution to the well-known equation system that includes: the material balance equations for the ionic components (superscripts "$+$" and "$-$" for cations and anions, respectively); the Nernst–Planck equation for the ion flows $\vec{j}^{\pm}$ under the action of the electric potential Φ, ionic concentration ($C^{\pm}$) gradients, and of the convection velocity of the solution $\vec{v}$; the Poisson equation, in which the bulk charge density is expressed in terms of the concentrations of cations and anions; and the Stokes equation, containing the volume force created by the action of electric field on the bulk charge density:

$$\nabla \cdot \vec{j}^{\pm} = \frac{\partial C^{\pm}}{\partial t} \tag{38}$$

$$\vec{j}^{\pm} = -D^{\pm}\nabla C^{\pm} \pm \frac{e}{kT}D^{\pm}C^{\pm}z^{\pm}\nabla\Phi + C^{\pm}\vec{v} \tag{39}$$

$$\nabla^2\Phi = -\frac{e(z^+C^+ - z^-C^-)}{\varepsilon_m} \tag{40}$$

$$\eta\nabla^2\vec{v} - \nabla P = (z^+C^+ - z^-C^-)e\nabla\Phi \tag{41}$$

This equation system can be simplified by using the principle of local equilibrium, which permits us to consider that every volume element that is sufficiently small is in equilibrium, despite the fact that the whole volume is in a nonequilibrium state. This principle makes it possible to characterize the distribution of nonequilibrium fields in the medium surrounding the polarized particle by the distribution of parameters in a so-called virtual electrolyte solution. This virtual solution can be defined for every point $\vec{r}$ of the "real" nonequilibrium system (including points inside the double layer), as an electroneutral solution with parameter values chosen in such a way that it is in equilibrium with the "real" solution at the considered point. It must be emphasized that, in general, the "real" solution is not electroneutral, for example, for points $\vec{r}$ inside a diffuse part of the EDL.

The parameters of the "real" solution are: electric potential $\Phi(\vec{r})$, cation and anion concentrations $C^{\pm}(\vec{r})$, and hydrostatic pressure $P(\vec{r})$. The parameters of the electroneutral virtual solution are: electric potential $\phi(\vec{r})$, electrolyte concentration $n(\vec{r})$, and hydrostatic pressure $p(\vec{r})$. Both sets of parameters differ from one another by the so-called quasiequilibrium potential $\Phi_{qeq}(\vec{r})$, the quasiequilibrium excesses of cation and anion concentrations $\gamma_{qeq}^{\pm}(\vec{r})$, and the quasiequilibrium excess of pressure Π:

$$\Phi = \phi + \Phi_{quq} \tag{42}$$

$$C^{\pm} = z^{\mp}n + \gamma_{qeq}^{\pm} \tag{43}$$

$$P = p + \Pi \tag{44}$$

The following relations between the values $n(\vec{r})$, $\Phi_{qeq}(\vec{r})$, $\gamma_{qeq}^{\pm}(\vec{r})$, and $\Pi(\vec{r})$ provide the thermodynamic and mechanical local equilibrium between any given macroscopically small volume of electrolyte solution at the point $\vec{r}$, and the corresponding virtual solution (see, e.g. Ref. 35):

$$\gamma_{qeq}^{\pm} = z^{\mp}n(\vec{r})\left[\exp\left(\mp\frac{z^{\pm}e\Phi_{qeq}}{kT}\right) - 1\right] \tag{45}$$

$$\Pi = RTn(\vec{r})\left\{z^{-}\left[\exp\left(-\frac{z^{+}e\Phi_{qeq}}{kT}\right) - 1\right] + z^{+}\left[\exp\left(\frac{z^{-}e\Phi_{qeq}}{kT}\right) - 1\right]\right\} \tag{46}$$

These expressions look like the well-known equilibrium relations for the excesses of ion concentrations and of hydrostatic pressure in the equilibrium EDL when the electrolyte concentration value outside the double layer is n. The difference between the equilibrium and the nonequilibrium systems is that in the latter the parameters of the virtual solution ϕ and n are position dependent, whereas for the former these parameters are constant throughout the system. This position dependence of the parameters of the virtual solution is a sufficient condition for nonequilibrium. Therefore, the equations of irreversible thermodynamics (particularly the Nernst–Planck and Navier–Stokes equations) can only contain gradients of these parameters and cannot include any gradients of quasiequilibrium distributions (particularly, no gradients of the field-induced component of Φ_{qeq}). Taking this into account, the substitution of Eqs (42)–(46) into Eqs (39)–(41) leads to the following form of the basic equation set of the problem:

$$\nabla \cdot \vec{j}^{\pm} = -\frac{\partial C^{\pm}}{\partial t} \tag{47}$$

$$j^{\pm} = (1 + g^{\pm})\left(-D^{\pm}\nabla n \mp z^{\pm}\frac{e}{kT}D^{\pm}n\nabla\phi + n\vec{v}\right) \tag{48}$$

$$\nabla^{2}(\phi + \Phi_{qeq}) = -\frac{e}{\varepsilon_{m}}nz^{+}z^{-}(g^{+} - g^{-}) \tag{49}$$

$$\eta\nabla^{2}\vec{v} - \nabla p - kT(z^{-}g^{+} - z^{+}g^{-})\vec{\nabla}n(\vec{r}) - ez^{+}z^{-}(g^{+} - g^{-})\nabla\phi = 0 \tag{50}$$

where

$$g^{\pm} = \exp\left(\mp\frac{ez^{\pm}\Phi_{\text{qeq}}}{kT}\right) - 1 \tag{51}$$

A very important simplification of this equation set arises in the case of a weak external field, using an expansion in successive powers of the field strength [35]. Since $\nabla\phi(\vec{r})$, $\nabla n(\vec{r})$, and $\nabla p(\vec{r})$ do not contain field-independent terms, restricting our consideration just to the first-order terms in the electric field strength, all the field-dependent terms of the coefficients of these gradients can be dropped. As a result, all the coefficients in the equation set only depend on the equilibrium potential distribution Φ_{eq}, and hence Eqs (47)–(50) should be solved using the following expression for $g^{\pm}$:

$$g^{\pm} \approx g_0^{\pm} = \exp\left(\mp\frac{ez^{\pm}\Phi_{\text{eq}}}{kT}\right) - 1 \tag{52}$$

In the electrolyte solution outside the double layer, the equilibrium potential vanishes: $\Phi_{\text{eq}} = 0$, so that the approximation of local electroneutrality applies even for an alternating external field [3]. Therefore, wherever $\Phi_{\text{eq}} = 0$, one may consider $C^{\pm} = z^{\mp}n$. Making the substitutions:

$$\Phi_{\text{eq}} = 0, \, C^{\pm} = z^{\mp}n \tag{53}$$

in Eqs (47), (48), and (52) leads, in view of Eqs (1) and (2), to the following equations for the dimensionless distributions of electric potential $\tilde{\phi}$ and electrolyte concentration $\tilde{n}$ outside the double layer:

$$\nabla^2\tilde{\phi} = \frac{i\omega}{D_{\text{ef}}}\Delta\delta\tilde{n} \tag{54}$$

$$\nabla^2\delta\tilde{n} = \frac{i\omega}{D_{\text{ef}}}\delta\tilde{n} \tag{55}$$

where

$$D_{\text{ef}} = \frac{D^+D^-(z^+ + z^-)}{D^+z^+ + D^-z^-} \tag{56}$$

$$\Delta = \frac{D^- - D^+}{D^+z^+ + D^-z^-} \tag{57}$$

Here, the field-induced (δn) and field-independent or equilibrium (N) parts of n are determined, as usual, by

$$n = N + \delta n$$

Equations (55) and (28) can be easily solved, leading to the following distributions of ϕ and δn outside the double layer:

$$\delta\tilde{n} = K_{\text{e}}\frac{\exp[-(1+i)W(r/a-1)]}{r^2/a^2}\frac{1+(1+i)W}{1+W+iW}\frac{r/a}{kT}\frac{aEa}{kT}\cos\theta \tag{58}$$

$$\tilde{\phi} = \left[\frac{K_{\text{d}}a^2}{r^2} - \frac{r}{a}\right]\frac{eEa}{kT}\cos\theta + \Delta\delta\tilde{n} \tag{59}$$

where

$$W = \sqrt{\frac{\omega a^2}{2D_{\text{ef}}}} \tag{60}$$

while K_c and K_d are frequency-dependent integration coefficients; K_c is proportional to the variation in electrolyte concentration on the surface of the particle, while K_d is proportional to its induced dipole moment (actually K_c is proportional to the variation in electrolyte concentration calculated using expressions valid outside the thin double layer and extrapolated up to the surface of the particle).

The coefficients K_e and K_d can be determined by integrating the continuity equation written for the difference between the total ion flows $\vec{j}^{\pm}$ and the flows $\vec{j}_l^{\pm}$ calculated using expressions which are only valid outside the double layer:

$$\int_a^{\infty} \frac{1}{r}\frac{\partial}{\partial r} r[\vec{j}^{\pm} - \vec{j}_l^{\pm}]_r \mathrm{d}r = - \int_a^{\infty} \frac{1}{r\sin\theta}\frac{\partial}{\partial\theta}\sin\theta[\vec{j}^{\pm} - \vec{j}_l^{\pm}]_\theta \mathrm{d}r - \int_a^{\infty}\frac{\partial}{\partial t}(C^{\pm} - z^{\mp}n)\mathrm{d}r \tag{61}$$

In this expression, the long-range (subscript "l") ion flows are

$$\vec{j}_l^{\pm} = -z^{\mp}ND^{\pm}\nabla\delta\tilde{\mu} + z^{\pm}N\vec{v} \tag{62}$$

The integrals can be analytically evaluated in the case when the double layer is thin as compared to the radius of the particle:

$$\chi a \gg 1 \tag{63}$$

The simplifications corresponding to this case, which were first used in Refs 27 and 28 for stationary fields, presented in detail in Ref. 1 for periodic fields, and whose validity was numerically verified [31] are:

- The curvature of the surface is only taken into account in the solution of the equations corresponding to the electroneutral electrolyte, while the equations inside the thin double layer are solved assuming a locally flat surface.
- Each portion of the double layer is considered to be in a state of local equilibrium, which means that the electrochemical potentials, as well as their tangential derivative, do not change across the double layer.

Using these simplifications Eq. (61) reduces to [36]

$$z^{\mp}ND\nabla_r\delta\tilde{\mu}^{\pm}\big|_a = -\frac{1}{a\sin\theta}\frac{\partial}{\partial\theta}\left\{\sin\theta\left[D^{\pm}G_0^{\pm}\nabla_\theta\delta\tilde{\mu}^{\pm}\big|_a + \int_a^{\infty}(z^{\mp}N - C_0^{\pm})\vec{v}\mathrm{d}r\right]\right\} \tag{64}$$

where $G_0^{\pm}$ are the adsorption coefficients for the two types of ions:

$$G_0^{\pm} = \int_a^{\infty}[C_0^{\pm} - C_0^{\pm}(\infty)]\mathrm{d}r \tag{65}$$

Equation (64) expresses the condition that ion flows arriving at the double layer spread out along the surface of the particle.

In all that follows we shall consider the case where $z^+ = z^- = z$, since it is the only situation when analytical results can be obtained. In this case, the adsorption coefficients become

$$G_{qeq}^{\pm} = \frac{2zn}{\chi_{qeq}}\left[\exp\left(\frac{z\zeta_{qeq}}{2}\right) - 1\right] \tag{66}$$

where ζ_{qeq} and χ_{qeq} are, respectively, the quasiequilibrium ζ potential of the particle and the local reciprocal thickness of the quasiequilibrium double layer. The local parameters of the thin quasiequilibrium double layer correspond to its local equilibrium with the virtual electrolyte solution of concentration $n(a, \theta)$ near the given point of the particle's surface. Note that the parameters of a fully equilibrium double layer correspond to an equilibrium electrolyte concentration, n_{eq}, identical for all the points.

In order to solve Eq. (50), the fluid velocity v_θ needs only be calculated inside the double layer, since the factor multiplying v_θ in this equation vanishes outside. For a thin double layer, and in view of restriction (63), the fluid velocity inside can be obtained by solving Eq. (50) for a plane interface under the action of tangential gradients of the electric potential ϕ and of the concentration n. The local equilibrium of the thin double layer means, in terms of the definition of the virtual solution, that the parameters of the latter, namely, ϕ, n, and p, do not change across the thin double layer:

$$\nabla_\theta \phi(r)\big|_{a \leq r \leq a + O\left(\frac{1}{\chi}\right)} \approx \text{const}; \ \nabla_\theta n(r)\big|_{a \leq r \leq a + O\left(\frac{1}{\chi}\right)} \approx \text{const}; \ \nabla_\theta p(r)\big|_{a \leq r \leq a + O\left(\frac{1}{\chi}\right)} \approx \text{const} \tag{67}$$

Consequently, the tangential components of the gradients of ϕ and n may be considered constant across the double layer. The tangential component of the gradient of p may be neglected, since the pressure gradient at the outer boundary of the double layer vanishes. This corresponds to the well-known feature of electrophoretic motion [37], when no total force acts on the electroneutral system composed of the particle together with its double layer:

$$\nabla_\theta p(r)\big|_{r=a+O\left(\frac{1}{\chi}\right)} = 0 \tag{68}$$

and, in accordance with Eq. (67):

$$\nabla_\theta p(z)\big|_{a \leq r \leq a + O\left(\frac{1}{\chi}\right)} = 0 \tag{69}$$

We consider a flat charged surface and choose an orthogonal coordinate system with the x-axis extending in the normal direction from the surface and towards the fluid. In accordance with Eqs (44), (46) and (69), the tangential component of Eq. (50) can be rewritten for the convection, considering a thin quasi-flat diffuse layer, as

$$\eta \frac{d^2 v_\theta}{dx^2} - kT(z^- g^+ + z^+ g^-)\nabla_\theta n(a, \theta) - ez^+ z^- (g^+ - g^-)\nabla_\theta \phi(a, \theta) = 0 \tag{70}$$

The surface conditions for this equation at the particle surface ($x = 0$) and at the outer boundary of the EDL ($x \to \infty$) are

$$v_\theta\big|_{x=0} = 0; \quad \frac{dv_\theta}{dx}\bigg|_{x \to \infty} = 0 \tag{71}$$

Integrating Eq. (70) and using condition (71) leads to the following distribution of v_θ:

$$v_\theta = kTz\nabla_\theta n(a,\theta) \int_{x'}^{\infty} dx' \int_0^{x'} (g^+ + g^-)dx'' + ez^2\nabla_\theta\phi(a,\theta) \int_{x'}^{\infty} dx' \int_0^{x'} (g^+ - g^-)dx'' \tag{72}$$

The integrals can be analytically calculated in the case of a symmetrical electrolyte by using the usual procedure that consists in a change of the integration variable from the coordinate x to the quasiequilibrium potential. In quasiequilibrium, the Poisson equation (49), combined with Eq. (51), reduces to

$$\frac{d^2\Phi_{qeq}}{dx^2} = \frac{enz^2}{\varepsilon_m}\sinh\frac{ze\Phi_{qeq}}{kT}$$

which can be integrated once leading to

$$\frac{d\tilde{\Phi}_{qeq}}{dx} = -\frac{\chi_{qeq}}{z}\sinh\frac{z\tilde{\Phi}_{qeq}}{2} \tag{73}$$

Using this result, the right-hand side of Eq. (72) can be reduced to

$$v_\theta = kT\chi_{qeq}\nabla_\theta n(a,\theta) \int_0^{\Phi_{qeq}} d\Phi'_{qeq} \int_{\zeta_{qeq}}^{\Phi'_{qeq}} (g^+ + g-)\sinh\frac{z\Phi''_{qeq}}{2}d\Phi''_{qeq}$$

$$+ ez\chi_{qeq}\nabla_\theta\phi(a,\theta) \int_0^{\Phi_{qeq}} d\Phi'_{qeq} \int_{\zeta_{qeq}}^{\Phi'_{qeq}} (g^+ - g^-)\sinh\frac{z\Phi''_{qeq}}{2}d\Phi''_{qeq} \tag{74}$$

Substituting $g^\pm$, written for the case $z^+ = z^- = z$, Eq. (51) leads to the following analytical solution of Eq. (74):

$$v_\theta = \frac{4\varepsilon_m}{\eta}\left[\frac{kT}{ze}\right]^2 \ln\left[\frac{\cosh\dfrac{z\tilde{\Phi}_{qeq}}{4}}{\cosh\dfrac{z\tilde{\zeta}_{qeq}}{4}}\right]\nabla_\theta\tilde{n} - \left[\frac{kT}{e}\right]^2\frac{\varepsilon_m(\tilde{\zeta}_{qeq} - \tilde{\Phi}_{qeq})}{\eta}\nabla_\theta\tilde{\phi} \tag{75}$$

Here

$$\tilde{\Phi}_{qeq} = \frac{e}{kT}\Phi_{qeq}, \tilde{\zeta}_{qeq} = \frac{e}{kT}\zeta_{qeq}, \tilde{\phi}_{qeq} = \tilde{\Phi}_{eq} + \delta\tilde{\Phi}_{qeq}(\theta),$$

$$\tilde{\zeta}_{qeq} = \tilde{\Phi}_{qeq}|_{x=0} - \tilde{\zeta}_{eq} + \delta\tilde{\zeta}_{qeq}(\theta) \tag{76}$$

are the potential distribution and the ζ potential of the thin polarized quasiequilibrium double layer, which contain field-induced terms: $\delta\Phi_{qeq}$, and $\delta\zeta_{qeq}$, in addition to the field-independent equilibrium terms: Φ_{eq} and ζ_{eq}. It must be emphasized that the capillary-osmotic and the electroosmotic liquid velocity distributions in the thin polarized double layer [first and second addends in the right-hand size of Eq. (75)] were obtained without any restrictions concerning a linear approximation with respect to the applied field. Therefore, this equation may be used for the consideration of nonlinear effects of the double-layer polarization. In particular, the second (electro-osmotic) term may be combined with expressions (31)–(34) to obtain easily the electro-osmotic slip component that is second order in the applied field and, correspondingly, the electro-osmotic component of electrorotation.

For the case of the linear approximation, the quasiequilibrium potential distributions of $\tilde{\Phi}_{qeq} = \tilde{\Phi}_{eq} + \delta\tilde{\Phi}_{qeq}(\theta)$ and $\tilde{\zeta}_{qeq} = \tilde{\zeta}_{eq} + \delta\tilde{\zeta}_{qeq}(\theta)$, which contain field-dependent terms, must be replaced by expressions corresponding to the equilibrium double layer:

$$\tilde{\Phi}_{qeq}, \tilde{\zeta}_{qeq}, \chi_{qeq} \Rightarrow \tilde{\Phi}_{eq}, \tilde{\zeta}_{eq}, \chi_{eq} \tag{77}$$

For brevity, hereafter we omit the subscript "eq" in the designation of potentials and reciprocal thickness of the equilibrium double layer, bearing in mind that $\Phi_{eq} \equiv \Phi, \zeta_{eq} \equiv \zeta, \chi_{eq} \equiv \chi$.

The integral in Eq. (64) can now be solved using for the velocity the sum of the electro-osmotic and the capillary osmotic contributions, Eq. (75), and, for the linear approximation, making the replacement Eq. (77). This leads to

$$\int_a^\infty (zN - C_0^\pm)\vec{v}dr = \frac{3m^\pm z^2 ND^\pm \nabla_\theta \delta\tilde{\phi}}{\chi}\{\pm 2[\exp(\mp z\tilde{\zeta}/2) - 1] + z\tilde{\zeta}\}$$
$$+ 3\frac{3m^\pm zND^\pm \nabla_\theta \delta\tilde{n}}{\chi}\left\{2\left[\exp\left(\mp z\tilde{\zeta}/2\right) - 1\right] - 8\ln\left(\cosh\frac{z\tilde{\zeta}}{4}\right) \pm z\tilde{\zeta}\right\} \tag{78}$$

where

$$m^\pm = \frac{2\varepsilon_m}{3\eta D^\pm}\left(\frac{kT}{ze}\right)^2 \tag{79}$$

Equations (64), (65), and (78), together with Eqs (58) and (59), lead to the following expressions, from which the coefficients K_c and K_d can be obtained:

$$K_c[(1 = W)[1 \pm z\Delta)(R^\pm + 2) - U^\pm] + iW[(1 \pm z\Delta)(R^\pm + 2W + 2) - U^\pm]\pm$$
$$zK_d(R^\pm + 2)(1 + W + iW) = \pm z(R^\pm - 1)(1 + W + iW) \tag{80}$$

where

$$R^\pm = \frac{2G_0^\pm}{zan_0} + 6m^\pm \frac{G_0^\pm}{zan_0} \pm z\frac{z\tilde{\zeta}}{\chi a}\right] \tag{81}$$

$$U^\pm = \frac{48m^\pm}{\chi a}\ln\left[\cosh\frac{z\tilde{\zeta}}{4}\right] \tag{82}$$

The results are

$$K_c = \frac{3z(R^+ + R^-)}{2}\frac{1 + W + iW}{iW^2 A + (1 + W + iW)B} \tag{83}$$

$$K_d = K_{dm} - K_c H = \frac{\gamma^*}{a^3} \tag{84}$$

where A, B, K_{dm} and H are real, frequency-independent coefficients:

$$A = R^+ + R^- + 4 - z\Delta(R^+ - R^-) \tag{85}$$

$$B = (R^+ + 2)(R^- + 2) - U^+ - U^- - (U^+ R^- + U^- R^+)/2 \tag{86}$$

$$K_{dm} = \frac{R^+ + R^- - 2 - z\Delta(R^+ - R^-)}{R^+ + R^- + 4 - z\Delta(R^+ - R^-)} \tag{87}$$

$$H = \frac{(R^+ - R^-)(1 - z^2\Delta^2) - U^+ + U^- + z\Delta(U^+ + U^-)}{zA} \tag{88}$$

Equation (84) has been written in such a way that the first term on the right-hand side is proportional to the "fast" part of the dipole moment, which is free from the influence of field-induced electrolyte concentration variations $\delta n(\omega, \vec{r})$ and, therefore, is frequency independent and in phase with the applied field.

In the frame of the system, Eqs (58), (59) and (64), resulting from applying the approximation of local equilibrium of the thin double layer to the standard model, the process of formation of field-induced concentration variations $\delta n(\omega, \vec{r})$ (or so-called "volume diffusion mechanism") is the only slow factor, determining the frequency dependence of the induced dipole moment. Its frequency dependence arises due to that of $\delta n(\omega, \vec{r})$, expressed by Eq. (58), as well as by means of coupling of the concentration variations and electric fields, expressed by Eq. (64). The second term in Eq. (84), which represents the "slow" part of the induced dipole moment, is proportional (with frequency-independent coefficient H) to the concentration changes $\delta n(\omega, \vec{r})|_{r=a} = K_c$ around the particle, and, correspondingly, is in phase with K_c. Here, the terms "slow" and "fast" refer to the volume diffusion characteristic time of the low-frequency dispersion (α dispersion):

$$\tau_\alpha = a^2/2D_{ef} \tag{89}$$

At low frequencies, when the period of the field is much longer than this characteristic time, the "slow" part of the dipole moment shifts in phase with the field while its amplitude attains its maximum value. On the other hand, at high frequencies, it shifts progressively 90° out of phase, while its amplitude tends to zero.

The "fast" part of the dipole coefficient K_{dm}, being free from the influence of the volume diffusion mechanism and, hence, from the contribution of diffusion flows of ionic components, and from the contribution of capillary osmosis, is determined solely by the distribution of ion flows due to electro-migration and electro-osmotic convection – the factors which were taken into account by Bikerman in his theory of surface conductivity [38]. Correspondingly, K_{dm} may be rewritten in terms of Bikerman's surface conductivity of the diffuse part of the EDL, λ_B, and the electrolyte conductivity k_m:

$$K_{dm} = \lim_{\omega \to \infty} K_d = \frac{\dfrac{2\lambda_B}{a} - k_m}{\dfrac{2\lambda_B}{a} + 2k_m} \tag{90}$$

It is very important to note that this formula simultaneously corresponds to the low-frequency limit of the dipole coefficient obtained from the Maxwell–Wagner–O'Konski theory [Eqs (16) and (17)] by substituting Bikerman's surface conductivity (λ_B) for λ in Eq. (17):

$$K_{dm} = \lim_{\omega \to 0} \frac{\gamma_{MW}}{a^3} \tag{91}$$

$$\gamma_{MW} = a^3 \frac{\varepsilon^*_{pef} - \varepsilon^*_m}{2\varepsilon^*_m + \varepsilon^*_{pef}} \tag{92}$$

$$\varepsilon^*_m = \varepsilon_m - \frac{ik_m}{\omega} \tag{93}$$

$$\varepsilon_{pef}^* = \varepsilon_p - \frac{i}{\omega}\frac{2\lambda_B}{a} \equiv \varepsilon_p - \frac{ik_m}{\omega}2\,Du \tag{94}$$

In Eq. (94), we use the dimensionless number "Du" (Dukhin's number) which plays a fundamental role in the theory of double-layer polarization [39]. It is the ratio between the surface conductivity and the conductivity of the solution and particle radius. For the standard model:

$$Du = \frac{\lambda_B}{k_m a} \tag{95}$$

The following expressions can be used for calculation of the conductivities involved in Eqs (93)–(95):

$$\lambda_B = \frac{z^2 e^2 n_0 a}{kT}(D^+ R^+ + D^- R^-) =$$

$$\frac{ze^2 n_0}{kT}\frac{2}{\chi}\left[D^+(e^{-\tilde{\zeta}/2} - 1)(1 + 3m^+) + D^-(e^{\tilde{\zeta}/2} - 1)(1 + 3m^-) \right] \tag{96a}$$

$$k_m = \frac{z^2 e^2 n_0}{kT}(D^+ + D^-) \tag{96b}$$

B. Wide-Frequency Range Polarizability and Dielectric Dispersion

A general formula for the induced dipole moment will now be described that yields simultaneously the frequency dependence of both dispersion ranges, namely, the α dispersion and the Maxwell–Wagner–O'Konski dispersion. The treatment was recently elaborated, in Ref. 40, for conditions corresponding to the quasiequilibrium polarization of the Stern layer at the isoelectric point (in the absence of an equilibrium diffuse layer), but it has not been applied yet in the frame of an analytical theory of thin double-layer polarization. However, the coincidence [see Eq. (91)] of the high-frequency limit of the dipole coefficient obtained from the analytical theory of the α dispersion, with the low-frequency limit of dipole coefficient obtained by using the theory of Maxwell–Wagner–O'Konski, creates the necessary prerequisites for a simple procedure of derivation of the general formula for the dipole coefficient γ. In order to derive γ, a superposition approximation will be used, whereby we have simply to replace the frequency independent term K_{dm} of Eq. (84) by the Maxwell–Wagner–O'Konski dipole coefficient, $K_{dm \to \gamma_{MW}/a^3}$:

$$\frac{\gamma^*}{a^3} = -K_c H + \frac{\varepsilon_{pef}^* - \varepsilon_m^*}{2\varepsilon_m^* + \varepsilon_{pef}^*} \tag{97}$$

where K_c, H, ε_{pef}^*, and ε_m^* are given, respectively, by the expressions (83), (88), and (93)–(96). The asymptotic validity of the superposition approximation is determined by the strong inequality expressing the very small value of the relaxation time τ_{MW} corresponding to Maxwell–Wagner–O'Konski dispersion, in comparison with τ_α, corresponding to α dispersion, connected with the volume diffusion mechanism:

$$\tau_{MW} \ll \tau_\alpha \tag{98}$$

The superposition approximation does not take into account the factors of the mutual influence of the volume diffusion mechanism inherent in α dispersion and the interplay between displacement and conductivity currents in the process of single-particle polarization, inherent in the Maxwell–Wagner–O'Konski dispersion. However, if Eq. (98) holds, the frequencies at which the volume diffusion mechanism is essential for the particle's polarization ($\omega \approx 1/\tau_\alpha$) happen to be very small, $\omega \ll 1/\tau_{MW}$. Hence, the displacement currents are negligible with respect to conductivity currents. On the other hand, the frequencies at which the displacement currents are comparable with conductivity currents, $\omega \approx /1/\tau_{MW}$, are so large ($\omega \gg 1/\tau_\alpha$) that the volume diffusion mechanism has a negligible influence on the fields and currents. The value of τ_{MW} is, for the majority of typical systems, close to the value of the electrolyte relaxation time [Eq. (11)]:

$$\tau_{MW} = \frac{\varepsilon_p + 2\varepsilon_m}{k_p + 2k_m} = \frac{\varepsilon_m}{k_m} \left(\frac{1 + \dfrac{\varepsilon_p}{2\varepsilon_m}}{1 + \dfrac{k_p}{2k_m}} \right) \approx \frac{\varepsilon_m}{k_m} = \tau_e \tag{99}$$

From Eqs (89), (99), together with Eq. (96) for electrolyte conductivity, Eq. (56) for effective ion diffusivity, and the well-known expression for reciprocal Debye length χ:

$$\chi = \sqrt{\frac{2e^2 z^2 n}{\varepsilon_m kT}} \tag{100}$$

it follows that

$$\frac{\tau_\alpha}{\tau_{MW}} \approx (\chi a)^2 \tag{101}$$

This means that the strong inequality (98) that allows us to use the superposition approximation is closely related to another inequality (63), describing the small thickness of the Debye atmosphere as compared to the particle radius. In such a way, the superposition approximation, leading to the validity of the general formula (97) for the dipole coefficient, happens to be applicable just in the case of the thin double layer, for which both theories – that of the volume diffusion mechanism of α dispersion and that of Maxwell–Wagner–O'Konski dispersion – are applicable.

    To check the accuracy of the general formula, the frequency dependence of the dipole coefficient, calculated with Eq. (97), has been compared with numerical calculations [32]. The results of the comparison for the imaginary part of polarizability, $\text{Im}\{\gamma^*(\omega)\}$, and for the real part of dielectric increment $\text{Re}\{\delta\varepsilon^*(\omega)\}$, calculated by substitution of $\gamma^*(\omega)$ from Eq. (97) into Eq. (8), are represented in Fig. 2. As is shown in this figure, the analytical results are confirmed by numerical calculations, with a relative error of M10% for $\kappa a > 10$, and $< 3\%$ for $\kappa a > 25$ in a frequency range including both volume diffusion and Maxwell–Wagner relaxations.

C. Wide-Frequency Range Electrorotation Spectra

In order to apply our analytical results to the description of electrorotation spectra, in terms of the traditional approach, it is sufficient to substitute $\text{Im}\{\gamma^*(\omega)\}$ from Eq. (97) into Eq. (29), and obtain in such a way the frequency dependence of the angular speed of a particle's electrorotation. Considering the positive values of $\text{Im}\{\gamma^*(\omega)\}$ at low frequencies, the traditional approach leads to the prediction of counterfield low-frequency electrorotation, which contradicts most of the experimental data.

    The calculation of the contribution of the combined movement of the particle and its thin Debye atmosphere (i.e., the contribution of rotational electroosmoic slip) to electro-

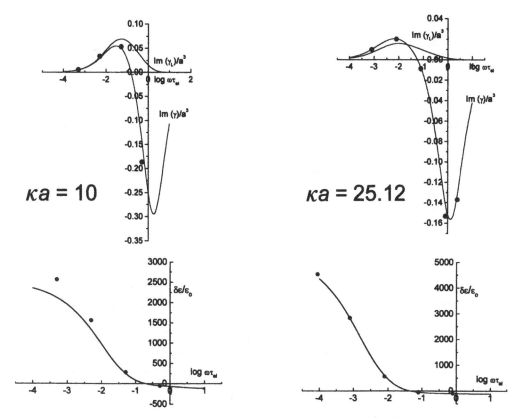

FIG. 2 Comparison between the results (particle polarizability and dielectric increment) of the wide-frequency range generalization [Eq. (97)] of the analytical theory of thin double-layer approximation (lines) with numerical calculations [29] (symbols), for $\kappa a = 10$ (left) and $\kappa a = 25.12$ (right). Polarizability data obtained with the LFDD theory [4] are also included as Im $(\gamma_L)/a^3$ curves. Parameters: $\varepsilon_m = 80\varepsilon_0$; $\varepsilon_p = 2\varepsilon_0$; $\zeta = 100$ mV.

rotation, Eq. (36), it is necessary to calculate the complex amplitude, K^*, of the field-induced variation of the ζ potential. As shown in Refs 4, 27, and 28, the main factor causing the variations in thin double-layer characteristics (for a dielectric particle with a thin double layer and an immobile dense part) is the variation in concentration of the electroneutral electrolyte solution along the particle's surface. This is the system under consideration, since the neutral salt concentration around the particle depends on the polar angle, Eq. (58), which causes an angular dependence of the double-layer thickness and, therefore, of the ζ potential. The dependence of ζ on the concentration n can be obtained from Eq. (73) with taking into account Eqs (76), (100) and the boundary condition expressing the approximate electro-neutrality of quasi-equilibrium EDL:

$$\frac{d\Phi_{qeq}}{dx}\bigg|_{x=0} = -\frac{\sigma}{\varepsilon_m} \tag{102}$$

where σ is the surface charge density of the dense part of the double layer, whereas $-d\Phi_{qeq}/dx|_{x=0}$ corresponds [Eq. (73)] to the surface charge density associated with the quasiequilibrium diffuse double layer.

Equation (102), together with (73) gives

$$\sinh\left[\frac{z\tilde{\zeta}_{qeq}}{2}\right] = \frac{e}{kT}\frac{z\sigma}{2\varepsilon\chi} \tag{103}$$

In this expression, χ can be replaced using Eq. (100), with the virtual system concentration near the particle surface, Eq. (58), $n = n_0(1 + \delta\tilde{n}|_{r=a}) = n_0(1 + K_c\cos\theta)$, while σ should be expressed as a function n using the appropriate isotherm. The simplest one corresponds to a constant surface charge, in which case Eq. (79) gives, to first order in the applied field:

$$\zeta_{qeq} = \zeta_O + \delta\zeta_{qeq} = \zeta_O - \frac{kT}{e}\frac{K_c}{Ea}\tanh\left[\frac{z\tilde{\zeta}_0}{2}\right]\cos\theta \tag{104}$$

From this result, and recalling the definition of K^*, it follows for the complex amplitude of the field-induced variation of the ζ potential:

$$-K^* = \frac{kT}{e}K_c\tanh\left[\frac{z\tilde{\zeta}_0}{2}\right] = \frac{kT}{e}\frac{3z(R^+ + R^-)}{2}\frac{1 + W + iW}{iW^2A + (1 + W + iW)B}\tanh\left[\frac{z\tilde{\zeta}_0}{2}\right] \tag{105}$$

This expression, together with Eqs (36) and (97), determines the contribution of the rotational electro-osmotic slip to the particle electrorotation. Recall that it has been obtained on the basis of the standard model with the simplest adsorption isotherm, based on the independence of the surface charge of the dense part of the EDL on the small variations in electrolyte concentration with respect to its equilibrium value.

Figure 3 shows the frequency dependence of both components of the electrorotation speed, Ω_S and Ω_M, as well as the total angular velocity Ω. It is obvious from the graphs, that the negative contribution of the rotational electro-osmotic speed Ω_S for not too high values of the ζ potential, predominates in the low-frequency range, determining low frequency cofield electrorotation, in spite of the negative sign of Ω_M in the low-frequency range. However, in the frequency range of Maxwell–Wagner dispersion, the contribution of Ω_S is negligible as compared to Ω_M, and the direction of electrorotation is determined in this range by the relationship between the Dukhin number $Du = \lambda_B/k_m a$ and the ratio $\varepsilon_p/\varepsilon_m$:

- Cofield rotation for

$$2Du > \frac{\varepsilon_p}{\varepsilon_m} \tag{106}$$

- Counterfield rotation for:

$$2Du < \frac{\varepsilon_p}{\varepsilon_m} \tag{107}$$

IV. INDUCED DIPOLE MOMENT AND AMPLITUDE OF FIELD-INDUCED ZETA POTENTIAL IN TERMS OF GENERALIZED SCHWARZ MODEL (SURFACE DIFFUSION MODEL)

A. Exact Solution of Electrodiffusion Problem for Particles Covered with Conducting Film in Electrolyte Solution

The first quantitative description of the low-frequency dielectric dispersion of suspensions in electrolyte solution was suggested by Schwarz [41] on the basis of a model now

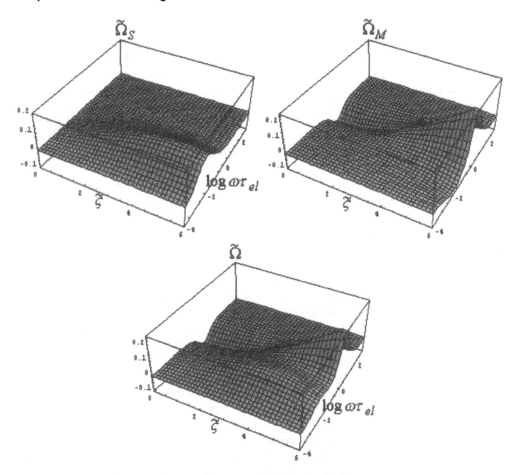

FIG. 3 Frequency and ζ-potential dependence of the angular velocity of electrorotation, Ω (in dimensionless form). Its components $\tilde{\Omega}_S$ and $\tilde{\Omega}_M$ are also shown. Parameters: $\varepsilon_m = 80\varepsilon_0$; $\varepsilon_p = 2\varepsilon_0$; $\zeta = 100$ mV; $a = 200$ nm; $D^+ = D^- = 2 \times 10^{-9}$ m^2/s; $\chi a = 26$; $\eta = 10^{-3}$ kg/m^{2}s.

known as the "surface diffusion model" (SDM). This model considered the colloidal particles covered by a thin surface layer of charges, which were capable of undergoing free migration and diffusion along the particle's surface, without any exchange in normal direction with adjoining electrolyte volume (they could be called *slipping charges*). Such behaviour was attributed [41] to counterions of the EDL on the basis of their high value of electrostatic attractive energy (as compared to thermal energy) near the surface. This makes charge exchange in the normal direction very difficult. However, a detailed consideration [1, 3, 4, 26–33] of the frequency dependence of the polarization of the diffuse part of the EDL, on the basis of the Nernst–Planck–Poisson–Stokes equation system, did not reveal any ionic behavior that could be described by the SDM. Moreover, such a detailed consideration confirmed the existence of a local equilibrium, and, correspondingly, of free exchange of ions between the thin diffuse double layer and the adjacent electrolyte volume. Hence, as was demonstrated in the previous section of this chapter, the field-induced variations of the ion in the diffuse double layer constitute the fast process, followed immediately by the slow one, namely, the field-induced variations in electrolyte concentration in the solution surrounding the particle. In view of the nature of the slow factor, this mechanism of low-frequency polarization was called the "volume diffusion model" (VDM).

The problem now arises of the predominant polarization mechanism in the dense part of the EDL – volume diffusion, surface diffusion, or some intermediate mechanism – and what conditions determine such a predominance. The problem remains open up to now, although some investigations of this complicated topic have been performed and are reviewed in Chapter 2 of this volume. It can be said that we can identify under what circumstances the SDM is not applicable (for instance, in the case of diffuse double-layer polarization), but it is difficult to establish, on a rigorous basis, when it is applicable.

Nevertheless, real systems exist for which the SDM is applicable without any doubt. Composite colloidal particles consisting of a dielectric core surrounded by one or more conducting shells have recently received growing interest [42–49]. Particles covered with either conducting polymers [46–49] or thin metal films [43–45] are of a special significance for applications in experiments and in technology. The promising way to fabricate particles with controlled shell dimensions in the nanometer range employs the stepwise electrostatically controlled adsorption of a wide range of materials, including metallic nanoparticles. The size and shape of these composite particles have been shown to be fully controlled by the original templates [42–49].

The characteristic feature of the polarization of such a system is the free displacement of charge carriers along the surface, and the difficult charge exchange in the normal direction, connected with a high resistance to electrochemical reaction. It is just such a feature that gives rise to the surface-diffusion polarization mechanism, but with several characteristic distinctions (see below), which require further generalizations of the Schwarz model. Application of the well-known patterns of current passing through metal–electrolyte interfaces facilitates the generalization of the model for taking into account the finite rate of charge exchange in the normal direction.

Future progress in nanoengineering of the surface of colloidal particles will be closely related to the parallel development of new thin-layer characterization techniques directly applicable to colloidal dispersions. For example, single-particle light scattering has been recently demonstrated to be capable of resolving the layer thickness growth with nanometer precision [44, 46, 47]. For the case of conducting colloidal particles it is rather attractive to employ electrorotation spectroscopy for the characterization of dielectric and conductive properties, and the nanostructures of the suspended colloidal particles [45–49]. However, the currently available simplified approach of electrorotation of particles with a shell structure does not properly take into account the polarization of a thin layer with electronic conductivity in electrolyte solution. Hence, it would be desirable to develop further the theory of electrorotation including a conducting layer with electronic conductivity to open new areas of application for electrorotation.

The order of magnitude of the characteristic time for both mechanisms – VDM and SDM – is given by Eq. (89). However, whereas in the former case D_{ef} denotes effective diffusivity of the electrolyte solution, determined by Eq. (56), for the latter case one has to replace D_{ef} by the effective diffusivity of slipping charges. For the case of a thin metallic layer, when the role of sleeping charges is played by electrons, then the characteristic time of their corresponding surface diffusion process may be several orders shorter than that characterizing the volume diffusion process.

Considering the range of frequencies exceeding the characteristic frequency provided by the volume diffusion mechanism, we will not take into account any field-induced variations in electrolyte concentration. Consequently, the Laplace equations in both the bulk electroneutral volume and in the dielectric inside the particle are sufficient to describe the volume field distribution:

$$\nabla^2\phi_m = 0, \quad \nabla^2\phi_p = 0 \tag{108}$$

If, however, a thin film with electronic conductivity is present at the particle surface, the high mobility of charges inside this film has to be taken into account when modeling the particle polarization. Generally, the high mobility of the charge carriers may give rise to concentration gradients in the conducting film. In addition, interfacial charge-transfer reactions between the film with electronic conductivity and the electrolyte solution, mediated by an electrochemical reaction, may also contribute to the particle polarization. In the limiting case, when the activation energy of the electrochemical reaction is very high, and, correspondingly, the value of the exchange current i_0 is very low, the tangential electric field induces electron migration along the film, but the normal component of the electrical current directed towards the electrolyte solution is negligibly small. In this case a net current in the film cannot exist, and the migration current has to be compensated for by diffusion flow in the opposite direction. Although in the general case of a finite value of the exchange current full compensation does not fully take place, the total electronic surface current in the film I_s) is assumed to consist of a migration ($-\lambda \cdot grad_\theta\phi_s$) and a diffusion component ($-D_{sef} \cdot grad_\theta\sigma_s$):

$$I_s = -\lambda \cdot grad_\theta\phi_s - Dsef \cdot grad_\theta\sigma_s \tag{109}$$

where λ is the surface conductivity of the film of sleeping charges σ_S is the surface charge density of slipping charge carriers (electrons) given by their number per unit film area,

$$D_{sef} = D_s \frac{\dfrac{\chi_s h}{2}}{\tanh\dfrac{\chi_s h}{2}} \tag{110}$$

is the effective diffusion coefficient of electrons (see Appendix), $D_S = kTu_s$ is the diffusion coefficient of electrons, with u_s as the mobility of electrons, h is the film thickness, $\chi_s = \dfrac{\lambda}{hD_s\varepsilon_s}$ (Debye length of the electronically conductive film), and ε_s is the dielectric constant of the film.

The coefficient:

$$\frac{\dfrac{\chi_s h}{2}}{\tanh\dfrac{\chi_s h}{2}}$$

in Eq. (110) takes into account that the density of electronic charges inside the film is inhomogeneous across the film's thickness due to screening. It gives rise to a locally higher field-induced density of electronic charges compared with its average value of σ_s/h and, correspondingly, D_{sef} is larger than the local value of the electron diffusivity, D_s.

If the exchange current is negligible, and the electronic current I_s is thus constrained within the conducting film, the surface divergence of I_s depends only on the change of σ_s with time. In the more general case, the electron flow due to an electrochemical reaction has to be included in the equation of electron conservation:

$$\text{div}_s \equiv \frac{1}{a\sin\theta}\frac{\partial}{\partial\theta}(\sin\theta \cdot I_s) = -\frac{\partial\sigma_s}{\partial t} - i_{elc} \tag{111}$$

Here, i_{elc} is the current density produced by an interfacial electrochemical reaction. Taking into account that there is no EDL in the dielectric underneath the conducting film, the electric potential at the surface of the core $(\phi_i|_{r=a})$ is equal to the potential of the inner surface of the film $(U_s|_i)$:

$$\phi_i|_{r=a} = U_s|_p \tag{112}$$

However, for writing the boundary conditions at the external surface of the film, one has to consider that a field-induced potential difference $\delta\Psi_1$ (overpotential) between the film and the bulk electrolyte may exist. This potential differnce is caused by possible field-induced charge distributions in the external Debye atmosphere:

$$U_s|_e + \delta\Psi_1 = \phi_m|_{r=a} \tag{113}$$

The potential difference $\delta\Psi_1$ is just the motive force for the electrochemical reaction at the surface permitting the current flow between the electronically conducting film and the ionic conductive solution. The current density of an electrochemical reaction taken in the linear approximation valid for small values of $\delta\Psi_1$ reads, according to the Tafel equation [50]:

$$i_{elc} = \frac{ez}{kT} i_0 \delta\Psi_1 \tag{114}$$

where, i_0, the exchange current density characterizes the effective reciprocal resistivity (Faradaic resistance) of the electrochemical reaction, i.e., in our case, the rate of charge exchange between the film of slipping charges and adjacent electrolyte solution.

The field-induced potential jump $\delta\Psi_1$ can be expressed through the field-induced surface charge density of the external ionic diffuse layer $\delta\sigma_{dif}$ and its differential capacity:

$$\delta\Psi_1 = -\frac{1}{C_{dif}} \delta\sigma_{dif} \tag{115}$$

In Eqs (112) and (113) ϕ_i and ϕ_e were both introduced for $r = a$. This simplification is possible due to the very small film thickness h compared with the particle radius:

$$h \ll a \tag{116}$$

and due to the smooth space dependence of the electrical potential resulting from the solution of the Laplace equations (108).

For the case of a very small film thickness the field-induced potential jump across the film may be also neglected:

$$U_s|_i \approx U_s|_e \tag{117}$$

and, correspondingly, from the latter equation, together with Eqs (112), (113), and (115), it follows:

$$\phi_p|_{r=a} - \phi_m|_{r=a} = -\frac{1}{C_d} \delta\sigma_{dif} \tag{118}$$

The validity of simplification [Eq. (117)] is obvious in the case when the electron concentration inside the film is very low, and the normal component of the field strength inside the thin film is comparable with that inside the dielectric core, with its much larger radius. However, upon increasing the electron concentration inside the film and, correspondingly, increasing the inside screening, the electric field strength inside the film may only reduce, and hence Eqs (117) and (118) are applicable both for small and large electron concentrations inside the thin conducting film.

The difference in electrostatic induction between the two bulk electroneutral phases (the inner dielectric core and the external bulk solution) is given by the sum of the electronic, $\delta\sigma_s$, and ionic, $\delta\sigma_{dif}$, charge densities:

$$\varepsilon_p \frac{\partial\phi_p}{\partial r}\Big|_{r=a} - \varepsilon_m \frac{\partial\phi_m}{\partial r}\Big|_{r=a} = \delta\sigma_{el} + \delta\sigma_{dif} \tag{119}$$

Here, we made specific use of the strong inequalities (30) and (116). The ionic component of the total surface charge may change as a result of the migration of ions, $k_m \frac{\partial\phi_m}{\partial r}\Big|_{r=a}$, and also due to the electrochemical reaction at the interface, i_{elc}. Hence, the condition of ionic charge conservation reads:

$$\frac{\partial\delta\sigma_{dif}}{\partial t} = k_m \frac{\partial\phi_m}{\partial r}\Big|_{r=a} + i_{elc} \tag{120}$$

In acocrdance with Eqs (2), (6), and (108), the electric field-induced distributions of the electric potentials in the particle interior and in the bulk, the field-induced electronic surface charge in the film, and the surface charge density of ionic diffuse layer take the following forms:

$$\phi_p \cdot e^{-i\omega t} = -E_p r \cos\theta \tag{121}$$

$$\phi_m \cdot e^{-i\omega t} = E r \cos\theta + \frac{\gamma^* E}{r^2}\cos\theta \tag{122}$$

$$\sigma_s \cdot e^{-i\omega t} = A_s \cos\theta \tag{123}$$

$$\sigma_{dif} \cdot e^{-i\omega t} = A_{dif} \cos\theta \tag{124}$$

The time- and space-independent coefficients, E_p, γ^*, A_s, and A_{dif}, can be found by substitution of ϕ_p, ϕ_m, $\delta\sigma_s$, and $\delta\sigma_{dif}$, from Eqs (121)–(124) into Eqs (111) and (118)–(120), and taking into account (114) and (115). This leads to the following system of linear algebraic equations:

$$\frac{2\lambda}{a}E_p - \frac{2D_{sef}}{a^2}A_s = i\omega A_s - \frac{ei_0}{kTC_{dif}}A_{dif} \tag{125}$$

$$E_p a - Ea\left(1 - \frac{\gamma^*}{a^3}\right) = \frac{1}{C_{dif}}A_{dif} \tag{126}$$

$$\varepsilon_m E\left(1 + \frac{2\gamma^*}{a^3}\right) - \varepsilon_p E_p = A_s + A_{dif} \tag{127}$$

$$k_m E\left(1 + \frac{2\gamma^*}{a^3}\right) = -i\omega A_{dif} - \frac{ei_0}{kTC_{dif}}A_{dif} \tag{128}$$

The solution of the latter system directly provides the value of γ^*, which determines the traditional (torque) component of electrorotation [Eq. (29)]. In order to determine the value of K^*, appearing in the electro-osmotic component of electrorotation [Eq. (36)], the well-known formula:

$$\tanh\left(\frac{e\zeta}{4kT}\right) = \tanh\left(\frac{\Psi_1}{4kT}\right)\exp(-\chi\delta) \tag{129}$$

is used, which connects the value of the ζ potential with the total potential jump Ψ_1 in the diffuse part of the external Debye layer. Here, χ is the reciprocal Debye length and δ is the thickness of the hydrodynamically immobile part of the diffuse layer. If the frequency sufficiently exceeds $2D_{ef}/a^2$, no concentration polarization occurs [see Eqs (58), (60) and (63)] and, hence, neither the electrolyte concentration nor the Debye length depend on the electric field. In these conditions, we may connect the field-induced variations of the ζ potential, $\delta\zeta$, and of Ψ_1 potential, $\delta\Psi_1$, by the following equation:

$$\tanh\left(\frac{e(\zeta eq + \delta\zeta)}{4kT}\right) = \tanh\left(\frac{e(\Psi_{1eq} + \delta\Psi_1)}{4kT}\right)\exp(-\chi\delta) \tag{130}$$

where ζ_{eq} and Ψ_{1eq} are the field-independent components of ζ and Ψ_1.

Using a linear approximation with respect to the field-induced variations of $\delta\zeta$ and $\delta\Psi_1$, we may rewrite Eq. (130) as

$$\delta\zeta = m\delta\Psi_1 \tag{131}$$

where

$$m = \exp(-\chi\delta)\frac{1 - \tanh^2\left(\dfrac{e\Phi_{1eq}}{4kT}\right)}{1 - \exp(-2\chi\delta)\tanh^2\left(\dfrac{e\Psi_{1eq}}{4kT}\right)} \tag{132}$$

and, in accordance with Eqs (33), (115), (124), and (131), one has for K^*:

$$K^* = -m\frac{1}{C_{dif}}\frac{A_{dif}}{Ea} \tag{133}$$

Next, A_{dif} is expressed in terms of γ^* by means of Eq. (128). For the case under consideration, i.e., when the volume diffusion-induced particle polarization is negligibly small, the electrorotation velocity, Eqs (36), (37), may then be expressed through the only characteristic parameter of particle polarization, namely, the Mosotti coefficient γ^*:

$$\Omega = -\frac{\varepsilon_m E^2}{2\eta}\left\{\frac{\mathrm{Im}\gamma^*}{a^3} - m\frac{k_e}{aC_{dif}}\mathrm{Im}\left[\frac{1}{\omega_{far} - i\omega}\left(1 + \frac{2\gamma^*}{a^3}\right)\left(1 - \frac{\gamma^*}{a^3}\right)\right]\right\} \tag{134}$$

Here and later we denote:

$$\omega_{far} = \frac{ezi_0}{kTC_{dif}}; \quad \tau = \frac{a^2}{2D_{sef}} \tag{135}$$

To obtain an expression for γ^* in a form similar to the simple and well-known formula of a "dielectric sphere in dielectric liquid", we will transform the system of euqations (125)–(128) into only two standard "electrotechnical" surface conditions, one describing the continuity of the normal components of the effective complex electrostatic induction, and the other providing the surface discontinuity of the electric potential. Instead of directly solving the system [Eqs (125)–(128] we will rewrite it in a way that it becomes formally equivalent to the description of a dielectric sphere in a bulk dielectric liquid.

As a first step of our transformation of Eqs (125)–(128), let us introduce a new unknown variable, the total field-induced surface charge δA:

$$\delta A = A_s + A_{dif} \tag{136}$$

δA and A_{dif} are substituted into Eq. (125) instead of A_s and A_{dif}. The latter quantity is excluded from Eq. (125) with the help of Eq. (128), and δA is substituted with the help of

Eq. (127). Taking into account Eqs (121) and (122), we obtain the transformed surface condition of the continuity of the normal components of the effective complex electrostatic induction:

$$\varepsilon_p^{*ef}\nabla_r\phi_i - \varepsilon_m^{*ef}\nabla_r\phi_m = 0 \tag{137}$$

where

$$\varepsilon_p^{*ef} = \frac{\dfrac{2\lambda}{a}\tau + (1 + i\omega\tau)\varepsilon_i}{1 + i\omega\tau + \omega_{far}\tau} \tag{138}$$

$$\varepsilon_{mj}^{*ef} = \frac{k_m}{i\omega + \omega_{far}} + \frac{(1 + i\omega\tau)\varepsilon_m}{1 + i\omega\tau + \omega_{far}\tau} \tag{139}$$

Relation (137) represents the surface condition of the continuity of normal components of the effective complex electrostatic induction of the internal medium, $\mathbf{D}_p^{*ef} = -\varepsilon_p^{*ef}\nabla_r\phi_i$ and the bulk medium $\mathbf{D}_m^{*ef} = -\varepsilon_m^{*ef}\nabla_r\phi_m$ written with, however, effective complex dielectric constants for the internal (ε_i^{*ef}) end bulk (ε_e^{*ef}) phases. The next step in the transformation of the system, Eqs (125)–(128), is directed towards finding an effective surface capacity in order to represent the equation for the field-induced surface jump of the potential in the simple notation:

$$(\phi_p - \phi_m)|_{r=a} = D_n^{*ef}\frac{1}{C_s^{*ef}} \tag{140}$$

Where D_n^{*ef} is the normal component of the effective complex electrostatic induction near the surface. In accordance with Eq. (137), D_n^{*ef} should be continuous at the surface:

$$D_n^{*ef} = -\varepsilon_p^{*ef}\nabla_p\phi_p|_{r=a} = -\varepsilon_m^{*ef}\nabla_n\phi_m|_{r=a} \tag{141}$$

Multiplying both sides of Eq. (128) by $\dfrac{\varepsilon_m^{*ef}}{k_m}$ and taking into account Eq. (122), we obtain:

$$D_n^{*ef} = -\varepsilon_m^{*ef}\frac{\omega_{far} + i\omega}{k_m}A_{dif} \tag{142}$$

Inserting Eq. (142) in Eq. (126), and comparing the result with Eq. (140) it follows:

$$C_s^{*ef} = \varepsilon_m^{*ef}\frac{\omega_{far} + i\omega}{k_m}C_{dif} \tag{143}$$

The system of the harmonic functions for the inner [Eq. (121)] and external [Eq. (122)] electric potential distributions together with the surface conditions [Eqs (137) and (140)] formally coincides with the description of the polarization in a homogeneous external field of a spherical particle with the dielectric constant ε_p^{*ef}, surrounded by an infinite thin film with a finite value of the capacity C_s^{*ef} per unit area of the particle surface immersed in a bulk medium with the dielectric constant ε_m^{*ef}. The effective parameters ε_p^{*ef}, ε_p^{*ef}, and C_s^{*ef} were obtained by means of an algebraic transformation of the system [Eqs (125)–(128)]. Hence, the more complicated initial model of a particle with a conducting thin layer (taking into account the diffusion of charge carriers inside it, with charge exchange between the film and electrolyte volume) has been brought to the well-known system of "dielectric in dielectric" with a thin layer of homogeneous surface capacity (see, e.g., Ref. 51). The existence of this formal analogy has the significant advantage that one can make use of the well-established procedures for the calculation of complex polarizability and complex amplitude of the field-induced variations in ζ potential. It is just sufficient to use the effective complex dielectric constants ε_p^{*ef}, ε_m^{*ef}, and C_s^{*ef}, the effective surface capacity, in the expression for the polarizability:

$$\frac{\gamma^*}{a^3} = -\frac{\left(\varepsilon_m^{*ef} - \varepsilon_p^{*ef}\right) + \dfrac{\varepsilon_m^{*ef}\varepsilon_p^{*ef}}{C_s^{*ef}a}}{\left(2\varepsilon_e^{*ef} + \varepsilon_i^{*ef}\right) + 2\dfrac{\varepsilon_m^{*ef}\varepsilon_p^{*ef}}{C_s^{*ef}a}} \tag{144}$$

B. Dispersion of Particle Polarizability, Suspension Admittance and Electrorotation

The characteristic examples of the frequency dependences of dimensionless angular velocity of electrorotation, given by the expression (134), of its traditional component, $-\text{Im}\{\gamma\}/a^3$, given by Eq (29) [the first term in brackets in Eq. (134)], and the frequency dependences of the disperse particles contribution to suspension permittivity [Eq. (8)] and conductivity [Eq. (9)], calculated with the expression (144) for the dipole coefficient, are represented in Figs 4–6.

Figures 4–6 show the frequency dependences for large values of diffusivity of slipping charge curriers (much larger than ionic diffusivity), which is characteristic of electronic conductivity. Apparently, from a comparison of the plots of the first lines of each of the figures, the difference between the full angular velocity of electrorotation (upper curve in every plot) and the contribution of its traditional component (lower curve) is significant only when the frequency is small as compared to the reciprocal electrolyte relaxation time, and when the equilibrium ζ potential is also small enough. This is true provided that the diffuse layer has a small capacitance, and, hence, given values of field-induced variations in the diffuse layer charge density σ_{dif} provoke large enough field-induced variations in the ζ potential. If the equilibrium ζ is large, and (or) the external field frequency is of the order of $1/\tau_{el}$ or larger, the contribution of the electro-osmotic component of electrorotation is small, and its traditional component accounts for the full angular velocity of electrorotation.

Note that the frequency dependence of $\text{Im}\{\gamma^*\}/a^3$ has a width of about three orders of magnitude for not very large values of film conductivity and, correspondingly, small and medium values of Du number (Figs 4–6). This frequency interval spreads through a significantly wider frequency range – about five orders of magnitude – under the condition of good conductive film, or large Du number (Figs 4 and 5). As a result, the frequency range, within the limits of which $\text{Im}\{\gamma^*\}/a^3$ has a noticeable value, may cover frequencies on both sides of $1/\tau_{el}$, far away from the latter (Figs 4 and 6). In accordance to the inequalities (23)–(26), this corresponds to the condition when both effects, LFDD and HFCD, manifest simultaneously (Figs 4 and 6).

The influence of the inner dielectric constant of the core particle can be ascertained by comparison of the plots presented in Fig. 4 ($\varepsilon_i = 2$), Fig. 5 ($\varepsilon_i = 25$), and Fig. 6 ($\varepsilon_i = 1000$). For small inner dielectric constant, when $\varepsilon_i/\varepsilon_e \ll 2$ Du, the frequency dependence of polarizability is represented by the almost antisymmetrical curve with maximum and minimum [see plots 11(lower curve), 12, and 13 in Fig. 4, and 11 and 12 in Fig. 5]. The left half of those antisymmetrical curves corresponds to surface diffusion relaxation: it provokes the increase in dipole coefficient (dimensionless polarizability) from $-1/2$ to 1 by decreasing the influence of diffusion of slipping carriers and increasing the surface current from zero to its maximum possible value. The right half of the curves corresponds to

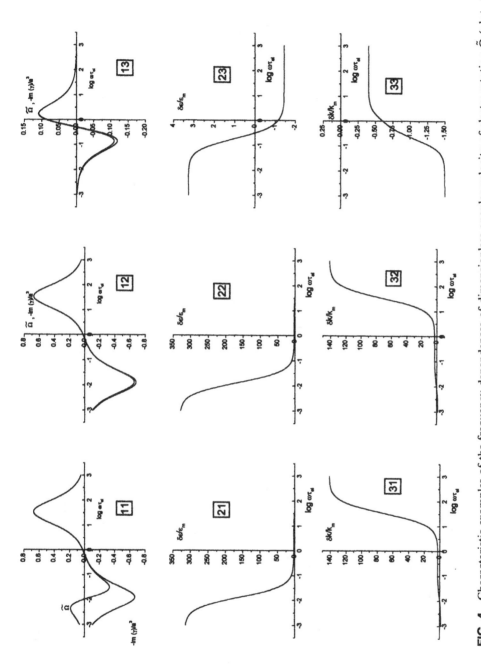

FIG. 4 Characteristic examples of the frequency dependences of dimensionless angular velocity of electrorotation $\tilde{\Omega}$ (plots 11, 12, 13) of its traditional component, $-\mathrm{Im}(\gamma)/a^3$ (same plots), and of the dispersed particles contributions to the permittivity (plots 21, 22, 23) and conductivity (plots 31, 32, 33) of the suspensions. Parameters: left column: $\Psi_1 = 25$ mV; $\lambda = 3.9 \times 10^{-7}$ S. Middle column: $\Psi_1 = 200$ mV; $\lambda = 3.9 \times 10^{-7}$ S. Right column: $\Psi_1 = 25$ mV; $\lambda = 3.0 \times 10^{-9}$ S. Common parameters: $\varepsilon_m = 80\varepsilon_0$; $\varepsilon_p = 2\varepsilon_0$; $a = 5\,\mu$m; $D^* = D^- = 2 \times 10^{-9}$ m^2/s; $D_s = 10^{-5}$ m^2/s; $\chi a = 237$; $\eta = 10^{-3}$ kg/m^{2}s; $\delta = 0$.

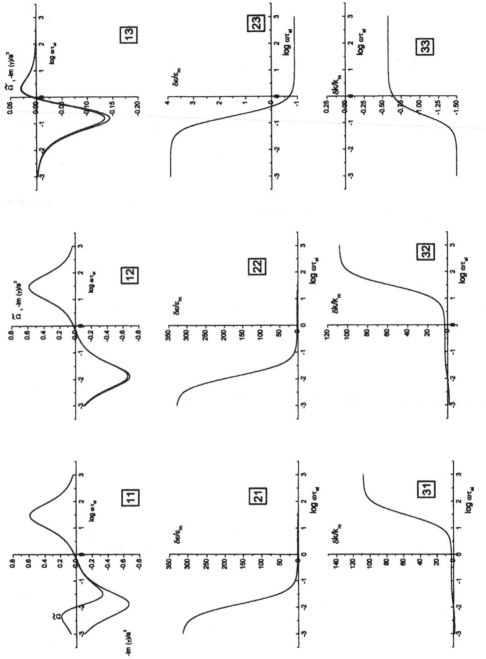

FIG. 5 Same as Fig. 4, but for $\varepsilon_p = 25\,\varepsilon_0$.

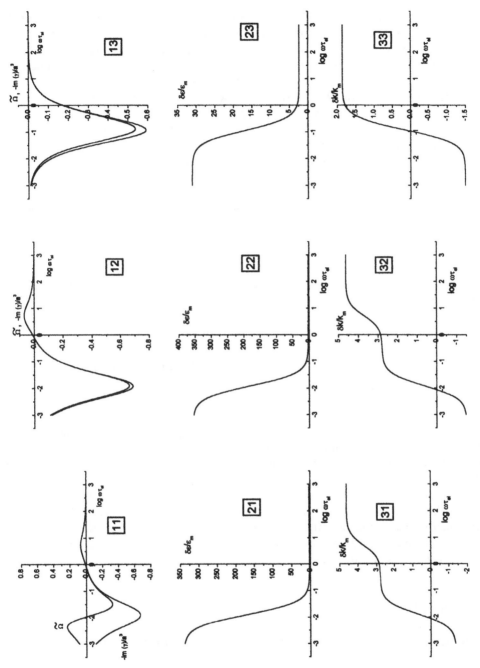

FIG. 6   Same as Fig. 4, but for $\varepsilon_p = 1000\ \varepsilon_0$.

Maxwell–Wagner relaxation: in this case, the dipole coefficient decreases because of the smaller influence of conductivity currents as compared to dielectric displacement currents. The decrease occurs also from 1 down to almost its minimum possible value, $-1/2$, because the very large ratio of the effective conductivities of particle and medium corresponds to a small ratio between their dielectric constants (under the condition $\varepsilon_i/\varepsilon_e \ll 2$ Du).

It follows from above that, if the dielectric constant of the particle increases, the amplitude of the right peak decreases, causing a characteristic distortion of the frequency dependence of polarizability, as at plot 13 of Fig. 5 and plots 11 and 12 of Fig. 6. Plot 13 of Fig. 6 demonstrates the complete disappearance of the right peak as a result of the combination of a small value of surface conductivity and a large value of particle permittivity.

The essential influence of the value of the exchange current density i_0, as is obvious from the expressions for the effective parameters, Eqs (138), (139), (142), and (143), is determined by the ratio between ω_{far} [see Eq. (135)] and the external field frequency:

$$\frac{\omega_{\text{far}}}{\omega} \geq 1$$

Substituting in Eq. (135) the well-known expression for the differential capacity of the diffuse layer:

$$C_{\text{dif}} = \varepsilon_m \chi \cosh\left[\frac{e\Psi_{10}}{2kT}\right]$$

and taking $\chi = 10^{-8}$ m^{-1}, $\varepsilon_m = 80\varepsilon_0$, $\Psi_{10} = 50$ mV, and a comparatively high value of $i_0 = 1$ A m^{-2}, the following estimation of ω_{far} may be given:

$$\omega_{\text{far}} \approx 380 \text{ Hz}$$

Such a small order of value for ω_{far} means that one may expect a slight influence of the finite exchange range between the slipping charges and electrolyte solution at least for the case in which the surface layer is a conductor of the first kind (charge carriers are electrons and holes).

The calculations show that the LFDD is the measurable effect most sensitive to the value of the exchange current density i_0. Figure 7 shows the influence of i_0 on the frequency dependence of the dielectric increment of LFDD. It follows from these results that, for the set of parameters used, a noticeable influence of the exchange processes on LFDD exists only if i_0 exceeds 10 A m^{-2}.

V. APPENDIX

Let us express the total tangential current (surface current) I_s in the thin electronic conductive film by integrating the local tangential electronic current density $i_{s\theta}$:

$$I_s = \int_a^{a+h} i_{s\theta} dr \tag{A.1}$$

The local tangential electronic current density includes two components, namely, migration and diffusion:

$$i_s = -k_s \nabla U - D_s \nabla \rho_s \tag{A.2}$$

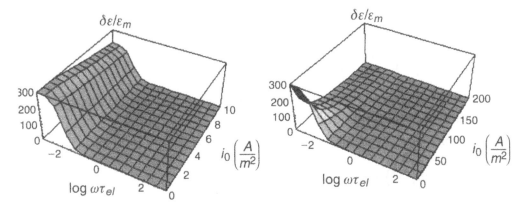

FIG. 7 Influence of the value of exchange current density i_0 on the dielectric dispersion in terms of the generalized Schwarz model. Parameters: see Fig. 4.

where k_s and D_s are the local values of the conductivity and the diffusion coefficient of electrons inside the film, and ρ_s is the local spatial density of electronic charges. Its associated charge density is $c_s = -\rho_s/e$.

For the single case described here, the field-independent electron concentration c_{seq}, the diffusion coefficient D_s, and the local conductivity are supposed to be constants throughout the film. Nevertheless, even for this model, the final field-induced spatial electronic charge density and, hence, the density of the local tangential electronic current due to screening from the adjacent double layer happens to be inhomogeneous in space. As a consequence, the integration of Eq. (A.1) is not a simple procedure.

A convenient way to consider the migration and diffusion current in the region of the film where the Debye screening mechanism influences the charge distribution (see, e.g., Ref. 10) is to represent the local electric potential as a superposition of

$$U_s = \phi_s + \Phi_s \tag{A.3}$$

where the second term, which may be called as a quasiequilibrium or Boltzmann term, is connected with the distribution of electronic charge according to Boltzmann's formula:

$$c_s = c_{s0} \exp \frac{e\Phi_s}{kT} \tag{A.4}$$

where c_{s0} is the concentration of electrons inn the neutral region of the film.

Assuming that, inside the electronic conductive film the only mobile charge carriers are the electrons and that the charge density of any fixed ionic groups inside the film is constant and equal to $\rho_{s0} = ec_{s0}$ (as required by electroneutrality), we may state that the Boltzmann prefactor c_{s0} is spatially constant and field independent.

Correspondingly, a weak field-induced electronic charge density leading to deviation from electroneutrality inside the film can be described in a linear approximation as

$$\delta\rho_s = \rho_s - \rho_{s0} \approx \rho_{s0} \frac{e\Phi_s}{kT} \tag{A.5}$$

By using Poisson's equation, Eq. (A.5) provides a Debye equation for the quasiequilibrium component of the electric potential Φ_s:

$$\nabla^2\Phi_s = \chi_s^2\Phi_s \tag{A.6}$$

where

$$\chi_s = \sqrt{\frac{e^2 c_{s0}}{kT\varepsilon_s}} = \sqrt{\frac{k_s}{\varepsilon_s D_s}} = \sqrt{\frac{\lambda}{\varepsilon_s D_s h}} \tag{A.7}$$

is the inverse Debye screening length of the electronically conducting film. Here, $\lambda = k_s h$, and k_s is the electronic conductivity ($k_s = \rho_s \dfrac{e}{kT} D_s$).

The solution of the Debye equation (A.6) inside the film is conveniently given by

$$\Phi_s = A \sinh\left(\frac{\chi_s x}{2}\right) + B \cosh\left(\frac{\chi_s x}{2}\right); \quad \frac{h}{2} \geq x \geq \frac{h}{2} \tag{A.8}$$

The first term in Eq. (A.8) is antisymmetric, and thus does not contribute to the integral of field-induced functions of type (A.1). Hence, for the calculation of I_s we need not take into account that term in Eq. (A.8):

$$\Phi_s = B \cosh\left(\frac{\chi_s x}{2}\right) \tag{A.9}$$

The integration constant B is connected with the surface density of the field-induced electronic charge σ_s by:

$$\sigma_s = \int_{-\frac{h}{2}}^{\frac{h}{2}} \delta\rho_s dx = \rho_s 0 \int_{-\frac{h}{2}}^{\frac{h}{2}} \frac{e\Phi}{kT} dx = 2B\rho_{s0} \frac{e}{kT} \frac{1}{\chi_s} \sinh\left(\frac{\chi_s h}{2}\right) = 2B\varepsilon_s \chi_s \sinh\left(\frac{\chi_s h}{2}\right) \tag{A.10}$$

Substituting of Eqs (A.3) and (A.4) into Eq. (A.2) transforms the equation for the electronic current density into a form which contains only the field-induced multiplier $\nabla\phi_s$:

$$i_s = -k_s \nabla\phi_s \tag{A.11}$$

The value of the current density i_s inside the film is restricted by the value of the current density in the external electrolyte solution. This is true even in the case where both the exchange current and the electron concentration are very large. From this it follows that $-\nabla\phi_s$ is of the order of the external field strength E. This means that, due to the small film thickness, we may neglect the variation of ϕ_{el} and, correspondingly, the variation in the value of $\nabla_\theta\phi_{el}$ across the thin film. It has to be noted that this variation is, however, not small along the surface of the large particle. The substitution of $i_{el\theta} = -kel\nabla_\theta\phi_{el}$ in Eq. (A.1) leads, upon taking into account the above considerations, to

$$I = -\lambda \nabla_\theta \phi_s \tag{A.12}$$

Using the continuity of the electric potential at the inner surface of the film, Eq. (15), and relations (A3), (A9), and (A10), we obtain:

$$I_s = -\lambda \cdot \mathrm{grad}_\theta \phi_s - D_s \frac{\dfrac{\chi_s h}{2}}{\tanh \dfrac{\chi_s h}{2}} \cdot \mathrm{grad}_\theta \sigma_s \tag{A.13}$$

VI. SUMMARY AND CONCLUSIONS

In this chapter we have described the classification of the spectrum of particle polariz-ability and the associated spectra of suspension admittance and particle electrorotation, for particles suspended in an electrolyte solution. This has been performed on the basis of a natural scale of frequencies connected with the reciprocal electrolyte relaxation time. The analysis, which is free of any model assumptions about the structure of the particle and its double layer, allows us, in particular, to predict the phenomenon of high-frequency dis-persion of suspension conductivity, which is characterized by a very large value for the specific increment of conductivity. The predicted phenomenon is in a certain sense the high-frequency analog of the well-known LFDD, characterized by a very large value for the specific increment of dielectric permittivity. Both phenomena are connected with the imaginary part of the induced dipole moment, and are located on both sides of the reciprocal relaxation time, far from it on a logarithmic scale.

The general expression for the angular velocity of electrorotation, including the traditional (torque) component and a new one (rotational electro-osmotic slip), was obtained by using a rather simple procedure.

The main steps and simplifications involved in the application of the standard model of the thin diffuse EDL to the solution of the problem of polarization of nonconducting charged spherical particles by uniform harmonic external fields have been described. The well-known results of the theory of thin diffuse double-layer polarization in low-frequency alternating fields have been generalized to describe simultaneously the low-frequency and Maxwell–Wagner frequency ranges. The generalized expression obtained for the induced dipole moment is confirmed by comparison with numerical results of the DeLacey and White approach [32]. Expressions have been derived describing the spectrum of dielectric dispersion of suspensions, and of electrorotation of a spherical particle with thin diffuse double layer for a wide frequency range. The standard model leads to the conclusion that the diffusion of ions adjacent to electroneutral electrolyte solution around the particle is the main slow process, controlling the low-frequency dependence of double-layer polar-ization (volume diffusion model).

A quite different approach is represented by the so-called surface diffusion model (SDM) (Schwarz model), which assumes the existence on the particle surface of special kinds of charges (slipping charges), which are capable of undergoing free migration and diffusion along the surface, whereas their migration and diffusion in the normal direc-tion is impossible or strongly hampered. This behavior is impossible for ions in the diffuse part of the EDL and its occurrence for ions in the dense part has never been confirmed up to now. The applicability of the SDM is no doubt for new interesting systems – particles covered by thin metallic films or thin polymeric films with electronic conductivity. In such systems, the role of slipping charges is played by electrons or holes. Hence, the SDM has been generalized to be applied to such new systems, by taking into account the inner screening in the layer of slipping charge, and the exchange current density, as a parameter determining the normal current between the film and enjoined solution. Analytical expressions have been obtained for the frequency dependences of particle polarizability, suspension admittance, and angular velocity of electrorotation over a broad frequency range, covering both low and high frequencies in the natural frequency scale. The existence of a high-frequency dispersion of suspension conductivity is predicted, and it is found to be characterized by a very large value for the specific conductivity increment when the mobility in slipping charges is comparable to that of electrons in metals.

ACKNOWLEDGMENTS

Financial support from INTAS (Proj. No. UA/95-0165) is gratefully acknowledged. One of us (V.N.S.) also acknowledges the Spanish Ministry of Education for his sabbatical fellowship SAB1998-0120.

REFERENCES

1. SS Dukhin, VN Shklov. Dielectric Phenomena and the Double Layer in Disperse Systems and Polyelectrolytes. New York: Wiley, 1974.
2. CT O'Konski. J Phys Chem 64: 605, 1960.
3. VN Shilov, SS Dukhin. Kolloidn Zh 32: 293, 1970.
4. VN Shilov, SS Dukhin. Kolloidn Zh 32: 117, 1970.
5. WM Arnold, HP Schwan, U Zimmermann. J Phys Chem 91: 5093, 1987.
6. C Grosse, KR Foster. J Phys Chem 91: 3073, 1987.
7. XB Wang, Y Huang, R Hölzel, JPH Burt, R Pethig. J Phys D. Appl Phys 26: 312, 1993.
8. WM Arnold, U Zimmermann. Z Naturforsch 37C: 908, 1982.
9. VPh Pastushenko, PI Kuzmin, YuA Chizmadshev. Stud Biophys 110: 51, 1985.
10. E Donath, VPh Pastushenko, M Egger. Bioelectrotech Bioenerg 23: 337, 1990.
11. T Müller, L Küchler, G Fuhr, T Schnelle, A Sokirko. Silvae Genet 42: 311, 1993.
12. XF Zhou, GH Marx, R Pethig, JM Eastwood Biochim Biophys Acta 1245: 85, 1995.
13. VL Sukhotukov, U Zimmerman J Membr Biol 153: 161, 1996.
14. J Gimsa, T Müller, T Schnelle, G Fuhr. Biophys J 71: 498, 1996.
15. JPH Brut, KL Chan, D Dawson, A Patron, R Pethig. Ann Biol Clin 54: 253, 1996.
16. XB Wang, Y Huang, PRC Gascoyne, FF Becker, R Hölzel, R Pethig. Biochim Biophys Acta 330: 1193, 1994.
17. M Egger, E Donath. Biophys J 68: 364, 1995.
18. Y Huang, XB Wang, R Hölzel, FF Becker, PRC Gascoyne. Phys Med Biol 40: 1789, 1995.
19. HP Schwan, G Schwarz, J Maczuk, H Pauly. J Phys Chem 66:2626, 1962.
20. R Barchini, DA Saville. J Colloid Interface Sci 173: 86, 1995.
21. AV Delgado, FJ Arroyo, F Conzalez-Caballero, VN Shilov, YB Borkovskaya. Colloids Surfaces A 140: 139, 1998.
22. H Maier. Biophys J 73: 1617, 1997.
23. R Georgieva, B Neu, VN Shilov, E Knippel, A Budde, R Latza, E Donath, H Kiesewetter, H Bäumler. Biophys J 74: 2114, 1998.
24. C Grosse, VN Shilov. J Phys Chem 100: 1771, 1996.
25. U Zimmerman, C Grosse, VN Shilov. J Colloid Interface Sci 159: 229, 1999.
26. JTG Overbeek. Kolloidchem Beih 59:287, 1943.
27. SS Dukhin. In: BV Derjguin, ed. Research in Surface Forces. New York: Plenum Press, 1971, pp 312–341.
28. SS Dukhin, VN Shilov. Kolloidn Zh 31: 706, 1969.
29. RW O'Brien, LR White. J Chem Soc, Faraday Trans 2 74: 1607, 1978.
30. SS Dukhin, NM Semenikhin. Kolloidn Zh 32: 366, 1970.
31. M Fixman. J Chem Phys 72: 5177, 1980.
32. EHB DeLacey, LR White. J Chem Soc, Faraday Trans 2 77: 2007, 1981.
33. WC Chew, PN Sen. J Chem Phys 77: 4683, 1982.
34. J Lyklema, SS Dukhin, VN Shilov. J Electroanal Chem Interfacial Electrochem 1: 143, 1983.
35. VN Shilov, OA Shramko, TS Simonova. Kolloidn Zh 54: 208, 1992.
36. VN Shilov, NI Zharkih, JB Borkovskaya. Kolloidn Zh 47: 927, 1985.
37. M Smoluchowski. In: E Graez ed. Handbuch der Electrizitat und der Magnetismus. vol. 2, Leipzig; Barth, p. 366.

38. JJ Bikerman. J Phys Chem 46: 725, 1942.
39. J Lyklema. Fundamentals of Interface and Colloid Science. New York and London: Academic Press, 1995, ch. 4.
40. C Grosse, M Tirado, W Pieper, R Pottel. J Colloid Interface Sci 205: 26, 1998.
41. G Schwarz. J Phys Chem 66: 2636, 1962.
42. JH Fendler, FC Meldrun. Adv Mater 7: 607, 1995.
43. SJ Oldenburg, RD Averitt, SL Westcott, NJ Halas. Chem Phys Lett 288: 243, 1998.
44. J Moriguchi, Y Teroaka, S Kagava, JH Fendler. Chem Mater 11: 1603, 1999.
45. MD Musick, CD Keating, MH Keefe, MJ Natan. Chem Mater 9: 1499, 1999.
46. MD Butterworth, SA Bell, SP Armes, AW Simpson. J Colloid Interface Sci 183: 91, 1996.
47. C Perruchot, MM Chehimi, M Delamar, F Fievet. Surface Interface Anal 6: 689, 1998.
48. MA Khan, SP Armes. Langmuir 15: 3469, 1999.
49. SF Lascelles, SP Armes. J Mater Chem 7: 1339, 1997.
50. PT Kissinger, CR Preddy, RE Shoup, WR Heineman. In: PT Kissinger, WR Heineman eds, Laboratory Techniques in Electroanalytical Chemistry. New York: Marcel Dekker, 1984, ch. 2.
51. T Hanai. In: P Sherman, ed. London and New York: Academic Press, 1998, pp. 313–415.

13

Characterization of Particles and Biological Cells by AC Electrokinetics

JAN GIMSA   Humboldt Universität zu Berlin, Berlin, Germany

I. INTRODUCTION

In the growing research into the interaction of AC electric fields with colloidal particles and biological cells, AC electrokinetic methods [1–15] are increasingly replacing classical impedance methods [5, 16–18]. The reason is the higher resolution of the electrokinetic methods for the electrical parameters of single objects, resulting from their different measuring principles. While impedance methods register the direct electric response of a suspension to an applied electric field, e.g., an alternating current, AC electrokinetic methods register force effects which arise from the interaction of the induced polarization charges with the inducing field. In the field of biology, interesting applications of electrokinetic methods are the determination, screening, or pursuit of changes in membrane capacitance, membrane conductance, and cytoplasmic properties [1, 9, 19–32].

In impedance the current through a suspension increases with frequency. This is caused by a decrease in the effective impedances of both particles and suspension medium due to the increasing current contributions of the permittivities of all the constituents. For a certain structure the switch from a current contribution predominantly based on its conductivity property at low frequencies to a contribution predominantly based on its permittivity property at higher frequencies occurs in a characteristic frequency range around the dispersion frequency. The different dispersion frequencies of the various constituents lead to a stepwise but continuous decrease in the suspension impedance with increasing frequency. The different properties of particle structures and the medium cause a different frequency dependence of the effective impedances of particle and medium. This results in a frequency-dependent redistribution of the current balance through and around the particles. Typically, for biological cells suspended in a medium of low conductivity, two structure-related redistribution processes occur in the kilohertz and the megahertz range, respectively. The first process is capacitive membrane bridging, i.e., changing the current from predominantly flowing around the cell to predominantly flowing through its highly conductive cytoplasm. Then, with increasing frequency the currents related to the bulk permittivities supersede the frequency-independent conductivity currents. This again leads to a predominant current flow around the cell due to the permittivity difference between the medium and cytoplasm. Although the latter process may cause a stronger current redistribution than the membrane bridging it is often neglected in impedance

---

Dedicated to my former teacher Prof. Dr. Roland Glaser on the event of his retirement in March, 2000.

modeling. One of the reasons is that the strength of a dispersion is related to the impedance change relative to its value at DC or at very low frequencies. For this reason, it seems to be more appropriate to relate the strength of a dispersion process to the height of the impedance step relative to zero, the impedance value at infinite frequency.

Whereas the impedance of a suspension depends, though in a complex way, on the sum of the current contributions through and around the suspended particles, the forces or torques in AC electrokinetics depend on the difference between particle and medium properties. Consequently, AC electrokinetic forces or torques may decrease as well as increase with frequency, thereby, reflecting the strength of the current redistributions. Summarizing these considerations, it can be concluded that particle and medium properties in impedance are qualitatively reflected in an integrative manner, whereas in the AC electrokinetic methods in a differential one. Nevertheless, impedance and the electrokinetic methods detect the same polarization processes and their dispersions and thus generally yield the same information on the dielectric particle properties.

There is a long-standing tradition in characterizing the dielectric properties of biological material. The dispersion processes leading to characteristic frequency-dependent properties of biological material are better understood than those of the various artificial colloidal particles. The impedance and permittivity of biological material drop over some frequency decades by several orders of magnitude [33, 34]. Two different approaches were chosen for the classification of the measured dispersions. They were either sorted according to their physical nature, e.g., as structural Maxwell–Wagner and molecular Debye dispersions, or according to the frequency range where they occur. A classification according to frequencies originally yielded a scaling where the dispersions were assigned to α, and β, and γ ranges with increasing frequencies. At that time it was assumed that well-defined processes were responsible in a certain dispersion range which could be assigned to certain biological structures [33–40]. Later, the ranges were subdivided, e.g., the β range into $\beta1$ and $\beta2$. Furthermore, processes based on a certain mechanism, which were originally assigned to a specific dispersion range, could be observed in another range, e.g., γ dispersions of large molecules or α dispersions of small molecules in the β-dispersion range.

Despite these difficulties, especially for suspended biological cells, a general classification is possible. Below a few kilohertz in the α-dispersion range many different processes, such as electrode processes, hydrodynamic relaxations of electro-osmotically induced convections within the measuring chamber and around the particles, as well as particle electrophoresis, may influence the measurements. These phenomena are hard to separate and concurrently influence the suspension medium and particles [2, 41–43]. In the β-dispersion range from a few kilohertz to some 100 MHz the main components are the aforementioned structural dispersions. A cell suspended in a medium of low conductivity may exhibit two extreme cases of polarizability. At low frequencies, when the membrane effectively insulates the cell, its polarizability is much lower than that of the medium. In the intermediate frequency band, after capacitive membrane bridging, conductivity effects dominate and the high cytoplasmic polarizability may strongly exceed that of the medium. Finally, at very high frequencies, permittivities dominate and the cell polarizability may again fall below that of the medium. The latter dispersion, resulting from a polarization according to the bulk conductivity properties to a polarization based on the bulk permittivity properties is qualitatively different from the dispersions of the polarization of the cytoplasmic membrane and that of internal membrane systems. Although some authors [44] consider the dispersions of the cytoplasmic and the internal membrane structures as $\beta1$ and $\beta2$ dispersions, respectively, it would be preferable to stress their different qualities by using $\beta1$ and $\beta2$ for the membrane and the bulk conductivity dispersions, respectively.

The different membrane structures can then be designated as $\beta 1'$, $\beta 1''$, etc. The structural dispersions can be further influenced by other processes such as the surface conductance [45–47] or membrane transport processes [48, 49]. In addition, Debye dispersions of cytoplasmatic proteins occur in the β range. For impedance, it has been known for a long time that these processes are masked by the stronger structural dispersions [50]. Only more recently has it become possible to resolve them by AC electrokinetic methods [8]. In the γ range, above some 100-MHz dissociation–association relaxations or Debye dispersions of small molecules or charged groups as well as the dispersion of bound water (in the range 1–10 GHz) and that of free water (above 10 GHz) occur [33, 51].

A number of reviews on AC electrokinetic phenomena is available [11, 31, 33, 52]. This chapter aims to provide an introduction to AC electrokinetics and its relation to classicial impedance. The interrelations of the various electrokinetic methods are described on the basis of a simple model. It allows for the derivation of expressions for electro-orientation, electrodeformation, dielectrophoresis (DP), and electrorotation (ER) and considers particles and cells of a general ellipsoidal shape covered by a thin, low conductive shell. A number of experimental results are presented, together with an introduction to the latest developments.

II. RELATION OF IMPEDANCE AND AC ELECTROKINETIC EFFECTS

A simple notion of the impedance of a suspension is presented by the scheme in Fig. 1.

This model permits a simplified but qualitatively correct comparison of impedance and AC electrokinetic effects. For simplicity, Fig. 1 only considers a small suspension element which consists of one-half of a single particle and the surrounding suspension. It is assumed that the particle is mirror symmetrical with respect to a plane oriented perpendicular to the field direction. For symmetry reasons such a plane is equipotential. An AC passing through the suspension may either flow around the particle, through the external solution elements ext 1 and ext 2, or pass through the external element ext 3 and the particle. In this model both current paths end at the equipotential plane which is at the reference potential Ψ_{ref}. The impedance of a complete particle and its surrounding solution is twice that of the circuit in Fig. 1, and the overall impedance of the suspension is then given by a Kirchhoff meshwork of a multitude of such circuits [53]. The sub-

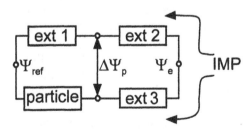

FIG. 1 Diagram for half a particle and its surrounding medium, demonstrating the different principles of impedance (IMP) and the AC electrokinetic effects. One of the elements describes the particle properties; three elements (ext 1, ext 2, and ext 3) represent the surrounding medium. Ψ_{ref}, Ψ_e, and $\Delta\Psi_p$ are the reference potential at the symmetry plane, the applied potential, and the potential difference in between the pole at the particle's surface and a reference in the external medium, respectively.

division of the external solution current path providing a reference in the external medium is assumed to be at a distance from the symmetry plane equal to the particle's radius. It is assumed that the potential at this point is frequency independent and exhibits no phase shift to the applied potential Ψ_e. $\Delta\Psi_p$ reflects the balance of medium and particle polarizability and is related to the mechanical work conducted by AC electrokinetic effects. $\Delta\Psi_p$ reflects the energy change of the particle–medium system, e.g., in DP when the particle is displaced by the external medium or vice versa. Thus, if the particle were to be replaced by the suspension medium or if it consisted of such a medium, $\Delta\Psi_p$ and accordingly all forces would vanish.

For the case of a homogeneous particle and a homogeneous external medium all elements of Fig. 1 can be described by parallel resistor–capacitor (RC) circuits [11, 53, 54]. These circuits reflect the dispersion relation [cf. Eq (A6)]. Since all external medium elements exhibit the same frequency behavior, the frequency behavior, the frequency dependence of the current redistribution between the top and the bottom current path is determined by the frequency dependence of the particle's effective impedance. It is noteworthy that the scheme of Fig. 1 resembles a Wheatstone bridge, the impedance of which is given by the voltage–current characteristics of its feed signal. In contrast, the AC electrokinetic effects are related to the bridge signal, $\Delta\Psi_p$. For a summary of the relationship between impedance and AC electrokinetics see Refs 54 and 55.

III. AC ELECTROKINETIC EFFECTS AND METHODS

The frequency and external conductivity dependence of orientation, deformation, dielectrophoretically or traveling wave-induced translational movement, aggregation, or rotation of the particles or cells can be exploited for particle characterization by AC electrokinetic methods. Different AC electrokinetic methods detect this dependence either on a suspension or on single objects. Apart from aggregation or pearl chain formation which are based on particle–particle interaction, each suspension method is complemented by a single-particle method. Generally, the various methods exhibit certain advantages and disadvantages. For example, the suspension methods detect many particles at once and may achieve a high statistical significance at short measuring times. Further, methods such as electro-optics can be applied to suspensions of submicroscopic objects. Electro-optics detects turbidity changes in a suspension arising from particle reorientations [5, 15]. In contrast, observations of the orientation of single particles, which can be held in place, e.g., by a laser tweezer [14], are restricted to the microscopic particle range. Nonetheless, a clear advantage of these methods is that the geometry and structural features of the object under observation are well known. This is a prerequisite for correct modeling and deduction of particle properties. On the other hand, when AC electrokinetic suspension methods or impedance detect, e.g., the broadening of a certain dispersion process, no unambiguous ascription to either the properties of each single particle or to the scattering of individual particle properties within a population is possible.

DP and ER analyze the translation and rotation of single cells in an inhomogeneous and rotating external field, respectively. The different motions in DP and ER depend on the different spatial and temporal properties of the field determining the interaction with the induced dipole moment. Like the deformational force in homogeneous AC fields, DP is proportional to the real part of the induced dipole moment [56–59]. Both effects can be employed for dielectric cell or particle characterization. Nevertheless, deformation is also used to explore the viscoelastic properties of cells [56, 58, 60]. Such experiments can be conducted by microscopic observation of a single cells [56] or by deformational diffracto-metry on suspensions [60]. The DP translation arises from the imbalance of the forces acting on the two hemiellipsoids in an inhomogeneous field. Consequently, particles or cells that would be elongated or compressed under certain conditions move towards or away from regions of high field [8, 11, 13, 20, 61–66]. Frequency-dependent changes in the DP force are mediated by dispersions of the particle's polarizability relative to that of the medium. In ER the rotating field induces an induced dipole moment which rotates at the angular frequency of the field. Any dispersion process causes a spatial phase shift of the external field vector and the dipole moment, giving rise to a torque which causes individual particle rotation. The torque, and therefore particle rotation, is at maximum if the relaxa-tion time of the dispersion process and external field frequency match. According to their mechanisms DP and ER reflect the sole real and imaginary parts of the induced dipole moment, respectively. Measurements of DP and ER can be conducted by direct micro-scopic observation or video processing [66–69]. Recently developed light-scattering meth-ods allow for the detection of DP (dielectrophoretic phase analysis light scattering, DPALS) and ER (ER light scattering, ERLS) in suspensions (see below; [70–72]).

The orientation of particles can be microscopically observed on individual particles in free suspension [3, 12, 73]. Alternatively, single particles can be held in place by a laser tweezer [14, 69]. When the turbidity or static light scattering of a suspension is registered, AC field-induced reorientations of nonspherical objects change the optical properties of the suspension. The respective methods are known as electro-optics or electrical birefrin-gence [5, 10, 74].

Induced mutual attraction of cells is exploited in biotechnology for their alignment in pearl chains before fusion [75]. Nevertheless, registration of the frequency dependence of the attraction force can also be used for cell characterization [76]. Microstructured electrode chambers allow for the generation of strongly inhomogeneous fields. In such fields even submicroscopic particles can be trapped or particle aggregates can be formed [72, 77, 78]. These effects cannot only be exploited for particle handling but also for characterization when the frequency dependence, e.g., of the trapping efficiency is ana-lyzed. Fluorescence markers may facilitate observation of submicroscopic particles. A special way to detect the frequency dependence of DP is to register turbidity changes in the gap between two electrode chips, each of which carries repetitive electrode structures [79]. In such a setup, positive DP attracting the objects towards the electrodes reduces the turbidity, whereas negative DP repelling the objects from the electrode surfaces leads to a turbidity increase. Flow chambers with a similar electrode setup operated at positive DP can also be used to filter out and analyze intact biological cells, e.g., from drinking water or homogenized meat [80–82].

Field-flow methods are chromatographic-like elution techniques in which a differ-ential retention of particles is caused by an external force field, e.g., medium cross flow, a thermal gradient, gravitation, etc. [83]. Recently, DP was introduced as a new kind of force field to the variety of fields already applied in field-flow fractionation [84]. Accordingly, DP field-flow fractionation does not mainly aim at dielectric property ana-lysis but at particle and cell separation according to their dielectric properties. The method is based on the DP positioning of the objects within the suspension flow velocity profile established through a special chamber. The chamber walls carry repetitive DP electrode structures and the mean particle distance to the walls or the strength of the particle interaction with the walls is determined by the DP force experienced by the objects.

Particle translation in traveling-wave DP (TWDP) is actually based on the ER effect in a repetitive, comb-shaped electrode structure driven by progressively phase-shifted signals [85, 86]. Besides the transition induced by the traveling field component, the particles experience a DP attraction or lift force evolving from the field inhomogeneities above the electrode structure. Besides microscopic observation and video processing a special laser Doppler method (TWDP-PALS) allows detection of the effect in suspensions (see below; [87]). The combination of TWDP with field-flow fractionation can also be considered. In this case even a two-dimensional separation of particle and cell populations seems possible when the field-induced traveling direction is oriented perpendicularly to the suspension flow.

IV. AC FIELD-INDUCED FORCE ACTING ON PARTICLES AND CELLS

Figure 2 presents a simple notion of how field-induced forces evolve. The two limiting cases of particle polarization can be respectively pictured by an air bubble in water (top) and a water droplet in air (middle). Interestingly, depending on frequency, biological cells may exhibit the two extreme cases of polarizability. Being insulated by the membrane at frequencies below the membrane dispersion, cell polarizability is usually very low (top). Above the membrane dispersion range, although still below the bulk con-

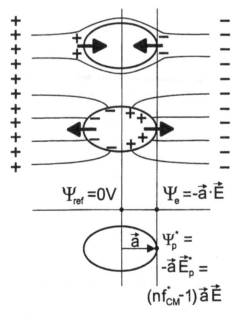

$$\Psi_{ref} = 0\,V \qquad \Psi_e = -\vec{a} \cdot \vec{E}$$

$$\vec{a} \qquad \Psi_p^* =$$

$$-\vec{a}\,\vec{E}_p^* =$$

$$(nf_{CM}^* - 1)\,\vec{a}\,\vec{E}$$

FIG. 2 Diagram of the field distributions and evolving forces for the two limiting cases of the polarization of an ellipsoidal particle. Charges are located at the particle/medium interface. Top: The particle's polarizability is much lower than that of the medium; Middle: the inverse case; Bottom: the induced potentials at three different sites: on the symmetry plane within the particle or in the external solution, a reference potential of $\Psi_{ref} = 0$ V is assumed. In the external solution at distance $|\vec{a}|$ from the symmetry plane the potential is Ψ_e. The complex, frequency-dependent potential Ψ_p^* at the pole site of the particle's surface is related to the effective local field $\vec{E}_p^*$ within the particle, which in turn can be expressed by the depolarizing coefficient n and the Clausius–Mossotti factor f_{CM}^*. (Please note that complex parameters are marked by an asterisk.)

ductivity dispersion, cell polarizability is dominated by the polarizability of the cyto-plasm. This can be much higher than the polarizability of the medium when cells are suspended at low conductivity. The charges depicted are an example of a DC or an instantaneous moment in an AC field. In AC mode the charges change their signs with the frequency of the field. At the bottom of Fig. 1 the induced potentials at three different sites are considered (see legend).

The reference potential at the symmetry plane within the particle or in the external solution would be 0 V, e.g., if the particle were to be localized in the center of two plane electrodes. Nevertheless, at any other place between such electrodes the offset potential would be the same for all points within the symmetry plane. Therefore, it is no limitation to assume Ψ_{ref} to be 0 V. In the external solution at distance $|\vec{a}|$ from the symmetry plane the potential is Ψ_e. The complex, frequency-dependent potential Ψ_p^* at the pole site of the particle's surface is related to the effective local field $\vec{E}_p$ within the particle which in turn can be expressed by the depolarizing coefficient n and the Clausius–Mossotti factor f_{CM}^* (see Appendix A). All AC electrokinetic effects are based on the difference between the potentials that would exist at the pole site of the particle's surface in the absence (Ψ_e) and the presence (Ψ_p^*) of the particle. This difference ($\Delta\Psi_p = \Psi_p^* - \Psi_e$) has already been intro-duced in Fig. 1. Please note that for a sphere the depolarizing coefficient $n = 1/3$ and the limiting values of the Clausius–Mossotti factor are $-1.5 \leq f_{CM^*} \leq 3$.

V. PARTICLES OF GENERAL ELLIPSOIDAL SHAPE WITH THIN, LOW CONDUCTANCE LAYER

In Fig. 2 "effective" values for the polarizability and the local field were assumed. These effective values are identical to the actual values for homogeneous particles. Nevertheless, colloidal and especially biological particles feature a strong compartmentalization and are covered by thin membrane layers of properties which are strikingly different from those of the bulk media. In impedance and in AC electrokinetics, multishell spherical, cylindrical, and ellipsoidal models were developed with the focus on biological cells [13, 18, 54, 63, 88–91]. A sphere or an ellipsoid covered by a single membraneous shell features the typical polarizability properties of a biological cell and can be considered as a standard model. The thin cytoplasmic membrane and its ion barrier function are the reasons for the two most striking frequency-dependent changes in cell polarizability which are expressed by two strong dispersions leading from one status to the other. These dispersions are reflected in the frequency dependence of the force effects. Curiously, the general solution for the polarization of the standard model of a biological cell, a single-shell ellipsoid, had already been derived before the biological work on the meteorological problem of dust particles carrying a water layer [92].

Generally, the calculation of the frequency-dependent polarization and the force effects on shelled spherical or ellipsoidal models starts with Maxwell's stress tensor [93, 94], or simpler with Laplace's equation [9, 54, 63, 90, 91, 95, 96]. In the Laplace solution a homogeneous ellipsoid or the homogeneous core of a shelled ellipsoid always exhibits a constant field. For an explicit solution of the single-shell ellipsoid a confocal shell has to be assumed to describe the ellipsoidal surfaces of the shell within a single co-ordinate system. This standard approach is in contradiction to the actual biological situation of a thin lipid membrane of constant thickness. Consequently, this model may lead to significant errors when the object deviates from a spherical shape [54]. Probably because of the shielding effects this error is reduced when the shell itself is highly polarizable, as for some artificial colloids or in the case of the meteorological problem mentioned above. For these cases, the reader may refer to the literature cited above.

A new alternative approach, which is more appropriate for the biological case, is presented in Appendix A. It assumes a body of general ellipsoidal shape consisting of a homogeneous core which is covered by a low-conductance layer of constant thickness. The model extends a notion of Maxwell that he developed for a shelled sphere to a shelled general ellipsoid (Fig. 3). According to Maxwell, for any given frequency for a shelled sphere a homogeneous equivalent sphere can be found that exhibits a constant local field and possesses the same external dipole field as the shelled one [92] (see also Refs 11, 54, and 97). Consequently, the shelled and the equivalent spheres experience the same induced forces. The derivation is based on three ideas: First, the existence of a Maxwellian equivalent body with a constant local field. Second, a shell impedance high enough to avoid significant lateral currents within the shell. Third, the introduction of influential radii which are equivalent to the depolarizing factors but allow for a special finite element "ansatz" [54].

The derivation starts from the Laplace solution for a homogeneous ellipsoid in a homogeneous field. To calculate the constant local field within a homogeneous or an equivalent ellipsoid, three field components along the three principal axes of the ellipsoid must be taken into account. These local field components can be obtained from the three principal axes and the respective induced potentials at the three poles of the ellipsoid's surface (see Fig. 3). A special finite element ansatz is employed to calculate the potentials [54]. Along each principal axis it assumes prismatic elements, consisting of the three media of core, shell, and exterior which possess infinitely small but constant cross-sectional areas (Fig. 4). The length of the core and shell elements are given by the object geometry. The lengths of the external medium elements are defined by the influential radii, a parameter previously introduced [53, 54]. The influential radii do not depend on frequency and are directly related to the electrostatic depolarizing factors. When they are normalized to their respective principal axis they reflect the maximum of the effective local field amplification factor (compared to a vacuum object). From the impedances of the finite elements the potentials at the ellipsoid's surface poles can easily be determined [Fig. 4; Eq. (A.5)]. The impedances of the three elements are given by their permittivities and conductivities [Eq. (A.6)].

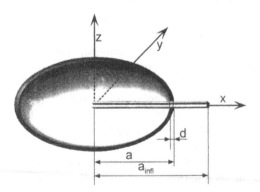

FIG. 3 Single-shell model of the general ellipsoidal shape with constant shell thickness d. The principal axes a, b, and c are oriented in the x, y, and z directions, respectively; a_{infl} is the influential radius along axis a.

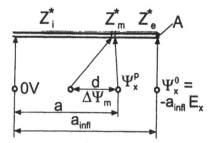

FIG. 4 Sketch of the prismatic volume element of Fig. 3. A, d, a, and a_{infl} stand for cross-sectional area, membrane thickness, the principal axis, and influential radius along the a axis. E_x, Ψ_x^p, Ψ_x^0 and $\Delta\Psi_m$ are the absolute value of the external field strength, and the actual potentials at the surface, the maximum potential at the surface, and the induced transmembrane potential, respectively. For symmetry reasons the induced potential in the center can be assumed as 0 V. The complex impedances Z^* of the internal medium, membrane layer, and external medium are designated by i, m, and e, respectively.

Finally, from the principal axes, the influential radii, and the impedances of the elements a new expression for the Clausius–Mossotti factor is derived [Eq. (A.9)] which immediately leads to the induced dipole moment [Eq. (A.1)]. For the description of a homogeneous ellipsoid or after capacitive bridging of the membrane layer of a single-shell particle the element Z_x^m of Eq. (A.9) can be canceled out. While the induced dipole moment obtained for a homogeneous ellipsoid is identical to the Laplace solution, in the single-shell case it is slightly different. A comparison between the new model with a constant shell thickness with the classical version with a confocal shell is given in Ref. 54 for single-shell spheroids. The comparison showed that the new model was superior for cell-sized objects with a low-conductive layer. On the other hand, it is clear that the calculation of the potential Ψ_x^p applying the finite element ansatz (Fig. 4) is only correct when the layer is "electrically thin" and tangential currents within the layer are negligible (for a consideration see Ref. (63)). Nevertheless, the introduction of the influential radius notion allows for the definition of the geometry of the impedance of the external medium element. This is a prerequisite for describing the frequency dependence of Ψ_x^p by Kirchhoff's rules using RC pairs for each impedance element of Fig. 4. By skipping certain elements from the RC model in a given range, simpler models are obtained which in turn can be used to derive simplified characteristic equations for characteristic spectra points (cf. Fig. 5; [54]).

VI. INTERRELATIONS OF THE AC ELECTROKINETIC EFFECTS

Regardless of the strong interrelations between impedance and diverse AC electrokinetic methods, one may gain the impression that these are often not fully recognized. One of the reasons might be the complexity of the existing theories which are often derived with the focus on a special experimental situation [9, 11, 17, 18, 55, 74, 90, 91, 93, 95, 96, 98, 99]. Nevertheless, several approaches towards a unified theory exist [11, 53–54, 63].

This chapter focusses on the AC electrokinetic effects and does not consider impedance. An overview of AC electrokinetic characterization method is given in Table 1. Expressions for electro-orientation, electro-deformation, DP, and ER have already been

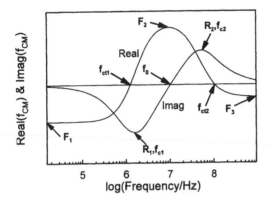

FIG. 5 Schematic DP and ER spectra of a biological cell consisting of a conductive core and a thin insulating membrane drawn as the real and imaginary parts of the Clausius–Mossotti factor (f_{CM}). Positive values of the DP curve indicate elongation or attraction to high field areas, negative values compression or repulsion from high field areas in electrodeformation and DP.

given in different forms and combinations by other authors [3, 5, 11, 12, 56, 66, 91, 95, 96, 99, 100]. Nevertheless, the combination of the dipole approach and component notation for field and induced dipole moment allows for an easy derivation of expressions for all phenomena (Appendix B). In the following, the phenomena and their interrelations will be discussed.

Electro-orientation of nonspherical single objects can either be microscopically observed in free suspension [3, 4, 12, 73, 101] or in a laser trap [14, 69] or can be detected as turbidity changes of a suspension (electro-optics) [5, 10, 15]. Two kinds of information can be obtained from electro-orientational spectra: the critical frequencies of turnover,

TABLE 1 AC Electrokinetic Characterization Methods

Effect	Direct observation, individual particles	Suspension methods
DP	Microscopic registration by eye or video processing	DP phase analysis light scattering (DPALS)
		DP field-flow fractionation
Deformation	Microscopic registration by eye or video processing	Electrodeformational diffractometry
ER	Microscopic registration by eye or video processing	ER light scattering (ERLS)
Orientation	Micro-observation in: • free suspension • laser trap	Electro-optics Electric birefringence
Collection and trapping	Micro-observation of: • pearl chains • (fluorescent) aggregate formation	Blocking or clearing of light path DP filtering, DP columns
Traveling-wave DP (TWDP)	Microscopic registration by eye or video processing	TWDP-PALS TWDP field-flow fractionation

and the preferred axis of orientation for the frequency bands within the turnover frequencies. Two different arguments are used to explain orientation, energy minimization, and the different torques induced around the three principal axes of the ellipsoid [4, 12]. In linear AC fields, the torque criterion results in orientation of the axis of the highest real part of the Clausius–Mossotti factor in the field direction [Appendix B, Eq. (B.6)].

The deformational force in linear AC fields is proportional to the real part of the Clausius–Mossotti factor [Eq (A.3) and (A.9)] [56–58]. In the context of the dipole ansatz, the expressions for electrodeformation in homogeneous fields and DP in weakly inhomogeneous fields are the same. The DP force results from an imbalance of the forces experienced by the two hemiellipsoids, leading to a particle translation towards areas of higher and lower field when the relative polarizability of the particle is higher or lower than that of the medium. In a homogeneous field the same forces elongate or compress the object.

In ER a circularly polarized, rotating field induces a rotating dipole moment. In the case of dispersion a time-independent spatial phase shift of the external field vector and the induced dipole moment occurs. Because of the angle between the two vectors a torque is generated which is proportional to the cross-product of the two vectors (B.2) and (B.11)]. The resulting individual rotation of particles of cells may occur in or against the sense of field rotation [1, 63, 102]. The frequency and external conductivity dependencies of DP and ER are especially suitable for recalculating dielectric properties since these methods independently access the sole real and imaginary parts of the induced dipole moment, respectively. Moreover, the mutual interdependency of the real and imaginary parts over the Kramers–Kronig relation can be exploited to test dielectric cell and particle models. Given a certain geometry, model parameters can be calculated from DP or ER data. Nevertheless, the model obtained can only be correct if it consistently explains DP and ER data [61, 63].

When experiments are conducted with one of the above methods the various effects cannnot usually be fully separated from one another. Generally, a frequency-dependent mutual attraction of the objects under investigation is observed as a disturbing effect. This attraction can be understood as the DP of one object in the local field inhomogeneity generated by the other object and vice versa. In ER chambers the field inhomogeneities in the vicinity of the electrodes also generate DP translations. Interestingly, ER was first observed in DP experiments on pairs of cells [103] when their connecting line had a particular orientation with respect to the field [104]. An orientation of 45° was ideal for one cell of the pair to experience the rotating field component generated by the relaxation of the polarization of the other cell and vice versa. As a result, both cells of a pair were spinning [104]. In both methods, DP and ER orientation and deformation may occur. To avoid complications in data interpretation of microscopic investigations the reorientation of cells can be hindered, e.g., hydrodynamically by the vicinity of the microscopic slide [8]. Nevertheless, for freely suspended cells, e.g., in ER light scattering, reorientation is a disturbing effect [70].

Traveling-wave DP of particles or cells was based on an idea analogous to the transition from a rotating to a linear motor [85, 105]. A traveling-wave field is generated over comb-shaped microelectrodes which are driven by progressively phase-shifted signals. In such a field, as in the rotating field of ER, the induced charges at the particles may exhibit a phase lag at certain propagation velocities of the field. Due to field propagation the resulting forces induce particle or cell translation. Analogous to ER, translation may be induced in or against the propagation direction of the field, depending on the dispersion process causing the phase lag [85, 86]. Similarly to traveling-wave DP, pumping of liquids can be induced when fluid interfaces or anisotropic media are polarized [106, 107].

Pumping is also possible for media which are homogeneous in the first place when the anisotropy is caused by temperature gradient generated by the applied field itself [87, 108]. Nevertheless, for traveling-wave electrode systems it is obvious that inhomogeneous field components must exist in the vicinity of the gaps between two neighboring electrodes. These components induce DP which may repel particles or cells from, or attract them to, the electrodes.

VII. SPECTRA AND THEIR REGISTRATION

Figure 5 depicts typical DP (or deformational force) and ER spectra for a spherical single-shell model of a biological cell suspended in a low-conductivity medium. Each of the two spectra is well described by five points: the DP spectrum is characterized by three different force plateaus (F_1–F_3) acting in different frequency ranges and by two critical frequencies (f_{ct1}, f_{ct2}) at which the DP force ceases. The two peaks of the ER spectrum are related to the two changes in the DP force from F_1 to F_2 and from F_2 to F_3, respectively. In the ER spectrum these dispersions are expressed as rotation peaks at the characteristic frequencies, f_{c1} and f_{c2}, respectively. At these frequencies the rotation speed of the object against or with the rotation direction of the field reaches the maxima R_1 and R_2, respectively. The rotation ceases at f_0.

It is clear that the spectra characteristics depend on the electrical and geometrical properties of the object as well as on the properties of the external medium (cf. Eq (A.9)]. At external conductivities higher than the internal one, the relative position of the DP plateaus F_2 and F_3 and the direction of the ER peak may be inverted. When the real and imaginary parts of the Clausius–Mossotti factor are plotted in the complex plane, the two dispersion processes result in two arcs [53, 109]. These arcs would have a perfect circular shape when the spectra consisted of well-separated Lorentzian functions. Nevertheless, in practice, the spectra are often modulated by underlying dispersion processes. For example, these processes may result from internal membrane systems [24, 28, 61, 110, 111], the dispersion of large cytoplasmatic molecules [8], lipid headgroup orientation [112], or transport processes within the plasma membrane [49, 113]. The spectra are further influenced by the orientation of a given object within the field (see Appendix B).

Several approaches have been tried to derive simplified characteristic equations for the characteristic points of the spectra [1, 54, 61]. This is a complex problem, especially for nonspherical objects. The correspondence of the finite element ansatz to a RC lumped model allows for simplifications at the RC level before the derivation of characteristic equations (cf. Appendix A). Different elements describing model properties with negligible contribution to the overall behavior in a given frequency range can be canceled out, e.g., the capacitors under DC or the resistors at very high frequencies (for details see Ref. 54). This approach, for the first time, allows for the derivation of all characteristic equations for a single-shell spheroid.

Figure 6 gives theoretical DP and ER spectra for the three orientations of typical ellipsoidal cells. For calculations, Eq. (B.7) and the parameters of chicken red cells were used. The model had a membrane thickness of 8 nm. Microscopically three principal axes of $a = 7.7$ μm, $b = 4$ μm, and $c = 1.85$ μm were measured. The conductivities and relative permittivities of the internal, membrane, and external medium were assumed to be 0.15, 0, and 0.01 S/m as well as 50, 10, and 80, respectively. Note that according to Eq. (B.7), cell orientation is such that the axis of highest polarizability is oriented in the field direction. Therefore, in a linear DP field a single axis is oriented in the field direction. In a circular

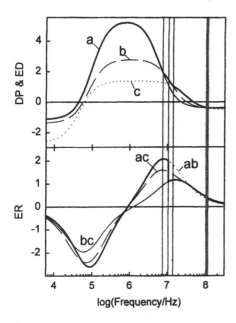

FIG. 6 DP, electrodeformation (ED), and ER spectra for chicken red cells given as real or imaginary parts of the Clausius–Mossotti factor. The bold lines are the predicted experimental spectra taking reorientation into account. Around 100-MHz orientation alters within a narrow frequency band (cf. Fig. 8). Top: three DP spectra for three different orientations; the principal axis orientated in the field direction is marked. Bottom: ER spectra for three pairs of axes laying in the plane of field rotation.

ER field the two axes of highest polarizability are oriented. From the theoretical spectra of Fig. 6 one can get the impression that, experimentally, for biological cells reorientation is a more disturbing effect in ER than in DP.

For the characterization of particle and cell properties it is not inevitable to measure complete frequency spectra. Actually, it is easier to determine characteristic points of the spectra, e.g., in dependence on the external conductivity or to follow their time-dependent changes. For this a variety of approaches exists. Most easy is the measurement of the critical frequencies of DP. It can be conducted by adjusting the DP field generator to a frequency at which the object's movement ceases [61]. A similar approach is very interesting for submicroscopic particles, such as viruses [77, 78, 114, 115]. For such objects the balance of DP accumulation at electrode surfaces or in niches of low field makes their microscopic observation feasible. Observability can be further facilitated by fluorescence markers.

Characteristic peak frequencies of ER can be determined by more sophisticated compensation methods [116]. A prerequisite for detecting the peak frequencies by compensation is a symmetrical shape of the peaks over a logarithmic frequency axis. To detect the peak frequency, two fields having different rotation senses and frequencies, e.g., of f and $f/4$, respectively, are alternately applied. At an alternation frequency of, e.g., 100 Hz and a key ratio of 1:1 the inertia of the particles results in a smooth rotation. The rotation ceases if the logarithmic center frequency $f/2$ and the peak frequency match. After determination of this frequency the peak rotation speed can be directly determined by switching the field to $f/2$. This method is especially useful for following rapid changes of cell or particle properties [23, 29, 30]. In principle, for two ER peaks of opposite signs but similar amplitude, it is also possible to determine the frequency of zero rotation by compensation [111]. To measure complete spectra by compensation a reference and a measuring field are alternately applied at a certain key ratio [117, 118]. Ceasing of rotation is obtained by variation of the key ratio. When the reference and the measuring frequency are located in the same peak or in different peaks of the same sign, the rotation sense of the field must be altered together with the frequency. The same principle can be applied in DP levitation [119]. Scanning of the ER spectrum by a compensation method seems to be especially interesting in ERLS for detecting the relative rotation sense of different ER peaks [118].

Of course, the results of compensation methods must not always be translated into DP or ER frequency spectra but can as well be directly interpreted, e.g., by characteristic equations. This is demonstrated in Fig. 7 for human red cells. The data were obtained by compensating the generator frequency to the critical frequencies of DP (top of Fig. 7) and by compensating for the first characteristic frequency of ER before measuring the rotation speed at that frequency (bottom of Fig. 7), respectively. The figure presents the dependence of the two critical frequencies of DP over medium conductivity and the changes of the antifield ER peak after addition of different ionophor concentrations (for details see Refs 8 and 30, repectively). The theoretical curves were calculated as outlined in the appendices. Membrane capacitance and specific conductance were assumed to be 10^{-2} F/m^2 and 480 S/m^2, respectively. The cells were modeled as spheroids with a radius of 3.6 μm and a third axis of 1.8 μm. To calculate the theoretical curve for the critical frequencies of DP, dispersing cytoplasmic properties were assumed. The dispersion frequency was 15 MHz at a distribution coefficient of 0.2. The dispersion mediated the transitions of the cytoplasmic permittivity and conductivity from 162 to 50 and from 0.4 to 0.535 S/m, respectively. To generate the "spider web" at the bottom of Fig. 7 the cytoplasmic dispersion was neglected. The external and internal permittivities were assumed to be 80 and 50, respectively. The cytoplasmic conductivity was varied as is shown in the figure.

For most cells, the critical frequencies of DP form a nose-shaped function which consists of two branches describing the conductivity dependence of the first (bottom branch) and the second (top branch) critical frequency, respectively. Above a certain external conductivity only negative DP is observed and the two branches join. For cells it was shown that this plot is especially sensitive to parameter variations [8]. In ER the Lorentzian peak generated by a single dispersion process is well defined by its characteristic frequency and amplitude. The idea of the spider-web like presentation is to allow for an efficient two-dimensional presentation of the dependence of the two well-separated ER peaks on two independent parameters [30, 31]. The presentation was obtained by plotting the trajectory of amplitude and frequency of the antifield peak by varying one parameter while keeping the other fixed. The "web" obtained is a helpful tool in interpreting, e.g., time-dependent cell parameter changes followed by the peak compensation method [29]. It can be seen that the membrane conductance has a value of about 480 S/m^2 and is fairly independent of the cytoplasmic conductivity.

When the axis orientation of a cell is plotted in dependence on the external conductivity, a function is obtained that is similar in shape to the critical DP frequency function (Fig. 7). Figure 8 shows an example for chicken erythrocytes which possesses a shape similar to a general ellipsoid. For calculations Eq. (B.6) was used (cf. Fig. 6). The parameters were the same as in Fig. 6.

Usually, cell or particle orientation is observed in linear fields. According to Eq. (B.6) the criterion for the orientation of a certain axis in parallel to the field is the max-

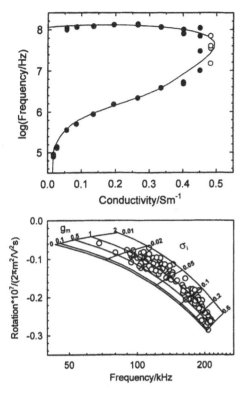

FIG. 7 Compensation methods applied to human red cells. For parameters of the theoretical curves see text. Top: critical frequencies of DP over medium conductivity; frequencies at which only part of the cell population exhibited a transition from positive to negative DP are marked by hollow points. Bottom: frequency and rotation speed at the antifield ER peak of red cells under the influence of the ionophor nystatin at a medium conductivity of 12.5 mS/m. A spider-web like presentation of the theoretical peak dependence on cytoplasmic conductivity (σ_i in S/m) and specific membrane conductance (g_m in kS/m^2) was fitted to measured points by changing the friction coefficient.

imum of the real part of the Clausius–Mossotti factor. To interpret linear field orientation, it is sufficient to consider alterations of the maximum axis in dependence on the external conductivity. Nevertheless, Fig. 8 gives a complete picture of the orientation phenomenon, presenting frequency curves of maxima and minima alterations. The inclusion of minima alterations also makes sense when the orientation in circular fields is considered. While a linear field orients the axis of the maximum of the real part of the Clausius–Mossotti factor, a rotating field orientates two axes, those of the largest and the second largest real part of the Clausius–Mossotti factor. The reason is that a rotating field can be considered as a superposition of two linear fields, perpendicularly applied within a field plane. As a result, the third axis of the lowest real part of the Clausius–Mossotti factor, when oriented perpendicularly to the field plane, exhibits an exceptional orientation.

In detail, the curves of Fig. 8 are different from those given by other authors [11, 12]. These differences arise from the ill-defined layer thickness of the common Laplace approach. The equatorial region of a polarized object oriented in field direction most strongly contributes to the induced forces. In the common Laplace model, reorientation turns membrane areas of very different thickness, i.e., electrical properties, into this region,

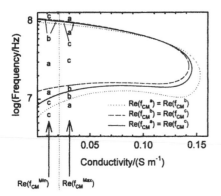

FIG. 8 Theoretical orientation of chicken erythrocytes over medium conductivity plotted as maxima, $Re(f_{CM}^{Max})$, and minima, $Re(f_{CM}^{Min})$, of the real part of the Clausius–Mossotti factor, respectively. The curves separate areas of a certain axis with maximal or minimal polarizability. While the maximum axis describes the orientation in a linear field, the minimum axis has an exceptional orientation in a circular field (see text). For cell parameters see Fig. 6.

thus changing the effective electrical properties of the membrane layer [54]. While the confocal model in cases where no reorientation occurred will have generated effective model parameters (e.g., Refs 8 and 9), in electro-orientation the application of a certain geometry to all orientations seems problematic (e.g., Ref. 12). The problem can probably be reduced by the application of different geometric models for each orientation, which is an intrinsic feature of the finite element model. The latter approach seems to be more reasonable for modeling ellipsoids with a low-conductivity membrane layer.

Figure 9 compares traveling-wave DP and ER measurements on Sephadex G15 particles. The traveling-wave induced particle velocity was measured over a comb-shaped, repetitive electrode structure which was driven by signals of progressive 90°-phase shifts (for details see Ref. 120).

The measurements are consistent with a homogeneous, spherical particle model possessing a dielectric constant and a conductivity of 40 and 0.9 mS/m, respectively [8,

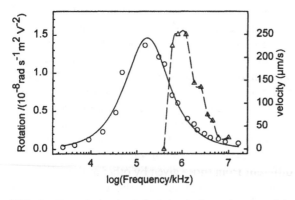

FIG. 9 Traveling-wave induced velocity (triangles, right ordinate) and rotation speed (circles, left ordinate) of Sephadex G15 particles at an external conductivity of 0.23 S/m in dependence on field frequency. Below 400 kHz the particles were attracted to the electrodes.

120]. The results of traveling-wave DP can qualitatively be explained by a superposition of a linearized ER and dielectrophoretic particle lifting or attraction by the inhomogeneous electrode field. Since attraction of the objects towards the electrode structure by positive DP hinders particle translation in the free suspension, it can only be observed in the frequency range of negative DP (cf. Fig. 5). Under the conditions used Sephadex particles exhibit a positive ER peak. With increasing frequency the positive DP changes at 400 kHz to negative DP. The positive ER peak corresponds to a translation in the direction of field propagation.

VIII. RECENT DEVELOPMENTS

A. Microstructured Chambers

Conventionally, AC electrokinetic measurements were carried out at low ionic strength to reduce heat production and medium convections. Such problems are minimized in micro-electrode chambers produced by microchip technology with an electrode size of the order of the particles themselves [8, 65, 77, 114, 121, 122]. The increased surface–volume ratio of the microchambers facilitates heat dissipation and reduces convections. Also, the necessary driving voltages decrease with the electrode spacing. Moreover, microelectrodes possess a reduced impedance and impose a lower electrical load on to the signal generators. Microchambers allow for an extension of the medium conductivity and the measuring frequency beyond the cell physiological ionic strength (about 1.5 S/m) and 1 GHz, respectively. Generally, at an increased medium conductivity all dispersions are shifted towards higher frequencies. For cells in ER the cofield peak becomes antifield for conductivities higher than that of the cytoplasmic core, which is usually of the order of 0.5 S/m. For shielded particles, like cells, the external medium and the core polarizability balance becomes delicate when their properties are similar. This allows for a more sensitive detection of the core properties [30].

Nevertheless, the introduction of microchambers and the extension of the measuring frequency range also caused unprecedented problems. In ER measurements on particles and cells above 30 MHz additional peaks of unexpectedly high and low rotation speed were observed. Figure 10 gives an example for the resonant properties of microchambers. The undistorted, theoretical curves (continuous) were calculated for the parameters used in Fig. 9. Around 180 MHz, the speed dramatically increased in a relatively narrow peak before a final drop. Checking the driving voltage at the terminating resistors (V_{drive}) revealed that the driving signal was frequency independent and that the reason for the ER resonance peaks must be located within the chamber structure itself. Analysis of the electrical chamber circuit properties showed that these effects are not due to object polarization but are instead caused by a resonance peak of the chamber field strength around 180 MHz. This chamber peak frequency was broadly independent of medium conductivity, but depended on the chamber and the chip carrier design.

For measurements square-topped fields were used. At 44 mS/m the inherent chamber resonance at 180 MHz was located far away at the high-frequency flank of the ER peak, causing only minor deviations from a Lorentzian shape. Increasing the conductivity to 370 mS/m shifted the ER peak to 66 MHz, resulting also in strong distortions at frequencies lower than 180 MHz. Similar distortions were observed at 44 mS/m when four inductances of 4.7 µH each were added in series to the electrode connectors. This drastic increase in inductance of the feed wires (see scheme at bottom of Fig. 10) shifted the resonance frequency and altered the spectrum at 44 mS/m (upright triangles), result-

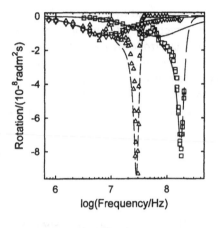

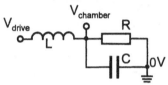

FIG. 10 Top: spectra of Sephadex G15 beads at two different external conductivities (44 mS/m – triangles; 370 mS/m – squares); the dashed curves consider resonance peaks at 180 and 28 MHz as well as the Fourier content of the square-topped field. Bottom: equivalent circuit for each of the four electrodes used to fit the resonance curves; L, R, and C represent the inductance of 0.807 μH for the bond wires in series with the electrode strips on the chip, the conductivity-dependent resistance of the filling solution, and a capacitance of 0.922 pF formed by each electrode.

ing in resonance distortions of the ER spectrum qualitatively identical to the chamber's inherent distortions.

How can such complex distortions be explained? For an ER generator producing two square-topped fields (in x and y directions) with duty cycles of 50%, the resulting field can be expressed in a Fourier series for the field. Consequently, Eq. (B.3) is transformed into

$$\vec{E} = \frac{4E_0}{\pi}\left(e^{j\omega t} - \frac{e^{-j3\omega t}}{3} + \frac{e^{j5\omega t}}{5} - + \cdots\right)\begin{pmatrix} 1 \\ j \\ 0 \end{pmatrix} \tag{1}$$

where E_0 stands for the peak value of the square-topped field. The field possesses only odd harmonics, the strength of which progressively decreases. Their spinning directions change with every component. The torque $\vec{N}$ induced on a particle can be developed in a series of torque components:

$$\vec{N} = \vec{N}(\omega) - \frac{\vec{N}(3\omega)}{9} + \frac{\vec{N}(5\omega)}{25} - + \cdots \tag{2}$$

In this series, the torque contributions of higher components decrease with the square of their harmonics and alternately change their signs, inducing torques with negative and positive spinning directions. Thus, if no resonances occur, ER spectra in square-topped fields of a duty cycle of 50% usually show only slight deviations from those in continuously rotating fields [123]. The situation changes when the chamber exhibits an inherent resonance. At a measuring frequency of, e.g., 60 MHz the third harmonic is at resonance, generating a strong negative contribution to the torque (cf. squares to theoretical curve in Fig. 10 and Eq. (2)]. Further evidence that the ER spectra distortions were caused by the chamber properties was that all spectra could be fitted by changing the resistor (scheme at bottom of Fig. 10) using the same friction coefficient (for details see Ref. 8).

Nevertheless, chamber resonances do not only cause problems. When the harmonic content of the drive signal is known it is possible to separate the narrow resonance distortions from the Lorentzian peaks generated by cell or particle polarization. Moreover, we found that fitting the distorted spectra of probe particles was much more sensitive in determining the effective electrical chamber parameters than direct impedance characterization of the chambers [8]. The electrorotating particles were used as field probes within the liquid dielectric of a capacitor, which was formed by the chamber. Since alterations in the ER torque are proportional to the square of the field strength, they are very sensitive to field strength variations. Furthermore, resonances can be applied for the local increase of forces in media, particle, or cell manipulation [8, 87] or in microsystem technology [124].

It is advantageous to avoid resonance in most cases of cell and particle characterization. To this end, as a rule, the chamber chip material should possess no conductivity and a low dielectric constant. Therefore, glass chips are superior to silicon chips. Further, the terminating resistors should be close to the electrode structure which may, unfortunately, cause temperature problems due to the resistors' heat dissipation. Despite these problems, a linear frequency behavior of microchambers up to the low gigahertz range can easily be achieved.

B. Dielectrophoretic Field Trapping

Planar or three-dimensional microelectrode chambers, e.g., with four electrodes as used in ER or with a special, intercastellated design allow for the generation of strongly inhomogeneous fields with well-defined positions of the field minima and maxima [65, 72, 77, 78, 121, 125]. Due to the strong gradients and the small distances between the sites of minimum and maximum field, particle collection at a certain site and the frequency dependent alteration of the collection sites can be observed even for submicroscopic particles. Usually, dielectrophoretic field-trapping experiments are theoretically interpreted by the real part of the induced dipole moment of the particles. Nevertheless, only the translation of single particles within the field gradient towards the site of collection is proportional to the real part. Fortunately, the frequency dependence of the polarizabilities of small aggregates is usually very similar to that of single particles. Moreover, the deviations of the collecting force acting in between the single particles of an electrically formed aggregate from the real part of the induced dipole moment are comparatively small. This is despite the deviation caused by the contribution of the imaginary part of the induced dipole moments and the differences in the polarizability of single particles and aggregates. For these reasons it is largely justified to make field-trapping results synonymous for the real part of the induced dipole moment.

In order to keep the experiments as simple as possible often only critical frequencies and their medium-conductivity dependence are registered (Figs 5 and 7) [72, 126]. For submicroscopic particles, e.g., viruses, fluorescence markers may facilitate observability. An interesting point is that, with decreasing size, the particle properties are strongly dominated by their surface properties, and the double-layer related α dispersions are shifted beyond 10 kHz into the β-dispersion range. As a result, the DP and ER behavior of, e.g., artificial particles becomes very sensitive to surface property alterations, e.g., by specific adsorption or binding of charged molecules, antibodies, etc. These features can be employed in screening systems, etc. [127].

C. Electrorotational Light Scattering (ERLS)

Microscopic registration of individual particle kinetics is tedious and restricted to the microscopic size range. Dynamic light-scattering techniques access the diameter range from about 5 nm up to 5 μm and are commonly used to characterize the size and surface charge of colloidal or biological particles. Recently developed methods such as ERLS and DPALS aim to combine the advantages of AC electrokinetics and dynamic light scattering. The methods are based on the simultaneous, computerized registration of many individual particles of a population and yield statistical significance at short measuring times.

ERLS is a homodyne dynamic light-scattering method with a single beam and a single detector [70, 118]. A scheme of the experimental setup is presented in Fig. 11.

For ERLS two principle designs for the measuring chamber were tested. The conventional setup consisted of four platinum-wire electrodes placed in a cylindrical drilled hole in a Plexiglas body. Alternatively, two microelectrode glass chips carrying the electrode structures were glued to a spacer with a drill hole, forming the measuring volume. With no field, the ERLS setup comprises a conventional dynamic light-scattering device. When the particles possess a deviation from a perfect optical rotational symmetry, the field frequency-dependent ER of the particles generates field frequency-dependent autocorrelation functions of the scattered light intensity. Figure 12 (top) gives an example for the change in autocorrelation function measured on human red cells.

The inverse of the decay time of the autocorrelation functions to 75% of its initial value ($1/\tau_{75}$) was introduced as a measure of the induced rotation speed [70]. The $1/\tau_{75}$ spectra allow for the determination of the frequency dependence of the particle rotation. As an example measurements on TiO_2 particles are presented in Fig. 12 (bottom). The TiO_2 particles were filtered by a micropore filter and had an average diameter of 320 nm. Nevertheless, since the criteria do not only depend on the rotation speed, but also on particle shape and additional particle motion, limits for clear detection of particle rotation exist [70]. Another problem of the method was its inability to detect the rotation sense of particles. This problem could partly be solved by a compensation method allowing for the detection of the relative sense of all ER peaks of a spectrum (compare with above) [118].

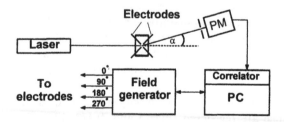

FIG. 11 Principal ERLS setup. A laser beam illuminates the measuring volume in the center of a special ER chamber. Scattered light intensity is detected by a photon-multiplier (PM) at the detection angle, α. The PM signal is fed into the correlator of a computer (PC) which also drives the four-phase field generator.

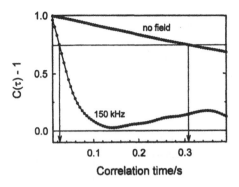

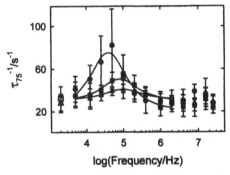

FIG. 12 ERLS measurements on human red cells (top) and TiO_2 particles (bottom). ER results in a steeper decay of the normalized autocorrelations function $[C(\tau) - 1]$; the inverse times necessary for the function to decay to 75% of its initial value $(1/\tau_{75})$ is marked by arrows for the case of no ER field and 150 kHz, respectively. TiO_2 particle spectra were measured at three media conductivities of 0.16, 1, and 2 mS/m, respectively. The fitted curves are single Lorentzian functions.

D. Dielectrophoretic Phase Analysis Light Scattering

To take advantage of the Kramers–Kronig relationship between DP and ER, e.g., for testing certain particle models, DP-induced translations also need to be registered. Commonly, heterodyne dual-beam laser–Doppler setups are applied to detect particle translations [128–130]. For DP measurements, several advantages favor a dual-beam setup. A striking advantage is that the measuring volume is restricted to the crossing region of the two beams and can thus be adjusted to an area of known field distribution. To detect translation, no optical anisotropy of the particles is required. This, in principle, allows for the detection of particles smaller than 5 nm. Since the classical laser–Doppler setups are not sensitive to very small particle displacements, we applied phase analysis light scattering (PALS). In PALS, Bragg cells introduce a small optical frequency difference into two laser beams to create an interference region with a moving fringe pattern (Fig. 13) [129]. The intensity of the scattered light of a particle, stationary within the crossing region, varies with the frequency difference between the two laser beams.

To detect particle translation, the intensity of the scattered light of a particle is compared to the difference frequency of the Bragg cell drive signals. For this, the difference frequency obtained from a mixer (equivalent to the optical difference frequency of the beams) is used as the reference for the lock-in amplifier. Phase demodulation of the light scattered by a single particle directly yields the particle velocity perpendicular to the fringe

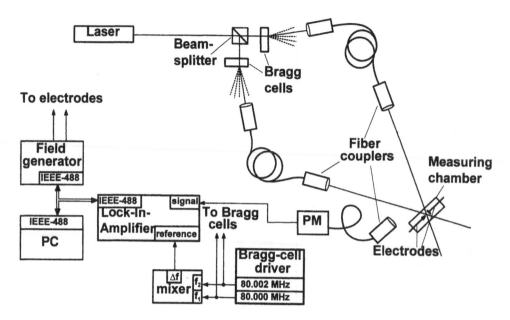

FIG. 13 Optical and electronic DPALS setup. The measuring chamber is illuminated by two focussed laser beams that path through Bragg cells. The frequency difference between the Bragg cell drivers, which is identical to the optical frequency difference between the two first-order beams, is used as the reference signal of the lock-in-amplifier. The scattered light intensity is detected by a PM through an optical fiber. A computer drives the lock-in amplifier and the field generator via an IEEE bus.

pattern. For diffusing particles the method registers the translational component perpendicular to the fringe pattern (Fig. 14, top). Amplitude weighting of the scattered light intensity yields phase differences in angular degrees. For pure diffusion the average shift vanishes. For moving ensembles of particles the shift obtained by amplitude weighting is directly related to the mean velocity of the ensemble [129].

Latex particles exhibit a single Lorentzian transition of their DP force. PALS registration of DP was possible despite an offset velocity of the particles of about 100 μm/s (for details see Ref. 71). This velocity is due to medium convections and gives another example for the superposition of different AC electrokinetic effects. With microchambers medium convections can only be reduced, and the construction of a convection-free DP chamber is still an unsolved problem.

IX. SUMMARY

Over the last two decades, a number of AC electrokinetic characterization techniques have been rapidly developed in biotechnology, but especially in the field of basic research. Prerequisites for this development were the improvement of the generators and the chamber connectors, the introduction of chip technology for the design of electrodes, video processing, light-scattering techniques, and last but not least, the development of adequate theories. Many AC electrokinetic effects used for characterization can also be applied to the manipulation of particles. Biological research in this field is aimed at the development

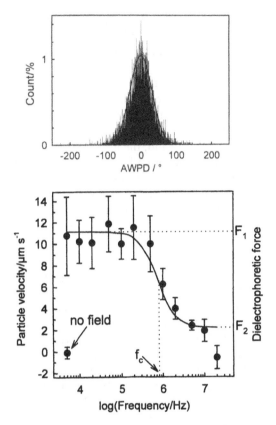

FIG. 14 PALS measurements of the diffusion (132 nm; top) and of the DP translation (15 μm, external conductivity 3.4 mS/m; bottom) of latex particles. In the case of diffusion amplitude weighting of the measured phase difference yields a Gaussian distribution with a center of 0°. The width depends on the beam geometry and the diffusion coefficient. Frequency-dependent DP translation shifts the center of the distribution. The shift can be recalculated into particle velocity (bottom). The solid line is a fit of the theoretical spectrum for a homogeneous particle with a permittivity of 2.6 and a conductivity of 0.75 mS/m

of "Bio-factories on a chip" based on electrode chips with channels and cavities for sorting, separation, collection, and different forms of electrical, biochemical, and genetic manipulations as well as for the characterization of dielectric properties [131–133]. In characterization, the development aimed at separating the different AC electrokinetic effects to simplify the interpretation of experimental data. An example is the development of ER. Whereas the first ER experiments were conducted on pairs of cells in linear fields [104] the introduction of rotating fields by Arnold and Zimmermann [1] based on an idea of H. Schwan (personal communication, 1998) allowed for the observation of ER on single objects. Nonetheless, in most methods for a correct interpretation of measurements, e.g., DP or ER spectra, it is still essential to take into account the interrelations of the different AC electrokinetic effects, e.g., cell reorientation, deformation, or even the distortion of membrane integrity by the induced transmembrane potential. The model presented here considers such interrelations. Despite its simplicity, the model is not only relevant for microscopic particles but also for submicroscopic biological particles, such as viruses or cell organelles and even molecular objects, e.g., for the electric birefringence of macro-molecular suspensions. Nevertheless, especially for submicroscopic objects, it is necessary to expand the simple structural model by additional dispersion processes, like those related to the double-layer properties [41, 43, 47, 72, 127, 134, 135].

Although the methodology for most of the AC electrokinetic methods was developed in biology, in the future colloid science may emerge as the main field of application. In colloid science the properties of the suspension medium and the different particle structures, such as their bulks, layers, and surfaces may vary much more than those of the relatively "standardized" biological structures. On the other hand, the molecular structure and composition of artificial particles are less complex and better understood than those of biological objects. Consequently, it can be expected that a variety of different new dispersion processes and their understanding will emerge in colloid science, rather than in biology. The improved models for the α dispersion can be considered as a first step in this development, and DP-screening systems based on antigen–antibody reactions at the surface of artificial colloids are its first applications. The new possibilities in characterizing particles will also find applications in the development of artificial particles for drug targeting, etc.

Progress in colloid science will improve our understanding of the interaction of biological material with electromagnetic fields, an acute problem given the current concern over "electro-smog." On the other hand, the characterization of individual microscopic and submicroscopic biological objects potentially has direct applications in the screening of microbial activity in fermentation, e.g., in breweries, where impedance methods are already established [33]. Knowledge of the dielectric properties of single objects such as yeast, bacteria, or cancer cells can finally be applied to improve the impedance models of suspensions in fermentation and of tissues in hyperthermal cancer therapy, respectively [136]. Today, having emerged from (and being inspired by) impedance techniques, AC electrokinetics has become an independent field of research that may help improve our understanding of impedance phenomena.

ACKNOWLEDGMENTS

I am grateful to Ms Ch. Mrosek, Mr D. Wachner, Mr S. Lippert, and Dr B. Prüger for their technical assistance and to Dr U. Gimsa and Dr R.F. Sleigh for help with the manuscript. Dr T. Müller, Dr R. Georgieva, and Dr P. Wust (from the SFB273) are acknowledged for fruitful discussions.

APPENDIX A

Induced Dipole Moment of Single-Shell Ellipsoid

In the following an expression for the general single-shell ellipsoid is derived. For the derivation it is assumed that the three principal axes of the general ellipsoid are oriented in the directions of the axes of the co-ordinate system. Axes a, b, and c should be oriented in the x, y, and z directions, respectively. For an ellipsoidal particle, assumed to be homogeneous (this assumption holds for an actually homogeneous ellipsoid as well as for the homogeneous equivalent ellipsoid), the x component of the induced dipole moment is given by the product of the particle volume V and the effective polarizability in the x direction $\bar{P}_x$:

$$\vec{m}_x = V\vec{P}_x = V\varepsilon_e\varepsilon_0 f_{CM}^x \vec{E}_x \quad \text{with} \quad V = \frac{4\pi abc}{3} \tag{A.1}$$

In Eq. (A.1), V, ε_e, ε_0, f_{CM}^X, and $\vec{E}_x$ stand for the particle volume, the permittivities of external medium and vacuum, the Clausius–Mossotti factor, and the external field in the x direction, respectively. As can be seen from Eq. (A.1), the Clausius–Mossotti factor stands for the frequency-dependent part of the induced dipole moment. The dipole moment can also be expressed using the constant local field $\vec{E}_x^p$ within the ellipsoid [137]:

$$\vec{m}_x = V\frac{\varepsilon_e\varepsilon_0}{n_x}(\vec{E}_x - \vec{E}_x^p) \tag{A.2}$$

Here, n_x stands for the depolarizing factor in the x direction (for depolarizing factors see Appendix C). When the Clausius–Mossotti factor is formally introduced, as in Eq. (A.1), it deviates from the conventional value by a factor of 3. The reason is that the factor of 3 in the denominator of the volume equation is conventionally cancelled out with the depolarizing factor of a sphere of 1/3. This cancellation was avoided here, since it is not reasonable for nonspherical objects with depolarizing factors different from 1/3.

The term $\vec{E}_x^p$ can be determined from the potential Ψ_x^p at the particle's surface at the pole of half-axis a (see Figs 2–4):

$$\vec{E}_x^p = \frac{\Psi_x^p}{a}\vec{i} = \frac{\Psi_x^p}{\Psi_x^e}\vec{E}_x \tag{A.2a}$$

For a homogeneous ellipsoidal particle the frequency-dependent part of the induced dipole moment is

$$f_{CM}^x = \frac{\varepsilon_i - \varepsilon_e}{\varepsilon_e + (\varepsilon_i - \varepsilon_e)n_x} \tag{A.3}$$

[137]. ε_i, ε_e, and n_x are the complex permittivities of the particle and external medium as well as the depolarizing factor in the x direction, respectively. Please note that due to the definition of Eqs (A.1) and (A.3) the numerical values of the Clausius–Mossotti factor are three times those usually used in the literature. The reason is that the factors "3" appearing in the volume and the depolarizing factors of a sphere are not canceled out (for discussion see Ref. 54). A comparison of Eqs (A.1)–(A.3) results in an expression for the Clausius–Mossotti factor in the x direction which depends on a normalized potential difference (cf. Figs 1 and 2):

$$f_{CM}^x = \frac{1}{n_x}\left(\frac{\Psi_x^e - \Psi_x^p}{\Psi_x^e}\right) \tag{A.4}$$

According to Fig. 4 the potential Ψ_x^p is determined by the voltage divider properties of the finite element model consisting of three impedance elements Z_x:

$$\Psi_x^p = \frac{Z_x^i + Z_x^m}{Z_x^i + Z_x^m + Z_x^e}\vec{E}_x\vec{a}_{infl} \tag{A.5}$$

The impedance elements for the internal, membrane, and external medium are designated by the indices i, m, and e, respectively. All elements possess the same cross-sectional area A_j but differ in lengths l_j. Element j has the impedance:

$$Z^j = \frac{1}{\sigma_j + j\omega\varepsilon_j\varepsilon_0}\frac{l_j}{A_j} \tag{A.6}$$

where $j, \omega, \sigma_j, \varepsilon_j$, and ε_0 stand for $(-1)^{0.5}$, the angular frequency of the field, the DC conductivity, the relative permittivity, and the permittivity of vacuum, respectively. For a given field strength the maximum of Ψ_x^p in Eq. (A.5) is determined by the influential radius $\vec{a}_{\text{infl}}$ (Fig. 4):

$$\text{Max}(\Psi_x^p) = \Psi_x^0 = \vec{a}_{\text{infl}} \vec{E}_x \tag{A.7}$$

Its minimum is 0 V (e.g., for a metallic body). Keeping in mind that the minimum of the Clausius–Mossotti factor [Eq. (A.3)] corresponds to the maximum of the local field, the combination of Eqs (A.3), (A.4), and (A.5) yields the relationship between n_x and the influential radius $\vec{a}_{\text{infl}}$ [53, 54]:

$$\frac{a}{a_{\text{infl}}} = 1 - n_x \tag{A.8}$$

Finally, Eqs (A.4), (A.5), and (A.8) can be combined to obtain a new expression for the Clausius–Mossotti factor in the x direction:

$$f_{\text{CM}}^x = \frac{a_{\text{infl}}}{a_{\text{infl}} - a}\left(1 - \frac{Z_x^i + Z_x^m}{Z_x^i + Z_x^m + Z_x^e}\frac{a_{\text{infl}}}{a}\right) \tag{A.9}$$

Analogous expressions are valid for the other two axes.

APPENDIX B

AC Electrokinetic Effects

In this appendix the dipole approximation is used to derive general expressions for the behavior of ellipsoidal objects in homogeneous, inhomogeneous, and rotating AC fields. The electrical properties of the objects are given by the frequency-dependent induced dipole moment along the three principal axes [Eq. (A.1)]. In weakly inhomogeneous fields the time-averaged dielectrophoretic force $\langle \vec{F} \rangle$ acting on a homogeneous particle of general ellipsoidal shape can be expressed by the product of the induced dipole moment $\vec{m}$ and the gradient of the conjugated, complex external field $\vec{E}^*$:

$$\langle \vec{F} \rangle = \frac{1}{2}\Re[\vec{m}\nabla\vec{E}^*] \tag{B.1}$$

The time-averaged torque $\langle \vec{N} \rangle$ in a rotating field is given by the cross-product of the induced dipole moment and the conjugated field:

$$\langle \vec{N} \rangle = \frac{1}{2} \, Re[\vec{m} \times \vec{E}^*] \tag{B.2}$$

In component notation the external AC field can generally be written as:

$$\vec{E} = \begin{pmatrix} E_x \\ E_y \\ E_z \end{pmatrix} = E_0 e^{j\omega t}\begin{pmatrix} e_x \\ e_y \\ e_z \end{pmatrix} \tag{B.3}$$

where E_0, ω, j, and t stand for the amplitude of the field, circular frequency, $(-1)^{-0.5}$, and time, respectively. The components of the field vector E_x, E_y, E_z are parallel to the ortho-normal base vectors $\vec{i}, \vec{j}, \vec{k}$, respectively, which are the vectors of an orthonormal base system. The induced dipole moment $\vec{m}$ is proportional to the external field, the external permittivity $\varepsilon_0\varepsilon_e$, and the volume V of the ellipsoid:

$$\vec{m} = \begin{pmatrix} m_x \\ m_y \\ m_z \end{pmatrix} = \varepsilon_0 \varepsilon_e V \begin{pmatrix} f_{CM}^{x\Re} + if_{CM}^{x\mathcal{I}} \\ f_{CM}^{y\Re} + if_{CM}^{y\mathcal{I}} \\ f_{CM}^{z\Re} + if_{CM}^{z\mathcal{I}} \end{pmatrix} \vec{E} \tag{B.4}$$

The frequency-dependent part of each component (index x, y, or z) along the three principal axes of the ellipsoid is given by three components of the Clausius–Mossotti factor. Each consists of a real (index $\Re$) and an imaginary part (index $\mathcal{I}$), which are in phase and out of phase with the inducing field, respectively. Introducing the expressions (B.3) and (B.4) into Eq. (B.2) one obtains:

$$\langle \vec{N} \rangle = \frac{1}{2} \Re \begin{pmatrix} m_y E_z^* - m_z E_y^* \\ m_z E_x^* - m_x E_z^* \\ m_x E_y^* - m_y E_x^* \end{pmatrix} \tag{B.5}$$

In the following, special cases for the external field shall be considered. For $e_x = e_y = e_z = 1$, Eq. (B.3) describes a linear AC field as it is applied in electro-orientation. If we assume the three principal axes a, b, c of the ellipsoid to be oriented parallel to the three base vectors the same magnitude of the external field applies along each principal axis. Therefore, the field-induced torque around a given axis depends on the induced dipole components along the other two axes. This relationship is obvious in Eq. (B.6) which can be directly obtained from Eq. (B.5) [12]:

$$\langle \vec{N} \rangle = \frac{1}{2} \varepsilon_e \varepsilon_0 V E_0^2 \begin{pmatrix} f_{CM}^{y\Re} - f_{CM}^{z\Re} \\ f_{CM}^{z\Re} - f_{CM}^{x\Re} \\ f_{CM}^{x\Re} - f_{CM}^{y\Re} \end{pmatrix} \tag{B.6}$$

For any given frequency and external conductivity, Eq. (B.6) can be used to obtain the two largest torque components that are induced around two of the three principal axes. Orientation would then occur along the third axis of weakest torque. These relationships were reflected by complicated tables of the signs of the components of Eq. (B.6) [11, 12]. The same result is obtained by looking up the maximum of the three components $f_{CM}^{x\Re}, f_{CM}^{y\Re}, f_{CM}^{z\Re}$.

Orientation will also be observed when DP spectra are detected on freely suspended particles. Particle reorientation and a subsequent change in its effective properties will result in discontinuous DP spectra. Neglecting thermal motion, one of the principal axes of the particle will always be oriented in parallel to the field. For a field in the x direction and a parallel orientation of a certain principal axis (e.g., axis a, index x), introduction of Eq. (B.4) into Eq. (B.1) then yields the DP force:

$$\langle \vec{F}_x \rangle = \frac{1}{2} \varepsilon_0 \varepsilon_e V \Re[(f_{CM}^{x\Re} + if_{CM}^{x\mathcal{I}}) E_x \nabla E_x^*] \tag{B.7}$$

A field that is weakly inhomogeneous in the x direction can be approximated by Eq. (B.3) for $e_x = 1 + \gamma x$ and $e_y = e_z = 0$; γ describes the small field inhomogeneity. For such a field one obtains:

$$\langle \vec{F}_x \rangle = \varepsilon_0 \varepsilon_e V f_{CM}^{x\Re} E_0^2 \frac{\gamma}{2} \vec{i} \tag{B.8}$$

Equation (B.8) shows that the frequency dependence of the DP force is described by the real part of the Clausius–Mossotti factor along the axis oriented parallel to the field. For $\gamma = 0$

the external field is homogeneous and the DP force vanishes. Nevertheless, an electrode-formational force will still be induced, leading to compression or elongation of the object.

To consider ER a field circulating at constant amplitude in the x plane can be assumed. Such a field is given by $e_x = 1$, $e_y = j$, and $e_z = 0$. It follows that:

$$\vec{E}_y = j\vec{E}_x \quad \vec{E}_y^* = -j\vec{E}_x^* \tag{B.9}$$

For simplicity, in the following only cases are considered in which one principal axis is oriented perpendicular to the field plane. Since the z component of the external field is zero for the oriented ellipsoid, the component of the induced dipole moment along axis c also vanishes and Eq. (B.5) becomes

$$\langle \vec{N}_z \rangle = \frac{1}{2}\varepsilon_0\varepsilon_e V\Re[(f_{CM}^{x\Re} + jf_{CM}^{x\mathcal{I}})E_xE_y^* - (f_{CM}^{y\Re} + jf_{CM}^{y\mathcal{I}})E_yE_x^*]\vec{k} \tag{B.10}$$

which can be simplified to

$$\langle \vec{N}_z \rangle = \varepsilon_0\varepsilon_e VE_0^2 \frac{f_{CM}^{x\mathcal{I}} + f_{CM}^{y\mathcal{I}}}{2}\vec{k} \tag{B.11}$$

Obviously only the out-of-phase parts of the induced dipole moment contribute to the torque. For the three possible orientations of the general ellipsoid the three different combinations of the Clausius–Mossotti factors apply and three different ER spectra will be obtained. In practice, at certain field frequencies reorientation of the particles will be observed, resulting in discontinuous ER spectra.

APPENDIX C

Depolarizing Factors

In Appendix A, Eq. (A.2) depolarizing factors for the general ellipsoids were introduced which were then related to the influential radii. To our knowledge explicit expressions for the depolarizing factors were first derived by Stille [138] for spheroids and for the general ellipsoid independently by Osborn [139] and Stoner [140]. For the general ellipsoid and principal axes with $a > b > c$ the depolarizing factors n_a, n_b, and n_c are given by

$$n_a = \frac{\beta\delta}{\sqrt{1 - \delta^2}(1 - \beta^2)}(LF(k, \Psi) - LE(k, \Psi))$$

$$n_b = n_a + \frac{\beta\delta}{\sqrt{1 - \delta^2}(\beta^2 - \delta^2)}LE(k, \Psi) - \frac{\delta^2}{\beta^2 - \delta^2} \tag{C.1}$$

$$n_c = -\frac{\beta\delta}{\sqrt{1 - \delta^2}(\beta^2 - \delta^2)}LE(k, \Psi) + \frac{\beta^2}{\beta^2 - \delta^2}$$

where β and δ are the axis ratios $\beta = b/a$ and $\delta = c/a$, and LF and LE are elliptical integrals that are functions of k and Ψ; k and Ψ also depend on the axis ratios according to

$$k = \sqrt{\frac{1 - \beta^2}{1 - \delta^2}} \quad \text{and} \quad \psi = \arccos(\delta) \tag{C.2}$$

The elliptical integrals are then

$$LF(k, \Psi) = \int_0^{\Psi} \frac{1}{\sqrt{1 - k^2 \sin^2 \varphi}} d\varphi$$

$$LE(k, \Psi) = \int_0^{\Psi} \sqrt{1 - k^2 \sin^2} d\varphi$$

(C.3)

For spheroids with two equal axes ($a = b$) it is possible to obtain explicit expressions [138]. For the oblate case ($a > c$) along an axis perpendicular to the symmetry axis one obtains:

$$n_a = \frac{1}{2}\left[1 - \frac{1 + e^2}{e^3}(e - \arctan e)\right] \quad \text{with } e = \sqrt{\left(\frac{a}{c}\right)^2 - 1}$$

(C.4)

and for the prolate case ($a < c$):

$$n_a \frac{1}{2}\left[1 - \frac{1 - e^2}{2e^3}\left(\ln\frac{1 + e}{1 - e} - 2e\right)\right] \quad \text{with } e = \sqrt{1 - \left(\frac{a}{c}\right)^2}$$

(C.5)

where e stands for the eccentricity of the spheroid. Since the sum of the depolarizing factors of the general ellipsoid is always unity ($n_a + n_b + n_c = 1$) the depolarizing factor along the symmetry axis of the spheroid is $n_c = 1 - 2n_a$.

REFERENCES

1. WM Arnold, U Zimmermann. Z Naturforsch 37c: 908, 1982.
2. H Maier. Biophys J 73: 1617, 1997.
3. M Saito, HP Schwan, G Schwarz. Biophys J 6: 313, 1966.
4. FJ Asencor, C Santamaria, FJ Iglesias, A Dominguez. Biophys J 64: 1626, 1993.
5. SP Stoylov. Biophys Chem 58: 165, 1996.
6. R Hölzel. Biophys J 73: 1103, 1997.
7. G Blum, H Maier, F Sauer, HP Schwan. J Phys Chem 99: 780, 1995.
8. J Gimsa, T Müller, T Schnelle, G Fuhr. Biophys J 71: 495, 1996.
9. T Müller, L Küchler, G Fuhr, T Schnelle. Silvae Genet 42: 311, 1993.
10. OV Ignatov, NA Khorkina, SY Shchyogolev, IN Singirtsev, VD Bunin, YA Tumaikina, VV Ignatov. FEMS Microbiol Lett 173: 453, 1999.
11. TB Jones (ed). Electromechanics of Particles. Cambridge, New York, Melbourne: Cambridge University Press, 1995.
12. RD Miller, TB Jones. Biophys J 64: 1588, 1993.
13. HA Pohl. Dielectrophoresis. The Behavior of Neutral Matter in Nonuniform Electric Fields. Cambridge, London, New York, Melbourne: Cambridge University Press, 1978.
14. M Nishioka, S Katsura, K Hirano, A Mizuno. IEEE Trans Ind Appl 33: 1381, 1997.
15. A Yu Ivanov, VG Gogvadze, AI Miroshnikov, BI Nedvedev. Biochim Biophys Acta 903: 241, 1985.
16. KR Foster, HP Schwan. In: C Polk, E Postow, eds. Biological Effects of Electromagnetic Fields. Boca Raton: CRC Press. 1996, pp 25–102.
17. K Asami, T Hanai, N Koizumi. Jap J Appl Phys 19: 359, 1980.
18. H Pauly, HP Schwan. Z Naturforsch 14b: 125, 1959.
19. U Zimmermann, WM Arnold. In: H Fröhlich, F Kremer, eds. Coherent excitations in biological systems. Berlin, Heidelberg: Springer-Verlag, 1983, pp 211–221.
20. KVIS Kaler, TB Jones. Biophys J 57: 173, 1990.
21. X Hu, WM Arnold, U Zimmermann. Biochim Biophys Acta 1021:191, 1990.

22. WM Arnold, BM Geier, B Wendt, U. Zimmermann. Biochim Biophys Acta, 885: 35, 1986.
23. R Georgiewa, E Donath, J Gimsa, U Löwe, R Glaser. Bioelectrochem Bioenerg 22: 255, 1989.
24. J Gimsa, G Fuhr, R Glaser. Stud Biophys 109: 5, 1985.
25. T. Müller, G Fuhr, F Geißler, R Hagedorn. In: M Markov, M Blank, eds. Electromagnetic Fields and Biomembranes. New York: Plenum Press, 1988, pp 197–182.
26. FF Becker, XB Wang, Y Huang, R Pethig, J Vykoukal, PRC Gascoyne. Proc Natl Acad Sci 92:860, 1995.
27. FF Becker, SB Wang, Y Huang, R Pethig, J Vykoukal, PRC Gascoyne, J Phys D: Appl Phys 27: 2659, 1994.
28. XB Wang, Y Huang, PRC Gascoyne, FF Becker, R Hölzel, R Pethig. Biochim Biophys Acta 1193: 330, 1994.
29. J Gimsa, C Pritzen, E Donath. Stud Biophys 130; 123, 1989.
30. J Gimsa, G Zechel, T Schnelle, R Glaser. Biophys J 66: 1244, 1994.
31. J Gimsa, R Glaser, G Fuhr. In: W Schütt, H Klinkmann, I Lamprecht, T Wilson, eds. Physical Characterisation of Biological Cells. Berlin: Verlag Gesundheitt, 1991, pp 295–323.
32. J Engel, E Donath, J Gimsa. Stud Biophys 125: 53, 1988.
33. R Pethig, DB Kell. Phys Med Biol 32: 933, 1987.
34. HP Schwan, S Takashima. Encycloped Appl Phys 5: 177, 1993.
35. H Fricke, HJ Curtis. J Phys Chem 40: 715, 1936.
36. CT O'Konski. J Phys Chem 64: 605, 1960.
37. C Ballario, A Bonincontro, C Cametti, A Rosi, L Sportelli. Z. Naturforsch C 39: 160, 1984.
38. JZ Bao, CC Davis, RE Schmukler. Biophys J 61: 1427, 1992.
39. R Lisin, BZ Ginzburg, M Schlesinger, YuD Feldman. Biochim Biophys Acta 1280: 34, 1996.
40. H Fricke. Nature 172: 731, 1953.
41. HP Schwan, G Schwarz, J Maczuk, H Pauly. J Phys Chem 66: 2626, 1962.
42. G Schwarz. J Phys Chem 66: 2636, 1962.
43. R Georgieva, B Neu, VM Shilov, E Knippel, A Budde, R Latza, E Donath, H Kiesewetter, H Bäumler. Biophys J 74: 2114, 1998.
44. K Asami, T Yonezawa. Biophys J 71: 2192, 1997.
45. G Fuhr, PI Kuzmin. Biophys J 50: 789, 1986.
46. R Paul, KVIS Kaler, TB Jones. J Phys Chem 97: 4745, 1993.
47. WM Arnold, HP Schwan, U Zimmermann. J Phys Chem 91: 5093, 1987.
48. E Donath, M Egger, VPh Pastushenko. Bioelectrochem Bioenerg 23: 337, 1990.
49. VL Sukhorukov, U Zimmermann. J Membr Biol 153: 161, 1996.
50. HP Schwan. Adv Biol Med 5: 147, 1957.
51. YuD Feldman, YuF Zuev, EA Polygalov, VD Fedotov. Colloid Polym Sci 270: 768, 1992.
52. G Fuhr, U Zimmermann, G Shirley. In: U Zimmermann, GA Neil, eds. Electromanipulation of Cells. CRC Press Inc. Boca Raton, New York, London, Tokyo, 1996, pp 259–328.
53. J Gimsa, D Wachner. Biophys J 75: 1107, 1998.
54. J Gimsa, D Wachner. Biophys J 77: 1316, 1999.
55. XB Wang, Y Huang, R Hölzel, JPH Burt, R Pethig. J Phys D: Appl Phys 26: 312, 1993.
56. H Engelhardt, H Gaub, E Sackmann. Nature 307: 378, 1984.
57. VL Sukhorukov, H Mussauer, U Zimmermann. J Membr Biol 163: 235, 1998.
58. M Krueger, F Thom. Biophys J 73: 2653, 1997.
59. P Pawlowski, I Szutowicz, P Marszalek, M Fikus. Biophys J 65:541, 1993.
60. HS Kage, H Engelhardt, E Sackmann. In: P Gaehtgens, ed. Biorheology. New York: Pergamon Press, 1990, pp 67–78.
61. J Gimsa, P Marszalek, U Löwe, TY Tsong. Biophys J 60: 749, 1988.
62. TB Jones, GW Bliss. J Appl Phys 48: 1412, 1977.
63. VPh Pastushenko, PI Kuzmin, YuA Chizmadshev. Stud Biophys 110; 51, 1985.
64. XB Wang, Y Huang, PRC Gascoyne, FF Becker. IEEE Trans Ind Appl 33: 660, 1997.

65. G Fuhr, WM Arnold, R Hagedorn, T Müller, W Benecke, B Wagner, U Zimmermann. Biochim Biophys Acta 1108: 215, 1992.
66. JM Cruz, FJ Garcia-Diego. J Phys D: Appl Phys 31: 1745, 1998.
67. R Hölzel. Biochim Biophys Acta 1525: 311, 1998.
68. A Budde, G Grümmer, E Knippel. Instrum Sci Technol 27: 59, 1999.
69. G Degasperis, XB Wang, J Yang, FF Becker, PRC Gascoyne. Meas Sci Technol 9: 518, 1998.
70. B Prüger, P Eppmann, E Donath, J Gimsa. Biophys J 72: 1414, 1997.
71. P Eppmann, B Prüger, J Gimsa. Colloids Surfaces A 149: 443, 1999.
72. J Gimsa. Ann NY Acad Sci 873:287, 1999.
73. JL Griffin. Exp Cell Res 61: 113, 1970.
74. H Funakoshi. Ann. Inst. Stat. Math. 5: 45, 1968.
75. U Zimmermann. Biochim Biophys Acta 694: 227, 1982.
76. DV Zhelev, PI Kuzmin, DS Dimitrov. Bioelectrochem Bioenerg 26: 193, 1991.
77. A Ramos, H Morgan, NG Green, A Castellanos. J Phys D: Appl Phys 31: 2338, 1998.
78. G Fuhr, T Schnelle, R Hagedorn, SG Shirley. Cell Eng 1: 47, 1995.
79. MS Talary, R Pethig. IEE Proc Sci Meas Technol 141: 395, 1994.
80. JP Burt, KL Chan, D Dawson, A Parton, R Pethig. Ann Biol Clin (Paris) 54:253, 1996.
81. GP Archer, WB Betts, T Haigh. Microbios 73: 165, 1993.
82. WB Betts. Trends Food Sci Technol 6: 51, 1995.
83. JC Giddings. Science 260: 1456, 1993.
84. Y Huang, XB Wang, FF Becker, PR Gascoyne. Biophys J 73: 1118, 1997.
85. R Hagedorn, G Fuhr, T Müller, J Gimsa. Electrophoresis 13: 49, 1992.
86. Y Huang, XB Wang, JA Tame, R Pethig. J Phys D: Appl Phys 26: 1528, 1993.
87. J Gimsa, P Eppmann, B Prüger. Biophys J 73: 3309, 1997.
88. HA Pohl, JS Crane. Biophys J 11: 711, 1971.
89. FA Sauer. In: H Fröhlich, F Kremer, eds. Coherent Excitations in Biological Systems. Berlin, Heidelberg: Springer Verlag, 1983, pp 134–143.
90. G Fuhr, R Glaser, R Hagedorn. Biophys J 49: 395, 1986.
91. AV Sokirko. Biol Membr 6/4: 587, 1992.
92. JC Maxwell. Treatise on Electricity and Magnetism. London: Oxford University Press, 1873.
93. FA Sauer, RW Schlögl. In: A Chiabrera, C Nicolini, HP Schwan, eds. Interactions Between Electromagnetic Fields and Cells. New York: Plenum Press, 1985, pp 203–251.
94. X Wang, X-B Wang, PRC Gascoyne. J Electrostat 39: 277, 1997.
95. T Kakutani, S Shibatani, M Sugai. Bioelectrochem Bioenerg 31: 131, 1993.
96. R Paul, M Otwinowski. J Theor Biol 148: 495, 1991.
97. H Pauly, HP Schwan. Biophys J 6: 621, 1966.
98. M Born. Z Phys 1: 221, 1920.
99. J Bernhardt, H Pauly. Biophysik 10: 89, 1973.
100. E Barnaby, G Bryant, EP George, J Wolfe. Stud Biophys 127: 45, 1988.
101. M Mischel, R Ackermann, R Hölzel, I Lamprecht. Bioelectrochem Bioenerg 27: 413, 1992.
102. R Glaser, G Fuhr, J Gimsa. Stud Biophys 96: 11, 1983.
103. AA Teixera-Pinto, LL Nejelski Jr, JL Cutler, JH Heller. Exp Cell Res 20: 548, 1960.
104. C Holzapfel, J Vienken, U Zimmerman. J Membr Biol 67: 13, 1982.
105. G Fuhr, R Hagedorn, R Glaser, J Gimsa. Elektrie. 43: 45, 1989. (In German.)
106. JR Melcher. Phys Fluids 9: 1548, 1966.
107. JR Melcher, MS Firebaugh. Phys Fluids 10: 1178, 1967.
108. T Müller, WM Arnold, T Schnelle, R Hagedorn, G Fuhr, U Zimmermann. Electrophoresis 14: 764, 1993.
109. G Fuhr, J Gimsa, R Glaser. Stud Biophys 108: 149, 1985.
110. KL Chan, PR Gascoyne, FF Becker, R Pethig. Biochim Biophys Acta 1349: 182, 1997.
111. H Ziervogel, R Glaser, D Schadow, S Heyman. Biosci Rep 6: 973, 1986.
112. B Klösgen, C Reichle, S Kohlsmann, KD Kramer. Biophys J 71: 3251, 1996.

113. E Donath, M Egger. In: E Bamberg, H Passow, eds. Progress in Cell Research. Elsevier Science, Amsterdam, 1992, pp 169–172.

114. G Fuhr, T Müller, T Schnelle, R Hagedorn, A Voigt, S Fiedler, WM Arnold, U Zimmermann, B Wagner, A Heuberger. Naturwissenschaften 81: 528, 1994.

115. T Schnelle, T Müller, S Fiedler, SG Shirley, K Ludwig, A Herrmann, G Fuhr, B Wagner, U Zimmermann. Naturwissenschaften 83: 172, 1996.

116. WM Arnold, U Zimmermann. Verfahren und Vorrichtung zur Unterscheidung von in einem Medium befindlichen Teilchen oder Partikel (1). [DE 3325843 A1], 1983.

117. J Gimsa, R Glaser, G Fuhr. Verfahren und Vorrichtung zur Messung des Rotationsspektrums dielektrischer Objekte. Patient [DD 256 192 A1], 1988.

118. B Prüger, P Eppmann, J Gimsa. Colloids Surfaces A 136: 199, 1966.

119. KVIS Kaler, JP Xie, TB Jones, R Paul. Biophys J 63: 58, 1992.

120. G Fuhr, S Fiedler, T Muller, T Schnelle, H Glasser, T Lisec, B Wagner. Sensors Actuators A 41: 230, 1994.

121. T Schnelle, R Hagedorn, G Fuhr, S Fiedler, T Müller. Biochim Biophys Acta 1157: 127, 1993.

122. R Pethig, Y Huang, XB Wang, JPH Burt. J Phys D: Appl Phys 25: 881, 1992.

123. J Gimsa, E Donath, R Glaser. Bioelectrochem Bioenerg 19: 389, 1988.

124. T Müller, J Gimsa, B Wagner, G Fuhr. Springer Microsystem Technol 3: 168, 1997.

125. S Fiedler, SG Shirley, T Schnelle, G Fuhr. Anal Chem 70: 1909, 1998.

126. H Morgan, NG Green. J Electrostatics 42: 279, 1997.

127. MP Hughes, H Morgan. Anal Chem 71: 3441, 1999.

128. BJ Berne, R Pecora. Dynamic Light Scattering. New York: Wiley, 1976.

129. K Schätzel, J Merz. J Chem Phys 81: 2482, 1984.

130. K Schätzel. Appl Phys B 42: 193, 1987.

131. MS Talary, JPH Burt, JA Tame, R Pethig. J Phys D: Appl Phys 29: 2198, 1996.

132. P Gascoyne, XB Wang, J Vykoukal, H Ackler, S Swierkowski, P Krulevitch. A microfluidic device combining dielectrophoresis and field flow fractionation for particle and cell discrimination. Solid-State Sensor and Actuator Workshop, Hilton Head, South Carolina, 1998, pp 37–38.

133. T Schnelle, T Müller, G Gradl, SG Shirley, G Fuhr. Electrophoresis 21: 66, 2000.

134. AV Delgado, F Gonzalez-Caballero, FJ Arroyo, F Carrique, SS Dukhin, IA Razilov. Colloids Surfaces A 131: 95, 1998.

135. C Grosse, VN Shilov.v J Phys Chem 100: 1771, 1996.

136. P Wust, R Felix, P Deuflhard. Spektrum Wiss 12: 78, 1999.

137. LD Landau, EM Lifschitz. Elektrodynamik der Kontinua. vol. 8. Berlin: Akademie Verlag, 1985. (In German.)

138. U Stille. Arch Elektrotech 38: 91, 1944. (In German.)

139. JA Osborn. Phys Rev 67: 351, 1945.

140. EC Stoner. Phil Mag 36: 308, 1945.

14

Electric Birefringence Spectroscopy: A New Electrokinetic Technique

TOMMASO BELLINI   Politecnico di Milano, Milan, Italy

FRANCESCO MANTEGAZZA   Università di Milano-Bicocca, Monza, Italy

I. INTRODUCTION

Aqueous solutions of proteins, nucleic acids, certain macromolecules, ionic micelles, colloids, and other polyelectrolyte systems respond to external electric fields in complicated ways. Mainly, the complexity stems from the presence of screened electric charge, since polyelectrolytes dissociate into large polyions and numerous small counterions. Because of the complexity and variety of the phenomena involved, the response of polyelectrolyte solutions to electric fields is still only partially understood. This chapter is devoted to a theoretical and experimental study of the electro-optic response of polyelectrolyte solutions.

The theoretical description of polyelectrolytes in the presence of an external electric field has to deal with the mobile charges in the diffuse atmosphere surrounding each particle. The most complete theoretical framework used to describe this matter is the so-called standard electrokinetic (SE) model [1–4]. The SE model embodies all the relevant electrical and hydrodynamic phenomena: charge conservation, description of hydrodynamics through the Navier–Stokes law, Poisson equation, and boundary conditions for the electric field, ionic diffusion, and conduction. Comprehensive numerical solutions of the SE model are available for spherical particles only [3]. In situations of thin double layer and/or low surface potential, approximate or asymptotic analytical solutions are available for spherical particles [5–10], for spheroidal particles at zero frequency [11], and for spheroidal particles at high frequency [12, 13]. Applications of the SE model allow interpretation of measurements of the electrophoretic mobility of individual particles and of the (complex) dielectric constant and conductivity of dilute suspensions. However, agreement between theory and experiment is poor. The analysis of such disagreement is made more difficult because few experimental techniques directly access the pertinent properties. Indeed, none of the classical electrokinetic methodologies (such as electrophoresis and dielectric spectroscopy) directly measure the particle's electric polarizability, α, which gauges the particle's reaction to an electric field in a direct way, and reflects the distribution of ions as determined by a combination of surface group ionization, diffusion, electromigration, and hydrodynamics. In this chapter we show that electric birefringence (EB) is a powerful tool providing direct information about α. Accordingly, this technique is more efficient, in many respects, than dielectric spectroscopy and conductivity measurements.

The EB experiment basically consists in biasing the orientational distribution of suspended anisotropic colloids by applying an external electric field, thus inducing in the sample a preferred orientation of the particles along the field. The amplitude of such a bias is measured by optical means, i.e., by detecting the birefringence of the suspension. Accordingly, EB is a technique that measures the electric torque on each particle. As such, it can be applied to electrically and optically anisotropic particles only, and it is highly sensitive to the magnitude of the anisotropy. This contrasts with the classical electrokinetic methodologies which, instead, typically measure orientationally averaged quantities, and are not particularly sensitive to anisotropies in either the particle shape, its internal structure, or its surface charge distribution.

EB is not a new technique [14, 15] and many measurements have been made on, e.g., tobacco mosaic virus, DNA, flexible polymer chains, and optically anisotropic colloids. Most experiments focus on the time decay of the induced birefringence after the electric field is switched off. The decay of the Kerr constant (the quantity measured by EB experiments) is controlled by the rotational correlation time, τ_R, which depends on particle geometry, solvent viscosity, and temperature [14]. Few studies are devoted to the steady-state amplitude of the Kerr constant in an a.c. field – the frequency-resolved electric birefringence (FREB) approach employed in our experiments [16, 17]. Especially important in this regard is the pioneering work of O'Konski, who disclosed the enhancement of the measured values of the Kerr constant with respect to those predicted from particle-shape anisotropy and bulk dielectric constants using the simple electrostatic theory [18]. As is now known, this dichotomy arises from transport processes in the diffuse double layer around each particle, i.e., it is an electrokinetic effect [19].

Only recently has the SE model been recognized as the correct environment, at least in principle, to describe the EB of polyelectrolytes. When O'Konski first observed the double layer-induced enhancement of the Kerr constant, little was known about electrokinetic phenomena and their interpretation. Thus, he proposed a Maxwell–Wagner (MW) model [1], where charged particles are treated as "equivalent" conductive particles [20]. As will be explained in this chapter, the MW approach is certainly more effective in describing the Kerr constant rather than the dielectric constant. In the form proposed by O'Konski, though, the model yielded unsatisfactory results: first, it failed to explain the low-frequency behavior and, second, it did not make contact with the physicochemical properties of the particles. Such a contact requires, in fact, connecting the MW model with the microscopic processes taking place on the particle surface. Even more recently, no serious attempt was made to connect results from EB experiments on polyelectrolytes with the modern knowledge of electrokinetic properties of charged colloids. This is in part due to the different environments of scientists working on electrokinetic phenomena and those performing EB experiments. This "cultural" misconnection has also a more specific justification since comprehensive solutions of the SE model exist only for spheres, but EB can be measured only with nonspherical particles.* The first use of approximate solutions of the SE model for nonspherical particles to analyze the low-frequency behavior of the Kerr constant of polyelectrolytes is quite recent [19].

In this chapter we present a theory for EB and a complete set of measurements interpreted with it. The experiments have been performed at various values of q, the charge density on the particle surface, controlled by adsorbing monolayers of cationic, anionic, or

---

* There are rare examples of spherical particles with internal anisotropic dielectric properties. In this case, the external electric field can indeed induce particle orientation.

nonionic surfactants on the particle surface. The amount of adsorbed surfactant molecules has been monitored by light-scattering techniques. In comparison with the values of q found in the literature, the range of surface charges here explored is rather large, extending from $q \approx 1 \ \mu C/cm^2$ (for particles coated by nonionic surfactants) to $q \approx 10 \ \mu C/cm^2$ (for particles coated by ionic surfactants).

For the sake of simplification, the structure of this chapter is divided into two parts: theory and experiments. Furthermore, since the relevance of the various processes contributing to the polarizability of the particles strongly depends on frequency, we will split our discussion into two main frequency regimes, distinguishing a low frequency (1 kHz < ν < 1 MHz) and high frequency (1 MHz < ν < 200 MHz) behavior, where different models apply.

In the high-frequency regime, the EB data display a richly structured frequency dependence, featuring two characteristic frequencies and (in some cases) a minimum before reaching the high-frequency asymptote. From the EB data we calculate the electric polarizability and interpret the results by using an extended Maxwell–Wagner (EMW) model, here described. By fitting the model to the experimental data, the effective particle surface charge is derived in a new and simple way. This is a relevant result since, despite its importance, ascertaining the effective electric charge of a colloidal particle remains an open question. The surface charge density is found to depend heavily on the strength of the interaction between the particle surface and the specific surfactant used in the experiments.

In the low-frequency regime, the EB data obtained with weakly charged particles are compatible with approximate electrokinetic models for the d.c. behavior of rod-like particles already available in the literature [11]. The same agreement is instead not found when studying highly charged particles, since in this case the measured EB is too large in comparison with the predictions. This disagreement between low-frequency EB data and theory may be a consequence of the approximations inherent in the available low-frequency models.

II. GENERAL THEORY OF ELECTRIC BIREFRINGENCE

The EB experiment involves measuring the induced optical anisotropy of a suspension. For a suspension of rod-like particles with no external forces, Brownian motion ensures that the dispersion is isotropic. Put another way, the time-averaged distribution of vectors describing the orientation of individual particles is uniform. Application of an electric field biases the orientation distribution due to the electric torque exerted on the particles. The optical anisotropy of the dispersion is reflected in the induced birefringence, defined as $\Delta n = n_\parallel - n_\perp$, where $n_\parallel$ and $n_\perp$ are the refractive indices detected by a light beam with a linear polarization parallel or perpendicular to the external electric field E_0. When E_0 is small, Δn is proportional to the square of the field. The Kerr constant, B, is defined as [15].

$$B = \frac{\Delta n}{\lambda E_0^2} \tag{1}$$

where λ is the wavelength of the incident light. For a dilute system, Δn is proportional to the particle number density N. In FREB experiments a sine-wave voltage of frequency ν is applied to the electrodes of a cell. Given in complex notation:

$$E_0 = \mathrm{Re}[E_{0\nu}e^{-i2\pi\nu t}] \tag{2}$$

After a starting transient, the signal consists of oscillatory and steady components, Δn_{ac} and Δn_{dc}. FREB measures the d.c. component of the induced birefringence as a function of frequency, $\Delta n_{dc}(\nu)$. The frequency-dependent d.c. value of the Kerr constant, $B(\nu)$ is

$$B(\nu) = \frac{\Delta n_{dc}(\nu)}{\lambda (E_0^2)_{dc}} \tag{3}$$

where $(E_0^2)_{dc} = E_{0\nu}^2/2$ is the mean-square magnitude of the applied electric field. A detailed description of the FREB experimental setup was given in an earlier paper [21].

Our treatment focuses on particles with no permanent dipole and whose polarizability tensor has a symmetry axis. Particle orientation can be specified by the angle, ϑ, between the applied field $\mathbf{E}_0$ and the particle axis. The optical polarizability tensor of a particle is denoted by $\underline{\underline{\alpha}}^o$. The diagonal elements of $\underline{\underline{\alpha}}^o$ are $\alpha_\parallel^o$ and $\alpha_\perp^o$; $\Delta\alpha^o = \alpha_\parallel^o - \alpha_\perp^o$. With these definitions it can be shown [14] that the Kerr constant $B(\nu)$ of the suspension, in SI units is

$$B(\nu) = \frac{N\Delta\alpha^o S_{dc}(\nu)}{\lambda n \varepsilon_0 E_{0\nu}^2} \tag{4}$$

Here, n is the refractive index of the solvent and ε_0 is the vacuum dielectric permittivity; S_{dc} denotes the d.c. component of the nonpolar orientational order parameter, defined as

$$S_{dc}(\nu) \equiv \int_0^\pi P_2(\cos\vartheta) f_{dc}(\vartheta) \sin\vartheta \, d\vartheta \tag{5}$$

where $f_{dc}(\vartheta)$ is the normalized d.c. component (or time-averaged) orientation distribution function, and $P_2(\cdot)$ is the Legendre polynomial of degree 2; $S = 1$ for perfect parallel alignment and $S = 0$ for isotropic alignment. The proportionality between $B(\nu)$ and S_{dc} holds for dilute systems where it is appropriate to neglect co-operative contributions to the optical field experienced by each particle. To calculate $B(\nu)$ it is necessary to know the angular distribution function, $f_{dc}(\vartheta)$, of the suspended rod-like particles in the presence of an oscillatory field. In the simplest case, i.e., when dealing with noncharged, nonconductive particles in a nonconductive medium, $f_{dc}(\vartheta)$ is proportional to the Boltzmann factor $\exp[-U_{dc}(\vartheta)/k_B T]$ where $U_{dc}(\vartheta)$ is the d.c. component of the electrostatic potential energy of the particle, and $k_B T$ is the thermal energy. This expression cannot be naively extended to include the situation of charged particles in aqueous solution because, in such a case, phenomena other than electrostatic coupling, such as ionic conduction, hydrodynamics, and ion diffusion, take place around the particle and, potentially, contribute to the particle alignment. Moreover, being electro-osmosis and ionic conduction dissipative phenomena, it is not clear, even in principle, if an energetic argument could be properly developed. For these reasons it appears to be more appropriate, and secure, to relate the Kerr constant to the forces acting on the particles rather than to their potential energy. We find [22] that $f_{dc}(\vartheta)$ depends on to the torque $\tau_{dc}(\vartheta)$ experienced by the particles as follows:

$$f_{dc}(\vartheta) = F \exp\left(\frac{1}{k_B T} \int_0^\theta \tau_{dc}(\vartheta) \cdot \mathbf{u}_\vartheta d\vartheta\right) \tag{6}$$

where F is a normalization constant and $\mathbf{u}_\vartheta$ represents the unit vector in the ϑ direction. Using Eq. (6) in Eqs (4) and (5) connects $B(\nu)$ with τ_{dc}. The separation of d.c. and a.c. components of the Kerr constant is a consequence of its linear response to the squared electric field [23]. Consequently, $B(\nu)$ can be considered separately for each frequency. It is worth noticing that Eq. (6) only holds for a stationary torque. The oscillatory component of the induced birefringence is computed using a Smoluchowski equation to establish $f_{ac}(\vartheta)$.

In principle, the orientational ordering of colloidal particles exposed to an electric field involves electric and hydrodynamic effects, since the total torque on the particle contains both, i.e., $\tau = \tau_E + \tau_H$. Both components of the torque can be obtained by evaluating the moment of the forces acting on the particle surface:

$$\tau_{E,H} = \int_S \mathbf{r} \times \left(\underline{\underline{\sigma}}_{E,H} \cdot \mathbf{n}\right) dS \tag{7}$$

where $\mathbf{r}$ is the vector identifying the position of dS (element of the particle surface S), $\mathbf{n}$ is the unit outer normal, $\underline{\underline{\sigma}}_H$ is the hydrodynamic stress tensor, accounting for viscous forces and pressure, and $\underline{\underline{\sigma}}_E$ is the Maxwell electric stress tensor. In the EB experiments discussed here, the Reynolds number is low and the hydrodynamics can be described by the Stokes equations with an extra body force to account for electrical effects [1]. The computation of Eq. (7) is simplified when it is recognized that, at equilibrium, the integral performed on any surface including the particle gives the same result as the integral performed on the particle surface [24].

Let us first evaluate the d.c. component of the electric torque, $\tau_{dc,E}$. For a dilute system, the electric field around a particle is the superposition of three terms, $\mathbf{E} = \mathbf{E}_0 + \mathbf{E}_1 + \mathbf{E}_2$, where $\mathbf{E}_0$ is the uniform external field (here assumed to be an a.c. field), $\mathbf{E}_1$ is the (d.c.) field within the unperturbed double layer, and $\mathbf{E}_2$ is the (a.c.) electric field due to the induced polarization of the particle. In the case of *spherical* particles, the latter contribution is a dipolar field oscillating at the same frequency ν of the applied external field. The amplitude of the associated dipolar potential $\Psi_2(\mathbf{r}) = \mathrm{Re}\lfloor \Psi_{2,\nu}(\mathbf{r})e^{-i2\pi\nu t}\rfloor$ is

$$\Psi_{2,\nu}(\mathbf{r}) = \frac{1}{4\pi\varepsilon_0\varepsilon_s}\frac{\alpha\mathbf{E}_{0\nu}\cdot\mathbf{r}}{r^3} \tag{8}$$

Here, $\mathbf{r}$ is the distance from the center of the sphere, ε_S is the relative dielectric constant of the solvent, and α indicates the electric polarizability of the particle accompanied by its counterion cloud; α can be obtained from the "standard" set of electrokinetic equations and boundary conditions (the SE model) either numerically [3] or through approximate analytical solutions [6,7,9]. It is important to realize that α is, even in the case of a d.c. field, a "kinetic" quantity, determined by a dynamic local balance of incoming and outgoing ions carried by hydrodynamic, diffusion, and electromigration flows. Thus, the electric polarizability of polyelectrolytes is, in general, a complex variable because of the finite mobility of ions and because of the finite velocity of the solvent flows.

For *nonspherical* particles, $\mathbf{E}_2$ contains multipolar terms of various orders. Thus, the associated electric potential Ψ_2 is

$$\Psi_{2,\nu}(\mathbf{r}) = \frac{(\underline{\underline{\alpha}}\mathbf{E}_{0,\nu})\cdot\mathbf{r}}{r^3} + \frac{1}{2}\sum_{i,j}Q_{ij}\frac{r_i r_j}{r^5} + \tag{9}$$

where $\underline{\underline{\alpha}}$ is the electric polarizability tensor while Q_{ij} is the quadrupolar tensor. The diagonal elements of $\underline{\underline{\alpha}}$, $\alpha_\parallel$ and $\alpha_\perp$, represent the electric polarizability of the particle along and perpendicular to its axis.

The electric stress tensor is given by [25]

$$\underline{\underline{\sigma}}_E = \varepsilon_0\varepsilon_s\left(\mathbf{EE} - \frac{1}{2}\mathbf{E}\cdot\mathbf{E}\underline{\underline{\delta}}\right) \tag{10}$$

where $\mathbf{EE}$ is a dyadic product and $\underline{\underline{\delta}}$ is the unit tensor. Note that Eq. (10) contains ε_s and not ε_e. This is because, in the electrokinetic model, counterions and coions are not considered part of the solvent, but rather described through their local densities.

We introduce the complex notation for $\mathbf{E}$ in Eq. (10): $\mathbf{E} = \mathrm{Re}[\mathbf{E}_\nu e^{-i2\pi\nu t}]$. By retaining only the nonoscillatory terms, i.e., by averaging $\underline{\underline{\sigma}}_E$ over a period, we obtain:

$$\underline{\underline{\sigma}}_{E,dc} = \frac{1}{2}\varepsilon_0\varepsilon_s\left(\mathrm{Re}(\mathbf{E}_\nu\mathbf{E}_\nu^*) - \frac{1}{2}\mathbf{E}_\nu\cdot\mathbf{E}_\nu^*\underline{\underline{\delta}}\right) \tag{11}$$

The asterisk indicates a complex conjugate. The nonoscillatory component of the torque is then

$$\tau_{dc,E} = \int_S \mathbf{r} \times \frac{1}{2}\varepsilon_0\varepsilon_s\left(\mathrm{Re}(\mathbf{E}_\nu\mathbf{E}_\nu^*) - \frac{1}{2}\mathbf{E}_\nu\cdot\mathbf{E}_\nu^*\underline{\underline{\delta}}\right)\cdot\mathbf{n}\,dS \tag{12}$$

By performing the surface integral in Eq. (12), we obtain:

$$\tau_{dc,E} = -\frac{1}{2}\Delta\alpha' E_{0\nu}^2 \sin\vartheta\cos\vartheta\mathbf{u}_\vartheta \tag{13}$$

Here, $\Delta\alpha' = \alpha'_\parallel - \alpha'_\perp$, where $\alpha'_\parallel$ and $\alpha'_\perp$ indicate, respectively, the real parts of $\alpha_\parallel$ and $\alpha_\perp$. It is worth noting that no contribution connected to multipolar terms in the electric potential higher than the dipole appears in Eq. (13). Hence, even if the nonspherical shape of the particle gives rise to a complicated pattern of electric fields, when it amounts to calculate the torque, only the dipolar term really matters. Indeed, the torque in Eq. (13) could be more simply expressed as $\tau_{dc,E} = \langle\mathbf{p}\times\mathbf{E}_0\rangle$, where the electric dipole $\mathbf{p} = \mathrm{Re}(\underline{\underline{\alpha}}\cdot\mathbf{E}_{0\nu}e^{-i2\pi\nu t})$.

The evaluation of the d.c. component of the torque $\tau_{dc,H}$ is more complicated. The basic problem is that, since the electric torque is proportional to E_0^2, we need to account for the hydrodynamics to the same extent. Calculations of the solvent flows to the E^2 level are not availbale even for the simplified geometry of a sphere. Thus, the only accessible way to tackle the problem is to use symmetry arguments. Along this line, Teubner [26] has shown that the hydrodynamic torque is zero for centrosymmetric, moderately charged particles with thin double layers. Furthermore, it is not implausible that the torque also vanishes for highly charged particles with thin double layers since expansions for the velocity, pressure, and electric fields for rigid-body motion proceed in *odd* powers of the applied field. However, no results exist for highly charged particles with thin double layers, the situation considered here. Accordingly, we cannot exclude, in general, a hydrodynamic contribution to the Kerr constant.

In order to give a more physical, although incomplete, picture of the issues related to the presence of hydrodynamic torque on the particles, let us consider the high-frequency situation, where the hydrodynamic penetration distance [1], $\sqrt{\eta_f/(2\pi\nu)}$, where η_f is the kinematic viscosity, is shorter than the size of the particle. Specifically, this condition, for a 0.5 μm radius particle, is obtained for $\nu > 1$ MHz. In such a case, the hydrodynamic flows, driven by surface electro-osmosis, are limited to a surface region whose thickness is less than the particle curvature and thus can be considered as local. Let us compare the forces due to hydrodynamic flow at two positions on the particle surface that are symmetric through the center. Given the centrosymmetry of the particle, the surface at either position has the same orientation. An inspection of the symmetry of the multipolar terms reveals also that E_0 and E_2 are the same at the two positions, while E_1 is reversed. Thus, the component of E tangential to the surface, $E_\parallel(r)$, is the same in the two positions, while the normal component, $E_\perp(r)$, differs. Therefore, to first order in E, the double layer is perturbed by $E_\perp$ in a different way in the two positions, while the electro-osmotic flow driven by $E_\parallel$, being equal in the two symmetry positions, gives rise to no torque. To second order in E, one should take into account the effects of the double-layer perturbation on the electro-osmotic flow. As the double-layer perturbation is noncentrosymmetric, the presence of a nonzero $\tau_{dc,H}$ to $O(E^2)$ cannot be ruled out using a pure symmetry argument.

In what follows we will simply assume, stretching the validity range of the Teubner argument [26], that the electric torque calculated by taking into account the effect of electrokinetics on the induced dipole [i.e., Eq. (13)] provides a good approximation. By inserting Eq. (13) into Eq. (6), and approximating $f_{dc}(\vartheta)$ by its first-order expansion in the electric field, from Eq. (4) we obtain:

$$B(\nu) = \frac{N \Delta \alpha^{o} \Delta \alpha'(\nu)}{15 \pi \lambda n \varepsilon_0 k_B T} \tag{14}$$

It follows that the problem of interpreting the Kerr constant in a system of rod-like charged particles has been reduced to calculating the parallel and perpendicular components of the real part of the electric polarizability as a function of frequency. Since the torque in Eq. (13) is formally equal to that expected for the simple case of noncharged particles in a nonconductive medium, Eq. (14) is similar to equations previously proposed. We want to stress, however, that Eq. (14) has been obtained for a dissipative system, α resulting from a kinetic balance rather than from bulk (molecular) polarizabilities. Theories for the calculation of α for charged particles will be discussed in the following sections.

III. HIGH-FREQUENCY THEORY OF ELECTRIC POLARIZATION OF SPHERES

Ideally, an EB theory should provide an understanding of the polarizability of a spheroid in terms of colloidal variables – size, shape, and charge of the particle, and the ionic strength of the suspending solution. This entails finding ways to deal with the relevant electrokinetic processes. Fortunately, this can be done using asymptotic methods in the high-frequency regime. To illustrate matters, we look first at the high-frequency electric polarizability of a sphere and show that in the megahertz range, the polarizability can be interpreted in electrokinetic terms by a simple extension of the MW model [1]. This is a key point, since MW-based calculations for ellipsoidal particles are straightforward. To establish the validity of the asymptotic approach, in this section we compare exact and approximate results for spherical particles.

The classical MW model represents a suspended polyelectrolyte as a conducting particle imbedded in a conductive medium. Polarization arises from the conductivity mismatch between the solvent and the particle. Due to the finite resistivity of the bodies involved, the induced dipole moment following a MW polarization process exhibits a phase lag with respect to the applied field, and the electric polarizability is thus a frequency-dependent quantity. The MW electric polarizability of a sphere, α_{MWS}, is a simple function of the complex dielectric constants of particle and suspending medium [25]:

$$\alpha_{MWS} = 4\pi\varepsilon_0\varepsilon_s R^3 \frac{\varepsilon_p - \varepsilon_e}{\varepsilon_p + 2\varepsilon_e} \tag{15}$$

Here, R is the particle radius, and ε_p is the complex dielectric constant of the particle, expressed as a function of its real dielectric constant, ε_p', and its volume conductivity, K_p:

$$\varepsilon_p = \varepsilon_p' + i\frac{K_p}{\omega\varepsilon_0} \tag{16}$$

Similarly, the complex dielectric constant of the solvent (water plus added electrolyte) is

$$\varepsilon_e = \varepsilon_s + i\frac{K_e}{\omega\varepsilon_0} \tag{17}$$

where K_e is the conductivity of electrolyte solution in the absence of colloidal particles. Here, ε_s and K_e are assumed to be real quantities with K_e proportional to the ionic strength of the solution; $\omega = 2\pi\nu$.

To show its Debye-like frequency dependence, the real part of α_{MWS} can be rearranged to give

$$\alpha_{MWS}' = \alpha^\infty + \frac{\alpha_{MWS}^0 - \alpha^\infty}{1 + \omega^2\tau_{MWS}^2} \tag{18}$$

where τ_{MWS} is the characteristic time, and α^∞ and α_{MWS}^0 are the MW electric polarizabilities at high and low frequencies, respectively. They can be expressed as*

$$\tau_{MWS} = \varepsilon_0 \frac{\varepsilon_p' + 2\varepsilon_s}{K_p + 2K_e} \tag{19}$$

$$\alpha^\infty = 4\pi\varepsilon_0\varepsilon_s R^3 \frac{\varepsilon_p' - \varepsilon_s}{\varepsilon_p' + 2\varepsilon_s} \tag{20}$$

$$\alpha_{MWS}^0 = 4\pi\varepsilon_0\varepsilon_s R^3 \frac{K_p - K_e}{K_p + 2K_e} \tag{21}$$

To calculate the polarizability of a charged spherical particle using the above equations, it is necessary to know the complex dielectric constant ε_p of the particle and, in particular, its conductivity K_p. Clearly, K_p has to be understood as an effective property of the particle which embodies different processes related to the double layer. This will be done in two steps: first, showing the equivalence between the polarizability of a sphere having a conductive surface versus a sphere having a conductive body, and second, mapping the electrokinetic processes into a surface conductivity. This combination of elements from the SE model [5,10] with the MW approach [1,18] constitutes an original model, which we will refer to as the EMW model. The EMW allows us to explain, unambiguously, the high-frequency structure of polarizability and EB. In this section we present an

---

* The real part of the polarizabilities at low and high frequency are indicated without the superscript (′), since at those frequencies the imaginary part is negligible.

EMW theory valid for spherical particles; the extension to spheroidal particles is more complicated and will be the subject of Section IV.

As shown by O'Konski [18], in his treatment of the MW polarizability of spherical particles, the volume conductivity K_p is equivalent to surface conductivity K_p^s in the following sense: a nonconducting spherical particle with surface conductivity K_p^s produces the same dipolar field as a particle with volume conductivity equal to

$$K_p = \frac{2K_p^s}{R} \tag{22}$$

Now, it turns out that for highly charged colloidal particles with thin diffuse layers, the electrokinetic features of the polarizability can be represented with an asymptotic theory by employing an appropriate value for surface conductivity. Building on the seminal work of Dukhin and Shilov [8], O'Brien and Rowlands [12] showed how this can be justified at high frequencies, i.e., where $D/R^2 \ll \omega \approx K_e/\varepsilon_0\varepsilon_s$, since the influence of the oscillating field on ion concentrations is negligible. Here, D is the diffusion coefficient of the free counterions. Thus, at high ionic strengths, high potentials, and high frequencies, the effects of processes around a sphere can be encapsulated in an equivalent particle surface conductivity $K_p^s(q, \kappa, K_e)$ expressed in terms of the electric surface charge of the particle, q, of the Debye length, κ^{-1}, and of the solvent conductivity, K_e. To relate the surface and volume conductivities to the particle charge we use Bikerman's expression [27] for the surface conductivity near a highly charged flat surface in a 1:1 electrolyte, viz.,

$$K_p^s = \kappa^{-1}\left[\exp\left(\left|\frac{e\zeta}{2k_B T}\right|\right) - 1\right](1 + 3m)K_e \tag{23}$$

In Equation (23) contributions from coions are neglected since their concentration is small near a surface with a large surface electric potential ζ. Here, e is the electron charge and m is a nondimensional ionic drag coefficient ($m = 0.18$ for KCl). Equation (23) is an exact expression for the incremental conductivity due to increased ion density and electro-osmosis along a flat surface. Moreover, at sufficiently high frequencies, electro-osmosis on a curved surface becomes a local process since the viscous penetration distance shortens. Thus, even for spheroidal particles or particles with slowly varying surface properties, the Bikerman expression provides the local surface conductivity at high frequency. This is further confirmed by the fact that Eq. (23) is identical to O'Brien's asymptotic formula for spheres [10] with thin double layers.

Next, because the ζ potential and surface charge density, q, for a 1 : 1 electrolyte are related as [1]

$$q = 2\frac{\varepsilon_s\varepsilon_0 k_B T}{e}\kappa \sinh\left(\frac{e\zeta}{2k_B T}\right) \tag{24}$$

when $\zeta \gg 2k_B T/e$, Eq. (23) can be expressed as

$$K_p^s(\mathbf{r}) \approx q(\mathbf{r})\frac{K_e}{zeI}(1 + 3m) \tag{25}$$

The **r**-position dependence has been added to emphasize the local nature of the Bikerman expression, i.e., the local conductivity depends on the local surface charge density. Here, I denotes the ionic strength of the solution.

It is interesting to note that we can obtain the relation between particle conductivity and charge in a much rougher way, i.e., by interpreting K_p as the conductivity given by an appropriate number of charges with their bulk mobility in an aqueous solution. Let us ideally replace the real particle with a droplet of water of the same size, containing a sufficient number of counterions to give the desired conductivity. In this way the particle charge (number of counterions in the droplet) becomes a simple function of the particle conductivity. The result is given by the following expression:

$$K_{\mathrm{p}}^{\mathrm{s}}(\mathbf{r}) = \frac{Rq(\mathbf{r})S}{2N_A Ve} \Lambda = \frac{3q(\mathbf{r})}{2N_A e} \Lambda \tag{26}$$

where V and S are the volume and surface of the particle, respectively, N_A is the Avogadro number, and Λ is the limiting ionic conductivity ($\Lambda \approx 75 \times 10^{-4} \Omega^{-1} \mathrm{m}^2 \mathrm{M}^{-1}$ for K^+). Since $K_e = 2 I \Lambda$, we observe that Eqs (25) and (26) yield, within 3%, the same value for K_p.

In synthesis, the EMW model construction involves several steps. First, the surface conductivity near a nonconducting particle, K_p^S, is computed with an asymptotic model that maps processes near the surface into a surface conductivity, Eq. (23). This is converted into an *equivalent* volume conductivity K_p of a sphere using the O'Konski relation [18], Eq. (22). Finally, using K_p in Eqs (15)–(17), the particle polarizability, α_{EMW}, is calculated according to the MW model. Basically, the EMW model relies on two main simplifications, i.e., on the notion that, at high enough frequency, ion diffusion can be neglected and hydrodynamic flows are local, their effect being reduced to a (minor) renormalization of the counterion conductivity in the double layer.

To demonstrate the validity of this approach we compare predictions computed from the EMW model just described, with analogous predictions of the complete SE model for spheres. Figure 1 shows $\alpha'(\nu)/(3\varepsilon_0\varepsilon_s V)$, the real part of the dimensionless particle electric polarizability, according to the SE model [3] for a specific choice of parameters (see figure caption). As evident, $\alpha'(\nu)$ has a rich frequency dependence over a wide range: $\alpha'(\nu)$ is an increasing function of ν up to 1 MHz and a decreasing function of ν at higher frequencies. It is interesting to note that the polarizability increases with ν when 1 kHz $< \nu <$ 1 MHz, suggesting that low-frequency processes (mainly ion diffusion) have the overall effect of reducing the particle polarizability. In Fig. 1 we also show (dashed line) the behavior of $\alpha'_{\mathrm{MWS}}(\nu)$, the real part of $\alpha_{\mathrm{MWS}}(\nu)$ for a sphere as given by Eq. (15), calculated from the same dielectric constants and solvent conductivity used in the numerical evaluation of the SE model, but with K_p "adjusted" as a free parameter to give an accurate representation of α as given by the SE model. As the figure demonstrates, when K_p is freely chosen, the polarizabilities calculated from the SE and MW models share the same high-frequency characteristics, supporting the notion that the high-frequency behavior of the complete SE model exhibits basic MW behavior. Figure 1 also shows (continuous line) the real part of $\alpha_{\mathrm{MWS}}(\nu)$ calculated (no free parameter) by using the EMW model, i.e., from Eq. (15) with K_p given by Eqs (22) and (23). Clearly, the EMW model accurately captures the high-frequency behavior of the SE model. The small differences between polarizabilities calculated from the SE model and the EMW model, stem from approximations inherent in the asymptotic formula for the conductivity. At frequencies below 1 MHz, the dipolar coefficient follows an "electrokinetic" regime where processes neglected in the EMW model are relevant. The common high frequency (above 200 MHz) asymptote of particle polarizability (α^∞) represents the polarizability of a dielectric sphere in a dielectric medium.

To test further the correspondence between the exact numerical solution of the SE model and our EMW model, the comparison shown in Fig. 1 was repeated for different choices of ζ potentials and ionic strengths. By fitting the high-frequency decay with K_p as free parameter, the "effective" particle volume conductivity was extracted; K_p was then also calculated, for the same condition, following the EMW prescriptions, from Eq. (23)

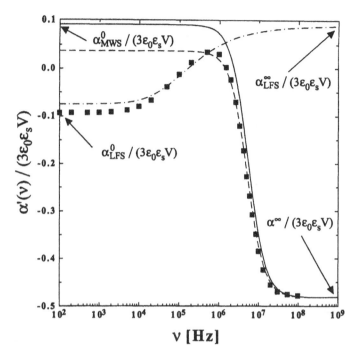

FIG. 1 Real part of the dimensionless particle electric polarizability, $\alpha'(\nu)/(3\varepsilon_0\varepsilon_s V)$, calculated for a sphere as a function of the frequency of the applied field, ν. Square: $\alpha'(\nu)$ calculated using numerical solutions of the SE model [3]; solid line: $\alpha'(\nu)$ calculated from the EMW model for spheres; dashed line: $\alpha'(\nu) = \alpha'_{MWS}(\nu)$ calculated using the MW model for spheres with K_p as an adjustable parameter to fit the SE model; dotted–dashed line: $\alpha'(\nu)$ calculated using Eqs (56) and (57). Parameters: particle radius $R = 135$ nm; ζ potential $= 100$ mV; ionic strength $I = 1$ mM KCl. The arrows indicate low- and high-frequency asymptotic values.

and the O'Konski relationship. The values of K_p obtained in the two ways are compared in Fig. 2 as a function of the ionic strength for different ζ potentials. The EMW model is found to be accurate at high potentials, almost independently of ionic strength.

IV. EXTENDED MAXWELL–WAGNER THEORY FOR HIGH-FREQUENCY POLARIZABILITY OF SPHEROIDS

To adapt the EMW approach to particles of spheroidal shape, different problems must be addressed. Following the sequence of the above paragraph, we first introduce the MW results for spheroids. In analogy with the O'Konski equation for spheres, we then construct relations mapping a spheroidal particle with conductive surface into an "effective" bulk-conductive particle having the same shape. Using Bikerman's expression [Eq. (23)] enables us to formulate the EMW model for spheroidal polyelectrolytes and to predict the frequency dispersion in the particle polarizability. It will be shown that an isotropic (independent from the particle orientation) surface conductivity K_p^s yields different particle volume conductivities along, $K_{p\parallel}$, and perpendicular, $K_{p\perp}$, to the particle axis due to geometrical factors alone. The consequences of nonuniform surface charge distribution are then discussed and it is shown how nonuniform charge introduces an additional

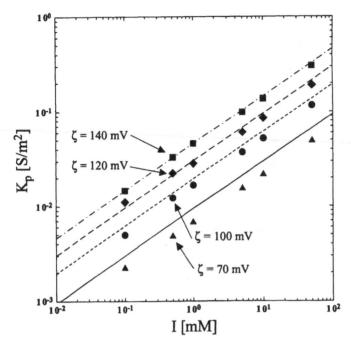

FIG. 2 Comparison between particle conductivities, K_p for spheres calculated in different ways. Symbols: K_p obtained by fitting results from the SE model with the MW model for spheres. Lines: K_p obtained with the EMW model for spheres using Eqs (22) and (23). Triangles and solid line: $\zeta = 70$ mV; circles and short dashed line: $\zeta = 100$ mV; diamonds and long dashed line: $\zeta = 120$ mV; squares and dot–dashed line: $\zeta = 140$ mV. Parameters: $\varepsilon_p' = 2$, $\varepsilon_s = 80$, and various KCl ionic strengths.

anisotropy, which accentuates (or depresses) the spheroid's already anisotropic volume conductivity. Finally, using the EMW model, particle charge can be readily calculated once the volume conductivities are extracted from the measured frequency dispersion of the Kerr constant. In order to clarify the treatment, we call EMW-1 the model developed assuming uniform surface charge, and EMW-2 the model obtained assuming nonuniform surface charge.

A. Maxwell–Wagner Polarizability

To develop explicit expressions we introduce the spheroidal coordinates ξ, χ, η, related to the Cartesian co-ordinates by

$$\begin{cases} x = c \sinh \xi \sin \chi \cos \eta \\ y = c \sinh \xi \sin \chi \sin \eta \\ z = c \cosh \xi \cos \chi \end{cases} \tag{27}$$

The particle surface is represented by a prolate spheroid aligned with the z axis. On the spheroid surface, the ξ co-ordinate is constant ($\xi = \xi_0$) and

$$\frac{z^2}{c^2 \cosh^2 \xi_0} + \frac{x^2 + y^2}{c^2 \sinh^2 \xi_0} = 1 \tag{28}$$

ξ_0 is chosen so that $\tanh \xi_0 = r$, r being the particle's aspect ratio, i.e., $r = a/b$. The particle volume is $V = 4\pi a^2 b/3$. The relationship between the short and long semiaxes, a and b, and the constant c is

$$c = \sqrt{b^2 - a^2} \tag{29}$$

Explicit expressions for the polarizabilities of ellipsoidal particles along and perpendicular to the external electric field follow from the standard electrostatic procedure [25]. These are

$$a_{\parallel,\perp} = 4\pi\varepsilon_0\varepsilon_s a^2 b \frac{\varepsilon_{p\parallel,\perp} - \varepsilon_e}{3(\varepsilon_e + (\varepsilon_{p\parallel,\perp} - \varepsilon_e)L_{\parallel,\perp})} \tag{30}$$

where $L_\parallel$ and $L_\perp$ are the ordinary geometrical depolarization factors for spheroids:

$$L_\parallel = 1 - 2L_\perp = \sinh^2 \xi_0 Q_1 \tag{31}$$

$$Q_1(z) = \frac{z}{2}\ln\left(\frac{z+1}{z-1}\right) - 1 \tag{32}$$

and $z = \cosh \xi_0$. In Eq. (30), the complex dielectric constants of the particle parallel ($\varepsilon_{p\parallel}$) or perpendicular ($\varepsilon_{p\perp}$) to the field are expressed as

$$\varepsilon_{p\parallel,\perp} = \varepsilon_p' + i\frac{K_{p\parallel,\perp}}{\omega\varepsilon_0} \tag{33}$$

where we have assumed that the real part of the dielectric constant of the particle, is the same when the particle is polarized parallel or perpendicular to its symmetry axis. On the other hand, the "effective" particle volume conductivities parallel and perpendicular to the symmetry axis, $K_{p\parallel}$ and $K_{p\perp}$, differ since they reflect particle geometry and transport in the double layer.

As in the case of spherical particles, the real part of the polarizability is a simple Debye function that can be expressed as a function of the high ($\alpha_{\parallel,\perp}^\infty$) and low ($\alpha_{\parallel,\perp}^0$) frequency polarizabilities as

$$\alpha_{\parallel,\perp}' = \alpha_{\parallel,\perp}^\infty + \frac{\alpha_{\parallel,\perp}^0 - \alpha_{\parallel,\perp}^\infty}{1 + \omega^2\tau_{\parallel,\perp}^2} \tag{34}$$

Here, $\tau_\parallel$ and $\tau_\perp$ are the reciprocals of the MW cut-off frequencies; $\alpha_{\parallel,\perp}^\infty$ and $\alpha_{\parallel,\perp}^0$ are the MW electric polarizabilities at high and low frequencies (respectively). They can be expressed as*

$$\tau_{\parallel,\perp} = \varepsilon_0 \frac{(1 - L_{\parallel,\perp})\varepsilon_s + L_{\parallel,\perp}\varepsilon_p'}{(1 - L_{\parallel,\perp})K_e + L_{\parallel,\perp}K_{p\parallel,\perp}} \tag{35}$$

$$\alpha_{\parallel,\perp}^\infty = 4\pi\varepsilon_0\varepsilon_s a^2 b \frac{\varepsilon_p' - \varepsilon_s}{3(\varepsilon_s + (\varepsilon_p' - \varepsilon_s)L_{\parallel,\perp})} \tag{36}$$

$$\alpha_{\parallel,\perp}^0 = 4\pi\varepsilon_0\varepsilon_s a^2 b \frac{K_{p\parallel,\perp} - K_e}{3(K_e + (K_{p\parallel,\perp} - K_e)L_{\parallel,\perp})} \tag{37}$$

---

* See footnote on p. 431.

B. O'Konski Relations for Spheroids

As in the case of spheres, particles having volume conductivity are "equivalent" to particles having a conductive surface, since in both cases the spatial distribution of electric field and charges is the same. Specifically, the electric field outside the particle is, in both cases, purely dipolar. Thus, provided that K_p^s and K_p are related as in Eq. (22), bulk-conducting and surface-conducting particles are electrically indistinguishable. This equivalence does not hold in the case of spheroidal particles: the field around a polarized spheroid is not purely dipolar, and has different multipolar contributions when the particle is a surface conductor and when it is a bulk conductor. A relevant simplification in this matter occurs, though, by recognizing that our present concern is only the dipolar coefficient, since, as detailed in the next paragraph, only the dipolar field contributes to the torque. When the focus is restricted to the dipolar level only, the O'Konski equivalence is extendable to spheroidal particles. To show this, let us first recall that the dipole moment $\mathbf{d}$ can be determined from the surface charged density $q(\mathbf{r})$ as an integral over the particle surface S:

$$d_w = \int_S w q(\mathbf{r}) dS \tag{38}$$

where d_w is the component of $\mathbf{d}$ along a generic co-ordinate w, that we take as directed along the external electric field E_0: when the particle's symmetry axis is parallel to E_0, then $w = z$; when E_0 is perpendicular to the particle axis, w lies in the xy plane.

At the surface of the spheroid we have the usual boundary condition for the surface charge $q(\mathbf{r})$ in some specific position $\mathbf{r}$:

$$q(\mathbf{r}) = \varepsilon_s \varepsilon_0 (\mathbf{E}(\mathbf{r}) \cdot \mathbf{n}(\mathbf{r}))_o - \varepsilon_p' \varepsilon_0 (\mathbf{E}(\mathbf{r}) \cdot \mathbf{n}(\mathbf{r}))_i \tag{39}$$

Here, the subscripts (o) and (i) denote outside and inside the interface, respectively, and $\mathbf{n}$ represents the normal to the surface. The surface charge density can be modified either as a consequence of transport of ions due to bulk conductivity of particle and solvent or due to surface conductivity. Surface charge conservation for the two limiting situations of particles having conductive surface and particles with bulk conductivity yields, respectively:

$$\frac{dq^s(\mathbf{r})}{dt} = -\nabla_s \cdot (K_p^s \mathbf{E}) - K_e (\mathbf{E} \cdot \mathbf{n})_o \tag{40}$$

$$\frac{dq_{\parallel,\perp}^v(\mathbf{r})}{dt} = K_{p\parallel,\perp} (\mathbf{E} \cdot \mathbf{n})_i - K_e (\mathbf{E} \cdot \mathbf{n})_o \tag{41}$$

where the bulk conductivity of the particle is $K_{p\parallel}$ or $K_{p\perp}$, depending on the relative orientation between the particle axis and $\mathbf{E}_0$. In Eq. (41) we let $K_{p\parallel}$ and $K_{p\perp}$ be different because, being "effective" quantities, they could reflect the surface conductivity in different ways; $\nabla_s \cdot (\ )$ represents the surface divergence, and $q_{\parallel,\perp}^v$ and q^s are the resulting surface charges considering the two limiting situations separately. Next, we equate the dipolar moments resulting from the surface charge densities q^s and $q_{\parallel,\perp}^v$:

$$\int_S w K_{p\parallel,\perp} (\mathbf{E} \cdot \mathbf{n})_i dS = -\int_S w \nabla_s \cdot (K_p^s \mathbf{E}) dS \tag{42}$$

This result is the same as the approximation introduced by Dukhin and Shilov [8], although here it has been obtained in a different and much simpler way, Equation (42) can be further simplified by introducing γ, the angle between $\mathbf{E}_0$ and $\mathbf{n}$, γ having different geometrical meanings in the parallel and perpendicular configurations. By noting that $V = \int_S w \cos \gamma \, dS$:

$$K_{p\parallel,\perp} V E = - \int_S w \nabla_s \cdot (K_p^s \mathbf{E}) dS \tag{43}$$

where we have made use of the fact that the electric field inside an insulating/conductive spheroid imbedded in an insulating/conductive medium, under the effect of an electric field, is uniform and, for the parallel and perpendicular geometries, parallel to $\mathbf{E}_0$. Upon integrating by parts, and since $(\nabla_s w)^2 = \sin^2 \gamma$, the integral for the parallel direction can be simplified to

$$K_{p\parallel} V = \int_S K_p^s \sin^2 \gamma \, dS = \int_S K_p^s [1 - (\mathbf{n} \cdot \mathbf{u}_E)^2] dS \tag{44}$$

In the second equality, $\mathbf{u}_E$ is a unit vector parallel to $\mathbf{E}_0$, which, in this geometry, is parallel to the particle's axis. A similar expression can be formed in the case of perpendicular geometry. Upon forming their average we have

$$\frac{K_{p\parallel} + 2K_{p\perp}}{3} = \frac{2}{3V} \int_S K_p^s dS \tag{45}$$

By combining Eq. (45) with the local Bikerman formula, Eq. (25), we obtain an explicit expression relating the charge to the components of the volume conductivity, viz.,

$$\Sigma = \frac{3\kappa^2 \varepsilon_s \varepsilon_0 k_B T}{2e(1 + 3m) K_e} V \frac{K_{p\parallel} + 2K_{p\perp}}{3} \tag{46}$$

Here, $\Sigma = q \, S$ is the total charge of the particle. Equation (46) is a powerful tool since it provides a simple relationship between the equivalent volume conductivities and the total charge of the particle. Moreover, Eqs (45) and (46) are generally valid since they do not depend on the specific χ and η dependence of q and K_p^s.

C. Spheroids with Uniform Surface Charge Density

Two formulations of the EMW model will be discussed. The first (EMW-1) is developed for uniformly charged particles and leads to explicit equations for the particle polarizability as a function of one parameter only, the surface charge density. According to the EMW-1 model, geometry places restrictions on the relative magnitudes of the volume conductivities $K_{p\parallel}$ and $K_{p\perp}$. The second (EMW-2) deals with a nonuniform surface charge distribution with an additional free parameter. In EMW-2, the relative magnitudes of the volume conductivities are no longer constrained by geometry.

When the surface conductivity is uniform, Eq. (44) can be simplified to

$$K_{p\parallel,\perp} V = K_p^s \int_s \sin^2 \gamma \, dS \tag{47}$$

The integral depends on particle geometry and when the electric field is parallel to the particle's axis:

$$K_{p\parallel} = \frac{3}{2a} K_p^s \cosh \xi_0 I_1 \tag{48}$$

with

$$I_1 = (1 - \sinh^2 \xi_0) \sin^{-1} \frac{1}{\cosh \xi_0} + \sinh \xi_0 \tag{49}$$

Similarly, evaluating the integral in Eq. (47) with the spheroid oriented perpendicular to the field yields:

$$K_{p\perp} = \frac{3}{4a} K_p^s \frac{\sinh^2 \xi_0}{\cosh \xi_0} I_2 \tag{50}$$

with

$$I_2 = \left(\frac{1}{\sinh \xi_0} - \sinh \xi_0 \right) + \frac{\cosh^4 \xi_0}{\sinh^2 \xi_0} \sin^{-1} \frac{1}{\cosh \xi_0} \tag{51}$$

Equations (48) and (50) show that the volume conductivities $K_{p\parallel}$ and $K_{p\perp}$ differ even when the surface conductivity is uniform. Their ratio:

$$\frac{K_{p\parallel}}{K_{p\perp}} = 2 \frac{\cosh^2 \xi_0}{\sinh^2 \xi_0} \frac{I_1}{I_2} \tag{52}$$

depends only on the axial ratio of the particle and, for uniformly charged, needle-shaped particles, $K_{p\parallel}/K_{p\perp}$ reaches a maximum value of 2.

Combining the expressions for $K_{p\parallel,\perp}$ with Eq. (30) provides explicit formulas for electric polarizability of uniformly charged spheroids according to the EMW model; $\alpha_{\parallel,\perp}$ are then related to properties of the bulk electrolyte, and shape and effective properties of the particle as

$$\alpha_\parallel = 4\pi\varepsilon_0\varepsilon_s a^2 b \frac{1}{3 \sinh^2 \xi_0 \cosh \xi_0} \frac{-\dfrac{2}{3}\left[1 + i\omega'\left(1 - \dfrac{\varepsilon_p'}{\varepsilon_s}\right)\right] + \lambda_s \cosh \xi_0 I_1}{\dfrac{2}{3}\partial Q_1(1 + i\omega') - \dfrac{2}{3}i\omega'\dfrac{\varepsilon_p'}{\varepsilon_s}\dfrac{Q_1}{\cosh \xi_0} - \lambda_1 I_1 Q_1} \tag{53}$$

$$\alpha_\perp = 4\pi\varepsilon_0\varepsilon_s a^2 b \frac{2}{3 \cosh \xi_0 \sinh^3 \xi_0} \frac{\dfrac{4}{3}\cosh \xi_0\left[1 + i\omega'\left(1 - \dfrac{\varepsilon_p'}{\varepsilon_s}\right)\right] + \lambda_s \sinh^2 \xi_0 I_2}{\dfrac{4}{3}\partial Q_1^1(1 + i\omega') - \dfrac{4}{3}i\omega'\dfrac{\varepsilon_p'}{\varepsilon_s}\dfrac{Q_1^1}{\sinh^2 \xi_0}\cosh \xi_0 - \lambda_1 Q_1^1 I_2} \tag{54}$$

Here, $\partial Q_1(z)$ and $\partial Q_1^1(z)$ are, respectively, the z derivatives of $Q_1(z)$ and $Q_1^1(z)$, defined as

$$Q_1^1(z) = \frac{\sqrt{z^2 - 1}}{2} \ln \frac{z + 1}{z - 1} - \frac{z}{\sqrt{z^2 - 1}} \tag{55}$$

with $z = \cosh \xi_0$. The normalized angular frequency, ω', is given by $\omega\varepsilon_0\varepsilon_s/K_e$, and the dimensionless conductivity ratio is $\lambda_s = K_p^s/(K_e a)$. Equations (53) and (54) constitute the essence of the EMW-1 model.

Figure 3 shows the real part of the dimensionless polarizabilities, $\alpha_\parallel'/(3\varepsilon_0\varepsilon_s V)$ and $\alpha_\perp'/(3\varepsilon_0\varepsilon_s V)$, calculated by using Eqs (53) and (54), for a uniformly charged spheroid with

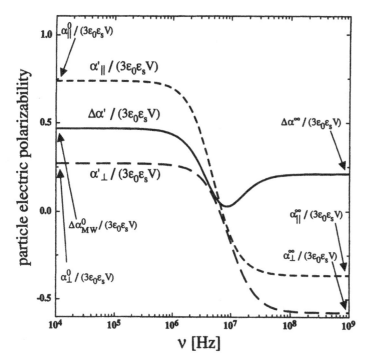

FIG. 3 Real part of the parallel $\alpha'_{\parallel}/(3\varepsilon_0\varepsilon_s V)$ (dashed line) and perpendicular $\alpha'_{\perp}/(3\varepsilon_0\varepsilon_s V)$ (long dashed line) dimensionless particle electric polarizability, for a prolate spheroid calculated with the EMW-1 model. The solid line represents the difference between the two quantities: $\Delta\alpha'/(3\varepsilon_0\varepsilon_s V) = (\alpha'_{\parallel} - \alpha'_{\perp})/(3\varepsilon_0\varepsilon_s V)$. Long semiaxis = 240 nm; $\zeta = 120$ mV; $I = 1$ mM KCl; axial ratio = 1/3. The arrows indicate the high- and low-frequency values of $\alpha'_{\parallel}/(3\varepsilon_0\varepsilon_s V)$, $\alpha'_{\perp}/(3\varepsilon_0\varepsilon_s V)$, and $\Delta\alpha/(3\varepsilon_0\varepsilon_s V)$.

an aspect ratio of 3, together with their difference, $\Delta\alpha'/(3\varepsilon_0\varepsilon_s V)$. In these examples, the EMW-1 polarizabilities $\alpha'_{\parallel}$ and $\alpha'_{\perp}$ have Debye-like frequency dispersions with different cut-off frequencies. The difference between the cut-off frequencies and asymptotic values of the two polarizabilities engenders the dip in $\Delta\alpha'$ which is then reflected in $B(\nu)$. This behavior is a striking signature of shape anisotropy.

The term $\Delta\alpha'$ depends on several variables, namely, a/b, $\varepsilon'_p/\varepsilon_s$, K_p^s/K, and $\omega\varepsilon_0\varepsilon_s/K_e$, in a complicated fashion. Figure 4 shows $\Delta\alpha$; $/(3\varepsilon_0\varepsilon_s V)$ calculated at constant ionic strengths, ζ potential, and surface charge. As the figure illustrates, both the frequency where $\Delta\alpha'$ is a minimum and the dip magnitude depend on the ionic strength and the ζ potential, but in different ways. From Fig. 4a we see that increasing the ionic strength diminishes the $\Delta\alpha'$ dip due to the lowered surface conductivity. At the same time, the dip frequency increases, due in part to the increase in $K_e/\varepsilon_0\varepsilon_s$. Figure 4b shows an increase in the magnitude of the dip and the dip frequency with increase in ζ potential. Evidently, the behavior of the dip frequency depends on both the bulk and surface conductivities. Figure 4c indicates the effects when working in terms of particle surface charge and ionic strength.

Figure 5 shows the (high-frequency) cut-off frequency, ν_{MW}, defined as the frequency of maximum negative slope of $B(\nu)$, as a function of the ionic strength for particles having either uniform charge or potential. The behavior in these situations is clearly different:

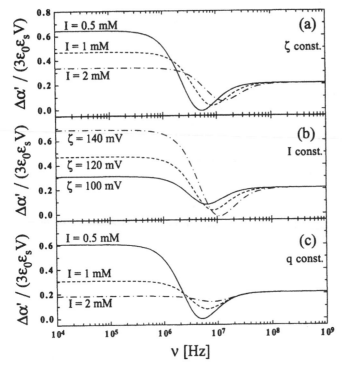

FIG. 4 Real part of the anisotropy in the dimensionless particle electric polarizability, $\Delta\alpha'/(3\varepsilon_0\varepsilon_s V) = (\alpha'_\parallel - \alpha'_\perp)/(3\varepsilon_0\varepsilon_s V)$, calculated using the EMW-1 model for various values of surface potential ζ, surface charge q, and ionic strength I. Parameters: long semiaxis = 240 nm; axial ratio = 1/3. (a) Behavior with $\zeta = 120$ mV at different KCl ionic strengths; (b) behavior with $I = 1$ mM KCl at different ζ potentials; (c) behavior with a surface charge $q = 1.271$ μC/cm^2 at different KCl ionic strengths. Solid line: I = 0.5 mM; dashed line: $I = 1$ mM; dot–dashed line: $I = 2$ mM.

ν_{MW} for constant potential particles has a stronger ionic strength dependence than for constant charge. This difference is particularly important since experimental data can encompass either situation, depending on the particle's surface characteristics.

D. Spheroids with Nonuniform Surface Charge Density

The theory for uniformly charged particles predicts, as the main feature of $\Delta\alpha$, a large, easily detectable minimum in $B(\nu)$. However, in many experiments the dip is less pronounced than allowed by the EMW-1 model, suggesting that $K_{p\parallel}/K_{p\perp}$ is larger than geometry alone permits, cf. Eq. (52). To illustrate this point, Fig. 6 shows $\Delta\alpha'/(3\varepsilon_0\varepsilon_s V)$ calculated from Eq. (30), treating $K_{p\parallel}$ and $K_{p\perp}$ independently. Note that the dip amplitude depends strongly on $K_{p\parallel}/K_{p\perp}$, becoming less pronounced as the anisotropy increases. One way this can arise is from a nonuniform surface charge, which alters the surface conductivity and therefore the dipole coefficient. For example, the charge density of some crystalline particles depends on the crystal plane exposed. Accordingly, in many instances the surface charge may be nonuniform. Inspection of Eq. (43) – taking into account Eq. (25) – shows that $K_{p\parallel}$ and $K_{p\perp}$ depend strongly on the charge distribution. For these reasons we introduce the EMW-2 formulation of the model, which involves fitting the $B(\nu)$ data with

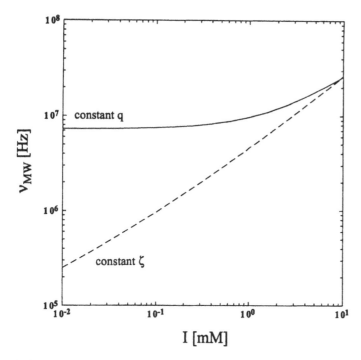

FIG. 5 Dependence of the MW characteristic frequency ν_{MW} on KCl ionic strength calculated from EMW-1 model at constant ζ potential or constant surface charge q. Constant charge – $q = 8.5\ \mu C/cm^2$ (solid line) and constant potential – $\zeta = 140\ mV$ (dashed line). Long semiaxis = 240 nm; axial ratio = 1/3.

the basic MW equations [Eq. (30)] to extract $K_{p\parallel}$ and $K_{p\perp}$ as independent quantities. In EMW-2, $K_{p\parallel}$ and $K_{p\perp}$ represent two different moments of the surface charge distribution, which combine in Eq. (46) to give the total charge. The main result is that one can extract Σ from $K_{p\parallel}$ and $K_{p\perp}$ without knowing the surface charge distribution.

E. Low-frequency Values of the Maxwell–Wagner Regime

According to the MW model, the high-frequency polarizabilities, $\alpha_\parallel^\infty$ and $\alpha_\perp^\infty$, depend on the real particle and solvent dielectric constants, whereas the low-frequency limits of the particle polarizabilities, $\alpha_\parallel^0$ and $\alpha_\perp^0$, depend on the particle and solvent conductivities [see Eqs (36) and (37)]. Since the low-frequency values of the Kerr constant in the MW regime are easily compared with experimental findings, it is worthwhile to observe in detail the related theoretical predictions. We define $\Delta\alpha^\infty = \alpha_\parallel^\infty - \alpha_\perp^\infty$ and $\Delta\alpha_{MW}^0 = \alpha_\parallel^0 - \alpha_\perp^0$. According to the EMW model, the parameter K_p in Eq. (37) depends on q or ζ as expressed by Eq. (23), and so we can calculate $\alpha_{\parallel,\perp}^0$ as a function of measurable quantities I and ζ. In Fig. 7, $\Delta\alpha_{MW}^0/\Delta\alpha^\infty$, the normalized value of the anisotropy of the real part of the low-frequency value of MW polarizability is plotted as a function of the ionic strength I for different values of the particle surface charge, as predicted by the EMW-1 model. From Fig. 7 we observe that the effect of the added salt is to decrease the ratio $\Delta\alpha_{MW}^0/\Delta\alpha^\infty$, whereas the effect of increasing ζ is to increase the ratio until an asymptotic value independent of I is reached.

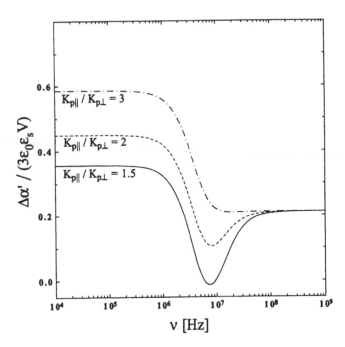

FIG. 6 Real part of the anisotropy in the dimensionless electric polarizability of the particles, $\Delta\alpha'/(3\varepsilon_0\varepsilon_s V) = (\alpha'_{\parallel} - \alpha'_{\perp})/(3\varepsilon_0\varepsilon_s V)$, calculated with the EMW-2 model for an anisotropic spheroid for various values of the ratio $K_{p\parallel}/K_{p\perp}$. Long semiaxis = 240 nm; axial ratio = 1/3; $K_{p\parallel} = 0.05$ S/m, $K_e = 0.015$ S/m (1 mM KCl).

V. LOW-FREQUENCY THEORY OF ELECTRIC POLARIZABILITY OF SPHERES

In the low-frequency regime (1 kHz $< \nu <$ 1 MHz), changes in ion densities beyond the double layer make the equations governing the particle polarization far more complex than in the high-frequency regime; this is in fact an expression of the fully electrokinetic character of the problem. In this regime, an exact numerical solution has been provided by DeLacey and White [3] in the case of spherical particles. When dealing with specific limiting cases, a number of approximate solutions are available. For instance, in situations of high ionic strength (thin double layer) and/or low surface potential, the following expression for the polarizability of spheres at low frequency can be used [5–7, 9]:

$$\alpha_{\text{LFS}} = -4\pi\varepsilon_0\varepsilon_s R^3 \frac{2K_e + K_p(\delta - 2)}{2(2K_e + K_p(\delta + 1))} \tag{56}$$

where

$$\delta = \frac{1 + \sqrt{\dfrac{i\omega R^2}{D}}}{1 + \sqrt{\dfrac{i\omega R^2}{D}} + \dfrac{1}{2}\dfrac{i\omega R^2}{D}} \tag{57}$$

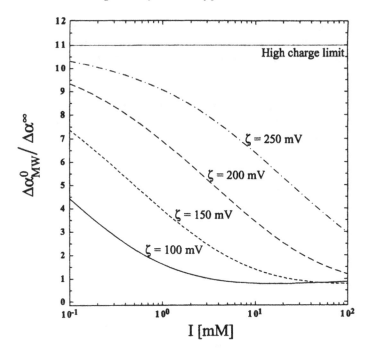

FIG. 7 Ionic strength dependence of the real part of the normalized low-frequency value of the anisotropy of polarizability $\Delta\alpha_{MW}^0/\Delta\alpha^\infty$ for spheroids. Data calculated for symmetric electrolyte by using the EMW-1 model for a spheroidal particle in the MW regime. Particle semiaxis = 240 nm; axial ratio = 1/3; $\Lambda = 75 \times 10^{-4}\Omega^{-1}$ m^{2}M^{-1}.

The definition of K_p used in Eq. (56) coincides with K_p as used in the high-frequency theory, i.e., it is computed by combining Eqs (22) and (25) [6]. The real and imaginary parts of δ are

$$\text{Re}[\delta] = \frac{1 + \sqrt{\omega\tau_s} + \omega\tau_s}{(1 + \sqrt{\omega\tau_s})(1 + \omega\tau_s)} \tag{58}$$

$$\text{Im}[\delta] = \frac{\omega\tau_s}{(1 + \sqrt{\omega\tau_s})(1 + \omega\tau_s)} \tag{59}$$

where $\tau_s = R^2/D$ ($D \approx 1.9 \times 10^{-9}$ m^2/s for KCl). According to Eqs (58) and (59), the frequency dispersion of δ is strongly non-Debye, its frequency decay being stretched over more than two decades.

In the case of thin double layers, assuming that $K_p \ll K_e$, Eq. (56) can be further approximated and α_{LFS} becomes

$$\alpha_{LFS} = -4\pi\varepsilon_0\varepsilon_s R^3\left[\frac{4K_e^2 - 6K_eK_p + 3K_p^2}{8K_e^2} + \frac{3}{8}\left(\frac{K_p}{K_e}\right)^2\delta\right] \tag{60}$$

In these conditions, the frequency dependence of α_{LFS} coincides with that of δ.

In Fig. 1 we have plotted the real part of the dimensionless polarizability for spheres, $\alpha_{LFS}/(3\varepsilon_0\varepsilon_s V)$, calculated with the Eqs (56) and (58) as a function of the frequency. As is apparent from this figure, the approximate expression is a good description of the low-frequency behavior of the polarizability. It is interesting to note in Fig. 1 that the low-frequency limit of the MW predictions for the polarizability (α_{MWS}^0) coincides with the high-frequency limit (α_{LFS}^∞) as expressed by Eqs (56–58).

In order to compare better and qualitatively the predictions of the approximate model with those of the SE approach, we have systematically calculated, with both models, the zero-frequency polarization at different values of I and ζ. At zero frequency, the α_{LFS} expression given by Eq. (56) becomes

$$\alpha_{LFS}^0 = 4\pi\varepsilon_0\varepsilon_s R^3 \frac{K_p - 2K_e}{4(K_p + K_e)} \tag{61}$$

In Fig. 8 we compare the normalized zero-frequency values of polarizability, $\alpha_{LFS}^0/|\alpha^\infty|$, calculated by using Eq. (61), with the predictions of the zero-frequency values of the SE model. As apparent, the agreement is good for high ionic strengths and low ζ potentials. Outside this limit, the approximate models for the low-frequency behavior of

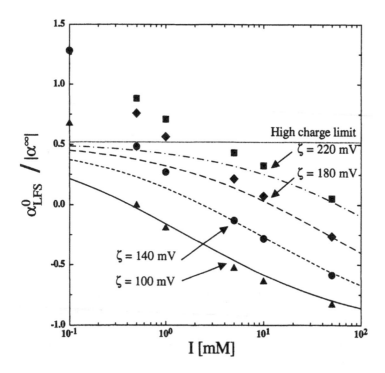

FIG. 8 Ionic strength dependence of the real part of the normalized zero-frequency values of the polarizability $\alpha_{LFS}^0/\alpha^\infty$ for spheres. Data calculated from the SE model for symmetric electrolyte for spherical particle, with particle radius $R = 135$ nm; $\varepsilon_p' = 2$; $\varepsilon_s = 80$; $\Lambda = 75 \times 10^{-4}$ $\Omega^{-1}\text{m}^2$ M^{-1}. Symbols: values calculated by using numerical solutions of SE model [3]; lines: values calculated by using the low-frequency theory for spheres [Eq. (61)]. Triangles and solid line: $\zeta = 100$ mV; Circles and short-dashed line: $\zeta = 140$ mV: diamonds and long-dashed line: $\zeta = 180$ mV: squares and dot–dashed line: $\zeta = 220$ mV; dotted line: high charge limit.

the electric polarizability of spherical polyelectrolytes can not be trusted, and it is necessary to take into account the complete numerical solution of the SE model.

VI. LOW-FREQUENCY THEORY OF ELECTRIC POLARIZABILITY OF SPHEROIDS

Having assessed the range of validity of the thin-double-layer approximation for calculating the polarizability of spheres, we now focus on analogous expressions for spheroids. When the simplifications inherent in the spherical shape are lost, all electrokinetic calculations become extremely difficult. As a result, not only no exact numerical solutions to the SE model for any nonspherical shape exists, but also the single available approximate theory for spheroids, by O'Brien and Ward [11], calculates the electric polarizability in the zero-frequency limit only. Following the O'Brien and Ward approach, the electric polarizability of spheroid at zero frequency is given, for a symmetric electrolyte, by the following expressions, along and perpendicularly to the applied field:

$$\alpha_{LF\parallel}^{0} = 4\pi\varepsilon_0\varepsilon_s a^2 b \frac{1}{3\sinh^2\xi_0\cosh\xi_0} \frac{-4\delta Q_1 + 3\cosh\xi_0\partial Q_1 I_1\lambda_s + 3Q_1 I_1\lambda_s}{2\partial Q_1(2\partial Q_1 - 3Q_1 I_1\lambda_s)} \tag{62}$$

$$\alpha_{LF\perp}^{0} = 4\pi\varepsilon_0\varepsilon_s a^2 b \frac{2}{3\cosh\xi_0\sinh^3\xi_0} \times$$

$$\frac{8\cosh\xi_0\partial Q_1^1 - 3\cosh\xi_0 I_2 Q_1^1\lambda_s + 3\partial Q_1^1 I_2\sinh^2\xi_0\lambda_s}{2\partial Q_1^1(4\partial Q_1^1 - 3I_2 Q_1^1\lambda_s)} \tag{63}$$

We define $\Delta\alpha_{LF}^{0} = \alpha_{LF\parallel}^{0} - \alpha_{LF\perp}^{0}$. In Fig. 9 we have plotted the normalized zero-frequency values of anisotropy of polarizability for spheroids, $\Delta\alpha_{LF}^{0}/\Delta\alpha^{\infty}$, calculated according to Eqs (62) and (63) as a function of the ionic strength at fixed ζ potential. Since the theoretical hypotheses in Eqs (62) and (63) are the same as those leading to Eq. (61), the range of validity of the predictions in Fig. 9 is probably the same as the one following from the analysis in Fig. 8. Thus, we argue that Eqs (62) and (63) are valid for I larger than 1 mM KCl and ζ potential lower than 140 mV.

By inspecting Figs 7 and 9 we observe that the predicted zero-frequency values of anisotropy of polarizability in the MW and in the low-frequency regimes are of the same order of magnitude. More specifically, at low ionic strength and low ζ potential $\Delta\alpha_{LF}^{0}/\Delta\alpha^{\infty} \approx \Delta\alpha_{MW}^{0}/\Delta\alpha^{\infty}$, but at higher ionic strength and ζ potential $\Delta\alpha_{LF}^{0}/\Delta\alpha^{\infty} < \Delta\alpha_{MW}^{0}/\Delta\alpha^{\infty}$. Thus, we expect that, at low ionic strength and low ζ potential the Kerr constant depends only mildly on frequency in the low frequency regime. At lower ζ potential, we instead anticipate $B(\nu)$ to be a decreasing function. This is sketched in Figs 10a and 10b, where we schematically report the expected behavior of the real part of $\Delta\alpha$ together with the consequent prediction for $B(\nu)$, and where we define the quantities introduced in the discussion above. Note the two distinct relaxations in the frequency dependence of $\Delta\alpha'$ (Debye-like at high frequency and less steep at low frequency) and the different levels attained at low and high frequency.

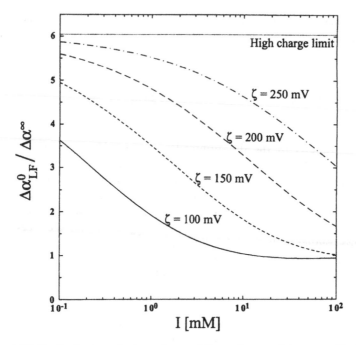

FIG. 9 Ionic strength dependence of the real part of the normalized zero-frequency values of the anisotropy of polarizability $\Delta\alpha_{LF}^0/\Delta\alpha^\infty$ for spheroids. Data calculated for symmetric electrolyte by using the O'Brien and Ward theory [Eqs (62), (63)] for spheroidal particle in the low-frequency regime. Particle semiaxis = 240 nm; axial ratio = 1/3; $\varepsilon_p' = 2$, $\varepsilon_s = 80$; $\Lambda = 75 \times 10^{-4}\Omega^{-1}$ m^2 M^{-1}.

VII. COMPARISON BETWEEN FREQUENCY DISPERSION OF KERR CONSTANT AND DIELECTRIC CONSTANT

In order to assess better the importance of FREB as a new experimental technique for measuring electrokinetic phenomena, it is useful to compare its predicted behavior with that of the dielectric constant. Here we want, in particular, to show that the MW polarization process, which is the dominating feature of FREB spectra, is instead hardly detectable by dielectric measurements.

Following the usual notation, the dielectric increment, $\Delta\varepsilon$, is defined as $\Delta\varepsilon = \varepsilon_d - \varepsilon_e$, where ε_d is the dielectric constant of the suspension. It is well known [3] that the dielectric increment is related to particle polarizability as

$$\Delta\varepsilon(\nu) = \frac{N\varepsilon_e}{\varepsilon_0\varepsilon_s}\alpha(\nu) = \frac{N}{\varepsilon_0\varepsilon_s}\left[\left(\varepsilon_s\alpha' - \frac{K_e\alpha''}{\omega\varepsilon_0}\right) + i\left(\varepsilon_s\alpha'' + \frac{K_e\alpha'}{\omega\varepsilon_0}\right)\right] \qquad (64)$$

In the above equation, $\Delta\varepsilon$ and ε_e are both complex quantities, and $\alpha(\nu)$ has different expressions in the low- and high-frequency regimes. In the latter, the dielectric increment can be evaluated from the MW polarization process. If we denote by $\Delta\varepsilon_{MW}^0$ and $\Delta\varepsilon_{MW}^\infty$ the dielectric increments at, respectively, the low- and high-frequency limits of the MW regime, and we further assume that $\varepsilon_p \ll \varepsilon_s$ (valid for most aqueous dispersions):

$$\frac{\Delta\varepsilon_{MW}^0 - \Delta\varepsilon_{MW}^\infty}{\varepsilon_s} = \frac{9}{2}\phi\frac{\lambda_s^2}{(1+\lambda_s)^2} \qquad (65)$$

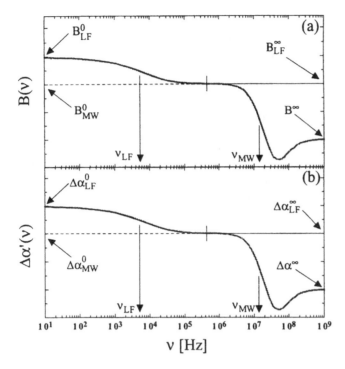

FIG. 10 (a) Thick line: schematic plot of the frequency dispersion of the real part of the anisotropy of the electric polarizability $\Delta\alpha'$ for spheroidal particles. (b) Thick line: schematic plot of frequency dispersion of the Kerr constant B for spheroidal particles. Solid lines: values obtained from the low-frequency model; dashed lines: values obtained from MW model. The arrows indicate low- and high-frequency values.

where the conductivity ratio λ_s in the case of the sphere becomes $\lambda_s = K_p/(2K_e)$. Equation (65) quantifies the signal $(\Delta\varepsilon_{MW}^0 - \Delta\varepsilon^\infty)$ to the background (ε_s) ratio for the MW feature as measured by dielectric spectroscopy, and reveals that such a ratio is always smaller than $9\phi/2$, which, for reasonable particle concentrations is, at most, 0.1. This value, large enough to be, at least in principle, experimentally detected, refers to the limit of large λ_s. Considering more realistic situations, λ_s has to be taken close to unity for low ionic strengths, and could be much smaller in the thin-double-layer limit. Thus, in the large majority of realistic experimental conditions, $(\Delta\varepsilon_{MW}^0 - \Delta\varepsilon_{MW}^\infty)/\varepsilon_s \leq 0.01$, small compared to typical resolutions in dielectric spectroscopy experiments. This explains why dielectric measurements have shown MW features only in very specific cases [28].

A similar approach can be adopted when considering the dielectric increment in the low-frequency regime. Let us call $\Delta\varepsilon_{LF}^0$ and $\Delta\varepsilon_{LF}^\infty$ the dielectric increments at the low- and high-frequency limit in the LF regime. We obtain:

$$\frac{\Delta\varepsilon_{LF}^0 - \Delta\varepsilon_{LF}^\infty}{\varepsilon_s} \approx \frac{9}{2}\phi\frac{\lambda_s^2 K_e \tau_s}{\varepsilon_0 \varepsilon_s (1 + 2\lambda_s)^2} \tag{66}$$

The right-hand side term in Eq. (66) is typically 10^2–10^3 times larger than the right-hand side term of Eq. (65), confirming the experimental fact that the low-frequency dielectric dispersion is largely predominant over the whole frequency range.

This brief analysis is summarized in Fig. 11 where we compare the frequency dispersion of the real part of electric polarizability (Fig. 11a) and of the real part of the dielectric enhancement (Fig. 11b, continuous line), both calculated for a suspension of charged spheres according to the low-frequency model in Eq. (56) combined with the EMW model at high frequency (the parameters for the calculation are given in the figure caption). As visible, the two quantities have dramatically different spectra, especially because $\Delta\varepsilon'/\phi$ does not show noticeable features around ν_{MW}. In order to explore the origin of this difference, we have plotted in Fig. 11b the two terms contributing to $\Delta\varepsilon'$ according to the last term in Eq. (64), one of which is simply proportional to α' (the dashed line in the figure). The aim of Fig. 11b is to show that, while the MW feature is the dominant relaxation in the α' spectrum, when $\Delta\varepsilon'$ is computed, the MW relaxation is basically absent because of an intriguing cancellation. Such a cancellation is better shown in Fig. 11c, which is a vertical zoom of Fig. 11b. We think that the comparison between Figs 11a and 11b is a powerful synthesis of the reason why the FREB technique is a very much better tool than dielectric spectroscopy for studying high-frequency electrokinetic phenomena.

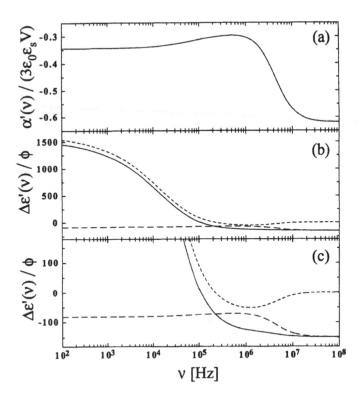

FIG. 11   (a) Dimensionless real part of the electric polarizability of a charged sphere. (b), (c) Solid lines: real part of the specific dielectric enhancement of a suspension of charged spheres, obtained, according to Eq. (64) as the sum of two terms: $\alpha'/(V\varepsilon_0)$ (short-dashed line) and $-K_e\alpha''/(V\omega\varepsilon_0^2\varepsilon_s)$ (long dashed line). Panel (c) is a vertical zoom of (b). Calculations were performed for a symmetric electrolyte according to the low-frequency model in Eq. (56) combined with the EMW model. $R = 135$ nm; $\varepsilon_p' = 2$, $\varepsilon_s = 80$, $\Lambda = 75 \times 10^{-4}\Omega^{-1}$ m^{2}M^{-1}; $\zeta = 70$ mV; $I = 0.5$ mM.

VIII. ELECTRIC BIREFRINGENCE EXPERIMENTS ON COLLOIDS OF VARIABLE CHARGE

In the following sections we present experimental results obtained at low and high frequency with colloids of variable charge, and we discuss in detail their comparison with the theoretical predictions. We first give a short description of the experimental technique and materials, followed by a presentation and discussion of the EB data.

A. Frequency-resolved Electric Birefringence

The EB experiment consisted of applying a voltage to the electrodes of the Kerr cell containing the suspension under study and measuring the induced birefringence, Δn. In FREB, the applied voltage was shaped in variable-frequency sine-wave pulses of zero average, having a duration long enough to reach a stationary value of the induced anisotropy. Our experiments employed pulses of about 40 ms, frequencies in the range 1–200 MHz, and field amplitudes between 1 and 10 V/mm. The time-dependent induced birefringence $\Delta n(t)$ was obtained by averaging over about 100 pulses. From the steady-state value of $\Delta n(t)$ measured at different frequencies of the applied field, we extracted $B(\nu)$.

In standard FREB experiments, frequencies range up to a few MHz [15]. In order to increase the maximum frequency, our experimental setup [21] was modified by devising a Kerr cell with specialized wiring and contacts to minimize inductive loops and avoid spurious responses. By making use of radio-frequency equipment to generate and amplify the applied electric field, the frequency range was then increased by two orders of magnitude. An automatic, computer-controlled setup was developed to make measurements over six decades in frequency within an interval less than the particle sedimentation time. Typically, with this setup we could measure B at 200 values of frequency (i.e., a typical FREB spectrum) in 2 h, about a factor 50 less than that of manual experiments.

B. Controlling the Particles' Charge Via Surfactant Adsorption

The measurements involve suspensions of elongated particles of single poly(tetrafluoroethylene) (PTFE) crystals having a 240 nm long semiaxis with an axial ratio close to 3. The polydispersity in the linear dimension is about 15%. These particles, kindly supplied by Ausimont (Milano, Italy), are shaped like spherocylinders, which, for the sake of comparison with theoretical models, will be approximated by prolate spheroids. PTFE particles were chosen for their special properties. First, their uniaxial internal crystal structure [29] allowed us to perform measurements on dilute dispersions and still maintain a favorable signal-to-noise ratio. The average refractive index of a PTFE particle is 1.38 and the optical anisotropy is $\Delta n = 0.04$, as measured by light-scattering techniques [29]. Second, the highly hydrophobic surface allows the surface charge to be controlled by adsorption of ionic and nonionic surfactants as described below. Third, since the refractive index of fluorinated compounds is quite close to that of water, the presence of an adsorbed layer of hydrogenated surfactant (with a much larger refractive index) can be detected by measuring the light scattering cross section. In fact, to assess surfactant adsorption, we measured the laser power transmitted by a suspension of PTFE particles of known concentration following the procedure outlined in Ref. [30]. When surfactant molecules are adsorbed on the particle surface, the scattering cross-section of the particle increases appreciably because the optical contrast between surfactant and water is much larger than that between PTFE and water. This effect allows measuring with accuracy better than 0.1 ng/mm^2 the quantity of surfactant adsorbed on particle surface. Thus, comparison between transmitted intensity data and $B(\nu)$ data enables comparing the number of adsorbed molecules with the number of ionized sites which yield particle charge.

In Ref. 30, we presented a more complete description of the PTFE particles. Here, we want to recall that, as prepared by emulsion polymerization, the original bare PTFE particles bear a negative surface charge due in part to adsorbed anionic surfactant (fluorinated carboxylate) and to the end groups of the polymer chains (fluorinated carboxyl ions). The amount of adsorbed fluorinated surfactant depends on the details of preparation, including a final dialysis to lower the ionic strength of the solution.

In this chapter we present data obtained with two different batches of PTFE. This induces only slight quantitative differences in the surface charge of the particles and, as a consequence, some slight differences in the Kerr constant. In the figure captions we have indicated the two different batches of particles with PTFE-1 and PTFE-2.

We have modified the particle charge by allowing the adsorption of various surfactants (nonionic, anionic, or cationic) on the particle surface. We have used three very common surfactants: Triton X-100 (BDH), 4-$(C_8H_{17})C_6H_4(OCH_2CH_2)_nOH$, n $\approx$ 10, $M_w \approx 650$, nonionic; AOT or sodium dioctylsulfosuccinate (BDH), $C_{20}H_{37}O_7SNa$, anionic, $M_w = 444.6$; HTAB or hexadecyltrimethylammonium bromide (Aldrich), $CH_3(CH_2)_{15}N(CH_3)_3Br$, cationic, $M_w = 364.5$.

In the past we have extensively studied the effect of adding nonionic surfactant to PTFE colloidal dispersions. In Ref. 30 we showed that, by controlling the amount of Triton X-100, we can induce competitive adsorption between the neutral and the ionic surfactants originally present on the particle surface, and thus modify the number of ionic surfactant molecules on the latter. When both Triton X-100 concentration, c_T, and PTFE volume fraction, ϕ, are large enough (in order to avoid desorption due to dialysis into the solvent), the number of adsorbed ionic surfactant molecules per particle is a function of c_T/ϕ alone. Overall, the effect of adding the Triton X-100 is to depress the particle charge without a noticeable increase in ionic strength. In this work we present part of the data already presented in Refs 19 and 30, together with new data obtained with various Triton X-100 concentrations ranging from $c_T/\phi = 10^{-3}$M to $c_T/\phi = 0.1$ M.

The effect of adding AOT to PTFE suspensions is instead to increment the (negative) charge of the particles. Our measurements involved two values of c_{AOT}/ϕ:

1. $c_{AOT}/\phi = 0.1$ M corresponds to almost complete coverage. Given the size and geometry of the particles, this corresponds to 2.3×10^5 adsorbed molecules per particle. If they are fully ionized, the charge density is 18.6 μC/cm^2.

2. $c_{AOT}/\phi = 0.01$ M corresponds to about 1/10 of complete coverage. This indicates about 3.8×10^4 adsorbed molecules per particle, corresponding to a surface charge density of 3.1 μC/cm^2. The actual number of ionizable groups per particle could be slightly larger due to the residual fluorinated surfactant molecules left from the PTFE synthesis.

In order to study the effect of adding cationic surfactant to PTFE suspension, we have performed some measurements with PTFE particles covered with HTAB. By adding HTAB to the PTFE suspension, the charge of the particles can be reversed. This effect is achieved by using HTAB concentrations (c_{HTAB}) larger than the point of zero charge, at which we have flocculation of samples. The measurements described in this chapter refer to $c_{HTAB}/\phi = 0.01$ M, for different ionic strengths. This corresponds to 3.8×10^4 adsorbed molecules per particle.

C. General Description of Experimental Results

In this section we describe $B(\nu)$ measurements performed with PTFE particles at different ionic strengths and different surfactant coverage; $B(\nu)$ results are presented in Figs 12–15.

Figure 12 shows $B(\nu)$ measured at various ionic strengths, for dispersions with no surfactant added. In Fig. 13 we show $B(\nu)$ measured at fixed ionic strength ($I = 1$mM KCl) at different Triton X-100 concentrations. Figure 14 shows data obtained with particles fully covered with AOT ($c_{AOT}/\phi = 0.1$ M) at different ionic strengths. In Fig. 15 we present $B(\nu)$ at various ionic strengths for PTFE particles in the presence of HTAB. As apparent from the figures, by adding different surfactants, the frequency dispersion of the Kerr constant dramatically changes. Note that in the cases of full AOT or HTAB coverage, we observe a dip in $B(\nu)$ in the 1–10 MHz range.

The following features are common to all the situations described: (1) at any given surfactant concentration and frequency, the Kerr constant decreases upon increasing the ionic strength, in agreement with previous work [21]; (2) for any given suspension, $B(\nu)$ is a decreasing function of ν in the kilohertz and megahertz regions (with the exception of the dip region). The dependence on I and ν clearly indicates that the Kerr constant is the result of double-layer polarization processes. The increment of dissolved ions increases the conductivity of the medium around the particles, which in turn more efficiently counteracts charge displacement. Moreover, kilohertz and megahertz regimes correspond to the time it takes for ions to diffuse across distances of about, respectively, the colloidal particle size

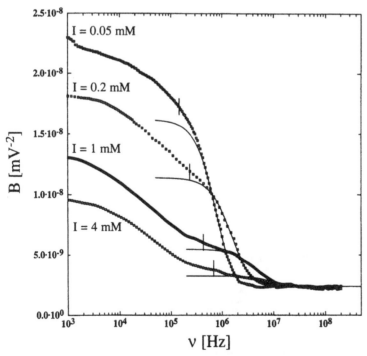

FIG. 12 Frequency dependence of the Kerr constant B measured in a $\phi = 10^{-3}$ suspension of PTFE particles where no surfactant has been added. The short vertical segments indicate, approximately, the transition frequency between the two regimes – low frequency and MW at high frequency. The solid lines represent a fit of the high-frequency data using the MW theory for a conducting spheroid (EMW-2).

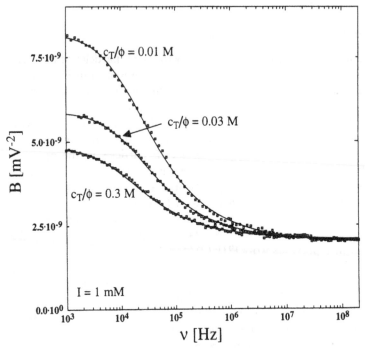

FIG. 13 Frequency dependence of the Kerr constant measured in a $\phi = 10^{-3}$ suspension of PTFE particles at the ionic strength 1 mM KCl. Curves refer to different concentrations of Triton X-100; the solid lines represent a fit with Eqs (56) and (57).

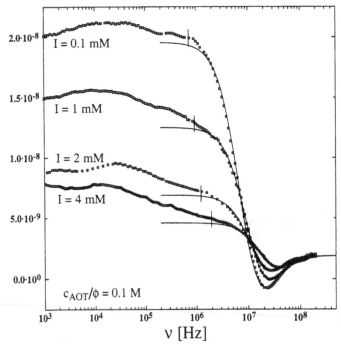

FIG. 14 Frequency dependence of the Kerr constant measured in a $\phi = 10^{-3}$ suspension of PTFE particles with 10^{-4} M AOT added. Curves refer to different ionic strengths; vertical and solid lines as in Fig. 12.

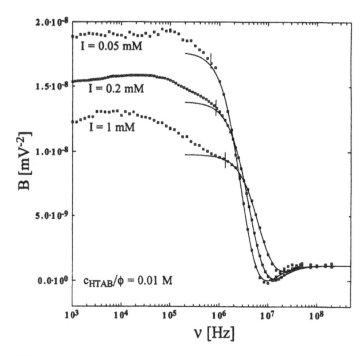

FIG. 15 Frequency dependence of the Kerr constant measured in a $\phi = 10^{-3}$ suspension of PTFE particles where 10^{-5} M HTAB has been added, for the indicated ionic strengths. Vertical and solid lines as in Fig. 12.

and the Debye length. Accordingly, the high-frequency ($\nu > 200$ MHz) asymptotic value of $B(\nu)$ appears to be independent of particle charge and ionic strength, in agreement with the notion that, when ion motions are negligible, $B(\nu)$ takes the value of uncharged (purely dielectric) PTFE particles in a purely dielectric medium.

To classify the behavior of $B(\nu)$, we will focus separately in the low-frequency (1 kHz–1 MHz) regime, where ionic diffusion and hydrodynamic motions are relevant, and in the high-frequency MW (1–200 MHz) regime where the EMW model applies. The intrinsic separation between the two regimes is particularly evident in some cases (see curve $I - 1$ mM in Fig. 12) where two distinct relaxations in different frequency intervals are evident, but it can be recognized in almost every situation. It should be remarked that the data presented here explore regimes of frequency and electric charge never considered before: all the previously published experiments [29–33] on the EB of polyelectrolyte solutions concerned particles of low charge, in relatively high ionic strength solutions, studied in the regime of low frequency ($\nu < 1$ MHz). Consequently, previous results were mainly similar to those represented by the low-frequency part of the curves in Fig. 13.*

---

* In our earlier works, in the absence of an electrokinetic theory for EB, the frequency dispersions of B have been fitted, in the low-frequency regime, using models developed for the real part of the dielectric constant. We now know that such fittings are not very meaningful, as the Kerr constant is proportional to the real part of the polarizability, instead. However, since in the low-charge limit the shape of the two frequency dispersions is similar, see Eq. (60), and given the data precision of those earlier measurements, the fitting yielded satisfactory results.

To highlight better the two frequency regimes, a short vertical segment separating them has been drawn in Figs 12–15. In either regime, a "cut-off frequency" corresponding to the frequency of maximum negative slope of $B(\nu)$ can be defined. The cut-off frequencies in the low-frequency and MW regimes are denoted by ν_{LF} and ν_{MW}, respectively (see Fig. 10). Measured values of ν_{MW} and ν_{LF} differ by about two orders of magnitude. The I dependence of ν_{MW} for bare or surfactant-covered particles is shown in Fig. 16; ν_{MW} changes by more than an order of magnitude over the range of ionic strength explored, whereas the measured values of ν_{LF} are almost independent of ionic strength. This could be expected since ν_{LF} is normally interpreted as the reciprocal of the polarization time of the double layer ($\nu_{LF} \approx D/a^2$), where a is colloid length scale. On the other hand, the strong dependence of ν_{MW} on I is due to mismatches in ionic conductivity, and thus in ion transport. In Fig. 16 we have also reported, in analogy with Fig. 5, the theoretical predictions for ν_{MW} according to the EMW model [see Eq. (35)] for constant q or constant ζ. Comparison of data with theory, as presented in Fig. 16, supports the notion that bare particles behave as constant ζ-potential entities, whereas AOT/HTAB covered particles behave as constant q particles. When particles are covered with Triton X-100 (i.e., low surface charge) the high-frequency dispersion is almost flat and thus ν_{MW} is hard to define.

In the analysis that follows, we will focus on the quantity $B(\nu)/B^\infty$, where B^∞ is the experimental value of B at $\nu > 200$ MHz. Since, according to Eq. (14),

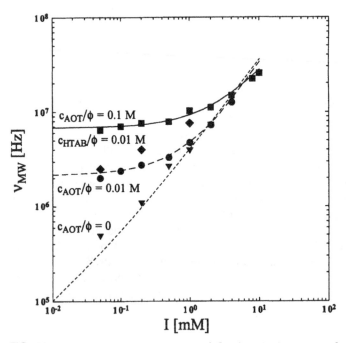

FIG. 16 Characteristic frequency ν_{MW} defined as the frequency of maximum negative slope of $B(\nu)$ measured for PTFE particles in presence of different surfactants. Symbols: $c_{AOT}/\phi = 0$ (triangles) $c_{AOT}/\phi = 0.01$ M (circles), $c_{AOT}/\phi = 0.1$ M (squares), and $c_{HTAB}/\phi = 0.01$ M (diamonds). Lines: EMW-2 model for constant $q = 8.0$ μC/cm^2 and $K_{p\parallel}/K_{p\perp} = 2.2$ (solid line), constant $q = 2.5$ μC/cm^2 and $K_{p\parallel}/K_{p\perp} = 3.5$ (long-dashed line), and constant potential $\zeta = -70$ mV and $K_{p\parallel}/K_{p\perp} = 6$ (short-dashed line).

$B(\nu)/B^\infty = \Delta\alpha'(\nu)/\Delta\alpha^\infty$, by dividing the Kerr constant by its asymptotic value, we can neglect the B dependence on all the nonelectrokinetic quantities while retaining all the relevant q and I dependences.

The data show different trends in the two regimes. For low frequency we observe that for particles with low charge (Figs 12 and 13), $B(\nu)$ decreases smoothly and monotonically, while it appears less dependent on ν for highly charged particles. In the MW regime $B(\nu)$ drops rather abruptly, in some cases (in the case of highly charged particles) shows a dip, and reaches its asymptotic value between 1 and 10 MHz.

IX. COMPARISON BETWEEN HIGH-FREQUENCY DATA AND M–W MODELS

In this section we focus on the data analysis limited to the high-frequency regime, where, as shown in Sections III and IV, simplifications of the theory are in order. We will make use of both the EMW-1 model, which assumes the charge to be uniformly distributed on the spheroid and involves fits with a single free parameter (the total charge Σ), and the EMW-2 model, which was developed to deal with nonuniformly charged particles and involves fits with two free parameters, $K_{\text{p}\parallel}$ and $K_{\text{p}\perp}$. As shown in the theory sections of this chapter, even though the charge distribution is unknown, Σ can be calculated from $K_{\text{p}\parallel}$ and $K_{\text{p}\perp}$ by using the "mixing rule" of Eq. (46).

The fits will be performed in the limited frequency range spanning from the crossover frequency (marked by vertical segments in Figs 12, 14, and 15) to the high-frequency limit. We will call B^0_{MW} the low-frequency value obtained in the fits. Accordingly $B^0_{\text{MW}}/B^\infty = \Delta\alpha^0_{\text{MW}}/\Delta\alpha^\infty$ has to be compared with the theoretical predictions shown in Fig. 7.

A. Data Analysis at High Frequency

Figures 17a and 17b show, respectively, curve $I = 0.1\text{mM}$ of Fig. 14 and curve $I = 1\text{mM}$ of Fig. 12 together with the best fit obtained using the EMW-1 model over the high-frequency region.

Figure 17a, which refers to particles fully covered with AOT and therefore highly charged, shows that the EMW-1 calculation captures all the main features of the data (amplitude, characteristic frequency, and dip). Despite a minor disagreement due to the EMW-1 model prediction of a larger dip, the one parameter fit is remarkable. For the data in Fig. 17a, the best fit yields $q = 8.1 \ \mu\text{C/cm}^2$ and $\zeta = 252$ mV. Data obtained with fully AOT-coated particles at various ionic strengths are approximated by the EMW-1 model with a quality similar to that found for the $I = 1$ mM data. The situation drastically worsens when the model is compared with data obtained for particles only partially covered by ionic surfactant, i.e., particles with lower surface charge, as shown in Fig. 17b. Here, although the EMW-1 model describes the frequency and amplitude of the $B(\nu)$ elbow, it fails to predict the disappearance of the dip. However, satisfactory fits to the FREB data can be obtained by using Eq. (30) and allowing $K_{\text{p}\parallel}$ and $K_{\text{p}\perp}$ to vary independently. This is shown in Figs 12, 13, and 15 which show, as continuous lines, the best fit obtained with the EMW-2 model.

Figure 18 shows best-fit values of $K_{\text{p}\parallel}$ as a function of the ionic strength for the various dispersions under study; $K_{\text{p}\parallel}$ is strongly dependent on the amount and the nature of the adsorbed surfactant, but only weakly changes upon changing I.

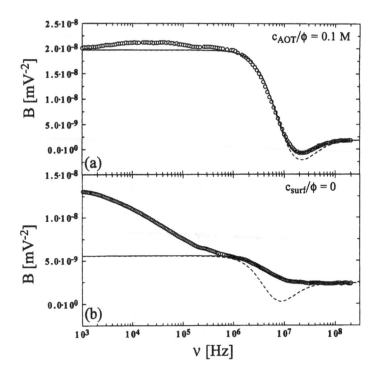

FIG. 17 Best fit on the high-frequency $B(\nu)$ with EMW-1 model (dashed line) and EMW-2 model (solid line) for particles of different AOT coverage. (a) PTFE particles fully covered with the anionic surfactant AOT (curve at $I = 0.1$ mM KCl of Fig. 14). Experimental conditions: $\phi = 10^{-3}$; $c_{AOT}/\phi = 0.1$ M; $I = 0.1$ mM KCl. From the fit: $\zeta = -252$ mV (EMW-1 model); $K_{p\parallel} = 0.26$ S/m and $K_{p\perp} = 0.12$ S/m (EMW-2 model). (b) PTFE particles with no added surfactant (curve at $I = 1$ mM KCl of Fig. 12). Experimental conditions: $\phi = 10^{-3}$; $c_{surf}/\phi = 0$; $I = 1$ mM KCl. From the fit; $\zeta = -138$ mV (EMW-1); $K_{p\parallel} = 2.5 \ 10^{-2}$ S/m and $K_{p\perp} = 5.5 \ 10^{-3}$ S/m (EMW-2 model). The fit with the EMW-2 model is almost indistinguishable from the data.

Figure 19 shows that, in some cases, the $K_{p\parallel}/K_{p\perp}$ ratio is significantly larger than the value allowed by the EMW-1 model ($K_{p\parallel}/K_{p\perp} = 1.72$, dashed line in Fig. 19). Other interesting features of these data are that $K_{p\parallel}/K_{p\perp}$ is larger with lower surfactant coverage [where the $B(\nu)$ dip is smaller] and that $K_{p\parallel}/K_{p\perp}$ is roughly independent of ionic strength for a given surfactant coverage. Since the $K_{p\parallel}/K_{p\perp}$ ratio as predicted by the EMW-1 model cannot exceed 2 [see Eq. (52)] even for the highest axial ratios, the disagreement between EMW-1 theory and EB experiment, shown in Fig. 17b and related to the value of the $K_{p\parallel}/K_{p\perp}$ ratio, cannot be the consequence of having adopted an inaccurate description of the particle shape (spheroids rather than spherocylinders), size, or polydispersity. The disagreement must be related to nongeometrical features, such as the charge distribution on the surface. A nonuniform surface charge density may originate from electrostatic interactions between ionic surfactant molecules (favoring surfaces at constant potential) or may be the consequence of the uneven chemical and physical properties of the surface of the PTFE particles. Indeed, the sides of PTFE rods have a flatter, more crystalline surface compared to their tips [34], which could favor (or disfavor) specific surfactant adsorption. That the local surface properties could be the source of the uneven charge distribution is also suggested by the weak dependence of $K_{p\parallel}$ and $K_{p\perp}$ on the ionic

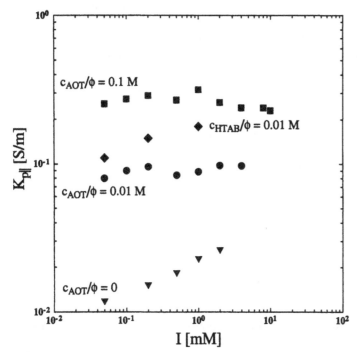

FIG. 18 Volume conductivity, $K_{p\parallel}$, as extracted from EMW-2 best fit on the high-frequency part of $B(\nu)$, for PTFE particles in presence of different surfactants: $c_{AOT}/\phi = 0$ (triangles), $c_{AOT}/\phi = 0.01$ M (circles), $c_{AOT}/\phi = 0.1$ M (squares), and $c_{HTAB}/\phi = 0.01$ M (diamonds).

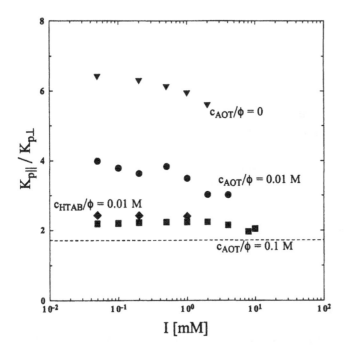

FIG. 19 $K_{p\parallel}/K_{p\perp}$ as extracted from EMW-2 best fit on the high-frequency part of $B(\nu)$, at different ionic strengths for PTFE particles in presence of different surfactants: $c_{AOT}/\phi = 0$ (triangles), $c_{AOT}/\phi = 0.01$ M (circles), $c_{AOT}/\phi = 0.1$ M (squares), and $c_{HTAB}/\phi = 0.01$ M (diamonds). The dashed line represents $K_{p\parallel}/K_{p\perp}$ predicted by the EMW-1 model ($K_{p\parallel}/K_{p\perp} = 1.72$).

strength. These considerations, together with the remarkably good fit to the data when using the MW formulas, indicate the EMW-2 model procedure as the most appropriate to interpret the data. When this is done we obtain, by virtue of the mixing rule [Eq. (46)], the I dependence of the surface charge presented in Fig. 20, where the strong dependence of q on the amount of added (and adsorbed) surfactant is evident. In the case of AOT, the increment in q is qualitatively proportional to c_{AOT}. Also notice that, at the same surfactant concentration $c_{AOT}/\phi = c_{HTAB}/\phi = 0.01$ M, the HTAB has a more conspicuous effect on the charge. In this case, the positive surfactant tends to be adsorbed on the surface and/or to be dissociated more efficiently with respect to the negative surfactant. Since the MW behavior is undetectable with Triton-covered particles, we cannot, in that case, extract q.

In Fig. 21 we have plotted the normalized values of the anisotropy of polarizability $\Delta\alpha^0_{MW}/\Delta\alpha^\infty$ as obtained from the EB data at low frequency in the MW regime. The data are taken for bare particles and for particles completely covered by AOT ($c_{AOT}/\phi = 0.1$ M). In Fig. 21 we also show the predictions of the MW theory for $\Delta\alpha^0_{MW}/\Delta\alpha^\infty$, as given by Eqs (36) and (37), assuming either ζ or q to be constant. In agreement of the behavior shown in Figs 16 and 20, bare particles seem to behave as constant-ζ entities, whereas AOT-coated particles behave as constant-q entities. While still in need of a microscopic explanation, this remarkably different behavior of surfactant coated and bare PTFE surfaces demonstrates that studies of FREB at the high-frequency limit enables one to access from a new perspective the electrokinetic properties of surfaces.

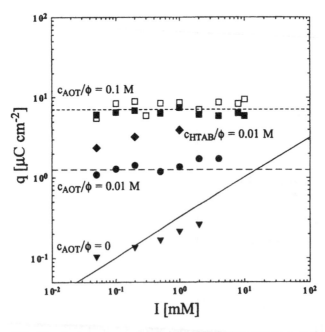

FIG. 20 Surface charge density, q, obtained from fitting $B(\nu)$ data using the EMW-1 (full symbols) and the EMW-2 models (open symbols) for PTFE particles in presence of different surfactants: $c_{AOT}/\phi = 0$ (triangles), $c_{AOT}/\phi = 0.01$ M (circles), $c_{AOT}/\phi = 0.1$ M (squares) $c_{HTAB}/\phi = 0.01$ M (diamonds). Lines indicate behavior at: $q = 8$ μC/cm^2 (short-dashed line), $q = 2.5$ μC/cm^2 (long-dashed line), and $\zeta = 40$ mV (solid line).

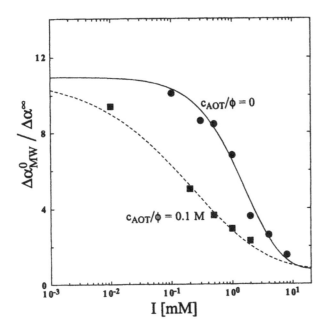

FIG. 21 Symbols: ionic strength dependence of the real part of the low-frequency normalized value of the polarizability $\Delta\alpha_{MW}^0/\Delta\alpha^\infty$ measured from $B(\nu)$ data at high frequency. Lines: $\Delta\alpha_{MW}^0/\Delta\alpha^\infty$ calculated as in Fig. 7; $\zeta = 130$ mV (solid line); $q = 6$ μC/cm^2 (dashed line).

X. COMPARISON BETWEEN LOW-FREQUENCY DATA AND O'BRIEN–WARD MODEL

In analogy to what was done in Section IX, we define here B_{LF}^0 as the measured low-frequency ($\nu = 1$kHz) values of the Kerr constant. In particular, we will focus on the ratio $B_{LF}^0/B^\infty = \Delta\alpha_{LF}^0/\Delta\alpha^\infty$ to compare it with the theoretical predictions (see Section VI), available only for the $\nu \to 0$ limit [11]. In Figs 22–25 we show $\Delta\alpha_{LF}^0/\Delta\alpha^\infty$ measured in suspensions of PTFE particles covered by different surfactants and at different ionic strengths. The comparison between Figs 22 and 23 shows the opposite effects of adding Triton X-100, which decreases the low-frequency value of B, and of adding AOT, which instead makes B_{LF}^0 grow. These effects are compared in Fig. 24, where the data referring to HTAB are also included. The difference between the B values of the uncovered particles presented in Figs 22–24 is probably a consequence of having used particles from different batches. A more detailed study of the behavior of $\Delta\alpha_{LF}^0/\Delta\alpha^\infty$ as a function of I is shown in Fig. 25 for different values of Triton X-100, AOT, and HTAB concentrations. It can be seen that the dependence of $\Delta\alpha_{LF}^0/\Delta\alpha^\infty$ on I is particularly strong at low ionic strength. The lines of Fig. 25 have been calculated according to the theory presented in Section VI and concern particles having different values for their surface ζ potential. The EB data in Fig. 25, obtained with PTFE particles either uncovered or covered with Triton X-100, nicely match with $\Delta\alpha_{LF}^0/\Delta\alpha^\infty$ calculated from the low-frequency model for constant ζ, which can therefore be determined from this comparison. The behavior of particles coated with AOT or HTAB cannot instead be compared with theory, since in such cases $\Delta\alpha_{LF}^0/\Delta\alpha^\infty$ takes values larger than the maximum obtainable from the theory in the

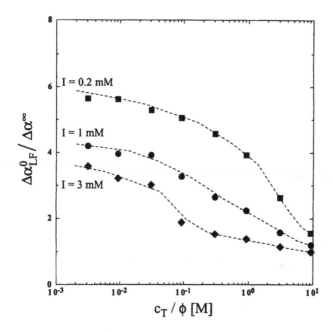

FIG. 22 Real part of the low-frequency value of the normalized anisotropy of particle polarizability $\Delta\alpha_{LF}^0/\Delta\alpha^\infty$, measured for PTFE-1 particles, as a function of the Triton X-100 concentration c_T/ϕ, at different NaCl ionic strengths. The data are taken at a fixed volume fraction $\phi = 0.5\%$. The dashed lines are intended to guide the eye through the data.

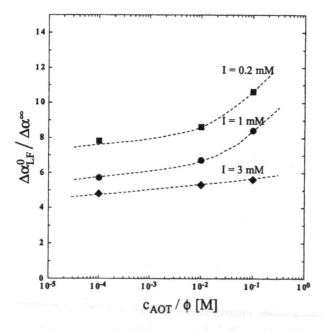

FIG. 23 Real part of the low-frequency value of the normalized anisotropy of particle polarizability $\Delta\alpha_{LF}^0/\Delta\alpha^\infty$, measured for PTFE-1 particles, as a function of the AOT concentration c_{AOT}/ϕ, at different NaCl ionic strengths. The data are taken at fixed volume fraction $\phi - 0.5\%$. The dashed lines are a guide to the eye through the data.

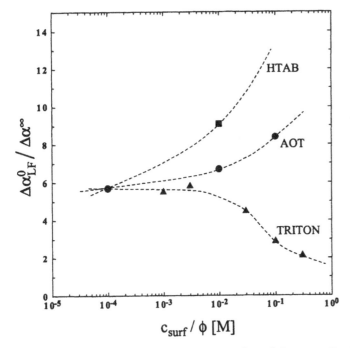

FIG. 24 Real part of the low-frequency value of the normalized anisotropy of particle polarizability $\Delta\alpha_{LF}^0/\Delta\alpha^\infty$, measured for PTFE-1 and PTFE-2 particles, for different surfactants as a function of surfactant concentration c_{surf}/ϕ, at fixed ionic strength $I = 1$ mM NaCl; c_{surf} is the
concentration of added surfactant. Triton X-100 (triangles), AOT (circles), and HTAB (squares).
The dashed lines are drawn to guide the eye through the data.

limit of high ζ. This is not too surprising when this behavior is compared to that shown in
Fig. 8, where a similar disagreement was found for spherical particles when comparing the
approximate theory to the exact solution of the SE model. If at high ζ (or low I) the
approximate theory for spheres underestimates the exact calculation, it seems reasonable
to conclude that the disagreement between theory and experiment in Fig. 25 could thus be
a consequence of the approximations inherent in the asymptotic theory of O'Brien and
Ward [11].

XI. CONCLUSIONS

Our experiments show that charged, anisotropic polyelectrolytes have a nonmonotonic
FREB response over a very wide frequency range. With respect to previous works, this
study covers a broader range of surface properties, from bare PTFE surfaces to surfaces
fully covered by surfactants, and extends the frequency span of experiments up to 200 MHz.

The EMW model provides a good description of the observed behavior at high
frequencies. Two versions of the EMW model were used to analyze data. The EMW-1
model, which assumes a uniformly distributed surface charge, accurately represents the
FREB data for particles whose surfaces are saturated with ionic surfactant. Bare or partially covered particles behave differently. To interpret their behavior we used the EMW-2

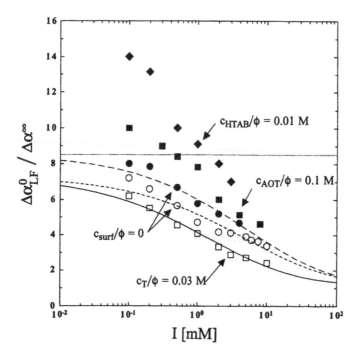

FIG. 25 Measured (symbols) and calculated (lines) real part of the low-frequency value of the normalized anisotropy of particle polarizability $\Delta\alpha_{LF}^0/\Delta\alpha^\infty$, plotted as a function of the ionic strength I, for PTFE particles in presence of different surfactants. PTFE-1: no surfactant (open circles); $c_T/\phi = 0.1$ M (open squares). PTFE-2: no surfactant (filled circles); $c_{AOT}/\phi = 0.1$ M (filled squares), $c_{HTAB}/\phi = 0.01$ M (filled diamonds). The lines represent calculations for NaCl at different ζ potentials: $\zeta = 140$ mV (solid line); $\zeta = 170$ mV (short-dashed line); $\zeta = 180$ mV (long-dashed line); high ζ limit (dotted line).

model, which allows for nonuniform surface distributions of charge. Overall, the success of EMW description provides an improved understanding of electrokinetic phenomena, offering a direct proof of MW processes in polyelectrolytes at high frequencies.

At low frequencies the comparison of EB data with the available theory is less conclusive. The low-frequency FREB data with highly charged particles are systematically underestimated by the available theories.

The rich behavior of $B(\nu)$ over such a wide frequency range, the direct access it provides to the electric polarizability of the particles, the success in interpreting the data in the high-frequency regime with a model greatly simplified with respect to the complete SE description, and the flexibility of the EMW model, allowing description of nonspherical particles suitable for EB experiments, make FREB a powerful technique for studying polyelectrolytes and characterizing their surface properties.

ACKNOWLEDGMENTS

A large part of the work presented here has been done in collaboration with D. A. Saville. We thank A. V. Delgado and M. Fixman for useful discussions.

REFERENCES

1. WB Russel, DA Saville, WR Schowalter. Colloidal Dispersions. Cambridge University Press, 1989.
2. J Lyklema. Fundamentals of Interface and Colloid Science. vol. II. Solid–Liquid Interfaces. London: Academic press, 1995.
3. EHB DeLacey, LR White. J Chem Soc, Faraday Trans 2 77: 2007, 1981.
4. M Mandel, T Odijk. Ann Rev Phys Chem 35: 75, 1984.
5. M Fixman. J Chem Phys 72: 5177, 1980.
6. EJ Hinch, JD Sherwood, WC Chen, PN Sen. J Chem Soc, Faraday Trans 2 80: 535, 1984.
7. WC Chew, PN Sen. J Chem Phys 77: 4683, 1982.
8. SS Dukhin, VN Shilov. Adv Colloid Interface Sci 80: 535, 1984.
9. RW O'Brien, J. Colloid Interface Sci 92: 204, 1983.
10. RW O'Brien. J Colloid Interface Sci 113: 81, 1986.
11. RW O'Brien, DN Ward. J Colloid Interface Sci 121: 402, 1988.
12. RW O'Brien and WN Rowlands. J Colloid Interface Sci 159: 471, 1993.
13. DA Saville, T Bellini, V Degiorgio, F Mantegazza. J Chem Phys 113: 6974, 2000.
14. CT O'Konski. Molecular Electro-optics. New York: Marcel Dekker, 1976.
15. E Fredericq, C Houssier. Electric Dichroism and Electric Birefringence. Oxford: Clarendon Press, 1973.
16. T Bellini, F Mantegazza, V Degiorgio, R Avallone, DA Saville. Phys Rev Lett 82: 5160, 1999
17. F Mantegazza, T Bellini, M Buscaglia, V Degiorgio, DA Saville. J Chem Phys 113: 6984, 2000.
18. CT O'Konski. J Phys Chem 60: 64, 1960.
19. F Mantegazza, T Bellini, V Degiorgio, AV Delgado, FJ Arroyo. J Chem Phys 109: 6905, 1998
20. CT O'Konski, S Krause. J Phys Chem 74: 3243, 1970.
21. AV Delgado, F Carrique, FJ Arroyo, T Bellini, F Mantegazza, ME Giardini, V Degiorgio. Colloids Surfaces A 140: 157, 1998.
22. T Bellini, V Degiorgio, F Mantegazza. Colloids Surfaces A 140: 103, 198.
23. H Watanabe, A Morita. Adv Chem Phys 56: 255, 1984.
24. M Fixman. Macromolecules 13: 711, 1980.
25. JA Stratton. Electromagnetic Theory. New York: McGraw-Hill, 1941.
26. M Teubner. J Chem Phys 76: 5564, 1982.
27. JJ Bikerman. Trans Faraday Soc 36: 154, 1940.
28. C Grosse, M Tirado, W Pieper, R Pottel. J Colloid Interface Sci 205: 26, 1998.
29. T Bellini, R Piazza, C Sozzi, V Degiorgio. Europhys Lett 7: 561, 1988.
30. T Bellini, V Degiorgio, F Mantegazza, F Ajmone-Marsan, C Scarnecchia. J Chem Phys 103: 8228, 1995.
31. N Ookubo, Y Hirai, K Ito, R Hayakawa. Macromolecules 22: 1539, 1989.
32. H Hoffman, U Kramer. In: DM Bloor, E Wyn-Jones, eds. The Structure, Dynamics and Equilibrium Properties of Colloidal Systems. NATO ASI Series, vol. 385–396, Amsterdam: Kluwer, 1990.
33. M Stoimenova, T Okubo. In: JA Schwarz, CI Contescu, eds. Surface of Nanoparticles and Porous Materials, vol. 103–124. New York: Marcel Dekker, 1999.
34. FJ Rahl, MA Evanco, RJ Fredericks, AC Reimschussel. J Polym Sci A-2 10: 1337, 1972.

15

Sedimentation Potential and Electric Conductivity of Suspensions of Polyelectrolyte-coated Particles

HUAN J. KEH   National Taiwan University, Taipei, Taiwan

I. INTRODUCTION

The transport phenomena in suspensions of colloidal particles have received quite an amount of attention in the past, due to their wide applications in the areas of chemical, biomedical, agricultural, and environmental engineering and science. The surface of a colloidal particle is generally not hard and smooth as assumed in many theoretical models. For instance, coating of colloidal particles with polymers plays an important role in the control of the stability/flocculation behavior of colloidal suspensions [1]. Even the surfaces of model colloids such as silica and polystyrene latexes are "hairy" with a gel-like polymeric layer extending from the bulk material inside the particle [2]. In particular, the surface of a biological cell is not a hard smooth wall, but rather is a permeable rough surface with various appendages ranging from protein molecules of the order of nanometers to cilia of the order of micrometers [3]. Such particles can be modeled as a composite particle having a central solid core and an outer porous shell [4].

The creeping flow of an incompressible Newtonian fluid past a composite sphere was solved by Masliyah et al. [5] using the Brinkman equation for the flow field inside the fluid-permeable surface layer and the Stokes equations for the flow field external to the particle. An analytical formula for the drag force experienced by the particle was derived as a function of the radius of the solid core, the thickness of the porous shell, and the permeability of the shell. They also measured the settling velocities of a solid sphere with attached threads and found that theoretical predictions for the composite sphere are in excellent agreement with the experimental results. On the other hand, the effects of a thin layer of adsorbed polymers on the motion of a spherical particle were determined by Anderson and Kim [6] using a method of matched asymptotic expansions. The result for the drag force produced by the fluid on the particle was expressed as the hydrodynamic thickness of the adsorbed polymer layer. Recently, these analyses for the motion of an isolated polymer-coated sphere have been extended to the cases of motion of a single polymer-coated sphere in the proximity of boundaries [2, 7, 8] and to the cases of simultaneous motion of two or more polymer-coated spheres [2, 9, 10]. Through the use of the "free-surface" and "zero-vorticity" cell models [11–13] for the creeping flow relative to an assemblage of identical polymer-coated spheres, analytical solutions for the dependence of the average drag force of this assemblage on the volume fraction of the particles were also obtained [14, 15].

The problems of the motion of charged particles in an electrolyte solution are more complex than those of uncharged particles. Theoretical investigations on the electrokinetic phenomena of colloidal particles covered by charged porous surface layers have been performed for many years [3, 16–18]. These investigations provided formulas for the electrophoretic mobility of such a polyelectrolyte-coated particle by introducing the modified Brinkman equation for the flow field inside the porous surface layer of the particle and assuming that the local radii of curvature of the particle are much larger than the thicknesses of the electric double layer and of the porous surface layer (i.e., the particle surface is planar and the applied electric field is parallel to it). Experimental results for the electrophoretic mobility of charged composite particles are also available for human erythrocytes [19], rat lymphocytes [20], and latex particles coated with poly(N-isopropylacrylamide) hydrogel layers [21]. Based on a formula derived from the theory of a planar particle surface, these experimental results could be used to calculate the fixed charge density and the hydrodynamic resistance parameter of the porous surface layer. Recently, analytical expressions for the electrophoretic mobility of a spherical polyelectrolyte-coated particle have been derived [22–24]. It has been found that the electrophoretic mobility of a charged composite sphere can be quite different from that of a "bare" rigid sphere.

In this chapter, the effects of particle charges on the sedimentation potential and electric conductivity of a suspension of polyelectrolyte-coated particles are presented. The densities of the fixed charges and of the hydrodynamic frictional segments are assumed to be uniform throughout the porous surface layer of each particle, but no assumption is made as to the thicknesses of the double layer and of the porous surface layer relative to the dimension of the particle. In Section II, we summarize the fundamental electrokinetic equations and boundary conditions which govern the electrolyte ion distributions, the electrostatic potential profile, and the fluid-flow field inside and outside the porous surface layer of a polyelectrolyte-coated particle migrating in an unbounded solution. These basic equations are linearized assuming that the ion concentrations, the electric potential, and the fluid pressure have only a slight deviation from equilibrium due to the motion of the particle. In Section III, the electric current density in a suspension of identical charged particles subjected to a uniformly imposed electric field is described. The suspension is sufficiently dilute that the suspended particles occupy only a small fraction of the total volume of the suspension and the double layer surrounding each particle does not overlap with the others. The average current density is expressed as an integral over a large surface enclosing a single particle and its adjacent double layer, and the effective conductivity of the suspension, which is the ratio of the average current density to the applied electric field, is related to the electrochemical potential energies of the ionic species.

A. Sedimentation Potential in a Suspension of Charged Particles

When charged colloidal particles are moving relative to an electrolyte solution, the electrical double layer surrounding each particle is distorted by the fluid flow around the particle. The deformation of the double layer resulting from the fluid motion is usually referred to as the relaxation (or polarization) effect and gives rise to an induced electric field. The sedimentation potential, which arises in a suspension of settling charged particles, was first reported by Dorn in 1878 and this phenomenon is often known by his name [25, 26]. The sedimentation potential gradient (which is of the order 1–10 V/m) not only alters the velocity and pressure distributions in the fluid due to its action on the electrolyte ions but also retards the settling of the particles by an electrophoretic effect.

An important contribution to the sedimentation theory for a dilute suspension of identical spherical particles with arbitrary double-layer thickness was made by Booth [25]. He solved a set of electrokinetic differential equations using a regular perturbation method to obtain formulas for the sedimentation velocity and sedimentation potential expressed as power series in the zeta potential (ζ) of the particle up to $O(\zeta^2)$ and $O(\zeta)$, respectively. Numerical results relieving the restriction of low surface potential in Booth's analysis were reported by Stigter [27] using a modification of the theory of electrophoresis developed by Wiersema et al. [28]. It was found that the Onsager reciprocal relation between the sedimentation potential and the electrophoretic mobility derived by de Groot et al. [29] is satisfied within good computational accuracy. Taking the double-layer distortion from equilibrium as a small perturbation, Ohshima et al. [30] obtained general expressions and presented numerical results for the sedimentation velocity and sedimentation potential of charged spheres over a broad range of zeta potential and double-layer thickness. Other than cases of spherical particles, the effect of the deformation on the ion cloud surrounding a charged cylinder on the sedimentation velocity of the particle has also been investigated semianalytically [31, 32].

Using the "zero-vorticity" cell model with the condition of zero net electric current, Levine et al. [33] derived analytical expressions for the sedimentation velocity and sedimentation potential in a suspension of identical charged spheres with low surface potential as functions of the fractional volume concentration of the particles, based on the assumption that the overlap of the double layers of adjacent particles is negligible on the outer (virtual) surface of the cell. In the limiting case of a single particle their result somewhat differs from that obtained by Booth [25], which is not subject to the constraint of zero net current. This analysis was also presented by Ohshima [34] to demonstrate the Onsager relation between the sedimentation potential and the electrophoretic mobility of charged spheres with low zeta potentials in concentrated suspensions. Recently, both the "free-surface" and the "zero-vorticity" cell models, allowing for the effects of overlap and polarization of the double layers, have been used to derive closed-form formulas for the sedimentation velocity and sedimentation potential in concentrated suspensions of charged spheres with arbitrary double-layer thickness [35].

Booth's perturbation analysis was also extended to the derivation of the sedimentation velocity and sedimentation potential in dilute suspensions of charged porous spheres [36] and charged composite spheres [23] with low densities of the fixed charges. In Section IV, the sedimentation phenomena in a dilute suspension of polyelectrolyte-coated spheres in the solution of a symmetric electrolyte are considered. First, the axisymmetric translational motion of a polyelectrolyte-coated sphere in an unbounded solution is analyzed. Using the Debye–Huckel approximation, we give the solution of the equilibrium electric potential and ion concentration distributions in Section IV.A; then, in Section IV.B, the linearized electrokinetic equations are transformed into a set of differential equations by a regular perturbation method with the fixed charge densities of the porous surface layer and of the rigid core surface as the small perturbation parameters. The perturbed ion concentration, electric potential, fluid velocity, and pressure profiles are determined by solving this set of differential equations subject to the appropriate boundary conditions. A closed-form expression for the settling velocity of the polyelectrolyte-coated sphere is obtained from a balance among its gravitational, electrostatic, and hydrodynamic forces. In Section IV.C, an explicit formula for the sedimentation potential is resulted from letting the net electric current in the suspension be zero. The Onsager relation between sedimentation and electrophoresis is found to be satisfied for the polyelectrolyte-coated spheres. Finally, in Section IV.D, analytical expressions in some limiting and special cases and typical numerical results for the sedimentation velocity and sedimentation potential for polyelectrolyte-coated spheres are presented.

B. Electric Conductivity of a Suspension of Charged Particles

When an external electric field is imposed on charged colloidal particles suspended in an electrolyte solution, the particles and the surrounding ions are driven to migrate. As a consequence, the fluid is dragged to flow by the motion of the particles and the ions, and there is an electric current through the suspension. To determine the current density distribution and transport properties such as the electric conductivity, it is necessary to find out not only the local electric potential but also the local ionic densities and fluid velocity. That is, one must first solve a set of coupled electrokinetic equations to obtain the distributions of electric potential, ionic concentrations, and fluid velocity in the electrolyte solution, and then compute the average electric current and effective conductivity in the suspension.

Dukhin and Derjaguin [37] derived a simple formula for the effective electric conductivity of a dilute suspension of impermeable charged particles by considering an infinite plane slab of suspension immersed in an infinite homogeneous electrolyte subjected to an electric field perpendicular to the plane of the slab. Extending this analysis, Saville [38] and O'Brien [39] assumed that the particles and their electrical double layers occupy only a small fraction of the total volume of the suspension to obtain approximate formulas for the conductivity using a regular perturbation method for particles with low zeta potential immersed in a symmetric electrolyte correct to $O(\zeta^2)$. Their results have some discrepancies with the experimental data reported by Watillon and Stone-Masui [40], who measured the surface conductances of a number of monodisperse polystyrene latexes over a range of particle volume fractions. Later, Saville [41] considered the effects of nonspecific adsorption, which alters the concentrations of ions in the solution outside the double layers, and of counterions derived from the particle charging processes, and obtained better agreement between theories and experiments.

The basic equations governing the electric conductivity of a dilute suspension of colloidal particles also describe the electrophoretic phenomena. O'Brien [42] derived analytical formulas for the electrophoretic mobility and the electric conductivity of a dilute suspension of dielectric spheres with thin but polarized double layers in a general electrolyte solution. Using a similar analysis, O'Brien and Ward [43] also determined the electrophoretic mobility and the effective conductivity of a dilute suspension of randomly oriented spheroids with thin polarized diffuse layers at the particle surfaces. On the other hand, approximate analytical expressions for the electrophoretic mobility and the effective conductivity of dilute suspensions of colloidal spheres in symmetric electrolytes were obtained by Ohshima et al. [44]. These expressions are correct to order $(\kappa a)^{-1}$, where κ is Debye–Huckel parameter [defined by Eq. (31)] and a is the particle radius. When the zeta potential of the particles is small, their reduced result is in agreement with O'Brien's [39].

The "zero-vorticity" cell model was also used to evaluate anlytically [45] and numerically [46] the electric conductivity of a concentrated suspension of identical charged spheres as a function of the fractional volume concentration of the particles based on the assumption that the overlap of the double layers of adjacent particles is negligible on the virtual surface of the cell. Recently, both the "free-surface" and the "zero-vorticity" cell models allowing for the effects of the overlap and polarization of the double layers have been used to derive closed-form formulas for the electric conductivity in concentrated suspensions of charged spheres with arbitrary double-layer thickness [47].

In many practical applications, the electric conductivity of a suspension is known from direct measurement and the zeta potential of dielectric particles in the suspension can be calculated. Similarly, one can also measure the electrophoretic mobility of a particle in order to obtain the zeta potential. O'Brien and Perrins [48] derived a formula for the electric conductivity of a porous plug composed of closely packed spheres and compared it with the conductivity data for dilute and concentrated dispersions of monodisperse polystyrene particles reported by Van der Put and Bijsterbosch [49]. They found significant differences between the zeta potentials evaluated from measurements of the electric conductivity and of the electrophoretic mobility. Similar differences were also found by another work [50] in which the conductivities and electrophoretic mobilities of polystyrene latex systems were measured. On the other hand, Stigter [51] developed a theory based on the concepts used to describe the conductivity of strong electrolyte solutions, in which the specific conductance of the suspension was computed by summing the individual contributions of the particle–ion interactions expressed in terms of equivalent conductances. The differences between the kinetic charges calculated from electrophoresis and from conductance in this theory were found to be small and within the errors of the experiments and the theoretical models [52].

Recently, analytical studies for the electric conductivity of dilute suspensions of charged porous spheres [53] and charged composite spheres [24] were performed under the assumption that the densities of the fixed charges are low. In Section V, we present the analysis for the effective conductivity of a suspension of polyelectrolyte-coated particles. The axisymmetric electrophoretic motion of a polyelectrolyte-coated sphere in an unbounded electrolyte solution is first considered in Section V.A. By use of the solution of the equilibrium electric potential distribution obtained in Section IV.A, the linearized electrokinetic equations are transformed into a set of differential equations by using the regular perturbation method with the fixed charge densities of the porous surface layer and of the rigid core surface as the small perturbation parameters. The perturbed electrochemical potentials of ions and the fluid velocity are determined by solving this set of differnetial equations subject to the appropriate boundary conditions. An analytical expression for the electric conductivity of a dilute suspension of identical polyelectrolyte-coated spheres is derived. Analytical expressions in some limiting and special cases and typical numerical results for the effects of the fixed charges of polyelectrolyte-coated spheres on the effective conductivity of the suspension are presented in Section V.B.

II. BASIC ELECTROKINETIC EQUATIONS FOR MOTION OF A POLYELECTROLYTE-COATED PARTICLE

In this section we consider the motion of a polyelectrolyte-coated particle of arbitrary shape in an unbounded liquid solution containing M ionic species when a constant gravitational field and/or electric field is applied. The polyelectrolyte-coated particle is modeled as a charged rigid particle core covered by a surface layer of charged porous substance in equilibrium with the surrounding electrolyte solution. The porous surface layer is treated as a solvent-permeable and ion-penetratable homogenous shell in which fixed-charged groups are assumed to distribute at a uniform density.

A. Governing Equations

In the electrolyte solution, conservation of all species, which do not react with one another, in the steady state requires that

$$\nabla \cdot \mathbf{J}_m = 0, \quad m = 1, 2, \ldots, M \tag{1}$$

where $\mathbf{J}_m(\mathbf{x})$ is the number flux of species m at position $\mathbf{x}$. If the solution is dilute, the flux is given by

$$\mathbf{J}_m = n_m \mathbf{u} - n_m \frac{D_m}{kT} \nabla \mu_m \tag{2}$$

with the electrochemical potential energy field of the mth species $\mu_m(\mathbf{x})$ defined as

$$\mu_m = \mu_m^0 + kT \ln n_m + z_m e \psi \tag{3}$$

Here, $n_m(\mathbf{x})$ and z_m are the concentration (number density) distribution and the valence of species m, respectively, D_m is the diffusion coefficient of species m which is assumed to be constant both inside and outside the porous surface layer, $\mathbf{u}(\mathbf{x})$ is the fluid velocity field relative to the particle, $\psi(\mathbf{x})$ is the electric potential distribution, μ_m^0 is a constant, e is the elementary electric charge, k is Boltzmann's constant, and T is the absolute temperature. The first term on the right-hand side of Eq. (2) represents the convection of the ionic species by the fluid and the second term denotes the diffusion and electrically induced migration of the species.

We assume that the Reynolds number of the fluid motion is vanishingly small, so the inertial effect on the fluid momentum balance can be neglected. The fluid flow is governed by a combination of the Stokes and Brinkman equations modified with the electrostatic effect:

$$\eta \nabla^2 \mathbf{u} - h(\mathbf{x}) f \mathbf{u} = \nabla p - \rho \mathbf{g} + \sum_{m=1}^{M} z_m e n_m \nabla \psi \tag{4}$$

$$\nabla \cdot \mathbf{u} = 0 \tag{5}$$

where η and ρ are the viscosity and density, respectively, of the fluid, f is the friction coefficient inside the porous surface layer of the particle per unit volume of the fluid, $p(\mathbf{x})$ is the fluid pressure distribution, $\mathbf{g}$ is the acceleration of gravity, and $h(\mathbf{x})$ is a unit step function which equals unity if $\mathbf{x}$ is inside the surface layer, and zero if $\mathbf{x}$ is outside the polyelectrolyte-coated particle; $\rho, \eta, f,$ and $\mathbf{g}$ are assumed to be constant. Note that f can be expressed as $6\pi \eta a_s N_s$, where N_s and a_s are the number density and the Stokes radius, respectively, of the hydrodynamic frictional segments in the surface layer.

The local electric potential ψ and the space charge density are related by Poisson's equation:

$$\nabla^2 \psi = -\frac{4\pi}{\varepsilon} \left[\sum_{m=1}^{M} z_m e n_m + h(\mathbf{x}) Q \right] \tag{6}$$

Here, Q is the fixed charge density inside the porous surface layer; $\varepsilon = 4\pi \varepsilon_0 \varepsilon_r$, where ε_r is the relative permittivity of the electrolyte solution which is assumed to be the same inside and outside the surface layer, and ε_0 is the permittivity of a vacuum. Note that the space charge density in the surface layer is the sum of the densities of the mobile ions and the fixed charges.

Because the governing equations are coupled nonlinear partial differential equations, it is a formidable task to find a general solution of them. Therefore, we shall assume that the Peclet number is sufficiently small so that the system is only slightly distorted from the equilibrium state where the particle and fluid are at rest and replace these nonlinear equations by approximate linear equations. One can write

$$p = p^{(eq)} + \delta p \tag{7a}$$

$$n_m = n_m^{(eq)} + \delta n_m \tag{7b}$$

$$\psi = \psi^{(eq)} + \delta \psi \tag{7c}$$

$$\mu_m = \mu_m^{(eq)} + \delta \mu_m \tag{7d}$$

where $p^{(eq)}(\mathbf{x})$, $n_m^{(eq)}(\mathbf{x})$, $\psi^{(eq)}(\mathbf{x})$, and $\mu_m^{(eq)}(\mathbf{x})$ are the equilibrium distributions of pressure, concentration of species m, electric potential, and electrochemical potential energy of species m respectively; $\delta p(\mathbf{x})$, $\delta n_m(\mathbf{x})$, $\delta \psi(\mathbf{x})$, and $\delta \mu_m(\mathbf{x})$ are the small perturbations to the equilibrium state (in which neither the gravitational field nor the electric field is imposed). The equilibrium concentration of any species is related to the equilibrium potential by the Boltzmann distribution:

$$n_m^{(eq)} = n_m^{\infty} \exp\left(-\frac{z_m e \psi^{(eq)}}{kT}\right) \tag{8}$$

where n_m^{∞} is the concentration of the type-m ions in the bulk (electrically neutral) solution where the equilibrium potential is set equal to zero. The perturbed quantities $\delta \mu_m$, δn_m, and $\delta \psi$ are linearly related:

$$\delta \mu_m = kT \frac{\delta n_m}{n_m^{(eq)}} + z_m e \delta \psi \tag{9}$$

Substituting Eq. (7) into Eqs (1), (4), and (6), canceling their equilibrium components, using Eqs (8) and (9), and neglecting the products of the small quantities $\mathbf{u}$, δn_m, $\delta \psi$ and δu_m, one obtains

$$\nabla^2 \delta \mu_m = \frac{z_m e}{kT}\left(\nabla \psi^{(eq)} \cdot \nabla \delta \mu_m - \frac{kT}{D_m}\nabla \psi^{(eq)} \cdot \mathbf{u}\right) \tag{10}$$

$$\nabla^2 \mathbf{u} - h(\mathbf{x})\lambda^2 \mathbf{u} = \frac{1}{\eta}\nabla \delta p - \frac{\varepsilon}{4\pi\eta}\left(\nabla^2 \psi^{(eq)}\nabla \delta \psi + \nabla^2 \delta \psi \nabla \psi^{(eq)}\right) - \frac{1}{\eta}h(\mathbf{x})Q\nabla \delta \psi \tag{11}$$

$$\nabla^2 \delta \psi = -\frac{4\pi}{\varepsilon}\sum_{m=1}^{M}\frac{z_m e n_m^{\infty}}{kT}\exp\left(-\frac{z_m e \psi^{(eq)}}{kT}\right)(\delta \mu_m - z_m e \delta \psi) \tag{12}$$

where $\lambda = (f/\eta)^{1/2}$. Note that the reciprocal of the parameter λ is the shielding length characterizing the extent of flow penetration inside the porous surface layer of the particle.

B. Boundary Conditions

The boundary conditions at the surface of the rigid particle core are

$$\mathbf{u} = \mathbf{0} \tag{13a}$$

$$\mathbf{n} \cdot \mathbf{J}_m = 0 \tag{13b}$$

$$\mathbf{n} \cdot \nabla \psi = -\frac{4\pi}{\varepsilon} \sigma \tag{13c}$$

where $\mathbf{n}$ is the unit normal outward from the particle surface and σ is the surface charge density of the "bare" particle core. In Eq. (13a), we have assumed that the shear plane coincides with the surface of the rigid core. Equations (13b) and (13c) state that no ions can penetrate into the rigid core and that the Gauss condition holds at the surface of the rigid core.

The conditions far from the particle are

$$\mathbf{u} \to -\mathbf{U} \tag{14a}$$

$$n_m \to n_m^\infty \tag{14b}$$

$$\psi \to -\mathbf{E}_\infty \cdot \mathbf{x} \tag{14c}$$

Here, $\mathbf{U}$ is the translational velocity of the particle and $\mathbf{E}_\infty$ is the uniform applied electric field. If no electric field is applied ($\mathbf{E}_\infty = \mathbf{0}$), the electric potential in the bulk solution is set equal to zero. Equation (14a) takes a reference frame where the particle is at rest and the velocity of the fluid at infinity is the particle velocity in the opposite direction.

The boundary conditions at the surface of the polyelectrolyte-coated particle (the boundary between the porous surface layer and the external solution) are

$$\mathbf{u}|_{s^+} = \mathbf{u}|_{s^-} \tag{15a}$$

$$\mathbf{n} \cdot \mathbf{\Pi}|_{s^+} = \mathbf{n} \cdot \mathbf{\Pi}|_{s^-} \tag{15b}$$

$$n_m|_{s^+} = n_m|_{s^-} \tag{15c}$$

$$\mathbf{J}_m|_{s^+} = \mathbf{J}_m|_{s^-} \tag{15d}$$

$$\psi|_{s^+} = \psi|_{s^-} \tag{15e}$$

$$\nabla\psi|_{s^+} = \nabla\psi|_{s^-} \tag{15f}$$

Here, $\mathbf{\Pi}$ is the hydrodynamic stress of the fluid given by

$$\mathbf{\Pi} = -p\mathbf{I} + \eta[\nabla\mathbf{u} + (\nabla\mathbf{u})^T] \tag{16}$$

where $\mathbf{I}$ is the unit dyadic. Equations (15a) and (15b) are the continuity requirement of the fluid velocity and stress tensor at the particle surface. Equations (15c) and (15d) state that the concentration and flux of species m must be continuous. Equations (15e) and (15f) indicate that the potential and electric field are also continuous. The continuity of the electric field results from the assumptions that the relative permittivity of the solution takes the same value both inside and outside the surface layer of the polyelectrolyte-coated particle.

From Eqs (7), (9), (13b), (13c), (14b), and (14c), the conditions for $\delta\mu_m$ and $\delta\psi$ can be obtained as

$$\mathbf{n} \cdot \nabla\delta\mu_m = 0 \tag{17a}$$

$$\mathbf{n} \cdot \nabla\delta\psi = 0 \tag{17b}$$

at the rigid core surface of the particle, and

$$\delta\mu_m = -z_m e\mathbf{E}_\infty \cdot \mathbf{x} \tag{18a}$$

$$\delta\psi = -\mathbf{E}_\infty \cdot \mathbf{x} \tag{18b}$$

at infinity. Equations (15c)–(15f) yield the following boundary conditions at the surface of the particle:

$$\delta\mu_m|_{s^+} = \delta\mu_m|_{s^-} \tag{19a}$$

$$\nabla\delta\mu_m|_{s^+} = \nabla\delta\mu_m|_{s^-} \tag{19b}$$

$$\delta\psi|_{s^+} = \delta\psi|_{s^-} \tag{19c}$$

$$\nabla\delta\psi|_{s^+} = \nabla\delta\psi|_{s^-} \tag{19d}$$

The fluid velocity $\mathbf{u}$ is a small perturbed quantity, and the boundary conditions for $\mathbf{u}$ have been given by Eqs (13a), (14a), (15a), and (15b).

III. AVERAGE CURRENT DENSITY IN A SUSPENSION OF CHARGED PARTICLES

In this section, we consider the electric current in a dilute suspension of identical charged particles immersed in a solution containing M ionic species. The particles may be impermeable to the fluid, porous, or composite. It is assumed that the suspension is statistically homogeneous and all effects of its boundaries are ignored. When the isotropic suspension is subjected to a uniform applied electric field $\mathbf{E}_\infty$, one has

$$\mathbf{E}_\infty = -\frac{1}{V}\int_V \nabla\delta\psi \, dV \tag{20}$$

where V denotes a sufficiently large volume of suspension to contain many particles. To obtain Eq. (20), we have used Eq. (7c) and the fact that the volume average of the gradient of the equilibrium electric potential is zero. There is a resulting volume-average current density, which is collinear with $\mathbf{E}_\infty$, defined by

$$\langle\mathbf{i}\rangle = \frac{1}{V}\int_V \mathbf{i}(\mathbf{x}) \, dV \tag{21}$$

where $\mathbf{i}(\mathbf{x})$ is the current density distribution. The effective electric conductivity Λ of the suspension can be assigned by the linear relation:

$$\langle\mathbf{i}\rangle = \Lambda\mathbf{E}_\infty \tag{22}$$

Since the measured electric field and current density are equal to $\mathbf{E}_\infty$ and $\langle\mathbf{i}\rangle$, respectively, Eq. (22) reduces to the usual experimental definition of electric conductivity, provided that the suspension is everywhere homogeneous.

The current density **i** can be written as

$$\mathbf{i} = \sum_{m=1}^{M} z_m e \mathbf{J}_m \tag{23}$$

Substituting Eqs (2), (3), and (7)–(9) into Eq. (23), using the fact that $\nabla \mu_m^{(eq)} = 0$, and neglecting products of the small perturbation quantities, one has

$$\mathbf{i} = \sum_{m=1}^{M} z_m e n_m^{(eq)} \left(\mathbf{u} - \frac{D_m}{kT} \nabla \delta \mu_m \right) \tag{24}$$

Far from any particle (beyond the double layer), $n_m^{(eq)} \to n_m^\infty$, and Eq. (24) becomes

$$\mathbf{i} \to - \sum_{m=1}^{M} z_m e D_m \left(\nabla \delta n_m + \frac{z_m e n_m^\infty}{kT} \nabla \delta \psi \right) \tag{25}$$

By adding and subtracting the current density given by the above equation in the integrand of Eq. (21), one obtains

$$\langle \mathbf{i} \rangle = - \sum_{m=1}^{M} \frac{z_m e D_m}{V} \int_V \left(\nabla \delta n_m + \frac{z_m e n_m^\infty}{kT} \nabla \delta \psi \right) dV$$
$$+ \frac{1}{V} \int_V \left[\mathbf{i} + \sum_{m=1}^{M} z_m e D_m \left(\nabla \delta n_m + \frac{z_m e n_m^\infty}{kT} \nabla \delta \psi \right) \right] dV \tag{26}$$

Note that the magnitude of **i** and D_m can be taken as zero inside the dielectric rigid core of each particle.

In a statistically homogeneous suspension with constant bulk ionic concentrations, the volume average of $\nabla \delta n_m$ is zero. According to the definition of Eq. (20) the first term on the right-hand side of Eq. (26) equals $\Lambda^\infty \mathbf{E}_\infty$, where

$$\Lambda^\infty = \frac{e^2}{kT} \sum_{m=1}^{\infty} z_m^2 n_m^\infty D_m \tag{27}$$

which is the electric conductivity of the electrolyte solution in the absence of the particles. The integral in the second term on the right-hand side of Eq. (26) can be calculated by first considering for a single particle as if the others were absent and then multiplying the result by the particle number N in volume V, since the integrand vanishes beyond the double layers surrounding the particles and the suspension is assumed to be sufficiently dilute that the double layers do not overlap with one another. Also, the volume integral can be transformed into a surface integral over a spherical boundary of infinite radius containing the single particle at its center. With this arrangement, the second term becomes

$$\frac{N}{V} \int_{r \to \infty} \left(\mathbf{n} \cdot \mathbf{i} \mathbf{r} + \sum_{m=1}^{M} \frac{z_m e n_m^\infty D_m}{kT} \delta \mu_m \mathbf{n} \right) dS =$$
$$- \frac{N}{V} \sum_{m=1}^{M} \frac{z_m e n_m^\infty D_m}{kT} \int_{r \to \infty} (\mathbf{n} \cdot \nabla \delta \mu_m \mathbf{r} - \delta \mu_m \mathbf{n}) dS \tag{28}$$

where **r** is the position vector relative to the particle center. To obtain Eq. (28), the requirement of the conservation of electric charges ($\nabla \cdot \mathbf{i} = 0$) and Eq. (25) have been used. Therefore, the average current density given by Eq. (26) can be expressed as

$$\langle \mathbf{i} \rangle = \Lambda^\infty \mathbf{E}_\infty - \frac{N}{V} \sum_{m=1}^{M} \frac{z_m e n_m^\infty D_m}{kT} \int_{r \to \infty} (\mathbf{r} \nabla \delta \mu_m \cdot \mathbf{n} - \delta \mu_m \mathbf{n}) \mathrm{d}S \qquad (29)$$

The determination of $\delta \mu_m$ in Eq. (29) is concerned with the solution of a set of basic electrokinetic equations for the electrolyte around a single particle. These electrokinetic equations for the case of a polyelectrolye-coated particle are described in the previous section and their analytical solutions for a spherical particle with low fixed charge densities in a gravitational field and in an electric field are presented in Sections IV and V, respectively. From these solutions we shall derive the sedimentation potential and effective conductivity of a dilute suspension of identical polyelectrolyte-coated spheres.

IV. SEDIMENTATION POTENTIAL IN A SUSPENSION OF POLYELECTROLYTE-COATED SPHERES

We first consider the axisymmetric motion of a polyelectrolyte-coated sphere of radius a in an unbounded solution of a symmetrically charged, binary electrolyte with a constant bulk concentration $n^\infty (M = 2, z_+ = -z_- = Z, n_+^\infty = n_-^\infty = n^\infty$, where subscripts $+$ and $-$ refer to the cation and anion, respectively). As illustrated in Fig. 1, the polyelectrolyte-coated particle has a porous surface layer of constant thickness d so that the radius of the rigid core is $r_0 = a - d$. The translational velocity of the particle $\mathbf{U} = U\mathbf{e}_z$, where $\mathbf{e}_z$ is the unit vector in the positive axial direction. The origin of the spherical co-ordinate system (r, θ, ϕ) is taken at the center of the particle.

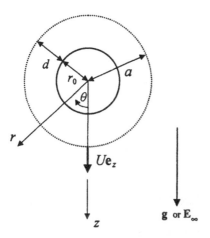

FIG. 1 Geometrical sketch for the motion of a polyelectrolyte-coated sphere driven by a gravitational field or an imposed electric field.

A. Equilibrium Electric Potential Distribution Around a Polyelectrolyte-coated Sphere

In this subsection we seek the solution of $\psi^{(eq)}$ which appears in the governing equations (10)–(12). Substituting the Boltzmann distribution [Eq. (8)] into Poisson's equation [Eq. (6)], one obtains the equilibrium Poisson–Boltzmann equation:

$$\nabla^2 \psi^{(eq)} = \frac{kT}{Ze}\kappa^2 \sinh\left(\frac{Ze\psi^{(eq)}}{kT}\right) - h(r)\frac{4\pi Q}{\varepsilon} \tag{30}$$

where $h(r)$ equals unity if $r_0 < r < a$ and is zero if $r > a$; κ is the Debye–Huckel parameter defined for symmetric electrolytes by

$$\kappa = \left(\frac{8\pi Z^2 e^2 n^\infty}{\varepsilon kT}\right)^{1/2} \tag{31}$$

The solution to Eq. (30) satisfying the boundary conditions (13c), (14c), (15e), and (15f) at equilbriium is

$$\psi^{(eq)} = \psi_{eq01}(r)\bar{\sigma} + \psi_{eq10}(r)\bar{Q} + O(\bar{\sigma}^3, \bar{\sigma}^2\bar{Q}, \bar{\sigma}\bar{Q}^2, \bar{Q}^3) \tag{32}$$

where

$$\bar{\sigma} = \frac{4\pi Ze\sigma}{\varepsilon \kappa kT} \tag{33a}$$

and

$$\bar{Q} = \frac{4\pi ZeQ}{\varepsilon \kappa^2 kT} \tag{33b}$$

which are the nondimensional charge densities of the rigid core surface and of the porous surface layer, respectively, of the polyelectrolyte-coated particle, and

$$\psi_{eq01} = \frac{kT}{Ze}\left(\frac{\kappa r_0}{\kappa r_0 + 1}\right)\frac{r_0}{r}e^{-\kappa(r-r_0)} \tag{34a}$$

$$\psi_{eq10} = \frac{kT}{Ze}\left\{1 - \left(1 + \frac{1}{\kappa a}\right)\frac{e^{-\kappa d}}{1 + \kappa r_0}[\kappa r_0 \cosh(\kappa r - \kappa r_0) + \sinh(\kappa r - \kappa r_0)]\frac{a}{r}\right\}, \text{ if } r_0 < r < a \tag{34b}$$

$$\psi_{eq10} = \frac{kT}{Ze}\left\{1 - \left(1 + \frac{1}{\kappa a}\right)\frac{e^{-\kappa d}}{1 + \kappa r_0}[\kappa r_0 \cosh(\kappa d) + \sinh(\kappa d)]\right\}\frac{a}{r}e^{-\kappa(r-a)}, \text{ if } r > a \tag{34c}$$

Note that $\psi^{(eq)}$ is a function of r only due to spherical symmetry, and the function $\psi_{eq01}(r)$ takes the same form in both regions of $r_0 < r < a$ and $r > a$.

Expression (32) for $\psi^{(eq)}$ as a power series in the fixed charge densities of the particle up to $O(\bar{\sigma}, \bar{Q})$ is the equilibrium solution for the linearized Poisson–Boltzmann equation that is valid for small values of the electric potential (the Debye–Huckel approximation). That is, the charge densities σ and Q of the particle must be small enough for the potential to remain small. Note that the contribution from the effects of $O(\bar{\sigma}^2, \bar{\sigma}\bar{Q}, \bar{Q}^2)$ to $\psi^{(eq)}$ in Eq. (32) disappears only for the case of symmetric electrolytes.

B. Sedimentation Velocity of a Polyelectrolyte-coated Sphere

To solve for the small quantities $\mathbf{u}$, δp, $\delta\mu_\pm$, and $\delta\psi$ for the sedimentation of a polyelectrolyte-coated sphere in a symmetric electrolyte in terms of the particle velocity U when the parameters $\bar{\sigma}$ and $\bar{Q}$ are small, these variables can be written as perturbation expansions in powers of $\bar{\sigma}$ and $\bar{Q}$:

$$\mathbf{u} = \mathbf{u}_{00} + \mathbf{u}_{01}\bar{\sigma} + \mathbf{u}_{10}\bar{Q} + \mathbf{u}_{02}\bar{\sigma}^2 + \mathbf{u}_{11}\bar{\sigma}\bar{Q} + \mathbf{u}_{20}\bar{Q}^2 + \cdots \tag{35a}$$

$$\delta p = p_{00} + p_{01}\bar{\sigma} + p_{10}\bar{Q} + p_{02}\bar{\sigma}^2 + p_{11}\bar{\sigma}\bar{Q} + p_{20}\bar{Q}^2 + \cdots \tag{35b}$$

$$\delta\mu_\pm = \mu_{00\pm} + \mu_{01\pm}\bar{\sigma} + \mu_{10\pm}\bar{Q} + \mu_{02\pm}\bar{\sigma}^2 + \mu_{11\pm}\bar{\sigma}\bar{Q} + \mu_{20\pm}\bar{Q}^2 + \cdots \tag{35c}$$

$$\delta\psi = \psi_{00} + \psi_{01}\bar{\sigma} + \psi_{10}\bar{Q} + \psi_{02}\bar{\sigma}^2 + \psi_{11}\bar{\sigma}\bar{Q} + \psi_{20}\bar{Q}^2 + \cdots \tag{35d}$$

$$U = U_{00} + U_{01}\bar{\sigma} + U_{10}\bar{Q} + U_{02}\bar{\sigma}^2 + U_{11}\bar{\sigma}\bar{Q} + U_{20}\bar{Q}^2 + \cdots \tag{35e}$$

where the functions $\mathbf{u}_{ij}$, p_{ij}, $\mu_{ij\pm}$, ψ_{ij}, and U_{ij} are independent of $\bar{\sigma}$ and $\bar{Q}$. Both $\mu_{00\pm}$ and ψ_{00} must equal zero as a result of not imposing the concentration gradient and electric field.

Substituting the expansions given by Eq. (35) and $\psi^{(eq)}$ and $n_\pm^{(eq)}$ given by Eqs (32) and (8) into the governing equations (10)–(12) and boundary conditions (13a), (14a), (15a), (15b), and (17)–(19) (taking $\mathbf{E}_\infty = 0$), and equating like powers of $\bar{\sigma}$ and $\bar{Q}$ on both sides of the respective equations, we can obtain a group of linear differential equations and boundary conditions for each set of the functions $\mathbf{u}_{ij}$, p_{ij}, μ_{ij} with i and j equal to 0, 1, 2, ... After solving this group of equations, the results for the r and θ components of $\mathbf{u}$, δp (to the orders of $\bar{\sigma}^2$, $\bar{\sigma}\bar{Q}$, and $\bar{Q}^2$), $\delta\mu_\pm$, and $\delta\psi$ [to the orders of $\bar{\sigma}$ and $\bar{Q}$, which will be sufficient for the calculation of the sedimentation velocity and potential to $O(\bar{\sigma}^2, \bar{\sigma}\bar{Q}, \bar{Q}^2)$] can be written as

$$\begin{aligned} u_r = \{ & U_{00}F_{00r}(r) + U_{01}F_{00r}(r)\bar{\sigma} + U_{10}F_{00r}(r)\bar{Q} + [U_{02}F_{00r}(r) + U_{00}F_{02r}(r)]\bar{\sigma}^2 \\ & + [U_{11}F_{00r}(r) + U_{00}F_{11r}(r)]\bar{\sigma}\bar{Q} + [U_{20}F_{00r}(r) + U_{00}F_{20r}(r)]\bar{Q}^2 \} \cos\theta \end{aligned} \tag{36a}$$

$$\begin{aligned} u_\theta = \{ & U_{00}F_{00\theta}(r) + U_{01}F_{00\theta}(r)\bar{\sigma} + U_{10}F_{00\theta}(r)\bar{Q} + [U_{02}F_{00\theta}(r) + U_{00}F_{02\theta}(r)]\bar{\sigma}^2 \\ & + [U_{11}F_{00\theta}(r) + U_{00}F_{11\theta}(r)]\bar{\sigma}\bar{Q} + [U_{20}F_{00\theta}(r) + U_{00}F_{20\theta}(r)]\bar{Q}^2 \} \cos\theta \end{aligned} \tag{36b}$$

$$\begin{aligned} \delta p = \frac{\eta}{a} \{ & U_{00}F_{P00}(r) + U_{01}F_{P00}(r)\bar{\sigma} + U_{10}F_{P00}(r)\bar{Q} + [U_{02}F_{P00}(r) + U_{00}F_{P02}(r) \\ & + \frac{a\varepsilon\kappa^2}{4\pi\eta}U_{00}\psi_{eq01}(r)F_{\psi01}(r)]\bar{\sigma}^2 + [U_{11}F_{P00}(r) + U_{00}F_{P11}(r) + \frac{a\varepsilon\kappa^2}{4\pi\eta}U_{00} \\ & \times (\psi_{eq01}(r)F_{\psi10}(r) + \psi_{eq10}(r)F_{\psi01}(r))]\bar{\sigma}\bar{Q} + [U_{20}F_{P00}(r) + U_{00}F_{P20}(r) \\ & + \frac{a\varepsilon\kappa^2}{4\pi\eta}U_{00}\psi_{eq10}(r)F_{\psi10}(r)]\bar{Q}^2 \} \cos\theta \end{aligned} \tag{36c}$$

$$\delta\mu_\pm = U_{00}[F_{01\pm}(r)\bar{\sigma} + F_{10\pm}(r)\bar{Q}]\cos\theta \tag{37}$$

$$\delta\psi = U_{00}[F_{\psi01}(r)\bar{\sigma} + F_{\psi10}(r)\bar{Q}]\cos\theta \tag{38}$$

Here, $F_{ijr}(r)$, $F_{ij\theta}(r)$, $F_{Pij}(r)$, $F_{ij\pm}(r)$, and $F_{\psi ij}(r)$ with i and j equal to 0, 1, and 2 are functions of r given by Eqs [A4], {A19}, [A20], and [A26] in Ref. 23.

The total force exerted on a charged sphere of radius a settling in an electrolyte solution can be expressed as

$$\mathbf{F}_{total} = \mathbf{F}_g + \mathbf{F}_e + \mathbf{F}_h \tag{39}$$

Here, $\mathbf{F}_g$ is the gravitational force (and buoyant force), $\mathbf{F}_e$ is the electric force, and $\mathbf{F}_h$ is the hydrodynamic drag force acting on the particle.

The gravitational force is given by

$$\mathbf{F}_g = V_t[\gamma_p(\rho_p - \rho) + \gamma_c(\rho_c - \rho)]g\mathbf{e}_z \tag{40}$$

where V_t is the dry volume of the polyelectrolyte-coated particle, ρ_p and ρ_c are, respectively, the true densities of the porous surface layer and the rigid core of the particle, γ_p and γ_c are the dry volume fractions of the surface layer and the rigid core, respectively, and $g\mathbf{e}_z(=\mathbf{g})$ is the gravitational acceleration.

The electric force acting on the charged particle can be represented by the integral of the electrostatic force density over the fluid volume outside the particle:

$$\mathbf{F}_e = -\frac{\varepsilon}{4\pi}\int_{r>a} \nabla\psi\nabla^2\psi \mathrm{d}V \tag{41}$$

As the net electric force acting on the particle at the equilibrium state is zero, the leading order of the electric force is given by

$$\mathbf{F}_e = -\frac{\varepsilon}{2}\int_0^\pi\int_a^\infty (\nabla\delta\psi\nabla^2\psi^{(eq)} + \nabla\psi^{(eq)}\nabla^2\delta\psi)r^2 \sin\theta \mathrm{d}r\mathrm{d}\theta \tag{42}$$

The hydrodynamic drag force acting on the spherical particle is given by the integral of the fluid pressure and viscous stress on the particle surface:

$$\mathbf{F}_h = 2\pi a^2\int_0^\pi \{-\delta p\mathbf{e}_r + \eta[\nabla\mathbf{u} + (\nabla\mathbf{u})^T]\cdot\mathbf{e}_r\} \sin\theta \, \mathrm{d}\theta \tag{43}$$

where $\mathbf{e}_r$ is the unit vector in the r direction.

At the steady state, the total force acting on the settling particle is zero. Using this constraint after the substitution of Eqs (40), (42), and (43) [with the help of Eqs (32), (36), and (38) for a symmetric electrolyte] into Eq. (39), one obtains

$$U_{00} = \frac{V_t[\gamma_p(\rho_p - \rho) + \gamma_c(\rho_c - \rho)]g}{4\pi\eta a C_{006}} \tag{44a}$$

$$U_{01} = 0 \tag{44b}$$

$$U_{10} = 0 \tag{44c}$$

$$U_{02} = \frac{U_{00}}{C_{006}}\left[-C_{026} + \int_a^\infty \left(\frac{r}{a}\right)^3 G_{02}(r)\mathrm{d}r\right] \tag{44d}$$

$$U_{11} = \frac{U_{00}}{C_{006}}\left[-C_{116} + \int_a^\infty \left(\frac{r}{a}\right)^3 G_{11}(r)\mathrm{d}r\right] \tag{44e}$$

$$u_{20} = \frac{U_{00}}{C_{006}}\left[-C_{206} + \int_a^\infty \left(\frac{r}{a}\right)^3 G_{20}(r)\mathrm{d}r\right] \tag{44f}$$

where functions $G_{02}(r)$, $G_{11}(r)$, and $G_{20}(r)$ and coefficients C_{006}, C_{026}, C_{116}, and C_{206} are defined by Eqs [41], [A6f], and [A27f] in Ref. 23. Note that U_{00} is the sedimentation velocity of an uncharged polymer-coated sphere, which was obtained by Masliyah et al. [5], and C_{006} is a function of parameters λa and r_0/a only. The definite integrals in the closed-form equations (44d), (44e), and (44f) as well as in coefficients C_{026}, C_{116}, and C_{206} can be performed numerically.

From Eqs (35e) and (44), the settling velocity of the polyelectrolyte-coated sphere can be expressed as

$$U = U_{00}[1 - (\kappa a)^2 H_1 \bar{\sigma}^2 - (\kappa a)^3 H_2 \bar{\sigma}Q - (\kappa a)^4 H_3 \bar{Q}^2 + O(\bar{\sigma}^3, \bar{\sigma}^2\bar{Q}, \bar{\sigma}\bar{Q}^2, \bar{Q}^3)] \qquad (45)$$

where the dimensionless coefficients:

$$H_1 = -\frac{U_{02}}{(\kappa a)^2 U_{00}} \qquad (46a)$$

$$H_2 = -\frac{U_{11}}{(\kappa a)^3 U_{00}} \qquad (46b)$$

$$H_3 = -\frac{U_{20}}{(\kappa a)^4 U_{00}} \qquad (46c)$$

and their typical numerical results calculated by using Eq. (44) will be given in Section IV.D. Note that $(\kappa a)^2\bar{\sigma}^2$, $(\kappa a)^3\bar{\sigma}Q$, $(\kappa a)^4\bar{Q}^2$ are independent of κ or n^∞. For a given electrolyte solution, coefficients H_1, H_2, and H_3 are functions of parameters κa, λa, and r_0/a only.

C. Sedimentation Potential

We now consider a dilute suspension of identical polyelectrolyte-coated spheres of radius a in the solution of a symmetric electrolyte. The electric fields around the individual particles superimpose to give a sedimentation field $\mathbf{E}_{SED}$ in the suspension. Since the suspension is statistically homogeneous, $\mathbf{E}_{SED}$ is uniform and can be regarded as the average of the gradient of electric potential over a sufficiently large volume V of the suspension to contain many particles. From Eq. (20):

$$\mathbf{E}_{SED} = -\frac{1}{V} \int_V \nabla \delta\psi \, dV \qquad (47)$$

In order to calculate $\mathbf{E}_{SED}$, we use the requirement that there exists no net current in the suspension. That is, the volume-average current density expressed by Eq. (21) or (29) must be zero.

For a symmetric electrolyte with the absolute value of valence Z, the average current density given by Eq. (29) becomes

$$\langle \mathbf{i} \rangle = \Lambda^\infty \mathbf{E}_{SED} - \frac{Z^2 e^2 n^\infty N}{kTV} \int_{r\to\infty} [D_+(\mathbf{r}\nabla\delta\mu_+ \cdot \mathbf{n} - \delta\mu_+\mathbf{n}) - D_-(\mathbf{r}\nabla\delta\mu_- \cdot \mathbf{n} - \delta\mu_-\mathbf{n})]dS$$

$$(48)$$

Substituting Eq. (37) into Eq. (48) and using the requirement that $\langle i \rangle = 0$, one has

$$
\mathbf{E}_{SED} = \lim_{r \to \infty} \frac{4\pi r^2 Z^2 e^2 n^\infty N U_{00}}{3kTV\Lambda^\infty} \left\{ \left[D_+ \left(r\frac{dF_{01+}}{dr} - F_{01+} \right) - D_- \left(r\frac{dF_{01-}}{dr} - F_{01-} \right) \right] \overline{\sigma} \right.
$$
$$
\left. + \left[D_+ \left(r\frac{dF_{10+}}{dr} - F_{10+} \right) - D_- \left(r\frac{dF_{10-}}{dr} - F_{10-} \right) \right] \overline{Q} + O\left(\overline{\sigma}^3, \overline{\sigma}^2\overline{Q}, \overline{\sigma}\overline{Q}^2, \overline{Q}^3 \right) \right\} \mathbf{e}_z
$$

(49)

After making relative calculations, the sedimentation potential field can be expressed as

$$
\mathbf{E}_{SED} = -\phi \left[\gamma_p(\rho_p - \rho) + \gamma_c(\rho_c - \rho) \right] \frac{\mu_E}{\Lambda^\infty} \mathbf{g}
$$

(50)

Here, $\phi = NV_t/V$ is the true volume fraction of the polyelectrolyte-coated particles and μ_E is the electrophoretic mobility of a polyelectrolyte-coated sphere of radius a at low electric potential:

$$
\mu_E = H_\sigma \frac{a\sigma}{\eta} + H_Q \frac{a^2 Q}{\eta} + O\left(\overline{\sigma}^3, \overline{\sigma}^2\overline{Q}, \overline{\sigma}\overline{Q}^2, \overline{Q}^3 \right)
$$

(51)

where the dimensionless coefficients H_σ and H_Q have the closed forms:

$$
H_\sigma = \frac{\kappa Z e}{6a^2 C_{006} kT} \int_{r_0}^{\infty} (2r^3 + r_0^3) F_{00r}(r) \frac{d\psi_{eq01}}{dr} dr
$$

(52a)

$$
H_Q = \frac{Ze}{6a^3 C_{006} kT} \int_{r_0}^{\infty} (2r^3 + r_0^3) F_{00r}(r) \frac{d\psi_{eq10}}{dr} dr
$$

(52b)

In Eq. (52), functions $\psi_{eq01}(r)$ and $\psi_{eq10}(r)$ are given by Eq. (34). Again, the contribution from the second-order effects of the fixed charge densities of the particles to $\mathbf{E}_{SED}$ and μ_E in Eqs (49) and (51) vanishes only for symmetric electrolytes. The numerical results for coefficients H_σ and H_Q calculated from Eq. (52) will be given in Section IV.D. For a specific electrolyte solution, these coefficients are functions of parameters κa, λa, and r_0/a only. It is understood that the result given by Eqs (50)–(52) is only valid with the requirements that $\phi \ll 1$ and $\kappa a \phi^{-1/3} \gg 1$.

Equation (50) is an Onsager reciprocal relation connecting the sedimentation potential with the electrophoretic mobility (correct to order ϕ) derived by de Groot et al. [29] on the basis of irreversible thermodynamics. For a dilute suspension of impermeable rigid spheres, this relation has also been demonstrated by Ohshima et al. [30]. For the situation of combined sedimentation and electrophoresis in a dilute suspension of colloidal particles, the average electric current density $\langle i \rangle$ and mass flux $\langle j \rangle$ can be expressed in terms of the Onsager transport coefficients b_{11}, b_{12}, b_{21}, and b_{22} as

$$
\langle i \rangle = b_{11} \mathbf{E}_\infty + b_{12} \mathbf{g}
$$

(53a)

$$
\langle j \rangle = b_{21} \mathbf{E}_\infty + b_{22} \mathbf{g}
$$

(53b)

with

$$
b_{11} = \Lambda
$$

(54a)

$$
b_{12} = b_{21} = \phi \left[\gamma_p(\rho_p - \rho) + \gamma_c(\rho_c - \rho) \right] \mu_E
$$

(54b)

$$b_{22} = \phi[\gamma_p(\rho_p - \rho) + \gamma_c(\rho_c - \rho)]\frac{U}{g} \tag{54c}$$

where Λ is the electric conductivity of the suspension defined by Eq. (22). For polyelectrolyte-coated spheres in a symmetric electrolyte, $\Lambda = \Lambda^{\infty} + O(\bar{\sigma}, \bar{Q})$ with $\Lambda^{\infty} = (D_+ + D_-)Z^2 e^2 n^{\infty}/kT$ as given by Eq. (27), μ_E is given by Eq. (51) to the first orders of $\bar{\sigma}$ and $\bar{Q}$, and U can be evaluated using Eq. (45) to the second orders. The details for the calculation of Λ will be presented in Section V.

D. Discussion

In this subsection, we first consider several limiting cases and a special case of the expressions (45) and (51) for the sedimentation velocity and electrophoretic mobility (or sedimentation potential), respectively, of polyelectrolyte-coated spheres. The correctness of these expressions may be confirmed by examining some of the limiting cases for which analytical solutions are already known. Numerical results of some typical cases are then presented.

When there is no permeable layer on the surface of the rigid particle core, one has $d = 0$, $r_0 = a$, $\gamma_p = 0$, $\gamma_c = 1$, $Q = 0$, and $\lambda = 0$. Equations (44a), (45), and (51) then reduce to

$$U_{00} = \frac{V_t(\rho_c - \rho)g}{6\pi\eta a} \tag{55a}$$

$$U = U_{00}\left[1 - \frac{\varepsilon\bar{\sigma}^2}{96\pi\eta}\left(\frac{1}{D_+} + \frac{1}{D_-}\right)\left(\frac{kT}{Ze}\right)^2\left(\frac{\kappa a}{\kappa a + 1}\right)^2 \Theta_1(\kappa a, 1) + O(\bar{\sigma}^3)\right] \tag{55b}$$

$$\mu_E = \frac{a\bar{\sigma}}{\eta}\left(\frac{1}{\kappa a + 1}\right)\Theta_2(\kappa a, 1) + O(\bar{\sigma}^3) \tag{55c}$$

where Θ_1 and Θ_2 are functions defined by

$$\Theta_1(x, y) = y^3 e^{2x}[5E_6(x) - 3E_4(x)]^2 - 8e^x[E_5(x) - E_3(x)] \\ + e^{2x}[7E_8(2x) - 3E_4(2x) - 4E_3(2x)] \tag{56a}$$

$$\Theta_2(x, y) = \frac{2}{3} + \frac{1}{3}y^3 - y^3 e^x[5E_7(x) - 2E_5(x)] \tag{56b}$$

and

$$E_n(x) = \int_1^{\infty} t^{-n}e^{-xt}dt \tag{57}$$

Equation (55a) is the result of Stokes' law. From Eq. (32), by letting $r_0 = a$, one can obtain the following relationship between the surface (zeta) potential ζ and the surface charge density of the rigid sphere at equilibrium:

$$\sigma = \frac{\varepsilon(\kappa a + 1)}{4\pi a}\zeta \tag{58}$$

Substituting the above equation into Eqs (55b) and (55c), these degenerated results are the same as those of a rigid sphere with a low surface potential [30].

When the particle is a homogeneous porous sphere, one has $r_0 = 0$, $d = a$, $\gamma_p = 1$, $\gamma_c = 0$, and $\sigma = 0$. In this limiting case, Eqs. (44a), (45), and (51) become

$$U_{00} = \frac{V_t(\rho_c - \rho)g}{6\pi\eta a}\left[\frac{2(\lambda a)^3 \cosh(\lambda a) + 3\alpha_1(\lambda a)}{2(\lambda a)^2 \alpha_1(\lambda a)}\right] \tag{59a}$$

$$\begin{aligned}
U = U_{00} &+ \frac{U_{00}\overline{Q}^2}{3(\lambda a)^2\alpha_1(\lambda a)}\left\{3\alpha_1(\lambda a)\int_0^a \left(\frac{r}{a}\right)^3 G_{20}(r)dr + 6\int_0^a \alpha_1(\lambda r)G_{20}(r)dr\right.\\
&+ [2(\lambda a)^3\cosh(\lambda a) + 3\alpha_1(\lambda a)]\int_a^\infty \left(\frac{r}{a}\right)^3 G_{20}(r)dr - 3(\lambda a)^2\alpha_1(\lambda a)\int_a^\infty \left(\frac{r}{a}\right)^2 G_{20}(r)dr\\
&+ \left.[(\lambda^2 a^2 + 6)\alpha_1(\lambda a) - 2(\lambda a)^2\sinh(\lambda a)]\int_a^\infty G_{20}(r)dr\right\} + O(\overline{Q}^3)
\end{aligned} \tag{59b}$$

$$\begin{aligned}
\mu_E = \frac{Q}{\eta\lambda^2}&\left\{1 + \frac{1}{3}\left(\frac{\lambda}{\kappa}\right)^2\left(1 + e^{-2\kappa a} - \frac{1 - e^{-2\kappa a}}{\kappa a}\right) + \frac{1}{3}\left(\frac{\lambda^2}{\lambda^2 - \kappa^2}\right)\left(1 + \frac{1}{\kappa a}\right)\right.\\
&\times \left.\left[\left(\frac{\lambda}{\kappa}\right)^2 \frac{\kappa a(1 + e^{-2\kappa a}) - 1 + e^{-2\kappa a}}{\lambda a \coth(\lambda a) - 1} - 1 + e^{-2\kappa a}\right]\right\} + O(\overline{Q}^3)
\end{aligned} \tag{59c}$$

where the function $\alpha_1(x) = x\cosh x - \sinh x$. Equations (59b) and (59c) are identical to the formulas for the sedimentation velocity and electrophoretic mobility, respectively, of a charged porous sphere derived previously [22, 36, 53, 54].

When $\lambda \to \infty$ (very high segment density in the surface layer of the polyelectrolyte-coated particle), the resistance to the fluid motion inside the porous surface layer is infinitely large and the velocity profile in the surface layer disappears. The ions can still penetrate the surface layer, and the equilibrium potential distributiont $\psi^{(eq)}$ is the same as given by Eq. (32). For this limiting case, Eqs (44a), (45), and (51) become

$$U_{00} = \frac{V_t[\gamma_p(\rho_p - \rho) + \gamma_c(\rho_c - \rho)]g}{6\pi\eta a} \tag{60a}$$

$$U = U_{00}\left\{1 - \frac{\varepsilon}{96\pi\eta}\left(\frac{1}{D_+} + \frac{1}{D_-}\right)\left[\psi^{(eq)}(a)\right]^2 \Theta_1\left(\kappa a, \frac{r_0}{a}\right) + O\left(\overline{\sigma}^3, \overline{\sigma}^2\overline{Q}, \overline{\sigma}\overline{Q}^2, \overline{Q}^3\right)\right\} \tag{60b}$$

$$\mu_E = \frac{\varepsilon\psi^{(eq)}(a)}{4\pi\eta}\Theta_2\left(\kappa a, \frac{r_0}{a}\right) + O\left(\overline{\sigma}^3, \overline{\sigma}^2\overline{Q}, \overline{\sigma}\overline{Q}^2, \overline{Q}^3\right) \tag{60c}$$

Note that in this case the mobile ions can penetrate the surface layer, and the fixed charges are distributed not only at the rigid core surface but also in the surface layer of the particle. Thus, the equilibrium potential distribution $\psi^{(eq)}(r)$, which is given by Eq. (32), and the results for the sedimentation velocity and sedimentation potential (or electrophoretic mobility) are generally different from those in the case of the true nonporous spheres of radius a.

When $\lambda \to 0$ (very low segment density), the surface layer of the polyelectrolyte-coated particle does not exert any resistance to the fluid motion. In this case, the function $F_{00r}(r)$ in Eq. (36a) for the r component of the fluid velocity reduces to

$$F_{00r}(r) = -\frac{1}{2}\left(\frac{r_0}{r}\right)^3 + \frac{3}{2}\left(\frac{r_0}{r}\right) - 1 \tag{61}$$

and Eqs (44a), (45), and (51) become

$$U_{00} = \frac{V_t[\gamma_p(\rho_p - \rho) + \gamma_c(\rho_c - \rho)]g}{6\pi\eta r_0} \tag{62a}$$

$$U = U_{00}\left\{1 - \frac{2a}{3r_0}\int_{r_0}^{\infty}\left(\frac{r}{a}\right)^3 F_{00r}(r)\left[G_{02}(r)\overline{\sigma}^2 + G_{11}(r)\overline{\sigma}\overline{Q} + G_{20}(r)\overline{Q}^2\right]dr \right.$$
$$\left. + O\left(\overline{\sigma}^3, \overline{\sigma}^2\overline{Q}, \overline{\sigma}\overline{Q}^2, \overline{Q}^3\right)\right\} \tag{62b}$$

$$\mu_E = \frac{Ze}{9r_0\eta kT}\int_{r_0}^{\infty}(2r^3 + r_0^3)F_{00r}(r)(\kappa\sigma\frac{d\psi_{eq01}}{dr} + Q\frac{d\psi_{eq10}}{dr})dr + O(\overline{\sigma}^3, \overline{\sigma}^2\overline{Q}, \overline{\sigma}\overline{Q}^2, \overline{Q}^3) \tag{62c}$$

To use the general expressions (44a), (45), and (51) for the sedimentation velocity and electrophoretic mobility and their simplified formulas (55), (59), (60), and (62) in the limiting cases, the parameters κa, λa, r_0/a, $\overline{\sigma}$, and $\overline{Q}$ of the colloidal system have to be determined. Experimental data for the surface layers of human erythrocytes [19] and rat lymphocytes [20] in electrolyte solutions indicate that the shielding length $1/\lambda$ has values of about 3 nm and the magnitude of Q ranges from quite low to about 1.6×10^6 C/m^3, depending on the pH value and ionic strength of the electrolyte solution. For some temperature-sensitive poly(N-isopropylacrylamide) hydrogel layers on latex particles in salt solutions [21], values of $1/\lambda$ were found to be about 1–50 nm and the magnitude of Q could be as high as 8.7×10^6 C/m^3. As to the surface charge density, an experimental study of the adsorption of poly(vinyl alcohol) on to AgI reported that the value of σ changes from 0 to -0.035 C/m^2 upon increasing the pAg from 5.6 to 11, while experimental data for a positively charged polystyrene latex used as the adsorbent for the polyelectrolyte poly(acrylic acid) showed that σ can have a value of 0.16 C/m^2 [55]. It is widely understood that the Debye length $1/\kappa$ is in the range from angstroms to about 1 µm, depending on the ionic strength of the solution. For a composite particle with $\sigma = 2 \times 10^{-3}$ C/m^2 and $Q = 2 \times 10^6$ C/m^3 in an aqueous solution with $1/\kappa = 1$ nm, one obtains the dimensionless charge densities $\overline{\sigma} \cong 0.1$ and $\overline{Q} \cong 0.1$.

According to Eqs (45) and (46), the sedimentation velocity of a charged polyelectrolyte-coated sphere in a given electrolyte solution can be calculated to the second orders σ^2, σQ, and Q^2. The corrections for the effects of the fixed distributed charges to U start from orders σ^2, σQ, and Q^2, instead of σ and Q. The reason is that these effects are due to the interaction between the particle charges and the local induced sedimentation potential gradient; both are of orders σ and Q, and thus the corrections are of orders σ^2, σQ, and Q^2. Figure 2 shows plots of numerical results for the dimensionless coefficients H_1, H_2, and H_3 in Eq. (46) as functions of the parameters κa and λa for the sedimentation velocity of a polyelectrolyte-coated sphere of $r_0/a = 0.5$ in an aqueous solution of KCl at room temperature. The values $\varepsilon k^2 T^2 / 4\pi\eta D_{\pm}Z^2 e^2 = 0.26$ are used in the calculations. It is found from the numerical results that H_1, H_2, and H_3 are always positive values and satisfy the relation $H_2 \leq 4H_1H_3$. With this relation of inequality, Eq. (45) illustrates that the presence of the particle charges would reduce the magnitude of the sedimentation velocity.

Figure 2 illustrates that, for specified values of r_0/a and the shielding ratio λa, the effects of the particle charges on the sedimentation velocity are maximal at some values

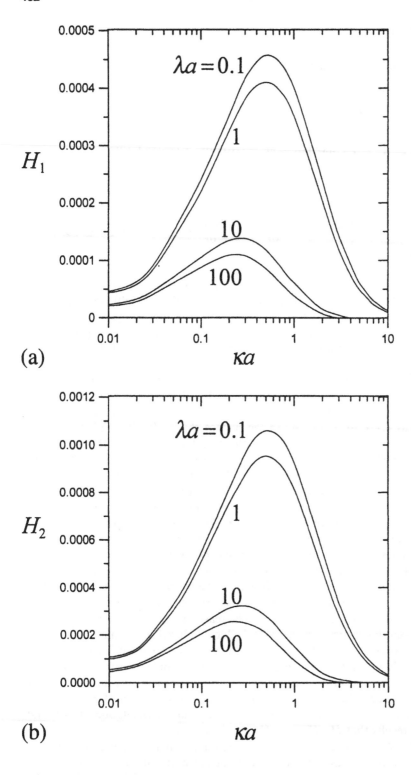

FIG. 2 The dimensionless coefficients H_1, (a); H_2, (b); and H_3, (c) in Eq. (45) for a sedimenting polyelectrolyte-coated sphere of $r_0/a = 0.5$ at various values of κa and λa.

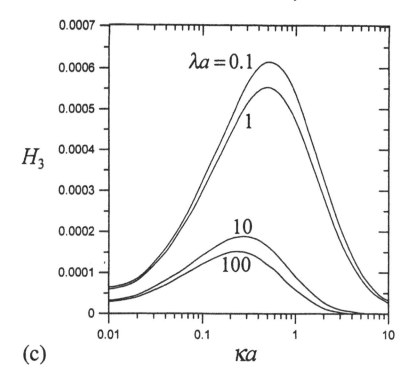

(c)

of κa and fade out when the value of κa is getting small or large. The reason for this behavior is obvious. The limit $\kappa a \to 0$ means that the effect of the presence of the counterions around the particle is negligible (viz, the particle is not affected by the electrostatic interaction with the surrounding counterions in hydrodynamic behavior), while the limit $\kappa a \to \infty$ indicates that the total charge density is zero everywhere and the total electric force on the particle vanishes. Both limits result in $H_1 = H_2 = H_3 = 0$ [as they are calculated from Eq. (46)] and the same sedimentation velocity as that of an uncharged polymer-coated particle, regardless of the values of r_0/a and λa. On the other hand, Fig. 2 also shows that H_1, H_2, and H_3 increase monotonically with decreasing λa for given values of κa and r_0/a. These dimensionless coefficients are sensitive functions of λa over the range $\lambda a = 1\text{--}10$.

The dimensionless coefficients H_σ and H_Q for the sedimentation potential or electrophoretic mobility in the KCl solution calculated from Eq. (52) are plotted as functions of the parameters κa and λa in Fig. 3. It can be seen that these coefficients, which are always positive, are monotonic decreasing functions of κa for given values of λa and r_0/a and are also monotonic decreasing functions of λa for fixed values of κa and r_0/a. When κa is small, H_σ and H_Q have the same order of magnitude; however, when κa is large, the magnitude of H_σ is much smaller than that of H_Q. Note that both H_σ and H_Q are not sensitive functions of λa when $\lambda a \le 1$. Also, these dimensionless coefficients are not sensitive functions of κa when $\kappa a \ll 1$, similar to the relevant results for charged impermeable spheres [30] and charged porous spheres [36].

Theoretical studies [56, 57] have predicted that a nonuniformly charged but "neutral" impermeable spherical particle (with zero area-averaged zeta potential) can translate

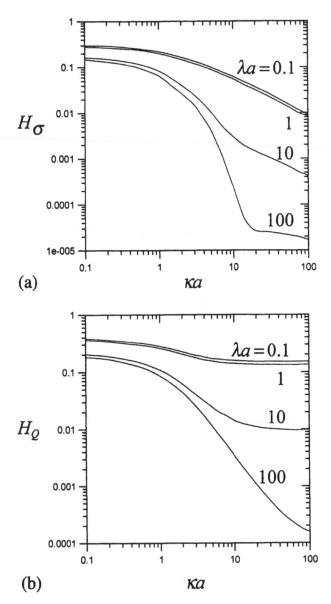

FIG. 3 The dimensionless coefficients H_σ (a) and H_Q (b) in Eq. (51) for a dilute suspension of identical polyelectrolyte-coated spheres of $r_0/a = 0.5$ at various values of κa and λa.

in electric fields. It would be of interest to know whether polyelectrolyte-coated spheres with zero net charge [$4\pi r_0^2 \sigma + (4\pi/3)(a^3 - r_0^3)Q = 0$] can undergo electrophoresis or produce sedimentation potential, and be retarded by this potential gradient during the sedimentation. For such spherically symmetric "neutral" polyelectrolyte-coated particles [with $\bar{\sigma} = -s\kappa a \bar{Q}/3$, where $s = (a/r_0)^2 - (r_0/a)$], Eqs (45) and (51) become

$$U = U_{00}[1 - (\kappa a)^4 H_0 \bar{Q}^2 + O(\bar{Q}^3)] \tag{63a}$$

$$\mu_E = H_\mu \frac{a^2 Q}{\eta} + O(\overline{Q}^3) \tag{63b}$$

where

$$H_0 = H_3 - \frac{s}{3} H_2 + \frac{s^2}{9} H_1 \tag{64a}$$

$$H_\mu = H_Q - \frac{s}{3} H_\sigma \tag{64b}$$

Figure 4 illustrates the numerical results for the dimensionless coefficients H_0 and H_μ associated with polyelectrolyte-coated spheres in KCl solution as functions of the parameters κa and λa. It can be seen that H_0 is positive and the presence of the fixed-charge distribution in the "neutral" particle would reduce its settling velocity. The tendency of the dependence of H_0 on κa is quite similar to that of the coefficient H_3 presented in Fig. 2c, except that the values of H_0 are about two orders of magnitude smaller and its maxima for fixed values of λa and r_0/a occur at values of κa about an order of magnitude larger. On the other hand, H_μ is always positive so that the "neutral" polyelectrolyte-coated spheres can experience electrophoresis or generate sedimentation potential. The direction of the electrophoretic velocity or the induced potential gradient is decided by the fixed charges in the porous surface layers (rather than the surface charges of the rigid cores) of the particles. Similar to the coefficient H_Q shown in Fig. 3b, H_μ is not necessarily a monotonic function of κa as λa is large. The trend of the dependence of H_μ on κa is quite different from that of H_Q. When κa is large, the coefficients H_μ and H_Q have the same order of magnitude. However, when κa is as low as 0.1, the values of H_μ are about three orders of magnitude lower than the values of H_Q.

V. ELECTRIC CONDUCTIVITY OF A SUSPENSION OF POLYELECTROLYTE-COATED SPHERES

In this section we present the effective electric conductivity of a dilute suspension of identical polyelectrolyte-coated spheres in the solution of a symmetrical electrolyte. The definition of the effective conductivity has been given in Section III. The linearized electrokinetic equations in Section II are solved for a polyelectrolyte-coated sphere in a uniformly applied electric field $\mathbf{E}_\infty$ using the regular perturbation method.

A. Analysis

We need to solve the small quantities $\delta\mu_\pm$, $\mathbf{u}$, and U (equal to $\mu_E \mathbf{E}_\infty$) in the form of perturbation expansions in powers of $\overline{\sigma}$ and $\overline{Q}$ given by Eqs (35a), (35c), and (35e) (in which $U_{ij} = \mu_{Eij} \mathbf{E}_\infty$ with i and j equal to 0, 1, 2, ...). Substituting the expansions given by Eq. (35) and $\psi^{(eq)}$ given by Eq. (32) into the linearized governing equations (10)–(12) and the boundary conditions (13a), (14a), (15a), (15b), and (17)–(19), and equating like powers of $\overline{\sigma}$ and $\overline{Q}$ on both sides of the respective equations, we obtain a group of differential equations and boundary conditions for each set of the functions $\mu_{ij\pm}$ and $\mathbf{u}_{ij}$ with i and j equal to 0, 1, 2, ... It is obvious that

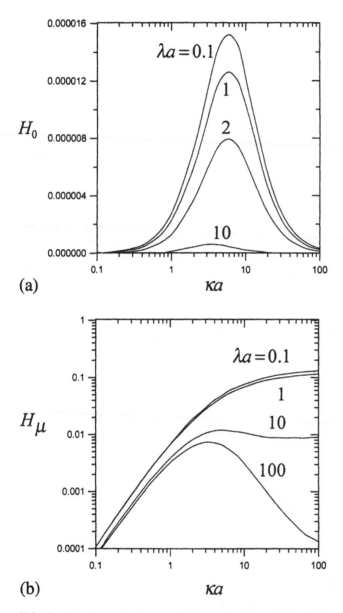

FIG. 4 The dimensionless coefficients H_0 (a) and H_μ (b) in Eq. (63) for the sedimentation of polyelectrolyte-coated spheres of $r_0/a = 0.5$ with zero net charge at various values of κa and λa.

$$\mu_{00\pm} = \mp ZeE_\infty \left(r + \frac{r_0^3}{2r^2} \right) \cos\theta \qquad (65)$$

The zeroth-order terms of $\mathbf{u}$ and μ_E disappear because an uncharged particle will not move by applying an electric field.

The first-order solutions for $\mu_{ij\pm}$ and the r and θ components of $\mathbf{u}_{ij}$ are

$$\mu_{ij\pm} = E_\infty F_{ij\pm}(r) \cos\theta \qquad (66)$$

$$u_{ijr} = E_\infty F_{ijr}(r)\cos\theta \tag{67a}$$

$$u_{ij\theta} = E_\infty F_{ij\theta}(r)\sin\theta \tag{67b}$$

where $F_{ij\pm}(r)$, $F_{ijr}(r)$, and $F_{ij\theta}(r)$ with $i+j=1$ are functions of r given by Eqs (40)–(43) in Ref. 24.

According to a characteristic of electrophoretic motion, the velocity field far from the particle (beyond the double layer) has the form [44, 58]:

$$\mathbf{u} \to -\mu_E E_\infty \mathbf{e}_z + O(r^{-2}) \tag{68}$$

Equation (68) satisfies the requirement that the net force on a large surface enclosing the particle and its adjacent double layer must be zero. Using this relationship and Eq. (67), one can obtain the first-order term for the electrophoretic mobility of a polyelectrolyte-coated sphere expressed as

$$\mu_{Eij}(r) = \frac{1}{C_{006}}\left[C_{001}\int_a^{r_0}\left(\frac{r}{a}\right)^3 G_{ij}(r)dr + C_{002}\int_a^{r_0} G_{ij}(r)dr \right.$$
$$+ C_{003}\int_a^{r_0}\alpha_1(\lambda r)G_{ij}(r)dr + C_{004}\int_a^{r_0}\beta_1(\lambda r)G_{ij}(r)dr \tag{69}$$
$$\left. - C_{005}\int_a^{\infty} G_{ij}(r)dr - C_{006}\int_a^{\infty}\left(\frac{r}{a}\right)^2 G_{ij}(r)dr + \int_a^{\infty}\left(\frac{r}{a}\right)^3 G_{ij}(r)dr \right]$$

where the function $\beta_1(x) = x\sinh x - \cosh x$; the functions $G_{ij}(r)$ and constants C_{001}, $C_{002}, \ldots$, and C_{006} are given by Eqs (42) and (A11)–(A16) in Ref. 24.

Among the second-order terms in the perturbation procedure, the only distributions we need in the following calculations are the electrochemical potential energies $\mu_{02\pm}$, $\mu_{11\pm}$, and $\mu_{20\pm}$. Their solution is

$$\mu_{ij\pm}(r) = \mp\frac{ZeE_\infty}{3kT}\left[\frac{r_0^3}{2r^2}\int_r^{\infty} K_{ij\pm}(r)dr + r\int_r^{\infty} K_{ij\pm}(r)dr + \frac{1}{r^2}\int_{r_0}^r r^3 K_{ij\pm}(r)dr\right]\cos\theta \tag{70}$$

where

$$K_{02\pm}(r) = \left(\frac{dF_{01\pm}}{dr} - \frac{kT}{D_\pm}F_{01r}\right)\frac{d\psi_{eq01}}{dr} \tag{71a}$$

$$K_{11\pm}(r) = \left(\frac{dF_{01\pm}}{dr} - \frac{kT}{D_\pm}F_{01r}\right)\frac{d\psi_{eq10}}{dr} + \left(\frac{dF_{10\pm}}{dr} - \frac{kT}{D_\pm}F_{10r}\right)\frac{d\psi_{eq01}}{dr} \tag{71b}$$

$$K_{20\pm}(r) = \left(\frac{dF_{10\pm}}{dr} - \frac{kT}{D_\pm}F_{10r}\right)\frac{d\psi_{eq10}}{dr} \tag{71c}$$

By substituting Eqs (35c), (65), (66), and (70) into Eq. (29), making relevant calculations, and then comparing the result with Eq. (22), the effective conductivity of a dilute suspension of identical polyelectrolyte-coated spheres is obtained as

$$\Lambda = \Lambda_\infty\left\{1 - \frac{1-\gamma_c\varepsilon_p}{1-\varepsilon_p}\phi\left[\frac{3}{2}\left(\frac{r_0}{a}\right)^3 + \Lambda_{01}\kappa a\bar{\sigma} + \Lambda_{10}(\kappa a)^2\overline{Q} + \Lambda_{02}(\kappa a)^2\bar{\sigma}^2\right.\right.$$
$$\left.\left. + \Lambda_{11}(\kappa a)^3\bar{\sigma}\overline{Q} + \Lambda_{20}(\kappa a)^4\overline{Q}^2 + O(\bar{\sigma}^3, \bar{\sigma}^2\overline{Q}, \bar{\sigma}\overline{Q}^2, \overline{Q}^3)\right]\right\} \tag{72}$$

Here, ε_p is the porosity of the surface layer of a polyelectrolyte-coated sphere, γ_c is the volume fraction of the rigid core of the particle, ϕ is the true volume fraction of the dry polyelectrolyte-coated particles, and

$$\Lambda_{ij} = \frac{Z(D_+ - D_-)}{D_+ + D_-} X_{ij}, \quad \text{if}(i, j) = (0, 1) \text{ or } (1, 0) \tag{73a}$$

$$\Lambda_{ij} = -\frac{\varepsilon k^2 T^2}{2\pi\eta(D_+ + D_-)e^2} Y_{ij} - Z^2 Z_{ij}, \quad \text{if } (i, j) = (0, 2), (1, 1), \text{ or } (2, 0) \tag{73b}$$

In Eq. (73):

$$X_{01} = \Phi_3\left(\kappa r_0, \frac{r_0}{a}\right) \tag{74a}$$

$$X_{10} = \frac{1}{(\kappa a)^2}\left\{1 - \frac{1}{2}\left(\frac{r_0}{a}\right)^3 - \frac{1}{2}\left(\frac{r_0}{a}\right)^6\right.$$
$$+ \frac{3}{2}\left(\frac{r_0}{a}\right)^6\left(1 + \frac{1}{\kappa a}\right)\frac{e^{-\kappa d}}{1 + \kappa r_0}\int_{r_0}^a \frac{a^4}{r^5}[\kappa r_0 \cosh(\kappa r - \kappa r_0) + \sinh(\kappa r - \kappa r_0)]dr \tag{74b}$$
$$\left. - \frac{3}{2}\left(\frac{r_0}{a}\right)^6 e^{\kappa a}E_5(\kappa a)\left[1 - \left(1 + \frac{1}{\kappa a}\right)\frac{e^{-kd}}{1 + \kappa r_0}(\kappa r_0 \cosh(\kappa d) + \sinh(\kappa d))\right]\right\}$$

$$Y_{02} = \frac{4\pi\eta e^2}{\varepsilon k^2 T^2}\frac{1}{(\kappa a)^2}\int_{r_0}^\infty W(r)F_{01r}(r)\frac{d\psi_{eq01}}{dr}dr \tag{75a}$$

$$Y_{11} = \frac{4\pi\eta e^2}{\varepsilon k^2 T^2}\frac{1}{(\kappa a)^3}\int_{r_0}^\infty W(r)\left[F_{01r}(r)\frac{d\psi_{eq10}}{dr} + F_{10r}(r)\frac{d\psi_{eq01}}{dr}\right]dr \tag{75b}$$

$$Y_{20} = \frac{4\pi\eta e^2}{\varepsilon k^2 T^2}\frac{1}{(\kappa a)^4}\int_{r_0}^\infty W(r)F_{10r}(r)\frac{d\psi_{eq10}}{dr}dr \tag{75c}$$

$$Z_{02} = -\frac{1}{Z^2 kT(\kappa a)^2}\int_{r_0}^\infty W(r)\frac{dF_{01+}}{dr}\frac{d\psi_{eq01}}{dr}dr \tag{75d}$$

$$Z_{11} = -\frac{1}{Z^2 kT(\kappa a)^3}\int_{r_0}^\infty W(r)\left(\frac{dF_{01+}}{dr}\frac{d\psi_{eq10}}{dr} + \frac{dF_{10+}}{dr}\frac{d\psi_{eq01}}{dr}\right)dr \tag{75e}$$

$$Z_{20} = -\frac{1}{Z^2 kT(\kappa a)^4}\int_{r_0}^\infty W(r)\frac{dF_{10+}}{dr}\frac{d\psi_{eq10}}{dr}dr \tag{75f}$$

where

$$W(r) = \left(\frac{r}{a}\right)^3 + \frac{1}{2}\left(\frac{r_0}{a}\right)^3 \tag{76a}$$

$$\Phi_3(x, y) = \frac{3y^4}{1 + x}\left[\frac{1}{x} + \frac{1}{x^2} + \frac{1}{2}e^x E_5(x)\right] \tag{76b}$$

The function E_n in the above equation has been defined by Eq. (57).

Note that Eq. (72) is correct to order ϕ. The parameters X_{ij} and Z_{ij} depend on the parameters κa and r_0/a only, while Y_{ij} are functions of κa. λa, and r_0/a. The coefficients Λ_{01} and Λ_{10}, which are independent of the shielding parameter λ, disappear for a symmetry electrolyte when the diffusivities of the cation and anion take the same value. As to the coefficients Λ_{02}, Λ_{11}, and Λ_{20}, for a symmetric electrolyte, the first term of the right-hand side of Eq. (73b) (which is a function of parameters κa and λa) denotes the effect due to the convection of the fluid, while the second term (which is independent of λa) represents the effect due to the deviations of the electrochemical potential distributions from their combined equilibrium and applied values. Although the parameters X_{ij}, Y_{ij}, Z_{ij} are positive values, the coefficients Λ_{01} and Λ_{10} can be either positive or negative, depending on the diffusion coefficients of the ionic species in the electrolyte solution, and the coefficients Λ_{02}, Λ_{11}, and Λ_{20} are always negative. It is understood that the result given by Eqs (72)–(75) is only valid with the requirements that $\phi \ll 1$ and $\kappa a \phi^{-1/3} \gg 1$.

B. Discussion

In this subsection, we first consider two limiting cases and a special case of the expression (72) for the effective conductivity of a dilute suspension of identical polyelectrolyte-coated spheres. The correctness of this expression may be confirmed by examining these limiting caes for which analytical solutions are already known. Then, numerical results of the dimensionless parameters X_{01}, X_{10}, Y_{ij}, and Z_{ij} (with $i+j=2$) in association with the coefficients Λ_{ij} by Eq. (73) for some typical cases will be presented.

When there is no permeable layer on the surface of each rigid particle core in the suspension, one has $d=0, r_0=a, \gamma_c=1, Q=0$, and $\lambda=0$. Equation (72) for a symmetric electrolyte then reduces to

$$\Lambda = \Lambda^\infty \left\{ 1 - \phi \left[\frac{3}{2} + \frac{Z(D_+ - D_-)}{D_+ + D_-} \kappa a \Phi_3(\kappa a, 1)\bar{\sigma} - \left(\frac{\varepsilon k^2 T^2}{2\pi\eta(D_+ + D_-)e^2} Y_{02} + Z^2 Z_{02} \right)(\kappa a)^2\bar{\sigma}^2 \right] \right\}$$

(77)

Here, function Φ_3 is defined by Eq. (76b);

$$Y_{02} = \frac{1}{(1+\kappa a)^2} \left\{ \frac{1}{2\kappa a} + \left[\frac{2}{\kappa a} + \frac{2}{(\kappa a)^2} + e^{\kappa a} E_5(\kappa a) \right] \left[1 + \frac{(\kappa a)^2}{4} e^{\kappa a}(E_3(\kappa a) - E_5(\kappa a)) \right] \right.$$
$$\left. - \left(3 + \frac{15}{2\kappa a} \right) e^{\kappa a} E_6(\kappa a) + e^{2\kappa a} \left[\frac{3\kappa a}{4} E_5(\kappa a)(E_4(\kappa a) - E_6(\kappa a)) + \frac{1}{2} E_6(2\kappa a) \right] \right\}$$

(78a)

and

$$Z_{02} = \frac{1}{(1+\kappa a)^2} \left\{ \frac{5}{4\kappa a} + \frac{3}{2(\kappa a)^2} - \left[\frac{6}{\kappa a} + \frac{6}{(\kappa a)^2} \right] e^{\kappa a} E_5(\kappa a) - e^{2\kappa a} \left[\frac{3}{2}(E_5(\kappa a))^2 + \frac{1}{4} E_6(2\kappa a) \right] \right\}$$

(78b)

where function $E_n(x)$ is given by Eq. (57). Substituting Eq. (58) into Eq. (77), this degenerated result is the same as that of a dilute suspension of identical rigid spheres with low surface (zeta) potential obtained by O'Brien [39]. Note that there is a typographical error in Eq. [5.34] of O'Brien's paper.

When the particles are homogeneous porous spheres, one has $r_0 = 0$, $d = a$, $\gamma_c = 0$, and $\sigma = 0$. In this limiting case, Eq. (72) for a symmetric electrolyte becomes

$$\Lambda = \Lambda^\infty \left\{ 1 - \frac{\phi}{1 - \varepsilon_p} \left[\frac{Z(D_+ - D_-)}{D_+ + D_-} \overline{Q} - \left(\frac{\varepsilon k^2 T^2}{2\pi\eta(D_+ + D_-)e^2} Y_{20} + Z^2 Z_{20} \right)(\kappa a)^4 \overline{Q}^2 \right] \right\}$$

(79)

where

$$Z_{20} = \frac{2(\kappa a)^2 + 3\kappa a + 3}{3(\kappa a)^7} \alpha_1(\kappa a) e^{-\kappa a} - \frac{\kappa a + 1}{3(\kappa a)^5} e^{-\kappa a} \sinh(\kappa a) - \frac{3}{4} \frac{\kappa a + 2}{(\kappa a)^8} [\alpha_1(\kappa a)]^2 e^{-2\kappa a}$$
$$- \frac{3}{4} \frac{(1 + \kappa a)^2}{(\kappa a)^8} e^{-2\kappa a} [\alpha_1(\kappa a) \sinh(\kappa a) - \sinh^2(\kappa a) + (\kappa a)^2]$$

(80a)

$$Y_{20} = \frac{1}{(\lambda a)^2 (\kappa a)^2} + \frac{2 + \kappa a}{(\kappa a)^8} [\alpha_1(\kappa a)]^2 e^{-2\kappa a} - \left(\frac{\kappa^2}{\lambda^2 - \kappa^2} \right) \frac{(1 + \kappa a)^2}{(\kappa a)^8} e^{-2\kappa a} [\alpha_1(\kappa a) \sinh(\kappa a)$$
$$- \sinh^2(\kappa a) + (\kappa a)^2] + \frac{2\lambda^2 \kappa^2}{(\lambda^2 - \kappa^2)^2} \frac{(1 + \kappa a)^2}{(\kappa a)^8} \alpha_1(\kappa a) e^{-2\kappa a} \left[\left(\frac{\lambda}{\kappa} \right)^2 \frac{\alpha_1(\kappa a)}{\alpha_1(\lambda a)} \sinh(\lambda a) - \sinh(\kappa a) \right]$$

(80b)

and function $\alpha_1(x) = x \cosh x - \sinh x$. Equation (79) is identical to the formula for the effective conductivity of a dilute suspension of charged porous spheres previously derived [53].

The dimensionless parameters X_{01}, X_{10}, Y_{ij}, and Z_{ij} (with $i + j = 2$) defined by Eq. (73) can be calculated for given values of the parameters κa, λa, and r_0/a using Eqs (74) and (75), and some typical results are plotted in Figs 5–7. It can be seen in Fig. 5 that both X_{01} and X_{10} (or the magnitudes of Λ_{01} and Λ_{10}) decrease monotonically with increasing κa for a given value of r_0/a. The magnitude of X_{10} can be quite large even for a particle with a relatively thin porous surface layer (say, with $r_0/a \cong 0.95$). Figure 6 indicates that the parameters Y_{02}, Y_{11}, and Y_{20} are monotonic decreasing functions of κa for given values of λa and r_0/a, and they decrease monotonically with increasing λa for given values of κa and r_0/a. Figure 7 illustrates that the parameters Z_{02}, Z_{11}, and Z_{20} decrease monotonically with increasing κa for a given value of r_0/a.

Equations (63) and (64) demonstrate that polyelectrolyte-coated spheres with zero net charge ($\bar{\sigma} = -s\kappa a \overline{Q}/3$) can undergo electrophoresis in an electric field and produce sedimentation potential under gravity. It would be of interest to know whether the electric conductivity of a dilute suspension of such "neutral" particles differs from that of a corresponding suspension of polymer-coated spheres with $\sigma = 0$ and $Q = 0$. For such "neutral" polyelectrolyte-coated spheres, Eq. (72) for a symmetric electrolyte becomes

$$\Lambda = \Lambda^\infty \left\{ 1 - \frac{1 - \gamma_c \varepsilon_p}{1 - \varepsilon_p} \phi \left[\frac{3}{2} \left(\frac{r_0}{a} \right)^3 - \frac{Z(D_+ - D_-)}{D_+ + D_0} X_0 (\kappa a)^2 \overline{Q} \right. \right.$$
$$\left. \left. - \left(\frac{\varepsilon k^2 T^2}{2\pi\eta(D_+ + D_-)e^2} Y_0 + Z^2 Z_0 \right)(\kappa a)^4 \overline{Q}^2 + O(\overline{Q}^3) \right] \right\}$$

(81)

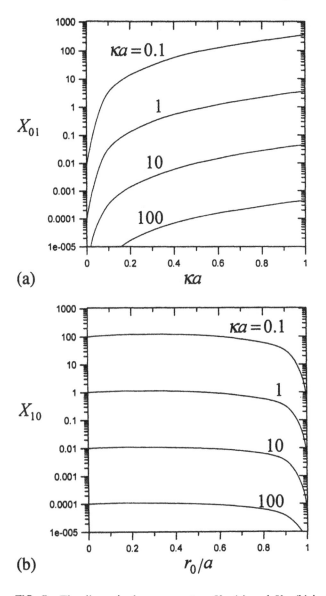

FIG. 5 The dimensionless parameters X_{01} (a) and X_{10} (b) in Eq. (73a) for the conductivity coefficients Λ_{01} and Λ_{10} at various values of κa and r_0/a.

where

$$X_0 = -X_{10} + \frac{s}{3} X_{01} \tag{82a}$$

$$Y_0 = Y_{20} - \frac{s}{3} Y_{11} + \frac{s^2}{9} Y_{02} \tag{82b}$$

$$Z_0 = Z_{20} - \frac{s}{3} Z_{11} + \frac{s^2}{9} Z_{02} \tag{82c}$$

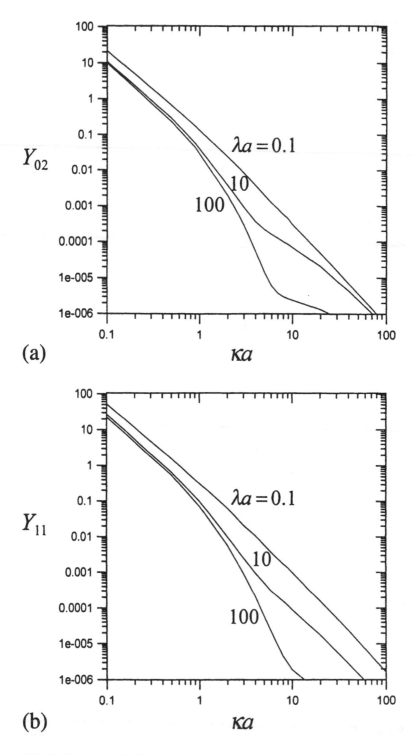

FIG. 6 The dimensionless parameters Y_{02}, (a); Y_{11}, (b); and Y_{20}, (c) in Eq. (73b) for the conductivity coefficients Λ_{02}, Λ_{11}, and Λ_{20} of a dilute suspension of identical polyelectrolyte-coated spheres of $r_0/a = 0.5$ at various values of κa and λa.

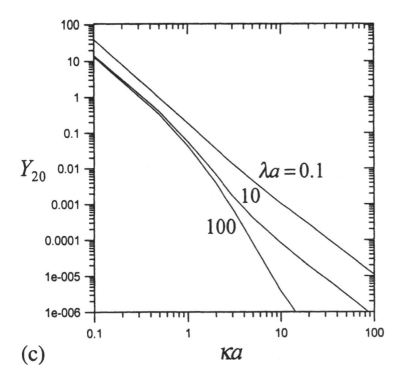

(c)

Numerical results for the dimensionless parameters X_0, Y_0, and Z_0 calculated using Eq. (82) are drawn as functions of the parameters κa and r_0/a in Fig. 8. It can be seen that X_0, Y_0, and Z_0 are all positive and the presence of the fixed-charge distribution in the "neutral" particles would influence the effective conductivity of the suspension. The direction of the influence is decided by the fixed charges in the porous surface layers (rather than the surface charges of the rigid cores) of the particles. The parameters X_0 and Z_0 decrease monotonically with the increase in κa for a given value of r_0/a, while Y_0 is not necessarily a monotonic function of κa for fixed values of λa and r_0/a. The parameters Y_0 and Z_0 decrease monotonically with increasing r_0/a for fixed values of κa and λa, while X_0 approaches zero in both limiting cases of $r_0/a = 0$ and $r_0/a = 1$ and has a maximum for a given value of κa. Further calculations also show that the parameter Y_0 is a monotonic decreasing function of λa for constant values of κa and r_0/a. In general, the trends of the dependence of X_0, Y_0, and Z_0 on κa and r_0/a are quite different from those of X_{10}, Y_{20}, and Z_{20} presented in Figs 5–7.

VI. CONCLUSIONS

The steady-state electrokinetic phenomena in a dilute suspension of polyelectrolyte-coated particles under the action of a gravitational field and/or an external electric field in an unbounded electrolyte solution are presented in this chapter. The porous surface layer of

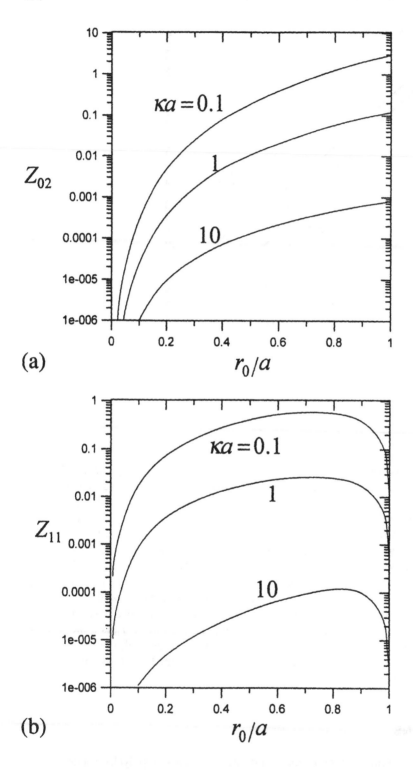

FIG. 7 The dimensionless parameters Z_{02}, (a); Z_{11}, (b); and Z_{20}, (c) in Eq. (73b) for the conductivity coefficients Λ_{02}, Λ_{11}, and Λ_{20} at various values of κa and r_0/a.

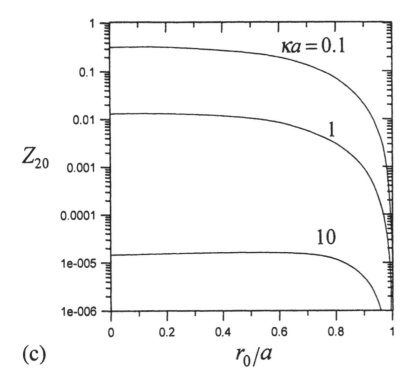

(c)

each particle is treated as a solvent-permeable and ion-penetrable shell in which fixed-charged groups and frictional segments are distributed at uniform densities. Solving the linearized continuity equations of ions, the Poisson–Boltzmann equation, and combined modified Stokes/Brinkman flow equations applicable in the system of an isolated polyelectrolyte-coated sphere by a regular perturbation method, we obtain the ion concentration (electrochemical potential energy) distributions, the electric potential profile, and the fluid flow field.

The sedimentation of a polyelectrolyte-coated sphere is considered in Section IV. Since the electric potential distribution differs from the equilibrium values, an electric force acting on the particle is induced. The total force exerted on the particle is the sum of the gravitational, electrostatic, and hydrodynamic forces, and the requirement that the total force is zero leads to an explicit formula, Eq. (45), for the settling velocity of the polyelectrolyte-coated sphere. The corrections for the effects of the fixed distributed charges to the settling velocity begin at the second orders (σ^2, σQ, and Q^2). Numerical results indicate that these effects have a maximum at a finite value of κa and disappear when κa approaches zero and infinity. The explicit formula, Eq. (50), for the sedimentation potential is derived by letting the net current in the suspension be zero. The Onsager reciprocal relation is satisfied between the sedimentation potential and the electrophoretic mobility. Expressions (45) and (50) for the sedimentation velocity and sedimentation potential (or electrophoretic mobility) in a dilute suspension of polyelectrolyte-coated spheres reduce to the corresponding formulas for the charged solid spheres and the charged porous spheres, respectively, in the limiting cases of $r_0/a = 1$ and $r_0/a = 0$.

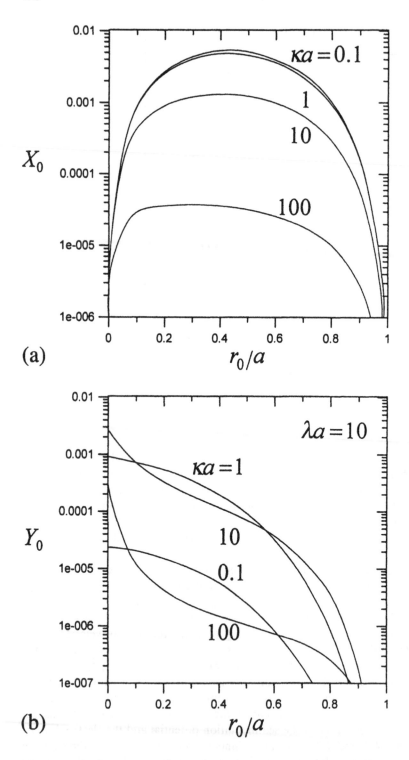

(a)

(b)

FIG. 8 The dimensionless parameters X_0, (a); Y_0, (b); and Z_0, (c) in Eq. (81) for a dilute suspension of identical polyelectrolyte-coated spheres with zero net charge at various values of κa and r_0/a.

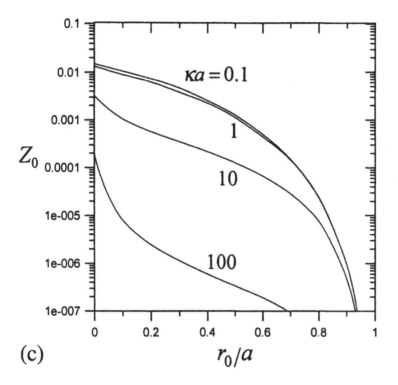

(c)

It is worth repeating that the Onsager relation between sedimentation and electrophoresis holds not only for impermeable charged particles but also for porous or composite ones. Another electrokinetic phenomenon in a circular capillary with the solution-permeable surface charge layer under the Debye–Huckel approximation has been generally studied [59], and similarly, the electro-osmotic flow rate and the streaming potential were shown to satisfy the Onsager reciprocity principle whether the capillary wall adsorbs a surface charge layer or not. Hence, different electrokinetic processes could reflect the same intrinsic phenomena, and measurements of one type of process can be used to check those on another. For instance, one can first measure the sedimentation potential in a dilute suspension of particles and then predict the electrophoretic mobility of the particles in the same system without any other measurements.

The effective electric conductivity of a dilute suspension of polyelectrolyte-coated spheres in an electrolyte solution is presented in Section V. The average electric current density passing through the suspension is expressed as an integral over a large spherical surface surrounding a single particle plus its double layer, and is related to the electrochemical potential energies of the electrolyte ions. An analytical expression, Eq. (72), is obtained for the effective electric conductivity of a dilute suspension of identical polyelectrolyte-coated spheres as a power series in the two fixed charge densities of the particles up to $O(\sigma^2, \sigma Q, Q^2)$. According to this formula, the presence of the fixed charges in the polyelectrolyte-coated particles can result in an increase or a decrease in the effective conductivity relative to that of a corresponding suspension of uncharged particles, depending on the diffusion coefficients of the electrolyte ions and the fixed-charge densities of the particles. Expressions (77) and (79) for the electric conductivity of a dilute suspension of identical spheres in a symmetric electrolyte in the limiting cases of $r_0/a = 1$ and $r_0/a = 0$ reduce to the corresponding formulas for the charged solid spheres and the charged porous spheres, respectively.

Equations (45), (50), and (72), with Eqs (46), (51), and (73), are obtained on the basis of the Debye–Huckel approximation for the equilibrium potential distribution around a polyelectrolyte-coated sphere. Similar formulas for the sedimentation velocity of an impermeable rigid sphere and for the electric conductivity of a dilute suspension of identical impermeable spheres with low zeta potential were shown to give good approximations for the case of reasonably high zeta potential (with errors less than 0.1% for $\zeta e/kT \leq 2$ in a KCl solution [30], of about 5% in another KCl solution, and less than 2% in a HClO$_4$ solution for the case of $\zeta e/kT = -2$ [39]). Therefore, the results of this chapter might be used tentatively for the situation of reasonably high electric potentials.

REFERENCES

1. DH Napper. Polymeric Stabilization of Colloidal Dispersions. London: Academic Press, 1983.
2. L Anderson, Y Solomentsev. Chem Eng Comm 148–150: 291, 1996.
3. RW Wunderlich. J Colloid Interface Sci 88: 385, 1982.
4. S Sasaki. Colloid Polym Sci 263: 935, 1985.
5. JH Masliyah, G Neale, K Malysa, TGM van de Ven. Chem Eng Sci 42: 245, 1987.
6. JL Anderson, J Kim. J Chem Phys 86: 5163, 1987.
7. HJ Keh, J Kuo. J Colloid Interface Sci 185: 411, 1997.
8. J Kuo, HJ Keh. J Colloid Interface Sci 210; 296, 1999.
9. J Kuo, HJ Keh. J Colloid Interface Sci 195: 353, 1997.
10. SB Chen. Phys Fluids 10: 1550, 1998.
11. J Happel. AIChE J 4: 197, 1958.
12. S Kuwabara. J Phys Soc Jpn 14: 527, 1959.
13. J Happel, H Brenner. Low Reynolds Number Hydrodynamics. The Hague, The Netherlands: Nijhoff, 1983.
14. D Prasad, KA Narayan, RP Chhabra. Int J Eng Sci 28: 215, 1990.
15. HJ Keh, J Kuo. Colloid Polym Sci 275: 661, 1997.
16. IS Jones. J Colloid Interface Sci 68; 451, 1979.
17. S Levine, M Levine, KA Sharp, DE Brooks. Biophys J 42: 127, 1983.
18. KA Sharp, DE Brooks. Biophys J 47: 563, 1985.
19. S Kawahata, H Ohshima, N Muramatsu, T Kondo, J Colloid Interface Sci 138: 182, 1990.
20. K Morita, N Muramatsu, H Ohshima, T Kondo. J Colloid Interface Sci 147: 457, 1991.
21. K Makino, S Yamamoto, K Fujimoto, H Kawaguchi, H Ohshima. J Colloid Interface Sci 166: 251, 1994.
22. H Ohshima. J Colloid Interface Sci 163: 474, 1994.
23. HJ Keh, YC Liu. J Colloid Interface Sci 195: 169, 1997.
24. YC Liu, HJ Keh. Langmuir 14: 1560, 1998.
25. F Booth. J Chem Phys 22: 1956, 1954.
26. DA Saville. Adv Colloid Interface Sci 16: 267, 1982.
27. D Stigter. J Phys Chem 84: 2758, 1980.
28. PH Wiersema, AL Leob, JTG Overbeek. J Colloid Interface Sci 22: 78, 1966.
29. SR de Groot, P Mazur, JTG Overbeek. J Chem Phys 20: 1825, 1952.
30. H Ohshima, TW Healy, LR White, RW O'Brien. J Chem Soc, Faraday Trans 2 80: 1299, 1984.
31. D Stigter. J Phys Chem 86: 3553, 1982.
32. SB Chen, DL Koch. J Colloid Interface Sci 180; 466, 1996.

33. S Levine, G Neale, N Epstein. J Colloid Interface Sci 57: 424, 1976.
34. H Ohshima. J Colloid Interface Sci 208: 295, 1998.
35. HJ Keh, JM Ding. J. Colloid Interface Sci 227: 540, 2000.
36. YC Liu, HJ Keh. Colloids Surfaces A 140; 245, 1998.
37. SS Dukhin, BV Derjaguin. In: E Matijevic, ed. Surface and Colloid Science. vol. 7. New York: Wiley, 1974.
38. DA Saville. J Colloid Interface Sci 71: 477, 1979.
39. RW O'Brien. J Colloid Interface Sci 81: 234, 1981.
40. A Watillon, J Stone-Masui. J Electroanal Chem 37: 143, 1972.
41. DA Saville. J Colloid Interface Sci 91: 34, 1983.
42. RW O'Brien. J Colloid Interface Sci 92: 204, 1983.
43. RW O'Brien, DN Ward. J Colloid Interface Sci 121: 402, 1988.
44. H Ohshima, TW Healy, LR White. J Chem Soc, Faraday Trans 2 79: 1613, 1983.
45. H Ohshima. J Colloid Interface Sci 212: 443, 1999.
46. TJ Johnson, EJ Davis. J Colloid Interface Sci 215: 397, 1999.
47. JM Ding, HJ Keh. J. Colloid Interface Sci 236: 180, 2001.
48. RW O'Brien, WT Perrins. J Colloid Interface Sci 99: 20, 1984.
49. AG Van Der Put, BH Bijsterbosch. J Colloid Interface Sci 75: 512, 1980.
50. CF Zukoski, DA Saville. J Colloid Interface Sci 107: 322, 1985.
51. D Stigter. J Phys Chem 83: 1670, 1979.
52. D Stigter. J Phys Chem 83: 1670, 1979.
53. YC Liu, HJ Keh. J Colloid Interface Sci 192: 375, 1997.
54. JJ Hermans, H Fujita. Proc Akad, Amsterdam B 58; 182, 1955.
55. J Blaakmeer, MR Bohmer, MA Cohen Stuart, GJ Fleer. Macromolecules 23: 2301, 1990.
56. JL Anderson. J Colloid Interface Sci 105: 45, 1985.
57. YE Solomentsev, Y Pawar, JL Anderson. J Colloid Interface Sci 158: 1, 1993.
58. RW O'Brien, RJ Hunter. Can J Chem 59: 1878, 1981.
59. HJ Keh, YC Liu. J Colloid Interface Sci 172: 222, 1995.

16

Sedimentation and Flotation Potential: Theory and Measurements

MASATAKA OZAKI   Yokohama City University, Yokohama, Japan

HIROSHI SASAKI   Waseda University, Tokyo, Japan

I. INTRODUCTION

The flow of charged colloidal particles through a liquid due to the gravitational force produces an electrical current developing an electrical potential difference in the liquid. This electrokinetic phenomenon is the sedimentation current or potential and it is known as the Dorn effect after his first experimental observations [1]. When the density of the dispersed material is lower than that of the liquid medium, as in the case of gas bubbles in a liquid, the related phenomenon of flotation potential can also occur. While the number of experimental determinations of sedimentation potential is rather scarce, the number of experimental studies on the electrokinetics of gas bubbles is even more limited. Thus, in some old studies, McTaggart [2], Alty [3], Komagata [4], and Bach and Gilman [5] employed the electrophoretic mobility technique, and Collins et al. [6] reported the zeta potential of gas bubbles in aqueous solutions obtained with a device by which the electrophoretic mobility of a small bubble could be evaluated at the stationary level. In the papers of Usui and coworkers [7–9], finely divided argon gas bubbles were allowed to float through an aqueous surfactant solution, and the Dorn potential (flotation potential for gas bubbles and sedimentation potential for solid particles in aqueous media) generated was used to calculate the zeta potential. Except for these investigations and those conducted by some researchers in the former Soviet Union [10, 11], little work has been devoted to flotation potential.

In this work, we show how sedimentation and flotation potentials can be measured, and how they can be applied to zeta potential estimations in hematite particles and gas bubbles, respectively. It must be mentioned that previous investigations on the latter topic have been restricted to aqueous solutions of surfactants, and no study has been carried out on solutions of inorganic electrolytes. Hence, we will focus our discussion on the flotation potential of gas bubbles in aqueous solutions.

II. THEORY

When a colloidal particle sediments at a constant speed v, the force due to gravitation is balanced by the frictional force. Therefore, the following equation holds:

$$6\pi\eta av = \frac{4}{3}\pi a^3(\rho - \rho_0)g \tag{1}$$

where a is the radius of the particle, η is the viscosity of the dispersing medium, ρ the density of the particle, ρ_0 is the density of the medium, and g is the acceleration due to gravity. The current density I which arises when the charged particles move is expressed by

$$I = nQv = \frac{2}{9}\frac{nQa^2(\rho - \rho_0)g}{\eta} \tag{2}$$

where n is the number of particles in a unit volume of the dispersion, and Q is the electrical charge on the particle. In electrophoresis, QE, the force acting on the particle is balanced by the frictional force, $6\pi a\eta v_e$. Therefore,

$$QE = 6\pi\eta av_e \tag{3}$$

holds, where E is the electric field, and v_e is the electrophoretic velocity. The electrophoretic mobility u is related to v_e by $u = v_e/E$, and u is related to ζ, the zeta potential, by

$$u = \frac{\varepsilon_r\varepsilon_0 f(\kappa a)}{\eta}\zeta \tag{4}$$

as described elsewhere in this book, where ε_r is the relative permittivity of the medium, ε_0 is the permittivity of vacuum, and κ is the reciprocal thickness of the electrical double layer. In the case when $\kappa a \gg 1$, that is, the size of the colloidal particle is far larger than the thickness of the electrical double layer, $f(\kappa a)$ becomes 1. In this case, Eq. (4) is called the Smoluchowski equation. From Eqs (3) and (4) we obtain:

$$Q = 6\pi a\varepsilon_r\varepsilon_0 f(\kappa a)\zeta \tag{5}$$

Therefore, I can be expressed by

$$I = \frac{4\pi\varepsilon_r\varepsilon_0 na^3 f(\kappa a)(\rho - \rho_0)g}{3\eta}\zeta \tag{6}$$

The electrical resistance, R of a liquid column of unit area with length l is expressed by

$$R = \frac{l}{\lambda} \tag{7}$$

where λ is the specific conductivity of the dispersion. Therefore, $\Delta\phi$, the potential difference (sedimentation potential) measured between two electrodes inserted at the ends of the column separated by l can be expressed by

$$\Delta\phi = RI = \frac{4\pi\varepsilon_r\varepsilon_0 na^3 lf(\kappa a)(\rho - \rho_0)g}{3\eta\lambda}\zeta \tag{8}$$

Replacing $\Delta\phi/l$ by E_s, and n by $\phi/(4\pi a^3/3)$, where ϕ is the volume fraction of the particles in the dispersion, if we then put $f(\kappa a) = 1$ we obtain the well-known Smoluchowski equation [12]:

$$E_s = \frac{\varepsilon_r\varepsilon_0\phi(\rho - \rho_0)g}{\eta\lambda}\zeta \tag{9}$$

where E_s is the sedimentation potential in volts per meter (or, more properly, the sedimentation field). This equation for the sedimentation potential is valid only when the fluid flow around the particle is laminar, the surface conductance is negligibly small, and the double-layer thickness is far less than the particle radius. Furthermore, it is also assumed that the particles are nonconducting, monodisperse, and spherical.

As the concentration of the colloids increases, particle–particle interactions occur, and Eq. (9) is no longer valid. Levine and Neale developed a theory useful for low zeta potential in a wide range of κa and ϕ, based on the Kuwabara cell model [13]. According to the Levine theory, correction factors must be applied to calculate zeta potential using sedimentation potential data. The model was further developed by Ohshima [14] to derive a simple and convenient expression for the sedimentation potential expressed as

$$E_s = \frac{\phi(1-\phi)(\rho-\rho_0)g}{(1+\phi/2)\lambda_0} u(\kappa a, \phi) \tag{10}$$

where $u(\kappa a, \phi)$ is the electrophoretic mobility depending both on κa and ϕ, and λ_0 is the specific conductivity of the dispersing medium without particles. The complete expression for $u(\kappa a, \phi)$ was derived by Ohshima [15]. However, this theory is only valid for small zeta potential.

In the case of flotation potential of gas bubbles, and according to the Dukhin–Derjaguin theory [16–18], i.e., the diffusion–electrical theory of the Dorn effect, the Dorn effect of bubbles in aqueous solutions is quite different from that of solid particles. This is mainly because the potential is strongly affected by such factors as Reynolds number, Peclet number, and surface activity of surfactants absorbed on bubble surfaces. Hence, the Smoluchowski equation holds only under limited conditions.

III. EXPERIMENTAL METHODS

A. Sedimentation Potential

For the estimation of the zeta potential, the sedimentation potential method requires only the concentration and the conductivity of the suspension, the density of the particle, and the potential difference developed between both ends of the column. It must also be emphasized that this method is applicable to concentrated systems. However, as above mentioned, the sedimentation method has been used only rarely [12, 19–23]. This may be partly because of the difficulty in measuring the sedimentation potential accurately, as the effects of drifts and/or lack of symmetry of the measuring electrodes are relatively large compared to the magnitude of the potential. This is so in spite of the relatively high values typically attained by the sedimentation potential (of the order of ~ 0.1–1 mV), which is not difficult to measure using a high-impedance digital millivolt meter. Among the measurements made so far, careful precautions were taken about the design of the apparatus for the sedimentation potential measurement. Quist and Washburn [19] developed an apparatus with which the colloidal dispersion was introduced into the column for the sedimentation potential measurement from another chamber connected through a stopcock, and the electric potential developed by the introduction of the colloidal dispersion was subsequently measured. Thus, the effects due to the ambiguity of the electrodes were minimized. Later, setups similar to that developed by Quist and Washburn were employed by Moza and Biswas [20] and Marlow and Rowell [21]. The latter authors found good agreement between zeta potentials obtained by the sedimentation potential and by electrophoresis in conditions where the Smoluchowski theory is valid. According to their conclusion, the sedimentation potential method is useful in aqueous dispersions of large particles below 1.8% volume concentration. They also showed the validity of sedimentation potential measurements for the determination of ζ at higher volume fractions, taking into account corrections due to both electrical and hydrodynamic interactions, based on the Levine's cell model. Hidalgo-Alvarez et al. [22] also measured the zeta potential of quartz particles by the sedimentation method and achieved good agreement with the results obtained by the streaming potential method.

It has been found that difficulties in accurate measurement of the sedimentation potential arising from the electrodes could be overcome by rotating the sedimentation column followed by measurement of the potential difference developed by the rotation [23]. A newly developed rotating-column method is described below, suggesting its applicability to concentrated colloidal dispersions.

A schematic drawing of the apparatus for the rotating-column method is shown in Fig. 1. The Pyrex glass column length is 60 cm with an inner diameter of 0.5–1 cm. The column is supported at its center on a rotatable axis, so that it can be rotated 360°. Ag–AgCl electrodes were prepared by electrolyzing silver wires in hydrochloric acid and were refreshed before each measurement. Test measurements with the apparatus were performed using hematite and silica dispersions prepared by the methods described in the literature [24, 25]. The sample dispersion was introduced into the column, taking care not to introduce air bubbles, after being adjusted to the desired solid content for a specific ionic concentration. The column was electrically shielded with aluminum foil in order to prevent introduction of any electrical noise. The electric potential differences developed on each rotation were measured with a high-impedance digital voltmeter, connected to a personal computer. The electric potential difference developed by a rotation of 180° is

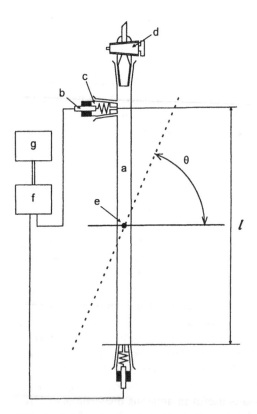

FIG. 1 Schematic drawing of rotating-column apparatus for measurement of the sedimentation potential: a, sample column; b, Ag–AgCl electrode; c, electrode chamber; d, Teflon stopcock; e, axis of rotation; f, high-impedance digital millivoltmeter; g, personal computer; θ, angle of rotation; l, length of the column.

equal to twice the sedimentation potential. Therefore, in this method, only the change in the potential difference is important, and the drifts and lack of symmetry of the electrodes are thus eliminated. The change in the measured potential difference on rotation is shown in Fig. 2. As observed in this figure, although the measured potential is drifting, the change in the potential difference can be read with high accuracy. The variation in the sedimentation potential as a function of $\sin \theta$ where θ is the rotating angle, is shown in Fig. 3. The sedimentation potential is proportional to $\sin \theta$, thus indicating that the measured potential is proportional to the effective length of the column. This is because the sedimentation velocity of the particles along the column is proportional to $\sin \theta$.

B. Flotation Potential

Figure 4 shows the experimental setup used for measuring the Dorn effect of bubbles. A side tube is connected to the bubbling cell for measuring the gas holdup, i.e., the increment of the liquid column height upon introduction of gas bubbles. That height was measured, as a function of time, by a pressure transducer which was connected to the side tube. When finely divided gas bubbles are allowed to rise through an aqueous solution, then the Dorn potential is created between the electrodes. The latter was measured by means of an electrometer through KCl–agar bridges and saturated calomel electrodes, and was recorded as a function of time. According to the Smoluchowski equation [Eq. (9)], the relation between the zeta potential and the Dorn potential $\Delta \phi$ is given by

$$\zeta = \frac{\lambda \eta \Delta \phi}{\phi(\rho - \rho_0)\varepsilon_r \varepsilon_0 g l} \tag{11}$$

By substitution of the relation:

$$\phi = \frac{\Delta V}{V + \Delta V} \tag{12}$$

into Eq. (11), we obtain:

$$\zeta = \frac{(V + \Delta V)\eta \lambda \Delta \phi}{\Delta V(\rho - \rho_0)\varepsilon_r \varepsilon_0 g l} \tag{13}$$

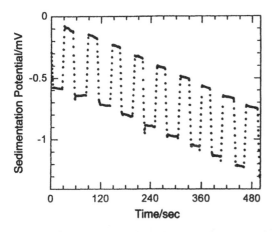

FIG. 2 Change and drift of the sedimentation potential for each rotation. The sedimentation potential changes with each rotation of the column.

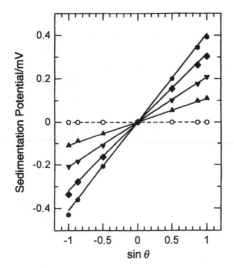

FIG. 3 Variation in the sedimentation potential as a function of sin θ, where θ is the angle of the rotation of the column.

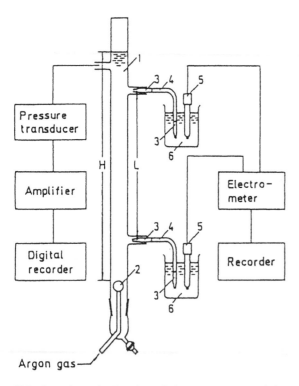

FIG. 4 Schematic drawing of the apparatus used for measuring the Dorn effect of bubbles: (1) Pyrex glass tube; (2) gas disperser; (3) 0.1 M KCl–agar bridge; (4) vinyl tube filled with 0.1 M KCl solution; (5) saturated calomel electrode; (6) 0.1 M KCl solution; $L = 480$ mm; $H = 660$ mm.

where V is the volume of the medium, and ΔV is the total volume of particles (bubbles) included in the medium. This equation was confirmed by Quist and Washburn [19] and Elton and Pearce [26], who used solid particles. If we neglect the term ΔV as compared to V, Eq. (13) can be rewritten as follows:

$$\zeta = \frac{H\eta\lambda}{(\rho - \rho_0)\varepsilon_r\varepsilon_0 g l}\frac{\Delta\phi}{\Delta H} \tag{14}$$

where H is the liquid column height, and ΔH is the increment of the latter upon introduction of gas bubbles. Inserting appropriate numerical values into Eq. (14), we can obtain the zeta potential in millivolts at 20°C:

$$\zeta = 1.98 \times 10^6 \lambda \frac{\Delta\phi}{\Delta H} \tag{15}$$

Hence, all we have to do in order to obtain the zeta potential is to measure λ (in S/cm) and $\Delta\phi/\Delta H$ (in mV/cm).

Figure 5 shows some examples of the Dorn potential recorded as a function of time with a 3×10^{-3} M NaBr aqueous solution. The solution contains 4.2×10^{-2} M ethanol in order to make the bubble size fine. Normally, the bubble diameters were less than 0.5 mm. If we introduce argon bubbles, the Dorn potential increases and reaches a stationary value. When we stop bubbling, the potential and the liquid column height decrease to their initial levels. Note also that if we increase the gas flow rate, the Dorn potential (and also ΔH) increases, as shown in the figure.

The Dorn potential obtained in this way is found to be proportional to ΔH in the range of low values of this quantity, although such a range increases in the presence of surfactants. From the slope $\Delta\phi/\Delta H$ and the specific conductance λ, we calculated the zeta potential by making use of Eq. (15). The mean diameter of the gas bubbles was assessed from the increment of the liquid column height, ΔH, upon introduction of the gas. The total volume of the liquid column ($\Delta H \times A$, where A is the area of the cross-section of the

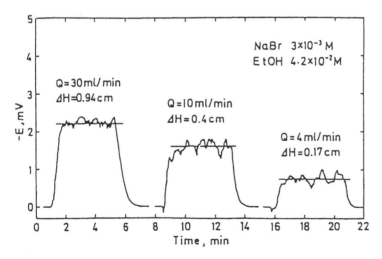

FIG. 5 Example of Dorn potential recorded as a function of time at various flow rates (Q) for 3×10^{-3} M NaBr, in the presence of 4.2×10^{-2} M ethanol.

tube) should be equal to the gas flow rate, Q, multiplied by the retention time of the bubbles:

$$\Delta H \times A = \frac{Q(H + \Delta H)}{U} \tag{16}$$

where U is the ascending velocity of the bubbles. Under experimental conditions, U was found to be proportional to the bubble diameter, d:

$$U = k \times d \tag{17}$$

where k is a proportionality constant, determined experimentally as approximately $125\ s^{-1}$ [27]. The procedure for determining d by means of Eqs (16) and (17) was called the $Q/\Delta H$ (this quantity is proportional to the bubble diameter, d) method by Usui et al. [8].

Figure 6 shows the zeta potential of bubbles as a function of their diameter, as determined by the $Q/\Delta H$ method. The solid symbols represent data obtained by averaging the zeta potential and d values, respectively. The values of negative zeta potentials increase with decreasing bubble size in each system. It is clearly seen that the smaller the bubble size the larger the zeta potential.

Classical electrokinetic theory does not predict any dependence of ζ on particle size. Thus, Fig. 7 shows the zeta potential of glass beads determined by the sedimentation method as a function of their mean diameter. Obviously, no appreciable variation in zeta potential with particle size is found. In the case of moving bubbles, a somewhat different situation must be considered: owing to the flow, the surface charges are displaced

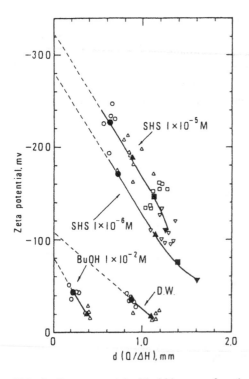

FIG. 6 Zeta potential of bubbles as a function of the mean diameter, determined by the $Q/\Delta H$ method. Bubbler number: ∇, No. 1; $\square$; No. 2; $\triangle$: No. 3; $\bigcirc$: No. 4. The filled symbols represent mean zeta potentials calculated by using the average of bubble diameters obtained by the $Q/\Delta H$ method.

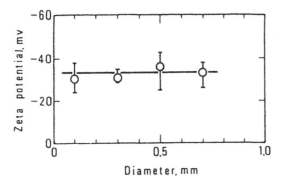

FIG. 7 Zeta potential of glass beads as a function of their diameter.

backwards resulting in a weaker electric dipole and, hence, a reduced Dorn potential. Thus, the zeta potential extrapolated to $d = 0$ should correspond to the value which the gas/water interface might have under static conditions.

IV. APPLICATIONS

A. Sedimentation Potential of Hematite Particles

Figure 8 shows the dependence of the zeta potential of a hematite dispersion, estimated using Eq. (9), on the solid concentration of the dispersion. Constant potentials were obtained in the range of the solid contents examined. The variation in zeta potential of a hematite dispersion with pH is shown in Fig. 9 together with the results obtained by an electrophoresis apparatus (the Pen-Kem System 3000), using the Smoluchowski equation relating electrophoretic mobility and ζ, for dilute suspensions. As observed, the zeta potentials of hematite particles obtained by sedimentation potential and electrophoresis agree well within experimental uncertainty.

B. Effects of Presence of Surfactants on Flotation Potential of Bubbles

Figure 10 shows the electrokinetically determined adsorption densities of Sodium Hexadecyl Sulfate (SHS) on bubble surfaces of different sizes in the presence of 1 and

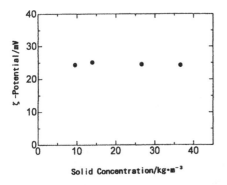

FIG. 8 Dependence of the zeta potential of hematite on the solid concentration at pH 4.

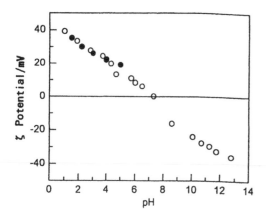

FIG. 9 Variation of the zeta potential of the hematite dispersion as a function of pH. (●) zeta potential measured by the sedimentation method; (○) zeta potential measured by electrophoresis.

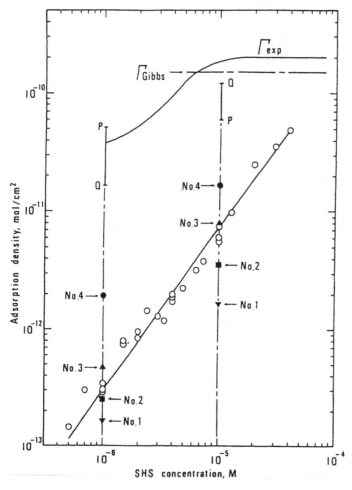

FIG. 10 Electrokinetic adsorption densities of SHS; Γ_{exp} and Γ_{Gibbs} represent the adsorption isotherms of SHS at the solution–air interface. (From Ref. 28.) (○) Reproduction of the electrokinetic adsorption density of SHS with bubbler No. 3. Arrows indicate the bubbler number; PQ represents the electrokinetic adsorption density extrapolated to $d = 0$.

10 μM SHS, as obtained using different bubblers. The adsorption densities of SHS obtained by Matsuura et al. [28], using radiotracer measurements (exp) and surface tension determinations (Gibbs), are also included in the figure. The data from Fig. 6, which were obtained with bubbler No. 3, are reproduced in Fig. 10 for comparison. The adsorption density was calculated using the Gouy–Chapman theory, i.e.,

$$Ne\Gamma = -\sigma_d = \left(\frac{2\varepsilon_r \varepsilon_0 kT}{\pi}\right)^{1/2} \times (n_0)^{1/2} \sinh\left(\frac{Ze\zeta}{2kT}\right) \qquad (18)$$

where Γ is the electrokinetic adsorption density in moles per square centimeter, N is Avogadro's number, e is the elementary charge, σ_d is the diffuse layer charge density per unit area, k is Boltzmann's constant, T is the absolute temperature, n_0 is the number concentration of ions, and Z is the valency of the ion. Under dynamic conditions, Γ_{SHS} increases almost linearly with surfactant concentration until an equilibrium value is reached. The electrokinetic adsorption densities, evaluated from the zeta potential at a bubble size of zero, are shown by the lines PQ, where P refers to the value obtained from the photographic method by camera [27], and Z from data in Fig. 6 ($Q/\Delta H$ method). If one takes the SHS concentration of 2×10^{-5} M as an example, a condition which yields an almost complete monolayer in the equilibrium state, the adsorption density in the dynamic state corresponds to only $\cong 10\%$ of a monolayer. The discrepancy in adsorption density between the equilibrium and dynamic (floating bubble) states tends to increase with decreasing concentration.

It is interesting to show the effect of inorganic electrolyte on the zeta potential of bubbles in the presence of a cationic surfactant. Figure 11 shows the zeta potential of bubbles as a function of NaCl concentration in the absence (○) and in the presence (●) of 10^{-5} M hexadecyltrimethylammonium bromide (HTAB). In the absence of NaCl, the zeta potential shows a positive value, i.e., $+0.3$ V. The zeta potential changes its sign from positive to negative with increasing NcCl concentration in the presence of HTAB. The profile of this curve with increasing NaCl concentration shows a similar trend in the case of the solution in the absence of HTAB.

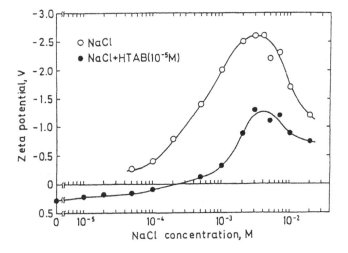

FIG. 11 Zeta potential of bubbles as a function of NaCl concentration in the absence (○) and in the presence (●) of 10^{-5} M HTAB. Ethanol: 4.2×10^{-2} M.

V. CONCLUSIONS

The examples discussed show that both the sedimentation and flotation potentials can be measured with high accuracy. The former is free of errors due to the drifts and/or the asymmetry of the measuring electrodes if the rotating-column method is used. It should also be emphasized that this method is applicable to concentrated systems for which many commercial electrophoresis apparatus are not useful. In the case of flotation potential, the experimental difficulties coexist with the lack of theoretical models. In this respect, it must be mentioned that Dukhin et al. [29] have reviewed the dynamic adsorption layer on buoyant bubbles and discussed the comparison between theory and experimental results.

REFERENCES

1. E Dorn. Ann Phys (Leipzig) 10: 46, 1880.
2. HA McTaggart. Phil Mag 27: 297, 1914; 28: 167, 1914; 44: 386, 1922.
3. T Alty. Proc Roy Soc London, Ser A 106: 3158, 1924; 112: 235, 1926.
4. S Komagata. Denki Kagaku 4: 380, 1936.
5. N Bach, A Gilman. Acta Physicochim URSS 9: 1, 1938.
6. GL Collins, M Motarjemi, G Jameson. J Colloid Interface Sci 63: 69, 1978.
7. S Usui, H Sasaki. J Colloid Interface Sci 65: 36, 1978.
8. S Usui, H Sasaki, H Matsukawa. J Colloid Interface Sci 81: 80, 1981.
9. H Sasaki, Q Dai, S Usui. Kolloidn Zh 48: 1097, 1986.
10. TZ Sotskova, YuF Bazhenov, LA Kul'skii. Kolloidn Zh 44: 989, 1982.
11. TZ Sotskova, VYa Poberezhnyi, YuF Bazhenov, LA Kul'skii. Kolloidn Zh 45: 108, 1983.
12. RJ Hunter. Zeta Potential in Colloid Science. New York: Academic Press, 1981, p 174.
13. S Levine, GH Neale, N Epstein. J Colloid Interface Sci 57: 424, 1976.
14. H Ohshima. J Colloid Interface Sci 208; 295, 1998.
15. H Ohshima. J Colloid Interface Sci 188: 491, 1998.
16. BV Derjaguin, SS Dukhin, Proceedings of the 3rd International Congress on Surface Activity, vol. 2, Universitatsdruckerei Meinz, Cologne, 1960, p 324.
17. SS Dukhin. In BV Derjaguin, eds., Research in Surface Forces, vol. 2. New York: Consultants Bureau, 1966. p 54.
18. SS Dukhin. Kolloidn Zh 45: 22, 1983.
19. JD Quist, ER Washburn. J Am Chem Soc 62: 3169, 1940.
20. AK Moza, AK Biswas. Colloid Polym Sci 254: 522, 1976.
21. BJ Marlow, RL Rowell. Langmuir 1: 83, 1985.
22. R Hidalgo-Alvarez, FJ de las Nieves, G Pardo. J Colloid Interface Sci 107: 295, 1985.
23. M Ozaki, T Ando, K Mizuno. Colloids Surfaces A: Physicochem Eng Aspects 159: 477, 1999.
24. E Matijevíc, P Scheiner. J Colloid Interface Sci 63: 509, 1978.
25. W Stöber, A Fink, E Bohn. J Colloid Interface Sci 26: 62, 1968.
26. GAH Elton, JB Peace. J Chem Soc 53: 22, 1956.
27. H Sasaki, H Matsukawa, S Usui, E Matijevíc. J Colloid Interface Sci 113: 500, 1986.
28. R Matsuura, H Kimizuka, S Miyamoto, R Shimozawa. J Chem Soc Jpn 313: 532, 1958.
29. SS Dukhin, G Kretzschmar, R Miller. Dynamics of Adsorption at Liquid Interfaces. Theory, Experiment, Application. Amsterdam: Elsevier, 1955, ch. 8.

17

Electroacoustic Phenomena in Concentrated Dispersions: Theory, Experiment, Applications

ANDREI S. DUKHIN and PHILIP J. GOETZ Dispersion Technology, Inc., Mount Kisco, New York

VLADIMIR N. SHILOV Institute of Biocolloid Chemistry, Ukrainian National Academy of Sciences, Kiev, Ukraine

HIROYUKI OHSHIMA Science University of Tokyo, Tokyo, Japan

I. INTRODUCTION

It is well known that when charged colloidal particles sediment due to gravity, a potential difference is produced between two vertically spaced electrodes. This electrokinetic phenomenon is called sedimentation potential or Dorn effect, and it is clear that if we externally short-circuit the electrodes, an electric current, the sedimentation current, will flow [1]. If we now assume that the particles are subjected to an applied acoustic field, the constant gravitational acceleration will be in fact replaced by a harmonically changing one provoked by the inertial forces associated with the sound wave. An acoustic potential or current will also be generated that will change harmonically with time: the phenomenon is now called colloid vibration potential (CVP) or colloid vibration current (CVI). Figure 1 illustrates schematically their origin: a particle with its double layer is moving relative to the liquid, and this motion involves ions of the double layer. In this case we consider the positive counterions opposing the negatively charged particle surface. The hydrodynamic surface current I_s reduces the number of positive ions near the right particle pole and enriches the double layer with extra ions near the left pole. As a result, the double layer shifts from the original equilibrium. A negative extra diffuse charge dominates at the right pole, whereas the positive one dominates at the left pole. The net result is that the motion has induced a dipole moment.

This induced dipole moment generates an electric field which is usually referred to as a CVP. This CVP is external to the particle double layer. It affects ions in the bulk of the electroneutral solution beyond the double layer, generating an electric current I_n. This electric current serves an important purpose: it compensates for the surface current I_s and makes the whole picture self-consistent. A related electrokinetic phenomenon is the electrokinetic sonic amplitude (ESA) effect: in this case, the externally applied field is an alternating electric field. The subsequent oscillation of the charged colloid units provokes small pressure disturbances around them, and if the density of the particle and the medium are different, a macroscopic harmonic sound wave will be produced [1,

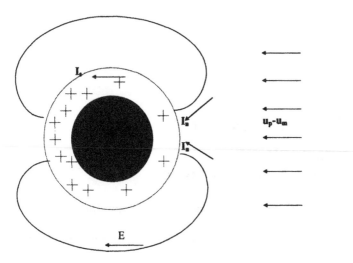

FIG. 1 Mechanism of double-layer polarization generating sedimentation current for a single particle.

2]. Hence, both CVP (or CVI) and ESA can be termed electroacoustic electrokinetic phenomena. There is growing interest in them in the colloid science community, mainly since commercial CVI (www.dispersion.com) and ESA (www.matec.com) devices became available in the 1980s.

This work is mainly devoted to CVI, based on the authors' experience and involvement in this electrokinetic phenomenon. A new theory of CVI in concentrated suspensions will be developed and experimental data with silica and rutile dispersions will be presented. The validity of the theoretical model will be discussed in terms of its agreement with experimental results and with exact low-frequency asymptotic theories, available for arbitrary suspension concentration, such as Smoluchowski formula and the fundamental Onsager reciprocity relations of the thermodynamics of irreversible processes.

II. THEORY

A. Background

There are two quite different approaches to deriving an electroacoustic theory. Historically, the first began with works by Booth and Enderby [3, 4]. They tried to solve a system of classical electrokinetic equations without using any thermodynamic relationships. It was very complex because they took into account surface conductivity effects without effective simplifications connected with small values of Debye screening length as compared to particle radius. Although this initial theory was valid only for dilute systems, the approach was later expanded by Marlow et al. [5], who tried to generalize it for concentrated systems using a Levine cell model [6]. This approach leads to somewhat complicated mathematical formulas. Perhaps this was the reason why it was abandoned.

An alternative approach to electroacoustic theory was suggested later by O'Brien [7, 8]. He introduced the concept of dynamic electrophoretic mobility μ_d, and derived a relationship between the quantity and the measured electroacoustic parameters such as CVI or ESA:

$$\text{ESA} = C_{\text{cal}} \sigma \phi \mu_{\text{d}} E$$

$$\text{CVI} = C_{\text{cal}} \sigma \phi \mu_{\text{d}} \nabla P \tag{1}$$

where C_{cal} is a cell constant, ϕ is the volume fraction of solids, P is the hydrodynamic pressure, and E is the external electric field strength. The quantity σ is a contrast between particle (ρ_{p}) and liquid (ρ_{m}) densities:

$$\sigma = \frac{\rho_{\text{p}} - \rho_{\text{m}}}{\rho_{\text{m}}} \tag{2}$$

According to O'Brien, a complete functional dependence of ESA (CVI) on the key parameters like ζ potential, particle size, and frequency is incorporated into dynamic electrophoretic mobility. The coefficient of proportionality between ESA (CVI) and μ_{d} is frequency independent, as well as independent of particle size and ζ potential. This peculiarity of Eq. (1) made dynamic electrophoretic mobility a central parameter of the electroacoustic theory.

The first theory of dynamic electrophoretic mobility, which relates this parameter to other properties of the dispersed system, was created initially by O'Brien for the dilute case only, neglecting particle–particle interactions. We can call this version the "dilute O'Brien's theory." Later he applied the Levine cell model trying to expand dynamic electrophoretic theory to concentrated systems [8]. This work was generalized by Ohshima [9], and can be termed the "O'Brien–Levine" theory. The last development of this approach was made recently by Dukhin and coworkers [10–12]. We used the Shilov–Zharkikh cell model [13] for dynamic electrophoretic mobility. We call the combination of O'Brien's relationship and our dynamic electrophoretic mobility theory the "hybrid O'Brien's theory." A block diagram of these various versions of the electroacoustic theory is presented in Fig. 2, which help us to understand this somewhat complicated situation.

Recently, the authors of this chapter returned to the original Enderby–Booth approach [3, 4] in order to generalize it for concentrates. We have done this using the Kuwabara cell model [14] for calculating the hydrodynamic drag coefficient, and the Shilov–Zharkikh cell model [13] for electrokinetics. In addition, we used a well-known "coupled-phase model" for describing the relative motion between the particles and the liquid in the concentrated system. The coupled-phase model allowed us to eliminate the assumption of superposition of hydrodynamic fields for incorporating particle polydispersity into the theory. In order to distinguish this new theory from other treatments, we suggest using the abbreviation "DSOG theory." The theory was initially developed in Ref. 15, assuming a negligible surface conductivity, reflected in the low values of the Dukhin number "Du" [16]:

$$\text{Du} = \frac{\kappa^{\sigma}}{K_{\text{m}} a} \ll 1 \tag{3}$$

where κ^{σ} is the surface conductivity, K_{m} is the bulk conductivity of the equilibrium medium, and a is the particle radius.

Later, we generalized it by including surface conductivity at frequencies below the Maxwell–Wagner frequency, when the displacement current is negligible as compared to that of the background of conductivity current [17]. We are presenting this most recent version in this chapter, which means that there is no restriction on the value of Du [16].

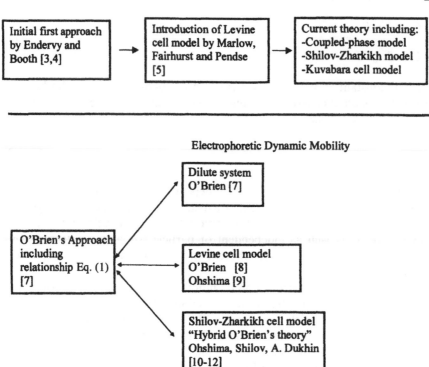

FIG. 2 Block diagram illustrating various versions of electroacoustic theory.

We restrict consideration to the simpler case of CVI and/or CVP, when a pressure gradient is the driving force generating the electroacoustic signal. We would like to be cautious concerning expanding this new theory to the ESA phenomenon. It turned out that the problem of frames of reference has different implications for these different electroacoustic effects.

As mentioned, we use a "coupled-phase model" [17–19] for describing the speed of the particle relative to the liquid. The Kuwabara cell model [14] yields the required hydrodynamic parameters such as the drag coefficient and we connect this parameter with the generated electric field by using the Shilov–Zharkikh cell model [13]. The DSOG theory is thus currently valid for polydisperse suspensions, without using the superposition assumption. In this work, we also describe the derivation of exact asymptotic solutions at the quasistationary limit, using just Onsager relationships and the Smoluchowski law [20], without any cell model. It is found that the DSOG theory satisfies the requirement of transition to the quasistationary limit at any volume fraction.

This situation resembles somewhat the problem of the sedimentation potential. There is a simple way to create the theory of sedimentation potential by using the Onsager reciprocal relationship. However, it turns out that a straight derivation rooted down to the basic physical equations is also very helpful. In the case of the sedimentation potential such a derivation, as performed by Ohshima [21], provides an important background and confirmation for Onsager-based theory.

B. Frames of Reference

When sound is the driving force (the case of CVI or CVP), the correct inertial frame is the laboratory frame of reference, since the acoustic wavelength is much shorter than the size of the sample chamber. Therefore, particles move with different phases inside the narrow sound beam. The chamber as an entity remains immobile.

The question of the frame of reference is more complicated in the case of the electric field as a driving force (ESA). The wavelength of the electric field is much longer, and as a result all particles move in phase. This motion exerts a certain force on the chamber. The motion of the chamber depends on its mass and that of the sample. Depending on the construction of the instrument, the inertial system might be related either to the chamber or to the center of mass, or to some intermediate case, depending on the chamber-to-sample mass ratio. It means that in the case of ESA, the final expression relating the measured ESA signal with properties of the dispersed system might contain a multiplier which depends on the mass of the chamber.

C. Coupled-phase Model

Let us consider an infinitesimal volume element in the field of the sound wave. There is an external force acting on this element, proportional to the pressure gradient of the sound wave, ∇P. This external force is applied to both the particles and liquid and is distributed between them according to the volume fraction ϕ.

Both particles and liquid move with an acceleration created by the sound-wave pressure gradient. In addition, because of inertia effects, the particles move relative to the liquid, which causes viscous friction forces to act between the particles and liquid.

The balance of these forces can be presented using the following system of equations written separately for particles and liquid:

$$-\phi\nabla P - \phi\rho_\text{p}\frac{\partial u_p}{\partial t} + \gamma(u_\text{p} - u_\text{m}) \tag{4}$$

$$-(1 - \phi)\nabla P = (1 = \phi)\rho_\text{m}\frac{\partial u_m}{\partial t}\gamma(u_\text{p} - u_\text{m}) \tag{5}$$

where u_m, and u_p are, respectively, the velocities of the medium and the particles in the laboratory frame of reference, t is the time, and γ is a friction coefficient which is proportional to the volume fraction and particle hydrodynamic drag coefficient Ω:

$$\gamma = \frac{9\eta\phi\Omega}{2a^2} F_f = 6\pi\eta a\Omega(u_\text{p} - u_\text{m}) \tag{6}$$

where η is dynamic viscosity, and a is the particle's radius.

The system of equations (4–6) is well known in the field of acoustics. It has been used in several papers [7–9] for calculating sound speed and acoustic attenuation. It is valid without any restriction on volume fraction. Importantly, it is known that this system of equations yields a correct transition to the dilute case.

This system of equations is normally referred to as the "coupled-phase model". The word "model" usually suggests the existence of some alternative formulation, but it is hard to imagine what one can change in this set of force balance equations, which essentially express Newton's second and third laws. Perhaps the word "model" is not suitable in this case.

This system of equations can be solved for the speed of the particle relative to the liquid. The time and space dependence of the unknowns u_m and u_p is assumed to be a monochromatic wave $Ae^{i(\omega t - kx)}$, where i is an imaginary unit, and k is a complex wavenumber. As a result, the system of equations (3)–(4) yields the following relationship between the pressure gradient and the speed of the particle relative to the fluid:

$$\gamma(u_p - u_m) = \frac{\phi(\rho_p - \rho_s)}{\rho_s + i\omega\phi(1-\phi)\dfrac{\rho_p\rho_m}{\gamma}}\nabla P \tag{7}$$

where $\rho_s = \phi\rho_p + (1-\phi)\rho_m$.

The model was generalized for polydisperse systems in Ref. 15. The velocity of the particles which belong to the ith size fraction $u_i - u_m$ is given by the following expression:

$$u_i - u_m = \frac{\left(\dfrac{\rho_p}{\rho_m} - 1\right)\nabla P}{\left(j\omega\rho_p + \dfrac{\gamma_i}{\phi_i}\right)\left(1 + \dfrac{\rho_p}{(1-\phi)\rho_m}\displaystyle\sum_{i=1}^{N}\dfrac{\gamma_i}{j\omega\rho_p + \dfrac{\gamma_i}{\phi_i}}\right)} \tag{8}$$

where

$$\gamma_i = \frac{9\eta\phi_i\Omega}{2a_i^2}$$

and the Stokes drag on the ith size fraction is

$$F^i_{Stokes} = 6\pi\eta a\Omega(u_i - u_m)$$

The relative motion of the particles disturbs the double layers surrounding them and as a consequence induces the electroacoustic phenomenon. The relationship between particle motion and the resulting electroacoustic signal is described next.

D. CVI as a Sedimentation Current

As already mentioned (Fig. 1), the analogy between CVI and sedimentation current is rather obvious. Simply put, charged particles sedimenting in a gravitational field will develop a measurable sedimentation current through externally short-circuited electrodes. The concept is extended to CVI by replacing the gravitational acceleration by a variable inertial one due to the action of the harmonic sound wave. The next step is to add a quantitative description to this simple qualitative picture. In order to do this, we must find a relationship between the CVP and the speed of the relative motion between particle and liquid ($\mu_p - u_m$). We did this under the assumption that the double-layer thickness must be much less than the particle radius a [23]:

$$\kappa a \gg 1 \tag{9}$$

where κ is reciprocal Debye length. It is possible to eliminate this restriction in the future following well known papers by Babchin and coworkers [24, 25].

We assume that the sound frequency ω is below the Maxwell–Wagner frequency ω_e [26]:

$$\omega \ll \omega_e = \frac{K_m}{\varepsilon\varepsilon_0} \tag{10}$$

where $\varepsilon\varepsilon_0$ is the dielectric permittivity of the medium. The strong inequality (10) permits us to neglect the contribution of displacement currents in comparison with that of conductivity currents.

The thin double-layer condition allows us to describe the distribution of the electric potential ϕ outside of the quasiflat double layer using the Laplace equation:

$$\Delta\phi = 0 \tag{11}$$

The general solution of this equation:

$$\phi = -Er\cos\theta + \frac{d}{r^2}\cos\theta \tag{12}$$

contains two unknown constants E and d. Two boundary conditions are required for calculating these constants.

The surface boundary condition reflects the continuity of the bulk current $I = -K_m \nabla_n\phi$ and the surface current I_s:

$$K_m \nabla_n\phi = \mathrm{div}_s I_s \tag{13}$$

where the surface current I_s contains two components, hydrodynamic (convective) and electromigration:

$$I_s = -\varepsilon\varepsilon_0\zeta\frac{\partial u_\theta}{\partial r}\bigg|_{r=1} - \kappa^\sigma\frac{1}{a}\frac{\partial\phi}{\partial\theta}\bigg|_{r=a} \tag{14}$$

where ζ is the electrokinetic potential. Substitution of Eq. (14) into Eq. (13) yields the first boundary condition:

$$\left(\frac{2\kappa\sigma}{a} - K_m\right)E - \left(\frac{2\kappa^\sigma}{a} + 2K_m\right)\frac{3}{a^3} = -\frac{1}{a}\varepsilon\varepsilon_0\zeta\frac{1}{\sin\theta}\frac{\partial u_\theta}{\partial r}\bigg|_{r=a} \tag{15}$$

We use the cell model concept for deriving the second boundary condition. According to this concept, we redistribute liquid equally beween particles, and assume that the liquid associated with each particle creates a spherical cell of radius b. This radius is related with the particle radius according to the following expression:

$$b^3 = \frac{a^3}{\phi}$$

We prefer to use the Shilov–Zharkikh cell model [13] over the Levine one [6]. All arguments for this decision are given in Ref. 10. These two cell models yield different expressions for the macroscopic electric field $\langle E \rangle$:

$$\langle E \rangle_{\text{Levine}} = -\frac{1}{\cos\theta}\frac{\partial\phi}{\partial r}\bigg|_{r=b} \tag{16}$$

$$\langle E \rangle_{\text{Shilov}} = \frac{\phi}{b\cos\theta}\bigg|_{r=b} \tag{17}$$

The cell boundary condition corresponding to CVI (external short circuit) specifies the zero value of $\langle E \rangle$ in terms of the condition on the cell surface. In the Shilov–Zharkikh cell model it becomes

$$\frac{\partial\phi}{\partial r}\bigg|_{r=b} = \frac{\phi_{r=b}}{b\cos\theta} = -E + \frac{d}{b^3} = 0 \tag{18}$$

In order to find the CVP we should calculate the unknown constants E and d using Eqs (15) and (18) and substitute this result into the following expression for the CVI

$$\text{CVI} = \langle I \rangle|_{(E)=0} = \frac{j_r|_{r=b}}{\cos\theta} = -\frac{K_m \frac{\partial\phi}{\partial r}\Big|_{r=b}}{\cos\theta} = K_m\left(E + \frac{2d}{b^3}\right) \tag{19}$$

As a result of Eqs (15), (18) and (19) we obtain the following relationship between CVI and the tangential fluid velocity:

$$\text{CVI} = \frac{3\phi}{\left(\dfrac{\kappa^\sigma}{K_m a} + 1\right) - \left(\dfrac{\kappa^\sigma}{K_m a} - 0.5\right)\phi} \frac{\varepsilon\varepsilon_0\zeta}{a} \frac{1}{\sin\theta} \frac{\partial u_\theta}{\partial r}\Big|_{r=a} \tag{20}$$

where K_s is the complex conductivity of the system r and θ are the spherical co-ordinates associated with the particle center, and u_r and u_θ are the radial and tangential velocities of the liquid motion relative to the particle.

The next step in the development of this CVI theory is the calculation of the hydrodynamic field $\frac{\partial u_\theta}{\partial r}_{r=a}$ assuming that the speed of the particle with respect to the liquid is given by expression (7). This is done in the next section.

E. Calculation of Hydrodynamic Field Using Nonstationary Kuwabara Cell Model

In this section we calculate the particle drag coefficient γ and the tangential velocity of the liquid u_θ which is a part of Eq. (20). We perform this calculation assuming that the liquid is incompressible. This condition is valid only when the wavelength λ is much larger than the particle size:

$$\lambda \gg a \tag{21}$$

This is the so-called long-wavelength requirement. It allows us to use, in the scale of the particle (cell) size, the hydrodynamic equations for incompressible liquids:

$$\rho_m \frac{d\vec{u}}{dt} = n\nabla \times \nabla \times \vec{u} + \nabla P \tag{22}$$

$$\nabla \cdot \vec{u} = 0 \tag{23}$$

This system of equations has been solved by Dukhin and Goetz [19] for a Happel cell model. Here, we suggest another solution, using the Kuwabara cell model [14]. Both models apply the same boundary conditions at the surface of the particle:

$$u_r(r = a) = u_p - u_m \tag{24}$$

$$u_\theta(r = a) = -(u_p - u_m) \tag{25}$$

However, the boundary conditions at the surface of the cell are quite different for the Kuwabara cell model and are given by the following equations:

$$\text{rot } u_{r=b} = 0 \tag{26}$$

$$u_r(r = b) = 0 \tag{27}$$

The general solution for the velocity field contains three unknown constants C, C_1, and C_2:

$$u_r(r) = C\left(1 - \frac{b^3}{r^3}\right) + 1.5\int_r^b \left(1 - \frac{x^3}{r^3}\right)h(x)dx \tag{28}$$

$$u_\theta(r) = -C\left(1 + \frac{b^3}{2r^3}\right) - 1.5\int_r^b \left(1 + \frac{x^3}{2r^3}\right)h(x)dx \tag{29}$$

$$h(x) = C_1 h_1(x) + C_2 h_x(x) \tag{30}$$

The values of these constants and special functions are given in the appendix. The final expressions for the drag coefficient and tangential velocity are

$$\gamma = \omega\rho_m\phi\left[\frac{3}{4I}\left(\frac{dh}{dx} + \frac{h}{x}\right)_{x=\alpha} - i\right] \tag{31}$$

$$\frac{1}{\sin\theta}\left(\frac{du_\theta}{dr}\right)_{r=a} = \frac{3(u_p - u_m)h(\alpha)}{2I} \tag{32}$$

where $\alpha = a\sqrt{\omega/2v}$, $\beta = b\alpha/a$, v being the kinematic viscosity, and η the dynamic viscosity.

F. CVI for Polydisperse System with Surface Conductivity

Let us assume now that we have a polydisperse system with N size fractions. Each fraction of particles has particle radius a_i, volume fraction ϕ_i, drag coefficient γ_i, and particle velocity u_i in the laboratory frame of reference. We assume that the density of the particles is the same for all fractions, as well as their surface conductivity, κ^σ, and ζ potential. The total volume fraction of the dispersed phase is ϕ. Generalization of the cell model concept for a polydisperse system, which is described in Ref. 19, yields the following relationship between fractional particle radius and radius of the shell b_i:

$$b_i^3 = \frac{a_i^3}{\phi} \tag{33}$$

or, equivalently,

$$\phi_i = \phi \tag{34}$$

The VI generated by the ith fraction equals:

$$\mathrm{CVI}_i = \frac{3\phi}{\left(\dfrac{\kappa^\sigma}{K_m a_i} + 1\right) - \left(\dfrac{\kappa^\sigma}{K_m a_i} - 0.5\right)\phi}\frac{\varepsilon\varepsilon_0\zeta}{a_i}\frac{1}{\sin\theta}\frac{\partial u_{i\theta}}{\partial r}\bigg|_{r=a} \tag{35}$$

The radial derivative of the tangential velocity has been calculated in our previous paper [15]:

$$\frac{1}{\sin\theta}\frac{\partial u_{i\theta}}{\partial r}\bigg|_{r=a} = \frac{3(u_{ip} - u_m)h(\alpha_i)}{I(\alpha_i)} \tag{36}$$

where h and I are special functions given in the appendix.

We can use either Eq. (7) or (8) for calculating the speed of the liquid relative to the particle surface only if we neglect the electro-osmotic flow caused by CVP. This electro-osmotic flow is a secondary effect, but it would still need justification for being neglected. This is done in Ref. 27 where we show that the electro-osmotic flow is reciprocally proportional to $(\kappa a)^2$.

The final expression for CVI can be obtained as a sum of the fractional currents generated by the fractional dipole moments. This expression is

$$
\text{CVI} = \frac{9\varepsilon\varepsilon_o\zeta(\rho_p - \rho_m)\nabla P}{4\eta} \cdot \frac{\displaystyle\sum_{i=1}^{N} \frac{1}{(\text{Du}_i + 1) - (\text{Du}_i - 0.5)\phi} j\alpha_i I(\alpha_i)\left[\rho_p - \rho_m\left(\frac{3H_i}{2I_i} + 1\right)\right]}{1 - \frac{\rho_p}{1-\phi}\displaystyle\sum_{i=1}^{N} \frac{\phi_i\left(\frac{3H_i}{2I_i} + 1\right)}{\rho_p - \rho_m\left(\frac{3H_i}{2I_i} + 1\right)}}
\tag{37}
$$

where $\alpha_i = a_i\sqrt{\omega/2\nu}$, $\beta_i = b_i\alpha_i/a$, $H_i = H(\alpha_i)$, $I_i = I(\alpha_i)$, and the special functions h, H, and I are given in the Appendix.

According to the assumption of monodisperse surface conductivity, the value of the fractional Dukhin number equals:

$$
\text{Du}_i = \frac{\kappa^\sigma}{K_m a_i}
\tag{38}
$$

Expression (37) for CVI is rather complex. It is important to have some reference point for testing its validity. We have managed to create a simplified version of the electroacoustic theory using quasistationary low-frequency limit. It turns out that upon restricting the frequency one can derive an electroacoustic theory for concentrates without using any cell model at all. This simple theory is described in the next section.

G. Low-frequency Asymptotic Limit

The Onsager reciprocity relationship follows from the time reversibility of the equations of motion on the molecular level. It links together various kinetic coefficients. This relationship is certainly valid in the stationary case. Much less is known about its validity in the case of alternating fields. This means that we can use this relationship only in the quasistationary limiting case of low frequency, i.e., when ω is much lower than the characteristic hydrodynamic, ω_{hd}, and electrodynamic (Maxwell–Wagner), ω_{ed}, frequencies:

$$
\omega \ll \omega_{hd} = \frac{\nu}{a^2}
\tag{39}
$$

$$
\omega \ll \omega_{ed} = \frac{K_m}{\varepsilon\varepsilon_m}
\tag{40}
$$

The well-known Saxen–Onsager reciprocal relation provides the following link [28]:

$$
\left.\frac{\langle V \rangle}{\langle I \rangle}\right|_{\langle \nabla P_{rel}\rangle = 0} = \left.\frac{\langle CVP_{\omega \to 0}\rangle}{\langle \nabla P_{rel}\rangle}\right|_{\langle I \rangle = 0}
\tag{41}
$$

between two kinetic coefficients:

1. *Electro-osmotic coefficient*: ratio between the electro-osmotic flow $\langle V \rangle$ and the macroscopic electric current density $\langle I \rangle$, flowing under the action of an external electric field, applied to the suspension and causing the electro-osmotic flow of liquid under the condition of zero macroscopic pressure gradient.
2. *Streaming potential coefficient*: ratio between streaming potential (or in our case, quasistationary CVP) and effective pressure gradient $\langle \nabla P_{rel} \rangle$, which moves the liquid relative to the particles and causes the CVP under the condition of zero macroscopic electric current.

In order to use this relationship in the case of CVI we need to know the effective pressure gradient ∇P_{rel}. This parameter can be easily obtained following the "coupled-phase model" [17–19, 22] for characterizing particle motion in the sound field for concentrated systems. The total friction force exerted on a particle equals $-\gamma(u_p - u_m)$. A force of the same value but opposite direction exerted on the liquid is due to of the pressure gradient which moves liquid relative to the particle. In the extreme case of low frequency, Eq. (7) leads to the following expression for this effective pressure gradient:

$$\nabla P_{rel}|_{\omega \to o} = \gamma(u_p - u_m)|_{\omega \to 0} = \frac{\phi(\rho_p - \rho_s)}{\rho_s} \nabla P \tag{42}$$

In addition, we can also use the fact that the CVP equals the CVI divided by the conductivity of the system K_s. As a result, from Eqs (41) and (42), and taking into account the definition of the static mobility $(\mu = \langle V \rangle / \langle E \rangle|_{\nabla P=0} = K_S \langle V \rangle / \langle I \rangle|_{\nabla P=0})$, we obtain the following expression for CVI:

$$CVI_{\omega \to 0} = CVP^*_{\omega \to 0} K_s = \left. \frac{\langle V \rangle}{\langle I \rangle} \right|_{\langle \nabla P \rangle = 0}^{*} K_s \frac{\phi(\rho_p - \rho_s)}{\rho_s} \nabla P \tag{43}$$

This is the equation for CVI at the low-frequency limit. Using the Smoluchowski law for electro-osmosis [10, 20], it is possible to obtain the following equation:

$$\left. \frac{\langle V \rangle}{\langle I \rangle} \right|_{(\nabla P)=0} = -\frac{\varepsilon \varepsilon_0 \zeta}{\eta} \frac{1}{K_m} \tag{44}$$

which would be explicitly asymptotically valid in concentrated systems, and would simultaneously be free from the assumptions of any electrokinetic cell model. Note that conditions which restrict the applicability of the Smoluchowski equation have been used, together with the assumption that the duble layer is quasiflat [10] and that the Dukhin number [Eq. (38)] is very small, that is, the surface conductivity is negligible [14]. The small Du condition allows us to apply the Maxwell–Wagner theory [26] for expressing the conductivity ratio through the volume fraction:

$$\frac{K_s}{K_m} = \frac{1 - \phi}{1 + 0.5\phi} \tag{45}$$

As a result we obtain the following equation for the asymptotic value of CVI at low frequency:

$$CVI_{\omega \to 0} = \frac{\varepsilon \varepsilon_0 \zeta (1 - \phi) \phi (\rho_p - \rho_s)}{\eta (1 + 0.5\phi)} \frac{(\rho_p - \rho_s)}{\rho_s} \nabla P \tag{46}$$

This is a very important result because it provides a test for electroacoustic theory. Comparing Eqs (37) and (46), it becomes clear that the DSOG theory satisfies this test because the ratio of the special functions I/H goes to 0 at low frequency (see Appendix).

If O'Brien's theory is used, the application of relation (1) to quasistationary CVI requires the low-frequency asymptotic value of electrophoretic mobility μ_d. Substitution into Eq. (1) of the Smoluchowski law for stationary electrophoresis [10, 20] in the form of Eq. (44) (valid in concentrated systems) leads to the following expression for CVI:

$$\text{CVI}_{\omega \to 0} = \frac{\varepsilon \varepsilon_0 \zeta (1 - \phi) \phi}{\eta (1 + 0.5\phi)} \frac{(\rho_p - \rho_m)}{\rho_m} \nabla P \tag{47}$$

The discrepancy between Eqs (47) and (46) comes from the different density contrasts. O'Brien's theory operates with particle–media density contrast, whereas DSOG theory replaces it with particle–dispersion density contrast. The following qualitative analysis helps us to understand the origin of this discrepancy between the two theories.

H. Qualitative Analysis

We think that it is helpful to get some heuristic understanding of the physical phenomena that take place when an ultrasound pulse passes through a dispersed system. This description provides answers to some general questions. For instance, why do we need density contrast in the case of ESA when the particles already move relative to the liquid under the influence of an electric field? Why do we need a density contrast to generate CVI at low frequency when the particles already move in phase with the liquid?

As far as we know, there are no simple published answers to these questions. In order to find these answers we utilize the analogy between sedimentation potential and electroacoustic phenomena. Marlow et al. have used this analogy before [5] and we will give further justification for this approach.

Let us consider a volume element of the concentrated dispersed system in the presence of the sound wave (Fig. 3). The size of this element is selected such that it is much larger than the particle size and also larger than the average distance between particles. As a result, the element contains many particles and at the same time is much smaller than the wavelength.

This dispersion element moves with a certain velocity and acceleration in response to the gradient of acoustic pressure. As a result, an inertia force is applied to the element. At this point we can use the principle of equivalence between inertia and gravity. The effect of the inertia force created by the sound wave is equivalent to the effect of the gravity force. Gravity acts on both the particle and liquid inside the dispersion element. The densities of the particles and liquid are different and the forces are different as well. Force acting on the particles depends on the ratio of the densities.

The question arises to what density should we take into account. To answer this question let us consider the forces acting on a given particle in the gravity field. The first force is the weight of the particle, which is proportional to its density ρ_p. This force will be partially balanced by the pressure of the surrounding liquid and other particles. The pressure is equivalent to that generated by an effective medium with density equal to the density of the dispersed system. This becomes more clear when one considers a larger particle surrounded by smaller ones as shown in Fig. 3.

We are coming to the well-known conclusion that this force is proportional to the density difference between particle ρ_p and dispersed system ρ_s. From one point of view (representing the medium and surrounding particle as a continuous medium) such a conclusion may come directly from Archimedes law, as the liquid being expelled by our

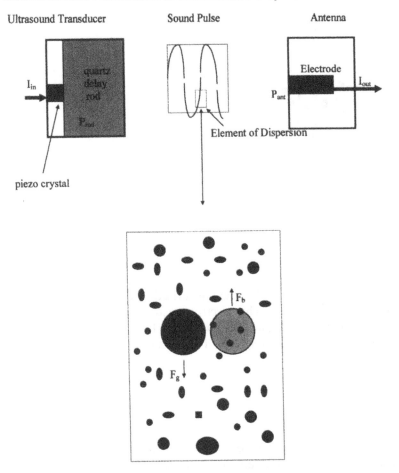

FIG. 3 Scheme illustrating the design of the electroacoustic sensor, and definition of the elements of the dispersion.

particle is the suspension, not only dispersion medium, and so the buoyant force is proportional to the difference between particle, ρ_p, and suspension, ρ_m, densities.

From another point of view (representing separately dispersion medium and particles surrounding our isolated particle), this conclusion follows from the following considerations, illustrated in Fig. 4. This figure shows sedimentation of a small spherical cloud of particles in the liquid. Case 1 corresponds to the situation when the cloud settles as one entity, and liquid envelops this settling cloud from outside. Case 2 corresponds to the different situation when liquid is forced to move through the cloud. There will be a difference between forces exerted on the particles within the cloud. There is an additional force in case 2, caused by liquid being pushed through the array of particles.

Electroacoustic phenomena correspond to case 2. It happens because the width of the sound pulse W is much larger than the wavelength λ at high ultrasound frequencies:

$$W \gg \lambda \tag{48}$$

The balance of forces exerted on the particles in a given element of the dispersion consists of the effective gravity force, the buoyancy force, and the friction force related to

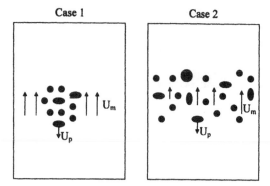

FIG. 4 Two possible scenarios for particle sedimentation (for explanation, see text).

the filtration of the liquid through the array of particles. As a result, particles move relative to the liquid with a speed $u_p - u_m$, which is proportional to the density difference between particle and system $\rho_p - \rho_s$.

The motion of the particles relative to the liquid disturbs their double layers, and as a result generates an electroacoustic signal. This electroacoustic signal is zero when the speed of the particle equals the speed of the medium, which happens when the density of the particle equals that of the dispersed system. This means that the electroacoustic signal must be proportional to $(\rho_p - \rho_s)$.

III. EXPERIMENTAL

The main goal of the experiments presented below is to test the validity of the DSOG theory in concentrated systems. Equilibrium dilution is the logical experimental protocol for achieving this goal because it provides a simple criterion of the theory. Equilibrium dilution maintains the same chemical composition of the dispersion medium for all volume fractions. As a result, parameters which are sensitive to the chemistry must be the same for all volume fractions. It means that the ζ potential calculated from CVI is supposed to remain the same for all volume fractions, as long as double layers do not overlap. Variation in ζ potential with volume fraction is the indication that the theory does not reflect volume fraction dependence properly.

We perform the dilution test with two different dispersions: silica Ludox and rutile R-746 produced by Dupont. We use two different techniques for producing the equilibrium dispersion medium for dilution: dialysis and centrifugation. Electroacoustic experiments were performed with the acoustic and electroacoustic spectrometer DT-1200 [29, 30]. The next section describes a method of CVI measurement employed by this instrument.

A. CVI Measurement

The electroacoustic spectrometer consists of two parts: electronic part and sensor part. All electronics are placed on two special-purpose boards (signal processor and interface). It requires also a conventional data acquisition card (DAC). The signal processor board and DAC are placed inside a personal computer which operates by interfacing with the user, using Windows 95 based software.

The electroacoustic sensor has two parts: a piezoelectric transducer with a critical frequency of 10 MHz and an electroacoustic antenna (Fig. 4). There is another design where the sensing electrode is placed on the surface of the transducers. We call this design "electroacoustic probe".

The antenna is designed as two coaxial electrodes separated by a nonconducting rigid ceramic insert. Internal electrical impedance between these electrodes can be selected, depending on the conductivity range of the samples by means of an internal transformer. The transformer is selected such that the input impedance is significantly less than the external impedance of the sample, so that the resulting signal is proportional to the short-circuit current. This transformer is located just behind the central electrode in order to minimize the stray capacitance.

The transmitting transducer and the receiving antenna are mounted in the opposite walls of the sample chamber such that the gap between the faces is 5 mm.

The signal processor generates the transmit gate which defines the 1-W pulse generated in the interface module as well as the necessary signals to set the frequency. Electroacoustic measurements can be performed either for one frequency or for the chosen set of frequencies from 1 to 100 MHz. The transducer converts these pulses into sound pulses with some certain efficiency. The sound pulse propagates through the quartz delay rod and eventually through the sample. An acoustic pulse propagating through the sample excites particles, disturbing their double layers. Particles gain dipole moments because of this excitation. These dipole moments generate an electric field that changes the electric potential of the central electrode of the electroacoustic antenna, and the difference between the electric potentials of the central and external reference electrodes gives rise to an electric current, registered as a CVI.

The value of this current is very low. The system averages at least 800 pulses in order to achieve a high signal-to-noise ratio, although the number of pulses depends on the properties of colloid. Thus, measurement of CVI in low-conducting oil-based systems requires averaging millions of pulses. In principle, this method makes it possible to measure any low-energy signals.

We suggest interpreting this measurement as propagation of the pulse through the transmission line with certain energy losses at different points. This approach allows us to eliminate measurement of the absolute powers. We simply compare pulse intensity before and after transmission and take into account all internal energy losses. This idea is accomplished as described below.

At the beginning of each measurement, the interface routes the pulses to a reference attenuator channel consisting of a fixed 40-dB attenuator, and similarly routes the output of this precision attenuator to the input section of the signal processor. Since the precision attenuator has a known response over the entire frequency range, this step allows us to characterize all energy losses in the measuring circuits at each frequency. The next step in the measurement is to determine the losses in the electroacoustic sensor. The signal processor now commands the interface to substitute the electroacoustic sensor for the reference attenuator. The 1-W pulses are now sent to the transmitting transducer which converts these electric pulses into sound pulses. We have certain energy losses at this point. These losses depend on the transducer efficiency and are pretty much constant.

The sound pulses propagate through the quartz delay rod (see Fig. 4) and eventually reach the surface of the transducer which faces the dispersion. It loses some energy at this point because of the reflection caused by mismatch of the acoustic impedances of the delay rod (Z_{tr}) and dispersed system (Z_s).

Some part of the pulse passes into the gap between the transducers, which is filled with the dispersion under test, and propagates through it. It loses energy during propagation due to the attenuation. Finally, this sound pulse reaches the electroacoustic antenna which converts it back into an electric signal. This conversion also involves energy losses.

The final electric pulse is routed through the interface to the input signal port on the signal processor where the signal level of the acoustic sensor output is measured. Comparison of the amplitude and phase of the electroacoustic sensor output pulse with that of the reference channel output pulse allows the program to calculate precisely the overall loss in the sensor at each frequency.

The experimental output of the electroacoustic sensor S_{exp} is the ratio of the intensity of the input electric pulse to the transducer, I_m, to the output electric pulse in the antenna I_{out} (Fig. 4):

$$S_{exp} = \frac{I_{out}}{I_{in}} \tag{49}$$

The intensity of the input electric pulse is related to the intensity of the sound pulse in the delay rod through some constant C_{tr}, which is a measure of the transducer efficiency and energy losses at this point:

$$I_{rod} = C_{tr} I_{in} \tag{50}$$

The intensity of the sound in the delay rod is proportional to the square of the sound pressure here, P_{rod}:

$$P_{rod} = \sqrt{2 \rho_{rod} c_{rod} C_{tr} I_{in}} \tag{51}$$

where ρ_{rod} and c are the density and sound speed in the rod material.

At the other end, we can use the definition of the electric pulse intensity as the square ot the electric current in the antenna, which is the CVI:

$$I_{out} = (\text{CVI})^2 C_{ant} \tag{52}$$

where the constant C_{ant} depends on geometrical factors of the CVI space distribution in the vicinity of the antenna, and on the electric properties of the antenna only for the proper ratio of the electric impedances of the antenna and dispersed system.

Substituting Eq. (51) into the Eq. (49) and taking into account Eq. (52), we obtain the following expression relating CVI with the measured parameter S_{exp}:

$$\frac{\text{CVI}}{P_{rod}} = \sqrt{S \exp / C_{tr} C_{ant} 2 \rho_{rod} c_{rod}} \tag{53}$$

The value of CVI depends on the pressure near the antenna surface P_{ant}. This pressure is lower than the pressure in the rod, P_{rod}, because of the reflection losses on the rod surface and attenuation of the pulse in the dispersion. There are two ways to take these effects into account. We can either measure corresponding losses using reflected pulses or we can calculate these losses. If we choose the second way we should use the following corrections:

$$P_{ant} = P_{rod} \frac{2 Z_s}{Z_s + Z_{rod}} \exp\left(-\frac{\alpha L}{2}\right) \tag{54}$$

where α is attenuation of the sound intensity expressed in neper/cm, and L is the distance between the transducer and antenna in centimeters. These corrections lead to the following expression for CVI:

$$\frac{\text{CVI}}{P_{\text{ant}}} = \sqrt{S_{\text{exp}}/2C_{\text{ant}}C_{\text{tr}}\rho_{\text{rod}}c_{\text{rod}}} \frac{Z_s + Z}{2Z_s} \exp\left(\frac{\alpha L}{2}\right) \tag{55}$$

The pressure gradient ∇P in Eqs (37) and (46) for CVI equals P_{ant}. Using this fact, we obtain the following equation, relating properties of the dispersion with the measured parameter S_{exp}:

$$\frac{3\varepsilon\varepsilon_0\zeta(1-\phi)\phi(\rho_p - \rho_s)}{2\eta(1 + 0.5\phi)}\frac{}{\rho_s}G(a,\phi) = \frac{cC_{\text{cal}}}{f}\left(1 - i\frac{\alpha c}{2\omega}\right)\sqrt{S_{\text{exp}}}\frac{Z_s + Z_{\text{rod}}}{2Z_s}\exp\left(\frac{\alpha L}{2}\right) \tag{56}$$

where c is the speed of sound in the dispersion, f is frequency in Hz, and the function G is defined in Eq. (47).

Equation (56) contains an unknown calibration constant C_{cal}, which is independent of the properties of the dispersion. This constant can be calibrated out using calibration with a known colloid. We use for this purpose silica Ludox at 10 wt% diluted with KCl 10^{-2} mol/L. These silica particles have a ζ potential of -38 mV at pH 9.3.

Equation (56) can be used for calculating either ζ potential only, in the case of a single frequency measurement, or both ζ potential and particle size in the case of multiple frequencies.

DSOG theory yields a new range of frequencies. This theory predicts that the critical frequency becomes higher with increasing volume fraction. Computer calculations show that this shift is about one order of magnitude for a volume fraction of 40%. It means that the optimum frequency range according to the new theory is

$$v/a^2 < \omega < 40v/a^2 \tag{57}$$

if we want to cover volume fractions up to 40%.

B. Materials and Experimental Protocol

We used silica Ludox and rutile R-746 from Dupont for this experiment. Selection of the silica Ludox is related to the small size of these particles. It allows us to eliminate any particle size dependence in Eq. (56) because $G(a,\phi) = 1$ for small particles. Using small particles gives one a more simplifying advantage: it eliminates the contribution of attenuation because small particles do not attenuate sound at low frequency. It means that the choice of small particles allows us to test volume fraction dependence only. This is important because this dependence is the most pronounced difference between different theories.

Silica Ludox TM satisfies all specified conditions because its nominal particle size on an area basis reported by DuPont is about 22 nm. We measured the size using acoustics, and obtained the particle size distribution (PSD) on a weight basis. It is quite close to the nominal value, as will be shown below, taking into account the difference in PSD basis. At the same time, the particle size should not be too small for the given ionic strength in order to satisfy the thin-double-layer restriction [Eq. (9)]. Silica Ludox meets this requirement because of the relatively high ionic strength of about 0.1 mol/L. Otherwise we would have to generalize the theory removing the thin-double-layer restriction according Babchin and coworkers [24, 25].

The selection of rutile as the second dispersion gives us an opportunity to test particle size dependence and enhance the density contrast contribution. We used rutile R-746 produced by DuPont. This product was a concentrated stable dispersion with a weight fraction of solids of 76.8%. We took 100 ml of this dispersion and weighed it. This weight was 234 g which, yields particles' material average density of 3.9 g/cm^3. This density was somewhat lower than that of the regular rutile, perhaps because of the stabilizing additives.

The equilibrium dilution protocol requires a pure solvent which is identical to the medium of the given dispersed system. In principle, one can try to separate the dispersed phase and dispersion medium using either sedimentation or centrifugation. This method does not work for silica Ludox because the particle size is too small.

The other way to create an equilibrium solution for small Ludox is dialysis. We used this method. Dialysis allows us to equilibrate the dispersion medium with the external solution over some period of time. We used regenerated cellulose tubular membrane Cell*Sept4 with pore size 12–14 kDa, the external solution being KCl (0.1 mol/L) adjusted to pH 9.5 using hydrochloric acid. Membrane filled with silica Ludox was placed in the KCl solution, which was continuously mixed with a magnetic stirrer. We prepared two samples in order to check reproducibility.

In addition, we prepared another setup using KCl solution of pH 3. This setup allowed us to estimate the equilibration time. The initial pH of the silica Ludox was about 9 at 23°C. We monitored the change of pH in the external solution, and the corresponding kinetic curve is shown in Fig. 5. It is seen that the pH becomes 8.6 after 3h of equilibration, this value being close to the final pH value of 8.7 after 12 days of equilibration. We waited 12 days because the equilibration time depends on the diffusion coefficient, which is highest for H$^+$ ions. The higher the diffusion coefficient, the lower the equilibration time.

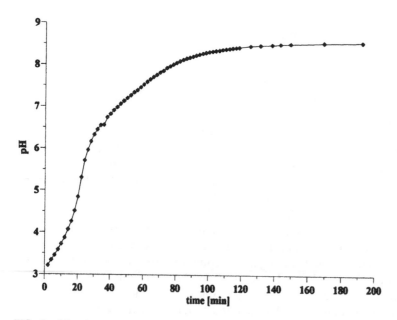

FIG. 5   Kinetic curve describing the variation of pH in the external dialysis solution versus time for silica Ludox.

Before starting dilution we checked again the weight fraction of the silica Ludox using a pyncometer. We were concerned about losing silica particles through the membrane pores into the solution. The weight fraction remained unchanged which means that the pores were too small for silica particles.

We had two sets of 50% silica with their corresponding equilibrium solutions. This allowed us to check two ways of dilution, as we used on set for diluting from the high weight fraction down. We did this by adding solution to the dispersed system. We used the opposite procedure with the other sample. We added dispersed system to the solution.

In the case of rutile, we used centrifugation of the initial 76.8 wt% dispersion in order to create equilibrium supernatant. We used this supernatant for preparing equilibrium 1.1 v/v% rutile dispersion, diluting the initial dispersion. After making measurements with this dilute system, we added more initial dispersion for preparing the next volume fraction: 3.2 v/v%. We proceeded this way, preparing more and more concentrated systems. All together 11 different volume fractions from 1.1 to 45.9 v/v% were tested (see Fig. 7).

For each volume fraction, we measured the attenuation spectra, sound speed, pH, conductivity, temperature and magnitude and phase of CVI. Attenuation spectra were measured within the frequency range 3–100 MHz, sound speed at 10 MHz, conductivity at 3 MHz, magnitude of CVI at 3 MHz, and phase of CVI at 1.5 MHz. Some of the results are discussed below.

IV. RESULTS AND DISCUSSION

The measured attenuation spectra are shown in Figs 6 and 7. It is seen that attenuation for silica Ludox is much lower than for rutile. This happens because of the smaller size and lower density contrast for silica. The attenuation spectra of silica become almost indistinguishable at volume fractions above 9%. This reflects a nonlinear dependence of the attenuation on the volume fraction. This nonlinearity appears because of the particle–particle interaction that shifts the critical frequency to higher values.

This peculiarity of the attenuation spectra was known before [19]. It is even more pronounced for rutile (Fig. 7). Attenuation at low frequency decreases with increasing volume fraction above 16.6 vol%. It is exactly the same effect which makes the attenuation constant for silica.

The described theory takes into account this nonlinear effect. As a result, the particle size calculated from this attenuation spectra is almost constant for all volume fractions for both silica and rutile (Fig. 8). The slight increase at high volume fraction can be caused by aggregation. It is important to mention here that dilute case theory would yield a size decreasing dramatically with volume fraction.

It is seen that our size is somewhat larger than nominal, due perhaps to the different technique applied by Dupont for characterizing the size of these particles. It is also clear that the nominal size corresponds to the dilute system whereas we measured size for the concentrated one.

It is seen (Fig. 6) that the attenuation for silica at 3 MHz is indeed negligible. It means that our expectations for eliminating this contribution to the CVI measurement using small particles were true. At the same time we have appreciable attenuation for rutile at 3 MHz. This gives us a chance to verify the way we correct CVI for sound attenuation [Eq. (56)].

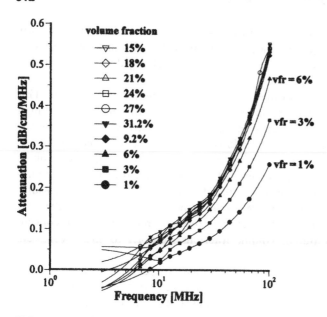

FIG. 6 Attenuation spectra measured for silica Ludox TM at different volume fractions.

The sound speed of the silica Ludox dispersion varies only within 2% for weight fractions changing from 1 to 50% (see Fig. 9), thus eliminating the contribution from the change in acoustic impedance to the measured CVI for silica as well.

Figures 10 and 11 illustrate the ζ potential values, calculated from the measured CVI using various theoretical models. One can see that the DSOG theory yields a ζ potential, which remains almost the same within the complete volume fraction range. Variations do

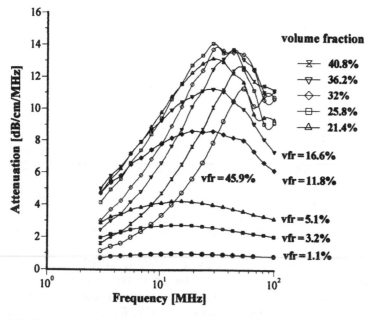

FIG. 7 Attenuation spectra measured for rutile R-746 at different volume fractions.

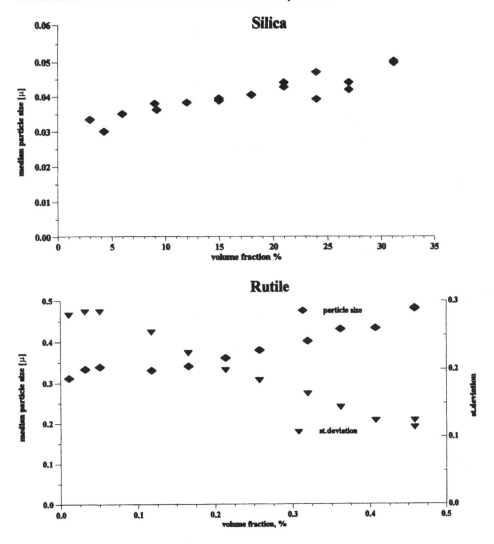

FIG. 8 Median particle size of silica and rutile calculated from the attenuation spectra of Figs. 6 and 7.

not exceed 10%, and its evident from the figures that this level of accuracy could not be achieved by any of the other theories.

V. APPLICATIONS

Electroacoustics can provide very fast single-point measurements of the ζ potential. For instance, typical measurement in aqueous systems using DT instruments [29, 30] takes about 15 s. These measurements are very precise, usually it is about 0.1–0.3 mV. Sample handling and preparation is very simple because there is no need to dilute the sample. However, the greatest advantage of electroacoustics is the possibility of performing fast

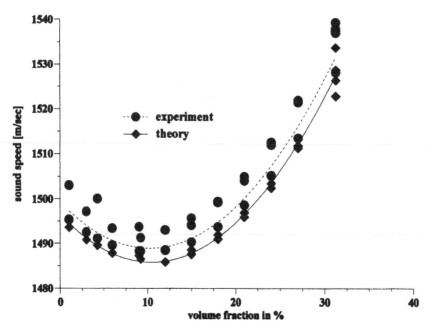

FIG. 9 Measured and calculated sound speed of the silica Ludox TM versus volume fraction.

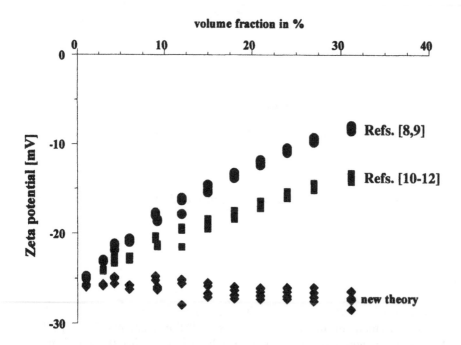

FIG. 10 Electrokinetic ζ potential calculated from the measured CVI at various volume fractions using different electroacoustic theories for silica Ludox.

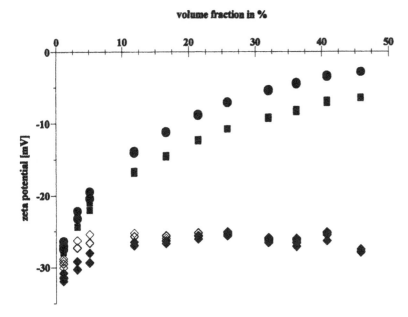

FIG. 11 Electrokinetic ζ potential calculated from the measured CVI at various volume fractions using different electroacoustic theories, for rutile R-746 from Dupont. (●) From Refs. 8 and 9; (■) from Refs. 10–12 (◇) new theory, monodisperse; (◆) new theory, polydisperse.

and precise automatic titrations. We show here several different titrations performed with DT instruments.

Figure 12 shows results of the typical pH titration performed with colloidal silica Ludox diluted down to 10% wt using KCl (0.01 mol/L). One can see that electroacoustics allows us to measure very low ζ potentials, even below 1 mV, with very high precision. It is thus very useful for determining isoelectric points.

Control of pH alone is not sufficient in many cases. For instance, preparation of a table precipitated calcium carbonate (PCC) slurry at 3 v/v% requires a special arrangement, since the ζ potential right after dispersing the solids in distilled water is very low: 1.3 mV. One can use sodium hexametaphosphate in order to increase the surface charge and improve the aggregative stability of this slurry. Electroacoustics allows us to determine the optimum dose of hexametaphosphate. The results of the corresponding surfactant titration are shown in Fig. 13. It is seen that the ζ potential reaches saturation at a hexametaphosphate concentration of about 0.5% by weight relative to the weight of the PCC solid phase. This illustrates the electroacoustic capabilities of equilibrium titrations.

Electroacoustics might be useful for kinetic experiments as well. One of the most interesting observations we have ever made was done with zirconia slurry. Zirconia is known as a complex material for electrokinetic characterization. We used to have problems running equilibrium titrations using electroacoustics. It turned out that these problems are related to the very long equilibration time of zirconia surfaces. Figure 14 shows the results of continuous ζ-potential measurements for zirconia slurry. This slurry was prepared at 3 vol% by adding the powder to a 0.01 mol/L KCl solution, adjusted initially to pH 4 in order to provide a significant ζ potential. It is

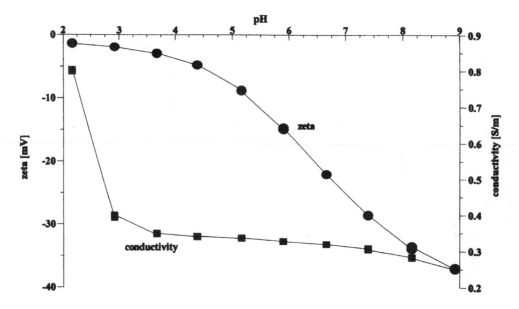

FIG. 12   pH titration of Ludox diluted with 0.01 M down to 10 wt%.

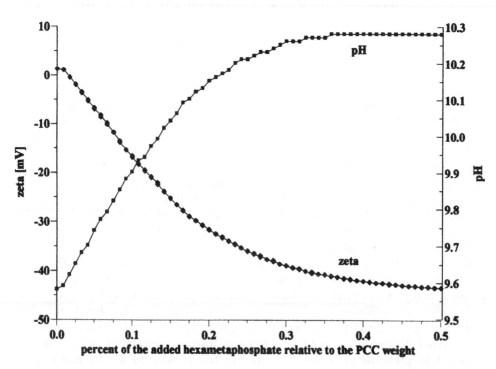

FIG. 13   Surfactant titration of PCC slurry with 0.1 g/g hexametaphosphate solution.

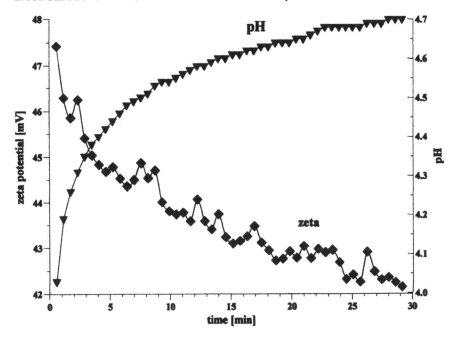

FIG. 14 Equilibration of 3 vol% zirconia slurry prepared in 0.01 M KCl, with pH adjusted initially to 4. It is seen that equilibration takes about 2 h.

seen that zirconia required about 2 h for the ζ potential and pH to equilibrate. It is not surprising now that equilibrium titrations with this material are not successful.

These various examples illustrate the potential use of electroacoustic techniques for the characterization of electrical surface properties.

VI. CONCLUSIONS

We have presented the principles of a new electroacoustic theory (DSOG), which was derived without using any relationship between electroacoustic signal and dynamic electrophoretic mobility. This theory is based on the coupled-phase model and the cell model concept. We managed to extend this theory to polydisperse systems without using a superposition assumption for the hydrodynamic part of the problem. The DSOG theory is considered to be valid for polydisperse concentrated dispersions with a thin double layer and any surface conductivity.

We have shown the results of the equilibrium dilution experimental test. The test with silica Ludox TM confirmed that DSOG theory gives correct volume fraction dependence within the whole available range of the volume fraction up to 30%.

Equilibrium dilution tests with a stable rutile dispersion proved that DSOG theory gives the correct particle size dependence within the volume fraction range from 1.1% up to 45.9%, as well as volume fraction dependence. We have shown that this new theory yields almost constant ζ potential ($\pm 10\%$ variation) within the whole volume fraction range. Polydispersity of the rutile sample was not a significant factor, at least compared with volume fraction and particle size.

We would like to stress that DSOG electroacoustic theory has been created so far for CVI only. It is not clear yet how to apply it to ESA effects. Required modifications will depend on the design of the instrument including the ratio of masses of the chamber and the sample. This ratio determines an appropriate frame of reference. At the same time, the basic physical framework should work for ESA effects as well as for CVI.

We have shown several applications of the electroacoustic technique for characterizing the electrical surface properties of various slurries. These examples illustrate the importance of electroacoustics for colloid science.

VII. APPENDIX SPECIAL FUNCTIONS

There are several special functions used in this theory. They are specified in this section.

$$H(\alpha) = \frac{ih(\alpha)}{2\alpha} - \frac{idh(x)}{2dx}\bigg|_{x=\alpha}$$

$$h(x) = h_z(x)h_2(\beta) - h_1(\beta)h_2(x)$$

$$I = I(\beta) - I(\alpha)$$

$$I(x) = I_1(x) - I_2(x)$$

$$I_1(x) = -h_1(\beta)\exp[x(1+i)]\left[\frac{3(1+x)}{2\beta^3} + i\left(\frac{x^2}{\beta^3} - \frac{3x}{2\beta^3} - \frac{1}{x}\right)\right]$$

$$I_2(x) = -h_2(\beta)\exp[-x(1+i)]\left[\frac{3(1+x)}{2\beta^3} + i\left(\frac{x^2}{\beta^3} + \frac{3x}{2\beta^3} - \frac{1}{x}\right)\right]$$

$$h_1(x) = \frac{\exp(-x)}{x}\left[\frac{x+1}{x}\sin x - \cos x + i\left(\frac{x+1}{x}\cos x + \sin x\right)\right]$$

$$h_2(x) = \frac{\exp(x)}{x}\left[\frac{x-1}{x}\sin x + \cos x + i\left(\frac{1-x}{x}\cos x + \sin x\right)\right]$$

REFERENCES

1. S Takeda, N Tobori, H Sugawara, K. Furusawa. In: H Ohshima, K Furusawa, eds. Electrical Phenomena at Interfaces. Fundamentals, Measurements, and Applications. New York: Marcel Dekker, 1998, ch. 13.
2. RJ Hunter. Colloids Surfaces 141: 37–65, 1998.
3. F Booth, J Enderby. Proc Am Phys Soc 208 A: 32, 1952.
4. JA Enderby. Proc Roy Soc, London A207: 329–342, 1951.
5. BJ Marlow, D Fairhurst, HP Pendse. Langmuir 4: 611–626, 1983.
6. S Levine, GH Neale. J Colloid Interface Sci 47: 520–532, 1974.
7. RW O'Brien. J Fluid Mech 190: 71–86, 1988.
8. RW O'Brien. Determination of particle size and electric charge. US Patent 5059909, 1991.
9. H Ohshima. J Colloid Interface Sci 195: 137–148, 1997.
10. AS Dukhin, VN Shilov, Yu Borkovskaya. Langmuir 15: 3452–3457, 1999.
11. AS Dukhin, H Ohshima, VN Shilov, and PJ Goetz. Langmuir 15 (20): 6692–6702, 1999.
12. H Ohshima, A Dukhin. J Colloid Interface Sci 212: 449–452, 1999.
13. VN Shilov, NI Zharkikh, YuB Borkovskaya. Colloid J 43: 434–438, 1981.
14. S Kuwabara. J Phys Soc Jpn 14: 527–532, 1959.

15. AS Dukhin, VN Shilov, H Ohshima, PJ Goetz. Langmuir 15: 3445–3451, 1999.
16. J Lyklema. J Fundamentals of Interface and Colloid Science. vol. 1. Academic Press, 1993.
17. AH Harker, JAG Temple. J Phys D: Appl Phys 21: 1576–1588, 1988.
18. RL Gibson, MN Toksoz. J Acoust Soc Am 85: 1925–1934, 1989.
19. AS Dukhin, PJ Goetz. Langmuir 12: 4987–4997, 1996.
20. HR Kruyt. HR Colloid Science. vol. 1. Irreversible Systems. Elsevier; 1952.
21. H Ohshima. J Colloid Interface Sci 208: 295–301, 1998.
22. AS Ahuja. J Appl Phys 44: 4863–4868, 1973.
23. SS Dukhin, BV Derjaguin. In: E Matijevíc, ed. Surface and Colloid Science. vol. 7. New York: John Wiley, 1974.
24. AJ Babchin, RS Chow, RP Sawatzky. Adv Colloid Interface Sci 30: 111, 1989.
25. RP Sawatzky, AJ Babchin. J Fluid Mech 246: 321–334, 1993.
26. SS Dukhin and VN Shilov. "Dielectric Phenomena and the Double Layer in Disperse Systems and Polyelectrolytes", John Wiley & Sons, NY, 1974.
27. AS Dukhin, VN Shilov, H Ohshima, PJ Goetz. Langmuir, in press.
28. I Prigogine. Introduction to Thermodynamics of Irreversible Processes. New York: Wiley-Interscience, 1967.
29. Dispersion Technology Inc., Website: www.dispersion.com.
30. AS Dukhin, PJ Goetz. Method and device for characterizing particle size distribution and zeta potential in a concentrated system by means of acoustic and electroacoustic spectroscopy. US Patent, 09,108,072.

18

Electrokinetics in Porous Media

PIERRE M. ADLER and S. BÉKRI Institut de Physique du Globe de Paris, Paris, France

JEAN-FRANÇOIS THOVERT Laboratoire de Combustion et Détonique, Poitiers, France

I. INTRODUCTION

Electro-osmotic effects are important in many situations of practical interest which are characterized by very different length scales. These phenomena are generated by the flow of electrolytes through or around charged solid surfaces. Polyelectrolytes and colloidal particles on the one hand, and porous media on the other, offer examples of such solid media. In the first case, the basic length scale is submicrometer while in the second manifestations of the local phenomena which occur at the pore scale can be found at much larger length scales.

For the sake of brevity, this review is good to be focused on porous media. An interesting and unusual application of these developments is the analysis of the anomalous electric and magnetic fields which are observed before earthquakes and volcanic activity; it has been proposed that these fields could be generated by the electrokinetic effect induced by water flow resulting from internal stresses, thermal buoyancy effects, and meteoritic waters.

A major difficulty is the determination of the relevant coefficients to be used in such macroscopic descriptions which necessitate at least two successive changes of scale (cf. Fig. 1). The first one goes from the plug scale l_1 to the core scale l_2; starting from the local equations in the pore [1], macroscopic coefficients at the plug scale could be derived. The same upward scaling was done for a single fracture [2].

Another increase in scale is necessary since all real porous media are heterogeneous and since their local properties such as porosity depend upon space. It necessitates a methodology to perform this second change of scale from the plug scale l_2 to the field scale l_3 and it includes semiempirical laws derived from the first change of scale (cf. [3]).

The first change of scale from l_1 to l_2 and its physical background can be summarized as follows. Consider an electrolyte flowing through a fracture with electrically charged walls. Far from the surfaces, the solute may be considered neutral. However, near the surfaces, the ion distribution within the fluid is disturbed. For example, a negatively charged face attracts positive ions from the solution and repels negative ones. Thus, in the region adjacent to the fracture surfaces, the electrolyte is charged. This zone is called the Debye–Hückel sheath layer. Its thickness κ^{-1} may vary from several Angströms up to a few tens of nanometers.

Because of this electrical perturbation within the solution, if the medium is embodied in an external electric field E, the ions are set in motion; this creates an electric current I

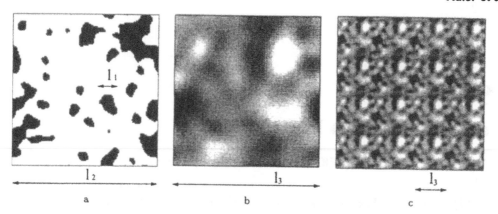

FIG. 1 Illustration of the length scales involved in the problem: (a) displays pores (in black) of a scale l_1 and the plug scale l_2 where porosity can be defined; (b) displays the porosity variations represented by gray levels at the field scale l_3. When the porosity variations are statistically homogeneous, the random porous medium can be replaced by a spatially periodic medium (c) where computations are performed.

passing through the material. In addition, a nonzero volumetric body force acts on the electrolyte, in the double layer, due to its net electric charge. Therefore, the fluid flows in the Debye layer and, as a result of viscous friction, in all the bulk although the solute experiences no body force in this region. This implies an interstitial flow with a certain seepage velocity U, in the absence of any macroscopic pressure gradient applied to the medium. Such a process is called electro-osmosis. Inversely, in the presence of a macroscopic pressure gradient ∇P, the fluid percolates through the fracture with a Darcy seepage velocity U. Additionally, the electrolyte motion within the double layer affects the equilibrium ion distribution and entails macroscopic electric current density I, occurring in the absence of any external electric field. These coupling phenomena are the basis of our study.

When the electric field and the pressure gradients are not too large, the fluxes I and U are linearly related to them by the following relations:

$$I = \sigma \cdot E - \alpha \cdot \nabla P \tag{1a}$$

$$U = \beta \cdot E - \frac{K}{\mu} \cdot \nabla P \tag{1b}$$

where σ and K are the conductivity and permeability tensors, respectively, and α and β are the electro-osmotic coupling tensors (with $\alpha = \beta^t$ by virtue of Onsager's irreversible thermodynamic theorem [4]). This situation has been analysed for porous media [1], for aggregates [5], and for fractures [2]. Note that it has been reviewed in Ref. 6.

Conservation of mass and current requires that U and I satisfy the continuity equations:

$$\nabla \cdot I = 0, \nabla \cdot U = 0 \tag{2}$$

The second change of scale goes from l_2 to l_3 (cf. [3]); the overbar denotes quantities defined at the field scale l_3. The tensors K, σ, α, and β are random or deterministic functions of space; I and U are given by Eq. (1) and they are solutions of Eq. (2). One has to determine the average seepage velocity $\overline{U}$ and the average current density $\overline{I}$ when the medium is submitted to an average pressure gradient $\overline{\nabla}P$ and an average electric potential gradient $\overline{\nabla}\psi$ at the scale l_3. Because of the linearity of the previous set of equations, the fluxes $\overline{U}$ and $\overline{I}$ are expected to be linearly related to the gradients $\overline{\nabla}P$ and $\nabla\psi$ [cf. Eq. (1)]. Specifically,

$$\overline{U} = -\frac{\overline{K}}{\mu} \cdot \overline{\nabla}P - \overline{\beta} \cdot \overline{\nabla}\psi \tag{3a}$$

$$\overline{I} = -\overline{\alpha} \cdot \overline{\nabla}P - \overline{\sigma} \cdot \overline{\nabla}\psi \tag{3b}$$

where the second-order tensors $\overline{K}$, $\overline{\sigma}$, and $\overline{\alpha}$ represent the macroscopic permeability, conductivity, and electro-osmotic coupling tensors, respectively. Again it can be proved that $\overline{\beta} = \overline{\alpha}^t$, where t denotes the transposition operator.

We tried to keep notations as compact as possible. On the microscopic scale, all fields are denoted by lower case letters such as the velocity u and the pressure p. On the plug scale l_2, they are denoted by upper case letters such as the seepage velocity U, the pressure P, and the electric current I. On the field scale l_3, they are denoted by upper case letters with overbars, such as $\overline{U}$ and $\overline{I}$.

This review is organized as follows. The general microscopic electrokinetic equations which govern the flow of an electrolyte are presented in Section II as well as the relevant boundary conditions. Section III is devoted to the presentation of the change of scale from l_1 to l_2. The macroscopic cofficients which appear in Eq. (1) are derived in a general way. The porous media and the fractures for which these coefficients have been systematically evaluated are then described. Finally, two semiempirical laws are derived for the electro-osmotic coupling coefficients in porous media and in fractures.

Section IV presents the derivation of the coefficients on the field scale l_3 when the coefficients at the plug scale l_2 are known. Various theoretical considerations are given, including a theoretical expansion for lognormal media. Results of numerical studies for stratified and fractured media are summarized.

This review is ended by some preliminary results summarized in Section V which show that the electrokinetic effect provides correct orders of magnitude for the electromagnetic measurements performed at La Fournaise volcano.

II. ELECTROKINETIC EQUATIONS AT THE MICROSCOPIC SCALE

All the relevant microscopic equations which govern electro-osmotic phenomena are given in this section, as well as a general description of the spatially periodic porous media.

A. Flux of the Ionic Species

Consider an N-component Newtonian electrolyte of density ρ_f, dynamic viscosity $\mu =$ constant, and dielectric constant ϵ_{el}, flowing with velocity $u(R, t)$ in interstices of a porous material. Let $\psi(R, t)$ be the electric potential prevailing within the solute. The flux j_i of each ith ion species, composing the solute, is given by the following constitutive equation [7]:

$$j_i = -D_i\nabla_R n_i - ez_ib_in_i\nabla_R\psi + n_iu, i = 1, 2, \ldots, N \tag{4}$$

where $\nabla_R = \partial/\partial R$, n_i is the ion's concentration in molecules per volume, ψ is the electric potential; D_i and b_i are the ion's diffusivity and electric mobility, related by the Stokes–Einstein equation:

$$D_i = b_i kT \tag{5}$$

Furthermore, z_i is the ion's algebraic valency ($> $ or < 0) and e is the electron charge ($e > 0$); j_i and the concentration n_i obey the continuity (species conservation) equation:

$$\frac{\partial n_i}{\partial t} + \nabla_R \cdot j_i = 0 \tag{6}$$

B. Flow Velocity Field

We will be concerned with dense porous materials, wherein the percolation flow velocity is normally small and the characteristic Reynolds number is much less than unity. In these circumstances the (generally nonsteady) flow velocity field is governed by the Stokes equations:

$$\mu \nabla_R^2 u - \rho_f \frac{\partial u}{\partial t} = \nabla_R p + f \ , \quad \nabla \cdot u = 0 \tag{7}$$

where ρ_f and μ are the density and the constant dynamic viscosity of the electrolyte, respectively, p is the pressure, and f is the electric volumetric force density:

$$f = -\rho \nabla_R \psi = - \sum e n_i z_i \nabla_R \psi \tag{8}$$

with

$$\rho = e \sum_{i=1}^{N} n_i z_i \tag{9}$$

being the electric charge density.

C. Electric Potential

The electric potential is given by the Poisson equation:

$$\nabla_R^2 \psi = -\frac{\rho}{\epsilon} = -\frac{e}{\epsilon_{el}} \sum_{i=1}^{N} n_i z_i \tag{10}$$

It is assumed that for any nonsteady process of ion transport, transient phenomena associated with the electric potential occur so fast that one can use Eq. (4) in a quasisteady approximation, i.e., assuming that $\psi = \psi(R, t)$ satisfies Eq. (10) with time t being a parameter. This means that the electromagnetic wave propagation characteristic time L/c is much smaller than the characteristic time L^2/D_i, associated with the diffusive species' transport, L and c being the characteristic interstitial scale and velocity of light, respectively.

D. Boundary Conditions

Equations (6–10) have to be solved subject to the following boundary conditions on the liquid–solid interface S:

$$\mathbf{v} \cdot \mathbf{j}_i = 0 \tag{11a}$$

$$\mathbf{u} = 0 \tag{11b}$$

$$\psi = \zeta \text{ or } \mathbf{v} \cdot \nabla_R \psi = -\frac{\sigma_s}{\epsilon_{el}} \tag{11c}$$

where $\mathbf{v}$ is the outer normal to S, σ_s its surface charge density, and ζ its zeta potential. The two boundary conditions, namely, the Dirichlet or Neuman condition on the electric field, are met in the literature (see [7], [18]).

III. MACROSCOPIC PROPERTIES AT THE PLUG SCALE

A. General Analysis

The previous system of equations and boundary conditions is by no means easy to solve in general. For porous media and fractures schematized as spatially periodic media, the system at equilibrium is analyzed first. The general system is then linearized close to equilibrium and made dimensionless.

1. Equilibrium Ion Density Distribution

Suppose that neither an external electric field nor a pressure (or concentration) gradient is imposed on the porous medium. One can then rewrite Eq. (6) in the following form:

$$\nabla_r \cdot \mathbf{j}_i^o = 0 \quad i = 1, 2, \dots, N \tag{12}$$

with the obvious zero-flux solution which, with the help of Eqs (4–5), may be written as

$$\mathbf{j}_i^o = 0 = -D_i \nabla_R n_i^o - \frac{e z_i}{kT} D_i n_i^o \nabla_R \psi^o \quad i = 1, 2, \dots, N \tag{13}$$

In the above, the superscript 0, appearing on n_i and ψ, refers to equilibrium conditions. From Eq. (13), one obtains the familiar Boltzmann distribution for n_i^o:

$$n_i^o = n_i^\infty \left(-\frac{e z_i}{kT} \psi^o \right) \tag{14}$$

where n_i^∞ is a certain value of n_i^o. To obtain the equilibrium potential distribution $\psi^o(R)$, introduction of Eq. (14) into Eq. (10) yields the Poisson–Boltzmann equation:

$$\nabla_R^2 \psi^o = -\frac{e}{\epsilon_{el}} \sum_{i=1}^N n_i^\infty z_i \exp\left(-\frac{e z_i}{kT} \psi^o \right) \tag{15}$$

This equation is to be solved subject to the equilibrium boundary condition [cf. Eq. (11c)]:

$$\psi^o = \zeta \quad \text{or} \quad \mathbf{v} \cdot \nabla_R \psi^o = -\frac{\sigma_s}{\epsilon_{el}} \tag{16}$$

Usually, the surface potential ζ is taken to be constant on S.

With the help of Eqs (8) and (14), the flow velocity field equation (7) is reduced to

$$0 = \nabla_R p^o + \nabla_R \psi^o \sum_{i=1}^N n_i^\infty e z_i \exp\left(-\frac{e z_i}{kT} \psi^o \right) \tag{17}$$

2. Equilibrium Ion Density Distribution in Spatially Periodic Porous Media

In order to complete the problem formulation, one should specify the porous microstructure of the material. Several types of ordered and disordered microstructures will be considered here, which are all characterized by the spatial periodicity property [8]. This periodicity will hold for three dimensions for porous media and for two dimensions for fractures; for the sake of brevity, only the three-dimensional (3D) periodicity is presented here. The porous medium is assumed to consist of a multitude of identical unit cells τ, indefinitely reproduced within an infinite space (see Fig. 2). Therefore, the solution of Eqs (6)–(11) is to be sought in the 3D space $\mathcal{R}^3$ with

$$R = r + R_n = r + n_1 l_1 + n_2 l_2 + n_3 l_3 \tag{18}$$

where l_1, l_2, l_3 are the three basic vectors characterizing the unit cell of the porous medium and the trio of integers $n = (n_1, n_2, n_3)$ belongs to $\mathcal{Z}^3$ [8, 9]; r is the local position vector within the unit cell.

At equilibrium, all properties, appearing in Eqs (14)–(15) are spatially periodic functions, i.e., only depend upon the intracell local position vector r:

$$n_i^o(R) = n_i^o(r) \ (i = 1, 2, \ldots, N), p^o(R) = p^o(r) \quad \text{and} \quad \psi^o(R) = \psi^o(r) \tag{19}$$

and may be considered only within the unit cell τ, on the external faces of which the periodicity conditions should be imposed. These will be formulated in the form [9]:

$$[|\psi^o|]_j = 0, \quad [|\nabla_r \psi^o|]_j = 0 \tag{20}$$

where $[|\ldots|]_j$ denotes the difference between the values of a function at the opposite points $(r, r + l_j)$, lying at the corresponding unit cell boundaries.

A last condition should be added to this set of conditions, which is called the macroscopic electroneutrality condition. The sum of all the charges within the unit cell is equal to zero:

$$\sum_{i=1}^{N} \left(\int_\tau n_i^o z_i dv + \int_S \sigma_i^o dS \right) = 0 \tag{21}$$

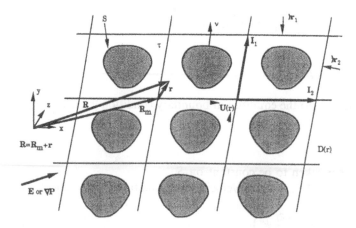

FIG. 2 Schematic view of a spatially periodic medium.

3. Dimensionless Linearized Transport Equations

(a) Small Perturbations. The resolution of the conservation equations (6), (7), and (10) presents some major difficulties. A simplification is possible if the perturbations of concentration, potential, pressure, and velocity fields, generated by the introduction of an electrical or pressure macroscopic gradient, are assumed to be small in comparison with their equilibrium values. This approximation enables us to replace these equations by linearized equations written in terms of the small perturbations $\delta n_i(R, t), \delta\psi(R, t),$ $\delta p(R, t)$, and $\delta u(R, t)$:

$$n_i = n_i^o(r) + \delta n_i(R, t) \tag{22a}$$

$$\psi = \psi^o(r) + \delta\psi(R, t) \text{ and } [|\delta\psi|]_j = -E \cdot l_j \tag{22b}$$

$$p = p^o(r) + \delta p(R, t) \text{ and } p[|\delta p|]_j = \nabla P \cdot l_j \tag{22c}$$

$$u = \delta u(r, t) \text{ and } [|\delta u|] = 0 \tag{22d}$$

where E and ∇P are the macroscopic electric field and the macroscopic pressure gradient on the plug scale l_2.

At this stage, it is convenient to introduce the ionic potentials ϕ_i, instead of the concentrations n_i, which one defines as

$$n_i = n_i^\infty\left[-\frac{ez_i}{kT}(\psi^o + \delta\psi + \phi_i)\right] = n_i^o \exp\left[-\frac{ez_i}{kT}(\delta\psi + \phi_i)\right] \tag{23a}$$

where n_i^o is given by

$$n_i^o = \exp\left[-\frac{ez_i}{kT}\psi^o\right] \tag{23b}$$

(b) Dimensionless Formulation. Finally, every dimensional quality X may be normalized by a scaling quantity X^*. The corresponding dimensionless variable X' is given by

$$X' = \frac{X}{X^*} \tag{24}$$

For instance, the position vector r is normalized by the Debye–Hückel length so that

$$r^* = \kappa^{-1} = \sqrt{\frac{\epsilon_{el}kT}{e^2n^*}}, \text{ with } n^* = \sum_{i=1}^{N} z_i^2 n_i^\infty \tag{25}$$

Table 1 provides the various scaling quantities and the dimensionless quantities. Note that, in order to linearize relation (23b), the normalized surface potential $\zeta' = \frac{e\zeta}{kT}$ is supposed to be small with respect to 1.

Using the above scaling relationships, one can write the following set of non-dimensionless equations:

$$\nabla' \cdot j_i = 0 \ , \ j_i = n_i^{0'} z_i D_i' \nabla' \phi_i' + n_i^{0'} + n_i^{0'} u' \tag{26a}$$

$$\nabla'^2 \psi^{0'} = -\sum_{i=1}^{N} n_i^{\infty'} z_i \exp(-z_i\psi^{0'}) \tag{26b}$$

$$\nabla'^2 \phi_i' = \nabla'\psi^{0'} \cdot \left(z_i\nabla'\phi_i' + \frac{1}{D_i'}u'\right) \tag{26c}$$

$$\nabla'^2 u' = \nabla'p' - \sum_{i=1}^{N} n_i^{\infty'} z_i \exp(-z_i\psi^{0'})\nabla'\phi_i' \ , \ \nabla' \cdot u' = 0 \tag{26d}$$

TABLE 1　Definitions of the Scaling Quantities

n^*	$\sum_{i=1}^{N} n_i^{\infty} z_i^2$
$\psi^* = \phi^* = \zeta^*$	kT/e
σ_s^*	$\epsilon_{el} \, \kappa \psi^* = \epsilon_{el} \, \kappa(kT/e)$
p^*	$n^* kT = \epsilon_{el} \, (kT/e)^2 \kappa^2$
u^*	$p^*/\kappa\mu = (\epsilon_{el} \, \mu)(kT/e)^2 \kappa$
D^*	$u^*/\kappa = (\epsilon_{el} \, /\mu)(kT/e)^2$
σ^*	$\epsilon_{el} \, kTn^*/\mu = (1/\mu)(\epsilon_{el} \, kT\kappa/e)^2$
$\alpha^* = \beta^*$	$\epsilon_{el} \, kT/\mu e$
K^*	κ^{-2}

The system (26) is to be solved subject to the boundary conditions formulated on the surface S of bed elements and to the periodicity conditions. Nondimensional forms of these conditions are

$$\text{on } S \begin{cases} \psi' & = \zeta' \text{ or } \boldsymbol{v}' \cdot \nabla' \psi_i' = -\sigma_s' \\ \boldsymbol{v}' \cdot \nabla' \phi_i' & = 0 \\ \boldsymbol{u}' & = 0 \end{cases} \tag{27a}$$

$$[|\boldsymbol{u}'| = 0 \quad , \quad [|\phi_i'|]_j = \boldsymbol{E}' \cdot \boldsymbol{l}_j' \quad , \quad [|\delta p'|]_j = \nabla' P' \cdot \boldsymbol{l}_j' \tag{27b}$$

It is now possible to solve the system of equations (26), subject to the conditions (27) as will be detailed in Section III.C.

B.　Macroscopic Coefficients at the Plug Scale; Circular Tube and Plane Channel

1.　Definitions

Consider first an external electric field of intensity $\boldsymbol{E}$ applied to a porous material. In its dimensionless form, the flux of ionic species is given by

$$\boldsymbol{j}_i' = n_i^0 z_i D_i' \nabla \phi_i' + n_i^{0'} \boldsymbol{u}' \tag{28}$$

The electric current density flowing through the porous medium, resulting from application of $\boldsymbol{E}'$, is

$$I' = \frac{1}{\tau} \int_{\tau} \sum_{i=1}^{N} z_i \boldsymbol{j}' d\tau = \frac{1}{\tau} \int_{\tau} \sum_{i=1}^{N} n_i^{0'} z_i^2 D_i' \nabla \phi_i' d\tau + \frac{1}{\tau} \int_{\tau} \sum_{i=1}^{N} n_i^{0'} \boldsymbol{u}' d\tau \tag{29}$$

The electro-osmotic velocity U' may be calculated by integrating the velocity $\boldsymbol{u}'$ throughout the unit cell volume:

$$U' = \frac{1}{\tau} \int_{\tau} \boldsymbol{u}' d\tau \tag{30}$$

Because of the linearity of the previous system of equations and boundary conditions, all solutions should be proportional to $\boldsymbol{E}$. The conductivity tensor σ of the medium is defined by

$$I' = \boldsymbol{\sigma} \cdot \boldsymbol{E}' , \quad \boldsymbol{\sigma} = \sigma^* \boldsymbol{\sigma}' \tag{31}$$

and the electro-osmosis tensor $\boldsymbol{\beta}$ by

$$U' = \beta' \cdot E' , \quad \beta = \beta^* \beta' \tag{32}$$

where σ^* and β^* are defined in Table 1. In order to follow common notations, σ, α, and β are lower case letters in contradiction with our rule stated at the end of Section I. In the following, the conductivity σ is sometimes normalized by the conductivity σ^∞ of the undisturbed fluid, instead of σ^*, as is customary for porous media. These two quantities are related by

$$\sigma^\infty = \sigma^* \sum_{i=1}^{N} n_i^{\infty'} z_i^2 D_i' \tag{33}$$

Let us consider now an external pressure gradient ∇P applied to the same porous material. Again, the solvent and the charge fluxes are proportional to the driving force ∇P. Thus, the permeability tensor K and the electro-osmosis tensor α are defined by

$$U' = -K' \cdot \nabla P', K = K^* K' \tag{34a}$$
$$I' = -\alpha' \cdot \nabla P' , \quad \alpha = \alpha^* \alpha' \tag{34b}$$

where K^* and α^* are defined in Table 1. Since both problems considered in this section are linear, a superposition of the two generalized forces, i.e., simultaneous application of the pressure gradient and of the electric field, leads to the relationship (1) with the electro-osmotic tensors σ, α, β and K. This is true only under the assumption that the ion distribution is slightly distorted, either by application of ∇P or E. The condition imposed on the latter quantity is obviously

$$E' \ll \zeta' \tag{35}$$

The comparable condition to be imposed on ∇P may also be obtained. Application of ∇P results in a force which must be much smaller than the electric forces near the walls $(en^*\kappa\zeta)$. Therefore, the restriction on ∇P is

$$\|\nabla P'\| \ll \zeta' \tag{36}$$

We are now ready to present elementary analytical results.

2. Circular Tube

Consider a tube of diameter $2a$. The unit cell is formed by the region $0 < r < a, 0 < z < L$, where L is an arbitrary length. The external field E and the pressure gradient ∇P are applied parallel to the tube axis. The local equations (26) can be analytically integrated [1], and the four global transport coefficients are given by

$$\sigma' = \frac{D_1' + D_2'}{2} + \zeta' \frac{I_1(\kappa a)}{\kappa a I_0(\kappa a)}(D_1' - D_2') \tag{37a}$$

$$\alpha' = \beta' = \zeta'\left[\frac{2I_1(\kappa)}{\kappa a I_0(\kappa a)} - 1\right] \tag{37b}$$

$$K' = (\kappa a)^2/8 \tag{37c}$$

3. Plane Channel

Assume that we consider a plane channel of height $2h$. The unit cell region is a parallelepipedic region of height $2h$, of width w and of length L; again the width and the length are arbitrary. The local equations can be analytically integrated, and the four global transport coefficients are given by

$$\sigma'_\parallel = \frac{D'_1 + D'_2}{2} + \frac{1}{2}\zeta'(D'_1 - D'_2)\frac{\tanh(\kappa h)}{\kappa h} + \frac{\zeta'^2}{2}\left[\frac{\tanh(\kappa h)}{\kappa h} - \frac{1}{\cosh^2(\kappa h)}\right] \tag{38a}$$

$$\alpha' = \beta' = \zeta'\left[\frac{\tanh(\kappa h)}{\kappa h} - 1\right] \tag{38b}$$

$$K' = \frac{(\kappa h)^3}{3} \tag{38c}$$

C. Numerical Simulation of Electro-osmotic Processes

The resolution of the linearized dimensionless electrokinetic equations requires three steps (cf. [1]). First, the equilibrium potential $\psi^{o\prime}$ is calculated by solving Eq. (26b). This operation is fast, though $\psi^{o\prime}$ is solved with a high accuracy in order to make possible the determination of u' and ϕ'_i. Equation (26c) and (26d) are then solved iteratively, until convergence of I and U. These computations usually converge in a few iterations. The required accuracy is of the order of 10^{-4} for the velocity and the ionic potentials, whereas it is of the order of 10^{-12} for the electrical potential. The numerical routine has been thoroughly tested [1, 10].

D. Porous Media

The purpose of this short section is to present the major situations where computations were performed [1]. Porous media can be classified as nonconsolidated and consolidated materials. Soils, clays, and packings of various sorts belong to the first class, while rocks such as sandstones and limestones belong to the second one. Both classes have been investigated.

A thorough study of regular packings has been carried out [1]. Computations were conducted for simple cubic arrays of spheres and ellipsoids for various solid concentrations in an aqueous solution of $HClO_4$. Of course, the conductivity and permeability of cubic sphere packings is isotropic, as may be reasoned from linear superposition. Computations were also performed for simple cubic arrays of oblate ellipsoids of revolution with aspect ratio of 5/1 and with solid concentrations varying from 0.02 up to 0.524; the double-layer thickness was kept equal to $\kappa R = 2.81$. Finally, Ref. 1 provides detailed results for orthorhombic arrays of ellipsoids of revolution with an aspect ratio of 5/1, in an attempt to describe the porous geometric structure of clays, which are composed of plate-like particles.

Random packings were also thoroughly studied [1]; they were built by simulating the sequential deposition of particles in a parallelepipedic cell with periodicity conditions along the two horizontal directions. After a particle is introduced at a random location from above the cell, it settles vertically along the z axis until it touches a grain already in place. It then rolls and glides until it reaches an equilibrium position, where it sits at least on three supporting points. While reaching its equilibrium position, it undergoes rotations and translations (without bouncing) along the bed surface until it reaches the lowest position. The numerical procedure is described in detail in Ref. 11. An example of such a packing of ellipsoids with an aspect ratio 5 is displayed in Fig. 3. Flat ellipsoids tend to settle with their largest section parallel to the xy plane. The average solid fraction is equal

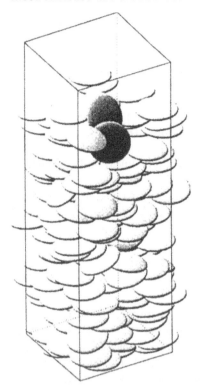

FIG. 3 Examples of a bed of $5 \times 5 \times 1$ ellipsoids obtained by sequential deposition.

to 0.595 for spheres and to 0.596 for ellipsoids. The transport properties of these random beds were investigated along the two horizontal x and y directions since the packings obtained in the above procedure are statistically isotropic in the horizontal plane.

Finally, electro-osmotic transport coefficients for reconstructed porous media, based on Fontainebleau sandstones, were calculated. These media are generated randomly, according to statistical geometrical characteristics measured on real materials. The reconstruction procedure has been described elsewhere [8, 12] and only the main points are given here. First, measurements are performed on a thin section of a given porous medium. A phase function $Z(r)$ is introduced:

$$Z(r) = \begin{cases} 1 & \text{if } r \text{ belongs to phase 1} \\ 0 & \text{if } r \text{ belongs to phase 2} \end{cases} \tag{39}$$

where r denotes the spatial position with respect to an arbitrary origin. Phase 1 corresponds to the pore space and phase 2 to the solid space. The porosity ϵ and the correlation function R_u) can be defined by the statistical averages (denoted by brackets):

$$\epsilon = < Z(r) > \tag{40a}$$

$$R_z(u) = \frac{< [Z(r) - \epsilon][Z(r + u) - \epsilon] >}{< (Z(r) - \epsilon)^2 >} \tag{40b}$$

or equivalently, because $Z^2(r) = Z(r)$:

$$R_z(u) = \frac{<[Z(r)-\epsilon][Z(r+u)-\epsilon]>}{\epsilon(1-\epsilon)} \tag{40c}$$

On a given sample, these measurements were performed using image analysis [8, 12] in a single, arbitrary chosen plane, since the considered materials were isotropic. Hence, $\boldsymbol{u}$ can be replaced by its absolute value u. The process of reconstruction of a 3D homogeneous and isotropic random medium of a given porosity ϵ and a given correlation function $R_z(u)$ is equivalent to generating a random discrete function $Z(r)$ which satisfies the properties specified in Eq. (40). It can be shown that such $Z(r)$ can be derived from a Gaussian field $X(r)$ when the latter is successively passed through a linear and a nonlinear filter [8].

For the present investigation, porous media have been reconstructed using the data measured on two samples of Fontainebleau sandstones, with porosities $\epsilon = 0.21$ and $\epsilon = 0.31$, respectively. The length scale $\mathcal{L}$ is the correlation length deduced from the correlation function by

$$\mathcal{L} = \int_0^\infty R_Z(u)du \tag{41}$$

An example of a reconstructed sample with $\epsilon = 0.31$ is displayed in Fig. 4. The correlation length scale $\mathcal{L}$, as measured by image analysis was about 20 μm, which exceeds the Debye layer thickness by at least two orders of magnitude. Nevertheless, we performed calculations with the dimensionless parameter $\kappa\mathcal{L} = 1/\sqrt{10}, 1, \sqrt{10}$. This corresponds to the situation where the geometry of the reconstructed samples was scaled down about 100 times.

E. Fractures

Let us present briefly fracture geometry and the cases which were analyzed. Fractures are relatively plane rock discontinuities which can be viewed as a void volume located between two solid walls (Fig. 5). Usually, the upper and lower surfaces are described by their heights above an arbitrary reference plane $z = 0$ [13] such as

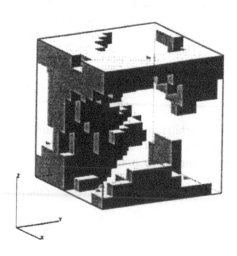

FIG. 4 Example of a reconstructed sample with $\epsilon = 0.31$.

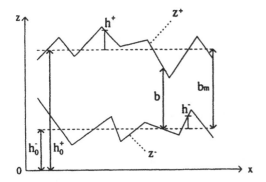

FIG. 5 Geometry of random fractures in the xz plane; conventions and notations.

$$z^{\pm}(x, y) = h_0^{\pm} + h^{\pm}(x, y)$$ (42)

where $h_0^{\pm}$ are the mean planes of each surface and $h^{\pm}$ are functions of the co-ordinates x and y. The surfaces of fractures may be characterized by the probability densities $\phi(F)$ which are often assumed to be Gaussian [14]:

$$\phi(F) = \frac{1}{\sqrt{2\pi}\sigma_F} \exp\left[-\frac{(F - <F>)^2}{2\sigma_F^2}\right], \quad \text{with } F = h^+, h^-, w$$ (43)

where $<F>$ is the statistical average, σ_F^2 the variance and w the difference between z^+ and z^-. At a given point (x, y), the aperture of a fracture is defined as $b = w$ if $w \geq 0$ and $b = 0$ if $w < 0$, and the mean aperture as $b_m = h_0^+ - h_0^-$.

The upper and lower surfaces can be correlated with one another or not; this is expressed by the dimensionless parameter θ ranging between 0, for uncorrelated surfaces, and 1, for perfectly correlated surfaces:

$$\theta = 1 - \frac{\sigma_w^2}{2\sigma_h^2}$$ (44)

Statistical properties of fractures in the xy plane are defined by means of a covariance function $C_F(r)$ of the $z^{\pm}$ fields, so that

$$r = \sqrt{\Delta x^2 + \Delta y^2}$$

$$C_F(r) = < \{F(x, y) - <F>\}\{F(x + \Delta x, y + \Delta y) - <F>\} >$$ (45)

with $F = z^{\pm}$. Two types of random fractures were considered. First, Gaussian fractures are described by a Gaussian covariance function, with the correlation length l:

$$C_F(r) = \sigma_F^2 \exp\left[-\left(\frac{r}{l}\right)^2\right]$$ (46)

Second, self-affine fractures are described by

$$C_F(r) = \sigma_F^2 \exp\left[-\left(\frac{r}{l}\right)^{2H}\right]$$ (47)

where H is the roughness or Hurst exponent, ranging between 0 and 1.

Deterministic fractures were studied in Ref. (2), including the plane channel comprised between the two solid walls $z^* \pm = 0$. Sinusoidal fractures were generated with limiting surfaces given by

$$h^{*\pm}(x, y) = \pm \sin\left(2\pi\frac{x}{l}\right) \sin\left(2\pi\frac{y}{l}\right) \tag{48}$$

The method of Fourier transforms can then be used to generate random fractures [13]. The functions $h^{*\pm}$ are Gaussian and the two surfaces are supposed to have the same spatial correlation length l.

The geometry of a fracture is fully described by its type (for instance deterministic, Gaussian, or self-affine) and by the three parameters $\dfrac{b_m}{\sigma_h}$, $\dfrac{l}{\sigma_h}$, and θ. It is equivalent to consider σ_h as the length unit. According to Refs 14–16, these quantities were measured in real samples and found to belong to the following intervals

$$0 < \frac{b_m}{\sigma_h} < 2.6 \text{ and } 1 < \frac{l}{\sigma_h} < 7 \tag{49}$$

Moreover, an additional parameter is essential to define self-affine fractures; the exponent H is often found close to 0.875 [17] for real fractures.

To summarize, examples of reconstructed fractures are gathered in Fig. 6. The sinusoidal fracture was built according to (48). The Gaussian and self-affine ones were generated from the same sequence of random numbers; thus, they differ only by the surface texture, which is related to the exponent H.

F. Local Coefficient α for Porous Media and Fractures

In the past, electro-osmotic phenomena have been mostly studied in the limit of vanishing thicknesses, for which calculations may be made analytically with the help of the Overbeek equation [18]:

$$\frac{\beta}{\zeta} = \frac{\epsilon_{el}\sigma}{\mu\sigma^\infty} \tag{50}$$

where σ^∞ is the electrolyte conductivity.

The thick double-layer limits can be easily deduced from Eqs (37) and (38) for circular and plane channels:

$$\frac{\mu\beta}{\epsilon_{el}\zeta} \to -\kappa^2 K \text{ or } \frac{\beta'}{\zeta'} \to -K', \kappa \to 0 \tag{51}$$

This result can be generalized to any porous medium. It is more convenient to consider a macroscopic pressure gradient ∇P applied to the porous medium, thus aiming to obtain the comparable limit of the coupling coefficient α, instead of the equivalent coefficient β. Suppose that ∇P induces a flow with local velocity u. The convected electric current is

$$I = -\alpha \cdot \nabla P = \frac{1}{\tau}\int \rho u dv \tag{52}$$

Using the same hypotheses and approximations as before, namely, that the equilibrium concentrations n_i^0 are slightly distorted by the flow, that the electrolyte is symmetric, and that the surface potential is small enough to linearize the Poisson–Boltzmann equation, Eq. (52) can be recast into

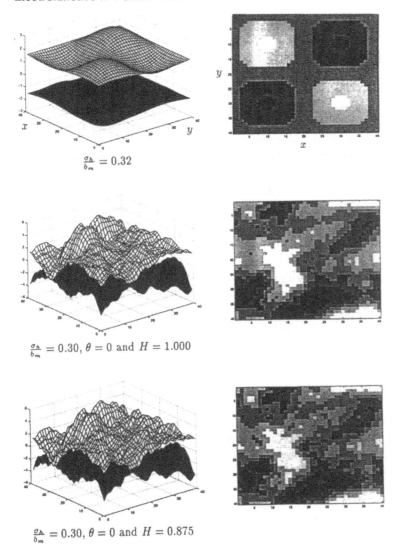

$\frac{\sigma_h}{b_m} = 0.32$

$\frac{\sigma_h}{b_m} = 0.30, \theta = 0$ and $H = 1.000$

$\frac{\sigma_h}{b_m} = 0.30, \theta = 0$ and $H = 0.875$

FIG. 6 Examples of sinusoidal, Gaussian, and self-affine fractures (top to bottom): 3D visualizations (left) and aperture maps (right). The mean aperture b_m is equal to 20 elementary cubes in all cases. For the 3D visualizations, the vertical z axis is graduated in σ_h units. For the maps, black areas correspond to the smallest distances between upper and lower surfaces, including contact zones, whereas white areas represent the largest apertures.

$$\boldsymbol{\alpha} \cdot \nabla P = \epsilon_{el} \kappa^2 \frac{1}{\tau} \int \psi^o \boldsymbol{u} \mathrm{d}v \tag{53}$$

Moreover, for thick double layers, with κ^{-1} larger than the characteristic pore size, $\psi^{0'}$ is nearly uniform and approximately equal to the surface potential ζ'. Therefore, replacing ψ^o by ζ, and using the definition of permeability, one obtains:

$$\boldsymbol{\alpha} \cdot \nabla P \approx \zeta \epsilon_{el} \kappa^2 \frac{1}{\tau} \int \boldsymbol{u} \mathrm{d}v = \frac{\zeta \epsilon_{el} \kappa^2}{\mu} \frac{\mu}{\tau} \int \boldsymbol{u} \mathrm{d}v = -\frac{\zeta \epsilon_{el} \kappa^2}{\mu} \boldsymbol{K} \cdot \nabla P, \kappa L \to 0 \tag{54}$$

Since ∇P is arbitrary, one obtains:

$$\alpha \approx -\frac{\zeta \epsilon_{el} \kappa^2}{\mu} K \text{ or } \alpha' \approx -\zeta' K' , \quad \kappa L \rightarrow 0 \tag{55}$$

Hence, this result suggests plotting the ratios $\beta'/\zeta' K'$ against a normalized double-layer thickness; for this purpose, a length scale applicable to all the directions and geometries has to be defined. As already pointed out above, and also shown by Eq. (53), the criterion for convergence toward Eq. (54) corresponds to the covering of the pore space by the electrical double layer, rather than its thickness relative to a typical grain size. Thus, an adequate length scale should be a measure of the characteristic pore size. Such a length scale Λ has been introduced [19]. It is essentially a pore volume-to-pore surface ratio, with a measure weighted by the local value of the electric field $E(r)$ in the conduction process:

$$\Lambda = 2 \frac{\int_{\tau_f} E^2(r) d\tau_f}{\int_S E^2(r) dS} \tag{56}$$

where τ_f is the pore volume inside the unit cell.

This length scale Λ is derived from the correction to σ due to the interfacial conductivity in thin layers. An electrokinetic method [20] was proposed to measure Λ, as an alternative way to determine the permeability via

$$K = \frac{\sigma^o}{\sigma^\infty} \frac{\Lambda^2}{8} \tag{57}$$

This relation was shown [20, 21] to provide excellent predictions for various models or real porous media. Note that for a plane channel, Eqs. (56) and (57) yield slightly different results, namely $\Lambda = 2h$ and $\Lambda = 2h\sqrt{2/3}$, respectively. For a circular channel, both expressions give $\Lambda = R$. For all the media considered here, Λ was evaluated by solving the Laplace equation in the pore space of the unit cell with insulating boundary conditions.

All the electro-osmotic data from our simulations are thus gathered in Fig. 7 and compared with the theoretical results [Eqs (37) and (38)] for channels. For all anisotropic configurations, Λ was evaluated along each direction. All the data cluster around a single curve, with very little dispersion considering the variety of geometrical configurations. As

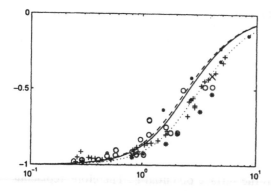

FIG. 7 Reduced coupling Ref. coefficients $\beta'/\zeta' K'$ as functions of $\kappa \Lambda$ for all the configurations studied in (1). Data are for simple cubic regular packings ($\bullet$), orthorhombic packings ($\bigcirc$), random packings (*), and reconstructed sandstones (+). Theoretical results [Eq. (38)] with $\Lambda = \sqrt{8/3h}$ (solid line) and Eq. (37b) with $\Lambda = R$ (broken line), and with Eq. (58) (dotted line).

expected, $\beta'/\zeta'K'$ tends to -1 for vanishing $\kappa\Lambda$. For intermediate Debye thicknesses, $\beta'/\zeta'K'$ is very well approximated by Eq. (37c) valid for a circular channel.

It is tempting to modify Eq. (56) to obtain an even better agreement with the numerical data. A least-square fit yields a modified expression for Λ:

$$\Lambda = 1.544 \frac{\int_{\tau_f} E^2(r)\mathrm{d}\tau_f}{\int_S E^2(r)\mathrm{d}S} \tag{58}$$

The prediction (37c) with R replaced by Λ in Eq. (58) is also plotted in Fig. 7. For an easier evaluation, expression for a plane channel [Eq. (38)] could be substituted into Eq. (37c). Alternatively, β can be deduced from the primary coefficients σ^o and K measured along the same direction by

$$\beta = \frac{8\epsilon_{el}\zeta K}{\mu \Lambda_e^2} \left[\frac{2I_1(\kappa\Lambda_e')}{\kappa\Lambda_e'I_0(\kappa\Lambda_e')} - 1 \right] \tag{59a}$$

$$\Lambda' = (12.5K\sigma^\infty) \tag{59b}$$

where I_0 and I_1 are modified Bessel functions of the first kind.

These results for porous media can be easily transposed to fractures by using a form derived from (38). In other words, Λ was systematically computed according to (56) and the data were compared to the following relationship for β'

$$\frac{\beta'}{\zeta'K'} = \frac{3}{\kappa^2\left(\dfrac{\Lambda}{2}\right)^2} \left[\frac{\tanh\left(\kappa\dfrac{\Lambda}{2}\right)}{\kappa\dfrac{\Lambda}{2}} - 1 \right] \tag{60}$$

A satisfactory representation of the electro-osmotic coupling coefficients of fractures is given by Eq. (60) as shown in Ref. 2. Nevertheless, it is again tempting to modify Eq. (56) in order to obtain an even better agreement with the numerical data. A least-square fit yields the following modified expression for Λ'' for all the predictions:

$$\Lambda'' = 2.201 \frac{\int_{\tau_f} E^2(r)\mathrm{d}\tau_f}{\int_S E^2(r)\mathrm{d}S} \tag{61}$$

This is displayed in Fig. 8 where the points cluster around the curve with a very small dispersion, considering the variety of geometrical configurations.

G. Local Coefficients K, σ, and α

It might be useful to summarize here the local coefficients for porous media in view of the second change of scale performed in Section IV. The permeability of Fontainebleau sandstones can be correlated by a power law as a function of the local porosity ϵ [22]. Such a correlation was later confirmed by computations made on reconstructed samples of rocks such as sandstones and chalk [23, 24]. Hence, a general expression for local permeability can be given as

$$K = k\mathcal{L}_c^2\epsilon^m \tag{62}$$

where k and m are dimensionless constants; $\mathcal{L}_c$ is the correlation length of the phase function (41). Typical values for sandstones are

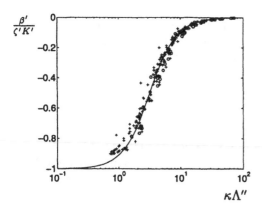

FIG. 8 The ratio $\dfrac{\beta'}{\zeta' K'}$ as a function of $\kappa \Lambda''$ for fractures. On this graph are gathered all the results for sinusoidal ($\bigcirc$) and self-affine ($+$) fractures. The solid line is derived from Eq. (60) where Λ has been substituted by Λ''.

$$k \simeq 0.2, m \simeq 3.75 \tag{63}$$

The macroscopic conductivity of a real porous medium is generally well correlated by the classical Archie law [8]:

$$\sigma = s\sigma_f \epsilon^n \tag{64}$$

where σ_f is the fluid conductivity, and s and n are dimensionless coefficients with the following typical values:

$$s \simeq 1.3 \; , \; n \simeq 2.5 \tag{65}$$

It should be noticed that surface conductivity is assumed to be negligible.

Finally, the local coefficient β is given by Eq. (59a). When the permeability and the conductivity of a porous medium are known, a length scale Λ' can be expressed by (59b); note that Λ is proportional to $\mathcal{L}_c$ according to (62).

Variations of these quantities as functions of the porosity ϵ are displayed in Fig. 9. When ϵ varies from 0.01 to 0.5, K, σ, and α vary by several orders of magnitude. For the range of values which is chosen here, the product $\kappa \Lambda$ is usually very large and the ratio of the two Bessel functions is equal to 1. Hence, α can be simplified as

$$\alpha = -0.64 \frac{\epsilon_{el} \zeta}{\mu} \frac{\sigma}{\sigma_o} \tag{66}$$

which corresponds within a factor 0.64 to the classical Overbeek relation (50).

A few parameters are going to be kept constant in this study. The characteristic coefficients of permeability and conductivity are always given by Eqs (63) and (65). Moreover, the fluid is water with electrical permittivity and viscosity given in SI units:

$$\epsilon_{el} = 7.10^{-10} \; , \; \mu = 10^{-3} \tag{67}$$

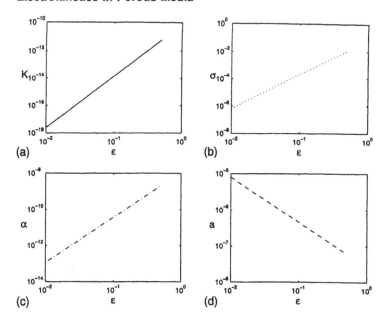

FIG. 9 Physical properties as functions of porosity ϵ: (a) permeability K; (b) conductivity σ; (c) dimensionless coupling coefficient α; (d) dimensionless coupling parameter a. ϵ ranges from 0.01 to 0.5. Data are for: $\mathcal{L}_c = 20$ μm; $\sigma_f = 0.05$ S/m; $\kappa = 10^8$ m^{-1}; $\zeta = -20$ mV; Eqs (63), (65), and (67).

IV. MACROSCOPIC PROPERTIES AT THE FIELD SCALE

The major purpose of this section is to determine the electro-osmotic coefficient $\bar{\sigma}$ by solving the divergence equations (2) for various physical situations, i.e., for various heterogeneous porous media.

A. Description of the Media

The media which are going to be systematically investigated belong to several classes, illustrated in Fig. 10; they can be listed as follows: deterministic media such as laminated materials, correlated media, and two-phase media.

Deterministic media are mostly of interest for verification of the numerical code. The simplest example of such a medium is an elementary one-dimensional medium composed of n identical cells. Each cell consists of two slices of two different materials 1 and 2 whose heights are h_1 and h_2; the physical properties are denoted by $K_i, \sigma_i (i = 1, 2)$.

It is often useful to use dimensionless quantities. Let K_o, σ_o, and α_o be some characteristic values of the local fields K, σ, and α; dimensionless local fields denoted by a prime can be defined by

$$K' = \frac{K}{K_o} \ , \quad \sigma' = \frac{\sigma}{\sigma_o} \ , \quad \alpha' = \frac{\alpha}{\alpha_o} \tag{68}$$

It is important to note that the dimensionless quantities defined in this section are different from the ones defined in Section III.

The second class of materials consists of correlated media. The properties of these media are assumed to be all lognormally distributed. More precisely, the random field $X(r)$

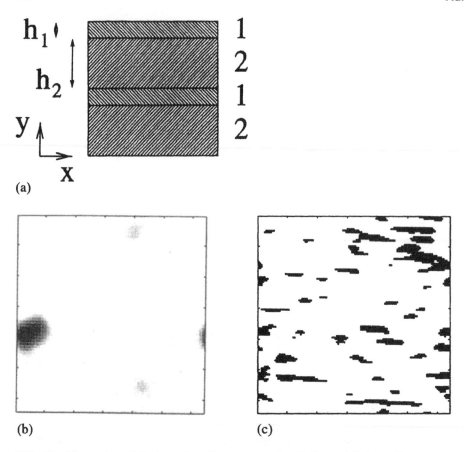

FIG. 10 Illustration of the investigated porous media: (a) deterministic medium composed of two different materials; (b) correlated medium (the displayed property is porosity which is lognormally distributed; this corresponds to the exponential of the field displayed in Fig. 1b), (c) two-phase medium (phase 1 is white, and phase 2 black). Data are for: $N_c = 128$; $P_1 = 0.1$; $\lambda_z = \lambda_y = 32$; $\lambda_s = 8$.

(which stands for the dimensionless fields K', σ', or α') can be expressed in terms of the field $Y_X(r)$:

$$X(r) = \exp(m_X + Y_X(r)) \tag{69}$$

where $Y_X(r)$ is a Gaussian field with zero mean and a variance denoted by $< Y_X^2 >$. The statistical average is denoted by $\langle \cdot \rangle$. Moreover, the Gaussian field $Y_X(r)$ is correlated; since it is assumed to be a stationary random field, its covariance is only a function of the vector $u = r - r'$. When $Y_X(r)$ is isotropic, C_{Y_X} only depends on the modulus u of u, i.e., $C_{Y_X}(u)$. Very often a Gaussian correlation is chosen:

$$C_{Y_X}(r, r') = < Y_X(r) Y_X(r') > = C_{Y_X}(r - r') = C_{Y_X}(u) = C_{Y_X}(u) = \exp\left(-\frac{u^2}{\lambda^2}\right) \tag{70}$$

where λ is the correlation length.

The three random fields $K, \sigma,$ or α can be assumed either independent or dependent, and it will be seen in Section IV.D that this has important consequences on the average properties.

The most important subclass of correlated media is called real correlated media and is defined by the property that the porosity ϵ is lognormally distributed; hence, Eq. (69) implies

$$\epsilon(r) = \exp(m_\epsilon + Y_\epsilon) \tag{71}$$

For simplicity, these media are restricted to be isotropic with a Gaussian correlation (70). Finally, the local physical properties $K, \sigma,$ and α are related to ϵ by Eqs (62), (64), and (66). Hence, in realistic applications, $K, \sigma,$ and α are lognormally distributed and inter-correlated. Reference 3 showed that

$$Y_K = mY_\epsilon, \ Y_\sigma = nY_\epsilon, \ Y_\alpha = nY_\epsilon \tag{72}$$

Hence,

$$< Y_\sigma Y_K > = < Y_\alpha Y_K > = nm < Y_\epsilon^2 > \tag{73a}$$

$$< Y_\sigma Y_\alpha > = < Y_\alpha^2 > = n^2 < Y_\epsilon^2 > \tag{73b}$$

The third class of materials is called two-phase media. Such a medium is composed of two phases called 1 and 2 which have different physical properties K_1, σ_1, α_1 and K_2, σ_2, α_2. These properties are assumed to be constant in each phase, but they could also be random. The distribution of the phases in space is characterized by a phase function $Z(r)$ defined analogously to Eq. (39). Such two-phase media are very similar to porous materials [8] and all the knowledge developed for these media can be applied here. To summarize, random two-phase media can be generated with a given proportion P_1 of phase 1 and a given phase correlation $R_z(u)$; $Z(r)$ is obtained by thresholding a standard Gaussian field $Y(r)$ with a correlation function $R_Y(u)$. The general form of $R_Y(u)$ which is used here also has a Gaussian form [cf. Eq. (70)], which is often anisotropic:

$$R_Y(u) = \exp\left[-\left(\frac{u_x}{\lambda_z}\right)^2 - \left(\frac{u_y}{\lambda_y}\right)^2 - \left(\frac{u_z}{\lambda_z}\right)^2\right] \tag{74}$$

In the important case of stratified media, two correlation lengths are of the same order of magnitude, but much larger than the third one; this can be schematized by

$$\lambda_x = \lambda_y \gg \lambda_z \tag{75}$$

Two major cases will be considered. In the first case of stratified media, we shall consider a medium 1 such as sandstones in which are embedded thin layers with low permeability. Stratified media which are random should not be confused with laminated media which have been introduced at the beginning of this subsection and which are deterministic. In the second case of fractured media, the same thin layers have a permeability much larger than that of medium 1; these layers correspond to fractures, or more precisely to fractured zones.

B. Dimensional Analysis and Expansion

Let us for instance consider the case where a constant but otherwise arbitrary macroscopic gradient $\overline{\nabla}P$ is imposed and the resulting fluxes are computed, thereby $\bar{\alpha}$ and $\overline{K}$. Let l be a characteristic local length over which the local coefficients K, σ, and α vary; l is assumed to be of the order of l_3 (cf. Fig. 1). The most natural definition for the dimensionless pressure P' is, therefore,

$$P = l \overline{\nabla P} P' \tag{76}$$

The electric potential Ψ is primarily induced through the coupling coefficient α since there is no overall superimposed electric field. The most natural definition for the dimensionless potential Ψ' is, therefore,

$$\Psi = \frac{\alpha_o}{\sigma_o} l \overline{\nabla P} \Psi' \tag{77}$$

Hence, the general system (2) is easily reduced to the dimensionless system:

$$\nabla' \cdot (K' \nabla' P' + a \alpha' \nabla' \Psi') = 0 \tag{78a}$$

$$\nabla' \cdot (\alpha' \nabla' P' + \sigma' \nabla' \Psi') = 0 \tag{78b}$$

where $r = l r'$ and $\nabla' = l \nabla$.

The dimensionless coupling parameter a is expressed as

$$a = \mu \frac{\alpha_o^2}{K_o \sigma_o} \tag{79}$$

This coefficient will play a central role in the following analysis. Typical orders of magnitude (see, e.g. Ref. 3) of the local coefficients are (in SI units)

$$K_o = 10^{-12}, \mu = 10^{-3}, \alpha_o = 310^{-9}, \sigma_o = 0.01 \tag{80a}$$

The implies that

$$a = 9 \times 10^{-7} \tag{80b}$$

Hence, because of the nature of the coupling coefficient, a is generally very small with respect to 1:

$$a \ll 1 \tag{81}$$

This statement is the basis of the systematic expansion summarized below.

A direct solution of the system (78) for P' and Ψ' turned out to be very difficult because standard and robust techniques, such as the conjugate gradient, were found to be unstable when realistic values of the coefficients K, α, and σ were used.

First consider that a macroscopic pressure gradient is imposed. The resulting fields P' and Ψ' verify the system (78). These two quantities can be expanded as

$$P' = \sum_{i=0}^{\infty} a^i P_i' \quad , \quad \Psi' = \sum_{i=0}^{\infty} a^i \Psi_i' \tag{82}$$

These expansions can be introduced into Eqs (78) which can be rearranged according to the powers a^i; then it is a standard procedure for such perturbation expansions to equate the terms of order a^i to zero:

$$i = 0: \nabla' \cdot (K' \nabla' P_o') = 0 \quad , \quad \nabla' \cdot (\sigma' \nabla' \Psi_o') = -\nabla' \cdot (\alpha' \nabla' P_o') \tag{83a}$$

$$i = 1: \nabla' \cdot (K' \nabla' P_1') = -\nabla' \cdot (\alpha' \nabla' \Psi_o') \quad , \quad \nabla' \cdot (\sigma' \nabla' \Psi_1') = -\nabla' \cdot (\alpha' \nabla' P_1') \tag{83b}$$

and for the term of order i:

$$i: \nabla' \cdot (K' \nabla' P'_i) = -\nabla' \cdot (\alpha' \nabla' \Psi'_{i-1}) \quad , \quad \nabla' \cdot (\sigma' \nabla' \Psi'_i) = -\nabla' \cdot (\alpha' \nabla' P'_i) \tag{83c}$$

Clearly, these equations can be solved recursively until convergence of the expansion is obtained. The right-hand side of Eq. (83c) is always known from the previous steps. These equations are elliptical partial differential equations with a known second member; their numerical resolution does not generate any difficulty and any standard conjugate gradient technique is efficient; no instability was ever found.

Precisely the same developments can be done when an external macroscopic electric gradient $\overline{\nabla}\Psi$ is imposed.

C. Theoretical Expansion for Lognormal Media

When the fluctuations of the local quantities K, σ, and α are not large, a small perturbation expansion in terms of these fluctuations can be applied to the coupled equations (2). There is a vast literature (see, e.g., Refs 25–29) on the determination of the macroscopic permeability from the Darcy equation alone (which corresponds to $\alpha = 0, \sigma = 9$) and we do not intend to review it here.

Let us assume that the three fields K, σ, and α are lognormally distributed and that their dimensionless counterparts are expressed by Eq. (69). The three constants K_o, σ_o and α_o are defined by

$$K_o = \exp m_K \quad , \quad \sigma_o = \exp m_\sigma \quad , \quad \alpha_o = \exp m_\alpha \tag{84}$$

The assumptions of small perturbation is equivalent to assuming that the variances $< Y_K^2 >, < Y_\sigma^2 >$, and $< Y_\alpha^2 >$ are small with respect to 1.

In order to have a tractable system, it is further assumed that the dimensionless coupling parameter a is much less than 1 as discussed in Section IV.B and that it is also much smaller than the previous variances:

$$a \ll < Y_K^2 > \quad , \quad < Y_\sigma^2 > \quad , \quad < Y_\alpha^2 > \ll 1 \tag{85}$$

This assumption, which corresponds to most cases of practical interest, implies that the expansions in terms of a can be limited to the zeroth order. For instance, let us consider a prescribed macroscopic pressure gradient $\overline{\nabla}P$; the relevant equations are thus reduced to Eq. (83a). All the detailed calculations can be found in Ref. 3. When the statistical fields K, σ, and α are isotropic, the resulting macroscopic permeability $< K >$, conductivity $< \sigma >$, and electro-osmotic coefficient $< \alpha >$ are given by

$$\frac{< K >}{K_o} = 1 + \frac{< Y_K^2 >}{6} \tag{86a}$$

$$\frac{< \sigma >}{\sigma_o} = 1 + \frac{< Y_\sigma^2 >}{6} \tag{86b}$$

$$\frac{< \alpha >}{\alpha_o} = 1 + \frac{< Y_\alpha^2 >}{2} + \frac{1}{3} < Y_\sigma Y_K - Y_\sigma Y_\alpha - Y_K Y_\alpha > \tag{86c}$$

These expressions can be commented on as follows. The first two equations are a mere repetition of the classical expression recalled by Ref. 30. The third expression is more interesting. First, Y_σ and Y_K play the same role; this is related to Onsager's symmetry of the tensors [3]. Second, various expressions are obtained when the fields are correlated or not. For instance,

1. All the fields are uncorrelated:

$$\frac{<\alpha>}{\alpha_o} = 1 + \frac{1}{2} < Y_\alpha^2 > \tag{87a}$$

2. K and σ are correlated, but they are not correlated to α:

$$\frac{<\alpha>}{\alpha_o} = 1 + \frac{1}{2} < Y_\alpha^2 > + \frac{1}{3} < Y_\sigma Y_K > \tag{87b}$$

3. σ and α are correlated, but they are not correlated to K:

$$\frac{<\alpha>}{\alpha_o} = 1 + \frac{1}{2} < Y_\alpha^2 > - \frac{1}{3} < Y_\sigma Y_\alpha > \tag{87c}$$

4. K and α are correlated, but they are not correlated to σ:

$$\frac{<\alpha>}{\alpha_o} = 1 + \frac{1}{2} < Y_\alpha^2 > - \frac{1}{3} < Y_K Y_\alpha > \tag{87d}$$

Of course, the most interesting result is obtained for real correlated media as defined in Section IV.A by Eq. (72). Because of the relations (73), one obtains:

$$\frac{<K>}{K_o} = 1 + \frac{m^2}{6} < Y_\epsilon^2 > \tag{88a}$$

$$\frac{<\alpha>}{\alpha_o} = 1 + \frac{n^2}{6} < Y_\epsilon^2 > \tag{88b}$$

A similar calculation, made for a prescribed overall potential gradient, yields

$$\frac{<\sigma>}{\sigma_o} = 1 + \frac{n^2}{6} < Y_\epsilon^2 > \tag{89}$$

This result which shows that $\dfrac{<\sigma>}{\sigma_o}$ is identical to $\dfrac{<\alpha>}{\alpha_o}$ is *a priori* puzzling, but it can be exactly demonstrated when an overall potential gradient is imposed [3].

D. Results

The full numerical solution of Eq. (2) is based on a few steps which can be summarized as follows. First, the porous medium is replaced by a spatially periodic pattern of unit cells with the same content as illustrated in Fig. 1c. The unit cell consists of $N_{cx} \times N_{cy} \times N_{cz}$ elementary volumes of sizes $\Delta_x \times \Delta_y \times \Delta_z$. In most cases, these quantities are equal to N_c and Δ. The *box integration method* is used to derive the difference equation.

This general methodology is applied to the couples of partial differential equations obtained by the perturbation expansions described in Section IV.B. When a macroscopic pressure gradient $\overline{\nabla P}$ is prescribed over the unit cell, the couples [Eqs (83a), (83b), (83c), ...] are solved sequentially.

It may be useful to provide the boundary conditions in this particular case. The field P_o' can be decomposed as [8]

$$P_o' = \check{P}_o' + r' \cdot \overline{\nabla' P_o'} \tag{90}$$

where the small inverted hat denotes a spatially periodic function. Likewise, all the fields $P_i'(i \geq 1)$ and $\Psi_i'(i \geq 0)$ are spatially periodic.

Similar equations and boundary conditions are obtained when a macroscopic potential gradient $\overline{\Psi}$ is applied.

1. Laminated Media

The first interesting case to consider is the one of laminated media which are displayed in Fig. 10a. It will prove convenient to schematize the relations (1) by the matrix formula:

$$\begin{pmatrix} U_i \\ I_i \end{pmatrix} = -M_i \cdot \begin{pmatrix} \nabla P_i \\ \nabla \Psi_i \end{pmatrix} \quad , \quad i = 1, 2 \tag{91}$$

where M_i is a 2×2 matrix with an obvious correspondence.

It is standard to study the so-called parallel and series cases. When the macroscopic gradients are parallel to the slices, i.e., perpendicular to the z axis in Fig. 10, the gradients are identical in each slice and the total fluxes add one to another. Without any loss in generality, it can be assumed that the fluxes are parallel to the x axis. Hence, the macroscopic properties are given by

$$\begin{pmatrix} U_x \\ I_x \end{pmatrix} = -M_x \cdot \begin{pmatrix} \frac{\partial P}{\partial x} \\ \frac{\partial \Psi}{\partial x} \end{pmatrix} \quad \text{with} \quad M_x = \frac{h_1 M_1 + h_2 M_2}{h_1 + h_2} \tag{92}$$

The series formula applies when the gradients are parallel to the z axis. In such a situation, the fluxes are conserved and it is preferable to use Eq. (91) in its inverse form. Hence,

$$\begin{pmatrix} U_z \\ I_z \end{pmatrix} = -M_z \cdot \begin{pmatrix} \frac{\partial P}{\partial z} \\ \frac{\partial \Psi}{\partial z} \end{pmatrix} \quad \text{with} \quad M_z = \left(\frac{h_1 M_1^{-1} + h_2 M_2^{-1}}{h_1 + h_2} \right)^{-1} \tag{93}$$

The macroscopic properties of the series case are, therefore, given by a much more complex expression.

These formulas were used to verify the numerical routines; the numerical data obtained for such laminated media were found to be in perfect agreement with Eqs. (92) and (93).

This offers us the opportunity to generalize to the present situation of the bounds which were first given [31] for elementary diffusive processes in a random medium. The proof closely follows the one recalled in Ref. 8. The result can be summarized by the two inequalities:

$$< M^{-1} >^{-1} \leq \overline{M} \leq < M > \tag{94}$$

In other words, the right-hand inequality means that the tensor $< M > -\overline{M}$ is positive definite; an analogous property holds for the left-hand inequality. Note that the tensor $\overline{M}$ is the tensor composed of the overbarred elements in Eq. (3).

2. Correlated Media

One of the major purposes of these media is to check further the numerical routine by comparing the numerical data and the analytical predictions [Eqs (86)] systematically [3].

A first series of checks was performed when the three fields $K, \sigma,$ and α are lognormally distributed. The average physical parameters were selected in order to be representative of a Fontainebleau sandstone. The variances of the fields Y [cf. Eq. (69)] are given by

$$< Y_K^2 >= 0.16 \quad , \quad < Y_\sigma^2 >= 0.25 \quad , \quad < Y_\alpha^2 >= 0.09 \tag{95}$$

In order to check the influence of the intercorrelation between the properties, a few samples were generated in the following way. A Gaussian standard field $Y(r)$ is generated; σ is said to be correlated with K when these two fields are expressed by

$$Y_K(r) =< Y_K^2 >^{1/2} Y(r) \quad , \quad Y_\sigma(r) =< Y_\sigma^2 >^{1/2} Y(r) \tag{96}$$

Moreover, the field $Y_\alpha(r)$ is statistically independent of $Y(r)$. Hence, the coupling coefficient is readily deduced from Eq. (87b) as

$$\frac{<\alpha>}{\alpha_o} = 1 + \frac{1}{2} < Y_\alpha^2 > + \frac{1}{3} < Y_K^2 >^{1/2} < Y_\sigma^2 >^{1/2} \tag{97}$$

The other cases are generated in the same manner.

In all cases, the results show that the routine was working satisfactorily. The theoretical predictions [Eq. (87)] were all verified numerically. The agreement between the numerical calculations and the analytical predictions (Eqs. (88) and (89)] is seen to be very good up to $< Y_\epsilon^2 >^{1/2} \leq 0.3$. The higher-order terms then become important; moreover, the fluctuations for permeability, for instance, are quite large.

3. Stratified Media

As defined in Section IV.A, these media are composed of two media denoted by the indices 1 and 2. Medium 1 has properties which are close to sandstones, while 2 is much less porous, but with a larger zeta potential. The standard case is summarized by

$$
\begin{aligned}
&\kappa = 10^8 , \ \sigma_f = 0.05 \\
&P_1 = 0.9 , \ \lambda_x = \lambda_y = 16\Delta , \ \lambda_z = 4\Delta , \ N_c = 64 \\
&\epsilon_1 = 0.3 , \ \mathcal{L}_{c1} = 2.10^{-5} , \ \zeta_1 = -0.02 \\
&\epsilon_2 = 0.15 , \ \mathcal{L}_{c2} = 2.10^{-6} , \ \zeta_2 = -0.05
\end{aligned} \tag{98}
$$

to which Eqs (63), (65), and (67) should be added. Everything is expressed in SI units; the scale Δ is arbitrary. The major differences between the two media are the porosities, the local correlation lengths, and the zeta potentials. The average properties for the media are given by

$$
\begin{aligned}
&\Lambda_1 = 0.131 \ 10^{-4} , \ K_1 = 0.876 \ 10^{-12} , \ \sigma_1 = 0.320 \ 10^{-2} \\
&\alpha_1 = 0.573 \ 10^{-9} , \ a_1 = 0.117 \ 10^{-6} \\
&\Lambda_2 = 0.847 \ 10^{-6} , \ K_2 = 0.651 \ 10^{-15} , \ \sigma_2 = 0.566 \ 10^{-3} \\
&\alpha_2 = 0.248 \ 10^{-9} , \ a_2 = 0.167 \ 10^{-3}
\end{aligned} \tag{99}
$$

The reference values used for the dimensionless fields are the ones of medium 1:

$$K_o = K_1 \quad , \quad \sigma_o = \sigma_1 \quad , \quad \alpha_o = \alpha_1 \tag{100}$$

Hence, the permeability K_2 of medium 2 is about three orders of magnitude smaller than the one K_1 of the sandstones, and the dimensionless coupling parameters a_2 and a_1 are in the opposite ratio.

Two large scale correlation lengths λ_x and λ_y are equal and different from λ_z. This symmetry property implies that, for the large scale tensor A, which stands for the three macroscopic tensors $< K >$, $< \sigma >$, and $< \alpha >$:

$$A_{xx} = A_{yy} \neq A_{zz} \text{ for } A = < K >, < \sigma >, \text{ or } < \alpha > \tag{101}$$

The influence of the parameters (98) was studied. These parameters can be divided into two groups, namely, the electrical parameters (such as the fluid conductivity, the Debye–Hückel length κ^{-1}, the zeta potentials ζ_1 and ζ_2) and the geometrical parameters [such as the large-scale correlations length $(\lambda_x, \lambda_y, \lambda_z)$ and the probability P_1 of phase 1]. Note that this list is not complete and that the local correlation lengths $\mathcal{L}_{c1}$ and $\mathcal{L}_{c2}$ have been studied only for fractured media. It should also be noticed that this study was by no means systematic; one parameter at a time was varied around the standard case provided by Eqs (98). Moreover, many numerical results suggested that analytical calculations could be usefully performed in several limiting cases.

The first general property which must be noticed is that the dimensionless results do not depend on the fluid conductivity σ_f, when it is assumed to vary independently of the other parameters. This is because only the ratio σ/σ_f appears in Eqs (64) and (59b). No detailed result is provided here for this verification.

The influence of the two zeta potentials was systematically studied numerically. Some results are illustrated in Figs 11 and 12. The variations in the dimensional coupling coefficient $< \alpha >$ are seen to be linear functions of the zeta potentials ζ_1 and ζ_2. This is expected since these effects are generally speaking proportional to ζ as shown in Ref. 1. Here, the situation is slightly different since there are two zeta potentials ζ_1 anad ζ_2 which cause the coupling phenomenon. Hence, $< \alpha >$ is expected to vary as

$$< \alpha >= \zeta_1 < \alpha_1 > + \zeta_2 < \alpha_2 > \tag{102}$$

where the tensors $< \alpha_1 >$ and $< \alpha_2 >$ are functions of all the other properties listed in Eqs (98). This also explains why the curve does not go to zero when ζ_2 tends towards 0 in Fig 12.

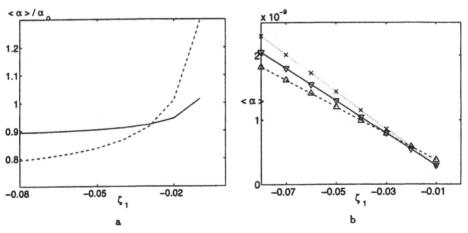

FIG. 11 Influence of the zeta potential ζ_1. Data are for the properties listed in Eq. (98). (a) Dimensionless components of the coupling tensor $< \alpha >$ as functions of ζ_1: $\dfrac{< \alpha_{xx} >}{\alpha_0}$ (solid line), $\dfrac{< \alpha_{zz} >}{\alpha_0}$ (dashed line); (b) dimensional coupling coefficients: $\alpha_0(\times)$, $< \alpha_{xx} > (\nabla)$, $< \alpha_{zz} > (\Delta)$. The lines are linear regression.

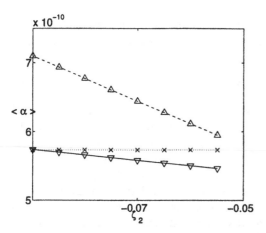

FIG. 12   Influence of the zeta potential ζ_2 on $\alpha_o(\times)$, $<\alpha_{xx}>(\nabla)$, and $<\alpha_{zz}>(\triangle)$. Data are for the properties listed in Eq. (98). The lines are linear regression lines.

The influence of κ was also investigated. Starting from the standard case $\kappa = 10^8$ (cf 98), κ was decreased down to 0.2×10^8; this corresponds to double-layer thicknesses ranging from 10 to 50 nm. The corresponding variations in the macroscopic properties are small; α_o which is related to medium 1 varies between $0.573\ 10^{-9}$ and $0.570\ 10^{-9}$, $\dfrac{<\alpha_{xx}>}{\alpha_o}$ between 0.946 and 0.940, and $\dfrac{<\alpha_{zz}>}{\alpha_o}$ between 1.01 and 0.984. Hence, the macroscopic coupling coefficients slightly increase with κ; this sense of variation is in agreement with the one obtained at the local scale in Ref. 1.

The geometrical parameters were then studied. The influence of anisotropy was analyzed in the following way: the two correlation lengths λ_x and λ_y were kept fixed and equal to 16Δ (in arbitrary units); λ_z was varied between 4Δ (which corresponds to a disk-like shape of the inclusions as illustrated in Fig. 10c and to the standard case, Eq. (98)] and 16Δ (which corresponds to an isotropic situation). Since the physical properties of each medium are kept constant, the reference values K_o, σ_o, and α_o are constant. Generally speaking, for each property A (cf. Eq. (101)], the values A_{xx} and A_{zz} become equal when the medium is isotropic as they should; one of them increases with λ_z while the other one decreases. The largest relative variations are obtained for $<K_{zz}>$; $\dfrac{<\alpha_{xx}>}{a_o}$ is almost constant, while the variations of $\dfrac{<\alpha_{xx}>}{\alpha_o}$ are relatively more important.

Finally, the influence of the probability P_1 of presence of phase 1 was systematically studied for the standard case [Eq. (98)]. Here again, the reference values K_o, σ_o and α_o are constant. The physical properties correspond to Eq. (98). It is remarkable that all the properties vary linearly with p_1. The regression lines are given by

$$\frac{<\alpha_{xx}>}{\alpha_0} = 1 - 0.54(1 - P_1) \ , \quad \frac{<\alpha_{zz}>}{\alpha_o} = 1 + 0.07(1 - P_1) \tag{103a}$$

$$\frac{<K_{xx}>}{K_o} = 1 - 1.24(1 - P_1) \ , \quad \frac{<K_{zz}>}{K_o} = 1 - 2.75(1 - P_1) \tag{103b}$$

$$\frac{<\sigma_{xx}>}{\sigma_o} = 1 - 0.95(1 - P_1) \ , \quad \frac{<\sigma_{zz}>}{\sigma_o} = 1 - 1.60(1 - P_1) \tag{103c}$$

These linear functions of $(1 - P_1)$ are of the same nature as the expansions which can be found in the literature for properties such as conductivity (see Ref. 8 for a general survey). The explanation of the linear dependences of all these properties of P_1 would necessitate a systematical theoretical effort as already mentioned in Ref. 3.

4. Fractured Media

These media are composed of two media denoted by the indices 1 and 2; 1 is the porous medium which corresponds again to a typical Fontainebleau sandstone [cf. Eq. (98)], while 2 corresponds to a fractured zone. The standard case can be summarized by

$$
\begin{aligned}
&\kappa = 10^8 \ , \quad \sigma_f = 0.05 \\
&P_1 = 0.8 \ , \quad \lambda_x = \lambda_y = 16\Delta \ , \quad \lambda_z = 4\Delta \ , \quad N_c = 64 \\
&\epsilon_1 = 0.3 \ , \quad \mathcal{L}_{c1} = 2.10^{-5}, \quad \zeta_1 = -0.02 \\
&\epsilon_2 = 0.35 \ , \quad \mathcal{L}_{c2} = 2.10^{-4} \ , \quad \zeta_2 = -0.05
\end{aligned}
\tag{104}
$$

to which Eqs (63), (65), and (67) should be added.

The major distinction between the two media is the correlation length $\mathcal{L}_{ci}(i = 1, 2)$ [cf. Eq. (62)], with $\mathcal{L}_{c2}$ 10 times larger than $\mathcal{L}_{c1}$. Except for conductivity, this implies large differences in the properties of the media:

$$
\begin{aligned}
&\Lambda_1 = 0.131 \cdot 10^{-4} \ , \quad <K_1> = 0.876 \cdot 10^{-12} \ , \quad <\sigma_1> = 0.320 \cdot 10^{-2} \\
&\alpha_1 = 0.573 \cdot 10^{-9} \ , \quad a_1 = 0.117 \cdot 10^{-6} \\
&\Lambda_2 = 0.144 \cdot 10^{-3} \ , \quad <K_2> = 0.156 \cdot 10^{-9} \ , \quad <\sigma_2> = 0.471 \cdot 10^{-2} \\
&\alpha_2 = 0.211 \cdot 10^{-8} a_2 = 0.606 \cdot 10^{-8}
\end{aligned}
\tag{105}
$$

Again the reference values K_o, σ_o, and α_o are related to medium 1.

In this preliminary study, it was found useless to make a full analysis of all the possible parameters and partly to duplicate what has been done for stratified media. Instead, a few preliminary and complementary checks were done on the influence of the local correlation lengths $\mathcal{L}_{ci}(i = 1, 2)$.

When the local correlation length $\mathcal{L}_{c1}$ is doubled, the permeability is multiplied by a factor of 4, the conductivity is unchanged, and the coupling coefficients are only very slightly modified; α_o varies from $0.5733 \ 10^{-9}$ to $0.5738 \ 10^{-9}$, $\dfrac{<\alpha_{xx}>}{\alpha_o}$ from 1.224 to 1.289, and $\dfrac{<\alpha_{zz}>}{\alpha_q}$ from 1.101 to 1.124.

A similar set of computations was performed with a doubling of $\mathcal{L}_{c2}$ from 200 to 400 μm. The probability P_2 of phase 2 is changed to 0.1. All the characteristics relative to medium 1 are of course unchanged, including α_o. The permeability K_2 is multiplied by a factor of 4. Again, the coupling coefficients are only slightly modified; $\dfrac{<\alpha_{xx}>}{a_o}$ varies from 1.078 to 1.059, and $\dfrac{<\alpha_{zz}>}{\alpha_o}$ from 1.044 to 1.040.

V. APPLICATION TO LA FOURNAISE VOLCANO

Let us present in this section a first large-scale application of the previous developments, which is contained in Ref. 32.

La Fournaise volcano (Réunion Island), 2630 m high, is one of the most active basaltic volcanoes in the world with one eruption every 18 months. It is characterized by heavy rainfalls reaching 6 m per year and large anomalies in the electric potential, which may be of the order of 2 V [33]. A large fracture zone, about 500 m wide, cuts across the whole volcano. As mentioned in Ref. 34, this zone is likely to play an important part in the water circulation through the massif.

For this study, we adopt the simplified structure shown in Fig. 13 which presents a cross-section of the volcano perpendicular to the fracture zone; the topography is simplified so that the cross-section is of a constant height $L \sim 2$ km, of the order of the height of the volcano; the top corresponds roughly to the top of the volcano and the bottom to sea level. This cross-section consists of three zones; the two external zones are assumed to be a porous medium with identical properties surrounding a fractured zone of width $2h \approx 500$ m with different properties. For simplicity, the structure is assumed to be translationally invariant along the z axis which is perpendicular to the plane of the figure.

Meteoritic water is assumed to flow through this structure. The upper and lower boundaries of the porous medium ($x = 0$ and $L.|y| > h$) are assumed to be impermeable; hence, water flows through the central zone and drags along water contained in the porous medium. We investigate the influence of this flow on the electric potential and the magnetic field which can be measured at the volcano surface ($x = 0$).

According to (1), the seepage velocity U and the current density I are given by

$$I_i = -\sigma_i \nabla \Psi_i - \alpha_i \nabla P_i \tag{106a}$$

$$U_i = -\alpha_i \nabla \Psi_i - \frac{K_i}{\mu} \nabla P_i \tag{106b}$$

where the subscript i stands for the porous medium ($i = p$) or for the fractured medium ($i = f$).

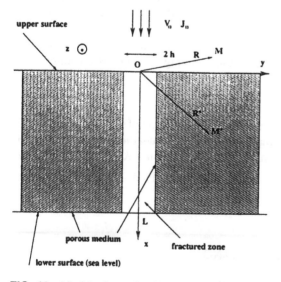

FIG. 13   Model schematic of La Fournaise volcano. V_0 and J_0 are defined in Eq. (110). The x and y axes are vertical and horizontal, respectively.

Conservation of mass and current requires that U_i and I_i satisfy the continuity equations, analogous to Eq. (2), in each region:

$$\nabla \cdot I_i = 0, \nabla \cdot U_i = 0, \text{for } i = p, f \tag{107}$$

This system has to be supplemented by boundary conditions. At $|y| = h, x \geq 0$, pressure, potential, and normal fluxes are all continuous:

$$P_p = P_f, \Psi_p = \Psi_f, n \cdot I_p = n \cdot I_f, n \cdot U_p = n \cdot U_f \tag{108}$$

where n denotes the unit normal to the discontinuity.

At $x = 0$ and L, for $|y| \geq h$, no flux crosses the surface. For $|y| < h$, the local seepage velocity V_0 and local current density J_0 are assumed not to vary with y. To summarize:

$$|y| \geq h, x = 0 \text{ and } L : u_{px} = 0, I_{px} = 0 \tag{109}$$

$$|y| \leq h, x = 0 \text{ and } L : u_{fx} = V_0, I_{fx} = J_0 \tag{110}$$

Other boundary conditions could have been chosen, such as $\psi = constant$ at $x = L$; however, measurements are performed at $x \sim 0$, and are insensitive to the conditions at $x = L$. With this set of boundary conditions, an analytical solution could be easily derived [32]. Once the solution is obtained, the magnetic field $B(R)$ at point M (cf. Fig. 13) is deduced by using the Biot–Savart law:

$$B(R) = \frac{\mu_0}{4\pi} \int \int \int_{x' \geq 0} I(R') \times \frac{R - R'}{|R - R'|^3} dR' \tag{111}$$

where μ_0 is the magnetic permittivity ($= 4\pi \times 10^{-7}$). The integration is performed over the whole medium (i.e., for $x' > 0$). $B(R)$ has a single horizontal component B_z. The observation point M is usually located slightly above the ground level (here, calculations are made for 1 m). The integration over z' can be performed analytically because of the translational symmetry along the z axis; the domain can also be reduced to the quarter plane $x > 0$, $y \geq 0, z = 0$ because of the symmetry with respect to the x axis.

The many parameters which govern this situation can be divided into several groups. The first group is the geometry of the layers; since it is imposed by the geometry of the volcano, it is kept fixed:

$$h = 260 \text{ m}, \quad L = 2000 \text{ m} \tag{112}$$

The second group includes the ingoing (and thus outgoing) fluxes. Usually, J_0 is taken to the equal to zero. An order of magnitude of the seepage velocity is estimated by assuming that when water is provided to the system, it falls under its own weight in the fractured zone:

$$V_0 = \rho g \frac{K_f}{\mu} \tag{113}$$

The third group of parameters consists of the local properties K_i, α_i, and σ_i. For a given medium and fluid, they can be either measured or calculated given certain assumptions [1]. Some of their properties are not intrinsic and also depend on water salinity and other physicochemical conditions. Hence, all these quantities can vary independently, at least to a first approximation.

The studied parameters are summarized in Table 2. With present knowledge, it was found preferable to make a sensitivity study around a set of central values, given as case c1.

TABLE 2 Range of Studied Parameters

Case	σ_p	α_p	K_p	σ_f	α_f	K_f
c1	0.01	3×10^{-9}	10^{-12}	0.1	3×10^{-9}	10^{-10}
c2	0.1	3×10^{-9}	10^{-10}	0.1	3×10^{-9}	10^{-10}
c3	0.001	3×10^{-9}	10^{-12}	0.01	3×10^{-9}	10^{-10}
c4	0.01	10^{-8}	10^{-12}	0.1	10^{-8}	10^{-10}
c5	0.01	3×10^{-9}	10^{-11}	0.1	3×10^{-9}	10^{-9}

All quantities are in SI units: σ (S), α (m^2 s^{-1}V^{-1}), K (m^2). Data are for $J_0 = 0$.

The electromagnetic fields predicted for ground level are displayed in Figs 14 and 15 for all the cases listed in Table 2. First of all, for case c1, it should be emphasized that the order of magnitude of $\Psi (\approx -1$ V) is consistent with the observations made [33]. These fields present a remarkable angular point at $|y| = h$ which is the limit of the fractured zone; this is consistent with the boundary conditions (108) where the normal components of the fluxes are continuous, but not the derivatives of the fields. Note that Ψ tends to zero far from the fractured zone. For $|y| < 1000$ m, B does not depend on the size of the integration domain of Eq. (111) along the y axis; computations were performed for sizes equal to 4000 and 10,000 m without any significant influence.

The influence of the heterogeneous character of the medium is demonstrated in these figures by the homogeneous situation c2 where the permeability of the porous medium is equal to that of the fractured zone. First, the kinks disappear from the curves as they should; second, Ψ and B_z are reduced by a factor 10. Hence, as is well known, the heterogeneous character of the underground medium is an essential feature for produducing large fields. The residual fields which remain in c2 are due to the no-flux condition (109) at the surface of the porous medium.

The influence of the conductivities was analyzed in case c3 where they were both divided by a factor of 10 (cf. Table 2). This parameter may vary widely; it depends on the local structure of the media and of the chemical content of the water. It is seen in Figs 14 and 15 that Ψ is multiplied by a factor of 10, while B_z is superposed with case c1 and thus

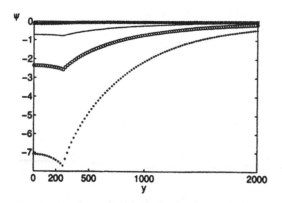

FIG. 14 The potential field (volts) on the ground ($x = 0$) as a function of the distance y (m) to the symmetry axis. Data are for: c1 (solid line); c2 (*); c3 ($\cdots$); c4 ($\bigcirc$); c5 ($- \cdot -$); c5 is overprinted by c1.

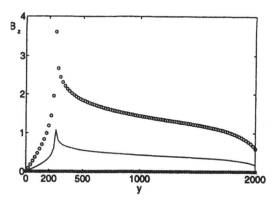

FIG. 15 Magnetic field B_z (nT) at 1 m above the ground as a function of the distance y (m) to the symmetry axis. Data are for: c1 (solid line); c2 (∗), c3 (···); c4 (○), c5 (– · –); c1 is overprinted by c3 and c5.

does not appear clearly in Fig. 15. Local water measurements are necessary to fix this important parameter.

The influence of larger coupling factors α_i was also clearly demonstrated by case c4 where these two factors are multiplied by a factor of 3 with respect to c1; Ψ and B_z are both roughly multiplied by a factor of 3. These orders of magnitude are the ones which fit the best the measured values [34]; note that the proposed variation is very small when compared with variations measured on plugs [35]. Hence, this factor should also be measured on samples extracted at La Fournaise.

Stronger permeabilities of the medium were studied in case c5 where both permeabilities were multiplied by a factor of 10 with respect to c1. U is modified accordingly and its influence on Ψ and B_z is negligible.

VI. CONCLUDING REMARKS

At the local scale, electrokinetic phenomena in porous media and in isolated fractures have been analyzed in the linear limit. The numerical code that has been developed gives reliable results for Debye–Hückel length comparable to the pore radius. Hence, the results are applicable mostly to finely dispersed media such as clays. In the opposite limit, the classical Overbeek formula can be used.

At the plug scale, the equations need to undergo a second change of scale. Physically, porosity may have some strong variations which imply changes in the transport properties of the medium. This change of scale has been done both numerically and theoretically; major results are summarized.

Finally, a preliminary example of application to a volcano is given.

Extensions are presently performed for all these scales. At the pore scale, it is interesting to derive values for multiphase flows through porous media and fractures. The same is true at the field scale where the most important effects are likely to take place in the nonsaturated zone. Finally, at the site scale, a complete 3D simulation is presently being performed of the La Fournaise volcano where the two-phase nature of the flow is taken into account and where the real geometry of the volcano is used.

REFERENCES

1. D Coelho, M Shapiro, JF Thovert, PM Adler. J Colloid Interface Sci 181: 169, 1996.
2. S Marino, D Coelho, S Békri, PM Adler. J Colloid Interface Sci 223: 292, 2000.
3. PM Adler. Math Geol 33: 63, 2001.
4. SR de Groot, P Mazur. Non-equilibrium Thermodynamics. Amsterdam: North-Holland, 1969.
5. D Coelho, R Thouy, JF Thovert, PM Adler. Fractals 5: 507, 1997.
6. PM Adler, D Coelho, JF Thovert, M Shapiro. In: JA Schwarz, CI Contescu, eds. Surfaces of Nanoparticles and Porous Materials. Surfactant Science Series, vol. 78, New York: Marcel Dekker, 1999, p 211.
7. VG Levich. Physico-chemical Hydrodynamics. Englewood Cliffs, NJ: Prentice-Hall, 1962.
8. PM Adler. Porous Media: Geometry and Transports. Stoneham, MA: Butterworth/ Heinemann, 1992.
9. H Brenner. Phil Trans Roy Soc London 297: 81, 1980.
10. D Coelho. Thesis, "Génération, géométrie et propriétés de transport des milieux granulaires", Université de Poitiers, Poitiers, France, 1996.
11. D Coelho, JF Thovert, PM Adler. Phys Rev E 55: 1959, 1997.
12. PM Adler, CG Jacquin, JA Quiblier. Int J Multiphase Flow 16: 691, 1990.
13. PM Adler, JF Thovert. Fractures and Fracture Networks. Dordrecht: Kluwer, p 429, 1999.
14. SR Brown, CH Scholz. J Geophys Res 91: 4939, 1986.
15. S Gentier. Thesis, "Morphologic et comportement hydromécanique d'une fracture naturelle dans le granit sous contrainte normale", Université d'Orléans, Orléans, France, 1986.
16. SR Brown, RL Kranz, BP Bonner. Geophys Res Lett 13: 1430, 1986.
17. KJ Maloy, A Hansen, EL Hinrichsen, S Roux. Phys Rev Lett 68: 213, 1993.
18. JTG Overbeek. In: HR Krugt, ed. Colloid Science. vol. 1. Amsterdam: Elsevier, pp 115–195, 1952.
19. DL Johnson, J Koplik, L Schwartz. Phys Rev Lett 57: 2564, 1986.
20. DB Pengra, S Li, SX Li, PZ Wong. In: JM Drake, J Klafter, R Kopelman, SM Troian, eds. Dynamics in Small Confining Systems 11. Mat Res Soc Symp Proc. vol. 366, 1995.
21. N Martys, EJ Garboczi. Phys Rev B 46: 6080, 1992.
22. CG Jacquin. Rev Inst Français Pétrole 19: 921, 1964.
23. J Yao, J-F Thovert, PM Adler, CD Tsariroglou, VN Burganos, AC Payatakes, J-C Moulu, F Kalaydijian. Rev Inst Français Pétrole 52: 3, 1997.
24. S Békri, K Xu, F Yousefian, PM Adler, J-F Thovert, J Muller, K Iden, A Psyllos, AK Stubos, MA Ioannidis. J. Petroleum Sci and Engineering 25: 107, 2000.
25. MJ Beran. Statistical Continuum Theories. New York: Interscience, 1968.
26. R Landauer. In: JC Garland, DB Tanner, eds. Electrical Transport and Optical Properties of Inhomogeneous Media. AID Conference Proceedings No. 40, American Institute of Physics, 1978.
27. AL Gutjahr, LW Gelhar, AA Bakr, JR McMillan. Wat Resour Res 14: 953, 1978.
28. P Indelman, B Abramovich. Wat Resour Res 30: 1857, 1994.
29. A de Wit. Phys Fluids 7: 2553, 1995.
30. G Dagan. Flow and Transport in Porous Formation. Berlin: Springer-Verlag, pp. 465, 1989.
31. O Wiener. Abhandl. d.K.S. Gesselsch. d. Wissensch. MathPhys 32:509, 1912.
32. PM Adler, J-L Le Mouel, J Zlotnicki. Geophys Res Lett 26: 795, 1999.
33. S Michel, J Zlotnicki. J Geophys Res B 103: 845, 1998.
34. J Zlotnicki, JL Le Mouel. Nature 343: 633, 1990.
35. T Ishido, H Mizutani. J Geophys Res 86: 1763, 1981.

19

Electrokinetic Effects on Pressure-driven Liquid Flow in Microchannels

DONGQING LI[*]   University of Alberta, Edmonton, Alberta, Canada

I. INTRODUCTION

Interfacial electrokinetic phenomena such as electro-osmosis, electrophoresis, and electro-viscous effects have been well known to colloidal and interfacial sciences for many years. However, the effects of these phenomena on transport processes (such as liquid flow and mixing in fine capillaries) are generally less well understood. Partially this is because in colloidal sciences electrokinetic phenomena are usually studied for closed systems, while in studies of transport phenomena the systems are open and involve complicated boundary conditions. Recently, the demands to understand interfacial electrokinetic phenomena in liquid transport processes increase significantly as various microelectromechanical systems (MEMS) and microfluidic devices involve liquid transport processes.

Just as rapid advances in microelectronics have revolutionized computers, appliances, communication systems, and many other devices, MEMS and microfluidic technologies will revolutionize many aspects of applied sciences and engineering, such as heat exchangers, pumps, combustors, gas absorbers, solvent extractors, fuel processors, and on-chip biomedical and biochemical analysis instruments. These lightweight, compact and high-performance microsystems will have many important applications in transportation, buildings, military, environmental restoration, space exploration, environmental management, and biochemical and other industrial chemical processing. Generally, the key advantages of MEMS and microfluidic include increases rates of heat, mass transfer, chemical reactions, and significantly reduced quantity of samples.

A fundamental understanding of liquid flow in microchannels is critical to the design and process control of various MEMS (e.g., micro pump and micro flow sensors) and modern on-chip instruments used in chemical analysis and biomedical diagnostics (e.g., miniaturized total chemical analysis system). However, many phenomena of liquid flow in microchannels, such as unusually high flow resistance, cannot be explained by the conventional theories of fluid mechanics. These are largely due to the significant influences of interfacial phenomena such as electroviscous effects at the micrometer scale.

In various processes involved in MEMS and microfluidic devices, a desired amount of a liquid is forced to flow through microchannels from one location to another. Depending on the specific structures of the MEMS or the microfluidic devices, the

---

[*]*Current affiliation*: University of Toronto, Toronto, Ontario, Canada.

shape of the cross-section of the microchannels varies. Two typical cases are the slit microchannels and the trapezoidal (may be approximated as rectangular) microchannels. This chapter tries to show how to model and to evaluate the interfacial electrokinetic effects on liquid flow through these microchannels.

II. ELECTRICAL DOUBLE LAYERS AND ELECTROVISCOUS EFFECTS

It is well known that most solid surfaces carry electrostatic charges, i.e., an electrical surface potential. If the liquid contains a certain amount of ions (for instance, an electrolyte solution or a liquid with impurities), the electrostatic charges on the solid surface will attract the counterions in the liquid. The rearrangement of the charges on the solid surface and the balancing charges in the liquid is called the electrical double layer (EDL) [1, 2], as illustrated in Fig. 1. Immediately next to the solid surface, there is a layer of ions which are strongly attracted to the solid surface and are immobile. This layer is called the compact layer, normally about several ångstroms (Å) thick. Because of the electrostatic attraction, the counterion concentration near the solid surface is higher than that in the bulk liquid far away from the solid surface. The coion concentration near the surface, however, is lower than that in the bulk liquid, due to the electrical repulsion. So there is a net charge in the region close to the surface. From the compact layer to the uniform bulk liquid, the net charge density gradually reduces to zero. Ions in this region are less affected by electrostatic interaction and are mobile. This region is called the diffuse layer of the EDL. The thickness of the diffuse layer is dependent on the bulk ionic concentration and electrical properties of the liquid, usually ranging from several nanometers for high ionic concentration solutions up to several micrometers for pure water and pure organic liquids. The boundary between the compact layer and the diffuse layer is usually referred to as the shear plane. The electrical potential at the solid–liquid surface is difficult to measure directly. The electrical potential at the shear plane, however, is called the zeta potential, ζ, and can be measured experimentally [1, 2].

According to the theory of electrostatics, the relationship between the electrical potential ψ and the local net charge density per unit volume ρ_e at any point in the solution is described by the Poisson equation:

$$\nabla^2 \psi = -\frac{\rho_e}{\varepsilon} \tag{1}$$

where ε is the dielectric constant of the solution.

Assuming that the equilibrium Boltzmann distribution equation is applicable, which implies a uniform dielectric constant, the number concentration of the type-i ion in a symmetric electrolyte solution is of the form:

$$n_i = n_{io} \exp\left(-\frac{z_i e\psi}{k_B T}\right) \tag{2}$$

where n_{io} and z_i are the bulk ionic concentration and the valence of type-i ions, respectively, e is the charge of a proton, k_B is the Boltzmann constant, and T is the absolute temperature. In the simple case of a symmetric electrolyte of valence z, the net volume charge density ρ_e is proportional to the concentration difference between cations and anions via

$$\rho_e = ze(n_+ - n_-) = -2zen_0 \sinh\left(\frac{ze\psi}{k_B T}\right) \tag{3}$$

where n_0 is the bulk number concentration of each ion.

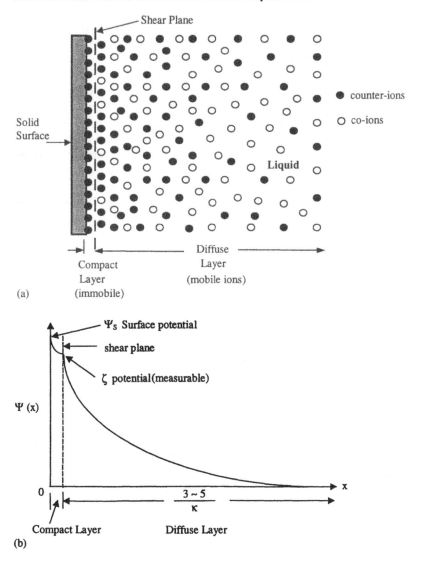

FIG. 1 Illustration of an EDL near a flat solid–liquid interface: (a) ion distribution; (b) electrical potential distribution.

Substituting Eq. (3) into the Poisson equation leads to the well-known Poisson–Boltzmann equation.

$$\nabla^2 \psi = \frac{2zen_0}{\varepsilon} \sinh\left(\frac{ze\psi}{k_B T}\right) \tag{4}$$

By defining the Debye–Hückel parameter $k^2 = 2z^2 e^2 n_0 / \varepsilon k_B T$ and the nondimensional electrical potential $\overline{\psi} = ze\psi / k_B T$, the Poisson–Boltzmann equation can be rewritten as

$$\nabla^2 \overline{\psi} = k^2 \sinh \overline{\psi} \tag{5}$$

Generally, solving this equation with appropriate boundary conditions, the electrical potential distribution ψ of the EDL can be obtained, and the local charge density distribution ρ_e can then be determined from Eq. (3).

It should be noted that the Debye–Hückel parameter is independent of the solid surface properties and is determined by the liquid properties (such as the electrolyte's valence and the bulk ionic concentration) only; $1/k$ is normally referred to as the characteristic thickness of the EDL and is a function of the electrolyte concentration. Values of $1/k$ range, for example, from 9.6 nm at 10^{-3} M to 304.0 nm at 10^{-6} M for a KCl solution. The thickness of the diffuse layer usually is about three to five times $1/k$, and hence may be as large as a few micrometers for pure water and pure organic liquids. Generally, for macrochannel flow the EDL effects can be safely neglected, as the thickness of the EDL is very small compared with the hydraulic diameter of channels. However, for microchannel flow the thickness of the EDL is often comparable with the characteristic size of flow channels. Thus, the EDL effects originated from the electrostatic interaction between ions in liquid and the charged solid (flow channel) surface may play an important role in microchannel flow and heat transfer.

When a liquid is forced through a microchannel under an applied hydrostatic pressure, the counterions in the diffuse layer (mobile part) of the EDL are carried towards the downstream end, resulting in an electrical current in the pressure-driven flow direction. This current is called the streaming current. Corresponding to this streaming current, there is an electrokinetic potential called the streaming potential. This flow-induced streaming potential is a potential difference that builds up along a microchannel. The streaming potential acts to drive the counterions in the diffuse layer of the EDL to move in the direction opposite to the streaming current, i.e., opposite to the pressure-driven flow direction. The action of the streaming potential will generate an electrical current called the conduction current, as illustrated in Fig. 2. It is obvious that when ions move in a liquid, they will pull the liquid molecules to move with them. Therefore, the conduction current will produce a liquid flow in the opposite direction to the pressure-driven flow. The overall result is a reduced flow rate in the pressure drop direction. If the reduced flow rate is compared with the flow rate predicted by conventional fluid mechanics theory without

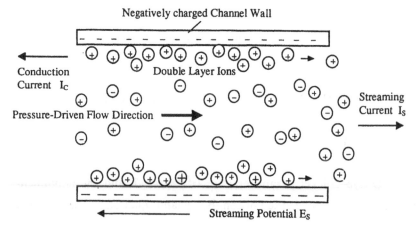

FIG. 2 Illustration of the flow-induced electrokinetic field in a microchannel (assume negatively charged channel wall surface). Steady state: $I_{net} = 0$, i.e., $I_s = I_c$.

considering the presence of the EDL, it seems that the liquid would have a higher viscosity. This is usually referred to as the electroviscous effect [1].

III. ELECTROKINETIC FLOWS IN SLIT MICROCHANNELS

In this section, we consider the liquid flow through a microchannel with a slit-shaped cross-section [3, 4], such as a channel formed between two parallel plates, as illustrated in Fig. 3. We assume that the length, L, and the width, W, of the slit channel are much larger than the height, $H = 2a$, of the channel, so that both the EDL field and the flow field can be considered as one-dimensional (i.e., with variation in the channel height direction only).

A. Poisson–Boltzmann Equation

Consider a liquid containing simple symmetric ions, i.e., the valences of the ions are the same, $z = z^+ = z^-$. For the slit microchannel, we consider only the EDL fields near the top and the bottom plates. The one-dimensional EDL field for a flat surface is described by the following form of the Poisson–Boltzmann equation:

$$\frac{d^2\psi}{dX^2} = \frac{2n_0 ze}{\varepsilon_0 \varepsilon} \sinh\left(\frac{ze\psi}{k_B T}\right) \tag{6}$$

The local net charge density in the liquid is given by Eq. (3). By nondimensionalizing eqs (2) and (6) via

$$\overline{X} = \frac{X}{a}, \quad \overline{\psi} = \frac{ze\psi}{k_B T}, \quad \overline{\rho} = \frac{\rho}{n_0 ze} \tag{7}$$

we obtain the nondimensional form of the Poisson–Boltzmann equation as

$$\frac{d^2\overline{\psi}}{d\overline{X}^2} = \kappa^2 \sinh(\overline{\psi}) = -\frac{\kappa^2}{2}\overline{\rho} \tag{8}$$

$$\overline{\rho} = -2\sinh(\overline{\psi}) \tag{9}$$

where $\kappa = a \times k = a/1/k$ is the electrokinetic separation distance or the ratio of the half-channel's height to the EDL thickness. Therefore, the parameter κ can be understood as the relative channel's height with respect to the EDL thickness.

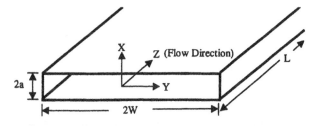

FIG. 3 A slit microchannel; the channel height is much less than the channel width and length.

B. Solution of the Poisson–Boltzmann Equation

If the electrical potential is small compared to the thermal energy of the ions, i.e., ($|ze\psi| < |k_BT|$) so that the $\sinh(\overline{\psi})$ function in Eq. (8) can be approximated by $\sinh(\overline{\psi}) \cong \overline{\psi}$, Equation (8) is transformed into

$$\frac{d^2\overline{\psi}}{d\overline{X}^2} = \kappa^2\overline{\psi} \tag{10}$$

This treatment is usually called the Debye–Hückel linear approximation [1, 5, 6]. The solution of Eq. (10) can be easily obtained. As illustrated in Fig. 3, if the separation distance between the two plates is sufficiently large so that the EDL fields near the two plates are not overlapped, the appropriate boundary conditions for the EDL fields are:

At the center of the slit channel: $\overline{X} = 0,$ $\overline{\psi} = 0$
At the solid surfaces: $\overline{X} = \pm 1,$ $\overline{\psi} = \overline{\zeta} = (ze\zeta/k_BT).$

With these boundary conditions Eq. (6) can be solved and the solution is given by

$$\overline{\psi} = \frac{\overline{\zeta}}{\sinh(\kappa)} |\sinh(\kappa\overline{X})| \tag{11}$$

Equation (11) allows us to plot the nondimensional EDL potential field in the slit channel. Due to the symmetry, we plot only the nondimensional EDL potential field from one surface to the center of the channel, as shown in Fig. 4. In this figure, the zeta potential is assumed to be 50 mV. For a given electrolyte, a large κ implies either a large separation distance between the two plates or a small EDL thickness. If the separation distance ($2a$) is given, increase in the bulk ionic concentration n_0 will increase the value of the Debye–Hückel parameter k [recall that $k = (2n_0z^2e^2/\varepsilon\varepsilon_0k_BT)^{1/2}$], the double-layer thickness $1/k$

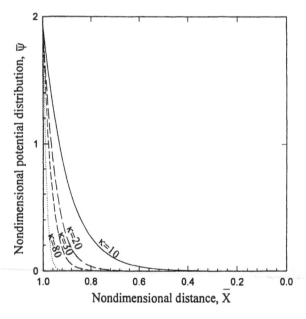

FIG. 4 Nondimensional electrical potential distribution near the channel wall for $\zeta = 50$ mV; $\overline{X} = 0$, center of the channel and $\overline{X} = 1$, the channel wall. (From Ref. 3.)

is reduced (or the double layer is "compressed"), and the electrokinetic separation distance $\kappa = a \times k = a/1/k$ is increased. It can be seen from Fig. 4 that as κ increases, the double-layer field (the $\overline{\psi} \neq 0$ region) exists only in the region close to the channel wall. For example, appreciable EDL potential exists only in a region less than a few per cent of the channel cross-sectional area for $\kappa = 80$. However, for dilute solutions such as pure water (infinitely dilute solution), the value of the electrokinetic separation distance $\kappa = a \times k$ is small, and hence the EDL field (the $\overline{\psi} \neq 0$ region) may affect a significant portion of the flow channel, as shown in Fig. 4.

It should be pointed out that Fig. 4 is the result of using the Debye–Hückel linear approximation, Eq. (11), and that the linear approximation is valid for small surface potential situations (i.e., $\psi \leq 2.5$ mV). However, for large surface potentials, it has been found that, in comparison with the exact solution of the Poisson–Boltzmann equation, Eq. (8), the linearized solution predicts slightly lower values of the potential in the region near the wall. After a short distance from the wall, the difference between the linear solution and the exact solution diminishes. Discussions of these two solutions can be found elsewhere [1, 5, 6].

C. Equation of Motion

Consider a one-dimensional, fully-developed, steady-state, laminar flow through the slit microchannel shown in Fig. 3. The forces acting on an element of fluid include the pressure force, the viscous force, and the electric body force generated by the flow-induced electrokinetic field (i.e., the streaming potential). The equation of motion is the Z-directional momentum equation:

$$\mu \frac{d^2 V_z}{dX^2} + P_z + E_z \rho(x) = 0 \tag{12}$$

Where $E_z = E_s/L$ is the electrokinetic potential gradient in the Z direction, and $E_z \rho(x)$ is the electrical body force; $P_z = -\Delta P/L$ is the pressure gradient in the Z direction. Nondimensionalizing Eq. (12) with $\overline{E}_s = E_s/\zeta_0$, $\overline{V}_z = V_z/V_0$ (V_0 is the fluid velocity at the center of the slit), and replacing $\overline{\rho}$ by Eq. (8), we obtain

$$\frac{d^2 \overline{V}_z}{d\overline{X}^2} + G_1 - \frac{2G_2 \overline{E}_s}{\kappa^2} \frac{d^2 \overline{\psi}}{dX^2} = 0 \tag{13}$$

where the two nondimensional numbers are given by

$$G_1 = \frac{a^2 P_z}{\mu V_0} \quad \text{and} \quad G_2 = \frac{n_0 z e a^2 \zeta_0}{\mu V_0 L} \tag{14}$$

Integrating Eq. (13) twice and employing the appropriate boundary conditions, we obtain the nondimensional velocity distribution in the slit microchannel as follows:

$$\overline{V}_z = \frac{G_1}{2}(1 - \overline{X}^2) - \frac{2G_2 \overline{E}_s \overline{\zeta}}{\kappa^2}\left[1 - \frac{\overline{\psi}}{\zeta}\right] \tag{15}$$

Substituting $\overline{\psi}$ in Eq. (15) by Eq. (11) yields:

$$\overline{V}_z = \frac{G_1}{2}(1 - \overline{X}^2) - \frac{2G_2 \overline{E}_z \overline{\zeta}}{\kappa^2}\left[1 - \left|\frac{\sinh(\kappa \overline{X})|}{\sinh(\kappa)}\right|\right] \tag{16}$$

From the definitions of G_1 and G_2 [Eq. (14)], one can clearly see that the first term in Eq. (16) represents the contribution of the applied pressure gradient, and the second term in Eq. (16) represents the EDL's contribution to the velocity. If there is no EDL effect on the flow, i.e., $G_2 = 0$, $\overline{E}_s = 0$, and $\overline{\zeta} = 0$, Eq. (12) will be reduced to $\overline{V}_z = G_1/2(1 - \overline{X}^2)$, which is the conventional Poiseuille flow velocity profile between two parallel plates.

D. Streaming Potential

As seen from Eq. (16), the velocity distribution can be calculated only if the streaming potential $\overline{E}$, is known. In the absence of an applied electric field, when a liquid is forced through a channel under hydrostatic pressure, the excess counterions in the diffuse layer of the EDL are carried by the liquid to flow downstream, forming an electrical current. This current resulting from the transport of charges by the liquid flow, called the streaming current, is given by

$$I_s = \int_{A_c} V_z \rho(X) \mathrm{d}A_c \tag{17}$$

Nondimensionalizing V_z and ρ, we obtain the nondimensional streaming current as

$$\overline{I}_s = \frac{I_s}{2n_0 V_0 zea} = \int_0^1 \overline{V}_z \overline{\rho} \mathrm{d}\overline{X} \tag{18}$$

Substituting $\overline{\rho}$ by using Eq. (8), the nondimensional streaming current becomes

$$\overline{I}_s = -\frac{2}{\kappa^2} \int_0^1 \overline{V}_z \mathrm{d}\left(\frac{\mathrm{d}\overline{\psi}}{\mathrm{d}X}\right) = -\frac{2}{\kappa^2}\left[\overline{V}_z \frac{\mathrm{d}\overline{\psi}}{\mathrm{d}X}\Big|_0^1 - \int_0^1 \frac{\mathrm{d}\overline{\psi}}{\mathrm{d}\overline{X}} \mathrm{d}\overline{V}_z\right] \tag{19}$$

Using the boundary conditions:

$$\overline{X} = 1, \quad \overline{V}_z = 0$$

$$\overline{X} = 0, \quad \frac{\mathrm{d}\overline{\psi}}{\mathrm{d}\overline{X}} = 0$$

one can easily see that the first term on the righthand side of Eq. (19) becomes zero. Therefore, the streaming current reduces to

$$\overline{I}_s = \frac{2}{\kappa^2} \int_0^1 \frac{\mathrm{d}\overline{\psi}}{\mathrm{d}\overline{X}} \mathrm{d}\overline{V}_z \tag{20}$$

By using Eqs (6), (11), and (13), we can show that the nondimensional streaming current is given by

$$\overline{I}_s = -\frac{2G_1 \overline{\zeta}\beta_1}{\kappa^2} + 4G_2 \overline{E}_s \beta_2 \left(\frac{\overline{\zeta}}{\kappa \sinh(\kappa)}\right)^2 \tag{21}$$

where

$$\beta_1 = 1 - \frac{\cosh(\kappa) - 1}{\kappa \sinh(\kappa)} \quad \text{and} \quad \beta_2 = \frac{\sinh(\kappa)\cosh(\kappa)}{2\kappa} + \frac{1}{2} \tag{21a}$$

The flow-induced streaming potential will drive the counterions in the diffuse layer of the EDL to move back in the opposite direction to the pressure-driven flow, and produce an electrical conduction current. The conduction current in the microchannel has two components, one is due to bulk electrical conductivity of the liquid, and the other corresponds to the electrical conductivity of the solid–liquid interface, as given by

$$I_c = \frac{\lambda_b E_s A_c}{L} + \frac{\lambda_s E_s P_s}{L} \tag{22}$$

where λ_b and λ_s are the bulk and surface conductivity, respectively, E_s/L is the streaming potential gradient, A_c is the cross-sectional area of the channel, and P_s is the wetting perimeter of the channel. The conduction current I_c can be rewritten as

$$I_c = \frac{E_s A_c \lambda_T}{L} \quad \text{where} \quad \lambda_T = \lambda_b + \frac{\lambda_s P_s}{A_c} \tag{23}$$

Nondimensionalizing Eq. (23) by using $\bar{L} = L/a$, $\bar{E}_s = E_s/\zeta_0$, and $\bar{A}_c = A_c/a^2$, the non-dimensional conduction current is given by

$$\bar{I}_c = \frac{I_c}{\zeta_0 \lambda_T a} = \frac{\bar{E}_s \bar{A}_c}{\bar{L}} \tag{24}$$

When the flow in the microchannel reaches a steady state, there will be no net electrical current in the flow, i.e., $I_c + I_s = 0$. Using I_c from Eq. (24) and I_s from Eq. (21), we obtain the streaming potential from the steady-state condition:

$$\bar{E}_s = \frac{2G_1 G_3 \bar{\zeta} \beta_1}{\kappa^2 + 4\beta_2 G_2 G_3 (\bar{\zeta}/\sinh(\kappa))^2} \tag{25}$$

where the nondimensional factor $G_3 = V_0 n_0 z e L / \zeta \lambda_T$.

In the classical theory of electrokinetic flow, the effect of the EDL on liquid flow and the effects of surface conductance are not considered. The streaming potential is related to the zeta potential and liquid properties through the following equation [1, 2]:

$$\frac{E_s}{\Delta P} = \frac{\varepsilon \varepsilon_0 \zeta_0}{\mu \lambda_b} \tag{26}$$

Equation (25) can be rearranged in a form similar to Eq. (26) by translating it in terms of dimensional parameters, i.e.,

$$\frac{E_s}{\Delta P} = \frac{\varepsilon \varepsilon_0 \zeta_0}{\mu \lambda_T} \Phi \tag{27}$$

where the correction factor Φ to the classical theory, Eq. (26), is given as

$$\Phi = \frac{\beta_1}{1 + \dfrac{\beta_2 \kappa^2 \varepsilon^2 \varepsilon_0^2 \zeta_0^2}{\sinh^2(\kappa) a^2 \mu \lambda_T}} \tag{28}$$

If the EDL effect on liquid flow is not considered, $\beta_1 = 1$ [see Eq. (21a)] and the second term in the denominator of Eq. (28) is zero; hence, $\Phi = 1$. Furthermore, if surface conductance is also not considered, i.e., $\lambda_T + \lambda_b$, Eq. (27) becomes the classical equation, Eq. (26).

Generally, the zeta potential, the surface conductance, the pressure gradient, and the liquid properties (such as the bulk ionic concentration, the dielectric constant, and the bulk conductivity) can be measured [1, 2]. By the above analyses, we can calculate the velocity field in the slit microchannel and the flow rate. Figure 5 shows the nondimensional velocity distribution of pure water in a slit silicon microchannel (20 mm (length) × 10 mm (width) × 20 μm (height)) for two sets of total pressure drops and zeta potentials. Clearly, by comparing the case without EDL field (i.e., $\zeta = 0$, we see that the EDL or the electroviscous effect reduces the velocity of the liquid appreciably.

E. Volume Flow Rate and Apparent Viscosity

The volume flow rate through the parallel plates can be obtained by integrating the velocity distribution over the cross-sectional area, as

$$Q = \int_{A_c} V_z dA_c \tag{29}$$

Using Eq. (16), the volume flow rate in nondimensional form is

$$\overline{Q} = \frac{2G_1}{3} - \frac{4G_2\overline{E}_s\zeta}{\kappa^2} - \frac{4G_2\overline{E}_s\zeta}{\kappa^3}\frac{[1 - \cosh(\kappa)]}{\sinh(\kappa)} \tag{30}$$

For steady flow under an applied pressure gradient, the volume flow rate is given by Eq. (30). The first term in this equation is the volume flow rate without the EDL effect. The other two terms clearly reflect the EDL effect on the flow rate and contribute to reduce the net flow in the pressure drop direction. This reduced flow rate seems to suggest that the liquid has a higher viscosity. If we define an apparent viscosity μ_a ($> \mu$), the classical Poiseuille volume flow rate (without considering the EDL effect) for flow between two parallel plates separated by a distance $2a$ is then given by

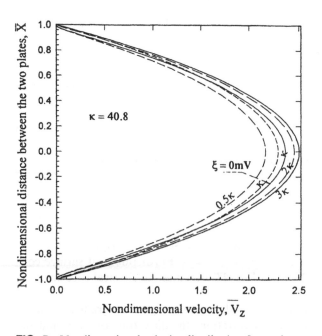

FIG. 5 Nondimensional velocity distribution for various κ values with $\zeta = 50$ mV. (From Ref. 3.)

$$Q_p = \frac{2P_z a^3}{3\mu_a} \tag{31}$$

Nondimensionalizing Eq. (31), the volume flow rate is

$$\overline{Q}_p = \frac{2G_1 \mu}{3\mu_a} \tag{32}$$

Equalizing Eq. (30) with Eq. (32), i.e., $\overline{Q} = \overline{Q}_p$, we obtain the ratio of the apparent viscosity to the true bulk viscosity:

$$\frac{\mu_a}{\mu} = \frac{\kappa^3 G_1}{\kappa^3 G_1 - 6G_2 \overline{E}_s \zeta \kappa - 6G_2 \overline{E}_s \zeta \left\{ \dfrac{1 - \cosh(\kappa)}{\sinh(\kappa)} \right\}} \tag{33}$$

Obviously, if there is no EDL or electrokinetic effect, the second and the third terms in the denominator of Eq. (33) will be zero, and hence $\mu_a = \mu$. Generally, because the second and the third terms in the denominator of Eq. (33) make it smaller than the numerator, it follows that $\mu_a > \mu$. This is usually referred to as the electroviscous effect. From Eq. (33), such an electroviscous effect depends, in addition to the pressure gradient or the flow rate via the parameter G_1, on the ionic properties of the liquid via the Debye–Hückel parameter k, the channel height, a, via the electrokinetic separation distance $\kappa = a \times k$, and the zeta potential ζ.

As explained above, the flow-induced streaming potential drives the counterions to move in the direction opposite to the pressure drop. These moving ions drag the surrounding liquid molecules with them. This generates a flow opposite to the pressure-driven flow and hence a reduced flow rate in the pressure-drop direction. According to the classical Poiseuille flow equation, Eq. (31), this reduced flow rate seems to suggest that the liquid has a higher viscosity. As an example, the ratio of the apparent viscosity to the bulk viscosity, μ_a/μ, is plotted by using Eq. (33) as a function of nondimensional electrokinetic separation distance κ in Fig. 6. It is observed that, for $\zeta = 50$ mV, the ratio μ_a/μ is approximately 2.75 when $\kappa = 2$ and then decreases as κ increases, approaching a constant value equal to unity for very large values of κ. For lower values of ζ the trend is the same except that the value of the ratio is lower. Generally, the higher the zeta potential ζ, the higher the ratio μ_a/μ.

IV. ELECTROKINETIC FLOWS IN RECTANGULAR MICROCHANNELS

Generally, most studies of electrokinetic flow phenomena deal with a one-dimensional EDL field, which holds only for simple geometric channels, such as cylinders and slit-shaped channels. However, in practice, the cross-section of microchannels made by modern micromachining technology is close to a rectangular (trapezoidal, more exactly) shape. As the EDL field depends on the geometry of the solid–liquid interfaces, the EDL field will be two-dimensional in a rectangular micro channel. In such a situation, the two-dimensional Poisson–Boltzmann (P–B) equation is required to describe the electrical potential distribution in the rectangular channel, and the corner of the channel may have a particular contribution to the EDL field and subsequently to the fluid flow field [7, 8].

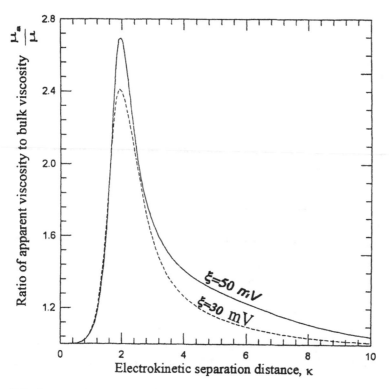

FIG. 6 Variation of the ratio of apparent viscosity to the bulk viscosity with the electrokinetic separation distance κ. (From Ref. 3.)

A. EDL Field in a Rectangular Microchannel

In order to consider the electroviscous effects on liquid flow in rectangular microchannels, we must evaluate distributions of the electrical potential and the net charge density, which can be determined by the theory of the EDL. Consider a rectangular microchannel of width $2W$, height $2H$, and length L as illustrated in Fig. 7. According to the theory of electrostatics, the relationship between the electrical potential ψ and the net charge density per unit volume ρ_e at any point in the solution is described by the two-dimensional Poisson equation [Eq. (1)]:

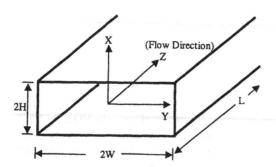

FIG. 7 Liquid flow in a rectangular microchannel (height $2H$, width $2W$).

$$\frac{\partial^2 \psi}{\partial y^2} + \frac{\partial^2 \psi}{\partial z^2} = -\frac{\rho_e}{\varepsilon} \tag{34}$$

Assuming as before that the equilibrium Boltzmann distribution [Eq. (2)] is applicable, and that the electrolyte is z–z valent, the following form of the P–B equation is reached:

$$\frac{\partial^2 \psi}{\partial y^2} + \frac{\partial^2 \psi}{\partial z^2} = \frac{2zen_0}{\varepsilon} \sinh\left(\frac{ze\psi}{k_B T}\right) \tag{35}$$

By using the Debye–Hückel parameter k, and the hydraulic diameter of the rectangular microchannel $D_h = 4HW/(H+W)$, and introducing the dimensionless groups: $Y = y/D_h$, $Z = z/D_h$, $K = kD_h$, and $\Psi = ze\psi/k_B T$, the above equation can be nondimensionalized as

$$\frac{\partial^2 \Psi}{\partial Y^2} + \frac{\partial^2 \Psi}{\partial Z^2} = K^2 \sinh \Psi \tag{36}$$

Due to symmetry, Eq. (36) is subjected to the following boundary conditions in a quarter of the rectangular cross-section:

$$Y = 0 \quad \frac{\partial \Psi}{\partial Y} = 0 \quad Y = \frac{H}{D_h} \quad \Psi = \bar{\zeta} \tag{37a}$$

$$Z = 0 \quad \frac{\partial \Psi}{\partial Z} = 0 \quad Z = \frac{W}{D_h} \quad \Psi = \bar{\zeta} \tag{37b}$$

where $\bar{\zeta}$, the nondimensional zeta potential at the channel wall, was defined in Section III.

For small values of ψ (Debye–Hückel approximation), the P–B equation (36) can be linearized as

$$\frac{\partial^2 \Psi}{\partial Y^2} + \frac{\partial^2 \Psi}{\partial Z^2} = K^2 \Psi \tag{38}$$

By using the method of separation of variables, the solution to the linearized P–B equation (38) can be obtained. Therefore, the electrical potential distribution in the rectangular microchannel is of the form:

$$
\Psi(Y,Z) = 4\bar{\zeta} \sum_{m=1}^{\infty} \frac{(-1)^{m+1} \cosh\left[\sqrt{1 + \dfrac{(2m-1)^2 \pi^2 D_h^2}{4K^2 W^2}} KY\right]}{(2m-1)\pi \cosh\left[\sqrt{1 + \dfrac{(2m-1)^2 \pi^2 D_h^2}{4K^2 W^2}} \dfrac{KH}{D_h}\right]} \cos\left[\frac{(2n-1)\pi D_h}{2W} Z\right]
$$

$$
+ 4\bar{\zeta} \sum_{n=1}^{\infty} \frac{(-1)^{n+1} \cosh\left[\sqrt{1 + \dfrac{(2n-1)^2 \pi^2 D_h^2}{4K^2 H^2}} KZ\right]}{(2n-1)\pi \cosh\left[\sqrt{1 + \dfrac{(2n-1)^2 \pi^2 D_h^2}{4K^2 H^2}} \dfrac{KW}{D_h}\right]} \cos\left[\frac{(2n-1)\pi D_h}{2H} Y\right]
\tag{39}
$$

For large values of ψ, the linear approximation is no longer valid. The EDL field has to be determined by solving the full Eq. (36). In order to solve this nonlinear, two-dimensional differential equation, a numerical finite-difference scheme may be introduced to divide the equation into discrete algebraic equations by integrating the governing differential equation over a control volume surrounding a typical grid point. The nonlinear source term is linearized as

$$\sinh \Psi_{n+1} = \sinh \Psi_n + (\Psi_{n+1} - \Psi_n)\cosh \Psi_n \tag{40}$$

where the subscripts $(n+1)$ and n represent the $(n+1)$th and the nth iterative value, respectively. The derived discrete, algebraic equations can be solved by using the Gauss–Seidel iterative procedure. The solution of the linearized P–B equation with the same boundary conditions may be chosen as the first guess value for the iterative calculation. The under-relaxation technique is employed to make this iterative process converge rapidly. The details of how to obtain the numerical solutions of Eq. (36) can be found elsewhere [8, 9].

After the electrical potential distribution inside the rectangular microchannel is computed, the local net charge density can be obtained from Eq. (3) as

$$\rho_e(Y, Z) = -2zen_0 \sinh \Psi(Y, Z) \tag{41}$$

Figure 8 shows a comparison of the EDL field in a rectangular microchannel predicted by the linear solution and the complete numerical solution. In these calculations, the liquid is a dilute aqueous 1:1 electrolyte solution (concentration is 1×10^{-6} M) at

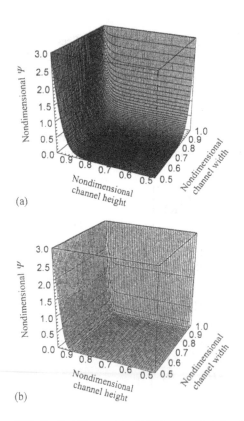

(a)

(b)

FIG. 8 Nondimensional EDL potential distribution in one-quarter of the microchannel, with $\zeta = 75$ mV. (a) Linear solution; (b) complete numerical solution. (From Ref. 9.)

18°C. The rectangular microchannel has a cross-section 30 μm × 20 μm with a zeta potential of 75 mV. Because of the symmetry, the EDL field is plotted only in one-quarter of the microchannel. With such a relatively high zeta potential, the linear approximation is obviously not good. It can be seen that there is a very steep decrease in the potential in the case of the complete solution, while the linear solution predicts a more gradual decay of the potential. The linear solution predicts a thicker layer from the wall of the liquid that has an appreciable nonzero EDL field. In order to see more clearly the differences between the linear solution and the complete solution, the EDL potential profiles (viewed from the channel-height side) are plotted in Fig. 9 for different zeta potentials. Clearly, the linear solution will result in large errors in the region close to the channel wall.

B. Flow Field in a Rectangular Microchannel

Consider the case of a forced, two-dimensional laminar flow through the rectangular microchannel, illustrated in Fig. 7. The equation of motion for an incompressible liquid is given by

$$\rho_f \frac{\partial V}{\partial t} + \rho_f (\underline{V} \cdot \underline{\nabla})\underline{V} = -\underline{\nabla}P + \underline{F} + \mu_f \underline{\nabla}^2 \underline{V} \tag{42}$$

In this equation, ρ_f and μ_f are the density and viscosity of the liquid, respectively. For a steady-state, fully developed flow, the components of velocity $\underline{V}$ (namely, u, v, w) satisfy $u = u(y, -z)$ and $v = w = 0$ in terms of Cartesian coordinates. Thus, both the time term $\partial \underline{V}/\partial t$ and the inertia term $(\underline{V} \cdot \underline{\nabla})\underline{V}$ vanish. Also, the hydraulic pressure P is a function of x only and the pressure gradient dP/dx is constant. If the gravity effect is negligible, the body force $\underline{F}$ is only caused by the action of an induced electrical field E_x (see explanation

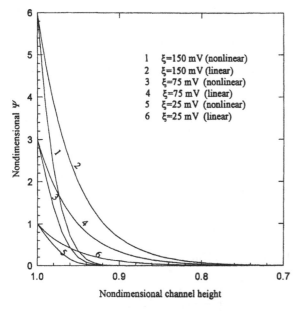

FIG. 9 Comparison of the profiles of the linear and the nonlinear EDL potential distributions for three zeta potential values. (From Ref. 9.)

in the electrokinetic potential section) on the net charge density $\rho_e(Y, Z)$ in the EDL region, i.e., $F_x = E_x\rho_e$. With these considerations, Eq. (42) is reduced to

$$\frac{\partial^2 u}{\partial y^2} + \frac{\partial^2 u}{\partial z^2} = \frac{1}{\mu_f}\frac{dP}{dx} = \frac{1}{\mu_f}E_x\rho_e(y, z) \qquad (43)$$

Defining the reference Reynolds number $\mathrm{Re}_0 = \rho_f D_h U/\mu_f$ and nondimensionalizing Eq. (43) via the following dimensionless parameters:

$$\bar{u} = \frac{u}{U} \qquad\qquad \bar{P} = \frac{P - P_0}{\rho_f U^2} \qquad\qquad X = \frac{x}{D_h \mathrm{Re}_0} \qquad (44a)$$

$$\frac{d\bar{P}}{dX} = \frac{D_h \mathrm{Re}_0}{\rho_f U^2}\frac{dP}{dx} \qquad\qquad \bar{E}_x = \frac{E_x D_h \mathrm{Re}_0}{\zeta_0} \qquad (44b)$$

(where U is a reference velocity, P_0 is a reference pressure, and ζ_0 is a reference electric potential), one can obtain the nondimensional equation of motion:

$$\frac{\partial^2 \bar{u}}{\partial Y^2} + \frac{\partial^2 \bar{u}}{\partial Z^2} = \frac{d\bar{P}}{dX} - \frac{\psi_c}{\rho_f U^2}\bar{E}_x\rho_e(Y, Z) \qquad (45)$$

Substituting $\rho_e(Y, Z)$ by Eq. (41), and defining a new dimensionless number $\bar{G}_1 = 2zen_0\zeta_0/\rho_f U^2$, the equation of motion may, therefore, be written as

$$\frac{\partial^2 \bar{u}}{\partial Y^2} + \frac{\partial^2 \bar{u}}{\partial Z^2} = \frac{d\bar{P}}{dX} + \bar{G}_1\bar{E}_x \sinh \Psi(Y, Z) \qquad (46)$$

The boundary conditions that apply for the velocity $\bar{u}$ are

$$Y = 0 \qquad \frac{\partial \bar{u}}{\partial Y} = 0 \qquad Z = 0 \qquad \frac{\partial \bar{u}}{\partial Y} = 0 \qquad (47a)$$

$$Y = \frac{H}{D_h} \qquad \bar{u} = 0 \qquad Z = \frac{W}{D_h} \qquad \bar{u} = 0 \qquad (47b)$$

Here, Eq. (47a) is the symmetric condition and Eq. (47b) is the nonslip condition at the walls of the microchannel.

By using Green's function formulation, the solution of Eq. (46) subject to the above boundary conditions is

$$\bar{u}(Y, Z) = -\lim_{t\to\infty}\int_{\tau=0} d\tau \int_{Y'=0}^{H/D_h}\int_{Z'=0}^{W/D_h} G(Y, Z, t|Y', Z', \tau)\left[\frac{d\bar{P}}{dX} + \bar{G}_1\bar{E}_x \sinh \Psi(Y', Z')\right]dY'dZ'$$

$$(48)$$

Here, $G(Y, Z, t|Y', Z', \tau)$ is the Green function which may be found by using the method of separation of variables [10]. The expression for $G(Y, Z, t|Y', Z', \tau)$ is

$$G(Y, Z, t|Y', Z', \tau) = \frac{4D_h^2}{WH}\sum_{m=1}^{\infty}\sum_{n=1}^{\infty}\exp\left\{\frac{-\pi^2 D_h^2}{4}\left[\frac{(2m-1)^2}{H^2} + \frac{(2n-1)^2}{W^2}\right](t - \tau)\right\}$$

$$\times \cos\left[\frac{(2m-1)\pi D_h}{2H}Y\right]\cos\left[\frac{(2m-1)\pi D_h}{2H}Y'\right]$$

$$\times \cos\left[\frac{(2n-1)\pi D_h}{2W}Z\right]\cos\left[\frac{(2n-1)\pi D_h}{2W}Z'\right]$$

$$(49)$$

Substituting Eq. (49) into Eq. (48) and rearranging, one can obtain the non-dimensional fluid velocity profile in the microchannel as follows:

$$
\bar{u}(Y, Z) = -\frac{64}{\pi^4 D_h^2} \frac{d\bar{P}}{dX} \sum_{m=1}^{\infty} \sum_{n=1}^{\infty} \frac{(-1)^{m+n} \cos\left[\dfrac{(2m-1)\pi D_h}{2H} Y\right] \cos\left[\dfrac{(2n-1)\pi D_h}{2W} Z\right]}{(2m-1)(2n-1)\left[\dfrac{(2m-1)^2}{H^2} + \dfrac{(2n-1)^2}{W^2}\right]}
$$

$$
-\frac{16}{\pi^2 HW} \bar{G}_1 \bar{E}_x \sum_{m=1}^{\infty} \sum_{n=1}^{\infty} \frac{\cos\left[\dfrac{(2m-1)\pi D_h}{2H} Y\right] \cos\left[\dfrac{(2n-1)\pi D_h}{2W} Z\right]}{\dfrac{(2m-1)^2}{H^2} + \dfrac{(2n-1)^2}{W^2}}
$$

$$
\times \int_{Y'=0}^{H/D_h} \int_{Z'=0}^{W/D_h} \cos\left[\frac{(2m-1)\pi D_h}{2H} Y'\right] \cos\left[\frac{(2n-1)\pi D_h}{2W} Z'\right] \sinh \Psi(Y', Z') dY' dZ
$$

(50)

If there is no electrostatic interaction, the second term on the right-hand side of Eq. (50) vanishes. The fluid velocity reduces to

$$
\bar{u}_0(Y, Z) = -\frac{64}{\pi^4 D_h^2} \frac{d\bar{P}}{dX} \sum_{m=1}^{\infty} \sum_{n=1}^{\infty} \frac{(-1)^{m+n} \cos\left[\dfrac{(2m-1)\pi D_h}{2H} Y\right] \cos\left[\dfrac{(2n-1)\pi D_h}{2W} Z\right]}{(2m-1)(2n-1)\left[\dfrac{(2m-1)^2}{H^2} + \dfrac{(2n-1)^2}{W^2}\right]}
$$

(51)

which is the well-known Poiseuille flow velocity profile through a rectangular channel.

Using Eqs (50) and (51), the mean velocity with and without the consideration of the effects of the EDL may be written, respectively, as

$$
\bar{u}_{ave} = -\frac{256 HW}{\pi^6 D_h^4} \frac{d\bar{P}}{dX} \sum_{m=1}^{\infty} \sum_{n=1}^{\infty} \frac{1}{(2m-1)^2(2n-1)^2\left[\dfrac{(2m-1)^2}{H^2} + \dfrac{(2n-1)^2}{W^2}\right]}
$$

$$
-\frac{64}{\pi^4 D_h^2} \bar{G}_1 \bar{E}_x \sum_{m=1}^{\infty} \sum_{n=1}^{\infty} \frac{(-1)^{m+n}}{(2m-1)(2n-1)\left[\dfrac{(2m-1)^2}{H^2} + \dfrac{(2n-1)^2}{W^2}\right]}
$$

$$
\times \int_{Y'=0}^{H/D_h} \int_{Z'=0}^{W/D_h} \cos\left[\frac{(2m-1)\pi D_h}{2H} Y'\right] \cos\left[\frac{(2n-1)\pi D_h}{2W} Z'\right] \sinh \Psi(Y', Z') dY' dZ'
$$

(52)

and

$$
\bar{u}_{0_{ave}} = -\frac{256 HW}{\pi^6 D_h^4} \frac{d\bar{P}}{dX} \sum_{m=1}^{\infty} \sum_{n=1}^{\infty} \frac{1}{(2m-1)^2(2n-1)^2\left[\dfrac{(2m-1)^2}{H^2} + \dfrac{(2n-1)^2}{W^2}\right]}
$$

(53)

Thus, the nondimensional volumetric flow rate through the rectangular microchannel, defined by $\bar{Q}_v = Q_v/4HWU$, is given by

$$\overline{Q}_v = \overline{u}_{ave} \tag{54}$$

Correspondingly, in the absence of the EDL, the nondimensional volumetric flow rate is expressed as

$$\overline{Q}_{ov} = \overline{u}_{oave} \tag{55}$$

In order to calculate the fluid velocity distribution, the analytical solution [Eq. (50)] for the velocity is used to obtain the "exact" solution" which, in practice, usually means an error of 0.01% or less. As seen from Eq. (50), the velocity distribution is finally expressed by two infinite series. Therefore, usually a very large number of terms in series is needed to achieve this error criteria. To reduce the computation time, the Aitken's procedure [10] may be employed for accelerating series.

C. Electrokinetic Field in a Rectangular Microchannel

As seen from Eq. (50), the local and the mean velocity can be calculated only when the nondimensional induced electrical field strength or the electrokinetic potential, $\overline{E}_x$, is known. Recall that the net electrical current, I, flowing in the axial direction of the microchannel, is the algebraic summation of the electrical convection current (i.e., streaming current) I_s and the electrical conduction current I_c. In a steady-state situation, this net electrical current should be zero:

$$I = I_s + I_c = 0 \tag{56}$$

Due to symmetry of the rectangular microchannel, the electrical streaming current is of the form:

$$I_s = 4D_h^2 U \int_{Y=0}^{H/D_h} \int_{Z=0}^{W/D_h} \overline{u}(Y, Z)\rho_e(Y, Z)\mathrm{d}Y\mathrm{d}Z \tag{57}$$

The electrical conduction current in the microchannel consists of two parts [7]: one is due to the conductance of the bulk liquid; the other is due to the surface conductance of the EDL. This electrical conduction current can be expressed as

$$I_c = I_{bc} + I_{sc} = \lambda_t E_x A_c = \frac{4\lambda_t H W \zeta_0}{D_h \mathrm{Re}_0} \overline{E}_x \tag{58}$$

where I_{bc} and I_{sc} are the bulk and the surface electrical conductance currents, respectively; λ_t is the total electrical conductivity and can be calculated by $\lambda_t = \lambda_b + \dfrac{\lambda_s P_s}{A_c}$ [7]. Here, P_s and A_c are the wetting perimeter and the cross-sectional area of the channel, respectively, λ_b is the bulk conductivity of the solution, and λ_s is the surface conductivity, which may be determined by experiment [4].

Substituting Eq. (41) for $\rho_e(Y, Z)$ into Eq. (57) and employing Eq. (56), the non-dimensional induced field strength can be expressed as

$$\overline{E}_x = \frac{D_h^2}{HW} \overline{G}_2 \mathrm{Re}_0 \int_{Y=0}^{H/D_h} \int_{Z=0}^{H/D_h} \overline{u}(Y, Z) \sinh \Psi(Y, Z)\mathrm{d}Y\mathrm{d}Z \tag{59}$$

Here, the nondimensional number $\overline{G}_2 = 2zen_0 D_h U/\lambda_t \zeta_0$.

The substitution of $\overline{u}(Y, Z)$ from Eq. (50) into Eq. (59) finally gives the dimensionless induced field strength as

$$\overline{E}_x = \frac{\dfrac{D_h^2}{HW}\overline{G}_2 \mathrm{Re}_0 \displaystyle\int_{Y=0}^{H/D_h}\int_{Z=0}^{W/D_h} \overline{u}_0(Y,Z)\sinh\Psi(Y,Z)\,dY\,dZ}{1 + \dfrac{4D_h^4}{H^2 W^2}\overline{G}_1\overline{G}_2\mathrm{Re}_0 \displaystyle\int_{Y=0}^{H/D_h}\int_{Z=0}^{W/D_h} C_o \sinh\Psi(Y,Z)\,dY\,dZ} \tag{60}$$

where

$$C_0 = \frac{4}{\pi^2 D_h^2}\sum_{m=1}^{\infty}\sum_{n=1}^{\infty} \frac{\cos\left[\dfrac{(2m-1)\pi D_h}{2H}Y\right]\cos\left[\dfrac{(2n-1)\pi D_h}{2W}Z\right]}{\dfrac{(2m-1)^2}{H^2}+\dfrac{(2n-1)^2}{W^2}}$$

$$\times \int_{Y'=0}^{H/D_h}\int_{Z'=0}^{W/D_h}\cos\left[\frac{(2m-1)\pi D_h}{2H}Y'\right]\cos\left[\frac{(2n-1)\pi D_h}{2W}Z'\right]\sinh\Psi(Y',Z')\,dY'\,dZ'$$

Consider a fully developed, laminar flow of a diluted aqueous 1:1 electrolyte (e.g., KCl) solution through a rectangular microchannel with a height of 20 μm, width 30 μm, and length 1 cm. At a typical room temperature $T = 298$ K, the physical and electrical properties of the liquid are $\varepsilon = 80$, $n_\infty = 6.023 \times 10^{19}$ (m^{-3}), $\mu_f = 0.90 \times 10^{-3}$ (kg/ms), and $\zeta = 75$ mV. In calculations the total electrical conductivity λ_t was chosen from experimental results [4]. With these base values, it is quite straightforward to calculate the parameters characterizing the flow, such as velocity distribution and streaming potential by the equations developed above. Figure 10 shows the nondimensional streaming potential as a function of the nondimensional applied pressure drop over the microchannel with different zeta potentials. As explained before, in the absence of an externally applied electrical field, when a fluid is forced to flow through a channel under a hydrostatic pressure difference, the mobile charges in the EDL are carried to the downstream end to form a

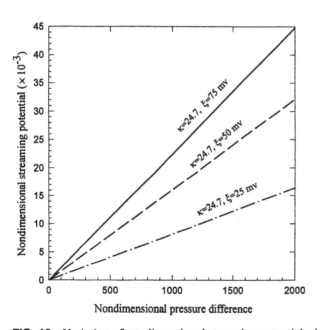

FIG. 10 Variation of nondimensional streaming potential with nondimensional pressure difference for different zeta potentials. (From Ref. 7.)

streaming current. Therefore, a larger pressure difference generates a larger volume transport and hence more ions are carried to the end of the channel, which results in a larger streaming current. Correspondingly, a stronger streaming potential may be produced as the pressure difference (drop) increases. This is clearly demonstrated in Figure 10. Also, it is shown that for a given pressure difference the streaming potential increases with an increase in zeta potential. It is also shown that for a given pressure difference the induced field strength decreases with an increase in ionic concentration of the aqueous solution. This may be understood as follows: if the ionic concentration is higher, which implies a larger Debye–Hückel length, i.e., a smaller EDL thickness, the effect of the EDL is less. Therefore, fewer ions are carried downstream with the flow and hence lower charge accumulation at the ends of the channel occurs.

It should be noted that the present model modifies the simple proportionality relationship between zeta potential and streaming potential in the classical electrokinetic theory, by considering the EDL effects on the liquid flow.

D. Electrokinetic Effects on Velocity Field

As seen from Eq. (50), the velocity field in a rectangular microchannel depends on the EDL field. Since the bulk ionic concentration and the shape of the channel's cross-section will affect the EDL field, these factors will in turn influence the velocity field. In this section, we wish to examine these effects for the same conditions described in Section IV.C.

The computation of nondimensional velocity distribution according to Eq. (50) is carried out for a fixed external pressure difference. In Fig. 11a,b, the distribution of nondimensional velocity is plotted for the channel of 20 μm × 30 μm with and without consideration of the EDL effects. As seen in Fig. 11a, the EDL field exhibits significant effects on the flow pattern. The maximum velocity in the center of the channel is lower when the EDL field effects are considered. The flow velocity near the channel wall approaches zero due to the action of the EDL field and the streaming potential. Moreover, the flow around the channel corner greatly deviates from the classical Poisseuille flow pattern, as shown in Fig. 11b.

It should be pointed out that the surface conduction current plays an important role in the total electrical conduction current for a dilute solution flow in microchannels. It makes a significant contribution to the streaming potential and therefore to the flow field. This is clearly demonstrated in Fig. 11c, which is the nondimensional velocity distribution with the same hydrodynamic and electrokinetic conditions, but without consideration of the surface condition current. One may overestimate the electrokinetic effects on the microchannel flow if the surface conduction current is not included.

Figure 12 displays the distribution of nondimensional velocity as a function of the geometric ratio of height to width with the same hydraulic diameter as the channel of 20 μm × 30 μm. For a fixed hydraulic diameter, a small geometric ratio represents a smaller channel height but a larger channel width. As shown in Fig. 12, it is obvious that the channel shape has a significant influence on the flow across microchannels because of EDL effects. The general pattern is the smaller the channel size, such as in Fig. 12a, the stronger the EDL effects and the larger portion of the flow field is affected. The explanation for this is that as the channel size decreases, the EDL thickness becomes relatively larger and hence the EDL effects are stronger.

In Fig. 13, the distribution of nondimensional velocity is plotted for the channel of 20 μm × 30 μm for two different bulk ionic concentrations. It is generally known

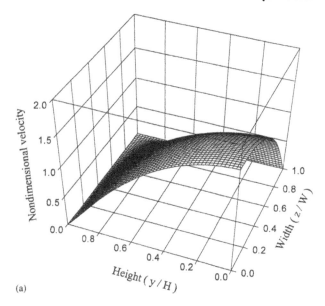

(a)

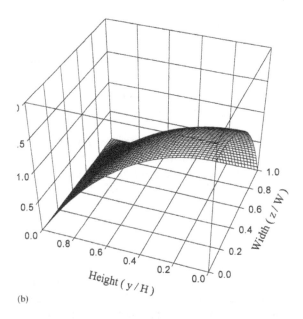

(b)

FIG. 11 EDL effects on nondimensional velocity distribution in one-quarter of a rectangular microchannel (geometric ratio of height to width = 2/3). (a) With EDL effects (nondimensional electrokinetic diameter $K = 24.7$); (b) without EDL effects; (c) with EDL effects (nondimensional electrokinetic diameter $K = 24.7$), but without consideration of the surface conduction. (From Ref. 7.)

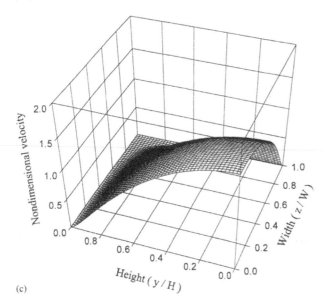

(c)

FIG. 11   (*continued*)

that the zeta potential has a dependence on the bulk ionic concentration. Recently, Mala et al. [4] reported that the zeta potential changes from 100 to 200 mV while the ionic concentration varies from 10^{-4} to 10^{-6} M for p-type silicon microchannels. Such an experimentally determined correlation between zeta potential and ionic concentration is used in our calculation. As seen, the EDL exhibits significantly stronger effects on the flow pattern for the dilute solution in Fig. 13b than that in the higher concentration in Fig. 13a.

In Fig. 14, the nondimensional volumetric flow rate is plotted as a function of nondimensional pressure difference for different concentrations of the aqueous solution and zeta potentials. As expected, the flow rate is reduced because of the electrokinetic effects. Basically, Fig. 14 shows that the flow rate exhibits the same EDL dependence as the streaming potential does.

E.   Electroviscous Effects

It is apparent from the previous analysis that the presence of an EDL exerts electrical forces on the ions in the liquid, and hence has a profound influence on the flow behavior. As discussed above, the streaming potential will produce a liquid flow in the direction opposite to the pressure-driven flow. The liquid thus appears to exhibit an enhanced viscosity if its flow rate is compared to that in the absence of the EDL effects (here, the viscosity is assumed to be independent of electrolyte concentration [1]).

As we have already shown, the nondimensional flow rate through the microchannel with and without the consideration of the EDL effects is given by Eqs (54) and (55), respectively. Equalizing Eq. (54) with Eqs. (55), i.e., $\overline{Q}_v = \overline{Q}_{ov}$, and using expressions for $\bar{u}_{ave}$ and $\bar{u}_{oave}$ in Eqs (52) and Eq. (53), one may obtain the ratio of the apparent viscosity to the bulk viscosity as follows:

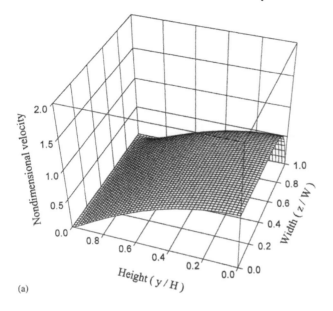

(a)

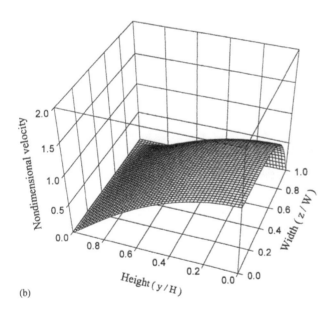

(b)

FIG. 12 Aspect ratio effects on nondimensional velocity distribution as a function of geometric ratio of height to width (nondimensional electrokinetic diameter $K = 24.7$). (a) Height/width $= 1/8$; (b) height/width $= 1/4$; (c) height/width $= 1/2$; (d) height/width $= 1/1$. (From Ref. 7.)

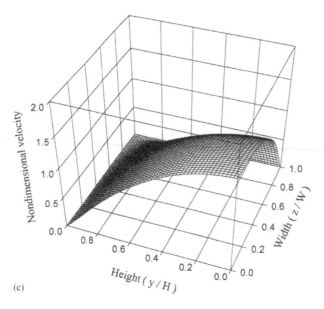

(c)

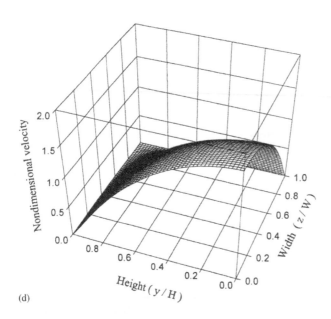

(d)

FIG. 12 (*continued*)

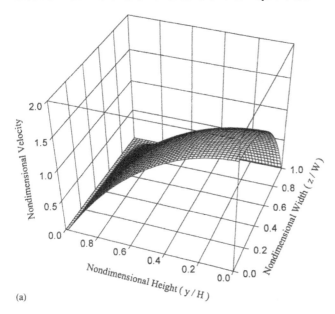

(a)

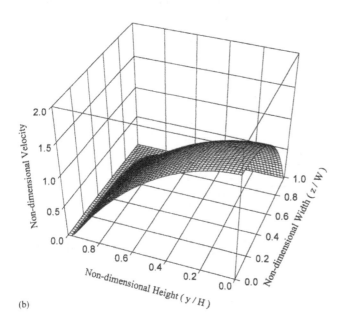

(b)

FIG. 13 Ionic concentration effects on nondimensional velocity distribution in one-quarter of a rectangular microchannel (geometric ratio of height to width = 2/3). (a) $n_\infty = 10^{-5}$ M, $\zeta = 150$ mV; (b) $n_\infty = 10^{-6}$ M, $\zeta = 200$ mV. (From Ref. 8.)

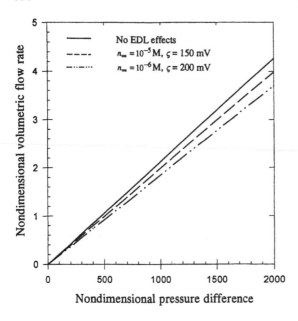

FIG. 14 Variation of nondimensional volume flow rate with nondimensional pressure difference for different concentrations and zeta potentials. (From Ref. 8.)

$$\frac{\mu_{af}}{\mu_f} = \frac{1}{1 + \dfrac{\pi^2 D_h^2 \overline{G}_1 \overline{E}_x}{4HW \, dP/dX} \dfrac{C_1}{C_2}} \tag{61}$$

where

$$C_1 = \sum_{m=1}^{\infty} \sum_{n=1}^{\infty} \frac{(-1)^{m+n}}{(2m-1)(2n-1)\left[\dfrac{(2m-1)^2}{H^2} + \dfrac{(2n-1)^2}{W^2}\right]}$$

$$\times \int_{Y'=0}^{H/D_h} \int_{Z'=0}^{W/D_h} \cos\left[\frac{(2m-1)\pi D_h}{2H} Y'\right] \cos\left[\frac{(2n-1)\pi D_h}{2W} Z'\right] \sinh \Psi(Y', Z') \, dY' dZ'$$

and

$$C_2 = \sum_{m=1}^{\infty} \sum_{n=1}^{\infty} \frac{1}{(2m-1)^2(2n-1)^2\left[\dfrac{(2m-1)^2}{H^2} + \dfrac{(2n-1)^2}{W^2}\right]}$$

Since the dimensionless pressure gradient is negative, and both C_1 and C_2 are greater than zero, it is easy to show that this ratio is greater than 1, which is the electroviscous effect.

Using Eq. (61), the ratio of the apparent viscosity to the bulk viscosity, μ_{af}/μ_f, is plotted as function of the nondimensional electrokinetic diameter for different values of the zeta potential of the solid surface in Fig. 15. It is seen from this figure that μ_{af}/μ_f is

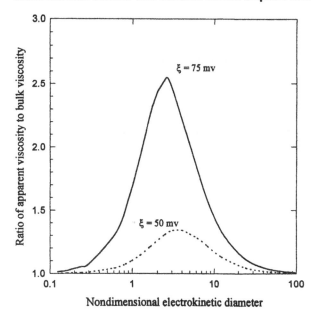

FIG. 15 Variation of ratio of apparent viscosity to bulk viscosity with nondimensional electrokinetic diameter for different zeta potentials. (From Ref. 7.)

strongly dependent on the strength of the EDL effects. This indicates that the higher the zeta potential, the larger the values of the ratio μ_{af}/μ_f. In addition, of particular interest is the prediction of a maximum in μ_{af}/u_f with respect to the nondimensional electrokinetic diameter. A similar pattern has been reported by Rice and Whitehead [5] and Levine et al. [6]. Unfortunately, no explanation was given in the literature before. We believe that there is a critical value for the nondimensional electrokinetic diameter K, corresponding to the strongest EDL effect, at which the largest reduction in the flow and hence the maximum ratio of μ_{af}/μ_f occur in rectangular microchannels. When K increases from this critical value, which implies either a larger hydraulic diameter or a thinner EDL thickness (i.e., a higher electrolyte concentration), the EDL effects become weaker. Thus, a smaller reduction in the flow and hence a lower ratio of μ_{af}/μ_f can be observed. On the other hand, for the case that K is less than this critical value, this means either a smaller hydraulic diameter or a larger EDL thickness (i.e., a lower electrolyte concentration). However, as discussed in the preceding theory, the EDL effects on the fluid flow in rectangular microchannels is considered by introducing an additional body force in the conventional equation of motion. By definition this additional body force is proportional to both the streaming potential and the net charge density, which in fact relate to the fluid velocity profile and the electrolyte ionic concentration. If K becomes very small, we may have either extremely narrow microchannels (which allow very little flow) or a relatively diluted electrolyte. In both cases, the streaming potential and the ionic net charge density decrease. Therefore, the EDL effects again become weaker and hence the smaller ratio of μ_{af}/μ_f is shown in Fig. 15. In conclusion, there is no monotonic relationship among the EDL effects on the fluid flow in rectangular microchannels and the channel size and electrolyte ionic concentration, because the streaming potential, the ionic net charge density, and the fluid velocity depend on a large number of basic parameters.

REFERENCES

1. RJ Hunter. Zeta Potential in Colloid Science, Principles and Applications. New York: Academic Press, 1981.
2. J Lyklema. Fundamentals of Interface and Colloid Science, vol. II. New York: Academic Press, 1995.
3. M Mala, D Li, JD Dale. Heat Mass Transfer 40 (13): 3079–3088, 1997.
4. M Mala, D Li, C Werner, H Jacobasch, Y Ning. Int J Heat Fluid Flow 18: 489–496, 1997.
5. CL Rice, R Whitehead. J Phys Chem 69: 4017, 1965.
6. S Levine, JR Marriott, G Neale, N Epstein. J Colloid Interface Sci 52: 136, 1975.
7. C Yang, D Li. Colloids Surfaces A 143: 339–353, 1998.
8. C Yang, D Li. J Colloid Interface Sci 194: 95–107, 1997.
9. M Mala, C Yang, D Li. Colloids Surfaces A 139: 109–116, 1998.
10. JV Beck. Heat Conduction Using Green's Functions. London: Hemisphere, 1992.

20

Electrokinetic Methods Employed in the Characterization of Microfiltration and Ultrafiltration Membranes

LAURENCE RICQ, ANTHONY SZYMCZYK, and PATRICK FIEVET Laboratoire de Chimie des Materiaux et Interfaces, Besançon, France

I. INTRODUCTION

Membrane separation processes have been in full expansion since the 1970s and have given rise to a great number of investigations. Indeed, besides their technological interest, membrane separation processes appear to be more economically competitive than conventional separation technologies such as distillation, crystallization, or solvent extraction.

A simplified working definition of a membrane can be conveniently stated as a selective barrier which, under a driving force, permits preferential passage of one selected species of a mixture [1]. Separation membrane processes are used for solution concentration, purification, or fractionation. The development of membrane technology has allowed us to extend the field of membrane applications (e.g., treatment of industrial effluents, drinking water production, biotechnology, food industry, pharmaceutical industry, textile industry).

Tangential micro-, ultra-, and nano-filtration are separation membrane processes in the liquid phase for which a pressure gradient acts as driving force. Microfiltration (MF) is used for selective separation of species having a size in the region of micrometers (yeast cells, bacteria, etc.), ultrafiltration (UF) is used for separation of species from ~ 2 to 100 nm (peptides, proteins, viruses), and nanofiltration is used for separation of species of less than 2 nm (sugars, dyes, salts).

There are a great variety of commercial organic (e.g., polyamide, polycarbonate, polysulfone) or inorganic (e.g., alumina, zirconia) membranes. Ceramic membranes, introduced in recent years, are currently in full expansion. They typically exhibit stability at high temperatures and extreme pH conditions. Furthermore, their mechanical resistance allows treatment of solutions with high viscosity, and their chemical resistance permits the use of effective and yet corrosive cleaning procedures and chemicals.

Up to now, most commercial membranes had a tubular geometry. The first plane ceramic membranes made their inroads to the market in recent years. The plane geometry is likely to provide higher shear rates compared with the tubular concept and so it is of great interest to limit membranes fouling (the fouling phenomenon, that is to say the accumulation of species on to the membrane surface, being the chief drawback to membrane separation processes).

It is now well acknowledged that the selectivity of a membrane depends on both steric effects and electrostatic interactions occurring between charged species (ions or molecules) and the membrane surface which is generally charged as well. From then on, the determination of parameters representing membrane–solution interactions, such as the zeta potential, is of the utmost importance to understand and predict the filtration performances of a membrane. The optimization of the filtration performances of a membrane requires, therefore, a preliminary study of its electric and electrokinetic properties.

This chapter lies within the scope of the characterization of the electrokinetic properties of MF and UF membranes by means of various experimental methods.

II. GENERALITIES

The aim of this section is to present, first, filtration membranes, and second, the membrane–solution interface. Electrokinetic phenomena are then developed since they allow us to characterize the interface.

A. Membranes and Filtration

1. Filtration Process

In general, a membrane plays the role of a molecular sifter with constituents which have to be separated. The membrane is tangentially swept by the liquid in which the constituents have to be separated (Fig. 1). The porous material separates two distinct aqueous phases called the retentate and permeate. This latter is constituted of solvent (with or without solute) which passes through the membrane because of a transmembrane pressure difference.

Both UF and MF then allow extraction of solvent and ionic solute from a solution which contains macrosolute. The filtration mechanism does not prevent physicochemical interaction between the species and membrane, at the surface and inside pore.

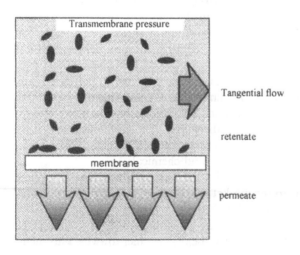

FIG. 1 Filtration principle.

2. Membrane Characteristics

Marketed membranes are generally defined by three characteristics: membrane material, permeability, and cut-off values. The user defines the membrane process efficiency by the permeate flux and solute retention or transmission.

(a) Membrane Material. Membranes used in UF have an asymmetric structure. They are composed of a macroporous support, which gives mechanical resistance, and a thinner filtering layer, which determines membrane selectivity properties. Among the materials used for the membrane filtering layer, there are mineral oxides (ZrO_2, Al_2O_3, etc.) and organic materials [cellulose acetate, polyacrylonitrile, sulfonated polysulfone, poly(ether sulfone), etc.].

(b) Membrane Permeability. The permeability, L_p, is an intrinsic characteristic. It defines the slope of solvent flux versus transmembrane pressure.

According to Darcy's law, all permeability variations can be performed by a membrane hydraulic resistance variation:

$$J_s = L_p \Delta P = \Delta P / \mu R_m \tag{1}$$

where J_s is solvent flux density (m s^{-1}), μ is solvent dynamic viscosity (Pa s), R_m is membrane hydraulic resistance m^{-1}), and ΔP is transmembrane pressure difference (Pa).

In the case where Poiseuille's law can be applied, the permeability depends on membrane porosity and pore geometric characteristics:

$$J_s = n_p \frac{\Delta P \pi r^4}{8 \mu l} \tag{2}$$

where n_p is pore number per unit area (m^{-2}), r is membrane pore radius (m), l is effective length of pore (m), and ΔP is transmembrane pressure (Pa).

When the solution contains species which can be retained by the membrane, the flux is less than that of the solvent in the same conditions, except at low membrane pressure. Beyond a transmembrane pressure value, we can generally observe that the flux does not vary with pressure, it tends toward a constant value called "limit flux" (Fig. 2).

(c) Pore Size and Cut-off Value. A membrane plays the role of a molecular sieve with constituents which have to be separated. A main characteristic of the membrane is, then, the pore size. In UF, this is evaluated by a molecular mass equivalent: the cut-off

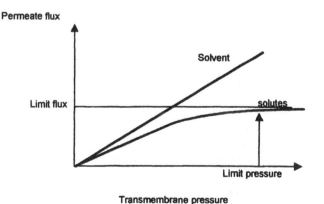

FIG. 2 Permeate flux versus transmembrane pressure for pure solvent and solutes.

value. It corresponds to a quasitotal retention of a given reference; in most cases, it comprises between 0.9 and 1 of a standard molecule in fixed conditions. It is expressed in g mol^{-1} (daltons). The cut-off value is not an intrinsic membrane characteristic since it depends on the standard molecules used and operating conditions.

(d) Retention. Retention R (or Transmission T_r) describes the ability of a membrane to prevent a compound from passing through a porous material. It defines the process selectivity:

$$R = (1 - C_p/C_r) = 1 - T_r \tag{3}$$

where R is retention, T_r is transmission, C_p is solute concentration in the permeate (mol L^{-1}), and C_r is solute concentration in the retentate (mol L^{-1}). A separation is called nonselective concerning a molecule when the solute retention is 0 and totally selective when the retention is equal to 1.

As far as a solute mixture is concerned, the separation process selectivity can be defined as the transmission ratio:

$$s = \frac{(C_{p1}/C_{r1})}{(C_{p2}/C_{r2})} \tag{4}$$

where C_{p1} is concentration of solute 1 in the permeate, C_{r1} is concentration of solute 1 in the retentate, C_{p2} is concentration of solute 2 in the permeate, and C_{r2} is concentration of solute 2 in the retentate.

3. Phenomena Limiting Solvent Transfer

In the presence of a macromolecular solute, the permeation flux is less than that measured with the pure solvent, in the same filtration conditions. Moreover, above a particular transmembrane pressure, the curve of flux versus transmembrane pressure wanders from linearity, and the permeate flux tends to reach a limiting value that does not depend on applied pressure (Fig. 2). In other respects, the measured UF permeate flux of a solution, under a constant transmembrane pressure, decreases with time.

This behavior can be expressed by the following equation:

$$J = \frac{\Delta P}{\mu(R_m + R_{pol} + R_{ads} + R_d)} \tag{5}$$

where R_{pol} is hydraulic resistance due to concentration polarization (m^{-1}), R_{ads} is hydraulic resistance due to solute adsorption (m^{-1}), R_d is hydraulic resistance due to particle deposit (m^{-1}), and μ is permeate dynamic viscosity (Pa s). The various hydraulic resistances are linked to interfacial phenomena which can occur during the solute filtration.

(a) Concentration Polarization R_{pol}. During macromolecular solute filtration, species are carried away by convection from the solution toward the membrane surface where they accumulate. The polarization layer is then the liquid layer near the surface where a macromolecular concentration gradient is established. The concentration polarization is a reversible phenomenon. It takes place in the first moment of filtration [2]. An increase in the fluid tangential velocity or the use of low pressure allows us to minimize this phenomenon.

(b) Adsorption R_{ads}. The contact of a solution with a membrane leads to electrostatic and hydrophobic interactions between species. These interactions induce adsorption of molecules or ions at the membrane surface, which limits transfer through the

membrane. Hence, solute transfer during UF depends strongly on physicochemical conditions.

(c) Solute Deposit R_d. The solute accumulation close to the membrane can lead to the formation of a particle deposit at the surface, which limits the solvent transfer. Fouling results from all phenomena which contribute, in a reversible or irreversible way, to modify membrane filtration properties. These modifications can be performed as pore size distribution variations, as the presence of a surface solute deposit, or as a mechanical pore blocking, resulting in a permeate flux decrease.

4. Prediction of Solute Transfer Through Membrane

(a) Steric Exclusion. Ferry [3] developed a theory which allowed correlation of solute retention or transmission with its steric dimensions. Ferry's model is based on the fact that solute within pores induces no-flow modification and that friction strength between the molecule and pore wall is negligible:

$$T_r = 1 - R = 1 - [1 - (1 - \lambda)^2]^2 \quad \text{for } 0 < \lambda < 1 \tag{6}$$

$$T_r = 0 \qquad\qquad\qquad \text{if } \lambda > 1$$

where λ is R_d/r, R_d is solute radius, supposed as spherical (m), and r pore radius (m). This model is only based on a steric exclusion mechanism.

(b) Steric and Ionic Exclusion. When a charged molecule, assumed to be a sphere with radius r_s, is filtered through a membrane (pore radius r), charges carried by both surfaces lead to repulsive or attractive electrostatic interactions. A possible consequence of an attraction is the solute adsorption on to the membrane surface and then a decrease in transmission. Several models take into account electrostatic interactions.

A first model considers the apparent size of solute r_{sa}, which depends on the double-layer thickness κ^{-1} : $r_{sa} = r_s + \alpha\kappa^{-1}$, where α is an empirical corrective term and r_s the solute radius. On this basis, the ISCR (ionic strength controlled retention) model gives the retention expression of a small molecule as a function of $I^{-1/2}$ (with I, the ionic strength) [4]:

$$R = R_0 + aI^{-1/2} \tag{7}$$

where $R_0 = \lambda'^2(2 - \lambda')^2$, $a = 4A\lambda'(2 - \lambda')(1 - \lambda')/r$, $A = \kappa^{-1}/I^{-1/2}$, $\lambda' = r_s/r$, and I is ionic strength $(\text{mol} \cdot \text{L}^{-1})$.

A retention model based on mixed interactions (mixed interaction retention, MIR) was developed by Millesime [4]. This one takes into account the steric exclusion, macromolecule – membrane interactions but also hydrophobic interactions. In a simple way, the model can be written:

$$R = a_0 + a_1 I^{-1/2} + a_1 BI^{+1/2} \tag{8}$$

where a_0 and a_1 are coefficients which depend only on r, r_s, A and B; B represents an interaction parameter between membrane and molecule: B is positive when the macromolecule develops hydrophobic interactions with the membrane, and B is weakly negative in the case of hydrophilic membranes.

At low ionic strength, the term $I^{-1/2}$ will then be predominant; it corresponds to an ionic retention phenomenon, whereas at high ionic strength, the term with $I^{+1/2}$, which corresponds to hydrophobic interactions, contributes to the increase in retention.

B. Origin of Surface Charge and Electrokinetic Phenomena

1. Origin of Surface Charge

Most membranes acquire a surface electric charge when brought into contact with an aqueous medium. The origin of this surface charge depends on the nature of the material. It can be due to defects in the crystal structure (e.g., aluminosilicates), surface groups (e.g., $-COOH$, $-OSO_3H$, or $-NH_2$), which can react with either acid or base to provide stabilizing charges, or surface groups, which are amphoteric in nature and can become either positively or negatively charged depending on pH (metal oxides) [5].

For instance, a mineral oxide in contact with an aqueous solution develops an electrical surface charge due to its amphoteric behavior. Depending on the solution pH, the implied reactions in the surface charge are:

$$\text{In acidic media:} \quad M - (OH)_{surf} + H_{aq}^+ \quad \leftrightarrow \quad M - OH_{2\,surf}^+$$

$$\text{In basic media:} \quad M - (OH)_{surf} + OH_{(aq)}^- \quad \leftrightarrow \quad M - O_{surf}^- + H_2O$$

In order to maintain the electroneutrality of the solution, ions rearrange at the solid–liquid interface. Ions which have a charge opposite to the surface (counterions) are attracted, whereas ions which have the same charge as the surface (coions) are repelled. The potential varies progressively from the solid surface to the bulk solution in a zone called the "electrical double layer", composed of a compact layer and a diffuse layer. The shear plane is located approximately between the compact and diffuse layers. Its potential is the electrokinetic potential or zeta potential; it can be determined by electrokinetic measurements.

2. Electrokinetic Phenomena

Electrokinetic phenomena which occur at the mineral oxide–solution interface are due to the electric charge distribution in the electrochemical double layer. They are observed when static equilibrium conditions of a charged surface are modified: this can occur when one phase (liquid or solid) tangentially moves with respect to the other phase. The diffuse layer then slips with respect to the compact layer along a plane usually called the "shear plane."

Electrokinetic methods can be distinguished by the driving force involved (mechanical or electrical) and by the nature of the mobile phase (solid or liquid) (Table 1). All these phenomena are linked to the electrokinetic potential or zeta potential.

3. Zeta-potential Determination

The zeta-potential determination can be made on the membrane itself by streaming potential or electro-osmosis measurements. Also, electrophoretic measurements realized on the

TABLE 1 Electrokinetic Phenomena

Liquid phase	Driving force		Solid phase
	Mechanical	Electrical	
Mobile	Streaming potential	Electro-osmosis	Stationary
Stationary	Sedimentation potential	Electrophoresis	Mobile

membrane material can provide information on the surface charge density. However, some recent works clearly show that the surface properties of a membrane can significantly differ from those of the crushed-membrane powder [6–9].

- When a pressure gradient acts through a membrane, the *streaming potential* (SP) can be defined as the electrical potential difference ($\Delta\phi$) arising between both sides of the membrane. The zeta potential (ζ) determination from the streaming potential can be done using the Helmholtz–Smoluchowski relationship, provided that there is no double-layer overlapping inside pores (i.e. $\kappa r \gg 1$, r being the pore radius) [10] that is in the case of a large pore radius and also at low zeta potential.

$$SP = \frac{\Delta\phi}{\Delta P} = \frac{\varepsilon_0 \varepsilon_r \zeta}{\eta \lambda_0} \tag{9}$$

where λ_0 is the conductivity of the electrolyte in the bulk, ε_0 is the vacuum permittivity, ε_r is the relative dielectric constant of the solvent, and SP is the measured streaming potential.

This relation assumes straight and independent pores.

- *Electro-osmosis* (EO) consists in applying a constant current through the membrane and measuring an electro-osmotic flow induced by the excess of counterions within the electrical double layer.

According to the theoretical analysis presented by Levine et al. [11], the electro-osmotic flow rate may be related to the zeta potential by means of the following equation:

$$V = \frac{I \varepsilon_0 \varepsilon_r \zeta}{\eta \lambda_0} f \tag{10}$$

where V is the electro-osmotic flow rate ($m^3 \cdot D^{-1}$), I is the applied current (A) and f is a function of κr and ζ.

For large pores and high ionic strength, $\kappa r \gg 1$ and $f = 1$, then the previous equation reduces to the Smoluchowski equation:

$$V = \frac{I \varepsilon_0 \varepsilon_r \zeta}{\eta \lambda_0} \tag{11}$$

- The determination of the zeta potential from particle *electrophoretic mobility* μ_e has been studied in several works [10, 12, 13]. There are two cases depending on the κa value, where κ is the reciprocal Debye length and a is the mean radius of the particle:

$$ka > 100 \qquad \mu_e = \frac{\varepsilon_0 \varepsilon_r \zeta}{\eta} \text{ (Smoluchowski's relation)} \tag{12}$$

$$\kappa a < 1 \qquad \mu_e = \frac{2\varepsilon_0 \varepsilon_r \zeta}{3\eta} \text{ (Huckel's relation)} \tag{13}$$

III. USE OF STREAMING POTENTIAL TO CHARACTERIZE UF AND MF MEMBRANES

Streaming potential measurement is one of the most used techniques for characterizing the electrokinetic properties of filtration membranes [14–20]. The streaming potential method allows the nonstop characterization of a membrane during the filtration process since the driving force involved is a pressure gradient. Thus, streaming potential can be used to study the influence of fouling phenomena on membrane surface properties [21, 22], to check the efficiency of cleaning treatments [23], or to investigate the effect of aging on membrane electrokinetic properties [24].

A. Streaming Potential Phenomenon

Let us consider a charged porous membrane brought into contact with a liquid containing charged species (Fig. 3a). A charge excess takes place in the electrical double layer that forms at the solid–liquid interface. When a pressure gradient is applied through the membrane pores, the charges in the mobile diffuse layer are carried towards the low-pressure compartment. This constitutes a streaming current (Fig. 3b). The accumulation

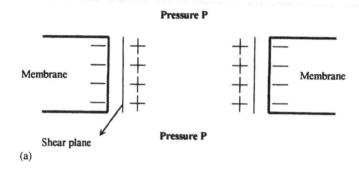

(a)

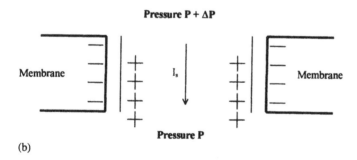

(b)

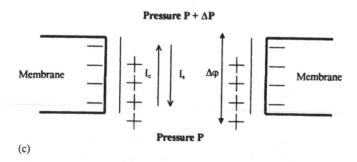

(c)

FIG. 3 Origin of the streaming potential. (See text for details.)

of charge at one end sets up an electric field which causes a current flow in the opposite direction. When this latter conduction current, I_c, is equal to the streaming current, a steady state is achieved [25]. The resulting potential difference between the pore ends ($\Delta\phi$), per unit pressure, is the streaming potential (Fig. 3c) [26, 27]. The streaming potential must be measured with a high-impedance millivoltmeter so that the global current is zero and polarization is possible.

Instead of allowing the streaming current to establish a potential difference across pores it is also possible to measure the streaming current directly by drawing it off through a low-impedance path which short-circuits the return path through the conducting liquid [25]. In such a case, the measurement must be made under conditions of essentially zero potential difference between the measuring electrodes, so that no current flows back through the pores [5, 28].

1. Description of Potential Signal

Figure 4 shows the evolution of the transmembrane electrical potential difference (resulting from a pressure pulse) as a function of time. Five zones can be distinguished:

(a–b): Base line. It represents the signal which corresponds to the initial transmembrane pressure difference (ΔP_0). The initial electrical potential difference may

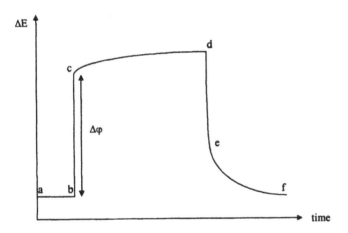

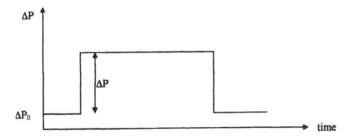

FIG. 4 Evolution of the transmembrane electrical potential difference, ΔE (resulting from a pressure pulse) versus time.

differ from zero although $\Delta P_0 = 0$ owing to electrode asymmetry and membrane heterogeneity.

(b–c): Sharp increase in $\Delta\phi$. The increase in electrical potential difference results from the establishment of the overpressure (ΔP) in the upstream compartment; $\Delta\phi$ allows the determination of the streaming potential.

(c–d): Evolution of the electrical potential difference caused by diffusion phenomena through the membrane. The potential difference due to these phenomena adds to that resulting from the streaming potential process. As a result, it may be difficult to distinguish what is due to diffusion from what is due to streaming potential.

(d–e): The potential difference sharply falls. This drop corresponds to the membrane depolarization which results from the restoration of the initial transmembrane pressure difference (ΔP_0).

(e–f): The signal falls gradually (concentration homogenization) to the base line.

B. Streaming Potential Measurements

The experimental measurement consists in recording the instantaneous electrical potential difference $\Delta\phi$ resulting from an overpressure ΔP applied on one side of the membrane. The streaming potential (SP) can then be defined as:

$$SP = \left.\frac{\Delta\phi}{\Delta P}\right|_{I=0} \tag{14}$$

The SP value can be determined either from the slope of $\Delta\phi = f(\Delta P)$ [29] or from a series of $\Delta\phi$ measurements performed at a constant overpressure [27, 30]. This latter method, called the "pulse method," has been used in several works. It entails carrying out a preliminary study so as to check the linear variation of $\Delta\phi$ versus ΔP (see Fig. 5). Remember that the SP method requires a pair of electrodes placed on both sides of the membrane and a high-impedance millvoltmeter to measure the transmembrane electrical potential difference.

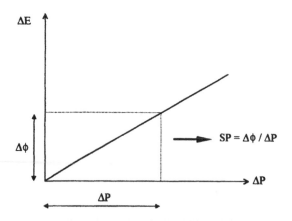

FIG. 5 Electrical potential difference (ΔE) versus overpressure (ΔP).

C. Streaming Potential Units

The filtration unit has to be adapted according to the membrane geometry. Generally, two types of unit are used. Figure 6 represents the apparatus used to perform SP measurements on plane membranes whereas Fig. 7 shows a filtration unit adapted to tubular membranes.

In both devices a pump allows circulation of the solution inside the filtration unit. Pressure in the retentate side is measured by means of a captor. The overpressure is applied in the retentate compartment by means of a bypass circuit. The transmembrane electrical potential difference ΔE is measured with a pair of electrodes on both sides of the membrane.

Ag/AgCl electrodes are commonly used since their asymmetry potential difference is very low (generally less than 0.1 mV) and they exhibit good chemical stability provided that the pH is about 8. To carry out SP measurements at higher pH, Ag/AgCl electrodes might be replaced by platinum electrodes although the asymmetric potential difference between platinum electrodes is much higher than between Ag/AgCl electrodes [27].

D. Influence of Various Parameters on Streaming Potential Measurements

1. Influence of pH

The theoretical analysis of the existence of a membrane net charge due to the ion distribution close to the surface, which is responsible for the streaming potential value

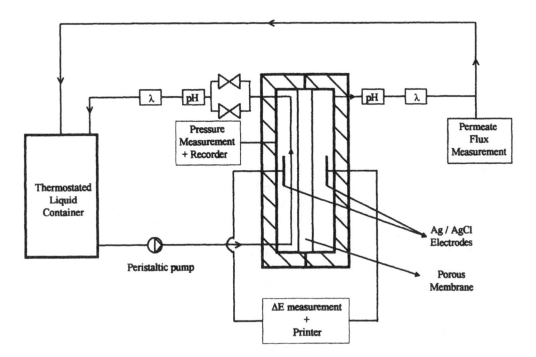

FIG. 6 Plane module. (From Ref. 41.)

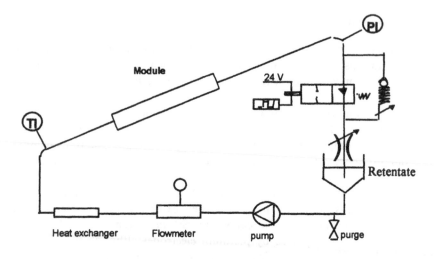

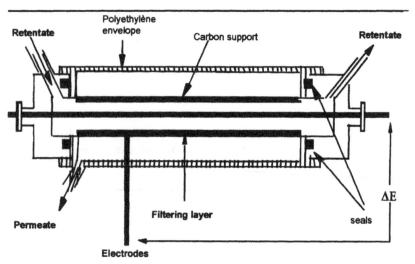

FIG. 7 Tubular module. (From Ref. 27.)

[30, 31], can be confirmed by experiment (Fig. 8) on an inorganic membrane (M1, Carbosep, Orelis, France) in a sodium chloride medium. The amphoteric character of the mineral oxide give rise to the negative or positive values of the SP, depending on the pH value: SP potential variations with pH allow determination of the net charge sign of the filtering layer.

The pH for which the SP is zero corresponds to the isoelectric point (i.e.p.) of the system [32]. The i.e.p. of the studied M1 membrane in NaCl medium is about 3.7 ± 0.2. It is a characteristic of the membrane – solution system [33]. At lower pH values, the membrane net charge is positive and at higher pH values, it is negative.

The slight decrease in SP value (absolute value) at pH higher than 8, may be due to an effect of ionic strength: actually, while at pH 7 the measured solution conductivity is about 0.11 mS cm^{-1}; at pH 9, it is about 0.20 mS cm^{-1} because of the addition of NaOH for pH adjusting.

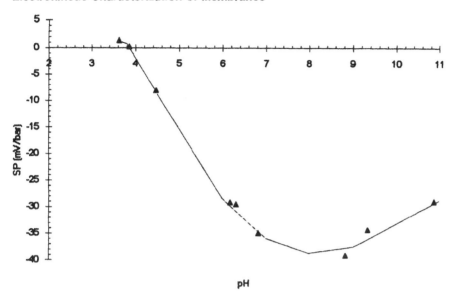

FIG. 8 Streaming potential of an M1 Carbosep (150 kD) membrane versus pH in 1 mM NaCl medium; $T = 50°C$. (From Ref. 31.)

The calculation of the zeta potential, using the Smoluchowski relationship allows us to determine only an apparent zeta potential value [34]. Nevertheless, this calculation takes into account the bulk solution conductivity. Figure 9 gives a representation of the apparent zeta potential versus pH value. The minimum previously observed at pH 8 disappears and the curve has a classical shape.

Similar experiments have been also realized on organic membranes such as polyether sulfonated and polysulfonated membranes [31, 35, 36]: i.e.p. values have been determined. In these cases, the negative net charge (measured with KCl electrolyte) observed in a wide range of pH is attributed to chloride anion adsorption [35, 36].

2. Influence of Filtered Electrolyte Solution

For most UF membranes, the permeate flux measured during filtration does not depend either on the nature of the electrolyte (KNO$_3$, KCl, NaCl, CACl$_2$, MgCl$_2$) or on pH [37].

Streaming potential variations versus pH allow determination of i.e.p. values, through which a dependence of SP on the nature of the ions can be measured. Actually, we can distinguish two types of behavior (Table 2). In the first type, monovalent electrolytes such as KNO$_3$, KCl, and NaCl lead to identical i.e.p., equal to 5.0 in the given example. In the second type, divalent electrolytes CaCl$_2$ and MgCl$_2$ lead to higher i.e.p. (5.5). This is due to the specific adsorption of calcium or magnesium, which displace the i.e.p. toward higher pH values, due to their valence and affinity for the surface [32, 33].

At pH 6.5 for example, the membrane surface sites are essentially ZrO$^-$. The SP depends on the net charge, mainly due to ZrO$^-$...K$^+$, ZrO$^-$...Na$^+$ or ZrO$^-$...Ca^{2+}, ZrO$^-$...Mg^{2+} complexes. Subsequently, the SP depends not only on the nature of the counterions [30, 38], but also on their mobility. This is because the SP measures a potential difference, under a pressure difference, which displaces the species. For

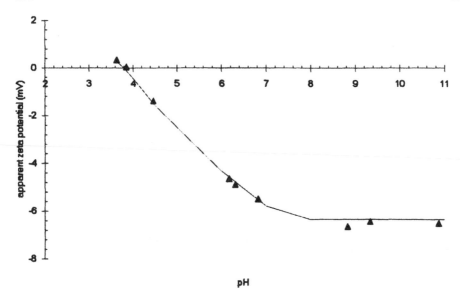

FIG. 9 Apparent zeta potential calculated from streaming potential measurements, using the Smoluchowski relationship, versus pH. M1 Carbosep (150 kD) membrane in 1 mM NaCl medium; $T = 50°C$. (From Ref. 31.)

example, during KNO_3 and KCl filtration with a negatively charged membrane, the SP is very similar for both electrolytes (at a given pH and ionic strength) because it depends mainly on K^+ cations (which are common for both electrolytes). However, SP nominal values in NaCl media are higher than those in KCl media. This is consistent with the conductivity values of the respective counterions [30]. Moreover, the apparent zeta-potential values (Table 2), calculated in NaCl and KCl media, are different: this can be explained by the greater affinity of K^+ cations for the surface than Na^+ cations; this can especially be due to the invalidity of the used relationship, which does not take into account the surface conductivity [36]. Therefore, the surface conductivity in KCl or KNO_3 solutions is beyond any doubt higher than in NaCl medium because of the respective conductivity of K^+ and Na^+ counterions.

TABLE 2 Isolectric Point (i.e.p.), Solution Conductivity (λ_0), Streaming Potential (SP) at pH 6.4, and Apparent Zeta Potential (ζ_{app}) Calculated at pH 6.5; M5 Carbosep Membrane; $I = 1m$ M; $T = 50°C$

Electrolyte	i.e.p. (±0.2)	λ_0 (mS cm^{-1})	SP (mV bar^{-1})	ζ_{app} (mV)
KNO_3	5.0	0.12	−6.9	−0.73
KCl	5.0	0.12	−7.0	−0.74
NaCl	5.0	0.10	−12.0	−1.06
$CaCl_2$	5.5	0.12	−5.0	−0.53
$MgCl_2$	5.5	0.12	−6.0	−0.64

Source: Ref. 31.

In CaCl$_2$ solution, authors have observed a lower SP because calcium is specifically adsorbed and partly neutralizes the surface charge [32, 39]. In this case of divalent ions, the role of counterions' conductivity is equally observed.

Similarly, Fig. 10 shows the influence of phosphate ions on a mineral membrane. The SP is reported versus the pH of the solution in various media. Measurements made in NaCl medium give an i.e.p. of about 3.7. Because of the nature of phosphate, the presence of these ions in solution leads to a shifting of the curve towards more negative values of SP and to a significant displacement of the i.e.p. equal to 2.4 in this case. It has also been observed that, even after several chemical cleanings of the membrane (according to the classical membrane conditioning protocol), the measured i.e.p. in NaCl medium does not recover its initial value. The interface has been then irreversibly modified.

Using a hydraulic criterion as a characterization tool, that is to say the permeate flux, does not allow us to put into the fore the membrane surface-state modification. Actually, it is the same in the three cases. As for the SP measurement, it allows characterization of the membrane state from a charge point of view.

To achieve an optimum characterization of the membrane, it will be useful to have two criteria: a hydraulic criterion determined by permeate flux and hydraulic resistance measurements, and an electrical criterion determined by the SP measurements and zeta-potential calculations. As well be observed below, modification in the electric charge distribution can influence significantly the filtration performances and the membrane selectivity.

3. Influence of Ionic Strength

In the case of indifferent electrolytes (i.e., electrolytes for which ion-surface interactions are purely electrostatic), the zeta potential calculated from SP measurements decreases as the salt concentration increases due to the phenomenon of double-layer

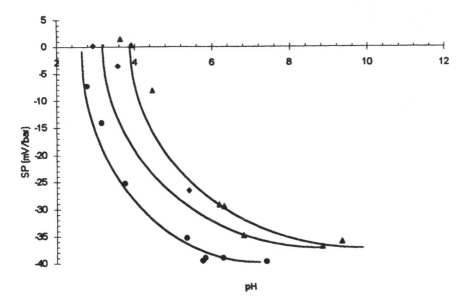

FIG. 10 Streaming potential versus pH in various media at $I = 1$ mM on a M1 membrane: (▲) NaCl; (●) H$_3$PO$_4$, (◆) NaCl after several membrane rinsings.

compression: the double-layer thickness is reduced (the surface charge is screened at a shorter distance) when the ionic concentration is raised. In some cases, when the effects of the inner or compact layer are important, more counterions can be adsorbed on it when the ionic strength is increased: as a consequence, the charge density in the diffuse layer will be reduced, as shown schematically in Fig. 11. Thus, the zeta potential (and thus the SP) decreases in nominal value as the ionic strength becomes greater, whatever the pH is [29, 30, 31, 40, 41]. With an electrolyte such as KCl the ionic strength does not modify the i.e.p. of the membrane, and the zeta potential tends to be zero at high salt concentration (Fig. 12). This is typical for an indifferent electrolyte.

Figure 13 [41] shows results obtained with Na_2SO_4 solutions under the same experimental conditions are in Fig. 12 (i.e., at pH 3.2, a pH lower than the membrane i.e.p.). In that case we can observe that the increase in ionic strength leads to the reversal of the zeta potential sign. The net surface charge of the membrane is then positive at low ionic strength and becomes negative when the salt concentration is high enough. This is due to the specific adsorption of the divalent anions, in this example, SO_4^{2-}. Sulfate ions can penetrate into the compact layer to reach the internal Helmholtz plane by losing part of their hydration sphere. Their interactions with the surface are not only electrostatic but also of a chemical nature. It must be noted that if experiments had been carried out at ionic strengths higher than 0.1 M, the zeta potential determined with Na_2SO_4 should tend to zero as for indifferent electrolytes.

E. Zeta Potential Determination

1. Domain of Validity of Helmholtz–Smoluchowski Relation

Zeta potential can be related to SP using the Helmholtz–Smoluchowski relationship, when there is no double-layer overlapping ($\kappa r \gg 1$, r being the pore radius) [10].

Christoforou et al. [42] have shown that the use of this equation requires a κr larger than 10 and low charge density. Westermann-Clark and Anderson [43] point out that for $\kappa r \leq 3$, the SP only allows us to indicate the sign of the surface net charge.

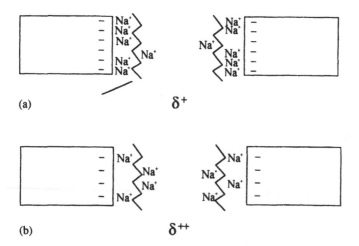

(a) δ^+

(b) δ^{++}

FIG. 11   Charge distribution in the double layer: (a) high ionic strength; (b) low ionic strength.

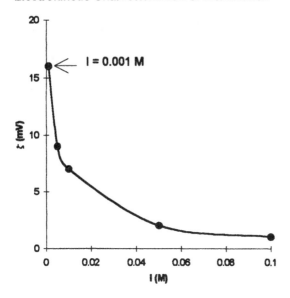

FIG. 12 Effect of ionic strength on zeta potential (ζ); KCl (pH 3.2), mixed alumina–titania–silica MF membrane. (From Ref. 41.)

In the case of a low capillary section, the contribution of the surface conductivity of the membrane material can be taken into account by replacing λ_0 by λ_p, the conductivity inside the pore [44]:

$$SP = \frac{\Delta E}{\Delta P} = \frac{\varepsilon_0 \varepsilon_r \zeta}{\eta \lambda_p} \tag{15}$$

with $\lambda_p = \lambda_0 + \frac{2\lambda_s}{r}$

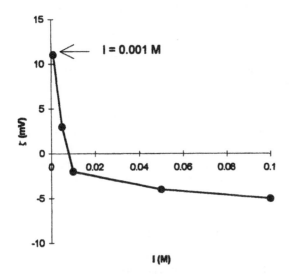

FIG. 13 Effect of ionic strength on zeta potential (ζ); Na$_2$SO$_4$ (pH 3.2), mixed alumina–titania–silica MF membrane. (From Ref. 41.)

Burgreen and Nakache [45] have shown that relation (15) can be corrected by a factor F which takes into account all the phenomena that the relation (15) does not consider: double-layer overlapping, zeta potential value, convection, and conduction current [25]:

$$SP = \frac{\Delta E}{\Delta P} = \frac{\varepsilon_0 \varepsilon_r \zeta}{\eta \lambda_0} F(\kappa r, \zeta) \tag{16}$$

The problem of taking into account double-layer overlapping has been equally treated by Hildreth [46] who proposed to correct it by a factor $G = \dfrac{\tanh(\kappa r)}{\kappa r}$:

$$SP = \frac{\Delta E}{\Delta P} = \frac{\varepsilon_0 \varepsilon_r \zeta}{\eta \lambda_p}(1 - G) \tag{17}$$

By correcting the zeta potential for surface conductivity, Agerbaek and Keiding [47] have shown that it decreases with ionic strength, in good agreement with the triple-layer model [25], whereas Eq. (15) gives a constant zeta potential when the ionic strength increases. It appears then that the zeta potential calculation from SP experimental measurements needs at least to take into account the surface conductivity and the double-layer overlapping when it exists.

The F factor has been theoretically calculated by Hildreth [46]. It has been experimentally confirmed [34]: the determination of the $F(\kappa r, \zeta)$ factor of Eq. (16) has been realized by dividing the zeta potential calculated from the SP by the one calculated from the electrophoretic mobility, the latter being considered as the true zeta potential:

$$F = \frac{\zeta_{app}}{\zeta_{true}} \tag{18}$$

where ζ_{app} is the zeta potential calculated from the streaming potential measurement and the use of Smoluchowski's equation, and ζ_{true} is the zeta potential calculated from electrophoretic mobility measurements.

This implies that the zeta potential found from electrophoretic mobility and SP measurements must be the same; that is, that the friction in a pore is the same as the one on a particle. Moreover, the possible influence of roughness of scraped particles needs to be taken into account.

Figure 14 [34] gives the values of F versus κr for the studied membranes. In the range of ionic strengths and membrane pore studied, κr varies from 0.2 to 110 since κ^{-1} decreases when the ionic strength increases. For high values of κr ($\kappa r > 20$). F tends toward 0.8 and we approach the validity domain of the Helmholtz–Smoluchowski's equation [relation (9)]. There is no double-layer overlapping, as F only takes into account the conductivity difference between the solution inside the pore and in the bulk; F does not reach unity since the location of the shear plane is different for both electrokinetic phenomena [34, 48].

For lower values of κr, the double-layer overlapping adds to the effect of conductivity; this leads to a correction coefficient F very different from 1. The experimental factor F found is similar to the one calculated by Hildreth [46]: it varies from 0 to 1, depending on κr and ζ.

2. Electrolyte Conductivity in the Pore and Space Charge Model

Measuring the streaming potential is among the most convenient techniques for the electrokinetic study of porous membranes [20, 42]. However, the determination of the zeta potential from a single SP measurement may be ambiguous [49, 50]. Figure 15 shows the zeta potential (ζ) dependence of the SP determined from the space charge model for

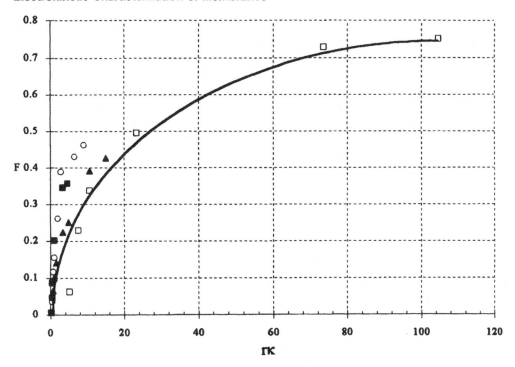

FIG. 14 Calculated correction factor F versus κr. $T = 50°C$; NaCl (pH $= 6.5$); ■: M5, ○: M4, ▲: M1, □: M14 Membranes Carbosep^{TM}, Orelis. (From Ref. 34.)

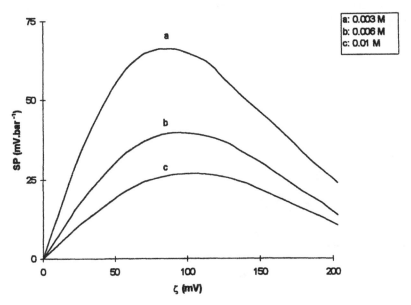

FIG. 15 Zeta potential (ζ) dependence of the streaming potential (SP) evaluated from the space charge model for various concentrations of KCl solutions: (a) 0.003 M; (b) 0.006 M: (c) 0.01 M; $r = 75$ nm. (From Ref. 49.)

various KCl concentrations. The SP has a maximum with respect to the zeta potential at fixed concentration. The downward trend of the SP for sufficiently high zeta potentials results from the excess conductance in the region near the pore walls. During the SP process, the pressure gradient acting through the membrane creates an electric field in the opposite direction of the fluid flow. This electric field produces a backflow of counterions by the electro-osmotic effect [25]. The surface conductance increases exponentially with the zeta potential at fixed electrolyte concentration [50]. The higher the surface conductance, the greater the counterion backflow is, which in turn causes the SP to decrease as zeta potential increases in magnitude.

It can be seen from Fig. 15 that one SP value may be associated with two different values of zeta potential. To remove this ambiguity, the SP can be studied as a function of the pH of the KCl solutions. Indeed, the zeta potential of inorganic membranes is affected by the pH of the solution flowing through the membrane pores. The plot SP $= f(pH)$ might then show one particular SP value associated with two different pH. For instance, Fig. 16 shows the variation of SP versus pH for the same KCl concentrations as in Fig. 15. The minimum pH value for each electrolyte concentration was chosen in such a way that the concentration of protons can be neglected with respect to that of K^+ ions. As expected, the curves present a maximum with respect to the pH of the solutions. For pH less than ~ 9.5, each SP value is associated with two distinct pH values. The i.e.p. of the membrane, that is, the pH for which SP $= 0$, is about 10.3. This allows us to determine the correct zeta potential corresponding to one SP value obtained at a known pH. Indeed, the farther the pH from the i.e.p., the higher the zeta potential is. The SP value corresponding to the pH furthest from the i.e.p. can then be associated with the highest of the two possible zeta potentials obtained from the plot SP $= f(\zeta)$ (see Fig. 15). This procedure has been repeated for various electrolyte concentrations in order to determine the correct value of the zeta potential from each SP.

Figure 17 presents the concentration dependence of the zeta potential (ζ) estimated from the space charge model and from the approximated Helmholtz–Smoluchowski relation [Eq. (9)]. The Helmholtz–Smoluchowski equation neglects the surface conduction

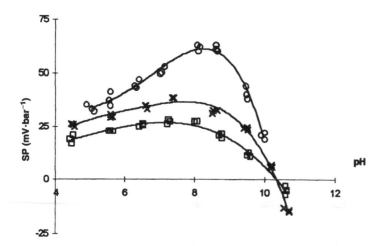

FIG. 16 pH dependence of the experimental streaming potential for various KCl concentrations–membrane Al$_2$O$_3$, $r = 75$ nm. ($\bigcirc$) 0.003 M; ($\times$) 0.006 M; ($\square$) 0.01 M. (From Ref. 49.)

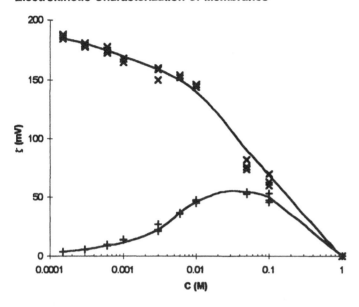

FIG. 17 Concentration dependence of the zeta potential (ζ)–membrane Al_2O_3, $r = 75$ nm. (×) From space charge model; (+) from Helmholtz–Smoluchowski equation [Eq. (9)]. (From Ref. 49.)

phenomenon and considers that the conductivity inside the pores is the same as the bulk solution conductivity (λ_0). It leads to an apparent value of the zeta potential also called the equivalent zeta potential [51].

Numerical calculations show that the membrane is strongly charged in a wide range of electrolyte concentrations. As expected, the zeta potential decreases as the electrolyte concentration increases. On the other hand, the curve obtained from Eq. (9) presents a maximum with respect to the electrolyte concentration. It appears that Eq. (9) can lead to a substantial underestimation of the true zeta potential, particularly at low concentration and for a strongly charged surface.

F. Determination of Filtering Layer SP in the Case of Multilayer Membrane

Ceramic membranes are commonly used in MF and UF to separate a wide range of products from undesirable compounds. Many of these membranes exhibit a multilayer composite microstructure with each underlying layer having progressively larger pore size and thickness. The predominantly thick layer, called the support layer, provides the necessary mechanical strength to the filtering layer, which has the smallest pore size. The support layer can be made of a different material from the filtering one. Both layers can consist of different metal oxides (Membralox^{TM}, SCT US Filter, and Tami membranes, for example) or the support layer can be a nonceramic material such as carbon (Carbosep^{TM} membranes, Orelis for example).

As an example, Table 3 gives the characteristics of a ceramic membrane consisting of three different metal oxide layers.

In the case of composite membranes for which the support layer as well as the intermediate support layers and the filtering one are made of metal oxides, the SP may

TABLE 3 Multilayer Tami^{TM} Membrane Characteristics

	Support	Intermediate layer	Filtering layer
Nature	Al_2O_3-TiO_2	TiO_2	ZrO_2
Thickness	2 mm	10 μm	5–6 μm
Mean pore radius	0.5 μm	0.2 μm	40 nm
Porosity	35%	40%	–

Source: Ref. 52.

not reflect only the surface properties of the filtering layer [52]. Indeed, the pore walls of each layer develop an electrical charge when such a membrane is put in contact with an aqueous medium. Because of its important thickness with respect to the separating layer, the support can influence the SP value measured across the membrane. In the same way, because the intermediate support layer pores are successively smaller than that of the bulk support, their contribution to the SP value may be not negligible. In other words, each oxide layer can play a significant role in the global SP measured across the whole membrane. Users have to access to the electrokinetic properties relative to each layer and especially to the filtering one.

Figure 18 illustrates the different structures (a), their respective permeation fluxes (b), and electrical potential differences (c) versus the transmembrane pressure. Curves giving the permeation flux variation as a function of the working pressure allow knowledge of, at a given flux J_x, the pressure drop across the support (ΔP_{sup}, the intermediate layer ($\Delta P_{i.l.}$), and the filtering layer ($\Delta P_{f.l.}$). Indeed, the pressure difference between both sides of the two-layered membrane ($\Delta P_{sup+i.l.}$) is equal to the sum of the pressure drop across the support and the intermediate layer:

$$\Delta P_{sup+i.l.} = \Delta P_{sup} + \Delta P_{i.l.} \tag{19a}$$

For the three-layered membrane, we have

$$\Delta P_{sup+i.l+f.l.} = \Delta P_{sup} + \Delta P_{i.l.} + \Delta P_{f.l.} \tag{19b}$$

In the same way, from curves showing the electrical potential difference versus the transmembrane pressure we can determine the potential drop across each layer (ΔE_{sup}, $\Delta E_{i.l.}$, and $\Delta E_{f.l.}$). Indeed, the electrical potential drop across the two-layered ($\Delta E_{sup+i.l.}$) and three-layered ($\Delta E_{sup+i.l.+f.l.}$) membranes can be also written as

$$\Delta E_{sup+i.l.} = \Delta E_{sup} + \Delta E_{i.l.} \tag{20a}$$

and

$$\Delta E_{sup+i.l.+f.l.} = \Delta E_{sup} + \Delta E_{i.l.} + \Delta E_{f.l.} \tag{20b}$$

If both the permeation flux and electrical potential difference measured experimentally are linear functions of the transmembrane pressure for the three structures, SP for both intermediate and filtering layers can be expressed as

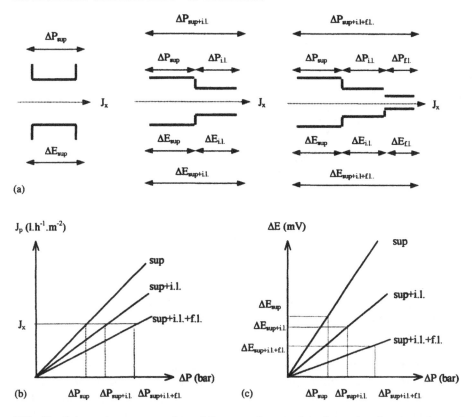

FIG. 18 Schematic representation of the procedure used to determine the streaming potential of the intermediate and filtering layer: (a) structures studied; (b) permeation flux J_p versus transmembrane pressure ΔP; (c) electrical potential difference ΔE versus pressure drop ΔP. (From Ref. 52.)

$$SP_{i.l.} = \frac{SP_{sup+i.l.} - SP_{sup} \cdot \dfrac{L_{P_{sup+i.l.}}}{L_{P_{sup}}}}{1 - \dfrac{L_{P_{sup+i.l.}}}{L_{P_{sup}}}} \qquad (21a)$$

and

$$SP_{f.l.} = \frac{SP_{sup+i.l.+f.l.} - SP_{sup+i.l.} \cdot \dfrac{L_{P_{sup+i.l.+f.l.}}}{L_{P_{sup+i.l.}}}}{1 - \dfrac{L_{P_{sup+i.l.+f.l.}}}{L_{P_{sup+i.l.}}}} \qquad (21b)$$

where $L_{P_{sup}}$, $L_{P_{sup+i.l.}}$, and $L_{P_{sup+i.l.+f.l.}}$ are the hydraulic permeabilities (i.e., the permeation fluxes per pressure unit) determined from flux measurements performed with the three structures.

Figure 19 shows SP variations versus pH for each membrane given in Table 3 in mM NaCl solutions. Isoelectric points found are very close for the three membranes (i.e., i.e.p. = 6.5 for the alumina support and the three-layered membrane; i.e.p. = 6.7 for the two-layered membrane consisting of the support and the intermediate titania layer). Jin and

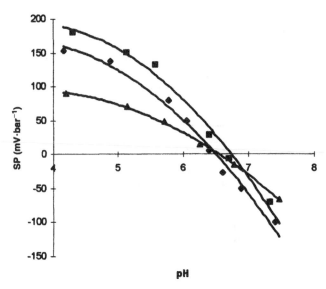

FIG. 19 pH dependence of streaming potential (SP) in 1 mM NaCl solution. (◆) Support; (■) two-layered membrane; (▲) three-layered membrane. (From Ref. 52.)

Sharma [53] showed that in the case of two tubes connected in series and having different radii and lengths, the resulting SP depends on the fraction of tube length and the ratio between the tubes' radii. In the present case, the support layer represents more than 99% of the total thickness. Moreover, pores of the intermediate and filtering layers are only about 2 and 12 times smaller than the support ones. Thus, it would be hazardous to conclude that measurements performed with the multilayered membranes allow characterization of the electrokinetic behavior of the separating layer.

Results shown in Fig. 19 and the membrane permeabilities allow calculation of SPs appropriate to titania and zirconia layers by using Eqs (21a) and (21b). The calculated SPs of the titania and zirconia layers are presented in Fig. 20 (continuous lines: curves a and d, respectively). It clearly appears that the properties of these layers differ significantly from results obtained with the two-layered membrane (dotted lines: curve b) and the three-layered one (dotted lines: curve c).

Each layer exerts an influence on the global SP measured across a multilayer membrane. The SP of such a membrane is a combination of the SP values of each layer composing the membrane. Isoelectric points determined for the different layers differ significantly from one another, whereas it does not clearly appear by considering the pH-dependent SP curves relative to the different structures studied (support, two-layered membrane, and three-layered membrane). The support layer seems to play a non-negligible role in the global SP due to its large thickness as compared to that of other layers.

G. Consequences of Surface Charge Density on Membrane Selectivity

The separation of molecules involves not only the steric effect exclusion, but ionic interactions between solute or ions and the membrane are also capable of modifying selectivity [27, 54, 55].

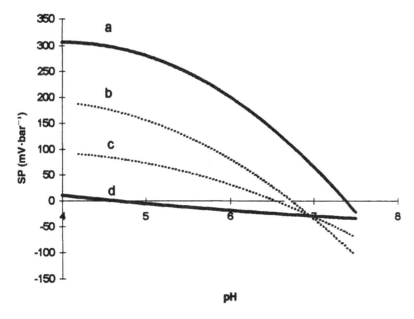

FIG. 20 pH dependence of streaming potential (SP) in 1 mM NaCl solution. Continuous lines: (a) SP determined for the titania layer; (d) SP determined for the filtering zirconia layer. Dotted lines: (b) experimental SP measured on the two-layered membrane; (c) experimental SP measured on the three-layered membrane. (From Ref. 52.)

Actually, the filtration of ionic solutions (NaCl, for example) at various pH and ionic strengths gives rise to a nontotal ion transmission even when the pore radius (2 nm) is larger than the ionic radius. The importance of such ionic effects has been quantified by SP measurements [56]. Thus, electric effects are observed on membrane selectivity. When the net charge is zero, that is, close to the i.e.p. (close to 5 in the example shown in Fig. 21), Cl^-, Na^+, and Ca^{2+} transmission are maximum, close to 1 (Fig. 22). This corresponds to the steric transmission forecast by Ferry [3] (Table 4).

At a pH far from the i.e.p., electrostatic repulsions between the membrane surface and ions in solution decrease the ion transmission. Notice that, even when the pore radius is widely larger than the ion size, transmission decreases by up to 40% at pH 4 in the case of $CaCl_2$ filtration and at pH 8 in the case of NaCl filtration. Actually, experiments were performed at 1 mM; at this ionic strength, the Debye length is about 10 nm whereas the pore radius is about 2 nm. There is, then, double-layer overlapping which leads to ionic exclusion.

At a pH lower than the i.e.p., the coions (Na^+ or Ca^{2+}) are rejected by the membrane. To maintain the fluid neutrality, Cl^- counterions are also rejected. That is why at a pH lower than 5, Cl^- transmission is lower than 1. The reverse phenomenon is observed at a pH higher than 5; that is, Na^+ and Ca^{2+} transmission are lower than 1. Hydrogen ions and OH^- also participate in the electrostatic equilibrium as has been mentioned in several studies [57, 58].

In summary, the higher the SP nominal value, the lower the ion transmission. This demonstrates the contribution of electric effects to membrane selectivity.

The study of electrolyte UF has been confirmed: Jeantet and Maubois [58] have investigated the nanofiltration selectivity mechanisms of ionic solutions. These are mainly determined by the Donnan effect. This has been more recently observed by Combe et al. [59].

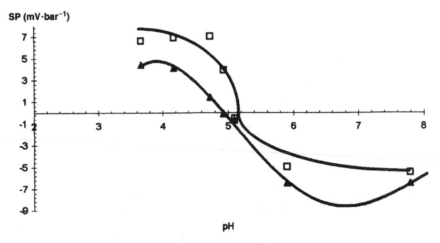

FIG. 21   Streaming potential (SP) versus pH of solution containing 10^{-3} M: (▲) NaCl; (□) CaCl$_2$. (From Ref. 56.)

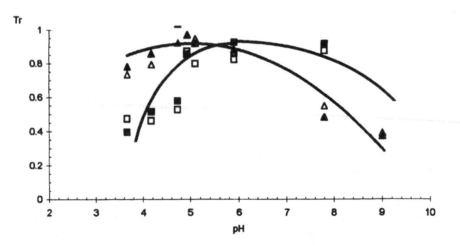

FIG. 22   Ion transmission Tr versus pH of solution containing 10^{-3} M of NaCl or CaCl$_2$: (▲) Na$^+$ NaCl); (■) Ca^{2+}(CaCl$_2$); (△) Cl$^-$ (NaCl); (□) Cl$^-$ (CaCl$_2$). (From Ref. 56.)

TABLE 4   Ionic Radius r_i and Steric Transmission Forecast by Ferry ($T_{r_{steric}}$)[a]

Ions	Ionic radius (nm) (r_i)	T_r steric
Na$^+$	0.095	0.991
Ca^{2+}	0.099	0.990
Cl$^-$	0.184	0.969

[a] Calculations were made with a membrane pore radius equal to 2 nm.
Source: Ref. 60.

For the same reasons, the efficient UF of macromolecules depends on electric and steric effects: experiments have been performed on single proteins of different sizes and charges as a function of pH, ionic strength, and nature of the salt. This has allowed evaluation of the filtration performance [60].

The presence of fouling species in solution induces an immediate membrane fouling and a decrease in the permeate flux down to a stable value. This value depends on the protein properties, especially on its solubility, but also on the sign of the membrane and of the fouling species' net charge. The permeate flux is lower when the pH is close to the protein i.e.p. At a given pH, if the protein and the membrane have the same net charge sign, the permeate flux is higher if the ionic strength variation resulting from the salt addition is high. On the other hand, it is lower when the protein and the membrane have opposite net charge.

Streaming potential measurements under various conditions of pH and ionic strength agree with the results of several authors [21, 51] according to which the proteins contribute to the net charge of the system. The adsorption of proteins onto the membrane surface then contributes to the filtration by its own selectivity [61].

Protein transfer through a mineral membrane is governed by ionic and steric exclusion phenomena. Single proteins have a higher transmission at a pH close to their i.e.p. and at high ionic strength, where the SP tends to zero.

The presence of ionic repulsion decreases the transmission. This depends on the SP which depends itself on the solution composition. However, no simple relation between solution and membrane characteristics and the SP exists.

A correlation (Fig. 23) which allows a correct estimation of transmission variation has been established. It can be exploited in order to choose the membrane, or operating conditions, when it is necessary to implement membrane separation operations. However, the correlation does not take into account interactions between proteins nor membrane

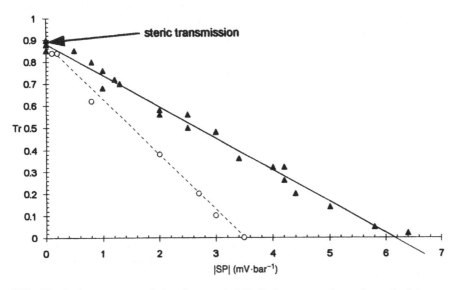

FIG. 23 Lysozyme transmission (inorganic M1 Carbosep membrane from Orelis) versus streaming potential nominal values (which vary with ionic strength added in salt). NaCl: (▲) pH = 7; (○) pH = 3.5. (From Ref. 46.)

fouling when proteins are in a mixture. The correlation makes clear the relationship between protein transmission and the measured SP.

All the results concerning the filtration of solutions, the composition of which has been adjusted (pH, ionic strength, solute nature), have allowed us to decompose the various phenomena which take place during filtration: steric effect and electric exclusion. This better understanding of filtration can be attributed to the use of SP as an electrical characterization tool.

IV. USE OF ELECTRO-OSMOSIS TO CHARACTERIZE UF AND MF MEMBRANES

Electro-osmosis is based on the movement of the liquid phase adjacent to a charged surface under the influence of an electric field applied parallel to the interface. It can be used to characterize the charge state of both organic [62–66] or inorganic membranes [67–70] as well as the electrokinetic properties of colloidal particles (after electrophoretic deposition of these particles on to membrane pores) in a concentration range inaccessible to microelectrophoresis [71].

A. Principle and Phenomenon of Electro-osmosis

The application of an electric field parallel to a charged pore filled with an electrolytic solution results in migration of ions situated beyond the shear plane (i.e., ions present in the diffuse layer as well as in the bulk solution) [72]. Ions move towards electrodes of opposite charge sign (Fig. 24), carrying solvent with them.

If both cations and anions have very similar mobilities, then the global flux in the bulk solution is zero and the solvent flow results only from the motion of ions present in the diffuse layer. This solvent flow is called the electro-osmotic flow rate.

The electro-osmosis experiment consists in performing the electrolysis of an electrolyte solution, the anode and the cathode being positioned on both sides of the membrane. Electrolysis can be carried out with a constant applied voltage or current. However, the latter option appears to be better than the former. Indeed, during constant-voltage electrolysis, electric current dramatically falls over the last minutes and it is not possible to maintain it at a constant value during the experiment [73].

The volume (v) of liquid electro-osmotically transported increases linearly with time (t). The electro-osmotic flow rate V is determined from the slope of the plot of v versus t, as shown in Fig. 25.

B. Electro-osmosis Measurements

Figure 26 depicts an experimental unit used for electro-osmosis experiments. The membrane is held at one end of a deep tube immersed in a thermostated electrolyte solution. Two platinum electrodes are used. The first one is a mesh electrode positioned behind the membrane, i.e., in the tube; the second one is a cylindrical electrode positioned outside the tube. A constant current is applied between the electrodes by means of a power supply, and a small capillary transfers the overflowed electrolyte to a precision balance.

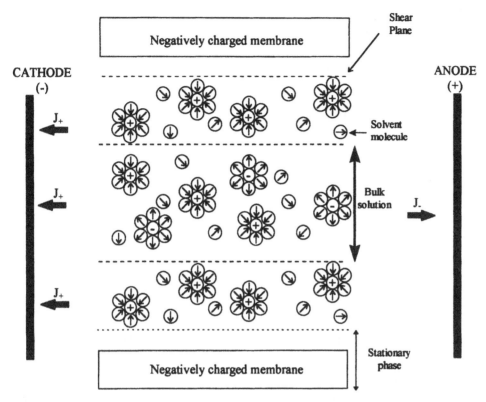

FIG. 24 Origin of the electro-osmosis phenomenon: (J_+) flux of cations; (J_-) flux of anions. (From Ref. 74.)

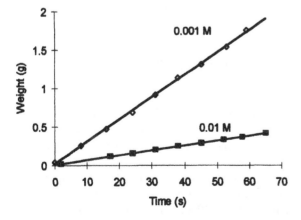

FIG. 25 Variation of electro-osmotically transported liquid volume as a function of time (t). (From Ref. 70.)

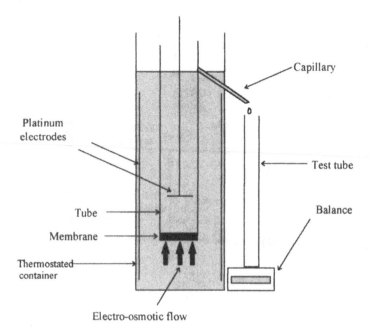

FIG. 26 Experimental unit for electro-osmosis experiments. (From Ref. 70.)

1. Operating Procedure

The membrane is first equilibrated with the measuring solution. Electro-osmotic experiments are carried out with a large volume of solution to prevent any change in pH (due to electrolysis of the electrolyte solution). At the beginning of the experiment, both liquid levels inside and outside the tube are made equal. A constant current (generally in the range 1–30 mA) applied between the two platinum electrodes induces an electro-osmotic flow of electrolyte in the tube.

The polarity of the electrodes is chosen so that the liquid electro-osmotically transported flows toward the capillary. The electro-osmotic flow rate is determined by weighting the fluid amount flowing through the membrane as a function of time. The measuring time is limited to prevent any change in pH (the pH has to be checked before and after each experiment to ensure that no change has occurred) [8, 70].

V. COMPARISON OF VARIOUS METHODS TO DETERMINE MEMBRANE ELECTROKINETIC CHARACTERISTICS

There are several procedures, including microelectrophoresis, SP measurements, and electro-osmosis, that allow determination of the zeta potential. The most widely used procedure, microelectrophoresis, is effective for studying powder dispersions, whereas SP and electro-osmosis can be applied to the membrane itself as we saw previously. So, electrophoresis measurements can be used to determine the zeta potential of a membrane sample after grinding, but the newly formed surface can differ considerably from the membrane surface.

The variation in zeta potential determined from SP and electro-osmosis in 0.001 and 0.01 M NaCl solutions in the case of a ceramic membrane (64% Al_2O_3, 27% TiO_2, 9% SiO_2, 0.9 μm pore size) is shown on Fig. 27.

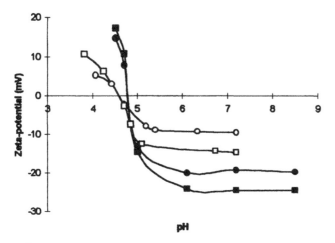

FIG. 27 pH dependence of zeta potential determined from streaming potential (SP) and electro-osmosis (EO) for NaCl solutions: (○) SP, 0.01 M; (●) EO, 0.01 M; (□) SP, 0.01M; (■) EO, 0.001 M. (From Ref. 75.)

The i.e.p. values obtained by both methods for NaCl solutions are very close: 4.7 ± 0.2 and 4.5 ± 0.2 from electro-osmosis and SP, respectively. The i.e.p. is independent of the ionic strength since NaCl is known to be an indifferent electrolyte.

For both techniques the profiles of the pH-dependent zeta-potential curves are similar. However, the numerical values of zeta potential differ from one method to another for pH values different from the i.e.p. Electro-osmosis gives higher absolute values and the gap increases as the pH moves to the i.e.p.

It is important to keep in mind that the zeta potential is the potential determined at the shearing plane between the liquid and the solid surface. Vernhet et al. [48] have suggested that the location of the shear plane depends on the electrokinetic method, as represented in Fig. 28. So, during electro-osmosis, the shear plane might be nearer the surface, leading to zeta potentials larger than obtained by SP measurements.

Figure 29 shows the ionic strength effect on the ratio of the zeta potentials calculated from the electro-osmotic flow rate and SP measurements for NaCl solutions at pH 7.2, corresponding to the plateau region on Fig. 27. It appears that the ratio ζ_{EO}/ζ_{SP} becomes higher as the ionic strength increases. The increase in ionic strength leads to a compression of the diffuse layer (the Debye length κ^{-1} decreases, see Fig. 30). These observed results agree with the assumption that the location of the shearing plane differs according to the electrokinetic methods used.

VI. CONCLUSION

The characterization of the membrane–solution interface by means of electrical and electrokinetic measurements is doubtless of the greatest interest for understanding and controlling the selectivity of a membrane. Indeed, as a rule, the filtration peformances of a membrane are governed not only by steric effects but also by the electrostatic interactions occurring between the species in solution and the membrane surface.

Several techniques providing information about the charge state of porous membranes can be used. Two electrokinetic methods, electro-osmosis and SP, can be applied to the study of UF and MF membranes.

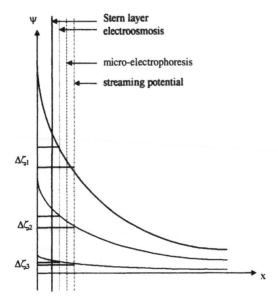

FIG. 28 Theoretical potential ψ evolution as a function of the distance with respect to the solid surface (x) within the electrical double layer: influence of the shear plane location on zeta potential values determined with different electrokinetic methods. $\Delta\zeta_i$ = variations between zeta potential values obtained with different electrokinetic methods at pH_i ($pH_1 < pH_2 < pH_3 <$ p.i.e.). (From Ref. 75.)

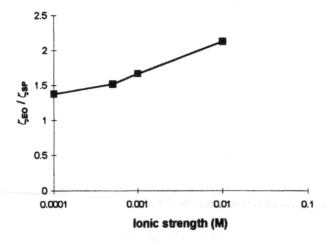

FIG. 29 Ionic strength effect on the ratio of zeta potentials calculated from the electroosmotic flow rate (ζ_{eo}) and streaming potential (ζ_{SP}) measurements. (From Ref. 75.)

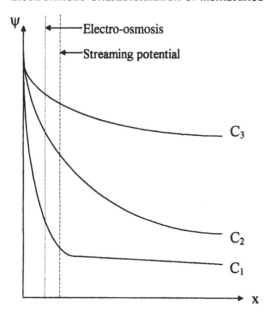

FIG. 30 Influence of the ionic strength (C) on the theoretical potential profile; $C_1 > C_2 > C_3$. (From Ref. 75.)

REFERENCES

1. HP Hsieh. Inorganic Membranes for Separation and Reactiton. Elsevier, Amsterdam, The Netherlands, 1996.
2. MW Chudacek, AG Fane. J Membr Sci 21: 145, 1984.
3. JD Ferry Chem Rev 18 (373): 373, 1936.
4. L. Millesime. Rôle des interactions sur la rétention de protéines par des membranes d'ultrafiltration. Cas particulier de membranes modifées par dépôt de polymère quarternisés. Thèse, Université Paris VI, 1993.
5. RJ Hunter. Introduction to Modern Colloid Science. Oxford Science Publications, New York, 1996.
6. JM Perrin. Propriétés de surface d'alumines: influence des impuretés minérales, du broyage et des traitements thermiques. Relation avec l'adsorption de polymères. Thèse, Mulhouse, 199.
7. L Dupont. Analyses théorique et expérimentale des interactions ioniques dans l'adsorption de poly(acide acrylique) sur plusieurs alumines. Application aux propriétés rhéologiques de suspensions concentrées de ces alumines. Thèse, Université de Franche-Comté, 1993.
8. M Mullet. Etude de membranes inorganiques: structure poreuse, propriétés, électrochimiques de surface et cinétique d'adsorption d'un peptide. Thèse, Université de Franche-Comté, 1996.
9. M Mullet, P Fievet, A Szymczyk, A Foissy, J-C Reggiani, J Pagetti. Desalination, 121: 41, 1999.
10. R O'Brien, LR White. Can J Chem 59: 1878, 1981.
11. S Levine, JR Marriot, G Neale, NJ Epstein. Colloid Interface Sci 52(1): 136, 1975.
12. JTG Overbeek, BH Bijsterbosch. Electrokinetic Separation Methods. Amsterdam: Elsevier, 1979, pp 1–32.
13. M Von Smoluchowski. Bull Akad Sci Cracovie, Classe Sci. Math Natur 1: 82, 1903.
14. KC Thomas, V Ramachandhran, BMJ Misra. J Appl Polym Sci 34: 2527, 1987.
15. E Staude, D Düputell, D, F Malejka, DJ Wyszynski. Dispers Sci Technol 12: 113, 1991.
16. C Werner, HJ Jacobasch, GJ Reichelt. Biomater Sci 7: 61, 1995.

17. J Benavente, G Jonsson. Colloids Surfaces A 138: 255, 1998.
18. R Blank, KH Muth, S Proske-Gerhards, E Staude. Colloids Surfaces A 140: 3, 1998.
19. M Minor, AJ van der Linde, HP van Leeuwen, J Lyklema. Colloids Surfaces A 142: 165, 1998.
20. D Möckel, E Staude, M Dal-Cin, K Darcovich, M J Guiver, Membr Sci. 145: 211, 1998.
21. M Nyström, A Pihlajamäki, N Ehsani. J Membr Sci 87: 245, 1994.
22. M Nyström, L Kaipia, SJ Luque. Membr Sci 98: 249, 1995.
23. M Pontie, X Chasseray, D Lemordant, JM Laine. J Membr Sci 129: 125, 1997.
24. MJ Pontie. J Membr Sci 154: 213, 1999.
25. RJ Hunter. Zeta potential in Colloid Science. Principles and Applications. Academic Press, London, 1981.
26. JM Reynard. Caractérisation d'une membrane échangeuse d'ions par l'étude de son potentiel d'écoulement. Thèse, Université Paris XII, 1987.
27. L Ricq. Caractérisation électrocinétique de membranes inorganiques au cours de la filtration de solutions non colmatantes (ions) et colmatantes (protéines). Thèse, Université de Franche-Comté, 1996.
28. J Garrido, JA Ibanez, Pellicier, J, AF Tejerina. J Electrostatics 12: 469, 1982.
29. M Nyström, M Lindström, E Matthiason. Colloids Surfaces 36: 297, 1989.
30. MGA Khedr, SM Abdelhaleem, A Baraka. J Electroanal Chem. 184: 161–169, 1985.
31. L Ricq, A Pierre, JC Reggiani, J Pagetti. Desalination 114: 101, 1997.
32. J Randon. Influence de l'interface oxyde–solution sur les performances des membranes d'ultrafiltration en zircone Thèse, Université de Montpellier, 1991.
33. Y Sun, GM Spencer. J Colloid Interface Sci 126 (1): 361, 1988.
34. L Ricq, A Pierre, JC Reggiani, J Pagetti, A Foissy. Colloids Surfaces A: Physicochem Eng Aspects 138 (2–3): 301, 1998.
35. R Takagi, M Nakagaki. J Membr Sci 111: 19, 1996.
36. R Hidalgo-Alvarez, F Gonzalez-Caballero, JM Bruque, G Pardo. J Colloids Interface Sci 82(1): 45, 1991.
37. L Ricq, A Pierre, JC Reggiani, S Zaragoza-Piqueras, J Pagetti, G Daufin. Membr Sci 114: 27, 1996.
38. KC Thomas, V Ramachandhran, BM Misba. J Appl Polym Sci 34: 2527, 1987.
39. JM Reynard. Caractérisation d'une membrane échangeuse d'ions par l'étude de son potentiel d'écoulement. Thèse, Université Paris VII, 1987.
40. JI Arribas, A Hernandez, A Martin, I Martinez, I Martinez, F. Tejerina. Some electrokinetic phenomena in microporous inorganic membranes. Proceedings of First International Conference on Inorganic Membranes, Montpellier, 1989, pp. 429–434.
41. A Szymczyk, P Fievet, JC Reggiani, J Pagetti. Desalination 115: 129, 1998.
42. CC Christoforou, GB Westermann-Clark, JL Anderson. J Colloid Interface Sci 106: 1, 1985.
43. GB Westermann-Clark, JL Anderson. Electrokin Membr 130(4): 839, 1983.
44. J Stock. Bull Int. Acad Sci Cracovie A. (1970) pp 635.
45. D Burgreen, FR Nakache. J Phys Chem 68: 1084, 1964.
46. DJ Hildreth. J Phys Chem 74: 2006, 1970.
47. ML Agerbaek, KJ Keiding. J Colloid Interface Sci 169: 342, 1995.
48. A. Vernhet, MN Bellon-Fontaine, A Doren. Phys 91 (11, 12): 1728, 1994.
49. A Szymczyk, P Fievet, B Aoubiza, C Simon, J Pagetti. J Membr Sci 161: 275, 1999.
50. A Szymczyk, B Aoubiza, P Fievet, J Pagetti. J Colloid Interface Sci 216: 285–296, 1999.
51. C Causserand, M Nystrom, PJ Aimar. J Membr Sci 88: 211, 1994.
52. A Szymczyk, P Fievet, JC Reggiani, J Pagetti. Desalination 116: 81, 1998.
53. M. Jin, MM Sharma. J Colloid Interface Sci. 142: 61, 1991.
54. L Millesime, J Dulieu, B Chaufer. J Membr Sci 108: 143, 1995.
55. EA Tsapiuk, MT Bryk, J Membr Sci 79: 227, 1993.
56. L Ricq, J Pagetti. J Membr Sci 115: 9, 1999.
57. L Ricq, A Pierre, JC Reggiani, J Pagetti. Desalination 114: 101, 1997.
58. R Jeantet, JL Maubois, Le Lait 75(6): 595, 1995.

59. C Combe, C Guizard, P Aimar, V Sanchez. J Membr Sci 129: 147, 1997.
60. L Ricq, S Narcon, JC Reggiani, J Pagetti. J Membr Sci 156: 81, 1999.
61. M Meireles, P Aimar, V Sanchez. Biotechnol Bioeng 38: 528, 1991.
62. M Tasaka, S Tamura, N Takemura. J Membr Sci 12: 169, 1982.
63. WR Bowen, RA Clark. J Colloid Interface Sci 97: 401, 1984.
64. HL Jacobasch and J Schurz. Progr Colloid Polym Sci 77: 40, 1988.
65. CK Lee, J Hong. Membr Sci 39: 79, 1988.
66. KJ Kim, AG Fane, M Nystrom, A Pihlajamaki, WR Bowen, H Mukhtar. J Membr Sci 116: 149, 1996.
67. W Koh, JL Anderson. AIChE J 21: 1176, 1975.
68. WR Bowen, DTJ Hughes. J Colloid Interface Sci 143: 252, 1991.
69. WR Bowen, H Mukhtar. Colloids Surfaces A 81: 93, 1993.
70. M Mullet, P Fievet, JC Reggiani, J Pagetti. J Membr Sci 123: 255, 1997.
71. WR Bowen, PM Jacobs. J Colloid Interface Sci 111: 223, 1986.
72. SS Dukhin, BV Derjaguin. In: E Matijevic, ed. Surface and Colloid Science, vol 7. New York and Toronto. Wiley, 1974.
73. M Bonnemay, J Royon. Electro-osmose, Techniques de l'ingénieur. D912 (1974) 1.
74. A Szymczyk. Etude des propriétés électriques et électrocinétiques de membranes foreuses, Thèse, Université de Franche-Comté, 1999.
75. A Szymczyk, P Fievet, M Mullet, JC Reggiani, J Pagetti. J Membr Sci 143: 189, 1998.

21

Electrokinetics and Surface Charges of Spherical Polystyrene Particles

HANS NIKOLAUS STEIN Eindhoven University of Technology, Eindhoven, The Netherlands

I. INTRODUCTION

Electrokinetic measurements are an important tool for learning something about the interfacial properties of dispersed particles, at least as long as the interface of such particles with the surrounding medium carries an electric charge which arises, e.g., by adsorption of ions from the surrounding solution, or desorption of ions from it into the solution. In the case of polystyrene latex particles, the surface frequently contains $-O-SO_3H$ groups dissociating into $-O-SO_3^-$ and H^+. Since the system as a whole must be uncharged, the surface charge (usually indicated as σ_0) must be compensated for by an excess of ions of opposite charge in the surroundings, arranged in the "electrical double layer." Charges and counterions will determine the behavior of the dispersions concerned (coagulation, sedimentation, rheology, etc).

However, when electrokinetic methods are applied to polystyrene latex particles, the data obtained show two unusual features:

1. At increasing electrolyte concentrations starting from very low values (0.1 mM or lower), the absolute value of the ζ potential first rises, then passes through a maximum, and at still higher concentration decreases again. A typical graph, based on data by Meijer et al. [1], is shown in Fig. 1, but similar findings have been reported by quite a few other authors, e.g., Elimelech and O'Melia [2], van der Put and Bijsterbosch [3], Goff and Luner [4], Brouwer and Zsom [5], Voegtli and Zukoski [6], and Verdegan and Anderson [7].
2. In the high electrolyte concentration region, the ζ potential retains nonzero values up to quite high concentrations. Thus, for polystyrene $\zeta \simeq -0.018\,\text{V}$ in 0.5 M $NaNO_3$ or KCl solutions.

On the other hand, polystyrene particles have been known for a long time to consist of equally sized, spherical particles that are perfectly smooth – at least on the scale of electron micrograph images. Thus, such particles closely approximate the systems which are at the background of most theoretical considerations on dispersions. Understanding the unusual features of electrokinetic behavior will then be a challenge for research workers interested in electrokinetics.

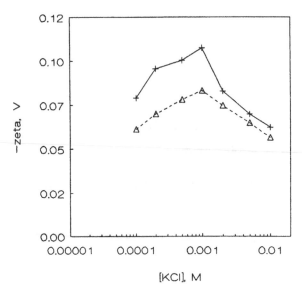

FIG. 1 Electrokinetic potentials of two types of polystyrene latex particles: (+) particle radius = 280 nm, $\sigma_0 = -0.0063\,C\,m^{-2}$; ($\Delta$) particle radius = 196 nm, $\sigma_0 = -0.027\,C\,m^{-2}$. (From Ref. 1.)

A first question to be asked is, whether the surface of the polystyrene latex particles is smooth not only on the scale of electron microscopy, but also on the molecular level. An answer to that question has been reported by Folkersma and Stein [8, 9], who found, on investigating polystyrene latex particles by atomic force microscopy, that the latex particles when dried were definitely rough, with maximum heights of protrusions being found of the order of up to 28 nm.

In the present contribution it is intended to review some of the data obtained by electrokinetics, to describe the picture of the interface polystyrene latex particle/aqueous solution that emerges from these data, and to mention an example of how this picture helps us to understand some data on coagulation of such latexes.

II. SOME BASIC NOTIONS OF ELECTROKINETICS

In interpreting electrokinetic observations, it should be borne in mind that the first layer of liquid near the solid/liquid interface is held fixed toward the solid (the "no slip" condition in hydrodynamics), and the liquid will be able to move towards the particle only at some (small) distance from the solid/liquid boundary itself. The place at which this transition occurs is idealized as a "slipping plane." Thus, the potential which plays a role in electrokinetics is not the potential at the solid/liquid boundary itself (usually indicated as ψ_0), but the potential at the slipping plane (indicated as ζ).

Thus, in the electrical double layer two boundaries should be conceptually discerned:

1. The real solid/liquid interface, separating the liquid medium with its double layer from the solid phase, and
2. The electrokinetic slipping plane – separating a layer of immobilized solution close to the interface from the solution at larger distance from the surface.

The charge compensating the surface charge will be partially present between the solid/liquid interface and the slipping plane, and in the space between the slipping plane and the bulk solution. These charges will be indicated here as σ_Δ and σ_ζ, respectively, and the requirement that the boundary as a whole must be electrically neutral can be expressed by

$$\sigma_0 + \sigma_\zeta + \sigma_\Delta = 0 \tag{1}$$

III. CALCULATING ZETA POTENTIALS FROM ELECTROKINETIC DATA WHEN CONDUCTION BY IONS BEHIND THE SLIPPING PLANE IS NEGLIGIBLE

A complicating factor is that the calculation of ζ potentials from electrokinetic data proves frequently to be quite difficult. The difficulties are of two kind: (1) the dependence of the relation between electrophoretic mobility and ζ potential, on the value of κa and (2) the influence of surface conductance.

The relation for calculating ζ potentials from electrokinetic data depends in a rather complicated way on the ratio between the length scales of the particle and of the double layer, κa (a = particle radius, and κ = the parameter of the Debye–Hückel theory, i.e., the reciprocal value of the extension of the double layer in a direction perpendicular to the interface).

This aspect is not really an insurmountable problem any more since the treatment of the problem by Wiersema et al. [10]. This covers the range of κa and ζ potential values of interest for electrokinetic work; in particular, the results they obtained by numerically solving the relevant equations are frequently referred to. These results are available as tabulated values of the dimensionless mobility E as a function of κa and the dimensionless ζ potential. The dimensionless potential is defined as $y_0 = ze_0\zeta/(k_{\mathrm{Bol}}T)$, and the dimensionless mobility is defined by:

$$E = \frac{3}{2} \frac{\eta e_0}{\epsilon_0 \epsilon_r k_{\mathrm{Bol}} T} \frac{U}{X} \tag{2}$$

where U is the velocity of the motion of a particle (m s^{-1}) in an electric field X (V m^{-1}), η is the viscosity of the liquid dispersion medium (Pa s), ϵ_0 is the permittivity of the vacuum (8.854×10^{-12} C V^{-1} m^{-1}), ϵ_r is the relative dielectric constant of the liquid medium, k_{Bol} is the Boltzmann constant (1.3806×10^{-23} J K^{-1}), T is the temperature (K), and e_0 is the proton electric charge (1.6022×10^{-19} C).

While the polarization of the double layer around the particles caused by the electric field (the so-called "relaxation" effect), and the retardation effect caused by the flow of (predominantly) counterions in the vicinity of a particle, are accounted for in the calculations by Wiersema et al. [10], one assumption remains: viz, that ions behind the slipping plane do not contribute to the electrical conductivity of the dispersion.

Figure 2 compares the results of Wiersema et al. with some of the equations presented in later years, for the case of y_0 (the dimensionless potential at the slipping plane) = 2 (i.e., $\zeta \simeq 0.05$ V). The equations concerned have been published by Semenikhin and Dukhin [11], O'Brien and White [12], O'Brien [13], and Ohshima et al. [14]. The equations are much easier to use, e.g., when calculating ζ potentials from electrokinetic data by a computer, than the tables of Wiersema et al. [10], which require some interpolation. It is seen that the agreement between the analytical equations and numerical

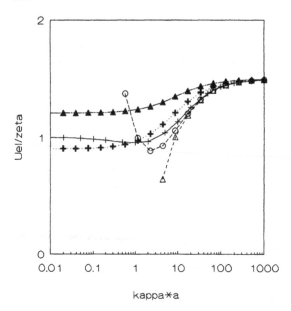

FIG. 2 Comparison of various electrokinetic mobility theories for $ze_0\zeta/(k_{Bol}T) = 2$. (++) Wiersema et al. [10]; (△) Semenikhin and Dukhin [11]; (○) Ohshima et al. [14]; (+) O'Brien and White [12]; (▲) O'Brien [13].

models of Wiersema et al. and those of O'Brien and White [12] is very good when $\kappa a > 100$, but that the situation is much less satisfactory for values of this parameter less than about 20. This is due to the neglect of terms of higher order in the parameter κa; most of the equations mentioned consist essentially of a power series in $1/\kappa a$. Thus, equations based on neglecting higher powers of $(1/\kappa a)$ than the first involve approximations which are not really serious at large values of this parameter, but become increasingly important for lower κa values.

For particles with a radius very large compared to the dimension of the double layer, i.e., if $\kappa a \gg 1$, the relationship between E and y_0 becomes

$$E = \frac{3}{2}y_0 \tag{3}$$

("von Smoluchowski" equation), while for $\kappa a \ll 1$:

$$E = y_0 \tag{4}$$

("Hückel" equation).

Although both the von Smoluchowski and the Hückel equations predict a linear dependence of electrophoretic mobility on the ζ potential, albeit with a different proportionality constant which depends on the κa value in the situation considered, in some cases of intermediate κa values the relationship between mobility and ζ potential becomes nonlinear to the extent that a maximum can be found. Such cases occur, especially at high ζ potentials, and with κa values around 1.

It will be noted that if the ratio:

$$\frac{E}{y_0} = \frac{\text{Dimensionless electrophoretic velocity}}{\text{Dimensionless } \zeta \text{ potential}} \tag{5}$$

is underestimated (e.g., by employing the Semenikhin and Dukhin equation when $\kappa a = 5$), the ζ potential calculated from the electrokinetic velocity will be too high; while the reverse is true if Eq. (5) is overestimated, e.g., by applying the von Smoluchowski equation (3) in a κa region in which the corrections to it are not really negligible.

IV. SPECIAL CASE OF POLYSTYRENE LATEX PARTICLES

As already indicated, in the case of polystyrene latex particles the ζ potential versus electrolyte concentration graphs show some unusual features: on increasing the concentration starting from low values, the ζ potential first starts to rise, passes through a maximum, and finally decreases again.

The unusual aspect of this behavior is especially the rise in the absolute value of the ζ potential with increasing concentration in the low concentration region, since a decrease in the absolute value with increasing concentration is what is usually observed, e.g., in the case of aqueous dispersions of TiO_2, and what is expected from theory ("compressing" action on the double layer, by excess electrolyte).

The following models have been proposed to account for the dependence of the ζ potential on concentration, in the low concentration region:

A. Passing Through a κa Region in which the ζ Potential Shows a Maximum with Increasing Electrophoretic Mobility [15]

This could occur only in the region where such a behavior is predicted by the results of Wiersema et al. [10]. However, the phenomenon has been reported to occur outside this region as well, so this cannot be an explanation which is valid in all cases. Thus, the data shown in Fig. 1 were obtained by calculating ζ potentials from electrophoretic mobilities by means of the values tabulated by Wiersema et al., and therefore the course of this graph cannot be due to the occurrence of a maximum in the ζ potential versus mobility curve at the κa value concerned.

B. Participation in Conductance by Ions Behind the Slipping Plane

This explanation through "anomalous conductance" by ions in the stagnant liquid behind the slipping plane has been advocated, among others, by van der Put and Bijsterbosch [3, 16].

Let us, before passing into the details of their argument, see how this effect might explain the ζ potential versus concentration graph found with polystyrene latex particles.

In many cases, the electric field strength in the experiments (X) is calculated from the quotient of the electric current (I) divided by the electrical conductivity of the solution studied (λ):

$$X = \frac{I}{\lambda} \tag{6}$$

This is applied generally if the geometry of the apparatus is too complicated for direct calculation of the field strength in the dispersion from the dimensions of the apparatus (as is usually the case). However, it entails the assumption that no stray currents occur. In the field of dispersions, the neglect of the so-called "surface conductivity" is particularly important.

By this is meant the following: the double layer around the particles contains an excess of counterions and a shortage of coions; however, the excess of counterions is larger than the shortage of coions. This can be easily seen from the expressions for a flat interface; in this case:

$$\sigma_+ = \sqrt{2\epsilon_0\epsilon_r n_\infty k_{\text{Bol}} T}\left(\exp\left(-\frac{y_0}{2}\right) - 1\right)$$
$$\sigma_- = \sqrt{2\epsilon_0\epsilon_r n_\infty k_{\text{Bol}} T}\left(\exp\left(\frac{y_0}{2}\right) - 1\right)$$

(7)

where n_∞ is the number concentration of each of the two ions in solution, σ_+ is the excess charge due to positive ions in the double layer further away from the solid/liquid interface than the slipping plane, and σ_- is that due to negative ions; y_0 is again the dimensionless value of the ζ potential. One of the σ values is a deficit rather than an excess; this is determined by the sign of y_0. Thus, if $y_0 < 0$, σ_+ will be an excess, while σ_- is a deficit; but in this case, the absolute value of $[\exp(-y_0/2) - 1]$ is larger than that of $[\exp(y_0/2 - 1)]$, and hence there will be a larger quantity of ions in the double layer than in an equal volume in the bulk liquid. Around a particle, then, there will be a larger conductivity than in the bulk (if the individual ions have similar mobilities), and part of the total current measured will not contribute to the building up of the overall field strength.

It is possible to take into account κ^σ, the conductivity by the ions involved in σ_+ and σ_-, i.e., by ions further away from the solid/liquid interface than the slipping plane, by means of an equation due to Bikerman [17]:

$$\kappa^\sigma = 1/F \times \left[\sigma_+\lambda_+ - \sigma_-\lambda_- + \sigma_+ \frac{2\epsilon_r\epsilon_0 RT}{\eta z_+} - \sigma_- \frac{2\epsilon_r\epsilon_0 RT}{\eta z_-}\right]$$

(8)

where λ_+ and λ_- are the ionic conductivities of the cation and anion, respectively, and R is the gas constant.

If we introduce into this equation the formulas for σ_+ and σ_- mentioned in Eq. (7), we obtain:

$$\kappa^\sigma = \frac{(2\epsilon_r\epsilon_0 RTc)^{\frac{1}{2}}}{F} \times \left(\left[\exp\left(-\frac{y_0}{2}\right) - 1\right] \times \left(\lambda_+ + \frac{2\epsilon_r\epsilon_0 RT}{\eta}\right)\right.$$
$$\left. + \left[\exp\left(\frac{y_0}{2}\right) - 1\right] \times \left(\lambda_- + \frac{2\epsilon_r\epsilon_0 RT}{\eta}\right)\right)$$

(9)

Use of this equation implies a knowledge of y_0, i.e., of the ζ potential. Van der Put and Bijsterbosch [16] compared this surface conductance with that calculated by means of a theoretical equation on the conductance of a dispersion of nonconducting particles derived in a way similar to that used in the derivation of the Bruggeman equation [18, 19]:

$$\kappa^\sigma = \frac{K_0 d\left[(1 - p) \times \left(\frac{K}{K_0}\right)^{\frac{1}{3}} - K/K_0\right]}{4 \times (1 - p) \times \left(\frac{K}{K_0}\right)^{\frac{1}{3}} - 4}$$

(10)

where p is the solid volume fraction in the plug, K is the conductivity of the dispersion, and K_0 is the conductivity of the liquid medium.

This formula appears to lead to values of the surface conductance of the particles which are *larger* than those calculated by the Bikerman equation. Thus, the existence of "anomalous Stern layer conductance" by ions in the space between the solid/liquid interface and the slipping plane was assumed. Neglect of this contribution leads to ζ potentials which are too low (in the absolute sense), since the real electrical field is smaller than that calculated from formula (6) using for I the total electrical current measured, instead of only that part which is not due to surface conductance.

Similarly, Zukoski and Saville [20] report that, on comparing ζ potentials calculated from mobility measurements with those calculated from measured values of the surface conductivity, the latter values were found to be inherently larger (in the absolute sense) than those calculated from mobilities. This is in line with the considerable role of surface conductivity, since if this is not take into account, the ζ potentials calculated from mobility values will be underestimated.

The existence of anomalous conductance may lead to an (apparent) rise in the ζ potential with increasing electrolyte concentration in the low concentration region, since at high electrolyte concentration the surface conductance will become negligible compared with the conduction through the bulk solution. Thus, the absolute value of the ζ potential will be underestimated, especially at very low concentrations. At larger concentrations, the underestimation will be less, and the ζ potential will rise to realistic values, while at still larger concentrations the compression of the double layer will lead to a further lowering of $|\zeta|$.

This is, according to van der Linde and Bijsterbosch [21], the reason for the finding of a $|\zeta|$ potential maximum: this was considered to be "either due to artifacts in measuring streaming currents and electrical conductance of plugs, or to anomalous surface conductance inherent to these systems not being accounted for in the theories relating electrophoretic mobilities to ζ potentials." Their main argument for this conclusion was that a polystyrene latex with negative σ_0, showing a maximum absolute value of the electrophoretic mobility in 0.01 M KBr solutions (in which the electrophoretic mobility is negative), could be converted into a latex showing a maximum value of the electrophoretic mobility of opposite sign at about the same concentration, by covering the latex with a poly-(L-lysine) layer. However, by hiding the original latex through adsorption of a polyelectrolyte, not only the sign of the surface charges will be changed, but also the hydrophilic/hydrophobic character of the polymer/aqueous solution interface. This may change the tendency of the K^+ and Br^- ions to be adsorbed.

Against the hypothesis of an important role of anomalous Stern layer conductance, the following arguments should be considered:

1. The assumption of the ions behind the slipping plane not participating in conductance appears to be fundamentally a consequence of the "nonslip" condition taken over from the basic science of hydrodynamics, for which there is wide experimental verification; one cannot easily imagine an ion moving through a stagnant liquid layer without disturbing the latter;

2. Midmore and Hunter confirmed doubts on the existence of anomalous Stern layer conductance by means of measurements of high-frequency dielectric response [15] on polystyrene dispersions: the number of ions responding to that method did not surpass more than a small fraction (3%) of the titratable surface charge, whereas if anomalous Stern layer conductance were to be really important, nearly all of the titratable surface charge would be compensated for by ions that are easily displaceable by an electric field. This was taken as an indication that anomalous surface conductance is not important ("Stern layer ions are those that are not displaced by a tangential electric field").

Thus, this technique appears to measure only the diffuse layer charge outside the slipping plane. It is remarkable that this result was found exactly on the same type of solids for which an important role of anomalous conductance had been anticipated. The only possibility for combining the measurements reported by Midmore and Hunter [15] with an important role of anomalous Stern layer conductance appears to the present author to be the following assumption: while ions embedded in a liquid layer, which is held fixed towards a solid/liquid interface, cannot respond to a high-frequency electric field as used by Midmore and Hunter, such ions might be able to respond to a low-frequency electric field or a direct-current field such as used in electrokinetic work.

C. The Surface of Polystyrene Particles is Covered by a "Hairy Layer"

In practice, polystyrene latex particles are prepared in most cases by emulsion polymerization [22] which may lead to the presence of various charged groups:

1. Initiator fragments, e.g., SO_4^{2-} from $K_2S_2O_8$, which react with part of the monomer and therefore are present as $-O-SO_3^-$ groups, frequently at polymer chain ends.
2. If surfactants with ionic head groups are present either during the polymerization reaction or in later usage, they may be adsorbed during the polymerization on to the micelles in which the polymerization reaction occurs, or on the final interface of the polystyrene particles, imparting ionic charges to that interface.
3. If a buffer solution is employed for proper pH control either during the polymerization reaction, or in later use, ions from this solution may be adsorbed as well.

In order to obtain polystyrene particles with a surface with easily surveyable surface properties, e.g., with only one type of ionic groups which are anchored by a chemical bond to the interface and thus may not easily be removed on subjecting the sample to a washing operation, surfactants and buffer solutions are frequently avoided during the polymerization (a "surfactant-free emulsion polymerization"). However, the ionic groups introduced by an initiator cannot easily be avoided. Initiators may either introduce negative surface groups into the polystyrene, or positive groups; both have been applied. As an example of the former, $K_2S_2O_8$ has already been mentioned; an example of the latter is 2,2'-azobis (2-amidinopropane) dihydrochloride [7].

The "hairy layer" model starts from the consideration that these ionic groups have a much higher affinity for an aqueous medium than for an apolar one, such as the inner part of a polystyrene particle. Therefore, they will tend to accumulate near the particle surface. In the model this is assumed to lead to the polystyrene particles being covered by polystyrene chains with ionic groups at their end. The polystyrene chains are kept extending into the aqueous solution by the hydrophilic character of their ionic end groups, and by the repulsion between these groups. The distance over which the liquid layer adhering to the solid particle extends is then thought to be determined by the length of extension of the polymer chains. In other words: the slipping plane is assumed to coincide approximately with the locus of the end groups of the polymer chains; the polymer chains extending into the solution are surrounded by an aqueous solution, but this solution is between these chains immobilized by them. Therefore, in this model the ζ potential is envisaged to be the potential at the end of the chains.

The repulsion between the ionic head groups is most pronounced at low electrolyte concentration, since at large concentrations the effect will prevail of the counterions preferentially accumulating around the ionic head groups, thus reducing the latter's mutual repulsion. Therefore, at very low ionic concentrations the polymer chains will extend for a relatively large distance into the solution. At that distance there will be only a small fraction of the original surface charge on the solid polystyrene which is still not compensated for by counterions, and therefore the ζ potential will be low.

According to the hairy layer model, two opposed effects on the ζ potential appear when the electrolyte concentration increases:

1. Reduction of the repulsion between the ionic head groups causing the polymer chains, which are extending into the solution, to collapse gradually; this would, if the potential versus distance decrease remains unchanged, lead to an increase in the ζ potential in the absolute sense.
2. Increasing neutralization of the surface charges on the polystyrene surface itself (by the "compressing" action on the double layer by the dissolved electrolyte), which would lead to a decrease in the ζ potential with increasing dissolved electrolyte concentration, if the distance of the slipping plane were to remain unchanged.

It will be noted that in both the cases mentioned, fundamentally the same effect is responsible for the compression of the surface layer, viz., the screening of electrostatic interaction by the crowding of oppositely charged ions around ionic groups; this refers in case 1 to the repulsion between equally charged ionic head groups of the polymer chain, and in case 2 to the electrical influence of the charged surface groups. Thus, it might be supposed that both contractions would occur with increasing electrolyte concentration to the same extent, which would lead to the expectation that the electric charge between solid/liquid interface and the slipping plane would be independent of the electrolyte concentration. However, this is too simple, since the ionic head groups, though more mobile than ionic groups on the interface itself, are less so than dissolved ions: the ionic head groups are restricted in their motions by the mobility of the polymer chain to which they are attached.

The model can explain the experimental ζ potential versus electrolyte concentration graphs such as Fig. 1, if effect 1 predominates at low electrolyte concentration, while effect 2 predominates at larger electrolyte concentrations. The predominance of effect 1 in the low concentration region is understandable on the basis of the restriction of motion of the ionic head groups by being bound to polystyrene chains, because the latter attract each other in an aqueous environment; this may lead to a pronounced collapse of the hairy layer especially in the low concentration region. In the high concentration region, where there is hardly any hairy layer left, the general compression of the double layer will become noticeable.

Against the hairy layer model, the argument has been raised [15] that such a layer is very unlikely on energetic grounds since the free energy required for the translation of one ethylbenzene residue from a hydrocarbon to an aqueous environment has been estimated by Tanford [23] to be 4×10^{-20} J (about 10 kT). Nevertheless, it has been reported by Masliyah et al. [24] that the presence of only a few such chains suffices to immobilize a volume of liquid around a spherical particle. It could be that the positive free energy change necessary for transferring these few chains into an aqueous environment is outweighed by the negative free energy effect through greater mobility of the ionic head groups, especially when the hydrophobicity effect of the chains is made less important by the adsorption of coions (see Section IV.D).

D. Coion Adsorption Model

This model has already been mentioned as one of the possible explanations of the electro-kinetic data by Meijer et al. [1]. It ascribes the remarkable features of the electrokinetics of polymer latex particles not so much to the presence of hairy layers, but to the adsorption of coions. Thus, with latex particles with a negative surface charge, arising, e.g., from their preparation with $K_2S_2O_8$ as initiator, in a $NaNO_3$ solution, NO_3^- ions are supposed to be chemisorbed on the polystyrene surface, in spite of the latter having a negative charge. This may lead, on increasing the $NaNO_3$ concentration from low values on, to a strong increase in the negative charge behind the slipping plane, and this is, according to the model, the reason for the increase in absolute values of the ζ potential with increasing concentrations at low concentration values.

In fact, electrostatically speaking, coions should be repelled from the surface rather than attracted. At higher concentrations, the "compressing" action of dissolved electrolyte ions then introduces again a decrease in the absolute value of the ζ potential with increasing concentration.

In the first instance this model does not seem to be very easy to accept, since the ions concerned (as an alternative to Cl^-, NO_3^- ions are frequently used) are not generally known as ions with a strong tendency to be chemisorbed. On the contrary, NO_3^- is often used as an "inert" ion, in studies involving AgI or oxides as solid substances in an aqueous medium.

Nevertheless, this model should not be too easily rejected. It is a question of whether or not our predominant experience with dispersions of inorganic solids in aqueous solution is really applicable in the case of polystyrene as a solid. The latter is a solid with a pronounced hydrophobic character, which is nevertheless kept dispersed by ionic groups. In this case, the chemisorption of such ions as are present may be a way to prevent direct contact (on an atomic scale) between hydrophobic material and surrounding aqueous solution.

This effect may be especially pronounced if a solid such as polystyrene latex is involved. If there really exists a hairy layer on its surface, there are quite large interfaces present, viz., the (atomically speaking) long polymer chains having a tendency to be kept in the aqueous environment by the behavior of the charged sulfate groups, but which will also have the tendency to collapse because of their hydrophobic character. Although ions such as Cl^- or NO_3^- themselves might be, with regard to their own thermodynamic free energy, better off in an aqueous solution than near an apolar material, the joint action of a preference for the aqueous solution by the sulfate groups and the reduction of the interface in contact with an aqueous medium, by adsorption of such ions, might be a deciding factor.

Important in this respect is the claim involved that the interaction of, say, Cl^- ions with a negatively charged apolar solid is strong enough to lead to pronounced adsorption at concentrations lower than, e.g., 1 mM (as implied by the rising branch of the ζ potential at low concentrations in Fig. 1), in spite of the electrostatic repulsion involved with the anionic $-OSO_3^-$ groups. For polystyrene latex particle with cationic surface groups, Verdegan and Anderson [7] describe the charge σ_Δ in the Stern layer due to adsorption of the coions i (viz., Na^+) by means of a Langmuir equation:

$$\sigma_{\Delta,i} = \frac{\Gamma_i K_i n_i}{1 + K_i n_i} \tag{11}$$

where $\sigma_{\Delta,i}$ is the contribution of i-type ions to the Stern-layer charge density, Γ_i is the amount of charge at a full surface occupation, K_i is an adsorption constant, and n_i is the bulk concentration of ions of type i.

The phenomenon can be described by assuming the adsorption constant K_i of an ion of type i to be given by the sum of an electrical term $(z_i e_0 \psi_0)/(k_{Bol} T)$ and a hydrophobicity related term $\Phi_i/(k_{Bol} T)$:

$$K_i = \exp\left(\frac{z_i e_0 \psi_0 + \Phi_i}{k_{Bol} T}\right) \tag{12}$$

Verdegan and Anderson [7] suggest some values for the constants in this equation, which are compatible with their observations. The main point of these constants is that the value of Φ_i must be such as to lead to nearly complete coverage of the polystyrene surface at very low concentrations. However, independent confirmation of such values appears yet to be lacking.

The coion adsorption hypothesis has been cast into doubt by Midmore and Hunter [15] because these authors did not find a significant different between different types of anions. However, Elimelech and O'Melia [2] certainly did find such differences, while Verdegan and Anderson report differences between anionic latexes and cationic latexes, polymerized with an anionic or a cationic type of initiator, respectively. In the former case, anions are the coions while in the latter case, cations are coions which are generally hydrated to a more pronounced degree than anions. Thus, the argument of Midmore and Hunter appears to be not decisive against the coion hypothesis.

V. ADDITIONAL DATA ON ELECTROKINETICS OF POLYSTYRENE LATEX PARTICLES

In view of the uncertainties expressed in the preceding paragraph, it might be welcome to focus attention on experiments with polystyrene latex particles of such a size that there is a reasonably broad concentration region available in which κa is $> \sim 100$, thus avoiding the pitfalls of uncertainties in the calculation of ζ potentials from electrokinetic data due to retardation and relaxation effects. The uncertainties arising from the contribution of an anomalous Stern-layer conductance to the electric current involved can be avoided by extending the concentration range up to large concentrations, where the influence of the surface conductance is negligible compared to the conduction through the bulk liquid. Thus, for particles with a radius of $1\zeta\,\mu$m in a 1 mM solution of a 1 : 1 electrolyte, $\kappa a = 103$, while in a 0.1 M solution (which is still a lower concentration than the critical coagulation concentration of a nonchemisorbing 1 : 1 electrolyte), $\kappa a = 1030$.

It was found possible to synthesize polystyrene latex particles with a radius of up to more than 1 μm, in a one-step surfactant-free emulsion polymerization reaction, using $K_2 S_2 O_8$ as initiator [25]. As indicated above, the absence of surfactants during the polymerization reaction is important with a view to obtaining a surface with only one type of ionic group; the restriction to a one-step process is important for avoiding new nucleation to occur during the second polymerization step, causing deviations from the monodispersity of the latex particles. The surface charges and (number averaged) particle radii of these latexes are reported in Table 1, together with the κa range valid during electrophoresis experiments, which on these latexes were performed (in NaNO$_3$ solutions with concentrations ranging from 0.0014 to 0.036 M). The surface charges are negative, in accordance with the initiator used during polymerization.

Electrophoretic mobilities were measured at pH 5.5 and converted into ζ potentials through the Ohshima equations, which closely follows the numerical results of Wiersema et al. [10] (Fig. 2) in the κa ranges concerned. From the ζ potentials and the value of σ_0, in

TABLE 1 Data on Polystyrene Latexes
Used in Electrophoresis

a (μm)	σ_0 (C m^{-2})	κa range
0.191	−0.0252	23–118
0.368	−0.0863	45–228
0.655	−0.0718	80–406
1.083	−0.0823	132–671
1.562	−0.00956	191–968

Source: Ref. 39.

all cases the charges were calculated between the slipping plane and the bulk solution (σ_ζ), using standard equations from the Gouy–Chapman theory. In addition, the charges between the solid/liquid interface and the slipping plane (σ_Δ) were calculated by using Eq. (1). It should be noted that such calculations imply only the following assumptions:

1. The surface charge at pH 5.5 is identical with the charge found by titration of the surface — OSO$_3$H groups; this comes down to the assumption that these groups are completely dissociated at pH 5.5.
2. The calculation of σ_ζ from the electrophoretic mobility is based solely on the validity of the Poisson equation in the region outside the slipping plane. The field strength is calculated neglecting anomalous Stern layer conduction. This may lead to too low ζ potential values, thus to too low σ_ζ values (both in the absolute sense), at electrolyte concentrations lower than about 5 mM.

Thus, the validity of the Gouy–Chapman equation is not claimed for the region between the solid/liquid interface and the slipping plane, and the reliability of the results is not influenced by the absence or presence of chemisorbed ions in that space.

Of special interest, in the context of the present contribution, is the dependence of the σ_Δ values thus calculated in σ_0. This is shown in Fig. 3; while the course of σ_Δ and σ_ζ as a function of electrolyte concentration are shown for two of the latexes concerned in Figs. 4 and 5.

The aspects of these data, which are of most concern to the present subject, are the following:

1. σ_Δ is, at constant NaNO$_3$ concentration, within experimental error a decreasing linear function of σ_0 (see Fig. 3), leading even to negative σ_Δ values at large NaNO$_3$ concentrations, and low $|\sigma_0|$.
2. The absolute value of $d\sigma_\Delta/d\sigma_0$ increases from a value of 0.89, found for latex with a low σ_0-value, to a value near 1 found for surfaces with a larger σ_0 value (see Fig. 6).
3. σ_Δ is, at constant σ_0, a decreasing function of the NaNO$_3$ concentration, but not a linear one; it is rather a linear function of log[NaNO$_3$] (see Fig. 7).
4. $d\sigma_\Delta/d(\log[\text{NaNO}_3])$ does not significantly depend on σ_0 (see Fig. 7).

These data are compatible with the following picture of the polystyrene/aqueous solution interface: in the absence of NaNO$_3$ in the solution, the whole charge between the phase boundary and the slipping plane consists of the counterions introduced during neutralization of the surface- O-SO$_3^-$ groups (in the present case, these were also Na$^+$

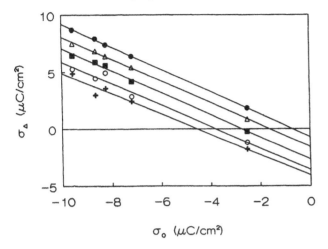

FIG. 3 σ_Δ, the charge in the space between the solid/liquid interface and the slipping layer, as a function of the surface charge σ_0, for polystyrene latex particles at pH 5.5. The latexes differ with regard to particle radius and surface charge (see Table 1). (●) 1.4×10^{-3} M NaNo$_3$; (△) 3.2×10^{-3} M NaNO$_3$; (■) 6.0×10^{-3} M NaNO$_3$; (○) 1.3×10^{-2} M NaNO$_3$; (+) 3.6×10^{-2} M NaNO$_3$.

ions). These counterions are present in the space between the phase boundary and the slipping plane such as nearly to compensate completely the charge of the surface groups ($d\sigma_\Delta/d\sigma_0$ approaches -1 for [NaNO$_3$] → 0). Thus, at low NaNO$_3$ concentrations, σ_Δ is positive, and $\approx |\sigma_0|$. Additional electrolyte in the solution introduces more ions into the space between the phase boundary and the slipping plane, but not only counterions. On the other hand, NO$_3^-$ ions are noticeably adsorbed on the polystyrene surface, in spite of the net negative charge present in these cases; thus, we are dealing with chemisorption of anions. This is the only way to account for σ_Δ becoming negative at large electrolyte

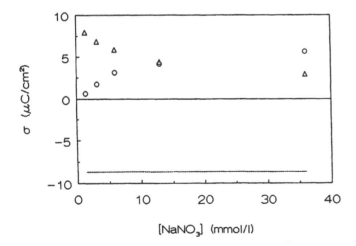

FIG. 4 Charge densities as function of [NaNO$_3$]; polystyrene sample L-78 ($a = 368$ nm, $\sigma_0 = 0.0863$ C m^{-2}) at pH 5.5. Dashed line: σ_0; ○: σ_ζ; △: σ_Δ. (From Ref. 39.)

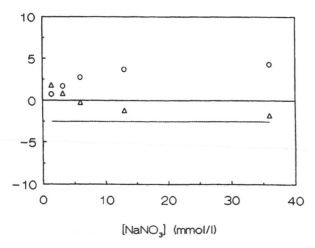

FIG. 5 As Fig. 4, for polystyrene sample L-80 ($a = 190\,\text{nm}$, $\sigma_0 = 0.0252\,\text{C}\,\text{m}^{-2}$). (From Ref. 39.)

concentration, without any change in the slope $d\sigma_\Delta/d\sigma_0$. This is conclusive evidence for chemisorption of NO_3^-, the coions in the present case.

It will have been remarked that, in the calculation of the ζ potential from the electrophoretic mobility, a possible role for anomalous Stern-layer conductance has been neglected. Any influence of anomalous Stern-layer conductance should be noticeable especially at the lowest $NaNO_3$ concentration, i.e., for the experiments presented by the uppermost two lines in Fig. 3. It is seen, however, that the slopes of these lines fit with those of the other lines (see also Fig. 6). This does not support the statement that surface conductance is an important factor at electrolyte concentrations of 1 mM and higher.

The chemisorption of NO_3^- ions is confirmed by the following arguments:

1. Negative values of $\sigma_\Delta 0$ are found especially at high $NaNO_3$ concentration; in this concentration region, the influence of surface conductivity becomes negligible.

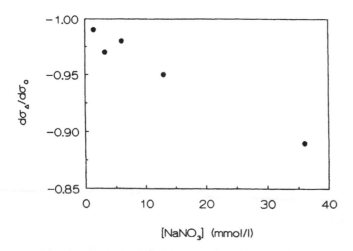

FIG. 6 $d\sigma_\Delta/d\sigma_0$ versus $NaNO_3$, polystyrene at pH 5.5, calculated from Fig. 3. (From Ref. 39.)

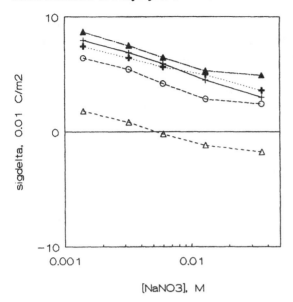

FIG. 7 Charge in the Δ layer at pH 5.5 for various polystyrene latexes: $(+)$ $a = 368\,\text{nm}$, $\sigma_0 = -0.0863\,\text{C}\,\text{m}^{-2}$; $(\triangle)$ $a = 190\,\text{nm}$, $\sigma_0 = -0.0252\,\text{C}\,\text{m}^{-2}$, diam. $0.381\,\mu\text{m}$; $\bigcirc$: $a = 612\,\text{nm}$, $\sigma_0 = -0.0718\,\text{C}\,\text{m}^{-2}$, diam. $1.224\,\mu\text{m}$; $+$: $a = 1009\,\text{nm}$, $\sigma_0 = -0.0823\,\text{C}\,\text{m}^{-2}$, diam. $2.018\,\mu\text{m}$; $\blacktriangle$: $a = 1583\,\text{nm}$, $\sigma_0 = -0.0956\,\text{C}\,\text{m}^{-2}$, diam. $3.166\,\mu\text{m}$.

2. Any role of surface conductivity would lead to the ζ potential values being underestimated (in the absolute sense); thus, the real ζ potentials should be even higher than those which have been used as the basis for our calculations. When the ζ potential is negative, this would lead to negative σ_Δ values, which are even larger in the absolute sense than those on which Figs 3–7 are based.

The NO_3^- ions are presumably adsorbed preferably at some distance from the charged-OSO_3^- groups, as shown by the slight effect of σ_0 on $d\sigma_\Delta/d\sigma_0$, and the absence of a distinct effect of σ_0 on $d\sigma_\Delta/d(\log[\text{NaNO}_3])$. It is to be expected that the Na^+ ions in the space between the slipping layer and the solid/liquid interface will be assembled preferably near the surface $-O\text{-}SO_3^-$ groups, since the Na^+ ions, which are hydrated to a greater extent than the NO_3^- ions, will be attracted both by the locally hydrophilic environment near the sulfate groups, and by their negative charges. Also, since the Na^+ ions screen the electrical influence of the negative charges of the sulfate groups, at least partially, the charge of the surface sulfate groups does not completely prevent the chemisorption of NO_3^- groups. This is similar to what has been found, both in the case of the adsorption of inorganic ions on $CaSiO_3$ surfaces [26, 27] and in that of the adsorption of short-chain tetra-alkylammonium ions on silica [28].

Figures 4 and 5 show the course of σ_Δ an σ_ζ as a function of electrolyte concentration; Fig. 4 the case of a low-surface charge sample of polystyrene, and the other for a high-surface charge sample. The surface charge σ_0 itself, i.e., the titratable charge of the $-OSO_3H$ groups on the surface, is of course independent of the electrolyte concentration.

It is seen from these graphs that σ_Δ is positive at low concentrations of electrolyte, and turns negative at higher concentrations only. The adsorption of coions therefore appears to occur, at least in the case of $NaNO_3$ as electrolyte, to a significant degree only at concentrations greater than 1 mM, and therefore cannot be held responsible for the rise in the $|\zeta|$ potential at low electrolyte concentration, calculated from electrophoresis data neglecting anomalous surface conductance, as shown in Fig. 1. This agrees with the statement in the preceding paragraph, that the coions are adsorbed at places remote from the $-SO_3^-$ groups, after the electrostatic effect of the latter is partially screened by Na^+ ions.

VI. SUMMARY OF PICTURE OF THE POLYSTYRENE/AQUEOUS SOLUTION INTERFACE

The electrokinetic data, together with data on the anomalous Stern-layer conductance, lead to the following picture of the interface of polystyrene latex particles.

The interface is not really the smooth surface of nearly ideal sphericity, which would correspond with the requirements of theory. In addition to having a surface roughness involving asperities of up to 28 nm, it is covered by a hairy layer, protruding into the surrounding solution, consisting of polystyrene chains with, at their end, charged ionic groups.

Water molecules between the hairy-layer polymer chains are kept at their places with regard to larger scale viscous motion, but the layer may accommodate small displacement of them, if the total volume of the hairy layer is not greatly disturbed. This is because the space between the particle interface and the ends of the polymer chains is filled only with a few chains per unit area: this suffices for immobilizing most of the liquid enclosed between the interface and the chain ends, but still allows some motion of dissolved ions, by interchange of places of a few water molecules within the layer. Thus, anomalous Stern-layer conductance becomes conceptually acceptable.

In addition, there is evidence for adsorption of coions; the clearest case of the latter appears to be that reported by Tuin et al. [25]. These data do not add new aspects to the question studied by van der Linde and Bijsterbosch [21], viz., whether or not the maximum in the ζ potential versus electrolyte concentration really exists, since the data reported by Tuin et al. refer to electrolyte solutions at high concentrations. However, the data obtained by Tuin et al. do not support an important role for anomalous surface conductance at electrolyte concentrations of 1 mM or higher.

Neither do these data support pronounced chemisorption of coions in the low concentration region, as implied by Verdegan and Anderson [7] through the values of the parameters in the adsorption constant [Eq. (12)] chosen by these authors.

Adsorption of coions at higher electrolyte concentrations may be an interesting background of the still appreciable ζ potential of polystyrene latex particles in electrolyte solutions of relatively high concentrations (-0.018 V in 0.5 M NaCl solution, see Ref. 7).

VII. APPLICATION OF INSIGHTS OBTAINED ON THE INTERFACE POLYSTYRENE/AQUEOUS SOLUTION: UNDERSTANDING THE INFLUENCE OF GRAVITY ON PERIKINETIC COAGULATION [29]

Experiments have been reported indicating that perikinetic coagulation is slowed down by the presence of a gravity field: under microgravity conditions, perikinetic coagulation is faster than under terrestrial gravity conditions [8, 30–32]. Results of a typical experiment

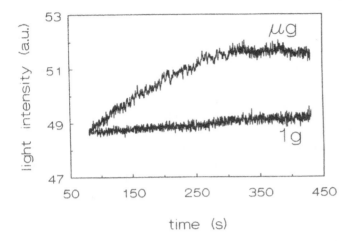

FIG. 8 Intensity of transmitted light during coagulation of polystyrene latex particles in 0.5 M NaCl, at 1 g and μg levels. (From Ref. 32.)

are shown in Fig. 8. The coagulation experiments concerned were performed using dispersions with low solid volume fractions (of the order of 0.0001), in 0.5 M NaCl solution. The retarding action by the gravity field was found for all suspensions investigated (dispersions of polystyrene latex, ground quartz, and amorphous silica particles), but the most pronounced effect was found in the case of polystyrene latex particles. The particle radius was about 1 μm in all cases; thus, the particles were relatively large compared with other colloidal dispersions.

That we are dealing here with a retardation by gravity rather than with an acceleration by the absence of gravity follows from the fact that the coagulation rate observed in the absence of gravity agrees with the value predicted by theory [33, 34], while that observed at terrestrial g values is slower than the theoretical value.

This retardation by gravity is not expected: theories on coagulation as treated in textbooks on colloid chemistry do not usually enter upon the influence of gravity; and, if anything, an accelerating effect by gravity rather than a retarding one would be expected, since in the presence of gravity there are, in addition to diffusion, a number of effects which may lead to "gravity induced" coagulation:

1. If the densities of the dispersed and continuous phases are not exactly matched, sedimentation or creaming will occur; two particles of not exactly the same size may meet because the larger will overtake the smaller particle;
2. If there are temperature gradients in the system, convection flows will occur, which may lead to some shear-induced coagulation.

These effects will be particularly pronounced in dispersions of large particles (such as discussed here), because in such dispersions diffusion of the particles will be relatively slow.

Both effects may, however, act in the opposite direction as well: sedimentation or creaming of the particles may disrupt pairs which are on the point of being formed, when the difference in sedimentation velocity is large compared to the velocity of approach due to mutual attraction; also, a shear driven by convection may disrupt a pair being formed when the viscous drag of the shear surpasses the attractive force. In general, both effects mentioned will lead to pair formation at high concentrations, and to pair separation at low ones; since in the latter case, the oncoming flow does not have sufficient particles to predominate over the pair-disrupting action. The transitional concentration between the two regimes will depend on the quotient of the forces in each particular case.

In order to estimate what this ratio is in the special case of polystyrene latex particles, the following observations and calculations were performed:

1. Coagulation under terrestrial gravity conditions, but with different density differences between the dispersed and continuous phases, around $\Delta\rho = 0$. The results are shown in Fig. 9; indeed there is a maximum coagulation rate in the vicinity of $\Delta\rho = 0$.

2. Visual inspection, by projecting a microscope image of the dispersion on a screen. In this case, convection flows could be observed and the approach of two particles or the disruption of some of the pairs during their formation could be followed.

The first impression, on following visually the approach or disruption of two particles, was that at the concentration concerned sedimentation and convection driven flows indeed tend to lead to fewer potential particle pairs becoming real pairs, rather than lead to addition pair formation. A typical experiment is shown in Fig. 10. In this figure, the number of projected distances of a certain length between two close particles are plotted as a function of time. First (stage A); in general, the particles approach each other independent of the presence or absence of a distinct density difference between particles and medium (indicated by open and closed circles, respectively). In stage B, the particle pairs indicated by closed circles remain in each other's vicinity, while the pairs indicated by open circles are disrupted, as shown by the large mutual distance with increasing time. Yet some 40 s later, the mutual distance between the pairs again decreases: at this stage the sedimenting particles frequently approach a new partner. In the absence of a distinct density difference between particles and medium, the interaction time between two particles is much longer; thus, there will be a greater number of potential pairs becoming real pairs in the absence than in the presence of a

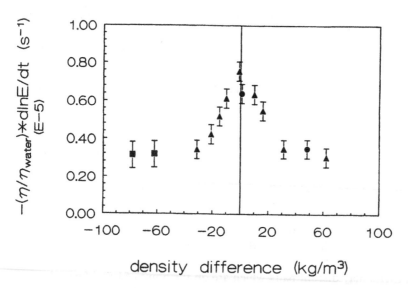

FIG. 9 Coagulation rate (measured as d ln E/dt, with E = intensity of transmitted light) at 1 g, corrected for viscosity changes by addition of substances for density matching, as a function of density difference between dispersed and continuous phases. Density of continuous phase adjusted with: (■) methanol; (▲) sucrose; (●) D_2O. (From Ref. 32.)

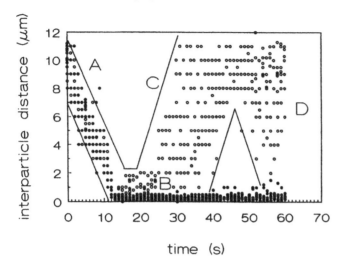

FIG. 10 Projected distance between two close particles as a function of time at 1 g. (A) particles approach each other; (B) particles interact during a certain period; (C) some particles move away from each other; (D) some particles approach new partners. (○) $\Delta\rho = -31\,\mathrm{kg\,m^{-3}}$; (●) $\Delta\rho = 1\,\mathrm{kg\,m^{-3}}$. (From Ref. 32.)

density difference. Gravity-induced coagulation indeed occurs also, but at the density differences between particles and medium investigated, only at higher gravity values (2–7 g).

3. Calculation of the forces involved due to the viscous drag exerted by convection flows. Such flows may occur at 1 g, but not at μg levels: under the latter conditions, the temperature gradients may induce local density changes, but in the absence of gravity these do not trigger convection flows. For this, a simple arrangement was considered in which one particle is caused to rotate by the convection flow around another particle (see Fig. 11). The shear rate due to convection could be estimated from the temperature gradients in the cell used for the perikinetic coagulation experiment, and from this the time necessary for one rotation could be estimated and compared with the time of approach of the two particles due to the mutual interaction.

For calculating the probability of separating a particle pair, we need the interaction force between the particles of a potential pair. This was calculated by the DLVO theory, adjusted for including the so-called "retardation" of the attractive forces as calculated by Clayfield et al. [35]. For the Hamaker constant, we took $9 \times 10^{-21}\,\mathrm{J}$ as reported by Gingell and Parsegian for polystyrene from spectroscopic data [36]. The repulsive forces were calculated on the assumption that the ζ potential of the particles in 0.5 M NaCl solution ($-0.018\,\mathrm{V}$) was equal to ψ_δ. It was assumed that the particles were in a secondary minimum, and the probability was calculated so that during a rotation the viscous drag by convection-driven shear could disrupt such pairs. The shear rate of such flows could be estimated from the temperature fluctuations during a coagulation experiment, affecting the outside of the coagulation cell to a greater extent than the latter's center.

On these premises, a pair in the secondary minimum had a reasonable probability of being disrupted by convection-driven flow, when the maximum net force between two particles outside of the secondary minimum was of the order of $10^{-14}\,N$.

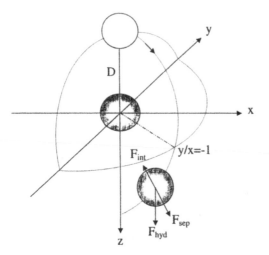

FIG. 11 Schematic representation of the forces acting on a doublet in simple shear flow, used for calculation of probability of disrupting the bond between two close particles by shear caused by convection. (*D*) Distance between the particles (thought to remain constant during the encounter); (F_{hyd}) hydrodynamic force; (F_{sep}) separation force; (F_{int}) interaction force.

However, the assumptions on the interaction forces mentioned led to a maximum attractive force (outside the secondary minimum) of the order of 10^{-12} to 10^{-11} N, at least if we started from perfectly smooth spherical particles.

The answer to the question how then to account for the observations, lies not in accepting a still lower value for the Hamaker constant, but in the realization that the polymer latex particles indeed approximate a sphere very closely, but not on a molecular scale. The existence of surface asperities has already been mentioned; these greatly diminish the attractive forces [37, 38]. In addition, there is a hairy layer, which keeps the particles still more at a distance with regard to their solid/liquid phase boundaries (where the Hamaker constant changes), while the ζ potential is measured at the end of the hairy layer. This together is the reason that the mutual attractive force between 1 μm radius latex particles is sufficiently weak, and the viscous drag sufficiently strong, to lead to the slowing down of the coagulation by convection flows which can occur at 1g, but not at 0g.

Neither the surface irregularities nor the hairy layer alone can be the reason for the retardation of the coagulation at 1g: a similar retardation is experienced by dispersions of quartz or amorphous silica, which are not covered by a hairy layer. However, in the case of polymer latexes the coagulation is slowed down to a greater extent, and this difference can be ascribed to the hairy layer.

REFERENCES

1. AEJ Meijer, WJ van Meegen, J Lyklema. J Colloid Interface Sci 66, 99, 1978.
2. M Elimelech, CR O'Melia. Colloids Surfaces 44: 165, 1990.
3. AG van der Put, BH Bijsterbosch. J Colloid Interface Sci 92: 499, 1983.
4. JB Goff, P Luner. J Colloid Interface Sci 99: 468, 1984.
5. WM Brower, RLJ Zsom. Colloids Surfaces 24: 195, 1987.

6. IP Voegtli, CF Zukoski IV. J Colloid Interface Sci 141: 92, 1991.
7. BM Verdegan, MA Anderson. J Colloid Interface Sci 158: 372, 1993.
8. R Folkersma. Microgravity Coagulation and Particle Gel Formation. PhD Thesis, Eindhoven University of Technology, 1997, p 114.
9. R Folkersma, HN Stein. J Colloid Interface Sci 206: 494, 1998.
10. PH Wiersema, AL Loeb, JTG Overbeek. J Colloid Interface Sci 22: 78, 1966.
11. NM Semenikhin, SS Dukhin. Kolloidn Zh 37: 1127, 1975.
12. RW O'Brien, LR White. J Chem Soc, Faraday Trans 2 74: 1607, 1978.
13. RW O'Brien. J Colloid Interface Sci 92: 204, 1983.
14. H Ohshima, TW Healy, LR White. J Chem Soc, Faraday Trans 2 79: 1613, 1983.
15. BR Midmore, RJ Hunter. J Colloid Interface Sci 122: 521, 1988.
16. AGJ van der Put, BH Bijsterbosch. J Colloid Interface Sci 75: 512, 1988.
17. JJ Bikerman. Z Phys. Chem A163: 378, 1933; Kolloid Z 72: 100, 1935; Trans Faraday Soc 36: 154, 1940.
18. DAG Bruggeman. Ann Phys 24: 636, 1935.
19. TJJ van den Hoven. Electrokinetic Processes and Conductance Relaxation of Polystyrene and Silver Iodide Plugs. PhD thesis, Wageningen University, 1984, p 65.
20. CF Zukoski IV, DA Saville. J Colloid Interface Sci 107: 322, 1985.
21. AJ van der Linde, BH Bijsterbosch. Croat Chem Acta 63: 455, 1990.
22. HN Stein. The Preparation of Dispersions in Liquids. New York: Marcel Dekker, 1995, pp 96–107.
23. C Tanford. The Hydrophobic Effect. New York: Wiley 1980.
24. JH Masliyah, G Neale, K Malysa, TGM van de Ven. Chem Eng Sci 42: 245, 1987.
25. G Tuin, ACIA Peters, AJG van Diemen, HN Stein. J Colloid Interface Sci 158: 508, 1993.
26. CAM Siskens, HN Stein, JM Stevels. J Colloid Interface Sci 52: 244, 1975; J Colloid Interface Sci 52: 251, 1975.
27. AJG van Diemen, HN Stein. Sci Ceram 9: 264, 1977.
28. JCJ van der Donck, GEJ Vaessen, HN Stein. Langmuir 9: 3553, 1993.
29. R Folkersma, AJG van Diemen, HN Stein. Adv Colloid Interface Sci 83: 71, 1999.
30. LLM Krutzer. The Influence of Flow Type, Particle Type and Gravity on Orthokinetic Coagulation. PhD Thesis, Eindhoven University, The Netherlands, 1993.
31. LLM Krutzer, R Folkersma, AJG van Diemen, HN Stein. Adv Colloid interface Sci 46: 59, 1993.
32. R Folkersma, AJG van Diemen, HN Stein. J Colloid Interface Sci 206: 482, 1998; J Colloid Interface Sci 206: 494, 1998.
33. M Von Smoluchowski. Phys Z 17: 557, 1916.
34. M Von Smoluchowski. Z Phys Chem 92: 129, 1917.
35. EJ Clayfield, EC Lumb, PH Mackey. J Colloid Interface Sci 37: 382, 1971.
36. D Gingell, VA Parsegian. J Colloid Interface Sci 44: 456, 1973.
37. SYu Shulepov, G Frens. J Colloid Interface Sci 170: 44, 1995.
38. SYu Shulepov, G Frens. J Colloid Interface Sci 182: 388, 1995.
39. G Tuin, JHJE Senders, HN Stein. J Colloid Interface Sci 179: 522, 1996.

22

Electrokinetics of Protein-coated Latex Particles

ANTONIO MARTÍN-RODRÍGUEZ, JUAN LUIS ORTEGA-VINUESA, and ROQUE HIDALGO-ÁLVAREZ   University of Granada, Granada, Spain

I. INTRODUCTION

Polymer colloids (latex) are usually considered as "model" particles, as they can possess a high monodispersity, smooth and homogeneous surfaces, and well-defined surface functional groups. In addition, these colloidal systems show a large surface/volume ratio, which makes them extremely suitable in adsorption studies. Due to these reasons, latex particles have often been used as ideal carriers for studying the adsorption of proteins and other macromolecules [1–16]. The interest in studying such systems lies in the fact that there exist numerous natural, technical, and industrial phenomena where protein adsorption processes take place on different surfaces. In most cases these adsorption processes contribute to the creation of new materials and systems. Consequently, a thorough characterization of the adsorbed protein layer becomes necessary, albeit it may become a difficult task due to the microscopic dimensions of this kind of structured interfaces. Fortunately, some important information can be obtained by using electrophoretic mobility, a simple technique based on a nonequilibrium process.

In particular, the immunolatex (antibody-coated latex particles) may be considered of biological interest. The reason lies in the fact that one of the most important applications of immunolatex is directed towards the development of particle-enhanced immunoassays used in clinical analysis laboratories. Presently, there are over 200 diseases that can be detected by commercially available latex agglutination immunoassays. In these immunoassays the colloidal aggregation of immunolatex particles must be only caused by the presence of the corresponding antigen. This makes it necessary to obtain immunolatex with a high colloidal stability, in order to avoid the unspecific aggregation of the system due to the physico-chemical conditions of the medium (pH, temperature, ionic strength, etc.). Information about such stability can be extracted from the electrokinetic behavior [17]. This statement is based on the strong relationship existing between the electrophoroetic mobility, μ_e and their ζ potentials, which can be considered responsible for the electrostatic repulsion between the particles.

Some of the topics that will be treated in this chapter are summarized below. First, we show that the isoelectric point of the protein-coated latex particles can be obtained from μ_e data, which allows setting of the pH ranges where latex–protein systems present positive and negative net charges. Electrophoretic studies can also be useful for obtaining information on the electrostatic interactions between the protein molecules and the latex surface, making it possible to estimate the participation of low molecular weight ions in the protein adsorption processes. This can be done by comparing the electrokinetic patterns of the latex particles before and after adsorption of the pure protein. On the other hand, there exist several factors affecting the electrokinetics of protein-coated latex particles, the effect caused by each of them being difficult to isolate. Actually, the electrokinetic behavior of these systems depends, at least, on (1) the amount of adsorbed protein, (2) the nature of the protein and adsorbent surfaces (3) the pH, and (4) the ionic strength of the medium. Due to this complexity there is almost no theoretical approach dealing with the electrophoretic mobility of protein-coated latex particles. In spite of that, Ohshima and Kondo have recently proposed a theory that could be applied to large colloidal particles with a surface-charge layer (as can be a protein layer). Details and references related to this theory are given in Sec. IV. Using this theory it is possible to obtain information on the adsorbed protein layer from μ_e data. Such information is related to the thickness, the charge density, and the frictional coefficient of the polypeptide layer.

II. ELECTROKINETIC BEHAVIOR OF PROTEIN-COATED LATEX PARTICLES

Upon adsorption, the electrical double layers (EDL) of the protein molecule and the adsorbent surface overlap each other, involving redistribution of charge in both of them. The latex-protein EDL so formed is a consequence of the interfacial behavior of protein under certain adsorption conditions. Therefore, it seems quite reasonable to compare electrokinetic properties (such as electrophoretic mobility and its conversion into electrokinetic charge or ζ potential) before and after adsorption, to understand better the various factors playing a role in protein adsorption. For this reason, during the past few decades most works dealing with adsorption of proteins on polymer colloids continued the study by using electrophoresis experiments. Under well-controlled conditions, and with well-defined systems, the effect of coverage degree, incubation, and dispersion conditions (pH and ionic strength and ion composition) on the electrophoretic mobility have been studied. These results and conclusions obtained will be presented in the next sections.

A. Degree of Coverage

Several authors [14, 16, 18–25] have studied the electrophoretic mobility of protein-coated latex particles as a function of the amount of adsorbed protein once the pH and the ionic strength are fixed.

The results obtained by adsorption of BSA (bovine serum albumin) [18–20], HSA [21], IgG [14, 21–24], and F(ab')$_2$ (amoiety of an IgG molecule) [16, 25] showed that upon increasing the amount of protein adsorbed (Γ) the absolute value of the mobility decreases to reach a plateau value. The decrease in the electrophoretic mobility is dependent on the pH, i.e., on the charge of the protein molecules, as shown in Fig. 1 [22].

The same μ_e–Γ dependence was found independently of the superficial groups, sulfonate, carboxyl, HEMA (2-hydroxyethyl methacrylate) sulfate, or amidine (cationic) latexes as can be seen in Fig. 2 [16]. The most striking result is that the μ_e decrease was obtained even when the protein molecules had a negative net charge. This can be justified by two nonexclusive phenomena [16]:

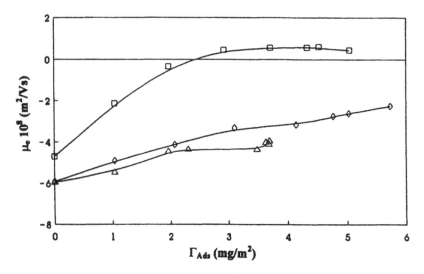

FIG. 1 Electrophoretic mobility versus adsorbed amount of rabbit IgG at three different adsorption (= dispersed) pH values; (□) pH = 5.0; (◇) pH = 7.0; (△) pH = 9.0; ionic strength 2 mM. (From Ref. 22.)

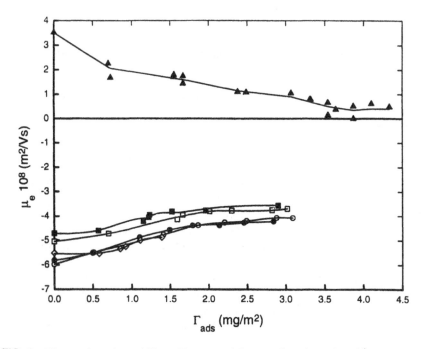

FIG. 2 Electrophoretic mobility of latex particles as a function of F(ab′)$_2$ coverage. Surface polar groups: (□, ■) sulfonate; (●, ○) carboxyl; (▲) amidine; (◇) hydroxyl and sulfate. Experimental conditions: pH = 7; ionic strength 2 mM. (From Ref. 16.)

1. A change in the electrophoretic mobility as a result of adsorption can be explained by a redistribution of the charged groups of the protein molecules. In this way, it is assumed that coadsorption of low molecular ions takes place in protein adsorption processes [18, 20, 24, 26]. This effect will be discussed in more detail in Section II.E.

2. One must keep in mind that μ_e reflects the ζ potential of the colloidal particles, but not the surface potential ψ_0 (directly related to σ_0). As a smooth polymer surface is being covered by macromolecules, it becomes more rigid and irregular. This shifts the slipping plane outward causing a decrease in ζ potential and thus decreasing the μ_e value.

B. Influence of Adsorption Conditions

Galisteo-González et al. [14] have studied the influence of the adsorption conditions on the electrokinetic behavior of the IgG-coated polystyrene (PS) particles. Two samples were selected with a similar degree of coverage ($3.3\,mg/m^2$), but obtained at different adsorption pH (7 and 5). The samples were dispersed at different pH (varying between 4 and 10) and at a constant ionic strength of 2 mM. They found that both immunolatexes displayed similar electrophoretic mobility. Similar conclusions were obtained for monomeric BSA, oligomeric BSA, and $F(ab')_2$ [19, 25].

Also, the effect of the ionic strength during adsorption experiments on the electrokinetic behavior of protein-coated latex particles has been studied. It was found that the ionic strength during such experiments does not affect the electrophoretic mobility of protein-coated latex [19].

The studies mentioned above have shown that the adsorption pH and ionic strength determine the amount of adsorbed protein on the latex surface, but the electrophoretic behavior of the sensitized polystyrene particles is controlled by the redispersion conditions.

C. Effect of Redispersion Conditions. Isoelectric Point of Protein-coated Latex Particles

Several works of electrophoresis experiments have been carried out (Table 1) with the aim of obtaining information about the electrical state of protein-coated latex particles. In these experiments, the particles, with different degrees of coverage, were selected and then the solid substrates were suspended under different pH and ionic strength conditions.

Some common features can be obtained from the studies shown in Table 1:

1. Desorption experiments gave negative results even when the redispersion conditions were more extreme (as to pH and ionic strength) than those used for the electrokinetic characterization of the complexes.

2. Even though the electrophoretic mobility of the bare latex was independent of pH (in the range 4–10) the protein-coated latex always presents an isoelectric point (i.e.p.).

3. The i.e.p. values of protein-coated anionic latex are lower than those of the dissolved protein. In contrast, the i.e.p. values of protein-coated cationic latex are greater than those of the dissolved protein. These results imply that the latex surface charge play an important role in the final EDL structure of the sensitized particle.

As an example, Fig. 3 [29] shows the electrophoretic mobility of bare and covered latex particles for cationic and anionic latex. The i.e.p. values of the immunolatex were 5 and 6

TABLE 1 Electrophoretic Studies Related to Dispersed Conditions of Different Latex–Protein Complexes

Latex	Protein	Amount adsorbed (mg/m^2)	Dispersed conditions	Reference
P(St/NaSS)	BSA	1.07, 3.26, 3.55	$I = 2$ mM pH 4–8	19
PS PS (+) P(St/HEMA) P(St/acryl. ac.)	BSA	Complete coverage	$I = 2$ mM pH 4–9	20
PS	IgG Monocl. IgG	Complete coverage	$I = 2$ mM, $I = 50$ mM pH = 4–10	22
PS	F(ab')$_2$	25% coverage Complete coverage	$I = 2$ mM pH = 3–9	25
P(St/NaSS) PS (+) P(St/HEMA) P(St/carboxyl)	F(ab')$_2$	50% coverage Complete coverage	$I = 2$ mM pH = 3–9	16
PS	HPA RNA	Complete coverage	$I = 10$ mM $I = 50$ mM pH 4–11	26
PMMA	BSA	Complete coverage	0, 2, 5 M Urea pH 4–10	27
PS	IgG	0.1–8	$I = 2$ mM pH 4–10	28
PS PS (+)	Monocl. IgG	Complete coverage	$I = 2$ mM pH 4–10	29
PS PS (+)	IgG	Complete coverage	$I = 2$ mM pH 4–10	14
P(St/carboxyl)	IgG (anti-HSA) Monocl. IgG	1, 1.8 1.3, 1.5	$I = 2$ mM pH 4–8	17
PS PS (+) PS–OH	Monocl. IgG F(ab')$_2$	From 0 to complete coverage	$I = 5, 30, 100$ mM pH 4–8	30

Key: P(St/NaSS) = polystyrene sulfonate latex; PS = polystyrene; PS(+) = cationic polystyrene; P(St/HEMA) = polystyrene–HEMA latex; P(St/acryl.\ ac.) = polystyrene–acrylic acid latex; PMMA = poly(methyl methacrylate) latex; P(St./carboxyl) = latex with carboxyl terminal groups on the particle; PS-OH = latex with hydroxyl terminal groups on the particle; BSA = bovine serum albumin; IgG = immunoglobulin G; F(ab\pri)_2 = Fab fragment of IgG; HPA = human plasma albumin; RNA = bovine pancreas ribonuclease; IgG (anti-HSA) = immunoglobulin G anti-human serum albumin.

for the anionic and cationic polystyrene particles. On the other hand, the i.e.p. values of the protein (monoclonal IgG) determined by isoelectric focusing were in the range 5.2–5.5.

    4. The i.e.p. values of immunolatexes are closer to those of the protein molecules the larger the adsorbed amounts of protein.

Figure 4 shows this phenomenon in the case of IgG adsorption on polystyrene latex particles [28].

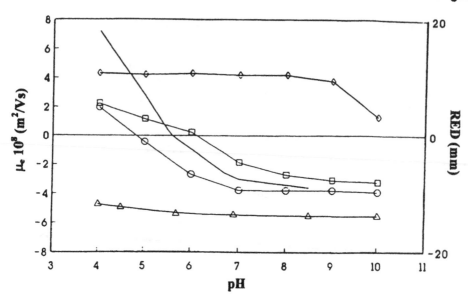

FIG. 3 Mobility of cationic ($\Diamond$) and anionic ($\triangle$) bare latex and covered latex particles ($\square$: cationic; $\bigcirc$: anionic) (left Y axis) together with the relative electrophoretic displacement (RED: right Y axis) of IgG Mab1 as obtained by two-dimensional isoelectric focusing (straight line as a function of pH). (From Ref. 29.)

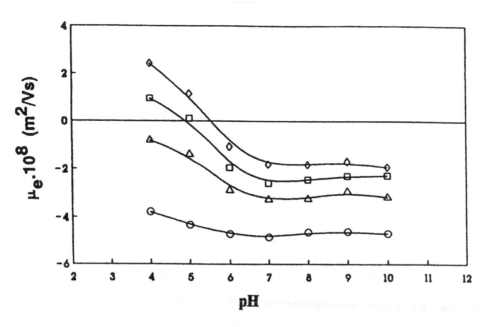

FIG. 4 Electrophoretic mobility versus dispersion pH for polystyrene bare latex ($\bigcirc$), and latex coated with IgG molecules at different degrees of coverage: ($\triangle$) 0.1; ($\square$) 2.2; ($\Diamond$) 8.0 mg/ IgG/m^2 PS. (From Ref. 28.)

5. Combining the results of mobility and adsorption, it possible to conclude that the maximum in adsorption appears closer to the i.e.p. of immunolatexes than to the average i.e.p. of the IgG in solution.

In Fig. 5 [24] this coincidence is illustrated for all monoclonal IgG and all latexes studied, as well as for BSA.

6. The i.e.p. of protein-coated latex particles shifts to a more acidic pH with increasing ionic strength. This result shows the pH shift of the maximum adsorption to an acidic side with an increase in ionic strength.

Finally, in relation to the i.e.p. of protein-coated latex particles, Shirahama et al. [31] have studied the adsorption and the electrokinetic properties of BSA adsorbed on to amphoteric latexes. These latexes have different i.e.p. values (3.6, 4.5, and 6.2) but much the same particle size were prepared by employing the Hoffmann reaction of styrene/acrylamide copolymer latexes treated with sodium hypochlorite and sodium hydroxide solutions. They also found that the i.e.p. values of BSA–latex complexes agree with those of pH values at maximum adsorption, and observed the shift of i.e.p. to a more alkaline (acid) pH, compared with i.e.p. of protein, for cationic (anionic) charged polystyrene latexes. They suggested that BSA molecules adsorbed on to positively charge surfaces undergo interfacial denaturing upon adsorption, as more basic amino acid residues are exposed to the aqueous phase.

D. Electrokinetic Characterization of Immunolatex of Interest for Particle-Enhanced Immunoassays

Latex particles have been used as a carrier for antigen and antibody reactions in agglutination tests where, for example, the presence of macroscopic agglomerates reveals that of

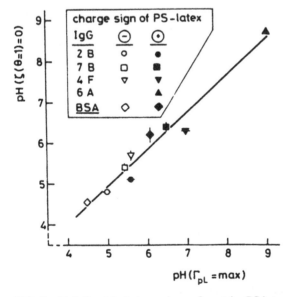

FIG. 5 Relationship between i.e.p of protein–PS latex complex, θ (= Γ/Γ_{PI}) = 1, and the pH where Γ_{PI} reaches a maximum value. BSA 2 B, 7 B, 4 F, and 6 A are monoclonal antibodies with different i.e.p. values. (From Ref. 24.)

the cross-linking antigen. In this way, Stoll et al. [32] used electrophoretic mobility to detect the agglutination of IgG-coated polystyrene latex in the presence of the antigen IgM in the solution. They found that the electrophoretic mobility decreases as a function of the IgM concentration towards a plateau value and they assumed that the ionic strength increases in the surface zone as the antibody–antigen reaction is nearing completion.

A substantial improvement in the use of synthetic polymer colloids for particle-enhanced immunoassays can be obtained by covalent coupling of antigens or antibodies with the surface of functional latex [33, 34]. A procedure for such coupling is the covalent coupling of IgG with carboxyl latexes by the carbodi-imide [1-ethyl-3-(3-dimethylamio-propyl)carbodi-imide chloride] method [35]. In this paper Ortega-Vinuesa et al. studied the electrokinetic behavior of physically adsorbed, and chemically bound, IgG to carboxyl latexes. The effect of carbodi-imide on the surface charge of latex, and therefore in the electrophoretic mobility, explains the electrokinetic properties of the latex compared to those of bare beads. Thus, the electrokinetic properties and the colloidal stability of the IgG–latex complex depend on the coupling mechanism of IgG with carboxyl latex.

Another important point in the latex agglutination immunoassay is to suppress nonspecific interactions of the complementary antigen with nonoccupied parts of the latex particles. A way to avoid this problem is to cover the latex surface with a second protein, for example, BSA. For this reason, sequential IgG and BSA adsorption has recently been of considerable interest in the development of diagnostic test systems. Electrophoretic mobility of IgG–BSA/PS complexes, obtained by sequential adsorption at pH 7, has been studied by Puig et al. [36]. The most striking features of this work are: first, the mobility shifts to more positive values with increasing preadsorbed amount of positively charged IgG; second, the i.e.p. of the protein–IgG complex shifts towards higher pH values with increasing amounts of preadsorbed BSA; and third, the PS surface charge must compensate, at least partially, for the charge on the protein since the i.e.p. of the protein–latex complexes is lower than the average i.e.p. of the dissolved protein. These features may be relevant for the development of particle-enhanced immunoassays.

Also in relation to the sequential adsorption, Ortega-Vinuesa et al. [15] have used the electrophoretic mobility to confirm the presence of monomeric BSA (m-BSA) on the surface of the F(ab')$_2$–cationic latex complex. The electrophoretic mobility of bare catio-nic latex was compared to that of particles sensitized with low and high degrees of coverage of F(ab')$_2$ and BSA on their surface. As we can see in Fig. 6 [15] the mobility of the bare cationic latex is practically independent of the pH in the range 4–9. The i.e.p. of F(ab')$_2$–cationic latex depends on the adsorbed protein, and it is around pH 8 for the F(ab')$_2$–full coated cationic latex particles. The most striking feature is the dramatic decrease in the mobility of the F(ab')$_2$–cationic latex when the amount of m-BSA adsorbed on the latex increases. The adsorption of m-BSA causes a displacement of the i.e.p. of the immunolatex toward lower pH values ($\simeq 6$). This displacement gives rise to negative mobility at pH 8 of the F(ab')$_2$–BSA/cationic latex and now it is possible to stabilize the reagent even at physiological ionic strength (170 mM). In fact, the mobility of this immunolatex is quite similar to that of the BSA–cationic latex, which reinforces the idea that the greater colloidal stability of the F(ab')$_2$–BSA/cationic latex is mainly due to the adsorbed BSA.

E. Adsorption of Low Molecular Weight Ions and Protein Layer

Another important aspect that can be inferred from electrokinetic characterization of protein-coated latex particles is the participation of ions in the protein adsorption process.

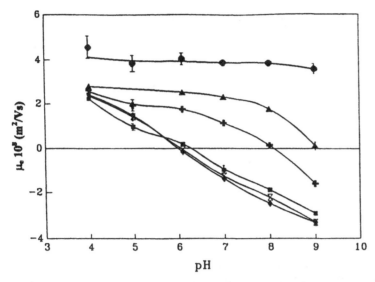

FIG. 6 Electrophoretic mobility versus dispersion pH of different complexes on the cationic latex. Bare cationic latex (●); low F(ab')$_2$ coverage (▲, 1.9 mg IgG/m^2); high F(ab')$_2$ coverage (+, 6.0 mg IgG/m^2); high F(ab')$_2$ coverage and BSA (■, 3.3 mg IgG/m^2 and 2.0 mg of BSA/m^2); low F(ab')$_2$ coverage and BSA (X, 1.3 mg IgG/m^2 and 2.2 mg of BSA/m^2); cationic latex totally covered by BSA (◆). (From Ref. 15.)

In general, proteins present high binding affinity for ions. For this reason, ions play an important role (in addition to hydrophobic dehydration of parts of the adsorbent and/ or protein, structural stability and size of the protein molecule, and electrostatic interaction between protein and surface) in the protein adsorption processes.

In an aqueous environment, charged surfaces and protein molecules are surrounded by counterions, which together with the surface charge, form an EDL. When the protein molecule and the surface approach each other, their EDLs overlap, which gives rise to a redistribution of the counterions.

An explanation for the coadsorption of low molecular weight ions can be found in Ref. 37. If the protein and the adsorbent surface have opposite charge signs, they attract each other, at least if the charge on the protein and the surface more or less compensate each other. If either of the two components has a large excess of charge, this would result in a consideration net amount of charge in the contact region between the protein layer and the adsorbent surface. This region has a low dielectric permittivity relative to that of water and, therefore, accumulation of charge in such an environment would cause the development of an extremely high electrostatic potential, which is energetically very unfavorable. A similar situation would result upon adsorption of a charged protein on a surface that has the same charge sign. Nevertheless, in many cases it is observed that, in spite of such adverse electrostatic conditions, proteins adsorb spontaneously. Based on a model [38] for the adsorbed protein layer it has been predicted that low molecular weight ions are transferred between the solution and the adsorbed layer to prevent accumulation of the net charge in the contact region between the protein and the adsorbent surface.

In this way, for example, the affinity of electrolyte anions on the adsorption of BSA on latex was studied by Shirahama and coworkers [27, 39]. The electrophoretic mobility of

BSA-coated PS particles as a function of pH, in three electrolyte anions (CH_3COO^-, Cl^-, and SCN^- used as Na salt) was determined. The results shown that in all anionic media, the i.e.p. of BSA–latex shifts to a more acidic pH in the order CH_3COO^-, Cl^-, and SCN^-. This indicates that the binding affinity of small ions to BSA molecules increases in the order $CH_3COO^- < Cl^- < SCN^-$. The results agree with the adsorption of BSA in the three media, since it was found that pH at maximum adsorption shifts to a more acidic pH in the order SCN^-, Cl^-, CH_3COO^-.

The number of ions transferred may be deduced from electrokinetic measurements. Because of the requirement of overall electroneutrality, the change in electrokinetic charge per unit surface area ($\Delta_{ads}\sigma_{Ek}$) upon transferring a protein molecule from solution to the latex surface can be expressed as

$$\Delta_{ads}\sigma_{Ek} = \sigma_{Ek}^{PS/Prot} - \sigma\frac{PS}{EK} - \sigma_{Ek}^{Prot}\Gamma_{ads}A \tag{1}$$

where σ_{Ek} is the charge density per unit surface area at the slipping layer of the covered protein (PS/Prot), bare latex (PS), and dissolved protein molecules (Prot); A is the surface area of protein per unit mass.

To calculate σ_{Ek} the electrophoretic mobilities can be converted into ζ potential values and, by using diffuse double-layer theory, into electrokinetic charge:

$$\sigma_{Ek} = \frac{4n_0ze}{\kappa}\sinh\left(\frac{ze\zeta}{2kT}\right) \tag{2}$$

where n_0 is the bulk concentration of ions, z is the ion valence, e is the elementary charge, κ^{-1} is the double-layer thickness, k is the Boltzmann constant, and T is the absolute temperature.

The conversion of electrophoretic mobility into ζ potential is still under discussion. Norde and coworkers [18, 26] converted μ_e into ζ potential values using the procedure of O'Brien and White [40]. A different approach has been used by Galisteo-González et al. [41] to the conversion of μ_e into ζ potential. In this paper they have used the Dukhin–Semenikhin theory [42]. Calculation of ζ potential values according to the theoretical treatment of Dukhin and Semenikhin requires knowledge of surface charge density, and it changes substantially with the pH of the medium in the case of a protein-covered latex surface. To obtain these data they performed a potentiometric titration, and compared the results with those obtained in the titration of a similar solution without latex or complex particles. For any pH value, the electric charge in the bare latex or complex surface can be estimated from the difference in titrant reactive volume, with respect to the solution without particles.

On the other hand, by using a moving-boundary technique, and knowing the dimension of the protein [26], the electrokinetic charge of dissolved protein molecules can be obtained.

By way of example, Fig. 7 [37] shows the change in electrokinetic charge per unit area of adsorbent surface, due an adsorbed layer of lysozyme and α-lactalbumin on a negatively charged PS surface. The data indicate that a pH < 8, positively charged ions are required Accordingly, coadsorption of positively charged ions accompanies the adsorption of α-lactalbumin (i.e.p. 4.3) on the negatively charged PS surface over the entire pH region considered.

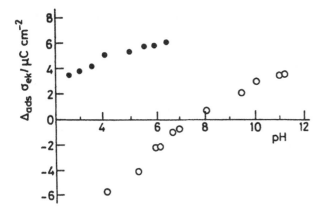

FIG. 7 Ion coadsorption, as reflected by the overall change in the electrokinetic charge density, accompanying the adsorption of lysozyme (○) and α-lactalbumin (●) on negatively charged polystyrene latex; 0.05 M KCl; 25°C. (From Ref. 37.)

Some come features can be obtained from these electrokinetic studies [16, 18–20, 26]:

1. The ion coadsorption increases as the charge on the latex decreases.
2. The uptake of ions seems to increase as the amount of adsorbed protein increases.
3. At the i.e.p. of the protein, the ion coadsorption does not depend on the amount of protein adsorbed.
4. The maximum ion adsorption occurs at pH values where particle and protein have the same sign charge.

III. RELATION BETWEEN COLLOIDAL STABILITY AND ELECTROPHORETIC MOBILITY OF PROTEIN-COATED LATEX

According to the theory elaborated by Derjaguin and Landau [43] and, independently by Verwey and Overbeek [44, 45], or DLVO theory, the stability of colloidal systems is related to EDL repulsion. Therefore, the ζ potential [46] plays a very important role in the colloidal stability of any mesoscopic system. In this way, the electrophoretic mobility of protein-coated surfaces can be suitable for predicting their electrostatic stability. In general, the electrokinetically mobilized charge, as well as the amount and type of adsorbed protein, will be the factor controlling the colloidal stability of protein-coated latex particles.

Tamai et al. [47] determined the ζ potential of bare and BSA-coated latex particles as a function of NaCl concentration. They found that the ζ potential decreases with the increasing amount adsorbed, and hence that the repulsive interaction between BSA–latex coated particles is lower than that between bare particles.

Peula and de las Nieves [13] used the electrophoretic mobility of monomeric (m) and oligomeric (o) albumin–sulfonate latex complexes to detect the existence or nonexistence of electrostatic repulsion between the coated particles. They found that the colloidal stability of the m-BSA latex complexes is mainly controlled by the electrostatic state of the adsorbed protein. The use of sulfonate latex results in a higher adsorption of protein and a very high colloidal stability at intermediate degrees of coverage in comparison with conventional sulfate latex. On the other hand, the colloidal stability of o-BSA–latex complexes is also controlled by the electrostatic state of the protein, but at the i.e.p. of the protein the complexes are stable owing to an additional steric stability by the adsorbed protein molecules.

Tamai et al. [48] studied the colloidal stability of polystyrene (PS) and styrene/ acrylamide copolymer (PS/Aam) latexes coated with human serum albumin (HSA) at high surface coverage. The results of stability obtained by light scattering were compared with those obtained by electrophoresis. The ζ potential values indicated that the high colloidal stability of HSA-coated PS latex in NaCl solution is a result of the adsorbed HSA. In addition, the ζ potential of coated latex is lower than that of bare latex in $MgCl_2$ and $CaCl_2$ solutions, and therefore it seems reasonable to suggest that HSA-coated PS latex flocculates due to HSA–HSA linkage by divalent cations.

On the other hand, the ζ potential of HSA-coated PS/Aam latex was almost the same as that of bare PS/Aam latex in $MgCl_2$ and $CaCl_2$ solutions. A possible explanation is that HSA molecules adsorbed on this latex might be incorporated to some extent into the inner part of the water-soluble polymer layer, and HSA–HSA linkages by divalent cations do not occur.

An explanation of the extremely low colloidal stability of the polyclonal IgG-coated surfaces was provided by Galisteo-González [41]. The results are related to the structure of the EDL of protein-coated polystyrene, and the main conclusions of this study are:

1. That ions in the EDL surrounding the IgG polymer surface (especially those under the hydrodynamic slipping plane) have a greater ionic mobility when the electric charge in the protein molecules has the same sign as the electric groups in the particle surface.
2. That the anomalous surface conduction mechanism is more pronounced in this case in the surface charge region.

Recently, Molina-Bolívar and coworkers [49, 50] have analyzed a new stabilization mechanism in hydrophilic colloidal particles. Actually, they worked with polystyrene beads covered by globular proteins, the surface of which is mostly hydrophilic. The proposed stabilization mechanism is explained by considering the adsorption of hydrated ions on to the adhered protein layer. An overlap of the ordered hydrated counterion layers near two mutually approaching surfaces creates a repulsive force. Its origin lies in the partial dehydration of the adsorbed ions and/or the surface charged groups, which will lead to an increase in the system energy. As a matter of fact, cations such as Na^+ and Ca^{2+} can stabilize negatively charged hydrophilic surfaces by means of these "hydration forces," stabilization becoming higher for increasing electrolyte concentration. Electrophoretic mobility measurements can be very useful for detecting the specific adsorption of counterions on latex–protein complexes. Specifically, the mobility of a polystyrene latex totally covered by antibody fragments [namely, $F(ab')_2$ fragments] was measured as a function of both the pH and the electrolyte concentration. The results obtained with $CaCl_2$ and NaCl are show in Figs 8 and 9, respectively. The solutions were not buffered in order to avoid the presence of foreign ions coming from the buffers. Hence, the pH was only controlled by adding NaOH or HCl. As can be seen in Fig. 8, for low Ca^{2+} concentrations (i.e., 2×10^{-3} M) there is an i.e.p. which disappears when the Ca^{2+} concentration increases to 38×10^{-3} M. It should be noted that the i.e.p. of such an antibody fragment is around 4.7, so the protein layer must be negatively charged in the pH range 5–8. However, the mobility is positive in that range. This interesting result indicates that Ca^{2+} ions adsorb on the negatively charged protein layer, provoking a change in the expected μ_e sign. As can be seen, these μ_e measurements give important

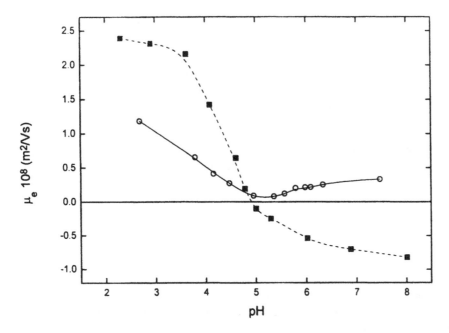

FIG. 8 Electrophoretic mobility of F(ab')$_2$-latex complex as a function of pH, for different CaCl$_2$ concentrations: 2.0×10^{-3} M (■) and 3.8×10^{-2} M (○). (Courtesy of Dr. J. A. Molina Bolivar.)

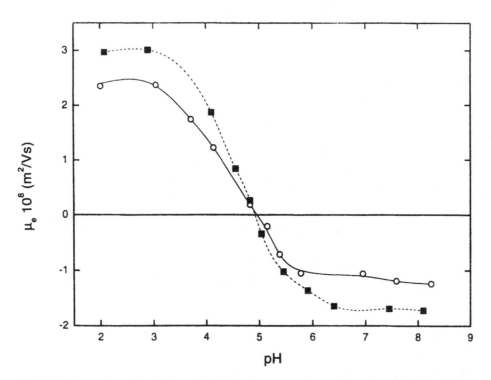

FIG. 9 Electrophoretic mobility of a F(ab')$_2$-latex complex as a function of pH, for different NaCl concentrations: 2.0×10^{-3} M (■) and 1.0×10^{-2} M (○). (Courtesy of Dr. J. A. Molina Bolivar.)

information on the adsorption of counterions on the adsorbed protein layer. These data firmly support the stability mechanism proposed by Molina-Bolivar and coworkers [49–50], which was based on hydration forces. However, taking Fig. 9 into account, the Na^+ ions do not attain inversion of the μ_e sign in the pH 5–8 range. It should be noted than this result does not mean a complete absence of Na^+ adsorption on to the protein layer. This monovalent cation could be adsorbed on the surface, although its surface concentration would not be high enough to change the ζ potential sign. Measurements of μ_e at higher NaCl concentration would be advised in order to make this clear. However, obtaining μ_e data at higher ionic strengths is a difficult task, due some experimental problems arising from the devices used for measuring electrophoretic mobility.

IV. THEORETICAL BACKGROUND

There exist few theoretical studies on the electrophoresis of colloidal particles with structured surfaces, which can be formed by adsorbing polyelectrolyte or protein molecules [50–56]. These studies revealed that the ζ potential loses its meaning for such particles, since the electrophoretic mobility becomes insensitive to the precise position of the slipping plane. This means that the electrokinetic behavior of latex–protein complexes does not only depend on the ζ potential, but also on other parameters, which are listed below: the hydrodynamic resistance of the protein layer to the flow of electrolyte solution through it, the charge density and the thickness of such a layer, and the electric potential distribution across it. In particular, Ohshima and Kondo have studied in detail the electrophoretic mobility of colloidal particles coated by surface-charged layers [50–60]. Their theory will just be summarized below, and subsequently, some fits of theoretical data and experimental results will be presented.

A. Theoretical Aspects

Consider a colloidal particle having a surface charge density σ_0 with a polyelectrolyte layer adsorbed on its surface. The charged groups of this surface layer of thickness d are distributed at a uniform density N. These charged groups have a valence z. The particle surface is assumed to be planar with the applied electric field parallel to it. The x axis is perpendicular to the surface and its origin is fixed at the boundary between the surface layer and solution. The charged colloidal particles are moving in a liquid containing a symmetrical electrolyte of valence v and bulk concentration n. In these cases, and using two Poisson–Boltzmann equations and hydrodynamic considerations, Ohshima and Kondo found an analytical expression of the electrophoretic mobility of such a system, which is given in Ref. 57. Nevertheless, it is quite difficult to calculate μ_e theoretically through such equation, as there is an integral that seldom has an analytical solution. Despite this problem, Ohshima and Kondo managed to obtain simpler expressions for the mobility of structured interfaces. Two general case can be considered:

1. $N \neq 0$ and $\sigma_0 = 0$

It should be noted that there is no charge density on the bare latex surface, but there is a net charge density due to the polyelectrolyte layer. For cases like that, these authors derived a mobility formula applicable whatever the potential value. They considered that as d tends to ∞, $\psi(-d)$ would tend to the Donnan potential ψ_{DON}, which corresponds to the potential distribution across a charged layer with a large thickness. A schematic representation of the potential distribution across a negatively charged surface layer of thickness d is shown in Fig. 10. Considering a negatively charged polyelectrolyte layer ($z = -1$), the electrophoretic mobility of a structured interface can be written as [57]

$$\mu_e = -\frac{\epsilon_r \epsilon_0 kT}{\nu e \eta}\left[\ln\left(\frac{|\sigma|}{2\nu e n d} + \left[\left(\frac{\sigma}{2\nu e n d}\right)^2 + 1\right]^{\frac{1}{2}}\right)\right] + \frac{\epsilon_r \epsilon_0 kT}{\nu e \eta}$$

$$\times \left(\left[\left(\frac{2\nu e n d}{\sigma}\right)^2 + 1\right]^{\frac{1}{2}} - \frac{2\nu e n d}{|\sigma|}\right)\frac{(\kappa/\lambda)[1 + (\sigma/2\nu e n d)^2]^{1/4}\tanh\lambda d - 1}{(\kappa/\lambda)^2[1 + (\sigma/2\nu e n d)^2]^{1/2} - 1} \qquad (3)$$

$$-\frac{|\sigma|}{\eta d\lambda^2}\left(1 - \frac{1}{\cosh\lambda d}\right)$$

where ϵ_r is the dielectric constant of the electrolyte solution (it is assumed that the surface-charged layer possesses the same ϵ_r value), ϵ_0 is the permittivity of vacuum, k is the Boltzmann constant, T is the absolute temperature, σ is the amount of fixed charges contained in the surface charged layer per unit area ($\sigma = zeNd$), and λ is a hydrodynamic parameter whose reciprocal has the dimension of length and gives information on the depth of the flow penetration in the surface region: λ is defined by the following relation:

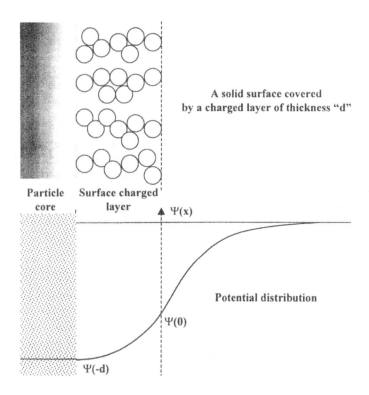

A solid surface covered
by a charged layer of thickness "d"

Particle   Surface charged
core   layer

$\Psi(x)$

Potential distribution

$\Psi(0)$

$\Psi(-d)$

FIG. 10 Schematic representation of the potential distribution across a negatively charged layer of thickness d adsorbed on an uncharged surface. (From Ref. 57.)

$$\lambda = \left(\frac{\gamma}{\eta}\right)^{\frac{1}{2}} \tag{4}$$

with η and γ being the viscosity of the fluid and the frictional coefficient of the adsorbed polyelectrolyte layer, and κ is the reciprocal of double-layer thickness:

$$\kappa = \left(\frac{2nv^2e^2}{\epsilon_r\epsilon_0 kT}\right)^{1/2} \tag{5}$$

Although Eq. (3) is valid for a thick polymer layer ($d \to \infty$), Ohshima and Kondo have demonstrated [57] that it remains an excellent approximation for $d \geq 10$ Å. This limit is lower than the mean size of any protein molecule, so it could be applied to latex–protein systems. Nevertheless, the same authors obtained a much simpler analytic formula for mobility if $d \geq 1/\kappa$ or $d \geq 1/\lambda$ [58]:

$$\mu = \frac{\epsilon_r\epsilon_0}{\eta}\frac{\Psi(0)/\kappa_m + \Psi_{DON}/\lambda}{1/\kappa_m + 1/\lambda} + \frac{zeN\,N_{Av}}{\eta\lambda^2} \tag{6}$$

where $\Psi(0)$ is the potential at the boundary between the polymer layer and the surrounding electrolyte solution, N_{Av} is the Avogadro number, and κ_m is the effective Debye–Hückel parameter of the polymer layer, which is related to κ through

$$\kappa_m = \kappa\left[1 + \left(\frac{zN}{2vn}\right)^2\right]^{\frac{1}{4}} \tag{7}$$

Expressions of Ψ_{DON} and $\Psi(0)$ are given by [60]

$$\Psi_{DON} = \frac{kT}{ve}\ln\left[\frac{-N}{2vn} + \left\{\left(\frac{-N}{2vn}\right)^2 + 1\right\}^{1/2}\right] \tag{8}$$

$$\Psi(0) = \Psi_{DON} - \frac{kT}{ve}\tanh\frac{ve\Psi_{DON}}{2kT} \tag{9}$$

Equation (6) consists of two terms: the first one is a weighted average of the Donnan potential and the polyelectrolyte layer/solution boundary potential, and thus it depends on the potential distribution across such a layer. The second one is directly related to the polymer fixed charges and is not subjected to shielding effects of electrolyte ions.

The dependence of the theoretical μ_e on the surface layer thickness d, on the total fixed charges contained in such a layer $\sigma = f(N)$, and on the electrolyte concentration n is shown in Figures 11, 12, and 13, respectively.

2. $N \neq 0$ and $\sigma_0 \neq 0$

The main problem of the above theoretical approach is that it would be only applicable to uncharged colloid particles covered by negatively charged proteins. However, most of the colloidal particles used to adsorb proteins present a surface charge density different from zero. The "$N \neq 0$ and $\sigma_0 \neq 0$" case represents a more general and realistic situation where a charged bare particle is coated with a charged polymer layer. For such systems an approximate expression for the theoretical electrophoretic mobility [60] can be obtained:

$$\mu = \frac{\epsilon_r\epsilon_0}{\eta}\frac{\Psi(0)/\kappa_m + \Psi_{DON}/\lambda}{1/\kappa_m + 1/\lambda} + \frac{zeN\,N_{Av}}{\eta\lambda^2} + \frac{8\epsilon_r\epsilon_0 kT}{\eta\lambda ve}\frac{\lambda}{1/\lambda^2 - 1/\kappa_m^2}\frac{\kappa_m}{}\tanh\frac{ve\zeta}{4kT} \tag{10}$$

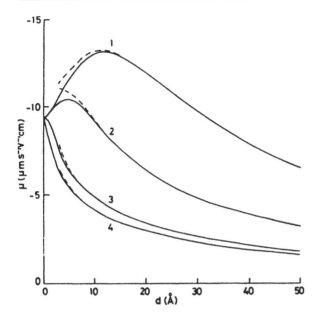

FIG. 11 Electrophoretic mobility as a function of the surface-layer thickness for several values of λ. Curve 1, $\lambda = 1 \times 10^9\,\text{m}^{-1}$; curve 2, $\lambda = 2 \times 10^9\,\text{m}^{-1}$; curve 3, $\lambda = 1 \times 10^{10}\,\text{m}^{-1}$; curve 4, $\lambda = \infty$. Calculated with $T = 298\,\text{K}$, $\epsilon_r = 78.5$, $n = 0.1\,\text{M}$, $\eta = 0.891 \times 10^{-3}\,\text{N}\,\text{s}\,\text{m}^{-2}$, and $\sigma = -0.19\,\text{C}\,\text{m}^{-2}$. Dashed lines: approximation given by Eq. (3); solid lines: exact results. (From Ref. 57.)

Now the potential distribution inside the polymer layer is different from the case $\sigma_0 = 0$. A scheme of such distribution is shown in Fig. 14. In this case, $\Psi(0)$ is different from that shown in Eq. (9), and is equal to

$$\Psi(0) = \Psi_{\text{DON}} - \frac{kT}{ve}\tanh\frac{ve\Psi_{\text{DON}}}{2kT} + \frac{4kT}{ve}e^{-\kappa_m d}\tanh\frac{ve\zeta}{4kT} \tag{11}$$

It should be noted that the third term of Eqs. (10) and (11) includes the ζ potential of the bare particle, and hence, these terms take into account the influence of the surface charge density of the adsorbent particles.

As can be seen, one can calculate the theoretical electrophoretic mobility of a colloidal ample if the potential distribution [$\Psi(x)$] across the adsorbed layer is known. For unchanged particles the problem would be solved knowing three different characteristics of the surface charged layer, namely, N, d, and λ. However, if the carrier particles are charged a fourth parameter is needed: the ζ potential of the bare surface.

B. Comparing Theoretical Predictions with Experimental Data

Ohshima and coworkers were the first researchers who tried to validate their own theory measuring the electrophoretic mobility of colloidal particles covered by protein layers: the electrophoretic mobility of particles coated with human serum albumin (HSA) and immunoglobulin G (IgG) was measured at different pH and electrolyte concentrations [61]. In order to avoid the use of Eq. (10), where there are four fitting parameters (N, d, λ, and ζ), the above authors fixed the proteins on the surface of carboxyl-latex particles through chemical bonding. The -COO⁻ groups from the polymer surface reacted with -NH₃⁺

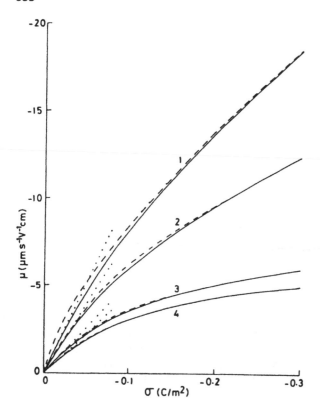

FIG. 12 Electrophoretic mobility as a function of the amount of total fixed charges contained in the surface layer (σ) for several values of λ. Curve 1, $\lambda = 1 \times 10^9 \, m^{-1}$; curve 2, $\lambda = 2 \times 10^9 \, m^{-1}$; curve 3, $\lambda = 1 \times 10^{10} \, m^{-1}$; curve 4, $\lambda = \infty$. Calculated with $T = 298 \, K$, $\epsilon_r = 78.5$, $n = 0.1 \, M$, $\eta = 0.891 \times 10^{-3} \, N \, s \, m^{-2}$, and $d = 10 \, \text{Å}$. Dashed lines: approximation given by Eq. (3); solid lines: exact results. (From Ref. 57.)

groups of the protein molecules by means of the carbodi-imine (CDI) method, which is described in detail in Refs. 35 and 62. So, the original surface charge density of the particles was reduced to zero after the chemical coupling. Therefore, their HSA– and IgG–latex complexes fulfilled the $N \neq 0$, $\sigma_0 = 0$ condition, and Eq. (6), which contains three undetermined quantities (N, d, and λ), could be used. In order to check the validity of such a formula, μ_e data were obtained at different pH and ionic strengths. The results for the HSA–latex (2.4 mg/m^2) and the IgG–latex (2.4 mg/m^2) complexes are shown in Figs 15 and 16, respectively, where the experimental data are depicted by symbols and the theoretical predictions by lines. The best fitting values are show in Table 2. A discussion of the results can be summarized as follows.

The mobility decreased in absolute value as the ionic strength of the medium increased, due to the screening effect of the electrolyte. The molecular sizes of HSA and IgG determined by other methods, including viscometry and dynamic light scattering, are reported to range from 40 to 150 Å [63] and from 44 to 243 Å [1], respectively, depending on the pH. Hence, the values of d obtained by μ_e data fall in the above ranges. In view of this, mobility measurements on protein-coated latex particles may be used to estimate the molecular size of proteins. The N values obtained seem to agree with the charge nature of

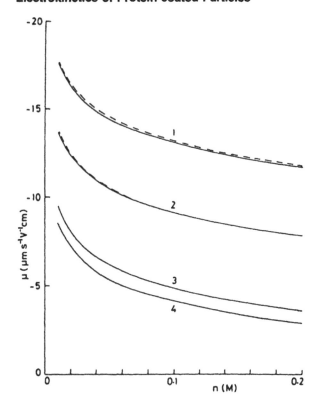

FIG. 13 Electrophoretic mobility as a function of the electrolyte concentration for several values of λ. Curve 1, $\lambda = 1 \times 10^9 \, \text{m}^{-1}$; curve 2, $\lambda = 2 \times 10^9 \, \text{m}^{-1}$; curve 3, $\lambda = 1 \times 10^{10} \, \text{m}^{-1}$; curve 4, $\lambda = \infty$. Calculated with $T = 298 \, \text{K}$, $\epsilon_r = 78.5$, $\eta = 0.891 \times 10^{-3} \, \text{N s m}^{-2}$, $\sigma = -0.19 \, \text{C m}^{-2}$, and $d = 10 \, \text{Å}$. Dashed lines: approximation given by Eq. (3); solid lines: exact results. (From Ref. 57.)

both proteins. Albumin is a highly charged protein, provided that the pH does not coincide with its i.e.p. (4.7), and its N is higher than that of IgG. In addition, the N values for the IgG molecules are different at pH 6.5 and pH 8.5, because its i.e.p. is more basic than that of HSA. Finally, the above authors were not able to explain adequately these results obtained at pH 2.5. The problem arose from the impossibility of fitting the experimental μ_e data with constant values of N, d, and λ, a controversial point for which no satisfactory explanation appears to have been given. On the other hand, the main drawback of obtaining information on the protein layer from μ_e data via the Ohshima and Kondo theory lies in the fact that one has three simultaneous fitting parameters. So, the value attained for one of them will depend on the values given to the others. Although it would not be an accurate method for obtaining detailed information on the adsorbed protein layer, it allows one to obtain approximate values on the charge density, thickness, and friction coefficient. Moreover, the accuracy would significantly improve, if, at least, one of the fitting parameters could be previously obtained by means of other techniques, thus reducing the number of parameters to be fitted.

Other authors who checked the validity of this theory were Ortega-Vinuesa and Hidalgo-Álvarez [64]. In their work, the chemical and physical adsorption of $F(ab')_2$ fragments were carried out on three carboxyl-latexes (JL3, JL4, and JL7), which differed in surface charge density. The main characteristics of these latexes are shown in Table 3. It should be noted that the chemical immobilization was also performed by the CDI method.

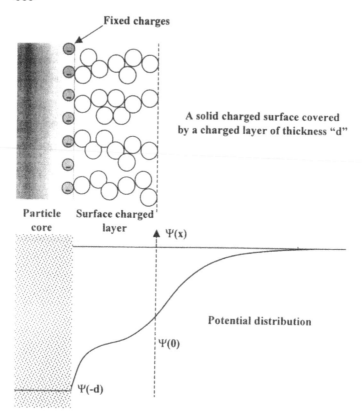

FIG. 14 Schematic representation of the potential distribution across a negatively charged layer of thickness *d* adsorbed on a negatively charged surface.

Both the bare latex and the adsorbed $F(ab')_2$ layer were characterized by electrophoretic mobility measurements. These authors demonstrated that the most important parameters affecting the μ_e data were not only those related to the protein layer (N, d, and λ), but also the charge density of the polymer particle.

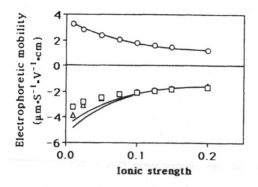

FIG. 15 Electrophoretic mobility of HSA–latex complexes as a function of ionic strength at different pH: 2.5 (○); 6.5 (△); 8.5 (□). The lines correspond to the best fitting of the data to the Ohshima and Kondo equation (6). (From Ref. 61.)

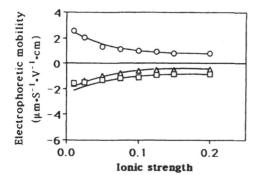

FIG. 16 Same as Fig. 15, but for IgG–latex complexes. (From Ref. 61.)

The main problem of applying the μ_e theory to those latex complexes formed by physical adsorption is that four fitting parameters would be needed, as it is a $N \neq 0, \sigma_0 = 0$ case. In order to reduce the number of adjustable parameters, one of them (the ζ potential of the bare latex particles) was previously calculated from electrophoretic mobility data. The μ_e data were converted into ζ potentials according to three different electrokinetic theoretical treatments [65]: (1) Smoluchowski (classical equation), (2) O'Brien and White (which takes into account the polarization of the mobile part of the EDL only, and (3) Dukhin and Semenikhin (which also takes into account polarization in the stagnant part of the double layer).

The three latexes showed similar electrokinetic behavior, and the calculated ζ potentials were also very similar. This is why only the values obtained for one of them are shown (see Fig. 17). Only the Dukhin–Semenikhin theory provides a continuous decrease in ζ potential for increasing salt concentration, that is, there is no ζ potential maximum. Therefore, Ortega-Vinuesa and Hidalgo-Álvarez [64] chose this theory to obtain the ζ potential values to be included in the Ohshima and Kondo theory [see Eqs. (10) and (11)]. On the other hand, the ζ potential of the bare latexes also depended on pH. This dependence is shown in Fig. 18, which was also obtained from μ_e measurements and then applying the Dukhin–Semenikhin conversion.

Subsequently, the μ_e measurements as a function of ionic strength at two pH values (4 and 8) were carried out for the following $F(ab')_2$ complexes:

- $F(ab')_2$–JL3 complex (2.6 mg/m², physical adsorption)
- $F(ab')_2$–JL3 complex (2.4 mg/m², covalently bounded)
- $F(ab')_2$–JL4 complex (2.3 mg/m², physical adsorption)

TABLE 2 Best Fitting Parameters of Eq. (6) to IgG–Latex and HSA–Latex Mobilities at Different pH (see Fig. 15)

	IgG–Latex			HSA–Latex		
pH	N (M)	λ^{-1} (nm)	d (Å)	N (M)	λ^{-1} (nm)	d (Å)
6.5	0.010	2.0	100	0.025	2.0	60
8.5	0.018	2.0	100	0.025	2.0	60

N = charge density; λ^{-1} = flow penetration depth; d = protein layer thickness.

TABLE 3 Particle Diameters Obtained by Photon Correlation
Spectroscopy (PCS) and Transmission Electron Microscopy (TEM)
and Surface Charge Density of the Latexes

Latex	Diameter, PCS (nm)	Diameter, TEM (nm)	Surface charge density ($\mu C/cm^2$)
JL 3	273 ± 5	264 ± 8	16.2 ± 0.3
JL 4	342 ± 7	331 ± 9	12.1 ± 0.2
JL 7	342 ± 5	333 ± 8	19.0 ± 0.4

Source: Ref. 64.

- $F(ab')_2$–JL4 complex (2.5 mg/m^2, covalently bounded)
- $F(ab')_2$–JL7 complex (2.0 mg/m^2, physical adsorption)
- $F(ab')_2$–JL7 complex (2.2 mg/m^2, covalently bounded)

Results are shown in Figs 19–21, and the most relevant aspects can be summarized as
follows:

1. Results are very similar for all three latex carriers. Therefore, the experimental
    μ_e response of $F(ab')_2$–latex complexes seems to be rather (although not totally)
    insensitive to the bare latex σ_0. Only JL7 latex with physically adsorbed $F(ab')_2$
    shows a slightly higher μ_e (in absolute value) at pH 8 and lower at pH 4 than the

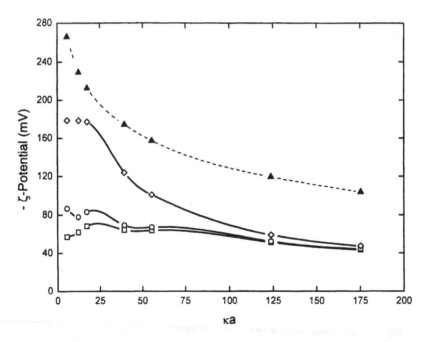

FIG. 17 Zeta potential (open symbols) and surface potential (▲) (obtained from the surface
charge density [64] as a function of the electrokinetic radius for the JL4 latex; μ_e data were converted into ζ potential values using the Smoluchowski (□), O'Brien–White (○), and Dukhin–
Semenikhin (◇) theories. (From Ref. 64.)

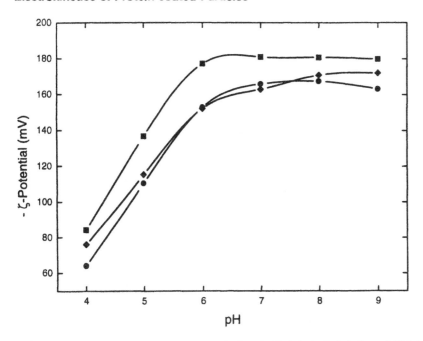

FIG. 18 Zeta potential dependence on pH for the JL3 (■), JL4 (●), and JL7 (◆) latexes. (From Ref. 64.)

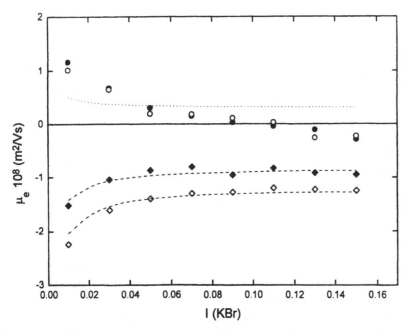

FIG. 19 Electrophoretic mobility of the following complexes: $F(ab')_2$–JL3 (2.6 mg/m^2, physical adsorption) at pH 4 (○) and pH 8 (◇); $F(ab')_2$–JL3 (2.4 mg/m^2, covalent coupling) at pH 4 (●) and pH 8 (◆). Dashed lines represent the theoretical μ_e obtained from Eq. (1) (the fitting parameters are shown in Table 4). Dotted line is the same theoretical mobility for $N = 0.003$ M, $\lambda = 0.3$ nm^{-1}, and $d = 80$ Å. (From Ref. 64.)

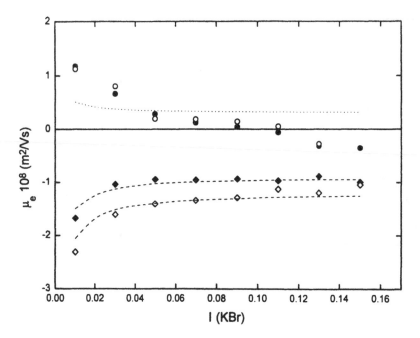

FIG. 20 Electrophoretic mobility of the following complexes: F(ab$'$)$_2$–JL4 (2.3 mg/m^2, physical adsorption) at pH 4 (○) and pH 8 (◇); F(ab$'$)$_2$–JL4 (2.5 mg/m^2, covalent coupling) at pH 4 (●) and pH 8 (◆). Dashed lines represent the theoretical μ_e obtained form Eq. (3) (the fitting parameters are shown in Table 4). Dotted line is the same theoretical mobility for $N = 0.003$ M, $\lambda = 0.3$ nm^{-1}, and $d = 80$ Å. (From Ref. 64.)

other two latexes. This could be explained by taking into account that JL7 is the most highly charged latex. However, one should also keep in mind that the protein coverage of the former latex is lower than that on JL3 and JL4 samples. These features could explain these differences.

2. At pH 8, there are clear differences between the samples with physically adsorbed and covalently bound F(ab$'$)$_2$.

3. At pH4 there is a KBr concentration where the mobility changes sign. This is caused by a flux of counterions through the adsorbed protein layer. In fact, at low electrolyte concentration μ_e is quite positive, as can be seen in last figures. So these experiments make clear the difference between zero charge point and i.e.p.

The discussion is summarized below.

1. First, the authors carried out a preliminary study of the equations to use, that is, Eqs. (3), (6), and (10), in order to test which of them works best. Ohshima and Kondo claimed that Eqs (3) and (6) were quite similar. However, the second is totally insensitive to the polyelectrolyte thickness d, which renders it useless for determination of this parameter via μ_e measurements. In addition the influence of d on the electrophoretic mobility [Eq. (3)] is important enough, and it cannot be considered. For this reason, Ortega-Vinuesa and Hidalgo-Álvarez [64] advised not to employ Eq. (6). On the other hand, these theoretical equations are not in a strict sense applicable to fit the μ_e data of the F(ab$'$)$_2$–latex systems, since these complexes do not obey the condition $\sigma_0 = 0$. Even those

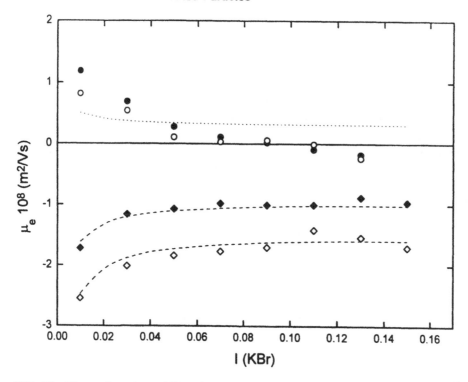

FIG. 21 Electrophoretic mobility of the following complexes: F(ab′)$_2$–JL7 (2.0 mg/m^2, physical adsorption) at pH 4 (○) and pH 8 (◇); F(ab′)$_2$–JL7 (2.2 mg/m^2, covalent coupling) at pH 4 (●) and pH 8 (◆). Dashed lines represent the theoretical μ_e obtained from Eq. (3) (the fitting parameters are shown in Table 4). Dotted line is the same theoretical mobility for $N = 0.003$ M, $\lambda = 0.3$ nm^{-1}, and $d = 80$ Å.

samples with chemically bound protein comply with $\sigma_0 \neq 0$, as the carboxyl groups activated by CDI that did not react with F(ab′)$_2$ amine groups were again transformed into carboxyl groups by adding glycine. This is why Eq. (10) should be used to analyze the mobility results. This equation does take into account the adsorbed layer thickness, not only in the third term but also through the $\Psi(0)$ expression [Eq. (11)]. Now, the most influential parameter is the ζ potential (or the surface potential Ψ_0) of bare latex particles.

2. Second, the authors made an exhaustive study of d, N, and λ parameters that best fit the theoretical mobility curves, given by Eq. (10) to their experimental μ_e data. They reached the following conclusions:

- It is impossible to fit the experimental mobility data using the ζ potential or Ψ_0 values calculated previously for their bare latexes (Fig.8).
- In addition, it is very difficult to know the real surface potential at $x = -d$ after protein adsorption, since during this process there is also coadsorption of low weight (LW) ions that remain enclosed within the polystyrene/F(ab′)$_2$ interface [66, 67].
- Moreover, there are different possibilities leading to different results: Ψ_0 or ζ potential obtained by the Smoluchowski, O'Brien–White, or Dukhin–Semenikhin theories. As the effect of the ζ potential in Eq. (10) is quite important, one could also obtain different behaviors for the theoretical μ_e. Due to this

problem of ambiguous solution, they decide to make the ζ potential (or Ψ_0) a fitting parameter as well. So, one would have again four fitting parameters for μ_e: N, d, λ, and ζ.

Because of the above stated reasons the authors advised to reject Eq. (10).

3. The best results are obtained using Eq. (3). In this case the protein-covered particles must comply with $\sigma_0 = 0$, although their systems do not obey this condition. Even so, one could apply Eq. (3) if the following reasoning is taken into account. As latex particles do not have a completely smooth surface and the carboxyl groups are probably heterogenously distributed, one can suppose that colloidal particles do not have surface charge but that all the charge of the protein–latex complexes is located in a layer of thickness d containing:

1. charges due to the carboxyl groups of the polymer surface;
2. charges due to the LW ions enclosed as the polymer/F(ab$'$)$_2$ interface;
3. charges that come from the protein molecules.

Figure 22 shows a description of this model. Consequently, one could use Eq. (3) without error, considering that N is not the charge density of the protein layer, but is the total average charge density given by merging the above three contributions.

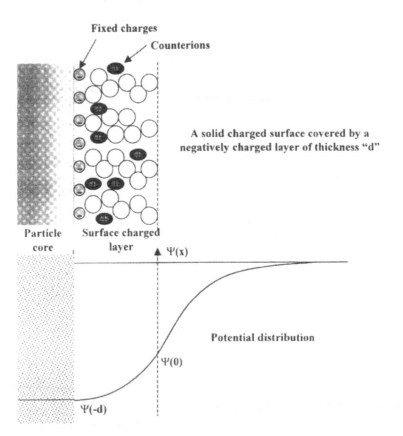

FIG. 22 Scheme of a new model, where Eq. (3) is applicable although the solid polymer particles that act as protein carriers were charged ($\sigma_0 \neq 0$). (From Ref. 64.)

The d, N, and λ parameters that best fit the experimental data for pH 8 (Figs 19–21) are given in Table 4. These results can be independently analyzed as follows.

The net charge density of the external layer (N) is lower for covalently bound $F(ab')_2$ than for physically adsorbed $F(ab')_2$. This is a surprising result. Theoretically both should have the same N, since in the first case some $-COO^-$ groups of the polymer surface react with some $-NH_3^+$ groups of the protein molecule to give a peptide link ($-CO-NH-$) and in the second case these groups are electrically neutralized. The explanation of this odd result might be found in the activation process of the carboxyl particles with CDI. It is worth commenting that this carbodi-imide is a positively charge molecule. Therefore, when a $-COO^-$ group reacts with a CDI molecule there is an exchange of negatively charged groups for positive ones. Ortega-Vinuesa et al. [17, 35] demonstrated that not all of the carboxyl groups react with CDI, and that even later a great part of the activated groups do not react with $-NH_3^+$ groups of either protein or glycine molecules, some CDI groups remaining on the latex surface. Therefore, these particles have positive and negative chemical groups linked to the polystyrene surface. This is why the mobility of such samples of pH 8 is lower (in absolute value) than that of the complexes with physically adsorbed $F(ab')_2$. For the same reason, μ_e data at pH 4 are higher for the former than for the latter. This difference is more pronounced at low ionic strength.

At pH 4, $F(ab')_2$ molecules are positively charged. However, this charge must be small since this pH is rather close to the i.e.p. (5.3 ± 0.6). Therefore, the N parameter should be small ($N \leq 5$ mM). However, it remains difficult (if not impossible) to fit these mobility data properly since there is a change of sign for [Kbr] ≈ 100 mM due to a counterion flux. This flux is not taken into account by the Ohshima and Kondo theory leading to Eq. (3). It should be noted that these authors only used this fitting method for pH values far from the i.e.p. of the adsorbed protein, and thus for a high surface charge density of the immobilized layer [61].

In addition, one can obtain information about another parameter related to the protein layer: its thickness. The best d value for fitting the experimental μ_e data is 8 nm. This is in good agreement with the $F(ab')_2$ dimensions ($14 \times 4 \times 4$ nm^3) obtained from the structures of crystallized IgG molecules by Sarma et al. [68]. This value also agrees with the results obtained by Nakamura et al. discussed above [61]: $d = 100$ Å for IgG (Mw $= 150$ kD) and $d = 60$ Å for BSA (MW $= 66$ kD). These last authors state that they used Eq. (6) to obtain information about d. They also could estimate that λ^{-1} (depth of the flow penetration) was 3.3 nm.

Finally, in another work, Ortega-Vinuesa eta l. [17] use the Ohshima and Kondo theory to calculate the surface charge density (N) of different IgG–latex complexes in

TABLE 4 Parameters that Best Fit the Theoretical Electrophoretic Mobility [Eq. (3)] to the Experimental μ_e Data Obtained at pH 8

Sample	N (M)	λ^{-1} (nm)	d (Å)
JL3 (covalently bound)	0.009	0.3	70
JL3 (physical adsorption)	0.013	0.3	70
JL4 (covalently bound)	0.009	0.3	80
JL4 (physical adsorption)	0.012	0.3	80
JL7 (covalently bound)	0.010	0.3	75
JL7 (physical adsorption	0.015	0.3	80

order to explain further colloidal stability data, as these are related somehow with the surface charge o the complexes. There were two different IgG samples: one of them was a polyclonal antibody (from goat), and in this work it will be referred to as "Pab." The second one was a monoclonal sample (from mouse), referred to as "Mab." They differed in their i.e.p. values. The i.e.p. of the Mab was lower (5.4 ± 0.1) than that of the Pab (6.9 ± 0.9). Both proteins were attached to the JL3 latex surface. Once more, physical and chemical (mediated by the CDI method) adsorption was carried out, and μ_e was measured. The results are shown in Fig. 23, where the protein loads are also given. The experimental data were again fitted by using Eq. (4). The choice of this expression has been justified previously. In this work, only N was chosen as the fitting parameter, as d and λ remained at constant values (obtained from previous work by the same authors [64, 35]). They used a value of 3.3 nm of the λ^{-1} parameter, and d was set at 100 Å, as reported by Nakamura et al. [61]. As can be seen from Fig. 23, the N values that best fit the experimental μ_e data are 25 mM (for the Pab physically adsorbed), 15 mM (Pab chemically bound, 0.5 mg/m^2), 12 mM (Pab chemically bound, 1.1 mg/m^2), 10 mM (Pab chemically bound, 1.8 mg/m^2), and 14 mM (Mab chemically bond, 1.5 mg/m^2). The great difference between the "physically adsorbed" and the "chemically bound" complexes is caused by the presence of attached CDI molecules, which drastically reduce the change density of the external layer of the latter particle complexes. It is worth noting that the N values decrease with increasing protein loading, a result that has also been found elsewhere [35]. These data are consistent with the idea of polyclonal IgG molecules having a low charge at pH 8. In addition, comparing the N values for the Mab samples and the Pab samples (mainly with those having 1.8 mg/m^2, as this coverage is similar to that of the Mab), one can conclude that the Mab-conjugated particles are more highly charged than the Pab species at pH 8. It should be noticed that this information, supplied by a electrokinetic technique, perfectly agrees with the i.e.p. values of both antibody samples

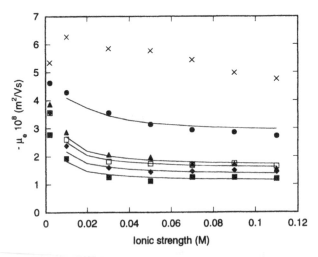

FIG. 23 Electrophoretic mobility versus ionic strength for the bare latex ($\times$); Pab (physical adsorption) (1.0 mg/m^2) ($\bullet$); Pab (covalent coupling) (0.5 mg/m^2) ($\blacktriangle$); Pab (covalent coupling) (1.1 mg/m^2) ($\blacklozenge$); Pab (covalent coupling) (1.8 mg/m^2) ($\blacksquare$); and Mab (covalent coupling) (1.5 mg/m^2) ($\square$). Solid lines represent the theoretical μ_e obtained from Eq. (3) using the following fitting values: $d = 100$ Å, $\lambda^{-1} = 3.3$ nm, and $N = 25$, 15, 12, 10, and 14 mM, respectively. (From Ref. 17.)

(obtained by isoelectric focusing), since the i.e.p. of the Mab IgG is more acidic than that of the Pab sample.

ACKNOWLEDGMENTS

In the past years a large number of people have been helpful in discussing the electrokinetics of protein-coated latex particles. We particularly want to thank W. Norde, Ch. Pichot, F. J. de las Nieves, F. Galisteo, J. M. Peula, J. A. Molina, and D. Bastos.

Financial support from "Comisión Interministerial de Ciencia y Technología," Project No. MAT 99-0662-C03-02, is gratefully acknowledged.

REFERENCES

1. P Bagchi, SM Bimbau. J Colloid Interface Sci 83: 460, 1981.
2. S Oku, A Kondo, K Higashitani. Biotech Bioeng 37: 537, 1991.
3. W Norde, F MacRitchie, G Nowicka, J Lyklema. J Colloid Interface Sci 112: 447, 1986.
4. JS Tan, PA Martic. J Colloid Interface Sci 136: 415, 1990.
5. DR Lu, K Park. J Colloid Interface Sci 144: 271, 1991.
6. MD Bale, SJ Danielson, JL Daiss, KE Goppert, RC Sutton. J Colloid Interface Sci 132: 176, 1989.
7. JWT Lichtenbelt, WJM Heuvelsland, ME Oldenzeel, RLJ Zsom. Colloids Surfaces B 1: 76, 1993.
8. T Basinska, S Slomkowski. J Biomater Sci Polym Edn 2: 1, 1991.
9. M Okubo, Y Yamamoto, M Uno, S Kamei, T Matsumoto. Colloid Polym Sci 265: 1061, 1987.
10. J Jozefowicz, M Jozefowicz. J Biomater Sci Polym Edn 3: 147, 1990.
11. F Betton, A Theretz, A. Elaissari, C. Pichot. Colloids Surfaces B 1: 97, 1993.
12. J Buijs, JWT Lichtenbelt, W. Norde, J Lyklema. Colloids Surfaces B 5: 11, 1995.
13. JM Peula, FJ de las Nieves. Colloids Surfaces A 90: 55, 1994.
14. F Galisteo-González, J Puig, A Martín-Rodríguez, J Serra-Domènech, R. Hidalgo-Álvarez. Colloids Surfaces B 2: 435, 1994.
15. JL Ortega-Vinuesa, R Hidalgo-Álvarez. Biotech Bioeng 47: 633, 1995.
16. JL Ortega-Vinuesa, MJ Gálvez-Ruiz, R Hidalgo-Álvarez. Langmuir 12: 3211, 1996.
17. JL Ortega-Vinuesa, R Hidalgo-Álvarez, FJ de las Nieves, CL Davey, DJ Newman, CP Price. J Colloid Interface Sci 204: 300, 1998.
18. V Elgersma, RLJ Zsom, W Norde, J Lyklema. J Colloid Interface Sci 138: 145, 1990.
19. JM Peula, J Callejas, FJ de las Nieves. In: R West, G Batts, eds. Surface Properties of Biomaterials. London: Butterworth-Heinemann, 1994, pp 61–69.
20. A Martín-Rodríguez, MA Cabrerizo-Vílchez, R Hidalgo-Álvarez. Colloids Surfaces A 92: 113, 1994.
21. W Mientus, E Knippel. J Biomater Sci Polym Edn 7: 401, 1995.
22. J Serra, J Puig, A Martín, F Galisteo, MJ Gálvez, R Hidalgo-Álvarez. Colloid Polym Sci 270: 574, 1992.
23. R Hidalgo-Álvarez, F Galisteo-González. Hetrogen Chem Rev 2: 249, 1995.
24. V Elgersma, RLJ Zsom, W Norde, J Lyklema. Colloids Surfaces 54: 89, 1991.
25. JL Ortega-Vinuesa, R Hidalgo-Álvarez. Colloids Surfaces B 1: 265, 1993.
26. W Norde, J Lyklema. J Colloid Interface Sci 66: 277, 1978.
27. T Suzawa, H Shirahama. Adv Colloid Interface Sci 35: 139, 1991.
28. A Martín, J Puig, F Galisteo, J Serra, R Hidalgo-Álvarez. J Dispers Sci Technol 13: 399, 1992.
29. F Galisteo-González, A Martín-Rodríguez, R Hidalgo-Álvarez. Colloid Polym Sci 272: 252, 1994.

30. J Buijs, JWWT Lichtenbelt, W Norde, J Lyklema. Colloids Surfaces B 5: 11, 1995.
31. H Shirahama, H Ohno, T Suzawa. Colloids Surf. 60: 1, 1991.
32. S Stoll, V Lanet, E Pefferkorn. J Colloid Interface Sci 157: 302, 1993.
33. C-H Suen, H Moratewz. Makromol Chem 186: 225, 1985.
34. JV Staros, RW Wrigth, DM Swingle. Anal Biochem 156: 220, 1986.
35. JL Ortega-Vinuesa, D Bastos-González, R Hidalgo-Álvarez. J Colloid Interface Sci 176: 240, 1995.
36. J Puig, Fernández-Barbero, D Bastos-González, J Serra-Domenech, R Hidalgo-Álvarez. In: R West, G Batts, eds. Surface Properties of Biomaterials. London: Butterworth-Heinemann, 1994, pp 39–48.
37. W Norde. Cells Mater 5: 97, 1995.
38. W Norde, J Lyklema. J Colloid Interface Sci 66: 285, 1978.
39. H Shirahama, K Takeda, T Suzawa. J Colloid Interface Sci 109: 86, 1986.
40. RW O'Brien, LR White. J Chem Soc, Faraday Trans 2 74: 1607, 1978.
41. F Galisteo-González, JA Moleón-Baca, R Hidalgo-Álvarez. J Biomater Sci Polym Edn 4: 631, 1993.
42. SS Dukhin, NM Semenikhin. Kolloidn Zh 31: 36, 1970.
43. V Derjaguin, LD Landau. Act Physicochem URSS 14: 633, 1941.
44. EJW Verwey, JTG Overbeek. Trans Faraday Soc 42B: 117, 1946.
45. EJW Verwey, JTG Overbeek. Theory of the Stability of Lyophobic Colloids. Amsterdam: Elsevier, 1948.
46. R Hidalgo-Álvarez, A Martín, A Fernández, D Bastos, F Martínez, FJ de las Nieves. Adv Colloid Interface Sci 67: 1, 1996.
47. H Tamai, M Hasegawa, T Suzawa. Colloids Surfaces 51: 271, 1990.
48. H Tamai, T Oyanagi, T Suzawa. Colloids Surfaces 57: 115, 1991.
49. JA Molina-Bolivar, JL Ortega-Vinuesa. Langmuir 15: 2644, 1999.
50. JA Molina-Bolivar, F Galisteo-González, R Hidalgo-Álvarez. Colloids Surfaces B 14: 3, 1999.
51. E Donath, V Pastushenko. Bioelectrochem Bioenerg 6: 543, 1979.
52. IS Jones. J Colloid Interface Sci 68: 451, 1979.
53. RW Wunderlich. J Colloid Interface Sci 88: 385, 1982.
54. S Levine, M Levine, KA Sharp, DE Brooks. Biophys J 42: 127, 1983.
55. KA Sharp, DE Brooks. Biophys J 47: 563, 1985.
56. H Ohshima, T Kondo. Colloid & Polymer Sci 264: 1080, 1986.
57. H Ohshima, T Kondo. J Colloid Interface Sci 116: 305, 1987.
58. H Ohshima, T Kondo. J Colloid Interface Sci 130: 281, 1989.
59. H Ohshima, T Kondo. Biophys Chem 39: 191, 1990.
60. H Ohshima, M Nakamura, T Kondo. Colloid Polym Sci 270: 873, 1992.
61. M Nakamura, H Ohshima, T Kondo. J Colloid Interface Sci 149: 241, 1992.
62. D Bastos-González, JL Ortega-Vinuesa, FJ de las Nieves, R Hidalgo-Álvarez. J Colloid Interface Sci 176: 232, 1995.
63. T Peters Jr. In: FW Putnam, ed. The Plasma Proteins. New York: Academic Press, 1975, p 147.
64. JL Ortega-Vinuesa, R Hidalgo-Álvarez. J Non-Equilib Thermodyn 21: 339, 1996.
65. R Hidalgo-Álvarez. Adv Colloid Interface Sci 34: 217, 1991.
66. W Norde. Adv Colloid Interface Sci 25: 267, 1986.
67. W Norde, J Lyklema. J Biomater Sci, Polym Edn 2: 183, 1991.
68. VR Sarma, EW Silverton, DR Davies, WD Terry. J Biol Chem 216: 3753, 1971.

23

Electrophoresis of Biological Cells: Models

ERIC LEE, FONG-YUH YEN, and JYH-PING HSU National Taiwan University, Taipei, Taiwan

I. INTRODUCTION

Electrophoresis, the movement of a charged entity as a response to an applied electric field, is characterized by the electrophoretic mobility [1], μ_E, defined as $\mu_E = U/E$, U being the magnitude of the terminal velocity of the entity and E the strength of the applied electric field. Smoluchowski [2] was able to derive the expression:

$$\mu_E = \frac{\epsilon_r \epsilon_0 \zeta}{\eta} \tag{1}$$

where ζ is the zeta potential, ϵ_r and η are, respectively, the relative permittivity and the viscosity of the liquid phase, and ϵ_0 is the permittivity of a vacuum. Due to its simplicity, this expression has been used widely in various fields. It should be pointed out, however, that the derivation of Eq. (1) is based on several assumptions, which include: (1) the entity is rigid and nonconducting, (2) its linear size is larger than the Debye length, (3) the surrounding fluid is unbounded, and (4) the surface properties are uniform over the entity surface. The analysis of Smoluchowski was extended to a more general case to take into account effects such as thick double layer, double-layer polarization and relaxation, and the presence of a boundary by several workers [3–16[. A review of the boundary effects and particle interactions in electrophoresis was provided by Chen and Keh [17]. A pseudospectral method was applied recently by Lee et al. [14–16] to the case of a sphere in a spherical cavity and to a concentrated spherical dispersion. In these studies, the effects of double-layer polarization, the presence of a rigid boundary, and the interaction of adjacent double layers were considered for the case of an arbitrary electrical potential and double-layer thickness.

Electrophoresis is one of the powerful analytical tools in biochemistry and biomedical engineering. It is often used to characterize the surface properties of biological cells. For example, the peripheral zone of human erythrocyte contains a glycoprotein layer about 15 nm thick that possesses some ionogenic groups and forms the outer boundary of the lipid layer. Since the dissociation of these groups leads to a charged layer, the surface structure of the erythrocyte can be estimate by electrophoresis, which has the merit that the destruction of cell structure can be avoided [18, 19]. Hashimoto et al. [20] compared the electrophoretic mobility of erythrocytes of healthy donors and those of patients. It was found that the mean mobility of the erythrocyte of a patient suffering systemic lupus erythematosus is smaller than that of a normal person. Makino et al. [21]

analyzed the electrophoretic behavior of malignant lymphosarcoma cells, Raw 117-P and Raw 117-H10 (variant of Raw 117-P). The results obtained justified the fact that different cells have different frictional resistance to surrounding fluid, and therefore, different electrophoretic velocities. These reveal the significance of electrophoresis and its potential applications in cell biology and biomedical engineering. For this reason, relevant phenomena have drawn the attention of researchers in various areas.

Although reported results for cell electrophoresis are ample in the literature, most of them, however, applied the classical theory of Smoluchowski [2]. As pointed out above, its derivation is based upon rigid entities, and can be unrealistic for biocolloids. For example, the cell surface often contains protein molecules' appendage and dissociable functional groups. The charge density on the cell surface is found to be a function of the degree of dissociation of the latter [22–27]. In addition, the cell surface is nonrigid, and can be penetrable to ions. Therefore, the classical theory of electrophoresis needs to be modified accordingly to reflect more closely the nature of the cell surface. Several attempts have been made to take the nonrigid property of a biocolloid into account [28–31]. Here, it is simulated by an entity comprising a rigid, uncharged core and a charged surface layer, which is ion penetrable and carries dissociable functional groups. It was concluded that the surface layer plays an important role in the determination of the electrophoretic behavior of an entity. In a study of the electrical interaction between two ion-penetrable charged membranes, Ohshima and coworkers [32–35] pointed out that the dissociation of the functional groups in the membrane leads to a fixed charge distribution. In this case, the classical Poisson–Boltzmann equation describing the electrical field needs to be modified to take this factor into account. In their studies the flow field in the membrane was assumed to be governed by the Debye–Bueche theory [36]. It should be pointed out that this theory becomes inapplicable if the Debye length is comparable or larger than the linear size of a particle. It is also inappropriate if the ionic strength is high [28]. Hsu and coworkers [37–41] and Ohshima and coworkers [42–46] have conducted a series of studies on the electrophoretic behavior of biocolloids. Similar to the treatment of Smoluchowski [2], the Debye length is assumed to be much smaller than the local curvature of the surface of a particle, i.e., the particle surface can be assumed planar. In their treatment, a particle comprises an inner rigid, uncharged region and an ion-penetrable surface layer, which can be a membrane or polymer chain. The governing equations for the liquid phase are the same as those for the case of rigid particles, and a modified set of governing equations are adopted for the surface layer. The results obtained provide the necessary information for the elaboration of the experiment data. In a recent study, Chen and Ye [47] derived Faxen's law for the case of a rigid sphere covered by a permeable layer of arbitrary thickness. Saville [48] considered the case of a rigid sphere covered by a polymer layer in which the hydrodynamic resistance of the latter is a function of position. Rasmusson et al. [49] discussed the electrophoresis of porous particles.

A thorough review of the literature reveals that a general theory for cell electrophoresis has not been established. Most of the available results are limited to low electrical potential, simple geometry, and thin Debye length. This is mainly because the governing equation for the electrical field and that for the flow field are both nonlinear and coupled, and solving these equations analytically is nontrivial, if not impossible. In this case, choosing a reliable numerical approach seems to be inevitable in order to obtain an overview of the problem under consideration. Also, the effects of the presence of a rigid boundary, the concentration of particles, and double-layer polarization and relaxation on the electrophoretic behavior of a particle are of practical significance, and should be considered in figure studies.

II. ISOLATED ENTITY

We begin by considering the simplest case of the electrophoresis of an isolated entity in an infinite fluid. In this case, since the presence of other objects such as a rigid boundary and nearby entities needs not to be considered, the problem under consideration can be simplified drastically. The behavior of a dilute dispersion, which neglects boundary effects, for instance, can be simulated by that of an isolated entity. Intuitively, the electrophoretic behavior of the entity is a function of its surface condition, the physical properties of the surrounding fluid, and the applied electrical field.

A. Electrokinetic Equations

The description of the electrokinetic phenomena in a colloidal dispersion, including electrophoresis, conductivity, electro-osmosis, and sedimentation, involves solving simultaneously a set of partial differential equations, called electrokinetic equations. These include the Poisson equation describing the electric field, the Navier–Stokes equation describing the flow field, and the ion conservation equation describing the concentration field. The equations are coupled and nonlinear, and in practice, analytical results are derived under drastic assumptions such as simple geometry, very thin or very thick double layers, and low surface potentials [2–5, 10–13]. In general, a numerical scheme is necessary [7, 14–16].

1. Poisson Equation
Under some restrictions, the spatial variation in electrical potential is governed by the Poisson equation, which can be derived by employing Gauss's law. We have

$$\nabla^2 \phi = -\frac{\rho_e}{\epsilon} \tag{2}$$

where ϕ, ρ_e, and ϵ are, respectively, the electrical potential, the space charge density, and the permeability. For an electrolyte solution containing N charged species:

$$\rho_e = \sum_{j=1}^{N} z_j e n_j \tag{3}$$

where z_j and n_j are, respectively, the valence and the number concentration of the jth ionic species, and e is the elementary charge. For the case of an ion-penetrable surface layer, the space charge comprises the ionic species in the liquid phase and the fixed charge in the surface layer, ρ_{fix}, i.e.,

$$\rho_e = \sum_{j=1}^{N} z_j e n_j + \rho_{fix} \tag{4}$$

2. Navier–Stokes Equation
The flow field of the problem under consideration can be described by the continuity equation:

$$\nabla \cdot \mathbf{v} = 0 \tag{5}$$

and the Navier–Stokes equation:

$$\rho_f \frac{\partial \mathbf{v}}{\partial t} = \eta \nabla^2 \mathbf{v} - \nabla p - \rho_e \nabla \phi \tag{6}$$

In these expressions, t is time, and $\mathbf{v}$, ρ_f, and η are, respectively, the velocity, the density, and the viscosity of the liquid phase, and p is the pressure. Here, we assume that $\mathbf{v}$ is in the creeping flow regime, that is, the Reynolds number is much less than unity, and the inertial term can be neglected. The last term on the right-hand side of Eq. (6) denotes the body force exerted on the liquid phase due to the presence of the electrical force. The flow field in the surface layer can be expressed on the basis of Debye–Bueche theory, in which a body force term, $-\gamma v$, is added to the right-hand side of Eq. (6) to reflect the resistance arises from the presence of the surface layer. We have

$$\rho_f \frac{\partial v}{\partial t} = \eta \nabla^2 v - \nabla p - \rho_e \nabla \phi - \gamma v \tag{7}$$

where γ is a frictional coefficient. Here, the surface layer is simulated by attached polymer segments [50], and each polymer segment is viewed as the center of resistance. If a polymer segment is treated as a sphere of radius r_p, then according to Stoke's law, the resultant friction force is $6\pi\eta r_p v$. If there are N_p polymer segments, then

$$\gamma = 6\pi\eta r_p N_p \tag{8}$$

If $\gamma \to \infty$, the surface layer becomes a rigid phase, and the radius of the particle needs to be adjusted accordingly. On the other hand, if $\gamma \to 0$, the surface layer does not exist.

3. Ion Conservation Equation

Suppose that the transport of ions in the liquid phase can be described by the Nernst–Planck equation [1]:

$$\frac{\partial n_j}{\partial t} = -\nabla \cdot f_j \tag{9}$$

$$f_j = -D_j \left[\nabla n_j + \frac{z_j e n_j}{k_B T} \nabla \phi \right] + n_j v \tag{10}$$

In these expressions f_j and D_j are, respectively, the local ion flux and the diffusivity of ionic species j, and k_B and T are the Boltzmann constant and the absolute temperature, respectively. The first term on the right-hand side of Eq. (10) denotes the contribution of Brownian motion, the second term arises from the local electrical field, and the third term arises from the flow of fluid. At steady state, the left-hand side of Eq. (9) vanishes. In this case, substituting Eq. (10) into Eq. (9) leads to the ion-conservation equation.

4. Double-layer Polarization

When a charged entity is placed in an electrolyte solution, the electrostatic interaction between the entity and nearby ions yields an ionic cloud surrounding the particle. If an electrical field is applied, the relative motion between the entity and the ions in the cloud leads to a distortion in the cloud, the so-called double-layer polarization. Let us consider first the interaction between a spherical particle and the surrounding ions in a constant electric field [7]. At a point far away from the particle, counterions move at constant velocity, which can be evaluated if the applied electrical field is known. The presence of the charged particle establishes an electrical field near its surface. The counterions that

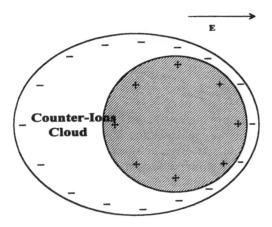

FIG. 1 Schematic representation of anion cloud surrounding a positively charged particle in an electric field **E**.

move toward the surface are accelerated, and those that move toward the bulk liquid phase are decelerated. This leads to the distribution of the ionic cloud of counterions shown in Fig. 1. The effect of the ionic cloud of coions on the electrophoretic behavior of the particle is more complex than that of the counterions. However, since the concentration of coions is much less than that of counterions, their influence is less significant than that of the counterions.

The effect of double-layer polarization on the electrophoretic behavior of a particle can be elaborated by referring to Fig. 1. The ionic cloud formed by the presence of counterions leads to an electrical field, which is in the opposite direction to that of the applied electrical field, and the electric force experienced by the particle is reduced. The movement of counterions also yields a flow of fluid, and the flow field near the particle is distorted. This increases the viscous force exerted on the particle surface. Therefore, the electrophoretic velocity of a particle in an electrolyte solution is slower than that in an electrolyte-free liquid.

At equilibrium, both f_j and $\mathbf{v}$ vanish, and integrating Eq. (10) leads to the Boltzmann distribution:

$$n_j = n_{j0} \exp\left(-\frac{z_j e\phi}{k_B T}\right) \tag{11}$$

Combining this expression with Eqs (2) and (3) gives the so-called Poisson–Boltzmann equation. The derivation of Eq. (11) suggests that it is applicable if the disturbance of the flow field arising from the applied electrical field is significant. Since the concentration field and the flow field are coupled, this implies that the degree of double-layer polarization is negligible. In general, double-layer polarization is insignificant if a double layer is thin, the surface potential is low, or the applied electrical field is weak.

O'Brien and White [7] suggested using the following expression of ionic concentration:

$$n_j = n_{j0} \exp\left[-\frac{z_j e(\phi + g_j)}{k_B T}\right] \tag{12}$$

Note that the inclusion of function g_j is arbitrary. Substituting Eq. (12) into Eqs (10) and (3) yields the governing equations for the case where the effect of double-layer polarization is significant.

5. Solution Procedure

As suggested by Lee et al. [14–16], the pseudospectral method based on Chebyshev polynomials [51] is readily applicable to the problem under consideration. This method has the merits of having a high order of convergence, and convergent properties that are independent of the boundary conditions. Since the computational domain is two-dimensional, the pseudospectral method is applied in both the r and θ domains. If $f_{NM}(r, \theta)$ is an unknown function, its Nth-order $\times$ Mth-order approximation can be expressed as

$$f_{NM}(r, \theta) = \sum_{i=0}^{N} \sum_{j=0}^{M} f_{NM}(r_i, \theta_j)g_i(r)g_j(\theta) \tag{13}$$

where $f_{NM}(r_i, \theta_j)$ is the value of f_{NM} at the kth, $k = [(N - 1)i + j]$, collocation point. The interpolation polynomials $g_i(r)$ and $g_j(\theta)$ depend on the collocation points, which are determined by mapping the computational domain on to the square $[-1, 1] \times [-1, 1]$ by

$$r = \frac{b - a}{2} \times y + \frac{b + a}{2}$$

$$\theta = \frac{\pi}{2} \times (x + 1) \tag{14}$$

In these expressions a and b are the radius of the outer and inner spheres, respectively. The $N + 1$ interpolation points in the interval $[-1, 1]$ are chosen to be the extreme values of an Nth-order Chebyshev polynomial $T_N(y)$:

$$y_j = \cos\left(\frac{\pi j}{N}\right), \quad j = 0, 1, \ldots, N \tag{15}$$

The corresponding interpolation polynomial $g_i(y)$ is

$$g_j(y) = \frac{(-1)^{j+1}(1 - y^2)dT_N(y)/dy}{c_j N^2 (y - y_j)}, \quad j = 0, 1, \ldots, N \tag{16}$$

where c_j is defined by

$$c_j = \begin{cases} 2, & j = 0, N \\ 1, & 1 \leq j \leq N - 1 \end{cases} \tag{17}$$

Both the partial derivative and the integration of $f_{NM}(r, \theta)$ can be estimated on the basis of Eq. (13). The axisymmetric nature of the problem under consideration suggests that only half of the physical domain needs to be considered in the numerical calculations. Therefore, we conduct calculations in half of the domain to obtain a high accuracy in stream function solutions, although the variables in our system are either symmetric or antisymmetric about the plane $\theta = \pi/2$ when the equations are linearized. The corresponding nonlinear problem is then solved by a Newton–Raphson iteration scheme. Double precision is used throughout the computation. Grid independence is checked to ensure that the mesh used is fine enough.

B. Surface Conditions

The evaluation of the electrophoretic mobility of a particle involves solving the spatial variation of the equilibrium electrical potential due to the presence of the charged particle. This potential is a function of the charged conditions of the surface of the particle. We consider several typical types of surface conditions in practice.

1. Constant Potential/Charge Surface

For a rigid entity, we assume that it is impenetrable to ions, the surface charges are homogeneously distributed, and the surface potential can be described by the Gouy–Chapman model. Two types of surface condition are often adopted, namely, constant surface potential and constant surface charge. The corresponding boundary conditions at the particle surface are, respectively,

$$\phi_e = \zeta \tag{18}$$

and

$$-\epsilon \nabla \phi_e \cdot \hat{n} = \sigma_0 \tag{19}$$

In these circumstances, ϕ_e is the equilibrium electrical potential, ζ is the zeta potential, σ_0 is the surface charge density, a constant, and $\hat{n}$ is the unit normal vector of the entity surface.

2. Charge-regulated Surface

The surface of biocolloids contains dissociable functional groups, and their degree of dissociation needs to be determined by chemical equilibrium [1]. In this case, the surface condition is a function of solution pH, dissociation constant, and the concentration of functional groups.

(a) One-site Model. Suppose that the dissociation reaction below occurs on the particle surface [52]:

$$AH \Leftrightarrow A^- + H^+ \tag{20}$$

Let K_a be the dissociation constant; then

$$K_a = \frac{[A^-][H^+]_s}{[AH]} \tag{21}$$

where a term in square bracket denotes concentration, and subscript "s" represents surface property. Assuming Boltzmann distribution for $[H^+]$, we have

$$[H^+]_s = [H^+]_b \exp\left(\frac{-e\phi_e}{k_B T}\right) \tag{22}$$

The total number of surface sites can be expressed as

$$N_s = [A^-] + [AH] \tag{23}$$

Equations (21)–(23) lead to

$$[A^-] = \frac{N_s}{1 + \frac{[H^+]_b}{K_a} \exp\left(-\frac{e\phi_e}{k_B T}\right)} \tag{24}$$

The surface charge, σ_0 is given by

$$\sigma_0 = -e[\text{A}^-] \tag{25}$$

Substituting Eq. (24) into this expression yields:

$$\sigma_0 = -\frac{eN_s}{1 + \dfrac{[\text{H}^+]_b}{K_a}\exp\left(-\dfrac{e\phi}{k_B T}\right)} \tag{26}$$

Suppose that the relative permittivity of the solid phase is much smaller than that of the liquid phase; then

$$\sigma = -\epsilon \nabla \phi_e \cdot \hat{n} \tag{27}$$

Combining Eqs (26) and (27) yields boundary condition at the particle surface:

$$-\epsilon \nabla \phi_e \cdot \hat{n} = -\frac{eN_s}{1 + \dfrac{[\text{H}^+]_b}{K_a}\exp\left(-\dfrac{e\phi}{k_B T}\right)} \tag{28}$$

(b)   Two-site Model.   Suppose that the dissociation reaction below occurs on the particle surface, which is an amphoteric one [52]:

$$\text{AH}_2^+ \Leftrightarrow \text{AH} + \text{H}^+ \tag{29}$$

$$\text{AH} \Leftrightarrow \text{A}^- + \text{H}^+ \tag{30}$$

Let K^+ and K^- be, respectively, the corresponding dissociation constants of the above two reactions. Following the same approach mentioned in the one-site model, we obtain:

$$\sigma_0 = eN_s \frac{[\text{AH}_2^+] - [\text{A}^-]}{[\text{AH}] + [\text{AH}_2^+] + [\text{A}^-]} = eN_s \frac{\delta \sinh[e(\phi_N - \phi_e)/k_B T]}{1 + \delta \cosh[e(\phi_N - \phi_e)/k_B T]} \tag{31}$$

where

$$\delta = 2\left(K^+/K^-\right)^{1/2} \tag{31a}$$

$$\phi_N = 2.303(k_B T/e)\left[\frac{1}{2}\left(pK^+ - pK^-\right) - pH\right] \tag{31b}$$

Therefore, the boundary condition at the particle surface becomes

$$-\epsilon \nabla \phi_e \cdot \hat{n} = eN_s \frac{\delta \sinh[e(\phi_N - \phi_e)/k_B T]}{1 + \delta \cosh[e(\phi_N - \phi_e)/k_B T]} \tag{32}$$

3.   Ion-penetrable Surface Layer

For biocolloids, charges carried by a particle may distribute in an ion-penetratable surface layer rather than over a rigid surface. In this case, the fixed charges are distributed in a finite volume in space, and the governing equations for the case of rigid charged surfaces need to be modified accordingly. Apparently, the idea of zeta potential becomes meaningless, and the distribution of fixed charge is of main concern in the calculation of the spatial variation of electrical potential. Two classes of fixed charge distribution are often considered, namely, uniform and nonuniform fixed charge distributions [37–46, 50]. The former can be expressed as

$$\rho_{\text{fix}} = \text{constant in the surface layer} \tag{33}$$

or

$$\rho_{\text{fix}} = e[X] \text{ in the surface layer} \tag{34}$$

where $[X]$ denotes the concentration of functional groups, which is assumed to distribute uniformly in the surface layer. The degree of dissociation of X is governed by the chemical equilibrium. In the latter, a function is assumed to describe the spatial variation of fixed charge. Typical examples includes linear and exponential functions. The condition that the degree of dissociation of the functional groups is a function of the conditions of the liquid phase may also lead to a nonlinear fixed charge distribution.

III. CONCENTRATED DISPERSION

In the previous section, the behavior of an isolated entity in an infinite solution was discussed. This is realistic for the case of a dilute dispersion. If the concentration of the dispersed phase is appreciable, the interaction between adjacent entities becomes significant, and a more rigorous treatment is necessary. For uniformly dispersed spherical particles, the so-called unit cell model is applicable [53]. In this model, the behavior of a dispersion is simulated by a particle surrounded by a concentric spherical liquid shell, and the concentration of the dispersion is measured by the relative magnitude of the particle and the liquid shell. Two types of boundary condition at the outer surface of the liquid shell, the virtual surface, have been proposed for the resolution of the governing equations for flow field. The first is suggested by Kuwabara [54], which assumes zero vorticity at the virtual surface, and the second is used by Happel [55], which assumes zero shear at the virtual surface. In the mathematical expression, we have

$$\nabla \times v = 0 \tag{35}$$

for the former, and

$$\left(\sigma^{h} \cdot \hat{n}\right) \times \hat{n} = 0 \tag{36}$$

for the latter. The first boundary condition is preferred since the result of Smoluchowski [2] can be recovered as the concentration of a dispersion becomes infinitely dilute [53]. On the other hand, the second boundary condition does not lead to this result.

IV. EXAMPLES

A. Entity Covered by an Ion-Penetrable Membrane

Let us consider the case shown in Fig. 2 where a planar particle, comprising a rigid, uncharged core and an ion-penetrable charged membrane of thickness d, is immersed in a symmetric electrolyte solution having a bulk a concentration n_0. The membrane contains a uniformly distributed fixed charge of density N_{fix}. An electric field $\mathbf{E}$ parallel to the surface of the particle is applied. Without loss of generality, we assume that the fixed charge is negative, and the particle moves in the opposite direction of $\mathbf{E}$.

Assuming Boltzmann distribution for ions, then the spatial variation in electrical potential in the membrane phase can described by

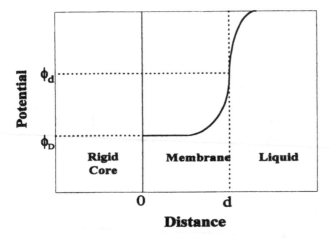

FIG. 2 Biocolloid simulated by a rigid core covered by an ion-penetrable, charged membrane layer of thickness d; ϕ_d and ϕ_D denote the electrical potential at the membrane–liquid interface and the Donnan potential.

$$\epsilon_1 \frac{d^2\phi}{dx^2} = 2n_0 z_1 e \, \sinh\left(\frac{z_1 e\phi}{kT}\right) + eN_{\text{fix}}, \quad 0 < x < d \tag{37}$$

Similarly, that in the liquid phase can be expressed as

$$\epsilon_2 \frac{d^2\phi}{dx^2} = 2n_0 z_1 e \, \sin\left(\frac{z_1 e\phi}{kT}\right), \quad d < x < \infty \tag{38}$$

The boundary conditions associated with these equations are assumed as

$$\phi \to 0 \text{ and } (d\phi/dx) \to 0 \text{ as } x \to \infty \tag{39}$$

$$\phi|_{x=d^-} = \phi|_{x=d^+} \tag{40}$$

$$\epsilon_1 \left(\frac{d\phi}{dx}\right)\bigg|_{x=d^-} = \epsilon_2 \left(\frac{d\phi}{dx}\right)\bigg|_{x=d^+} \tag{41}$$

$$\frac{d^2\phi}{dx^2} = \frac{d\phi}{dx} = 0 \text{ and } \phi = \phi_D \text{ at } x = 0 \tag{42}$$

where ϕ_D is the Donnan potential. Here, we assume that the membrane is sufficiently thick so that the Donnan potential is reached at the core–membrane interface. Equation (40) suggests that the electrical potential is continuous at the liquid–membrane interface. Equation (41) assumes that no net charges accumulate at the liquid–membrane interface.

The flow fields can be described by the Navier–Stokes equation (44). We have, in the membrane phase:

$$\eta \frac{d^2 v}{dx^2} - \gamma v + \rho_e(x)E = 0, \quad 0 < x < d \tag{43}$$

and in the liquid phase:

$$\eta \frac{d^2 v}{dx^2} + \rho_e(x)E = 0, \quad d < x < \infty \tag{44}$$

where E is the strength of electric field. The boundary conditions associated with Eqs. (43) and (44) are assumed to be

$$v \rightarrow -U \text{ and } (dv/dx) \rightarrow 0 \text{ as } x \rightarrow \infty \tag{45}$$

$$v|_{x=d^-} = v|_{x=d^+} \tag{46}$$

$$\left(\frac{dv}{dx}\right)\bigg|_{x=d^-} = \left(\frac{dv}{dx}\right)\bigg|_{x=d^+} \tag{47}$$

$$v|_{x=0} = 0 \tag{48}$$

Equation (45) implies that the magnitude of liquid velocity far away from the particle surface is U, and its direction is parallel to the applied electric field. Equations (46) and (47) state that both the liquid velocity and the tangential component of the stress tensor are continuous at the liquid–membrane interface, and Eq. (48) represents the no-slipping condition.

Solving the governing equations for the electric fields and those for the flow field simultaneously subject to the boundary conditions assumed and employing the definition $\mu = U/E$, the electrophoretic mobility can be evaluated. Detailed derivations can be found in the literature [37–41].

B. Entity in a Cavity

This problem has been analyzed by Lee et al. [15, 16] where the motion of a rigid non-conducting spherical particle of radius a in a concentric spherical cavity of radius b is considered. As illustrated in Fig. 3, a uniform electric field $\mathbf{E}$ is applied in the z direction. Following the treatment of O'Brien and White [7], the total electrical potential ϕ is expressed as the sum of the electrical potential that would exist in the absence of the applied electric field ϕ_1 (or the equilibrium potential) and the electrical potential arising from the applied electric field ϕ_2. The effect of double-layer polarization is taken into account by defining the function g_j. We have

$$\nabla^2 \phi_1 = -\frac{\rho_1}{\epsilon} - \sum_{j=1}^{N} \frac{z_j e n_{j0}}{\epsilon} \exp\left(-\frac{z_j e \phi_1}{k_B T}\right) \tag{49}$$

$$n_j = n_{j0} \exp\left(-\frac{z_j e (\phi_1 + \phi_2 + g_j)}{k_B T}\right) \tag{50}$$

$$\nabla^2 \phi_2 = -\frac{\rho_2}{\epsilon} = -\left(\begin{array}{c} \sum_{j=1}^{N} \frac{z_j e n_{j0}}{\epsilon} \exp\left(-\frac{z_j e (\phi_1 + \phi_2 + g_j)}{k_B T}\right) \\ -\sum_{j=1}^{N} \frac{z_j e n_{j0}}{\epsilon} \exp\left(-\frac{z_j e \phi_1}{k_B T}\right) \end{array}\right) \tag{51}$$

$$\phi = \phi_1 + \phi_2 \tag{52}$$

$$\rho = \rho_1 + \rho_2 \tag{53}$$

$$\nabla^2 g_j - \frac{z_j e}{k_B T} \nabla \phi_1 \cdot \nabla g_j = \frac{1}{D_j} u \cdot \nabla \phi + \frac{1}{D_j} u \cdot \nabla g_j + \frac{z_j e}{k_B T} \nabla \phi_2 \cdot \nabla g_j + \frac{z_j e}{k_B T} \nabla g_j \cdot \nabla g_j \tag{54}$$

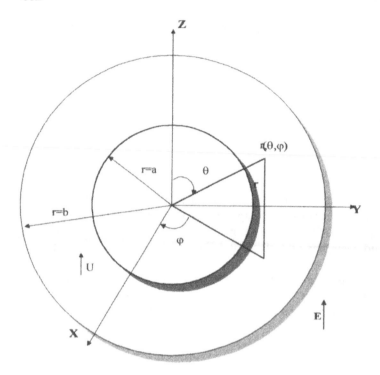

FIG. 3 Sphere of radius a in a spherical coordinates. E and U are, respectively, the applied electric field and the velocity of the particle.

In these expressions n_{j0} is the bulk concentration of ionic species j. To reduce the number of variables in the hydrodynamic equations the stream function formulation is adopted. It can be shown that the stream function ψ satisfies

$$E^4\psi = -\frac{1}{\eta}\sin\theta \cdot \nabla \times [\rho\nabla(\phi_1 + \phi_2)] \tag{55}$$

where $E^4 = E^2E^2$ with

$$E^2 = \frac{\partial}{\partial r^2} + \frac{\sin\theta}{r^2}\frac{\partial}{\partial\theta}\left(\frac{1}{\sin\theta}\frac{\partial}{\partial\theta}\right) \tag{55a}$$

The r and the θ components of the particle velocity, v_r and v_θ, can be expressed respectively, as

$$v_r = -\frac{1}{r^2\sin\theta}\frac{\partial\psi}{\partial\theta} \tag{55b}$$

and

$$v_\theta = \frac{1}{r\sin\theta}\frac{\partial\psi}{\partial r} \tag{55c}$$

Equations (49)–(55) need to be solved simultaneously.

For illustration, we assume that both the particle and the cavity are nonconducting and their surfaces remain at a constant electrical potential characterized by the corresponding zeta potentials, ζ_a and ζ_b, respectively. In this case, the boundary conditions for ϕ_1 are

$$\phi_1 = \zeta_a, \quad r = a \tag{56a}$$

$$\phi_1 = \zeta_b, \quad r = b \tag{56b}$$

The boundary conditions for ϕ_2 are assumed to be

$$(\partial\phi_2/\partial r) = 0, \quad r = a \tag{57a}$$

$$(\partial\phi_2/\partial r) = -E_z \cos\theta, \quad r = b \tag{57b}$$

Suppose that the concentration of ions reaches the equilibrium value at the cavity surface and that the surface of the particle is nonpenetrable to ions. Therefore, the boundary conditions for g_j are

$$g_j = -\phi_2, \quad r = b \tag{58a}$$

$$(\partial g_j/\partial r) = 0, \quad r = a \tag{58b}$$

Since the magnitude of the velocity of the particle is U, we have the following boundary conditions:

$$v_r = U\cos\theta \text{ and } v_\theta = -U\sin\theta, \quad r = a \tag{59a}$$

$$v_r = 0 \text{ and } v_\theta = 0, \quad r = b \tag{59b}$$

The corresponding boundary conditions for the stream function ψ are

$$\psi = -\frac{1}{2}Ur^2\sin\theta \text{ and } \frac{\partial\psi}{\partial r} = -Ur\sin^2\theta, \quad r = a \tag{60a}$$

$$\psi = \frac{\partial\psi}{\partial r} = 0, \quad r = b \tag{60b}$$

The symmetric nature of the problem under consideration suggests the following conditions:

$$\frac{\partial\phi_1}{\partial\theta} = \frac{\partial\phi_2}{\partial\theta} = \frac{\partial g_1}{\partial\theta} = \frac{\partial g_2}{\partial\theta} = \psi = \frac{\partial\psi}{\partial\theta} = 0, \quad \theta = 0 \text{ and } \theta = \pi \tag{61}$$

For a simpler mathematical treatment, the governing equations and the corresponding boundary conditions are rewritten in scaled forms. The zeta potential ζ_a, (or ζ_b for an uncharged particle), the radius of the particle a, and the bulk density of cations, n_{10}, are selected as the scaling quantities. For $z_1 : z_2$ electrolyte the condition of electroneutrality in the bulk liquid phase requires that $n_{20} = n_{10}/\alpha$, where $\alpha = (-z_2/z_1 n_{10})$ and n_{20} are, respectively, the bulk concentrations of cation and anion. The reciprocal Debye length κ can be expressed as

$$\kappa = \left(\frac{\epsilon k_B T}{\sum n_{j0}(ez_j)^2}\right)^{-1/2} \tag{62}$$

Since $n_{10}z_1 = (\kappa a)^2\epsilon k_B T/(1+\alpha)e^2 a^2 z_1$, $\phi_1 = \phi_1^*\zeta_a$ (or $\phi_1^*\zeta_b$), $n_j = n_j^* n_{j0}$, and $\nabla^* = a\nabla$, the scaled form of Eq. (49) is

$$\nabla^{*2}\phi_1^* = -\frac{1}{(1+\alpha)}\frac{(\kappa a)^2}{\phi_r} - (n_1^* - n_2^*) \tag{63}$$

where $n_1^* = \exp[-\phi_r(\phi_1^* + \phi_2^* + g_1^*)]$ and $n_2^* = \exp[\alpha\phi_r(\phi_1^* + \phi_2^* + g_2^*)]$. The corresponding scaled boundary conditions become

$$\phi_1^* = 1 \text{ at } r^* = 1 \tag{64a}$$

$$\phi_1^* = \zeta_b/\zeta_a, \quad r^* = 1/\lambda \tag{64b}$$

or for an uncharged particle:

$$\phi_1^* = 0, \quad r^* = 1 \tag{65a}$$

$$\phi^* = 1, \quad r^* = 1/\lambda \tag{65b}$$

In these expressions, $\lambda = a/b$ and $\phi_r = \zeta_a z_1 e/k_B T$ (or $\zeta_b z_1 e/k_B T$). Note that the Debye–Hückel approximation is applicable if $\phi_r \ll 1$. Similarly, the scaled form of Eq. (50) is

$$\nabla^{*2}\phi_2^* = -\frac{1}{(1+\alpha)} \frac{(\kappa a)^2}{\phi_r} \{(n_1^* - n_2^*) - [\exp(-\phi_r\phi_1^*) - \exp(\alpha\phi_r\phi_1^*)]\} \tag{66}$$

The corresponding scaled boundary conditions are

$$\frac{\partial\phi_2^*}{\partial r} = 0, \quad r^* = 1 \tag{67a}$$

$$\frac{\partial\phi_2^*}{\partial r} = -E_z^* \cos\theta, \quad r^* = 1/\lambda \tag{67b}$$

where $E_z^* = E_z a/\zeta_a$. The scaled form of Eq. (54) for $j = 1$ is

$$\nabla^{*2}g_1^* - \phi_r\nabla^*\phi_1^* \cdot \nabla^*g_1^*$$
$$= Pe_1 u^* \cdot \nabla^*\phi_1^* + Pe_1 u^* \cdot \nabla^*\phi_2^* + Pe_1 u^* \cdot \nabla^*g_1^* + \phi_r\nabla^*g_1^* + \phi_r\nabla^*g_1^* \cdot \nabla^*g_1^* \tag{68}$$

The corresponding scaled boundary conditions are

$$\frac{\partial g_1^*}{\partial r^*} = 0, \quad r^* = 1 \tag{69a}$$

$$g_1^* = -\phi_2^*, \quad r^* = 1/\lambda \tag{69b}$$

Similarly, the scale form of Eq. (54) for $j = 2$ is

$$\nabla^{*2}g_2^* + \alpha\phi_r\nabla^*\phi_1^* \cdot \nabla^*g_2^*$$
$$= Pe_2 u^* \cdot \nabla^*\phi_1^* + Pe_2 u^* \cdot \nabla^*\phi_2^* + Pe_2 u^* \cdot \nabla^*g_2^* - \alpha\phi_r\nabla^*\phi_2^* \cdot \nabla^*g_2^* \tag{70}$$
$$- \alpha\phi_r\nabla^*g_2^* \cdot \nabla^*g_2^*$$

The corresponding scaled boundary conditions are

$$\frac{\partial g_2^*}{\partial r^*} = 0, \quad r^* = 1 \tag{71a}$$

$$g_2^* = -\phi_2^*, \quad r^* = 1/\lambda \tag{71b}$$

where $Pe_j = U_E a/D_j$ are the electric Peclet numbers for ionic species j, $U_E = \epsilon\zeta_a^2/\eta a$ being the characteristic velocity. Note that U_E is the electrophoretic velocity based on Smoluchowski's theory if an electric field of magnitude ζ_a/a is applied. The electrostatic force exerted on the particle can be calculated by $F_{Ez} = \int\int_S \sigma(-\nabla\phi)dA$, where σ is the surface charge density and is related to ϕ_1 by Gauss's theorem. In spherical co-ordinates we have

$$F_{Ez} = 2\pi\epsilon\zeta_a^2 \int_0^\pi \left(\frac{\partial\phi_1^*}{\partial r^*}\right)_{r^*=1} \left(\frac{\partial(\phi_1^* + \phi_2^*)}{\partial r^*}\cos\theta - \frac{1}{r^*}\frac{\partial(\phi_1^* + \phi_2^*)}{\partial\theta}\sin\theta\right)_{r^*=1} r^{*2}\sin\theta\, d\theta \quad (72)$$

For convenience, Eq. (69) is rewritten as

$$F_{Ez} = 2\pi\zeta_a^2 K^* K_E \quad (73)$$

where $E^* K_E$ is the definite integral in Eq. (72).

We define $\psi/U_E a^2$ as the scaled stream function. The space charge density can be expressed by $\rho = \rho_0\rho^*$ with

$$\rho_0 = \epsilon\kappa^2 \frac{\zeta_a}{(1+\alpha)\phi_r} \quad (74)$$

$$\rho^* = n_1^* - n_2^* \quad (75)$$

It can be shown that the scaled form of Eq. [55] is

$$E^{*4}\psi^* = -\frac{(\kappa a)^2}{(1+\alpha)}\left[\left(\frac{\partial g_1^*}{\partial r^*}n_1^* + \frac{\partial g_2^*}{\partial r^*}\alpha n_2^*\right)\frac{\partial\phi^*}{\partial\theta} - \left(\frac{\partial g_1^*}{\partial\theta}n_1^* + \frac{\partial g_2^*}{\partial\theta}\alpha n_2^*\right)\frac{\partial\phi^*}{\partial r^*}\right]\sin\theta \quad (76)$$

The corresponding scaled boundary conditions are

$$\psi^* = -\frac{1}{2}U^* r^{*2}\sin\theta, \quad \frac{\partial\psi^*}{\partial r^*} = -U^* r^*\sin^2\theta, \quad r^* = 1 \quad (76a)$$

$$\psi^* = \frac{\partial\psi^*}{\partial r^*} = 0, \quad r = 1/\lambda \quad (76b)$$

Where $U^* = U/U_E$ is the magnitude of the scaled terminal velocity of the particle.

According to Happel and Brenner [56], the drag force acting on the particle surface in the z direction can be evaluated by

$$F_{Dz} = \eta\pi \int_0^\pi \left(r^4\sin^3\theta\frac{\partial}{\partial r}\frac{E^2\psi}{r^2\sin^2\theta}\right)_{r=a} d\theta - \pi\int_0^\pi\left(r^2\sin^2\theta\rho\frac{\partial\phi}{\partial\theta}\right)_{r=a} d\theta \quad (77)$$

In terms of scaled variables, we have

$$F_{Dz} = \pi\epsilon\zeta_a^2 \int_0^\pi \left(r^{*4}\sin^3\theta\frac{\partial}{\partial r^*}\frac{E^{*2}\psi^*}{r^{*2}\sin^2\theta}\right)_{r^*=1} d\theta$$

$$- \pi\epsilon\zeta_a^2\frac{(\kappa a)^2}{(1+\alpha)\phi_r}\int_0^\pi\left(r^{*2}\sin^2\theta(n_1^* - n_2^*)\frac{\partial\phi_2^*}{\partial\theta}\right)_{r^*=1} d\theta \quad (78)$$

$$= \pi\epsilon\zeta_a^2\left(U^* K_{Df} - \frac{(\kappa a)^2}{(1+\alpha)\phi_r}E^* K_{De}\right)$$

where

$$U^* K_{Df} = \int_0^\pi\left(r^{*4}\sin^3\theta\frac{\partial}{\partial r^*}\frac{E^{*2}}{r^{*2}}\frac{\psi^*}{\sin^2\theta}\right)_{r^*=1} d\theta \quad (79a)$$

$$E^* K_{De} = \int_0^\pi\left(r^{*2}\sin^2\theta(n_1^* - n_2^*)\frac{\partial\phi_2^*}{\partial\theta}\right)_{r^*=1} d\theta \quad (79b)$$

The mobility of the particle can be calculated on the basis that the net force exerted on it vanishes at the steady state. We have $F_{Dz} + F_{Ez} = 0$, F_{Dz} and F_{Ez} being the drag force and the electric force, respectively. Equations (73) and (78) yield:

$$U_m^* = \frac{\eta U}{\epsilon \zeta_a E} = \frac{U^*}{E^*} = \frac{[(\kappa a)^2 K_{De}/(1+\alpha)\phi_r] - 2K_E}{K_{Df}} \tag{80}$$

Note that since both $K_E K_{De}$ and K_{Df} in this expression are functions of the applied field and the scaled terminal velocity of the particle, the estimation of U_m^* involves an iterative procedure. For a specified scaled electric field E^*, and initial guess for U^*, U_i^*, is assumed which is used to calculate K_E by Eq. (73) and K_{Df} and K_{De} by Eq. (79). Equation (80) is then used to calculate the value of U^* in the next stage, U_{i+1}^*, by Eq. (80). This procedure is continued until a convergent U^* is obtained.

If the applied electrical field is low, the ion density, the electrical field, and the flow field are only slighted distorted. In this case the nonlinear electrokinetic equations can be approximated by the corresponding linearized forms. It can be shown that

$$\nabla^{*2}\phi_1^* = -\frac{1}{(1+\alpha)}\frac{(\kappa a)^2}{\phi_r}[\exp(-\phi_r\phi_1^*) - \exp(\alpha\phi_r\phi_1^*)] \tag{81}$$

$$n_1^* = \exp(-\phi_r\phi_1^*)\lfloor 1 - \phi_r(\phi_2^* + g_1^*)\rfloor \tag{82}$$

$$n_2^* = \exp(\alpha\phi_r\phi_1^*)\lfloor 1 + \alpha\phi_r(\phi_2^* + g_2^*)\rfloor \tag{83}$$

$$\nabla^{*2}\phi_2^* - \frac{(\kappa a)^2}{(1+\alpha)}[\exp(-\phi_r\phi_1^*) + \alpha\exp(\alpha\phi_r\phi_1^*)]\phi_2^*$$

$$= \frac{(\kappa a)^2}{(1+\alpha)}\{[\exp(-\phi_r\phi_1^*)g_1^* + \exp(\alpha\phi_r\phi_1^*)\alpha g_2^*]\} \tag{84}$$

$$\nabla^{*2}g_1^* - \phi_r\nabla^*\phi_1^* \cdot \nabla^*g_1^* = Pe_1 u^* \cdot \nabla^*\phi_1^* \tag{85}$$

$$\nabla^{*2}g_2^* + \alpha\phi_r\nabla^*\phi_1^* \cdot \nabla^*g_2^* = Pe_2 u^* \cdot \nabla^*\phi_1 \tag{86}$$

and

$$E^{*4}\psi^* = \frac{(\kappa a)^2}{(1+\alpha)}\left[\left(\frac{\partial g_1^*}{\partial\theta}n_1^* + \frac{\partial g_2^*}{\partial\theta}\alpha n_2^*\right)\frac{\partial\phi_1^*}{\partial r^*}\right]\sin\theta \tag{87}$$

According to O'Brien and White [7], the problem described by Eqs (81)–(87) can be decomposed in two problems. In the first problem the particle moves with a uniform velocity U with no applied field. We have

$$u_r^* = U^*\cos\theta, \quad r^* = 1 \tag{88a}$$

$$u_\theta^* = -U^*\sin\theta, \quad r^* = 1 \tag{88b}$$

$$\frac{\partial\phi_2^*}{\partial r} = 0, \qquad r^* = 1/\lambda \tag{88c}$$

In the second problem the particle is held fixed in an applied electric field E, and the nonslip condition is assumed at the cavity surface. We have

$$u_r^* = u_\theta^* = 0, \quad r^* = 1 \tag{89a}$$

$$\frac{\partial\phi_2^*}{\partial r} = -E^*\cos\theta, \quad r^* = 1/\lambda \tag{89b}$$

As a result, the force required to move the particle with uniform U^* in problem 1, f_1, is proportional to U^*. Similarly, the force exerted on the particle in problem 2, f_2, is proportional to E^*. The fact that the sum of forces acting on the particle vanishes leads to $U_m^* = f_2/f_1$.

C. Concentrated Dispersions

Let us consider the case of a concentrated colloidal dispersion [14]. Referring to Fig. 4, the cell model of Kuwabara [54] introduced in Section III is adopted. Here, the assumptions and the parameters used are the same as those of Section IV.B, except for the boundary conditions. For illustration, the linearized problem of Section IV is considered. The governing equations can be expressed by Eqs (81)–(87), and the corresponding boundary conditions are

$$\phi_1^* =, \quad r^* = 1 \tag{90a}$$

$$\frac{d\phi_1^*}{dr^*} = 0, \quad r^* = 1/\lambda \tag{90b}$$

$$\frac{\partial \phi_2^*}{\partial r} = 0, \quad r^* = 1 \tag{91a}$$

$$\frac{\partial \phi_2^*}{\partial r} = -E_z^* \cos\theta, \quad r^* = 1/\lambda \tag{91b}$$

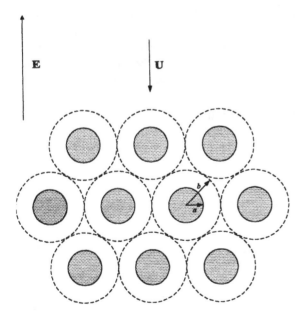

FIG. 4 The cell model of Kuwabara [54] used to simulate a concentrated spherical dispersion where a is the radius of a particle and b is a spherical shell of thickness $(b - a)$. E and U are, respectively, the applied electric field and the velocity of fluid.

$$\frac{\partial g_1^*}{\partial r^*} = 0, \quad r^* = 1 \tag{92a}$$

$$g_1^* = -\phi_2^*, \quad r^* = 1\lambda \tag{92b}$$

$$\frac{\partial g_2^*}{\partial r^*} = 0, \quad r^* = 1 \tag{93a}$$

$$g_2^* = -\phi_2^*, \quad r^* = 1/\lambda \tag{93b}$$

$$\psi^* = -\frac{1}{2} U^* r^{*2} \sin\theta, \quad \frac{\partial \psi^*}{\partial r^*} = -U^* r^* \sin^2\theta, \quad r^* = 1 \tag{94a}$$

$$\psi^* = \left[\frac{1}{r^*}\frac{d^2}{dr^{*2}} - \frac{2}{r^{*3}}\right]\psi^* = 0, \quad r^* = 1/\lambda \tag{94b}$$

Note that the boundary conditions for ϕ_2^* and ψ^* are different from those for the case of a sphere in a spherical cavity discussed in Section IV.B. To take into account the effect of double-layer overlapping, and the fact that a unit cell as a whole is electroneutral, we must have $d\phi_1^*/dr^* = 0$ at the virtual surface ($r^* = 1/\lambda$). For flow field, the vorticity vanishes at the virtual surface, that is, $\nabla \times \vec{v} = 0$. In terms of the stream function we have

$$\left[\frac{1}{r^*}\frac{d^2}{dr^{*2}} - \frac{2}{r^{*3}}\right]\psi = 0$$

Following the treatment introduced in Section IV.B, the mobility can be evaluated on the basis of a force balance.

D. Porous Entity

Let us consider the case of a spherical particle in an infinite fluid. The former comprises a rigid core of radius a and an ion-penetrable surface layer of thickness d. The particle moves at a constant velocity U as a response to an applied electric field E. Following an approach similar to that presented in Section IV.B, and taking the nature of the surface layer into account, the governing equations for the liquid phase in scaled forms are

$$\nabla^{*2}\phi_1^* = -\frac{1}{(1+\alpha)}\frac{(\kappa a)^2}{\phi_r}[\exp(-\phi_r\phi_1^*) - \exp(\alpha\phi_r\phi_1^*)] \tag{95}$$

$$n_1^* = \exp(-\phi_r\phi_1^*)\lfloor 1 - \phi_r(\phi_2^* + g_1^*)\rfloor \tag{96}$$

$$n_2^* = \exp(\alpha\phi_r\phi_1^*)\lfloor 1 + \alpha\phi_r(\phi_2^* + g_2^*)\rfloor \tag{97}$$

$$\nabla^{*2}\phi_2^* - \frac{(\kappa a)^2}{(1+\alpha)}[\exp(-\phi_r\phi_1^*) + \alpha\exp(\alpha\phi_r\phi_1^*)]\phi_2^*$$

$$= \frac{(\kappa a)^2}{(1+\alpha)}\{[\exp(-\phi_r\phi_1^*)g_1^* + \exp(\alpha\phi_r\phi_1^*)g_2^*]\} \tag{98}$$

$$\nabla^{*2}g_1^* - \phi_r\nabla^*\phi_1^* \cdot \nabla^*g_1^* = Pe_1 u^*\nabla^*\phi_1^* \tag{99}$$

$$\nabla^{*2}g_2^* + \alpha\phi_r\nabla^*\phi_1^* \cdot \nabla^*g_2^* = Pe_2 u^* \cdot \nabla^*\phi_1^* \tag{100}$$

$$E^{*4}\psi^* = \frac{(\kappa a)^2}{(1+\alpha)}\left[\left(\frac{\partial g_1^*}{\partial\theta}n_1^* + \frac{\partial g_2^*}{\partial\theta}\alpha n_2^*\right)\frac{\partial\phi_1^*}{\partial r^*}\right]\sin\theta \tag{101}$$

Similarly, the governing equations for the surface layer in scaled forms are

$$\nabla^{*2}\phi_1^* = -\frac{1}{(1+\alpha)}\frac{(\kappa a)^2}{\phi_r}[\exp(-\phi_r\phi_1^*) - \exp(\alpha\phi_r\phi_1^*) + q^*] \tag{102}$$

$$n_1^* = \exp(-\phi_r\phi_1^*)\left[1 - \phi_r(\phi_2^* + g_1^*)\right] \tag{103}$$

$$n_2^* = \exp(\alpha\phi_r\phi_1^*)\left[1 + \alpha\phi_r(\phi_2^* + g_2^*)\right] \tag{104}$$

$$\nabla^{*2}\phi_2^* - \frac{(\kappa a)^2}{(1+\alpha)}[\exp(-\phi_r\phi_1^*) + \alpha\exp(\alpha\phi_r\phi_1^*)]\phi_2^*$$

$$= \frac{(\kappa a)^2}{(1+\alpha)}\left\{[\exp(-\phi_r\phi_1^*)g_1^* + \exp(\alpha\phi_r\phi_1^*)g_2^*]\right\} \tag{105}$$

$$\nabla^{*2}g_1^* - \phi_r\nabla^*\phi_1^* \cdot \nabla^*g_1^* = \chi_1 Pe_1 u^* \nabla^*\phi_1^* \tag{106}$$

$$\nabla^{*2}g_2^* + \alpha\phi_r\nabla^*\phi_1^* \cdot \nabla^*g_2^* = \chi_2 Pe_2 u^* \cdot \nabla^*\phi_1^* \tag{107}$$

$$E^{*4}\psi^* - (\gamma a)^2 E^{*2}\psi^* = \frac{(\kappa a)^2}{(1+\alpha)}\left[\left(\frac{\partial g_1^*}{\partial\theta}n_1^* + \frac{\partial g_2^*}{\partial\theta}\alpha n_2^*\right)\frac{\partial\phi_1^*}{\partial r^*}\right]\sin\theta \tag{108}$$

In these expressions q^* denotes the scaled fixed charge distribution, and χ_1 and χ_2 are modified terms, which reflect the difference between the diffusivity of the surface layer and that of the liquid phase. Solving the above equations subject to the assigned boundary conditions yields the spatial variations in the electric field and flow field. The mobility of a particle can then be calculated on the basis of a force balance similar to that introduced in Section IV.B.

NOTATION

a	radius of a particle (m)
b	radius of spherical cavity (m)
d	thickness of surface layer (m)
D	diffusivity of ions (m^2/s)
e	elementary charge (1.6×10^{-19} coul)
$\mathbf{E}$	applied electric field (volt/m)
E	magnitude of $\mathbf{E}$ (volt/m)
f	molar flux of ions (mol/m^2/s)
g	potential function for the concentration of electrolytes (volt)
k_B	Boltzmann constant (1.38×10^{-23} J/K)
n	molar density of ions (mol/m^3)
p	pressure (N/m^2)
r	r co-ordinate of spherical co-ordinates
t	time (s)
$\mathbf{U}$	terminal velocity (m/s)
U	magnitude of $\mathbf{U}$ (m/s)
$\mathbf{v}$	velocity of $\mathbf{U}$ (m/s)
x, y, z	Cartesian co-ordinates

Greek Letters

α $= -z_2/z_1$

γ drag coefficient

ϵ permeability (Coul/volt $\cdot$ m)

ζ zeta potential (volt)

η viscosity of fluid (kg/m $\cdot$ s)

θ θ co-ordinate of spherical co-ordinates

κ inverse Debye length (1/m)

ρ space charge density in liquid phase (coul/m^3)

σ surface charge density (coul/m^2)

ϕ electric potential (volt)

χ a correction factor for diffusivity

ψ stream function

Symbols

∇ gradient operator

∇^2 Laplace operator

$\nabla\cdot$ divergence operator

E^2 $= \frac{\partial}{\partial r^2} + \frac{\sin\theta}{r^2}\frac{\partial}{\partial\theta}\left(\frac{1}{\sin\theta}\frac{\partial}{\partial\theta}\right)$

E^4 $= E^2 E^2$

Superscripts

$*$ scaled quantity

Subscripts

0   equilibrium state

1   cations

2   anions

j ion species

m   mobility

r r co-ordinate

θ θ co-ordinate

REFERENCES

1. RJ Hunter. Foundations of Colloid Science. vol. 1. Oxford: Oxford University Press, 1989.
2. M Smoluchowski. Z Phys Chem 92: 129, 1918.
3. E Huckel. Phys Z 25: 204, 1924.
4. DC Henry. Proc R Soc London, Ser A 133: 106, 1931.
5. F Booth. Proc R Soc London, Ser A 203: 514, 1950.
6. PH Wiersema, AL Loeb, JTG Overbeek. J Colloid Interface Sci 22: 78, 1966.
7. RW O'Brien, LR White. J Chem Soc, Faraday Trans 2 74: 1607, 1978.
8. RW O'Brien, RJ Hunter. Can J Chem 59: 1878, 1981.
9. H Ohshima, TW Healy, LR White. J Chem Soc, Faraday Trans 2 79: 1613, 1983.
10. S Levine, GH Neale. J Colloid Interface Sci 47: 520, 1974.

11. MW Kozak, EJ Davis. J Colloid Interface Sci 127: 497, 1989.
12. MW Kozak, EJ Davis. J Colloid Interface Sci 129: 166, 1989.
13. H Ohshima. J Colloid Interface Sci 188: 481, 1997.
14. E Lee, JW Chu, JP Hsu. J Colloid Interface Sci 209: 240, 1999.
15. E Lee, JW Chu, JP Hsu. J Colloid Interface Sci 196: 316, 1997.
16. E Lee, JW Chu, JP Hsu. J Colloid Interface Sci 205: 65, 1998.
17. SB Chen HJ Keh. In: JP Hsu, ed. Interfacial Forces and Fields: Theory and Applications. New York: Marcel Dekker, 1999.
18. S Kawahata, H Ohshima, N Muramatsu, T Kondo. J Colloid Interface Sci 138: 182, 1990.
19. GV Seaman. In: DM Sergenor, ed. The Red Blood Cells. vol. 2. New York: Academic Press, 1975, p 1136.
20. N. Hashimoto, S Fujita, T Yokoyama, Y Ozawa, I Kingetsu, D Kurosaka, D Sabolovic, W Schuett. Electrophoresis 19: 1227, 1998.
21. K Makino, T Taki, M Ogura, S Handa, M Nakajima, T Kondo, H Ohshima. Biophys Chem 47: 261, 1993.
22. BW Ninham, VA Parsegian. J Theor Biol 21: 405, 1971.
23. DYC Chan, TW Healy, JW Perran, LR White. J Chem Soc, Faraday Trans 1 70: 1046, 1975.
24. TW Healy, LR White. Adv Colloid Interface Sci 9: 303, 1978.
25. YI Chang. J Theor Biol 139: 561, 1989.
26. YI Chang. Colloid Surfaces 41: 245, 1989.
27. YI Chang, JP Hsu. J Theor Biol 147: 509, 1990.
28. RW Wunderlich. J Colloid Interface Sci 88: 385, 1982.
29. E Donath, V Pastushenko. Bioelectrochem Bioenerg 6: 543, 1979.
30. S Levine, M Levine, KA Sharp, DE Brooks. Biophys J 42: 127, 1983.
31. KA Sharp, DE Brooks. Biophys J 47: 563, 1985.
32. H Ohshima, S Ohki. Biophys J 47: 673, 1985.
33. H Ohshima, S Ohki. Bioelectrochem Bioenerg 15: 173, 1986.
34. H Ohshima, T Kondo. J Theor Biol 124: 191, 1987.
35. H Ohshima, T Kondo. J Theor Biol 128: 187, 1987.
36. P Debye, A Bueche. J Chem Phys 16: 573, 1948.
37. JP Hsu, WC Hsu, YI Chang. J Colloid Interface Sci 155: 1, 1993.
38. JP Hsu, WC Hsu, YI Chang. Colloid Polym Sci 272: 251, 1994.
39. JP Hsu, JP Fan. J Colloid Interface Sci 172: 230, 1995.
40. JP Hsu, SH Lin, S Tseng. J Theor Biol 182: 137, 1996.
41. S Tseng, SH Lin, JP Hsu. Colloids Surfaces B 13: 277, 1999.
42. H Ohshima, T Kondo. J Colloid Interface Sci 130: 261, 1989.
43. H Ohshima. J Colloid Interface Sci 163: 474, 1994.
44. H Ohshima, T Kondo. J Colloid Interface Sci 116: 305, 1987.
45. H Ohshima, M Nakamura, T Kondo. J Colloid Interface Sci 270: 873, 1992.
46. H Ohshima, T Kondo. Biophys Chem 39: 191, 1991.
47. SB Chen, X Ye. J Colloid Interface Sci 222: 137, 2000.
49. M Rasmusson, B Vincent, N Marston. Colloid Polym Sci 278: 253, 2000.
50. H Ohshima. Adv Colloid Interface Sci 62 : 189, 1995.
51. C Canuto, MY Hussaini, A Quarteroni, TA Zang. Spectral Methods in Fluid Dynamics. New York: Springer-Verlag, 1986.
52. RJ Hunter. Foundations of Colloid Science. vol 2. Oxford: Oxford University Press, 1989.
53. S Levine, GH Neale. J Colloid Interface Sci 47: 520, 1974.
54. S Kuwabara. J Phys Soc Jpn 14: 527, 1959.
55. J Happel. Am Inst Chem Eng J 4: 197, 1958.
56. J Happel, H Brenner. Low Reynolds Number Hydrodynamics. Martinns Nighoff: Boston, 1983.

24

Electrophoretic Fingerprinting and Multiparameter Analysis of Cells and Particles

EBERHARD KNIPPEL   Consultant, Langenfeld, Germany

AXEL BUDDE   HaSo Tec Hard- and Software Technology GmbH, Rostock, Germany

I. INTRODUCTION

Most substances acquire a surface electric charge when brought into contact with a polar, normally aqueous medium. The underlying charging mechanisms, leading to the formation of an electrical double layer, are dissociation, ion adsorption, and ion dissolution. One of the electrokinetic phenomena which arises when attempts are made to shear off the mobile part of the electrical double layer from a charged surface is the movement of a charged particle surface relative to stationary liquid by an applied electric field (particle electrophoresis). The mobility of a particle under the influence of an electric field appears to depend entirely on the nature of the particle surface, and is independent of the size or the nature of the particle itself. Reuss may be considered to have provided the first demonstration of electrophoretic phenomena in 1809 [1]

Particle electrophoresis has an important practical applicability for the characterization and control of the surface behavior of particles in fields such as colloid and polymer chemistry, biotechnology, and biological applications. Thus, considerable efforts have been undertaken to develop and improve this surface characterization technique. A remarkable advance in particle electrophoresis was the electrophoretic fingerprinting introduced by Marlow and coworkers [2–16]. The electrophoretic mobility has been shown to be a function of the measurable quantities pH and specific conductance of the medium [17, 18]. Thus, the electrophoretic fingerprint is a three-dimensional diagram of these experimental data which can be used to characterize an unknown system of particles or to explore the nature of the electrochemical surface.

However, the concept of zeta potential, developed for smooth surfaces, was often in contradiction to the behavior of complex (real) surfaces, for which reason it was improved experimentally as well as theoretically. For a comprehensive electrophoretic analysis of "hairy" layers, local charge density isotherms have been combined with hydrodynamic flow penetration into the layer to provide an iterative numerical procedure for the calculation of the electrophoretic mobility as a function of pH and salt concentration based on the linearized Poisson–Boltzmann equation [19]. The consideration of highly charged surfaces led to the development of a numerical procedure providing a solution to the nonlinear electrostatic problem including dissociation, adsorption, and association [20]. This procedure makes it possible to fit the experimental fingerprint data with a theoretical fingerprint calculated by variation of a number of physiochemical parameters.

Another way to improve the analysis of electrophoresis data is the electrophoretic multiparameter analysis of cells and particles which is the simultaneous measurement and combination of cell (particle) characterizing parameters such as electrophoretic mobility, size, density (sedimentation), and shape [21]. The final data are represented as histograms and two- or three-dimensional (2D or 3D) plots or compared by using statistical analysis. This method can be used to characterize complex processes on surfaces and to detect small differences between particles or cells including pathological changes in biomedical and clinical research.

The focus of this chapter is to demonstrate the usefulness of electrophoretic fingerprinting and multiparameter analysis in analyzing biological and nonbiological surfaces. The presented work covers the theoretical background as well as experimental aspects and applications.

II. INSTRUMENTATION

Owing to the important practical applicability of particle electrophoresis many attempts have been done to improve electrophoretic devices. In the last years it became clear that the comprehensive analysis and prediction of reactions on cell and particle surfaces and the detection of small differences between normal and pathological cells require a modern electrophoretic instrumentation capable of measuring several parameters simultaneously and processing a large number of data. Most commercially available electrophoretic devices have these properties to a high degree.

One of the modern devices with a very high accuracy is the electrophor (Fig. 1). This apparatus, which has been described in detail elsewhere [22], relies on the conception of single-cell electrophoresis and should serve here as typical for a modern instrument measuring several parameters simultaneously. Essentially, the observation of particles in the electrophoresis chamber is realized with a microscope and a video camera. A personal computer with a frame grabber tracks each particle in the video image on-line. A feature extraction determines the evident parameters for the particles, such as size and co-ordinates of the center of mass. From these data, the tracing procedure calculates the whole set of parameters for each particle. The on-line measurement gives the following parameters: (1) electrophoretic mobility, (2) sedimentation velocity (density), (3) particle size, (4) intensity, and (5) shape.

A set of parameters for the tracking procedure (e.g., a size range) defines which particles will be considered in the measurement. This allows one to select cells from a cell mixture and to eliminate the influence of dirt and particles which are out of focus. The particle diameter range extends from some 1 nm (darkfield illumination) up to 50 μm. This is optimal for biological, medical, and biophysical purposes. The construction of the apparatus allows use of the spectrum of available microscopic methods, e.g., darkfield, phase contrast, and fluorescence. This minimizes the preparation procedure for the particles and, thus, irreversible changes to the particle surface by the preparation. The electrophoretic chamber, which has a flat profile, and the stationary layer at the center, are separated from the electrodes by a semipermeable membrane. Separate electrolyte flow circuits at both platinum electrodes protect particles from electrolysis products. The tem-

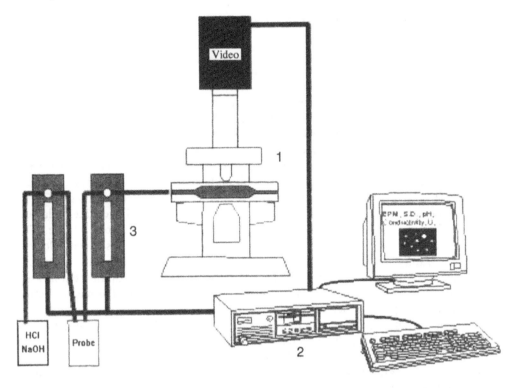

FIG. 1 Schematic representation of the Electrophor Electrophoretic Analyzer: (1) microscope, video camera, and electrophoresis system; (2) computer; and (3) automated titration system.

perature of the electrophoretic chamber, and of the pH-electrode and conductivity cell, is stabilized to exclude the influence of changing viscosity on the electrophoretic mobility of particles. Automatic sample injection and sample movement between cycles of the electrophoretic measurement simplify the handling and assure the tracking of different cells at each cycle for the calculation of mean mobility values and mobility distributions.

The apparatus allows one simultaneously to measure pH and specific conductance which strongly influence the particle mobility and surface charge. This and the automatic titration allow the measurement of the pH and/or conductance dependence of the surface charge (electrophoretic fingerprints).

From the measurements one can save different data for further use: (1) snapshots from the microscopic image of the particles, (2) a file with the complete data set (11 parameters) of each measured particle for statistical calculations, comparison and plotting (MES file), and (3) in the case of automated titration, a file with mean values of the particle parameters at each titration step (TIT file).

A program for statistical calculations on MES files allows one to: (1) generate and statistically compare histograms of all measured parameters, (2) calculate mean values and standard deviation, (3) compare mean values using the t-test, (4) generate different types of 2D and 3D plots, (5) compare different classes of measurements (variance analysis), (6) calculate a density coefficient from sedimentation and size, and (7) calculate the portion of different populations in a histogram.

III. ELECTROPHORETIC FINGERPRINTING

A. Introduction

The zeta potential and its directly measured quantity, electrophoretic mobility, have been widely used as a characterizing parameter in colloid science. The idea of representing the electrophoretic mobility of colloidal particles as function of pH and the logarithm of the conductance was developed at Pen Kem, Inc., and first applied to latex surfaces at the University of Massachusetts by Marlow and coworkers [2–5]. The term "electrophoretic fingerprinting" was coined in the first journal publication of the method [8]. The electrophoretic fingerprinting is based on a theoretical point of view, since all theories that treat the charging of colloidal surfaces in aqueous medium have as variables potential, conductivity, and pH (when hydrogen ions and hydroxyl ions are potential determining) [17, 18].

The electrophoretic mobility may be represented as a function of two state variables: the pH and the $p\lambda$, which is the negative logarithm of the specific conductance. The representation of electrophoretic mobility as a function of pH and $p\lambda$ is called "electrophoretic topography" and yields a 3D template or "fingerprint" which is the corresponding isomobility contour plot. This is the definition by Marlow et al. [2–5]. In the following we define the electrophoretic fingerprint as the 3D representation of the mean electrophoretic mobility of particles versus pH and conductivity. The first electrophoretic fingerprints were measured by Marlow and coworkers [2, 4, 5] using the Pen Kem System 3000 Automated electrokinetics analyzer which has been described elsewhere [23, 24]. The colloid was sampled automatically and the electrophoretic mobility, pH, and conductivity were measured. The sample was automatically dosed with the appropriate acid or base from a digital buret by using back and forth titration. The mobility data were represented as a 3D plot, where pH and conductivity represented the independent variables and mobility the dependent variable, giving a third space axis. For a fingerprint, typically 100–400 individual spatial points were used. The data were first gridded to produce regularly spaced data in the x–y plane required for contour mapping and 3D display. The gridding procedure used Kringing, a technique that utilizes a regional variable theory and assumes an underlying linear variogram [25]. The gridded data were then smoothed by using a cubic spline. Some studies of the dependence of the fingerprint pattern on the number and distribution of experimental points have been carried out [10]. It became clear that of the order of 70 data points are required to produce meaningful isomobility contour plots. However, a much larger number would be preferable for defining the finer details of these plots. We have used a similar procedure for data plotting [21]. Electrophoretic titration data, measured in solutions with different sodium chloride concentrations, are not regularly distributed in the pH–conductivity plane. Therefore, values for an equally spaced grid have been calculated from the measured data using an inversed distance method followed by splining. The experimental fingerprints can be represented in a system of pH–conductivity–electrophoretic mobility or pH–inverse Debye radius–apparent charge co-ordinates. The procedure mentioned above is important for making fingerprints comparable, presenting the data as 3D plots or 2D isomobility contour plots and calculating gradients and differences between fingerprints.

In the course of application the electrophoretic fingerprint was shown to be a characteristic property of a system of charged colloidal particles which may be used to characterize an unknown system of particles or to explore the nature of the electrochemical surface by comparison with theoretical models. The electrophoretic fingerprint seems to be a characteristic of the surface electrical state of a system. Fingerprints for different systems are quite different and should be reproducible in different laboratories.

Electrophoretic fingerprints have been used as a sensitive probe of the surface electrochemical state of particle surfaces and a measure of colloid stability. However, the fingerprint technique has also been applied to biological problems. The idea of detecting pathological changes in human body fluids in contact with colloidal indicator particles was conceived at the University of Rostock in 1987 [26]. Clinical studies have revealed that the electrophoretic mobility of indicator colloids can be used to diagnose and follow the treatment of disorders such a cystic fibrosis, fetal lung maturity, meningitis, and respiratory distress syndrome in newborns. The disease-induced changes in body fluids are manifest by a shift, relative to healthy persons, in the electrophoretic mobility of indicator particles that have been contacted with the body fluid. Interestingly, it has been found that different batches of the same type of particles may or may not respond in this simple test. The objective of the collaboration with Marlow et al. [8] was to describe the differences in surface chemistry between responsive and unresponsive polystyrene particles and thus to demonstrate the usefulness of electrophoretic fingerprinting in analyzing biological active versus nonactive surfaces. From the electrophoretic isomobility contour plots cuts have been taken at true constant conductivity and pH, for comparison with theoretical calculations. It has been shown that the unresponsive particles had a surface which could be described by a single acid site dissociation model [17, 18], the acid site being characterized by a pK similar to that of carboxylic acid. The surface charge density was low and there was no evidence of more than one type of acidic group, nor was there evidence to suggest an expandable layer at the surface [3–5]. This type of surface had no isoelectric point but approached a zero mobility at low pH. There was evidence provided that Na^+ and Cl^- adsorption does not occur at a pH of 7.4 and that specific adsorption of cations occurs, but only at low pH values.

In contrast, the polystyrene latex particles, responsive to body fluid, had a surface which could be described by a two-site dissociation model [8, 17, 18]. At low ionic strength the responsive particles showed characteristics of an amphoteric surface, the surface group being more than likely hydroxyl. Such a group can be easily produced through hydrolysis of sulfate groups. With an increased conductivity this surface undergoes significant uptake of Cl^- ions. There was also no evidence to suggest that the responsive polystyrene particles had an expandable surface. Summarizing, one can conclude that the unresponsive particles had predominantly negative electron-donor sites whereas the responsive microspheres had both negative donor sites, positive acceptor sites, and, in addition, sites containing adsorbed Cl^- ions. These studies have clearly shown that surface chemistry plays a key role in the biological activity of indicator particles and that electrophoretic fingerprinting is a powerful method for characterizing biologically active and nonactive surfaces.

Later, some attempts have been made to explore the nature of the electrochemical surface by comparison with theoretical models [12]. Electrokinetic data on polymer colloids can only be quantitatively explained by using models involving drastically different physical interpretations, e.g., anomalous surface conductance or ion mobility of a bound layer of ions within a shear plane [27–29], expansion or contraction of surface layers under changing electrochemical stresses [12, 29, 30], and variation in surface charge density through preferential ion adsorption [12, 27–29, 31]. It has been demonstrated how electrophoretic fingerprinting can sort out which of these phenomena is occurring for a model polymer colloid in a given electrochemical state. Furthermore, it was shown that discrepancies between zeta potential calculated from electrophoretic mobility measurements and relatively low-frequency conductivity under similar thermodynamic conditions can be understood by making electrophoretic fingerprints. The surfaces were fingerprinted by laser Doppler electrophoresis. The electrophoretic mobilities were calculated using a single acid dissociation model and the theoretical fingerprints fitted to the experimental data. These studies have revealed that there are several mechanisms operative concerning the surface layers structures and accompanying double layer of the polymer colloid investigated, which have different degrees of significance under different electrochemical states. They include electrophoretic relaxation, shear plane expansion, and conductance within the shear plane. It has been demonstrated how the electrophoretic fingerprinting can be used as an "in situ surface analysis" tool and can sort out the electrochemical conditions under which different mechanisms controlling the chemistry and structure of the surface and accompanying double layer are operative.

Later, Prescott et al. [16] have reported on hydrodynamic fingerprinting where the hydrodynamic fingerprint was the isothermal contour diagram of the hydrodynamic size as a function of pH and conductance. It was shown that the hydrodynamic fingerprint, like the electrophoretic fingerprint, is a characteristic pattern of a particular colloidal system.

Paulke et al. [32] have reported on electrophoretic fingerprinting of latex particles with divergent characteristics such as charge, surface morphology, and functional groups. The surface of these microspheres was more complex than the colloids investigated by Marlow et al. [8] The fingerprints revealed two types of latex particles: (1) the electrophoretic mobilities depend on the conductivity of the medium, and the influence of pH is negligible, and (2) the electrophoretic mobility depends strongly on pH, and the influence of conductivity is slight. The first group concerns polystyrene particles with hydroxyl as well as ammonium and amidinium groups and polymethacrylate particles with hydroxyl groups, and the second group polystyrene particles with the combination of sulfate/hydroxyl groups and ammonium/hydroxyl groups on the surface. The electrophoretic behavior of these two types of latex particles has been discussed using independent charge density data.

B. Smooth Surfaces

For a better characterization of dissociation and adsorption processes on particle surfaces, the pK values of surface chemical groups and their concentration have been calculated by use of the electrophoretic fingerprint [22]. Briefly, the corresponding program uses a model of a smooth surface with charges originating from the dissociable surface group, considering monovalent ions from the NaOH/HCl electrolyte. For large particles, the zeta potential, ζ, may be calculated from the electrophoretic mobility, μ, by means of the Smoluchovsky equation:

$$\zeta = \frac{\mu \eta}{\epsilon_r \epsilon_0} \tag{1}$$

where ϵ_r and ϵ_0 are the relative and absolute dielectric permeability, respectively, and η is the dynamic viscosity of the solvent. For smooth surfaces the zeta potential does not differ essentially from the electrostatic surface potential and the surface charge density σ can be calculated by using the Gouy–Chapman equation:

$$\sigma = \frac{2\epsilon_r \epsilon_0 \kappa k T}{\epsilon_0} \sinh \frac{e_0 \zeta}{2kT} \tag{2}$$

when κ^{-1} is the Debye screening radius:

$$\kappa^{-1} = \sqrt{\frac{\epsilon_r \epsilon_0 kT}{e_0^2 N_A \sum_i z_i^2 c_i}} \tag{3}$$

where e_0 and N_A are the electronic charge and the Avogadro number, respectively, c_i and z_i are the concentration and valency of the ith ion species, and k and T are the Boltzmann constant and absolute temperature, respectively.

The surface charge density, σ, calculated from Eq. (2), results essentially from two processes on the particle surface: (1) dissociation of surface chemical groups depending on pH of the solution, and (2) adsorption of the four different ions (Na^+, Cl^-, H^+, OH^-) on to the surface. The surface charge density, σ_i^{Diss}, which is caused by the dissociation of molecules of the sort i, may be described by a chemical equilibrium:

$$\sigma_i^{Diss} = \frac{z_i c_i^{Diss}}{1 + 10^{z_i(pH_{Surf} - pK_i)}} \tag{4}$$

where c_i^{Diss} is the surface density of dissociable sites, pK_i is the decadic logarithm of the dissociation constant, and pH_{surf} is the pH on the particle surface. We have to consider that the ion concentration near the surface c_{surf} differs form the bulk concentration c_0 owing to the surface potential. This is described by the Boltzmann relation:

$$c_{surf} = c_0 \exp\left[-\frac{ze_0\zeta}{kT}\right] \tag{5}$$

The surface charge density σ_i^{Ads} caused by ion adsorption at the surface can be described by a Langmuir adsorption isotherm:

$$\sigma_i^{Ads} = \frac{z_i c_i^{Ads}}{1 + c_i^{50}/c_i^{surf}} \tag{6}$$

where c_i^{50} is the ion concentration where half of the adsorption sites are occupied, c_i^{Ads} is the total density of adsorption sites on the surface, and c_i^{Surf} is the concentration of sort i at the surface given by Eq. (5). Using the measured electrophoretic mobility (zeta potential), pH, and the specific conductance, the equations may be now solved, resulting in the parameters c_i^{Diss}, pK_i, c_i^{Ads}, and c_i^{50}.

The solution of the set of nonlinear equations is calculated with a modified Monte Carlo algorithm which searches the global minimum of the deviation of measured data from electrophoretic mobilities from pK and concentration values [33].

C. Nonideal Surfaces

The model mentioned above has been developed for smooth surfaces and provides reasonable results for liposomes and simple latex microspheres. More complex surfaces such as "hairy surfaces" with a spatial charge distribution may lead to inaccurate results. This was the starting point for our efforts to improve the theory of electrophoretic fingerprinting which considers the hairy character of structured real surfaces [19, 20]. In the case of a complex distribution of the fixed charges at the surface/solution interface the electrostatic as well as hydrodynamic properties may significantly differ from those of smooth surfaces where the fixed charges are assumed to be homogeneously distributed in a plane. Therefore, theories have been developed to describe the case of a charged layer penetrable for hydrodynamic flow [34–47] and applied to different classes of hairy layer particles such as cells, polymer micelles, and microgels [48–52]. Although providing considerable information about electrostatic and hydrodynamic properties [42, 47, 53] electrokinetic methods are rarely used for studies of the structure and dynamics of charged surface layers. This could be explained by the exclusion of the mechanism of fixed charge generation from the hairy layer electrokinetic theories. For this reason we have added an iterative numerical procedure to the charged layer electrophoretic concept which considers for the first time the additional dependence of the surface change density on the electrical potential provided by the shift in the local pH as compared to the bulk pH [19]. This shift depends on the distance of the charge from the surface and represents the fundamental difference from the classic case of charges distributed in a plane. The contribution of fixed surface charges, located at different distances from the particle surface, to the electrophoretic mobility is controlled by the Debye length. If one can assume a flow through the charged layer, more deeply located fixed charges also determine the electrophoretic mobility. Thus, varying the Debye length can provide information on the spatial distribution of the fixed charges. Therefore, a variation in ionic strength as well as pH, which is realized in the technique of electrophoretic fingerprinting, leads to a comprehensive interpretation of electrophoretic results.

In the following we will describe briefly the feature of our attempt to combine local charge density isotherms with the hydrodynamic flow penetration into the layer in order to develop an iterative procedure to calculate the electrophoretic mobility as a function of pH and salt concentration based on the linearized Poisson–Boltzmann equation. Details are described elsewhere [19].

We consider a particle whose radius is much larger than the Debye length κ^{-1}. The fixed surface charges are spatially distributed in a layer extending from the particle surface toward the bulk solution. This layer is considered to be uniform in the tangential direction. The thickness of the layer is assumed to be much smaller than the particle radius. Perpendicular to the surface, arbitrary distribution functions of m dissociable groups describe the potential spatial charge density of the dissociable sites $\rho_m(x)$. Here x denotes a co-ordinate axis perpendicular to the surface with the origin located at the particle surface. The m different dissociable sites are described by their respective pK_m and z_m values; z_m is $+1$ for a proton being released or, respectively, -1 for a proton generating a positively charged site by means of association. In the case of monovalent groups the actual fixed charge density ρ^m produced by the mth group is given by

$$\rho^m = \frac{\rho_m(x)}{1 + 10^{z_m\left(pK_m - pH - \frac{e_0\psi(x)}{kT\ln(10)}\right)}} \tag{7}$$

Here, $\psi(x)$ denotes the electrical potential. A corresponding relationship could be used to treat ion adsorption or association. $\psi(x)$ is given by the solution of the linearized Poisson–Boltzmann equation:

$$\psi'' - \kappa^2\psi = -\frac{1}{\epsilon\epsilon_0}\sum_m \frac{\rho_m(x)}{1 + 10^{z_m\left(pK_m - pH - \frac{e_0\psi(x)}{kT\ln(10)}\right)}} = f[x, \psi(x)] \tag{8}$$

where ϵ and ϵ_0 are the dielectric constant and the permittivity of a vacuum, respectively.

As we have shown previously [54], when the right-hand side of Eq. (8) does not depend on ψ, an analytical solution is possible. This was given in terms of Green function integrals over the net fixed charge density. This is not possible with Eq. (8) here. A further linearization of Eq. (8) is not meaningful, since this would lead to a linear differential equation with variable but arbitrary coefficients. Therefore, an iterative numerical procedure was executed, starting from the analytical solution of Eq. (8), but with

$$f[x, \psi(x)] = -\frac{1}{\epsilon \epsilon_0} \sum_m \rho_m(x) \tag{9}$$

The calculated intermediate electric potential distribution was used to find, successively, the respective intermediate fixed charge density distributions.

We will now discuss some curves resulting from the theoretical considerations. The first plot (Fig. 2) demonstrates the effective charge density of a particle covered with a charged layer of different thickness as a function of pH and ionic strength. This is the simple case of a spatial charge distribution of one single, negatively charged site, homogeneously distributed in the charged layer, and without flow penetration into the layer. In case of a 5 nm layer thickness it has been found that the electrophoretic effective charge is much less than the net surface charge density. The curves show that even the saturation value of σ_{eff}, far from the pK, does not approach the theoretical limit for plane surface charges ($-0.022\, C/m^2$). As one can expect, the difference between the theoretical limit for plane surface charges and the plateau values in the case of spatially distributed charges is reduced when the layer thickness decreases compared to the Debye length (see curves for a layer thickness of 0.5 nm). It becomes clear that the distribution of surface charges in a layer is responsible for the decreased σ_{eff} with increasing ionic strength. The dependence of σ_{eff} on ionic strength can be explained as follows. In the case of a short Debye length compared to the layer thickness, most fixed charges are screened by counterions within the layer and, consequently, do not contribute to the electrophoretic mobility, provided that the flow does not penetrate into the charged layer.

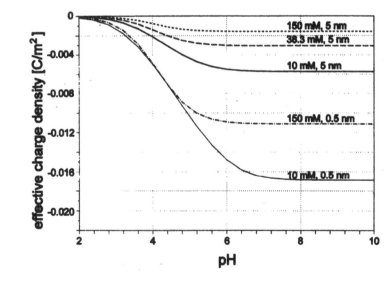

FIG. 2 Theoretical plot of the effective charge density $\sigma_{eff} = \eta \kappa \mu$ of a particle covered with a charged layer of different thicknesses as a function of pH and ionic strength. The spatial charge density in the layer was $-0.022\, C\, m^{-2}$, p$K = 4$. No flow inside the layer. (Reprinted with permission from Ref. 19. Copyright 1996 American Chemical Society.)

Exploring the influence of a moderate flow penetration into the layer and the presence of an additional, positively charged site, one finds an increase of the absolute value of σ_{eff} compared to the situation where no flow occurs. This may be explained by a more effective contribution from more deeply located fixed charges to the particle electrophoretic mobility. The point of zero σ_{eff} does not depend on ionic strength, which is attributed to the underlying assumption of the identical, uniform distribution for both dissociable sites.

Another aspect can be drawn from Fig. 3 where the comparison of the effective charge densities as a function of the ionic strength in the presence and absence of hydrodynamic flow penetration is shown. The increase of σ_{eff} mentioned above is more pronounced for larger ionic strength, leading to a nontrivial, nonmonotonous dependence of σ_{eff} on the Debye length.

Information about the fixed charge distribution can be obtained from the dependence of the zero point of electrophoretic mobility on ionic strength (Fig. 4). For this purpose a theoretical calculation of the effective charge density as a function of pH has been carried out; instead of a homogeneous charge density distribution a linearly increasing site density distribution function together with one decreasing toward the outer edge of the hairy layer were assumed. From this graph we can conclude that (1) in this way the plateau value of σ_{eff} far from the pK of the dominating group is modified and the charged site, which increases toward the bulk, contributes more efficiently to σ_{eff}, and (2) this gives rise to a dependence of the zero σ_{eff} point on ionic strength.

In Fig. 4 the point of zero mobility is plotted as a function of the salt concentration. It clearly shows that a qualitatively different behavior of the zero point of mobility as a function of salt concentration occurs when the charge distributions are reversed. The influence of charging site distribution asymmetry is discussed in detail in Ref. 19, where we also considered structural changes in the surface charge distribution, structural changes in the surface charge distribution function, and the theoretical and experimental change in σ_{eff} around the zero point of mobility.

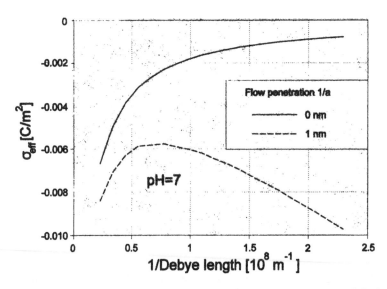

FIG. 3 Comparison of the effective charge densities as a function of the ionic strength expressed in inverse Debye length in the absence and presence of hydrodynamic flow penetration. (Reprinted with permission from Ref. 19. Copyright 1996 American Chemical Society.)

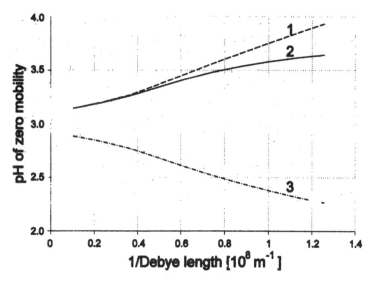

FIG. 4 Zero mobility point as a function of ionic strength for the two cases of charge distribution. Curves 1 and 2 have the positive charges preferentially located toward the bulk (outside) whereas in curve 3 the negative charges increase toward the outside. In curve 2 a flow penetration depth of 1 nm is additionally taken into account. (Reprinted with permission from Ref. 19. Copyright 1996 American Chemical Society.)

D. Application to Red Blood Cells

The above theoretical considerations have been compared to experimental electrophoretic fingerprints of the often measured, but not fully understood, surface of human red blood cells. The experimental plot of the effective charge density of erythrocytes as a function of pH and ionic strength shows, down to pH 5, a negative electrophoretic mobility (sialic acid and some carboxyl groups) whereas in the range pH 3–5 a charge reversal occurs, clearly demonstrating the presence of positive groups (amino groups of the surface glycoproteins). A comparison of the fingerprints of normal erythrocytes, neuraminidase-treated erythrocytes (removal of sialic acid sites), and glutaraldehyde-treated erythrocytes (removal of the majority of positive sites) has revealed the following results. In the case of neuraminidase-treated red blood cells the point of zero σ_{eff} has significantly shifted to their pH, reflecting the increased relative contribution of positive charges to the net effect. Furthermore, a gradual reversal of the order of the pH shift as a function of ionic strength takes place. Obviously, the removal of negative charges occurs preferentially in the outer regions of the charged layer. The theoretical curves reproduce the pH shift in the zero point of mobility when the remaining positive charges are preferentially distributed to the outer border of the layer, signifying the removal of sialic acid by neuraminidase predominantly in the outer region of the glycocalyx. As in the case of control red blood cells, the theoretical results are significantly lower than the experimental data at low pH. Obviously, structural alterations to the surface layer and consequently a more condensed layer has to be considered to explain this result [19].

In the case of glutaraldehyde-treated red blood cells it is clearly seen that the point of zero σ_{eff} is strongly shifted to lower pH values. This point does not depend on ionic strength, demonstrating that the different distribution functions of positive and negative groups are responsible for the pH shift.

Summarizing, the results revealed that negative groups on the red blood cell surface occur preferentially in the outer regions of the layer. To assume a penetration depth of 1 nm into the 3.5 nm thick hairy layer seems to be reasonable. Upon sialic acid removal, alterations to the glycocalyx structure occur, resulting in possible changes in the fixed charge distribution. One can conclude that intramolecular as well as intermolecular electrostatic interactions are important for the structure of the hairy layer. No evidence for the adsorption of small ions has been found. The comprehensive interpretation of theoretical and experimental fingerprints in the case of human red blood cells on the basis of the charged layer concept of electrophoresis provides an improved understanding of the electrophoretic behavior of the erythrocyte surface, a weakly charged grafted polyelectrolyte layer. The sensitivity of electrophoretic fingerprinting to the charged layer properties in the short-nanometer range has been impressively demonstrated.

E. Highly Charged Surfaces

However, there are some restrictions of the theoretical approach based on the linearized Poisson–Boltzmann equation, particularly at low ionic strength and highly charged surfaces. In the case of highly charged surfaces such as polyelectrolytes the equations obtained with the linearization of the Poisson–Boltzmann equation do not describe comprehensively the electrophoretic fingerprint data because the influence of surface conductance has to be considered [55]. Additionally, one has to account for steric small-ion exclusion and layer polarity.

These considerations led to the development of a numerical procedure providing the solution of the nonlinear electrostatic problem including dissociation, adsorption, and association [20]. This procedure enables us to fit the experimental fingerprint data with a theoretical fingerprint calculated by variation of various physicochemical parameters such as thickness of the charged layer, depth of hydrodynamic flow, association constants of ions, surface conductivity, charge distribution, dissociation constants of surface chemical groups, and dielectric permittivity. From this fitting one can obtain information about the real structure of the charged hairy layer. The procedure has been applied to the consecutive layer-by-layer adsorption of polyelectrolytes on colloidal charged latex particles. In the following we will describe the characteristics of this theoretical approach. Details are described in Ref. 20.

It is now well accepted that the electrophoretic mobility μ of a particle with a hairy surface layer cannot be described by means of the Helmholtz–Smoluchowsky equation valid only for smooth surfaces with fixed charges arranged in a plane.

The Smoluchovsky equation essentially provides the definition of the zeta potential ζ. In the case of a smooth surface the zeta potential is the electric potential difference between the plane of shear and infinity. In the absence of a well-defined plane of shear, which is the case for particles covered with hairy layers, the definition of the zeta potential is more complex. A new description of the electrophoretic properties of surfaces with hairy layers is required.

To solve the problem of the electrophoretic mobility of a particle covered with a charged hairy layer, we make use of Onsager's law in order to replace the problem of calculating the electrophoretic mobility by the calculation of the streaming current near a hypothetical hairy surface in a rectangular chamber [56]. This approach has the advantage of conveniently separating the hydrodynamic and the electrostatic part of the problem. Indeed, the magnitude of the electrophoretic velocity of a particle is equal to the electroosmotic velocity of the flow at infinite distance from the particle surface. On the other hand, the electro-osmotic velocity coefficient represents one of the cross coefficients on Onsager's flow matrix, L_{12}

$$J = L_{11}\Delta p + L_{12}\Delta \psi$$
$$I = L_{21}\Delta p + L_{22}\Delta \psi \tag{10}$$

where L_{11} has the meaning of the hydrodynamic conductance of the hypothetic rectangular chamber, L_{21} is the streaming current coefficient, and L_{22} represents the electric conductance of the system; Δp is the hydrodynamic pressure difference, and J and I are charge and volume flow, respectively. Onsager's law $L_{12} = L_{21}$ proves that the streaming current is, except for a normalizing coefficient, identical to the electrophoretic mobility.

The electrostatic part consists of calculating the electrical potential $\psi(x)$ as a function of distance from the particle surface; x denotes the co-ordinate perpendicular to the interface with the origin placed at the particle surface. Hydrodynamics provide the convective flow velocity $v(x)$ in the hypothetical rectangular chamber driven by the pressure difference Δp applied along the chamber [57]. The streaming current is given by the flow of ions transported by a convective flow parallel to the surface under study. Thus, the streaming current per unit width of the chamber is equal to

$$I = \int_0^\infty \rho_{\text{mob}}(x) n y(x) \mathrm{d}x \tag{11}$$

where $\rho_{\text{mob}}(x)$ is the equilibrium charge density given by the Boltzmann distribution of mobile ions according to the electric potential profile $\psi(x)$. The electrophoretic mobility is obtained by the streaming current divided by the pressure gradient and the chamber width.

Summarizing, Eqs (10)–(12) will not be linearized but solved numerically. Having finally obtained the electrical potential $\psi(x)$, the spatial charge density due to the distribution of the electrolyte ions inside the layer is given by

$$\rho_{\text{mob}}(x) = N_A e_0 \sum_i c_i z_i X_i e^{-\frac{z_i e_0 \psi(x)}{kT}} \tag{12}$$

where N_A, e_0, c_i, z_i, X_i, k, and T are Avogadro's constant, elementary charge, concentration of the ith ionic species of the electrolyte, charge number, steric exclusion factor with $X_i = 1$ for $x > \delta$, Boltzmann constant, and absolute temperature. This charge density provides a contribution to the surface conductivity.

If thick and highly charged layers are considered, it becomes necessary to correct the mobility for surface conductivity. The surface conductivity consists of the migrative and convective part. The migration contribution, K_m, is given by the excess spatial density of the mobile ions in the double layer compared to the bulk:

$$K_m = N_A \sum_i z_i e_0 c_i \int_0^\infty \left(u_i X_i e^{-\frac{z_i e_0 \psi(x)}{kT}} - u_i^{\text{bulk}} \right) \mathrm{d}x \tag{13}$$

Here, u_i denotes the mobility of the ith ionic species. As a first approximation we shall equate the ionic mobility within the layer to its bulk value, u^{bulk}. The convective part of the surface conductivity K_c is given by

$$K_c = \frac{1}{E} \int_0^\infty v(x) \rho_{\text{mob}}(x) \mathrm{d}x \tag{14}$$

Here, $v(x)$ is the electro-osmotic flow velocity and is given by the solution of

$$v(x)'' - a^2 v(x) = E\rho_{mob}(x) \tag{15}$$

where a^{-1} is the Brinkman length, with the boundary conditions:

$$v(0) = 0$$

$$v'(\delta) = \text{continuous}$$

$$v(\infty) = \text{bounded}$$

The mobility is corrected for the surface conductivity as follows:

$$\mu_{corr} = \mu \left[\frac{1 + \text{Rel}(1 - \Theta)\frac{v_1 + v_2}{v_1}}{1 + \text{Rel}\frac{v_1 + v_2}{v_1}} \right] \tag{16}$$

where Rel is the dimensionless criterion:

$$\text{Rel} = \frac{2K_s}{K d} \tag{17}$$

where K_s, K, and d are surface conductivity, bulk conductivity, and particle diameter, respectively; Θ is a dimensionless coefficient describing the influence of the Rel criterion on the mobility. It expresses the ratio between the noncorrected mobility calculated by means of using the counterion excess charge density instead of the total mobile ion charge density, over the conventional noncorrected particle mobility; u_1 and u_2 are counterion and coion mobility, respectively. Equation (17) is an extension of the Henry–Booth concept [58, 59] of the correction of the electrophoretic mobility for surface conduction.

This theoretical approach results in new aspects introduced by considering the nonlinearity of the Poisson–Boltzmann equation. In Fig. 5 the influence of this nonlinearity on the apparent charge density and, in addition, the effect of the surface conductivity correction [Eq. (17)] is demonstrated. The linear approximation is compared to the nonlinear solution which is derived with and without the correction for the surface conductivity. Under conditions of high dissociation the linear solution yields a much larger apparent charge density of the particle than the nonlinear one. The extremum of the linear solution in the region of low concentrations is a result of an increasingly large shift in the apparent pK due to unrealistically high electrical potentials. The extremum of the apparent charge density in the case of the nonlinear solution is basically a result of the increasing importance of nonlinearity toward low ionic strength. The correction for the surface conductivity at low ionic strength is quite important and reduces the apparent charge density at 1 mM by almost 50%. This gives rise to a maximum of the mobility itself. At a smaller degree of dissociation at pH $=$ pK the surface conductivity correction is less pronounced. Here, only the nonlinearity is of some importance, yield a reduction the mobility of the order of 15–30%. This effect can be partly explained by the influence of the potential. Equally important is the indirect effect of nonlinearity on the actual charge density. At pH $=$ pK the effect of the electric potential on the state of dissociation is most pronounced because the slope of the titration curve is at minimum at pH $=$ pK. For this reason relatively small changes in the potential results in a pronounced effect upon the group dissociation equilibrium. To summarize Fig. 5, it can be concluded that nonlinearity as well as surface conductivity provide an important influence on the mobility (apparent charge density) of particles with charged hairy layers.

The theoretical plot of the electrophoretic apparent charge density as a function of pH for a particle covered with a charged layer revealed a small but well-pronounced shift in the half-saturation value of the mobility toward higher pH with decreasing ionic

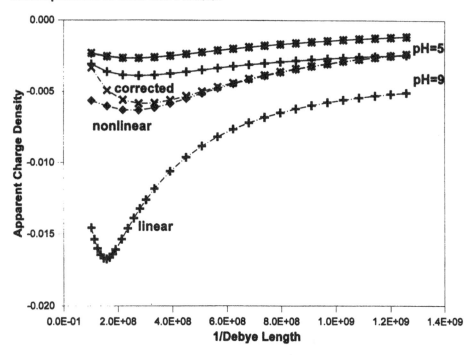

FIG. 5 Effect of nonlinearity and surface conductivity on the apparent charge density as a function of ionic strength. Layer parameters were as follows: thickness 4 nm; density of charged sites 0.03 $C\,m^{-2}$; $pK = 5$; zero flow penetration; relative permittivity 80. Particle radius 1 μm. No steric exclusion effects and no surface conductivity correction. (Reprinted with permission from Ref. 20. Copyright 1997 American Chemical Society.)

strength [20]. This was caused by the dependence of the apparent pH on the electrical potential. Furthermore, concerning the effect of the hydrodynamic flow inside the hairy layer on the apparent charge density, it was shown that a characteristic flow penetration depth of a few angstroms results in a large effect on the particle mobility.

A further theoretical plot has demonstrated the effect of steric mobile ion exclusion on the apparent charge density as a function of ionic strength, which was not considered before in the context of electrophoresis. The complete or gradual exclusion of a cation, for instance, results in a remarkable increase in the apparent charge density.

In order to provide a theoretically based description of the polyion-covered latex particle surface we compared theoretical with experimental fingerprints. For this reason a number of physicochemically based assumptions including molecular considerations have been made to achieve a good agreement between theory and experiment. However, one cannot expect a good fit over the whole range of fingerprints because the structure of the charged layer most probably changes with pH and conductivity.

In the following we will shortly discuss two typical 3D graphs representing the comparison between the theoretical electrophoretic fingerprint and the corresponding experimental data. The objects of investigation are consecutively adsorbed poly(allylamine hydrochloride) (PAH) and poly(styrenesulfonate) (PSS) layers on polystyrene latex beads. Figure 6 shows the comparison of the experimental fingerprint of the fist adsorbed PAH layer (filled circles) with the corresponding theoretical fingerprint (grid). There is a good coincidence between experiment and theory (except at low pH) if a 1 nm thick adsorbed

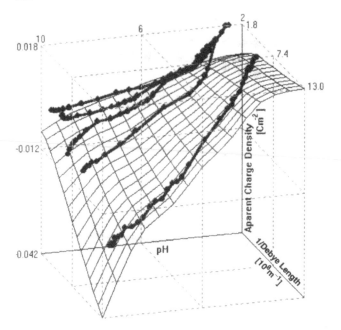

FIG. 6 3D comparison in terms of apparent charge density between the theoretical fingerprint (grid) and the experimental data of the PAH covered sulfate latex (filled circles). Parameters for the theoretical surface: latex surface – charged sites, density $0.04\,C\,m^{-2}$, $pK = 4.5$, and density $0.05\,C\,m^{-2}$, $pK = 6.5$, homogeneous site distribution and no adsorption of ions are assumed; PAH charges: density $0.03\,C\,m^{-2}$, $pK = 9$, and density $0.03\,C\,m^{-2}$, $pK = 10$, both distributions homogeneous. PAH layer thickness 1 nm. Adsorption sites in the PAH layer, density $0.04\,C\,m^{-2}$, anion adsorption constant 100/M, homogeneous distribution. Flow penetration depth is 1.34 nm; 85% ion accessible volume within the layer; layer hairs 0.5 nm radius. (Reprinted with permission from Ref. 20. Copyright 1997 American Chemical Society.)

hairy layer is assumed, which presumably consists of PAH. The divergence at low pH that has been observed at any ionic strength can be explained considering that the underneath latex charges (carboxyl groups) are not dissociated and thus release some PAH charges which have been immobilized by ion-pair formation with the latex charged groups at high pH. The PAH largely replaces the adsorbed cations from the latex surface; the PAH layer is too thin to be capable of hydrodynamically screening the latex surface charges. About one-third of the PAH charges form ion pairs with the polystyrene carboxyl groups and become electrophoretically active only after protonation of these groups. The coverage of the latex surface by adsorbed PAH seems to be probably not complete. Figure 7 demonstrates the influence of the second consecutively adsorbed PAH layer. Again one observes the characteristic release of some apparent extra charges at low pH, indicating the ion-pair formation of approximately one-third of the polyion charges and again the fingerprint is consistent with the assumption of 1 nm layer thickness. It is worth noting that the latex charge forms a significant part of the electrophoretic effect charge concluded from the broad pK dependence. We favor a picture of polymer-free spots at the surface which might be too small in diameter to provide a possibility for further ion-pair formation between polyanions and polycations, especially since the binding energy per ion pair is only of the order of kT [60]. The interpretation of the fingerprint does not require interpenetration of the charged polymer layers. Otherwise, it cannot be ruled out.

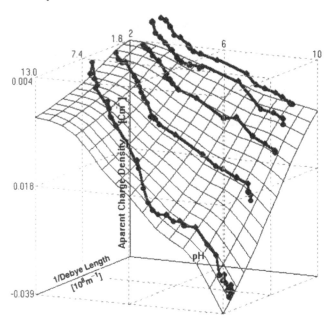

FIG. 7 3D comparison in terms of apparent charge density between the theoretical fingerprint (grid) an the experimental data on PAH–PSS–PAH covered sulfate latex. Parameters for the theoretical surface: latex surface same as in Fig. 6. First PAH layer has no charges. PSS charges: density $0.005\,C\,m^{-2}$, $pK = 3.5$, homogeneous distribution, layer thickness 1 nm. No adsorption to PSS. Outermost PAH layer charges: density $0.02\,C\,m^{-2}$, $pK = 9$; density $0.02\,C/m^2$, $pK = 10$, both distributions homogeneous. Layer thickness 1 nm. Anion adsorption sites in the outermost PAH layer: density $0.05\,C\,m^{-2}$, adsorption constant 100/M, homogeneous distribution. Flow penetration depth is 1.34 nm; 85% ion accessible volume within the layer; layer hairs 0.5 nm radius. (Reprinted with permission from Ref. 20. Copyright 1997 American Chemical Society.)

Summarizing, each layer deposition is accompanied by charge overcompensation. Not only the top layer but also the underneath layers and the naked latex surface contribute to the particle mobility which can be interpreted as an incomplete coverage or a polyelectrolyte interpenetration. The thickness of the top adsorbed hairy layer is of the order of 1 nm. About one-third of the charged groups of the top layer forms ion pairs with the underneath charges. Counterion adsorption to the charged groups of the top layer can be observed. These conclusions regarding the consecutively adsorbed polyion layers on a latex support underline the usefulness of electrophoretic fingerprinting.

The large number of unknown parameters in the procedure mentioned above may produce a certain arbitrariness. However, the results are mostly satisfactory, particularly if further molecular and structural information is available. We hope that the comprehensive interpretation of electrophoretic fingerprints described here give rise to further applications of electrokinetics to hairy particles.

IV. ELECTROPHORETIC MULTIPARAMETER ANALYSIS

In the last years it became clear that many problems in research and practice cannot be solved by simple electrophoretic investigations and that the additional, simultaneous measurement of several particle and cell characterizing parameters such as size, density, and shape are necessary to improve the interpretation of electrophoretic results.

The simultaneous measurement of several parameters characterizing the particle and cell may be successful for the discrimination between normal and pathological cells, particularly in cases where conventional methods fail to work. Figure 8 shows typical graphs of particle-characterizing parameters of human erythrocytes. The combination of the measured parameters electrophoretic mobility, density (sedimentation), and size provides 2D graphs with distinct spots which are characterized by shape, localization, and standard deviation. The corresponding 3D plot represents electrophoretic mobilities of the cells versus cell size and density (Fig. 9).

Furthermore, electrophoretic fingerprints obtained by measuring the electrophoretic mobility of particles as a function of pH and the specific conductance can be practically used to detect very small differences and changes in surface properties. The aim of the following examples is to demonstrate the usefulness of the multiparameter analysis for the discrimination of particles and cells and for the detection of small changes on surfaces.

A typical example for the detection of cellular alterations in the course of a disease is shown in Fig. 10 a–c [21]. The figure demonstrates the simultaneous measurement of electrophoretic mobility and size of individual cells in a cell mixture which was prepared from the peritoneal dialysate of a patient in different stages of peritonitis. The cells were independently identified using monoclonal antibodies. In the acute state, the electrophoretic mobility of macrophages is decreased as expected for activated cells [61]. Six weeks after drug treatment a normalization of number and composition of cells was observed, accompanied by an increased macrophage mobility.

Moreover, the sensitive measurement of the particle/cell density (via sedimentation rate) using the electrophoretic multiparameter analysis is a simple alternative to time-consuming standard methods. The density is an important parameter for cell characterization, particularly the restorage of granula, the organelle content, and heterogeneity of circulating platelet populations. Density and volume/density relations play an important but not completely understood role in hematological diseases with pathological platelet populations [62]. The combination of density and surface charge, which is known to be characteristic for blood cells, opens a powerful method in the diagnosis of hematological diseases.

For the detection of structural alterations on surfaces one has to analyze exactly where the changes in the 3D graph of an experimental electrophoretic fingerprint took place. One possibility is the subtraction of the electrophoretic mobilities of two measured fingerprints. Their difference was estimated by calculating the electrophoretic mobilities in the same PH – pλ (logarithm of the specific conductance) space, followed by subtraction of the corresponding grid points. Differences below the noise range, which was defined by reproducibility measurements of erythrocytes, have been eliminated. Such differences in the case of two fingerprints representing human red cell concentrates are shown in Fig. 11 [21]. Leucocyte-depleted red cell concentrates are generally considered to be the standard product to administer to immunocomprised patients [63]. Their advantages are the prevention of nonhemolytic febrile transfusion reactions, reduction of alloimmunization, and prevention of cytomegalie virus transmission [64, 65]. The depletion of leucocytes by means of filtration of red cell concentrates reduces the risk that the degradation of leucocytes during storage leads to the release of leucocyte-associated virus particles and humoral factors.

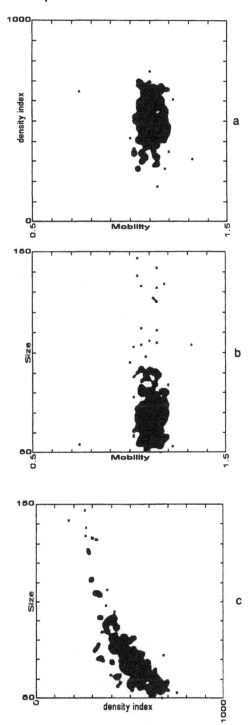

FIG. 8 Typical 2D graphs representing human erythrocytes: (a) electrophoretic mobility $(10^{-8} \, m^2 \, s^{-1} \, V^{-1})$ versus density index (arbitrary units), (b) electrophoretic mobility versus size (arbitrary units), and (c) density versus size. (From Ref. 22.)

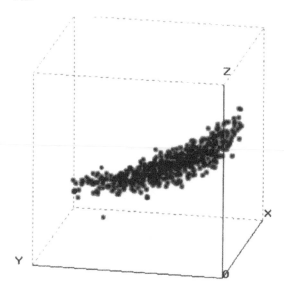

FIG. 9 3D graph of human erythrocytes. Electrophoretic mobility in $10^{-8}\,m^2\,s^{-1}\,V^{-1}$ (x axis) versus size (y axis) and density index (z axis), both in arbitrary units. (From Ref. 22.)

To investigate the influence of filtration on the erythrocyte membrane, red cell concentrates were stored for five weeks after filtration and then compared with nonfiltered batches using the electrophoretic fingerprint technique. The measured fingerprints showed a considerable difference between filtered and nonfiltered cells one week after filtration. This difference became smaller in the course of storage and disappeared nearly after five weeks. This demonstrates a storage-dependent effect of filtration on the red cell membrane [21].

In conclusion, the combination of several, simultaneously measured parameters such as size, density, and shape with the electrophoretic mobility offers new potential fields of application in nonbiological and biological research and clinical medicine. The 2D and 3D graphs make it possible to detect small changes in small portions of cells.

V. CONCLUSIONS

Particle electrophoresis has become an increasingly effective method for characterizing the surface electrostatic properties of colloidal particles and cells. An important step was the introduction and theoretical treatment of electrophoretic fingerprinting. The measurement of the electrophoretic mobility of a single particle under varying conditions (pH and conductance of the medium) connected with a comprehensive analysis of experimental fingerprints by using calculated fingerprints makes it possible to obtain a better insight into charge-generating processes and their dynamics at particle surfaces and leads to more information about the surface electrical structure. Using the simultaneous measurement and combination of electrophoretic mobility with particle size, density, and shape one can detect small differences between normal and pathological cells and thus can complete diagnosis and prognosis of diseases.

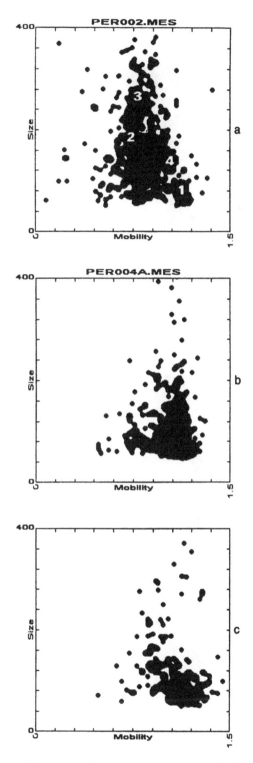

FIG. 10 Electrophoretic mobilities ($10^{-8}\,\mathrm{m^2\,s^{-1}\,V^{-1}}$) versus size (arbitrary units) of cells obtained from the peritoneal dialysis solution of a patient in different stages of peritonitis: (a) acute state, (b) 2 weeks later, (c) 6 weeks later, 1 – erythrocytes; 2 – granulocytes, 3 – macrophages, 4 – lymphocytes. (From Ref. 22.)

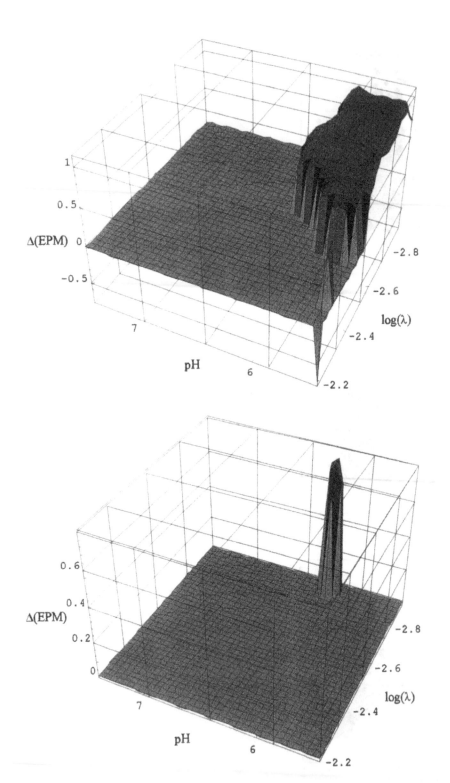

FIG. 11 Difference between experimental fingerprints of filtered and nonfiltered red blood cells after storage for 1 week (top) and 5 weeks (bottom). The difference between the electrophoretic mobilities (EPM in $10^{-8}\,\mathrm{m^2\,s^{-1}\,V^{-1}}$) is plotted versus pH and the decadic logarithm of conductivity (in S/m). (From Ref. 21.)

ACKNOWLEDGMENTS

We would like to thank all collaborators who are mentioned in the common papers. We are particularly indebted to Dr. Edwin Donath for his fundamental preliminary theoretical work, stimulating discussions, and his considerable contribution to the theory of electrophoretic fingerprinting.

REFERENCES

1. FF Reuss. Memoires de la Societe Imperiales de Naturalistes de Moskou 2: 327, 1809.
2. BJ Marlow, D Fairhurst, T Oja. Proceedings of 60th Colloid and Surface Science Symposium, Atlanta, GA, 1986.
3. A Morfesis. PhD thesis, University of Massachusetts, Amherst, MA, 1986.
4. BJ Marlow, RL Rowell, A Morfesis. Proceedings of the Gordon Research Conference on Polymer Colloids, Tilton, NH, 1987.
5. BJ Marlow, RL Rowell, A Morfesis. Proceedings of Fall ACS Meeting, New Orleans, 1987.
6. BJ Marlow, RL Rowell. J Energy Fuels 2: 125, 1988.
7. RE Marganski, RL Rowell. J Energy Fuels 2: 132, 1988.
8. BJ Marlow, D Fairhurst, W Schütt. Langmuir 4: 776, 1988.
9. A Morfesis, RL Rowell. Langmuir 6: 1088, 1990.
10. S-J Shiau. PhD thesis, University of Massachusetts, Amherst, MA, 1989.
11. RL Rowell. In: F Candeau, RH Ottewill, eds. Scientific Methods for the Study of Polymer Colloid. Dordrecht: Kluwer Academic, 1990, pp 187–208.
12. BJ Marlow, RL Rowell. Langmuir 7: 2970, 1991.
13. BJ Marlow, RL Rowell. Energy Fuels 4: 125, 1988.
14. RL Rowell, S-J Shiau, BJ Marlow. ACS/PMSE Proc 62: 52, 1990.
15. RL Rowell, S-J Shiau, BJ Marlow. In: T Provder, ed. Particle Assessment and Characteriation. ACS Symposium Series, Washington, DC, 1991, pp. 326–336.
16. JH Prescott, S-J Shiau, RL Rowell. Langmuir 9: 2071, 1993.
17. TW Healy, LR White. Adv Colloid Interface Sci 9: 303, 1978.
18. RJ Hunter. Zeta Potential in Colloid Science. New York: Academic Press, 1981, ch. 7.
19. E Donath, A Budde, E Knippel, H Bäumler. Langmuir 12: 4832, 1996.
20. E Donath, D Walther, VN Shilov, E Knippel, A Budde, K Lowack, CA Helm, H Möhwald. Langmuir 13: 5294, 1997.
21. A Budde, E Knippel, G Grümmer, J Treichler, H Brockmann, E Donath, H Bäumler. electrophoresis 17: 507, 1996.
22. G Grümmer, E Knippel, A Budde, H Brockmann, J Treichler. Instrum Sci Technol 23: 265, 1995.
23. P Goetz. US Patent 4 154 669, 1979.
24. P Goetz. In W Schütt, H Klinkmann, eds. Cell Electrophoresis. Berlin: Walter de Gruyter, 1985, pp 41–53.
25. BD Ripley. Spatial Statistics. New York: Wiley-Interscience, 1981.
26. E Knippel, H Wolf. Z Klin Med 42: 1621, 1987.
27. CF Zukowski, DA Saville. J Colloid Interface Sci 114: 32, 1986.
28. CF Zukowski, DA Saville. J Colloid Interface Sci 114: 45, 1986.
29. BR Midmore, RJ Hunter. J Colloid Interface Sci 122: 521, 1988.
30. AEJ Meijer, WJ van Meegan, L Lyklema. J Colloid Interface Sci 66: 99, 1978.
31. AG van der Put. PhD thesis, University of Wageningen, The Netherlands, 1980.
32. B-R Paulke, P-M Möglich, E Knippel, A Budde, R Nitzsche, RH Müller. Langmuir 11: 70, 1995.
33. A Budde, E Knippel, G Grümmer, R Nitzsche. Electrophoresis 15: 577, 1994.

34. V Pastushenko, E Donath. Stud Biophys 56: 7, 1976.
35. E Donath, V Pastushenko. Bioelectrochem Bioenerg 6: 543, 1979.
36. IS Jones. J Colloid Interface Sci 68: 451, 1979.
37. RW Wunderlich. J Colloid Interface Sci 88: 385, 1982.
38. S Levine, M Levine, KA Sharp, D Brooks. Biophys J 42: 127, 1983.
39. P Snabre, P Mills. Colloid Polym Sci 263: 494, 1985.
40. AE Jaroshchuk. Colloid J 47: 369, 1985.
41. KA Sharp, DE Brooks. Biophys J 47: 563, 1985.
42. E Donath, A Voigt. J Colloid Interface Sci 109: 122, 1986.
43. E Donath, A Voigt. Biophys J 49: 493, 1986.
44. H Ohshima, T Kondo. J Colloid Interface Sci 116: 305, 1987.
45. VM Starov, YE Solomentsev. J Colloid Interface Sci 158: 159, 193.
46. VM Starov, YE Solomentsev. J Colloid Interface Sci 158: 166, 1993.
47. H Ohshima. Adv Colloid Interface Sci 62: 189, 1995.
48. E Donath, D Lerche. Bioelectrochem Bioenerg 7: 41, 1980.
49. RV McDaniel, A McLaughlin, AP Winiski, M Eisenberg, S McLaughlin. Biochemistry 23: 4618, 1984.
50. C Holt, DG Dalgleish. J Colloid Sci 114: 513, 1986.
51. S McLaughlin. Annu Rev Biophys Chem 18: 113, 1989.
52. H Ohshima, K Makino, T Kato, K Fujimoto, T Kondo, H Kawaguchi. J Colloid Interface Sci 159: 512, 1993.
53. E Donath, A Krabi, G Allan, B Vincent. Langmuir 12: 3425, 1996.
54. V Patushenko, E Donath. Stud Biophys 56: 7, 1976.
55. E Donath, A Voigt. Biophys J 49: 493, 1986.
56. A Voigt, E Donath, G Kretzschmar. Colloids Surfaces 47: 23, 1990.
57. E Donath, P Kuzmin, A Krabi, A Voigt. Colloid Polym Sci 271: 930, 1993.
58. DC Henry. Trans Faraday Soc 44: 1021, 1948.
59. F Booth. Trans Faraday Soc 44: 955, 1948.
60. K Lowack, CA Helm. Submitted for publication in Macromolecules.
61. E Knippel, J Rychly. Thesis, University of Rostock, Rostock, Germany, 1989.
62. A Wehm. Hämostasestörungen bei hämatologischen Erkrankungen und malignen Tumoren. Lengerich: Wolfgang Pabst Verlag, 1990, pp 93–144.
63. G Matthes. In: CF Högmann, ed. Leucocyte Depletion of Blood Components. Amsterdam: VU University Press, 1994, pp 115–124.
64. G Sirchia. In: SR Krzt, ML Baldwin, G Sirchia, eds. Controversies in Transfusion Medicine. Arlington, American Association of Blood Banks, 1990, pp 1–12.
65. RN Pietersz. Transfus Med Rev 7: 17, 1993.

25

Electrophoresis of Polymer-coated Particles

KIMIKO MAKINO Science University of Tokyo, Tokyo, Japan

I. INTRODUCTION

Polymer coating produces a new surface on the original surface of the material. The surface properties and structure change by this modification. That is, if the layer produced on the surface by the modification is thick enough, the surface shows the properties of the polymer layer used to coat. It is well known that PEGylation, i.e., modification with poly(ethyleneglycol), makes various polymer surfaces more hydrophilic [1, 2]. Therefore, PEGylation using block copolymer, by surface immobilization or other methods, has been reported to reduce the levels of cell adhesion and protein adsorption, the effects of which have been applied to stealth liposomes and microspheres in the field of drug delivery systems. Such changes in surface properties are especially observed in charge density originating from various groups fixed in the surface area and in surface softness, which can be predicted by electrokinetic measurements on polymer particles and from analysis of the data [3, 4].

Electrokinetic measurements on small particles give us information on the structure of their surface layers. The properties of the surfaces are known to change markedly in the direction of the depth from the surfaces, which is more obvious in the surfaces of the smaller particles than of the larger ones, because the relative surface area is much larger in the former. It is well known that the interfacial region between two phases has a finite thickness and in this narrow region properties of the system change over microscopic dimensions. Due to such changes of the properties, it is often observed that the charged groups do not distribute uniformly in the surface layer. In particular, in microspheres prepared by polymerization of aqueous monomer molecules proceedings in water/oil (w/o) emulsion, the charged groups distribute nonuniformly in a single microsphere, as will be discussed later. Also, the membrane of microcapsules prepared by an interfacial polymerization method at the interface between the oil phase and water phase has an asymmetrical structure in the normal direction to the membrane surface.

In this chapter, it will be discussed that the structure of the particles is dependent on the particle size, that is, the smaller particles have higher surface charge density than larger ones, if they are produced by the polymerization of monomers from the w/o emulsion process, even though the initial monomer composition before polymerization is kept constant. This phenomenon is especially observed in small systems and not in bulk polymerization. The relationship between distribution of charged groups in microcapsule membranes and the membrane swellability will also be discussed. The changes in polymer surface properties produced by PEGylation of surfaces and the effects of PEGylation upon the interaction between the polymer surface and biological cell surfaces will then be discussed.

II. PREPARATION OF MONODISPERSE POLY(ACRYLAMIDE-*CO*-ACRYLIC ACID) HYDROGEL MICROSPHERES BY MEMBRANE EMULSIFICATION TECHNIQUE AND THEIR SIZE-DEPENDENT SURFACE PROPERTIES

When microspheres are used as drug devices, their size and surface properties are primarily important. Highly monodisperse microspheres of less than 1 μm are applicable to the targeting of drugs to various organs such as the lung, liver, kidney, and especially the brain [5–8]. Since the pioneering work by Nakashima et al. [9] on the preparation of monodisperse microspheres using SPG (Sirasu porous glass) membranes, various kinds of monodisperse microspheres with a narrow size distribution have been developed [10–13]. The electrical surface properties are also important when they are used as drug devices, because the device surface interacts with various kinds of biological cell surfaces carrying different electrical charges after the microspheres are administered and before they reach the target organ [14–18]. For the same reasons, the softness and hydrophilicity of the surfaces are also required for the microspheres used as drug devices. Hydrogels have been considered to be useful as a material for drug devices, because of their high biocompatibility and softness [9–22]. In view of these points, we have prepared monodisperse poly(acrylamide-*co*-acrylic acid) hydrogel microspheres with different polymer compositions (0–10 mol% acrylic acid and 90–100 mol% acrylamide) and with different sizes from w/o emulsion using SPG membranes of various pore sizes [23–25]. By measuring the electrophoretic mobility of microspheres and by analyzing the data with an electrokinetic theory for "soft" surfaces, it was found that the microsphere surface became more negatively charged and softer by the addition of acrylic acid, as will be described later. It was also suggested that the microspheres with smaller sizes have higher surface charge density than those with larger sizes, although the microspheres were prepared from the solution with the same monomer composition. From this, it is found that copolymerization of acrylamide monomers and acrylic acid monomers does not proceed homogeneously within a single microsphere. The density of polymer networks composed of both monomers is higher in the core region than in the surface layer of the microspheres. The accumulation of polymers in the core region of a microsphere explains the decrease in microsphere size with the increase in acrylic acid concentration observed in smaller microspheres.

A. Preparation of Microspheres

Three kinds of monomer solutions containing different concentrations of acrylamide and acrylic acid were prepared. The composition of the monomer solutions is shown in Table 1. The total monomer concentration was kept at 4.0 M in each solution. A 15-ml portion of the aqueous monomer solution was dispersed in an oil phase composed of 500 ml of cyclohexane containing 0.06% (w/v), 2,2'-azobis(isobutyronitorile), 1.0% (v/v) Sunsoft818H (surfactant) to prepare a w/o emulsion by the use of a hydrophobic MPG (microporous glass) membrane apparatus (Ise Chemical Corp.). The average pore sizes of the respective MPG membranes used in this process were 0.33, 0.73, 1.15, and 1.70 μm. The emulsion prepared was stirred at 365 r.p.m. at 70°C under nitrogen atmosphere, and 10 mL of cyclohexane containing 1.0% (v/v) Sunsoft818H and 1.5% (w/v) 2,2'-azobiis (isobutyronitorile) was added every 5 min. After adding a total of 300 mL of the cyclohexane solution, the reaction was continued at 70°C for 3 h. The microspheres prepared were centrifuged for 10 min at 1000 r.p.m. and washed with cyclohexane, isopropanol, ethanol and methanol, and finally with distilled water.

TABLE 1 Feed Composition of Aqueous Solution Phases for Preparing Microspheres

	Acrylamide (mol/L)	Acrylic acid (mol/L)	N,N'-methylene-bisacrylamide (mol/L)	N, N, N', N'-tetramethyl ethylenediamine (% v/v)	NaCl (mol/L)
Poly (acrylamide) microsphere	4.0	0	0.4	0.2	0.75
Poly[acrylamide-*co*-acrylic acid (5 mol%)] microsphere	3.8	0.2	0.4	0.2	0.75
Poly[acrylamide-*co*-acrylic acid (10 mol%)] microsphere	3.6	0.4	0.4	0.2	0.75

B. Measurements of Particle Size

The volume-averaged diameter of the microspheres and their size distribution were measured with a light scattering particle sizer (Malvern 3601, Master sizer/E/Malvern, Inc.). The size distribution was evaluated with the Span value defined as follows.

$$\text{Span} = \frac{D_{90\%} - D_{10\%}}{D_{50\%}} \tag{1}$$

Here ,$D_{N\%}$ (N = 10, 50, 90) means that the volume percentage of microspheres with diameters up to $D_{N\%}$ is equal to $N\%$. A smaller span value indicates a narrower size distribution. The results are summarized in Table 2. A broader size distribution was observed in the microspheres composed of hydrogel containing more acrylic acid. The size distribution was not affected by the membrane pore size. The microsphere size, on the other hand, was dependent on both the membrane pore size and the polymer composition. As the membrane with a larger pore size was used, larger microspheres were obtained. When the membrane with a pore size of $1.70 \, \mu m$ was used, the microsphere size clearly increased as acrylic acid concentration in the monomer mixture increased, while it decreased as acrylic acid concentration increased when the membranes with smaller pore sizes were used. It will be discussed in Section II.E how the pore size of the SPG membrane and the monomer composition affects the microsphere size and distribution, with their effects on the structure of the microsphere.

C. Measurements of Electrophoretic Mobility of Microspheres

The surface properties of the microspheres have been studied from their electrophoretic mobility values. The electrophoretic mobility of microspheres was measured in pH 7.4 buffer solutions of various ionic strengths by using an automated electrokinetic analyzer (Pen Kem System 3000) at 37°C. The ionic strength was adjusted by dilution of the buffer solution (with ionic strength 0.154) with distilled water. The values obtained for the electrophoretic mobility of microspheres containing no arylamide are plotted against the ionic strength of the dispersing media in Fig. 1. The electrophoretic mobility values for the poly(acrylamide) microspheres were almost zero or slightly negative in solutions at pH 7.4 with ionic strengths between 0.005 and 0.154, and no size dependence was observed.

Figures 2 and 3 show the electrophoretic mobility of poly(acrylamide-*co*-acrylic acid) microspheres containing 5 and 10 mol% acrylic acid, respectively. The electrophoretic mobility value of the microspheres were negative at all ionic strengths, implying that the surfaces of those microspheres have net negative charges, which originate from acrylic acid. Also, the electrophoretic mobility values became more negative as the acrylic acid concentration increased, with two exceptions. Only two types of poly(acrylamide-*co*-acrylic acid) microspheres (samples 11 and 12 in Table 2) prepared with the membrane of largest pore size ($1.70 \, \mu m$) showed almost the same electrophoretic mobility values (-1.5 to $-0.4 \, \mu m \, s^{-1} \, V^{-1} \, cm$) at all ionic strengths although they were composed of hydrogels of different acrylic acid concentrations (5 and 10 mol%). It should be emphasized here that the dependence of the electrophoretic mobility upon acrylic acid concentration in the hydrogel constituting the microspheres is not clear when the microsphere size is large. Interestingly, the electrophoretic mobility values for the microspheres containing acrylic acid have shown clear size dependence. More negative mobility values were obtained with smaller microspheres than with the larger ones, as shown in Figs 2 and 3.

TABLE 2 Relationship Between Pore Size and Microsphere Diameter with Size Distribution

Feeding composition									
				Dispersion phase					
		4.0 M Acrylamide			3.8 Acrylamide +0.2 M Acrylic acid			3.6 M Acrylamide +0.4 M Acrylic acid	
Membrane poor size (μm)	Sample #	Mean diameter (μm)	Span	Sample #	Mean diameter (μm)	Span	Sample #	Mean diameter (μm)	Span
0.33	1	1.45	0.65	2	1.47	0.72	3	1.35	0.80
0.73	4	2.59	0.35	5	2.37	0.42	6	2.33	0.42
1.15	7	3.75	0.65	8	3.42	0.73	9	3.09	0.77
1.70	10	6.29	0.49	11	6.64	0.64	12	7.43	0.98

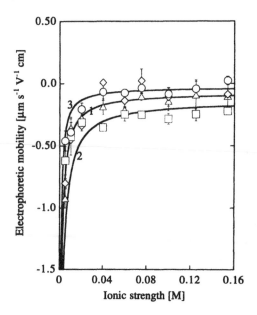

FIG. 1 Electrophoretic mobility of poly(acrylamide) microspheres. Symbols are experimental data measured as a function of the ionic strength in the suspending medium at pH 7.4 and 37°C: ($\triangle$) sample 1; ($\diamond$) sample 4; ($\square$) sample 7; ($\bigcirc$) sample 10. Solid curves are theoretical ones calculated with $zN = -0.007$ M and $1/\lambda = 0.83$ nm (curve 1); $zN = -0.012$ M and $1/\lambda = 0.89$ nm (curve 2); $zN = -0.004$ M and $1/\lambda = 0.7$ nm (curve 3).

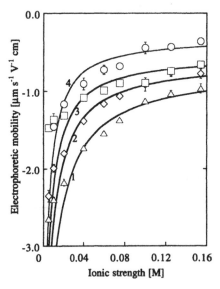

FIG. 2 Electrophoretic mobility of poly[acrylamide-*co*-acrylic acid (5 mol%)] microspheres. Symbols are experimental data measured as a function of ionic strength in the suspending medium at pH 7.4 and 37°C: ($\triangle$) sample 2; ($\diamond$) sample 5; ($\square$) sample 8; ($\bigcirc$) sample 11. Solid curves are theoretical ones calculated with $zN = -0.078$ M and $1/\lambda = 0.80$ nm (curve 1); $zN = -0.053$ M and $1/\lambda = 0.90$ nm (curve 2); $zN = -0.039$ M and $1/\lambda = 0.98$ nm (curve 3); $zN = -0.023$ M and $1/\lambda = 1.05$ nm (curve 4).

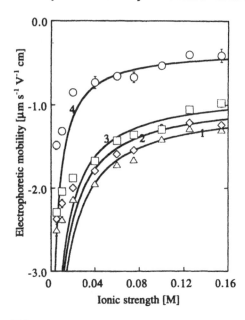

FIG. 3 Electrophoretic mobility of poly[acrylamide-*co*-acrylic acid (10 mol%)] microspheres. Symbols are experimental data measured as a function of ionic strength in the suspending medium at pH 7.4 and 37°C: (△) sample 3; (◇) sample 6; (□) sample 9; (○) sample 12. Solid curves are theoretical ones calculated with $zN = -0.080$ M and $1/\lambda = 0.90$ nm (curve 1); $zN = -0.066$ M and $1/\lambda = 0.99$ nm (curve 2); $zN = -0.053$ M and $1/\lambda = 1.02$ nm (curve 3); $zN = -0.023$ M, and $1/\lambda = 1.05$ nm (curve 4).

The microspheres composed of 3.8 M acrylamide and 0.2 M acrylic acid (5 mol% acrylic acid) showed changes in electrophoretic mobility values from -2.6 to $-1.0\ \mu\text{m s}^{-1}$ V^{-1} cm in solutions with ionic strengths between 0.005 and 0.154 in sample 2 (smaller size), while they have shown changes from -1.5 to $-0.4\ \mu\text{m s}^{-1}\ \text{V}^{-1}$ cm in the same solutions in sample 11 (largest size). The microspheres composed of 3.6 M acrylamide and 0.4 M acrylic acid (10 mol% acrylic acid) displayed similar size-dependent changes in their electrophoretic mobility. That is, the electrophoretic mobility of sample 3 changed from -2.6 to $-1.5\ \mu\text{m s}^{-1}\ \text{V}^{-1}$ cm in solutions with ionics strengths between 0.005 and 0.154, while that of sample 12 changed from -1.5 to -0.4 in the same solutions. The size dependence of the electrophoretic mobility is unexpected because the electrophoretic mobility is independent of the particle size, if the latter is much larger than the Debye length, $1/\kappa$ [where κ is given later in Eq. (8)]. This phenomenon will be explained by the size dependence on the structure of microspheres. In order to obtain information on the structure of the microspheres, the electrophoretic mobility data were analyzed with Ohshima's electrokinetic theory for "soft" particles [26].

The electrophoretic mobility of samples 1–12 is negative at all ionic strengths, and it becomes less negative as the ionic strength becomes higher in each sample. Also, the electrophoretic mobility values tend to a nonzero value even in the solution with an ionic strength as high as 0.154. This phenomenon suggest that the surface of this particle is "soft" and the data obtained can be discussed in terms of Ohshima's theory [26]. In this model, ionized groups of valency z are uniformly distributed at a number density of $N(\text{m}^{-3})$ in the surface layer. Suppose that the particle moves in a liquid containing a symmetrical electrolyte of valency v in the applied field, and n (m^{-3}) is the bulk concentration of symmetrical electrolytes in the dispelling medium. The electrophoretic mobility μ is then expressed as [26]

$$\mu = \frac{\epsilon_r \epsilon_0}{\eta} \frac{\psi_0/\kappa_m + \psi_{DON}/\lambda}{1/\kappa_m + 1/\lambda} f\left(\frac{d}{a}\right) + \frac{zeN}{\eta\lambda^2} \tag{2}$$

with

$$f\left(\frac{d}{a}\right) = \frac{2}{3}\left[1 + \frac{1}{2\left(1+\frac{d}{a}\right)^3}\right] \tag{3}$$

$$\psi_{DON} = \frac{kT}{ve}\ln\left[\frac{zN}{2vn} + \left\{\left(\frac{zN}{2vn}\right)^2 + 1\right\}^{1/2}\right] \tag{4}$$

$$\psi_0 = \frac{kT}{ve}\left(\ln\left[\frac{zN}{2vn} + \left\{\left(\frac{zN}{2vn}\right)^2 + 1\right\}^{1/2}\right] + \frac{2vn}{zN}\left[1 - \left\{\left(\frac{zN}{2vn}\right)^2 + 1\right\}^{1/2}\right]\right) \tag{5}$$

$$\lambda = (\gamma/\eta)^{1/2} \tag{6}$$

$$\kappa_m = \kappa\left[1 + \left(\frac{zN}{2vn}\right)^2\right]^{1/4} \tag{7}$$

$$\kappa = \left(\frac{2ne^2v^2}{\epsilon_r\epsilon_0 kT}\right)^{1/2} \tag{8}$$

Here, a is the particle radius, d is the thickness of the ion-penetrable surface layer, η is the viscosity, γ is the frictional coefficient of the surface layer, ϵ_r is the relative permittivity of the solution, ϵ_0 is the permittivity of a vacuum, ψ_{DON} is the Donnan potential of the surface layer, ψ_0 is the potential at the boundary between the surface layer and the surrounding solution, and κ is the Debye-Hückel parameter. We call ψ_0 the surface potential of a "soft" particle and ψ_m can be interpreted as the Debye-Hückel parameter in the surface layer. The parameter λ characterizes the degree of friction exerted on the liquid flow in the surface layer, and zN represents the number density of the fixed charges in the surface layer. The reciprocal of λ, i.e., $1/\lambda$, has the dimension of length and can be considered to be a "softness" parameter, since at the limit $1/\lambda \to 0$, the surface layer becomes rigid. Although Eq. (2) assumes symmetrical electrolytes, the electrolytes used are not symmetrical (cations are univalent but anions are not); the value of v is set approximately equal to unity ($v = 1$), since anions are less important for negatively charged particles. Note that in typical "soft" particles having core particles [14–18], the factor $f(d/a)$ in Eqs. (2) and (3) can practically be set equal to 1, because $d \ll a$. However, microspheres in this section may be regarded as spherical polyelectrolytes with no core ($a = 0$) so that f may be put equal to 2/3. Equation (2) involves two unknown parameters, N (m^{-3}) and $1/\lambda$ (nm), which now represent the fixed charge density in the microsphere and its softness, respectively. By a curve-fitting procedure [15–18], zN and $1/\lambda$ were determined. The values of μ calculated via Eq. (2) were plotted again the ionic strength in comparison with the experimental data in Figs 1–3 (solid lines). We found it possible to draw a curve with a pair of single values each of zN and $1/\lambda$ that is in good agreement with the experimental data over a wide range of ionic strengths (0.04–0.154) except for very low ionic strength. This means that the hydrogel microsphere surface can be considered as a "soft surface" described by Eq. (2) at ionic strengths between 0.04 and 0.154. Thus, this agreement enables us to estimate the values of the unknown parameters zN and $1/\lambda$ by a curve-fitting procedure. In the calculation, we used the value of the relative permittivity ϵ_r of water ($\epsilon_r = 74.39$ at 37°C).

The best-fit values for the charge density (zN) and the softness parameter ($1/\lambda$) of the surface of each sample are presented in Table 3. In this table, the size dependencies of both charge density (zN) and the softness parameter ($1/\lambda$) are clearly observed in poly(acrylamide-*co*-acrylic acid) microspheres, while they are not in poly(acrylamide) microspheres. As the size of the poly (acrylamide-*co*-acrylic acid) microspheres increases, the negative charge density decreases and the softness parameter ($1/\lambda$) slightly increase, although the microspheres of each mean diameter were all prepared from monomer mixture solutions with a constant composition. The increase in the negative charge density with decreasing particle size can be caused by the concentration of acrylic acid at the surface layer of the microsphere, or by the decrease in concentration of acrylic acid incorporated in the polymerization reaction. Both mechanisms may be available to decrease the charge density.

D. Determination of Total Amount of Carboxylic Acid Groups Located in Microspheres

Poly(acrylamide-*co*-acrylic acid) microspheres were redispersed in 0.01 M NaOH and were titrated potentiometrically with 0.01M HCl under a nitrogen atmosphere. By means of this titration, it was confirmed that constant concentration of acrylic acid is incorporated in the polymerization reaction with acrylamide, independent of the emulsion size.

E. Size-dependent Structure of Microspheres

In the electrokinetic experiments, the fixed charges that are located over the depth of order $1/\kappa_m$ ($\approx 1/\kappa$) measured inward from the surface contribute to the values of the electrophoretic mobility. The parameter N can be considered to be an average density over the depth $1/\kappa$. The increase in zN with the decrease in particle size is probably caused by the increase in the density of acrylic acid in the surface layer. That is, in the polymerization process, acrylic acid monomer or oligomer containing acrylic acid are considered to move toward the emulsion core, escaping from the w/o interfaces, because the microspheres were prepared from w/o emulsion, as noted in Section II.A. Therefore, the more charged polymer chains are stable in the core of the microspheres, whereas the surface region of the microspheres is composed of the less charged polymer chains. In a microsphere with a large diameter, charged polymer chains can be localized in the microsphere core, but with decreasing diameter of the microsphere, charged polymer chains cannot exist in the core escaping from the w/o interface. For example, when two particles (particles A) with a radius of $0.25\,\mu\mathrm{m}$ exist in a particle (particle B) with a radius of $0.5\,\mu\mathrm{m}$, the surfaces of particles A are in contact with the surface of particle B, and the volume fraction of particles A in particle B is 12.5%. If the radius of particle B is $3.5\,\mu\mathrm{m}$ and includes 12.5% (v/v) of particles A with a radius of $0.25\,\mu\mathrm{m}$, then particles A can be localized in the core region within a radius of $2.2\,\mu\mathrm{m}$, assuming that the porosity of particles A to be 47.6% in particle B. This may cause changes in zN, depending on the particle size, as shown in Table 3.

TABLE 3 Dependence of Surface Charge Density (zN) and Softness Parameter ($1/\lambda$) on Size and Composition of Microspheres

| Feeding composition | Dispersion phase | | | | | | | | | | | |
| | 4.0 M Acrylamide | | | | 3.8 M Acrylamide +0.2 M Acrylic acid | | | | 3.6 M Acrylamide +0.4 M Acrylic Acid | | | |
Membrane pore size (μm)	Sample #	Mean diameter (μm)	zN (M)	$1/\lambda$ (nm)	Sample #	Mean diameter (μm)	zN (M)	$1/\lambda$ (nm)	Sample #	Mean diameter (μm)	zN (M)	$1/\lambda$ (nm)
0.33	1	1.45	−0.007	0.83	2	1.47	−0.078	0.80	3	1.35	−0.080	0.90
0.73	4	2.59	−0.007	0.83	5	2.37	−0.053	0.90	6	2.33	−0.066	0.99
1.15	7	3.75	−0.012	0.89	8	3.42	−0.039	0.98	9	3.09	−0.053	1.02
1.70	10	6.29	−0.004	0.70	11	6.64	−0.023	1.05	12	7.43	−0.023	1.05

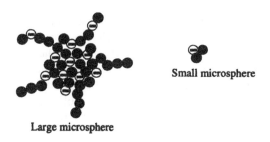

FIG. 4 Polymerization of acrylamide and acrylic acid in w/o emulsion: (●) acrylamide (non-charged monomer); (⊖) acrylic acid (negatively charged monomer).

Figure 4 shows schematically the possible structure of poly(acrylamide-*co*-acrylic acid) microspheres of different sizes. Particle A in the above discussion corresponds to the more charged polymer chains, and particle B to the microsphere as a whole. The concentration of acrylic acid monomers and charged oligomers is higher in the core region than in the surface layer of the microspheres, and the surface charge density in the surface layer increases as the size of microsphere decreases. At the same time, acrylamide monomer concentration should be higher in the core region of the microsphere in order to reduce electrical repulsion acting between negative charges. From these, not only the density of acrylic acid but also that of acrylamide should be lower in the surface region than in the microsphere core. On the other hand, when the hydrogel microspheres are composed of noncharged homopolymer such as poly(acrylamide), the microsphere structure is almost homogeneous, because such a heterogeneity of polymerization of monomers does not occur.

In Table 3, it is also clear that the microsphere surface becomes softer by the addition of acrylic acid, with two exceptions. That is, by the addition of acrylic acid, the softness parameter ($1/\lambda$) of the microspheres prepared using a SPG membrane with a pore size of $0.33 \mu m$ increases from 0.83 nm (sample 1) to 0.9 nm (sample 3). These increases in magnitude of zN and $1/\lambda$ by the addition of acrylic acid are observed also in microspheres prepared with the SPG membranes of pore sizes 0.73 and 1.15 m. Two exceptions are samples 11 and 12, which are prepared using a SPG membrane of pore size 1.7 μm. These microspheres are composed of 95 mol% acrylamide with 5 mol% acrylic acid, and 90 mol% acrylamide with 10 mol% acrylic acid, respectively. The same values of zN and $1/\lambda$ were, however, obtained for these two samples, the reasons for which will be discussed later.

Also in Table 2, it is clearly observed that the size distribution of microspheres becomes broader as the ratio of acrylic acid to acrylamide increases, independent of the pore size of SPG membranes, by comparison between samples 1–3, samples 4-6, samples 7–9, and samples 10–12, as noted before (Section II.B). The mean diameter clearly increases as acrylic acid concentration increases when SPG membranes with a pore size of $1.70 \mu m$ are used. When SPG membranes with smaller pore sizes are used, such dependence of microsphere size upon acrylic acid concentration is not observed, but the microsphere size decreases as acrylic acid concentration increases. This phenomenon would be explained by the size-dependent microsphere structure, as discussed above. Acrylic acid concentration is higher in the core region than in the surface layer of the microspheres, and acrylamide monomers should accumulate in the core region of the microspheres to reduce electric repulsion acting between acrylic acid molecules. Such an accumulation of polymers in the core region would explain the decrease in microsphere size with the increase in acrylic acid concentration observed in microspheres prepared with the SPG membranes of pore sizes of 0.33, 0.73, and 1.15 μm. In these cases, the polymer networks in the core region become more dense as the acrylic acid concentration increases, although the surface layer is composed of a loose polymer network. On the other hand, when the microsphere size is relatively large (samples 10–12), sufficient volume is available for 10 mol% acrylic acid inside the microspheres, so that the polymer network become looser. Therefore, the microsphere size increases as the acrylic acid concentration increases, as observed in samples 10–12. This volume increase explains the same electrophoretic mobility values observed in two types of large microspheres containing different concentrations of acrylic acid (5 and 10 mol%) and the same values of zN and $1/\lambda$ for both samples reported in Table 3. We have also found that poly(acrylamide-co-acrylic acid) microspheres containing 15 mol% could not be obtained from w/o emulsions with this system, because the hydrogel was composed of a loose polymer network and was mechanically weak.

III. EFFECTS OF PEGYLATION ON SURFACE PROPERTIES OF MICROCAPSULES

A. Two-layer Structure of Microcapsule Membrane as Predicted from Electrophoretic Studies

In this section, we discuss the preparation and properties of two types of microcapsule membranes with a two-sublayer structure, each of which has a different charge density and softness [27]. By PEGylation of these microcapsule surfaces [1, 2, 28–30], it will be shown that the charge density in the outer sublayer decreases, while that in the inner sublayer is not affected. Also, the surface layer became softer by PEGylation. The effects of membrane composition upon the swellability of the microcapsule membrane will be also discussed [27, 31]. That is, the asymmetrical structure of microcapsule membranes causes the nonuniform distribution of charges and then the swellability of the microcapsule changes, depending on the permittivity of the dispersing medium.

1. Preparation of Microcapsules

Four types of hydrophilic gel microcapsules containing water have been prepared by an interfacial polymerization method. Each type of microcapsule has membranes of different composition. Using three kinds of monomers, N,N-dimethylacrylamide (DMMAm), 4-aminomethylstyrene (AmSt), and N,N-dimethylaminopropylacrylamide (DMAPAA), two types of water-soluble copolymers with different compositions having primary and tertiary amino groups were synthesized. Two more types of copolymers were also synthesized by copolymerization of α-acryloxypoly(ethyleneglycol) (a-PEG) with the above two kinds of monomer mixture. Structural formulas of the copolymers are shown in Fig. 5. Polymerization conditions are summarized in Table 4; copolymers 1 and 2 represent poly(DMAAm-co-DMAPAA-co-AmSt) and copolymers 3 and 4 represent poly-(DMAAm-co-DMAPAA-co-AmSt-co-PEG). These four types of copolymer (copolymers 1–4) were polymerized with terephthaloy dichloride at the w/o interface to prepare four types of microcapsules (MCs 1–4) containing water. A mixed organic solvent was prepared from a mixture of cyclohexane and chloroform (3 : 1, v/v). Triethylamine (0.5 ml), sorbitan trioleate (SO-30) (0.5 ml), and 2-propanol (1.5 ml) were dissolved in 100 ml of the mixed organic solvent. One of the copolymers 1–4 (Table 4) was dissolved in 10 ml of distilled

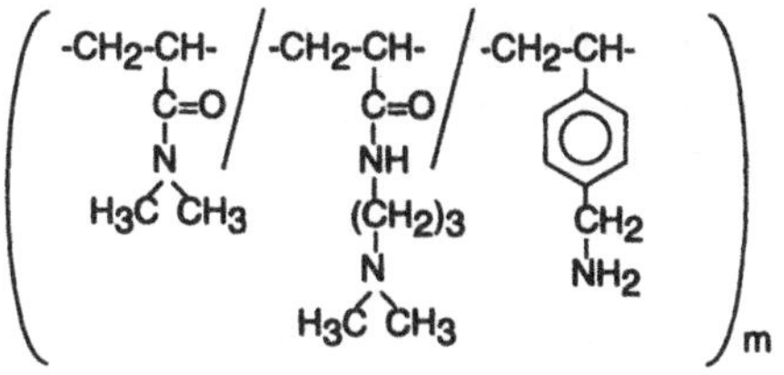

poly(DMAAm-*co*-DMAPAA-*co*-AmSt)

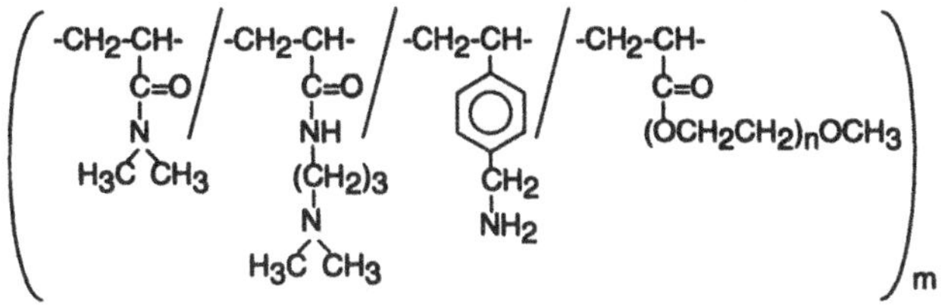

poly(DMAAm-*co*-DMAPAA-*co*-AmSt-*co*-PEG)

FIG. 5 Structures of poly(DMAAm-*co*-DMAPAA-*co*-AmSt) and poly(DMAAm-*co*-DMAPAA-*co*-AmSt-*co*-PEG).

water to a concentration of 2.0% (w/v). The copolymer solution was added to the above organic solution and the mixture was stirred at 848 r.p.m. for 20 min at room temperature. Terephthaloyl dichloride was dissolved in 100 ml of the mixed organic solvent to a concentration of 0.03 M and the solution was then added to the above emulsion without stopping the stirring. After stirring for 90 min at 4°C, cyclohexane was added in order to terminate the reaction. Microcapsules were centrifuged at 2000 r.p.m. for 15 min and were washed in cyclohexane, in 2-propanol, in ethanol, in methanol, and finally in distilled water. Microcapsules thus prepared were kept in distilled water at 4°C, before use for the following experiments. These microcapsules were termed MCs 1–4 corresponding to the kind of copolymers used.

TABLE 4 Synthesis of Copolymers Having Primary and Tertiary Amino Groups

Samples	Feed composition (mol%)				Yield %
	DMAAm	AmSt	DMAPAA	a-PEG[a]	
1	85.00	5.00	10.0	0.0	53.5
2	80.00	5.00	15.0	0.0	55.0
3	84.80	4.99	9.98	0.23	68.4
4	79.81	4.99	14.96	0.24	54.8

Solvent: ethanol (100 mL), initiator 2,2'-azobis (2,4-dimethylvaleronitrile) V-65 (10 mM), reaction temp.: 45°C, reaction time: 24 h.
[a] a-PEG: α-acryloxylpoly(ethylene glycol).

2. Total Amount of Fixed Charges Located in Microcapsule Membranes

To determine the number of carboxylic acid groups fixed in microcapsule membranes, a certain volume of microcapsules was redispersed in 0.01 M NaOH and then the dispersion medium was titrated with 0.01 M HCl under a nitrogen atmosphere and the titration was monitored potentiometrically. For the determination of the number of amino groups fixed in microcapsule membranes, the microcapsules were redispersed in 0.01 M HCl and the dispersion medium was titrated with 0.01 M NaOH.

The total amounts of amino groups and carboxylic acid groups located in the microcapsule membranes measured in above process are summarized in Table 5. The values show the number of charges which 1 ml of microcapsules carries. The values were calculated by assuming that microcapsules are spherical and that 25% of their sedimentation volume is the void volume. Negative and positive charges originate from carboxylic acid groups and amino groups, respectively. It is clear that MCs 1–4 have both positive and negative charges in their membranes, but the numbers of negative charges are larger than those of positive ones. Therefore, the net charges are negative in the membranes of MCs 1–4. If the charges are uniformly distributed in the membranes, these values can be converted into charge density in the membranes by using the values of microcapsule size and the membrane thickness. If the microcapsule membrane has an asymmetrical structure in the normal direction to the membrane surface, however, the charge density is not uniform in the membrane.

3. Membrane Structure of Microcapsules Predicted from Electrophoretic Studies

Further information on the surface properties of hydrogel microcapsules and charge distribution in the membranes is available from the electrophoretic mobility measurement of the microcapsules [32–34]. The electrophoretic mobility of microcapsules was measured in Dulbecco's buffer solutions at pH 7.4 with various ionic strengths by using an automated electrokinetic analyzer (Pen Kem System 3000) at 37°C. The ionic strength was adjusted by dilution of Dulbecco's buffer solution (ionic strength 0.154 M) with distilled water.

The measured values of electrophoretic mobility of MCs 1–4 are plotted against the ionic strength of the dispersion medium in Figs 6 and 7.

Two kinds of microcapsules (MC 1 and MC 3) exhibit negative mobility values in all solutions at pH 7.4 with ionic strengths between 0.005 and 0.154 M, as shown in Fig. 6, implying that the surfaces of these microcapsules have net negative charges which originate from rich carboxylic acid groups. Unexpectedly, the electrophoretic mobility values show a minimum at an ionic strength of 0.01. That is, the electrophoretic mobility

TABLE 5 Total Fixed Charges Located in Membranes of MCs 1–4

Samples	Negative charges (mmol)	Positive charges (mmol)	Net charges (mmol)
MC 1	0.121	0.021	−0.100
MC 2	0.089	0.018	−0.071
MC 3	0.101	0.028	−0.072
MC 4	0.091	0.035	−0.056

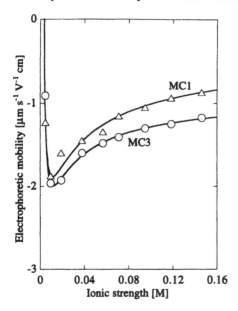

FIG. 6 Electrophoretic mobility of MC 1 and MC 3. Symbols are experimental data with MC 1 ($\triangle$) and MC 3 ($\bigcirc$) measured as a function of ionic strength in the suspending medium at pH 7.4 and 37°C. Solid curves are theoretical ones calculated with $z_1 N_1 = -0.111$ M, $z_2 N_2 = 0.041$ M, $1/\lambda = 0.505$ nm, $d = 1.706$ nm (MC 1), and with $z_1 N_1 = -0.048$ M, $z_2 N_2 = 0.048$ M, $1/\lambda = 1.230$ nm, $d = 3.758$ nm (MC 3).

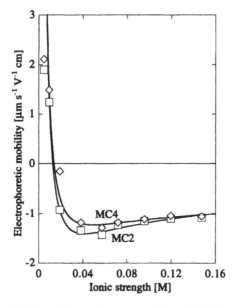

FIG. 7 Electrophoretic mobility of MC 2 and MC 4. Symbols are experimental data with MC 2 ($\square$) and MC 4 ($\diamondsuit$) measured as a function of ionic strength in the suspending medium at pH 7.4 and 37°C. Solid curves are theoretical ones calculated with $z_1 N_1 = -0.140$ M, $z_2 N_2 = 0.166$ M, $1/\lambda = 0.471$ nm, $d = 2.090$ nm (MC 2), and with $z_1 N_1 = -0.084$ M, $z_2 N_2 = 0.159$ M, $1/\lambda = 0.778$ nm, $d = 2.751$ nm (MC 4).

increases in magnitude from about -1.0 to $-2.0\,\mu\mathrm{m\,s^{-1}\,V^{-1}}$ cm between ionic strengths of 0.005 and 0.01 M. In solutions with ionic strengths higher than 0.01 M, the mobility becomes less negative as the ionic strength rises and reaches nonzero values.

On the other hand, the electrophoretic mobility values of MC 2 and MC 4 alter their sign from positive to negative as the ionic strength rises at ionic strengths between 0.01 and 0.02 M, as shown in Fig. 7. They also show a minimum mobility for an ionic strength around 0.06 M and reach nonzero negative values as the ionic strength rises. In all cases (Figs 6 and 7), the electrophoretic mobility tends to nonzero values even in solutions with ionic strengths as high as 0.154. This means that the surfaces of MCs 1–4 are "soft" and their surface properties can be analyzed by the electrokinetic theory for soft surfaces [26]. This theory provides us with information on the softness and charge density of a surface layer, as has already been seen in the previous section. As described below, the presence of the mobility minimum observed in MCs 1–4 suggests that the surface layer of each of the microcapsules is composed of at least two sublayers, each of which has a different charge density. In particular, in the case of MC 2 and MC 4, the surface layers are considered to be composed of two oppositely charged sublayers, the outer sublayer of which is negatively charged and the inner one is positively charged. The data obtained can be discussed with Ohshima's electrokinetic theory for a "soft" particle with a nonuniformly charged surface layer [35].

Suppose that a colloidal particle covered with an ion-penetrable surface charge layer consisting of two charged sublayers moves in a liquid containing a symmetrical electrolyte of valency v and bulk concentration n in an applied electric field. The particle radius and the thickness of the surface charge layer are assumed to be much larger than $1/\kappa$. An x axis is taken perpendicular to the particle surface with its origin at the front surface of the surface charge layer (Fig. 8). The outer sublayer (sublayer 1) of the particle surfaces carries ionized groups of valency z_1 and number density N_1, while the inner sublayer (sublayer 2) carries ionized groups of valency z_2 and number density N_2. Let the thickness of sublayer 1 be d. We then have

$$\rho_{\mathrm{fix}}(x) = \rho_1 = ez_1 N_1, \quad -d < x \le 0 \tag{9}$$

$$\rho_{\mathrm{fix}}(x) = \rho_2 = ez_2 N_2, \quad x \le -d \tag{10}$$

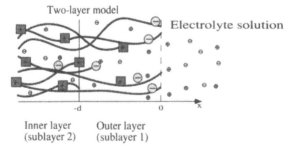

Charged polymer layer

Two-layer model

Electrolyte solution

Inner layer (sublayer 2)   Outer layer (sublayer 1)

FIG. 8 Schematic representation of a soft surface covered by an ion-penetrable surface charge layer consisting of two oppositely charged sublayers 1 and 2. Fixed charges are represented by large squares and circles with plus and minus signs, while electrolyte ions are by small circles with plus and minus signs. The thickness of the outer sublayer (sublayer 1) is d. The slipping plane is located inside at $x < -d$. It is assumed that $\kappa d \gg 1$.

where ρ_1 and ρ_2 are, respectively, the densities of fixed charges in sublayers 1 and 2, and e is the elementary electric charge. Since the thickness of the inner sublayer is assumed to be much larger than $1/\kappa$, then it can practically be set equal to infinity for mathematical convenience. Thus, deep inside sublayer 2, the potential is equal to the Donnan potential in that sublayer, which will be shown later in Eq. (16). The electrophoretic mobility μ is then expressed by Eq. (11) [35]:

$$\mu = \frac{e}{\eta\lambda^2}\left[z_1N_1\left\{1+\left(\frac{\lambda}{\kappa}\right)^2\frac{1+\frac{\lambda}{2\kappa}}{1+\frac{\lambda}{\kappa}}\right\}+(z_1N_1-z_2N_2)\right.$$
$$\left.\left\{\left(\frac{\lambda}{\kappa}\right)^2\frac{\lambda}{2(\kappa-\lambda)}e^{-\kappa d}-\frac{\kappa^2}{\kappa^2-\lambda^2}e^{-\lambda d}\right\}\right]$$

(11)

where λ and κ are, respectively, given by Eqs (6) and (8); z_1N_1 and z_2N_2 represent the number density of the fixed charges in sublayers 1 and 2, respectively.

By a curve-fitting procedure, z_1N_1, z_2N_2, $1/\lambda$, and d were determined. The values of μ calculated via Eq. (11) were plotted against the ionic strength together with the experimental data in Figs 6 and 7 (solid lines, MCs 1–4). We found it possible to draw a curve with a combination of single values for each of z_1N_1, z_2N_2, $1/\lambda$, and d that is in good agreement with the experimental data over a wide range of ionic strengths (0.005–0.154 M). This means that the hydrogel microcapsule surface can be considered as a "soft surface" composed of two sublayers described by Eq. (11). Thus, this agreement enables us to estimate the values of the unknown parameters z_1N_1, z_2N_2, $1/\lambda$, and d by the curve-fitting procedure. In the calculation, we used the value of the relative permittivity ϵ_r of water ($\epsilon_r = 74.39$ at 37°C).

The best-fit values for the charge density of sublayers 1 and 2 (z_1N_1 and z_2N_2), the softness parameter ($1/\lambda$) of the surface, and the thickness of sublayer 1 (d) of each type of microcapsule are shown in Table 6. As summarized in this table, it was found that PEGylation causes the surface charge density of the outer sublayer (z_1N_1) to decrease, while that of inner sublayer (z_2N_2) is little affected. That is, the value of z_1N_1 for MC 3 is less negative than that for MC 1, and that for MC 4 is less negative than that for MC 2. On the other hand, the values of z_2N_2 for MC 1 and MC 3 are almost the same, and those for MC 2 and MC 4 are also almost identical. The softness parameter ($1/\lambda$) for MC 3 is larger than that for MC 1, implying that the surface of MC 3 is softer than that of MC 1. The parameter $1/\lambda$ for MC 4 is also larger than that for MC 2. From this, it is clear that the surface becomes softer and its charge density is reduced by PEGylation.

The potential distribution across the surface layer as predicted from the present two-sublayer model is shown in Fig. 9. We take an x axis perpendicular to the microcapsule surface with its origin 0 at the front surface of the outer sublayer, as in Fig. 8. The region

TABLE 6 Values of z_1N_1, z_2N_2, $1/\lambda$, and d of MCs 1–4

Samples	z_1N_1 (M)	z_2N_2 (M)	$1/\lambda$ (nm)	d (nm)
MC 1	−0.111	0.041	0.505	1.706
MC 2	−0.140	0.166	0.471	2.090
MC 3	−0.048	0.048	1.230	3.758
MC 4	−0.084	0.159	0.778	2.751

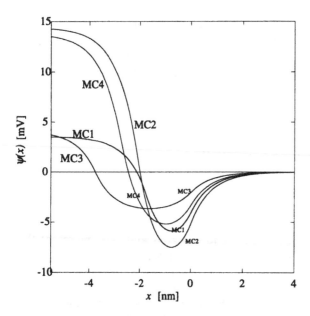

FIG. 9 Potential distribution across the membranes of microcapsules redispersed in an aqueous solution with ionic strength 0.154 at 37°C.

$x < 0$ shows the microcapsule membrane and the region $x \geq 0$ shows the bulk solution phase. These potential curves are for MCs 1–4 suspended in a solution with an ionic strength of 0.154 at 37°C. The potential distribution $\psi(x)$ was calculated from Eqs (12)–(14) [35]:

$$\psi(x) = \frac{1}{2}\psi_{\mathrm{DON},1}\left[e^{-\kappa x} - e^{-\kappa(x+d)}\right] + \frac{1}{2}\psi_{\mathrm{DON},2}e^{-\kappa(x+d)}, \quad x \geq 0 \tag{12}$$

$$\psi(x) = \frac{1}{2}\psi_{\mathrm{DON},1}\left[1 - \frac{1}{2}\left\{e^{\kappa x} + e^{-\kappa(x+d)}\right\}\right] + \frac{1}{2}\psi_{\mathrm{DON},2}e^{-\kappa(x+d)}, \quad -d \leq x < 0 \tag{13}$$

$$\psi(x) = \frac{1}{2}\psi_{\mathrm{DON},1}\left[e^{\kappa(x+d)} - e^{\kappa x}\right] + \psi_{\mathrm{DON},2}\left[1 - \frac{1}{2}e^{\kappa(x+d)}\right], \quad x < -d \tag{14}$$

with

$$\psi_{\mathrm{DON},1} = \frac{ez_1 N_1}{\epsilon_r \epsilon_0 \kappa^2} \tag{15}$$

$$\psi_{\mathrm{DON},2} = \frac{ez_2 N_2}{\epsilon_r 0 \kappa^2} \tag{16}$$

where $\psi_{\mathrm{DON},1}$ and $\psi_{\mathrm{DON},2}$ are, respectively, the Donnan potentials of sublayers 1 and 2.

Figure 9 shows that the potential distribution in the membranes of MC 2 and MC 4 varies much more than that of MC 1 and MC 3 in the normal direction to the membrane surface. The potential values far inside the surface layers of MC 2 and MC 4 are more positive than those of MC 1 and MC 3, while the differences in potential values in the outer sublayers are relatively small. The differences in the charge distribution depending on the kind of microcapsules are explained by the differences in the partition coefficients of the copolymers in the preparation process, as follows. The amounts of total positive charges fixed in each type of microcapsules were little affected by the composition of copolymers used, as shown in Table 5. The content of DMAPAA in copolymer 2 used to prepare MC 2 is, however, larger than that in copolymer 1 used to prepare MC 1, as shown in Table 4. In the monomers used to prepare the copolymers 1 and 2, the amino group in DMAPAA is more easily protonated than that in DMAAm. Also, both of the water-soluble copolymers 1 and 2 diffuse from the inner aqueous phase to the outer oil phase of w/o emulsion followed by the reaction with terephthaloyl dichloride. Therefore, it is reasonable to assume that the dissociable and more hydrophilic moiety in the polymer chains has a tendency to remain in the inner side of the microcapsule membranes. Also, more concentrated carboxylic acid groups originating from terephthalic acid are considered to be fixed in the outer surface layer rather than in the inner one, because terephthaloyl dichloride diffuses from the outer oil phase to the inner aqueous one. This mechanism may lead to the lower value of $z_2 N_2$ for MC 1 than that for MC 2. This tendency is also observed when MC 3 and MC 4 are compared. That is, the value of $z_2 N_2$ for MC 3 is less than that for MC 4. Also, the degree of polymerization is considered to be relatively low, and that the microcapsule membranes have a soft structure.

4. Changes in Microcapsule Sizes Depending on Relative Permittivity of Dispersing Medium

The values for the average diameters of MCs 1–4 kept in water and in cyclohexane are summarized in Table 7. The average diameters of MCs 1–4 kept in distilled water are not affected by the membrane composition. When microcapsules are kept in cyclohexane, however, MC 1 and MC 3 have much larger average diameters than in distilled water, while the average diameters of MC 2 and MC 4 are almost the same as those in distilled water [27,31]. These differences are considered to be caused by those in the swelling ratio of the membranes, which may be due to the differences in charge density and charge distribution in the microcapsule membranes. That is, these phenomena can be explained by the dissociation of carboxylic acid and amino groups in the microcapsule membranes. When microcapsules are in water, the dissociable groups in membranes are easily dissociated and make an ion complex between anionic and cationic groups. The ion-complex formation leads the membrane to be in a shrunken state. Microcapsules in cyclohexane, however, cannot be in a shrunken state because the groups are little dissociated due to the low permittivity of cyclohexane. This would explain the changes in diameters of MC 1 and MC 3, depending on the permittivity of the solvents. On the other hand, the average diameters of MC 2 and MC 4 were almost the same in both water and cyclohexane. This suggests that ion-complex formation occurs less effectively to shrink the membrane of MC 2 and MC 4 when they are redispersed in water. It is of interest to observe the

TABLE 7 Average Diameter of MCs 1–4 in Distilled Water and in Cyclohexane

Samples	Average diameter in distilled water (μm)	Average diameter in cyclohexane (μm)
MC 1	18.12	38.41
MC 2	19.50	18.91
MC 3	19.43	26.76
MC 4	15.23	17.38

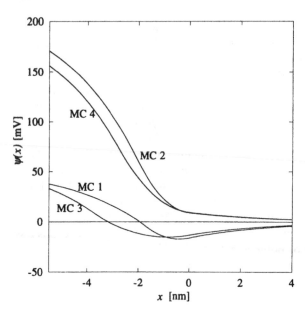

FIG. 10 Potential distribution across the membranes of microcapsules redispersed in an aqueous solution with ionic strength of 0.01 at 37°C.

potential distribution in microcapsules which are redispersed in a very diluted electrolyte solution. We have calculated the potential distribution of MCs 1–4 redispersed in an aqueous solution with ionic strength of 0.01 at 37°C (Fig. 10).

As clearly seen in Fig. 10, the outer sublayers of MC1 and MC 3 are negatively charged and the potential in the outer sublayers is negative, while the potential in the inner sublayers is positive. Therefore, the membranes of MC 1 and MC 3 can be shrunken by ion-complex formation (between positive charges and negative charges) in the normal direction to the membrane surface. On the other hand, the potential in the outer sublayers of MC 2 and MC 4 is positive and that in the inner sublayers is also positive. In these microcapsule membranes, the shrunken state is less available, because the number of negative charges is too small compared with that of positive charges in to allow formation of the ion complex. When MC 2 and MC 4 are kept in the aqueous solution with the higher ionic strengths, however, the potential in the outer sublayer is negative, as shown in Fig. 9.

The observed difference is the potential distributions between Figs 9 (ionic strength 0.154) and 10 (ionic strength 0.01) is caused by the difference in the Donnan potential of the surface layers. The Donnan potential is negative in the outer sublayers and is positive in the inner sublayers of MCs 1–4. The magnitude of the Donnan potential is higher in a solution with lower ionic strength. Also, the Donnan potential of the inner sublayers of MC 2 and MC 4 is much higher than that of MC 1 and MC 3, as is seen in Figs 9 and 10. The potential values for MC 2 and MC 4 are highly positive in the inner sublayers, and the potential values in the outer sublayers are affected not only by the Donnan potential in the outer sublayer but also the Donnan potential in the inner sublayers. Therefore, when the Donnan potential in the inner sublayer is highly positive, the potential in the outer sublayer becomes positive, even if the Donnan potential in the outer sublayer is negative. It was thus concluded that the asymmetrical structure of microcapsule membranes cause a nonuniform distribution of charges and affects the swellability of the membrane in solvents with different permittivities.

B. Interaction Between Charged Soft Microcapsules and Red Blood Cells

It has been reported that cationic surfaces are easily recognized as foreign materials and positively charged materials are eliminated from the biosystems, because the cell surfaces are usually negatively charged [3]. PEGylation of polymer surfaces has been reported to suppress effectively the interaction of the polymeric materials with biological cells, while the mechanisms are still obscure [1, 2, 28, 29]. In this section, it will be discussed how the surface properties such as softness and surface charge density of the microcapsules are related to the interaction of these membranes with those of biological cells such as RBC (red blood cells).

As has been discussed in Section III.A, the membranes of both of poly(DMAAm-*co*-DMAPAA-*co*-AmSt-*alt*-terephthalic acid) microcapsules (denoted by MC 1 and MC 2) and poly(DMAAm-*co*-DMAPAA-*co*-Amst-*co*-PEG-*alt*-terephthalic acid) microcapsules (MC 3 and MC 4) have a two-layer structure, the outer sublayer of which is negatively charged and the inner one positively charged. By PEGylation the surface charge density of the outer sublayer becomes less negative, while that of inner sublayer is not affected, and the surface layers become softer.

In addition to the microcapsules (MCs 1–4) noted in Section III.A, two more kinds of microcapsules, the membranes of which are positively charged, have been prepared by the same method as mentioned in the previous section. A monomer, 2-[(methacryloxy) ethyl] trimethyl ammonium chloride (METAC), which has one ammonium group in the structure, was used instead of DMAPAA to synthesize an aqueous copolymer, poly(DMAAm-*co*-METAC-*co*-AmSt). Another type of copolymer, poly(DMAAm-*co*-METAC-*co*-AmSt-*co*-PEG) was also synthesized by the combination of a-PEG with DMAAm, METAC, and AmSt. The structure of those copolymers is shown in Fig. 11. Using these two aqueous copolymers, poly(DMAAm-*co*-METAC-*co*-AmSt-*alt*-terephthalic acid) microcapsules (MC 5) and poly(DMAAm-*co*-METAC-*co*-AmSt-*co*-PEG-*alt*-terephthalic acid) microcapsules (MC 6) were prepared.

Four types of hydrophilic gel microcapsules containing water (MC 2 and MC 4 with MC 5 and MC 6) are used to study the interaction between their surfaces and RBC surfaces in this section. The compositions of the copolymers used to prepare each type of microsphere are summarized in Table 8.

1. Membrane Structure of Microcapsules Predicted from Electrophoretic Studies

The structure of the membranes of MC 5 and MC 6 has been analyzed by the same method as mentioned in Section III.A. As shown in Fig. 12, the electrophoretic mobilities of MC 5 and MC 6 are positive in all solutions at pH 7.4 with ionic strengths between 0.005 and 0.154, implying that the surfaces of those microcapsules have net positive charges, which originate from quaternary ammonium groups and primary amino groups. As the ionic strength rises, the electrophoretic mobility decreases, because the shielding effects of electrolyte ions in the medium increase as the ionic strength rises. The electrophoretic mobility reaches a nonzero value at an ionic strength as high as 0.154, which means that the surfaces are soft and the surface properties can be discussed in terms of the electrokinetic theory for "soft" particles. The electrophoretic mobilities of MC 2 and MC

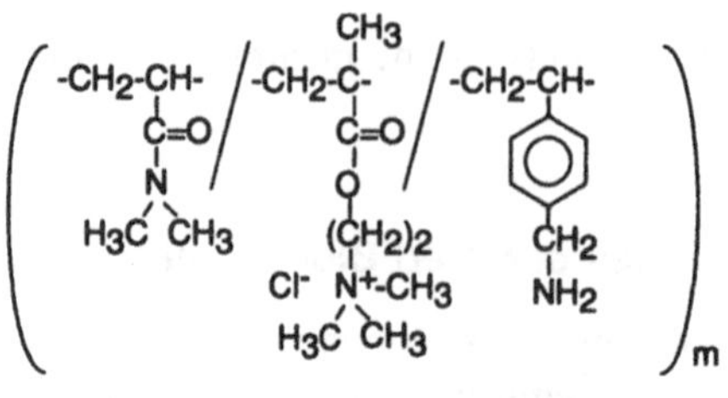

poly(DMAAm-*co*-METAC-*co*-AmSt)

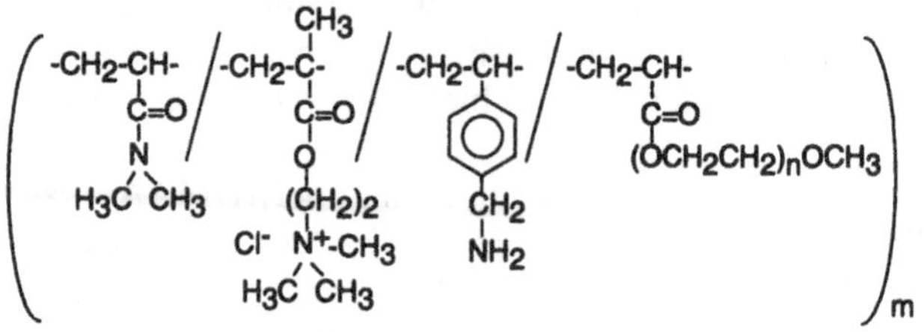

poly(DMAAm-*co*-METAC-*co*-AmSt-*co*-PEG)

FIG. 11 Structure of poly(DMAAm-*co*-METAC-*co*-AmSt) and poly(DMAAm-*co*-METAC-*co*-AmSt-*co*-PEG).

4 are also plotted in Fig. 12 as a function of the ionic strength. By a curve-fitting procedure, $z_1 N_1$, $z_2 N_2$, $1/\lambda$, and d were determined. Theoretical values of μ calculated via Eq. (11) are plotted against the ionic strength (solid lines) together with the experimental data (symbols) in Fig. 12. The best-fit values of $z_1 N_1$, $z_2 N_2$, $1/\lambda$, and d of each type of microcapsule are given in Table 9.

As summarized in Table 9, it was found that the values of charge densities in the outer sublayers ($z_1 N_1$) of MC 5 and MC 6 are positive and those in the inner sublayers ($z_2 N_2$) are almost zero. The values for charge densities in the outer sublayers ($z_1 N_1$) of MC 2 and MC 4 are negative and those in the inner sublayers ($Z_2 N_2$) are positive, as mentioned before. The softness parameters ($1/\lambda$) for MC 5 and MC 6 are larger than those for MC 2 and MC 4, implying that the surfaces of MC 5 and MC 6 are softer than those of MC 2 and MC 4. In softer surfaces, water moves more easily than in harder surfaces. In Table

TABLE 8 Synthesis of Copolymers Having Primary and Tertiary Amino Groups,[a] Copolymers 2 and 4, and Those Having Primary and Quaternary Ammonium Groups, Copolymers 5 and 6

Samples	Chemical composition in feed (mol%)					Yield %
	DMAAm	AmSt	DMAPAA	METAC	a-PEG[b]	
Copolymer 2	80.00	5.00	15.00	—	—	55.0
Copolymer 4	79.81	4.99	14.96	—	0.24	54.8
Copolymer 5	80.00	5.00	—	15.0	—	57.0
Copolymer 6	79.80	4.99	—	14.97	0.25	35.6

[a] Solvent: ethanol (100 mL), initiator: V-65 (10 mM), reaction temp.: 45°C, reaction time: 24 h.

[b] a-PEG: α-acryloxypoly(ethylene glycol).

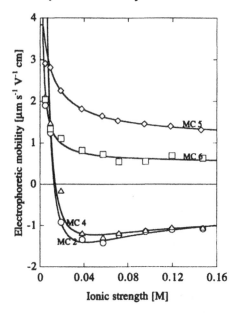

FIG. 12 Electrophoretic mobility of MC 2, MC 4, MC 5, and MC 6. Symbols are experimental data for MC 2 ($\bigcirc$), MC 4 ($\triangle$), MC 5 ($\diamond$), and MC 6 ($\square$) measured as a function of the ionic strength in the suspending medium at pH 7.4 and at 37°C. Solid curves are theoretical ones calculated with $z_1 N_1 = -0.140$ M, $z_2 N_2 = 0.166$ M, $1/\lambda = 0.471$ nm, $d = 2.090$ nm (MC 2), with $z_1 N_1 = -0.084$ M, $z_2 N_2 = 0.159$ M, $1/\lambda = 0.778$ nm, $d = 2.751$ nm (MC 4), with $z_1 N_1 = 0.049$ M, $z_2 N_2 = -0.012$ M, $1/\lambda = 1.336$ nm, $d = 3.000$ nm (MC 5), and with $z_1 N_1 = 0.010$ M, $z_2 N_2 = 0.006$ M, $1/\lambda = 1.916$ nm, $d = 10$ nm (MC 6).

10, the average diameter of each of the microcapsules redispersed in distilled water is shown. As clearly observed, MC 5 and MC 6 are much larger than MC 2 and MC 4, even though they were prepared under the same conditions except for their materials. Also, from this observation, it is apparent that the membranes of MC 5 and MC 6 are softer and more hydrophilic than those of MC 2 and MC 4.

As for the effects of PEGylation, it was found that the surface charge density of outer sublayer ($z_1 N_1$) decreases, while that of inner sublayer ($z_2 N_2$) was little affected by the PEGylation. That is, the value of $z_1 N_1$ for MC 4 is less negative than that for MC 2, and that for MC 6 is less positive than that for MC 5. On the other hand, the values of $z_2 N_2$ for MC 2 and MC 4 are almost the same. Also, the values of $z_2 N_2$ for MC 5 and MC

TABLE 9 Surface Charge Density, Softness Parameter, and Thickness of Outer Sublayers of MCs 1–4 and RBC

Samples	$z_1 N_1$ (M)	$z_2 N_2$ (M)	$1/\lambda$ (nm)	d (nm)
MC 2	−0.140	+0.166	0.471	2.090
MC 4	−0.084	+0.159	0.778	2.751
MC 5	+0.049	−0.012	1.336	3.000
MC 6	+0.010	+0.006	1.916	10.00
RBC	−0.830	+0.194	0.841	1.000

TABLE 10 Average Diameter of Each
Microcapsule Redispersed in Distilled Water

Samples	Diameter (μm)
MC 2	19.50
MC 4	15.23
MC 5	28.98
MC 6	29.22

6 are very small, while the sign is different. The softness parameter $(1/\lambda)$ for MC 4 is larger than that for MC 2, implying that the surface of MC 4 is softer than that of MC 2. The softness parameter $(1/\lambda)$ for MC 6 is also larger than that for MC 5. From these data, it is clear that the surface becomes softer by PEGylation.

2. Membrane Structure of RBC Predicted from Electrophoretic Studies

Sheep blood was centrifuged at 3000 r.p.m. for 10 min to separate RBC from plasma. The separated blood cells were dispersed in isotonic saline solution. The surface structure of RBC was also studied by electrophoretic mobility measurements in the same way as above mentioned. To adjust the ionic strength, Dulbecco's buffer solution at pH 7.4 was diluted with distilled water and then sucrose was added to the solution to adjust the osmotic pressure. As shown in Fig. 13, the electrophoretic mobility of RBC is negative in all solutions at pH 7.4 with ionic strengths between 0.005 and 0.154, implying that the surface has net negative charges. As the ionic strength rises the electrophoretic mobility becomes less negative and reaches a nonzero value. By a curve-fitting procedure the best-fit values of $z_1 N_1$, $z_2 N_2$, $1/\lambda$, and d were determined and are shown in Table 9. The theoretical value of μ calculated with the values via Eq. (11) were plotted against the ionic strength (a solid line) for comparison with the experimental data (closed circles) in Fig. 13. It was thus found that the membrane of RBC is composed of two sublayers, the inner one of which is positively charged, while the outer one is negatively charged, and that the surface of RBC is soft.

3. Interaction of Microcapsules with RBC

The interactions between microcapsule surfaces and RBC surfaces were evaluated by the hemolysis of RBC and their adsorption on microcapsule surfaces. Each kind of microcapsule was resuspended so that the final concentration was 10% (v/v) in saline solution and was kept at 37°C. Sheep RBC (50 μL) were added to 3 mL of the microcapsule suspension. The mixture was incubated at 37°C for 15 min. The size distributions of the mixture were measured with a light-scattering particle sizer (Malvern 3601, Master sizer/E/ Malvern, Inc.). The size distributions of microcapsules and RBC redispersed in isotonic saline solution were also measured.

The changes in size distribution of microcapsules and RBC before and after they were mixed and incubated were studied. The size distributions of MCs 2, 4, 5, and 6 are shown in Figs 14a–17a, respectively, together with that of RBC before mixing, and those after mixing in Figs 14b–17b. The sizes of MC 2 and MC 4 before mixing with RBC are distributed between about 4 and 40 μm with the maximum frequency value at the fraction

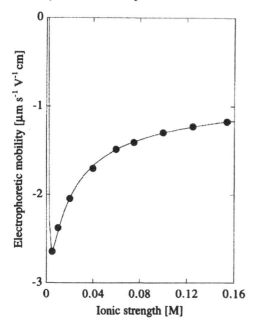

FIG. 13 Electrophoretic mobility of RBC. Symbols are experimental data measured as a function of ionic strength in the suspending medium at pH 7.4 and at 37°C. The solid curve is a theoretical one calculated with $z_1 N_1 = -0.830$ M, $z_2 N_2 = 0.194$ M, $1/\lambda = 0.841$ nm, $d = 1.000$ nm.

with a diameter of about $12.5 \, \mu$m. The sizes of MC 5 and MC 6 before mixing with RBC are distributed between 7 and $84 \, \mu$m with the maximum frequency value at the fraction wit ha diameter of about $25 \, \mu$m. The size of RBC is distributed between 2 and $7 \, \mu$m. In Figs 14a–17a, the sum of the frequencies in the distributions of RBC and microcapsules is adjusted to 100% in each figure. This adjustment was done as follows. The microcapsules and RBC are mixed under the condition that the volume fractions of microcapsules and RBC are, respectively, 9.84 and 1.64% (v/v). The size distributions of microcapsules and RBC in Figs 14a–17a were normalized by taking into account the above volume fraction ratio.

The mixtures of RBC and each type of microcapsule suspension were incubated for 15 min at 37°C. Figures 14b–17b show the size distribution of the mixture after incubation. By comparison of Fig. 14a with Fig. 14b, it is clearly seen that the fraction of RBC disappears and the fractions between 40 and $70 \, \mu$m newly appear when RBC are incubated with MC 2. This change in size distribution is considered to be caused by the simultaneous adsorption of a certain portion of RBC on two microcapsule surfaces. This type of adsorption of RBC on microcapsule surfaces bridges between two microcapsules and then produces new fractions with larger diameters. This possibility is also confirmed by the observation that the frequency of the fractions of microcapsules with diameters between 10 and $20 \, \mu$m decreases by the incubation. Also, it was found that MC 4 interacts with RBC, although the interaction is weaker than that observed in the case of MC 2, as shown in Fig. 15. After incubation, the fractions of RBC decrease and new fractions with diameters between 40 and $70 \, \mu$m appear. From the comparison of Fig. 14 with Fig. 15, it was found that the interaction of microcapsule with RBC is suppressed by PEGylation of the microcapsule surfaces.

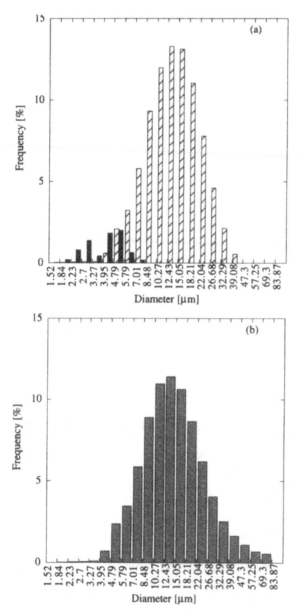

FIG. 14 Size distribution of MC 2, RBC, and their mixture. (a) MC 2 and RBC; MC 2 is shown with striped columns. (b) The mixture of MC 2 and RBC after incubation.

On the other hand, no changes in the size distribution of RBC were observed by the incubation of RBC with MC 5, although the frequency of the fractions of RBC increases, as shown in Fig. 16. Also, the frequency of the fractions of MC 5 decrease after incubation. The increased frequency of RBC fractions can be explained by the differences in refractive index between RBC and MC 5. The refractive index of MC 5 seems to be less than that of RBC, because the microcapsule membrane may be more hydrophilic than the RBC membrane. Also, the size distribution of MC 5 did not change

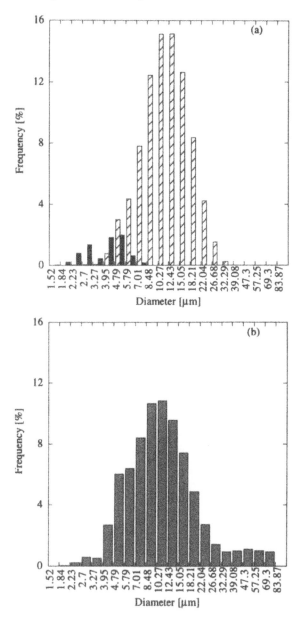

FIG. 15 Size distribution of MC 4, RBC, and their mixture. (a) MC4 and RBC; MC 4 is shown with striped columns. (b) The mixture of MC 4 and RBC after incubation.

with incubation and the new fraction which was observed in Figs 14b and 15b did not appear after incubation.

The mixture of MC 6 and RBC also shows the same tendency, as observed in Fig. 17. From Figs 16 and 17, it is clear that the surfaces of both MC5 and MC 6 doe not interact with the RBC surfaces.

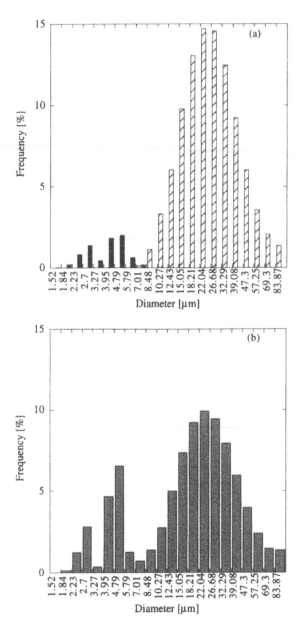

FIG. 16 Size distribution of MC 5, RBC, and their mixture. (a) MC 5 and RBC; MC 5 is shown with striped columns. (b) The mixture of MC 5 and RBC after incubation.

A certain portion of RBC may disappear by hemolysis, when they are incubated with MC 2 or MC 4. The data on hemolysis by measuring the absorbance at 540 nm were not available, because it was confirmed by the following experiments that the hemoprotein adsorbs on the microcapsule. RBC 50 μm was redispersed in 3 mL of distilled water to obtain complete hemolysis, and then each type of microcapsule was added to the solution. After incubation at 37°C for 15 min, the suspension was centrifuged and the absorbance of

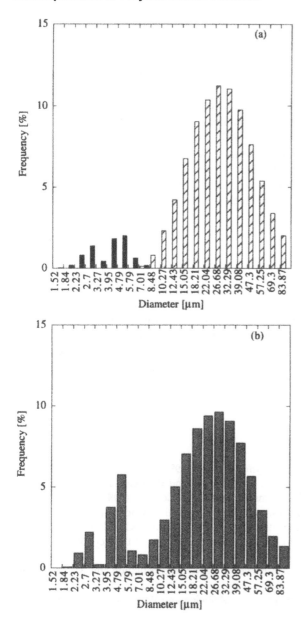

FIG. 17 Size distribution of MC 6, RBC, and their mixture. (a) MC 6 and RBC; MC 6 is shown with striped columns. (b) The mixture of MC 6 and RBC after incubation.

the supernatant was measured spectrophotometrically at 540 nm. In all cases, the absorption at 540 nm was almost zero, and the microcapsules obtained at the bottom after centrifugation were observed to be colored brown, while they were almost transparent or white before being added to the solution obtained by the hemolysis of RBC.

From Figs 14–17, it is clear that the surfaces of MC 2 and MC 4 interact with RBC surfaces, although those of MC 5 and MC 6 do not. This result was unexpected, because the outer surfaces of MC 2 and MC 4 are negatively charged, while the surfaces of MC 5 and MC 6 are positively charged. These results suggest that the surface charges do not control the interaction of the microcapsule surfaces with RBC surfaces.

It should be emphasized that microcapsules with soft surfaces (MC 5 and MC 6) do not interact with RBC, even though the microcapsule surfaces are positively charged and the surface of RBC is negatively charged. On the other hand, microcapsules with negatively charged but harder surfaces (MC 2) interact with RBC to introduce hemolysis. The membrane surface of MC 4, which is obtained by PEGylation of MC 2, becomes softer than that of MC 1 so that the interaction with RBC is weakly suppressed. From these observations, we have concluded that the dominant factor controlling the interaction between RBC and the synthetic polymer is not their surface charges but the softness of the surfaces. Therefore, it is considered that synthetic polymers are not recognized by biological cells as foreign materials, if the polymer surfaces are soft and hydrophilic, even though their surfaces are positively charged.

ACKNOWLEDGMENTS

This work was supported in part by the Kumagai Foundation for Science and Technology and by a Grant-in-Aid (10680805) from the Ministry of Education, Science, and Culture, Japan. I wish to thank Professor Dr. H. Ohshima, Professor Dr. Y. Sakurai, Professor Dr. T. Okano, Professor Dr. T. Tsukamoto, Dr. A. Kikuchi, and Dr. S. Ando, my coworkers, for their valuable discussion, and Mr. S. Nagashima, Ms. M. Umetsu, and Ms. Y. Goto, my former students, for their help.

REFERENCES

1. GR Lannos, MV Sefton. J Biomater Sci, Polym Ed 4: 381, 1993.
2. H Milton, S Salipsky, eds. Poly(ethylene Glycol):Chemistry and Biological Applications. ACS Symposium Series 680. Washington, DC: American Chemical Society, 1998.
3. K Makino. In: H Ohshima, K Furusawa, eds. Electrical Phenomena at Interfaces: Fundamentals, Measurements, and Application. New York: Marcel Dekker, 1998, pp 583–593.
4. K Makino. In: N Kallay, ed. Interfacial Dynamics. New York: Marcel Dekker, 2000, pp 591–619.
5. PB Deasy. Drugs and Pharmaceutical Sciences. vol. 20. Microencapsulation and Related Drug Processes. New York: Marcel Dekker, 1984.
6. M Donbrow. Microcapsules and Nanoparticles in Medicine and Pharmacy. CRC Press, FL. U.S.A. 1992.
7. J Kreuter. Drugs and Pharmaceutical Sciences. vol. 66. Colloidal Drug Delivery Systems. New York: Marcel Dekker, 1994.
8. AJ Hickey. Inhalation Aerosols: Physical and Biological Basis for Therapy. New York: Marcel Dekker, 1996.
9. T Nakashima, M Shimizu, M Kukizaki. Membrane Emulsification Operation Manual. Miyazaki: Industrial Research Institute, 1991, pp 1–24.
10. S Omi, K Katami, A Yamamoto, M Iso. J Appl Polym Sci 51: 1, 1994.
11. S Omi, K Katami, T Taguchi, K Kaneko, M Iso. J Appl Polym Sci 57: 1013, 1995.
12. S Omi. Colloids Surfaces A 109: 97, 1996.
13. K Shiga, N Muramatsu, T Kondo. J Pharm Pharmacol, 48: 891, 1996.
14. Y Nakano, K Makino, H Ohshima, T Kondo. Biophys Chem 50: 249, 1994.
15. K Makino, M Ikekita, T Kondo, S Tanuma, H Ohshima. Colloid Polym Sci 272: 487, 1994.

16. K Makino, F Fukai, S Hirata, H Ohshima. Colloids Surfaces B 7: 235, 1996.
17. K Makino, S Yamamoto, K Fujimoto, H Kawaguchi, H Ohshima. J Colloid Interface Sci 166: 251, 1994.
18. K Makino, K Suzuki, Y Sakurai, T Okano, H Ohshima. J Colloid Interface Sci 174: 400, 1995.
19. N Ogata, SW Kim, J Feijen, T Okano, eds. Advanced Biomaterials in Biomedical Engineering and Drug Delivery Systems. Tokyo: Springer-Verlag, 1996.
20. T Akaike, T Okam, M Akashi, M Terano, N Yui, eds. Advances in Polymeric Biomaterials Science. Tokyo: CMC, 1997.
21. K Park, WSW Shalaby, H Park. Biodegradable Hydrogels For Drug Delivery. Lancaster: Technomic, PA, U.S.A. 1993.
22. R Gurny, HE Junginger, NA Peppas, eds. Pulsatile Drug Delivery, Current Applications and Future Trends. Stuttgart: Wissenschaftliche Verlagsgessellschaft, 1993.
23. S Nagashima, S Ando, K Makino, T Tsukamoto, H Ohshima. J Colloid Interface Sci 197: 377, 1998.
24. S Nagashima, S Ando, T Tsukamoto, H Ohshima, K Makino. Colloids Surfaces B: Biointerfaces, 11: 47, 1998.
25. S Nagashima, M Koide, S Ando, K Makino, T Tsukamoto, H Ohshima. Colloids Surfaces A: PhysicoChem Aspects, 153: 221, 1999.
26. H Ohshima. J Colloid Interface Sci 163: 474, 1994.
27. K Makino, M Umetsu, Y Goto, A Kikuchi, Y Sakurai, T Okano, H Ohshima. J Colloid Interface Sci 218:275, 1999.
28. L Illum, IM Hunneyball, SS Davis. Int J Pharm. 29: 53, 1986.
29. D O'Mullane, K Patrak, E Tomlinson. Biomaterials 9:203, 1988.
30. K Makino, M Umetsu, Y Goto, A Nakayama, T Suhara, J Tsujii, A Kikuchi, H Ohshima, Y Sakurai, T Okano. Colloids Surfaces B: Biointerfaces 13: 287, 1999.
31. K Makino, M Umetsu, A Kikuchi, H Ohshima, Y Sakurai, T Okano. Colloids Surfaces B: Biointerfaces 12: 97, 1998.
32. E Miyauchi, Y Togawa, K Makino, H Ohshima, T Kondo. J Microencaps 9: 329, 1992.
33. N Ueda, R Oht, Y Togawa, K Makino, T Kondo. Polym Gels Networks 3: 135, 1995.
34. T Dobashi, T Narita, J Masuda, K Makino, T Mogi, H Ohshima, M Takenaka, B Chu. Langmuir 14: 745, 1998.
35. H Ohshima, T Kondo. Biophys Chem 46: 145, 1993.

26

Electrical Surface Properties and Electrokinetics of Colloidal Particles: Theory and Applications

RYSZARD SPRYCHA   Sun Chemical Corporation, Carlstadt, New Jersey

I. INTRODUCTION

Colloidal systems containing finely dispersed particles of solids in liquids are commonly met in everyday life, e.g., paints printing inks, polymer dispersions, photographic emulsions, pharmaceuticals, insecticides and pesticides, cosmetics, etc. Dispersion of solids can be prepared in different ways, e.g., by precipitation and stabilization of colloidal size particles or by comminution of large particles by grinding or milling. Whenever a new solid/liquid interface is formed some rearrangement of electrical charges in the interfacial region takes place so that the surface of the solid becomes charged.

The surface charge plays a very important role in stabilizing of colloidal dispersions. In addition, systems containing charged particles are used in practice, e.g., in electrophoretic painting and coating or in printing inks (liquid toners). The electrical charge on the surface of a solid can affect significantly such processes as adsorption at the solid/liquid interface or the kinetics of surface reactions. The mechanism of surface charge formation and distribution of charges in the interfacial region are important and interesting from both theoretical and practical points of view.

The purpose of the present chapter is to discuss briefly: the origin of the electrical charge at the solid/liquid interface in different systems; the structure of the electrical interfacial layer (EIL); and electrokinetic properties of such systems. Section II outlines the mechanisms of surface charge formation at different types of interfaces and experimental methods for investigation of the EIL at the solid/liquid interface. The methods of evaluation of some parameters of the EIL from electrokinetic data are discussed in Section III. Finally, practical applications of electrokinetic measurements in selected areas are overviewed in Section IV with special emphasis on the printing ink industry.

II. ORIGIN OF SURFACE CHARGE AT THE SOLID/LIQUID INTERFACE: THE ELECTRICAL INTERFACIAL LAYER

In most cases, solid particles dispersed in a liquid carry on their surface net electric charge. Because the entire system is electroneutral the surface charge is counterbalanced by an equal and opposite charge in the solution. The surface charge on the particle and counter-charge in the solution form the so-called electrical interfacial layer (EIL). The mechanism by which the surface of a solid particle acquires electrical charge depends on the type of both the solid and the solution. For practical purposes, colloidal dispersions can be divided into aqueous systems (water is a solvent of high dielectric constant, ~ 80) and nonaqueous systems (mostly containing solvents of low dielectric constant, e.g., hydrocarbons). The structure and properties of the electrical interfacial layer in different systems have been extensively studied [1–40].

A. Aqueous Systems

The surface charge at the solid/electrolyte interface can be formed mainly as a result of:

- dissociation of surface functional groups
- adsorption of ions from solution
- ion exchange (clays)

Examples of different charging mechanisms in water are presented below. More details on this subject can be found in the literature [1–11, 17–20].

1. Metal/Electrolyte Interface

Metal electrodes may behave as perfectly polarizable (no charge transfer across the interface) or reversible (charges are transferred across the interface until the system reaches equilibrium) electrodes. The mercury electrode is a good example of a polarizable electrode. The surface of the metal may acquire a positive or negative charge, depending on the externally applied potential. The mercury/electrolyte system has been widely used to study the properties of the EIL because for this metal direct measurements of capacitance as well as of surface tension of the externally polarized liquid metal are possible [21–23].

2. Ionic Crystal/Electrolyte Interface

The surfaces of ionic crystals (e.g., AgCl) cannot be charged by an externally applied potential. For such systems, electrical surface charge can be formed by adsorption of potential determining ions (constituents of crystal lattice). Thus, AgCl can carry positive or negative net surface charge depending on the concentration of Ag^+ and Cl^- ions in the aqueous solution. The surface potential of AgCl is described by the Nernst equation and changes by about 59.2 mV for every 10-fold increase in the concentration of potential determining ions [22, 24–26].

3. Metal (Hydr)oxide/Electrolyte System

Potential determining ions for metal oxides and hydroxides are H^+ and OH^- ions [2, 7, 8, 20. 27, 28]. This results from the fact that the surfaces of metal oxides are covered with metal hydroxyl groups (-MOH) which have amphoteric properties and may accept or donate protons in water, depending on the pH of the solution:

$$-MOH + H^+ \leftrightarrow -MOH_2^+ \tag{1}$$

$$-MOH \leftrightarrow -MO^- + H^+ \tag{2}$$

As -MO⁻ and -MOH$_2^+$ charged groups are not constituents of the crystal lattice of the metal oxide their concentrations may vary significantly when the pH of the electrolyte changes. Thus, the surface potential of oxides cannot be described by the Nernst equation [3, 9, 15, 29, 30]. The deviation of surface potential from Nernst's value depends on the oxide type and extent of ionization of surface–MOH groups [17, 29].

B. Nonaqueous Systems

Solid particles dispersed in low-conductivity nonaqueous solvents can carry the electrical surface charge. However, the mechanism of charging in such systems is different from that in aqueous dispersions [31]. For example, particles of carbon black dispersed in benzene containing calcium alkylsalicylate were positively charged [32]. Assuming adsorption of alkylsalicylate ions on the particle (such a mechanism operates in aqueous systems) the surface should carry a net negative charge.

Though the mechanism of surface charge formation in nonaqueous systems is less understood than that in aqueous dispersions it is generally accepted that acid–base (Brönsted) interactions between the surface of the particle and solvent/solute molecules are responsible for surface charge development in such systems. In the system consisting of both protic particles (PH) and solvents (SH) only, Lyklema [33] proposed the following mechanism of particle charging:

$$PH + SH \leftrightarrow P^- + SH_2^+ \tag{3}$$

or

$$PH + SH \leftrightarrow PH_2^+ + S^- \tag{4}$$

In the first case, the particle gains negative charge by donating protons to the solvent. The particle can gain positive charge by accepting protons from the solvent, Eq. (4). Thus, for a given system the sign of the surface charge depends on the acid–base properties of the solid and solvent.

A similar approach regarding the mechanism of charging in low-conductivity media (e.g., hydrocarbons) was proposed by Fowkes and coworkers [34–36]. In aprotic solvents, particles can be charged by addition to the system of special chemicals that can be adsorbed at the surface of the particle via acid–base interactions. The process of charging takes place in three steps. For example, for acidic particles (AH) and basic additive (B), in the first step basic additive adsorbs on to acidic sites of the solid particle:

$$AH + B \leftrightarrow AH \cdots B \tag{5}$$

In the second step proton transfer from acid solid to the basic additive takes place:

$$AH \cdots B \to A \cdots HB \tag{6}$$

Finally, desorption of basic additive carrying the proton from the surface to the solution takes place, leaving a negatively charged solid particle:

$$A \cdots HB \to A^- + HB^+ \tag{7}$$

The electrical properties of nonaqueous systems can be affected very strongly by even a small amount of water. Because it is very difficult to remove water from a nonaqueous system completely its presence should be taken into account in many practical applications. Dissociation of water (H⁺ and OH⁻ ions) can contribute to the charging in totally aprotic systems.

For aproptic particle (P) and solvent (S) the following mechanism of charging has been proposed [37]:

$$P + S + H^+ + OH^- \leftrightarrow PH^+ + SOH^- \tag{8}$$

or

$$P + S + H^+ + OH^- \leftrightarrow POH^- + SH^+ \tag{9}$$

Thus, if surface of particle is more basic than solvent molecule reaction (8) dominates and is responsible for particle charging. If solvent is more basic than particle surface than reaction (9) will dominate in the system.

C. Structure of the Electrical Interfacial Layer

To describe the distribution of the electrical charges in the interfacial region early theories used the double-layer (DL) models. The compact layer (flat condenser) model was proposed by Helmholtz [21] and a diffuse-layer model by Gouy and Chapman [21–23], who assumed that the surface charge is counterbalanced by a purely diffuse atmosphere of charges. Due to the very limited applicability of such models more complicated models have been developed [1–22, 38–44].

The triple-layer model (TLM) was first proposed by Stern [38] and developed by Grahame [21] and others [1–30, 39–44]. Later, a four-layer model was proposed to take into account the very strong specific adsorption of ions which were placed closer to the surface than to the inner Helmholtz plane (IHP) [39]. Other authors also used multilayer models for a more accurate description of the EIL structure [40–44].

Though the term "electrical double layer" is commonly used to describe the charge distribution at the interfaces it seems that using the term "electrical interfacial layer" is more appropriate. The latter term is very general and does not implicate any specific distribution of the charges at the interface. Different parameters of the EIL can be determined by use of appropriate experimental techniques.

1. Surface Charge

The surface charge (σ_0) of a mercury electrode can be calculated from the theory of electrocapillarity by using the Lippman equation [21–23]. For ionic crystals and metal oxides the surface charge can be determined by a potentiometric titration method [12, 20, 25, 28, 45–46]. The surface charge density is calculated from the difference between the amount of potential determining ions added to the dispersion and the amount of those ions remaining in the solution. The potentiometric titration experiments are performed at a constant concentration of so-called "indifferent electrolyte." For sparingly soluble oxides, potential determining ions (H^+ and OH^-) can be consumed in the process of oxide dissolution. Therefore, modified procedures of potentiometric titration are used to take this effect into account [47, 48].

The charge density of particles dispersed in low-conductivity nonaqueous media can be determined directly by separation of the particles from the liquid. Because the EIL in low-conductivity solution is very thick [37] it is easy to remove particles by a simple "filtration" process. The liquid with entrained countercharges is collected in the container connected directly to an electrometer [49]. More details on the determination of surface change in nonaqueous low-conductivity media can be found in the literature [37, 49–53] and in Section IV of this chapter.

2. Surface Potential

The potential of metal electrodes can be measured directly. Such measurements for oxides are rather difficult. Recently, directly measurements of the surface potential (ψ_0) were performed for oxides using the ion-sensitive field effect transistor [54] and other techniques [28, 55]. More detailed discussion on the surface potential can be found elsewhere [12, 22, 28, 56].

3. Electrokinetic Potential

The zeta (electrokinetic) potential, (ζ) i.e., the potential at the effective shear plane between the moving and stationary parts of the EIL, cannot be measured directly. However, the zeta potential can be calculated from electrokinetic measurements by using the appropriate theory describing the given phenomenon. All electrokinetic phenomena are due to relative motion between the charged solid surface (including the compact immobile part of the EIL) and the solution containing the diffuse layer. This motion may result, e.g., from an applied electric field (electrophoresis), force liquid flow (streaming potential) or other factors [18, 22, 56–62].

Though electrokinetic effects can be measured relatively easily the conversion of experimental electrokinetic data into zeta potential is not a trivial matter. An excellent review paper on this subject has been published by Hidalgo-Alvarez [58]. More details on different experimental techniques for zeta-potential measurements can be found in the literature [56–60] as well as in other chapters of this book.

Microelectrophoresis, electrophoretic light scattering, and electroacoustic methods are at present most frequently used to measure zeta potential in aqueous as well as non-aqueous systems [59, 60]. It is, however, well known that electrokinetic measurements in low conductivity and low dielectric constant media are difficult [60–63].

Zeta-potential measurements can be used for many purposes in fundamental research [64–101] as well as industrial research [102–117]. Examples of the application of electrokinetic data in basic research will be discussed in Section III. Practical applications of zeta-potential measurements in different industrial areas will focus mainly on those related to the printing ink industry (see Section IV).

III. EVALUATION OF EIL PARAMETERS FROM ELECTROKINETIC DATA

Electrokinetic data can be used to evaluate certain parameters of the EIL at the solid/solution interface. For example, one of the important parameters in the surface complexation model (SCM) of the EIL [2, 7, 9, 18] is the surface ionization constant. These constants can be determined graphically [1, 7, 9] or numerically [17, 118, 119] from potentiometric titration data, or graphically from electrokinetic data [120–122].

A. Estimation of Surface Ionization Constants

According to the SCM model of the EIL for oxides [2, 7, 9, 18] the following four reactions are responsible for the development of surface charge:

$$-MOH_2^+ \leftrightarrow -MOH + H_S^+ \tag{10}$$

$$-MOH \leftrightarrow -MO^- + H_S^+ \tag{11}$$

$$-MOH_2^+ X^- \leftrightarrow -MOH + H_S^+ + X_\beta^- \tag{12}$$

$$-MOH + Y_\beta^+ \leftrightarrow -MO^- Y^+ + H_S^+ \tag{13}$$

Reactions (10) and (11) describe the process of dissociation of surface metal hydroxyl groups (-MOH). Reactions (12) and (13) represent the process of surface complex formation between surface hydroxyl groups and anions (X^-) or cations (Y^+) of the supporting 1 : 1 electrolyte YX – subscripts "s" and β denote the surface and IHP, respectively [2, 7].

Assuming that the concentration of H^+ ions within the EIL can be described using the Boltzman equation [1, 2, 7] the ionization constants in the logarithmic form for reactions (10) and (11), respectively, can be expressed as follows:

$$pK_{a_1}^{int} = pH + \log\left[-MOH_2^+\right] - \log\left[-MOH\right] + \frac{e\psi_0}{2.3kT} \tag{14}$$

$$pK_{a_2}^{int} = pH - \log\left[-MO^-\right] + \log -\left[MOH\right] + \frac{e\psi_0}{2.3kT} \tag{15}$$

where k is the Boltzman constant and T denotes absolute temperature. If [-MOH. [-MO$^-$], and [-MO$^-$] components in Eqs (14) and (15) are known, then the acidity constants can be obtained by extrapolation of the respective acidity quotients:

$$pQ_{a_1} = pH + \log\left[-MOH_2^+\right] - \log\left[-MOH\right] \tag{16}$$

$$pQ_{a_2} = pH - \log\left[-MO^-\right] + \log\left[-MOH\right] \tag{17}$$

to the point of zero charge (p.z.c.) where surface potential ψ_0 is equal to zero. For indifferent electrolytes the values of the p.z.c. and isoelectric points (i.e.p.) coincide [2, 7, 14, 24].

According to the SCM model of the electrical interfacial layer the diffuse-layer charge (σ_d) is determined by [-MO$^-$] and [-MOH$_2^+$] components [2, 7, 120]:

$$\sigma_d = B\left(\left[-MOH_2^+\right] - \left[-MO^-\right]\right) \tag{18}$$

where B is a conversion factor from mol dm^{-3} to $C\,m^{-2}$.

Assuming that, for low ionic strength of the electrolyte solution ($c \leq 10^{-2}$ mol dm^{-3}) the potential at the onset of the diffuse layer (σ_d) is equal to the electrokinetic potential [123–126], then the diffuse-layer charge can be calculated from Gouy–Chapman theory [18, 21, 22].

$$\sigma_d = -11.74\sqrt{c}\sinh\frac{ze\zeta}{2kT} \tag{19}$$

where c is electrolyte concentration. The concentrations of [-MOH$_2^+$] and [-MO$^-$] groups are comparable only in the vicinity of the p.z.c. [127, 128] and at a certain distance from the p.z.c. one can assume that

$$\sigma_d \approx -B[-MO^-] \tag{20}$$

for pH > pH$_{pzc}$, and

$$\sigma_{d+} \approx B[-MOH_2^+] \tag{21}$$

for pH < pH_{pzc}. For low electrolyte concentrations one can use as a first approximation that [-MOH] $\approx N_s$ (total number of surface sites available per unit area [7, 9]. The value of N_s can be evaluated experimentally for a given solid [2, 18, 127]. Thus, from Eqs (16)–(21) the acidity quotients can be expressed as follows:

$$pQ_{a_1} \approx pH + \log\frac{\sigma_{d+}}{B} - \log N_s \tag{22}$$

$$pQ_{a_2} \approx pH - \log\frac{|\sigma_{d-}|}{B} + \log N_s \tag{23}$$

These acidity quotients can be calculated for different values of pH and electrolyte concentration, using electrokinetic data, and then plotted as a function of pH. To eliminate the effect of electrolyte concentration on the surface ionization constants the so-called double extrapolation technique [2, 7, 9] can be applied (extrapolation to pH_{pzc} and to $c = 0$). An example of such straight-line extrapolation for an alumina/KNO$_3$ system is presented in Fig. 1.

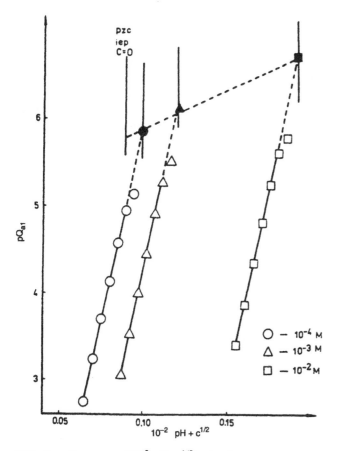

FIG. 1 pQ_{a1} versus $(10^{-2} pH + c^{1/2})$ plot for Al$_2$O$_3$/KNO$_3$ system. The dashed lines are extrapolated ones. (From Ref. 120; copyright © 1984 Academic Press, Inc.)

From Eqs (14) and (15) and (22) and (23) the value of the surface potential can be evaluated; for pH < pH$_{pzc(iep)}$:

$$\frac{e\psi_0}{2.3kT} \approx pK_{a_1}^{int} - pQ_{a_1} \tag{24}$$

and for pH > pH$_{pzc(iep)}$:

$$\frac{e\psi_0}{2.3kT} \approx pK_{a_2}^{int} - pQ_{a_2} \tag{25}$$

It was found (from the electrokinetic data for different oxides) that the surface potential changed by about 53 mV for SiO_2, 56 mV for TiO_2, 56 mV for Al_2O_3, and 54 mV for FeOOH per pH unit 10, 28, 120, 122. Moreover, far enough from the p.z.c.(i.e.p.) those values were almost independent of the electrolyte concentration.

For more accurate evaluation of the surface potentials, especially in the vicinity of the i.e.p., the so-called "curvilinear" interpolation method [121] can be used instead of the straight-line extrapolation technique described above. An example of such inter-polation for the alumina/KNO$_3$ system is shown in Fig. 2. As seen for 0.01 mol dm^{-3} electrolyte solution all experimental points lie on the curve. However, for diluted electrolyte solution (0.1 mmol dm^{-3}) the experimental points in the pH range 8–9 deviate from the interpolated curve because in this range the assumptions expressed by Eqs (20) and (21) are not valid [121]. Using the data from Fig. 2 the surface potentials of alumina for different ionic strengths were evaluated [121]. The results are presented in Fig. 3. As observed, far enough from the i.e.p. the slope of the curves does not depend on electrolyte concentration. However, in the vicinity of the i.e.p. the surface potential change for higher electrolyte concentration is much less than that observed for diluted electrolyte solution. This is in agreement with literature data for oxides [15–29].

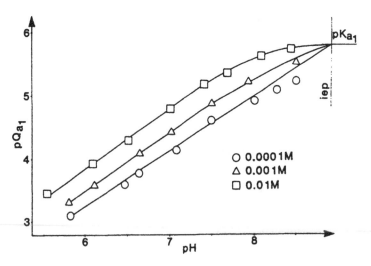

FIG. 2 pQ_{a1} versus pH plot for Al_2O_3/KNO_3 system obtained by curvilinear interpolation method. (From Ref. 121; copyright © 1987 Academic Press, Inc.)

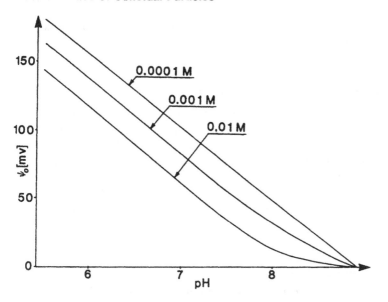

FIG. 3 Surface potential of alumina versus pH for Al_2O_3/KNO_3 system. (From Ref. 121; copyright © 1987 Academic Press, Inc.)

Other investigators used the approach presented in this chapter for different systems. They found that such analysis of zeta-potential data could be useful in evaluation of surface ionization constants as well as for obtaining information on the surface potential versus pH for oxides [122]. The plot of calculated surface potential for a fused silica sample versus pH for different electrolyte concentrations is presented in Fig. 4. As seen, the

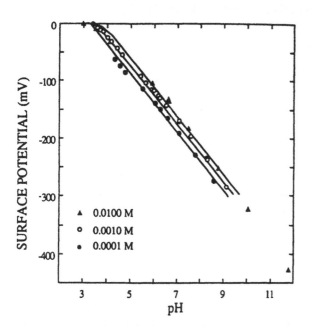

FIG. 4 Surface potential for a fused silica versus pH for SiO_2/KCl system. (From Ref. 122; copyright © 1992 American Chemical Society.)

surface potential changed by $\sim 54\,mV$ per pH unit, and the shape of the curves was in good agreement with model calculations for silica [29].

B. Evaluation of Shear-Plane Separation Distance

The location of the shear plane within the EIL is not known and cannot be measured directly. The evaluation of the slipping plane location, though possible, is not an easy task and is based on some reasonable assumptions. Some authors concluded that this plane coincides with the outer Helmholtz plane (OHP), i.e., $\psi_d = \zeta$ [2, 120, 123–126] while others assumed the slipping plane to be farther out from the OHP [19].

Eversole and coworkers [129, 130] were the first who evaluated slipping-plane separation distance for glass samples using the Gouy–Chapman theory of the diffuse electrical layer. The values obtained by them ranged from 24 to 110 Å, depending on the type of electrolyte used. Later on it was shown that those values were considerably overestimated. Eversole and coworkers [129, 130] assumed applicability of the diffuse-layer theory in the entire interfacial region, and independence of surface potential and separation distance (x) from the electrolyte concentration. They used data obtained for diluted solutions, but high pH values (high zeta potentials). Such an approach resulted in an underestimation of surface potential.

Sprycha and Matijevic [131] used the same approach to evaluate the slipping-plane separation distance for silver iodide ($c \leq 0.01$ mol dm^{-3}) and metal oxides ($c \leq 0.001$ mol dm^{-3}) by analyzing electrokinetic data in the close vicinity of the i.e.p. Under such conditions the entire EIL (low electrolyte concentrations and low surface potentials) most closely corresponds to the diffuse-layer model. In addition, for silver iodide and some metal oxides, independent methods of surface potential determination can be used as a test of validity of this approach.

According to the Gouy–Chapman theory of the EIL [18, 21, 22]:

$$\ln \tanh\left(\frac{ze\zeta}{4kT}\right) = \ln \tanh\left(\frac{ze\psi_0}{4kT}\right) - \kappa x \tag{26}$$

where κ denotes the reciprocal of the Debye length [21, 22], which can be expressed as follows:

$$\kappa = \sqrt{\frac{8\pi e^2 n z^2}{\epsilon kT}} \tag{27}$$

where e is an electronic charge, n denotes concentration of ions per unit volume, z is the valency of the ions, and ϵ denotes dielectric constant of the solvent. At room temperature ($T = 298$ K) and assuming $\epsilon = 78.8$ for water in the bulk, Eq. (26) can be written as [131]

$$\log \tanh \frac{\zeta}{102.8} = \log \tanh \frac{\psi_0}{102.8} - 0.142 x\sqrt{c} \tag{28}$$

If ψ_0, x, and ϵ in the region between the solid surface and the shear plane are indeed independent of the electrolyte concentration the plot of the left-hand term in Eq. (28) versus $\sqrt{c}$ should yield straight lines. In such a case the separation distance x can be determined from the slope of the straight line, and the surface potential value from the intercept of the straight line with the y axis.

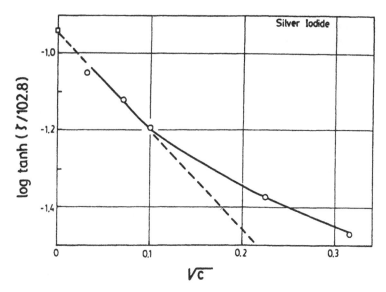

FIG. 5 Plot of log tanh ($\zeta/102.8$) versus $c^{1/2}$ for AgI/KNO$_3$ system. (From Ref. 131; copyright © 1989 American Chemical Society.)

Examples of such plots obtained for silver iodide and for chromium hydroxide and titania (rutile) are presented in Figs 5 and 6, respectively. Dashed lines represent extrapolated parts of the curves. As seen in Fig. 5, in the concentration range 0–10^{-2} mol dm^{-3} the EIL can be satisfactorily described by the Gouy–Chapman model in the vicinity of the p.z.c. (a straight line). For higher electrolyte concentrations and far from

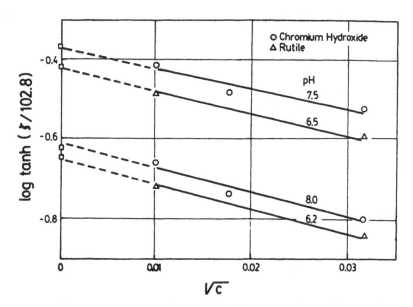

FIG. 6 Plot of log tanh ($\zeta/102.8$) versus $c^{1/2}$ for chromium hydroxide and titania (rutile) for two different pH values. (From Ref. 131, copyright © 1989 American Chemical Society.)

the i.e.p. the presence of the Stern layer causes deviations from the linear dependence of the plot. The shear plane separation distance for AgI calculated from Fig. 5 is $\sim 17\,\text{Å}$. This value differs from the ~ 6 Å value commonly accepted for AgI [24, 126]. The discrepancy arises from the value of the dielectric constant for water used in Eq. (28). Due to the ordering of water molecules at the AgI/water interface the value of the dielectric constant for the ordered layer is about 5–10 [24, 126]. When this lower value of ϵ is substituted in Eq. (27) the calculated separation distance is ~ 5–6 Å, in agreement with literature data [24, 126].

Using data from Fig. 6 the slipping-plane separation distance calculated for chromium hydroxide and rutile was ~ 40 Å for both oxides. For oxides, in addition to a strongly bound first layer of water molecules ($\epsilon \approx 6$) a secondary layer of water is also present with an estimated ϵ of ≈ 20–40 [59]. Assuming an average value of $\epsilon = 15$–20 in the immobile region between the surface and the shear plane, the value of the slipping-plane separation distance calculated from Fig. 6 is about 17–20 Å. This value is similar to that evaluated for colloidal hematite from adsorption data [11] and in line with the value used in the TLM model of the EIL [2].

The shear-plane separation distance is sensitive to the value of the solvent dielectric constant used in the calculations, Eq. (26). To visualize how ϵ affects the value of the separation distance for AgI, TiO_2, and $Cr(OH)_3$ the plot of x versus ϵ is presented in Fig. 7. As seen, the slipping plane for metal oxides is located considerably farther from the surface than that for AgI.

Analysis of electrokinetic data for oxides in the vicinity of the i.e.p. can furnish some information regarding the relationship between ψ_d and ζ potentials. Using experimental electrokinetic data for chromium hydroxide and following the analysis by Smith [132] it was shown that the assumption used in the literature that $\psi_d = \zeta$ is justified for low potentials and low electrolyte concentrations [131].

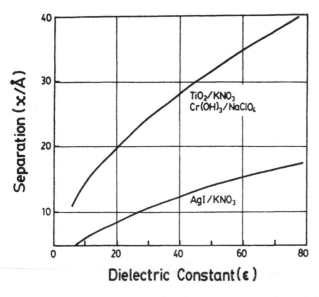

FIG. 7 Slipping-plane separation distance as a function of the dielectric constant of water in the compact part of the EIL for chromium hydroxide, titania, and silver iodide. (From Ref. 131; copyright © 1989 American Chemical Society.)

IV. APPLICATION OF ELECTROKINETIC MEASUREMENTS IN THE PRINTING INK INDUSTRY

Electrical properties of colloidal systems play a very important role in numerous practical applications. They contribute significantly to the stability and rheology of colloidal dispersions, emulsions, or foams. Zeta-potential measurements are widely used in the industry and industrial research related to the recovery of mineral ores by flotation [115, 131–135], paper making [108, 116, 136–138], pigments, and printing inks [106, 109, 114, 139–145], to mention a few. In this section a few examples of the application of electrokinetic measurements related to the printing ink industry will be briefly discussed.

A. Stability of Pigment Dispersions and Inks

Pigment dispersions and printing inks are examples of concentrated and very complex colloidal systems. Printing ink consists of pigment, resins, solvents, and different additives. Properties of inks and dispersions may vary considerably. Many commercial products contain organic solvents while others are water based. Some of these solvents may be toxic and harmful for humans, e.g., toluene. Thus, there is growing pressure on the graphic arts industry to reduce or eliminate the use of toxic organic solvents and switch to water-based formulations. Depending on the area of application the printing inks may have relatively low concentrations of pigments and binders (low viscosity), e.g., ink-jet ink or gravure inks, or very high concentrations of pigments and resins, e.g., lithographic paste inks (high viscosity). The stability of inks is a key issue in achieving good printability and color development. In addition, it determines the shelf-life of the product.

Colloidal dispersions can be stabilized by different mechanisms such as electrostatic or steric stabilization. If colloidal particles carry high surface charge then electrostatic repulsion between them may be strong enough to prevent aggregation. The same can be achieved by adsorption of a thick layer of polymeric material (polymers, surfactants). The most stable systems are formed when both mechanisms of stabilization are involved, so-called electrosteric stabilization [57, 146–150].

The first theory that described the stability of lyophobic colloidal systems in terms of attraction–repulsion forces (Van der Waals' attraction forces – electrostatic repulsive forces) was the DLVO theory derived by Deriaguin, London, Vervey, and Overbeek [149, 150]. The DLVO theory can predict the effect of electrolyte concentration on the stability of dispersions (Shultze–Hardy rule) and can explain the phenomenon of weak reversible flocculation at the "secondary minimum" [22, 56, 149, 150]. The existence of this minimum is very important from a practical point of view. Colloidal particles flocculated in secondary (very shallow) minimum can be very easily redispersed. On the other hand, the structure formed is strong enough to prevent particle sedimentation and keep the system in the dispersed state for a long time (paints, inks). Though the DVLO theory was one of the most important achievements of colloidal science in the twentieth century and was successful in the description of lyophobic colloids, it failed to predict the stability of lyophilic colloidal dispersions and the effect of surfactant addition on the stability of lyophobic systems [57, 149, 150].

Different dispersants and dispersing techniques are used to prepare pigment dispersions, depending on the chemistry of pigment and solvent used [60]. Oxide-type pigments, e.g., TiO_2, $CaCO_3$, or Fe_2O_3 can be successfully dispersed in water using inorganic (e.g., polyphosphates) or organic (e.g., polyacrylate) polyions [60]. In the first case, electrostatic charge and high zeta potential are responsible for dispersion stabilization. The latter case is an example of electrosteric stabilization. For the pigment dispersion to be electrostatically stabilized the pigment particles must be charged. The mechanisms of solid surface charging in aqueous and nonaqueous systems were briefly discussed in Section II.

The mechanism by which pigment dispersion is stabilized has to be carefully selected. Most water-based inks are formulated using anionic polymers (acrylates, maleates, sulfonated polyesters, etc.), therefore pigment dispersions designed for such systems must be stabilized by anionic or nonionic polymers (surfactants) to achieve good compatibility. Polymers and pigments which carry opposite charges cannot be used together to formulate ink.

The sign and value of the electrical charge of pigment particles can be determined by one of the electrokinetic methods. Because inks and pigment dispersions are concentrated systems the experimental technique that can handle such a system without dilution is desired and recommended for use, e.g., electrokinetic sonic amplitude [60].

Electrokinetic properties of pigments depend strongly on surface treatment [139, 140, 151, 152]. Many pigments are treated to ease their dispersibility and improve ink stability or change other properties such as chemical resistivity, weathering, durability, etc.. Some pigments may be abrasive and cause excessive wear of printing plates or the gravure printing cylinder. Larger pigment aggregates can accelerate wear processes very significantly. Thus, keeping such pigments well dispersed in the submicrometer size range is very important for printers. Treatment pigments showing high affinity towards polymers (binders) used in printing inks will usually exhibit a very high degree of dispersion (lack of aggregates) not only in the liquid ink but also in the dry ink film. Good pigment dispersion is very important for achieving high gloss and color strength of the print. Different approaches were used to describe the polymer/pigment interactions in practical systems. Some of them considered pigment/polymer interactions and the process of pigment dispersion in terms of surface energies of the pigments and polymers used [60, 153].

Inorganic pigments are very often coated to change their surface properties, e.g., with silica or aluminum oxide. In such a case, the electrokinetic properties of coated pigments change and approach that characteristic for a given coating [139]. For example, the i.e.p. of silica ($pH_{iep} = 2$) is shifted to the higher pH values when silica is coated with alumina. The shift depends on the coverage of the silica surface with aluminum oxide. When coverage is complete the i.e.p. of such particles is the same as that for alumina ($pH_{iep} = 9$). If basic particles ($pH_{iep} > 7$) are coated with a more acidic coating than the i.e.p. of coated material will be shifted towards lower pH values. This technique is commonly used for surface treatment of titania and iron oxide pigments [153–156]. The effect of surface coating of the values of the zeta potential and i.e.p. of the particles of chromium hydroxide and hematite is presented in Figs 8 and 9, respectively. As seen in Fig. 8, contamination of the chromium hydroxide surface with silica shifted the i.e.p. from the original value ($pH_{iep} = 8.4$) to lower pH values ($pH_{iep} = 7.2$). Similarly, the i.e.p. of hematite particles ($pH_{iep} = 6.5$) coated with chromium hydrous oxide was shifted to the higher pH values ($pH_{iep} = 7.5$). The latter value for coated particles is the same as that for pure chromium hydrous oxide [69]. Detailed information on the objectives of surface treatment of pigments can be found in the literature [60, 154].

To improve the dispersibility of organic pigments their surfaces can be treated with substances which anchor with one end of the molecule attached to the pigment surface and expose functional groups (e.g., carboxylic, amino) to the air. Surface functional groups are easily accessible for interactions with solvent and can render surface charge to the particle. Electrokinetic properties of coated organic pigments depend on the chemistry of treatment applied [60, 152]. Surface treatment can be performed at different stages of pigment

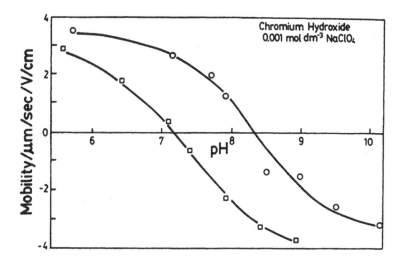

FIG. 8 Electrophoretic mobilities of chromium hydroxide particles versus pH: pure sample (○); sample contaminated with silica (□). (From Ref. 131; copyright © 1989 American Chemical Society.)

manufacture. There is a vast patient literature available on surface treatment of pigments, e.g. [157–159].

There is a special class of pigment dispersants called "hyperdispersants" [160, 161]. These substances are polymeric materials which were designed specifically to anchor very strongly on the surface of a given pigment and form a very effective steric layer protecting pigment particles against flocculation. Examples of hyperdispersants are fatty polyesters containing carboxylic functionality, fatty polyureas, or polyurethanes. Hyperdispersants

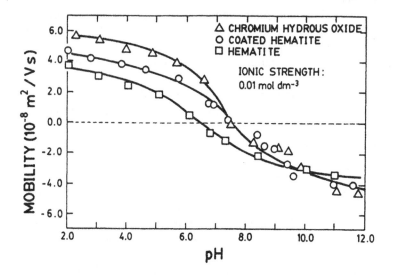

FIG. 9 Electrophoretic mobilities of hematite, hematite coated with chromium hydrous oxide, and pure chromium hydrous oxide particles versus pH. (From Ref. 69; copyright © 1988 American Chemical Society.)

are widely used in the printing ink industry to make high-strength pigment bases of the desired rheological properties. They allow for reduction of processing costs and can increase the capacity of the plant [160, 161].

B. Liquid Toners

Liquid toner is a special type of ink designed to be used to in electrophotographic printing as an alternative to dry toning ("xerography"). The process of liquid development of a latent image was first proposed by Metcalfe [162]. In the printing devices using liquid toners a latent image is created on the photoconductive insulating surface of a cylinder as a pattern of electrostatic charges. The process of image development is based on electrophoretic transfer of charged pigmented toner particles to the charged sites of the cylinder. The image on the cylinder and the colloidal toner particles have opposite charges. The developed image from the cylinder is then transferred to paper. Liquid immersion development has some advantages over the dry toning process. For example, high resolution can be achieved by use of fine colloidal toner particles. Detailed information on liquid immersion development processes can be found elsewhere [51–53, 163, 164].

Liquid toner consists of pigment particles dispersed in low-conductivity solvent containing polymer (for stabilization, fusing), dispersant, and charge-control agent. The materials used to make ink must meet special requirements [51].

The solvent has to be an insulating liquid of very low conductivity (resistivity greater than $10^9 \, \Omega/cm^{-1}$ to prevent the latent image from discharging. In addition, the solvent should have very low viscosity (to allow fast migration of toner particles to the surface), to be safe for users (high flash point, nontoxic, odorless), and to be inexpensive. Some aliphatic hydrocarbons meet these criteria and are used in liquid toners.

Different types of pigments are used in liquid toners, depending on the color. they may be inorganic or organic pigments. The way of their manufacture is protected with patents. Pigments also have to meet special requirements regarding color, lightfastness, resistance, low bleed, and others.

A polymer (dispersant) used in liquid toner should ensure good pigment dispersion, protection of particles against flocculation, and fixing of the toner particles to the substrate. These materials may be natural or synthetic and have to adsorb very strongly on the pigment surface. Many of the synthetic materials are alkyl methylacrylate-based copolymers.

The main purpose of using "charge-control agents" in liquid toners is to give the toner particles electrical charge so that the electrophoretic deposition of particles is possible. A variety of different charging agents can be used such as metal soaps, metal resinates, surfactants, etc. The mechanism of charging in nonaqueous systems was briefly discussed in Section II of this chapter.

There are three charged species in the liquid toner: toner particles, coions, and counterions. For effective image development it is important that the number of coions is relatively low so that mostly pigment particles are electrophoretically transferred to the surface. The charge density on pigment particles is a very important parameter characterizing liquid toners. Usually, the so-called Q/m (charge-to-mass ratio) is used to characterize toners because this value can be easily measured [51–53].

The electrokinetic charge on the toner particle can be expressed as follows [51]:

$$Q = 4\pi\epsilon\zeta r \tag{29}$$

where, r is the radius of the toner particle. The mass of that particle is

$$m = \frac{4}{3}\pi r^3 \rho \qquad (30)$$

where ρ is the toner particle density. By combining Eqs (29) and (30) one obtains:

$$\frac{Q}{m} = \frac{3\epsilon\zeta}{r^2\rho} \qquad (31)$$

Equation (31) clearly shows that for a given system (r = constant, ρ = constant, and ϵ = constant) the Q/m value is a measure of the zeta potential of the particles. For fast development of the latent image the Q/m should be high, but some limitations also apply [51].

One of the methods for measuring Q/m values for toners is the plate-out technique [51–53] in which the electrical current versus time is measured during the deposition of toner particles in a special cell. This allows for calculation of the total charge that passes through the system. The amount of toner particles deposited on the electrode is determined gravimetrically. Having these two values the Q/m parameter can be easily calculated. Though the above method of electrokinetic characterization of toner particles seems to be very simple the relationships between particle charge, mobility in the electric field, and zeta potential in low-conductivity media are rather complicated [37, 50–52]. Detailed studies showed that the electrophoretic mobility of charged particles in nonaqueous media depended on the time the measurement was performed after the electric field was applied [61]. The results obtained for carbon-black particles and two different concentrations of the charge-control agent [poly(isobutylenesuccinimide), OLOA 1200, by Chevron] are presented in Fig. 10. According to the authors [61], to make a claim that the particles in nonaqueous media carry the net electrical charge, their measured mobility has to be constant versus time or its value extrapolated to zero time should be greater than zero. They also observed that the values of extrapolated zeta potential may depend on the cell design [61]. The plots of zeta potentials versus concentration of charge-control agent for two different cell designs are presented in Fig. 11.

It is noteworthy that the EIL that develops in low-conductivity solvents is significantly different from that in aqueous solutions. Because in a nonaqueous system the

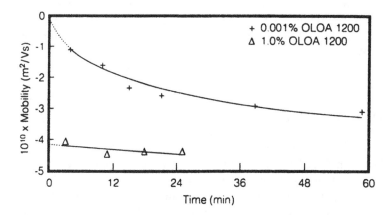

FIG. 10 Electrophoretic mobility of carbon black particles versus time from filling of the cell for two suspensions containing different concentrations of the charge-control agent (OLOA 1200). (From Ref. 61; copyright © 1992 American Chemical Society.)

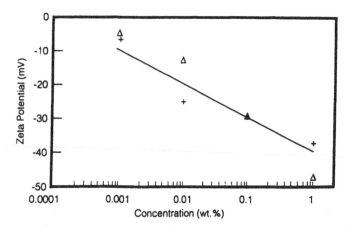

FIG. 11 Zeta potential of carbon black particles versus concentration of charge-control agent (OLOA 1200) measured in a parallel-plate cell (+) and using Delsa 440 instrument cell ($\triangle$). (From Ref. 61; copyright © 1992 American Chemical Society.)

dissociation of electrolyte is very limited, the number of ions n in Eq. (27) is very low. The lowering of n in low-conductivity media considerably surpasses the lowering of the dielectric constant so that the value of κ in Eq. (27) is significantly lower than that for aqueous systems. Due to the low value of κ the "thickness" ($1/\kappa$) of the EIL [21–27, 56, 57] in low-conductivity media is very large.

C. Lithographic Printing

Lithography is one of the major printing processes. The principle of the process is based on the difference in surface properties between image and nonimage areas of the printing plate. In conventional lithography the image area is oleophilic and thus it shows very high affinity towards lithographic inks (oil based) and very low affinity towards water (water repellent). The nonimage area is hydrophilic (high affinity towards water) and oil repellent when wet. The image and nonimage areas are in the same plane on the printing plate. During printing the plate before inking has to be damped with water (fountaining/dampening solution) to form a thin film of water on the nonimage area. The wet nonimage area is now ink repellent. Water film applied by the damping system to the image area would break almost immediately into tiny droplets of water, leaving uncovered ink-accepting surface. When ink film is applied to the damped plate by the inking roller only image areas of the plate will be covered with the ink. Ink from the image area can be transferred directly to the substrate in the nip of the printing press. The process of direct lithographic printing (plate in direct contact with substrate) is nowadays rather rare [164, 165]. Most printers use offset process in which the image from the printing plate is at first "offset" to the blanket roller and then transferred from the rubber blanket to the paper. Different types of printing presses and damping systems are available on the market [164, 165].

For good printability a proper ink/fountain solution balance has to be achieved. The tolerance of a given system for water, the so-called water window, depends on the chemistry of the lithographic ink and fountain solution used [165–168]. The main function of the dampening solution is to keep nonimage areas of the printing plate clean and ink repellent. New plates are covered with a thin layer of highly hydrophilic desensitizer, e.g., gum arabic. The thin layer of gum arabic would be worn very quickly during printing. Thus, all fountain solutions contain desensitizer as a main component. Other additives can also be used in the fountain solutions, depending on the particular task.

A fountain solution is not necessary in the "waterless" lithographic process. Formulas for waterless inks are different from the conventional ones. Also, the printing plate is designed and manufactured differently. The image area of the plate is receptive to waterless ink, but the nonimage area covered with special silicone polymer is ink repellent. Thus, lithographic printing without dampening solution is possible. Both conventional and waterless processes have some advantages and disadvantages. More details on lithographic printing can be found in the literature [164–173].

Due to the complexity of the lithographic systems it is very difficult to predict the performance of a given ink on the press from laboratory tests [165–173]. Among different experimental techniques used to characterize the interactions between the ink, plate, and fountain solution, electrokinetic methods also caught the attention of researchers [112, 169, 170].

Virtanen and coworkers [169, 170] used zeta-potential measurements in a rectangular cell [174] to study the effect of pH and the concentration of surfactants and calcium ions in the fountain solution on ink performance. They observed "charge reversal" at higher concentrations of calcium ions in the ink/fountain solution system. Calcium ions can migrate from the coated papers (coating contains calcium carbonate) to the acidic fountain solution. They can be subsequently adsorbed on different offset material surfaces (including ink and plate) and change their surface properties. The authors claim that the streaming potential technique they used [169, 170] yielded valuable information about interactions in the ink/plate/fountain solution system.

Brito et al. [112] and Kuehn [175] used electrokinetic measurements (streaming potential technique in a rectangular cell) to investigate the surface properties of printing lithographic plates. They found that the zeta-potential measurements could provide a sensitive measure of the nature of the surface groups on the surface of the printing plate [112] and a better understanding of the interactions between the printing plate and fountain solution [175].

REFERENCES

1. DE Yates, S Levine, TW Healy. J Chem Soc, Faraday Trans 1 70: 1807, 1974.
2. JA Davis, RO James, JO Leckie. J Colloid Interface Sci 63: 480, 1978.
3. L Bousse, NF de Roij, P Bergveld. Surface Sci 135: 479, 1983.
4. T Hiemstra, JC de Vit, WH van Riemsdijk. J Colloid Interface Sci 133: 91, 1989.
5. N Kallay, M Tomic. Langmuir 4: 449, 1988.
6. M Tomic, N Kallay. Langmuir 4: 565, 1988.
7. R Sprycha. J Colloid Interface Sci 102: 173, 1984.
8. LK Koopal, WH van Riemsdijk, MG Roofey. J Colloid Interface Sci 118: 117, 1987.
9. R Sprycha. J Colloid Interface Sci 127: 1, 1989.
10. R Sprycha. J Colloid Interface Sci 127: 12, 1989.
11. R Torres, N Kallay, E Matijevic. Langmuir 4: 706, 1988.
12. N Kallay, T Preocanin, S Zalac. In: N Kallay, ed. Interfacial Dynamics. Marcel Dekker Inc. New York, 2000, pp 255–247.
13. M Kosmulski, R Sprycha, J Szczypa. In: N Kallay, ed. Interfacial Dynamics. Marcel Dekker Inc., New York, 2000, pp 163–223.
14. R Sprycha, M Kosmulski, J Szczypa. J Colloid Interface Sci 128: 88, 1989.

15. TW Healy, LR White. Adv Colloid Interface Sci 9: 303, 1978.
16. P Hesleitner, D Babic, N Kallay, E Matijevic. Langmuir 3: 815, 1987.
17. J Westall, H Hohl. Adv Colloid Interface Sci 12: 265, 1980.
18. JA Davis, DB Kent. In: MF Hochella, AF White, eds. Reviews in Mineralogy. vol. 23. Mineralogical Society of America, Washington D.C., 1990, pp 177–260.
19. MA Blesa, N Kallay. Adv Colloid Interface Sci 28: 111, 1988.
20. GA Parks, PL de Bruyn. J Phys Chem 66: 967, 1962.
21. DC Grahame. Chem Rev 41: 441, 1947.
22. RJ Hunter. Foundations of Colloid Science. vol. 1. Oxford: Oxford University Press, 1987.
23. P Delahay. Double Layer and Electrode Kinetics. New York: Wiley-Interscience, 1965.
24. BH Bijsterbosch, J Lyklema. Adv Colloid Interface Sci 9: 147, 1978.
25. J Lyklema, TG Overbeek. J Colloid Interface Sci 16: 595, 1961.
26. J Lyklema. Disc Faraday Soc 42: 81, 1966.
27. GA Parks. Chem Rev 65: 177, 1965.
28. N Kallay, R Sprycha, M Tonic, S Zalac, Z Torbic. Croat Chem Acta 63: 467, 1990.
29. GR Wiese, RO James, DE Yates, TW Healy. In: JO'M Bockris, ed. International Review of Science. Phys Chem Ser 2, vol. 6. Butterworths, London, 1976.
30. S Levine, AL Smith. Disc Faraday Soc 52: 290, 1971.
31. M Kosmulski. In: N Kallay, ed. Interfacial Dynamics. New York, 2000, pp 273–312.
32. JH van der Minne, PHG Hermanie. J Colloid Sci 7: 600, 1952.
33. J Lyklema. Adv Colloid Interface Sci 2: 65, 1968.
34. FM Fowkes, H. Junai, M.A. Mostefe, FW Anderson and RY Moore In: M Hair, MD Croucher, ed. Colloids and Surfaces in Reprographic Technology. ACS, Washington D.C., 1982, pp 307–324.
35. FM Fowkes, Disc Faraday Soc 42: 246, 1966.
36. FM Fowkes, TB Lloyd, WJ Chen, GW Heebner. Proc SPIE 1253: 52–63, 1990.
37. A Kitahara. In: A Kitahara, A Watanabe, eds. Electrical Phenomena at Interfaces. New York: Marcel Dekker, 1984, pp 119–143.
38. O Stern. Z Electrochem 30: 508, 1924.
39. NJ Barrow, JW Bowden, AM Posner, JP quirk. Austr J Soil Res 18: 37, 1980.
40. L Bousse, JD Meindl. In: JA Davis, KF Hayes, eds. Geochemical Processes at Mineral Surfaces. ACS Symp Ser 323, Washington, DC: American Chemical Society, 1986, pp 79–98.
41. W Smit. J Colloid Interface Sci 109: 295, 1986.
42. W Smit. J Colloid Interface Sci 113: 288, 1986.
43. W Rudzinski, R Charmas, S Partyka. Langmuir 7: 354, 1991.
44. W Rudzinski, R Charmas, S Partyka, F Thomas, IY Bottero. Langmuir 8: 1154, 1992.
45. J Lyklema. J Electroanal Chem 37: 53, 1972.
46. N Kallay, D Babic, E Matijevic. Colloids Surfaces 19: 375, 1986.
47. L Blok, PL de Bruyn. J Colloid Interface Sci 32: 51, 1970.
48. R Sprycha. Colloids Surfaces 5: 147, 1982.
49. A Klinkenberg, JL van der Minne. Electrostatics in the Petroleum Industry. New York: Elsevier, 1958.
50. I Chen, J Mort, MA Machonkin. Langmuir 13: 5036, 1997.
51. MD Croucher, S Drappel, J Duff, K Lok, RW Wong. Colloids Surfaces 11: 303, 1984.
52. V Novotny, ML Hair. J Colloid Interface Sci 71: 273, 1979.
53. V Novotny. Colloids Surfaces 2: 373, 1981.
54. REG van Hal, JCT Eijkel, P Bergveld. Adv Colloid Interface Sci 69: 3, 1996.
55. NHG Penners, LK Koopal, J Lyklema. Colloids Surfaces, 21: 457, 1986.
56. J Lyklema. Fundamentals of Interface and Colloid Science. vol. II. Academic Press, London, 1995.
57. S Ross, ID Morrison. Colloidal Systems and Interfaces. New York: Wiley, 1988.
58. R Hidalgo-Alvarez. Adv Colloid Interface Sci 34: 217, 1991.
59. R Hunter. Zeta Potential in Colloid Science. New York: Academic Press, 1981.

60. E Kissa. Dispersion. New York: Marcel Dekker, 1999.
61. RE Kornbrekke, ID Morrison, T Oja. Langmuir 8: 121, 1992.
62. V Novotny. J Appl Phys 50: 324, 1979.
63. ER Arndt, KL Kreuz. J Colloid Interface Sci 123: 230, 1988.
64. ME Labib, R Williams. J Colloid Interface Sci 97: 356, 1984.
65. A Delgado, F Gonzalez-Caballero, JM Bruque. Colloid Polym Sci 264: 435, 1986.
66. O El-Gholabzopuri, MA Cabrierizo, R Hidalgo-Alvarez. Colloid Surfaces A 159: 449, 1999.
67. BJ Marlow, RL Rowell. Langmuir 1: 83, 1985.
68. J Salcedo, A Delgado, F Gonzalezz-Caballero. Prog Colloid Polym Sci 79: 64, 1989.
69. A Garg, E Matijevic. Langmuir 4: 38, 1988.
70. A Bismarck, MT Kumru, J Springer. J Colloid Interface Sci 217: 377, 1999.
71. T Jimbo, A Tanioka, N Minoura. Colloids Surfaces A 159: 459, 1999.
72. RA van Wagenen, JD Andrade. J Colloid Interface Sci 76: 305, 1980.
73. AJ van der Linde, BH Bijsterbosch. Croat Chem Acta 63: 455, 1990.
74. A Wiacek, E Chibowski. Colloids Surfaces A 159: 253, 1999.
75. AJG van Diemen, HN Stein. J Colloid Interface Sci 67: 213, 1978.
76. M Alfridsson, M Rassmussen, S Wall. Colloid Surfaces A 159: 413, 1999.
77. J Salcedo, A Delgado, F Gonzalez-Caballero. J Colloid Interface Sci 138: 10, 1990.
78. J Salcedo, A Delgado, F Gonzalez-Caballero. J Colloid Interface Sci 133: 278, 1989.
79. A Delgado, F Gonzalez-Caballero, J Salcedo, MA Cabrerizo. J Dispers Sci Technol 10: 107, 1989.
80. J Lyklema, JTG Overbeek. J Colloid Sci 16: 501, 1961.
81. R Hidalgo-Alvarez, FJ de las Nieves, AJ van der Linde, BH Bijsterbosch. Colloids Surfaces 21: 259, 1986.
82. M Colic, DW Fuerstenau, N Kallay, E Matijevic. Colloids Surfaces 59: 169, 1991.
83. R Sprycha, J Jablonski, E Matijevic. J Colloid Interface Sci 149: 56, 1992.
84. M Kosmulski, E Matijevic. Colloids Surfaces 64: 57, 1992.
85. A Delgado, F Gonzalez-Caballero, JM Bruque. J Colloid Interface Sci 113: 203, 1986.
86. A Delgado, F Gonzalez-Caballero, J Salcedo, M Cabrerizo. Mater Chem Phys 19: 327, 1988.
87. R Hidalgo-Alvarez, A Delgado, J Callejas, F Gonzalez-Caballero. Colloid Polym Sci 263: 941, 1985.
88. CF Zukoski IV, DA Saville. J Colloid Interface Sci 114: 45, 1986.
89. FJ de las Nieves, ES Daniels, MS El-Aasser. Colloids Surfaces 60: 107, 1991.
90. M Kosmulski, E Matijevic. Langmuir 7: 2066, 1991.
91. RO James, J Texter, PJ Scales. Langmuir 7: 1993, 1991.
92. M Kosmulski. Colloids Surfaces A 159: 277, 1999.
93. S Chen, K Kiruna. Langmuir 15: 1075, 1999.
94. R Xu. Langmuir 14: 2593, 1998.
95. RG Alargova, IY Vakarelsky, VN Panczow, SD Stoyanov, PA Kralchevsky, A Mehreteab, G Broze. Langmuir 14: 1996, 1998.
96. L Millesime, C Amiel, F Michel, B Chaufer. Langmuir 12: 3377, 1996.
97. M James, RJ Hunter, RW O'Brien. Langmuir 8: 420, 1992.
98. M Colic, DW Fuerstenau. Langmuir 13: 6644, 1997.
99. PG Hartley, I Larson, JP Scales. Langmuir 13: 1220, 1997.
100. EA Barringer, HK Bowen. Langmuir 1: 420, 1985.
101. EE Remsen, KC Thurmond, KL Wooley. Macromolecules 32: 3685, 1999.
102. C Wuertz, J Springer. Prog Org Coat 37: 117, 1999.
103. A Bismarck, J Springer. Colloids Surfaces A 159: 341, 1999.
104. I Leitschkis, W Rammensee. Colloids Surfaces A 159: 40, 1999.
105. A Borner, R Herbig. Colloids Surfaces A 159: 439, 1999.
106. H Giesche, E Matijevic. Dyes Pigments 17: 323, 1991.
107. Z Uhlberg, LR Marochko. Colloids Surfaces A 159: 513, 1999.
108. BW Greene, AS Reder. Tappi J 57: 101, 1974.

109. CM Hansen. Surface Coat Int 8: 386, 1997.
110. C Werner. Colloids Surfaces 159: 519, 1999.
111. T Jimbo, M Higa, N Minoura, A Tanioka. Macromolecules 31: 1277, 1998.
112. PSD Brito, IMGCS Paira, CAC Sequeira. NATO Asi Ser, Ser C 1993, pp 413–421.
113. Z Kovac, CJ Sambucetti. In OJ Murphy, S Srinivasan, BE Conway, eds. Electrochemistry in Transition. New York: Plenum Press, 1992, pp 39–49.
114. J Winkler. Eur Coat J 1: 38, 1997.
115. J Leja. Surface Chemistry of Froth Flotation. New York: Plenum Press, 1982.
116. YC Huang, FM Fowkes, ND Sanders, TB Lloyd. Tappi Proc (Coating), 1995, p 123.
117. C Werner, U Konig, A Augsburg, C Arnhold, H Korber, R Zimmermann, HJ Jacobasch. Colloids Surfaces A 159: 519, 1999.
118. J Westall. In: MC Kavanagh, JO Leckie, eds. Particulates in Water. Advances in Chemistry Series, No. 189, ACS, Washington D.C., 1980, pp 33–44.
119. J Westall. In W Stumm, ed. Aquatic Surface Chemistry. New York: John WIley, 1987, pp 3–32.
120. R Sprycha, J Szczypa. J Colloid Interface Sci 102: 288, 1984.
121. R Sprycha, J Szczypa. J Colloid Interface Sci 115: 590, 1987.
122. PJ Scales, F Grieser, TW Healy, LR White, DYC Chan. Langmuir 8: 965, 1992.
123. HC Li, PL de Bruyn. Surface Sci 5: 203, 1966.
124. RJ Hunter, HJL Wright. J Colloid Interface Sci 37: 564, 1971.
125. AG van der Put, BH Bijsterbosch. J Colloid Interface Sci 92: 499, 1983.
126. AL Smith. In: GD Parfitt, ed. Dispersions of Powders in Liquids. London: Applied Science, 1973 pp. 86–131.
127. HP Boehm. Disc Faraday Soc 52: 264, 1971.
128. PW Schindler, H Gamsjager. Dis Faraday Soc 52: 286, 1971.
129. WG Eversole, PH Lahr. J Chem Phys 9: 530, 1941.
130. WG Eversole, WW Boardman. J Chem Phys 9: 798, 1941.
131. R Sprycha, E Matijevic. Langmuir. 5: 479, 1989.
132. AL Smith. J Colloid Interface Sci 55: 525, 1976.
133. MH Jones, JT Woodcook, eds Principles of Mineral Flotation. Symposie Series No. 40, Australasian Institute of Mining and Metallurgy, Carlton, Australia, 1984.
134. SH Castro, J Alvarez, eds. Froth Flotation. New York: Elsevier, 1985.
135. JS Laskowski, J Ralston, eds. Colloid Chemistry in Mineral Processing. New York: Elsevier, 1992.
136. E Strazdinis. Tappi 545: 1691, 1972.
137. MJ Jaycock, JL Pearson. J Colloid Interface Sci 55: 181, 1976.
138. H Tanaka. In M Ohshima, K Furusawa, eds. Electrical Phenomena at Interfaces. New York: Marcel Dekker, 1998, pp 389–405.
139. DN Furlong, KSW Sing, BGD Parfitt. J Colloid Interface Sci 69: 409, 1979.
140. M Ishimori, M Asada, M Hosaka. J Jpn Soc Colour Mater 65: 155, 1992.
141. GMS El-Shafei. J Colloid Interface Sci 182: 249, 1996.
142. T Lossai. J Coat Technol 61: 57, 1989.
143. YC Huang, FM Fowkes, TB Lloyd. J Adhes Sci Technol 5: 39, 1991.
144. JE Hall. J Coat Technol 55: 41, 1983.
145. MJB Franklin, K Goldsborough, GD Parfitt, J Peacock. J Paint Technol 42: 740, 1970.
146. DH Napper. Polymeric Stabilization of Colloidal Dispersion. London: Academic Press, 1983.
147. RJ Pugh. In: RJ Pugh, L Bergstrom, eds. Surface and Colloidal Chemistry in Advanced Ceramics Processing. New York; Marcel Dekker, 1994 pp 195–223.
148. BV Velamakanni, DW Fuerstenau. Powder Technol 75: 1, 1993.
149. AW Adamson, AP Gast. Physical Chemistry of Surfaces. New York: Wiley-Interscience, 1997.
150. D Meyers. Surfaces, Interfaces and Colloids. Wiley-VCH, New York, 1999.
151. T Kobayashi, T Terada, S Ikeda. J Oil Colour Chem Assoc 73: 252, 1990.

152. T Kobayashi. In: M Ohshima, K Furusawa, eds. Electrical Phenomena at Interfaces. New York: Marcel Dekker, 198, p 405.
153. GD Cheever, JC Ulicny. J Coat Technol 55: 53, 1983.
154. J Schroeder. Prog Org Coat 16: 3, 1988.
155. H Sander. Farbe Lack 92: 286, 1986.
156. H Sander. Farbe Lack 93: 257, 1987.
157. F Babler. US Patent 5 931 997, 1999.
158. R Langley, WG Warwick. US Patent 3 775 149, 1973.
159. R Langley. US Patent 3 764 360, 1973.
160. TG Vernardakis. Modern Paints Coat 75: 32, 1985.
161. ACD Cowley. Am Ink Maker 71: 26, 1993.
162. KA Metcalfe. J Sci Instrum 32: 79, 1955.
163. RM Schaffert. Electrophotography. London: Focal Press, 1965.
164. RH Leach, RJ Pierce, eds. The Printing Ink Manual. London: Blueprint (imprint of Chapman and Hall), 1993.
165. J McPhee. Fundamentals of Lithographic Printing. Pittsburgh, PA: GATF Press, 1998.
166. RW Bassemir, R Kirshnan. TAGA Proc, 1989, pp 240–256.
167. RW Bassemir, R Krishnan. TAG Proc 1987, pp 560–573.
168. RW Bassemir, R Krishnan. TAGA Proc 1988, pp 339–353.
169. J Virtanen, A Karttunen, R Trauzeddel, R Tosch, U Lindquist. Advances in Printing Science and Technology. Pentech Press, London, 1989, pp. 41–61.
170. R Tosch, R Trauzeddel, U Lindquist, J Virtanen, A Karttunen. Am Ink Maker 69: 16, 1991.
171. RW Bassemir, F Shubert. TAGA Proc 1982, pp 290–310.
172. R Krishnan, D Klein. TAGA Proc 1991, pp 478–489.
173. JF Padday. Print Technol 13: 1, 1969.
174. VA de Palma. Rev Sci Instrum 51: 1390, 1980.
175. N Kuehn. TAGA proc 1998, pp 272–278.

27

Electrokinetic Investigations of Clay Mineral Particles

IVAN SONDI and VELIMIR PRAVDIĆ   Ruđer Bošković Institute, Zagreb, Croatia

I. INTRODUCTION

The term clay (or clays) is used in soil science to describe any naturally occurring inorganic materials composed primarily of fine-grained minerals of particle size less than $2\,\mu$m. The term clay minerals refers to a specific group of layer-type aluminosilicate minerals, which contain structural hydroxyl groups and belong to the general class of phylosilicates [1]. They are very fine particles or crystals, often colloidal in size, and usually plate-like in shape, less commonly tabular or scroll shaped. Because of their fineness they have the surface chemical properties of colloids. Due to their overwhelming diversity, their structural properties, chemical composition, and related unique surface chemical properties, clay minerals are fascinating colloid systems, which have been studied extensively from various aspects and for various purposes.

Some typical properties of clay minerals such as plasticity, swelling (lattice expansion), ion exchange, high specific surface area, and adsorption of organic and inorganic components, make them suitable for many practical applications. Therefore, they are widely used in many industrial products and processes [2–4].

Clay minerals are the main components of sediments and suspended matter in natural waters. Chemical processes in soils and recent sediments are largely determined by reactions at the surface of their mineral components [5–8]. In terms of adsorption of organic and inorganic compounds, clay minerals are the most active inorganic constituents of soils, sediments, and suspended matter. Therefore, they play a significant role in the chemical–biological transformation, transport, and deposition of contaminants in the natural environment [9–12].

Furthermore, layer aluminosilicates, such as clay minerals, catalyze numerous organic reactions [13–16]. Accordingly, the role of clay minerals in prebiotic chemistry was significant as chemical microreactors in selectively adsorbing and catalyzing reactions of amino acids [17–19]. Alvarez et al. [20] have shown that the adsorption of DNA on clay surfaces provides protection against biodegradation, but does not eliminate the ability to amplify DNA by the polymerase chain reaction. These observations have significant paleontological, archeological, and anthropological implications for the detection of ancient DNA. Bishop and Philip [21] have proposed a role for clays in kerogen formation. According to them, following mineral dissolution of sediments, organic material adsorbed on clays may give rise to the formation of an insoluble polymer, with many of the characteristics of amorphous kerogen. All these findings indicate the importance of surface properties and, indeed, of charge formation at the clay particle surfaces.

The aim of this chapter is to discuss the surface characteristics and colloidal behavior of clay mineral particles, in view of their electrokinetic phenomenology, physicochemical processes at clay solid/liquid interfaces, and finally, their role in the adsorption, transfer, and deposition of organic and inorganic compounds in natural environments. The review is limited to naturally occurring clays, clay minerals extracted from these, and the factors that regulate their colloid stability in dilute suspensions. The presentation of electrokinetic data for model clay minerals in the presence of fulvic acids (FAs) and some artificial polymers such as poly(acrylic acids) (PAAs) describes the structure and properties of the clay/aqueous electrolyte solution interface in a natural aquatic environment.

II. CHEMISTRY AND STRUCTURE OF CLAY MINERALS

Clay minerals are hydrous aluminosilicates with magnesium- and iron-substituting aluminum in varying degrees, and with exchangeable alkali and alkaline earth elements as potentially essential constituents. These substitutions in the basic aluminosilicate structure cause a wide diversity in chemical composition and structural characteristics of clay minerals. It is almost impossible to find in nature two clay specimens that are absolutely identical in structure and chemical composition. Structural parameters and chemical compositions make distinctions among clay minerals. Basically, the crystalline clay minerals are composed of continuous two-dimensional layers of silica tetrahedra arranged in a hexagonal form (siloxane layer), condensed with two-dimensional aluminum or magnesium hydroxyl octahedral layers, well known as gibbsite and brucite layers. In most clay minerals these structural units are superimposed in different ways. For additional information and a detailed discussion on the structures and chemistry of crystalline clay minerals the reader is referred to specialized literature [22-24]. We will consider only the basic structures of most common clay minerals, which describe the clay mineral systems that have been most extensively used as models in colloid chemistry.

Several characteristics are important for the classification of clay minerals: (1) structural arrangement of layers, (2) chemical composition of the octahedral layer, and (3) type of chemical bonding between the layers [25]. According to Brown et al. [24] there are six main structural groups of layer silicate mineral: (1) kaolinite–serpentine, (2) pyrophyllite–talc, (3) mica, (4) smectite–vermiculite, (5) palygorskite–sepiolite, and (6) the chlorite group.

According to the structural arrangements and numbers of tetrahedral and octahedral layers combined, phyllosilicates can be classified into three layer types. In the case when a tetrahedral siloxane layer is linked to one dioctahedral (gibbsite) or trioctahedral (brucite) layer, the structure is known as a 1 : 1 layer silicate. The term dioctahedral means that only two of the three possible positions in the octahedral layer are occupied by trivalent aluminum ions. In trioctahedral clay minerals, all three possible positions in the octahedral layer are occupied by magnesium in order to achieve the charge balance. The layers are uncharged and bonded together partly through van der Waals' forces and partly through hydrogen bonds from the hydroxyls of the octahedral sheet to the oxygens of the next silica sheet. The most important species of this kaolinite–serpentine group is kaolinite, a dioctahedral nonswelling clay mineral. The extent of atom substitution in the kaolinite lattice is relatively small and the chemical composition of kaolinite can be expressed by the ideal formula $Al_4Si_4O_{10}(OH)_8$.

When two tetrahedral layers are linked to one central octahedral layer, a 2:1 layer structure is formed. The most important groups in this structure are the smectite–vermiculite and the mica groups. Smectites are derived structurally from pyrophyllite or talc and the variety of minerals of this group arises from different types of isomorphous substitutions and of cation vacancies in the octahedral layer. Therefore, the surface charge density is mainly determined by the substitution of Si^{4+} by Al^{3+} in the tetrahedral layer, and by the amount of divalent and trivalent cations within the octahedral layer. The excess of negative charge in the structure of smectite is compensated for by interlayer cations (commonly Na^+, K^+, Ca^{2+}, and Mg^{2+}), located between the 2:1 structures, and whose hydration states vary with humidity. Furthermore, in water the compensating cations can be easily exchanged by other cations if available in proper concentrations. Montmorillonite and beidellite are the most important dioctahedral minerals of the smectite group. The commonest trioctahedral smectite is saponite.

The stability of the structure, which is determined by the type of chemical bonding between layers, is particularly reflected in swelling properties. The significant difference between the 1:1 and the 2:1 minerals is that there is no possibility of hydrogen bonding between triple layers in the structure of 2:1 minerals, and oxygen planes can interact with each other only by van der Waals' forces. Accordingly, they can be easily cleaved along this plane, exposing newly formed basal surfaces to the aqueous medium.

The mica structure is essentially the same as that of smectites. The only significant difference is that the small excess of negative charge between the silicate layers is balanced by potassium ions. The introduction of potassium stabilizes the structure, and the uptake of water molecules is prevented, causing formation of a nonswelling structure. Clay micas are particles with a size less than $2\ \mu m$ and contain less potassium and more water than the coarser grained mica minerals. Illite is the most widely used group name for hydrous clay micas present in soil and sediments.

As shown previously [5] and in Table 1, the chemical formula of the unit cell for a 2:1 layer type clay mineral can be written as

$$M_x[Si_aAl_{8-a}](Al_bFe(III)_cFE(II)_{c'}Mg_{4-b-c-c'})O_{20}(OH)_4 \qquad (1)$$

Chlorites belong to the third type of phyllosilicates (2:1 + 1), a group of nonswelling clay minerals, very common in sediments and soils. They are structurally related to the 2:1 clays in which the charge-compensating cations between mica-type unit layers are replaced by a brucite layer (octahedral magnesium hydroxide). Because of some replacement of MG^{2+} by Al^{3+}, this layer has a net positive charge and therefore compensates the net negative charge of the unit layers. Therefore, the presence of the 2:1 layer and one

TABLE 1 Layer Structures and Chemical Formula [Eq. (1)] of Clay Mineral Groups [5]

Group	Layer type	Layer charge	a	b	c	c'
			Chemical formula coefficients			
Kaolinite	1:1	< 0.01	–	–	–	–
Mica (illite)	2:1	1.4–2.0	6.8	3	$c+c' \approx 0.25$	
Vermiculite	2:1	1.2–1.8	7	3	0.5	–
Smectite	2:1	0.5–1.2	8	3.2	$c+c' \approx 0.25$	
Chlorite	2:1 + 1	Variable	2.4	8.4	0.5	1.5

layer of hydroxide structure are characteristic of chlorite clay minerals. The general unit cell chemical formula for dioctahedral chlorite can be written [5]:

$$[Si_aAl_{8-a}](Al_bFe(III)_cMg_{c'})O_{20}(OH)_{16} \qquad (2)$$

Sometimes natural clays contain clay mineral particles in which unit layers of different types of clay minerals are stacked together in either a regular or irregular manner (interstratification). They are known as mixed-layer clays or interstratified minerals. These clays, particularly illite–smectite, are very common minerals and illustrate the transitional nature of the 2:1 layered silicates.

III. THE ORIGIN OF CLAYS

Weathering and related reactions have dominated the formation of clay minerals during geological time. The weathering of rocks and soils is a complex interaction with the atmosphere, the hydrosphere, and the biosphere. Generally, clay minerals originate through several processes: (1) by hydrolysis and hydration of silicate, (2) by dissolution of a soluble rock containing relatively insoluble clay minerals, (3) by slaking and weathering of clay-rich sedimentary rocks such as shales, (4) by bacterial and other organic activity, and (5) by diagenetic processes following sedimentation [26–29].

Natural clays are rarely "pure," and are difficult to work with, while synthetic clay minerals permit control of the composition and the history of the product. Previously reported investigations [30–32] have described the synthesis, structure, and properties of smectite minerals, as well as kaolinite [33]. Syntheses of clay minerals offer a possibility for obtaining clay minerals with optimized properties for particular applications by regulating the nature and degree of isomorphous substitution, structural properties, catalytic activity, etc.

IV. STRUCTURAL ELEMENTS OF SURFACE CHARGING IN CLAY MINERALS

A. Typology of Surfaces and Surface Reactions

As colloids, clay minerals attract attention because their surfaces are structurally and chemically heterogeneous. Only by understanding the relationship between structural elements and surface charging can electrokinetic data be brought to use as a source of information in colloid science. One of the most instructive examples is the heterogeneity of the kaolinite mineral surface, originally proposed by van Olphen [34], confirmed and elaborated further in many subsequent and recent investigations [35–41]. According to van Olphen, the kaolinite surface exhibits three morphological planes of different structure, chemical composition, and consequently different surface properties. The kaolinite surface consists of a siloxane surface (silicon oxide) basal layer, a gibbsite (aluminum hydroxide) basal layer, and an edge surface.

Generally, the basal (siloxane) surfaces carry a constant negative charge, attributed to the isomorphous substitution of Si^{4+} by Al^{3+} in the tetrahedral layer. At the solid/aqueous solution interface this negative charge is pH independent. The model is known as the constant basal surface charge model [37]. In addition, Sposito [5] has described the structural features on the siloxane layer, the siloxane ditrigonal cavities that have properties of soft Lewis bases.

The isomorphic substitution of Al^{3+} by Fe^{2+} and/or Mg^{2+} in the octahedral (gibbsite layer) results in a net negative surface charge. The charge of the edge surface ("broken bonds") depends on pH and shows amphoteric character. It arises from adsorption and desorption of potential determining ions, H^+, and OH^-, and can be described by surface proton (acid–base) equilibria:

$$S\text{-}OH + H^+ \Leftrightarrow S\text{-}OH_2^+ \tag{3}$$

$$S\text{-}OH(+OH^-) \Leftrightarrow S\text{-}O^-(+H_2O) \tag{4}$$

The hydrous oxide groups at the edges are considered to be major reactive sites in clay surfaces [42, 43]. They can bind metal ions and ligands. The following equations [7, 44] are examples of surface metal binding, Eqs (5)–(7), of ligand exchange, Eqs (8) and (9), and of ternary surface complex-formation equilibria, Eqs (10) and (11):

$$S\text{-}OH + M^{z+} \Leftrightarrow S\text{-}OM^{(z-1)} + H^+ \tag{5}$$

$$2S\text{-}OH + M^{z+} \Leftrightarrow (S\text{-}O)_2M^{(z-2)} + 2H^+ \tag{6}$$

$$S\text{-}OH + M^{z+} + H_2O \Leftrightarrow S\text{-}OMOH^{(z-2)+} + 2H^+ \tag{7}$$

$$S\text{-}OH + L^- \Leftrightarrow S\text{-}L + OH^- \tag{8}$$

$$2S\text{-}OH + L^- \Leftrightarrow S_2\text{-}L^+ + 2OH^- \tag{9}$$

$$S\text{-}OH + L^- + M^{z+} \Leftrightarrow 2\text{-}L\text{-}M^{z+} + OH^- \tag{10}$$

$$S\text{-}OH + L^- + M^{z+} \Leftrightarrow S\text{-}OM\text{-}L^{(z-2)+} + H^+ \tag{11}$$

Because the siloxane basal surface is permanently negatively charged, the fraction of the positively charged edge surfaces in the total surface is determined by the net surface charge. The area of edge surfaces varies, depending on the type of the clay mineral, and in turn depends on the degree of physical disintegration of the particular clay mineral crystal [40, 41, 45].

B. Surface Areas and Cation-exchange Capacities of Clays

Specific surface areas (SSAs, surface areas per unit mass) of clays are largely due to their size and structural characteristics. In smectite minerals there are two types of surfaces, external and internal. The definition of external or internal surface depends to a large extent on the method and the technique of measurement. All of the techniques are based on adsorption of some nonreactive adsorbate at the accessible surface.

One of the most widely used techniques for measuring the external surface area of clays is sorption of nonpolar gases at liquid nitrogen temperature. Nitrogen and argon are used for medium and large SSAs ($\simeq$ from 1 to several hundred $m^2\,g^{-1}$), while krypton is used for very low SSA solids ($< 0.1\,m^2\,g^{-1}$). The procedure of calculation derives from the multilayer model of adsorption of the Brunauer, Emmett, and Teller theory [46]. The method of calculation is based on the statistically determined number of adsorbed molecules in the monolayer coverage, with the independently chosen surface area occupied by a single molecule (or atom) of the adsorbate. The choice is between the measured cross-sectional are of the adsorbate gas, and the one calculated from the area of exclusion based on repulsion forces. These two numbers differ by more than 20%. The adsorption has to proceed on a "clean" surface, that is, on one from which any other adsorbates, and, in particular water molecules, have been removed. The removal of water films from the surface of clay particles by degassing in vacuum at elevated temperatures often causes collapse of the expanded clay lattice and a diminution of the surface area. Even the cleaning of the surface in a helium gas stream with subsequent nitrogen adsorption cannot prevent these problems. Such experimental treatment of clay minerals, even under the most carefully chosen experimental conditions, is the cause of erroneously reported external SSAs.

The interlayer areas are mostly inaccessible to the adsorbate gas molecules. The cross-section of pores, a low multiple of ionic diameters, can cause capillary condensation of adsorbate gases, and thus invalidate the adsorption model. Therefore, gas adsorption methods are inadequate for measuring the internal surface areas of clay minerals.

The use of adsorption of polar molecules from a liquid, mostly from aqueous media, is experimentally a simple method for measurements of surface areas, and has significantly improved measurements of internal surface areas of 2 : 1 expanding clay minerals. As polar adsorbates, ethylene glycol [47], N-cetylpyridinium bromide [48], and glycerol [49] have been used. While these techniques have helped much in the understanding of the adsorption properties of clays, they still involve assumptions of the existence of interlayer voids and of the ability of the adsorbate to penetrate these.

Besides surface area, the density of the charge of the clay mineral surfaces is very important for their colloid chemical behavior in natural waters. As described previously, the excess of negative lattice charge, originating from isomorphic replacement in the interior of the crystal lattice, is compensated for by the adsorption of cations on the layer surfaces, which are too large to be incorporated in the interior of the clay lattice. In aqueous solutions these cations can be easily exchanged by other cations, if present. For more information about cation-exchange equilibria in clays the reader is referred to Laudelout [50].

The cation-exchange capacity (CEC) of the negative double layer can be defined as the excess of counterions that can be exchanged for other cations under given conditions of temperature, pressure, solution composition, and soil–solution mass ratio [5]. It can be determined analytically by replacement of native exchangeable ions by "standard" cations and it is conventionally expressed in milliequivalents per gram (meq g^{-1}). Therefore, the number of exchanged cations (i.e., the CEC) is directly related to the degree of isomorphous substitution that has occurred in the clay mineral lattice. Many different methods can be applied and usually cations such as Na^+ and NH_4^+ are used in exchange processes and in the determination of CEC.

Table 2 shows a compilation of the results in the measurements of CEC by using ammonia exchange monitored by an ammonia-selective electrode [51] and of external SSA

TABLE 2 Specific Surface Area and Cation-exchange Capacity of Some Clay Samples

Sample	SSA ($m^2 g^{-1}$)	CEC (meq/100 g)
Ripidolite	2.6	1.5
Kaolinite	11.7	12.0
Montmorillonite	71.1	142.3
Beidellite	61.0	107.0
Illite	44.3	24.6

determined by single-point nitrogen adsorption for some typical clay minerals [52]. The results indicate large external SSA and high CEC for beidellite and "Otay" montmorillonite (smectites), while illite, kaolinite, and ripidolite (chlorite) have considerably lower values.

C. Electrical Double Layer of Clay Particles

In the case when clay mineral particles are immersed in an electrolyte solution, an electrostatic field created by the charged surface of the particle governs near-surface distribution of the electrolyte anions and cations, forming an electrical double layer (EDL). The colloid stability of clay mineral suspensions is influenced by the structure of this layer. In order to describe the behavior of the EDL, many physicochemical models have been proposed and discussed. The purpose of this paragraph is to review some basic information about the models applied. Detailed discussion can be found in several monographs and papers [5, 7, 53–55].

Mostly, the simple Gouy–Chapman model has been considered as sufficient for the interpretation of electrokinetic data for clay surfaces. Since the physical reality of the EDL on clay mineral particles cannot be satisfactorily explained with this model, many modifications have been attempted with more or less success.

The simple model that is commonly used to describe the hard-sphere ion and inner-potential distribution in the diffuse part of the EDL is the modified Gouy–Chapman (MGC) theory. This theory is applied to clay minerals when the clay particles are considered equivalent to a uniformly charged, infinite planar surface [5]. As discussed above, the clay minerals are thin plate-like particles with negatively charged basal planes and positively charged edge surfaces. Therefore, the disk model, which can accommodate spatial differentiation of constant and pH-dependent charge, is closer to reality in description of the EDL and of the surface reactions of clay mineral particles in electrolyte solutions than the infinite-plane model [56]. The charge at disk-shaped clay mineral particles originates from the ratio between the negatively charged basal planes and the positively charged edge surfaces.

Secor and Radke [56] have shown that by using the disk model, the electrostatic field near the basal planes may "spill over" to dominate the positive edge surface for 2:1 clay mineral particles suspended in an electrolyte solution at concentrations below 5×10^{-3} mol dm^{-3}. Accordingly, the potential can always be negative, including the edge surface. Furthermore, in order to study the effect of finite particle size on the EDL of 2:1 clay minerals in 1:1 electrolyte solutions, Chang and Sposito [57, 58] applied the MGC theory to the disk model and successfully solved the Poisson–Boltzmann equation using a self-adaptive finite-element method.

V. CLAY MINERALS IN ENVIRONMENTAL STUDIES

The most important characteristics of clay minerals in the aquatic environment are their colloid stability and adsorptive properties. Colloid stability and interaction of clay mineral particles with organic and inorganic compounds and with other mineral particles depend, besides their mineralogical, chemical, and surface properties, on environmental characteristics, such a electrolyte concentration, pH [7, 59–62], presence of organic matter [63–66], and the relationship between the density of particles and the hydrodynamic conditions in the aquatic environment [67–69].

   Investigation of the clay mineral–solution interactions through electrokinetic properties is one way to study the complex material exchange processes in natural environments.
Charge formation, charge density, and compensation of charge due to adsorption of
dissolved substances are reflected in electrophoretic mobilities [41, 45, 65].

A. Measurements of Electrophoretic Mobilities of Suspended Particles

Suspended particulate matter, an important component present in all natural waters,
consisting predominantly of clays, is implicated in many of the biogeochemical processes
occurring in natural waters. It presents the most important vehicle in the transport and
deposition of inorganic and organic compounds in natural aquatic environments. This
material is heterogeneous in particle size and chemical composition, containing various
minerals such as aluminosilicates, carbonates, oxides, partially coated with organic matter
and organic macromolecules, biological debris, etc. [70].

   Clay minerals are the most important inorganic component of suspended matter in
natural aquatic environments. They surpass carbonates and many metal oxides in their
chemical persistence, colloid stability, and adsorption capacities. Biogeochemical processes govern the land/sea interaction and effectively regulate the adsorption and deposition phenomena in estuaries, in the transition between soft river water and seawater [9].

   Hunter and Liss [71] measured surface electrical charge on suspended particles in
estuaries and established that particles were negatively charged. Recently, the influence of
electrokinetic potential on flocculation and sedimentation dynamics of suspended matter
was investigated in detail in the Raša River estuary, Croatia [68]. The small Raša River
and its estuary, represent a relatively simple natural system, which offers a chance of
recognition of basic processes regulating the kinetics of clay mineral transport and deposition. Stability versus transport of suspended matter was found to coincide with changes in
the magnitude of the negative electrokinetic potential.

   Rapid flocculation and sedimentation of fine-grained particles, mostly clay minerals,
characterize the investigated natural environment. In addition, segregation of clay minerals was also observed within the estuary. Such mineral segregation in the natural environment has been previously observed [64, 72] and attributed to different rates of flocculation
[73, 74]. At the Raša River mouth, where a rapid sedimentation of clay minerals was
observed, the concentration of suspended matter is high and a strong salinity gradient
is present. Therefore, flocculation is enhanced. Electrokinetic measurements and particle
size analyses provide a significant indicator for such processes, indeed there is some predictive potential in these.

   Figure 1 shows electrokinetic potentials of suspended particles in original samples of
water in dependence on salinity. The results showed negative ζ-potential values, decreasing
with increasing salinity* from roughly -25 to $-10\,\mathrm{mV}$. The observed negative electrokinetic potential of particles in the Raša River and its estuary confirms literature data on the
uniquely negative charge of suspended mineral particles in riverine, estuarine, and marine
environment, indeed from ionic strengths of 10^{-3} to almost 0.56 mol dm^{-3}.

---

*In oceanography the measure of salt concentration is expressed as

   salinity $= 0.03 + 1.805 \times$ chlorinity

Chlorinity is the mass in grams of "atomic mass silver" just necessary to precipitate all the halogens (chloride,
bromide, iodide, and fluoride) in 0.3285 kg of a seawater sample. For an approximate measure, a salinity of 38‰
corresponds to an ionic strength of 0.56 M.

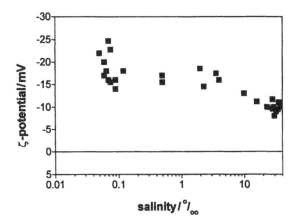

FIG. 1 Zeta potential of suspended particles of the Raša River estuary in dependence on salinity.

There is experimental evidence that fine-grained suspended inorganic particles, entering seawater, are unstable as single colloidal particles, and flocculate into aggregates with settling rates many times faster than those of the constituent grains [67–69, 75]. The surface charge density of the suspended mineral particles has influence on their sedimentation rates and on the sorption capacity for environmental contaminants. In clays, the most important factors affecting the charge density are the presence of divalent cations, the pH of the medium, and the adsorption of organic and inorganic compounds. The observed decrease in the electrokinetic potential with increasing salinity is significant, causing flocculation induced by charge neutralization. In addition, the hydraulics of turbulence and the resulting mechanical particle collisions also enhance flocculation. This phenomenon is related to particle concentration [76]. It was found that particles from (the optical) turbidity maximum in the water stream were more flocculated and settled faster than particles at lower concentration but higher salinity (ionic strength) [67].

B. Model Approach

Natural waters are complex systems of electrolytes, dissolved and particulate organic matter and inorganic particles, composed of clay minerals, quartz, carbonates, and a variety of metal oxides and hydroxides (most frequently of Al, Fe, and Mn).

A reductionistic approach to the understanding of processes involving clays (and other particulate matter) is to identify single processes, critical parameters governing the kinetics of these, and the consequences, mostly for the stability of colloidal systems. Laboratory experiments done with heterogeneous particulate matter, or an average clay of complex composition, mostly fail. This is also true for studies of rivers with a high concentration of suspended matter of nondescript origin. Electrokinetic measurements of such dense suspensions give an average potential, which obfuscates the single and important surface processes involving charge formation, distribution, and interaction. Accordingly, the use of model minerals and simple electrolyte solutions in carefully designed experiments appears to be the way out of these problems. The next step is the most intriguing one: does the understanding of a single process reveal the mechanisms of complex phenomena in nature? The answer is that electrokinetics helps us to understanding complex phenomena.

C. Preparation of Clay Minerals for Colloid Chemical Studies and Electrokinetic Measurements

The sources of clay minerals are natural clays. The main difficulty is the separation of clay minerals from clays always containing several different minerals such as quartz, carbonates, oxides, hydroxides, allophanes, and many diverse organic impurities. In all colloidal systems surface properties can be radically influenced and modified by the presence of small amounts of compounds adsorbed at, or covering the surface. The surface adsorbed oxyhydroxides may have a significant effect on the electrokinetic potential and on the isoelectric point (i.e.p.) (or the point of zero charge) of mineral particles [77]. A shift in the pH of the i.e.p. is towards that of the adsorbate on the clay surface. In addition, these adsorbates significantly participate in ligand-exchange reactions with anions and organic compounds present in the aquatic medium. The consequence is a change in surface and electrokinetic properties of the mineral surface and in collidal stability [77]. Therefore, in order to obtain results of electrokinetic measurements at a significant fundamental and reproducible level, it is absolutely necessary to remove all surface impurities. Sometimes, relatively simple electrokinetic techniques have to be corroborated by sophisticated physical. spectroscopic, and chemical analytical techniques. With these meticulous requirements, it comes as no surprise that reported colloid, surface, and electrokinetic properties of the same clay minerals obtained form various sources might be different [52].

Experimental procedures, which are necessary to separate the clay minerals from natural clays and to remove organic and inorganic impurities from the clay mineral surfaces, are well established. In order to separate clay minerals from natural soils and sediments it is necessary to employ fractionation techniques used in sedimentology [78]. The clay minerals can be treated with dilute Na acetate-buffered H_2O_2 (oxidation and decomposition), to remove residual organic compound from the surface. Other reactants and procedures can also be used in the decomposition of organic matter [79–82]. The surface-adsorbed crystalline and amorphous iron oxides and oxyhydroxides can be removed by citrate–bicarbonate–dithionate solution [79]. Carbonates and amorphous silica can be decomposed and removed by acids, or by boiling in Na_2CO_3 solution, respectively [83–85]. Finally, the monoionic forms of clay minerals can be prepared by ion exchange, following the method of Karen and Shainberg [86].

D. Clay Mineral Suspensions in Single Electrolyte Solutions

Previous investigations on the electrokinetic properties* of different types of clay minerals [35, 38-41, 60] indicated differences between various minerals in their pH dependence.

Figure 2 shows ζ-potential curves versus pH in a 10^{-3} mol dm^{-3} NaCl solution for three smectite clay minerals: "Otay" montmorillonite and beidellite (dioctahedral smectitates), and saponite, as a representative of a trioctahedral smectite. The most important

---

* *A note on methods and materials*

Electrokinetic measurements, on which the figures are based, were made by an automated electrophoresis instrument (S 3000, Pen-Kem, Bedford Hills). Zeta potentials were calculated from the measured electrophoretic mobilities (EPMs) by a preprogrammed method. The transformation of EPM into ζ potential involves many assumptions [54]. In these studies the ζ potential was calculated from the measured EPM data according to the Henry equation [54]:

$$u_e = (2\epsilon\zeta 13\eta) \cdot f_1(\kappa a)$$

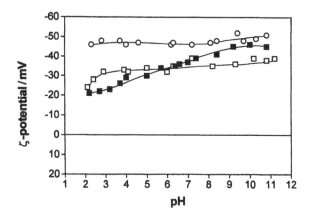

FIG. 2 Zeta potential of smectite clay minerals suspended in 1×10^{-3} mol dm^{-3} NaCl solution as a function of pH: (○) saponite; (■) beidellite; (□) "Otay" montmorillonite.

observation is that these natural minerals do not exhibit i.e.p. values in the pH range 2–11. In addition, the electrokinetic potentials of saponite and montmorillonite are almost insensitive to changes in pH, while beidellite shows distinct differences in the values of negative potential with increasing pH.

Figure 3 shows ζ-potential curves versus pH in a 10^{-3} mol dm^{-3} NaCl solution for kaolinite (kaolinite group) and ripidolite (trioctahedral chlorite). No i.e.p. values were observed for these minerals either, but, in contrast to the smectites, they showed a decrease in ζ potential, from -50 to -10 mV, with a decrease in pH from 10 to 2. Accordingly, changes in the magnitude of the ζ potential with pH for kaolinite and ripidolite are higher than for smectite clay minerals.

Results indicate that the constant charge of the basal faces of clay minerals significantly influences their electrokinetic properties, particularly the EPMs. The pH-dependent changes are a consequence of the existence of hydroxyl groups at the edge surfaces, the major reactive sites in clay surfaces. Indeed, the fraction of the edge surfaces and, thus, of reactive hydroxyl groups, in the total surface area vary, depending on the type of clay minerals. For kaolinite, the fraction of edge surface is 12% [87] to 14% [88], whereas for smectite clay minerals it is no more than 1% [47, 89]. It has been also shown that mechanical disintegration of clay mineral particles, e.g., by milling, increases the fraction of edge surfaces and accordingly changes their electrokinetic properties [39, 40].

(*continued*)
where ϵ is the permittivity of the medium (F m^{-1}), η is viscosity (N m^{-2} s^{-1}), ζ is electrokinetic (zeta) potential, u_e is EPM (m^{2}V^{-1} s^{-1}), a is particle radius (m), κ is the reciprocal double-layer thickness (m^{-1}), and f_1 (κa) is a function dependent on particle size and shape. In these measurements a high ratio between particle radius and the double-layer thickness was assumed, giving $\kappa a > 1$ and using the limiting value for $f_1(\kappa a)$ of 3/2. Accordingly, the ratio between ζ potential and EPM becomes:

$$\zeta = 12.8 \times 10^8 \cdot u_e$$

For sources and characterizations of minerals and chemicals the reader is referred to literature references [39–41].

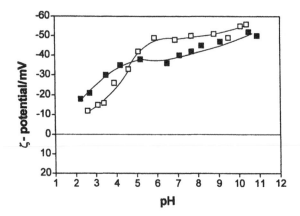

FIG. 3 Zeta potential of ripidolite (□) and kaolinite (■) suspended in 1×10^{-3} mol dm^{-3} NaCl solution as a function of pH.

Figures 4 and 5 show ζ-potential curves versus pH in a 10^{-3} mol dm^{-3} NaCl solution for natural and milled ripidolite and beidellite. The most important observation is the appearance of an i.e.p. for both of the milled minerals, for ripidolite at pH 6 and for beidellite at pH 3. In addition, changes in the ζ potential with pH for milled ripidolite are more pronounced than those for beidellite.

The effects of milling on the SSA and CEC of ripidolite and beidellite clays are shown in Table 3. For ripidolite the effect of milling causes a 12.3-fold increase in the SSA and a three-fold increase in CEC. For beidellite, the increase in SSA is a 1.5-fold and the CEC increases by 20%. Milling, that is, the break up of larger particles, creates new edges, increasing their contribution to the total surface area (Fig. 6). The reported findings show that the electrokinetic and surface properties depend also on the degree of physical "weathering" of the source clay minerals, a process responsible for enervation of small particles in nature. This is an important observation for predicting the interaction of clay mineral particles with organic and inorganic compounds in natural aquatic environments,

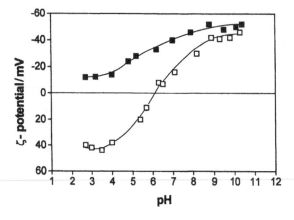

FIG. 4 Zeta potential of ripidolite suspended in 1×10^{-3} mol dm^{-3} NaCl solution as function of pH: (■) natural and (□) milled samples.

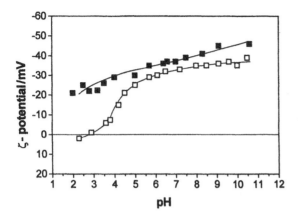

FIG. 5 Zeta potential of beidellite suspended in 1×10^{-3} mol dm^{-3} NaCl solution as function of pH: (■) natural and (□) milled samples.

allowing the conclusion that the extent of mechanical wear is more decisive for the electrokinetic properties than the type of clay mineral [39–41].

E. Influence of Adsorption of Organic Matter on Electrokinetic Properties of Clay Minerals

Investigations have shown that adsorbed organic matter masks, to some extent, the physiochemical properties of underlying surfaces and dominates surface interactions of mineral particles with the aquatic medium [59, 90, 91]. In natural aquatic media, humic substances are omnipresent, and thus they are the most important contributors to organic films covering the surfaces of mineral particles, particularly clay minerals [41, 45, 65, 66]. Accordingly, adsorbed humic materials influence dissolution, particle formation, adsorption, colloid stability, coagulation, and sedimentation of particulate and colloid matter in natural aquatic environments [65, 66, 92, 93].

Humic substances, mixtures of biomacromolecules generated by degradation of lignine, are mostly acidic due to the presence of phenolic groups. They are present in a range of molecular mass, and these macromolecules can appear also as charged colloidal particles. Their charge is due to, entirely or partially, deprotonated carboxylic and phenolic groups, which are responsible for proton binding of hydrated multivalent cations [7, 94]. Metal-ion binding to humic substances is a significant phenomenon regulating the concentration and the mobility of metals in natural waters and in soils [95–97]. Therefore,

TABLE 3 Effect of Milling on Specific Surface Area and Cation-exchange Capacity of Clay Samples

Sample	SSA (m^2 g^{-1})		CEC (meq/100 g)	
	Natural	Milled	Natural	Milled
Ripidolite	2.6	32	1.5	4.3
Beidellite	61	92	107	132

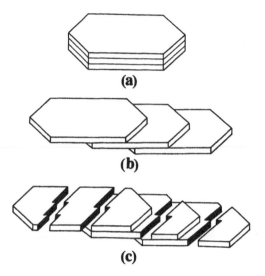

FIG. 6 Model for the disintegration of clay particles during milling: (a) model of the stack of clay minerals; (b) cleavage along the basal planes; (c) break up of the particles normal on to the basal plane.

interaction between clay mineral particles and these natural organic compounds is the most important process determining the transport, distribution, and removal of inorganic and organic pollutants in the natural environment.

The high sensitivity of electrokinetic measurements to solute concentration allows studies at low concentrations of polymeric materials. Accurate in situ measurements of small adsorbed amounts require sophisticated techniques, and are exceedingly difficult to perform and interpret. Thus, the investigation of the clay surface/solution interactions through electrokinetic properties is, and will possibly remain, an important way to study the complex exchange processes.

An example of the influence of changing concentration of FAs on the electrokinetic behavior of natural clay minerals is shown in Figs 7 and 8. In these figures the results of titration of original and disintegrated (milled) ripidolite and beidellite minerals with a solution of FA are shown. Both natural clays show no influence of the presence of FA, while the milled samples show a significant increase in the negative ζ potential with increasing concentration of FA. The changes in ζ potential of milled samples are caused by adsorption of FA on newly generated edge surfaces.

Zhou et al. [98] claimed that the amount of adsorbed humic substances varies with clay mineral type, particle size, and the SSA. These results show that natural samples of ripidolite and beidellite, two minerals with different clay structures, particle size, SSA, and CEC (Table 3), carefully prepared and free from surface impurities, show the same electrokinetic response, and do not interact with FA. However, mechanical wear (milling) a process that correspond to natural physical weathering, creates new edge surfaces, which are responsible for the interaction of clays. Accordingly, neither do particle size, clay mineral type, or SSA determine the character of adsorption, but the creation of new amphoteric (edge) surfaces [40, 41].

The adsorption of FA by solids is affected by the type of surfaces and by the pH of the aqueous media [99]. According to Schnitzer and Khan [94] —COOH is the major

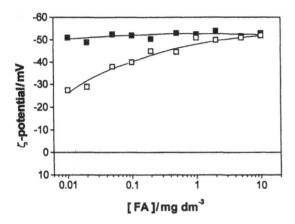

FIG. 7 Zeta potential of ripidolite in dependence on the concentration of fulvic acid (FA) in 1×10^{-3} mol dm^{-3} NaCl solution, pH 6.5: (■) natural and (□) milled samples.

functional group in FA active in biogeochemical processes. In the adsorption process this group forms complexes with surface hydroxyls, as shown in the case of goethite [100]. The same process is operative in the case discussed here. The disintegration of the clay particles by milling leaves the basic lattice structure intact, but increases the number of oxyhydroxy sites with interact with —COOH groups. Several authors [45, 65, 101] claimed that adsorbed organic matter predominantly determines the surface charge density of suspended particles in natural waters.

Oxide minerals, such as goethite [65, 100, 101], hematite [91], manganese oxides [90], and alumina [99, 102–105] have all been used in experiments with humic materials to represent mineral surfaces of particulate matter in natural waters. An understanding of these is best achieved by experiments on these oxides where hydroxyl groups are present on all exposed crystallographic faces. In all of these papers the common conclusion was that the adsorption of humic acids modifies the mineral surface toward a charge-average type. Clay minerals, either in native form or in the form of disintegrated particles, have

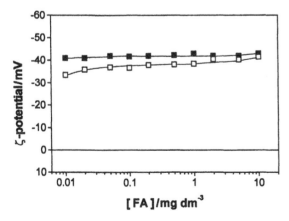

FIG. 8 Zeta potential of beidellite in dependence on the concentration of FA in 1×10^{-3} mol dm^{-3} NaCl solution, pH 6.5: (■) natural and (□) milled samples.

been less studied within this context. In order to study the adsorption of substituted nitrobenzenes and nitrophenols on clays, Haderlein and Schwarzenbach [106] have used model minerals such as silica, γ-Al$_2$O$_3$ and gibbsite as representatives of major types of surfaces met in clay minerals. The studies of adsorption and of the influence of FA on the electrokinetic potential of various types of clay surfaces have produced positive evidence on the assumed functional importance of surface type [40, 41, 45, 106].

In these investigations, ζ-potential curves of model minerals versus pH in a 10^{-3} mol dm^{-3} NaCl were determined (Fig. 9). Gibbsite and γ-Al$_2$O$_3$ show i.e.p. values at pH 5.0 and 9.5, respectively. Previous investigation [77] shows the existence of an i.e.p. for silica between pH 0.5 and 2.0. These extreme pH values are of no interest for the present discussion. Furthermore, these results indicate the validity of Al$_2$O$_3$ and gibbsite surfaces as models for hydrophilic oxyhydroxy surfaces, and of silica as a model for hydrophobic siloxane surfaces.

Figure 10 shows the next step in the investigation of the influence of humic substances on electrokinetic potentials of the same model minerals at a constant pH of 6.5. No changes were registered for silica in the FA gradient from 0.01 to 10 mg dm^{-3}. For gibbsite, for the same concentration range of FA, there is an increase in negative potential, leveling off at approximately -60 mV. For γ-Al$_2$O$_3$ there is a threshold for the influence, possibly adsorption, of FA at approximately 0.3 mg dm^{-3}, initiating a steep slope from positive to negative values with a reversal of charge at 3 mg dm^{-3}.

Hydroxylated edge surfaces of clay minerals, which can be represented by the γ-Al$_2$O$_3$ mineral surface, are the sites of maximum interaction with these polymers. Previous investigations by Lockhart [107] have also shown that the complexation of humic substances with kaolinite occurs at the edges, where Lewis acid aluminol groups are exposed. In solid samples, Lockhart found than the basal plane of kaolinite was free from organic substances in a humate–clay association, with less than 6% of the organic carbon in the solid phase.

Is the functional adsorption of FA on hydrophilic sites due to carboxylic ($-$COOH) groups? The answer was sought by using PAAs as models for active carboxylic groups, which are structurally well defined, and as models for studying the behavior of organic materials in natural waters [103, 104]. Figure 11 shows the answer to the question of

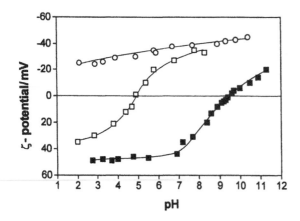

FIG. 9 Zeta potential of amorphous SiO$_2$ (O), gibbsite ($\square$), and γ-Al$_2$O$_3$ (■) suspended in 1×10^{-3} mol dm^{-3} NaCl solution as a function of pH.

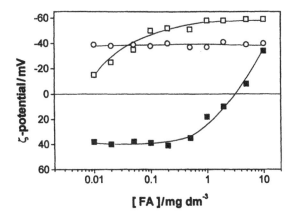

FIG. 10 Zeta potential of amorphous SiO_2 (○), gibbsite (□), and γ-Al_2O_3 (■) as a function of concentration of FA in 1×10^{-3} mol dm^{-3} NaCl solution, pH 6.5 ± 0.2.

interaction of two PAAs, of different molecular masses, with hydroxylated γ-Al_2O_3. The trend, showing first a threshold and than a steep change and reversal of electrokinetic potentials, is similar to the response of the same surface with FA (Fig. 10).

In view of the results discussed above, PAAs show a significant influence on the electrokinetic potentials of milled ripidolite, a clay mineral in which the incidence of edge surfaces was significantly increased. Data in Fig. 12 show the similarity of the electro-kinetic responses of FA and PAA on ripidolite: a monotonic increase in negative potential with increasing concentration of organic materials. A conclusion is near that the similarity of the responses indicates that the same functional group, carboxyls, are responsible for the interaction and, consequently, for the changes in electrokinetic potentials.

High-resolution transmission electron-microscopic analysis of the milled samples indicates that ripidolite and beidellite particles are randomly oriented, showing both (hk0) and (00L) planes in the same area (Fig. 13). Edge surfaces, to which much electro-kinetic influence is associated, are presented with the (hk0) planes with the lattice fringes

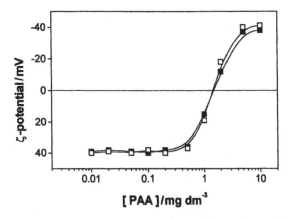

FIG. 11 Zeta potential of γ-Al_2O_3 suspended in 1×10^{-3} mol dm^{-3} NaCl solution as a function of concentration of PAAs: (□) MW 2000, (■) MW 5000, pH 6.5 ± 0.2

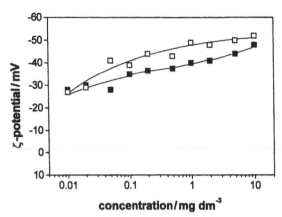

FIG. 12 Zeta potential of milled ripidolite suspended in 1×10^{-3} mol dm^{-3} NaCl solution as a function of the concentration of PAA (MW 5000) (■) and FA (□), pH 6.5 ± 0.2.

(basal unit repeat) of 1.4 and 1.1 nm for ripidolite and beidellite, respectively. These planes represent active hydroxyl sites where Lewis acid metallic groups are exposed. Surfaces with constant basal surface charge are presented as (00L) planes.

Kumert and Stumm [108] and Ochs et al. [103] claim that co-ordinate adsorption (ligand exchange) of humic acids on the oxide surfaces is predominant. Stumm [7] postulated that adsorption of humic substances at the Al$_2$O$_3$ surfaces is dominated by co-ordinate interaction between acidic functional groups of the humic substances and the metal centers of the oxide mineral surfaces. Accordingly, exposed metal centers of the edge surfaces of clay minerals present active sites in interaction with humic substances, which is dominated by formation of mononuclear bidentate and/or polynuclear surface complexes [103, 108]. Exposed metallic groups at the clay mineral surface edges are not always aluminol as encountered in clay minerals such as kaolinite and beidellite. Structural and chemical characteristics of clay minerals also indicate the possibility that exposed metallic groups can be represented by magnesium (e.g., in saponite and in clays from the chlorite group). Figure 14 shows an idealized structure, projection [010], of chlorite clay minerals. Edge surfaces of clay minerals may contain several metallic groups (mostly Al, Mg, and Fe), depending on the type of clay and on isomorphous substitution. All of these modify the surface charge densities and the concentration dependence of the electrokinetic potentials.

F. Influence of Humic Materials on Electrokinetic Properties of Clay Minerals in Complex Electrolyte Solution

In natural waters, clay mineral particles are exposed to a variety of organic compounds as well as to different physicochemical conditions such as temperature, oxygen, pH, and ionic strength [7, 65, 66, 101]. Investigations have shown that natural organic compounds, mostly humic acids, are dominant factors determining the colloidal stability of mineral particles in the natural aquatic environment [41, 65, 66, 109–112]. Accordingly, the processes of flocculation and sedimentation of colloid material in a natural aquatic environment are, besides the other influences, also governed by organic surface modification. According to Hunter [113] and Hunter et al. [114], the term flocculation is used to describe the formation of particle aggregates by bridging of polymer molecules from one particle to

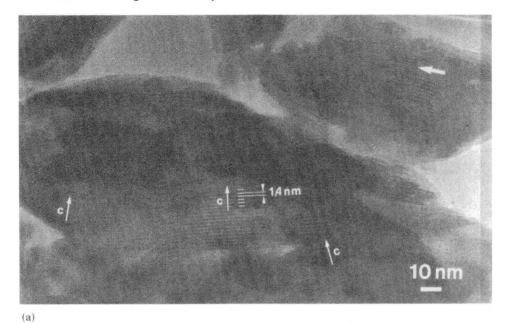

(a)

(b)

FIG. 13 High-resolution transmission electron micrographs of milled ripidolite (a) and beidellite (b) clays.

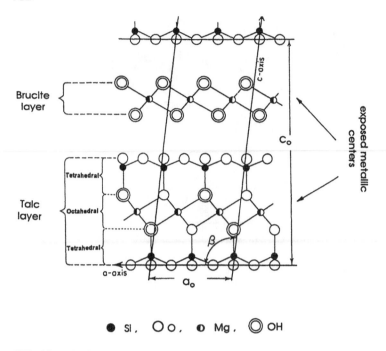

FIG. 14 Idealized structure of chlorite clay minerals, projection [010]. (From Ref. 22.)

another as distinct from coagulation in which the particles come into close contact as a result of changes in the EDL around the particles.

This part describes results of a series of experiments in which the electrokinetic potential has been used to determine the surface charge behavior of clay particle suspensions in the presence of FA in artificial seawater (ASW) at three different salinities 1, 15, and 37‰), corresponding to an estuarine salinity gradient.

The influence of FA on the ζ potential of the chlorite clay for salinities 1, 15, and 37 is shown in Fig. 15. At salinity 1, addition of FA from 0.01 to 0.1 mg dm^{-3} causes an

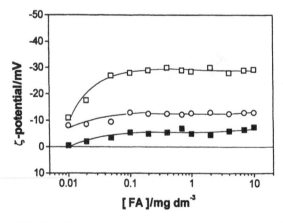

FIG. 15 Zeta potential of chlorite clay mineral dependence on the concentration of FA in ASW of different salinities: (□) salinity 1, (○) salinity 15, and (■) salinity 37.

increase in the negative values of the potential, from approximately -10 to $-30\,mV$. Accordingly, the influence of FA on the electrokinetic potential of chlorite in ASW of low salinity is similar to one described above for the single $1:1$ electrolyte system. It is obvious that the positive edge charge of chlorite clay mineral particles is neutralized by adsorption of FA, making the excess negative charge higher and accordingly the electro-kinetic potential more negative. At higher salinities the effect is smaller, most probably due to the presence of divalent cations, particularly Ca^{2+}, in the ASW solution.

These results support investigations [65, 101], which have shown that the most important factor determining the surface charge density of particles in a natural aquatic environment is the adsorbed layer of natural organic matter, mostly humic substances. The adsorption of these compounds on colloid particles occurs at its low concentration, and was found to reach equilibrium in seconds, under conditions similar to those found in natural waters [101. Tombacz et al. [115] have shown that a relatively small amount of organic anions, adsorbed on the oxyhydrous surfaces of minerals, significantly increases their colloidal stability. One may also speculate that an increase in the negative charge, resulting from FA adsorption on the clay edge surface, might stabilize the suspended clay particles in the aqueous medium [42]. The adsorption of natural organic compounds such as FA on the most reactive edge surface smear out the heterogeneity of the surface charge of clay minerals, and the particles become of uniformly negative charge. Therefore, mutual interaction of clay particles through the edge–edge or edge–face type cannot initiate the flocculation process, and the adsorbed layer of FA might provide steric and electrostatic stabilization for clay minerals.

On the other hand, increase in the concentration of divalent cations, such as Mg^{2+} and particularly Ca^{2+}, with higher salinities significantly influence the electrokinetic poten-tial of the clay minerals. Day et al. [101] have shown that the adsorption of natural organic matter on to the surface of goethite can be enhanced in the presence of Ca^{2+}. Furthermore, Jarnstrom and Stenius [116] have shown that the adsorption of sodium polyacrylate on kaolinite is sharply enhanced in the presence of Ca^{2+} ions. Two mechanisms were pro-posed. First, the presence of divalent cations might govern the increase in the number of the active surface sites due to the formation of oxide–cations' bridges through coulombic interactions of the cations with the edge surface of the clay mineral and FA functional groups. Second, the complexation of some functional groups of FA with Ca^{2+} causes reduction in the number of contact sites the FA molecules have with clay edge surfaces. Therefore, by increasing the number of molecules that can bind per unit area an increased adsorption of FA might result.

According to Day et al. [101] Ca^{2+} is able to complex negative functional groups of the adsorbed FA, resulting in a decrease in electrokinetic potential. As the charge density of the clay minerals is reduced, electrostatic repulsion between the particles also decreases. If the particles can approach to a shorter distance, van der Waals' attraction force will induce their flocculation. It was shown that Ca^{2+} ions induced aggregation of kaolinite suspensions due to delocalized binding of the ions to the PAA molecules, reducing the charge of the PAA [117].

Some authors [65, 112] pointed out that bridging flocculation of the adsorbed organic compounds, caused by divalent cation complexation, might be one of the most significant processes in destabilization, flocculation, and deposition of colloidal particles in hard water and in an estuarine environment. Ruehrwein and Ward [118] have proposed a bridging mechanisms for flocculation. It has been described as an adsorption of the seg-ments of individual polymeric molecules on to the surfaces of more than one particle, causing the formation of aggregates which sediment rapidly. The conformational characteristics of adsorbed organic polymers depend on ionic strength, pH, and molecular weight, and directly affect flocculation [44, 113, 119]. Because the loops of the polymer molecules have to be able to bridge the distance over which the electrostatic repulsion between particles is operative, it was shown that adsorption of high molecular weight polymers is the most effective flocculant. In addition, an increase in ionic strength decreases the thickness of the EDLs and, therefore, lower molecular weight polymers can act as flocculants.

As a final note on recent investigations of the electrokinetics of clay minerals it has been recognized that many diverse factors are in play. Electrokinetic investigations give answers only to bulk structure/surface charge related phenomena. Combined with other techniques and methodologies of research, electrokinetic techniques significantly contribute to the understanding of the colloidal behavior of clay minerals in aquatic environments.

VI. A LOOK INTO THE FUTURE OF RESEARCH IN ELECTROKINETICS OF CLAYS

Electrokinetic investigations have established the interaction of surfaces with single components of the aquatic medium. From this essentially reductionistic mode of experimentation, one is tempted to synthesize the probable behavior of clay particulate matter in the natural environment.

Another area in which much information is missing is the mutual electrokinetic influence of particles, a phenomenon known as particle–particle charge interaction. This process is operative in natural aquatic systems where changes in the hydrodynamic regime often favor coalescence of particles. The main obstacles to a breakthrough in this field remain in the limitations imposed by electrokinetic techniques, limited to laboratory conditions. It seems that the technique, such as electrophoresis, in which electric fields are imposed on particles and the resulting movements are measured, have reached their experimental limits. More is expected of acoustic or ultrasound (mechanical) perturbations, with the measurement of the generated alternating electric fields. Whether these techniques will be "gentle" enough for clay surfaces, sensitive enough in measuring small responses, and allow straightforward interpretation, remains a worthy question.

REFERENCES

1. SW Bailey. Clays Clay Miner 28: 73, 1980.
2. P Sennett. In: M-H Grant, ed. Encyclopedia of Chemical Technology. vol. 6. New York: Wiley, 1993, pp 405–423.
3. RE Grim. Applied Mineralogy. New York: McGraw Hill, 1962, p 422.
4. HH Murray. Clay Miner 34: 39, 1999.
5. G Sposito. The Surface Chemistry of Soils. New York: Oxford University Press, 1984, p 234.
6. DJ Greenland, CJB Mott. In: DJ Greenland, MHB Hayes, eds. The Chemistry of the Soil Constituents. New York; Wiley, 1978, pp 321–350.
7. W Stumm. The Chemistry of Solid–Water Interfaces. New York: Wiley, 1992, p 428.
8. S Yariv, H Cross. Geochemistry of Colloid Systems. Berlin: Springer-Verlag, 1979, p 450.
9. W Salomons, U Förstner. Metals in Hydrocycle. Berlin: Springer-Verlag, 1984, p 349.
10. I Sondi, M Juračić, E Prohić, V Pravdić. Sci Tot Environ 1555: 173, 1994.

11. D Eisma, GC Cadee. In: ET Degens, S Kempe, JE Richey, eds. Biogeochemistry of Major World Rivers. New York: Wiley, 1991, pp 283–296.
12. WH van Riemsdijk, LK Koopal. In: J Buffle, HP van Leeuwen, eds. Environmental Particles. Boca Raton, FL: Lewis, 1992, pp 455–495.
13. JP Rupert, WT Granquist, TJ Pinnavaia. In: ACD Newman, ed. Chemistry of Clays and Clay Minerals. New York: Wiley, 1987, pp 275–319.
14. TJ Pinnavaia. Science 220: 365, 1983.
15. P Laszlo. Science 235: 1473, 1987.
16. ET Degens, J Matheja, TA Jackson. Nature 227: 492, 1970.
17. TA Jackson. Chem Geol 7: 275, 1971.
18. N Lahav, D White, S Chang. Science 201: 67, 1978.
19. B Siffert, A Naidja. Clay Miner 27: 109, 1992.
20. AJ Alvarez, M Khanna, GA Toranzos, G Stotzky. Mol Ecol 7: 775, 1998.
21. AN Bishop, RP Philip. Energy Fuels 8: 1494, 1994.
22. RE Grim. Clay Mineralogy. New York: McGraw Hill, 1968, p 596.
23. ACD Newman, G Brown. In: ACD Newman, ed. Chemistry of Clays and Clay Minerals. New York; Wiley, 1987, pp 1–128.
24. G Brown, ACD Newman, JH Rayner, AH Weir. In: DJ Greenland, MHB Hayes, eds. The Chemistry of Soil Constituents. New York: Wiley, 1978, pp 29–178.
25. VA Franck-Kemenetsky. Clay Miner Bull 4: 161, 1960.
26. B Velde, A Meunier. In: ACD Newman, ed. Chemistry of Clays and Clay Minerals. New York: Wiley, 1987, pp 423–458.
27. H Chamley. Clay Sedimentology. Berlin: Springer-Verlag, 1989, p 623.
28. D Righi, A Meunier. In: B Velde, ed. Origin and Mineralogy of Clays. Berlin: Springer-Verlag, 1995, pp 43–157.
29. MJ Willson. Clay Miner 34: 7, 1999.
30. WT Granquist, SS Pollack. Am Miner 52: 212, 1967.
31. AC Wright, WT Granquist, JV Kennedy. J Catal 25: 65, 1972.
32. H Yamada, H Nakazawa, H Hashizume. Clays Clay Miner 42: 674, 1994.
33. S Fiore, FJ Huertas, F Huertas, J Linares. Clays Clay Miner 43: 353, 1995.
34. H van Olphen. Disc Faraday Soc 11: 82, 1951.
35. DJA Williams, KP Williams. J Colloid Interface Sci 65: 79, 1978.
36. RK Schofield, HR Samson. Disc Faraday Soc 18: 135, 1954.
37. Z Zhou, WD Gunter. Clays Clay Miner 40: 365, 1992.
38. GM Beene, R Bryant, DJA Williams. J Colloid Interface Sci 147: 358, 1991.
39. I Sondi, M Stubičar, V Pravdić. Colloids Surfaces 127: 141, 1997.
40. I Sondi, V Pravdić. J Colloid Interface Sci 181: 463, 1996.
41. I Sondi, O Milat, V Pravdić. J Colloid Interface Sci 189: 66, 1997.
42. HD Morris, B Shelton, PD Ellis. J Phys Chem 94: 3121, 1990.
43. KF Hayes, G Redden, W Ela, JO Leckie. J Colloid Interface Sci 142: 448, 1991.
44. W Stumm. Colloids Surfaces 73: 1, 1993.
45. I Sondi, V Pravdić. Croat Chem Acta 71: 1061, 1998.
46. SJ Greg, KSW Sing. Adsorption, Surface Area and Porosity. London: Academic Press, 1967, p 371
47. RS Dayal, SB Hendricks. Soil Sci 62: 421, 1950.
48. DJ Greenland, JP Quirk. J Soil Sci 15: 178, 1964.
49. S Diamond, EB Kinter. Clays Clay Miner 5: 334, 1958.
50. H Laudelout. In: ACD Newman, ed. Chemistry of Clays and Clay Minerals. New York: Wiley, 1978, pp 225–236.
51. E Brusenberg, CV Clamency. Clays Clay Miner 21: 213, 1973.
52. I Sondi. Flocculation of five-grained suspended matter and sedimentation processes in karstic estuaries of the Adriatic region. PhD Thesis, University of Zagreb, 1995, p 137.

53. H van Olphen. In: ACD Newman, ed. Chemistry of Clays and Clay Minerals. New York: Wiley, 1987, pp 203–224.
54. RJ Hunter. Zeta Potential in Colloid Science. London: Academic Press, 1981, p 386.
55. PC Hiemenze. Principles of Colloid and Surface Chemistry. New York: Marcel Dekker, 1986, p 815.
56. RB Secor, CJ Radke. J Colloid Interface Sci 103: 237, 1985.
57. FC Chang, G Sposito. J Colloid Interface Sci 163: 19, 1994.
58. FC Chang, G Sposito. J Colloid Interface Sci 178: 555, 1996.
59. HH Hahn, W Stumm. Am J Sci 268: 354, 1970
60. I Sondi, J Bišćan, V Pravdić. J Colloid Interface Sci 178: 514, 1996.
61. LS Swartzen-Allen, E Matijević. J Colloid Interface Sci 56: 159, 1976.
62. R Perkins, R Brace, E Matijević. J Colloid Interface Sci 48: 417, 1974.
63. JI Hedges. Geochim Cosmochim Acta 41: 1119, 1977.
64. RJ Gibbs. J Sediment Petrol 53: 1193, 1983.
65. R Beckett, NP Lee. Colloids Surfaces 44: 35, 1990.
66. C O'Melia. Colloids Surfaces, 39: 255, 1989.
67. K Kranck. Sedimentology 28: 107, 1981.
68. I Sondi, M Juračić, V Pravdić. Sedimentology 42: 769, 1995.
69. JPM Syvitski. In: RH Bennet, WR Bryant, MH Hulbert, eds. Microstructure of Fine-Grained Sediments: from Mud to Shell. New York: Springer-Verlag, 1991, pp 31–137.
70. GG Leppard. In; J Buffle, HP van Leeuwen, eds. Environmental Particles. Boca Raton, FL: Lewis, 1992, pp 231–289.
71. KA Hunter, PS Liss. Limnol Oceanogr 27: 322, 1982.
72. RJ Gibbs. J Sediment Petrol 47: 237, 1977.
73. JK Edzwald, JB Upchurch, CR O'Melia. Environ Sci Technol 8: 58, 1974.
74. JK Edzwald, CR O'Melia. Clays Clay Miner 23: 39, 1975.
75. K Kranck. Sedimentology 22: 111, 1975.
76. IN McCave. In: DAV Stow, DJW Piper, eds. Fine-Grained Sediments: Deep-Water Processes and Facies. London: Blackwell, 1984, pp 35–69.
77. GA Parks. Chem Rev 65: 177, 1965.
78. R Hardy, M Tucker. In: M Tucker, ed. Techniques in Sedimentology. Oxford: Blackwell, 1988, pp 191–229.
79. OP Mehra, ML Jackson. Clay Clays Miner 7: 317, 1960.
80. LP Meier, AP Menegatti. Clay Miner 32: 557, 1997.
81. AP Menegatti, GL Früh-Green, P Stille. Clay Miner 34: 247, 1999.
82. MS Stul, L van Leemput. Clay Miner 17: 209, 1982.
83. JK Anderson. Clay Clays Miner 10: 380, 1963.
84. BD Mitchell, BFL Smith. J Soil Sci 25: 239, 1974.
85. RJ Cook. Clay Miner 27: 73, 1992.
86. R Karen, I Shainberg. Clay Clays Miner 23: 193, 1975.
87. AP Ferris, WB Jepson. J Colloid Interface Sci 51: 245, 1975.
88. AE James, DJA Williams. Adv Colloid Interface Sci 17: 219, 1982.
89. R Greene-Kelly. Clay Miner Bull 5: 392, 1964.
90. E Tipping, MJ Heaton. Geochim Cosmochim Acta 47: 1913, 1983.
91. E Tipping, DC Higgins. Colloids Surfaces 5: 85, 1982.
92. DJ Greenland. Soil Sci 111: 34, 1971.
93. TD Waite, IC Wrigley, R Szymczak. Environ Sci Technol 22: 778, 1988.
94. M Schnitzer, SU Khan. Humic Substances in the Environment. New York: Marcel Dekker, 1972, p 327.
95. MDG Kinniburgh, WH van Riemsdijk, LK Koopal, M Borkovec, MF Benedetti, MJ Avena. Colloids Surfaces, 151: 147, 1999.
96. JP Pinherio, AM Mota, HP van Leeuwen. Colloids Surfaces 151: 181, 1999.

97. EM Thurman. Organic Geochemistry of Natural Waters, Dordrecht: Nijhoff/Junk, 1985, p. 497.
98. JL Zhou, S Rowland, RFC Mantoura. Water Res 28: 571, 1994
99. CP Schulthess, CP Huang. Soil Sci Soc Am J 55: 34, 1991..
100. E Tipping, D Cook. Geochim Cosmochim Acta 46: 75, 1982.
101. G McD Day, BT Hart, ID McKelvie, R Beckett. Colloids Surfaces 89: 1, 1994.
102. M Buleva, I Petkanchin. Colloids Surface 151: 225, 1999.
103. M Ochs, B Ćosović, W Stumm. Geochim Cosmochim Acta 58: 639, 1994.
104. M Plavšić, B Ćosović, S Rodić. Colloid Polym Sci 274: 548, 1996.
105. JT Kunjappu, P Somasundaran, K Sivadasan. Colloids Surfaces 97: 101, 1995.
106. SB Haderlein, RP Schwarzenbach. Environ Sci Technol 27: 316, 1993.
107. NC Lockhart. Clays Clay Miner 29: 413, 1981.
108. R Kumert, W Stumm. J Colloid Interface Sci 75: 377, 1980.
109. J Bišćan, D Dragčević. Mar Chem 43: 127, 1993.
110. J Bišćan, I Rhebergen, M Juračić, J-M Martin, J-M Mouchel. Mar Chem 32: 235, 1991.
111. RJ Gibbs. Environ Sci Technol 17: 237, 1983.
112. JA Davis. Geochim Cosmochim Acta 46: 2381, 1982.
113. RJ Hunter. Foundation of Colloid Science. vol. 1. Oxford: Oxford University Press, 1986, p 673.
114. RJ Hunter, R Matarese, DH Napper. Colloids Surfaces 7: 1, 1983.
115. E Tombacz, G Filipcsei, M Szekeres, Z Gingl. Colloids Surfaces 151: 233, 1999.
116. L Jarnstrom, P Stenius. Colloids Surfaces 50: 47, 1990.
117. P Stenius, L Jarnstrom, M Rigdahl. Colloids Surfaces 51: 219, 1990.
118. RA Ruehrwein, DW Ward. Soil Sci 73: 485, 1952.
119. A Yokoyama, KR Srivinasan, HS Foyler. Langmuir 5: 534, 1989.

28
Electrokinetics of Mineral Particles

KALYAN K. DAS Tata Research Development and Design Centre, Pune, India

I. INTRODUCTION

The electrokinetics of mineral particles is a topic of great interest not only from the fundamental surface chemistry angle but also from the mineral beneficiation point of view. Researchers have looked into the myriad of aspects of electrokinetics of mineral particles. While some of the studies were motivated by the applied chemistry and engineering perspectives, there were many others which concentrated on the development of theory and validation thereof. Interfacial phenomena at mineral–water interfaces are usually controlled by the electrical double layer (EDL) forces. It is, therefore, necessary to understand the behavior of the ions that adsorb as counterions to maintain electroneutrality as well as those adsorbing specifically on the surface. Excellent literature articles on the electrokinetics of mineral particles are available [1–7].

II. ELECTRICAL DOUBLE LAYER AT THE MINERAL/WATER INTERFACE

It is well known from classical literature that when a solid is brought into contact with an aqueous solution, charged species are transferred across the solid/liquid interface until an equilibrium is established and this condition is characterized by a potential ψ_0 and a surface charge density σ_s. The interface between the solid and solution may be treated as a semipermeable membrane which allows only the charged species common to both the solid and the solution to pass through. These species are called the potential-determining ions. As a consequence of the relative motion between the charged dispersed phase and the bulk liquid (continuum), the EDL is sheared. The potential at this shear plane is called the electrokinetic or zeta (ζ) potential. For detailed treatises on EDL theories, readers may refer to the books by Shaw [8], Adamson [9], Hunter [10], and Heimenz and Rajagopalan [11].

A. Electrokinetic Phenomena

There are four main kinds of electrokinetic methods, which measure the potential at the shear plane. These are: (1) electrophoresis, (2) streaming potential, (3) sedimentation potential, and (4) electro-osmosis. For studies concerning the mineral–solution interface, the most appropriate electrokinetic technique is electrophoresis. This technique involves setting up a potential gradient in a suspension containing charged particles and determining their velocity. The movement of such colloidal particles under an applied electric field is known as electrophoresis. Electrophoretic mobility (velocity per unit potential gradient) is often measured using one of the commercially available instruments, mostly based on either microscopic (visual) determinations or on electrophoretic light scattering. Since the internal surfaces of the electrophoresis cell are also charged, there is an electro-osmotic flow of liquid near the cell walls together with a compensating return flow of liquid with maximum velocity at the center of the cell. The true electrophoretic velocity is measured at locations in the cell where the electro-osmotic flow and the return flow of liquid cancel each other. This location is called the "stationary level." For a cylindrical cell, this level is located at 0.146 of the internal diameter from the cell wall. For a flat cell, the stationary levels are located at fractions of about 0.2 and 0.8 of the total depth, the exact locations depending on the width/depth ratio of the cell [8].

B. Point of Zero Charge and Isoelectric Point

The single most important parameter that describes the EDL of a mineral in an aqueous suspension is the point of zero surface charge (p.z.c.). The p.z.c. is expressed as the condition in the aqueous solution at which the surface charge σ_0 is zero, and this is determined by a particular value of the activity of the potential-determining ion (a^{+z}M,pzc or a^{-z}A,pzc). Assuming that potential differences due to dipoles, etc., remain constant, the surface potential is also considered to be zero at the p.z.c. The value of the surface potential at any activity of potential-determining electrolyte, a_M^{+z} is given by

$$\psi_0 = \frac{RT}{z\mathrm{F}} \ln \frac{a_{M^{+z}}}{(a_{M^{+z}})_{pzc}} \tag{1}$$

where R is the gas constant, T is the absolute temperature, z is the valence of the potential-determining cation, and F is the Farady constant.

The importance of the p.z.c. is that the sign of the surface charge has a major effect on the adsorption of all other ions and, in particular, those ions charged oppositely to the surface, because these ions function as the counterions to maintain electroneutrality. Determinations of the p.z.c. are normally carried out by conductometric or potentiometric titration: some property of the suspension is measured as a function of the concentration of the potential-determining ions, for different concentrations of an "indifferent" electrolyte. If a common intersection of all the curves corresponding to the various electrolyte concentrations is found, the activity of the potential-determining ion at that intersection will be the p.z.c. If, alternatively, one measures the zeta potential as a function of the activity of the potential-determining ions, at different ionic strengths, all the zeta-potential curves cross at a single point, which is called the point of charge reversal (CR) or "isoelectric point" (i.e.p.). The i.e.p. would be same as the p.z.c. if the electrophoretic measurements are carried out as a function of the concentration of potential-determining ions in the presence of an indifferent electrolyte.

C. Conversion of Electrophoretic Mobility into Zeta Potential

The electrophoresis technique has been widely used in studying the mineral/solution interface. The measured electrophoretic mobilities (μ_e) can be converted into zeta potentials at the shear plane, using a suitable model. For large particles with radius a ($\kappa a \gg 1$, κ being the reciprocal Debye length) and a thin EDL, Smoluchowski [12] derived an equation relating the electrophoretic mobility to ζ:

$$\zeta = \frac{\eta}{\epsilon\epsilon_0}\mu_e \tag{2}$$

where η is the viscosity of the dispersion medium and $\epsilon\epsilon_0$ is its dielectric permittivity. For small particles, where $\kappa a \ll 1$ (a condition seldom fulfilled by real disperse systems), Hückel derived the following equation:

$$\zeta = \frac{3\eta}{2\epsilon\epsilon_0}\mu_e \tag{3}$$

For most particulate systems of practical interest in our field, the Smoluchowski equation has been found to be adequate, at least as a semiquantitative follow-up of the behavior of the solid/liquid interface for different physicochemical conditions of the dispersion medium. The following practical formula can be useful for the conversion of mobility into ζ in aqueous systems at 25°C.

$$\zeta\,(\text{mV}) = 12.83 \times \mu_e\left(\mu\text{m s}^{-1}/\text{V cm}^{-1}\right) \tag{4}$$

Without going into further theoretical details (see, e.g., Refs 10–12), we will rather focus on the advances made in the field of mineral particles. The electrokinetics of mineral particles has been researched quite well. In discussing the progress made in the field, the minerals have been classified into four categories, namely, (1) oxide minerals, (2) sulfide minerals, (3) salt-type minerals, and (4) siliceous and other minerals.

III. OXIDE MINERALS

Oxides are industrially important minerals and are also the starting materials for many specialized applications. Classical papers by de Bruyn and Agar [2], James and Parks [3], Fuerstenau [4], and Parfitt [7] have dealt rigorously with the electrokinetics of oxide minerals. In addition to the potential-determining H^+ and OH^- ions, the presence of metal ions and their hydroxy species in the suspension affect the interfacial electrokinetics of an oxide mineral and this aspect is delineated below.

A. Adsorption of Hydrolyzable Metal Ions on Oxide Surfaces

The electrokinetics of mineral–solution interfaces is substantially influenced by the local environment in which the mineral resides in the suspension, as well as by the solubility characteristics of the mineral itself. Depending on the solution chemistry conditions and the solubility of the minerals in question, there may be a number of ions present in the vicinity of the interface. Under favorable thermodynamic conditions, these ions can adsorb on to the mineral surfaces and control their surface charge characteristics. Some major features of electrokinetic potential–pH isotherms for surface nucleated hydroxy species on different types of colloidal mineral particles have been rigorously investigated by James and Healy [13]. These authors have studied the adsorption of hydrolyzable metal ions such as Co(II), Fe(III), Cr(III), and Ca(II) on SiO_2, and Th(IV) on SiO_2 and TiO_2. It has been argued that the hydrolysis products of multivalent cations are adsorbed more rapidly at the mineral–water interface. James and Healy [13] have identified three characteristic charge reversals for the adsorption of Co^{2+} on SiO_2. The salient characteristics of the three charge reversals are:

1. The first point of charge reversal CR1 is the p.z.c. of the mineral, since H^+ and OH^- are potential-determining ions.
2. The third charge reversal CR3 observed in the presence of hydrolyzable metal ions reflects a partial or complete coating of the metal hydroxide on the colloidal substrate. If sufficient metal ions are present to coat completely the surface with adsorbed metal hydroxide, then CR3 will coincide with the p.z.c. of the metal hydroxide. Due to incomplete coating at lower concentrations of metal ion, CR3 occurs at pH values below the p.z.c. of metal hydroxide.
3. The most interesting point, however, is CR2, the point of zeta-potential reversal in between the p.z.c. of the substrate and the p.z.c. of the metal hydroxide coating. It has been suggested that CR2 could indicate specific adsorption or surface precipitation. CR2 occurs usually at a pH below that of the pH of bulk precipitation.

James and Healy's model of the adsorption of hydrolyzable metal ions on the mineral surface is based on the concept that the electric field at the surface induces precipitation in the interface, usually 0.5–1.0 pH unit below the bulk precipitation edge. CR2 corresponds to the onset of the surface precipitation. Several researchers have investigated the adsorption of hydrolyzable metal ions on different mineral substrates; for example, Pugh and Bergstrom [14] have studied the adsorption of Mg(II) on ultrafine α-silicon carbide and α-alumina. For both systems, strong adsorption of Mg(II) occurred well below the bulk precipitation threshold on the negatively charged SiC surface and positively and negatively charged α-Al_2O_3 surface. The uptake of Mg(II) on these surfaces could be conveniently interpreted in terms of James and Healy's cation hydrolysis/adsorption model.

The electrokinetic investigations on some of the important oxide minerals are presented in the following sections.

B. Hematite

One of the widely studied oxide minerals is hematite (α-Fe_2O_3). The hematite/water interface has been examined by a number of researchers both in the presence and absence of reagents. Attempts have also been made to explore the possibility of using monodisperse hematite particles as model colloids. For possible high-technology applications, hematite has been used in the synthesis of core–shell particle systems as well. Some of the salient studies involving hematite are delineated in the following paragraphs.

In an important investigation, Delgado and González-Caballaro [15] have examined the possibility of using hematite (which can be prepared in a controlled and reproducible manner) as a model colloid. The effects of the size and shape on the electrokinetics were investigated as a function of NaCl and $CaCl_2$ concentrations for pseudospherical hematite samples of different diameters. The experiments were conducted at sufficiently alkaline conditions, i.e., at pH 10 in order to ensure enough surface charge (the p.z.c. of hematite is at pH 7–8). The authors have compared the electrokinetic behavior with the predictions of a rigorous theory for the mobility of spheres, assuming that both types of particles have equal zeta potentials. Interestingly, it was found that smaller particles show a considerable effect of anomalous (Stern-layer) ionic conduction, which was, however, less appreciable for particles larger in diameter. In the carefully chosen electrolyte concentration range (above 10^{-3} M), in which this surface contribution is masked by the bulk solution conductivity, a satisfactory agreement is found between measured and predicted mobility values. The effect of pH and indifferent electrolyte ($NaNO_3$) concentration on the electrophoretic mobility of two different size spherical samples has been compared to the predictions of the models by O'Brien and Ward [16] and Yoon and Kim [17]. For most experimental conditions involving both spheroidal and nonspheroidal geometries, there were good agreements with theory. Based on their findings, the authors have suggested the possibility of using hematite of controlled geometry (not necessarily spherical) as model colloids.

The possibility of using monodisperse oxide particles as model colloids was also examined by studying the effects of simple 1–1, 2–1, and 3–1 inorganic electrolytes on the electrokinetics of hematite particles [18]. Contrary to the expectation, a maximum in the electrophoretic mobility–electrolyte concentration curve was observed. In addition to inorganic electrolytes, the effects of five different amino acids (alanine, glycine, lysine, threonine, and glutamic acid) were investigated on the well-characterized hematite colloids. It was observed that the suspension pH was one of the most important parameters, which controlled the adsorption of amino acids on the hematite surface. Interestingly, a maximum in adsorption was found close to the i.e.p. of each amino acid. Based on the results of adsorption, the authors have argued that electrostatic interactions alone cannot explain the adsorption process completely; acid–base interactions may have to be considered.

Preparation of colloidal hematite particles is of significant interest among the researchers. In one such endeavor, Kandori et al. [19] have examined the zeta potential, X-ray fluorescence, and surface analysis (by X-ray photoelectron spectroscopy, XPS) of monodispersed cubic hematite prepared from $FeCl_3$ solution in relation to the number of washing cycles. No appreciable difference in zeta potentials was observed among four washing cycles. The i.e.p. of hematite was observed at pH 3, much lower than the reported literature value of pH 8.5. Since the chloride ions could not be completely eliminated by washing, it was speculated that trace amounts of chloride ions are accumulated in the surface layer of cubic hematite, thus inducing high surface acidity.

Owing to promising technological applications, the synthesis and characterization of core–shell particles have received considerable attention in the recent past. The stability and physicochemical characteristics of the core–shell composites are predominantly dictated by the shell particles, which are specialized coating materials, and are generally expensive. Plaza et al. [20] have investigated the chemical and electrokinetic surface characterization of a core–shell system containing hematite. Nearly spherical hematite particles formed the *core* coated by a thin layer of yttrium basic carbonate or yttrium oxide as the *shell* particles. The morphology and the surface characteristics of the particles could be controlled by modifying the initial yttrium nitrate concentration. The changes in i.e.p. for the different synthesis conditions were monitored through interfacial electrokinetics. Based on measurements in the presence of different NaCl concentrations, pure hematite showed an i.e.p. at pH 7.4–7.6. Electrophoretic measurements performed on different $Y(OH)CO_3$- and Y_2O_3-coated particles showed that the pH_{iep} of all the samples were above that of pure hematite and approached the values of yttrium carbonate and oxide (i.e., $\sim$ 1–1.5 pH unit above that of pure hematite).

In an investigation on the adsorption of salicylic acid on hematite, Kovacevic et al. [21] have attempted the interpretation of adsorption and electrokinetic data by introducing the slipping-plane separation as an adjustable parameter. In the experimental part, they have measured the adsorption of salicylic acid on hematite and the corresponding electrophoretic mobilities as a function of pH. Similar values of surface potentials were observed at different ionic strengths. The equilibrium in the system was defined by an adsorption isotherm, which considered the dissociation of the salicylic acid in the bulk of the solution and the electrostatic interactions between the surface and the adsorbed ions. The electrostatic interactions were calculated by taking into account the surface potential (obtained from electrokinetics using the Gouy–Chapman theory). Singly charged anions, which occupied a surface area of $87 \pm 15\,\text{Å}^2$, were concluded to be the absorbable species. From the hematite surface, the slipping-plane separation was found to be $15\,\text{Å}$.

In addition to the studies mentioned above, a notable investigation was conducted by Colic et al. [22] who have determined the lyotropic effect of different anions as well as the surface charge, electrokinetic, and coagulation kinetic measurements on aqueous positively charged hematite particles. From the surface charge and electrokinetic data, the intrinsic equilibrium constant for counterion association and inner layer thickness were found.

C. Cassiterite

The i.e.p. of cassiterite (SnO_2) has been determined by many investigators. In the case of a cassiterite sample from Australia, the p.z.c. determined by potentiometric titration and i.e.p. determined by microelectrophoresis gave similar values, $\sim$ pH 4.5 [23]. Although the natural sample contained some trace elements, mainly iron, the p.z.c./i.e.p. of the natural sample was similar to that obtained for the synthetic one. The similarity in the i.e.p. and p.z.c. values has been attributed to a similar kind of adsorption of organic and inorganic ions via an electrostatic mechanism.

In one of the early investigations [24], the i.e.p. of cassiterite was determined to be at pH 3. The authors had found that for the cassiterite–anionic collector system, bivalent cations had a moderate effect on the floatability and zeta potential, presumably due to the formation of sparingly soluble compounds in the slurry. However, trivalent cations, notably Fe^{3+}, strongly improved the floatability in the pH region 3–6.

The effect of collectors on the flotation of cassiterite has been delineated through interfacial electrokinetics by other researchers also. Kamaruddin and Khangaonkar [25] have observed that, for the cassiterite–sulfosuccinamate system, the collector substantially affected the zeta potential, and the i.e.p. was lowered to pH 2.5. The change in zeta potential in the presence of sulfosuccinamate indicates specific adsorption and/or chemical interaction. The flotation of cassiterite was optimum at pH 3 and ions like Fe^{3+} and Al^{3+} had significant effects on increasing the zeta potential to the positive side. Invoking the species distribution diagrams, the authors attributed the increase in zeta potential to the adsorption of cationic hydroxy complexes at the solid/liquid interface.

For a bimineral system containing cassiterite and quartz, the electrokinetic properties have been investigated in the presence of cationic, anionic, and nonionic surfactants by Taha et al. [26, 27]. Considerable variations in surface charge were noticed, depending on the pH and the solution chemistry conditions. The authors also studied the effect of different polyvalent electrolytes ($FeCl_3$, $AlCl_3$, $LaCl$, $CeCl$, and $ThCl_4$) on the adsorption of different surfactants (namely, primary amine hydrochloride, sodium oleate, sodium dodecyl sulfate, and Ph disodium orthophosphate) on cassiterite and quartz particles and observed significant changes in the zeta potentials and i.e.p. of the oxides.

D. Wolframite

The electrokinetics of wolframite ($FeMnWO_4$), an important tungsten-bearing mineral, has been well studied. The p.z.c. was found to be at pH 6.1 [28]. Externally added ions such as Fe^{2+}, Mn^{2+} and WO_4^{2-} controlled the zeta potential of the system and had an activating effect on the flotation of wolframite. The origin of the negative charge on the wolframite surface has been attributed to the selective dissolution of Mn^{2+} or Fe^{2+} rather than WO_4^{2-}. It has also been postulated that the electrokinetic characteristics of wolframite are closely related to the ratio of Mn to Fe in the mineral [29]. Some researchers [30] have suggested that ionic species such as Na^+, Cl^-, SO_4^{2-}, CO_3^{2-}, SiO_3^{2-}, and WO_4^{2-} are "surface nonactive" to wolframite. Addition of constituent metal ions, namely, Fe^{2+} and Mn^{2+} caused reversal of charge.

Interfacial electrokinetics has been extensively monitored in the flotation of wolframite with different collectors. In the case of a cationic collector, dodecyl ammonium acetate, the flotation response of wolframite was observed to be good between pH 5 and 10. Interestingly, Fe^{2+} (one of the constituent metal ions) was found to be a depressant for amine flotation. The hydrophobic behavior and optimum flotation conditions of wolframite have also been evaluated by electrokinetic studies in the presence of other flotation collectors, namely, oxalic acid, aminonitroparaffin, and naphthenehydroxamic acid [31]. It was found that the concentration of the flotation reagents had little effect on the electrokinetic potential.

E. Barite

The i.e.p. of barite ($BaSO_4$) has been measured to be at pH 4.2. The zeta potential varied from $+140\,mV$ at pH 1 to $-120\,mV$ at pH 10 [32]. The potential determining ions for barite are Ba^{2+}, SO_4^{2-}, H^+, and OH^- [33]. Sadowski and Smith [34] have investigated the effects of different metal cations like Ca^{2+}, Mg^{2+}, Fe^{2+}, and Al^{3+} on the stability and zeta potential of barite suspension. It was noticed that the i.e.p. shifted to more alkaline pH in the presence of Fe^{2+} and Al^{3+}. However, in the presence of Ca^{2+} and Mg^{2+}, the i.e.p. shifted to more negative values with increase in pH. In a study involving flotation, Taha et al. [35] have found that sodium oleate and primary amine collectors adsorb on barite and celestite surfaces. This has been confirmed by the correlation between maximum floatability and the zeta-potential values.

IV. SULFIDE MINERALS

The sulfide minerals of economical interest are those of copper (chalcopyrite, $CuFeS_2$), lead (galena, PbS), and zinc (sphalerite, ZnS). Secondary sulfide minerals of importance are mainly those of copper, namely, chalcocite (Cu_2S) and covellite (CuS). An important iron sulfide mineral is pyrite (FeS_2). The electrokinetics of the three base metal sulfides of copper, lead, and zinc have been extensively investigated. The presence and nature of adsorbed species in the aqueous sulfide mineral system have been studied by electrokinetics [1], electrochemical [36, 37], and surface spectroscopic techniques [38]. Two important factors affecting the electrokinetics of sulfide minerals are their oxidation characteristics and adsorption of hydrolyzable metal ions on sulfide surfaces.

A. Oxidation Characteristics of Sulfide Minerals

Sulfides are known to be thermodynamically unstable in the presence of oxygen. As a consequence, in ore bodies sulfide minerals are oxidized to varying degrees. Mechanisms of oxidation of sulfides have been studied for a long time and are a subject of immense controversy [39–45]. In aqueous medium a sulfide mineral represents an electrochemical system. The dissolved species from the suspended solid are adsorbed as metal hydroxy species or metal hydroxides on the surface, depending on the Eh–pH conditions and the concentration of ions in solution. The electrokinetic behavior of sulfide minerals is more complicated to interpret than that of oxide minerals. Metal cation(s), S^{2-}, HS^-, H^+, and OH^- anions, and H_2S may all be responsible for the charge on the sulfide mineral [2, 46]. According to Luttrell and Yoon [47], under limited oxidation conditions, in addition to OH^- and HS^-, a variety of polysulfide species may also contribute to the negative surface charge in alkaline solution.

Compared to oxides, the sulfide minerals show wide variations in the point of zero charge (pH_{pzc}) or isoelectric point (pH_{iep}) due to their proneness to oxidation. Depending on the pH values, solid concentration, and aging, the zeta potential–pH characteristics change dramatically as the nature and quantity of surface species vary. Since the sulfide ion is readily hydrolyzed in solution, the sulfur-bearing species, HS^- and H_2S also need to be considered. The observed low pH_{iep} in the case of most sulfide minerals has been attributed [48] to a surface coating of elemental sulfur produced by the reaction:

$$MS \rightarrow M^{2+} + S^0 + 2e \text{ (oxidation)} \tag{5}$$

and

$$1/2O_2 + 2H^{2+} + 2e \rightarrow H_2O \tag{6}$$

or

$$2H_2O + 2e \rightarrow H_2 + 2OH^- \text{ (reduction)} \tag{7}$$

The products of oxidation are both elemental sulfur and metal ion(s). Sulfur is thermodynamically stable up to pH $\sim$ 8. However, the oxidation of sulfide to elemental sulfur (S^0) increases as the pH decreases. With increase in pH, S^0 and S^{2-} are both increasingly unstable and oxidize to $S_xO_y^{n-}$ species:

$$S^0 + 4H_2O \rightarrow SO_4^{2-} + 8H^+ + 6e \tag{8}$$

Sulfur coating of the mineral surface is known to render self-induced floatability of many sulfides. Also, depending on the pH conditions, the metal ions may undergo hydrolysis and can be precipitate on the surface in the form of metal hydroxide, making the surface hydrophilic. The level of oxidation (and nature of oxidation products) determines whether the mineral surface would be hydrophobic or hydrophilic.

B. Adsorption of Hydrolyzable Metal Ions on Sulfide Minerals

As in the case of oxides, sulfide minerals are also influenced by the adsorption of hydrolyzable metal ions. Pugh [49] and Pugh and Tjus [50] in their studies on (1) Cu(II) hydroxy-coated zinc sulfide particles, and (2) uptake of Pb^{2+}/Zn^{2+} by PbS/ZnS have corroborated the findings of James and Healy [13] on oxide minerals. The adsorption of Zn(II) at the cinnabar (HgS)/H_2O interface was studied by James and Parks [3]. It has been observed that the adsorption of hydrolysis products in the EDL or ion exchange of free aqueous metal ions for surface protons can describe the uptake of the species. Das et al. [51] have investigated the role of constituent metal ions, namely, Cu^{2+} and Fe^{3+} on the oxidation of chalcopyrite. It was suggested that the ratio of Cu/Fe on the chalcopyrite surface governs the interfacial potential and this was reflected by a shift in pH_{iep} values for chalcopyrite samples oxidized to different extents. The effects of Cu^{2+}, Ni^{2+}, and Fe^{2+} addition on covellite, pyrrhotite, and millerite have been investigated by Acar and Somasundaran [52]. They observed a shift in i.e.p. of the metal sulfides in the presence of various concentrations of externally added metal ions. The shift in i.e.p. with the addition of a lower concentration of metal ions was attributed to the precipitation of metal hydroxide either in the bulk or on the surface. The authors have explained the change in zeta-potential values in the presence of metal ions as due to the conversion of the sulfide surfaces into that of the adsorbing metal sulfide surface. This was found to be true in the case of Cu^{2+} adsorption on pyrrhotite (FeS) and millerite (NiS), thus changing the surfaces to those of covellite (CuS). Similarly, surface conversion of pyrrhotite into millerite was observed in the presence of Ni(II) ions.

The important electrokinetic investigations on some of the sulfide minerals are described in the following subsections.

C. Galena

1. Determination of Isoelectric Point

Among the sulfides, the most extensively studied minerals are those of lead and zinc. Wide variations in the pH_{iep} values, as much as 5 pH units, have been reported for galena and sphalerite. According to Pugh [49] these variations could be explained by considering the complicating effects associated with the oxidation of sulfide surfaces. Although potential determining lattice ions such as M^{2+}, M^{3+}, and S^{2-} control the EDL, other Eh–pH dependent processes, for example oxidation, are also important. This can result in the formation of sulfoxides, basic thiosulfates, sulfates, basic carbonates, etc., which may enhance dissolution, leading to the release of significant quantities of soluble species. For a naturally occurring galena sample, Pugh [49] observed negative charge over pH 2–12 while for a synthetic sample, the pH_{iep} was found to be at pH $\sim$ 5. The effect of externally added Pb(II) (the constituent metal ion) on the galena surface was interesting, e.g., at low Pb(II) concentration, the galena surface showed a lowering in negative charge but at higher Pb(II) concentrations, charge reversals occurred; negative to positive at pH $\sim$ 4 and positive to negative at pH $\sim$ 9. In contrast to the pH_{iep} observed at pH $\sim$ 5 by Pugh [49], Gaudin and Sun [53] determined the i.e.p. of galena to be at pH 3, while studies by McGlashan et al. [54] have shown lead sulfides to be negatively charged over most of the pH range. Yarar and Kitchener [55] have measured the zeta potential of galena as a function of pH, Pb^{2+} concentration, and S^{2-} concentration and observed the i.e.p. at pH 3. It was interesting to note that using Pb^{2+} and S^{2-} as potential-determining ions at fixed pH, the zeta potential did not become zero at any concentration of the added ions. Yucesoy and Yarar [56] determined the zeta potential of lead sulfide as a function of pH, Pb^{2+} concentration, ethyl xanthate concentration, and $NaNO_3$ concentration. With respect to Pb^{2+}, the i.e.p. was observed at pH 4.0 at $10^{-5}\,M\,Pb^{2+}$. The negative zeta potential of Pb increased in magnitude (negative direction) on adsorption of ethylxanthate ion. Although nitrate ion is known to be indifferent, it appeared to be specifically adsorbing on galena at pH 5.5. Fuerstenau [57] has reported i.e.p. values for galena at pH 2.6 and 8 in the absence and presence of oxygen, respectively. Aplan et al. [58] observed that various galena samples including synthetic ones, yielded i.e.p. values at pH $\sim$ 2.1.

2. Oxidation of Galena

Like most sulfide minerals, galena is also prone to oxidation. Plante and Sutherland [59] have observed that oxidation of galena produced soluble cation and sulfate in neutral/

acidic solutions, but polythionate, sulfate, and thiosulfate in alkaline solutions. Lead was not found to be released. Contrary to this, Pugh and Bergstrom [14] detected significant amounts of lead ions dissolved from two galena samples during conditioning. The more highly oxidized sample released almost 10 times more lead ions into water at neutral pH.

An important electrokinetic investigation on the oxidation of galena has been reported by Fornasiero et al. [60]. The galena samples were conditioned in aqueous solutions at different pH values with different gases for varying amounts of time. To account for the changes in zeta potential occurring during surface oxidation, the authors proposed a mechanism involving the dissolution of Pb ions followed by their readsorption as lead hydroxide. The Gouy–Chapman–Stern double-layer model was invoked to calculate the zeta potential as a function of pH. It has been argued that the driving force for the interaction between the solution Pb species and the surface sites are electrical in nature.

3. Effect of Sample Preparation Technique and Conditioning

It is well known that the electrokinetic characteristics of minerals vary, depending on the method of sample preparation. For example, freshly wet ground galena in distilled water showed an i.e.p. at pH 6.8 while there was a gradual shift in zeta potential–pH curves to more negative values, especially in the acidic range [61], depending on whether the sample was dry ground (i.e.p. at pH 5.2), the dry-ground sample was pretreated in ammonium chloride solution (i.e.p. at pH 4.2), or the sample was dry ground and stored for about five months (i.e.p. at pH 2.2). Interestingly, a similarity in the zeta-potential behavior of a heavily oxidized galena sample with that of elemental sulfur (at acidic pH values) was also noticed. It is fascinating to note that chalcopyrite as well as sulfur sample obtained from galena itself showed an i.e.p. at pH 2.2. These results corroborated the findings of earlier researchers on the influence of sulfur in rendering the minerals hydrophobic. Accordingly, it has been suggested that the surface reactions are predominantly driven by sulfur. Based on the observation that the electrokinetic characteristics of heavily oxidized galena samples showed remarkable similarity to that of elemental sulfur, Kelebek and Smith [61] have proposed the following reaction in the acidic pH conditions for the formation of sulfur from galena:

$$3\,PbS + 4\,H_2SO_4 \rightarrow 3\,PbSO_4 + 4\,S^0 + 4\,H_2O \tag{9}$$

Like the sample preparation technique, equilibration under different conditions also influences zeta-potential measurements. Lee and Whang [62] have measured the i.e.p. of galena in aqueous solution at pH 3.4. Pre-equilibration at pH 3 and 6 shifted the i.e.p. to the alkaline region (i.e., pH 11.3 and 10.5, respectively) whereas pre-equilibration at pH 10 shifted the i.e.p. to acidic pH. The addition of Na_2S was found to have a depressing effect on the flotation of galena while at the same time it increased the negative charge on galena over the whole pH range. On the other hand, $Pb(NO_3)_2$ had an activating effect on the flotation of galena. Neville and Hunter [63] have noticed a relationship between equilibrating conditions and zeta potential. Depending on the equilibration conditions, the i.e.p. was found to vary between pH 2 and 4. When the mineral was conditioned at pH 4, two CRs were observed; negative to positive at pH 6 and positive to negative at pH 10.

4. Effect of Xanthate Collector

The widely used collector for flotation of galena is potassium ethyl xanthate. The interfacial electrokinetics of galena–ethyl xanthate system has been studied [56, 64]. As expected of an adsorption process, the negative zeta potential of galena was found to increase with xanthate concentration [56, 62, 64]. Espinosa-Jiménez et al. [64] measured the p.z.c. of galena at pH 4.8. The observed increase in zeta potential subsequent to the interaction between xanthate and the galena surface has been ascribed to the adsorption of xanthate on galena, leading to enhanced hydrophobicity of the mineral surface.

5. Aggregation Studies

In addition to several studies on the electrokinetics of minerals and its effect on flotation, there are also some reports of aggregation studies on sulfide minerals [65, 66]. It is well known that for the individual minerals, the pH of maximum settling occurs at their respective p.z.c. values. It was interesting to note that in the agglomeration study of galena, pyrite, and their mixtures, the effect of externally added Ca and Pb ions on the overall settling behavior was minor, but Fe(II) significantly lowered the settling rate of both galena and pyrite. In the case of mineral mixtures, the changes in zeta potential *vis-à-vis* galena alone were attributed to galvanic interactions between the two minerals, resulting in oxidation products on the galena surface.

D. Sphalerite

1. Determination of p.z.c. and Effect of Variables

The electrokinetics of sphalerite, the predominant zinc mineral, has been well studied. Pugh [49] has observed $pH_{iep} \sim 3$ for sphalerite and $pH_{iep} \sim 9$ for synthetic ZnS. Moignard et al. [48] have investigated the electrokinetics of ZnS in great detail. The effect of percentage solids, conditioning time, and conditioning pH on the zeta potential of synthetic and natural zinc sulfides have been studied. Increasing the percentage solids from 0.1 to 2% resulted in the shift in i.e.p. from pH 4 to 8.5. Similarly, the i.e.p. shifted from pH 5 to 8.5 as the pH of pre-equilibration was increased from pH 3 to 9.8. These phenomena were explained as due to the formation of oxidation and hydrolysis products at the ZnS surface. These authors postulated that, in acid solutions, the ZnS surface is oxidized by residual oxygen of the system to Zn^{2+} and S^0:

$$ZnS + 2H^+ + O_2 \rightarrow Zn^{2+} + S^0 + H_2O \qquad (10)$$

At alkaline pH, Zn^{2+} hydrolyzes in aqueous solution to form a surface coating of insoluble zinc oxide–hydroxide. Formation of both elemental sulfur and zinc oxide–hydroxide coatings on the ZnS surface was cited as responsible for changes in the double-layer characteristics.

William and Labib [67] have measured the zeta potentials of synthetic ZnS, ZnCdS, and ZnO in an aqueous solution as a function of pH. In the pH range 3–12, three distinct regions were observed within each of which the zeta potential exhibited different behavior. At low pH values, ZnS dissolved symmetrically due to the attack of H^+ ions, probably at sulfur sites. According to the authors, at high pH values the surface appeared to be covered with an oxide layer and, for pH > 9, a dissolution reaction took place, adding to the negative charge on the surface.

2. Electrokinetics of Sphalerite Depression and Activation

In the differential flotation of lead and zinc sulfides from a Pb–Zn ore, sphalerite is depressed by adding zinc sulfate, and galena is floated first. The depression of sphalerite with Zn(II) in neutral and alkaline solutions has been discussed in detail by Finkelstein and Allison [68]. The depression has been attributed to the adhesion of $Zn(OH)_2$ to the surface of sphalerite. The thick coating of hydroxide encapsulating the particles, together with the possibility of strong hydrogen bonding with localized water, renders the surface extremely hydrophilic.

Since sphalerite is sluggish to float, a metal ion (generally Cu^{2+}) is added to the flotation pulp to activate the surface. This enables the collector (normally xanthate) to adsorb on the mineral surface and render it floatable. Zeta-potential studies have been extensively used to delineate the surface chemical characteristics of this system. Hukki et al. [69] observed a large change in zeta potential after addition of 25 ppm of $CuSO_4 \cdot 5H_2O$ to the ZnS suspensions. The shift in CR (from pH 3 to 4.2) is explained as due to specific adsorption of Cu^{2+}. The other two CRs, negative to positive at $\sim$ pH 6.5 and positive to negative at $\sim$ pH 9, are attributed to hydroxide formation. According to Laskowski et al. [70], the activation of sphalerite is believed to be a two-stage process: the overall activation of sphalerite becomes somewhat obscured by the hydrolysis of Cu^{2+} and precipitation of $Cu(OH)_2$ in the alkaline pH conditions (under which flotation of sphalerite is generally carried out). The activation products formed on the surface of sphalerite in the near neutral/alkaline solution in the first rapid activation stage are "flotation inactive." These unstable layers subsequently transform into a "flotation-active" form in the second stage. However, in the presence of Pb^{2+} ions, the surface layers formed due to activation by Cu^{2+} are stable but do not improve the flotation rate.

Pugh and Tjus [50] have studied the electrokinetics of Cu(II) hydroxy-coated zinc sulfide particles. The uptake of hydroxylated metal ions on the mineral surface resulted in the reversal of zeta potentials and these have been discussed in terms of specific adsorption of positively charged hydroxy species and the interfacial precipitation of metal hydroxides [71–73]. Pugh and Tjus [50] also found that the coating of accumulated copper-hydroxy species on a ZnS surface depleted over a period of time. This was reflected by the reduction in positive zeta potential which tended to revert to its original negative value. Based on solubility–pH diagrams, these authors argued that for $Cu(OH)_2(s)$, the precipitation edge increases from pH 6.2 to 7.3. The instability of the coating has been attributed to the diffusion of Cu(II) into the ZnS lattice and subsequent release of Zn(II) and the formation of a surface film [74].

In an interesting investigation on the effect of Fe-lattice ions on the adsorption, electrokinetic, calorimetric, and flotation properties of sphalerite, Gigowski et al. [75] have observed that Fe-rich sphalerites adsorb xanthate with preference. The effect was noticed particularly in the case of sphalerite activated with Cu. It was observed that the adsorption led to an increase in negative potential whereas other factors had little effect. The electrokinetic studies suggested that the i.e.p. values of different sphalerite samples differed depending on their iron contents. Similarly, depending on the Fe content, the interaction of isopropylxanthate–sphalerite showed differences in electrokinetic behavior. From calorimetric measurements, the authors have argued that the activation of the sphalerite surface with Cu ions was more influenced by the Fe content than by the degree of oxidation.

It is well recognized that reagent chemistry plays an important role in the flotation recovery of minerals. In the flotation of sulfide minerals with some new collectors, it was observed that, for 2-mercaptobenzoxazole–chalcocite, 2-mercaptobenzothiazole–galena, and 2-aminothiophenol–sphalerite systems, the new collectors strongly improved the mineral recovery in the neutral pH range [76]. In comparison with xanthates, the new reagents required a lower concentration to achieve a similar flotation performance. Electrophoresis and sedimentation studies confirm the results obtained from flotation.

From the electrokinetic measurements, chalcocite was estimated to have an i.e.p. at pH 6–7. Though according to DLVO theory, this pH range should have been an unstable region, but in reality no instability of suspension was observed. This led the authors to suggest that repulsive solvation/hydration forces might be responsible for the observed suspension stability.

E. Chalcopyrite

Chalcopyrite is a ternary system involving copper, iron, and sulfur. Like most other sulfide minerals, chalcopyrite is also prone to oxidation. The electrokinetic behavior of the mineral surface reflects the changes in the EDL brought about as a consequence of the adsorption of different ions. The oxidation characteristics of chalcopyrite are complicated due to the presence of two metal atoms, namely, Cu^{2+} and Fe^{3+}. One or both of these ions may be released into the system and subsequently undergo hydrolysis, the end product being the respective hydroxides. The properties of the chalcopyrite/water interface are governed by the concentrations of Cu^{2+}, Fe^{3+}, S^{2+}, and H^+/OH^-. The oxidation characteristics of chalcopyrite have been a subject of controversy especially with respect to its influence on flotation [39–43]. In the past, attempts have been made to identify the species responsible for flotation with and without collectors, by different electrochemical methods suitably supplemented by surface spectroscopic techniques such as XPS and Auger electron spectroscopy [36, 39–45], but there is no broad agreement. However, it is now generally believed that a certain amount of oxidation (either by chemical or electrochemical control) assists the flotation of chalcopyrite in the absence of collectors. Heavy oxidation by strong oxidants like hydrogen peroxide could completely suppress the flotation of chalcopyrite whereas pretreatment with mild glacial acetic acid increased flotation [77]. Excellent literature articles are available on the above subject [40, 42, 44, 45, 78].

Electrokinetic studies on chalcopyrite are somewhat scarce. Previous studies [54, 79] on chalcopyrite indicate a low i.e.p. for this mineral (pH ~ 3). Kelebek and Smith [61] reported an i.e.p. of pH 2.2 for chalcopyrite and suggested that the pH dependence of zeta potential for chalcopyrite is similar to that of sulfur, both having a common i.e.p. at pH 2.2. McGlashan et al. [54] proposed that various ionic sites involving copper, iron, and sulfur atoms, created on fracture, would react with oxygen to produce negatively charged surface species. Freshly ground chalcopyrite samples at low pH values have zeta potentials similar to that of sulfur [48, 61]. In both cases, it is difficult to interpret the low i.e.p. and negative zeta-potential values since the specific surface groups capable of accepting protons or hydroxyl ions are unknown.

Depending on the level of oxidation, either due to geochemical reactions, aging, or deliberate oxidation, the surface characteristics of chalcopyrite undergo dramatic changes which are appropriately reflected by interfacial electrokinetics [51]. The authors observed that chalcopyrite samples from four different geographical locations showed differences in electrokinetic behavior which could be attributed to the changes in the concentrations of the constituent metal ions on the mineral surface (from XPS analysis). It has been suggested that the relative preponderance of one metal ion over another (e.g., Cu^{2+} over Fe^{3+}) as well as the solubility of the surface contributing the metal ions in solution (as a result of the oxidation reaction) control the electrokinetic behavior of chalcopyrite. The oxidation reaction could be the result of deliberate oxidation carried out in the laboratory (e.g., by hydrogen peroxide treatment), it could have occurred due to aging, or it could have taken place over geological periods of time. In essence, the ratio of Cu : Fe controlled the interfacial electrokinetics as a result of dissolution, hydrolysis, precipitation, and adsorption of the metal ions on the mineral surface. This hypothesis was confirmed by monitoring the interfacial electrokinetics of the particular metal-deficient chalcopyrite sample in the presence of externally added metal ions. It was also demonstrated that, depending on the solid/liquid ratio and "aging," the chalcopyrite surface can become copper- or iron-rich as reflected by a shift in pH_{iep}.

1. Chalcopyrite–Collector Interactions

Although chalcopyrite responds well to flotation by xanthate collectors, oxidized samples show reduced efficiency with xanthate. Electrokinetics has been used to study the effect of various nonxanthate collectors on chalcopyrite. Mangalam and Khangaonkar [80–82] have studied the Hallimond tube flotation of chalcopyrite with 8-hydroxyquinoline, cupferron, and sodium diethyldithiocarbamate. In the case of 8-hydroxyquinoline [80], correlation of zeta potential with flotation behavior was found to be poor. The authors have explained the lowering of floatability with respect to coagulation effects. In the case of cupferron [81], it was found that the floatability of chalcopyrite decreased with increasing adsorption as a result of coagulation and dissolution effects. The application of Modi–Fuerstenau and Stern–Grahame models indicated the predominantly electrostatic nature of the collector adsorption in the case of both cupferron and sodium diethyldithiocarbamate.

The importance of reagent chemistry on the flotation behavior of chalcopyrite has also been demonstrated by Fairthorne et al. [83]. These authors have investigated the adsorption of a commercially used collector, butyl ethoxycarbonyl thiourea (BECTU) on chalcopyrite as a function of pH in the presence of oxygen and nitrogen. An increase in the flotation recovery due to addition of BECTU was observed. The adsorption was monitored in solution by UV–visible spectroscopy and on the chalcopyrite surface by XPS, secondary-ion mass spectroscopy (SIMS), zeta potential, and flotation measurements. The chalcopyrite sample showed changes in zeta potential on addition of BECTU. It was noticed that the zeta potential – pH curves became less negative and even reversed sign at high BECTU concentrations. The reversal of charge has been attributed to the release of a proton from the BECTU molecule when forming the surface cuprous–BECTU complex. SIMS analysis provided strong evidence for the presence of a 1:1 Cu–BECTU surface complex.

F. Chalcocite

Lekki and Laskowski [84] have studied the electrokinetics of the secondary copper sulfide, chalcocite, to evaluate the effect of flotation reagents. Chalcocite was observed to be negatively charged from pH 3 to 10. Both ethyl xanthate (collector) and α-terpineol (frother) adsorbed at the chalcocite surface and thus increased the magnitude of the negative zeta potential. The i.e.p. for chalcocite was observed at pH 6.5 in the presence of α-terpineol while no shift in i.e.p. was observed in the presence of ethyl xanthate.

The surface oxidation of chalcocite has been studied by Oestreicher and McGlashan [85] as a function of conditioning time and equilibrium pH. Their results tend to indicate that copper hydroxide formation takes place when the sulfide is equilibrated at pH 3.4 where dissolution occurs with the release of high concentrations of Cu^{2+}. In the absence of significant oxidation, the chalcocite surface remains essentially negative over the pH range 4–11.

G. Covellite

The electrokinetics of another secondary copper sulfide mineral, covellite, was studied by Acar and Somasundaran [52] to find out the effect of dissolved mineral species. They have observed two i.e.p. values for covellite, at pH 6.3 and 9.5. Since CuO has an i.e.p. at pH 9.5, the second i.e.p. has been interpreted as that due to the formation of copper oxide on the surface of covellite. The lower i.e.p. at 6.3 has been explained as due to the formation of $Cu(OH)_2$ on the covellite surface.

In an electrokinetic study involving both sulfide and oxide copper minerals like covellite (CuS), cuprite (Cu_2O) and tenorite (CuO), Bhaskar Raju and Forsling [86] have investigated the role of potential-determining ions. In the case of CuS, additional CRs were noticed in pH conditions other than the regions of metal ion precipitation. The interaction of protons with the hydrous mineral surface, and the adsorption of desorbed or excess metal ions with protonated surface sites are explained by an ion-exchange mechanism.

H. Pyrite

The important electrokinetic studies on iron sulfide mineral, pyrite, are those by Fuerstenau et al. [87], Fuerstenau and Elgillani [88], and Fornasiero et al. [89]. In the first study [87], the adsorption of xanthate on pyrite was dealt with. It was observed that the changes in i.e.p. were dependent on the conditioning period: thus, the pH_{iep} was 6.2 after 5 min conditoning, and 6.9 after 15 min. Gaudin and Sun [53] had observed an i.e.p. value of pH 6.4. In the flotation of sulfide minerals, cyanides are generally used as depressants for pyrite. Fuerstenau and Elgillani [88] studied the effect of cyanide on the zeta-potential behavior of pyrite. They proposed that either CN^- or $Fe(CN)_6^{3-}$ decreased the zeta potential and moved the i.e.p. to acidic values, indicating specific adsorption of an anion. The depression of pyrite by cyanide was attributed to the formation of a complex $Fe[Fe(CN)_6]_y$ at the surface.

Fornasiero et al. [89] have studied the oxidation of pyrite as a function of pH under various pretreatment conditions: conditioning in Ar, N, air, and O atmospheres, at several conditioning pHs, and for several conditioning periods. The zeta potential of "virgin" pyrite was found negative over the pH range 3–10; the i.e.p. of pyrite under Ar was at pH 1.4 ± 0.4. However, exposure to O reversed the sign of the zeta potential from negative to positive at low pH values. It was also observed that the CR takes place more rapidly at lower conditioning pH. The results were explained on the basis of dissolution of Fe from the surface followed by the electrostatic adsorption of positively charged Fe hydroxide species on to negatively charged surface sites of pyrite. Good agreement was found between the experimental and the calculated zeta-potential data by using the EDL theory of Gouy–Chapman–Stern.

V. SALT-TYPE MINERALS

The electrokinetic investigations on salt-type minerals are varied and interesting. Some of the important ones are discussed below.

A. Apatite [(CaF)Ca$_4$(PO$_4$)$_3$ and (CaCl)Ca$_4$(PO$_4$)$_3$]

1. Determination of p.z.c. and Effect of Potential-Determining Ions

Apatite is an important phosphatic mineral. One of the early electrokinetic studies on apatite was conducted by Somasundaran [90]. The i.e.p. of apatite in aqueous solution was determined to be at pH 4, but with the passage of time, the i.e.p. shifted towards a final value of pH 6. In the case of apatite, the potential determining ions are H$^+$, OH$^-$, and (PO$_4$)$^{3-}$. Zeta-potential changes due to the addition of calcium and fluoride were found to be significant. The major effect of Ca was attributed to its specific adsorption characteristics. The pH-dependent hydrolysis of the surface species was responsible for the charge development at the apatite surface. Rao et al. [91] determined the i.e.p. of apatite at pH 4. Similar results were obtained by Smani et al. [92]. However, in the literature the i.e.p. of apatite has been reported to vary from pH 3.5 to 6.7, depending on the source of the mineral [93].

2. Apatite–Reagent Interactions

The interaction of oleate (a commonly used flotation collector) on apatite has been investigated by adsorption density measurements and electrokinetics [91]. It was found that the adsorption density at monolayer coverage corresponded to a condensed state of surfactant with a molecular area of 33 Å^2. The isotherm at pH 8 showed monolayer coverage followed by the precipitation step of Ca oleate. At pH values of 9–11, the isotherms revealed a bilayered structure of oleate on the surface before the precipitation step of Ca oleate. The zeta-potential studies on the interaction of oleate with apatite showed that at very low concentration (e.g., 3.55×10^{-7} M oleate), there was no change in zeta potential. With increase in the oleate concentration, the zeta potentials became more negative and even showed totally negative surface charges at 2.66×10^{-5} M oleate. The zeta-potential values were found to correlate well with the adsorption of oleate up to the bilayer formation.

In an interesting study of the bacterial adhesion on to apatite minerals, Yellojirao et al. [94] have examined the role of electrokinetic properties of *streptococcus sanguis* and *actinomyces naeslundii* in determining their adhesion to apatite minerals. The apatite sample exhibited an i.e.p. at pH 8. It was fascinating to note that bacterial adhesion on to the apatite surface could take place even when both the surfaces were negatively charged and the adhered layer was resistant to washing. After exposure to the bacteria, the mineral fines exhibited zeta-potential values intermediate between those of the mineral and the bacteria. The authors have argued that in the absence of salivary proteins, the inorganic species present in saliva (including Ca^{2+}, Mg^{2+}, K$^+$, and various phosphate species) alter the magnitude of the surface charge but do not affect the bacterial adhesion process. On the basis of the observed adsorption of negatively charged bacteria on to negatively charged minerals, it has been suggested that electrostatic interactions are not the primary factors responsible for adhesion. The authors have identified surface heterogeneity of bacteria as a possible reason for controlling adhesion of similarly charged particles. It was further argued that the dissolved species from the mineral could alter the surface charge of the bacteria so as to reduce repulsion and thus promote adhesion.

B. Calcite (CaCO$_3$)

Calcite (calcium carbonate) is one of the most important and cheap industrial minerals. Calcite exhibits complex behavior in aqueous medium due to its solubility characteristics.

The salient features of different investigations on the calcite electrokinetics are delineated below.

1. Determination of i.e.p.

The i.e.p. values of calcite samples have been reported to be widely different. Fuerstenau et al. [95] have determined the i.e.p. at pH 10.8, while Somasundaran and Agar [96] have reported a value between pH 8 and 9.5. Smani et al. [92] have found the i.e.p. to be at pH 5.4. It has also been observed that under identical conditions, synthetic and natural calcite samples show varying zeta-potential values [97]. For example, synthetic calcite in 10^{-3} mol dm^{-3} NaCl and at a pH of 8.6 was positively charged (+22 mV), whereas natural calcite under the same conditions showed a potential of -18 mV. The synthetic sample showed an i.e.p. at pH 9.5 whereas no clear i.e.p. was noticed for a natural sample. The zeta-potential values for synthetic calcite were significantly altered by the addition of natural seawater or organic solutes. However, no such change was observed for the natural sample. It has been suggested that the organic coating attached to the surface of natural calcite prevented calcite from reacting with ions and solutes from the solution.

2. Effect of Solids Concentration

The electrokinetic properties of calcite in water were found to vary, depending on the operating conditions [98]. The zeta potentials (ranging from negative to positive values) decreased due to vigorous shaking but were not much affected by atmospheric CO_2. However, the quantity of suspended $CaCO_3$ had a strong effect. In 0.01 M NaCl medium, the zeta potentials of calcite were measured at different pH values for various concentrations of dispersed calcite. The results showed a maximum in zeta potential at pH 9.1 ± 0.1. However, depending on the calcite concentrations; one, two, or no i.e.p. values were observed. This was ascribed to the concentration change of dehydrated calcium ions in the inner Helmholtz plane, due to variations in the dissolution rate of the solid. It has been argued that when the dispersed mass is small, the hydrated calcium ions concentrate largely in the outer Helmholtz plane while the less hydrated carbonate anions remain in contact with the surface in the inner Helmholtz plane, thus the zeta potential becomes negative. However, at a mass greater than 30 mg per 100 cm^3, the mineral surface in contact with water becomes large, thereby allowing certain surface calcium ions to be mobilized and become hydrated. The surface calcium ions are smaller than the carbonates and impart a positive charge to the surface.

3. Calcite–Reagent Interactions

Electrokinetic properties of the calcite/water interface have also been studied in the presence of magnesium and organic matter [99]. The zeta potentials of calcite suspended in water were negative and only slightly changed in the pH range 8.5–10.5. At constant Ca^{2+} concentration, pH did not affect the zeta-potential values. However, the zeta potential was rather dependent on pCa^{2+} or pCO_3^{2-}. The authors observed an i.e.p. at pCa^{2+} 2.7. The factors which significantly influenced the zeta potential of calcite were pH, Mg^{2+} concentration, and aging time. The findings involving various metal ions are interpreted in terms of the formation of a surface layer of Mg-bearing calcite, Mg^{2+}, Ca^{2+}, and CO_3^{2-} being the potential determining ions. With regard to the influence of organic matter, it is interesting to note that the adsorption of dodecyl sulfate on calcite produced more negative surface charge due to electrostatic interactions between Ca and dodecyl sulfate anions. In a somewhat similar study, Smallwood [100] has investigated the adsorption of Ca^{2+}, Mg^{2+}, CO_3^{2+}, and SO_4^{2-} on calcite and argonite and has observed that the ions are more strongly adsorbed by aragonite (orthorhombic calcium carbonate) than by calcite (rhombohedral structure). The adsorption of several organic substances (albumin, agar, Na alginate, chondroitin sulfate, and glucose) revealed that, except for glucose, other organic substances showed CRs indicating adsorption. With the exception of agar, the adsorbed organic compounds stabilized the mineral suspension by steric and double-layer repulsion.

Researchers have also investigated the adsorption of surfactants on a calcite surface. Rao et al. [101] have studied the mechanisms of oleate adsorption on calcite by electro-kinetics and adsorption density measurements. The zeta potentials were found to correspond to the adsorption of oleate until monolayer coverage. Somasundaran [102] has reported that the adsorption of oleate on calcite below its p.z.c. occurs due to electrostatic forces, while above the p.z.c. the adsorption takes place by chemical interactions. The adsorption of different surfactants, namely, dodecyltrimethylammonium chloride (DTAC), sodium dodecyl sulfonate (SDS), and Na oleate on calcite was investigated by Andersen et al. in the alkaline conditions (pH 9.8) [103]. The electrokinetic potential in DTAC solutions indicated that the adsorption was due to electrostatic attraction between negatively charged calcite particles and surfactant cations. Adsorption of SDS involved at least partial chemical interaction. There was a strong calcite–oleate interaction even at low oleate concentrations.

One of the classical problems in mineral beneficiation is the difficulty in flotation separation of scheelite from calcite due to the similarities in physicochemical character-istics of the two minerals. Ozcan and Bulutcu [104] have compared the electrokinetic and flotation behavior of calcite and scheelite in the presence of oleoyl sarcosine, oxine, alkyl oxine, and quebracho. Alkyl oxine and oxine are generally used as modifiers for the scheelite surface in the separation of calcite from scheelite. It was found that alkyl oxine increased the negative zeta-potential value of scheelite to a less negative value whereas the zeta potential of calcite did not change, thus suggesting that the reagents adsorbed chemically on the scheelite surface and not on calcite. From the flotation studies, the authors found that conditoning of calcite and scheelite minerals individually with alkyl oxine before quebracho addition increased the efficiency of oleoyl sarcosine in scheelite flotation, but had no effect on calcite flotation.

In the development of a process for the direct flotation of phosphates, Yaniv et al. [105] have measured the surface charges of francolite and calcite by microelectrophoresis at different pH values in the presence of amine collectors. The results suggested that the collector adsorbed predominantly on the negatively charged surface of francolite at pH 7–9. There was agreement between flotation recovery and electrokinetic studies; a primary amine was found to be a very effective collector for francolite.

C. Fluorite

Different i.e.p. values for fluorite (CaF_2) are reported in the literature, varying from pH 4.2 to pH 11.3 [106–110]. For fluorite, the potential-determining ions are H^+, OH^-, and F^-. In one study the p.z.c. was determined to be at pH 8.4 [111]. While univalent ions such as K^+, Cl^-, and NO_3^- were indifferent to the interface, some multivalent ions such as Al^{3+}, Th^{4+}, SO_4^{2-}, and $Fe(CN)_6^{3-}$ showed special affinity and changed the values of the zeta potential significantly. Similar observations were made by Morales et al. [112].

The fluorite–oleate system has been investigated by Rao et al. [113] also. It was observed that the zeta potential of the interface decreased continuously with increasing oleate concentration. Zeta-potential measurements on fluorite correlated with the adsorption of oleate up to the bilayer formation and with formation of calcium oleate on the mineral surface at high oleate concentrations. From a comparison of the theoretical chemical equilibrium with experimental data, the authors postulated that the infinite increase in the slope of the isotherms was due to the precipitation of calcium oleate.

Hu et al. [114] have found that, in a mixture of scheelite and fluorite, the fluorite surface is converted to that of scheelite (detected by AES). Fluorite in the supernatant of scheelite exhibited nearly identical electrokinetic properties and flotation behavior as those of scheelite. Based on their findings, the authors suggested that the selective flotation of fluorite from scheelite might be achieved by using a selective amphoteric collector for fluorite at pH < 4 to avoid surface conversion.

D. Magnesite and Dolomite

Electrokinetic investigations on some other salt-type minerals like magnesite ($MgCO_3$), dolomite ($CaCO_3$, $MgCO_3$) and smithsonite ($ZnCO_3$) are rather scarce. The zeta potential and flotation characteristics of magnesite in the presence of sodium lauryl sulfate have been studied by Mangalam et al. [115]. It was observed that, under neutral pH conditions, the zeta potential showed a reversal of charge at higher concentrations ($0.42\,gL^{-1}$) of the collector, whereas at pH 5.5 a reversal in sign was noted even at a very low concentration ($0.03\,gL^{-1}$) of sodium lauryl sulfate. Better floatability was obtained under weakly acidic conditions.

In a study on the flotation separation of magnesite and dolomite [116] using sodium oleate, it was observed that the signs of zeta potential became opposite (positive for magnesite and negative for dolomite) upon addition of $10^{-3}\,mol\,Na_2SiF_6$ to a solution containing $50\,mgL^{-1}$ sodium oleate. This resulted in a higher floatability for magnesite and a lower one for dolomite in the pH range 8–9.

VI. KAOLIN/SILICA MINERALS

The electrokinetic investigations on clay and siliceous materials have been quite extensive. Some aspects of the surface characteristics of the two widely studied systems, kaolinite and silica, are discussed below.

A. Kaolinite

The electrokinetic properties of minerals are important in determining the effectiveness of a flotation process. In the kaolin industry, for example, flotation is used extensively to remove anatase impurities from white kaolinite, generally by using oleate, which is selective towards anatase. Due to the presence of exchangeable metal ions in the crystal structure, kaolinite is significantly influenced by the aqueous environment as well as by the additives. Thus, it is of interest to learn about the interfacial electrokinetics of the kaolin–aqueous system.

1. Determination of i.e.p., Effect of Surface Charge, and Sample Preparation Technique

Several researchers [117–120] have measured the i.e.p. of kaolinite suspension. Smith and Narimatsu [119] have used a microelectrophoresis technique and observed the i.e.p. of kaolinite at pH ~ 2.2. However, by using the streaming potential method, the authors did not find any i.e.p. Cases et al. [120] have determined the i.e.p. of kaolinite at pH 3 and found that the kaolin sample spontaneously flocculated at pH < 3.5 and stabilized well at pH > 5. The zeta potential versus pH curves did not account for the change in surface charge of kaolinite near pH 3 – this has been attributed to the electronegative character of the (001) faces consisting of either OH^- ions or O^{2-} ions distributed in the hexagonal layer. It was, thus, observed that the particles migrated under an applied field even with zero surface charge. Using a single-particle microelectrophoresis technique, Williams and Williams [121] have studied the effects of preparation technique and pH and NaCl concentration on the electrophoretic mobility of Na kaolinite. Based on an estimation of the edge zero point of charge (p.z.c.), they developed theories regarding the surface electrical properties of kaolinite. The Gouy–Stern–Grahame model was found to fit the experimental data only after rather unrealistic adjustment of some model parameters.

In a study on kaolin clays from five different commercial deposits in Georgia, Yuan and Pruett [122] have attempted to understand the variations in zeta potential and related fundamental properties. They observed that i.e.p. values varied from pH 1.5 to pH 3.5 for different samples. In one case, the kaolinite sample was negatively charged all throughout. The differences in the zeta-potential values for different kaolinite samples under identical conditions were attributed to the differences in ionic characteristics of the clay samples. The authors suggested that higher negative zeta potential in the case of various clays could be attributed to the exchangeable or soluble calcium ions in the clays.

2. Rheology of Kaolinite Suspensions

Because of the uniqueness of their structure, clay minerals are also important from the rheology point of view. The rheological properties of kaolinite suspension in the presence and absence of polymeric reagents have been examined and correlations between the electrokinetics and rheological properties have been attempted. Nicol and Hunter [123] have studied the rheological behavior of kaolinite solutions at different pH conditions. The Bingham yield values of kaolinite suspensions in the presence of adsorbates showed a pronounced maximum at neutral pH. This was attributed to the presence of extended chains of particles attached by electrostatic interactions between edges and faces. At high pH values, the particles supposedly behaved as individuals and the yield value could be correlated with attractive and repulsive forces between them. It was also noticed that metal ions under different pH conditions affected the kaolinite rheological properties differently.

From the measurement of the electroacoustic zeta potential and shear yield stress (τ_y) properties of nondilute kaolinite suspensions over a wide range of pHs and Al(III) concentrations, Johnson et al. [124] have observed distinct trends for the different faces present in the system. At zero or low Al(III) concentration, the zeta-potential values were dominated by the silica-like kaolinite face, pH-dependent face, and edge interactions. With the increase in Al(III) concentration, the zeta-potential values became more positive and the i.e.p. values shifted towards higher pH. The zeta-potential data are supported by shear yield stress results. It was interesting to note that the τ_y versus pH properties systematically changed from kaolinite-like to alumina-like behavior with increase in Al(III) concentration.

The viscosity of mineral suspensions are known to be altered by polymeric additives. The addition of poly(vinyl alcohol) to a kaolinite suspension resulted in an increase in the viscosity [125]. The effects of electrolytes (NaCl, $CaCl_2$, and $AlCl_3$) and flocculants (e.g., polyacrylamide) on the zeta potential of kaolin have revealed that when the electrolytes alone were used, the zeta potential decreased. However an electrolyte–polyacrylamide mixture increased the zeta potential [126]. On the other hand, other researchers [127] have found that the adsorption of polyacrylamide actually decreases the electrophoretic mobility due to the shift in the slipping plane towards the bulk solution.

Based on the relationship between viscosity, zeta potential, and dispersant addition, Gerischer and Sanderson [128, 129] have suggested a quantitative stability criterion for the clay suspensions in the absence and presence of additives like polyphosphate and poly-acrylate. The authors have observed that the addition of dispersant alone could not inhibit the agglomeration of kaolin at high solids concentration. It was also necessary to increase the pH to a fairly alkaline condition (pH $\sim$ 11). Nakaishi and Kuroda [130] have studied the flocculation of the Na kaolinite/water system as a function of salt concentration at pH 10. It was found that the mobility of particles increased with salt concentration. However, the experimental results of flocculation did not agree with the theory of diffuse double layer. In the dispersion state, the relationship between zeta potential and salt concentration agreed with the Gouy–Chapman theory of a diffuse double layer under constant surface potential. A surface potential of 56 mV and the position of the slipping plane (2.3 nm) were obtained from the experimental data. Furthermore, the authors argued that it is necessary to know the particle size of a clay–water system when measuring the zeta potential because mobility is heavily dependent on the dispersion – flocculation of a clay–water system.

3. Decontamination of Clay Minerals

One of the direct applications of electrokinetics is in the decontamination of clays [131, 132]. The efficiency of electro-osmosis in removing heavy metals from contaminated soils has been examined by applying an electric field to kaolin suspensions. Due to the application of the field, an electro-osmotic flow of water is produced, which is superimposed on the hydraulic flow. Since the electro-osmotic flow is somewhat stronger in smaller pores, as in the case of clays, it can facilitate the displacement of contaminated solution by fluid convection. Furthermore, local changes in pH can affect adsorption equilibrium and the distribution of solutes between the solution and the solid surfaces [131]. In the past, attempts have been made to separate heavy metals like lead from porous kaolinite. It may be of interest to note that the electro-osmosis experiments were conducted by applying either a constant current or a constant voltage across 8 cm inside diameter, 25–30 cm long specimens of kaolinite clay with graphite electrodes [131]. Samples of the pore fluid and of the clay were analyzed to determine metal content, pH, and other conditions as functions of time and position. Results show that removal of heavy metals from soil can be accomplished effectively by electrokinetic treatment of sufficient duration. Approximately 95% of the lead was removed after 189 days. Other researchers, e.g., Coletta et al. [132] have used natural solutions containing clay extracts and synthetic solutions with varying concentrations of Al^{3+}, Ca^{2+}, and Na^+ as anodic flushing solutions for enhanced in situ removal of Pb from clay by electrokinetics.

B. Silicates and Silica

The electrokinetics of silica and silicate minerals have been studied both from the application (silica is the impurity in practically every flotation system) as well as fundamental understanding points of view. The effects of surfactants, polymers, and electrolyte concentrations on the electrokinetics of silica have been investigated. Generally, silica and silicate minerals show negative zeta potential over a wide range of pH.

1. Chrysocolla

For chrysocolla (a copper silicate mineral), Bowdish and Plouf [133] obtained an extrapolated i.e.p. value at pH 5.5. It was found that at pH $\sim$ 6, incipient leaching of Cu begins, leaving vacant sites on the surface which change the electrical potential of the particles and thus cause an overall increase in the negative charge. The incipient leaching range of pH is nearly identical to the pH range in which chrysocolla can be sulfidized and floated. At pH $\sim$ 6, complete flotation of chrysocolla has been obtained by using K octyl hydroxamate as chelating agent. Similarly, a high peak in the recovery of chrysocolla by ethyl xanthate has been found at a pH near that of the minimum zeta potential.

2. Forces at the Silica–Solution Interface

Scales et al. [134] have studied the silica–solution interface by using a flat plate streaming potential apparatus. The extrapolated i.e.p. was determined to be at pH 2.8 ± 0.2. The authors have analyzed the findings on the basis of the GCSG model for the EDL which could partially account for the experimental results.

Hartley et al. [135] have used atomic force microscopy to investigate the forces between a silica sphere in the colloidal size range and silica/mica flat surfaces as a function of distance of separation. It was interesting to note that, at low ionic strength and under identical conditions, zeta-potential measurements on the spheres and streaming potential measurements on the flat surfaces show excellent agreement with the diffuse double-layer potentials derived from the force data using DLVO theory. No short-range repulsion between the silica surfaces was observed here. However, at higher electrolyte concentration, the electrokinetically derived potentials were found to deviate from those derived from the fitted atomic-force microscopy data. A short-range steric type repulsion was observed between the surfaces, the magnitude of which increased with decreasing pH.

3. Adsorption of Polymers and Surfactants

The adsorption of polymers and surfactants on the silica surface has been studied [136, 137]. In an interesting work, Sidorova et al. have investigated the electrokinetic and adsorption properties of silica in solutions of weak polyelectrolytes, namely, polyethyleneimine, poly(acrylic acid), and an uncharged surfactant $(C_8H_{15}C_6H_4O)$ $(C_2H_4O)_m$ H, OP7 [137]. Electrokinetic measurements were performed on a plane-parallel fused quartz capillary while the adsorption measurements were carried out on Aerosil (specific surface area of $175\,m^2\,g^{-1}$). It was found that the adsorption of OP7 on silica did not change the surface charge and specific surface conductivity at neutral pH. The observed decrease in zeta-potential value in OP7 solution is attributed to the displacement of the slipping plane into the bulk of the solution. However, in the case of polyelectrolyte solution, specific interactions between the polycation and the silica surface were reflected by an increase in negative surface charge, a shifting of the i.e.p. and a decrease in the surface conductance.

4. Effect of Electrolyte Concentration

In an important article, Dunstan [138] has interpreted the electrokinetic potentials and surface charge densities of the silica surface from both electrophoretic mobility and electro-osmosis measurements over a range of KCl concentrations. The interpreted surface charge densities were found to increase, thereby indicating effective adsorption of the negatively charged chloride ions on to the negatively charged surface. The silica surface was, thus, postulated to be nonclassical in nature.

VII. CONCLUSIONS

The electrokinetics of mineral particles is a fascinating field, and has attracted the attention of researchers for quite a long time. The understanding gained from the excellent works of several investigators have led to a better insight into the fundamentals of the interactions in mineral–solution systems. The mineral–aqueous solution system may consist of different kinds of species, both organic and inorganic. Some of the species are contributed by the mineral itself due to its inherent solubility characteristics. The knowledge gained from such studies has also helped in choosing the appropriate operating conditions for flotation of mineral particles. In this chapter, an attempt has been made to acquaint the reader with the progress made in this field, particularly in the last few decades. It is hoped that further understanding of the electrokinetics of mineral particles will help us in designing effective strategies to solve some of the challenging problems in mineral beneficiation.

ACKNOWLEDGMENTS

The author gratefully acknowledges the support and encouragement of Professor Mathai Joseph, Executive Director, Tata Research Development and Design Centre. Sincere thanks are due to Drs EC Subbarao and Pradip for helpful discussions.

REFERENCES

1. TW Healy, MS Moignard. In: MC Fuerstenau, ed. Flotation. vol. 1. New York: SME-AIME, 1976, p 275.
2. PL de Bruyn, GE Agar. In: DW Fuerstenau, ed. Froth Flotation – 50th Anniversary Volume. New York: Rocky Mountain Fund Series AIME, 1962, p 91.
3. RO James, GA Parks. In: P Somasundaran, RB Grieves, eds. Advances in Interfacial Phenomena. AIChE Symp Ser 150(71); 157, 1975.
4. DW Fuerstenau. In: ML Hair, ed. The Chemistry of Biosurfaces, vol. 1. New York: Marcel Dekker, 1971, p 376.
5. DW Fuerstenau, T Wakamatsu. Faraday Disc Chem Soc 59: 157, 1975.
6. K Osseo-Asare, DW Fuerstenau. Croat Chem Act 45: 149, 1973.
7. GD Parfitt. Croat Chem Acta 45: 189, 1973.
8. DJ Shaw. Introduction to Colloid and Surface Chemistry. Oxford: Butterworth, 1970.
9. AW Adamson. Physical Chemistry of Surfaces. New York: Wiley Interscience, 1967.
10. RJ Hunter. Zeta Potential in Colloid Science. London: Academic Press, 1981.
11. PC Hiemenz, R Rajagopalan. Principles of Colloid and Surface Chemistry. 3rd ed. New York: Marcel Dekker, 1997.

12. JTG Overbeek. In: HR Kruyt, ed. Colloid Science. vol. 1. Amsterdam: Elsevier, 1952. Chaps 4 and 5.
13. RO James, TW Healy. J Colloid Interface Sci 40: 42, 53, 65, 1972.
14. RJ Pugh, L Bergstrom. Colloids Surfaces 19: 1, 1986.
15. AV Delgado, F González-Caballero. Croat Chem Acta 71: 1087, 1998.
16. RW O'Brien, DN Ward. J Colloid Interface Sci 121: 402, 1988.
17. BJ Yoon, S Kim. J Colloid Interface Sci 128: 275, 1989.
18. A Ben-Taleb, P Vera, AV Delgado, V Gallardo. Mater Chem Phys 37: 68, 1994.
19. K Kandori, Y Kawashima, T Ishikawa. J Mater Sci Lett 12: 288, 1993.
20. RC Plaza, JDG Durán, A Quirantes, MJ Ariza, AV Delgado. J Colloid Interface Sci 194: 398, 1997.
21. D Kovacevic, N Kallay, I Intol, A Pohlmeier, H Lewandowski, HD Narrers. Colloids Surfaces A 140: 261, 1998.
22. M Colic, DW Fuerstenau, N Kallay, E Matijevic. Colloids Surfaces 59: 169, 1991.
23. MR Houchin, LJ Warren. Colloids Surfaces 16: 117, 1985.
24. H Schubert, H Baldauf, W Raatz. Bergakademie 21: 550, 1969.
25. H Kamaruddin, PR Khangaonkar. EPD Congress 1994, Proceedings of Symposium of TMS Annual Meeting, Warrendale, PA: Minerals Metals & Materials Society, 1994, pp 141–151.
26. F Taha, MB Saleh, KME Attyia, MMR Khalaf. Adsorpt Sci Technol 11: 161, 1994.
27. F Taha, KME Attyia, M Saleh, MMR Khalaf. Adsorpt Sci Technol 12: 7, 1995.
28. JC Woo, KU Whang. Taehan Kwangsan Hakhoe Chi 23: 98, 1986.
29. D Wang, Y Hu. Youse Jinshu 39: 33, 1987.
30. S Itoh, T Okada. Tohoku Kogyo Gijutsu Shikenso Hokoku 11: 29, 1980.
31. TL Gak, AA Baishulakov, KI Omarova. Kompleksn Ispol'z Miner Syr'ya 4: 14, 1982.
32. KJ Young. Kwangsan Hakhoe Chi 7: 60, 1970.
33. J Stachurski, E Oruba. Arch Gorn 25: 377, 1980.
34. Z Sadowski, RW Smith. Miner Metall Process 4: 114, 1987.
35. F Taha, GV Illyuvieva, AA Megahed. Izv Vyssh Uchebn Zaved Tsvetn Metall 1: 8, 1985.
36. R Woods. Chem Austral 57: 392, 1990.
37. KK Das, A Briceno, S Chander. Miner Process Extractive Metall Rev 8: 229, 1992.
38. AN Buckley. Appl Surface Sci 27: 437, 1987.
39. GH Luttrell, RH Yoon. Proceedings of 112th SME-AIME Meeting, Atlanta, GA, 1983, preprint no. 83, p 196.
40. RH Yoon. Int J Miner Process 8: 31, 1981.
41. JR Gardner, R Woods. Int J Miner Process 6: 1, 1979.
42. PJ Guy, WJ Trahar. In: KSE Forssberg, ed. Flotation of Sulfide Minerals, Developments in Mineral Processing. vol. 6. Amsterdam: Elsevier, 1985, pp 91–110.
43. AN Buckley, IC Hamilton, R Woods. In: KSE Forssberg, ed. Flotation of Sulfide Minerals, Developments in Mineral Processing. vol. 6. Amsterdam: Elsevier, 1985, pp 41–60.
44. R Woods. In: MC Fuerstenau, ed. A. M. Gaudin Memorial Volume. vol. 1. New York: SME-AIME, 1976, pp 298–331.
45. GW Heyes, WJ Trahar. Int J Miner Process 4: 317, 1977.
46. TW Healy. Principles of mineral flotation. Proceedings of The Wark Symposium, Australasian Institute of Mining and Metallurgy, Parkville Victoria, 1984, p 43.
47. GH Luttrell, RH Yoon. Colloids Surfaces 12: 239, 1984.
48. MS Moignard, DR Dixon, TW Healy. Proc Aust IMM 263: 31, 1977.
49. RJ Pugh. In: E Forssberg, ed. Proceedings of the International Mineral Processing Congress. Amsterdam: Elsevier, 1988, p 751.
50. RJ Pugh, K Tjus. J Colloids Interface Sci 117: 231, 1987.
51. KK Das, Pradip, KA Natarajan. J Colloid Interface Sci 196: 1, 1997.
52. S Acar, P Somasundaran. Miner Eng 5: 27, 1992.
53. AM Gaudin, SC Sun. Trans AIME 169: 347, 1946.
54. D McGlashan, A Rovig, D Podobnik. Trans AIME 244: 446, 1969.

55. B Yarar, JA Kitchener. Trans IMM 79: 123, 1970.
56. A Yucesoy, B Yarar. Trans IMM 83: 96, 1974.
57. MC Fuerstenau. US Bur Mines Inf Circ 8818: 7, 1978.
58. FF Aplan, G Simkovich, EY Spearin, KC Thompson. US Bur Mines Inf Circ 8818: 25, 1978.
59. EC Plante, KL Sutherland. Trans AIME 183: 160, 1949.
60. D Fornasiero, F Li, J Ralston. J Colloid Interface Sci 164: 345, 1994.
61. S Kelebek, GW Smith. Colloids Surfaces 40: 137, 1989.
62. MY Lee, KU Whang. Taehan Kwangsan Hakhoe Chi 23: 403, 1986.
63. PC Neville, RJ Hunter. Proceedings of the 4th RACI Electrochemistry Conference, Adelaide, 1976.
64. M Espinosa-Jiménez, A Hayas-Barrú, F González-Caballero. Collect Czech Chem Commun 58: 259, 1993.
65. JM Vergouw, A Difeo, Z Xu, JA Finch. Miner Eng 11: 159, 1998.
66. JM Vergouw, A Difeo, Z Xu, JA Finch. Miner Eng 11: 605, 1998.
67. R Williams, ME Labib. J Colloid Interface Sci 106: 251, 1985.
68. NP Finkelstein, SA Allison. In MC Fuerstenau, ed. Flotation – A. M. Gaudin Memorial Volume. New York: SME-AIME, 1976, pp 414–457.
69. RT Hukki, A Palomaki, E Orivuori. Cited in Ref. 6 in TW Healy, MS Moignard. In: MC Fuerstenau, ed. Flotation. vol. 1. New York: SME, 1976, p 275.
70. JS Laskowski, Q Liu, Y Zhan. Miner Eng 10: 787, 1997.
71. MC Fuerstenau, BR Palmer. In MC Fuerstenau, ed. Flotation. vol. 1. New York: SME-AIME, 1976, p 148.
72. DJ Murray, TW Healy, DW Fuerstenau. Advances in Chemistry Series. No. 79, Adsorption from Aqueous Solutions, 1968, p 68.
73. KP Anantpadmanabhan, P Somasundaran. Colloids Surfaces 13: 151, 1985.
74. SRB Cooke. Adv Colloid Sci 3: 357, 1950.
75. B Gigowski, A Vogg, K Wierer, B Dobias. Int J Miner Process 33: 103, 1991.
76. GS Maier, X Qiu, B Dobias. Colloids Surfaces A 122: 207, 1997.
77. KK Das, Pradip. In: YA Attia, BM Moudgil, S Chander, eds. Proceedings, Interfacial Phenomena in Biotechnology and Materials Processing. Amsterdam: Elsevier, 1985, p 305.
78. S Chander. Miner Metall Process 5: 104, 1988.
79. D Salatic, S Pustric, D Djakovic. Proceedings of 11th International Mineral Processing Congress, Cagliari, Italy, 1975.
80. V Mangalam, PR Khangaonkar. Colloids Surfaces 7: 209, 1983.
81. V Mangalam, PR Khangaonkar. Trans Indian Inst Met 40: 49, 1987.
82. V Mangalam, PR Khangaonkar. Int J Miner Process 15: 269, 1985.
83. G Fairthorne, JS Brinen, D Fornasiero, DR Nagaraj, J Ralston. Int Miner Process 54: 147, 1998.
84. J Lekki, J Laskowski. Trans IMM 80: 174, 1971.
85. CA Oestreicher, DW McGlashan. Proceedings of AIME Annual Meeting, San Francisco, 1972.
86. G Bhaskar Raju, W Forsling. Bull Electrochem 8: 402, 1992.
87. MC Fuerstenau, MC Kuhn, D Elgillani. Trans AIME 241: 148, 1968.
88. MC Fuerstenau, D Elgillani. Trans AIME 241: 437, 1968.
89. D Fornasiero, V Eijt, J Ralston. Colloids Surfaces 62: 63, 1992.
90. P Somasundaran. J Colloid Interface Sci 27: 659, 1968.
91. K Hanumantha Rao, BM Antti, E Forssberg. Int J Miner Process 28: 59, 1990.
92. MS Smani, P Blazy, JM Cases. Trans AIME 258: 168, 1975.
93. SK Mishra. Int J Miner Process 5: 69, 1978.
94. MK Yellojirao, P Somasundaran, KM Schilling, B Carson, KP Ananthapadmanabhan. Colloids Surfaces A 79: 293, 1993.
95. MC Fuerstenau, G Gutiérrez, DA Elgillani. Trans AIME 241: 319, 1968.
96. P Somasundaran, GE Agar. J Colloid Interface Sci 24: 433, 1967.

97. N Vdovic, J Biscan. Colloids Surfaces A 137: 7–14, 1998.
98. B Siffert, P Fimbel. Colloids Surfaces 11: 377, 1984.
99. DS Cicerone, AE Regazzoni, MA Blesa. Colloid Interface Sci 154: 423, 1992.
100. PV Smallwood. Colloid Polym Sci 255: 881, 1977.
101. K Hanumantha Rao, BM Antti, E Forssberg. Colloids Surfaces 34: 227, 1989.
102. P Somasundaran. J Colloid Interface Sci 31: 557, 1969.
103. JB Andersen, SE El-Mty, P Somasundaran. Colloids Surfaces 55: 365, 1991.
104. O Ozcan, AN Bulutcu. Int J Miner Process 39: 275, 1993.
105. I Yaniv, R Dimitrova, O Ciobanu. Rev Chem Eng 9: 283, 1993.
106. KI Marinakis, HL Shergold. Int J Miner Process 14: 161, 1985.
107. RJ Pugh. Colloids Surfaces 18: 19, 1986.
108. JD Miller, JB Hiskey. J Colloid Interface Sci 41: 567, 1972.
109. MC Fuerstenau, DA Elgillani, G Gutiérrez. Trans AIME 241: 391, 1968.
110. RJ Pugh, P Stenius. Int Miner Process 15: 193, 1985.
111. F González-Caballero, G Pardo, JM Bruque. An Quim 72: 345, 1976.
112. J Morales, F González-Caballero, G Pardo Sánchez, R Perea. Afinidad 39: 24, 1982.
113. K Hanumantha Rao, JM Cases, P De Donato, KSE Forssberg. J Colloid Interface Sci 145: 314, 1991.
114. Y Hu, J Xu, G Qiu, D Wang. J Cent South Univ Technol (Engl Ed) 1: 63, 1994.
115. V Mangalam, V Mohan, PR Khangaonkar. Bull Electrochem 2: 605, 1986.
116. D Salatic, M Milic. Freiberg Forschungsh A 593: 205, 1978.
117. N Street, AS Buchanan. Aust J Chem 9: 450, 1956.
118. I Iwasaki, SRB Cooke, DH Harroway, HS Choi. Trans AIME 233: 97, 1962.
119. RW Smith, Y Narimatsu. Miner Eng 6: 753, 1993.
120. JM Cases, C Touret-Poinsignon, D Vestier. CR Acad Sci Ser C 272: 728, 1971.
121. DJA Williams, KP Williams. J Colloid Interface Sci 65: 79, 1978.
122. J Yuan, RJ Pruett. Miner Metall Process 15: 50, 1998.
123. SK Nicol, RJ Hunter. Aust J Chem 23: 2177, 1970.
124. SB Johnson, DR Dixon, PJ Scales. Colloids Surfaces A 146: 281, 1999.
125. W Schempp, A Gokalp, T Kashmoula, J Schurz. Papier (Darmstadt) 26: 558, 1972.
126. H Hoppe, W Troeger, F Winkler, E Worch. Z Chem 15: 412, 1975.
127. I Petkanchin, TS Radeva, R Varoqui. Izv Khim 24: 269, 1991.
128. GFR Gerischer, RD Sanderson. Pap Puu 63: 477, 483, 497, 1981.
129. GFR Gerischer, RD Sanderson. Pap Puu 63: 561, 566, 1981.
130. K Nakaishi, Y Kuroda. Nendo Kagaku 35: 56, 1995.
131. I-M Lu, S-C Yen, TW Chapman. In: RG Bautista, ed. Emerging Sep. Technol. Met. II, Proc. Symp. Warrendale, PA: Minerals, Metals & Materials Society, 1996, pp 349–361.
132. TF Coletta, CJ Bruell, DK Ryan, HI Inyang. J Environ Eng (Reston, Va) 123: 1227, 1997.
133. FW Bowdish, TM Plouf. SME-AIME 254: 66, 1973.
134. PJ Scales, F Grieser, TW Healy, LR White, DYC Chan. Langmuir 8: 965, 1992.
135. PG Hartley, I Larson, PJ Scales. Langmuir 13: 2207, 1997.
136. ML Malysheva, OD Rusina, LN Momot, TB Zheltonozhskaya, BV Eremenko. Kolloidn Zh 56: 544, 1994.
137. M Sidorova, T Golub, K Musabekov. Adv Colloid Interface Sci 43: 1993.
138. DE Dunstan. J Chem Soc, Faraday Trans 90: 1261, 1994.

29

Electrokinetics of Gas Bubbles

ALAIN GRACIAA, PATRICE CREUX, and JEAN LACHAISE   University of Pau, Pau, France

I. INTRODUCTION

The surface charge residing at a gas–liquid surface is a factor that acts upon a number of practical issues such as froth flotation, for example, a process which is commonly used in domains as different as mineral recovery, the paper-making industry, the farm-product industry, wastewater treatment, etc. In this separation process, air bubbles collect hydrophobic particles selectively and push them up to the surface of the pulp, leaving hydrophilic particles behind in the pulp. It is well known that among the many factors which affect the process, electrostatic charges of both the particles and the bubbles play an important role [1, 2].

Although there is now abundant information on the charges of particles, until recent years there has been a general lack of information on bubble charges. This lack was due to the difficulty in measuring the electrophoretic mobility of bubbles, which is often much lower than their rising mobility in the gravity field. However, for about 20 years, new methods have been developed to remove this difficulty. This chapter presents a review of these methods, discusses their advantages and inconveniences, and tries to give a brief synthesis of the results obtained through their implementation.

II. EXPERIMENTAL METHODS

The difficulties encountered in the measurement of the electrokinetic behavior of a gas bubble are numerous. Most of them come from the natural adsorption of foreign molecules at the gas/liquid surface and from the high rising velocity of the bubble in the gravity field.

It is known that the surface of a bubble can easily adsorb alcohol molecules, surfactant molecules, ions, and more generally all sorts of impurities or particles present in the immersing liquid. This adsorption can drastically modify the electromobility of the bubbles. So it is very important to purify carefully the liquid before introducing the bubble, to obtain valuable information on its intrinsic electric properties.

These surface-active agents are often present in very low concentrations in the liquid. Their presence at the surface cannot be detected by any diversion of the interfacial tension. So as a precaution, it is always useful to drain the liquid with masses of oxygen and hydrogen microbubbles generated directly in the liquid by electrolysis [3, 4]. This drainage collects the undesirable molecules at the surface of the liquid; it is then easy to remove them from the measurement cell. When coarser impurities are present as, for example, metal hydroxide precipitates, their elimination by ultracentrifugation is also possible [5].

Sometimes the presence of surfactants at the bubble's surface is highly desirable. As these molecules come by diffusion from the immersing liquid solution, their adsorption can require minutes or even hours. In order to reach equilibrium the bubble must be very stable, which is not easy to achieve.

Finally, since the immersing liquid is most often an aqueous solution, attention must also be paid to water, which could dissolve CO_2 from the atmosphere. As it is known that, at saturation, the pH can reach 5.6, the presence of a significant amount of carbonates and bicarbonates could affect the surface potential of bubbles. To avoid such problems, it is necessary to work under a neutral atmosphere or, at the very least, systematically to drain the water with nitrogen bubbles.

Bubbles can be produced individually or collectively. They are produced individually by means of a syringe, or by electrolysis followed by isolation from the other bubbles generated in the process, before introducing the selected bubble into the measurement cell. Bubbles can be produced collectively by circulating a gas through a porous glass frit or by depressurizing a liquid saturated in dissolved gas.

Bubbles generated by syringes have diameters of the order of 1 mm. Bubbles generated by the other processes are smaller; their diameters range from some micrometers to about 100 µm.

The high rising velocity of the bubbles is due to the large difference between gas and water densities, and also to the low viscosity of the immersing liquid. It is often much higher than the electromobility of the bubble. Consequently, it is impossible to measure the electromobility of a bubble with the ordinary electrophoretic method. Tricks have been used to overcome this major difficulty and they are at the basis of the differences between four principal methods which have been developed.

The first method cancels the rising movement of the bubble by trapping it in a spinning tube; it is the spinning-bubble method (Fig. 1a). On the other hand, the second method uses exclusively the rising movement of the bubbles, which creates an electric current and consequently a potential difference between two points; it is the rising-potential method (Fig. 1b). The third and fourth methods keep the rising movement of the bubbles, but they try to reduce it by using very small bubbles. They exploit the additional electrophoretic movement created, horizontally or vertically, by means of, respectively, an horizontal or a vertical electric field; they are the horizontal electrophoresis (Fig. 1c) and the vertical electrophoresis methods (Fig. 1d).

A. The Spinning Bubble

There have been numerous attempts to measure electromobility or ζ potential in the past. McTaggart [6–8] and Alty [9–11] trapped bubbles in water in the center of horizontal cylindrical tubes, closed at both ends with metal disks acting as electrodes. By rotating the tube, the bubble could be held on the axis, and the velocity could be measured under an applied potential difference. This method was strongly criticized by Bach and Gilman [12, 13] because it failed to take into account the electro-osmotic flow of the water in the closed tube. Bach and Gilan built a cell in which bubbles could be generated by electrolysis. The cell was somewhat irregular in shape and the electrodes were rather obtrusive. They went to some lengths to measure a "cell constant" to correct for electro-osmotic streaming. Nevertheless, the reliability of the results has been questioned by Samygin et al. [14].

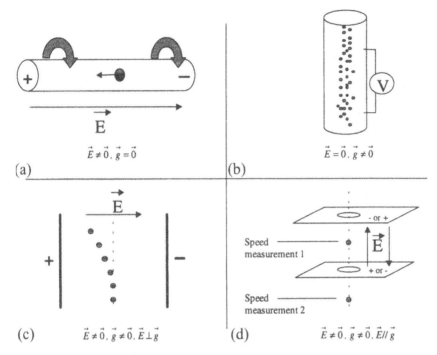

FIG. 1 Methods for measuring the ζ potential of gas bubbles: (a) the spinning bubble; (b) the rising potential; (c) horizontal electrophoresis; (d) vertical electrophoresis.

Cichos [15], Huddleston [16], Huddleston and Smith [17], and Sakai [18] employed the McTaggart technique but did not address the contribution of electro-osmosis, although they were aware of the difficulty. McShea and Callaghan neglected the phenomenon [19].

In closed systems electro-osmosis creates a flux of the liquid along the naturally charged surfaces of the tube and a reflux of the liquid towards the center. Thus, the electrophoretic mobility of the bubble is disrupted by the electro-osmotic movement of the liquid. The true electrophoretic mobility would be only obtained in "stationary levels" where the electro-osmotic flux is just opposed to the electro-osmotic reflux [20–24]. However, these stationary levels are difficult to locate with precision.

In open systems the electro-osmotic contribution to bubble mobility can be evaluated in an outer circuit, and then it would be possible to go back to true mobility [25], but the evaluation of the electro-osmotic contribution is not easy.

Another solution has been proposed more recently [26]. It consists in simultaneously neutralizing the surface charge and creating a highly viscous zone that impedes motion in the region near the surface. This is obtained by coating a thin layer of diethylaminoethyl-dextran, which is further cross-linked through reaction with 1,4-butanediglycidyl ether. The disappearance of the electro-osmotic flow is noted by observing a linear velocity profile for latex microspheres.

The electrophoretic mobility of a gas bubble in a rotating fluid has been expressed relative to the bubble ζ potential by Sherwood [27]. The required conditions for using this expression are:

- A perfectly horizontal rotating tube.
- A laminar fluid flow.
- A sufficiently high rotating speed (generally higher than 90 rad/s) for maintaining the bubble on the axis of the tube.
- A tube very much longer than the Taylor columns which appear both ahead of and behind the moving bubble [28].
- A bubble radius lower than 0.4 times the inner radius of the tube, to avoid disruptions from the Stewartson and Ekman layers [29, 30].

This method will be particularly useful for studying the kinetics of the adsorption of surfactant molecules at the air–water surface or at the oil–water interface [31], research which has undergone full development [32].

B. The Rising Potential

The rising bubbles create a streaming current in the immersing liquid, which was first analyzed theoretically by Derjaguin and Dukhin [33–35]. Since then, devices have been developed to measure the intensity of this current so as to obtain the ζ potential of the bubbles [36–40]. In these apparatuses, a high-precision variable resistance is placed in series with the two measurement electrodes; if the value of this resistance is negligible in comparison with the resistance of the solution between the electrodes, mainly all the charge displaced by the bubbles is returned to the lower electrode through the external circuit. The streaming current may then be calculated from the voltage developed across the known resistance.

As no electric field is applied, there is no electro-osmosis. Furthermore, as the length of the vertical tube in which the bubbles rise can be of the order of 50 cm, or even higher, the duration of the ascent of the bubbles is long enough to provide some information on the beginning of the adsorption of surfactants. Yet, the bubble streaming current can produce disruptive convection currents in larger tubes [36].

C. Horizontal Electrophoresis

The configuration of the devices based on this method is the most common [20–24, 40, 41]. The apparatus is generally composed of a parallelepipedic microelectrophoresis cell to avoid the possibility of convection currents observed in larger cells [42]. As soon as an electric potential is applied between electrodes located on two vertical and parallel walls of the cell, electro-osmotic flows are generated on the other walls. In order to avoid the influence of these disruptive flows the movement of the bubble must take place at a stationary level, which, here, is a vertical plane.

The bubble is directly formed at this level, following a burst of current which passes through a solution saturated in gas via two platinum wires judiciously placed in the upper and lower walls of the cell.

A microscope system is focused on the stationary level. As a bubble rises in the cell, its horizontal electrophoretic velocity is very much lower than its rising velocity. So it is necessary to move the cell vertically relative to the microscope objective in order to keep the bubble in the field of vision during a time sufficient to perform the measurement of the electrophoretic velocity. This is a difficult operation.

D. Vertical Electrophoresis

This is a recent method, which derives from the development of laser techniques [3, 4]. It consists in the use of a double-laser Doppler electrophoresis apparatus to determine bubble electrophoretic mobilities, by measurement of the difference in the bubble rise rates, with and without an electric field. This one is applied parallel to the bubble rise vector, causing either a decrease or an increase in the natural rise rates of the bubbles according to its direction.

A miniature camera fitted with a microscope lens is used to measure bubble diameters. The cylindrical cell has a diameter about 100 times that of the bubble diameter, so that effects of electro-osmotically induced flow at the cell walls on bubble velocities at the cell axis could be ignored.

This device is able to measure electrophoretic mobilities with a high precision. However, as the duration of a measurement is extremely short, it cannot be used for studying the kinetics of adsorption at the bubble's surface.

III. EXPERIMENTAL RESULTS

Each of the methods the principle of which has been described briefly in the preceding paragraphs presents advantages and disadvantages. They have been used by numerous researchers under various conditions. As for the methods, our presentation of the results will not be exhaustive. We shall report only trends and convergences, first on systems without surfactant, then on systems with surfactant(s).

In the two cases, bubbles were generated from different gases. The gases used were air [1, 19, 22, 24, 26, 31, 43, 44], oxygen [20, 21], hydrogen [3, 4, 38, 40], nitrogen [2, 23, 36, 41, 45–47], carbon dioxide [38], and chlorine [40]. In general, whatever the nature of the gas, the observed behaviors are similar, with perhaps an exception for gases which, like carbon dioxide, can dissolve easily in water [38].

A. Bubbles Without Surfactant

1. Bubbles in Pure Water

The earliest investigators found that air bubbles immersed in pure water presented a negative electrophoretic mobility [48]. Afterwards, this result was confirmed whatever the method used [4, 19, 26, 37, 49]. However, the evaluated ζ potentials are dispersed from some negative millivolt value [4] to $-100\,\text{mV}$ [37]. These variations could be attributed to the disparities of the methods of purification and to the difficulties of the measurements. In spite of these differences, the authors are in agreement on the origin of this electronegativity. It could be assigned to a preferential adsorption of OH^- ions, probably under the influence of an orientation of the water dipoles near the interface with their positive poles directed towards the solution [4].

The pH of water can be varied by adding hydrogen chloride or sodium hydroxide. McShea and Callaghan [19] have measured the electrophoretic mobility of air bubbles versus pH, while Kubota and Jameson [45] have performed the same study on nitrogen bubbles. They found that the magnitude of the electrophoretic mobilities of their bubbles decreased at the same time that the pH decreased and became zero for pH values close to 3 (Fig. 2). At very low pH, Kubota and Jameson [45] have claimed that the bubble could be positive charged, probably because of a preferential accumulation of H^+.

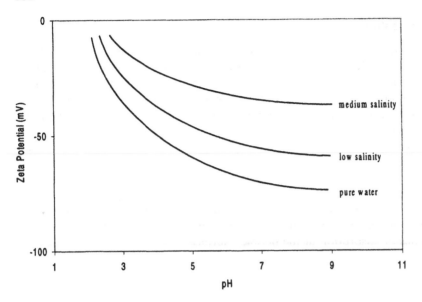

FIG. 2   Zeta potential of gas bubbles versus pH or salinity.

2.   Bubbles in Solutions of Electrolytes

Generally, whatever the nature of the gas or the type of salt, the magnitudes of electrophoretic mobilities are monotonically depressed by increasing the concentration of the mono- or di-valent salts, without reversing the sign of the mobilities: hydrogen bubbles in Na_2SO_4, $NaClO_4$, and $NaNO_3$ [40], and nitrogen bubbles in NaCl and $MgSO_4$ [45]. This behavior is attributed to the screening of the interactions between the applied electric field and the charged bubbles, induced by the salts used.

Studies on the influence of pH have been performed for bubbles of different gases immersed in various saline solutions: nitrogen bubbles in NaCl [17, 41], hydrogen, oxygen, or chlorine bubbles in Na_2SO_4 and KCl [40], and oxygen bubbles in $NaClO_4$ [4]. Whatever the gas or salt, the observed trends are the same (Fig. 2). The ζ potential (or electrophoretic mobility) magnitude decreases as the acidity of the immersing solutions increases. This decrease is more pronounced the lower the salinity. For all the authors, the isoelectric point ranges (or would range by extrapolation) from pH 1.5 to 3.

For a monovalent or divalent salts, the influence of the counterions has also been examined. Thus, Sakai [18] has found that the electrophoretic mobility of nitrogen bubbles decreases with increasing hydrated anionic radius in the order:

$$Br^-(3.30\,\text{Å}) > NO_3^-(3.35\,\text{Å}) > Cl^-(3.32\,\text{Å}) > ClO_4^-(3.38\,\text{Å}) > SO_4^{2-}(3.7\,\text{Å})$$

for cation Mg^{2+} (3.74 Å) in 4.92 mM aqueous MgX salt solutions. This classifying is in agreement with the one which had been previously found by Kelsall et al. [4] for oxygen bubbles and NO_3^-, ClO_4^-, and SO_4^{2-} ions in 10^{-4} M aqueous NaX salt solutions.

Sakai [18] has also found that the electrophoretic mobility increases with increasing hydrated cationic radius in the order:

$$Li^+(3.82\,\text{Å}) > Na^+(3.58\,\text{Å}) > K^+(3.31\,\text{Å}) > Rb^+(3.29\,\text{Å}) > Cs^+(3.29\,\text{Å})$$

for the anion Cl^- (3.32 Å) in 4.92 mM aqueous XCl salt solution.

These behaviors seem to show that, at the air/water surface, ionic adsorption would be dependent on the size of the ions.

In contrast to mono- or di-valent salts, the trivalent salts can reverse the sign of the mobility at a certain concentration: nitrogen bubbles in $AlCl_3$ [45] or in $Al_2(SO4)_3$ [6, 24, 43, 45]. No explanation is provided by the investigators who reported these results. However, it is likely that the three charges of the aluminum ion must play a prominent part in this explanation.

Li and Somasundaran have also observed a charge reversal relative to pH for nitrogen bubbles in 10^{-3} and 10^{-2} M $MgCl_2$ aqueous solutions [41], or in 5×10^{-6} M, 10^{-4} M, and 10^{-3} M $AlCl_3$ aqueous solutions [2], using 10^{-2} M NaCl as background electrolyte. In the first case, the observed charge reversal in the pH range 9–11 has been attributed to the precipitation of $Mg(OH)_{2(s)}$ at the gas–liquid interface, while the reduction in magnitude of the ζ potential of the bubble observed in the acidic pH range has been considered to be due to the specific adsorption of Mg^{2+}. In the second case, the reversal of bubble charge has been attributed to specific adsorption of Al^{3+} and its hydroxo complexes in the low pH range and to precipitation of aluminum hydroxide in the intermediate pH range.

B. Bubbles with Surfactant(s)

Surfactants are molecules characterized by the presence of hydrophilic and lipophilic moieties. As a consequence, they readily adsorb at surfaces or interfaces, and they form micelles above a critical concentration, etc. At the gas/liquid surface, the hydrophilic groups of the surfactant molecules are directed towards the liquid, and the lipophilic groups towards the gas. The surface activity of the surfactants causes a decrease in the surface tension relative to the concentration until a concentration, called the critical micelle concentration (c.m.c.), is reached. Above this concentration, the surface tension remains constant, and the surfactant molecules form micelles, which are multimolecular aggregates in kinetic equilibrium with monomer molecules.

Adsorption of surfactants at the gas/liquid surface can change its electric charge, especially if they are ionic species. So the study of the influence of surfactants on the electrical behavior of bubbles is of the greatest importance.

We consider blow the influence of nonionic surfactants, anionic surfactants, cationic surfactants, and mixtures of surfactants. In most of the works reported, the surfactant concentration ranges only from very low values to values close to the c.m.c., in order to avoid the presence of micelles prejudicial to electrokinetic measurements.

1. Nonionic Surfactants

The variation in the ζ potential of bubbles immersed in aqueous solutions of a nonionic surfactant according to the surfactant concentration depends on pH and also, but much less, on the nature of the surfactant. Thus, at neutral pH, in polyoxyethylene dodecyl ether (6.5 EO) solutions, Usui and Sasaki [37] have found that argon bubbles present a ζ potential nearly constant over the concentration range 10^{-5}–5×10^{-4} M of the surfactant. This tendency has also been observed by Okada et al. [1] for air bubbles in polyoxyethylene dodecyl ether (23 EO) solutions at acid pH over the concentration range 10^{-5}–10^{-2} M, and by Yoon and Yordan [23] for microbubbles in polyoxyethylene methyl ether solutions at neutral and acid pH over the concentration range 10^{-4}–10^{-2} M. At alkaline pH, the latter authors have found that negative ζ potential values increase in magnitude with surfactant concentration. A trend which has been observed also by

Saulnier et al. [31] at neutral pH for air bubbles in polyoxyethylene octylphenol (10 EO) solutions over the concentration range 10^{-6}–5×10^{-3} M. This increase in the electronegativity of bubbles with surfactant concentration could be attributed to the low cationic character of the surfactant used [50]. Furthermore, when this increase is important enough, a saturation seems to be observed close to the c.m.c. [31].

Yoon and Yordan [23] and Okada et al. [1] have shown a strong dependence of the ζ potential on pH. Clearly negative at alkaline pH, ζ tends to become positive in a pH zone between 2 and 3.

2. Anionic Surfactants

The most studied anionic surfactant has been sodium dodecylsulfate [1, 19, 22, 23, 40]. The others are sodium hexadecylsulfate [37], sodium dodecylbenzenesulfonate [22], and sodium hexadecylbenzene sulfonate [31, 44]. The trends observed with nonionic surfactants are found again, but more pronounced. Increase in surfactant concentration raises the electronegativity of the bubbles whatever their nature. At saturation point, which can be observed near to the c.m.c. ζ potential values can decrease at values more negative than $-100\,\text{mV}$ [22, 31, 37].

Kubota et al. [22] have compared their ζ potential values obtained with sodium dodecylsulfate with the ones obtained by Usui and Sasaki [37] with sodium hexadecylsulfate. They proposed that the observed discrepancies would be the result of differences in the adsorptions of the surfactant molecules. It would be interesting to test this idea by varying systematically the number of methylene groups per hydrocarbon chain in the molecules, as has been performed for c.m.c. and solubilities by Graciaa et al. [51].

The variation in ζ potential with pH has been studied for sodium dodecylsulfate and air bubbles by Yoon and Yordan [23] and Okada et al. [1]. They agree to say that the pH decreases the magnitude of the ζ potential in the highly acidic range, the isoelectric point being reached below pH 2.

3. Cationic Surfactants

The main cationic surfactants studied have been alkyltrimethylammonium bromides: $n = 10$ [45], $n = 12$ [40, 45], $n = 14$ [31, 45], and $n = 16$ [20, 21, 24, 37, 45]. Dodecyltrimethylammonium chloride [19], dodecylamine hydrochloride [1, 23, 45], and hexadecylpyridinium chloride [22] have been also examined.

The most important trend observed in these systems is that the electrophoretic mobility and the ζ potential become positive for a concentration or a pH which depend on the nature of the surfactant. Thus, at pH 5.6, Kubota and Jameson [45] find that the zero electrophoretic mobility is obtained for approximately 10^{-3} M for decyltrimethylammonium bromide and for approximately 10^{-6} M for hexadecyltrimethylammonium bromide. Yoon and Yordan [23] find that for hexadecylcetylpyridinium chloride the ζ potential is always positive whatever the pH, while for dedecylamine hydrochloride, the zero ζ potential is obtained in alkaline pH (pH 10 for a 2.5×10^{-4} M surfactant solution, a result which was confirmed by Okada et al. [1], pH 11 for 1.0×10^{-5} M).

At constant pH and constant surfactant concentration, the zero ζ potential can be obtained also by adding electrolytes to the immersing liquid. This effect was observed by Collins et al. [20], Fukui and Yuu [21], and Okada and Akagi [24] on oxygen bubbles in a flotation solution composed of hexadecyltrimethylammonium bromide at a concentration of 5×10^{-5} M and ethanol as a frother at 0.5% (v/v). Their corroborating measurements show that it would be reached for a concentration of the order of 1 M in Na_2SO_4.

The main trends which should be kept in mind from this brief literature survey are schematized in Fig. 3.

With a neutral pH, the ζ potential of gas bubbles in pure water is negative. When surfactant is added two distinct behaviors are observed, according to the surfactant concentration: anionic surfactants and nonionic surfactants increase the magnitude of the negative ζ potential (anionics being much more effective than nonionics), while cationic surfactants can change the sign of the ζ potential (Fig. 3a). For the three types of surfactants, saturation with the variation in ζ potential seems to be observed for concentrations close to the c.m.c.

The ζ potential of bubbles is very sensitive to the pH of the surfactant solutions. The two distinct behaviors of anionic surfactants and nonionic surfactants on the one hand and of cationic surfactant on the other are conserved (Fig. 3b). The pH can give a zero ζ potential with nonionic surfactants and anionic surfactants in the highly acidic range (a little more acidic for anionic than for nonionic); it gives zero ζ potential for cationic surfactants in a relatively alkaline range.

4. Mixtures of Surfactants

The case of mixtures of surfactants was first tackled by Usui and Sasaki [37] for ionic/nonionic surfactant mixtures; the ionic surfactant was hexadecyl sulfate (anionic) or hexadecyltrimethylammonium bromide (cationic), and the nonionic surfactant was polyoxyethylene dodecyl ether (6.5 EO). A more extensive study has been performed by Graciaa et al. [52] for anionic/nonionic surfactant mixtures, anionic/cationic surfactant mixtures, and fluorocarbon/hydrocarbon surfactant mixtures. In this study, the nonionic species was hexaethylene glycol dodecanol, the cationic surfactant was decyltrimethylammonium chloride, and the anionic surfactants were hexadecylbenzene sodium sulfonate, sodium decyl sulfonate, and a perfluorononylbenzene sodium sulfonate.

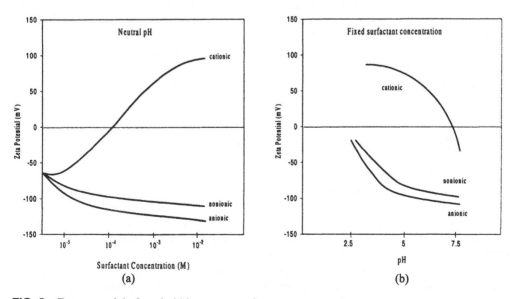

FIG. 3 Zeta potential of gas bubbles versus surfactant concentration at neutral pH or versus pH at fixed surfactant concentration.

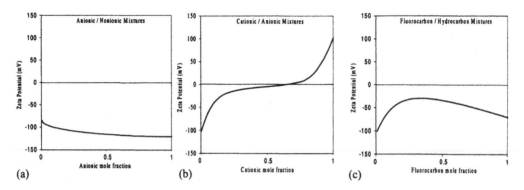

FIG. 4 Zeta potential of gas bubbles versus the composition of mixtures of surfactants.

The main results of these studies are schematized in Fig. 4, where for each type of mixtures, the variations in the ζ potential are reported according to the mole fraction of one of the two surfactants.

For the anionic/nonionic surfactant mixtures, the ζ potential decreases monotically from its value for the nonionic surfactant to its value for the anionic surfactant when the anionic mole fraction varies from 0 to 1 (Fig. 4a). This variation is not linear, which indicates that the two surfactants do not behave ideally at the bubble's surface, but deviation from ideality is low.

For the anionic/cationic surfactant mixtures, the ζ potential is close to zero, though not exactly zero, over the wide midrange composition where the c.m.c. corresponds to the catanionic species [52]. It is also worth noting that the isoelectric point does not take place exactly at a 50% molar proportion, but at a higher cationic amount, e.g., 60%, in accordance with the value reported in research on detergency [53].

For the fluorocarbon/hydrocarbon surfactant mixtures, the ζ potential undergoes a maximum, which is a minimum in absolute value, that indicates a maximum of incompatibility located near the equimolar mixture at which the electrostatic repulsion is enhanced by the lipophobic character of the fluorocarbon hydrophobic group. Since the ζ potential in the equimolar case is almost half the value attained with the pure surfactants, it may be conjectured that the adsorption density is considerably reduced by the extra repulsion provided by the lipophobic effect of the fluorocarbon group.

IV. CONCLUSIONS

The trends reported in the variations in the ζ potentials of gas bubbles are coherent whatever the parameter considered: pH, salinity, surfactant, etc. However, discrepancies remain in the absolute values. They range from the variety of the devices used for circumventing the problem posed by the natural rising velocity of the bubble in the gravity field, to the difficulties of the tuning of these apparatuses, and sometimes to the use of different relationships to work out ζ potentials.

Most of these problems would vanish if experiments could be performed in weightlessness within an orbital station. In the absence of any external electric field, bubbles would then be absolutely motionless and fine electrophoretic studies could be performed outside any disturbance by submitting them to an induced electric field.

However, while waiting for this ideal experiment, the reported results on gas/water surfaces and similar results observed on oil/water interfaces [54] are sufficiently convincing for initiating a theoretical research which promises to have an exciting future [55].

REFERENCES

1. K Okada, Y Akagi, M Kogure, N Yoshioka. Can J Chem Eng 68:393, 1990.
2. C Li, P Somasundaran. J Colloid Interface Sci 148:587, 1992.
3. GH Kelsall, S Tang, S Yurkadul, AL Smith. J Chem Soc, Faraday Trans 92:3879, 1996.
4. GH Kelsall, S Tang, S Yurkadul, AL Smith. J Chem Soc, Faraday Trans 92:3887, 1996.
5. C Li, P Somasundaran. J Colloid Interface Sci 146:215, 1991.
6. HA McTaggart. Phil Mag 27:297, 1914.
7. HA McTaggart. Phil Mag 28:367, 1914.
8. HA McTaggart. Phil Mag 44:386, 1922.
9. T Alty. Proc Roy Soc A 106:315, 1924.
10. T Alty. Proc Roy Soc A 110:178, 1926.
11. T Alty. Proc Roy Soc A 112:235, 1926.
12. N Bach, A Gilman. Acta Physicochim URSS 9:1, 1938.
13. A Gilman, N Bach. Acta Physicochim URSS 9:27, 1938.
14. VD Samygin, BV Derjaguin, SS Dukhin. Kolloidn Zh 26:424, 1964.
15. C Cichos. Freiberg Forschungsh A 513:7, 1973.
16. RW Huddleston. An electrokinetic study of the gas–aqueous solution interface. PhD dissertation, Unilever Research/Liverpool Polytechnic CNAA, 1974.
17. RW Huddleston, AL Smith. In: RJ Akers, etc. Foams. London: Academic Press, 1976, pp 163–177.
18. M Sakai. Prog Colloid Polym Sci 77:136, 1988.
19. JA McShea, IC Callaghan. Colloid Polym Sci 261:757, 1983.
20. GL Collins, M Motarjemi, GJ Jameson. J Colloid Interface Sci 63:69, 1978.
21. Y Fukui, S Yuu. AIChE J 28:866, 1982.
22. K Kubota, S Hayashi, M Inaoka. J Colloid Interface Sci 95:362, 1983.
23. RH Yoon, JL Yordan. J Colloid Interface Sci 113:430, 1986.
24. K Okada, Y Akagi. J Chem Eng Jpn 20:11, 1987.
25. RJ Hunter. Zeta Potential in Colloid Science. New York: Academic Press, 1981, pp 125–178.
26. A Graciaa, G Morel, P Saulnier, J Lachaise, RS Schechter. J Colloid Interface Sci 172:131, 1995.
27. JD Sherwood. J Fluid Mech 162:129, 1986.
28. GI Taylor. Proc Roy Soc London A 102:180, 1922.
29. DW Moore, PG Saffman. J Fluid Mech 31:365, 1968.
30. LM Hocking, DW Moore, IC Walton. J Fluid Mech 90:781, 1979.
31. P Saulnier, J Lachaise, G Morel, A Graciaa. J Colloid Interface Sci 182:395, 1996.
32. SS Dukhin, G Kretzschmar, R Miller. In: D Mobius, R Miller, eds. Dynamics of Adsorption at Liquid Interfaces. Amsterdam: Elsevier, 1995, pp 100–136.
33. BV Derjaguin, SS Dukhin. Proceedings of the 3rd International Congress on Surface Activity, Universitatsdruckerei Mainz GmbH, Cologne, 1960, p 324.
34. SS Dukhin. In BV Derjaguin, ed. *Research in Surface Forces*, vol. 2. New York: Consultants Bureau, 1966, p 54.
35. SS Dukhin. Kolloid Zh 45:22, 1983.
36. HP Dibbs, LL Sirois, R Bredin. Can Metal Quart 13:395, 1974.
37. S Usui, H Sasaki. J Colloid Interface Sci 65:36, 1978.
38. LA Kuznetsova, NY Kovarskii. Colloid J 57:657, 1995.

39. M Ozaki, H Sasaki. In: H Ohshima, K Furusawa, eds. Electrical Phenomena at Interfaces. New York: Marcel Dekker, 1998, pp 245–252.
40. NP Brandon, GH Kelsall, S Levine, AL Smith. J Appl Electrochem 15:485, 1985.
41. C Li, P Somasundaran. J Colloid Interface Sci 146:215, 1991.
42. LL Sirois, G Millar. Can Metal Quart 12:281, 1973.
43. K Okada, Y Akagi, N Yoshioka. Can J Chem Eng 66:276, 1988.
44. P Saulnier, J Lachaise, G Morel, A Graciaa. Colloid Polym Sci 273:1060, 1995.
45. K Kubota, GJ Jameson. J Chem Eng Jpn 26:7, 1993.
46. GL Collins, GJ Jameson. Chem Eng Sci 32:239, 1977.
47. K Kubota, T Harima, S Hayashi. Can J Chem Eng 68:608, 1990.
48. P Currie, T Alty. Proc Roy Soc A 122:622, 1928.
49. JJ Bikerman. Trans Faraday Soc 34:1268, 1938.
50. MJ Rosen, ZH Zhu. J Colloid Interface Sci 33:473, 1989.
51. A Graciaa, Y Barakat, M El-Emary, L Fortney, RS Schechter, S Yiv, WH Wade. J Colloid Interface Sci 89:209, 1982.
52. A Graciaa, P Creux, J Lachaise, JL Salager. J Ind Eng Chem Res 39: 2677, 2000.
53. DN Rubingh. In: DN Rubingh, PM Holland, eds. Cationic Surfactants: Physical Chemistry. New York: Marcel Dekker, 1990, pp 469–507.
54. KG Marinova, RG Alargova, ND Denkov, OD Velev, DN Petsev, IB Ivanov, RP Norwankar. Langmuir 12:2045, 1996.
55. V Bergeron. Curr Opin Colloid Interface Sci 4:249, 1999.

30

Electroviscoelasticity of Liquid/Liquid Interfaces

ALEKSANDAR M. SPASIC Institute for Technology of Nuclear and Other Mineral Raw Materials, Belgrade, Yugoslavia

I. INTRODUCTION

A. Brief Review of Various Approaches to Liquid/Liquid Interfaces in Emulsions

Following a classical deterministic approach, the phases that constitute a multiphase dispersed system are assumed to be a continuum, i.e., without discontinuities inside the entire phase, which is considered homogeneous and isotropic [1–5]. Therefore, the basic laws, e.g., conservation of mass, first and second Cauchy laws of motion, and first and second laws of thermodynamics, are applicable. Also, such concepts as heat, mass, and momentum transfer phenomena are of common use for the description of the related events. It is hydrodynamic, electrodynamic, and thermodynamic instabilities that occur at the interface, and the rheological properties of the interfacial layers that are responsible for the existence of droplets or droplet-film structures in fine dispersed systems [6, 7].

According to the classical approach, the behavior of liquid–liquid interfaces in fine dispersed systems is based on an interrelation between three forms of "instabilities." These are sedimentation, flocculation, and coalescence. These events, represented schematically in Fig. 1, can be understood as a kind of interaction between the liquid phases involved [1, 2, 5–25]. Furthermore, the forces responsible for sedimentation and flocculation are gravity and van der Waals' forces of attraction, respectively, but the forces responsible for coalescence are not well known [2, 25], although some suggestions have been made recently [6, 7].

A new, deterministic approach discusses the behavior of liquid–liquid interfaces in fine dispersed systems as an interrelation between three other forms of "instabilities". These are rigid, elastic, and plastic [6–11]. Figure 2 shows the events that are understood as interactions between the internal and external periodical physical fields. Since both electric/electromagnetic and mechanical physical fields are present in a droplet, they are considered as internal, whereas ultrasonic, temperature, or any other applied periodical physical fields are considered as external. Hereafter, the rigid form of instability comprises the possibility of two-way disturbance spreading, or dynamic equilibrium. This form of instability, when all forces involved are in equilibrium, permits a two-way disturbance sreading (propagation or transfer) or entities either by tunneling (low-energy dissipation) or by induction (medium- or high-energy dissipation). A classical particle or

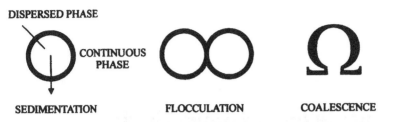

FIG. 1 Possible instabilities in liquid–liquid interfaces. Classical approach.

system could not penetrate a region in which its energy would be negative, that is, barrier regions in which the potential energy is greater than the system energy. In the real world, however, a wave function of significant amplitude may extend into and beyond such a region. If the wave function extends into another region of positive energy, then the barrier is crossed with some probability; this process is termed tunneling (since the barrier is penetrated rather than climbed). The elastic form of instability comprises the possibility of reversible disturbance spreading, with or without hysteresis. Finally, the plastic form of instability comprises the possibility of irreversible disturbance spreading with a low or high intensity of influence between two entities. Entity is the smallest indivisible element of matter that is related to the particular transfer phenomenon. The entity can be either a differential element of mass/demon or a phonon as quantum of acoustic energy, or an infon as quantum of information, or photon or electron.

Now, a disperse system consists of two phases, "continuous" and "dispersed." The continuous phase is modeled as an infinitely large number of harmonic electromechanical oscillators with low-strength interactions among them. Furthermore, the dispersed phase is a macrocollective consisting of a finite number of microcollectives/harmonic electromechanical oscillators (clusters) with strong interactions between them. The cluster can be defined as the smallest repetitive unit that has a character of integrity. Clusters appear in micro and nano dispersed systems. The microcollective consists of the following elements:

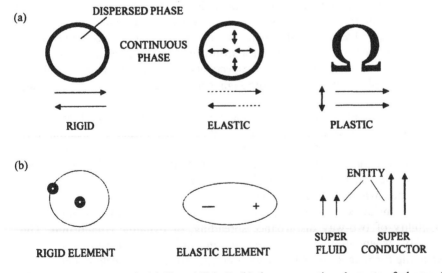

FIG. 2 A new approach: (a) "instabilities"; (b) the constructive elements of phases. (From Ref. 6, with permission from Academic Press.)

rigid elements (atoms or molecules), elastic elements (dipoles or ions that may be recombined), and entities (as the smallest elements) [6, 8, 10].

Validation of these theoretical predictions will be corroborated experimentally by means of electrical interfacial potential (EIP) measurements, and nuclear magnetic resonance (NMR) spectroscopy [6–11].

B. Electromechanical Analogy: Interfacial Tension-Electrical Interfacial Potential

A fundamental approach will be used to analyze a special stability and rupture problem of the droplet-film structure immersed in the droplet homophase continuum. Formation and rupture processes of the secondary liquid/liquid droplet-film structures will be discussed, considering mechanical and electrical principles. The analogy interfacial tension–EIP will be illustrated, considering the physical model of the processes appearing during the secondary separation of the droplet-film structure submerged in the droplet homophase continuum (double emulsion) on an inclined plate. Figure 3 shows the physical model of the processes involved: approach, rest, disturbance, rupture, and flow up. The generator pole is the origin/source of the disturbance, and the rupture pole is the point where the electrical and mechanical waves change the direction of travel (feed in/feed back).

The effect of an acting force on the rate of thinning of the film covering the secondary drops is to be discussed. Because of the pressure gradient associated with flow in the film and because a fluid–liquid interface must deform when there is a pressure difference across it, the film thickness varies with position, giving rise to the well-known dimple [15]. An applied or an acting force changes the distribution of the pressure in the film, and hence the variation in its thickness with position and time.

The film thickness of a droplet-film structure δ_0, at rest, is calculated from the relation given by [3]

$$\delta_0 = 0.70d\left(\frac{d^2\Delta\rho g}{\sigma_{\text{in}}}\right)^{0.5} \tag{1}$$

where d is the droplet-film structure diameter, $\Delta\rho$ is the density difference between the two liquids, σ_{in} is the interfacial tension, and g is the acceleration due to gravity.

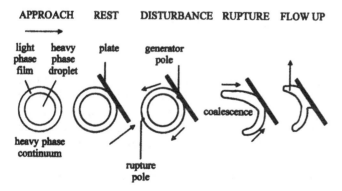

FIG. 3 Physical model of the processes involved in the secondary separation of a double emulsion on an inclined plate.

Coalescence of the primary liquid drops and gas bubbles with their homophase is determined by the rate of drainage of the intervening fluid film. For a uniform film of viscosity μ, the variation in film thickness δ with time t is given by [15]

$$-\frac{d\delta}{dt} = \frac{8\pi}{3n^2} \frac{\delta^3}{\mu} \frac{f_c}{a^2} \tag{2}$$

where f_c is the force pressing on the film of area a, and n is the number of immobile interfaces bounding the film. Thus, in the case of a double emulsion stopped on an inclined plate n is equal to 1, since there is one immobile interface.

Coalescence of the secondary droplets from the liquid/liquid droplet-film structures with their homophase is limited by the rate of rupture of the covering fluid film. When the droplet-film structure stops on an inclined plate, the film surface is in hydrodynamic equilibrium and the electric double layers (EDLs) are at rest. Since the droplet-film structures are very small, it is postulated that the electrical forces become dominant when compared with the viscous forces on the junction point between the droplet-film structure and the plate. Based on the hypothesis that the electrical forces are responsible for the droplet-film rupture, one can adopt Eq. (2) with necessary changes relevant to the nature of forces involved in the rupture process [10]. Now, the variation in film thickness δ with time t is given by

$$-\frac{d\delta}{dt} = \frac{8\pi}{3} \frac{\delta^3}{Z} \frac{f_e}{a^2} \tag{3}$$

where f_e is the electrical force acting on the droplet-film structure surface of area a, and Z is the complex resistance/ impedance of the equivalent electric circuit. Therefore, the rate constant in differential equation (3) is given by

$$C_m = \frac{8\pi}{3} \frac{1}{Z} \frac{f_e}{a^2} \tag{4}$$

Also, the electrical force f_e may be introduced using an analogous relation to

$$\frac{f_c}{a^2} = \frac{2\sigma_{in}}{v} \tag{5}$$

namely

$$\frac{f_e}{a^2} = \frac{2\sigma_{ep}}{v} \tag{6}$$

where v is the droplet-film structure volume, and σ_{ep} is the EIP.

1. Formation Process

The study of the spreading phenomenon and interfacial stability at the surface/interface boundary of two immiscible liquids has been the subject of few publications, in spite of the fact that such information has a great importance for the examination of the rupture mechanism. During the formation of the secondary liquid/liquid structures, viscous and electrical forces are predominantly involved. Therefore, hydrodynamic and electrodynamic equilibria have to be reached.

Existence of the secondary liquid/liquid droplet-film structure or its electroviscoelastic behavior is dependent on:

- droplet film particle size

- curvature of the droplet-film interface
- density difference between the phases
- viscosity ratio of the phases
- impedance ratio of the phases
- interfacial tension
- EIP
- temperature effects
- mechanical effects (vibration)
- third-phase presence
- mutual solubility
- external periodical physical fields (e.g., temperature, electric, magnetic, ultrasonic)
- internal periodical physical fields (mechanical, electric, and magnetic).

2. Stability

Figure 4 shows the graphical interpretation of a droplet-film structure stopped on the inclined plate, and an acting force with its components at the structure–plate junction point. Figure 5 shows the graphical interpretation of a droplet-film structure approach to the inclined plate, equilibrium and rupture using mechanical principles. F_x, F_y, and F_z are the component vectors of F_s, which is the resultant surface force vector in $3N$ dimensional configuration space.

According to Newton's second law, the general equation of fluid dynamics in differential form is given by

$$\rho \frac{D\tilde{u}}{Dt} = \sum_i \tilde{F}_i (dx\,dy\,dz) + d\tilde{F}_s \tag{7}$$

When a droplet-film structure rests on the inclined plate, the term on the left-hand side of Eq. (7) becomes equal to zero; further, the terms on the right-hand side represent the volume, F_i (gravitational F_g, buoyancy F_{bo}, and electromagnetic/Lorentz F_l), and the surface, F_s forces, respectively. The gravitational force is superimposed on the buoyancy force; therefore, the volume force term is equal to zero. The surface forces are supposed to be associated with the interface between the fluid continuum and the droplet, and between

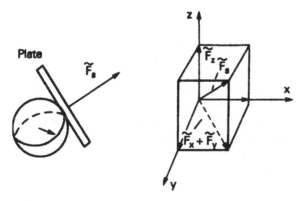

FIG. 4 Graphical representation of the contact point between a droplet-film structure and a plate. (From Ref. 10, with permission from Elsevier Science.)

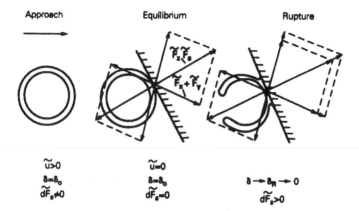

FIG. 5 Illustration of the mechanical forces involved in droplet-film structure approach, equilibrium, and rupture. (From Ref. 10, with permission from Elsevier Science.)

the latter and the plate. They can be calculated for by means of a stress tensor T_n as follows:

$$d\tilde{F}_s = \tilde{T}_n d\tilde{A} \tag{8}$$

where T_n is composed of two tensors given by

$$\tilde{T}_n = -p[\delta_j^i] + [\xi_j^i] \tag{9}$$

In the first, isotropic tensor, the hydrostatic pressure is dominant and the contribution of the other elements is neglected; further, in the second tension tensor, tangential elements are due to the interfacial tensions, and the normal elements are presumed to be of adhesive origin. The force balance at the junction point between the plate and the droplet-film structure isgiven by (Fig. 4):

$$d\tilde{F}_s - (d\tilde{F}_x + d\tilde{F}_y + d\tilde{F}_z) = 0 \tag{10}$$

The tangential elements τ in the second tension tensor of Eq. (9) are presumed to be identical; therefore, the mechanical equilibrium condition is given by

$$\tau = \frac{p - \left(\dfrac{\sigma_{in}}{d}\right)}{2} \tag{11}$$

Introducing the impedance Z instead of the viscosity μ and the electron flux density ϕ instead of the velocity u, an electrical analog of the droplet-film rupture mechanism on the inclined plate can be proposed, as shown in Fig. 6.

At first, the droplet-film structure surface is at rest, and then the structure–plate junction point becomes the generator pole (source or sink of the incident sinusoidal wave) and disturbs the EDL, changing the distribution of pressure in the film and, hence, causing the variation in the film thickness, which ends with the rupture. Now, the electrical equilibrium condition analogous to Eq. (11) is given by

$$\tau = \frac{U - \left(\dfrac{\sigma_{ep}}{d}\right)}{2} \tag{12}$$

where U is the electrostatic potential and σ_{ep} is the EIP.

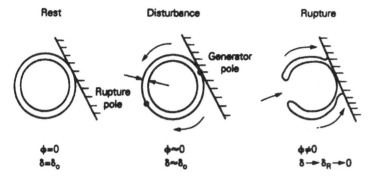

Rest **Disturbance** **Rupture**

Generator
pole

Rupture
pole

$\phi=0$ $\phi\sim0$ $\phi\neq0$
$\delta=\delta_0$ $\delta\sim\delta_0$ $\delta\to\delta_R\to0$

FIG. 6 Electrical interpretation of the processes of droplet-film structure rest, disturbance, and rupture. (From Ref. 10, with permission from Elsevier Science.)

For spherical polydispersions, the relation between the interfacial tension σ_{in} and the internal pressure p_{in} in the droplet-film structure is given by [26]

$$\frac{\sigma_{in}}{d} \approx p_{in} \tag{13}$$

and the relation between the EIP σ_{ep} and the electrical internal potential u is given by

$$\frac{\sigma_{ep}}{d} \approx u \tag{14}$$

So, a discussion of the formation and hydrodynamic and electrodynamic equilibra as well as rupture processes is facilitated in terms of the proposed electrical analogy.

3. Rupture Process

The rupture process is to be analyzed for the special case of the droplet-film structure on an inclined plate. Now, under the assumption that about the same interfacial jump potential appears during the droplet-film formation and rupture processes, it is postulated that the generator pole (junction point of the droplet-film structure and the inclined plate) is the source or sink of the incidental sinusoidal wave. Hence, the impedance of the structure consists of resistance and reactance terms. The surface of the droplet-film structure resting on the inclined plate is supposed to achieve initially hydrodynamic and electrodynamic equilibria.

Therefore, from Eq. (3) the process of the film thinning is given by

$$\delta = \frac{\delta}{(1 + 2\delta_0^2 C_m t)^{0.5}} \tag{15}$$

where C_m is given by Eq. (4). When time tends to the rupture time t_R the film thickness δ tends to zero, which may be represented by

$$\lim_{t \to t_R} \delta = \delta_R \to 0 \tag{16}$$

In general, after the rupture process of the light phase film is completed, the coalescence process of the heavy-phase droplet with its homophase occurs. Factors affecting the secondary coalescence time are [3]:

- droplet-film structure size
- distance of fall of droplet to the interface
- curvature of the droplet–side interface
- density difference between phases
- viscosity ratio of the phases
- interfacial tension effects
- temperature effects
- vibration and electrical effects
- presence of the EDLs
- solute transfer effects.

As can be seen in Fig. 3, the phases of the overall process during the secondary separation of a double emulsion on an inclined plate are: approach, rest, disturbance, rupture, and flow up. Besides the analysis of the wave propagation or disturbance spreading by the impedance model, the secondary separation process can be represented by successive time sequences as follows:

$$t_{SE} = t_A + t_{RE} + t_{DIST} + t_R + T_{FU} \tag{17}$$

Under the assumption that the rest t_{RE} and rupture t_R times are infinitely short, and since the two EDLs are to be destroyed, the disturbance time t_{DIST} is composed of two disturbance and two collapse subsequences, and may be written as

$$t_{DIST} = t_{distII} + t_{colII} + t_{distI} + t_{colI} \tag{18}$$

Hence, for this particular case after the rupture process is completed, the start of "flow up" (lifting of the separated light-phase film) takes place. Again the volume forces (gravitational, buoyancy, and electromagnetic) enter into the game [10, 21]. The "flow up" occurs when the following condition is satisfied:

$$\tilde{F}_{bo}dxdydz + \tilde{F}_1dxdydz \geq \tilde{F}_gdxdydz \tag{19}$$

C. Marangoni Instabilities of First and Second Order and Possible Electrical Analog

In a number of papers on Marangoni instability or the mechanism of Benard cell formation, it is supposed that the surface tension is a linear, monotonically decreasing function of temperature [27, 28]. This behavior is typical for a large class of fluids, e.g., water, silicone oil, water/benzene solutions, etc. There are exceptions, such as some alloys, molten salts, and liquid crystals that show a linear growth of surface tension with temperature. Also, there exists a third class of fluid systems characterized by a surface tension showing nonlinear dependence with respect to temperature. This behavior is representative of aqueous long-chain alcohol solutions and some binary metallic alloys [28].

The Marangoni instability of the first order was first elucidated and demonstrated theoretically by Pearson [29]. It was shown that if there was an adverse temperature gradient of sufficient magnitude across a thin film with a free surface, such a layer could become unstable and lead to cellular convection. This instaiblity mechanism is illustrated in Fig. 7.

A small disturbance is assumed to cause the film of initially uniform thickness to be heated locally at a point on the surface. This results in a decreased surface tension and a surface tension gradient that leads to an induced motion tangential to the surface away

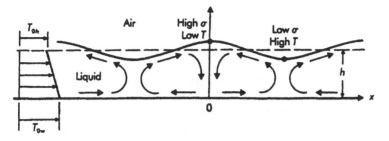

FIG. 7 Marangoni instability mechanism. (From Ref. 27, reprinted by permission of John Wiley.)

from the point of local heating. From mass conservation, this motion in turn induces a motion of the bulk phase towards the surface. The liquid coming from the heated region is warmer than the liquid/gas interface. The motion is thus reinforced, creating cellular convection patterns and will be maintained if the convection overcomes viscous shear and heat diffusivity [27]. It is appropriate to introduce the critical parameter termed the Marangoni number given by

$$\mathrm{Ma} = \frac{\sigma_T \beta_T \delta^2}{\mu} \tag{20}$$

where σ_T is the surface tension, β_T is the uniform temperature gradient, δ is the film thickness, and μ is the viscosity. This number can also be interpreted as a thermal Peclet number that represents a measure of the ratio between heat transport by convection due to surface tension gradients, and bulk heat transport by conduction. More detailed discussion on the derivation of the Marangoni number can be found in the literature, e.g., Refs 27, 28, and 30.

Now, it is possible to redefine the parameters with the changes relevant to the nature of the forces involved, and to obtain the relation for the electrical critical parameter given by

$$\mathrm{AD_R} = \frac{\sigma_E \gamma_T \delta^2}{Z} \tag{21}$$

where σ_E is the EIP, γ_T is the uniform potential gradient, δ is the film thickness, and Z is the impedance [8]. The subscript R indicates a rigid form of "instability," i.e., the stable existence of a droplet or droplet-film structures.

Using an analogous approach, where the buoyancy forces are neglected as in, e.g., a microgravity environment, a nonlinear dependence of the surface tension with respect to the temperature can give rise to the second-order Marangoni effect, with characteristic number:

$$\mathrm{Ma}'' = \frac{\left(\frac{\partial^2 \sigma_T}{\partial T^2}\right)(\Delta T)^2 \delta}{\rho \nu k} \tag{22}$$

where ΔT is the temperature drop between the lower and upper boundaries of the layer, δ is the distance between the boundaries of the layer, ρ is the constant density, k is the heat diffusivity, and ν is the kinematic viscosity [28].

Finally, the nonlinear electrical critical parameter is derived and expressed by

$$AD_E = \frac{\left(\dfrac{\partial^2 \sigma_E}{\partial y^2}\right)(\Delta y)^2 \delta}{Z} \qquad (23)$$

where Δy is the potential drop between the lower and upper boundaries of the layer, δ is the distance between the boundaries, and Z is the impedance. The subscript E corresponds to an elastic form of "instability," indicating formation, breathing, and destruction of a droplet or a droplet-film structure. It is suggested that the reader should consult Ref. 22 for other nonlinear interactions.

D. Electroviscosity/Electroviscoelasticity of Liquid/Liquid Interfaces

Electroviscosity and electroviscoelasticity are terms that may be broadly defined as dealing with fluid-flow effects on physical, chemical, and biochemical processes. The hydrodynamic and/or electrodynamic motion is considered in the presence of both potential (elastic forces) and nonpotential fields (resistance forces). The elastic forces are gravitational, buoyancy, and electrostatic/electrodynamic (Lorentz), and the resistance forces are continuum resistance/viscosity and electrical resistance/impedance.

According to the classical deterministic approach, the phases that constitute the multiphase dispersed systems are assumed to be a continuum; i.e., without discontinuities inside the one entire phase, homogeneous, and isotropic. The principles of conservation of momentum, energy, mass, and charge are used to define the state of a real fluid system quantitatively. In addition to the conservation equations, which are insufficient to define the system uniquely, statements on the material behavior are also required. These statements are termed constitutive relations, e.g., Newton's law, Fourier's law, Fick's Law, and Ohm's law.

In general, the constitutive equations are defined empirically, although the coefficients in these equations (e.g., viscosity coefficient, heat conduction coefficient, and complex resistance coefficient/impedance) may be determined at the molecular level. Often, these coeffficients are determined empirically from related phenomena; therefore, such a description of the fluid state is termed a phenomenological description or model.

Sometimes, particular modifications are needed when dealing with fine dispersed systems. An example is Einstein's modification of the Newtonian viscosity coefficient in dilute colloidal suspensions [27]. Later on, Smoluchowski's modification of the Einstein relation for particles carrying EDLs [30]. Finally, a recent more profound elaboration of the entropic effects [22].

Now, using the above described stability and electromechanical analogies, an approach to nonNewtonian behaviors and to electroviscoelasticity is to be introduced. When Eq. (7) is applied to the droplet when it is stopped, e.g., as a result of an interaction with some periodical physical field, the term on the left-hand side becomes equal to zero. Furthermore, if the droplet is in the state of "forced" levitation, and the volume forces balance each other, then the volume force term is also equal to zero [6–9]. It is assumed that the surface forces are, for the general case that includes the electroviscoelastic fluids, composed of interaction terms expressed by

$$d\tilde{F}_s = \tilde{T}^{ij} d\tilde{A} \qquad (24)$$

where the tensor T^{ij} is given by

$$T^{ij} = -\alpha_0 \delta^{ij} + \alpha_1 \delta^{ij} + \alpha_2 \zeta^{ij} + \alpha_3 \zeta_k^i \zeta^{kj} \qquad (25)$$

where T^{ij} is composed of four tensors, δ^{ij} is the Kronecker symbol, ζ^{ij} is the tension tensor, and $\zeta_k \zeta^{kj}$ is the tension coupling tensor. In the first isotropic tensor the potentiostatic pressure $\alpha_0 = \alpha_0(\rho, U)$ is dominant and the contribution of the other elements is neglected. Here, U represents hydrostatic or electrostatic potential. In the second isotropic tensor, the resistance $\alpha_1 = \alpha_1(\rho, U)$ is dominant and the contribution of the other elements is neglected. In the third tension tensor, its normal elements $\alpha_2 \sigma$ are due to the interfacial tensions and the tangential elements $\alpha_2 \tau$ are presumed to be of the same origin as the dominant physical field involved. In the fourth tension coupling tensor, there are normal, $\alpha_3 \sigma_k^i$, and $\alpha_3 \sigma^{kj}$ elements, and tangential $\alpha_3 \tau_k^i$ and $\alpha_3 \tau^{kj}$ elements, which are ascribed to the first two dominant periodical physical fields involved. Now, the general equilibrium condition for the dispersed system with two periodical phyiscal fields involved may be derived from Eq. (25), and may be expressed by

$$\tau_d = \frac{-\alpha_0 + \alpha_1 + \alpha\left(\frac{\sigma}{d}\right) + \alpha_3\left(\frac{\sigma}{d}\right)}{2(\alpha_2 + \alpha_3)} \tag{26}$$

where τ_d are the tangential elements of the same origin as those of the dominant periodical physical field involved. Figure 8 shows the schematic equilibrium of surface forces at any point of a stopped droplet-film structure while in interaction with some periodical physical field [6]. Note that for dispersed systems consisting of, or behaving as Newtonian fluids, $\alpha_3 = \alpha_3(\rho, U)$ is equal to zero.

The processes of formation/destruction of the droplet or droplet-film structure are nonlinear. Therefore, the viscosity coefficients $\mu_i(i = 0, 1, 2)$, where each consists of bulk, shear, and tensile components, when correlated with the tangential tensions of mechanical origin τ_v, can be written as

$$\tau_v = \mu_0 \frac{du}{dx} + \mu_1 \frac{d^2 u}{dx^2} + \mu_2 \left(\frac{du}{dx}\right)^2 \tag{27}$$

where u is the velocity, and x is one of the space co-ordinates.

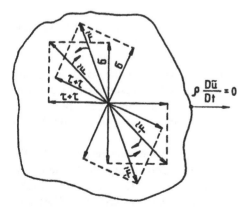

FIG. 8 Balance of surface forces at any point of a stopped droplet-film structure while in interaction with some periodic physical field (a two-dimensional projection). F represents the projection of the resultant surface forces' vector in three N dimensional configuration space, τ stands for the tangential, and σ for the normal components. (From Ref. 6, with permission from Academic Press.)

Using the electrical analog, the impedance coefficients $Z_i(i = 0, 1, 2)$, where each consists of ohmic, capacitive, and inductive components, will be correlated with the tangential tensions of electrical origin τ_e, as follows:

$$\tau_e = Z_0 \frac{d\phi_e}{dt} + Z_1 \frac{d^2\phi_e}{dt^2} + Z_2 \left(\frac{d\phi_e}{dt}\right)^2 \tag{28}$$

where ϕ_e is the electron flux density, and t is the time co-ordinate.

More detailed discussion about the derivation of these equations can be found in Refs 6–10.

II. THEORY

A. Previous Work

In normal viscous fluids, only the rate of deformation is of interest. In the absence of external and body forces, no stresses are developed and there is no means of distinguishing between a natural state and a deformed state [31]. It is rather disturbing to think of the very large overall deformations obtained in the flow of fluids being associated in any way with substances that have elasticity. The rationalization lies in realizing that, for the substances considered here, the behavior is essentially that of a fluid; although much translation and rotation may occur, the "elastic" distortion of the elementary volumes around any point is generally small. This "elastic" distortion or material's strain is nevertheless present and is a feature that cannot be neglected. It is responsible for the recovery of reverse flow after the removal of applied forces and for all the other nonNewtonian effects [31]. These distortions or strains are determined by the stress history of the fluid and cannot be specified kinematically in terms of the large overall movement of the fluid. Another way of looking at the situation is to say that the natural state of the fluid changes constantly in flow and tries to catch up with the instantaneous state or the deformed state. It never does quite succeed in doing so, and the lag is a measure of the memory or the elasticity. In elastic solids, the natural state does not change and there is perfect memory [31].

The entropy of elasticity of a droplet is a measure of the increase in the available volume in configuration space. This increase occurs with a transition from a rigid, regular structure to an ensemble of states that include many different structures. If the potential wells in the liquid state were as narrow as those in the solid state, and if each of those potential wells were equally populated and corresponded to a stable amorphous structure (and vice versa), then the entropy of elasticity would be a direct measure of the increase in the number of wells, or a direct measure of the number of available structures [6, 22].

In the last two centuries a lot of attempts and discussion have been performed for the elucidation and development of the various constitutive models of liquids. Some of the theoretical models that can be mentioned here are: Boltzmann, Maxwell (UCM, LCM, COM, IPM), Voight or Kelvin, Jeffrey's, Reiner–Rivelin, Newton, Oldroyd, Giesekus, graded fluids, composite fluids, retarded fluids with a strong backbone and fading memory, etc. Further and deeper knowledge related to the physical and mathematical consequences of the structural models of liquids and of the elasticity of liquids can be found in Ref. 32.

B. Electrified Interfaces: a New Constitutive Model of Liquids

The secondary liquid–liquid droplet or droplet-film structure is considered as a macroscopic system with internal structure determined by the way the molecules (ions) are tuned (structured) into the primary components of a cluster configuration. How the tuning/structuring occurs depends on the physical fields involved, both potential (elastic forces) and nonpotential (resistance forces). All these microelements of the primary structure can be considered as electromechanical oscillators assembled into groups, so that excitation by an external physical field may cause oscillations at the resonant/characteristic frequency of the system itself (coupling at the characteristic frequency) [6–8].

Figure 9 shows a series of graphical sequences that are supposed to facilitate the understanding of the proposed structural model of electroviscoelastic liquids. The electrical analog Fig. 9a consists of passive elements (R, L, and C) and an active element (emitter coupled oscillator W). Further on, the emitter-coupled oscillator is represented by the equivalent circuit as shown in Fig. 9b. Figure 9c, shows the electrical (oscillators j) and mechanical (structural volumes V_j) analogs when they are coupled with each other, e.g., in the droplet. Now, the droplet consists of a finite number of structural volumes or spaces/electromechanical oscillators (clusters) V_j, a finite number of excluded surface volumes or interspaces V_s, and a finite number of excluded bulk volumes or interspaces V_b. Furthermore, the interoscillator/cluster distance or internal separation S_i represents the equilibrium of all forces involved (electrostatic, solvation, van der Waals, and steric [33]). The external separation S_e, in introduced as a permitted distance when the droplet is in interaction with any external periodic physical field. The rigidity droplet boundary R presents a form of droplet instability when all forces involved are in equilibrium. Nevertheless, two-way disturbance spreading (propagation or transfer) of entities occur, either by tunneling (low-energy dissipation) or by induction (medium- or high-energy dissipation). The elasticity droplet boundary E represents a form of droplet instability when the equilibrium of all forces involved is disturbed by the action of any external periodic physical field, but the droplet still exists as a dispersed phase. In the region between the rigidity and elasticity droplet boundaries, a reversible disturbance spreading occurs. After the elasticity droplet boundary, the plasticity as a form of droplet instaiblity takes place; the electromechanical oscillators/clusters do not then exist any more and the beams of entities or atto-clusters appear. Atto-clusters are the entities that appear in the atto-dispersed systems. In this region, one-way propagation of entities occurs.

Considering all the arguments and comments presented, the probability density function (PDF) in general form can be expressed by

$$F_d(V) = F_d(V_j) + [F_d(V_s) + F_d(V_b)] \tag{29}$$

where the first term on the right-hand side of the equation is due to energy effects, and the second term (consisting of two subterms) is due to enytropic effects; subscript j is related to the structural volumes or energies, and subscripts s and b are related to the excluded surface and bulk volumes or energies. Consider an uncertain physical property and a corresponding space describing the range of values that the property can have (e.g., the configuration of a thermally excited N particle system and the corresponding $3N$ dimensional configuration space). The PDF associated with a property is defined over the corresponding space; its value at a particular point is the probability per unit volume that the property has a value in an infinitesimal region around that point [22].

(a)

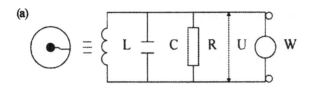

(b)

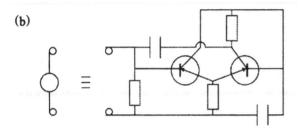

(c)

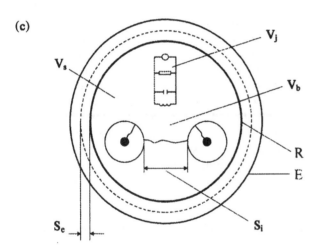

FIG. 9 Graphical interpretation of the structural model: (a) electrical and mechanical analog of the micro collective/cluster; (b) equivalent circuit for the emitter coupled oscillator; (c) the macro collective: a schematic cross-section of the droplet and its characteristics (V_j – structural volumes/clusters; V_s – excluded surface volumes/interspaces; V_b – excluded bulk volumes/interspaces; S_i – internal separation; S_e – external separation; R – rigidity droplet boundary; E – elasticity droplet boundary).

An alternative expression of this PDF, considering Fig. 9c, may be written as

$$\Delta_{1,2}(V) = \sum_{i=3}^{n}(V_j)_i + \Delta_{1,2}\left[\sum_{i=1}^{n-3}(V_b)_i + \sum_{i=3}^{n}(V_s)_i\right] \tag{30}$$

therefore, the number i of clusters V_j remains constant while the droplet passes through the rigid (e.g., the state related to the subscript 1) and elastic (e.g., the state related to the subscript 2) form of "instabilities." The integer summation index (i) takes values in the interval $[3, n]$ for the number of clusters V_j and for the number of excluded surface volumes V_s, and $[1, n-3]$ for the number of excluded bulk volumes; while integer n takes values in the interval $[0, \infty]$. The simplest case, used only as an example, is the droplet that contains three oscillators V_j $[(i) = 3; n = 0]$: then there is one excluded bulk volume $V_b[(i) = 1; n = 3]$ and three excluded surface volumes $V_s[(i) = 3; n = 0]$. Differences $\Delta_{1,2}$ in volumes or energies V_b and V_s occur only in the entropic part, i.e., the internal separation S_i (Fig. 9c) changes (increases or decreases) during the transition of the droplet from rigid to elastic or vice versa. Consequently, the external separation S_e decreases or increases, depending on the direction of transition.

1. Classical Assumptions for Interfacial Tension Structure and for Partition Function

1. The droplet is considered as a unique thermodynamic system that can be described by a characteristic free energy function expressed by

$$\Delta G = (\sigma_i + T\Delta S) = \left[\sigma_i - T\left(\frac{d\sigma_i}{dT}\right)_{\chi_i}\right] = -kT \ln Z_p \tag{31}$$

where χ_i correspond to the constant chemical potential, and Z_p is the partition function.

2. According to quantum mechanical principles, the droplet possesses vacancies or "free volumes" and the relation for interfacial tension can be writte as

$$\sigma_i = \frac{G^s - G^b}{\kappa^0} = \frac{1}{\kappa^0}\left[(\Phi^s - \Phi^b) + N^s kT \ln\left(\frac{V_f^b}{V_f^s}\right)\right] \tag{32}$$

where Φ^s and Φ^b are the overall energies of N heavy-phase molecules in their ideal positions, s on the surface and b in the bulk; N^s is the number of molecules on the surface, V_f^s and V_f^b are "free volumes" of the molecules on the surface and in the bulk, and κ^0 is the surface of "free surface" [6, 7, 24, 25, 34–67]. This means the droplet is a macro system with physicochemical properties that may be described with the help of different thermodynamic parameters.

The phenomenological meaning if the given interfacial tension structure is in agreement with the "free volume" fluid model. Hence, a fluid is a system with ideal or ordered neighbor elements' distribution and with discontinuities of the package density (boundaries of the subsystems or microcollectives/clusters as some particular physical systems) [6, 7].

Furthermore, the partition function indicates how molecules are distributed among the available energy levels. It is possible to separate various contributions (the sum of the translational, rotational, vibrational, and electronic energy terms) to the partition function [59].

3. Using the equivalency of the mean energies W, at the instant of equilibrium, a characteristic free energy function can be expressed by

$$\Delta G = \bar{w} = -kT \ln Z_p \tag{33}$$

where the partition function for this particular system is derived from

$$W = \frac{h\omega}{2\pi} \frac{\partial \ln Z_p}{\partial \theta} \tag{34}$$

and

$$Z_p = \frac{Q_N}{\lambda^{3N}} = \sum_{j=0}^{\infty} \exp\left[-\left(j+\frac{1}{2}\right)\Theta\right] = \frac{1}{2}\cosh\frac{\Theta}{2} \tag{35}$$

and j is the number of the identical oscillators, where each is given by

$$\Theta = \frac{h\omega}{2\pi kT} \tag{36}$$

λ is a free path between two collisions expressed by

$$\lambda = \frac{h^2}{2\pi mkT} \tag{37}$$

and Q_N is a configuration integral.

Further and more detailed discussion and derivation of the partition function can be found in, e.g., Refs 6, 7, and 59.

2. Postulated Assumptions for an Electrical Analog

1. The droplet is a macrosystem (collective of particles) consisting of structural elements that may be considered as electromechanical oscillators.
2. Droplets as microcollectives undergo tuning or coupling processes, and so build the droplet as a macrocollective.
3. The external physical fields (temperature, ultrasonic, electromagnetic, or any other periodic) cause the excitation of a macrosystem through the excitation of a microsystem at the resonant/characteristic frequency, where elastic and/or plastic deformations may occur.

Hence, the study of electromechanical oscillators is based on electromechanical and electrodynamic principles.

A nonhomogeneous nonlinear differential equation of the Van der Pol type represents the initial electromagnetic oscillation:

$$C\frac{dU}{dt} + \frac{U}{R} - \alpha U + \gamma U^3 + \frac{1}{L}\int U dt = 0 \tag{38}$$

where U is the overall potential difference at the junction point of the spherical capacitor C and the plate, L is the inductance caused by potential difference, and R is the ohmic resistance (resistance of the energy transformation, electromagnetic into mechanical or damping resistance); α and γ are constants determining the linear and nonlinear parts of the characteristic current and potential curves. U_0, the primary steady-state solution of this equation, is a sinusoid of frequency close to $\omega_0 = 1/(LC)^{0.5}$ and amplitude $A_0 = [(\alpha - 1)/R/3\gamma/4]^{0.5}$.

The noise in this system, due to linear amplification of the source noise (the electromagnetic force is assumed to be the incident external force, which initiates the mechanical disturbance), causes the oscillations of the "continuum" particle (molecule surrounding the droplet or droplet-film structure), which can be represented by the particular integral:

$$C\frac{dU}{dt} + \left(\frac{1}{R} - \alpha\right)U + \gamma U^3 + \frac{1}{L}\int U dt+ = -2A_n \cos \omega t \tag{39}$$

where ω is the frequency of the incident oscillations.

Finally, considering the droplet or droplet-film structure formation, "breathing," and/or destruction processes, and taking into account all the noise frequency components, which are included in the driving force, the corresponding equation is given by

$$C\frac{dU}{dt} + \left(\frac{1}{R} - \alpha\right)U + \frac{1}{L}\int U dt + \gamma U^3 = i(t) = \frac{1}{2\pi}\int_{-\infty}^{\infty} \exp(i\omega t)A_n(\omega)d\omega \qquad (40)$$

where $i(t)$ is the noise current and $A_n(\omega)$ is the spectral distribution of the noise current as a function of frequency.

In the case of nonlinear oscillators, however, the problem of determining the noise output is complicated by the fact that the output is fed back into the system, thus modifying in a complicated manner the effective noise input [6, 31, 41]. The noise output appears as an induced anisotropic effect.

C. Theory of Electroviscoelasticity

A number of theories that describe the behavior of liquid–liquid interfaces have been developed and applied to various dispersed systems: Stokes, Reiner–Rivelin, Ericksen, Einstein, Smoluchowski, Kinch, etc. [6, 7, 26, 42–55]. The reader is suggested to review some of the following topics for a better understanding of the present theory: potential energy ssurfaces (PES) (Dirac, Millikan, Feynman, Schwinger, Tomonaga, Born–Oppenheimer, etc.), quantum electrodynamics (QED) (Schrödinger), molecular mechanics (MM2-MM3-CSC) (Allinger, Lii, Cambridge Science Computation), and transient state theories TST (Wigner, etc.) [22, 32, 56–66].

Figure 10 shows a series of graphical sequences that may help in an understanding of the proposed theory of electroviscoelasticity. This theory describes the behavior of electrified liquid–liquid interfaces in fine dispersed systems, and is based on a new constitutive model of liquids [6–11]. If an incident periodic physical field (Fig. 10b), e.g., electromagnetic, is applied to the rigid droplet of Fig. 10a, then the resulting, equivalent electrical circuit can be represented as shown in Fig. 10c. The equivalent electrical circuit, rearranged under the influence of an applied physical field, is considered as a parallel resonant circuit coupled with another circuit, such as an antenna output circuit. Thus, in Fig. 10c, W_d, C_d, L_d, and R_d correspond to the circuit elements; active emitter coupled oscillator W_d and passive C_d, L_d, and R_d respectively. The subscript d is related to the particular droplet diameter, i.e., the droplet under consideration. Now again, the initial electromagnetic oscillation is repressneted by the differential equations, Eq. (39) and Eq. (40), and when the nonlinear terms are omitted and/or cancelled, the following linear equation is obtained:

$$C\frac{dU}{dt} + \left(\frac{1}{R} - \alpha\right)U + \frac{1}{L}\int U dt + = -2A_n\cos\omega t \qquad (41)$$

with a particular solution resulting in the following expression for the amplitude:

$$A = \frac{2\omega C A_n}{\left[4(\omega_0 - \omega)^2 + \left(\frac{1}{R} - \alpha\right)^2\right]^{0.5}} \qquad (42)$$

and for all the noise frequency components, the linear equation is given by

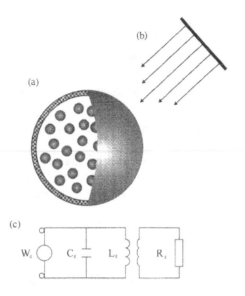

FIG. 10 Structural model of electroviscoelasticity: (a) rigid droplet; (b) incident physical field; (c) equivalent electrical circuit. W_d represents the emitter-coupled oscillator, C_d, L_d, and R_d are capacitive, inductive, and resistive elements of the equivalent electrical circuit, respectively. Subscript d is related to the particular diameter of the droplet under consideration.

$$C\frac{dU}{dt} + \left(\frac{1}{R} - \alpha\right)U + \frac{1}{L}\int U dt = i(t) = \frac{1}{2\pi}\int_{-\infty}^{\infty} \exp(i\omega t)A_n(\omega)d\omega \qquad (43)$$

with the particular solution expressed by

$$U_n = \frac{i\omega A_n \exp(i\omega t)}{C(\omega_0^2 - \omega^2) + i\left(\frac{1}{R} - \alpha\right)\omega} - \frac{i\omega A_n \exp(-i\omega t)}{C(\omega_0^2 - \omega^2) + i\left(\frac{1}{R} - \alpha\right)\omega} \qquad (44)$$

Furthermore, the electrical energy density e_e inside the capacitor is given by

$$w_e = \frac{1}{8\pi}\varepsilon E^2 \qquad (45)$$

where ε is the dielectric constant and E is the electric field, and the magnetic energy density w_m inside the capacitor is given by

$$w_m = \frac{1}{8\pi}\mu_e H^2 \qquad (46)$$

where μ_e is the magnetic permeability constant and H is the magnetic field; hence, the overall mean energy may be written as

$$\bar{w} = \frac{1}{8\pi}(\varepsilon E^2 + \mu_e H^2) \qquad (47)$$

The electromagnetic oscillation causes the tuning or structuring of the molecules (ions) in the "electric double layers." The structuring is realized by complex motions over the various degrees of freedom, whose energy contributions depend on the positions of the individual molecules in and around the stopped droplet-film structure under the action of some periodic physical field.

The hydrodynamic motion is considered to be the motion in the potential (elastic forces) and nonpotential (resistance forces) fields. There are several possible approaches for correlating the electromagnetic and mechanical oscillations; for example, this motion may be represented by the differential equation for the forced oscillation:

$$\frac{d^2\xi}{dt^2} + 2\beta\frac{d\xi}{dt} + \omega_0^2\xi = A\cos\omega t \tag{48}$$

with a solution representing the mechanical oscillation of the ordered group of molecules expressed by

$$\xi = \xi_0\exp(-\beta t)\cos(\omega t + \xi) \tag{49}$$

Now, if the electromagnetic force is assumed to be the incident (external) force which initiates the mechanical disturbance, then the oscillation of the "continuum" particles (molecules surrounding the droplet-film structure) is described by the differential equation (49) where ω is the frequency of the incident oscillations. After a certain time, the oscillations of the free oscillators ω_0 (molecules surrounding the droplet-film structure) tune with the incident oscillator frequency Ω. This process of tuning between free oscillations of the environmental oscillators and the incident oscillations of the electromechanical oscillator can be expressed by

$$\Omega = (\omega_0^2 - 2\beta^2)^{0.5} \tag{50}$$

Thereafter, there are two possibilities, depending on the energy appearing during the tuning process; the first leads towards the constant energy/rigid sphere, and the second leads towards the increasing energy/elastic sphere. For example, if a wave with high enough amplitude appears then the rupture of the droplet-film structure occurs [7].

During the interaction of the droplet or droplet-film structure with an incident periodic physical field at the instant of equilibrium, the mean electric, electromagnetic, and mechanical energies will be equal:

$$\bar{w} = \frac{1}{8\pi}(\varepsilon E^2 + \mu H^2) = \frac{1}{2}\rho\xi_0^2\omega^2 \tag{51}$$

and hence the frequency of the incident wave can be expressed by

$$\omega = \left[\frac{1}{4\pi\rho\xi^2}(\varepsilon E^2 + \mu H^2)\right]^{0.5} \tag{52}$$

Figure 11 shows the behavior of the circuit depicted in Fig. 10c, using the correlation impedance–frequency–arbitrary droplet diameter. If the electromagnetic oscillation causes the tuning or structuring of the molecules (ions) in the EDL, then the structuring is realized by complex motions over the various degrees of freedom. Since all events occur at the resonant/characteristic frequency, depending on the amount of coupling, the shape of the impedance–frequency curve is judged using the factor of merit or Q factor [67]. The

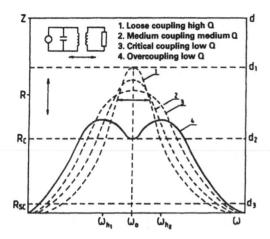

FIG. 11 Impedance of the equivalent electric circuit versus its frequency; ω_0 is the resonant frequency of this circuit, when the impedance is equal to the resistance. On the right axis arbitrary droplet diameters are plotted. (From Ref. 6, with permission from Academic Press.)

Q factor primarily determines the sharpness of resonance of a tuned circuit, and may be represented as the ratio of the reactance to the resistance, as follows:

$$Q = \frac{2\pi f L}{R} = \frac{\omega L}{R} \tag{53}$$

Furthermore, the impedance Z can be related to the factor of merit Q as given by the equations:

$$Z = \frac{(2\pi f L)^2}{R} = \frac{(\omega L)^2}{R} \tag{54}$$

and

$$Z = \omega L Q \tag{55}$$

From these equations and Fig. 11 it can be seen that the impedance of a circuit is directly proportional to its effective Q at resonance. Also, at the resonant frequency ω_0, the impedance Z is equal to the resistance R, R_c/critical, and R_{sc}/supercritical, respectively. These resistances and Z–ω curves correspond to the various levels of coupling (1 – loose coupling/high Q; 2 – medium coupling/medium Q; 3 – critical coupling/low Q; 4 – over coupling/low Q). The ω_{h1} and ω_{h2} represent the hump frequencies that appear during the overcoupling/curve no. 4. On the right axis of the Fig. 11 the corresponding critical diameters d_1, d_2, and d_3 are arbitrarily plotted.

The theory presented has been applied to the representative experimental system described in Section III.A of this chapter. Validation of the theoretical predictions was corroborated experimentally by means of EIP measurements and NMR spectroscopy. These methods and aparatus are briefly presented in Sections III.C and III.D. The experimental results obtained were in fair agreement with the postulated theory. Measured, calculated, and estimated data are presented in Sections I.V.A and IV.B [6, 7].

III. EXPERIMENTAL

A. Description of the System

The particular secondary liquid–liquid system which has been used to corroborate the validity of the theoretical predictions was a heavy-phase droplet/light-phase film structure immersed in a heavy-phase continuum (double emulsion). This system was the heavy-phase output from a "pump-mix" mixer–settler battery together with its entrained light phase. The battery is part of a pilot plant for extraction of uranium from wet phosphoric acid by the D2EHPA-TOPO process [6–11, 17, 21, 42, 68–83]. The heavy liquid was 5.6 M phosphoric acid, and the light liquid was a synergistic mixture of 0.5 M di(2-ethylhexyl) phosphoric acid (D2EHPA) and 0.125 M tri-n-octylphosphine oxide (TOPO) in dearomatized kerosene (DTK) [6–11, 21].

The structural formulas of the constituent liquids are

$$
\begin{array}{ccc}
\text{OH} & \text{OH} & \text{R}' \\
| & | & | \\
\text{HO}-\text{P}=\text{O} & \text{RO}-\text{P}=\text{O} & \text{R}'-\text{P}=\text{O} \\
| & | & | \\
\text{OH} & \text{OR} & \text{R}'
\end{array}
$$

where R and R' are given, respectively, by

$$
\text{R:} \qquad -\text{CH}_2-\underset{\underset{\text{C}_2\text{H}_5}{|}}{\text{CH}}(\text{CH}_2)_3\text{CH}_3 \qquad\qquad \text{R}': \qquad -(\text{CH}_2)_7\text{CH}_3
$$

Figure 12 shows the measured variations of the physical properties of the liquids (density, viscosity, and interfacial tension) with temperature.

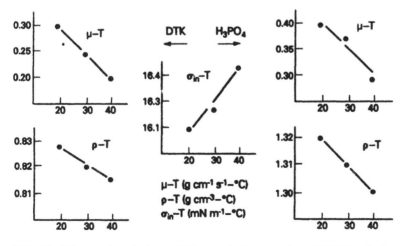

FIG. 12 Measured variations of the physical properties of DTK and phosphoric acid solutions with temperature, ρ: density, μ: viscosity, and σ_{in}: interfacial tesion. (From Ref. 10, with permission from Elsevier Science.)

B. Generation of Physical Model

A polydispersion was generated and the primary separation performed in a laboratory mixer for the batch studies under the conditions applied in the pilot plant [6, 21]. The selected hydrodynamic characteristics in the mixer unit were as follows:

- The ratio of the phases in the mixer, light/heavy, was equal to 1.1.
- The number of revolutions of the eight-blade double-shrouded impeller in the mixer was equal to $15\,s^{-1}$.
- The mixing criterion, $\rho n^3 D^2$, was equal to 13 (ρ is the density of the heavy phase, n is the number of revolutions of the impeller, and D is the diameter of the latter).

Thereafter, the sample of the heavy phase, together with the entrained light phase, was isolated and observed in situ with an optical microscope. Figure 13 shows in situ photographs of the examined liquid/liquid droplet-film structure immersed in the droplet homophase continuum. This isolated system is in hydrodynamic and electrodynamic equilibrium, i.e., the rigid sphere can be observed.

Further, Fig. 14 shows in situ photographs of the system examined at the junction point of the droplet-film structure and plate. The glass plate disturbs this system during observation, i.e., the elastic sphere can be observed. Finally, Fig. 15 shows in situ photographs where both rigid and elastic spheres may be observed.

C. Measuring Electrical Interfacial Potential

A method and apparatus were developed to monitor voltammetrically the EIP appearing during the formation of the EDL while the two-phase contact occurs [10, 21]. Measurements of the EIP have been performed during the processes of formation and transition of the electroviscoelastic sphere into the rigid sphere [6]. Figure 16

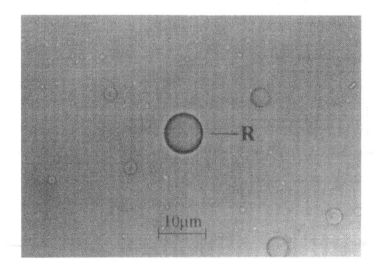

FIG. 13 In situ photograph of the liquid/liquid droplet-film structure immersed in the droplet homophase continuum; R, rigid sphere.

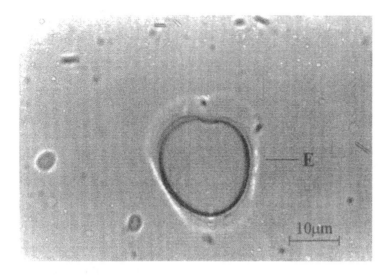

FIG. 14 Microscope image of the system at the point of contact between the droplet-film structure and a glass plate; E, elastic sphere, disturbed. (From Ref. 6, with permission from Academic Press.)

shows the developed liquid/liquid contact cell with its peripheral modulus for EIP measurements.

Measurements of EIP, during EDL formation at the interface boundary between two liquids, were performed by introducing the heavy liquid through a syringe, whose needle constituted one electrode, into a platinum vessel, which constituted the other electrode. The electrodes were connected via a sensor (high-input impedance instrumental amplifier) to an oscilloscope with memory. A Faraday cage, to avoid effects of the environment, surrounded the electrodes and the sensor.

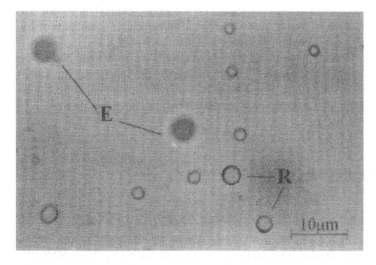

FIG. 15 In situ photograph of the system; R, rigid spheres, E, disturbed elastic spheres ("breathing" expansion/contraction). (From Ref. 6, with permission from Academic Press.)

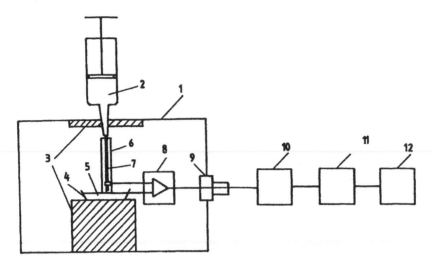

FIG. 16 Liquid/Liquid contact cell with its peripheral modulus: (1) Faraday cage; (2) syringe with the heavy phase; (3) insulators; (4) platinum vessel; (5) light phase; (6) plastic tube; (7) steel needle; (8) high-input impedance amplifier; (9) connector with coaxial cable; (10) oscilloscope with memory; (11) analog-to-digital converter; (12) data acquisition system.

During the liquid/liquid contact cell development, two important arguments were considered: first, concerning very low-energy levels appearing during EDL formation at the boundary surface of two immiscible liquids, and second, the possible influence of the cell construction elements on the impedance structure determination.

D. Measuring Resonant Frequency

In order to determine the resonant (characteristic) frequency, a NMR spectrometer was used as a reactor for the energetic analysis. The impedance Z at the resonant frequency ω_0 is equal to the resistance R. The resonant frequency of the electromechanical oscillator can be considered as some characteristic frequency within the vibrorotational spectrum of the molecular complex that builds the droplet-film structure [6].

All experiments were performed and all spectra acquired on a Bruker MSL 400 spectrometer with a 9.395 T magnet and at a ^{31}P frequency of 161.924 MHz. The transmitter was set at resonance frequency with phosphoric acid standard solution, and a sweep width of 15 kHz was employed. The swept region corresponded to the range between −10 and 90 ppm.

IV. RESULTS AND DISCUSSION

A. EIP Measurements

Figure 17 shows the measured change in EIP appearing during the introduction of the heavy-phase droplet into the light-phase continuum [6]. It can be seen in the figure that an interfacial jump potential peak appears during the formation of the EDL. Thereafter, the EIP decreases to a constant value. The lowering of the EIP in absolute value during the flow is due to the participation of cations that form the dense part of the EDL. The anions

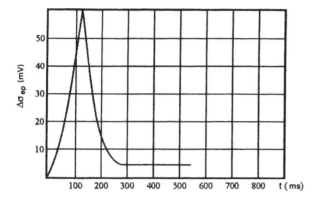

FIG. 17 Measured variations of EIP with time for the system phosphoric acid/D2EHPA–TOPO–kerosene at a spherical interface. (From Ref. 6, with permission from Academic Press.)

are the counterions in the diffuse part. Redistribution of these anions and cations between the region close to the surface and the surface layers of the heavy-phase define the kinetics of the EIP [6, 10].

Figure 18 shows the measured spontaneous oscillations of the EIP during the "breathing" period. After the EIP jump, which is in the millivolt–millisecond scale, the EIP continues to oscillate in the millivolt–minute range. Its damped oscillatory mode is (probably) due to the hydrodynamic instability of the interfacial surface, as a consequence of the local gradients of interfacial tension and density in mutual saturation processes of liquids [6, 10]. Other relevant interpretations of the EIP spontaneous oscillations may be expressed as follows: the electroviscoelastic sphere undergoes transformation into the rigid sphere. This transformation process can be understood as memory storage.

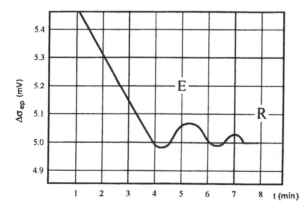

FIG. 18 Measured spontaneous oscillations of the EIP during the "breathing" period; transformation of the electroviscoelastic sphere into the rigid sphere. (From Ref. 6, with permission from Academic Press.)

B. ³¹P NMR Measurements

Figure 19 shows ³¹P NMR spectra of the molecular complex that builds the examined droplet-film structure. The impedance Z at the resonant frequency ω_0 is equal to the resistance R. The resonant frequency of the electromechanical oscillator can be considered as some characteristic frequency within the vibrorotational spectrum of the molecular complex that forms the droplet-film structure.

V. EXPERIMENTAL IMPLICATIONS

A. Application to Particular Entrainment Problem in Solvent Extraction: Breaking of Emulsions

This part of the chapter is an attempt to present some of the practical/engineering utility potentials of the discussed topics. Therefore, some of the electromechanical principles presented here have been used for elucidation of the secondary liquid–liquid phase separation problems, methods, equipment, and/or plant conception. The example given is related to the appearance of droplets/emulsions or droplet-film structures/double emulsions in solvent-extraction operations. Since these emulsions or double emulsions occur as an undesirable side effect, both nondestructive and destructive methods for their separation or elimination can be shown. Table 1 presents the two kinds of methods for secondary liquid–liquid phase separation [21].

The particular problem considered as the representative one was mechanical entrainment of one liquid phase by the other in the solvent-extraction operation. Experimental

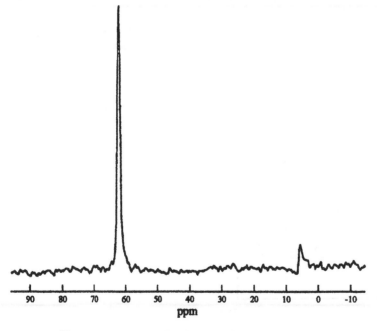

FIG. 19 ³¹P NMR spectrum of D2EHPA–TOPO–kerosene sample (phosphoric acid standard solution, sweep width of 15 kHz); the peak at 7 ppm corresponds to D2EHPA and that at 62 ppm corresponds to TOPO. (From Ref. 6, with permission from Academic Press.)

TABLE 1 Secondary Liquid–Liquid Phase Separation Methods

Considerations	Methods								
	Coalesc. lamellar	Coalesc. densely packed	Flotation aero	Flotation air induced	Electrostatic	Electrodynamic	Centrifugal	Dissolution	Demulsifier
Process									
Eff. (%)	60–70	90	–	70–80	80–95	–	–	–	–
Sens. (%)	0.01	0.002	–	0.005	0.001	–	–	–	–
Operat.									
Thro.	Large	Small	Small	Large	Small	–	Small	–	–
Flex.	Yes	No	No	Yes	No	–	No	Yes	Yes
Maint.	Easy	Diff.	Diff.	Easy	Diff.	–	Diff.	Easy	Easy
Specific	–	STC TSP SFC	STC TSP SFC	STC TSP SFC	TSP	–	–	–	–
Design									
Information patents	Ins.	Ins.	Ins.	Many	Ins.	No	Ins.	Ins.	Ins.
Economic									
C. costs	Low	High	Medium	V. high	V. high	V. high	Low	Low	Low
Rentability	Medium	High	–	Medium	Low	Low	High	High	Low

Coalesc. = Coalescence, Eff. = Efficiency, Sens. = Sensitivity, Operat. = Operational, Thro. = Throughput, Flex. = Flexibility, Maint. = Maintenance, Diff. = Difficult, STC = sensitive to temperature changes, TSP = sensitive to presence of third stable phase, SFC = sensitive to feed content of entrained phase, Ins. = Insufficient, C = Capital.

Source: From Ref. 21 (reprinted with permission from Elsevier Science).

results obtained in the pilot plant for uranium recovery from wet phosphoric acid were used as the comparable source [7, 10, 21, 67–79, 84–87]. In this plant, a secondary liquid–liquid phase separation loop has also been carried out [36–38]. The loop consisted of a lamellar coalescer and four flotation cells in series. The central equipment in the loop, relevant to this investigation, is the lamellar coalescer. The phase separation in this equipment is based on the action of external forces of mechanical and/or electrical origin, while adhesive processes at the inclined filling plates occur [10, 73, 88–92].

A complete evaluation of the details of this solvent-extraction procedure can be found in Ref. 21.

B. Other Applications

There are indications, experimentally corroborated, that a new constitutive model of liquids can be generalized for liquid–solid and liquid–gas systems (micro, nano, and atto). Micro- and nano-dispersed systems can behave either as elastic and/or rigid, while atto-dispersed systems behave only as plastic. Examples of micro- and nano-dispersed systems are suspensions, emulsions, fluosols, and foams. Finally, atto-dispersed systems are conductors of various kinds, including information streams, beams of phonons, photons, electrons, or ions, and combinations of the latter.

It might be of interest in the future to derive a rigorous mathematical formalism for the developed theory of electroviscoelasticity. This formalism could begin with the Hamilton and Onsager–Sedov approaches, which have been applied in the nonlinear theory of fluid-permeable elastic continua by Grinfeld and Norris [93], and thereafter more sophisticated mathematical instruments may be needed [20].

Also, further evaluation of the idea that the basic entity can be understood as an energetic ellipsoid (based on the model of the electron following Maxwell–Dirac isomorphism, MDI [94, 95] seems to be sensible. MDI states that the electron is an entity, at the same time quantum-mechanical (microscopic) and electrodynamic (macroscopic).

VI. CONCLUSIONS

A brief recapitulation of the theoretical and experimental contents of this chapter could be shown, using the developed theory of electroviscoelasticity, and its three forms of "instabilities," namely, rigid, elastic, and plastic. Figure 20 summarizes the correlation between the EIP and time, related to the formation of the electroviscoelastic sphere and its transition into the rigid sphere.

A new constitutive model of liquids is supposed to facilitate the understanding of liquid//liquid interfaces, and in particular the interactions and phenomena which occur at very developed surfaces, e.g., in emulsions or double emulsions.

The theory of electroviscoelasticity constitutes a new interdisciplinary approach to colloid and interface science. This theory can be helpful in solving entrainment problems in solvent extraction, for what this research was initiated fourteen years ago [21]. Furthermore, this knowledge can be used, e.g., in studies of fine dispersed systems (micro, nano, and atto; suspensions, emulsions, fluosols, foams, and beams/streams of entities), the physics of liquids, and biological systems (neurophysiology/infon transfer).

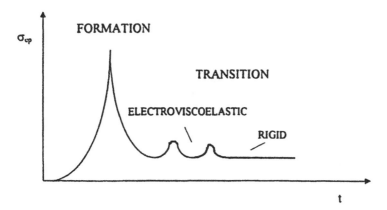

FIG. 20 Correlation of EIP with time, during formation of the electroviscoelstic sphere and its transition into the rigid sphere.

Finally, using the knowledge and/or some of the ideas presented and discussed, it will be possible to develop a general knowledge and database organization system related to the breaking of emulsions. This job can be done for both nondestructive and destructive methods for secondary liquid–liquid phase separation, as indicated in Table 1.

REFERENCES

1. IL Osipov. Surface Chemistry. New York: Reinhold, 1964, pp 295–340.
2. A Adamson. Physical Chemistry of Surfaces. New York: John Wiley, 1967, pp 505–533.
3. TC Lo, M Baird, C Hanson, eds. Handbook of Solvent Extraction. New York: John Wiley, 1983, pp 275–514.
4. J Jaric. Mehanika kontinuuma. G.K. Beograd, 1988, pp 117–159.
5. RJ Pugh. Adv. Colloids Interface Sci 64:67, 1996.
6. AM Spasic, V Jokanovic, DN Krstic. J Colloid Interface Sci 186:434, 1997.
7. AM Spasic, V Jokanovic. J Colloid Interface Sci 170:229, 1995.
8. AM Spasic, DN Krstic. ISE & ECS Joint Meeting Proceedings, Paris, 1997, p 415.
9. AM Spasic. Proceedings of the 12th CHISA, Prague, 1996, C1.6.
10. AM Spasic. Chem Eng Sci 47:3949, 1992.
11. AM Spasic, DN Krstic. Proceedings of 13th CHISA, Prague, 1998, CD 7.
12. J Godfrey, C Hanson, MJ Slater, S Thamalingam. AIChE Symp Ser 74:127, 1978.
13. JC Godfrey. The Formation of Liquid–Liquid Dispersions – Chemical and Engineering Aspects – Flow Phenomena of Liquid–Liquid Dispersions in Process Equipment. London: Institute of Chemical Engineers, 1984.
14. TF Tadros. Interfacial Aspects of Emulsification. London: Institute of Chemical Engineers, 1984.
15. S Hartland, SM Wood. AIChE J 19:810, 1983.
16. W Spisak. Nature 349:23, 1991.
17. AM Spasic, NN Djokovic, MD Babic, MM Marinko. 13th CHISA, Prague, 1998, CD 7.
18. JY Oldshue. Fluid Mixing Technology. New York: McGraw-Hill, 1983, pp 125–140.
19. LL Schramm. Emulsions-Fundamentals and Application in Petroleum Industry. Washington, DC: American Chemical Society, 1992, pp 131–170.
20. M Mitrovic, J Jaric. Lecture, Mathematical Faculty. University of Belgrade, Belgrade, 1997.
21. AM Spasic, NN Djokovic, MD Babic, MM Marinko, GN Jovanovic. Chem Eng Sci 52:657, 1997.

22. EK Drexler. Nanosystems; Molecular Machinery, Manufacturing and Computation. New York: John Wiley, 1992, pp 161–189.
23. SE Friberg, R Bothorel, eds. Microemulsions: Structure and Dynamics. Boca Raton, FL: CRC Press, 1987, pp 173–195.
24. EU Condon, H Odishaw, eds. Handbook of Physics. New York: McGraw-Hill, 1958, pp 4, 13.
25. I Prigogine. The Molecular Theory of Solutions. New York: North Holland, 1957.
26. T Misek. Proceedings of ISEC, '86, Munich, 1986, vol. III, p 71.
27. RF Probstein. Physicochemical Hydrodynamics. New York: John Wiley, 1994, pp 352–369.
28. A Cloot, G Lebon. J Phys 47:23, 1986.
29. JRA Pearson. J Fluid Mech 4:489, 1958.
30. HR Kruyt. Colloid Science. Vol I. Amsterdam, Houston, New York, London: Elsevier, 1952, pp 302–341.
31. YH Pao. J Appl Phys 28:352, 1957.
32. DD Joseph. Fluid Dynamics of Viscoelastic Liquids. New York: Springer-Verlag, 1990, pp 539–596.
33. JN Israelachvili. Intermolecular and Surface Forces. New York: Academic Press, 1992, pp 260–339.
34. CR Reid, JM Prausnitz, BE Pauling. The Properties of Gasses and Liquids. New York: McGraw-Hill, 1989, pp 632–655.
35. I Prigogine, R Defay. Chemical Thermodynamics. London: Longmans, 1965, pp 32–47; 437–449.
36. OF Devereaux. Topics in Metallurgical Thermodynamics. New York: John Wiley, 1983, pp 333–372.
37. M Karapetiantz. Thermodynamique Chimique. Moscow: Mir, 1978, pp 110–135.
38. ACD Rivet. The Phase Rule and the Study of Heterogeneous Equilibrium. Oxford: Clarendon Press, 1923, pp 25–29.
39. HJ Perry. Chemical Engineering Handbook. New York: McGraw-Hill, 1941, pp 1213–1268.
40. DC Hodgman, CR Weast, MS Selby. Handbook of Chemistry and Physics. 40th ed. Cleveland, OH: Chemical Rubber Co., 1958, pp 2512.
41. MA Garstens. J Appl Phys 28:352, 1957.
42. AM Spasic, ed. Multiphase Dispersed Systems. Belgrade: ITNMS, 1997, pp 1–46.
43. WM Rosenow, H Choi. Heat, Mass and Momentum Transfer. Englewood Cliffs, New Jersey: Prentice Hall, 1961, pp 24–29.
44. R Schuhmann Jr. Metallurgical Engineering. vol. I. Engineering Principles. Cambridge, MA: Addison Wesley, 1952, pp 143–146.
45. AG Turner. Heat and Concentration Waves. New York: Academic Press, 1972, pp 94–118.
46. SH Davis. J Appl Mech 50:977, 1983.
47. JL Ericksen. Isledovanie po mehanike splosnih sred. Moscow: Mir, 1977.
48. PT Andjelic. Tensorski racun. N.K. Beograd, 1967, pp 263–266.
49. M Plavsic. Mehanika viskoznih fluida. PMF, University of Belgrade, 1986, pp 33–44.
50. D Gallez, A De Wit, M Kaufman. J Colloid Interface SWci 180:524, 1996.
51. D Gallez, WT Coakley. Heterogenous Chemistry Reviews, 1996, pp 1–70 (to appear).
52. Nabile TM Eldabe. J Mater Phys 28:2791, 1987.
53. M Takashima, AK Gosh. J Phys Soc Jpn 47:1717, 1979.
54. MJ Pilling. Reaction Kinetics. Oxford: Clarendon Press, 1975, pp 37–48.
55. EN Yeremin. The Foundation of Chemical Kinetics. Moscow: Mir, 1979, pp 103–149.
56. LI Sif. Quantum Mechanics. New York: McGraw-Hill, 1955, pp 7–90.
57. AN Krall, WA Trivelpiece. Principles of Plasma Physics. New York: McGraw-Hill, 1973, pp 98–128.
58. A Sen. Modern Plasma Physics. Trieste course 1979, Vienna: IAEA, 1981, pp 249–273.
59. RPH Gasser, WG Richards. Entropy and Energy Levels. Oxford: Clarendon Press, 1974, pp 27–38.
60. JO Hirchfelder, CF Curtiss, RB Bird. Molecular Theory of Gases and Liquids. New York: John Wiley, 1954, pp 139.

61. S Ono, S Kondo. In: S Flugge, ed. Handbuch der Physik. vol. X. Structure of Liquids. Berlin: Springer-Verlag, 1960.
62. M Peric, B Engels, SD Peyerimhoff. J Mol Spectrosc 171:494, 1995.
63. M Peric, M Thummel, CM Marian, SD Peyerimhof. J Chem Phys 102:7142, 1995.
64. M Peric, SD Peyerimhof. J Chem Phys 102:3685, 1995.
65. M Peric, SD Peyerimhof, RJ Buenker. Mol Phys 55:1129, 1985.
66. RA Millikan. Electrons (+ and −), Protons, Neutrons, Mesotrons, and Cosmic Rays. 1934, Belgrade: Prosveta, pp 249–275 (1948 translation).
67. WI Orr. Radio Handbook. 17th ed. New Augusta, IN: Editors and Engineers, 1967, pp 60–65.
68. MD Babic. Proceedings of 9th CHISA, Prague, 1987, D 3.70.
69. MD Babic, AM Spasic, MM Marinko, NN Djokovic. Uranium in Phosphoric Fertilizers and Possibility of its Elimination. Belgrade: SANU, 1983, pp 41–55.
70. D Negoicic, N Djokovic, A Spasic, M Babic, A Tolic. Proceedings of 2nd Congress on Chemical Engineering and Process Technology, Durbrovnik, 1987, p 269.
71. A Spasic, N Djokovic, N Canic, M Babic, A Tolic. Proceedings of 9th CHISA, Prague, 1987, D3 72.
72. A Spasic, N Djokovic, N Canic, M Babic. Proceedings of ISEC '88, Moscow, 1988, vol. IV, 227 pp.
73. AM Spasic, NN Djokovic, MD Babic, GN Jovanovic. Chem Biochem Eng Q 5:35, 1991.
74. AM Spasic, NN Djokovic, MD Babic, GN Jovanovic. Proceedings of 11th CHISA, Prague, 1993, p D9.23.
75. AM Spasic, NN Djokovic, MD Babic, GN Jovanovic. Proceedings of 11th CHISA, Prague, 1993, p D9.24.
76. AM Spasic, V Jokanovic, DN Krstic. Proceedings of 187th ECS Meeting, Liquid/Liquid Interfaces, Reno, NV, 1995.
77. AM Spasic. Lecture in Symposium on Selforganization of Nonequilibrium Processes, Ecka, Yugoslavia, 1995.
78. AM Spasic. A Theory of Electroviscoelasticity: Structure and Stability of Fine Dispersed Systems, Lecture at PMF, University of Belgrade, 1995.
79. MD Babic, AM Spasic, MM Marinko, NN Djokovic. Proceedings of Geo Congress, Uranium-Mining and Hydrogeology, Bergakademie, Freiberg, 1995, 9 pp.
80. V Jokanovic, D Janackovic, AM Spasic, D Uskokovic. Mater Trans JIM 37:627, 1996.
81. AM Spasic, NN Djokovic, MD Babic, MM Marinko, DN Krstic. Physical Chemistry '98, Belgrade, 1998.
82. AM Spasic, MM Marinko, MD Babic, NN Djokovic, P Jovanic, V Jokanovic, NV Vunjak, GN Jovanovic, DN Krstic. Centennial of the Serbian Chemical Society, Belgrade, 1997, p 78.
83. AM Spasic, DN Krstic. Lecture at the University of Belgrade, Belgrade, 1999.
84. G Cordero, LG Jodra, JL Otero, JM Josa. Proceedings of Conference on Nuclear Power and Fuel Cycle, IAEA, 1977, 377 pp.
85. FJ Hurst. Ind Eng Chem Process Des Deve 11:122, 1972.
86. FJ Hurst WD Arnold. Hydrometallurgy 9:69, 1982.
87. D Petkovic. Nucl Technol 4:29, 1982.
88. V Hancil, V Rod, J Reznickova. Proceedings of ISEC'86, Munich 1986, vol. III, p 81.
89. JM Josa, A Moral. Proceedings of ISEC'80, Liege, 1980, p 80.
90. JD Scherwood, J Nitmann. J Phys 47:15, 1986.
91. W Jost. Diffusion. New York: Academic Press, 1952, pp 436–479.
92. L Kolar-Anic. Osnove Statisticke Termodinamike. Belgrade: University of Belgrade, 1995, pp 174–184.
93. MA Grinfeld, AN Norris. Int J Eng Sci 35:75, 1997.
94. D Koruga. Fulerenes & Nanotubes Review 1:75, 1997.
95. B Ostojic, M Peric, J Radic-Peric. Centennial of the Serbian Chemical Society, Belgrade, 1997, p 77.

31

Surfactant-Stabilized Emulsions from an Electrokinetic Perspective

O BOEN HO   Akzo Nobel Surface Chemistry, Deventer, The Netherlands

I. INTRODUCTION

Many products involve emulsification in the exploitation or manufacturing process. The applications range from bitumen emulsions and drilling fluids to food products, cleaning and personal care (including cosmetics), carrier media for pharmaceuticals, and many more. Industrial interest in emulsions and emulsification and in their physical chemistry is evident. Industrial research is, however, characterizied by a pragmatic approach and a strong focus on development and technology. The incentive of product or process development renders the industrial approach principally driven by performance. Academic basic science principally opens up new horizons. It may, therefore, focus on mechanistic details, theories, and tests. This does not necessarily mean that the two approaches cannot be mutually reinforcing. Nevertheless, starting points remain different, and one must be aware of some practical differences. For example, the industrial definition of emulsion stability or stability in a more general sense differs from the one academics tend to practice. Industrial product stability focuses on (shelf) lifetime, or product tenability. "Industrial" stability expresses the product status quo rather than some point in a phase diagram.

Industrial incentives and objectives seem to drive industrial emulsion research to "toolkit approaches." To illustrate this statement we discuss two well known examples:

- To select the appropriate set of surfactant ingredients, people in the field usually apply the so-called HLB method [1]. The HLB is an abbreviation of *hydrophile lipophile balance*. It assumes that all surfactants and surfactant mixtures can be characterized by an HLB value. The method couples the performance of an emulsion system with a match between a surfactant system and a substracted emulsion system. Performance optimization is reduced to two steps. First, the two (or poly) phase system has to be characterized in terms of an HLB. This is called the *required* HLB. Once this HLB is identified, then further optimization occurs by testing several surfactant types and sets of surfactant mixtures, all having an HLB equal to the required HLB.
- Within a processing method aimed at obtaining the smallest emulsion droplets the PIT method [2] is often applied, which involves the identification of a critical temperature. At this temperature, the dispersed phase and continuous phase interchange. This temperature is called the PIT (*phase inversion temperature*).

The final process contains a trajectory in which the emulsion system is brought to the PIT, dispersed and brought back to the basic temperature.

We will return to this "tools" aspect later on.

By definition, emulsions consist of at least two immiscible liquids, one dispersed in the other. However, emulsions consisting of only two components hardly occur. Usually, they consist of at least three components apportioned into two phases. The third component of the emulsion is the dispersing agent. The dispersing state can be maintained employing several means. We ignore the use of rheological modifiers such as thickeners, and focus on the surfactant (the majority of emulsions is surfactant based).

The literature shows a bias towards specific types of emulsion systems. The majority of references are dedicated to the so-called "oil-in-water" (o/w) systems. This is a constraint we will have to keep in mind. Therefore, our objective will be to consider the role of surfactants in aqueous emulsions in the context of electrical phenomena. We will discuss the electrokinetics of emulsions from two perspectives. First, the interfacial potential is treated from a phenomenological point of view. The role of the interfacial potential in emulsions is then considered. As detailed treatments of the basics of electrokinetics have preceded this contribution, no detailed attention will be given to these topics.

This chapter includes four sections. To begin with, a framework will be outlined. Aspects such as classes of surfactant, types of emulsions, and emulsification will be treated in more detail. This will be done in Section II. In this section, measuring techniques are also briefly discussed. The practical value of commonly employed techniques will be reviewed. Section III focuses on the origin and nature of the droplet potential. Section IV provides interpretation and application of electrokinetic data.

II. SURFACTANTS, EMULSIFICATION, AND DETECTING INTERFACIAL POTENTIALS

A. The Surfactant

There exists a bewildering variety of surface-active agents. At present, approximately 60,000 commercial products are available [3]. Evidently, this large variety clearly necessitates some sort of classification of these agents (usually abbreviated as *surfactants*). There are many ways to accomplish that. In most cases, classes are defined on the basis of the nature of the polar or charged *head* group. On the basis of the type of charge, four classes can then be distinguished:

- Anionic, including the carboxylates, sulfates, sulfonates, and the phosphates.
- Nonionic, for the ethoxylates and glycerides.
- Cationic, including the quaternary ammonium salts.
- Amphoteric when anionic and cationic species are combined into more complex structures.

This is the simplest and most obvious classification and for a long time it was considered satisfactory. Nowadays, there are also many hybrids, which render the above classification incomplete or even incorrect. Therefore, other ways of classification may be preferred. One obvious approach is by taking typical properties of a surfactant, e.g.:

- The number of elements within the molecule: monomers, dimers, up to polymers.
- Its origin: biosurfactant, animal, industrial, etc.
- Its function: foaming agents, demulsifiers, dispersants, etc.

Obviously, any classification has restraints. In fact, the quintessence of classifying is to differentiate between molecules based on a particular property or feature. Recalling the electrical aspect of the objective in mind, employing the difference in charge type evidently provides sufficient differentiation. In order to differentiate the class of nonionic surfactants, we characterize them in terms of the HLB. This value was calculated by using Griffin's expression [1]:

$$HLB = 20 \times (M_H/M_T) \tag{1}$$

where M_H and M_T are the molecular weights of the hydrophilic moiety of the surfactant and the whole surfactant, respectively. To avoid long chemical names, we will identify a number of surfactants by their trade names. Other surfactants will be denoted by abbreviations.

B. Types of Emulsion

There are several types of emulsions. Usually, they are differentiated in terms of the dispersion structure or the input required in their formation and the related issue of stability.

When the stability is considered, three types of emulsions are recognized. The so-called macroemulsion is fundamentally unstable and requires high-energy input. In time, this emulsion separates into the different liquid phases. To maintain the system in this state of high (free) energy a stabilizing agent must be added. As such, stability implies the slowing down of the destabilization process. Unlike macroemulsions, microemulsions are thermodynamically stable. No external force is needed to obtain such a system: it occurs spontaneously once all ingredients are present. Spontaneous emulsification is also observed in miniemulsion systems. Just like macroemulsions, miniemulsions are not thermodynamically stable. In time, they deteriorate. The spontaneous formation is a complex aspect, which is not fully understood. One prerequisite must be the low interfacial tension. At low interfacial tension, small mechanical perturbations are sufficient to generate small droplets. Although the droplet size is not the critical factor, miniemulsions differ from macroemulsions by their droplet size. Microemulsions do not have a specific range in which they are stable. In our considerations we propose to deal with macroemulsions.

When focusing on the nature of the liquid phases comprising the emulsion, a different division can be made. We know many types of liquids. Even under ambient conditions one can recognize a metallic type (mercury), organic liquids (hydrocarbons, esters, and ethers), polar liquids (like methanol or water), and a class including silicon and fluorinated organic substances. In practice, we limit our attention to two types of liquids: "oil" and "water." Within this limitation, we distinguish two basic types of emulsions. In Section I we mentioned the o/w type with the oil dispersed in water. The water-in-oil (w/o) emulsion comprises oil as the continuous phase and water as the dispersed phase. Finally, one may encounter multiple emulsions. In these emulsions at least two phases are dispersed: the one in the other.

C. Emulsification

The composition and evolution of an emulsion determine its properties. In Section I we briefly discussed the HLB concept for choosing ingredients. Another aspect, which needs some attention, is the emulsification process, i.e., the way the emulsion is made. In particular the dosing sequence of ingredients has a strong impact on the properties of the

TABLE 1   Properties of Two Emulsions Made Along Different Routes

Method applied	Zeta potential (mV)	Flocculation rate ($\mu m^3 s^{-1}$)	Number of droplets (mm^{-3})
A	83	0.045	28×10^5
B	65	0.082	14×10^5

emulsion. To illustrate this aspect we briefly discuss the observation by Stalidis *et al.* [4]. Two 0.5% emulsions of methyl cyclohexane in 0.01 M aqueous NaCl solution were prepared by mixing in two surfactants: polyoxyethylene (20 EO) sorbitan mono-oleate (Tween 80, 2.25 $\times 10^{-3}$%) and sodium lauryl sulfate (SLS, 1.28 $\times 10^{-3}$%). To obtain these emulsions, two different methods were applied. In method A, both surfactants were added together and before the sonification step; in method B, Tween 80 and SLS were added separately. Just as in method A, Tween 80 was added before sonification; SLS was, however, added afterwards. The properties of the emulsions obtained were different, as is illustrated by Table 1. The emulsion prepared by applying method A shows a more negative zeta potential, is more stable, and is more finely divided than the one prepared by applying method B.

D.   Techniques for Determining Potential

To explore interfacial charge structure, electrokinetic or titration techniques are commonly employed. This is just a matter of experimental convenience (since the other electrokinetic phenomena essentially yield the same information). In electrophoresis, determining the "potential" usually implies determining the "zeta potential" (ζ). Here, the dispersion is placed between electrodes in an electric field. When charge is present on the surface of the droplets, they will move in the electric field. The displacement of the dispersion boundary (per unit of time) is a measure of the zeta potential. Nowadays, sophisticated methods are available. If the periodicity of the electric field is considered, two main techniques can be distinguished. One group of techniques, the AC technique, applies a high-frequency alternating current. Dispersion mobility is derived from the acoustic which the dispersion generates. In addition to this "electrosonic amplitude" (ESA), option, modern electrophoresis equipment often provides a "streaming potential" option. In this complementary mode the droplets are moved by sound waves, giving rise to a "colloid vibration current." The high-frequency AC section then functions as a receiver. In the alternative technique a stationary current is imposed. This is often referred to as the DC technique of which classical electrophoresis is an example. The dispersion mobility is determined by means of optical techniques such as Doppler shifts. Finally, there are titration techniques. The classical method of titration uses a potential-determining ion as titrant. In general, this laborious technique is not appropriate to emulsions, as titration may modify the system. We limit our discussion to a special type of titration technique.

   The DC technique is almost synonymous with microelectrophoresis or optical electrophoresis. Two papers are illustrative of the possibilities of current microelectrophoresis techniques. Hantz et al. [5] applied so-called electrophoretic light scattering to study their emulsion system. In this technique, light scattering is combined with microelectrophoresis. In addition to these options it yields an apparent droplet charge. The setup employed by Pisárčik et al. [6] allows for monitoring of zeta-potential distribution. A drawback of DC electrophoresis is its requirement of dilution. "Best-guess" supernatant serum partly mimicks continuous phase conditions, but cannot compensate for the effect of dilution. Besides, the advantage of seeing the subject(s) must be questioned, as one tends to become subjective and to discriminate certain species of the polydisperse population.

The AC technique is founded on a firm theoretical basis. In a recent paper by Dukhin et al. [7] an "electroacoustic" theory is presented, covering systems containing up to 45% (by volume) of dispersed phase. By modulating the frequency with which the current is generated, droplet size distributions can also be obtained, as shown by Hunter and O'Brien [8]. Another advantage of the AC technique is its semiquantitative character. Sometimes results do not need to be quantitative as has been demonstrated by the author [9] as relative effects can be attributed to differences in surfactant systems. We will discuss this point in more detail later on. A disadvantage of the AC technique is the need for rather detailed knowledge of the studied system. Additionally, a density difference between the dispersed and continuous phase is required in order to produce an acoustic effect. Also, whereas the DC technique produces a moderate and stationary distortion of the droplets (but large deformation of the double layer), the (high frequency) AC technique keeps droplets in a permanent transient state. Investigation of this issue has led to the consensus that there is a negligible level of disturbance.

Users tends to prefer a specific type of equipment. However, the choice depends on the particular application one has in mind, since each technique focuses on certain properties and features, disregarding other effects. Table 2 provides a short (and necessarily incomplete) listing of features characterizing the two techniques.

We end this section with the discussion of a remarkable method for obtaining interfacial potentials. Grieser and Drummond [10] were able to determine so-called micellar "electrostatic interfacial potentials." Their method applies the shift in apparent acidity when indicator molecules experience an electric field. This shift obeys Nernst's law:

$$\psi_{\text{Nernst}} = 2.303(kT/e)(pK_a^0 - pK_a^{\text{obs}}) \tag{2}$$

where ψ_{Nernst} is the Nernst potential, and pK_a^0 and pK_a^{obs} are the intrinsic and the apparent acidity constants of the indicator, respectively. By "placing" indicator molecules in the neighborhood of the micelle surface, a shift is thus expected when the nature of the surfactant is ionic. The interpretation of the Nernst potential depends on the distance of the micellar interface to the average indicator molecule. A special selection of indicators

TABLE 2 Advantages of Both the AC and DC Techniques

DC Technique	AC Technique
Implies well-controlled experimental conditions	System remains under near-equilibrium conditions
Allows combinations with light scattering techniques	In situ measurement
Selective (no interference by air bubbles)	Dilution not necessary
Droplets can be made visible	Semiquantitative up to high-volume fraction of dispersed phase
Yields information about the "individual" droplets	Data allow for direct applications, e.g., in heteroflocculation and other complex experiments
Zeta potential determined with a minimum of information	
Requires small quantity of sample material	Other information (droplet size distribution) can be acquired

facilitates nearest approach, allowing the reasonable assumption that the potential obtained is equal to the "wall" potential [11]. Obtaining the potential requires spectroscopic means and thus the system to be transparent. Application to emulsions is, therefore, limited to near-transparent miniemulsion systems.

III. NATURE OF THE DROPLET POTENTIAL

A. Origin of the Droplet Potential

The best-known examples of charged emulsion droplets are no doubt those prepared with ionic surfactants. These surfactants are, however, not the only source of droplet charge. It has been known for a long time that one does not need (ionic) surfactant to cause droplets to move in an external electric field. The following systems will be discussed:

- Droplets of pure oil in water.
- Oil droplets stabilized by nonionic surfactant in water.
- Oil-in-water emulsions containing ionic surfactants or a mixture of ionic and nonionic surfactants.

A significant negative potential is built up around droplets of pure oil in pure water. Table 3 gives an impression of the zeta potential on such "naked" droplets. A number of studies on these "naked" droplet/water systems has been reported. On the origin of the zeta potential a number of suggestions has been made [12–14]:

- Presence of impurities in one of the two phases.
- Polarization of one of the two phases.
- Preferential adsorption or desorption of electrolyte ions.
- Adsorption of hydroxyl ions or desorption of hydrogen ions.

Carruthers [12] studied the aqueous electrophoretic mobility of droplets of several types of oil differing in polarity. The poor correlation of oil polarity to the oil droplet mobility indicates the minor role of the oil phase in creating an interfacial potential. This is supported by the observation of zeta potentials (in the same order of magnitude) at air bubbles in water [15]. Marinova et al. [16] ruled out polarization in the aqueous phase. They did so by comparing the zeta potential of oil droplets in aqueous systems with and without urea. The zeta potential was the same irrespective of the presence of this chaotropic (water-destructuring) agent. As potential buildup is also observed in pure water systems, preferential adsorption or desorption of electrolyte ions as a *main* source of charge can be ruled out as well. In support of this claim, Marinova et al. also observed pH dependence of the zeta potential. One does not expect such a dependence when preferential adsorption or desorption of electrolytes dominates the potential. (Although these sorption processes cannot be neglected, as has been demonstrated by means of surface tension measurements.) In some of their experiments extra care had to be taken to avoid contamination. Zeta potentials remained unaltered, pointing to the minor role of contamination. Led by an increasing magnitude of the zeta potential with pH (see Fig. 1), most authors believe the potential is due to adsorption of hydroxyl ions. Still, this observation cannot exclude the simultaneous contribution of desorption of hydrogen ions or orientation (polarization) of the water molecules in the vicinity of the interface.

A considerable number of papers discuss the electrokinetics of emulsions stabilized with nonionic surfactant [5, 16–25]. These droplets also generate a zeta potential. They exhibit a close similarity to the "naked" oil droplets, suggesting a similar mechanism of

TABLE 3 Room-Temperature Zeta Potentials of "Naked" Oil Droplets and of Oil Droplets Covered by Nonionic Surfactants in Water

Emulsion system	ζ (mV)	Source (references)
Olive oil/mixture of (ethoxylated) sorbitan esters, HLB = 9/water, $I = 10^{-6}$ M,[a] pH = 6[a]	−37	19
Ethyl laurate/water,[b] $I = 0.01$ M, pH = 6	−45	12
Liquid paraffin/ethoxylated oleyl alcohol, HLB[c] = 8.5/water, $I = 5\,10^{-4}$ M, pH = 5.4	−49	5
Mineral oil/mixture of (ethoxylated) sorbitan esters, HLB = 9–10/water, $I = 10^{-6}$ M,[a] pH = 6[a]	−40 to −45	22
Octadecane/water, $I = 0.01M$, pH = 5.7	−23	12
Chlorobenzene/polyoxyethylene (6) hexadecanol, HLB = 10.5/water, $I = 10^{-6}$ M,[a] pH = 6[a]	−53	23
n-Octyl bromide/water, $I = 0.01$ M, pH = 6	−10	12
Toluene/sorbitan monolaurate (Span 20), HLB = 8.6/water, $I = 0.01$ M, pH = 6	−75	25
Xylene/ethoxylated octylphenol, HLB = 13.7/water, $I = 0.01$ M, pH = 6	−43	20
Xylene/water, $I = 10^{-3}$ M, pH = 6	−61	17
Decalin/water, $I = 10^{-6}$ M,[a] pH = 6[a]	−71	14
Air	−55 to −65	15, 26

[a] In pure water.
[b] Data on the "naked" oil system are expressed in italics.
[c] The nonionic surfactants are also characterized by their HLB.

interfacial polarization and corroborating the above arguments. However, introducing nonionic surfactants at the interface adds a new dimension to the problem, as a surface potential buildup can now also result from adsorption of ions from the bulk solution through binding to the surfactant. The oil drop/water potential difference may therefore be considered to be a consequence of:

- Adsorption of hydroxyl ions or desorption of hydrogen ions.
- Preferential adsorption of electrolyte ions through binding to surfactant.

Figure 1 also reproduces zeta potentials versus pH data collected by Tajima et al. [18]. A progressively more negative zeta potential is observed with increasing pH. As discussed above, this trend can be understood in terms of hydroxyl adsorption or desorption of hydrogen ions. Comparison of zeta-potential data collected with both "naked" oil/water systems and in the presence of nonionic surfactant shows these potentials to be essentially similar (see Table 3). It is therefore tempting to oversimplify and assume the same mechanism for systems containing nonionic surfactants. However, it must be kept in mind that the introduction of surfactant molecules at the interface will shift the plane of shear. Assuming an unchanged charge structure this is expected to result in a decreasing zeta potential, as was indeed observed by Marinova et al. [16]. Then, in view of the similarity in behavior as is evident from Table 3, the zeta

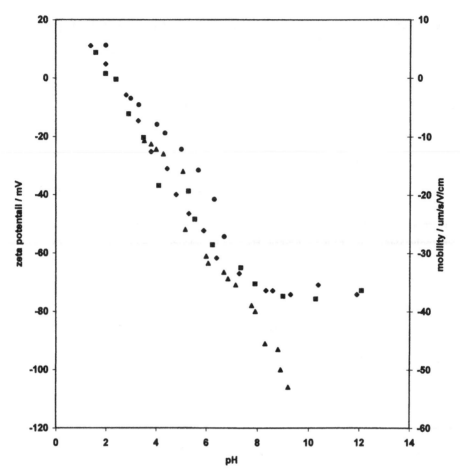

FIG. 1 Plots of zeta potential and electrophoretic mobility versus pH: (♦) Tajima et al. [18]: liquid paraffin/water/ethoxylated (3 EO) dodecanol; (■) Tajima et al. [18]: liquid paraffin/water/ethoxylated (6 EO) dodecanol; (▲) Marinova et al. [16]: xylene droplets in water; (●) Carruthers [12]: liquid *n*-octadecane/0.01 N NaCl.

potential must pass through a maximum as the surfactant concentration is increased. This indicates a change in the charge structure. The study of Becher and Tahara [22] shows that the zeta potential depends on the polyxoyethylene nonionic surfactant composition. In their view the zeta potential is built up by binding of hydroxyl ions to the ether groups.

The phenomenon of preferential adsorption has been studied by a number of authors. Yoshihara et al. [17] studied the effect of NaSCN and $Ca(SCN)_2$ on the zeta potential of a nonionic surfactant based *micro*emulsion. They observed a potential buildup with increasing SCN^- ion concentration. A tendency of the hydrophilic moiety of the surfactant to bind the anionic SCN^- ions more strongly than to the counterions (Na^+ and Ca^{2+}) was inferred. Rabinovich and Baran [27] studied the effect of chloride ions on the zeta potential. Figure 2 shows the zeta potential passing through an optimum as a function of the concentration of KCl (closed squares). Evidently, the buildup of the zeta potential must be associated with Cl^- adsorption.

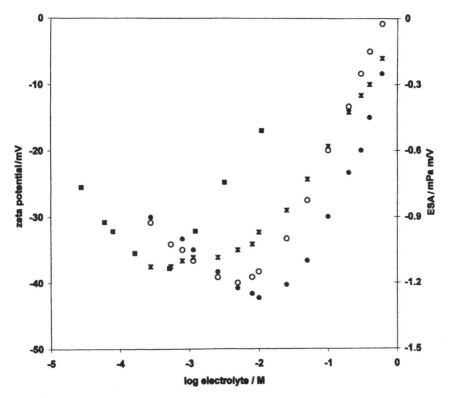

FIG. 2 Zeta potential or ESA versus electrolyte concentration: (■) Rabinovich and Baran [27]: butanol + KCl; (●) Goetz and El-Aasser [30]: SLS + CA; (○) Goetz and El-Aasser [30]: SLS + CA/NaOH; (∗) Goetz and El-Aasser [30]: SLS.

In the presence of ionic surfactants, zeta potentials are the result of a combination of dissociation or association and subsequent adsorption of the surface-active species:

- Dissociation, e.g., of SLS:

$$CH_3(CH_2)_{11}OSO_3Na \rightarrow CH_3(CH_2)_{11}OSO_3^- + Na^+$$

- The $CH_3(CH_2)_{11}OSO_3^-$ moiety is surface active and will adsorb on to the droplets.
- Association of, e.g., dodecylamine and acetic acid:

$$CH_3(CH_2)_{11}NH_2 + H{-}OOCH_3 \rightarrow CH_3(CH_2)_{11}CH_3^+ + CH_3COO^-$$

- The $CH_3(CH_2)_{11}NH_3^+$ moiety is surface active and will determine the sign of the potential.

A negative zeta potential is found in the presence of anionic surfactants, while cationic surfactants give rise to positive zeta potentials. As the "naked" oil droplet is negatively charged, the zeta potential passes through an isoelectric point (i.e.p.) as the concentration of cationic surfactant is increased. In the case of a decyl trimethylammonium bromide (DTAB) stabilized xylene emulsion in water, zeta potentials as high as 98 mV were observed [20]. In Table 4 a short list of zeta-potential data, illustrating the typical ranges observed, is presented.

TABLE 4 Illustrating Zeta Potentials (ζ) of Emulsion Droplets with a
Closed-Packed Arrangement of Ionic Surfactant Molecules

Emulsion System	ζ (mV)	Source (references)
Xylene/water/DTAB	98	20
Bitumen/water/DTAB	118	28
Bitumen/water/CTAB	94	28
Liquid paraffin/water/CTAB	93	29
Liquid paraffin/water/Na dodecyl benzene sulfonate	−93	29

According to the classical picture, the surface charge on the droplet is balanced by electrostatically adsorbed ions, which form the diffuse part of the electrical double layer. Hypothetically, these ions can bulk in the interface, giving rise to an i.e.p. and subsequent charge reversal. In the *micellar* system of bis(dodecyl)-1,4-butane quaternary ammonium bromide, micelles having a negative zeta potential ($\sim -130\,\text{mV}$) were observed [6]. Change of sign evidently occurs when potential-determining ions (of the opposite sign) are added. Figure 3 shows the dependence of the zeta potentials on the concentration of specific electrolytes of DTAB and sodium cholate stabilized emulsions.

The electrokinetics of o/w emulsions stabilized by a combination of nonionic and ionic surfactants exhibit interesting features. Goetz and El-Aasser [30] studied the effect of the electrolyte a concentration on the zeta potential in o/w emulsions stabilized by cetyl alcohol (CA) and SLS. In this system the zeta potential passes through a minimum as electrolyte is added. They also established the dependence of both position and depth of the minimum on the nature of the electrolyte, see Fig. 2 (open and closed circles). In the absence of the nonionic surfactant (Fig. 2, asterisks) the minimum is not formed. In their discussion they refuted a number of theories including displacement of H^+ ions in the inner Helmholtz layer by larger cations, ionic conduction, and surface roughness. Supported by their findings, they concluded that the origin of the optimum must be coion adsorption (in competition with the compression of the double layer). We have discussed the formation of the minimum in the context of emulsions stabilized by nonionic surfactants alone where the same conclusion was reached. Interestingly, in the presence of the anionic SLS, the position of the optimum shifts to higher electrolyte concentrations (Fig. 2, squares and circles). This shift must be attributed to the anionic character of SLS. By the same token, the zeta potential of a SLS-stabilized o/w emulsion does not pass through a minimum (Fig. 2, asterisks).

The zeta potential of emulsions stabilized by a mixture of nonionic or anionic surfactants in combination with a cationic surfactant depends on the surfactant ratio. When the "cations" dominate, the potential is positive. Domination of especially anionic surfactants provides negative zeta potentials. An i.e.p. is observed at some particular composition.

B. Dependence of Potential on Amount of Surfactant

The dependence of the zeta potential on the concentration of ionic surfactant shows an interesting feature. Figure 4 shows typical plots. Neglecting the very low and high con-

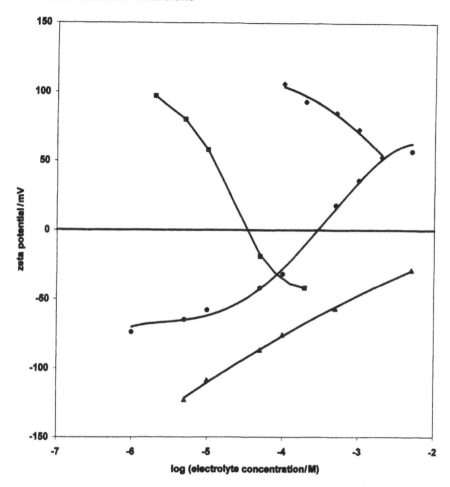

FIG. 3 Plots of zeta potential versus electrolyte concentration: (◆) KBr; xylene/water/DTAB; (■) K₃[Fe(CN)₆]; xylene/water/DTAB; (▲) KNO₃; xylene/water/Na cholate; (●) La(NO₃)₃; xylene/water/Na cholate. (From Ref. 20.)

centration regimes, for both anionic and cationic surfactants this dependence generally consists of two regimes. In the "low" concentration regime (left-hand side of the curves in Fig. 4) proportionality is observed of the zeta potential to the logarithm of the surfactant concentration. The higher concentration regime is characterized by a zeta potential insensitive to changes in the surfactant concentration. Sometimes, one of the regimes is missing (Fig. 4, closed circles).

An explanation for the different nature of the two regimes may well be found in the mechanism of surfactant adsorption. A constant zeta potential, insensitive to increase in surfactant concentration, indicates saturation. We discuss this saturation effect in more detail later on. With respect to the nature of the proportionality regime, Healy et al. [11] collected and analyzed data on *micellar* surface potential (ψ_0) as a function of surfactant activity. The main result is the following remarkable relationship:

$$d|\psi_0|/d\log a_s = 59.2\,\text{mV} \tag{3}$$

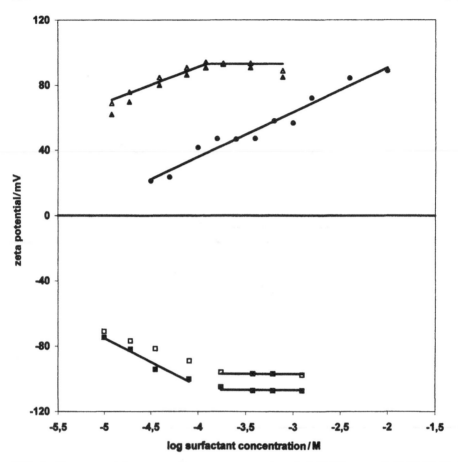

FIG. 4 Zeta potential versus surfactant concentration: (□) SLS/heptane/0.01 M NaCl; (■) SLS/
cyclohexane/0.01 M NaCl; (▲) CDBAC/cyclohexane/0.01 M NaCl; (△) CDBAC/heptane/0.01 M
NaCl; (●) CTAB/xylene/0.5 M NaCl. (From Refs. 31 and 32.)

where a_s is the surfactant activity in the bulk solution. They arrived at the conclusion that
the absorbing surfactant can be considered as a potential-determining ion. Our (macro)-
emulsion system slightly differs from the micellar system Healy et al. studied. Once it has
formed, its number of adsorption sites is more or less fixed and the droplets differ in
curvature. However, provided each shift in surfactant concentration is accompanied by
a redispersing step, our treatment of the surface potential data can be made to resemble
the approach of Healy et al. By assuming a constant ratio between the surface potential
and zeta potential, this implies that the observed dependence, as illustrated by Fig. 4 (left-
hand side), is the result of a potential buildup by potential-determining ions. It is interesting to note that such dependence differs from the usual picture of hyperbolic adsorption
on a macroscopic interface [33].

C. Constraint of Potential Buildup

Figure 1 illustrates that the interfacial potential lies between certain limits. For the
"naked" oil droplets and the oil droplets covered by nonionic surfactants, the zeta poten-

tial decreases with pH, levels off, and reaches its minimum at pH 8–9 [12, 18]. At low pH a sign reversal occurs. The i.e.p. occurs at pH 2–3. Thus, the zeta potential ranges from 70 to 110 mV. (This range becomes narrower at higher ionic strengths). Similar maximum levels are observed for ionic surfacant-stabilized oil droplets (see Table 4).

Basically, the zeta potential depends on the concentration of the so-called potential-determining ions and ionic strength. The maximum potential depends on the capacity of the (o/w) interface to adsorb these potential-determining ions. As such, an estimate can be made of the zeta-potential maximum or minimum (depending on the sign of the potential). For this purpose a number of methods is available.

Following the approach of Haydon and Phillips [34], a closed-packed adsorption layer of ionic surfactants can be assumed, using the following equation for the wall potential ψ_0 (Gouy–Chapman, see e.g., Davies and Rideal [35, Ch. 2]):

$$\psi_0 \approx (ikT/e)\sinh^{-1}(139/Ac^{1/2}) \tag{4}$$

where k is Boltzmann's constant, T is the absolute temperature, e is the elementary charge, A is the area per surfactant molecule, and c is the surfactant concentration. For an "A–B" surfactant $i = 1$ or 2, depending on the electrolyte concentration. To obtain the zeta potential a correction factor must be introduced for which Davies and Rideal [35, Ch. 3] suggest a value of approximately 0.50.

The "naked" oil droplet system does not allow for such an estimate. Although a good guess can be made as to the nature of the adsorbing species, viz., the hydroxyl ion, assuming that close packing of the adsorbed layer is not realistic. For oil droplets covered by nonionic surfactants similar arguments are valid. For systems containing an ionic surfactant the data of Haydon and Phillips [34] predict −95 mV for SLS in the system petroleum ether/water. [Data on dodecyl trimethylammonium bromide (DoTAB) in the same system were also provided. Due to the size of the hydrophilic moiety of this surfactant, calculation of ζ requires a more complex expression accounting for its dipole moment.]

In the case of ionic surfactants, micellar data may be used. Micelles may be a suitable model, because they are densely packed. However, in working out this concept one needs to exercise some caution. Johnson et al. [36] compared the charge structure of cetyl trimethylammonium chloride (CTAC) micelles and planar adsorbed CTAC. By employing a site-binding model and classical electrical double layer theory they arrive at the conclusion that the counterion dissociates stronger from a planar adsorbed layer than from a strongly curved surface like that of a micelle. We collected micellar zeta potential data (see Table 5). This table shows these potentials are essentially similar to the ones observed in

TABLE 5 Listing Room-Temperature Zeta Potentials of Micelles

Micellar System	ζ (mV)	Reference
Bis(tridecyl)-1,4-butane quarternary ammonium bromide	25	6
Bis(hexadecyl)-1,4-butane quarternary ammonium bromide	40	6
Dodecyl trimethylammonium chloride	~ 92	11
CTAC	90–100	36
Sodium decyl sulfate	~ −74	11
Sodium lauryl sulfate (SLS)	−77 to −101	(ch. 3 of 35), 37

emulsion systems (Table 4). Overall, the indications are that in emulsion systems stabilized by nonionic or monovalent surfactant, zeta potentials occur in the range of -120 to $+120 \, \mathrm{mV}$.

IV.   ROLE OF POTENTIAL IN STABILITY OF EMULSIONS

Four processes may account for isothermal emulsion breakdown:

- Ostwald ripening
- Sedimentaiton and subsequent creaming or settling
- Flocculation
- Coalescence.

Ostwald ripening drives the larger droplets to grow at the expense of the smaller ones. It is associated with the difference in chemical potential for droplets of different size. Size differences thus become ever more pronounced, until one (large) drop is left. Sedimentation is driven by gravity and occurs when the droplets and surrounding medium differ in density. Flocculation (or coagulation) is the formation of aggregates. It occurs when attractive forces dominate. The identity of the individual droplets is retained until coalescence occurs. Droplets flocculate either into a deep, primary energy minimum or in a relatively shallow secondary minimum. Flocculation velocities are often evaluated by applying the well known von Smoluchowski equation:

$$1/N(t) = 1/N(t = 0) + K_F t \tag{5}$$

where $N(t)$ is the number of droplets (per unit of volume) at time t, and K_F is the so-called flocculation rate constant. Entry into the deep primary minimum is generally followed by coalescence. In this process, smaller droplets merge into larger ones as droplet surfaces touch. This breakdown process is easily identified by a characteristic droplet number decay time. Coalescence studies apply the exponential expression according to Van den Tempel:

$$N(t) = N(t = 0) \exp(-K_c t) \tag{6}$$

where K_C is the coalescence rate constant. Just as with K_F, a "high" K_C implies that coalescence occurs rapidly.

In general, the different destabilization processes occur simultaneously, although within certain periods one process may be found to prevail. Needless to say, coalescence is preceded by flocculation or sedimentation.

A.   Emulsion Stabilization and Electrostatic Repulsion

In the struggle for the extension of emulsion shelf-life most approaches are directed to the inhibition of the above-mentioned destabilization processes. Adding surfactants is one approach. As is shown in previous sections, part of the input of the "surfactant" compo-nent is the buildup of the interfacial (or wall) potential. According to the well known DLVO theory, lyophobic colloidal systems derive stability from the electrostatic repulsion between droplets. Interestingly, as argued by Elworthy and Florence [38], another effect of the added surfactants is the increase in the attractive forces which accompanies the for-mation of the adsorbed film.

In the overall picture of stabilization, attractive and repulsive forces determine the interaction between emulsion droplets. A potential V can be defined which is a resultant of the attraction (V_A) and the repulsion (V_R). On mutual approach, V_A and V_R change differently with interdroplet distance. When a net positive result occurs, certain stabilization is imparted to the emulsion. The DLVO theory predicts a repulsive interaction between two identical diffuse double layers. The repulsive potential of mean force is approximately given by the expression:

$$V_R = \tfrac{1}{2}\varepsilon\varepsilon_0 R\psi_0^2 \ln[1 + \exp(-\kappa H)] \tag{7}$$

where $\varepsilon\varepsilon_0$ represent the dielectric constant of the medium, R is the droplet radius, κ is the reciprocal Debye length, and H is the distance between droplet surfaces.

It is well known that adding sufficient amounts of electrolyte destabilizes emulsions. Destabilization proceeds more rapidly when the emulsion is exposed to electrolyte containing counterions of high valency. If the DLVO theory is valid (or the result of Van der Waals' attraction and electrostatic repulsion dictates stabilization), the Schulze–Hardy rule applies. A good example is found in the emulsion of xylene in water stabilized by sodium cholate or DTAB [20]. In the case of the cholate-stabilized emulsion, coagulation was initiated by adding nitrate salts of Pb^{2+}, La^{3+} and Th^{4+}. The DTAB-stabilized emulsion was coagulated with potassium salts of $Fe(CN)_6^{3-}$ and $Fe(CN)_6^{4-}$. In Table 6 the critical coagulation concentration (c.c.c.) as well as the concentrations predicted by the Schulze–Hardy rule (in parentheses) are listed. This table shows good agreement between theoretical prediction and experiment (last rows). Similar efficacy gains are achieved when added counterions adsorb preferentially. As expected, near the i.e.p. destabilization occurs [20, 21, 25, 39, 40].

However, dissonant observations were also reported. As is also shown by Table 6, in many cases the Schulze–Hardy rule strongly underestimates experimental values of c.c.c.

TABLE 6 List of Emulsion Systems and Their Properties with Respect to Coagulation: Zeta Potentials at Which Coagulation Occurs and c.c.c. for the Different Electrolytes Added

System (and source)	Electrolyte	ζ (mV)	c.c.c. (mM)
n-Decane/water [33]	NaNO$_3$	−33	13 (111)[a]
	Ba(NO$_3$)$_2$	−24	3.5 (1.7)
	La(NO$_3$)$_3$	−25	0.65 (0.15)
Kerosene/butanol/water [27]	KCl	−16	50 (286)
	CaCl$_2$	−5	20 (4.5)
	AlCl$_3$	−4	0.5 (0.39)
Chlorobenzene/polyoxyethylene (6)	NaNO$_3$	−20	6.2 (10.6)
cetylalcohol/water [23]	Ca(NO$_3$)$_2$	−17	0.72 (0.17)
	Al(NO$_3$)$_3$	−44	0.0057 (0.015)
Xylene/Ca soap/water [21]	Pb(NO$_3$)$_2$	< 0	—
	Cr(NO$_3$)$_3$	i.e.p.	0.30 (0.14)
	Zr(NO$_3$)$_4$	i.e.p.	0.012 (0.025)
Xylene/Na cholate/water [20]	Pb(NO$_3$)$_2$	i.e.p.	3.4 (3.1)
	La(NO$_3$)$_3$	i.e.p.	0.32 (0.27)
	Th(NO$_3$)$_4$	i.e.p.	0.036 (0.048)
DTAB [20]	K$_3$Fe(CN)$_6$	i.e.p.	0.067 (0.07)
	K$_4$Fe(CN)$_6$	i.e.p.	0.013 (0.012)

[a] The c.c.c. values predicted by the Schulze–Hardy rule are given in parentheses.

for the higher valency electrolytes. Violation of this rule may be due to two aspects influencing the efficacy of the electrolytes. One aspect is hydrolysis to which higher valency ions are particularly prone. Due to this process the actual electrolyte concentration is reduced. Also, as shown by Table 6, the zeta potential tends to break down more quickly with higher electrolyte valency. Such behavior must be associated with specific adsorption.

The other aspect is the correlation between emulsion stability and the zeta potential. DLVO theory predicts a second-order relationship [see Eq. (7)]. The studies of Elworthy et al. [23] showed an emulsion breakdown to occur (long) before the zeta potential had reached the zero level (see Fig. 5). In another study [41] they examined the effect of *n*-hexadecanol on the stability of the same emulsion system. Coalescence rates dropped by almost a factor of 5 when the concentration of *n*-hexadecanol was increased from 0 to 5%. Strikingly, during this procedure the zeta potential hardly changed. Becher and Tahara [22] and Becher et al. [42] studied the effect of electrolyte on the properties of cottonseed and mineral oil emulsions in water. These nonionic surfactant-stabilized emulsions show a critical electrolyte concentration behond which the zeta potential is rapidly broken down to almost zero. Also, this concentration is reached with less electrolyte when higher valency electrolyte is used, just as is predicted by the Schulze–Hardy rule. Surprisingly, emulsion stability is not affected by addition of electrolyte. The following conclusion may be drawn: the stability of (o/w) emulsions is not exclusively attributable to electrostatic repulsion. We believe that more effects should be taken into account. Table 7 compiles literature data on emulsion systems and the nature of the stability. Evidently, emulsions

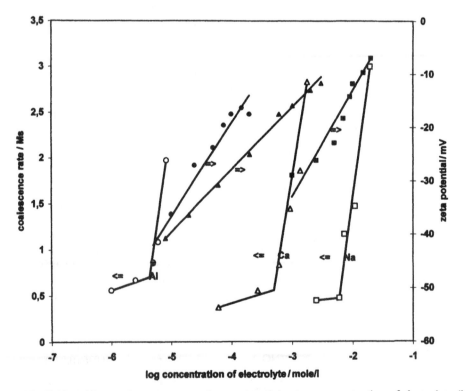

FIG. 5 Emulsion coalescence rate and zeta potential versus concentration of electrolyte (log scale): (□) $NaNO_3$; (△) $Ca(NO_3)_2$; (○) $Al(NO_3)_3$; (■) $NaNO_3$; (▲) $Ca(NO_3)_2$; (●) $Al(NO_3)_3$. (From Ref. 23.)

TABLE 7 Overview of Features of Emulsions Prepared from Data Found in Literature: Oil Type, Type of Surfactant, and Suggested Nature of the Stability[a]

Oil type (and source)	Surfactant and class	Nature of the stability
Chlorobenzene [41]	n-Hexadecanol + polyoxyethylene (6) cetylalcohol (nonionic, HLB[b] = 6.4)	Interfacial viscosity + steric hindrance
Light mineral oil [42]	Sorbitan mono-oleate (Span 80) + Tween 80 (nonionic, HLB = 5–11.5)	Steric hindrance
Light paraffin [5]	Polyoxyethylene (2 + 10) oleyl alcohol (nonionic, HLB = 8.5)	Steric hindrance + electrostatic repulsion
Toluene [25]	Span 20 (nonionic, HLB = 8.6)	Electrostatic repulsion
Dodecyl benzene [43]	Polyoxyethylene (6.5) nonylphenols + (2) tetradecanol (1:1) (or higher, HLB > 8.6)	Interfacial viscosity
Liquid paraffin [18]	Polyoxyethylene dodecanols (1–3) (nonionic, HLB < 9.7)	Electrostatic repulsion
Liquid paraffin [18]	Polyoxyethylene nonylphenols (2–4) (nonionic, HLB < 9.7)	Electrostatic repulsion
Chlorobenzene [23]	Polyoxyethylene (6) cetyl alcohol (nonionic, HLB = 9.7)	Steric hindrance
Liquid paraffin [18]	Polyoxyethylene dodecanols (4–8) (nonionic, HLB > 9.7)	Steric hindrance + electrostatic repulsion
Liquid paraffin [18]	Polyoxyethylene nonylphenols (5–12) (nonionic, HLB > 9.7)	Steric hindrance + electrostatic repulsion
Xylene [20]	Polyoxyethylene (9.5) octylphenol (Triton X-100, nonionic, HLB = 13.2)	Steric hindrance + electrostatic repulsion
Methyl cyclohexane [4]	Tween 80 (nonionic, HLB = 15)	Electrostatic repulsion
n-Decane [33]	SLS (anionic)	Electrostatic repulsion
Methyl cyclohexane [4]	SLS (anionic)	Electrostatic repulsion
Xylene [20]	Na cholate (anionic)	Electrostatic repulsion
Xylene [20]	DTAB (cationic)	Electrostatic repulsion
Xylene [39]	Cationic + anionic	Electrostatic repulsion
Xylene [21]	Ca soap (anionic)	Electrostatic repulsion
Xylene [31]	CTAB (cationic) + cetyl alcohol (nonionic)	Interfacial viscosity + electrostatic repulsion
Xylene [40]	Dodecyl benzene sulfonate (anionic)	Electrostatic repulsion

[a] The list is compiled on the basis of type of surfactant.
[b] When the surfactant is nonionic the HLB is used as a guide.

stabilized by ionic surfactant are stable because of electrostatic repulsion. The input of nonionic surfactants is a combination of electrostatic and steric components. Sometimes, the electrostatic component is outweighed. A third component of the stability is the interfacial viscosity (and elasticity). The role of this component is difficult to assess, because evaluation is not easy. Few studies have included viscosity measurements and stated their role. In mixed ionic–nonionic surfactant systems a balance is reached between the individual contributions. Overall, the predictive value of zeta potentials depends on the class of surfactant involved. Zeta potentials predict ionic surfactant-based emulsion stability well. Nonionic surfactant-stabilized emulsions require a different approach.

B. Surfactant Composition and Zeta Potential

It is common knowledge that a well-balanced composition and amount of surfactants are a prerequisite for optimal emulsion stability. Therefore, as the next item we evaluate the role of surfactant composition on emulsion stability.

Pithayanukul and Pilpel [43] studied the emulsion system dodecyl benzene/water, stabilized by a mixture of polyoxyethylene (6.5) nonylphenols (HLB = 11.3) and polyoxyethylene (2) tetradecanol (HLB = 5.9). These authors examined the effect of concentration and ratio of these nonionic surfactants on the stability of the emulsion. Three regions of surfactant concentration could be distinguished. At very low concentrations (< 0.25% on oil phase) emulsions separated immediately. At high concentration (> 10% on oil phase) the emulsions were stable. In the intermediate concentration range the emulsions initially showed improved stability towards coalescence and in a second stage also towards flocculation. The occurrence of the transition between these stages was determined by the ratio between the two types of nonionic surfactants applied. When the polyoxyethylene (2) tetradecanol dominated, more surfactant was needed to obtain a certain stability level. However, the shift in ratio could not always be compensated for. Below a "critical" ratio, emulsions always coalesced when they flocculated. We tested these findings in a wider context by collecting references on destabilization experiments in the concentration range Pithayanukul and Pilpel found to be crucial ($\sim$ 2–8% surfactant on oil phase). The diversity in types of nonionic surfactant was tackled by characterizing these surfactants in terms of the HLB value. In Table 8 the results of this exercise are shown. In principle the same trends are found. Nonionic surfactants characterized by an HLB $\geq$ 9.7 or ionic surfactants are favorable with respect to coalescence. Use of nonionic surfactant characterized by a low HLB ($\leq$ 8.6) gives an emulsion that coalesces when it flocculates. In an intermediate HLB zone there is a transition region. This zone is not as narrow as the one observed by Pithayanukul and Pilpel [43], most probably due to the number of oil types involved in our collection. This reminds us of the HLB concept we discussed in Section I. The impact of the required HLB seems to appear.

The belief in a "critical" surfactant mix (linked to an oil type) is supported by earlier studies by Becher and Tahara [22] and Becher et al. [42]. These authors studied the effect of surfactant composition on emulsion properties. They prepared aqueous emulsions of light mineral oil and of cottonseed oil. The surfactant system was a mixture of nonionic surfactants [sorbitan monostearate (Span 60), Span 80, polyoxyethylene (20) sorbitan monostearate (Tween 60), and Tween 80] in different ratios, spanning a range of HLB = 5–12. They identified oil systems, which inclined to certain combinations of nonionic surfactants. At certain combinations, emulsion stability was found to pass through a maximum. When expressed in terms of an HLB value, maximum stability occurred at specific HLB values (HLB = 9–10 for light mineral oil and HLB = 6–7 for cottonseed oil), independent of surfactant types composing the mixture. This is clearly in support of the notion of a "required" HLB. They also determined the zeta potentials of the emulsions. Strikingly, the zeta potential passed through a minimum when the HLB passed the required values (see Fig. 6). The conclusion must be drawn that zeta potentials do express balanced conditions with respect to the nonionic surfactant mix.

In a recent study this conclusion is combined with the predictive value of zeta potentials in the case of ionic surfactants. The present author [9] studied the electrophore-

TABLE 8 List of Features of Emulsions (Oil and Surfactant Type) and Corresponding Breaking Process

Oil phase (and source)	Surfactant	Breakdown process
n-decane [33]	None	Flocculation + coalescence
Kerosene [27]	Butanol (HLB $\approx$ 0)	Flocculation + coalescence
Soybean oil [24]	Glycerides (HLB < 4.1)	Flocculation + coalescence
Light paraffin oil [5]	Polyoxyethylene (2) oleyl alcohol (HLB = 8.5)	Flocculation + coalescence
Dodecyl benzene [43]	Polyoxyethylene (6.5) nonylphenols + idem (2) tetradecanol (HLB < 8.5)	Flocculation + coalescence
Toluene [25]	Span 20 (HLB = 8.6)	Flocculation
Dodecyl benzene [43]	Polyoxyethylene (6.5) nonylphenols + idem (2) tetradecanol (HLB > 8.5)	Flocculation
Liquid paraffin [18]	Polyoxyethylene (1–3) dodecanols (HLB < 9.7)	Flocculation + coalescence
Liquid paraffin [18]	Polyoxyethylene (2, 4) nonylphenols (HLB < 9.6)	Flocculation + coalescence
Chlorobenzene [23]	Polyoxyethylene (6) cetylalcohol (HLB = 9.7)	Flocculation + coalescence
Xylene [27]	Triton X-100 (HLB = 13.2)	Flocculation
Methyl cyclohexane [32]	Tween 80 (HLB = 15)	Flocculation
n-Decane [33]	SLS	Flocculation
Xylene [27]	Na cholate	Flocculation
Xylene [27]	DTAB	Flocculation
Xylene [39]	Na dioctyl sulfosuccinate	Flocculation
Xylene [39]	CTAB	Flocculation
Xylene [21]	Ca soap	Flocculation
Xylene [40]	Na dodecyl benzene sulfonate	Flocculation

tic mobility of hexane emulsions in water stabilized by ionic surfactants. Surprisingly, when related to the molecular structure of the included surfactants, the electrophoretic mobility could be split into contributions corresponding to well-defined molecular segments. This shows strong similarities with the approach taken by Davies in designing his incremental HLB concept [44]. Therefore, as a next step, the identified increments were recalculated using the HLB increments of anionic surfactants (salts of the fatty acids, sulfates, and sulfonates) listed by Davies. A list of HLB increments is obtained in this way which also includes many cationic segments. In Table 9 a number of segments and their HLB increments are shown. This table shows that increments due to ionic segments are limited to 20–25 HLB units. This is much *less* than the figure often quoted by many authors for the sulfate group, i.e., 39 HLB units. One may have doubts about the HLB system in general [38]. However, experience has provided numerous good examples. As support, the reliability of Table 9 is illustrated. In Table 10, predictions of optimal emulsion stability are compared with experimental data. The table clearly shows the excellent

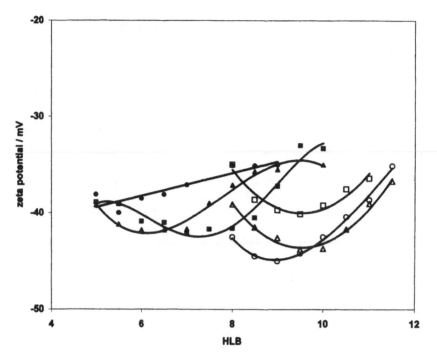

FIG. 6 Zeta potential versus HLB. Mineral oil: (○) Span 80/Tween 80; (△) Span 60/Tween 60; (□) Span 80/Tween 60. Cottonseed oil: (●) Span 80/Tween 80; (▲) Span 60/Tween 60; (■) Span 80/Tween 60. (From Ref. 22.)

agreement between prediction and experiment. Note that such a table cannot be used in predicting optimal conditions in the case where anionic and cationic surfactants are combined. However, there is no advantage of combining anionic and cationic surfactants. The net effect of combining them is a reduction in the (zeta) potential, thus nullifying the stabilizing effect of the individual surfactants [39, 45].

TABLE 9 HLB Increments of a Number of Molecular Segments

Molecular segments	HLB increment
$-COOH$	2.1
$-OH$	1.9
$-O-$	1.3
CH_3-, $-CH_2-$, $-CH=$	-0.475
Benzyl$-$	-1.66
$-OSO_3^- Na^+$	20.8
$-SO_3^- Na^+$	20.7
$-COO^- Na^+$	19.1
$-N(CH_3)_3^+ Cl^-$	22.0
$=N(CH_3)_2^+ Cl^-$	22.5
$-N(C_2H_4OH)_2 CH_3^+ Cl^-$	24.9
$-NH_3^+ Cl^-$	20.0

TABLE 10 Comparison of Calculated HLB values for a Number of Ionic Surfactants and Determined Values[a]

	HLB value	
Surfactant	Calculated	Determined
CTAB	24.2	24 [46]
Tetradecyl trimethylammonium bromide	25.8	27 [46]
DoTAB	27.2	26 [19]
Decylamine, acidified to pH 3 by adding HCl	22.3	23 [46]
Cetyl ethyl morpholinium ethylsulfate	25.8	25–35 [1]
SLS	22.1	21 [46]
Sodium decanoate	21.8	22 [46]

[a] The calculations apply Table 9; the determinations make use of the optimal emulsion stability concept.

C. Miscellaneous Aspects of Surfactants

Surfactants are also known to promote Ostwald ripening, especially at high concentrations [47, 48]. In the presence of surfactant micelles, oil molecules are solubilized and also transported by micelle diffusion [43]. In addition, they change the rheology of the continuous phase and the density of the participating phases, which in turn affect the sedimentation velocity. These aspects of surfactants fall outside the scope of this review, and will be ignored.

One important aspect should be mentioned, however. It concerns the effect of droplet potential on the sedimentation velocity. The higher this potential, the longer it takes before an emulsion settles. The following equation describes the retardation of the settling velocity (v) as a result of zeta potential:

$$v/v_0 = 1 - (\varepsilon\varepsilon_0\zeta/4\pi a)^2/(\eta\Lambda) \tag{8}$$

where η and Λ are, respectively, the viscosity and specific conductance of the continuous phase; a is the droplet radius, and v_0 is the settling velocity at zero zeta potential. Use of this equation is limited to the case of thin double layers ($\kappa a \gg 1$) [35, 49, 50].

V. CONCLUSION

We have overviewed the electrokinetics of o/w emulsion stabilized by surfactants. The droplets of virtually every o/w emulsion are charged, irrespective of the type of surfactant used. Therefore, electrophoresis is a powerful tool in exploring and developing emulsion systems. However, the predictive value of zeta potentials depends on the class of surfactant involved. In emulsion systems containing nonionic surfactants, DLVO theory does not apply. In this case, zeta potentials indicate when stability is maximal as a function of the surfactant mix. Electrokinetics is, then, a tool to explore emulsion systems and to determine required HLB values. Emulsions stabilized by only ionic surfactants obey the Schulze–Hardy rule; then, zeta potentials predict the stability well. Electrokinetics can be used to develop and identify stable emulsions. Evidently, the correct application of electrokinetics requires information about the type of surfactant involved.

We conclude with some suggestions for further research. We believe that the possibilities of acoustophoresis must be better exploited, as this technique enables us to examine many systems and processes. A system, which becomes accessible, is the w/o system [51, 52]. Interesting processes to study comprise the uptake of "slow" ions by oil droplets or vesicles in water. The link to drug research makes these kinds of studies even more exciting.

ACKNOWLEDGMENTS

The author wishes to acknowledge Dr. W.H. Mulder (Dept. Chem., Univ. West Indies, Jamaica), Dr. B. Parr (Akzo Nobel Surface Chemistry, Stennungsund, Sweden), and Dr. E. Putman (RIVM, Bilthoven, The Netherlands) for their stimulating discussions.

SURFACTANT INDEX

Abbreviation or trade name	Chemical formulas
CDBAC	Cetyldimethylbenzylammonium chloride
CTAB	Cetyl trimethylammonium bromide
CTAC	Cetyl trimethylammonium chloride
DoTAB	Dodecyl trimethylammonium bromide
DTAB	Decyl trimethylammonium bromide
SLS	Sodium lauryl sulfate
Span 20	Sorbitan monolaurate
Span 60	Sorbitan monostearate
Span 80	Sorbitan mono-oleate
Tween 60	Polyoxyethylene (20) sorbitan monostearate
Tween 80	Polyoxyethylene (20) sorbitan mono-oleate
Triton X-100	Polyoxyethylene (9.5) octylphenol

REFERENCES

1. WC Griffin. J Soc Cosmet Chem 1:311, 1949; WC Griffin. J Soc Cosmet Chem 5:4, 1954.
2. T Förster, F Schamibl, H Tesmann. Int J Cosmet Sci 12:217, 1990; T Förster, W Von Rybinski, H Tesmann, A Wdale. Int J Cosmet Sci 16:84, 1994.
3. McCutcheon's Emulsifiers & Detergents, Glen Rock, NJ: The Manufacturing Confectioner (appears annually); Ash & Ash: Handbook of Industrial Surfactants: An International Guide to More Than 1600 Products by Trade Name, Application, Composition & Manufacturer. Aldershot Gower Technical, 1993.
4. G Stalidis, A Avranas, D Jannakoudakis. J Colloid Interface Sci 135(2):313, 1990.
5. E Hantz, A Cao, P Depraetere, E Taillandier. J Phys Chem 89(26):5832, 1985.
6. M Pisárčik, M Dubničková, F Devínsky, I Lacko, J Škvarla. Colloids Surfaces 143(1):69, 1998.
7. AS Dukhin, VN Shilov, H Ohshima, PJ Goetz. Langmuir 15(20):6692, 1999.
8. RJ Hunter, RW O'Brien. Colloids Surfaces A 126(2–3):123, 1997.
9. O Boen Ho. J Colloid Interface Sci 198(2):249, 1998.
10. F Grieser, CJ Drummond. J Phys Chem 92:5580, 1988.

11. TW Healy, CJ Drummond, F Grieser, BS Murray. Langmuir 6:506, 1990.
12. JC Carruthers. Trans Faraday Soc 34:300, 1938.
13. W Dickinson. Trans Faraday Soc 37:140, 1941.
14. AJ Taylor, FW Wood. Trans Faraday Soc 53:523, 1957.
15. HA McTaggart. Phil Mag 27:29, 1914.
16. KG Marinova, RG Alargova, ND Denkov, OD Velev, DN Petsev, IB Ivanov, RP Borwankar. Langmuir 12:2045, 1996.
17. K Yoshihara, H Ohshima, N Momozawa, H Sakai, M Abe. Langmuir 11:2979, 1995.
18. K Tajima, M Koshinuma, A Nakamura. Colloid Polym Sci 270:759, 1992.
19. BJ Floy, JL While, SL Han. J Colloid Interface Sci 125(1):23, 1988.
20. SP Singh, MISS Madhuri, P Bahadur. Rev Roum Chim 7:803, 1982.
21. RP Varna, P Bahadur, P Bahadur. Cellul Chem Technol 9(4):381, 1975.
22. P Becher, S Tahara. Phys Chem Anwendungstech. Grenzflächenaktiver Stoffe, Ber Int Kongr 1972, 6th (1973), teil 2, 519.
23. PH Elworthy, At Florence, JA Rogers. J Colloid Interface Sci 35(10):23, 1971.
24. M Kako, S Kondo. J Colloid Interface Scie 69:163, 1978.
25. M Sharma, P Bahadur, SP Jain. Rev Roum Chim 24(5):747, 1979.
26. A Graciaa, G Morel, P Saulner, J Lachaise, RS Schechter. J Colloid Interface Sci 172:131, 1995.
27. YaI Rabinovich, AA Baran. Colloids Surfaces 59:47, 1991.
28. AR lane, RH Ottewill. In: AL Smith, ed. Theory and Practice of Emulsion Technology. Proc Symp Soc Chem Ind 1974, London: Academic Press, 1976, ch. 9.
29. TF Tadros. In: AL Smith, ed. Theory and Practice of Emulsion Technology. Proc Symp Soc Chem Ind 1974, London: Academic Press, 1976, ch. 17.
30. RJ Goetz, MS El-Aasser. J Colloid Interface Sci 142(2):317, 1991.
31. TF Tadros. Colloids Surfaces 1:3, 1980.
32. A Avranas, G Stalidis. J Colloid Interface Sci 143(1):180, 1991; J Colloid Interface Sci, Chim Chron (new series) 20:129, 1991.
33. S Usui, Y Imamura, E Barouch. J Dispers Sci Technol 8(4):359, 1987.
34. DA Haydon, JN Phillips. Trans Faraday Sco 54:698, 1957.
35. JT Davies, EK Rideal. Interfacial Phenomena. 2nd ed. New York: Academic Press, 1963.
36. SB Johnson, CJ Drummond, PJ Scales, S Nishimura. Colloids Surfaces a 103(3):195, 1995.
37. F Tokiwa, K Ohdi. Kolloid Z Z Polym 239:687, 1970; K Kameyama, T Takagi. J Colloid Interface Sci 140(2):517, 1990.
38. PH Elworthy, AT Florence. J Pharm Pharmacol 21 (suppl):79S, 1969.
39. AK Goswami, P Bahadur. Ann Chim 69:45, 1979.
40. MK Sharma, G Chandra, SK Jha, SN Srivastava. Progr Colloid Polym Sci 63:56, 1978.
41. PH Elworthy, AT Florence, JA Rogers. J Colloid Interface Sci 35(10):34, 1971.
42. P Becher, SE Trifiletti, Y Machida. Theory and practice of emulsion technology. Proceedings of Symposium, 1974, 271 (1976).
43. P Pithayanukul, N Pilpel. J Colloid Interface Sci 89(2):494, 1982.
44. JT Davies. Proceedings of 2nd International Congress on Surface Activity. vol. 1, London: Butterworth, 1957, p 477.
45. AK Goswami, P Bahadur. Progr Colloid Polym Sci 63:27, 1978.
46. BH O. Proceedings of 4th World Surfactant Congress, Barcelona, 1996, vol. 2, p 451.
47. M Laradji, H Guo, M Grant, MJ Zuckerman. Phys Rev A 44:8184, 1991.
48. Y De Smet, L Deiemaeker, R Finsy. Langmuir 15:6745, 1999.
49. M Smoluchowski. Graetz Handbuch der Elektrizität u. des Magnetismus II. Leipzig, 1914, p 366.
50. F Booth. J Chem Phys 22:1956, 1954.
51. VG Bedenko, AV Pertsov. Kolloidn Zh 50(2):335, 1988.
52. EE Isaacs, H Huang, AJ Babchin, RS Chow. Colloids Surfaces 46:177, 1990.

32

Electrokinetics of *n*-Alkane Oil-in-Water Emulsions

EMIL CHIBOWSKI and AGNIESZKA WIĄCEK   Maria-Curie Sklodowska University, Lublin, Poland

I. INTRODUCTION

Emulsion systems still attract many researchers. This is because of their tremendous practical application in many fields of human activity and occurrence as natural systems [1].

Nevertheless, many aspects of emulsion properties (stability/instability) are not well known yet. Mostly emulsions are required to be stable, and their stability is dependent on their concentration and is also a function of the oil droplet size. The smaller the diameter the more stable the emulsion is [2]. Considering an oil-in-water (o/w) type emulsion, the Laplace pressure ΔP inside small oil droplets (say, below 1 µm) is sufficiently high to prevent their deformation in most practical conditions, so the droplets behave like rigid spheres:

$$\Delta P = \frac{2\gamma}{r} \tag{1}$$

where γ is the interfacial oil/water tension and r is the droplet radius. It should be stressed that because thermodynamically the o/w system is unstable, as long as there are no factors that hinder decrease in the total surface of the droplets, it will tend to coalesce very fast. Emulsion instability can occur via creaming (sedimentation), aggregation, and coalescence [2, 3], and in the case of a pure oil phase dispersed in water these processes will cause separation of the phases. Therefore, to prolong the life of the dispersed state of the oil phase a third component is added, which is called an emulsifier. It collects at the oil/water interface, thus forming an adsorbed film on the oil droplet surface. This causes a decrease in the interfacial tension, and usually some changes in the electric potential (charge) take place at the interface. Applying the DLVO theory, the balance of attractive dispersion and repulsive electric interactions between the oil droplet–oil droplet and the water phase can be evaluated. However, at present it is known that the classical DLVO theory in many systems is not sufficient, so the extended theory is needed to describe the interactions correctly [4, 5].

This chapter deals with some electrical aspects of oil/water interface, mostly with the electrokinetic potential, i.e., zeta potential, of the oil droplets suspended in the water phase. *n*-Alkanes were used as the oil phase. In some of the emulsion systems discussed, metal cations, alcohol, and/or protein were present. The authors wish to stress that this is not a review article on the electrokinetics of emulsions in general, but rather a summary of the results obtained with n-alkane emulsions during the last years. Some results of other authors will also be recalled.

II. OIL/WATER INTERFACE

A. Galvani Potential Across the Interface

As was mentioned above, the oil submicro-size droplets suspended in the water phase behave as hard spheres, and the system may be considered interesting for studies of electrical properties at the oil/water interface. However, if such a system has reached thermodynamic equilibrium, actually no electric potential difference should appear between two reversible and identical electrodes dipped in the two phases; nevertheless, a partition between the phases takes place if some ionic species are present in the system. However, in most practical emulsion systems a potential difference between the oil surface and the bulk of the continuous phase (aqueous solution) will apear due to nonequilibrium (metastable) states of the system.

First, let us consider, after Kruyt [6] an oil/water system containing a 1:1 electrolyte. At equilibrium the distribution of ions in the oil (o) and water (w) phases is determined by their electrochemical potentials, $\bar{\mu}_+$ and $\bar{\mu}_-$, in both phases, respectively:

$$\bar{\mu}_{+(o)} = \bar{\mu}_{+(w)} \quad \text{and} \quad \bar{\mu}_{-(o)} = \bar{\mu}_{-(w)} \tag{2}$$

Hence,

$$\bar{\mu}_{+(o)} = \mu^o_{+(o)} + kT \ln a_{+(o)} + e\varphi_{(o)} = \mu^o_{+(w)} + kT \ln a_{+(w)} + e\varphi_{(w)} = \bar{\mu}_{+(w)} \tag{3}$$

and

$$\bar{\mu}_{-(o)} = \mu^o_{-(o)} + kT \ln a_{-(o)} - e\varphi_{(o)} = \mu^o_{+(w)} + kT \ln a_{-(w)} - e\varphi_{(w)} = \bar{\mu}_{-(w)} \tag{4}$$

where μ^o_+, μ^o_- are the standard chemical potentials of the cation and anion, respectively, k is Boltzman's constant, a_+ and a_- are the activities of the cation or anion in oil (o) or water (w) phase, e is the elemental charge, and $\varphi_{(o)}$ and $\varphi_{(w)}$ are the Galvani potentials in bulk oil and water phases, respectively. In the bulk phases the condition of electroneutrality has to be fulfilled, and the difference in Galvani potential between the oil and water phase reads [6]:

$$\Delta\phi = \varphi_{(o)} - \varphi_{(w)} = \tfrac{1}{2}e[(\mu^o_{+(w)} - \mu^o_{+(w)}) - (\mu^o_{+(o)} - \mu^o_{-(o)})] \tag{5}$$

Here, the Galvani potential difference does not depend on the electrolyte concentration but only on the specificity of the electrolyte. If the solution contains several electrolytes, $\Delta\phi$ will depend on the relative amounts of the ions, but not on their concentration [6]. However, as is well known, there is no means to determine the Galvani potential difference, $\Delta\phi$, across the interface. Formally, it consists of a potential jump χ (surface potential), which is due to the orientation of permanent dipoles at the interface, and a double-layer potential ψ^o resulting from an excess of ionic charge at the interface. Because of the different electric permittivities of the oil $\varepsilon_{(o)}$ and water $\varepsilon_{(w)}$ at the interface the ψ potential will show a discontinuity as a function of distance x. Therefore, a relationship holds:

$$\varepsilon_{(o)} \left(\frac{d\psi}{dx}\right)_{(o)} = \varepsilon_{(w)} \left(\frac{d\psi}{dx}\right)_{(w)} \tag{6}$$

Considering now a double diffuse electrical layer model at the oil/water interface, it results that for small values of the double-layer potential ψ° (say, below 50 mV) its partition is governed by the ratio of products of the electric permittivity and ion concentration in both phases [6]. Taking the Gouy–Chapman relationship for $d\psi/dx$, from Eq. (6) it results that:

$$\sqrt{n_{(w)}\varepsilon_{(w)}} \sinh \frac{ze\psi_{(w)}}{2kT} = \sqrt{n_{(o)}\varepsilon_{(o)}} \sinh \frac{ze\psi_{(o)}}{2kT} \qquad (7)$$

Also

$$\Psi^\circ = \psi_{(o)} + \psi_{(w)}, \qquad n_{(o)}\varepsilon_{(o)}/n_{(w)}\varepsilon_{(w)} = \alpha \qquad \text{and} \qquad \sqrt{\alpha} = \psi_{(w)}/\psi_{(o)} \qquad (8)$$

Taking, for example, the *n*-dodecane/water interface at 20°C, $\varepsilon_{(o)} \approx 2$ and $\varepsilon_{(w)} \approx 80$, thus $\varepsilon_{(w)} = 40\varepsilon_{(o)}$. Accordingly, although the number of ions will be much smaller in the oil phase, the greater drop in the double diffuse-layer potential will occur in the oil phase. Because the energy of hydration for small anions is less than for cations, there should be an excess of the former in the oil phase, thus giving a negative potential drop in the oil phase in a 1:1 electrolyte.

The double-layer structure in the water phase is weak; therefore, pure hydrocarbon emulsion in pure water is unstable [6]. Figure 1 shows an example of the double-layer potential partition (150 mV is assumed) between the oil (dodecane) and water phases for different values of α [Eq. (8)]. From this figure it can be found that for equal partition of the double-layer potentials to occur, $\psi_{(o)} = \psi_{(w)} = 75$ mV; i.e., $\log \alpha = 0$, the ion distribution should be $n_{(o)} = 40n_{(w)}$, which is simply impossible. On the other hand, for $\log \alpha = 3$, $n_{(w)} = 25n_{(o)}$ (which is a more reasonable partition of the ions) the potential distribution would be $\psi_{(o)} = 145$ mV in dodecane phase and only $\psi_{(w)} = 5$ mV in water. So, if this model works, one can expect that in a "pure emulsion" system in 1:1 electrolyte, the double diffuse double-layer potential is practically situated in the oil phase. However, it appears that in real emulsion systems in pure water the measured zeta potentials may

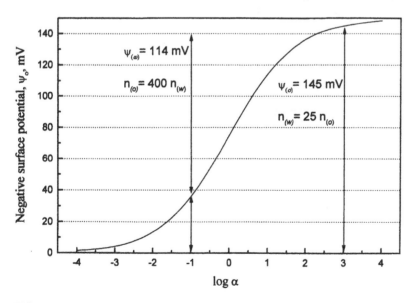

FIG. 1 Distribution of the total potential drop (150 mV assumed) between *n*-dodecane and water phases, as a function of the parameter α[Eq. (8)].

amount to 50 mV or more. This might be an effect of water dipoles structuring at the interface (χ potentials) as well as unequal numbers of cations and anions "dissolving" in such a layer (see below). Also, if an ionic surface-active substance is present in the system, it will concentrate at the interface, and the potential drop in the water phase will now be much greater.

B. Zeta Potential and Charge Origin at Hydrocarbon/Water Interface

Stachurski and Michalek [7] have determined electrophoretically zeta potentials for a series of *n*-alkane emulsions in 10^{-3} M NaCl. The emulsions were obtained by mechanical dispersing of the oil phase (0.5v/v%) in a mixer at 15,000 r.p.m. without any addition of emulsifier. They found a strong pH dependence of the negative zeta potentials for the alkanes having C_9–C_{16} atoms in the chain, while for C_6–C_8 *n*-alkanes the zeta potential changed much less and the isoelectric point (i.e.p.) had not been reached up to pH 3 (Fig. 2). Considering the classical diffuse double-layer theory, these results show that OH^- ions are potential determining ones and are preferentially "adsorbed" relative to H^+ ones. The authors [7] have not obtained positive values of zeta potentials in the pH range tested, and the extrapolated i.e.p. appears at pH 2.8 (Fig. 2). Parreira and Schulman [8], over 40 years ago, obtained a positive 10-mV zeta potential for solid paraffin in strongly acidic solution by applying the streaming potential method. A question arises why *n*-hexane to *n*-octane alkanes behave in a different way from higher chain-length *n*-alkanes and why their negative zeta potentials are about half those of the rest of the *n*-alkanes in the pH range 6–11. Moreover, below pH 6 for *n*-tridecane, Stachurski and Michalek [7] have

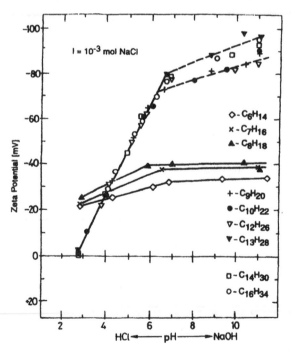

FIG. 2 Zeta potential of *n*-alkanes (C_6–C_{16}) droplets as a function of pH at constant ionic strength ($I = 10^{-3}$ mol of NaCl) of aqueous solutions. (From Ref. 7.)

found a straight-line correlation between zeta potential and emulsion stability. They observed that the faster the emulsion decomposition the lower was the zeta potential (Fig. 3). The same trend has been observed for hexane and octane. Very similar changes in zeta potential for *n*-tetradecane in 1 mM NaCl and pure water were obtained by the present authors, which will be discussed later.

It is interesting to learn whether the measured negative zeta potentials are really due to an excess adsorption of OH⁻ ions at the oil surface and what kind of force would be responsible for their adsorption. Then, if so, where are they located at the oil/water interface during the droplet movement in the electric field? Marinova et al. [9] conducted very careful studies on the mechanism of the oil/water charging. Using water very well purified from surface-active substances, as well as hexadecane and xylene, they investigated, besides OH⁻ adsorption, several possible mechanisms by which the droplets could gain their negative charge, such as adsorption of Cl⁻ and/or HCO_3^- ions, negative adsorption of positive ions (e.g., H_3O^+), orientation of water dipoles at the interface, and adsorption of some surface-active contamination being present in the water or oil phase. Conducting measurements in 2.28×10^{-3} M NaCl or 10^{-3} M Na_2CO_3, and at the same pH (9.8), they found that the zeta potential of xylene droplets was practically the same (-120 to -122 mV). Moreover, at pH 6 in 10^{-3} M NaCl they found that the zeta potential of xylene, dodecane, hexadecane, and perfluoromethyldecalin was also about the same ($\simeq -55 \pm 9$ mV). Hence, they concluded that the negative charge of the droplets could

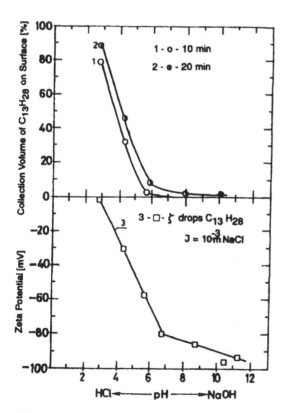

FIG. 3 Changes in zeta potential of $C_{13}H_{28}$ droplets with the pH of aqueous solutions at different times. (From Ref. 7.)

not be explained by specific adsorption of these negative ions. Their results on the electrophoretic mobility and zeta potentials for xylene as a function of pH and NaCl concentration are shown in Figs 4 and 5. Also, no positive values of the mobility were obtained here (even at pH 3.5) and a maximum appeared on the mobility curve as a function of NaCl concentration, which seems to be a characteristic feature (see later). Eliminating the above-mentioned reasons, Marinova et al. [9] arrived at the conclusion that the negative surface charge at the oil/water interface results from specific adsorption of OH^- ions. As the most probable mechanism for this specific adsorption the authors picked out formation of hydrogen bonding between the OH^- ions and water molecules in the boundary layers. Finally, they generalized the statement that the negative potential is due to the adsorption of hydroxyl ions and it is an inherent property of the oil–water interface. Though that is the case, the problem is still open as to the shear-plane location at the interface while the droplet is moving in an electric field. Despite the general statement that the oil/water interface shows a negative electrophoretic mobility (zeta potential), the general agreement that it is solely due to OH^- ions true adsorption at this interface is a debatable issue.

In a series of papers by Dunstan [10, 11] and Dunstan and Saville [12, 13] the authors postulate that there is a redistribution of ions close to a hydrophobic surface. Although both types of ions (cation and anion) are repelled from the adjacent layer of the solution, yet, for example, in the case of KCl as the supporting electrolyte Cl^- ions are preferentially "soluble" in this region, giving rise to the observed negative mobility of the particles. This difference in "solubility" results from different values of the solubilization entropy of K^+ and Cl^- ions. In consequence the water structure is also modified in this region and its entropy is increased. Moreover, the authors [10–13] also conclude that HCO_3^- and OH^- ions are also preferentially solubilized in comparison with H^+ ions. They conducted experiments on docosane ($C_{22}H_{46}$ n-alkane) particles (0.5 µm average radius) in different electrolyte solutions, applying various "cleaning" procedures for the water, docosane, and the reagents used prior to the measurements [10–13].

Figure 6 presents the results for zeta potential as calculated from the measured mobility, using the O'Brien and White theory [12]. In the presence of Al^{3+} cations, at 10^{-5} M and higher concentrations, the zeta potentials are positive, while in HCl solutions above 2×10^{-3} M and up to 3×10^{-2} M (the largest concentration used, pH $\approx$ 2) the

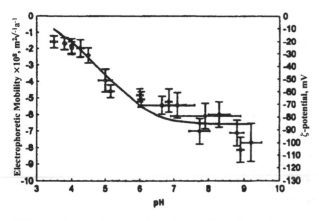

FIG. 4 Measured electrophoretic mobility, U_E, and zeta potential of xylene droplets as a function of pH at fixed ionic strength, $I = 10^{-3}$ M. (From Ref. 9.)

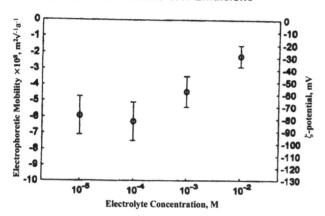

FIG. 5 Measured electrophoretic mobility, U_E, and zeta potential of xylene droplets as a function of NaCl concentration at pH 6. (From Ref. 9.)

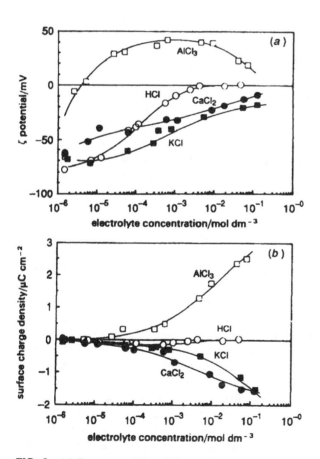

FIG. 6 (a) Zeta potential and (b) electrokinetic charges, σ, for docosane particles versus electrolyte concentration using the theory of O'Brien and White; the electrolytes are as indicated in the figures. (From Ref. 12.)

reversal of negative zeta potential sign could not be obtained, although it was close to zero. According to the classical double-layer theory such a behavior as that observed in Fig. 6 can point to a specific (excess) adsorption of ions; the authors [10–13] explain it by preferential solubility of the ions in the vicinity of the hydrophobic surface, as was discussed above. The specific adsorption of an ion on a solid surface is considered to require at least partial dehydration of the ion, a process which is not spontaneous in the case of hydrophobic surfaces like that of paraffin. A proof that no "true" charging (specific ion adsorption) of the alkane (hydrophobic) surface takes place is direct measurement of the forces with the apparatus for surface force measurements, as well as measuring the so-called negative adsorption of ions [14, 15]. No electrostatic repulsion, but a long-range attraction was measured between two hydrocarbon surfaces as a function of the concentration of the electrolyte solutions [14, 15]. These facts prompted the authors' conclusion about the "preferential solubility" mechanism of the observed negative electrophoretic mobility of the hydrophobic particles in an electrolyte solution. This mechanism in fact depends on negative adsorption in the interfacial layer. This "excess" solubility is due to differences in the entropy of solubilization of such ions as H^+, K^+, Ca^{+2}, Al^{+3}, and Cl^-. The authors assumed that the solubilization enthalpies of ions are constant throughout the suspension because the degree of hydration is probably the same in the bulk and vicinal water [10, 11].

However, there is a weak point in their approach because the entropy of Cl^- and K^+ ions is practically the same, while at all studied KCl concentrations (10^{-6}–10^{-1} M) the zeta potential was negative and the absolute values were higher than in $CaCl_2$ and HCl solutions (Fig. 6). The entropies ΔS_{hydr} for the ions [relative values to ΔS_{hydr} (H^+) = 0], in J/ K mol were [16]: K^+ −93, Ca^{+2} −271, Al^{+3} −557, Cl^- −94, and OH^- −180. The authors [10,11] have not considered any role for OH^- ions in the negative charge creation at the interface. However, accepting this approach and looking at the entropy for OH^- ions, which is equal to −180 J/K mol, it may be concluded that the ions play an essential role in the negative charge creation at the alkane/water interface. Dunstan [10] has found that removing dissolved CO_2 from docosane suspension in water (pH = 5.8, 1.5×10^{-6} M H_2CO_3) caused a mobility increase from −2.9 to −4.0 μm cm/s V, with pH increasing to 7. It means that OH^- concentrations increased from 6.31×10^{-9} to 10^{-7} M. The entropy ΔS_{hydr} for HCO_3^- is 156 J/K mol, and it is lower than ΔS_{hydr} (OH^-) = 180 J/K mol. Applying the approach of Dunstan and Saville [10–13] the increase in negative mobility may result from preferential solubility of OH^- ions in the interfacial region.

The observed extreme on the mobility/concentration curves may result from combined effects, increasing charge with increasing electrolyte concentration and retardation effect as a result of "very mobile" electrokinetic surface charge in such a system [10–13]. The retardation effect can dominate at a higher concentration of the electrolyte because the thickness of the restructured vicinal water decreases. However, if the charge is very mobile, one would expect the electrophoretic mobility to change with changing potential gradient across the measuring cell. However, this appears not to be the case, as Dunstan and Saville [12, 13] found that "no observed variation in the electrophoretic mobility" in water and in 10^{-3} M KCl took place, while the field strength was changed from 80 to 400 V/cm. It indicates that the charge is rather "fixed" to the solid particle of docosane. What is surprising, as the authors stated, was that the suspension was very stable [10, 13] up to 6 months. Marinova et al. [9] found that the xylene droplet size increased from the initial 100–200 nm to "micrometer values after 5–10 min," and the emulsion did not break out during 20 min. The content of the oil phase was below 0.5v/v%. A very similar size (average diameter 0.8–1 μm) was that of the docosane suspension at a very low volume fraction [11]. What is common for both suspensions is the way in which they are prepared.

Marinova et al. [9] and Dunstan [11] heated the mixtures up to a temperature of 60°C for 1 h (xylene) [9] and 50°C (docosane, melting temp. 44°C), then sonicated during cooling ($\sim$ 5 min) [11]. Such a procedure could cause some chemical processes (e.g., surface oxidation) or water molecule inclusion because of London dispersion interactions. The surface free energy of xylene (28.4–30.1 mJ/m^2) and docosane (about the same as that of xylene) totally originates from the dispersion interaction (if no oxidation takes place) and it provides work of adhesion with water, $W_a = 2\sqrt{(30 \times 21.8)} = 51$ mJ/m^2, which is not so much lower than the cohesion work of xylene or docosane molecules ($\sim$ 60 mJ/m^2), but obviously much less than the water cohesion work (145.6 mJ/m^2). If some water molecules were occluded in the oil phase during the cooling process, it could give rise to a hydrogen-bonding interaction with OH$^-$ ions and water molecules.

However, similarly to the findings of Dunstan [10] and Dunstan and Saville [13] that no specific adsorption of the charge takes place on the docosane surface, Jabloński et al. [17] found no adsorption of OH$^-$/H$^+$ by applying the potentiometric titration technique to a suspension of octadecane, which was also obtained by melting the alkane in water at 80°C and stirring mechanically, then ultrasonicating for 10 min before cooling to room temperature. The resulting suspension (5 g/200 mL of water) possessed particles of average 0.7 μm diameter, which gave 10.5 m^2/g for specific surface area. The particles showed negative electrophoretic mobilities, both pH and NaCl concentration dependent. These results are shown in Fig. 7. Again, up to pH 2.8 no positive values of the mobility appeared, and a maximum of the negative values as a function of pH occurred at the highest NaCl concentration, i.e., 0.1 M. However, maybe more interesting is the maximum in the electrophoretic mobility as a function of NaCl concentration appearing at 10^{-3} M. As was discussed above, according to Dunstan and Saville [10–13] the reason for the presence of an increase in the mobility versus electrolyte concentration curve is due to simultaneously occurring preferential solubility of the ions in the restructured region and retardation effect. The most important result of this study [17] is a direct finding that

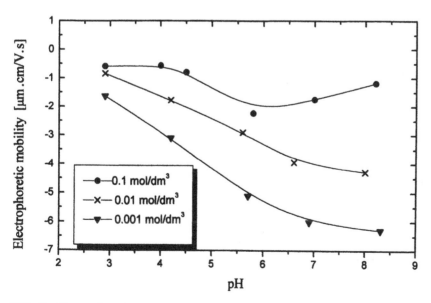

FIG. 7 Electrophoretic mobility of octadecane particles as a function of pH and NaCl concentration. (With permission of the authors: Ref. 17.)

neither H^+ nor OH^- ions adsorb (specifically) on to the octadecane surface. The blank and the suspension titration curves totally overlapped [17], while from the observed mobilities the surface charge should be $\sim 1 \mu C/cm^2$. Hence, taking into account the specific surface value ($10.5 \, m^2/g$) of the octadecane particles in the suspension, the consumption of 0.1 M NaOH should be about 0.2 mL, which is not the case. The authors [17] also postulate an indirect charging mechanism by the effect of H^+ and OH^-, as well as of electrolyte ions, on the vicinal water structure.

In the light of the above discussed results, the conclusion must be that H^+/OH^- are not charge-creating ions on the paraffin surface, although OH^- ions are mobility (zeta potential) determining ions. Furthermore, the structure of the electrical double layer cannot be well understood on the basis of the classical double-layer theories. Moreoever, the distance at which the slipping plane is located is still an open problem. However, because both from the electrophoresis and electro-osmosis [13] methods negative zeta potentials are obtained for hydrocarbon/water (electrolyte solution), it is strongly suggested that the negative charge must be in some way relatively strongly fixed to the hydrocarbon surface. Moreover, over 20 years ago, Chibowski and coworkers [18–23] measured the zeta potential of thick n-alkane films (from hexane to hexadecane) deposited on the surface of sulfur and other solids by the streaming potential method. While the zeta potential of the bare sulfur surface was $-25.6 \pm 0.5 \, mV$, the zeta potentials of the film-covered surface oscillated between ~ -98 and $-122 \, mV$. These zeta-potential changes are shown in Fig. 8, while Fig. 9 presents the zeta-potential changes of sulfur as a function of statistical monolayers of n-heptane deposited on sulfur. Comparing the results from Figs. 8 and 9, it may be concluded that the thickness of n-alkane films from Fig. 8 corresponds to about 15 statistical monolayers, assuming vertical orientation of the molecules on the surface [18]. However, the maximum thickness of such films on Teflon was found to be about four statistical monolayers [19]. If the orientation of the n-alkane molecules were not vertical, there should be no difference between zeta potentials for odd- and even-

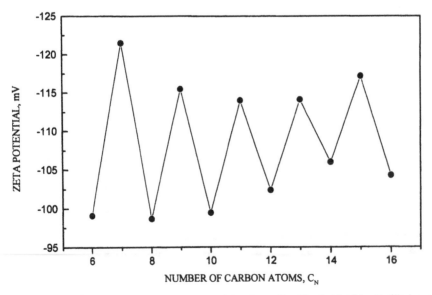

FIG. 8 Relationship between zeta-potential values of sulfur in doubly distilled water and number of carbon atoms in the n-alkane deposited on the sulfur surface. (From Ref. 18.)

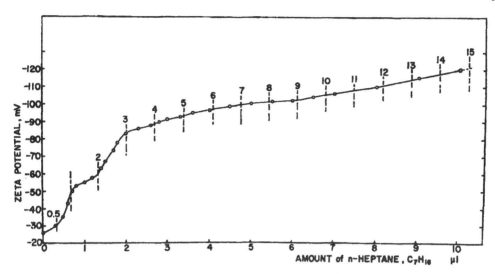

FIG. 9 Zeta potential of sulfur in water as a function of *n*-heptane volume used for wetting the sulfur surface. Vertical dashes denote the number of statistical monolayers; vertical orientation is assumed. (From Ref. 18.)

numbered *n*-alkane chains (see also Ref. 24). The observed individual changes of zeta potentials for "odd" and "even" alkanes are probably connected with different solid-state structure of the molecules [23]. Figure 10 shows schematically such a difference [25]. Among other things, the difference in the structure demonstrates also in the changes of the alkanes melting temperatures, which go in a similar way as the zeta potential changes presented in Fig. 8. Moreover, Stachurski and Michalek [7] found that at pH > 6.5 in 10^{-3} M NaCl, the zeta potential of *n*-alkane droplets also changed in a different way for "even" and "odd" alkanes (Fig. 2). Based on these results it may be concluded that the structure of the outermost layer of *n*-alkane droplets differs slightly for these two groups. The present authors obtained larger effective diameters of odd-alkane droplets than those of even alkanes in 1 M *n*-propanol emulsion (natural pH 6.8) for two independent series of the emulsion. The emulsions were prepared by dissolving the alkane in propanol, then adding water (to obtain 1 M solution) and homogenizing in an ultrasonic bath for 15 min. These results are shown in Figs 11a and 11b (effective diameters and zeta potentials, respectively) for 5-min-old emulsion, and in Fig. 12 for 2-h-old emulsion [26]. However, no clearly visible changes were observed for zeta potential, although in the first series (5 min old emulsion) this tendency could be observed.

Returning to the problem the origin of zeta potential in the systems discussed, Chibowski and Waksmundzki [19] postulated in 1978 that "essential contribution of the water dipoles in the double layer structure may be expected" (see also Refs 20–23). The results recently published in our papers [26–30], as well as those discussed above seem to support the idea, which actually is not in contradiction with the mechanism proposed by Dunstan and Saville [10–12] or by Marinova et al. [9]. Assuming that water dipoles are immobilized and ordered at the oil surface, the slipping plane might be located next to them. The role of OH^- and other ions may be structure making or breaking and possibly they may be fixed to the dipole layer (hydrated). This model is consistent with the suggestion of Israelachvili and Wennerström that "the non-slip plane is located no further than one water layer from these surfaces," which concerned a silica or mica surface [15].

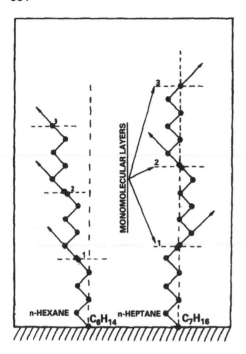

FIG. 10 Schematic representation of "odd" and "even" n-alkane successive monolayers in the solid state. (From Ref. 23.)

Because in a broad pH range of water the zeta potential at the oil surface is negative, the dipoles should be oriented with their negative pole $\text{>O}^-\text{O}$ towards the water phase. The zeta potential of n-alkane emulsions in different n-alcohol solutions (0.1–1.0 M), which will be discussed next, led us to conclude about the essential role of water and/or alcohol dipoles in the zeta-potential generation at the oil/water (solution) interface.

III. n-ALKANE/ALCOHOL SOLUTION EMULSION

A. Zeta Potential of Droplets of n-Alkanes in Alcohol Solutions

Many investigations of zeta potential and droplet size of n-alkane emulsions (0.1 v/v%) in methanol, ethanol, and propanol solutions have been conducted by the present authors [26–30]. The results in Figs 11 and 12 show that not only in-water but also in 1 M propanol the zeta potential of n-alkane droplets is negative. The multimodal size distribution of droplets in the n-alkane emulsions is shown in Fig. 13a (for 5-min-old emulsions) and in Fig. 13b (2-h-old emulsions). The most stable and reproducible emulsions in this study were obtained for decane and tetradecane [26]. Figure 14a presents the effective diameter and Fig. 14b the zeta potential for two series of tetradecane emulsion in 1 M propanol (natural pH 6.8). The values were determined with the ZetaPlus instrument of Brookhaven, which applies dynamic light scattering. Emulsion samples were poured into polyacrylic measuring cells and were left without any shaking. The particular values of the diameter and zeta potential concern the same sample. The effective diameter, which can also be called 'equivalent sphere diameter," is that which a sphere would have in order to diffuse at the same rate as the particle being measured [30, 31]. In a polydisperse system

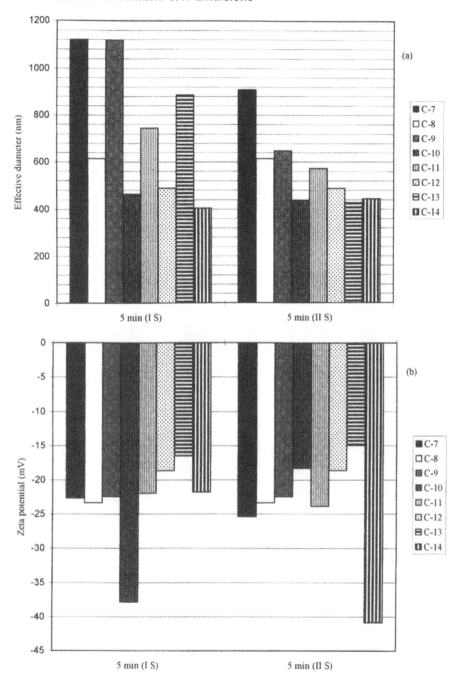

FIG. 11 (a) Effective diameter and (b) zeta potential of _n_-alkane/water/_n_-propanol (1 M) emulsion versus _n_-alkane chain length 5 min after emulsion preparation (two series). (From Ref. 26.)

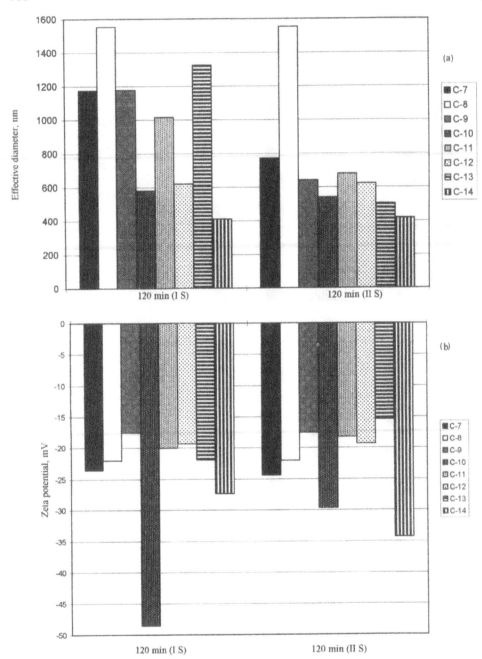

FIG. 12 (a) Effective diameter and (b) zeta potential of n-alkane/water/n-propanol (1 M) emulsion versus n-alkane chain length 120 min affer emulsion preparation (two series). (From Ref. 26.)

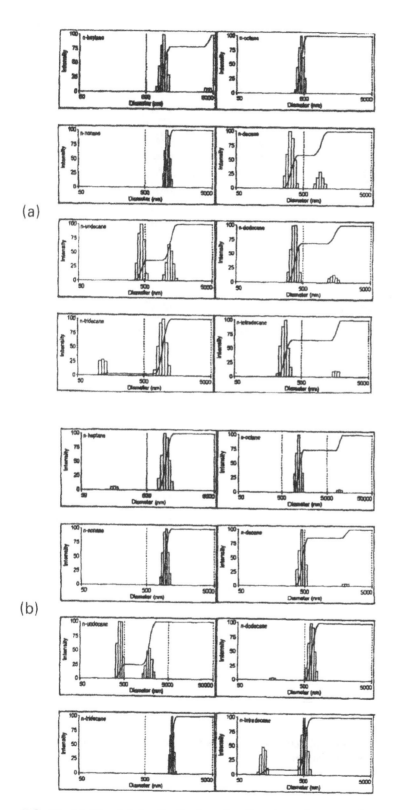

FIG. 13 Multimodal size distribution of *n*-alkane/water/*n*-propanol (1 M): (a) 5 min after emulsion preparation; (b) 120 min after emulsion preparation. (From Ref. 26.)

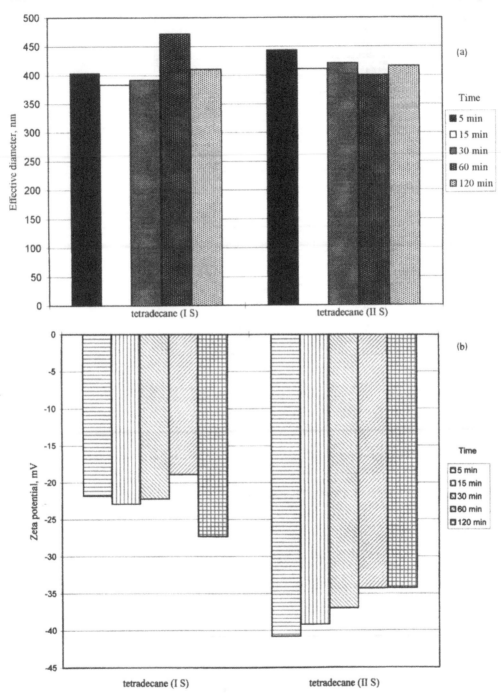

FIG. 14 (a) Effective diameter and (b) zeta potential of n-tetradecane/water/n-propanol (1 M) emulsion versus time (two series). (From Ref. 26.)

the effective diameter is an average value obtained by the averaged intensity of the light scattered by each particle (here, alkane droplets). As is seen from Fig. 13, in most cases only one principal population of droplets is present in the emulsions. If a decrease in effective diameter as a function of time is detected by dynamic light scattering in the sample, which has not been mixed in the meantime, it means that sedimentation (floating to the emulsion surface) of larger droplets took place. On the other hand, an increase in the effective diameter most probably results from coalescence of small droplets in the emulsion, the droplets still being in the emulsion phase. Obviously, both processes may occur simultaneously. The coalescence process, for example, took place in octane emulsion during 2 h (Fig. 13a). In the case of *n*-tetradecane in propanol at natural pH 6.8, the effective diameter (~ 400 nm) of the droplets was well reproducible and stable during 2 h (Fig. 14a, two series), while the negative zeta potentials were different, being higher by ~ 15–20 mV in the second series (Fig.14b). It is likely that in these diluted emulsions 0.1 $\frac{v}{v}$%) 20 mv of zeta potential is enough to keep the emulsions stable. This will be discussed later in detail. It is worth noticing that on the time scale the zeta potential is more or less contant in each series, but is is not a general feature of all *n*-alkane emulsions.

The pH effect on the effective diameter and zeta potentials of tetradecane emulsion in propanol is depicted in Fig. 15a and 15b, respectively. Although the absolute values of zeta potentials at pH 4 are less in comparison with those at pH 6.8 (natural) and 11, they are still negative and decrease slightly (from -35 to -20 mV) during the first hour (Fig. 15b). This decreasing trend is also seen at natural pH, but being only about 5 mV. The highest negative zeta potentials (average -45 mV) of tetradecane droplets were obtained at pH 11. In general, these zeta potentials in alcohol solutions are much lower than those obtained by Stachurski and Michalek [7] (Fig. 2, ~ -100 mV) in 10^{-3} M NaCl, as well as those of dodecane and hexadecane in 10^{-3} M NaCl at pH 6, as determined by Marinova et al. [9] and others for analogous systems [10–13, 17]. However, at pH 4 the zeta potentials in Fig. 15b are comparable to those in the literature [7, 9, 10–13, 17]. Such a behavior leads to the conclusion that the alcohol dipoles adsorbed on the alkane surface play a role here. Because the dipole moment of *n*-propanol is 1.69 D (1 D = 3.336×10^{-30} C m), while that for water is 1.84 D, water replacement by propanol dipoles should decrease the zeta potential, if the immobilized and oriented dipoles are really responsible for zeta-potential creation [15]. This case is believed to be true. A support for this idea is the zeta potential of *n*-hexadecane emulsion in a 1 M solution of methanol, ethanol, propanol, and 2-butanol shown in Fig. 16. The average values of zeta potential in alcohol solutions behave practically the same way as their dipole moments. Zeta-potential fluctuations in the case of methanol probably result from low stability of the emulsion. This is shown in Fig. 17 for dodecane emulsion in 1 M methanol, where the emulsion breaks after 1 h and it is accompanied by large zeta-potential fluctuations, as it was in the case of hexadecane. In ethanol and propanol solutions the emulsion is stable during one day.

The role of OH$^-$ ions in zeta potential setting may depend on hydrogen bonding formation with the dipoles immobilized at the alkane surface. More hydrogen bonds may occur in the case of water than propanol dipoles. This could explain the smaller negative zeta potentials in the presence of alcohol than in pure water. It is to be noted that the zeta potentials of the alkanes deposited on a sulfur surface in water were also over -100 mV [18].

Zeta potentials of decane droplets in propanol solution were also determined in the presence of NaCl, CaCl$_2$, or LaCl$_3$ salts in the 10^{-5} to 10^{-2} M concentration range [26]. In general, the presence of the cations increased the effective diameter of the emulsion

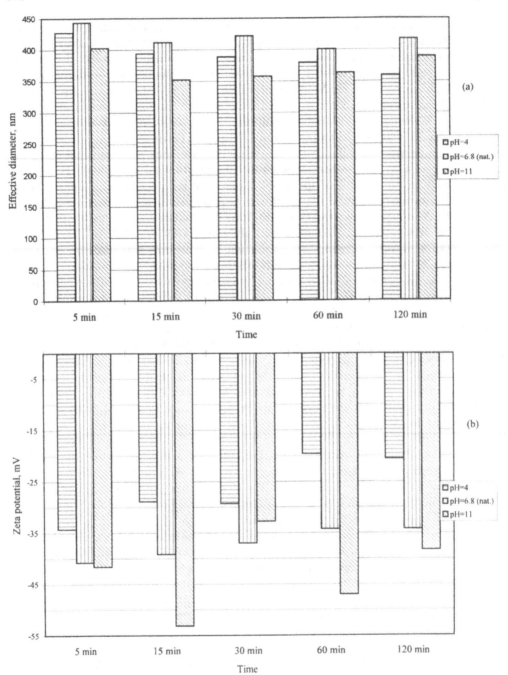

FIG. 15 (a) Effective diameter and (b) zeta potential of *n*-tetradecane/water/*n*-propanol (1 M) emulsion versus pH. (From Ref. 26.)

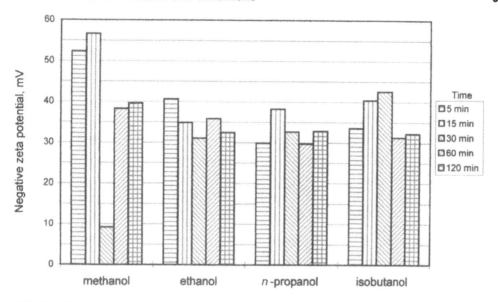

FIG. 16 Zeta potential of n-hexadecane/water/alcohol (1 M) emulsion versus alcohol chain length.

droplets, especially at the higher concentrations and in La^{3+} solution. An example is given in Fig. 18 for LA^{3+}. The effect of the presence of cations on the zeta potentials of decane droplets in 1 M propanol is shown in Figs 19–21, for Na^+, Ca^{2+}, and La^{3+}, together with zeta potentials in the reference emulsion, i.e., without those cations. The zeta potentials freshly prepared emulsions vary much in the presence of the cations (Figs 19–21) and generally the absolute zeta potential values are reduced in 10^{-2} M solutions relative to those in water. It looks as if the valency of the cations plays an essential role. While in the presence of mono- and bi-valent cations (Na^+, Ca^{2+}) at their higher concentrations (10^{-3} and 10^{-2} M) the zeta potentials remain negative, so in the presence of trivalent La^{3+} they are close to zero or even reverse the sign to positive in 1- and 2-day-old emulsions (10^{-2} M). Also, a positive zeta potential appeared in a very fresh emulsion (Fig. 21). The observed zeta-potential fluctuations, especially in the presence of cations, mean that the structure of the water and alcohol dipoles changes on the droplet surface due to "adsorption–desorption" (hydration) processes of cations. However, there is no strong relationship between the effective diameter (stability) and zeta potential of the emulsion, although in many systems for small absolute zeta potential values the breaking of the emulsion occurred faster. Figure 22a presents the changes in zeta potentials for decane emulsions in 1 M propanol and in solutions of the discussed cations (10^{-5}–10^{-2} M), and Fig. 22b shows the corresponding effective diameters for 15-min-old emulsions. In solutions of Na^+ and La^{3+} cations with increasing concentration, decreasing zeta potentials are accompanied by increasing diameters of the droplets. However, just a reverse relationship can be seen for Ca^{2+}. The reasons for such behavior may be manifold and more experimental evidence (using other cations) would be needed to draw a conclusion. It can only be stated that in a diluted emulsion any linear relationship between the magnitude of zeta potential and the effective diameter of the droplets actually does not exist, at least if the absolute value of zeta potential is higher than, say, 20 mV. The reason for this will be discussed later in terms of the extended DLVO theory [4], which considers the acid–base interactions.

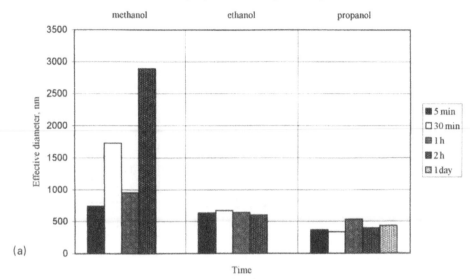

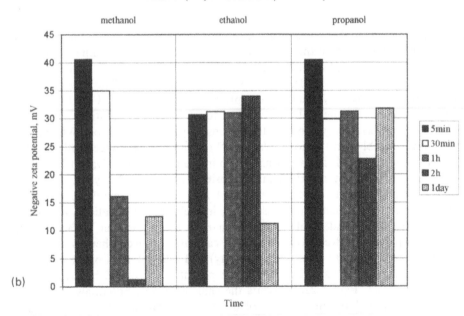

FIG. 17 (a) Effective diameters and (b) zeta potentials of *n*-dodecane droplets in emulsions prepared in 1 M methanol, ethanol, or propanol for different times after preparation of the emulsion. (From Ref. 28.)

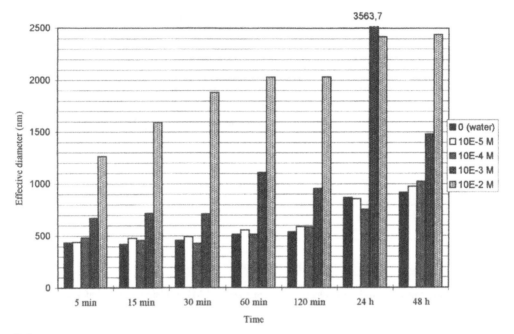

FIG. 18 Effective diameter of *n*-decane/electrolyte/*n*-propanol (1 M) emulsion versus LaCl₃ concentration. (From Ref. 26.)

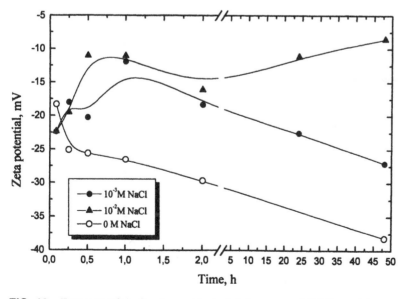

FIG. 19 Zeta potential of *n*-decane/electrolyte/*n*-propanol (1 M) emulsion versus NaCl concentration.

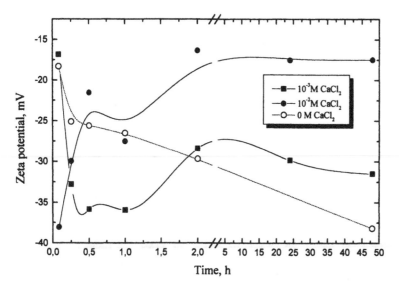

FIG. 20 Zeta potential of *n*-decane/electrolyte/*n*-propanol (1 M) emulsion versus CaCl₂ concentration.

B. Stability of *n*-Alkane/Alcohol Solution Emulsion in Terms of Extended DLVO Theory

It is well known that the classical DLVO theory, balancing attractive London dispersion and repulsive electrostatic forces only, may be applied to a limited number of dispersed systems to calculate the total interactions between two particles suspended in a liquid medium. Derjaguin and Churajev [32] suggested that it can be applied to moderately

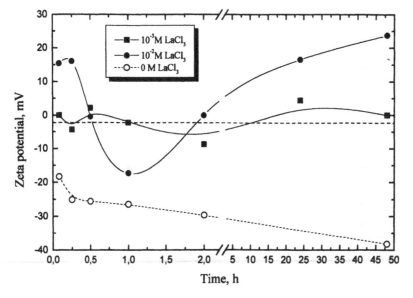

FIG. 21 Zeta potential of *n*-decane/electrolyte/*n*-propanol (1 M) emulsion versus LaCl₃ concentration.

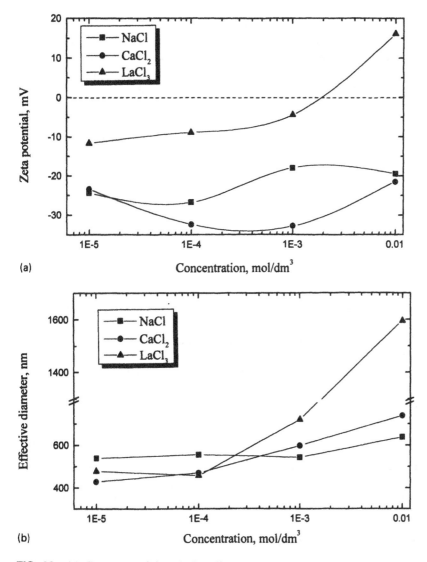

FIG. 22 (a) Zeta potential and (b) effective diameter of *n*-decane/electrolyte/*n*-propanol (1 M) versus electrolyte concentration 15 min after emulsion preparation.

hydrophobic/hydrophilic surfaces characterized by contact angles of water in the 15°–64° range. If the contact angle is higher than 64°, hydrophobic forces [33] contribute significantly, and if it is less than 15°, hydration forces operate across the interface. Van Oss [4, 34–36] were among the first who proposed a quantitative evaluation of the hydration forces, which in their approach are due to Lewis' acid–base electron-donor and electron-acceptor interactions. In most systems the forces are simply hydrogen bonding. The total interaction ΔG_{131}^{TOT} in the extended DLVO theory can be written:

$$\Delta G_{131}^{TOT} = \Delta G_{131}^{LW} + \Delta G_{131}^{AB} + \Delta G_{131}^{EL} \tag{9}$$

where ΔG_{131}^{LW} is the apolar Lifshitz–van der Waals' interaction, ΔG_{131}^{AB} is the acid–base (electron donor and electron acceptor) interaction, and ΔG_{131}^{EL} is the electrostatic interaction between the same two material particles (droplets) (1) dispersed in a liquid (water) (3) [4, 34–36]. Obviuolsy, if ΔG_{131}^{TOT} is negative, an attraction between the particles will occur, and the surface of the particles is hydrophobic if they are dispersed in water [37].

In the case of two identical flat surfaces and at a minimum equilibrium distance d_o $(0.158 \pm 0.008\,\mathrm{nm})$ the free energy ΔG_{131}^{LW} of apolar interaction reads [34]:

$$\Delta G_{131}^{LW} = -2\gamma_{13}^{LW} \tag{10}$$

For the case of interest here, two identical spheres (the oil droplets) having radius R, and if $R \gg d$, the suitable equation for the interaction as a function of the interparticle distance d is

$$\Delta G_{131}^{LW} = -\frac{A_{131}R}{12d} \tag{11}$$

where A_{131} is the Hamaker constant, which can be determined from ΔG_{131}^{LW} value at d_o distance:

$$-\Delta G_{131,d_o}^{LW} = \frac{A_{131}}{12\pi d_o^2} \tag{12}$$

From Eqs (11) and (12) it results that the apolar Lifshitz–van der Waals' interaction can be determined if the droplet radius is known and interfacial tension is determined for the oil/water interface. If any third substance is present (e.g., an alcohol) the situation is more complicated because adsorption of the alcohol molecules takes place on the oil surface and polar interaction appears. An attempt to evaluate the interactions in such a system will be discussed below.

The second term in Eq. (9) deals with polar acid–base interactions, which for two flat parallel surfaces at the equilibrium distance d_o can be expressed as [36]

$$\Delta G_{131}^{AB} = -4[(\gamma_1^+ \gamma_1^-)^{1/2} + (\gamma_3^+ \gamma_3^-) - (\gamma_1^+ \gamma_3^-)^{1/2} - (\gamma_1^- \gamma_3^+)^{1/2}] \tag{13}$$

where subscripts 1 and 3 mean oil and water (solution), respectively, and the sign "$+$" is for electron acceptor, and "$-$" is for electron donor interactions.

In the case of two spheres having radius R and interacting at a distance d, the acid–base free energy of interaction reads [4]:

$$\Delta G_{131}^{AB} = \pi \lambda R \Delta G_{131,d_o}^{AB} \exp[(d_o - d)/\lambda] \tag{14}$$

where $\Delta G_{131,d_o}^{AB}$ is the same as in Eq. (13) and λ is the decay length [38, 39], which for the discussed systems may be assumed to be equal to 1 nm [4]. From Eq. (13) it is clear that to calculate the ΔG_{131}^{AB} energy the surface free-energy components γ_i^- and γ_i^+ for the n-alkane droplet and water (alcohols solution) should be known. Note that for a "bare surface" of the alkane droplet in water no polar interactions appear. Nevertheless, the acid–base free energy of the interaction for oil–water–oil will be negative and equal to $-4(25.5 \times 25.5)^{1/2} = -102\,\mathrm{mJ/m^2}$, because the second term in Eq. (13) describes water acid–base interaction. In consequence, the two oil droplets attract each other and the emulsion breaks, if the repulsive electrostatic interaction does not exceed the acid–base one, which is not the case in a pure oil–water–oil system. However, in the presence of an alcohol solution its polar molecules are adsorbed on alkane droplets and no term in Eq. (13) is zero. This system will be discussed later.

First, let us describe the electrostatic term in Eq. (9). In order to evaluate the electrostatic repulsion energy the electric potential Ψ^o at the droplet surface should be known. Because it is hardly possible to determine this potential experimentally, usually the zeta potential is taken for the calculations. In the case of these emulsions this seems quite justified. Among many equations describing electrostatic repulsion between surfaces of plates and spheres for different magnitudes of the electrical potential Ψ° and κR parameter (Debye–Hückel reciprocal length times the sphere radius) [40], an approximate simple equation for two identical spheres having radius R and zeta potential, say up to $50\,\text{mV}$, reads [41]:

$$\Delta G_{131}^{\text{EL}} = 0.5\varepsilon R\zeta^2 \ln[1 + \exp(-\kappa d)] \tag{15}$$

where ε is the dielectric constant of the liquid medium, and d is the distance between the spheres.

Thus, the evaluation of the total interaction using the extended DLVO theory needs solution of Eqs (11)–(15). This is possible if one knows the surface free-energy components of the droplet and the liquid (solution) as well as the droplet zeta potentials. Such evaluation has been conducted for n-dodecane emulsion in 1 M solutions of methanol, ethanol, and propanol [28], the zeta potentials of which are shown in Fig. 17. The problem was with determination of the free-energy components of alkane droplets with the adsorbed alcohol molecules to evaluate $\Delta G_{131}^{\text{LW}}$ and $\Delta G_{131}^{\text{AB}}$ free energies. The interfacial free energy (interfacial tensions) were taken from Jańczuk et al.'s paper [42]. They used the ring and sessile drop methods and determined the total dispersion (Lifshitz–van der Waals) and polar (acid–base) components of the surface tension. The polar components were 6.49, 3.32, and $2.79\,\text{mJ/m}^2$ for methanol, ethanol, and propanol, respectively. Although from these data individual values of the electron donor, γ_i^-, and the electron acceptor, γ_i^+, parameters cannot be obtained, the true value of the acid–base interaction with the solution can be calculated [28]. This is possible because γ_i^- and γ_i^+ are interrelated through the relationship $\gamma_i^{\text{AB}} = 2(\gamma_1^+ \gamma_1^-)^{1/2}$. The problem was how to evaluate the acid–base interaction for dodecane droplets in the alcohol solutions. Two variants were considered [28]. The first, a closed-packed monolayer of the alcohol molecules on the alkane droplet surface, because the emulsions were produced by adding water to the alkane solution in the alcohol. The second, the amount of alcohol molecules on the alkane droplet surface was calculated from the zeta-potential value, assuming that it totally resulted from the oriented alcohol dipoles present on the surface in this electrolyte-free system.

The details of the calculations can be found in our paper [28]. In Table 1 there are listed the calculated free energy of interactions in the alcohol solutions at the minimum equilibrium distance d_0 as calculated from both variants. Also, the zeta potentials, effective diameters, and maximum distance between the droplets are shown. Because the Debye parameter κR was in the range 0.28–3.31, the measured zeta potentials (Fig. 17b), which were determined with the zetameter from the Smoluchowski equation, were recalculated using Henry's equation, and these values are presented in Table 1. Next, because in ethanol and propanol solutions the changes in zeta potential and effective diameter during the first two hours were relatively small, they were averaged for the calculations. In the case of methanol the free energy of interaction was calculated for 5 and 30 min and 1 and 2 h-old emulsions.

As seen in Table 1, the extended DLVO gives a dominant interaction for the acid-base type independently of the variant used for the calculations. The assumption of a close-packed layer of the alcohol dipoles on the alkane droplet surface (variant 1) gives ~ 2.5 times higher acid–base interaction than that calculated from the zeta potential (variant 2). To compare the free energy of interaction for a droplet–solution–droplet in different alcohol solutions, normalized values of the interactions were calculated (ΔG_{131} divided by radius R, Table 1). It can be easily found that the total energy of interaction at

TABLE 1 Interfacial Parameters for n-Dodecane (0.1 v/v%) in Alcohol Solution (1 M) Emulsion at Natural pH

			Dispersing phase			
				Methanol		
Parameter	Propanol	Ethanol	5 min	30 min	1 h	2 h
pH	6.8	7.0		6.3		
Zeta potential (mV)	−46.2	−47.1	−60.8	−35.0	−23.8	−1.3
X_{max} (μm)	3.9	5.9	6.7	15.6	8.8	26.1
ΔG_{131}^{LW} (kT)	−9.6	−10.3	−12.0	−27.9	−15.7	−46.0
ΔG_{131}^{EL} (kT)	340	540	1020	780	74	190
$V.1^a$ ΔG_{131}^{AB} (kT)	−5160	−17,000	−23,000	−54,000	−30,400	−87,900
$V.2$ ΔG_{131}^{AB} (kT)	−1700	−7800	−8600	−19,900	−11,200	−32,800
$\Delta G_{131}^{LW}/R$ (kT/nm)	−0.04	−0.03	−0.03	−0.03	−0.03	−0.03
ΔG_{131}^{EL} (kT/nm)	1.63	1.57	2.71	0.90	0.15	0.13
$V.1$ $\Delta G_{131}^{AB}/R$ (kT/nm)	−23.9	−51.5	−61.5	−62.1	−62.2	−61.4

[a] V = variant.

minimum equilibrium distance d_0 is negative, because of high negative values of ΔG_{131}^{AB}. Nevertheless, the emulsions were stable for 2 h, except for that in methanol solution (Fig. 17b). This is mainly caused by a large droplet-to-droplet distance (X_{max}, Table 1). Obviously, the larger the droplets (from the same total volume of the alkane) the larger the maximum distance and the lower is the emulsion stability [28]. The lowest attractive interactions (apolar Lifshitz–van der Waals and polar acid–base) appear in propanol solution, and in accordance with the experimental results, this emulsion is the most stable, even during 1 day (Fig. 17). Note that although in methanol solution the droplet effective diameter varies much during the first 2 h (Table 1), the normalized apolar ΔG_{131}^{LW} and polar acid–base ΔG_{131}^{AB} interactions are constant during 2 h.

The changes in total interaction free energy ΔG_{131}^{TOT} in the discussed emulsions as a function of multiples of d_0 are shown in Fig. 23 for 5 min-old emulsions. The calculation showed [28] that at a distance $50d_0$ the total interaction energy is positive: 149, 283, and 214 kT for propanol, ethanol, and methanol, respectively, and the same is true at $100d_0$ (Fig. 23). However, in methanol solution this energy drops down fast and after 1 h it is only 38.6 kT. The appearance of the positive total interaction energy at a distance less than the maximum drop–drop distance explains why these diluted emulsions are relatively stable, even if the assumed models may be considered as debatable.

C. Zeta Potential of *n*-Alkane Droplets in Alcohol (Protein) Solution

Application of natural or natural origin emulsifiers and/or stabilizers, like proteins, is of great interest because of the practical use of emulsion systems in many agricultural, pharmaceutical, and cosmetic products. Many papers have been published on protein adsorption and behavior at different interfaces, too many to quote all of them here, but some examples of those more recently published should be mentioned [2, 43–57]. However,

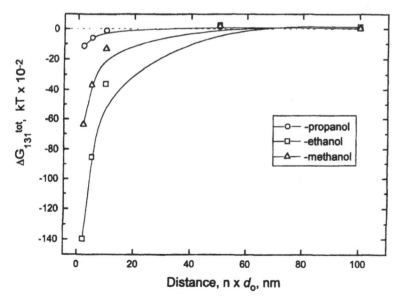

FIG. 23 Changes in total free energies of interaction as a function of distance expressed in multiples of equilibrium distance $d_0 = 0.158$ nm. (From Ref. 28.)

many aspects of protein behavior in the adsorbed state are still far from well understood. In this chapter some results will be presented for n-alkane emulsions in alcohol solution in which lysozyme or lysine hydrochloride were present. Lysine is an amino acid having the chemical formula $H_2NCH_2(CH_2)_3CH(NH_2)COOH$, for which the pK_a of 1-COOH, 1-NH_3^+, and 5-NH_2 are 2.2, 9.2, and 10.8, respectively [43]. Its i.e.p. occurs at pH 9.72, while that for lysozyme molecules lies at pH 10.7–11.1 [43, 54]. Lysozyme is one of the best charactereized proteins. It is a small globular protein (MW 14,603) with 18 amino acid (cationic) and 12 carboxyl (anionic) residues [57]. Proteins adsorb at various interfaces, including oil/water, thus stabilizing, for example, food emulsions [52]. The hydrophobic parts of the molecules tend to the oil phase while at least some polar -NH_3^+ and -COO^- groups are directed toward the water phase. Protein adsorption is considered to be irreversible, but small surfactant molecules may desorb them [2, 52]. Obviously, temperature, pH, and ionic strength may affect the adsorotpion [2]. In the case of globular proteins (e.g., lysozyme) the adsorbed molecules may possess some transition forms between native and denaturated [52, 54], among them the so-called "molten globule" which can also exist in the solution phase [52]. The hen's egg-white lysozyme has dimensions 4.6 nm $\times$ 3.0 nm $\times$3.0 nm and 59% of its surface is apolar [54]. During its adsorption process some structural rearrangements always take place [54]. Computer images show a tight packing of the atoms with an average density of 75%, which is much larger than that for water (44%) at 25°C and 1013 hPa [54]. The polar and apolar residues are nearly evenly distributed on the globular compact surface of the protein. From the changes in free energy ΔG of the denaturation process as a function of temperature it results that the whole globular protein molecule could unfold if disruption of an essential part of the molecule took place. At 25°C lysozyme exists in solution as a monomer, but dimers and/or higher numbers of oligomers may also appear [54]. It is a rigid and stable molecule over a broad range of temperatures (from ambient up to 75°C) and pH (2–11) [58]. The adsorption plateau is reached at 3–4 mg/m^2 [2, 54] and it is higher at a pH close to its i.e.p. [2]. According to Haynes and Norde [54] adsorption of globular protein to an apolar surface (polystyrene) must cause removal or neutralization of the electrical charge between the molecule and the surface. This may be achieved by ion-pair formation (protein molecule–surface), protonation/deprotonation of the ionizable residual groups, and coadsorption of small ions in the contact layer. Lateral interactions between the adsorbed molecules may obviously be attractive or repuslive, depending on the kind and magnitude of the electric charge of the residues.

Figure 24 shows zeta potentials of n-tetradecane droplets in 1 M ethanol as a function of pH, for the emulsions in which 10^{-3} M lysine hydrochloride, 4 mg/100 mL Lysozyme, or 10^{-3} M KCl was present. For comparison there are also shown zeta potentials of emulsion in 1 M ethanol alone and in pure water [29]. Emulsions in 1 M ethanol were prepared by dissolving 0.1 mL of tetradecane in an appropriate volume of ethanol, then water was added to yield 100-mL mixtures which were next homogenzied in an ultrasonic bath for 15 min. In the case of emulsions prepared in pure water or 10^{-3} M KCl, they were obtained by mechanical stirring (10,000 r.p.m. for 3 min). The emulsion pH was achieved by adding a suitable volume of concentrated HCl or KOH solution. As is seen in Fig. 24 the i.e.p. of the droplets (freshly prepared emulsion) in the presence of lysozyme occurs at pH $\simeq$ 9.4. It is shifted by about 1.5 pH unit toward neutral pH value in comparison with the i.e.p. of the native molecule. The conclusion might be that more residual -COO^- groups are directed toward the water phase in the adsorbed state of the molecule, as a result of conformational changes upon the adsorption process. In the absence of lysozyme, the zeta potentials of the tetradecane droplets in 1 M ethanol are

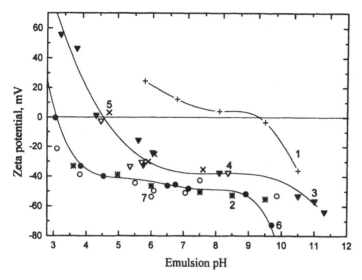

FIG. 24 Zeta potential of *n*-tetradecane droplets in emulsions (0.1% v/v) as a function of pH: (1) in 1 M ethanol + lysozyme (4 mg 100 ml^{-1}); (2) in 1 M ethanol + 1 mM lysine hydrochloride; (3) in 1 M ethanol (5 min after preparation); (4) as (3) after 1 h; (5) as (3) after 24 h; (6) in pure water; (7) in 1 mM KCl. (From Ref. 29.)

negative up to pH 4.5 and are practically constant within 24 h (Fig. 24, points 3, 4, and 5). The values deal with the same emulsion sample the pH of which was decreased by subsequent addition of HCl to the already prepared emulsion. Moreover, as will be discussed later, if the emulsion was prepared by adding water (with dissolved lyzosyme 4 mg/100-mL and the solution obtained by dilution of a stock, more concentrated lysozyme solution, 20 mg/100 mL) whose pH had been already regulated to 4, negative zeta potentials of the droplets in 1 M ethanol appeared up to 24 h and in the next 24 h the zeta potentials changed the sign to positive. However, this is not the case in 0.5 M ethanol or 0.5 and 1 M propanol solutions [27, 30].

In pure water or KCl solutions the potentials are also negative up to pH 3 (Fig. 24, points 6 and 7), and the changes are parallel to those in alcohol solution. In the pH range 4–9, no effect of KCl on zeta potential appears relative to the values in pure water. Only at pH 3 and 9.7 are the negative zeta potentials about 20 mV lower and higher, respectively. It is hardly possible to explain this behavior based on these results, except for an essential increase in the ionic strength at both pH values because of the presence of KCl. Maybe more interesting is the i.e.p. occurrence in water at about 1.5 pH unit lower than it happens in the alcohol solution, pH 3 and 4.5, respectively. Assuming water and the alcohol dipoles to be primarily responsible for the zeta potentials of the droplets, the observed i.e.p. shift suggests a weaker negative charge originating from the adsorbed and oriented alcohol dipoles on the droplet surface, and this seems reasonable. Also, in 10^{-3} M lysine hydrochloride the zeta potentials are practically constant, $\simeq -40 \pm 5$ mV in the pH range 4–9.5, and they fall into the values obtained in water and KCl (Fig. 24). This might suggest that lysine has no effect on the droplet zeta potentials. If the molecules would affect the zeta potential of the tetradecane droplets, one might expect the i.e.p. occurrence at a pH somewhere close to 9.72 (lysine i.e.p.). Some effect of the presence of lysine on the zeta potential was observed, but only in 0.5 M ethanol [30]. Finally, it should be noted that the values presented in Fig. 24 were obtained from the Smoluchowski equation, and in some cases they should be corrected by using Henry's equation. It would cause an increase in the values by a factor of 1.005–1.175, depending on the pH and the salt present. The details can be found in Ref. 29.

Zeta potentials of the emulsion prepared in 1 M ethanol, using water whose pH was regulated to 4 or 11 and at which the lysozyme solution was added (4 mg/100-mL) from its stock solution (20 mg/100 mL), are shown Fig. 25a. In this figure there are also plotted zeta potentials in analogous emulsions, but lysozyme free. As is seen, even at pH 4 the zeta potentials are negative during the first 24 h, although they are reduced in relation to those without lysozyme. Also, at pH 11 the negative zeta potentials in solutions with lysozyme are lower than those in lysozyme-free solutions. Figure 25b shows effective diameters of the discussed emulsions. The presence of lysozyme at pH 4 permits much more stable droplets than in the alcohol alone and the droplets are on average 120 nm smaller, whereas at natural pH 6.8 and 11 lysozyme rather destabilizies the emulsion (Fig. 25b). The results in Fig. 25 show again that the magnitude of the zeta potential is not a principal factor in the stability of these emulsions. It is worth paying attention to the positive zeta potential at pH 4 in 48-h-old emulsion. It indicates that the structure formation process of the adsorbed lysozyme is slow. If first the emulsion was prepared and next its pH was regulated, positive zeta potentials were obtained up to pH 9 for freshly prepared emulsion (Fig. 24). In investigations of the emulsion properties it should be remembered that protein adsorption and conformational processes occurring at the oil/alcohol solution interface are very slow. It may happen that emulsions prepared in similar conditions display different zeta potentials. Burns et al. [56] investigated the adsorption of lysozyme on plasma-polymerized 1,2-diaminocylcohexane on polystyrene. The surface changed its acid-to-base properties with time, probably as a result of the reaction between the surface amino groups and CO_2 from the atmosphere, resulting in formation of carbamic acid [56]. After 64 days the surface was almost neutral with an acid/base ratio of 1.02. They found decreasing adsorption of lysozyme both at pH 5 and 11 with duration of the experiment. Although at pH 5 both the surface and lyzozyme were positively charged, the adsorption was higher at this pH than at pH 11, where the surface was negatively charged and the molecule was close to the i.e.p. The authors [56] concluded that lysozyme was "pH insensitive" and had a strong tendency to adsorb on positively charged surfaces. They quoted the reuslts of similar behavior on different surfaces as obtained by other authors. This of course does not explain the mechanism of such a behavior. In our opinion this may be due to hydrogen-bonding formation between the residual molecules and the surface groups, which are stronger than the electrostatic repulsion between not too many charged groups on the surface and protein molecules. On the other hand, a typically electrostatic mechanism of protein adsorption (serum albumin and others) was suggested by Kato et al. [55] on ionic surfaces.

The results in Fig. 25a indicate that positively charged lysozyme molecules in a solution probably adsorb very slowly on an apolar oil surface, especially if alcohol molecules are already present on it. It took 2 days to reverse the zeta-potential sign, probably by removing the ethanol dipoles or creating hydrogen bonds with them (the net effect is the same). Dickinson and Matsumura [52] recorded changes in the interfacial tension in n-tetradecane/water (from 52 to $\simeq$ 28 mN/m) during up to 2 h in the presence of α-lactalbumin (10^{-4} wt%) at pH 7.5 and 40°C. This throws some light on the possibility of differently charged n-alkane droplets at the same pH in freshly prepared emulsion, depending on the method of preparation (Figs 24 and 25a); however, this needs further investigation. According to these authors [52] molecular mechanisms contributing to the free energy of the adsorption process are as follows: dehydration of the hydrophobic parts of the mole-

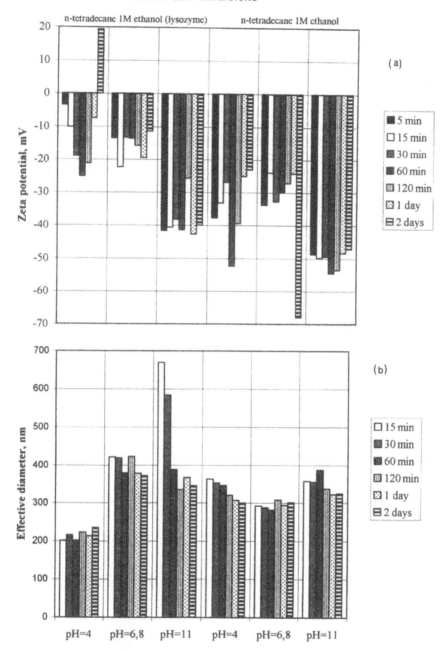

FIG. 25 (a) Zeta potentials and (b) effective diameter of *n*-tetradecane droplets in 1 M ethanol + 4 mg/100 mL lysozyme solution in pH 4, 6.8 (natural), and 11 for different times after emulsion preparation. (From Ref. 30.)

cule, its unfolding at the surface, charge redistribution, changes in pK on adsorption, and restructuring of van der Waals' interactions. The principal driving force for protein adsorption is removal of its hydrophobic amino acid chains from polar environment, i.e., bulk water phase [52]. Dilution of the water phase does not cause desorption of protein (irreversible adsorption). However, there are always small areas among protein train segments (which occupy only about one-third of the surface area) which are accessible to small surfactant molecules, such as alcohols [51, 52]. Ethanol already acts as a denaturating agent at 20°C [51, 59]. Note, however, that 1 M ethanol is only $\simeq 5$ wt% solution. Obviously, pH and temperature are also factors affecting the native state of the protein, causing a transition to "molten globule" structure. In general, "protein adsorption is characterized by time dependence in protein conformational structure" [52]. From the above cited conclusions and statements it results that much is unclear and debatable as regards protein behavior at the oil/water (alcohol) interface. Also, our results lead to the same conclusion. It should be mentioned here that in other series of experiments, positive zeta potentials (not presented here) of tetradecane emulsion at natural pH and at pH 4 were obtained in 1 M ethanol, despite the fact that they were prepared in the same way as those which exhibited the negative values. More careful investigations are being conducted to clarify this point.

The effect of ethanol concentration on lyzosyme adsorption on a tetradecane droplet surface as characterized by zeta potential is shown in Fig. 26, where the zeta potentials are presented for the emulsion prepared in 0.5 M ethanol, while Fig. 25a shows the potentials in 1 M ethanol. At this lower concentration, at pH 4, the zeta potentials oscillate around zero (± 3 mV) on the time scale (up to 2 days). Similar oscillations appear at pH 6.8; however, within a larger range (± 10 mV). These emulsions were unstable, while in the absence of lysozyme they were stable, especially at pH 4 [30].

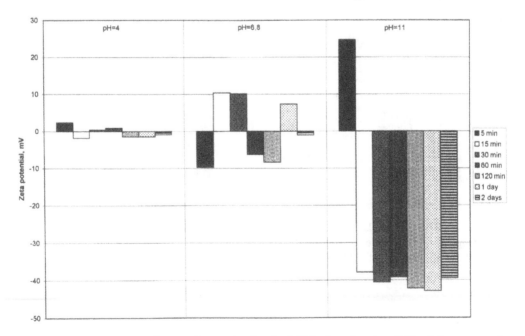

FIG. 26 Zeta potentials of n-tetradecane droplets in 0.5 M ethanol $+$ 4 mg/100 mL lysozyme solution in pH 4, 6.8 (natural), and 11 for different times after the emulsion preparation. (From Ref. 30.)

Using *n*-propanol instead of ethanol, stable emulsions were obtained [27]. Even in 0.1 M propanol solution the emulsion was stable during 2 days at pH 4, 6.8, and 11, having practically the same average diameter of 290 nm. Without lysozyme, however, the emulsion was stable only at pH 4. These results are presented in Fig. 27a, and Fig. 27b shows the corresponding zeta potentials. As can be seen, the zeta potentials in propanol were positive at pH 4 and 6.8. At pH 11 the potentials were negative and very stable during 2 days. If lysozyme was not present in the emulsion, its zeta potential was negative at the pH values discussed and in 0.1, 0.5, and 1 M propanol solution, similarly to what it was in 0.5 and 1 M ethanol solutions (Fig. 25a). As an example, Fig. 28a and 28b show the effective diameters and zeta potentials, respectively, for tetradecane emulsion at natural pH and in 0.1, 0,5 and 1 M propanol. These diluted emulsions (0.1 v/v%) were stable during 2 days (without any shaking). The effective diameter increased with increasing propanol concentration, especially in 1 M solution (Fig. 28a) and it was accompanied by decreasing zeta potentials (Fig. 28b). The emulsion was still stable (Fig. 28a), although in 1 M solution the zeta potential dropped very soon to 10–12 mV after preparation. It may be concluded that the droplets are being stabilized by other than electrostatic repulsive forces. Hydrogen bonding interactions between the residues of the adsorbed protein molecules and water (alcohol) molecules in the bulk phase may play here an essential role [26–30]. Actually, the importance of hydrogen bonds was discussed above for emulsions in the alcohol situations alone (Table 1).

The zeta-potential changes point to a competitive adsorption taking place between the alcohol dipoles and protein molecules. Alcohol concentration and its polar interaction (Lewis acid–base interaction) [42] seem to determine protein adsorption, followed by restructuring of the molecule, and in consequence the zeta potential of the alkane droplet. Similar competition between ethanol (0.5 and 1 M) and lysine (10^{-3} M) molecules was observed via the zeta-potential changes in tetradecane droplets [30]. Relatively strong acidic -COOH groups ($pK_a = 2.2$) caused a clearly visible decrease in the negative zeta potentials of the emulsion only in 0.5 M ethanol (Fig. 29), while in 1 M ethanol the effect was not so clearly visible (see Fig. 24). The 10^{-3} M lysine solution in most investigated systems destabilized the emulsion, except for one in 0.5 M ethanol at natural pH, where the emulsion was much more stable than in lysine-free emulsion [30]. This amino acid could be expected to increase the emulsion stability, but probably its molecular structure with polar residues on both sides of the molecule hinders its adsorption on the apolar alkane surface. On the other hand, hydrogen bonding between the adsorbed alcohol molecules, which are weak electron donors, would be formed if the lysine molecule had a high electron acceptor (proton donor) ability, which does not seem to be the case. The behavior also suggests that the electrostatic mechanism of lysine adsorption does not play any important role here either. At pH 4 and 6.8 the lysine molecule is positively charged [43] and alkane droplets show a negative zeta potential in alcohol solution. This supports the conclusion that the zeta potential of *n*-alkane droplets does not originate from excess ionic charge adsorbed but rather oriented dipoles.

Finally, the temperature effect on the emulsion zeta potential in the presence of lysozyme was investigated. The impact of both parameters, ethanol concentration (0.5 and 1 M) and temperature at natural pH, on the zeta potentials of tetradecane are plotted in Figs 30 and 31. In both solutions (0.5 and 1 M) the temperature increase from 20° to 30°C causes a reversal of zeta potential from negative to positive. Further temperature increase does not practically affect the zeta-potential value. At these higher temperatures, zeta potentials are practically the same in both ethanol solutions (0.5 and 1 M) and relatively stable. As was mentioned above, in one series positive zeta potentials were

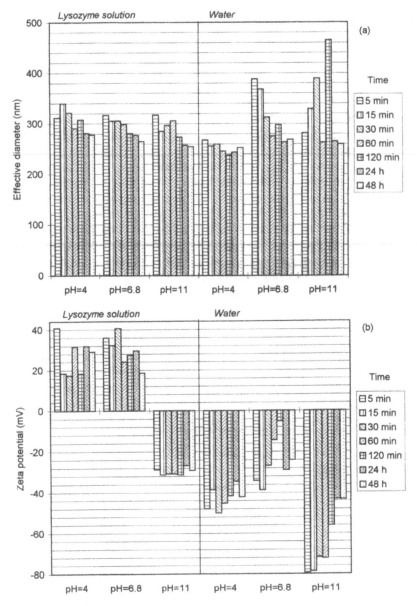

FIG. 27 (a) Effective diameter and (b) zeta potential of n-tetradecane/n-propanol (0.1 M)/lysozyme (4 mg/100 mL) and n-tetradecane/water/n-propanol (0.1 M) emulsions versus pH. (From Ref. 27.)

already observed at 20°C and they were in the range of the values at 30° and 50°C. High oscillations of zeta potential (−10 to +10 mV) in 0.5 M ethanol occurred together with large changes in the droplet size of the emulsion (4–10 µm) (Fig. 32). This must be the consequence of a coalescence process taking place in the emulsion and resulting in the adsorption/desorption of lysozyme, thus leading to the observed zeta-potential sign reversal. The temperature increase caused stabilization of the droplets at their drastically smaller effective diameter 400–600 nm (Fig. 32). Obviously, the temperature increase affects the protein adsorption kinetics and its structure (see above, e.g., the molten globule concept),

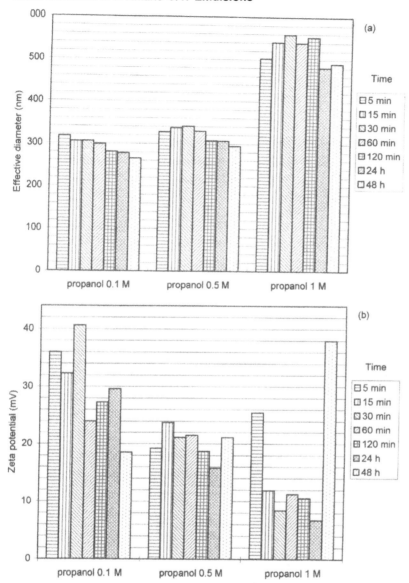

FIG. 28 (a) Effective diameter and (b) zeta potential of *n*-tetradecane/*n*-propanol/lysozyme (4 mg/ 100 mL) emulsion at pH 6.8 (natural) versus *n*-propanol concentration in water. (From Ref. 27.)

leading to its unfolding (denaturation) [2, 52, 59] and finally a faster establishment of the adsorption equilibrium.

IV. SUMMARY

Emulsion systems are very interesting for electrokinetic studies both for theoretical and practical purposes. Although in true thermodynamic equilibrium conditions no electric potential drop should be observed across the oil/aqueous solution interface, in most emulsion systems including those in pure water or simple inorganic electrolyte solution,

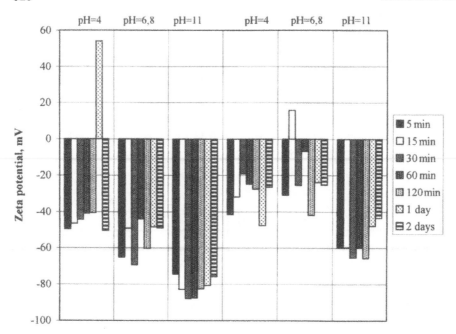

FIG. 29   Zeta potentials of *n*-tetradecane droplets in (0.5 M ethanol and in 0.5 M ethanol + 1 mM lysine monohydrochloride) solutions in pH 4, 6.8 (natural), and 11 for different times after emulsion preparation. (From Ref. 30.)

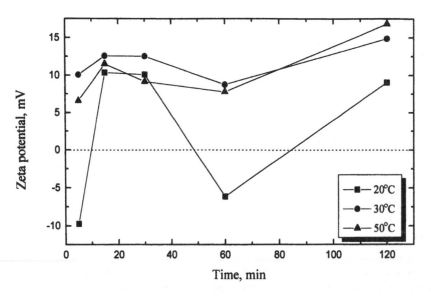

FIG. 30   Zeta potentials of *n*-tetradecane droplets in 0.5 M ethanol + 4 mg/100 mL lysozyme at 20°, 30°, and 50°C as a function of time.

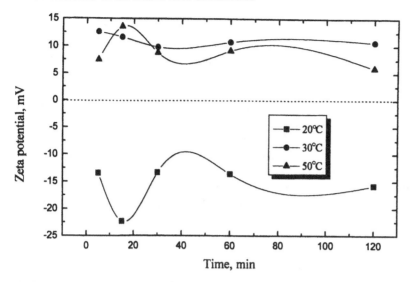

FIG. 31 Zeta potentials of *n*-tetradecane droplets in 1 M ethanol + 4 mg/100 mL lysozyme at 20°, 30°, and 50°C as a function of time.

electrokinetic phenomena are observed. This is due to some metastable states inhibiting thermodynamic equilibrium as well as to the nonionic origin of the interfacial potential, which is most probably due to permanent dipoles, like water or alcohol, immobilized and preferentially oriented with their negative poles towards the water phase. Therefore, practically all reported experimental results show negative zeta potentials of an oil droplet in water or alcohol solution in a broad pH range, say 4–11. Small inorganic ions, especially multivalent cations (La^{3+} or Al^{3+}), affect the zeta potential and even cause its reversal to positive values at increasing concentration. The mechanism of the zeta-potential forma-

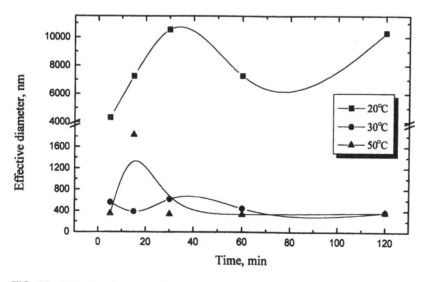

FIG. 32 Effective diameter of *n*-tetradecane droplets in 0.5 M ethanol + 4 mg/100 mL lysozyme at 20°, 30°, and 50°C as a function of time.

tion probably relies on "attachment" of ions to the first structured and immobilized layer of water (alcohol) dipoles. In another concept, it is due to preferential (competitive) "solubility" of ions in the vicinal layer of water. The present authors believe that hydrogen-bonding interactions between water and alcohol dipoles play an important role in stabilizing/destabilizing the emulsion systems. In fact, the model calculation showed that this type of interaction predominates in comparison with the attractive London dispersion and repuslive electrostatic interactions.

The intriguing electrokinetic behavior of oil droplets in alcohol solution in the presence of protein (lysozyme) needs more experiments to clarify the observed, sometimes unexpected, results. It is believed that in such very complex systems, Lewis acid–base interactions (electron donor and electron acceptor, hydrogen bonding) play an important role. The mechanism of zeta-potential generation actually consists of two competing processes, as in adsorption/desorption of alcohol and protein molecules, hydrogen bond formation between the molecules, conformational changes in the adsorbed protein molecules (unfolding, denaturation), ionization of amino and carboxyl residues, and others. These processes, first of all depend on time, but also on the emulsion temperature, pH (ionic strength), and the kind and concentration of alcohol solution. To obtain more experimetnal evidence on the effect of protein on the electrokinetic behavior of emulsions and their stability, experiments are under way in which other proteins (bovine serum albumin, α-lactalbumin, and β-casein) are used.

REFERENCES

1. Proceedings of the Second World Congress on Emulsion. vols 1–4, Bordeaux, France, 1997.
2. P Wlastra. In: P Becher, ed. Encyclopedia of Emulsion Technology. New York: Marcel Dekker, 1996, pp. 1–61.
3. W Adamson. Physical Chemistry of Surfaces. 5th ed. New York: John Wiley, 1990.
4. CJ Van Oss. Interfacial Forces in Aqueous Media. New York: Marcel Dekker, 1994.
5. H Yotsumoto, RH Yoon. J Colloid Int Sci 157:426, 1993.
6. HR Kruyt. Colloid Science. vol. 1. Amsterdam: Elsevier, 1952.
7. J Stachurski, M Michalek. J Colloid Int Sci 184:433, 1996.
8. HC Parreira, IH Schulman. Advances in Chemistry Sciences No. 33. Washington, DC: American Chemical Society, 1961, p 160.
9. KG Marinova, RG Alargova, ND Denkov, OD Velev, DN Petsev, IB Ivanov, RP Borwankar. Langmuir 12:2045, 1996.
10. DE Dunstan. Langmuir 12:2045, 1996.
11. DE Dusntan. J Phys Chem 97:11143, 1993.
12. DE Dunstan, DA Saville. J Chem Soc, Faraday Trans 88:2031, 1992.
13. DE Dunstan, DA Saville. J Chem Soc, Faraday Trans 89:527, 1993.
14. JN Israelachvili. Intermolecular and Surface Forces. New York: Academic Press, 1985.
15. J Israelachvili, H Wennerström. Nature 379:219, 1996.
16. Y Marcus. Biophys Chem 51:111, 1994.
17. J Jabloński, W Janusz, J Szczypa. J Dispers Sci Technol 20(1&2):165, 1999.
18. E Chibowski, A Waksmundzki. J Colloid Interface Sci 64:380, 1978.
19. E Chibowski, A Waksmundzki. J Colloid Interface Sci 66:213, 1978.
20. E Chibowski, L Holysz. J Colloid Interface Sci 81:8, 1981.
21. E Chibowski. J Colloid Interface Sci 69:326, 1979.
22. E Chibowski, L Holysz. J Colloid Interface Sci 77:37, 1980.
23. E Chibowski, L Holsyz. J Colloid Interface Sci 127:377, 1989.
24. RJ Hunter. Zeta Potential in Colloid Science. London, San Francisco: Academic Press, 1981.

25. R Macy. Organic Chemistry Simplified. Warsaw: PWN, 1960 (in Polish).
26. E Chibowski, , S Soltys, M Lazarz. Progr Colloid Polym Sci 105:260, 1997.
27. A Soltys, M Lazarz, E Chibowski. Colloids Surfaces A: Physico Chem Eng Aspects 127:163, 1997.
28. A Wiącek, E Chibowski. Colloids Surfaces B: Biointefaces 14:19, 1999.
29. A Wiącek, E Chibowski. Colloids Surfaces A: Physicochem Eng Aspects 159:253, 1999.
30. A Wiącek, E Chibowski. Colloids Surfaces B: Biointerfaces 17:175, 2000.
31. Manual Instruction for BI-TOQELS. Holtsville, NY: Brookhaven Institute, 1995.
32. BV Derjaguin, NV Churajev. Colloids Surfaces 41:223, 1989.
33. HK Christenson. In: ME Schrader, G Loeb, eds. Modern Approaches to Wettability: Theory and Applications. New York: Plenum Press, 1992, pp 29–51.
34. CJ van Oss, MK Chaudhury, RJ Good. Chem Rev 88:927, 1988.
35. CJ van Oss, RJ Good, MK Chaudhury. Langmuir 4: 884, 1988.
36. CJ van Oss, RF Giese, PM Constanzo. Clays Clay Miner 38:151, 1990.
37. CJ van Oss, RF Giese. Clays Clay Miner 43:474, 1995.
38. DYC Chan, DJ Mitchell, BW Ninham, BA Pailthorpe. In: F Franks, ed. Water. vol. 6. New York: Plenum Press, 1979.
39. H Christensen, PHM Claesson. Science 239:390, 1988.
40. M Elimelech, J Gregory, X Jie, R Williams. Particle Deposition and Aggregation Measurements, Modeling and Simulation. Butterworth, Oxford 1994, ch. 3.
41. R Hogg, TW Healy, DW Feuerstenau. Trans Faraday Soc 62:1638, 1966.
42. B Jańczuk, T Bialopiotrowicz, W Wójcik. Colloids Surfaces 36:391, 1989.
43. L Stryer. Biochemistry. Warsaw: PWN, 1986 (in Polish).
44. DG Dalgleish. Colloids Surfaces 46:141, 1990.
45. AR Mackie, J Mingins, AN North. J Chem Soc, Faraday Trans 87:3043, 1991.
46. M Paulsson, D Dejmek. J Colloid Interface Sci 150:394, 1992.
47. M Subirade, J Gueguen, KD Schwenke. J Colloid Interface Sci 152:442, 1992.
48. T Cosgrove, JS Philips, RM Richardson. Colloids Surfaces 62:199, 1992.
49. E Dickinson. J Chem Soc, Faraday Trans 20:2973, 1992.
50. S Welin-Klinström, A Askendal, H Elwing. J Colloid Interface Sci 158:188, 1993.
51. A Dussaud, GB Han, L Ter Minassian-Saraga, M Vignes-Adler. J Colloid Interface Sci 167:247, 1994.
52. E Dickinson, Y Matsumura. Colloids Surfaces B: Biointerfaces 3:1, 1994.
53. FK Hansen, R. Myrvold. J Colloid Interface Sci 176:408, 1995.
54. CA Haynes, W Norde. J Colloid Inteface Sci 169:313, 1995.
55. K Kato, S Sano, Y Ikada. Colloids Surfaces B: Biointerfaces 4:221, 1995.
56. NL Burns, K Holmberg, C Brink. J Colloid Interface Sci 178:116, 1996.
57. F Sarmiento, JM Ruso, G Prieto, V Mosquera. Langmuir 14:5725, 1998.
58. K Murakami. Langmuir 15:4270, 1999.
59. A Bonincontro, A De Francesco, M Matzeu, G Onori. A Santucci. Colloids Surfaces B: Biointerfaces 10:105, 1997.

33

Electrophoretic Studies of Liposomes

BRUNO DE MEULENAER, PAUL VAN DER MEEREN,
and JAN VANDERDEELEN   Ghent University, Ghent, Belgium

I. INTRODUCTION

Liposomes or lipid vesicles are spherical, self-closed structures composed of curved lipid bilayers, which entrap part of the solvent, in which they freely float, into their interior [1, 2]. As shown in Fig. 1, they may consist of one or several concentric membranes and hence are referred to as either unilamellar or multilamellar vesicles. Their sizes range from 20 nm for small unilamellar vesicles (SUV) up to several micrometers for large unilamellar vesicles (LUV) or multilamellar vesicles (MLV). The thickness of a single lipid bilayer membrane is about 4 nm.

Liposomes are interesting structures that have rapidly come into widespread use as models of biological membranes on the one hand and as delivery systems, where encapsulation and protection of hydrophilic and lipophilic substances are required, on the other [3–8].

In membrane research, it is known that electrostatic phenomena play a crucial role in many biological processes taking place at the membrane surface. For example, they bring about a change in the value of the pH at the surface, resulting in a modulation of the enzyme activity near the membrane surface. Also, electrostatic interactions can lead to an accumulation of charged species near the membrane surface, which then acts as a catalytic site. In addition, the electrostatic potential is important for many, if not all, types of interactions with membranes. For example, calcium ion binding to membranes has a major role in the initiation of many processes including pharmacological action, exocytosis, messengers, and the binding of substrates with receptors. Other examples relate to the binding of charged drugs to membranes and interactions with surfaces occurring during, for example, virus fusion or phagocytosis. Moreover, a change in the membrane surface potential can affect adhesion between cells as well as fusion of cell membranes through the contribution of the electrostatic force to the total adhesion process.

Investigating the applicability of liposomes as drug delivery systems, the control and prediction of the liposomal stability are important since the dispersions have to be stored for a long time after preparation. According to classical DLVO theory, the electrostatic potential is of paramount importance for the colloidal stability of these liposomal suspensions [9–11]. Besides the stability, the electrostatic potential will also affect the interaction of the liposomes with membranes, drugs, or other vesicles. Thus, in vivo methods have shown that the surface charge density exerts an influence over the distribution of liposomes [12,13]. As a further consequence, measuring and controlling the electrostatic properties of lipid vesicles is crucial for the basic understanding and practical applications of liposomes.

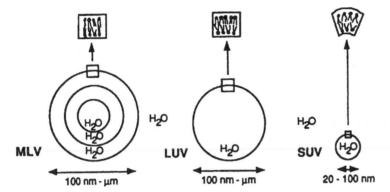

FIG. 1 Schematic representation of different types of liposomes; both multilamellar (MLV) and large (LUV) as well as small unilamellar vesicles (SUV) are shown.

Historically, numerous electrophoretic studies on liposomal dispersions were done by microelectrophoresis. Nowadays, laser Doppler velocimetry, based on dynamic laser light scattering by the lipid suspensions, is probably the best tool for a fast and/or routine determination of the electrostatic potential of liposomes.

This chapter will start with a thorough discussion of the effects of various electrolytes on the electrophoretic properties of liposomes. Whenever possible, mathematical models will be presented, allowing us to describe experimental data. Hereby, the effect of experimental conditions, such as temperature and particle size, will be highlighted. Subsequently, it will be shown that electrophoretic experiments may provide useful information on the encapsulation properties of small molecular weight organic molecules. Finally, the interaction with polymeric substances will be reviewed. This interaction may be due to electrostatic or hydrophobic forces, which applies for proteins and polynucleotides, or may involve chemical bonding to the liposomal surface; polyethyleneglycol-coated liposomes, which are also referred to as Stealth liposomes [14,15] are typical examples of the latter category.

II. LIPOSOMES IN INDIFFERENT ELECTROLYTES

It is generally known that the zeta potential (ζ) of dispersed particles is not solely determined by the characteristics of the surface of the particles, but also depends on the composition and content of the dispersion liquid. Thus, numerous authors have investigated the effect of the electrolyte concentration on the zeta potential of liposomal dispersions. A typical example is shown in Fig. 2, which demonstrates the effect of the concentration of NaCl, NaBr, and Na_2SO_4 on the zeta potential of liposomes containing both zwitterionic phosphatidylcholine (PC)* and anionic phosphatidic acid (PA) at pH 7.2; the PC:PA molar ratio varied from 10:0 to 8:2. As expected from the classical electrical double layer (EDL) theory developed by Gouy and Chapman, the absolute value of the zeta potential decreases as the electrolyte concentration increases. This behavior can be explained by assuming that either the surface potential ψ_0 or the surface charge density σ_0 remain constant.

---

*The chemical structures of some phospholipids are shown in Fig. 3.

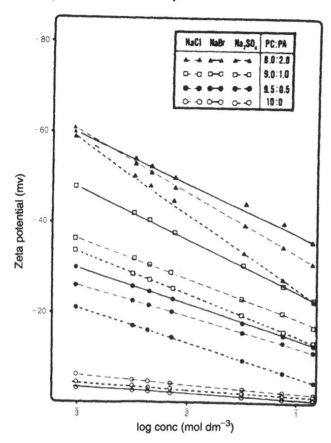

FIG. 2 Zeta potential of mixed PC:PA liposomes as a function of log concentration of either NaCl, NaBr, or Na₂SO₄. (From Ref. 10.)

A. Constant Surface Potential Model

Assuming that the surface potential remains constant as the electrolyte concentration increases, the absolute value of the zeta potential, i.e., the potential at a limited distance from the surface, decreases because the thickness of the double layer decreases. Mathematically, the zeta potential may be calculated from the Stern potential ψ_S using the Eversole and Boardman equation. The latter is based on the Gouy–Chapman theory, which describes the variation of the electrostatic potential ψ in the aqueous phase as a function of the distance x from the surface:

$$\ln\left(\tanh\left(\frac{Ze\zeta}{4kT}\right)\right) = \ln\left(\tanh\left(\frac{Ze\psi_S}{4kT}\right)\right) - \kappa\Delta x_{sp} \tag{1}$$

with

$$\kappa = \sqrt{\frac{2e^2 C N_A Z^2}{\varepsilon_0 D k T}}$$

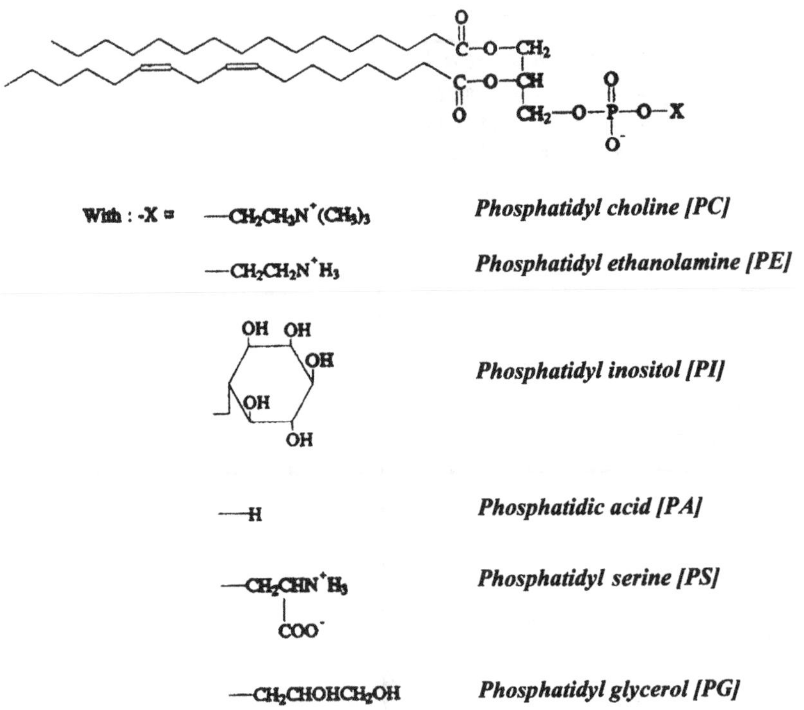

FIG. 3 Chemical structures of some phospholipids.

Here, Z is the valency of the ion (a symmetric Z-valent electrolyte is assumed as dispersion medium). Equation (1) allows fitting of zeta-potential values at different electrolyte concentrations, and hence at different values of the Debye–Hückel parameter, to a linear model: the thickness of the plane of shear Δx_{sp} follows from the slope, and the surface potential may be deduced from the intercept. According to this model, the surface potential remains constant when the electrolyte concentration is changed.

The Gouy equation indicates the relationship between the overall charge density in the Stern plane σ_S, the Stern potential ψ_S, and the electrolyte concentration C. Here, the overall charge density in the Stern plane, which will be referred to as the surface charge density in further discussion, is the summation of the original surface charge density σ_o (i.e., without adsorbed ions) and the charge density of adsorbed ions σ_{ads}.

$$\sinh\left(\frac{Ze\psi_S}{2kT}\right) = \frac{A\sigma_S}{\sqrt{C}} \tag{2}$$

with

$$A = \frac{1}{\sqrt{8N_A\varepsilon kT}}$$

From Eq. (2) it follows that the charge density is proportional to the square root of the electrolyte concentration if the surface potential remains constant. Hence, a roughly threefold increase in surface charge density is expected as the salt concentration is increased by a factor of 10. Carrión et al. [10] analyzed their experimental data (Fig. 2) obtained for mixed PC/PA liposomes, and the parameters of the curve-fitting procedure indicated that the plane of shear was located at least 0.36 and at most 2.1 nm from the Stern layer. At 5% anionic lipid content, a (constant) surface potential of about -25 mV was obtained for each of the three electrolytes studied. Based on the Gouy–Chapman theory it follows that this surface potential corresponds to one exchange site per 85 nm^2 at 1 mM electrolyte and to one negative charge per 7 nm^2 at 150 mM. Based on the composition, however, one negative charge should occupy 12 nm^2 if the molecular projected areas of both PC and PA are assumed to be 0.6 nm^2. Hence, the surface charge density values derived from the curve-fitting procedure seem unrealistic. Similar derivations also apply for the estimation of the position of the plane of shear. Comparing experimentally determined values of both the zeta potential and the surface potential (obtained by fluorescent surface probes), Eisenberg et al. [16] found that the plane of shear is located at about 0.2 nm from the Stern plane. Hence, the constant surface potential model seems to be physically unrealistic. Its main advantage is that the influence of the electrolyte concentration on the zeta potential can be easily described using only two parameters.

B. Constant Surface Charge Density

According to the double-layer theory, monovalent metal cations are electrostatically attracted toward a negatively charged surface and thus reduce the membrane surface potential of constant surface charge density colloidal particles. This so-called screening effect is purely electrostatic in nature and hence not specific. It may be adequately described by the Gouy–Chapman EDL theory.

1. Surface Charge Density

The net electrical charge density σ_0 on a liposomal surface is a function of both the lipid molecular projected area pa_1 and degree of ionization α:

$$\sigma_0 = -\frac{\alpha e}{pa_1} = -eN_{bs}^- \tag{3}$$

This is sometimes also referred to as the maximum surface charge density because no counterion binding is taken into account.

   The degree of dissociation is assumed to be 100% for acidic phospholipids such as PA, phosphatidyglycerol (PG), phosphatidylserine (PS), and phosphatidylinositol (PI), and zero for zwitterionic phospholipids such as PC and phosphatidylethanolamine (PE). For the sake of completeness, it has to be mentioned that numerous authors have observed nonzero potentials for PC liposomes over a wide range of ionic strengths, despite the fact that PC is zwitterionic over a wide pH range. According to Cevc [17], the phospholipid projected area is typically 0.40–0.55 nm^2 in the gel phase and 0.60–0.80 nm^2 in the liquid crystalline phase. For vesicles made entirely of singly charged phospholipids the surface charge density is thus expected to be between -0.29 and -0.40 C/m^2 below the chain melting phase transition temperature and between -0.20 and -0.27 C/m^2 above the phase-transition temperature. The overall charge density on a typical biological membrane is not high, however: it seldom exceeds -0.05 C/m^2. In modeling the electrokinetic behavior of the liposomes, McLaughlin and coworkers [16,18–20] assumed that the projected area of the dialkyl phospholipids used (PC and PS) was 0.7 nm^2, whereas Egorova and coworkers [21–23] usually considered the molecular projected area of anionic phospholipids, such as PS and PG, to be $0.6\,nm^2$.

Considering mixed phospholipids, the degree of ionization is approximated by the mass fraction of anionic phospholipids. Actually, this simplification assumes that both acidic and neutral phospholipids have the same molar mass, as well as the same projected area. In addition, this simple rule is only valid provided that the composition of the surface layer is equal to the overall composition, i.e. preferential accumulation or depletion in the surface layer is not taken into account. In liposomes containing PC and stearylamine (STE), though, the latter seems to be preferentially located in the outer monolayer. In addition, part of this monoalkyl surfactant exists in micellar form. It follows that, when a predetermined percentage of positive charge is required, it is evident that STE containing PC liposomes do not ensure the presence of the theoretical charge in the vesicle [24]. Incorporation of PA, on the other hand, seems to yield electrophoretic mobilities that are in line with theoretical predictions.

2. Surface Potential and Zeta Potential

When the electrolyte ions do not adsorb on to the surface, the Gouy equation, Eq. (2), adequately describes the surface potential as a function of the electrolyte concentration.

The zeta potential is related to the surface potential by the Eversole and Boardman equation, Eq. (1), assuming a specified position of the plane of shear. Most researchers assume that the plane of shear is at 0.2 nm from the surface. The latter assumption was corroborated by independent measurements of the surface potential using fluorescent probes [16]. According to Cevc [17], the uncertainty associated with the position of the plane of shear is the main pitfall of surface-potential estimations from electrophoretic studies. In fact, probably the actual location may change with the membrane or solvent composition and with temperature. Combining Eqs (1), (2), and (3), Eisenberg et al. [16] obtained very nice fits of the predicted zeta potentials to the experimentally determined values considering vesicles formed from mixtures of zwitterionic PC and anionic PS in 3.1, 15, and 105 mM tetramethylammonium chloride (TMACl), respectively. In Fig. 4 the PS:PC ratio is expressed as the number of PS molecules per unit surface area (N_{bs}^-). In the theoretical analysis, the molecular projected area of both phospholipids was assumed to be $0.7\,nm^2$, whereas the plane of shear was supposed to be located 0.2 nm from the surface of the liposomes.

According to the Gouy equation, a linear relationship should be found between the (hyperbolic sine function of the) surface potential and the inverse of the square root of the electrolyte concentration. From the slope of this linearized equation the surface charge density may be calculated.

III. COUNTERION BINDING

Biological membranes are usually exposed to alkaline and alkaline earth cations. In numerous experiments, it was observed that the surface potential of a phospholipid bilayer membrane exposed to sodium, potassium, calcium, or magnesium ions is not as negative as predicted by Gouy–Chapman theory; the latter assumes that the surface charge density (σ_o) is solely determined by the surface concentration of negative lipids (N_{bs}^-). The parsimonious interpretation is that the counterions are bound to the surface as well as exerting a screening effect. Thus, calcium and other alkaline earth cations change the electrostatic

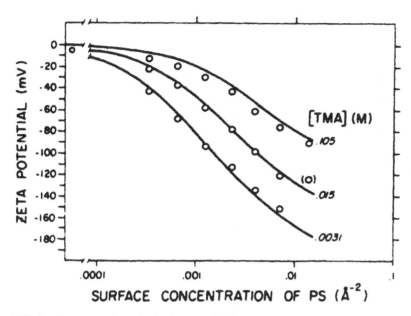

FIG. 4 Zeta potentials of mixed PC:PS liposomes in either 3.1, 15, or 105 mM TMACl at 25°C. The 15 and 105 mM TMACl electrolyte also included 1 mM MOPS buffer (pH 7.5) and 1 mM EDTA as the TMA$^+$ salt, whereas the 3.1 mM TMACl electrolyte also contained 0.1 mM MOPS and 0.5 mM EDTA. Solid lines are the predictions of the Gouy–Chapman theory, assuming that the plane of shear is 0.2 nm from the surface of the membrane. (From Ref. 16.)

potential adjacent to negatively charged bilayer membranes both by accumulating in the aqueous diffuse double layer adjacent to the membrane (as described by Gouy–Chapman theory) and by adsorbing to the phospholipids. Similarly, metal cations are found to (partly) neutralize the negative surface potential of membranes formed from acidic lipids by binding to the ionized phosphate groups and/or carboxyl groups of these lipids.

The Gouy–Chapman theory also predicts that all monovalent cations should exert identical effects on the surface potential of bilayer membranes containing negative lipids. However, Fig. 5 reveals that the magnitude of the zeta potential of multilamellar PS liposomes in 0.1 M chloride solutions depends on the nature of the cation; the electrolyte solutions were buffered to pH 7.5 with 3-(N-morpholino)propanesulfonic acid (MOPS) with a molar concentration 100 times lower than that of the monovalent salt. Figure 5 demonstrates that the zeta potential of PS vesicles in 0.1 M NaCl and 0.1 M TMACl is −62 and −91 mV, respectively. As the type of salt influences the value of the zeta potential, the simple electrolyte screening effect, which is a purely physical effect, is not sufficient to model the experimental data. Hence, the authors concluded that monovalent cations adhere to the surface of negatively charged liposomes.

Ion binding evokes structural changes both at the level of single molecules and the whole membranes, thus largely affecting membrane properties. Lipid headgroups with bound ions, for example, may turn out of the surface plane if the complex is sufficiently hydrophilic and charged, as observed by phosphorus nuclear magnetic resonance [25]. Ion binding, furthermore, affects the membrane phase-transition temperature. The salt-induced phase temperature shift is in most cases towards higher temperatures [26]. In diluted electrolytes, the magnitude of the shift increases with the ion valency in the sequence $La^{3+} \gg Ca^{2+} > Cd^{2+} > Co^{2+} \gg Cs^+ > Rb^+ > K^+ \approx Na^+$. Moreover, the per-

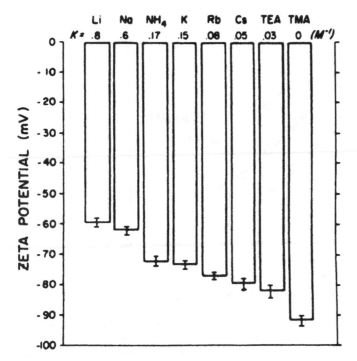

FIG. 5 Zeta potential of PS liposomes in 100 mM chloride solutions of Li^+, Na^+, NH_4^+, K^+, Rb^+, Cs^+, tetraethylamine (TEA$^+$), and tetramethylamine (TMA$^+$) at 25°C. The solutions also contained 0.1 mM EDTA as well as 1 mM Tris buffer (pH 7.5). Values for the intrinsic binding constant K derived from Gouy–Chapman–Stern theory are included in the graph. (From Ref. 16.)

meability of membranes with bound ions differs from the leakiness of the corresponding virgin membranes in dilute salt solution. Polyvalent ions may induce severe membrane perturbations and may even release the content of the membrane vesicles, probably by inducing membrane dehydration and lateral phase separation [27].

A. Monovalent Ions

Eisenberg et al. [16] demonstrated that the electrostatic potential at the surface of a membrane exposed to alkali metal cations could be satisfactorily described by assuming that the simplest form of the diffuse double layer theory is valid and that some specific adsorption of the alkali metal cations also occurs:

$$(PL^-)_s + (Me^{z+})_s \rightarrow (PL - Me^{(z-1)+})_s$$

$$K_{int} = \frac{(PL - Me^{(z-1)+})_s}{(PL^-)_s (Me^{z+})_s} \tag{4}$$

The equilibrium constant K_{int} of this adsorption reaction is referred to as the intrinsic binding constant. Here, PL$^-$ denotes the phospholipid site, Me^{z+} is a cation of valency $+Z$, and the subscript s refers to surface concentration. The concentration of counterions of valency Z in the aqueous phase at the membrane solution interface C_s is related to the bulk concentration C through the Boltzmann equation:

$$C_s = C \exp\left(-\frac{Ze\psi_S}{kT}\right) \tag{5}$$

In general, counterions adsorb on the surface of liposomes giving rise to a Stern layer so that the double layer must be described by an extended version of the Gouy–Chapman theory, which is referred to as the Gouy–Chapman–Stern theory. Herein, ion adsorption in the Stern layer is represented by a Langmuirian isotherm. Assuming a 1/1 binding stoichiometry of monovalent cations to anionic lipids, it follows that the surface charge will approach zero when all surface sites are occupied:

$$\sigma_S = \sigma_0 - e(\text{PL} - \text{Me}^{0+})_s = \sigma_0 - \sigma_0\frac{K_{int}C_s}{1 + K_{int}C_s} = \frac{\sigma_0}{1 + K_{int}C_s} \tag{6}$$

In general, the change in surface charge density due to adsorption of an ion of valency Z at a bulk concentration C is given by the Stern equation, which is an extension of the classical Langmuir adsorption isotherm:

$$\sigma_s = \sigma_0 - ZeN_{bs}\frac{K_{app}C}{1 + K_{app}C} \tag{7}$$

with N_{bs} representing the number of binding sites per unit area and K_{app} the apparent binding constant of the cation–phospholipid complex. The latter constant depends on the electrostatic potential in the aqueous phase adjacent to the surface of the membrane. The intrinsic constant K_{int} takes this concentrating effect into account and should be independent of the surface potential. As $K_{app}C$ equals $K_{int}C_s$, the relationship between the apparent and the intrinsic binding constant may be deduced from Eq. (5):

$$K_{int} = K_{app} \exp\left(\frac{Ze\psi_S}{kT}\right) \tag{8}$$

In Fig. 5, the zeta potential of multilamellar PS vesicles, formed in 0.1 M chloride solutions of various monovalent cations, is shown. Based on the Gouy–Chapman–Stern theory, Eisenberg et al. [16] concluded from these data that, except for TMACl, cation binding always occurs, thus leading to a smaller (absolute) value of the zeta potential than expected based on the simple Gouy–Chapman theory. As indicated in Fig. 5, the intrinsic association constants deduced for the adsorption of the alkali metal cations to PS decreased in the lyotropic sequence $\text{Li}^+ > \text{Na}^+ > \text{K}^+ > \text{Rb}^+ > \text{Cs}^+$. Graham et al. [28] derived similar values for the 1/1 binding constant of Na^+ and TMA^+ from surface potential measurements on monolayers.

In a second series of experiments, the surface charge density was varied by mixing anionic PS with zwitterionic PC in PC/PS ratios from 40/1 to 1/1. In this case too, a very good fit of the Gouy–Chapman–Stern model to the experimental data obtained at different concentrations of NaCl was observed, assuming the plane of shear to be at about 0.2 nm from the surface of the membrane. For the sake of completeness, it should be mentioned that electrokinetic experiments as such do not allow the determination of the value of the position of the plane of shear. In fact, Eisenberg et al. [16] mention that very good fits to the experimental data of PC/PS liposomes in NaCl may be obtained by different combinations of the intrinsic binding constant and the position of the plane of shear, such as 0.6 M^{-1} and 0.2 nm, 1.5 M^{-1} and 0.1 nm, as well as 3 M^{-1} and 0.0 nm. Thus, independent information is required, which may be obtained from measurements made with a fluorescent probe such as 2-(p-toluidinyl)naphthalenesulfonate (TNS), which responds to the potential at the membrane–solution interface.

McLaughlin [20] demonstrated that the zeta potential of mixtures of the zwitterionic PC with either anionic or cationic lipids in both 0.01 and 0.1 M NaCl containing 0–100% of charged lipids could be very well described by the Gouy–Chapman–Stern theory if the intrinsic association constant of counterions was assumed to be $1 \, M^{-1}$ (coions are assumed not to bind) and the plane of shear was 0.2 nm from the liposomal surface.

For the sake of completeness, it has to be mentioned that Cevc [17], as well as McLaughlin [20], stated that the Gouy–Chapman–Stern model is actually too simplistic. Thus, the relative permittivity and polarity in the interfacial region are assumed to remain constant and equal to the bulk characteristics. In reality, however, both characteristics vary smoothly between the values for the membrane interior and for the bulk solution. The effective dielectric constant can, therefore, vary between 2 and 78. Nevertheless, for most liposomal dispersions, the Gouy–Chapman–Stern theory can describe the experimental observations very accurately, despite the fact that it dramatically oversimplifies the reality.

In the above-mentioned studies concerning the effects of the electrolyte concentration on the surface properties of (partly) anionic liposomes, only cation adsorption was taken into account. For purely zwitterionic PC dispersions, it is generally accepted that inorganic monovalent cations normally do not bind to the liposomal surface, with the exception of Li^+. On the other hand, Tatulian [29] observed large differences in the effect of a range of potassium salts on the electrophoretic mobility of dimyristoyl-PC liposomes, which was explained from differences in the affinity of the anions for the liposomal surface. From his experimental results, as represented in Fig. 6, Tatulian [29] deduced that the affinity for the PC liposomal surface seemed to follow the sequence:

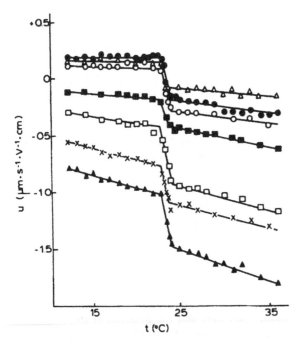

FIG. 6 Electrophoretic mobility of DMPC liposomes in 10 mM potassium salt solution (■ KBr; ● KCl; ○ KNO$_3$; □ KSCN; × KI; ▲ KClO$_4$; △ K$_2$SO$_4$) and 5 mM Tris–HCl buffer (pH 7.4) as a function of the temperature. (From Ref. 29.)

$ClO_4^- > I^- > SCN^- > Br^- > NO_3^- > Cl^- \approx SO_4^{2-}$. This statement was supported by Akeson et al. [30]; comparing the electrophoretic mobility of PC liposomes in both 0.12 M KNO_3 and 0.12 M KCl, he also observed that NO_3^- ions were binding more strongly than Cl^- ions to PC vesicles and hence gave rise to a more negative zeta potential.

B. Divalent Ions

Many fundamental studies have investigated the properties of PS liposomes in the presence of Ca^{2+} ions. In these model studies, PS is preferred as it is the most abundant anionic phospholipid in natural membranes. The huge interest in the effects of Ca^{2+} ions originates from the fact that this divalent ion is known to be important in many physiological processes. Besides, calcium can induce phase transitions and separation of the lipid components or aggregation and fusion of membranes.

Divalent transition elements always bind better than alkali earth elements: divalent ions normally bind to charged membranes, whereas monovalent ions can but need not always do so. For divalent ions, surface charge neutralization as well as charge reversal are frequently observed.

As opposed to all inorganic divalent cations, ethane-bis(trimethylammonium), a divalent organic cation, does not adsorb specifically to negatively charged bilayer membranes. A possible explanation may be found in the fact that this compound actually consists of two separate monovalent quaternary ammonium sites: one single dimethonium ion is hence very similar to two individual tetramethylammonium ions.

The intrinsic 1/1 binding constant of inorganic divalent cations with anionic phospholipid dispersions may be determined from the concentration at which the vesicles reverse charge and become positive. Two approaches have been proposed using either the charge reversal concentration as such or the slope of the zeta potential versus electrolyte concentration at the charge reversal concentration.

1. Charge Reversal Concentration

In the first approach, the intrinsic association constant is equal to the reciprocal of the divalent cation concentration at which the mobility of anionic phospholipid containing liposomes reverses sign. This follows from the definition of the intrinsic binding constant, Eq. (4), for divalent counterions:

$$K_{int} = \frac{(PL - Me^+)_s}{(PL^-)_s (Me^{2+})_s} \tag{4a}$$

At the charge reversal concentration, the number of (negatively charged) free binding sites is exactly equal to the number of (positively charged) occupied binding sites. Moreover, the electrolyte concentration at the surface is exactly equal to the bulk concentration because electrostatic effects are absent. Hence, in this special case, the intrinsic binding constant is equal to the apparent binding constant.

Although the binding constant of divalent cation–anionic phospholipid complexes may be deduced from the zeta potential at any divalent cation concentration, still the charge reversal concentration is by far preferable, since some uncertainties in the above-described Gouy–Chapman–Stern model disappear at this particular concentration. First, 1/1 complexes of monovalent cations, as well as electroneutral 1/2 complexes of divalent cations with acidic surface groups do not interfere as they do not impose any charge effects on the surface. In addition, the location of the plane of shear becomes irrelevant, as both the surface and the zeta potential are equal to zero. Moreover, the molecular projected area pa_1 is not incorporated. On the other hand, the fit at any other electrolyte concentration depends on the 1/1 association constant for monovalent ions, on the number of 2/1 complexes with divalent cations, on the location of the plane of shear, and on the estimated value of the molecular projected area.

The major limitation of the charge-reversal concentration approach is that it is strictly limited to liposomal dispersions containing only anionic phospholipids. Figure 7 reveals that the zeta potential of PS vesicles reverses sign at 0.08 M Ca^{2+}, which implies that the intrinsic association constant for the 1/1 complex is $12 M^{-1}$. Similarly, the intrinsic 1/1 association constants of Ni^{2+}, Co^{2+}, Mn^{2+}, Ba^{2+}, Sr^{2+}, and Mg^{2+} with PS were found to be 40, 28, 25, 20, 14, and $8 M^{-1}$, respectively. These data were supported by Minami et al. [11] based on an experimental study of the effect of external addition of Mg^{2+}, Ca^{2+}, Sr^{2+}, and Ba^{2+} salts on the initial aggregation kinetics of dimyristoylphosphatidylglycerol (DMPG) liposomes. Graham et al. [28] provided additional support for the values of 1/1 binding constant of Ca^{2+} and Mg^{2+}, based on fitting a theoretical model

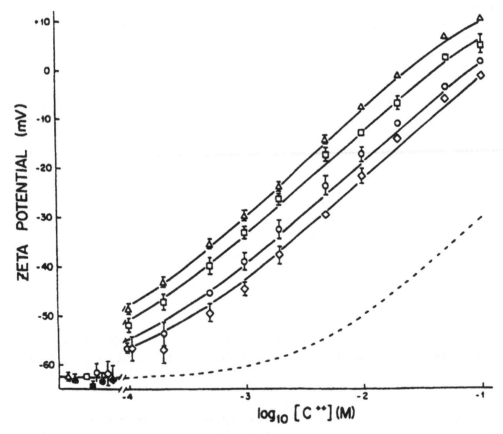

FIG. 7   Zeta potential of PS liposomes in 0.1 M NaCl containing 1 mM MOPS (pH 7.4) and either 0.1 mM EDTA (filled symbols) or the indicated concentration of divalent cation ($\triangle$ Ni^{2+}; $\bigcirc$ Ca^{2+}; $\square$ Mn^{2+}; $\diamond$ Mg^{2+}). Errors bars of the experimental points were derived from measurements on at least 20 liposomes. The lines represent the best fit of the Gouy–Chapman–Stern model with a PS–Na^+ intrinsic binding constant of $0.6 M^{-1}$ and a distance between the plane of shear and the Stern plane of 0.2 nm. (From Ref. 18.)

for surface-potential measurements on monolayers. Under identical conditions, the selectivity sequence for PC is $Mn^{2+} > Ca^{2+} \approx Co^{2+} \approx Mg^{2+} \approx Ni^{2+} > Sr^{2+} > Ba^{2+}$.

Additional proof for the above-mentioned value of the intrinsic association constant of the Ca^{2+}–PS complex follows from the fact that all the microelectrophoretic data are consistent with the simple Gouy–Chapman–Stern model, assuming that calcium forms mainly 1/1 complexes with PS and that the intrinsic association constant is about $10\,M^{-1}$. In this analysis, though, several assumptions are included. First, the plane of shear is assumed to be at 0.2 nm from the liposomal surface. Also, it is assumed that monovalent cations form 1/1 complexes with anionic, but not with zwitterionic lipids. For divalent counterions, on the other hand, 1/1 complexes with both anionic and zwitterionic lipids are considered, albeit with a different intrinsic association constant; thus, adsorption to mixed PS/PC liposomes is accurately described, assuming the binding constant of Ca^{2+} and Mg^{2+} to zwitterionic PC to be 3 and $2\,M^{-1}$, respectively. Moreover, adsorption of anions is completely ignored. The molecular projected area of each phospholipid is assumed to be $0.7\,nm^2$ in the membrane. Finally, it is assumed that phase separation does not occur. Figure 7 reveals that a good fit may be obtained over the whole concentration range studied. In order to allow evaluation of the effect of divalent cation adsorption, the dotted line in Fig. 7 represents the hypothetical zeta potential that would be obtained if sodium ions adsorbed to the PS liposomes (with an intrinsic association constant of $0.6\,M^{-1}$), whereas the divalent ions do not adsorb. It follows that even very small amounts of divalent cations highly influence the zeta potential of negatively charged liposomes. As a further consequence, trace amounts of divalent cationic impurities may largely affect the zeta potential in a monovalent electrolyte.

Eisenberg et al. [16] observed that the addition of small amounts of EDTA sometimes led to a largely increased electrophoretic mobility. According to the authors, this behavior is most probably due to the presence of divalent or trivalent cationic contaminants, which are known to be very efficiently bound by EDTA. Indeed, in a separate experiment they demonstrated that EDTA (both in 0.1 and 1 mM concentration) itself had no significant effect on the electrophoretic mobility of vesicles formed from specially purified PS in 0.1 M monovalent cation chloride and 1 mM Tris at pH 7.5. Hence, trace amounts of multivalent ions may be responsible for differences in electrophoretic mobility of phospholipids in monovalent electrolytes observed by various authors.

In order to overcome this possible effect of multivalent ions, McLaughlin and coworkers [16,18,19] suggested including small amounts of EDTA in the liquid phase as this compound is known to have a very strong binding capacity for di- and multi-valent ions. As Fig. 7 indicates that EDTA addition does not affect the zeta potential of PS liposomes in the absence of divalent cations, it may be deduced that no multivalent cationic contaminants prevailed in the lipid sample used. In the absence of divalent cations, the zeta potential of the PS liposomes in 0.1 M NaCl at pH 7.4 (which was fixed by incorporation of 1 mM MOPS buffer) was about $-62.5\,mV$ (Fig. 7). This value corresponds to the Gouy–Chapman–Stern prediction if the association constant K_{Na} is assumed to be $0.6\,M^{-1}$, whereas the plane of shear is supposed to be located 0.2 nm from the liposomal surface.

In a more extended set of experiments, both PS and PC/PS (5/1) liposomes were used in 10 and 100 mM of both NaCl and CsCl with varying concentrations of Ca^{2+}. The intrinsic association constants of Na^+ and Cs^+ with PS were assumed to be 0.6 and $0.1\,M^{-1}$, whereas the 1:1 association constant of Ca^{2+} with PS and PC was 12 and $3\,M^{-1}$, respectively, except at the lowest salt concentration studied (i.e., 10 mM), whereby the Ca^{2+}–PS and Ca^{2+}–PC binding constants of 36 and $5\,M^{-1}$ seemed to give better fits.

According to the authors, this apparent concentration effect is most probably not realistic and might be due to the fact that concentrations, rather than activities, are used in their model. According to Akeson et al. [30], the activity coefficients of divalent and mono-valent ions in 0.12 M monovalent electrolyte are 0.36 and 0.77, based on the Davies equation.

Independent studies of Ca^{2+} and Mg^{2+} adsorption were done by Reboiras [31] using ion-exchange membrane electrodes to investigate the residual concentration of cations. These experiments yielded a 1/1 association constant PS/Ca^{2+} of $11.4 \pm 2.7\,M^{-1}$, with no significant variation between pH 6.5 and 7.5. Hence, the results agreed with zeta-potential data. Also, these experiments clearly demonstrated the higher affinity of Ca^{2+}, compared with that of Mg^{2+}, to the surface of mixed dipalmitoyl-PC:PI sonicated liposomes.

Ermakov et al. [32] also found a very good fit of the Gouy–Chapman–Stern theory to experimental data for dipalmitoylphosphatidylcholine (DPPC) liposomes in 0.1 M KCl and 20 mM Tris–HCl (pH 7.4) in the presence of differrent $BeSO_4$ concentrations, assuming an intrinsic association constant of $400\,M^{-1}$. Hence, Be^{2+} is about 100 times more effective in comparison with other divalent cations. The high affinity of Be^{2+} as compared to Ca^{2+} or Mg^{2+} is proved by differential scanning calorimetry measurements, indicating that the phase-transition temperature is shifted by cation adsorption [33]. This effect is much more pronounced for Be^{2+} and it is observed at about 100 times lower concentration as compared to that for Ca^{2+} or Mg^{2+}.

2. Change in Zeta Potential at Charge Reversal Concentration

An alternative approach was described by Reboiras and Jones [34]. They performed electrophoretic mobility measurements on pure PC, pure PI, and mixed PC/PI liposomal dispersions in a pH 6.6 buffer prepared from 20 mM imidazole ($pK_a = 6.953$) plus hydro-chloric acid, giving an ionic strength of 0.0138 M of 1/1 electrolyte. The authors assumed that the surface potential was equal to the zeta potential, that the molecular area of both phospholipids was $60\,\text{Å}^2$, and that no specific adsorption of the buffer ions occurred. However, these assumptions led to the conclusion that the degree of ionization of PI was only 23% for pure PI vesicles, which seems to indicate that buffer ion adsorption was actually taking place. The theoretical analysis proposed by Reboiras and Jones [34] is based on the linear relationship between the surface charge density and the surface poten-tial, which is valid at low values of the surface potential. However, this simplified formula will produce a highly underestimated value for the surface charge density at higher surface potential values. For a spherical particle this relationship corresponds to

$$-\sigma_S = \frac{\varepsilon_o D(1 + \kappa a)\psi_S}{a} \qquad \text{if} \qquad |\psi_S| < 25\,\text{mV} \tag{9}$$

where D is the dielectric constant of the medium. Assuming that the zeta potential is nearly the same as the Stern potential and that κa is much larger than unity, it follows from Eqs (7) and (9) that

$$\zeta \approx \psi_S = \frac{-\sigma_S}{\varepsilon_o D\kappa} = \frac{-\sigma_o}{\varepsilon_o D\kappa} + \frac{ZeN_{bs}}{\varepsilon_o D\kappa} \frac{K_{int}C_s}{1 + K_{int}C_s} \tag{10}$$

As long as Eqs (9) and (10) are only applied close to the isoelectric region, this approach is justified. Thus, differentiation yields the slope of the zeta potential versus $\log(C)$ curve at the charge reversal concentration:

$$\left(\frac{d\zeta}{d\log C_s}\right)_{\zeta=0} = 2.303\frac{ZeN_{bs}}{\varepsilon_o D\kappa}\left(\frac{K_{int}C_s}{(1+K_{int}C_s)^2}\right) \tag{11}$$

This equation may be used to derive an average intrinsic association constant for mixed anionic–zwitterionic phospholipid liposomes. However, in their theoretical analysis, Reboiras and Jones [34] extended the use of the linear relationship between surface potential and surface charge density to the zeta (which is assumed to be the same as the surface) potential ζ_0 in the absence of adsorbing ions, from which they deduced an alternative expression for the electrolyte concentration at which the liposomes become electrically neutral:

$$\zeta = \zeta_o + \frac{ZeN_{bs}}{\varepsilon_o D\kappa}\frac{K_{int}C_s}{1+K_{int}C_s} \tag{12}$$

$$\frac{1}{C_{\zeta=0}} = K_{int}\left(\frac{ZeN_{bs}}{\varepsilon_o D\kappa\zeta_o} - 1\right) \tag{12a}$$

Using Eq. (12a), Eq. (11) may be rewritten as a function of the zeta potential in the absence of adsorbing ions:

$$\left(\frac{d\zeta}{d\log C_s}\right)_{\zeta=0} = 2.303\zeta_o\left(\frac{\varepsilon_o D\kappa\zeta_o}{ZeN_{bs}} - 1\right) \tag{13}$$

According to the authors [34], both the number of binding sites N_{bs} and the intrinsic binding constant K_{int} can be determined by solving Eqs (12a) and (13) simultaneously. The use of this set of equations is, however, not justified at all, as Eq. (9) has been used outside the isoelectric region in replacing the contribution $-\sigma_0/(\varepsilon_o D\kappa)$ in Eq. (10) by ζ_o in Eq. (12). As a further consequence, unrealistic values for the intrinsic binding constant, i.e., $1.14 \times 10^6\,M^{-1}$ and $3.28 \times 10^5\,M^{-1}$ for Ca^{2+} and Mg^{2+} adsorption to mixed DPPC/PI (75/25) liposomes at pH 6.6, are obtained. This is clearly an unrealistic result as the Langmuir equation indicates that in this special case more than 99% of the binding sites would be occupied at the charge reversal concentration. This is in contradiction with the experimental data shown in Fig. 8, since an additional large increase in zeta potential is observed following charge neutralization. On the other hand, an intrinsic binding constant of $3.0\,M^{-1}$ is found by applying Eq. (11) to the experimental results in Fig. 8, assuming the number of binding sites per unit surface area (N_{bs}) to be equal to the number of phospholipid molecules per unit surface area. The latter value of the intrinsic binding constant is very similar to the results obtained by McLaughlin et al. (15): they found an intrinsic binding constant of $3\,M^{-1}$ for zwitterionic and $12\,M^{-1}$ for anionic phospholipids. Similarly, Akeson et al. [30] found a Ca^{2+}–PC intrinsic association constant of about $3\,M^{-1}$ from the slope of the zeta potential versus electrolyte concentration behavior by applying Eq. (11).

Besides the binding constant, Reboiras and Jones [34] claim to deduce information about the number of binding sites. According to the parameters of their model, the number of divalent cation binding sites is $6.55 \times 10^{16}\,m^{-2}$ and $14.8 \times 10^{16}\,m^{-2}$ for Ca^{2+} and Mg^{2+}, respectively, which is less than half the number of anionic phospholipids present per unit surface area. According to these data, the surface should remain negatively charged even at complete coverage of all binding sites. This result is clearly in contradiction with the experimentally determined charge reversal, which is an additional proof of the incorrect use of the EDL theory.

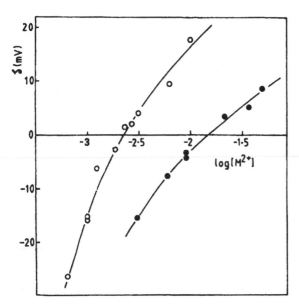

FIG. 8 Zeta potential of mixed DPPC:PI (75:25 wt%) liposomes in either $CaCl_2$ or $MgCl_2$ solution at 25°C. (From Ref. 35.)

C. Trivalent Ions

In general, metal ions bind to the phosphate group in phospholipids with stability constants ranked as trivalent > divalent > monovalent metals. Akeson et al. [30] studied Al^{3+} adsorption in both 0.12 M KNO_3 and 0.12 M KCl. The 1/1 binding constant of the trivalent ion to PC was estimated from the change in zeta potential with electrolyte concentration in the isoelectric region. It follows that the Al–PC binding constant is about 1500 M^{-1}. The authors conclude that the affinity of Al^{3+} is about 500 times larger as compared to that of Ca^{2+}. According to Akeson et al. [30], the Stern model underestimates surface bound Al^{3+} when the shear plane is at the liposomal surface or at 0.2 nm from the surface: the predicted adsorbed amount is only half the experimentally determined value. According to the authors, this may be due to an underestimation of the distance between the plane of shear and the lipid membrane surface.

Ermakov et al. [32] observed that the adsorption of Gd^{3+} at the membranes made from both PC and PS is suitable for quantitative description by the Gouy–Chapman–Stern theory. The binding constant was 1000 and 50,000 M^{-1} for PC and PS, respectively.

IV. PHASE TRANSITION

Tatulian [29] demonstrated that the electrophoretic mobility of zwitterionic DMPC vesicles depended on temperature with a discontinuous behavior at the phase-transition temperature (Fig. 6). Similarly, Ermakov et al. [32] observed that the zeta potential of positively charged Be^{2+}-coated DPPC liposomes was about 10–15 mV higher below the phase-transition temperature. In general, a downward shift of about 10 mV is observed at temperatures above the phase-transition temperature (T_g) both for positively charged, zwitterionic and anionic phospholipid liposomes. According to Ermakov et al. [32], the lower zeta potential of positively charged Be^{2+}-coated DPPC liposomes above the phase-transition temperature could be accurately described by the Gouy–Chapman–Stern theory if the parameters of adsorption were chosen to be about 20 times higher for lipids in the solid state. McLaughlin et al. [18] also reported increased binding constants in the gel as compared to the liquid crystalline state; in the latter case, describing Ca^{2+} and Mg^{2+} adsorption to PS liposomes, the difference was, however, only 40%. Similarly, the affinity of PS bilayers for Mg^{2+} has been reported to decrease with the degree of chain unsaturation and fluidity. Thus, more negative values of the zeta potential above the phase-transition temperature may be due to a partial release of bound cations.

However, this cannot explain that the same phenomenon is also observed in the absence of strongly adsorbing divalent cations. In this respect, Ermakov et al. [32] state that additional factors might be of importance, such as changed dielectric properties of the surface. The latter effect is completely ignored in the Gouy–Chapman theory, which assumes that the dielectric constant remains constant and equal to the dielectric constant of bulk water throughout the whole EDL. Alternatively, Cevc [17] suggests that the different electrokinetic behavior might be explained from the existence of an additional hydration potential in addition to the electrostatic potential. The experimentally observed increase in the membrane hydration at the lipid chain melting phase transition thus offers a possible explanation for the decrease in zeta potential.

Moreover, it is important to stress that the three parameters of the Gouy–Chapman–Stern theory, i.e., the maximum surface charge density, the binding constant, and the position of the slipping plane, are not independent of the fitting procedure. Very similar results may be obtained by either increasing the number of binding sites at constant binding affinity or by increasing the binding affinity at constant binding site density. The increased number of binding sites below the transition temperature could be a logical consequence of the decreased molecular projected area.

V. RELAXATION EFFECT

As illustrated by Fig. 7, as well as by Fig. 9, the classical Gouy–Chapman–Stern model is quite successful in predicting the zeta potential as a function of the (logarithm of the) electrolyte concentration for acidic lipid dispersions in monovalent electrolyte solutions at relatively high salt concentrations, i.e., above 20–50 mM. It follows that the three parameters of this model, i.e., the maximum surface charge density (which was assumed to be one negative charge per $0.6\,nm^2$), the cation binding constant K_{int}, and the location of the plane of shear, may be estimated from the best fit of the Gouy–Chapman–Stern theory to the experimental data obtained at high electrolyte concentration.

In the 1–50 mM range, though, the experimentally determined values of the zeta potential as a function of ionic strength are significantly less pronounced than the predictions of Gouy–Chapman–Stern theory. This is shown in Fig. 9 where the dashed line represents the calculated zeta potential based on the Gouy–Chapman–Stern theory, whereas the circles show the experimentally determined zeta potential for PS liposomes in KCl electrolyte. These anomalies are mainly because all commercial equipment used to determine the zeta potential of colloidal dispersions actually determines the electrophoretic mobility and calculates the corresponding zeta-potential value using the Helmholtz–Smoluchowski equation. This also holds for the experimental points included in Fig. 9, which were indeed derived from the electrophoretic mobility, using the Helmholtz–Smoluchowski equation. Hence, a poor fit of the Gouy–Chapman–Stern model to experi-

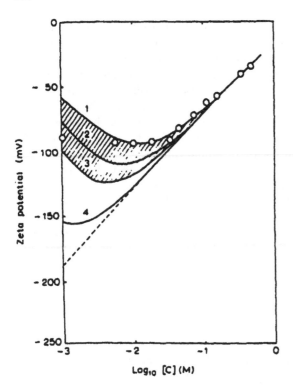

FIG. 9 Comparison of the experimentally determined values of zeta potential (○) of PS liposomes in KCl, as derived from their electrophoretic mobility using the Helmholtz–Smoluchowski equation, and theoretical (real) zeta-potential values (dashed line) calculated by the Gouy–Chapman–Stern model with $\Delta x_{sp} = 0.2\,\text{nm}$, $K_{int} = 0.25\,\text{M}^{-1}$, and $\sigma_o = -267\,\text{mC/m}^2$. The solid curves represent the apparent zeta potential that follows from the theoretical electrophoretic mobility using the Helmholtz–Smoluchowski equation. Hereby, the mobility was derived from the theoretical zeta potential using the Dukhin–Deryaguin model assuming the particle radius to be 250 nm (1), 500 nm (2), 1 μm (3), and 5 μm (4). (From Ref. 21.)

mentally determined zeta-potential values is obtained if the use of the Smoluchowski equation is not justified. This may be because the Smoluchowski equation ignores the effect of both surface conductivity and double-layer relaxation. The latter arises from the fact that the center of the ionic atmosphere lags slightly behind the center of the particle and hence slows down the electrophoretic mobility.

This effect has been taken into account in more advanced theories for the electrophoretic mobility as developed by Wiersema et al. [36], O'Brien and White [37], and Dukhin and Deryaguin [38]. All these models yield very similar results: the thicker and more mobile an interfacial region is, the more important is the relaxation effect. Decreasing vesicle size also increases the significance of this effect appreciably. According to Egorova and coworkers [21,22], this information may be expressed in a quantitative way in the following relaxation effect criterion.

Relaxation is negligible provided that: $\dfrac{\exp\left(\dfrac{e|\psi_s|}{2kT}\right)}{\kappa a} < 0.02$ \hfill (14)

From this relation, it follows that κa should be larger than 50 if the Stern potential is negligibly small, whereas a limiting value of 350 is obtained at a Stern potential of 100 mV. In quasiphysiological solutions, i.e., 100 mM of a 1/1 electrolyte, the thickness of the double layer is of the order of 1 nm, and hence the relationship between the mobility and the zeta potential according to the Dukhin–Deryaguin theory is nearly equal to the prediction of the simple Helmholtz–Smoluchowski equation provided that the particle diameter is at least 100 nm. Hence, the relaxation effect is negligible at salt concentrations of at least 0.1 M. At lower salt concentrations, on the other hand, the relaxation phenomenon gives rise to a much lower electrophoretic mobility than expected from the Helmholtz–Smoluchowski equation. As a further consequence, the experimentally determined zeta potential (which is actually a rescaled mobility using the Helmholtz–Smoluchowski equation) is much less than the effective zeta potential. This effect was clearly demonstrated by Roy et al. [39] who studied the influence of size on the electrokinetic behavior of mixed PC/PS liposomes in 1 mM phosphate buffer (pH 7.4) with 1 mM NaCl. The liposomes were prepared by extrusion through polycarbonate membrane filters, by sonication or by hydrating a thin film of phospholipids. Here, the size of the liposomes increased in the sequence from extrusion over sonication to MLV production by hydrating a thin phospholipid film. Roy et al. [39] observed that the experimentally determined zeta potential was not only affected by the percentage of PS present, but also by the preparation method, i.e., the particle size. This behavior could be ascribed to the small value of the Debye–Hückel parameter κ (i.e., 0.329 nm^{-1}), causing the relaxation criterion to be larger than 0.02, especially for the smaller liposomes that were prepared by extrusion. Although Roy et al. [39] introduced the Henry coefficient into the Smoluchowski equation in order to take account of this relaxation effect, a lower zeta potential was still found for smaller particles, thus indicating that the Henry correction could not solve the relaxation problem.

From a scientific point of view, the most logical procedure may seem to use a more complicated model to convert the experimentally determined electrophoretic mobility to a zeta potential. However, this process may give rise to severe errors since large changes in zeta potential may produce small changes in electrophoretic mobility. Besides, the calculations of O'Brien and White [37] revealed that the electrophoretic mobility as a function of the zeta potential has a maximum value in 1.2 mM 1/1 electrolyte when κa exceeds 3. In this case, two widely different zeta-potential values may give rise to the same mobility. It follows that mobility to potential transformation is not straightforward. Therefore, a different approach is mostly used (Fig. 10): following the calculation of the real zeta potential by the Gouy–Chapman–Stern model, the (real) electrophoretic mobility may be obtained by the Dukhin–Deryaguin model, from which the apparent zeta potential according to the Helmholtz–Smoluchowski equation may be obtained. Although the latter has no real physical meaning, it still allows direct comparison of the theoretical prediction with the experimental data. Thus, the continuous lines in Fig. 9 represent the apparent zeta potential as a function of the electrolyte concentration at four different values of the particle radius. Figure 9 indicates that the apparent zeta potential, which is basically a rescaled electrophoretic mobility, becomes less than the real zeta potential from the Gouy–Chapman–Stern theory, at electrolyte concentrations below 0.1 M. The smaller the particle, the larger the difference. As indicated by line 4 in Fig. 9, even for 10-µm particles, the difference becomes quite noticeable below 10 mM of 1/1 electrolyte. Hence, even for very large liposomes, a significant difference may be expected between the effective zeta potential and the Smoluchowski prediction based on the electrophoretic mobility in dilute electrolyte solutions.

THEORETICAL CALCULATION   ◄---►   EXPERIMENTAL DETERMINATION

1) In the absence of relaxation : $\zeta_{H-S} = \zeta$

$$\sigma_o \xrightarrow{G-C-S} \psi_S \xrightarrow{E-B} \zeta \qquad\qquad\qquad \zeta_{H-S} \xleftarrow{H-S} \mu$$

2) In the presence of relaxation : $|\zeta_{H-S}| < |\zeta|$

2.a) theoretical approach

$$\sigma_o \xrightarrow{G-C-S} \psi_S \xrightarrow{E-B} \zeta \qquad\qquad\qquad \zeta \xleftarrow{D-D} \mu$$

2.b) practical approach

$$\sigma_o \xrightarrow{G-C-S} \psi_S \xrightarrow{E-B} \zeta \xrightarrow{D-D} \mu \xrightarrow{H-S} \zeta_{H-S} \qquad\qquad \zeta_{H-S} \xleftarrow{H-S} \mu$$

FIG. 10 Schematic representation of the calculation procedure used to obtain the real and apparent values (also referred to as the Helmholtz–Smoluchowski approximations) of electrophoretic mobility and zeta potential, based on theoretical analyses on the one hand and experiments on the other. G–C–S = Gouy–Chapman–Stern theory; E–B = Eversole–Boardman equation; H–S = Helmholtz–Smoluchowski equation; D–D = Dukhin–Deryaguin theory.

As the experimental data are reasonably well described by the Dukhin–Deryaguin theory, it is concluded that the observed deviation of the experimentally determined zeta potential from the true zeta potential (as indicated by the dashed line in Fig. 9) is likely to be (at least partly) the result of the relaxation effect. Egorova et al. [22] found a very reasonable fit of the Dukhin–Deryaguin theory for cardiolipin in NaCl. For cardiolipin in KCl or PS in NaCl, however, the Dukhin–Deryaguin approach was by far superior, but nevertheless, systematic deviations between theory and experiment were observed. They concluded that although the relaxation effect is undoubtedly present at low salt concentration, it could hardly be regarded as the only reason for deviation between experiment and Smoluchowski theory.

For the sake of completeness, it should be mentioned that in all hitherto described experiments, the relaxation effect could be neglected thanks to the well-selected experimental conditions used. First, most studies were performed in quite concentrated electrolyte, such as 0.1 M of monovalent electrolyte. Even when studying the effect of low concentrations of divalent or trivalent cations, typically 0.1 M monovalent electrolyte was added, so that the Debye–Hückel parameter was quite large. In addition, microelectrophoresis was mostly used. As a further consequence of the need to monitor individual liposomes, the size of the studied particles was typically above 1 μm. Besides, the larger particles within the field of view can be selected for electrophoretic observations. Thus, Eisenberg et al. [16] mention that especially when the salt concentration was low and the surface potential high, measurements were made on large vesicles of at least 10 μm in diameter in order to circumvent the relaxation effect. Both the high electrolyte concentration and the large particle size contributed to a large value of κa, so that retardation effects were negligible. On the other hand, most recent studies prefer electrophoretic light scattering as this technique is much less labor intensive. In this technique, it is not required to visualize individual vesicles, and hence the size is generally much smaller. Considering a typical diameter of 100 nm in a 10 mM electrolyte, relaxation effects are very likely to occur.

VI. SURFACE DISSOCIATION

Since protons, just as any kind of counterions, are electrostatically attracted towards negatively charged surfaces, it follows that the surface pH will be lower than the bulk pH. As a further consequence, the negative logarithm of the apparent dissociation constant for any structural membrane residue, the so-called pK_{app}, is higher than the pK_a value for the same group in the bulk solution, provided that the membrane surface is negatively charged. Thus, Table 1 shows that the pK_a values for the carboxyl and amino groups of the water-soluble phosphorylserine are 2.65 and 9.9, respectively, whereas the corresponding pK_{app} values for these groups in PS liposomes in 100 mM of a monovalent electrolyte are shifted upward to 5.5 and 11.5. Similarly, the pK_a of water-soluble butyric acid is 4.8, whereas the apparent pK_{app} of long-chain carboxylic acids incorporated into PC liposomes is 7.3. Hence, the pK_{app} values for phospholipids and long-chain carboxylic acids inserted into PC liposomes in the presence of 100 mM of a monovalent electrolyte may be shifted by 1–3 units as compared to the pK_a values for simple water-soluble organic compounds. In general, the shift is larger for the more charged membranes.

As the electrolyte concentration largely affects the coulombic membrane potential, it follows that the pK_{app} value for surface-bound groups will be influenced by the presence of salts: at lower electrolyte concentrations, the difference between the pK_a value for dissolved groups and the pK_{app} value for membrane-bound groups will become more pronounced, whereas at higher salt concentrations this difference will tend to become zero. It

TABLE 1 Apparent Dissociation Constants, in 0.1 M of a Monovalent Salt, of the Ionizable Groups of Common Phospholipids and Long-Chain Fatty Acids Imbedded in a Liposomal Membrane[a]

Compound	Medium	$pK_{PO_4^-}$	$pK_{PO_4^{2-}}$	pK_{COO^-}	$pK_{NH_3^+}$
PS	Liposome	≤ 1		5.5	11.5
phosphorylserine	Water	≤ 1	5.8	2.65	9.9
PI	Liposome	2.7			
PG	Liposome	2.9			
PA	Liposome	3.5	9.5		
Glycerophosphoric acid	Water	1.4	6.2		
R–COOH in PC	Liposome			7.3	
Butyric acid	Water			4.8	

[a]To allow comparison, the dissociation constants of some similar water-soluble compounds are included.
Source: Ref. 40.

has to be stressed that, according to this discussion, it is the apparent, but not the intrinsic dissociation constant of surface-bound groups that is affected by the bulk electrolyte concentration.

In recent years, Egorova [23] presented some experimental data suggesting that the electrolyte concentration affects the degree of dissociation of surface-bound groups not only by affecting the pH at the surface of the membrane, and hence affecting the apparent dissociation constant, but also by changing the intrinsic pK_a value for the dissociating group itself. This hypothesis was derived from electrophoretic experiments using both dioleoyl PG and PS bilayer membranes: the intrinsinc dissociation constants of the phosphate group of dioleoyl-PG and of the carboxyl group of PS were found by fitting a theoretical model to the experimental dependencies of electrophoretic mobilities on ionic strength in solutions of monovalent salts. The model is based on a simple dissociation reaction for the acidic ionizable group. Assuming that specific ion binding is described by the Langmuir adsorption isotherm and expressing the surface ion concentrations through the Boltzmann relation, Eq. (5), the surface charge density for dioleoyl-PG and PS membranes in 1/1 electrolyte may be written as a function of the surface potential:

$$\sigma_S = \frac{\sigma_o}{1 + \left(\dfrac{[H^+]}{K_d} + K_{int}[C]\right) \exp\left(-\dfrac{Ze\psi_S}{kT}\right)} \tag{15}$$

In Eq. (15), K_d and K_{int} are the intrinsic surface dissociation constant of the ionizable group and the intrinsic binding constant of the cation, respectively. The latter was $0.13\,M^{-1}$ for dioleoyl-PG in KCl and $1\,M^{-1}$ for PS in NaCl. Assuming the molecular projected area pa_1 to be $0.6\,nm^2$, Eq. [3] reveals that σ_o amounted to $0.27\,C/m^2$. In order to obtain σ_S and ψ_S values as a function of the electrolyte concentration, Eq. (15) is combined with the Gouy equation (2) and the solution is searched by means of an optimization program.

Figure 11a shows that fitting with due regard for cation adsorption was successful using a fixed value of the intrinsic dissociation constant at intermediate to high salt concentrations for dioleoyl-PG in KCl at pH 7. The same behavior was also observed for PS in NaCl at pH 7. In the 1–20 mM range, however, a nearly constant value for the electrophoretic mobility was obtained. In this low concentration region, changes in intrinsic dissociation constant pK_a had to be introduced in order to obtain a good fit of theory to experiment (Fig. 11b). Thus, the experimental data for the electrophoretic mobility of dioleoyl-PG in KCl at pH 7 could be described nicely assuming that the intrinsic dissociation constant of the phosphate group increased from 1.54 in 1 M salt up to 5.75 in the presence of only 1 mM of KCl (Fig. 11c). From these values, it follows that dioleoyl-PG is completely dissociated at pH 7 in 0.1 up to 1 M KCl, whereas only 9% of dissociation would occur in 1 mM KCl. Although the highly improved fit to the experimental data provides a strong indication that the pK_a of charge-forming ionizable groups suffers changes at low ionic strength, Egorova [23] still mentions that it should be kept in mind that other sources of changes in charge density, such as changes in the molecular projected area or cation-binding constant, could not be rejected. Hence, additional experiments are needed to prove the effect of both the salt concentration and the pH on the intrinsic dissociation constant of surface-bound dissociating groups.

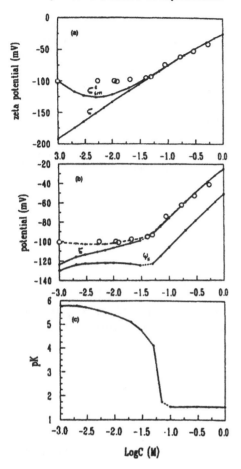

FIG. 11 Zeta potential of dioleoyl-PG liposomes in KCl solution at pH 7: (a) experimental values of zeta potential derived from Helmholtz–Smoluchowski, real theoretical values (cf. ζ in Fig. 10), and apparent theoretical values (indicated as ζ_{H-S} in Fig. 10) assuming $pK_a = 1.654$, $K_{int} = 0.13\,M^{-1}$, $\sigma_o = -267\,mC/m^2$, and $\Delta x_{sp} = 0.2\,nm$; (b) experimental values of zeta potential (○), real theoretical values, and apparent theoretical values taking the same values for the parameters as described in (a), except for pK_a, which was assumed to be dependent on KCl concentration; (c) pK_{int} versus KCl concentration profile that yields the best fit of apparent zeta potential to experimentally determined values, as shown in (b). (From Ref. 23.)

VII. INTERACTIONS WITH LOW MOLECULAR WEIGHT ORGANIC MOLECULES

Organic ions, such as pharmacologically significant molecules, hydrophobically adsorb on to or into the phospholipid membranes and decrease or increase the net membrane surface charge density [41]. Thus, Bermudez et al. [42] observed that the tuberculostatics ofloxacin and rifampicin both modify the surface charge of long-circulating and thermosensitive liposomes; the extent of drug–liposome interaction depended on the nature of both the bilayer constituents and the tuberculostatics. Similar phenomena were observed in the interaction of liposomes with some antibiotics, anesthetics and calcium-channel antagonists.

Whereas the antimitotics ethidium bromide, adriamycin, and celiptium were shown to be able to form specific stable complexes with the phospholipids by localized adsorption on to the membrane, others penetrate the membranes in a nonspecific manner. Thus, cholesterol is incorporated into liposomes as evidenced by the reduction in electrophoretic mobility observed in distilled water from -5.4×10^{-8} m^2/V/s for pure PS to -3.0×10^{-8} m^2/V/s for 25%PS/75% cholesterol [43].

In the latter case, the lipidic membrane will be considered as a separate phase, the hydrophobic phase, and a partition coefficient can be determined to characterize the passing of the drug from the aqueous phase to the lipidic phase. For imipramine, a tricyclic molecule having a neuroleptic activity, the partition coefficient was calculated through electrophoretic mobility measurements on liposomes using both anionic and neutral lipids. As the pK_a of the drug's amino group is near 8.6, it follows that the molecule is (at least partly) protonated at acidic to neutral pH conditions. Hence, drug incorporation into either anionic or zwitterionic liposomes induces a more positive value for the electrophoretic mobility. This effect is illustrated in Fig. 12 [44], which represents the electrophoretic mobility of egg PC MLV at pH 4.5 as a function of the logarithm of the imipramine concentration in 10 mM NaCl; at this particular pH condition, the drug may be assumed to be completely in its protonated form, which is represented as DH$^+$. From the change in zeta potential, the drug concentration in the lipid phase $X_{abs}^{(+)}$ (in mole fraction) may be deduced by using Eqs (1) and (2). The partition coefficient of the drug in the liposomes can then be calculated:

$$K_{DH} = \frac{(PL - DH^+)_s}{(PL)_s(DH^+)_s} = \frac{X_{abs}^{(+)}}{(DH^+)_s} \tag{16}$$

In Eq. (16), $(DH^+)_s$ represents the concentration of the (protonated form of the) drug in the close neighborhood of the surface of the liposomes, whereas $(PL - DH^+)_s$ represents the concentration of drug molecules incorporated into the liposomal surface. At pH 4.5,

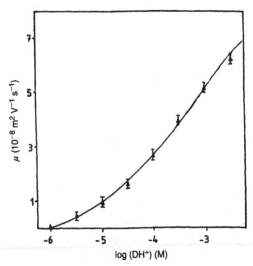

FIG. 12 Electrophoretic mobility of egg PC multilamellar liposomes in 10 mM NaCl at pH 4.5 as a function of log imipramine concentration. The solid line represents the theoretical profile if the partition coefficient of the drug is 2000 M^{-1}. (From Ref. 44.)

the interaction between egg PC and imipramine was nicely modeled by assuming a partition coefficient K_{DH} of 2000 M^{-1} and an anion-binding constant K_{Cl} of 0.5 M^{-1} (Fig. 12, Ref. 44). The latter parameter takes account of the adsorption of chloride anions to adsorbed positively charged drug molecules. In the case of PA at pH 4.5, the electrophoretic data corresponded to a K_{DH} of 600 M^{-1} and a K_{Na} of 0.5 M^{-1}.

At pH 8, the concentrations of the protonated (DH$^+$) and deprotonated (D) form of the imipramine drug must be calculated using the acidic dissociation constant in water. The parameters of the best-fitting model revealed that the partition coefficients of the protonated and the deprotonated drug were very similar. Also, similar values for the intrinsic partition coefficient were obtained for both zwitterionic and anionic phospholipid liposomes.

Independent determinations of the partition coefficient were also performed by measuring the increase in the area of a phospholipid film maintained at a constant surface pressure of 20 mN/m as drug molecules were penetrating into the film. For all lipids concerned, the partition coefficients derived from surface-pressure experiments were in close agreement with the values derived from electrophoretic mobility measurements.

In contrast to the intrinsic value, the apparent partition coefficient is largely affected by the state of ionization of both the membrane lipids and the drug, as shown by Kraemer et al. [45] who studied the interaction between oleic acid-containing PC membranes and propanolol, a β-receptor blocker. The highest values for the apparent partition coefficients were observed in the pH range from 7.5 to 9.5, i.e., in between the intrinsic pK_a of the fatty acid and the drug, respectively. In this region, electrostatic interactions highly favor the drug–liposome interaction. Similarly, Pohl et al. [46] observed an enhanced binding of both verapamil and propanolol in the presence of negatively charged phospholipids such as PS; as both drugs are positively charged at physiological pH values, it follows that drug–membrane interactions are not only affected by hydrophobic interactions, but also by electrostatic forces. On the other hand, the apparent partition coefficient in zwitterionic liposomes of the protonated form is about 10 times less than that of the neutral form of these drugs. In the former case, electrostatic repulsion between the drug-loaded membrane and free drug molecules hinders incorporation into the membrane.

VIII. PROTEIN–LIPOSOME INTERACTIONS

Protein–liposome interactions are mainly important in two types of investigations. If phospholipid vesicles are considered, protein–liposome interactions are studied in order to elucidate the in vivo interactions between proteins or enzyme and biological membranes (see, e.g., Ref. 47). If cationic liposomes, containing positively charged synthetic lipids, are considered, the main goal of these studies is to understand the relationship between the clearance behavior in vivo of these vesicles and the interaction between liposomes and proteins or cells [48].

Of course this section will not deal specifically with these topics. Only the use of zeta-potential measurements in these studies and their importance will be highlighted.

A. Phospholipid–Protein Interactions

Amino acids, the basic building blocks of proteins, only induce a pH variation in the solution, influencing the acid–base dissociation equilibra of the PC phospholipid membrane upon addition of lysine and glutamic acid [49].

Polypeptides, on the other hand, like poly(L-lysine) or poly(L-glutamic acid), were adsorbed to the PC liposomal surface, causing a significant change in electrophoretic mobility towards more positive values for poly(L-lysine). Addition of poly(L-glutamic acid) resulted in slightly negative values [49]. These observations are in correspondence with those obtained by Kim et al. [50] using negatively charged PS and PG vesicles. A poly(L-lysine)-dependent decrease in the absolute value of the zeta potential was observed for both kinds of vesicles. Increasing the polymerization degree of the polypeptide resulted in a similar decrease due to the larger number of positive charges per molecule. Consequently, charge neutralization occurred at higher lipid/polypeptide molar ratios for longer poly(L-lysine) molecules compared to the shorter ones.

Binding of *proteins* to phospholipid vesicles does not occur if both carry similar charges. De Meulenaer et al. [51] observed no change in electrophoretic mobility of soybean PC upon addition of positively charged cytochrome *c*. Bergers et al. [47] revealed, in a systematic study, that protein adsorption to negatively charged phospholipid liposomes was only observed at pH values where the number of positive charge moieties exceeded the number of negative charge moieties of the protein by at least 3 charge units. If binding occurs, this is reflected in a change of electrophoretic mobility or zeta potential. Bergers et al. [47] studied the binding of trypsin inhibitor, myoglobin, ribonuclease, and lysozyme to PC/PG vesicles. The electrophoretic mobility became less negative after absorption of positively charged proteins. Similar observations were made by Matsumura and Dimitrova [52] for serum albumin and cytochrome *c*.

With increasing protein-to-phospholipid ratios, the drop in the electrophoretic mobility leveled off, reaching a plateau. Protein adsorption profiles showed a similar shape [47]. De Meulenaer et al. [51] observed similar phenomena studying the interaction of cytochrome *c* with mixed dimyristoylphosphatidylcholine (DMPC)–dimyristoylphosphatidylglycerol (DMPG) vesicles, as shown in Fig. 13. The cytochrome *c*/phospholipid ratio at which the plateau was reached seemed to be dependent on the content of the negatively charged DMPG. Similarly, charge neutralization occurred at a fairly constant cytochrome *c*/DMPG molar ratio for all DMPG–DMPC vesicles studied.

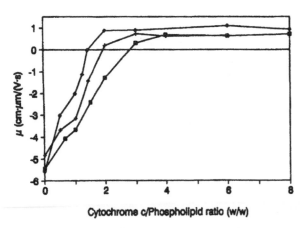

FIG. 13 Electrophoretic mobility of pure DMPG (□), 80:20 (w/w) DMPG:DMPC (+), and 60/40 (w/w) DMPG:DMPC (◆) dispersions in 5 mM TES buffer (pH 7.0) containing 2.5 mg phospholipids/ml as a function of cyctochrome *c*-to-phospholipid ratio (w/v). (From Ref. 51.)

All these data confirm that protein–phospholipid vesicle interactions are largely influenced by electrostatic interactions. However, other types of interaction forces, such as hydrophobic interactions, control the protein–phospholipid association as well. Thus, proteins such as immunoglobulin G (IgG) and albumin have a strong affinity for hydrophobic domains and consequently can interact hydrophobically with membrane lipids as well [48,52].

B. Cationic Lipid–Protein Interaction

Synthetic cationic lipids have been developed to complex negatively charged nucleic acids and deliver them to cells. The variability of in vivo gene delivery efficiency suggests the possibility of interaction between the cationic lipid–nucleic acid complex and blood components [48].

The identification of blood proteins that interact with cationic vesicles has been the subject of a recent study by Zelphati et al. [53]. The addition of such proteins as heparin, IgG, bovine serum albumin (BSA), and high-density lipoprotein (HDL) to cationic lipid–nucleic acid vesicle complexes induced a decrease in the number of positive charges on the complex. A more detailed study comparing the behavior of IgG and BSA revealed that BSA induced charge reversal and even formation of highly negative complexes, while IgG could only induce charge neutralization.

Zelphati et al. [53] demonstrated that serum-mediated inhibition of vesicle take up in vivo systems is coincident with changes in complex charge and size. Thus, interactions with the cell membrane are not solely based on electrostatic interactions, and internalization becomes limited with increasing complex size. Moreover, the presence of proteins such as BSA on the complex surface may act as a steric barrier to prevent endosomal destabilization in the cell, which is necessary for the gene therapeutic use of these complexes.

As a conclusion, it seems that serum components interact with cationic lipid nucleic acid complexes. Thus, interactions with the cell membrane (due to charge neutralization), internalization within the cell (due to size), and gene release in the target (due to steric hindrance) are altered. In these cases too, zeta-potential measurements can be a useful tool for studying some of these interactions.

IX. POLYNUCLEOTIDE–LIPOSOME INTERACTION

A large variety of cationic lipids have been selected to prepare positively charged liposomes in order to study their interaction with nucleotides and evaluate them with respect to their pharmaceutical use. It is not the intention of the authors to fully review the interaction of cationic liposomes with nucleotides and their use, because this will fall outside the scope of this book. However, for the sake of completeness and understanding, some basics will be discussed as a kind of introduction to this subject. Basically, the importance of zeta-potential measurements will be illustrated.

A. Cationic Liposomes and Their Use in Gene Therapy

Liposomes increasingly take an important place in gene therapy research, because they can act as a carrier for antisense oligodeoxynucleotides, plasmid DNA, or other nucleotides. Various liposomes are being investigated with regard to this objective: anionic, fusogenic, pH sensitive, immuno-, and cationic. In particular, the last are reported markedly to prevent nuclease degradation and enhance the rate of oligonucleotide uptake [54].

These positively charged liposomes consist of cationic lipids. The latter may contain a single or double lipid chain and a positively charged polar head. The first two important lipids used are the synthetic DODAB (dioctadecyl dimethylammonium bromide) and DOTAP (1,2-dioleolyl-3-trimethylammonium propane), both quarternary ammonium salts (Fig. 14). Following the success of these compounds in genes transfection, many other lipids have been synthesized. Cationic liposomes are mainly composed of (binary) mixtures of a cationic lipid and a neutral component, especially DOPE (dioeoyl phosphatidyl ethanolamine), which is known to be a strong destabilizer of lipid bilayers [55]. This destabilizing effect seems to be necessary to order to release the complexed nucleotides once the liposomes entered the target cell. Commercial kits for the preparation of cationic liposomes for gene transfer are available [56].

Cationic liposomes have intrinsic properties which make them attractive as a vehicle for gene delivery. First, they are not immunogenic and they are easy to standardize, because of their synthetic character. Due to their positive charge they are, moreover, likely to interact strongly with negatively charged nucleotides [57]. As natural nucleotides are rapidly degraded in biological fluids and cellular uptake is inefficient because of the restricted permeability characteristics of the cell membrane towards nucleotides, association with a carrier, giving rise to a so-called "lipoplex" is needed to resolve these problems [58].

A serious disadvantage of cationic liposomes is their intrinsic toxic and detergent character, excluding high-dose applications [59]. Therefore new, nontoxic cationic surfactants have been developed (e.g., Ref. 60). Apart from their possible toxicity, a high number of nonspecific interactions with serum components or cell surfaces may give rise to a lower efficiency of these gene carriers when applied to in vivo conditions [59,61], as briefly discussed in the preceding paragraph.

For more complete information about the use of cationic liposomes in gene transfer, we would like to recommend some excellent reviews [54,55,56,61].

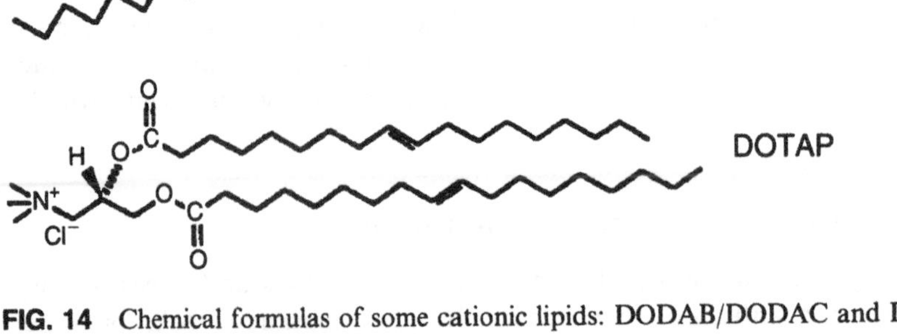

FIG. 14 Chemical formulas of some cationic lipids: DODAB/DODAC and DOTAP.

B. Role of Zeta-Potential Measurements in Research on Cationic Liposomes' Interaction with Nucleotides

Although charge interactions seem to play a very important role in the complex formation between cationic liposomes and nucleotides, it is only recently that zeta-potential measurements were involved.

As could be expected, cationic liposomes exhibit a highly positive zeta potential. By incorporation of zwitterionic surfactants such as PE or PC, a reduction in zeta potential can be observed. In these mixed zwitterionic–cationic liposomes the zeta potential remains positive if the cationic surfactant is present. Thus, Fig. 15 illustrates that the zeta potential of DOTAP is significantly higher than that of a DOTAP/PE (1/1) mixture [62]. By addition of the negatively charged nucleotides, a decrease in the zeta potential can be observed giving rise to neutral complexes, which exhibit charge reversal on further addition of nucleotides, as can be observed from Fig. 15. Similar observations have been made by other researchers for other lipoplex systems [57–59,63]. This decrease in zeta potential can be explained by the electrostatic attraction between the negatively charged nucleotides and the cationic liposomes. However, even after charge reversal, zeta potentials become increasingly more negative on addition of more nucleotides (Fig. 15). Similarly, Arima et al. [58] observed a decrease in the zeta potential of PC vesicles on addition of oligonucleotides down to $-30\,mV$ at an oligonucleotide/lipid ratio of 0.5. In the same study a less impressive but significant reduction in the zeta potential of negatively charged PS vesicles was observed. Hence, nucleotides bind to the vesicles, resulting in a reduction in the zeta potential, even if they are electrically neutral or negatively charged.

As can be observed from Fig. 15, charge inversion is observed at a 1/1 positive/ negative charge ratio. In other studies, however, even at a charge ratio equal to ±0.9 highly positive lipoplexes could be observed [57]. This seems to depend on the kind of lipids and nucleotides used, as well on the characteristics of the liposome, such as lamellarity.

Through a more in-depth investigation of the nature of the lipoplexes by other techniques, some remarkable observations were made, which help us to understand the kind of complexes between cationic liposomes and nucleotides are formed. At low nucleotide loads, lipids are supposed to enclose totally the negatively charged nucleotides [57]. According to Pires et al. [62], the neutralization of the positive charges results in a decrease in the effective size of the lipid polar headgroup. Consequently, the ability of the lipids to

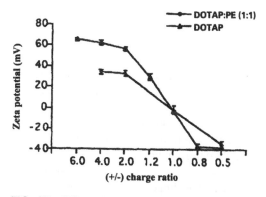

FIG. 15 Effect of DNA on zeta potential of DOTAP containing liposomes. Total lipid concentration 4.1 mM in a 10 mM HEPES, 150 mM NaCl, pH 7.4 buffer. (From Ref. 62.)

form a lipid bilayer is reduced. The fact that the nucleotides seem to be completely condensed with the lipid matrix, even at a positive/negative charge ratio equal to one (Fig. 15), indicates that the final structure of the lipoplexes is not spherical, but rather planar where the particle charges of both leaflets of the bilayer become available to interact with the nucleotides [62]. These observations seem to be in correspondence with those obtained by Gershon et al. [64] who proposed a model according to which cationic liposomes bind initially to nucleotides to form clusters of aggregated vesicles along the nucleic acids. At a critical liposome concentration, nucleotide-induced membrane fusion and liposome-induced nucleotide collapse would occur, giving rise to the formation of condensed structures. Zelphati and Szoka [61] mention as well that the lipoplexes cannot really be called liposomes because after their complexation with nucleotides they do not necessarily retain the classic bilayer form surrounding an aqueous layer. According to Koltover et al. [65], these synthetic cationic amphiphilic structures would be an inverted hexagonal phase.

Most researchers link the zeta-potential data to stability studies of these lipoplexes, by combining the electrophoretic data with particle size measurements. Generally, as could be expected, minimum stability of the complexes is observed when charge neutralization occurs. Due to the lack of charge repulsion, extensive aggregation or fusion can be observed [58,59,62].

Many recent investigations still assume an important relationship between the charge of the lipoplex and its encapsulation efficiency or its transfection efficiency, without confirming the observations with zeta-potential measurements (e.g., Refs 60 and 66). The relevance of such measurements is related to the large influence of electric charges in the deposition of macromolecules after intravenous injections [62]. Thus, it is observed that the charge ratio of a lipoplex is one of the parameters influencing the efficiency of nucleotide uptake. Moreover, it seems that the optimal charge ratio decreases due to the incorporation of neutral lipids. Positive zeta potentials are necessary for effective delivery of the nucleotides according to Zelphati and Szoka [54].

C. Conclusion

From this restricted review it can be concluded that zeta-potential measurements are an essential and convenient tool in studying the interaction of cationic liposomes and nucleotides. As the final charge of the lipoplex is of utmost importance with regard to its interaction with serum components, it is clear that zeta-potential measurements offer a quick and easy tool for partially predicting the efficiency of the nucleotide transfection.

X. POLYMER–LIPOSOME INTERACTIONS (BINDING + DEPLETION)

The Gouy–Chapman–Stern theory has been shown to describe successfully the electostatic potential of many bilayer model systems. Through combination with the Helmholz–Smoluchowski equation, the experimentally assessed electrophoretic mobility, μ, can be related to zeta potential [20,40]. However, several observations indicate that, for some systems such as red blood cells or phospholipid vesicles containing gangliosides [67], large discrepancies were found between the experimentally determined μ values (or zeta potential) and the actual surface potential. For vesicles containing terminally grafted hydrophilic polymers, similar observations were reported. The huge interest in this type of liposomes originates from the substantial increase in therapeutic activity of these systems as drug carriers. An example of such a sterically modified phospholipid is shown in Fig. 16.

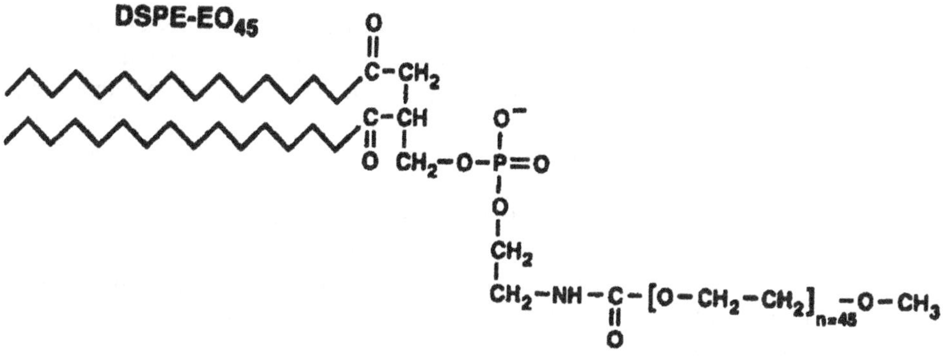

DSPE-EO$_{45}$

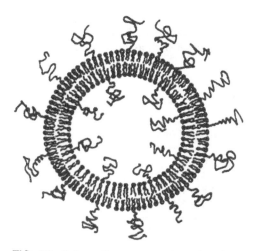

FIG. 16 Chemical structure of a sterically modified distearoyl-PE.

Typically these modifications have been achieved by incorporation of poly-(ethylenegly-col) (PEG) that is chemically bound to the polar headgroup of PE, which is also referred to as pegylated PE [68]. By incorporation of these sterically modified phospholipids, polymer-grafted liposomes can be produced as presented in Fig. 17.

Since it is only the intention of the authors to point out the importance of studying the zeta potential of sterically modified liposomes, reference is made to some excellent reviews for additional information on sterically stabilized liposomes and their applications [14,69–71].

A. "Reduction in Electrophoretic Mobility but not in Surface Potential" [72]

The first systematic study on the discrepancy between the electrophoretic mobility and the electrostatic surface potential of pegylated liposomes has been presented by Arnold et al. [73]. Actually, similar effects for bilayer membranes containing the ganglioside G_{M1} were previously reported by McDaniel et al. [67].

FIG. 17 Schematic representation of sterically stabilized liposomes with covalently attached PEG. (From Ref. 14.)

Woodle et al. [72] compared the zeta potential, calculated from the electrophoretic mobility using the Helmholz–Smolukowski equation, with the surface potential, as determined by a fluorescent probe, of various phospholipid vesicles as indicated in Table 2. As could be expected, zwitterionic PC vesicles were characterized by an electrophoretic mobility and zeta potential approaching zero. By incorporation of the negatively charged PG, a drastic increase in the absolute value of the ζ and surface potential is observed. If the pegylated PE is included in the bilayer structure of the PC vesicles in the same amounts as PG, a parallel increase in surface potential is observed. This is in correspondence with the anionic character of the pegylated PE. However, the change in surface potential is not reflected in a comparable increase in zeta potential or electrophoretic mobility, despite the fact that the pegylated PE carries a similar negative charge as PG.

Table 2 also reveals that, by decreasing the amount of polymer-grafted phospholipids in the PC vesicles, a less negative surface potential is observed, while this is again not reflected in the experimentally determined zeta potential. Sadzuka and Hirota [74] obtained similar results using pegylated PG containing DMPC/cholesterol/DMPG vesicles. Nonpegylated vesicles attained zeta potentials of $-57.2 \, \text{mV}$ in a sucrose-containing lactate buffer, while incorporation of the sterically modified phosphatide decreased the zeta potential to values of $-16.4 \, \text{mV}$. Varying concentrations of pegylated PG in the range 3.7 up to 5.4 mol % did not significantly influence the experimentally determined zeta potential. At a high concentration of 7 mol % pegylated surfactant, a further decrease in the zeta potential could be observed.

Kostarelos et al. [75] performed experiments with soybean vesicles to which an A–B–A copolymer of polyoxyethylene (99 units) and polyxoypropylene (67 units) was added. The block copolymer in this was not chemically bound to a phospholipid, but due to the presence of the hydrophobic polyoxypropylene moiety an interaction with the hydrophobic tails of the soybean phospholipids was possible, thus fixing the polymer in the membrane. Kostarelos et al. [75] observed a significant reduction in zeta potential: $-56.7 \, \text{mV}$ to $-20 \, \text{mV}$ after a 0.04 wt % addition of the polymer. In contrast to the previous observations of Woodle et al. [72] and Sadzuka and Hirota [74], a clear concentration dependence of the zeta potential could be observed up to a 0.04 wt % presence of the block copolymer. From the study of Kuhl et al. [76], the concentration dependence can be explained by the arrangement of the polymer at the vesicles surface as a function of polymer concentration, as shown in Fig. 18.

Even more sophisticated vesicles were designed recently by Kono et al. [77] by incorporation of temperature-sensitive hydrophilic polymers such as poly(N-acryloylpyr-

TABLE 2 Zeta Potential and Surface Potential for Various Phospholipid Vesicles

Lipid composition[a]	Zeta potential (mV)	Surface potential (mV)
PC	1.3 ± 0.2	0^{b}
PG:PC (7.5, 92.5)	-21.2 ± 0.8	-25.4 ± 5.8
PEG:PC (7.5, 92.5)	-5.4 ± 0.4	-24.8 ± 4.7
PEG:PC (5, 95)	-7.5 ± 2.6	-14.4 ± 2.4

[a] PC = phosphatidylcholine; PG = phosphatidylglycerol; PEG = poly(ethyleneglycol) linked to distearoylphosphatidylethanolamine (mol %).
[b] By definition.
Source: Ref. 71.

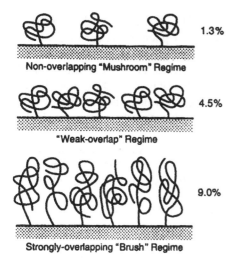

1.3%

Non-overlapping "Mushroom" Regime

4.5%

"Weak-overlap" Regime

9.0%

Strongly-overlapping "Brush" Regime

FIG. 18 Schematic representation of change in polymer layer thickness and density as a function of concentration in the exposed outer monolayer of the bilayer. (From Ref. 76.)

rolidine) and its derivatives. These polymers exhibit a so-called lower critical solution temperature (LCST). At temperatures below this LCST the water-soluble polymer, present as a hydrated coil, changes to a dehydrated globule and becomes water insoluble. The effects of temperature change are outlined in Fig. 19, indicating that as soon as the LCST is reached, polymers become hydrated, which results in a decrease in zeta potential.

Meyer et al. [59] used polymer-grafted surfactants in cationic liposomes and a similar drop in the zeta potential could be observed. For the effectiveness of sterically modified surfactants, the presence of a hydrophobic moiety, enabling the fixation of the molecule in the liposomal bilayer, is indispensable, as illustrated by Virden and Berg [78]. They observed no steric stabilization by addition of polyoxyethylene [47] to phospholipid vesi-

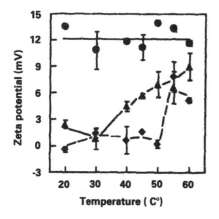

FIG. 19 Zeta potential of unmodified (●), poly(N-acryloylpyrrolidine)–2C$_{12}$ modified (◆), and poly(N-acryloylpyrrolidine-co-N,N-didodecylacrylamide)–2C$_{12}$ modified (▲) 3β-[N-(N',N'-dimethylaminoethane)carbamoyl]cholesterol DOPE (1:1 mol) liposomes as a function of temperature. (From Ref. 75.)

cles, while the addition of nonylphenol polyoxyethylene [52] to the same vesicles improved their stability significantly. Arnold et al. [73] observed similar phenomena for PEG added to PC-phosphatidic acid mixed vesicles. Krabi and Donath [79] observed an even higher than expected electrophoretic mobility by the addition of PEG and dextran to a liposomal solution. This was explained by an apparently reduced viscosity in the EDL region due to depletion and a subsequently decreased polymer-segment density near the interface. From electrophoretic measurements, the apparent thickness of this depletion layer could be calculated.

Arnold et al. [73] proposed that the decreased zeta potential was due to an increased friction at the liposome surface due to the presence of the hairy polymer tails. Similarly, Janzen et al. [80] postulated a model in which the polymer forms a surface layer subjected to a viscous drag.

Another approach to explain the observed reduction in ζ but not in surface potential is to assume a shift of the hydrodynamic plane of shear away from the charge-bearing vesicle surface caused by the hydrophilic polymer chains which are present in a brush-like formation (Fig. 18). This theory has already been presented by McDaniel et al. [67] but in their study using membranes containing ganglioside, the observed phenomena could also be explained by the fact that the charge of the ganglioside moiety was located in a plane at a definite distance from the vesicle surface. Woodle et al. [72] proved by using synthetic vesicles, in which all charges were present at the vesicle surface, that the shift in the hydrodynamic plane due to the presence of the polymer chains is indeed the reason for the observed discrepancies between the zeta potential and surface potential.

From the relationship:

$$\psi_x = \psi_0 e^{(-\kappa x)} \tag{17}$$

where ψ_x is the surface potential at a distance x from the vesicle surface, ψ_0 is the surface potential at the surface, and κ^{-1} is the Debye length, a realistic estimation of the PEG coating thickness was obtained: a shift from 21 to 5 mV corresponded to an increase in the hydrodynamic radius from 0.52 to 4.8 nm, which is very similar to other estimates of 6 nm for the extended length of the PEG from the surface by independent methods. Sadzuka and Hirota [74] obtained comparable results by comparing the zeta potentials at different salt concentrations. By linear regression of the obtained zeta potentials with respect to the Debye length, the position of the slipping plane or the thickness of the fixed aqueous layer around the vesicle could be calculated.

B. Consequences of Steric Stabilization of Liposomes

Due to the presence of polymer brushes at the liposomal surface, liposomes become sterically stabilized. Various examples illustrate this improved stability. Thus, Kostarelos et al. [75] observed an improvement in the vesicle stability towards aggregation caused by the cation-associated collapse of the surface electrostatic double layer. Similarly, an increased stabilizing effect was observed at higher molar PEG ratios for calcium-induced fusion of pegylated PE/PS liposomes in a study by Holland et al. [81]. At high electrolyte concentrations, however, the polymer chains may become insufficiently hydrated, resulting in an inadequate solubility. Consequently, the hairy vesicles' surface will acquire a less extended configuration, and the effectiveness of the steric barrier against flocculation will decrease with increasing electrolyte concentrations [75]. Similar observations were made by Virden and Berg [78] for nonylphenol polyoxyethylene [52] sterically stabilized phospholipid vesicles.

The main feature of these sterically stabilized liposomes is the improved blood circulation time, allowing the use of liposomes as an efficient drug delivery system. Due to the presence of the polymer brushes at the vesicle surface their approach to blood components and cellular surfaces in the body, is prevented [82]. The interaction with blood components and cellular surfaces is supposed to stimulate adhesion, fusion, lysis, or destruction by the immune system of the liposomes, which in turn results in their rapid removal from the vein system and their restricted efficiency. Due to a longer blood circulation time, an increased therapeutic efficiency of liposomal drug carrier systems is observed.

NOMENCLATURE

a	particle radius (m)
BSA	bovine serum albumin
C	bulk electrolyte concentration (mol/m^3)
C_s	electrolyte concentration at the surface (mol/m^3)
D	dielectric constant (−)
DH$^+$	protonated form of a drug
DMPC	dimyristoylphosphatidylcholine
DMPG	dimyristoylphosphatidylglycerol
DODAB	dioctadecyl dimethylammonium bromide
DOPE	dioleoyl phosphatidylethanolamine
DOTAP	1,2-dioleolyl-3-trimethylammonium propane
DPPC	dipalmitoylphosphatidylcholine
e	elementary charge unit (1.6×10^{-19} C/charge)
EDL	electrical double layer
HDL	high-density lipoproteins
IgG	immunoglobulin G
IP	imipramine
k	Boltzmann constant (1.38×10^{-23} J/K)
K_{app}	apparent binding constant (m^3/mol)
K_{int}	intrinsic binding constant (m^3/mol)
LCST	lower critical solution temperature
LUV	large unilamellar vesicles
Me	metal
MLV	multilamellar vesicles
MOPS	3-(N-morpholino)propanesulfonic acid
N_A	Avogadro's number (6.02×10^{23} /mol)
N_{bs}	number of binding sites per unit surface area (1/m^2)
N_{bs}^-	number of negatively charged binding sites per unit surface area (1/m^2)
pa_1	molecular projected area (m^2)
PA	phosphatidic acid
PC	phosphatidylcholine
PE	phosphatidylethanolamine
PEG	poly(ethyleneglycol)
PG	phosphatidylglycerol
PI	phosphatidylinositol
pK_a	− log (dissociation constant of acidic groups)

pK_{app}	$-\log$ (apparent dissociation constant)
PL	phospholipid
PS	phosphatidylserine
STE	stearylamine
SUV	small unilamellar vesicles
T	Kelvin temperature (K)
TMACl	tetramethylammonium chloride
TNS	2-(p-toluidinyl)naphthalenesulfonate
$X_{abs}^{(+)}$	drug concentration in the lipid phase
Z	valency

Δx_{sp}	distance between Stern plane and plane of shear (m)
ε_o	permittivity in vacuum ($8.85 \times 10^{-12}\, C^2/N/m^2$)
ζ	zeta potential [V]
ζ_0	zeta potential with no ion adsorption [V]
κ	Debye–Hückel parameter (1/m)
μ	electrophoretic mobility ($m^2/V/s$)
μ_{H-S}	apparent electrophoretic mobility as derived from the real zeta potential using the Helmholtz–Smoluchowski equation ($m^2/V/s$)
σ_{ads}	charge density of adsorbed species in the Stern plane (C/m^2)
σ_o	charge density at the surface of particle (without adsorbed ions) (C/m^2)
σ_S	charge density at the surface of the Stern plane (C/m^2)
ψ_o	surface potential (without adsorbed ions) [V]
ψ_S	Stern potential [V]
ψ_x	surface potential at distance x from surface [V]

REFERENCES

1. RRC New. Liposomes, a Practical Approach. Oxford: Oxford University Press, 1990.
2. MJ Ostro. Liposomes from Biophysics to Therapeutics. New York: Marcel Dekker, 1987.
3. G Blume, G Cevc. Biochim Biophys Acta 1029:91, 1990.
4. G Gregoriadis. Liposomes as Drug Carriers, Recent Trends and Progress. New York: John Wiley, 1988.
5. G Gregoriadis. TIBTECH 13:527, 1995.
6. MJ Ostro, PR Cullis. Am J Hosp Pharm 46:1576, 1989.
7. BF Haumann. Inform 3:1172, 1992.
8. T Sato, J Sunamoto. Prog Lipid Res 31:345, 1992.
9. AM Carmona-Ribeiro. J Phys Chem 93:2630, 1989.
10. FJ Carrión, A De La Maza, JL Parra. J Colloid Interface Sci 164:78, 1994.
11. H Minami, T Inoue, R Shimozawa. J Colloid Interface Sci 158:460, 1993.
12. RL Juliano, D Stamp. Biochem Biophys Res Co. 63:651, 1975.
13. A Gabizon, D Papahadjopoulos. Biochim Biophys Acta 1103:94, 1992.
14. MC Woodle, DD Lasic. Biochim Biophys Acta 1113:171, 1992.
15. DD Lasic, D Needham. Chem Rev 95:2601, 1995.
16. M Eisenberg, T Gresalfi, T Riccio, S McLaughlin. Biochemistry 18:5213, 1979.
17. G Cevc. Chem Phy Lipids 64:163, 1993.
18. S McLaughlin, N Mulrine, T Gresalfi, G Vaio, A McLaughlin. J Gen Physiol 77:445, 1981.
19. A McLaughlin, WK Eng, G Vaio, T Wilson, S McLauglin. J Membr Biol 76:183, 1983.
20. S McLaughlin. Annu Rev Biophys Biophys Chem 18:113, 1989.

21. EM Egorova, AS Dukhin, IE Svetlova. Biochim Biophys Acta 1104:102, 1992.
22. EM Egorova, LL Yakover, TF Svitova. Biochim Biophys Acta 1109:1, 1992.
23. EM Egorova. Colloids Surfaces 131:19, 1998.
24. E Casals, M Soler, M Gallardo, J Estelrich. Langmuir 14:7522, 1998.
25. O Söderman, G Arvidson, G Lindblom, K Fontell. Eur J Biochem 134:309, 1983.
26. G Cevc. Biochemistry 26:6305, 1987.
27. M Deleers, JP Servais, E Wuelfert. Biochim Biophys Acta 813:195, 1985.
28. IS Graham, JA Cohen, MJ Zuckerman. J Colloid Interface Sci 135:335, 1990.
29. SA Tatulian. Biochim Biophys Acta 736:189, 1983.
30. MA Akeson, DN Munns, RG Burau. Biochim Biophys Acta 986:33, 1989.
31. MD Reboiras. Bioelectrotech Bioenerg 39:101, 1996.
32. YA Ermakov, SS Makhmudova, AZ Averbakh. Colloids Surfaces A 140:13, 1998.
33. PWM Van Dijck, PHJ Ververgaert, AJ Verkleij, LLM Van Deenen, J de Gier. Biochim Biophys Acta 406:465, 1975.
34. MD Reboiras, MN Jones. Colloids Surfaces 15:239, 1985.
35. K Hammond, MD Reboiras, IG Lyle, MN Jones. Colloids Surfaces 10:143, 1984.
36. PH Wiersema, ALJ Loeb, JTG Overbeek. J Colloid Interface Sci 22:78, 1966.
37. RW O'Brien, LR White. J Chem Soc, Faraday Trans 2 74:1607, 1978.
38. SS Dukhin, BV Deryaguin. In: E Matijević, ed. Surface and Colloid Science. vol. 7. New York: John Wiley, 1974.
39. MT Roy, M Gallardo, J Estelrich. J Colloid Interface Sci 206:512, 1998.
40. G Cevc. Biochim Biophys Acta 1031–1033:311, 1990.
41. F Molina, C Llacer, AO Vila, A Puchol, J Figueruelo. Colloids Surfaces A 140:91, 1998.
42. M Bermudez, E Martinez, M Mora, ML Sagrista, MA de Madariaga. Colloids Surfaces 158:59, 1999.
43. DP Gregory, L Ginsberg. Biochim Biophys Acta 769:238, 1984.
44. S Banerjee, M Bennouna, J Ferreira-Marques, JM Ruysschaert, J Caspers. J Colloid Interface Sci 219:168, 1999.
45. SD Kraemer, S Jakits-Deiser, H Wunderli-Allenspach. Pharm Res 14:827, 1997.
46. EE Pohl, AV Krylov, M Block, P Pohl. Biochim Biophys Acta 1373:170, 1998.
47. JJ Bergers, MH Vingerhoeds, L Vanbloois, JN Herron, LHM Janssen, MJE Fischer, DJA Crommelin. Biochemistry 32:4641, 1993.
48. SC Semple, A Chonn, PR Cullis. Adv Drug Deliv Rev 32:3, 1998.
49. H Matsumura, F Mori, K Kawahara, C Obata, K Furasawa. Colloids Surfaces A 92:87, 1994.
50. J Kim, M Mosior, LA Chung, H Wung, S McLaughlin. Biophys J 60:135, 1991.
51. B De Meulenaer, P Van der Meeren, M De Cuyper, J Vanderdeelen, L Baert. J Colloid Interface Sci 189:254, 1997.
52. H Matsumura, M Dimitrova. Colloids Surfaces B 6:165, 1996.
53. O Zelphati, LS Uyechi, LG Barron, FC Szoka. Biochim Biophys Acta 1390:119, 1998.
54. O Zelphati, FC Szoka. Pharm Res 13:1367, 1996.
55. D Litzinger, L Huang. Biochim Biophys Acta 1113:201, 1992.
56. DD Lasic, NS Templeton. Adv Drug Deliv Rev 20:221, 1996.
57. JC Birchall, IW Kellaway, SN Mills. Int J Pharm 183:195, 1999.
58. H Arima, Y Armaki, S Tsuchiya. J Pharm Sci 86:438, 1997.
59. O Meyer, D Kirpotin, K Hong, B Sternberg, JW Park, MC Woodle, D Papahadjopoulos. J Biol Chem 273:15621, 1998.
60. C Puyal, P Mulhaud, A Bienvenüe, JR Philippot. Eur J Biochem 228:97, 1995.
61. O Zelphati, FC Szoka. J Control Release 41:99, 1996.
62. P Pires, S Simões, S Nir, R Gaspar, N Düzgünes, MC Pedrose de Lima. Biochim Biophys Acta 1418:71, 1999.
63. RI Mahoto, K Kawabata, T Nomura, Y Takakura, M Hashida. J Pharm Sci 84:1267, 1995.
64. H Gershon, R Ghirlando, SB Guttman, A Minsky. Biochemistry 32:7143, 1993.
65. I Koltover, T Salditt, JO Räder, CR Safinya. Science 281:78, 1998.

66. K Koltover, I Jääkeläinen, J Märkkönen, A Urtti. Biochim Biophys Acta 1195:115, 1994.
67. RV McDaniel, A McLaughlin, AP Winiski, M Eisenberg, S McLaughlin. Biochemistry 23:4618, 1984.
68. MS Webb, D Saxon, FMP Wong, HJ Lim, Z Wang, MB Bally, LSL Choi, PR Cullis, LD Mayer. Biochim Biophys Acta 1372:272, 1998.
69. MN Jones. Adv Colloid Interface Sci 54:93, 1995.
70. TM Allen. Adv Drug Deliv Rev 13:285, 1994.
71. TM Allen. J Liposome Res 2:289, 1992.
72. MC Woodle, LR Collins, E Sponsler, N Kossovsky, D Papahadjopoulos, FJ Martin. Biophys J 61:902, 1992.
73. K Arnold, O Zschoernig, D Marthel, W Herold. Biochim Biophys Acta 1022:303, 1990.
74. Y Sadzuka, S Hirota. Adv Drug Deliv Rev 24:257, 1997.
75. K Kostarelos, PF Luckham, TF Tadros. J Chem Soc, Faraday Trans 94:2159, 1998.
76. TL Kuhl, DE Leckband, DD Lasic, JN Israelachvili. Biophys J 66:1479, 1994.
77. K Kono, A Henmi, T Takagishi. Biochim Biophys Acta 1421:183, 1999.
78. JW Virden, JC Berg. J Colloid Interface Sci 153:411, 1992.
79. A Krabi, E Donath. Colloids Surfaces A 92:175, 1994.
80. J Janzen, X Song, DE Brooks. Biophys J 70:314, 1996.
81. JW Holland, C Hui, PR Cullis, TD Madden. Biochemistry 35:2618, 1996.
82. DD Lasic, FJ Martin, A Gabiezon, SK Huang, D Papahadjopoulos. Biochim Biophys Acta 1070:187, 1991.

34

Aggregation of Liposomes and Effects of Electric Field on It

HIDEO MATSUMURA National Institute of Advanced Industrial Science and Technology, Tsukuba, Japan

KUNIO FURUSAWA University of Tsukuba, Tsukuba, Ibaraki, Japan

I. INTRODUCTION

In biological worlds, many types of supermolecular systems have characteristic organized structures: cell membrane, protein assembly such as muscles, etc. The origins of forces to construct such organized structures are due to some specific interactions and/or nonspecific ones. So far, many studies have been conducted in various fields of biology concerning the interactions between biological molecules. However, even nonspecific interactions between biological supermolecular systems have not yet been understood. Elucidating such nonspecific interactions requires building the scientific foundation of self-organization of biological systems. This chapter shows one example of organization of model cell systems (liposome particles) by nonspecific internal interactions and also that under external electric fields.

II. LIPOSOME

A liposome is a spherical particle consisting of bilayer membranes of phospholipid molecules and dispersed in an aqueous solution. Because of its vesicular structure of membrane organization, it is sometimes called an "artificial cell." Lipid molecules consist of a hydrophilic head group and lipophilic hydrocarbon chains. Acidic lipids bear negative charges on the head group at neutral pH because of dissociation of H^+. Amphoteric lipids have a positive and a negative charge on the head groups. The outer surface of the liposome, consisting of natural lipids, usually bears negative charges on it because of the presence of acidic lipids. Even the liposomes of amphoteric lipids from nature bear negative charges, which can be easily seen by electrophoresis of the liposomes. The origin of these negative charges is not clear but acidic impurities and/or specific adsorption of anions (including OH^-) are typical candidates. The chemical structures of several phospholipids and the shape of a liposome are shown in Fig. 1. Further information on the chemical structure of lipids and the methods of preparation of liposomes can be found in Ref. 1.

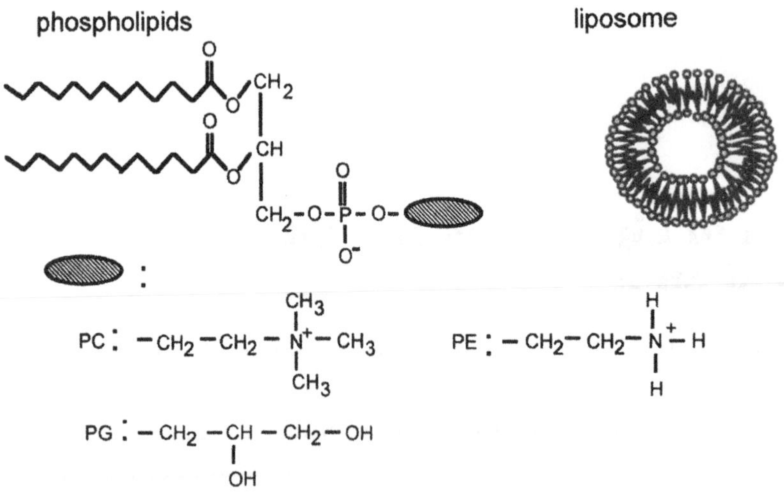

phospholipids   liposome

FIG. 1   Chemical structures for several kinds of phospholipids and diagram of a liposome.

III. AGGREGATION OF LIPOSOMES

A. Electrophoresis of Liposomes

Whether liposomes undergo aggregation or not depends on the interparticle interaction between them. The barrier in the interaction potential usually comes from the combination of electrostatic repulsion and van der Waals' attraction potentials. If the thermal fluctuation energy which liposome particles have is much larger than the potential barrier, the aggregation begins easily. To facilitate aggregation, we usually reduce the electrostatic repulsive force between liposomes by adding some ionic compounds such as metal salts to the medium of liposome dispersion. The general features of the change in the interfacial electrical potentials in relation to ion binding have been reported [2, 3]. Figure 2 shows zeta potentials of liposomes for each type of phospholipid: phosphatidylcholine (PC) from egg yolk, phosphatidylethanolamine (PE) from bovine braine, and phosphatidylglycerol (PG) from egg yolk, dispersed in various concentrations of $CaCl_2$ aqueous solution [4]. Even liposomes of PC and PE bear negative charges. The surface

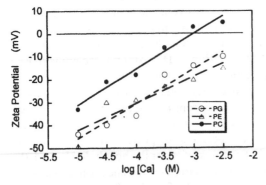

FIG. 2   Zeta potentials of liposomes made of each phospholipid (PG: phosphatidylglycerol; PE: phosphatidylethanolamine; PC: phosphatidylcholine) immersed in various concentrations of $CaCl_2$.

charges of liposomes of PC can be neutralized by the binding of Ca^{2+} ions (the iso-electric point of a PC liposome is around 1×10^{-3} M Ca^{2+} concentration), but the neutralization does not occur in the case of PE and PG in the experimental concentration range analyzed.

B. Coagulation Rate of Liposomes

The reduction of electrical repulsive forces causes an instability in liposome dispersion resulting in aggregation; in some cases the aggregation by Na^+ ions is reversible [5–7]. The coagulation behavior depends on the kinds of lipids and metal ions. The coagulation rate of liposomes (averaged diameter 200 nm) from each phospholipid (PC, PE, or PG) in 1×10^{-3} M $CaCl_2$ aqueous solution is shown in Fig. 3 [4]. It shows the change in the intensity of transmitted light through the dispersion sample after mixing with $CaCl_2$ solution (final concentration of Ca 1×10^{-3} M). DLVO theory suggests that the iso-electric point induces the rapid coagulation of colloid particles. This Ca concentration is the isoelectric point for PC liposomes (Fig. 2), but no change in the light signal occurs for these liposomes (curve a). On the other hand, PE and PG liposomes undergo coagulation at this Ca concentration even though they have some negative zeta potentials. This phenomenon is contrary to the concept of the theory. That is, the electrostatic repulsive potentials observed in PE and PG cases are not enough to prevent the Brownian coagulation, but the zero electrostatic potential in the PC case appears to prevent the coagulation. This means that some additional repulsive forces must exist between PC liposomes.

The van der Waals and electrostatic forces have a particle-size dependence, whereas the so-called structural forces such as the hydration force are short ranged and we can consider them as particle-size independent. So by conducting aggregation experiments with liposomes of different particles sizes, we can obtain some information on the mechanism of aggregation. The size dependence of PC liposome aggregation shows interesting behavior; the larger size liposomes produce an aggregation easily. Figure 4 shows the change in the intensity of transmitted light through the dispersion sample after mixing with $CaCl_2$ solution (final concentration of Ca 1×10^{-3} M) for each size of PC liposome.

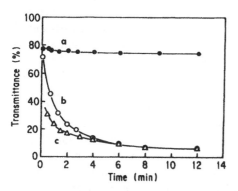

FIG. 3 Change in light transmittance of liposome dispersion (a: PC; b: PG; c: PE) after mixing with $CaCl_2$ aqueous solution. The final concentration of Ca^{2+} ions is 1×10^{-3} M.

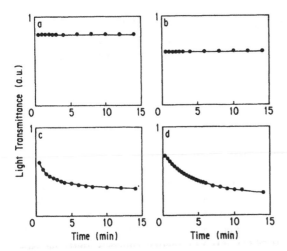

FIG. 4 Change in light transmittance after mixing with aqueous CaCl$_2$ solution (final concentration, 110^{-3} M) for various diameters of the PC liposomes: (a) 100 nm; (b) 200 nm; (c) 320 nm; (d) 2000 nm.

At this Ca^{2+} concentration, the electrostatic repulsion is negligible but the van der Waals attractive force increases with the increase in liposome size. The experimental results suggest that the interaction potential between PC liposomes consists of a short-range structural repulsive force and the van der Waals' attractive one. From the hygroscopic nature of the PC molecule, the short-range attractive force can be hydrophilic hydration. The combination of the van der Waals attractive potential and the short-range hydration potential has a potential minimum somewhat away from the particle surface [8, 9]; it is similar to the secondary minimum in the interaction potential between larger sized particles with larger Hamaker constant. In this case, aggregated flocs have a thin liquid layer between particles so that they are likely to be influenced by the change in the condition of surrounding medium. Furthermore, contrary to the case of coagulation in the primary minimum of the potential, the flocculated particles do not need to overcome any potential barrier to be redispersed into primary units. In fact, when one reduces the Ca concentration, PC liposomes easily separate into single particles (Fig. 5).

In the high ion concentration range, the thickness of the electrical double layer (EDL) on the particle surface is rather small and hence the aggregation is induced because of the near absence of repulsive forces. However, the repulsive forces from short-ranged hydration and electric ones prevent the coagulation at the deep primary potential minimum and only the flocculation at a shallow well of potential minimum somewhat away from the surface is possible. In the range of Ca concentration from 0.01 to 0.1 M, only doublets or triplets of PC liposomes are observable. This suggests that reversible flocculation prevails in these salt conditions. In the range of Ca concentration above 0.1 M, flocs consisting of several primary particles are observable and they grow to a dendritic structure (Fig. 6), which is similar to the structure formed by Brownian coagulation [10]. It means that the depth of the potential well increases with salt concentration and hence backward reaction to the separation decreases. However, those flocs are rather easily separated to form the primary particles by the dilution of salt concentration. It means that the liposome particles gather together somewhat away from the particle surface, as described before.

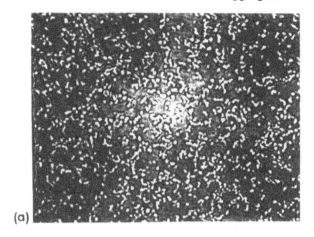

FIG. 5 Microsope images: (a) flocculated larger PC liposomes ($d = 1000–2000\,\text{nm}$) under the condition of $CaCl_2 = 1 \times 10^{-3}\,\text{M}$, 5 h after mixing; (b) redispersed liposomes obtained by dilution of $CaCl_2$ with pure water.

IV. EFFECTS OF ELECTRIC FIELD ON FLOCCULATION

A. Electrical Phenomena in Dispersion Samples

When we apply external electric fields to dispersion samples, various types of electrokinetic phenomena will occur, such as electro-osmosis, electrophoresis, ionic conduction, electric polarization, etc. These depend on the electrical properties of dispersion particles and they can affect the aggregation behavior of particles through the change in interparticle interaction.

Within the low-frequency range of electric fields ($< 1\,\text{MHz}$), the size of colloidal particles is much smaller than the wavelength of the alternating electric fields. In this range we can handle the electric field separately from the magnetic one and the usual electric-circuit treatment can be applied to understand the phenomena. In a homogeneous electrolyte solution, we can treat the sample as a combination of conductor and capacitor.

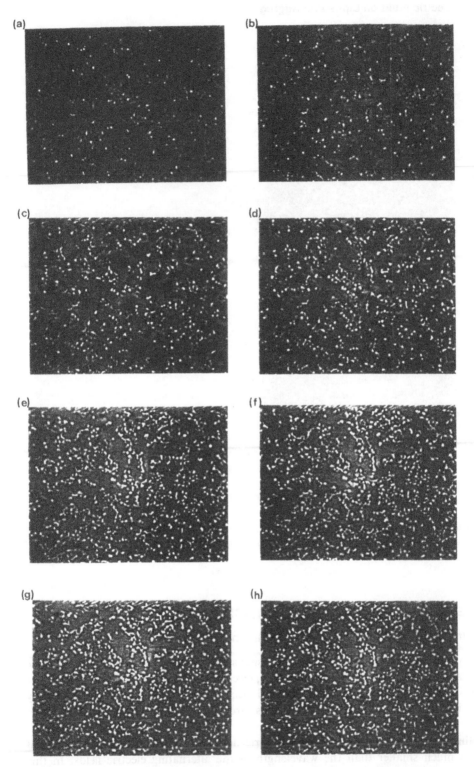

FIG. 6 Microscope image of liposome dispersion for different times elapsed after mixing with CaCl$_2$ aqueous solution (final concentration, 0.55 M): (a) 30 min; (b) 45 min; (c) 1 h; (d) 1 h 15 min; (e) 2 h; (f) 2 h 15 min; (g) 2 h 30 min; (h) 2 h 45 min.

However, in a heterogeneous sample like a colloidal dispersion, we must consider the interfacial polarization effect at the particle/solution interface [11, 12]. Therefore, the properties of the EDL at the interface are an important factor for electrical properties of the dispersion samples. We can express the electrical polarization by the difference in conductivity between the double layer and the bulk solution in the case of nonconducting particles. The quantity that expresses the ratio between the surface conductivity and bulk conductivity is the dominating factor for the surface polarization. It is called the Dukhin number: $Du = k_S/ak_b$, where k_S is the surface conductivity, k_b is the conductivity in the bulk, and a is the radius of the particle [13]. The electrical polarization at the surface of a particle is expressed by an induced dipole moment (d) on the particle. It is described by using the Du number. Similarly, the ratio between the conductance of the dispersion sample (K^*) and that of the pure medium (K) is described by using Du. These are mathematically expressed as [12–14]:

$$d = \varepsilon a^3[-1/2 + (3/2)Du/(Du + 1)]E \qquad (1)$$

$$K^*/K = 1 + (3/2)p(2Du - 1)/(Du + 1) \qquad (2)$$

where ε is the permitivity of the medium, E is the electric field strength, and p is the volume fraction of particles. Here, the relaxation phenomenon of electrical polarization by relaxation of concentration polarization through the diffusion mechanism is ignored. These values decrease when the relaxation occurs in the low-frequency range ($< D/a^2$ where D is the diffusivity of the counterion in the bulk); K^*/K is an experimentally determined quantity, so we can calculate the Du number if the volume fraction of the particles is known [Eq. (2)]. Hence, the induced dipole moment is estimated from Eq. (1).

Figure 7 shows K^*/K for PC liposome dispersions at various concentrations of Ca^{2+} ions [15]. The induced dipole moments on PC liposomes can be calculated using these values with the volume fraction of the samples (in these samples, $p = 0.1$) as shown in Fig. 8 [15]. The induced dipole moment has a similar tendency to that of K^*/K.

B. Aggregation of PC Liposomes by Induced Dipoles

When we apply electric fields to the dispersion sample in unstable conditions of the system, the aggregation behavior can be changed. As described in the preceding sec-

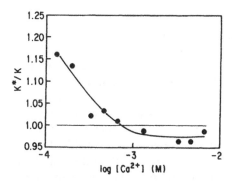

FIG. 7 Conductance ratio of liposome dispersion to medium versus concentration of $CaCl_2$ at frequency of 1 kHz.

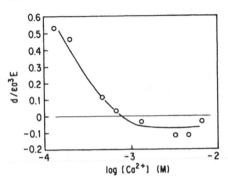

FIG. 8   Reduced induced dipole moments versus concentration of CaCl$_2$ at frequency of 1 kHz.

tion, PC liposomes have their isoelectric point around 1×10^{-3} M CaCl$_2$ aqueous solution. When we apply an electric field to the PC liposome sample around the iso-electric point, the formation of linear chains of liposome can be observed. Figure 9 shows linear aggregation of PC liposomes near a needle electrode induced by an alternating electric field of 15 V between two electrodes located 5 mm apart (an asym-metric "T" letter configuration of electrodes) in the salt condition of 3.6×10^{-3} M CaCl$_2$ [15]. As can be seen in the figure, the formtion of a linear chain of PC liposome has a frequency dependence. Electric fields of low frequency have low potentiality for inducing linear aggregations. At 10 and 100 kHz, we can see good linear alignment of the liposomes, but we see insufficient alignment of them at 100 Hz and 1 kHz. This feature may be explained by the magnitude of the induced dipole moment on the PC liposomes, which is described in the preceding section. That is, in the lower frequency range, the relaxation mechanism of concentration polarization provokes insufficient electric polarization around the liposomes.

Since the induced dipole moment is attributed to the nature of the EDL on the particle surface, it changes with the ion concentration in the solution. Figure 10 shows the Ca^{2+} ion concentration dependence of the linear alignment of PC liposomes in the concentration range containing the isoelectric point [15]. In the lower concentration range of Ca^{2+} ions, the linear alignment of the PC liposome is perfect, but in the higher concentration range, it cannot be observed. We can imagine it is caused by the induced dipole–induced dipole interaction, and it is truly attributable to the nature of the double layer. As is described in the preceding section, the induced dipole moment depends on the conductance difference between that in the double layer and that in the bulk solution. We can see in Fig. 9 that the magnitude of the dipole is larger in the lower concentration and decreases with increasing Ca^{2+} ion concentration. Around the isoelectric point $(1 \times 10^{-3}$ M), the aggregates formed are somewhat thicker, due to the additional contribution from the intrinsic liposome–liposome interaction. This aggrega-tion does not break to the original single liposome particles, whereas the linear aggre-gates in the lower concentration range separate into single particles when the electric field is switched off (Fig. 11). At a Ca concentration of 8×10^{-4} M, once the structure is obtained, it maintains its shape after switching off the applied electric fields. In this case, the intrinsic interparticle attractive interaction is dominant. On the other hand, at a Ca concentration of 4×10^{-4} M, after switching off the electric fields, each liposome particle begins Brownian motion as a single particle. In this case, the attractive force is mainly from induced dipole moments.

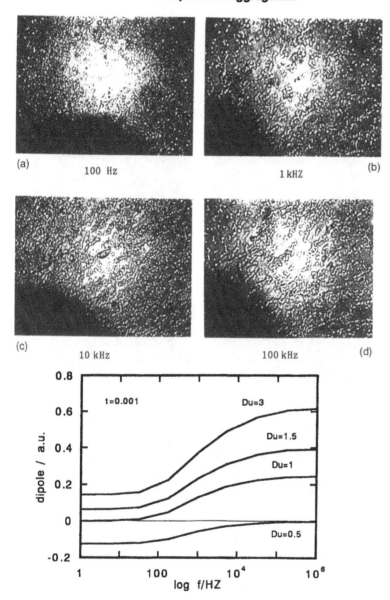

FIG. 9 Microscope images of liposome clusters near a Pt electrode in electric fields of various frequencies: (a) 100 Hz; (b) 1 kHz; (c) 10 kHz; (d) 100 kHz. $V_{pp} = 15 \text{V}/5 \text{mm}$; $CaCl_2$, 3.6×10^{-3} M. The frequency dependency of the induced dipole moments calculated by the equation in Ref. 13 (p 4.118) is also shown. In this figure, the relaxation time for concentration polarization is 1 ms.

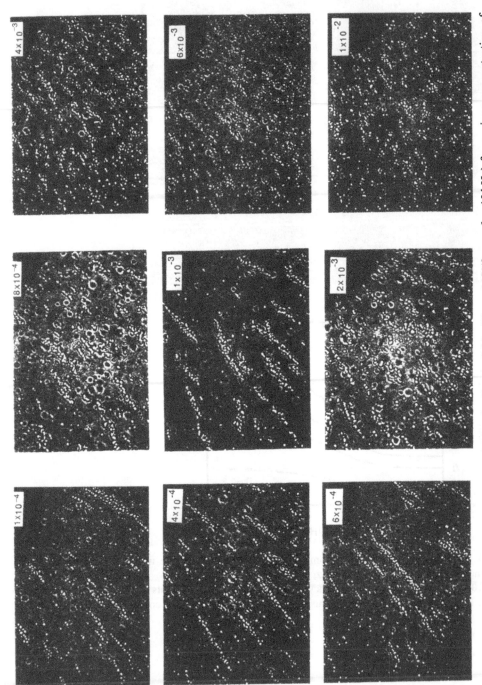

FIG. 10 Microscope images of liposome clusters in electric fields ($V_{pp} = 15\,\mathrm{V}/3\,\mathrm{mm}$; $f = 10\,\mathrm{kHz}$) for various concentration of $CaCl_2$. In the photos, the concentration of $CaCl_2$ is shown.

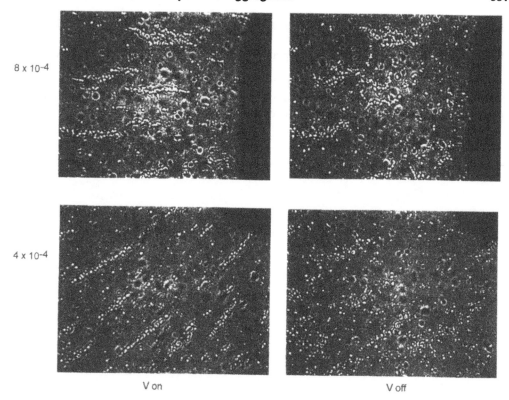

8×10^{-4}

4×10^{-4}

V on   V off

FIG. 11 Microscope images of the structure changes after switching off the applied electric fields ($V_{pp} = 15\,\text{V}/3\,\text{mm}; f = 10\,\text{kHz}$) for two Ca concentrations ($4 \times 10^{-4}\,\text{M}$ and $8 \times 10^{-4}\,\text{M}$).

The Ca^{2+} mobility in the surface region of the PC liposome is rather high [16, 17], which gives rise to a high Du number and hence a rather large induced dipole moment. This may be a specific character of PC liposomes. Thus, there remain interesting features to be clarified on the surface of liposome particles.

REFERENCES

1. RRC New. Liposomes, a Practical Approach. IRL Press, [at Oxford University Press, Oxford], 1990, ch. 1.
2. H Matsumura, K Furusawa. Adv Colloid Interface Sci 30:71, 1989.
3. H Matsumura, K Furusawa. In: H Ohshima, K Furasawa, eds. Electrical Phenomena at Interfaces. New York: Marcel Dekker, 1998, ch. 27.
4. H Matsumura, K Watanabe, K Furusawa. Colloids Surfaces A 98:175, 1995.
5. E Day, A Kwok, S Hark, J Ho, W Vail, J Bentz, S Nir. Proc Natl Acad Sci USA 77:4026, 1980.
6. T Yoshimura, K Aki. Biochim Biophys Acta 813:167, 1985.
7. J Virden, C Berg. Langmuir 8:1532, 1992.
8. J Mara, J Israelachvili. Biochemistry 24:4608, 1985.
9. J Israelachvili. Intermolecular & Surface Forces. London: Academic Press, 1991, ch. 17.
10. T Vicsek. Fractal Growth Phenomena. London: World Scientific, 1989.
11. S Takashima. Electrical Properties of Biopolymers and Membranes. Adam Hilger, 1989, ch. 6.
12. SS Dukhin, VN Shlov. Adv Colloid Interface Sci 13:153, 1980.

13. J Lyklema. Fundamentals of Interface and Colloid Science. vol. 2. London: Academic Press, 1995, p 3.208.
14. H Matsumura. In: H Ohshima, K Furasawa, eds. Electrical Phenomena at Interfaces. New York: Marcel Dekker, 1998, ch. 14.
15. H Matsumura. Colloids Surfaces A 104:343, 1995.
16. S Verbich, S Dukhin, H Matsumura. J Disp Sci Technol 20:83, 1999.
17. H Matsumura, S Verbich, S Dukhin. Colloids Surfaces A 159:271, 1999.

Index

Printed and bound by CPI Group (UK) Ltd, Croydon, CR0 4YY

23/10/2024

01778254-0017